Moments of Inertia of Common Geometric Shapes

Rectangle

$\bar{I}_{x'} = \frac{1}{12}bh^3$
$\bar{I}_{y'} = \frac{1}{12}b^3h$
$I_x = \frac{1}{3}bh^3$
$I_y = \frac{1}{3}b^3h$
$J_C = \frac{1}{12}bh(b^2 + h^2)$

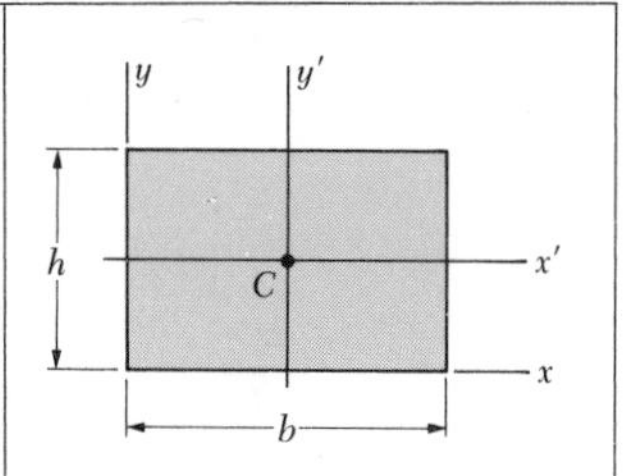

Triangle

$\bar{I}_{x'} = \frac{1}{36}bh^3$
$I_x = \frac{1}{12}bh^3$

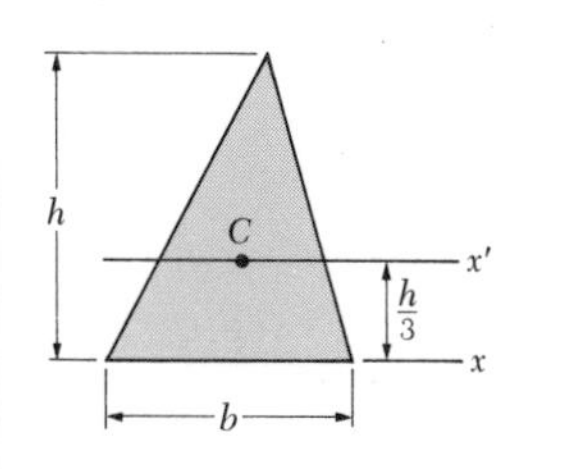

Circle

$\bar{I}_x = \bar{I}_y = \frac{1}{4}\pi r^4$
$J_O = \frac{1}{2}\pi r^4$

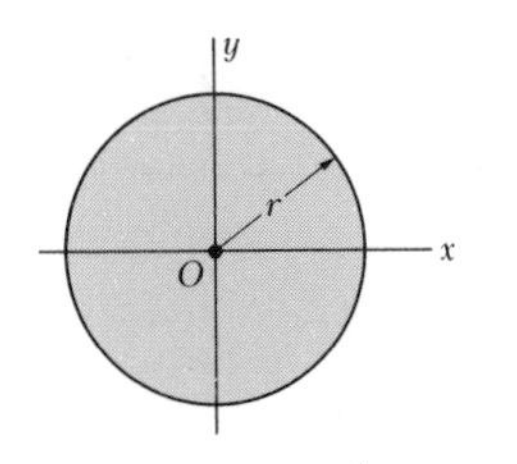

Semicircle

$I_x = I_y = \frac{1}{8}\pi r^4$
$J_O = \frac{1}{4}\pi r^4$

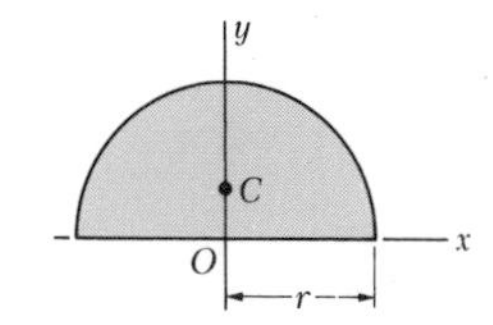

Quarter circle

$I_x = I_y = \frac{1}{16}\pi r^4$
$J_O = \frac{1}{8}\pi r^4$

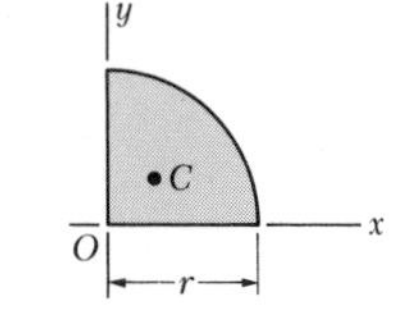

Ellipse

$\bar{I}_x = \frac{1}{4}\pi ab^3$
$\bar{I}_y = \frac{1}{4}\pi a^3b$
$J_O = \frac{1}{4}\pi ab(a^2 + b^2)$

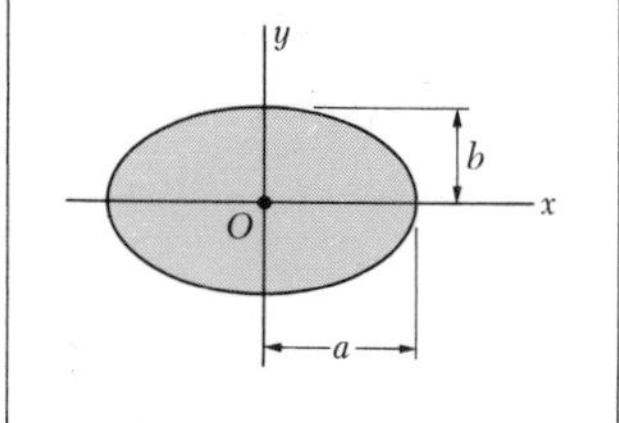

Mass Moments of Inertia of Common Geometric Shapes

Slender rod

$I_y = I_z = \frac{1}{12}mL^2$

Thin rectangular plate

$I_x = \frac{1}{12}m(b^2 + c^2)$
$I_y = \frac{1}{12}mc^2$
$I_z = \frac{1}{12}mb^2$

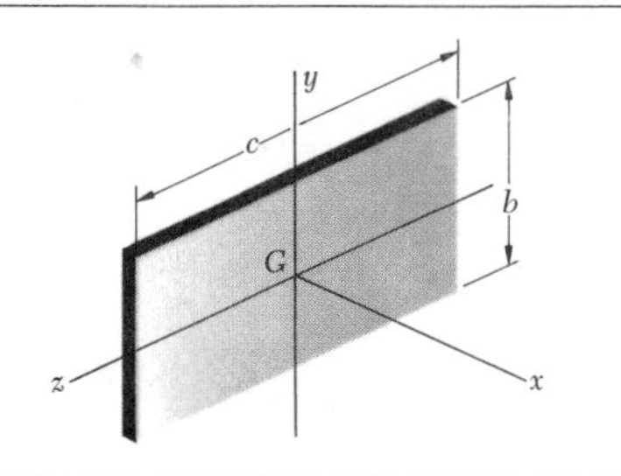

Rectangular prism

$I_x = \frac{1}{12}m(b^2 + c^2)$
$I_y = \frac{1}{12}m(c^2 + a^2)$
$I_z = \frac{1}{12}m(a^2 + b^2)$

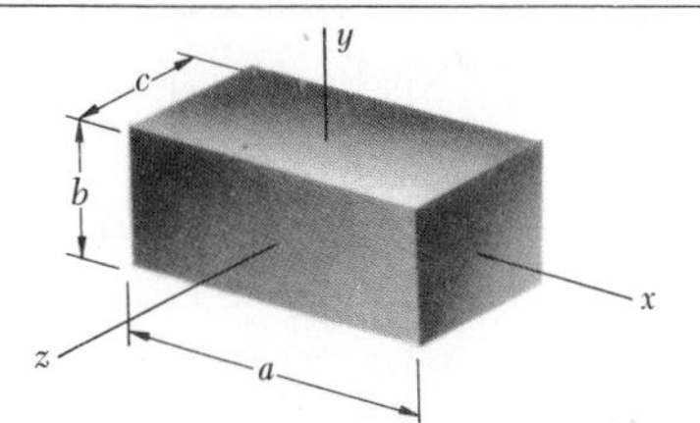

Thin disk

$I_x = \frac{1}{2}mr^2$
$I_y = I_z = \frac{1}{4}mr^2$

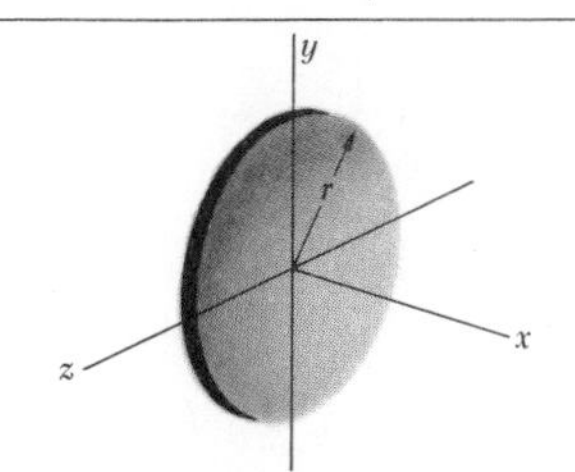

Circular cylinder

$I_x = \frac{1}{2}ma^2$
$I_y = I_z = \frac{1}{12}m(3a^2 + L^2)$

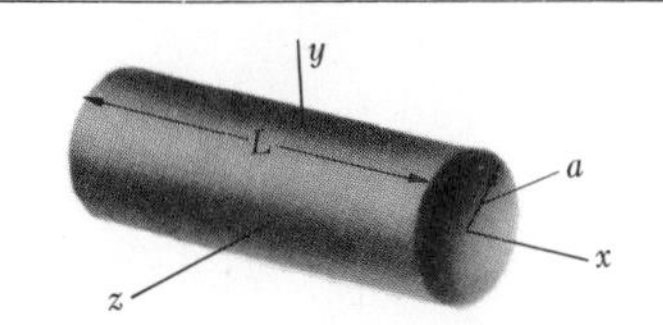

Circular cone

$I_x = \frac{3}{10}ma^2$
$I_y = I_z = \frac{3}{5}m(\frac{1}{4}a^2 + h^2)$

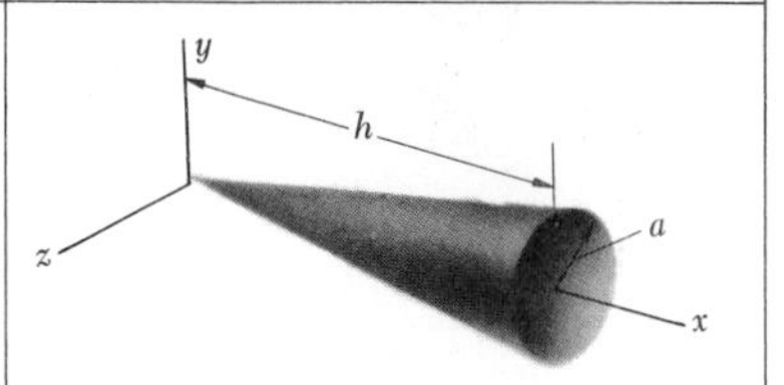

Sphere

$I_x = I_y = I_z = \frac{2}{5}ma^2$

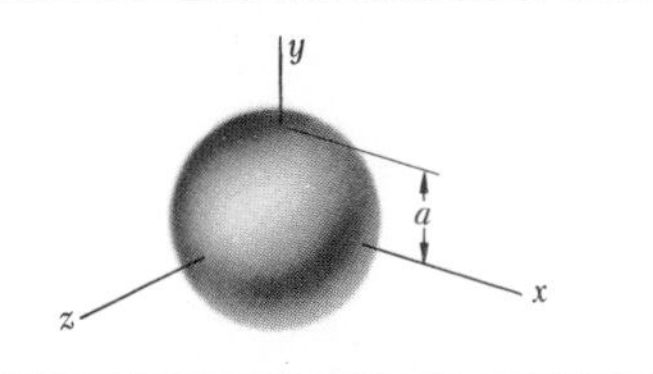

Mechanics for Engineers

DYNAMICS

Mechanics for Engineers

FOURTH EDITION

DYNAMICS

Ferdinand P. Beer
Lehigh University

E. Russell Johnston, Jr.
University of Connecticut

McGraw-Hill Book Company
New York St. Louis San Francisco Auckland Bogotá Hamburg
Johannesburg London Madrid Mexico Milan Montreal New Delhi
Panama Paris São Paulo Singapore Sydney Tokyo Toronto

MECHANICS FOR ENGINEERS: DYNAMICS

1 2 3 4 5 6 7 8 9 0 KGPKGP 8 9 4 3 2 1 0 9 8 7 6

ISBN 0-07-004582-8

This book was set in Laurel by York Graphic Services, Inc. The editors were Anne Duffy and David A. Damstra; the cover was designed by Caliber Design Planning, Inc.; the production supervisor was Leroy A. Young. The drawings were done by Felix Cooper. Arcata Graphics Kingsport was printer and binder.

Library of Congress Cataloging-in-Publication Data

Beer, Ferdinand Pierre.
Mechanics for engineers: dynamics.

1. Mechanics, Applied. 2. Dynamics. I. Johnston, Elwood Russell. II. Title.
TA350.B3544 1987 620.1′054 86-18633
ISBN 0-07-004582-8

Contents

Preface *ix*

List of Symbols *xiii*

11 Kinematics of Particles 435

11.1 Introduction to Dynamics 435

Rectilinear Motion of Particles *436*

11.2 Position, Velocity, and Acceleration 436
11.3 Determination of the Motion of a Particle 440
11.4 Uniform Rectilinear Motion 448
11.5 Uniformly Accelerated Rectilinear Motion 448
11.6 Motion of Several Particles 449
*11.7 Graphical Solution of Rectilinear-Motion Problems 456
*11.8 Other Graphical Methods 457

Curvilinear Motion of Particles *464*

11.9 Position Vector, Velocity, and Acceleration 464
11.10 Component Motions 467
11.11 Relative Motion 468
11.12 Tangential and Normal Components 479
11.13 Radial and Transverse Components 481
11.14 Review and Summary 491

12 Kinetics of Particles: Newton's Second Law 498

12.1 Introduction 498
12.2 Newton's Second Law of Motion 499
12.3 Systems of Units 500
12.4 Equations of Motion. Dynamic Equilibrium 503
12.5 Systems of Particles 504
12.6 Motion of the Mass Center of a System of Particles 506
12.7 Rectilinear Motion of a Particle 507

12.8 Curvilinear Motion of a Particle 515
12.9 Motion under a Central Force 517
12.10 Newton's Law of Gravitation 517
*12.11 Trajectory of a Particle under a Central Force 528
*12.12 Application to Space Mechanics 529
*12.13 Kepler's Laws of Planetary Motion 533
12.14 Review and Summary 538

13 Kinetics of Particles: Work and Energy 546

13.1 Introduction 546
13.2 Work of a Force 547
13.3 Kinetic Energy of a Particle. Principle of Work and Energy 550
13.4 Applications of the Principle of Work and Energy 552
13.5 Systems of Particles 553
13.6 Potential Energy 564
13.7 Conservation of Energy 566
13.8 Power and Efficiency 574
13.9 Review and Summary 578

14 Kinetics of Particles: Impulse and Momentum 585

14.1 Introduction 585
14.2 Principle of Impulse and Momentum 586
14.3 Impulsive Motion 588
14.4 Systems of Particles 589
14.5 Conservation of Momentum 590
14.6 Impact 598
14.7 Direct Central Impact 598
14.8 Oblique Central Impact 601
14.9 Problems Involving Energy and Momentum 611
14.10 Angular Momentum 619
14.11 Conservation of Angular Momentum 620
14.12 Motion under a Conservative Central Force. Application to Space Mechanics 620
14.13 Principle of Impulse and Momentum for a System of Particles 622
*14.14 Variable Systems of Particles 630
*14.15 Steady Stream of Particles 630
*14.16 Systems Gaining or Losing Mass 633
14.17 Review and Summary 645

15 Kinematics of Rigid Bodies 653

15.1 Introduction 653
15.2 Translation 655
15.3 Rotation 656
15.4 Linear and Angular Velocity, Linear and Angular Acceleration in Rotation 658
15.5 General Plane Motion 664
15.6 Absolute and Relative Velocity in Plane Motion 666
15.7 Instantaneous Center of Rotation in Plane Motion 674
15.8 Absolute and Relative Acceleration in Plane Motion 682
*15.9 Analysis of Plane Motion in Terms of a Parameter 684
*15.10 Particle Moving on a Slab in Translation 692
*15.11 Particle Moving on a Rotating Slab. Coriolis Acceleration 693
15.12 Review and Summary 701

16 Kinetics of Rigid Bodies: Forces and Accelerations 709

16.1 Introduction 709
16.2 Plane Motion of a Rigid Body. D'Alembert's Principle 710
16.3 Solution of Problems Involving the Plane Motion of a Rigid Body 714
16.4 Systems of Rigid Bodies 716
16.5 Constrained Plane Motion 733
16.6 Rotation of a Three-Dimensional Body about a Fixed Axis 753
16.7 Review and Summary 762

17 Kinetics of Rigid Bodies: Work and Energy 767

17.1 Introduction 767
17.2 Principle of Work and Energy for a Rigid Body 768
17.3 Work of Forces Acting on a Rigid Body 769
17.4 Kinetic Energy in Translation 770
17.5 Kinetic Energy in Rotation 771
17.6 Systems of Rigid Bodies 771
17.7 Kinetic Energy in Plane Motion 777
17.8 Conservation of Energy 778
17.9 Power 779
17.10 Review and Summary 788

18 Kinetics of Rigid Bodies: Impulse and Momentum 794

18.1 Introduction 794
18.2 Principle of Impulse and Momentum for a Rigid Body 795
18.3 Momentum of a Rigid Body in Plane Motion 796
18.4 Application of the Principle of Impulse and Momentum to the Analysis of the Plane Motion of a Rigid Body 798
18.5 Systems of Rigid Bodies 801
18.6 Conservation of Angular Momentum 801
18.7 Impulsive Motion 812
18.8 Eccentric Impact 812
*18.9 Gyroscopes 824
18.10 Review and Summary 832

19 Mechanical Vibrations 838

19.1 Introduction 838

Vibrations Without Damping 839

19.2 Free Vibrations of Particles. Simple Harmonic Motion 839
19.3 Simple Pendulum (Approximate Solution) 842
*19.4 Simple Pendulum (Exact Solution) 843
19.5 Free Vibrations of Rigid Bodies 849
19.6 Application of the Principle of Conservation of Energy 857
19.7 Forced Vibrations 863

Damped Vibrations 870

*19.8 Damped Free Vibrations 870
*19.9 Damped Forced Vibrations 873
*19.10 Electrical Analogues 874

Appendix Moments of Inertia of Masses 889

Index 905

Answers to Even-numbered Problems 915

Preface

The main objective of a first course in mechanics should be to develop in the engineering student the ability to analyze any problem in a simple and logical manner and to apply to its solution a few, well-understood, basic principles. It is hoped that this text, as well as the preceding volume, *Mechanics for Engineers: Statics,* will help the instructor achieve this goal.†

In this fourth edition, the vectorial character of mechanics has again been emphasized. The concept of vectors and the laws governing the addition and the resolution of vectors were discussed at the beginning of the volume on statics. Throughout the text, forces, velocities, accelerations, and other vector quantities have been clearly distinguished from scalar quantities through the use of **boldface** type. Products and derivatives of vectors, however, are not used in this text.‡

One of the characteristics of the approach used in these volumes is that the mechanics of *particles* has been clearly separated from the mechanics of *rigid bodies.* This approach makes it possible to consider simple practical applications at an early stage and to postpone the introduction of more difficult concepts. In the volume on statics, the statics of particles was treated first, and the principle of equilibrium was immediately applied to practical situations involving only concurrent forces. The statics of rigid bodies was considered later, at which time the principle of transmissibility and the associated concept of moment of a force were introduced. In this volume, the same division is observed. The basic concepts of force, mass, and acceleration, of work and energy, and of impulse and momentum are introduced and first applied to problems involving only particles. Thus the students may familiarize themselves with the three basic methods used in dynamics and learn their respective advantages before facing the difficulties associated with the motion of rigid bodies.

Although the authors strongly believe in the advantage of the approach they use, they do not wish to impose their choice on the instructor. Consequently, the various chapters have been written so that they

†Both texts are also available in a single volume, *Mechanics for Engineers: Statics and Dynamics.*

‡In a parallel text, *Vector Mechanics for Engineers: Dynamics,* vector analysis is used throughout the presentation of dynamics, making it possible to analyze more advanced problems in three-dimensional kinematics and kinetics.

may be taught in alternate sequences. For example, Chap. 15 may be taken immediately after Chap. 11 if the instructor prefers to keep kinematics apart from kinetics, and Chaps. 13 and 14 may be postponed if he or she wishes to teach the method of work and energy (Chaps. 13 and 17) and the method of impulse and momentum (Chaps. 14 and 18) as single units. Whatever the sequence used, the very fact that distinct chapters are devoted to the dynamics of particles and to the dynamics of rigid bodies should help the students organize their own thoughts.

Since this text is designed for a first course in dynamics, new concepts have been presented in simple terms and every step explained in detail. On the other hand, by discussing the broader aspects of the problems considered and by stressing methods of general applicability, a definite maturity of approach has been achieved. For example, the concepts of mass and weight are carefully distinguished, and the limitations of the methods used to analyze plane motion are indicated, so that the students will not be tempted to apply them to cases in which they are not valid.

The fact that mechanics is essentially a *deductive* science based on a few fundamental principles has been stressed. Derivations have been presented in their logical sequence and with all the rigor warranted at this level. However, the learning process being largely *inductive*, simple applications have been considered first. Thus the dynamics of particles precedes the dynamics of rigid bodies, and, in the latter, the emphasis has been placed on the study of plane motion.

Free-body diagrams were introduced early in statics. They were used not only to solve equilibrium problems but also to express the equivalence of two systems of forces or, more generally, of two systems of vectors. The advantage of this approach becomes apparent in the study of the dynamics of rigid bodies. By placing the emphasis on "free-body-diagram equations" rather than on the standard algebraic equations of motion, a more intuitive and more complete understanding of the fundamental pinciples of dynamics may be achieved. This approach, which was first introduced in 1962 in the first edition of *Vector Mechanics for Engineers,* has now gained wide acceptance among mechanics teachers in this country. It is, therefore, used again in preference to the method of dynamic equilibrium in the solution of all sample problems in this edition of *Mechanics for Engineers.*

Color has again been used in this edition to distinguish forces from other elements of the free-body diagrams. This makes it easier for the students to identify the forces acting on a given particle or rigid body and to follow the discussion of sample problems and other examples given in the text.

Because of the current trend among American engineers to adopt the international system of units (SI metric units), the SI units most frequently used in mechanics were introduced in Chap. 1 of *Statics.* They are discussed again in Chap. 12 of this volume and used throughout the text. Approximately half the sample problems and 60 percent of the problems to be assigned have been stated in these units, while the remainder retain U.S. customary units. The authors believe that this approach will best serve the needs of the students, who will be entering the engineering profession during the period of transition from one system of units to the other. It also should be recognized that the passage from one system to the other entails more than the use of conversion factors. Since the SI system of units is an absolute system based on the units of time, length, and mass, whereas the U.S. customary system is a gravitational system based on the units of time, length, and force, different approaches are required for the solution of many problems. For example, when SI units are used, a body is generally specified by its mass expressed in kilograms; in most problems of statics it was necessary to determine the weight of the body in newtons, and an additional calculation was required for this purpose. On the other hand, when U.S. customary units are used, a body is specified by its weight in pounds and, in dynamics problems, an additional calculation will be required to determine its mass in slugs (or $\text{lb} \cdot \text{s}^2/\text{ft}$). The authors, therefore, believe that problem assignments should include both systems of units. The actual distribution of assigned problems between the two systems of units, however, has been left to the instructor, and a sufficient number of problems of each type have been provided so that four complete lists of assignments may be selected with the proportion of problems stated in SI units set anywhere between 50 and 75 percent. If so desired, two complete lists of assignments may also be selected from problems stated in SI units only and two others from problems stated in U.S. customary units.

A number of optional sections have been included. These sections are indicated by asterisks and may thus easily be distinguished from those which form the core of the basic dynamics course. They may be omitted without prejudice to the understanding of the rest of the text. The topics covered in these additional sections include graphical methods for the solution of rectilinear-motion problems, the trajectory of a particle under a central force, the deflection of fluid streams, problems involving jet and rocket propulsion, Coriolis acceleration, gyroscopes, damped mechanical vibrations, and electrical analogues. These topics will be found of particular interest when dynamics is taught in the junior year.

The material presented in the text and most of the problems require no previous mathematical knowledge beyond algebra, trigonometry, and elementary calculus. However, special problems have been included, which make use of a more advanced knowledge of calculus, and certain sections, such as Secs. 19.8 and 19.9 on damped vibrations, should be assigned only if the students possess the proper mathematical background.

Each chapter begins with an introductory section setting the purpose and goals of the chapter and describing in simple terms the material to be covered and its application to the solution of engineering problems. The body of the text has been divided into units, each consisting of one or several theory sections, one or several sample problems, and a large number of problems to be assigned. Each unit corresponds to a well-defined topic and generally may be covered in one lesson. In a number of cases, however, the instructor will find it desirable to devote more than one lesson to a given topic. Each chapter ends with a review section summarizing the material covered in that chapter. Marginal notes have been added to help the students organize their review work, and cross-references have been included to help them find the portions of material requiring their special attention.

The sample problems have been set up in much the same form that students will use in solving the assigned problems. They thus serve the double purpose of amplifying the text and demonstrating the type of neat and orderly work that students should cultivate in their own solutions. Most of the problems to be assigned are of a practical nature and should appeal to engineering students. They are primarily designed, however, to illustrate the material presented in the text and to help students understand the basic principles of mechanics. The problems have been grouped according to the portions of material they illustrate and have been arranged in order of increasing difficulty. Problems requiring special attention have been indicated by asterisks. Answers to all even-numbered problems are given at the end of the book.

The introduction in the engineering curriculum of instruction in computer programming and the increasing availability of personal computers or mainframe terminals on most campuses make it now possible for engineering students to solve a number of challenging dynamics problems. Only a few years ago these problems would have been considered inappropriate for an undergraduate course because of the large number of computations their solutions require. In this new edition of *Mechanics for Engineers: Dynamics*, a group of four problems designed to be solved with a computer has been added to the review problems at the end of each chapter. These problems may involve the determination of the motion of a particle under various initial conditions, the kinematic or kinetic analysis of mechanisms in successive positions, or the numerical integration of various equations of motion. Developing the algorithm required to solve a given dynamics problem will benefit the students in two different ways: (1) it will help them gain a better understanding of the mechanics principles involved; (2) it will provide them with an opportunity to apply the skills acquired in their computer programming course to the solution of a meaningful engineering problem.

The authors wish to acknowledge gratefully the many helpful comments and suggestions offered by the users of the previous editions of *Mechanics for Engineers* and of *Vector Mechanics for Engineers*.

Ferdinand P. Beer
E. Russell Johnston, Jr.

List of Symbols

$\mathbf{a}$, a	Acceleration
a	Constant; radius; distance; semimajor axis of ellipse
$\bar{\mathbf{a}}$, $\bar{a}$	Acceleration of mass center
$\mathbf{a}_{B/A}$	Relative acceleration of B with respect to A
$\mathbf{a}_c$	Coriolis acceleration
$\mathbf{A}, \mathbf{B}, \mathbf{C}, \ldots$	Reactions at supports and connections
$A, B, C, \ldots$	Points
A	Area
b	Width; distance; semiminor axis of ellipse
c	Constant; coefficient of viscous damping
C	Centroid; instantaneous center of rotation; capacitance
d	Distance
e	Coefficient of restitution; base of natural logarithms
E	Total mechanical energy; voltage
f	Frequency; function
$\mathbf{F}$	Force; friction force
g	Acceleration of gravity
G	Center of gravity; mass center; constant of gravitation
h	Angular momentum per unit mass
H_O	Angular momentum about point O
i	Current
$I, I_x, \ldots$	Moment of inertia
$\bar{I}$	Centroidal moment of inertia
$I_{xy}, \ldots$	Product of inertia
J	Polar moment of inertia
k	Spring constant
k_x, k_y, k_O	Radius of gyration
$\bar{k}$	Centroidal radius of gyration
l	Length
L	Length; inductance
m	Mass
m'	Mass per unit length
$\mathbf{M}$	Couple
M_O	Moment about point O
M	Moment; mass of earth
n	Normal direction

$\mathbf{N}$	Normal component of reaction
O	Origin of coordinates
p	Circular frequency
$\mathbf{P}$	Force; vector
q	Mass rate of flow; electric charge
$\mathbf{Q}$	Force; vector
$\mathbf{r}$	Position vector
$\mathbf{r}_{B/A}$	Position vector of B relative to A
r	Radius; distance; polar coordinate
$\mathbf{R}$	Resultant force; resultant vector; reaction
R	Radius of earth; resistance
$\mathbf{s}$	Position vector
s	Length of arc
t	Time; thickness; tangential direction
$\mathbf{T}$	Force
T	Tension; kinetic energy
$\mathbf{u}$	Velocity
u	Variable
U	Work
$\mathbf{v}$, v	Velocity
v	Speed
$\overline{\mathbf{v}}$, $\overline{v}$	Velocity of mass center
$\mathbf{v}_{B/A}$	Relative velocity of B with respect to A
V	Volume; potential energy
w	Load per unit length
$\mathbf{W}$, W	Weight; load
x, y, z	Rectangular coordinates; distances
$\dot{x}$, $\dot{y}$, $\dot{z}$	Time derivatives of coordinates x, y, z
$\overline{x}$, $\overline{y}$, $\overline{z}$	Rectangular coordinates of centroid, center of gravity, or mass center
$\boldsymbol{\alpha}$, α	Angular acceleration
α, β, γ	Angles
γ	Specific weight
δ	Elongation
ε	Eccentricity of conic section or of orbit
η	Efficiency
θ	Angular coordinate; angle; polar coordinate
μ	Coefficient of friction
ρ	Density; radius of curvature
τ	Period; periodic time
ϕ	Angle of friction; phase angle; angle
φ	Phase difference
$\boldsymbol{\omega}$, ω	Angular velocity
ω	Circular frequency of forced vibration
Ω	Rate of precession

Mechanics for Engineers

DYNAMICS

CHAPTER ELEVEN
Kinematics of Particles

11.1. Introduction to Dynamics. Chapters 1 to 10 were devoted to *statics*, i.e., to the analysis of bodies at rest. We shall now begin the study of *dynamics*, which is the part of mechanics that deals with the analysis of bodies in motion.

While the study of statics goes back to the time of the Greek philosophers, the first significant contribution to dynamics was made by Galileo (1564–1642). Galileo's experiments on uniformly accelerated bodies led Newton (1642–1727) to formulate his fundamental laws of motion.

Dynamics is divided into two parts: (1) *Kinematics*, which is the study of the geometry of motion; kinematics is used to relate displacement, velocity, acceleration, and time, without reference to the cause of the motion. (2) *Kinetics*, which is the study of the relation existing between the forces acting on a body, the mass of the body, and the motion of the body; kinetics is used to predict the motion caused by given forces or to determine the forces required to produce a given motion.

Chapters 11 to 14 are devoted to the *dynamics of particles*, and Chap. 11 more particularly to the *kinematics of particles*. The use of the word particles does not imply that we shall restrict our study to that of small corpuscles; it rather indicates that in these first chapters we shall study the motion of bodies—possibly as large as cars, rockets, or airplanes—without regard to their size. By saying that the bodies are analyzed as particles, we mean that only their motion as an entire unit will be considered; any rotation about their own mass center will be neglected. There are cases, however, when such a rotation is not negligible; the bodies may not then be considered as particles. The analysis of such motions will be carried out in later chapters dealing with the *dynamics of rigid bodies*.

In the first part of Chap. 11, we shall analyze the rectilinear motion of a particle; that is, we shall determine at every instant the position, velocity, and acceleration of a particle as it moves along a straight line. After first studying the motion of a particle by general methods of analysis, we shall consider two important particular cases, namely, the uniform motion and the uniformly accelerated motion of a particle (Secs. 11.4 and 11.5). We shall then consider in Sec. 11.6 the simultaneous motion of several particles and introduce the concept of the relative motion of one particle with respect to another. The first part of this chapter concludes with a study of graphical methods of analysis and their application to the solution of various problems involving the rectilinear motion of particles (Secs. 11.7 and 11.8).

In the second part of this chapter, we shall analyze the motion of a particle as it moves along a curved path in a given plane. The position P of the particle at a given time will be defined by the *position vector* $\mathbf{r}$ joining the origin O of the coordinates and point P, and the velocity and acceleration of the particle will be represented, respectively, by the *vectors* $\mathbf{v}$ and $\mathbf{a}$ (Sec. 11.9). Using rectangular components of the vectors involved, we shall analyze the motion of a projectile (Sec. 11.10). In Sec. 11.11, we shall consider the motion of a particle relative to a reference frame in translation. Finally, we shall analyze the curvilinear motion of a particle in terms of components other than rectangular. In Sec. 11.12, we shall introduce the tangential and normal components of the velocity and acceleration of a particle; and in Sec. 11.13, the radial and transverse components of its velocity and acceleration.

RECTILINEAR MOTION OF PARTICLES

11.2. Position, Velocity, and Acceleration. A particle moving along a straight line is said to be in *rectilinear motion*. At any given instant t, the particle will occupy a certain position on the straight line. To define the position P of the particle, we choose a fixed origin O on the straight line and a positive direction along the line. We measure the distance x from O to P and record it with a plus or minus sign, according to whether P is reached from O by moving along the line in the positive or the negative direction. The distance x, with the appropriate sign, completely defines the position of the particle; it is called the *position coordinate* of the particle considered. For example, the position coordinate corresponding to P in Fig. 11.1a is $x = +5$ m, while the coordinate corresponding to P' in Fig. 11.1b is $x' = -2$ m.

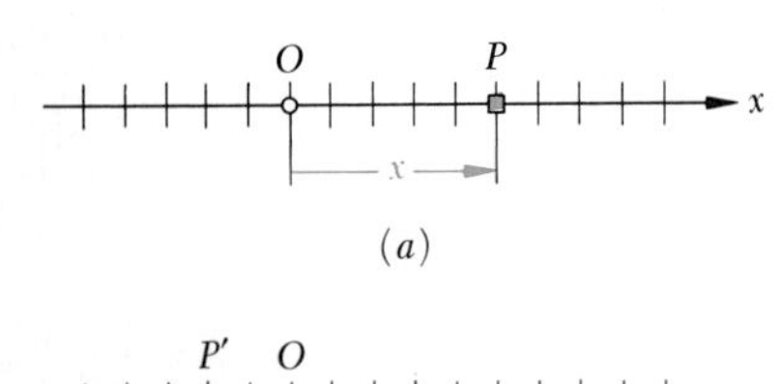

Fig. 11.1

When the position coordinate x of a particle is known for every value of time t, we say that the motion of the particle is known. The "timetable" of the motion may be given in the form of an equation in x and t, such as $x = 6t^2 - t^3$, or in the form of a graph of x versus t as shown in Fig. 11.6. The units most generally used to measure the position coordinate x are the meter (m) in the SI system of units,† and the foot (ft) in the U.S. customary system of units. Time t will generally be measured in seconds (s).

†Cf. Sec. 1.3.

Consider the position P occupied by the particle at time t and the corresponding coordinate x (Fig. 11.2). Consider also the position P' occupied by the particle at a later time $t + \Delta t$; the position coordinate of P' may be obtained by adding to the coordinate x of P the small displacement Δx, which will be positive or negative according to whether P' is to the right or to the left of P. The *average velocity* of the particle over the time interval Δt is defined as the quotient of the displacement Δx and the time interval Δt:

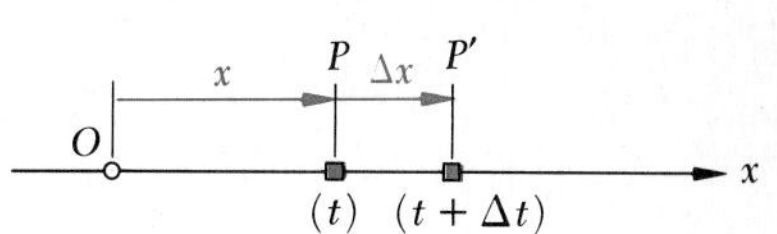

Fig. 11.2

$$\text{Average velocity} = \frac{\Delta x}{\Delta t}$$

If SI units are used, Δx is expressed in meters and Δt in seconds; the average velocity will thus be expressed in meters per second (m/s). If U.S. customary units are used, Δx is expressed in feet and Δt in seconds; the average velocity will then be expressed in feet per second (ft/s).

The *instantaneous velocity* v of the particle at the instant t is obtained from the average velocity by choosing shorter and shorter time intervals Δt and displacements Δx:

$$\text{Instantaneous velocity} = v = \lim_{\Delta t \to 0} \frac{\Delta x}{\Delta t}$$

The instantaneous velocity will also be expressed in m/s or ft/s. Observing that the limit of the quotient is equal, by definition, to the derivative of x with respect to t, we write

$$v = \frac{dx}{dt} \tag{11.1}$$

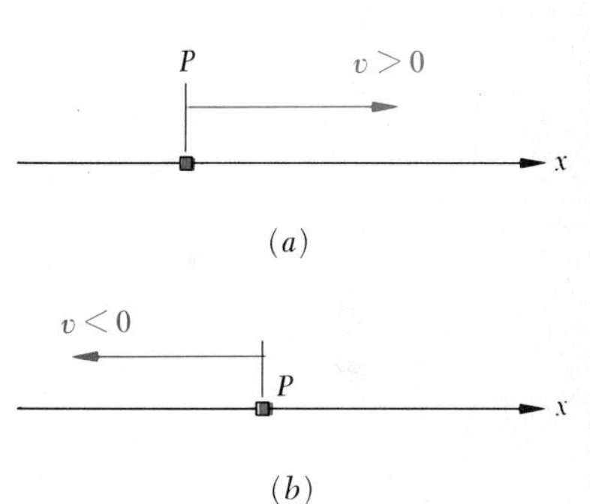

Fig. 11.3

The velocity v is represented by an algebraic number which may be positive or negative.† A positive value of v indicates that x increases, i.e., that the particle moves in the positive direction (Fig. 11.3*a*); a negative value of v indicates that x decreases, i.e., that the particle moves in the negative direction (Fig. 11.3*b*). The magnitude of v is known as the *speed* of the particle.

Consider the velocity v of the particle at time t and also its velocity $v + \Delta v$ at a later time $t + \Delta t$ (Fig. 11.4). The *average acceleration* of the particle over the time interval Δt is defined as the quotient of Δv and Δt:

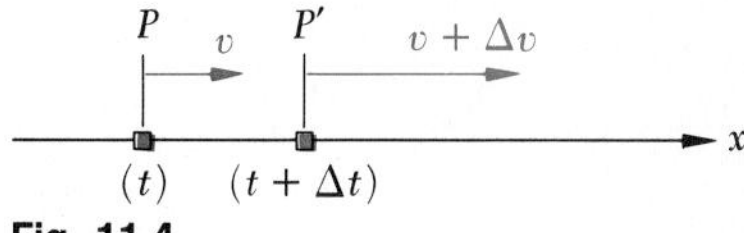

Fig. 11.4

$$\text{Average acceleration} = \frac{\Delta v}{\Delta t}$$

† As we shall see in Sec. 11.9, the velocity is actually a vector quantity. However, since we are considering here the rectilinear motion of a particle, where the velocity of the particle has a known and fixed direction, we need only specify the sense and magnitude of the velocity; this may be conveniently done by using a scalar quantity with a plus or minus sign. The same remark will apply to the acceleration of a particle in rectilinear motion.

If SI units are used, Δv is expressed in m/s and Δt in seconds; the average acceleration will thus be expressed in m/s^2. If U.S. customary units are used, Δv is expressed in ft/s and Δt in seconds; the average acceleration will then be expressed in ft/s^2.

The *instantaneous acceleration* a of the particle at the instant t is obtained from the average acceleration by choosing smaller and smaller values for Δt and Δv:

$$\text{Instantaneous acceleration} = a = \lim_{\Delta t \to 0} \frac{\Delta v}{\Delta t}$$

The instantaneous acceleration will also be expressed in m/s^2 or ft/s^2. The limit of the quotient is by definition the derivative of v with respect to t and measures the rate of change of the velocity. We write

$$a = \frac{dv}{dt} \tag{11.2}$$

or, substituting for v from (11.1),

$$a = \frac{d^2x}{dt^2} \tag{11.3}$$

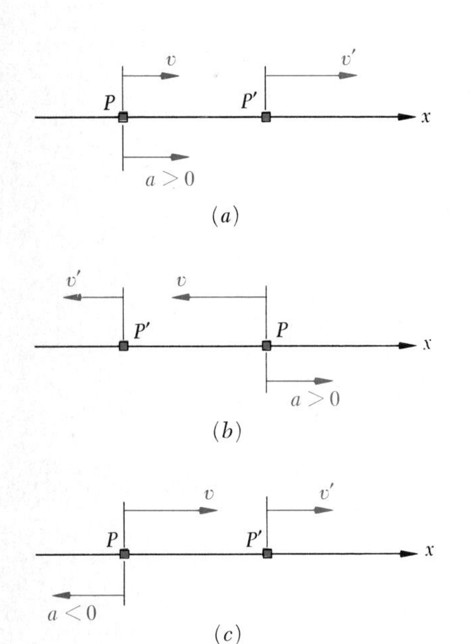

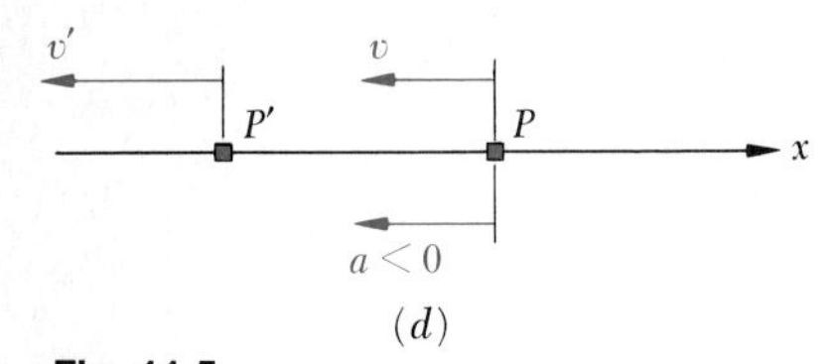

Fig. 11.5

The acceleration a is represented by an algebraic number which may be positive or negative.† A positive value of a indicates that the velocity (i.e., the algebraic number v) increases. This may mean that the particle is moving faster in the positive direction (Fig. 11.5*a*) or that it is moving more slowly in the negative direction (Fig. 11.5*b*); in both cases, Δv is positive. A negative value of a indicates that the velocity decreases; either the particle is moving more slowly in the positive direction (Fig. 11.5*c*) or it is moving faster in the negative direction (Fig. 11.5*d*).

The term *deceleration* is sometimes used to refer to a when the speed of the particle (i.e., the magnitude of v) decreases; the particle is then moving more slowly. For example, the particle of Fig. 11.5 is decelerated in parts b and c, while it is truly accelerated (i.e., moves faster) in parts a and d.

Another expression may be obtained for the acceleration by eliminating the differential dt in Eqs. (11.1) and (11.2). Solving (11.1) for dt, we obtain $dt = dx/v$; carrying into (11.2), we write

$$a = v\frac{dv}{dx} \tag{11.4}$$

† See footnote, page 437.

Example. Consider a particle moving in a straight line, and assume that its position is defined by the equation

$$x = 6t^2 - t^3$$

where t is expressed in seconds and x in meters. The velocity v at any time t is obtained by differentiating x with respect to t:

$$v = \frac{dx}{dt} = 12t - 3t^2$$

The acceleration a is obtained by differentiating again with respect to t:

$$a = \frac{dv}{dt} = 12 - 6t$$

The position coordinate, the velocity, and the acceleration have been plotted against t in Fig. 11.6. The curves obtained are known as *motion curves*. It should be kept in mind, however, that the particle does not move along any of these curves; the particle moves in a straight line. Since the derivative of a function measures the slope of the corresponding curve, the slope of the x–t curve at any given time is equal to the value of v at that time and the slope of the v–t curve is equal to the value of a. Since $a = 0$ at $t = 2$ s, the slope of the v–t curve must be zero at $t = 2$ s; the velocity reaches a maximum at this instant. Also, since $v = 0$ at $t = 0$ and at $t = 4$ s, the tangent to the x–t curve must be horizontal for both these values of t.

A study of the three motion curves of Fig. 11.6 shows that the motion of the particle from $t = 0$ to $t = \infty$ may be divided into four phases:

1. The particle starts from the origin, $x = 0$, with no velocity but with a positive acceleration. Under this acceleration, the particle gains a positive velocity and moves in the positive direction. From $t = 0$ to $t = 2$ s, x, v, and a are all positive.

2. At $t = 2$ s, the acceleration is zero; the velocity has reached its maximum value. From $t = 2$ s to $t = 4$ s, v is positive, but a is negative; the particle still moves in the positive direction but more and more slowly; the particle is decelerated.

3. At $t = 4$ s, the velocity is zero; the position coordinate x has reached its maximum value. From then on, both v and a are negative; the particle is accelerated and moves in the negative direction with increasing speed.

4. At $t = 6$ s, the particle passes through the origin; its coordinate x is then zero, while the total distance traveled since the beginning of the motion is 64 m. For values of t larger than 6 s, x, v, and a will all be negative. The particle keeps moving in the negative direction, away from O, faster and faster.

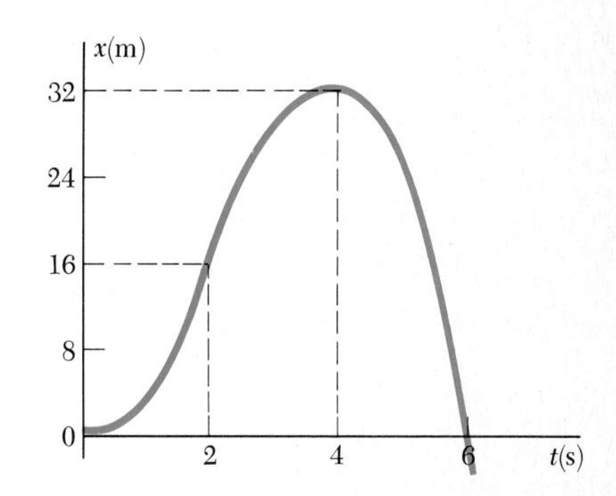

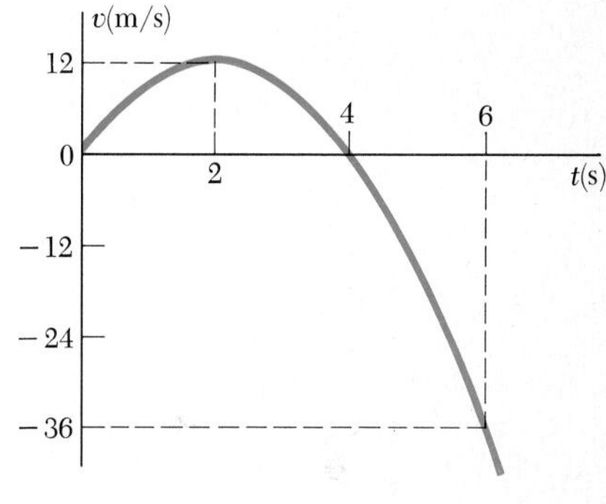

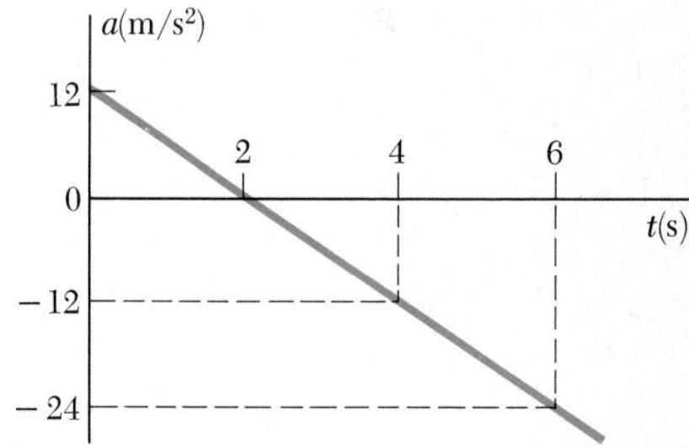

Fig. 11.6

11.3. Determination of the Motion of a Particle. We saw in the preceding section that the motion of a particle is said to be known if the position of the particle is known for every value of the time t. In practice, however, a motion is seldom defined by a relation between x and t. More often, the conditions of the motion will be specified by the type of acceleration that the particle possesses. For example, a freely falling body will have a constant acceleration, directed downward and equal to 9.81 m/s^2, or 32.2 ft/s^2; a mass attached to a spring which has been stretched will have an acceleration proportional to the instantaneous elongation of the spring measured from the equilibrium position; etc. In general, the acceleration of the particle may be expressed as a function of one or more of the variables x, v, and t. In order to determine the position coordinate x in terms of t, it will thus be necessary to perform two successive integrations.

We shall consider three common classes of motion:

1. $a = f(t)$. *The Acceleration Is a Given Function of t.* Solving (11.2) for dv and substituting $f(t)$ for a, we write

$$dv = a\,dt$$
$$dv = f(t)\,dt$$

Integrating both members, we obtain the equation

$$\int dv = \int f(t)\,dt$$

which defines v in terms of t. It should be noted, however, that an arbitrary constant will be introduced as a result of the integration. This is due to the fact that there are many motions which correspond to the given acceleration $a = f(t)$. In order to uniquely define the motion of the particle, it is necessary to specify the *initial conditions* of the motion, i.e., the value v_0 of the velocity and the value x_0 of the position coordinate at $t = 0$. Replacing the indefinite integrals by *definite integrals* with lower limits corresponding to the initial conditions $t = 0$ and $v = v_0$ and upper limits corresponding to $t = t$ and $v = v$, we write

$$\int_{v_0}^{v} dv = \int_0^t f(t)\,dt$$

$$v - v_0 = \int_0^t f(t)\,dt$$

which yields v in terms of t.

We shall now solve (11.1) for dx:

$$dx = v\,dt$$

and substitute for v the expression just obtained. Both members are then integrated, the left-hand member with respect to x from $x = x_0$ to $x = x$, and the right-hand member with respect to t from $t = 0$ to $t = t$. The position coordinate x is thus obtained in terms of t; the motion is completely determined.

Two important particular cases will be studied in greater detail in Secs. 11.4 and 11.5: the case when $a = 0$, corresponding to a *uniform motion*, and the case when $a =$ constant, corresponding to a *uniformly accelerated motion*.

2. $a = f(x)$. *The Acceleration Is a Given Function of* x. Rearranging Eq. (11.4) and substituting $f(x)$ for a, we write

$$v\,dv = a\,dx$$
$$v\,dv = f(x)\,dx$$

Since each member contains only one variable, we may integrate the equation. Denoting again by v_0 and x_0, respectively, the initial values of the velocity and of the position coordinate, we obtain

$$\int_{v_0}^{v} v\,dv = \int_{x_0}^{x} f(x)\,dx$$

$$\tfrac{1}{2}v^2 - \tfrac{1}{2}v_0^2 = \int_{x_0}^{x} f(x)\,dx$$

which yields v in terms of x. We now solve (11.1) for dt:

$$dt = \frac{dx}{v}$$

and substitute for v the expression just obtained. Both members may be integrated, and the desired relation between x and t is obtained.

3. $a = f(v)$. *The Acceleration Is a Given Function of* v. We may then substitute $f(v)$ for a either in (11.2) or in (11.4) to obtain either of the following relations:

$$f(v) = \frac{dv}{dt} \qquad f(v) = v\frac{dv}{dx}$$

$$dt = \frac{dv}{f(v)} \qquad dx = \frac{v\,dv}{f(v)}$$

Integration of the first equation will yield a relation between v and t; integration of the second equation will yield a relation between v and x. Either of these relations may be used in conjunction with Eq. (11.1) to obtain the relation between x and t which characterizes the motion of the particle.

SAMPLE PROBLEM 11.1

The position of a particle which moves along a straight line is defined by the relation $x = t^3 - 6t^2 - 15t + 40$, where x is expressed in feet and t in seconds. Determine (*a*) the time at which the velocity will be zero, (*b*) the position and distance traveled by the particle at that time, (*c*) the acceleration of the particle at that time, (*d*) the distance traveled by the particle from $t = 4$ s to $t = 6$ s.

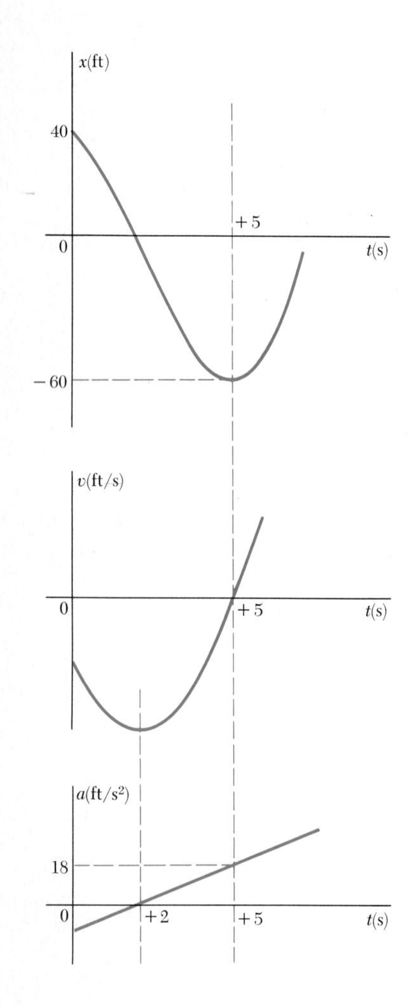

Solution. The equations of motion are

$$x = t^3 - 6t^2 - 15t + 40 \tag{1}$$

$$v = \frac{dx}{dt} = 3t^2 - 12t - 15 \tag{2}$$

$$a = \frac{dv}{dt} = 6t - 12 \tag{3}$$

a. Time at Which $v = 0$. We make $v = 0$ in (2):

$$3t^2 - 12t - 15 = 0 \qquad t = -1 \text{ s} \quad \text{and} \quad t = +5 \text{ s} \blacktriangleleft$$

Only the root $t = +5$ s corresponds to a time after the motion has begun: for $t < 5$ s, $v < 0$, the particle moves in the negative direction; for $t > 5$ s, $v > 0$, the particle moves in the positive direction.

b. Position and Distance Traveled When $v = 0$. Carrying $t = +5$ s into (1), we have

$$x_5 = (5)^3 - 6(5)^2 - 15(5) + 40 \qquad x_5 = -60 \text{ ft} \blacktriangleleft$$

The initial position at $t = 0$ was $x_0 = +40$ ft. Since $v \neq 0$ during the interval $t = 0$ to $t = 5$ s, we have

$$\text{Distance traveled} = x_5 - x_0 = -60 \text{ ft} - 40 \text{ ft} = -100 \text{ ft}$$

$$\text{Distance traveled} = 100 \text{ ft in the negative direction} \blacktriangleleft$$

c. Acceleration When $v = 0$. We carry $t = +5$ s into (3):

$$a_5 = 6(5) - 12 \qquad a_5 = +18 \text{ ft/s}^2 \blacktriangleleft$$

d. Distance Traveled from $t = 4$ s to $t = 6$ s. Since the particle moves in the negative direction from $t = 4$ s to $t = 5$ s and in the positive direction from $t = 5$ s to $t = 6$ s, we shall compute separately the distance traveled during each of these time intervals.

From $t = 4$ s to $t = 5$ s: $\quad x_5 = -60$ ft

$$x_4 = (4)^3 - 6(4)^2 - 15(4) + 40 = -52 \text{ ft}$$

$$\begin{aligned}\text{Distance traveled} &= x_5 - x_4 = -60 \text{ ft} - (-52 \text{ ft}) = -8 \text{ ft} \\ &= 8 \text{ ft in the negative direction}\end{aligned}$$

From $t = 5$ s to $t = 6$ s: $\quad x_5 = -60$ ft

$$x_6 = (6)^3 - 6(6)^2 - 15(6) + 40 = -50 \text{ ft}$$

$$\begin{aligned}\text{Distance traveled} &= x_6 - x_5 = -50 \text{ ft} - (-60 \text{ ft}) = +10 \text{ ft} \\ &= 10 \text{ ft in the positive direction}\end{aligned}$$

Total distance traveled from $t = 4$ s to $t = 6$ s is

$$8 \text{ ft} + 10 \text{ ft} = 18 \text{ ft} \blacktriangleleft$$

SAMPLE PROBLEM 11.2

A ball is tossed with a velocity of 10 m/s directed vertically upward from a window located 20 m above the ground. Knowing that the acceleration of the ball is constant and equal to 9.81 m/s² downward, determine (*a*) the velocity v and elevation y of the ball above the ground at any time t, (*b*) the highest elevation reached by the ball and the corresponding value of t, (*c*) the time when the ball will hit the ground and the corresponding velocity. Draw the v–t and y–t curves.

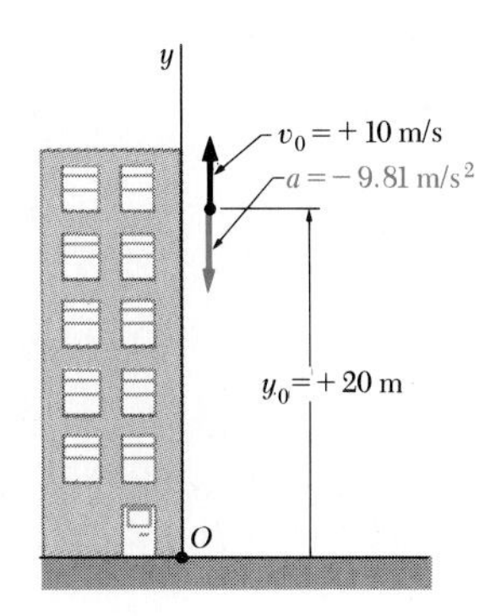

a. Velocity and Elevation. The y axis measuring the position coordinate (or elevation) is chosen with its origin O on the ground and its positive sense upward. The value of the acceleration and the initial values of v and y are as indicated. Substituting for a in $a = dv/dt$ and noting that at $t = 0$, $v_0 = +10$ m/s, we have

$$\frac{dv}{dt} = a = -9.81 \text{ m/s}^2$$

$$\int_{v_0=10}^{v} dv = -\int_0^t 9.81\, dt$$

$$[v]_{10}^{v} = -[9.81t]_0^t$$

$$v - 10 = -9.81t$$

$$v = 10 - 9.81t \quad (1)$$

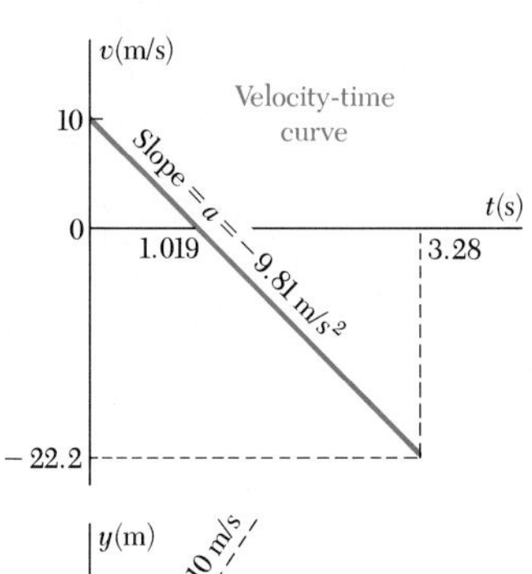

Substituting for v in $v = dy/dt$ and noting that at $t = 0$, $y_0 = 20$ m, we have

$$\frac{dy}{dt} = v = 10 - 9.81t$$

$$\int_{y_0=20}^{y} dy = \int_0^t (10 - 9.81t)\, dt$$

$$[y]_{20}^{y} = [10t - 4.905t^2]_0^t$$

$$y - 20 = 10t - 4.905t^2$$

$$y = 20 + 10t - 4.905t^2 \quad (2)$$

b. Highest Elevation. When the ball reaches its highest elevation, we have $v = 0$. Substituting into (1), we obtain

$$10 - 9.81t = 0 \qquad t = 1.019 \text{ s}$$

Carrying $t = 1.019$ s into (2), we have

$$y = 20 + 10(1.019) - 4.905(1.019)^2 \qquad y = 25.1 \text{ m}$$

c. Ball Hits the Ground. When the ball hits the ground, we have $y = 0$. Substituting into (2), we obtain

$$20 + 10t - 4.905t^2 = 0 \qquad t = -1.243 \text{ s} \quad \text{and} \quad t = +3.28 \text{ s}$$

Only the root $t = +3.28$ s corresponds to a time after the motion has begun. Carrying this value of t into (1), we have

$$v = 10 - 9.81(3.28) = -22.2 \text{ m/s} \qquad v = 22.2 \text{ m/s} \downarrow$$

SAMPLE PROBLEM 11.3

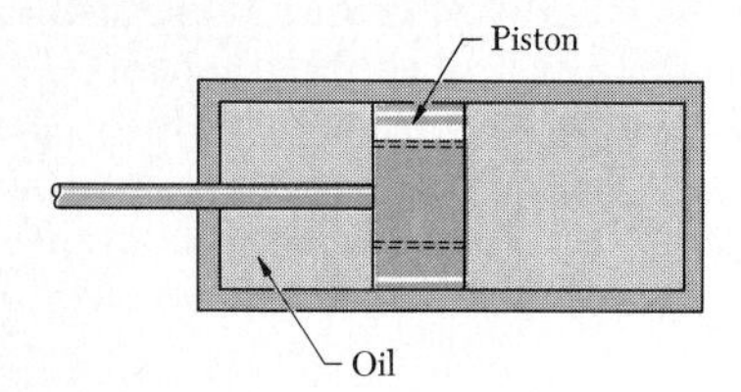

The brake mechanism used to reduce recoil in certain types of guns consists essentially of a piston which is attached to the barrel and may move in a fixed cylinder filled with oil. As the barrel recoils with an initial velocity v_0, the piston moves and oil is forced through orifices in the piston, causing the piston and the barrel to decelerate at a rate proportional to their velocity, that is, $a = -kv$. Express (*a*) v in terms of t, (*b*) x in terms of t, (*c*) v in terms of x. Draw the corresponding motion curves.

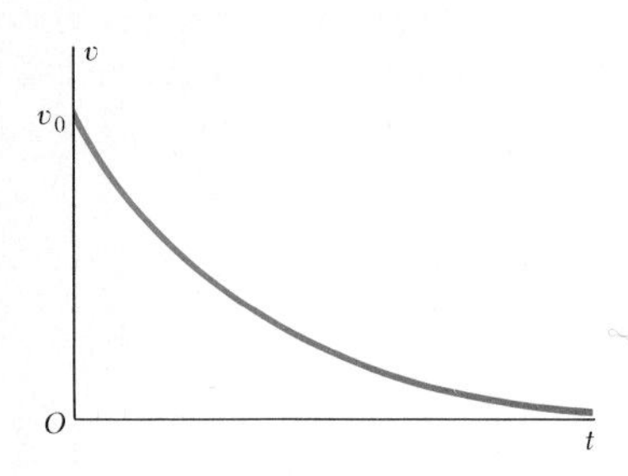

a. v in Terms of t. Substituting $-kv$ for a in the fundamental formula defining acceleration, $a = dv/dt$, we write

$$-kv = \frac{dv}{dt} \qquad \frac{dv}{v} = -k\,dt \qquad \int_{v_0}^{v} \frac{dv}{v} = -k\int_0^t dt$$

$$\ln\frac{v}{v_0} = -kt \qquad\qquad v = v_0 e^{-kt} \blacktriangleleft$$

b. x in Terms of t. Substituting the expression just obtained for v into $v = dx/dt$, we write

$$v_0 e^{-kt} = \frac{dx}{dt}$$

$$\int_0^x dx = v_0\int_0^t e^{-kt}\,dt$$

$$x = -\frac{v_0}{k}[e^{-kt}]_0^t = -\frac{v_0}{k}(e^{-kt} - 1)$$

$$x = \frac{v_0}{k}(1 - e^{-kt}) \blacktriangleleft$$

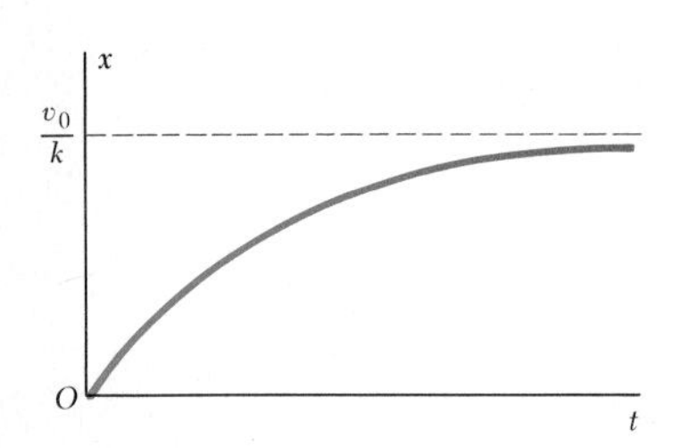

c. v in Terms of x. Substituting $-kv$ for a in $a = v\,dv/dx$, we write

$$-kv = v\frac{dv}{dx}$$

$$dv = -k\,dx$$

$$\int_{v_0}^{v} dv = -k\int_0^x dx$$

$$v - v_0 = -kx \qquad\qquad v = v_0 - kx \blacktriangleleft$$

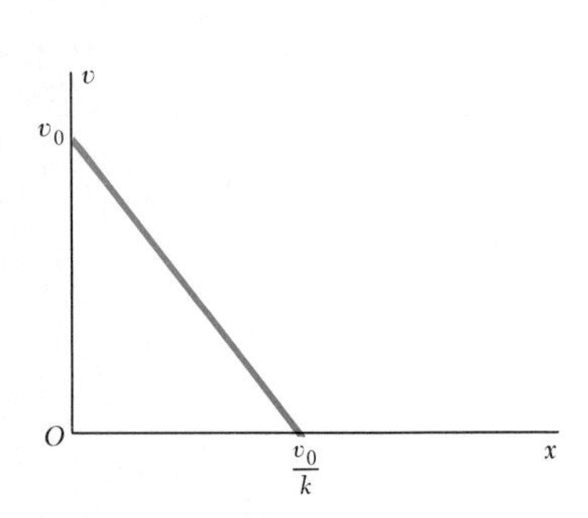

Check. Part *c* could have been solved by eliminating t from the answers obtained for parts *a* and *b*. This alternate method may be used as a check. From part *a* we obtain $e^{-kt} = v/v_0$; substituting in the answer of part *b*, we obtain

$$x = \frac{v_0}{k}(1 - e^{-kt}) = \frac{v_0}{k}\left(1 - \frac{v}{v_0}\right) \qquad v = v_0 - kx \qquad \text{(checks)}$$

Problems

11.1 The motion of a particle is defined by the relation $x = 2t^3 - 15t^2 + 36t - 10$, where x is expressed in meters and t in seconds. Determine the position, velocity, and acceleration when $t = 4$ s.

11.2 The motion of a particle is defined by the relation $x = t^4 - 3t^3 + t$, where x is expressed in inches and t in seconds. Determine the position, velocity, and acceleration when $t = 3$ s.

11.3 The motion of a particle is defined by the relation $x = t^3 - 3t^2 + 6$, where x is expressed in feet and t in seconds. Determine the time, position, and acceleration when $v = 0$.

11.4 The motion of a particle is defined by the relation $x = t^3 - 9t^2 + 15t + 18$, where x is expressed in meters and t in seconds. Determine the time, position, and acceleration when $v = 0$.

11.5 The motion of a particle is defined by the relation $x = t^3 - 12t^2 + 45t - 20$, where x is expressed in meters and t in seconds. Determine (*a*) when the velocity is zero, (*b*) the position, acceleration, and total distance traveled when $t = 5$ s.

11.6 The motion of a particle is defined by the relation $x = 2t^3 - 15t^2 + 24t + 4$, where x is expressed in meters and t in seconds. Determine (*a*) when the velocity is zero, (*b*) the position and the total distance traveled when the acceleration is zero.

11.7 The acceleration of a particle is defined by the relation $a = -5$ ft/s^2. If $v = +30$ ft/s and $x = 0$ when $t = 0$, determine the velocity, position, and total distance traveled when $t = 8$ s.

11.8 The acceleration of a particle is directly proportional to the time t. At $t = 0$, the velocity of the particle is $v = -9$ m/s. Knowing that $v = 0$ and $x = 12$ m when $t = 3$ s, write the equations of motion of the particle.

11.9 The acceleration of a particle is defined by the relation $a = 32 - 6t^2$. The particle starts at $t = 0$ with $v = 0$ and $x = 50$ m. Determine (*a*) the time when the velocity is again zero, (*b*) the position and velocity when $t = 6$ s, (*c*) the total distance traveled by the particle from $t = 0$ to $t = 6$ s.

11.10 The acceleration of a particle is defined by the relation $a = kt^2$. (*a*) Knowing that $v = -24$ ft/s when $t = 0$ and that $v = +40$ ft/s when $t = 4$ s, determine the constant k. (*b*) Write the equations of motion, knowing also that $x = 6$ ft when $t = 2$ s.

11.11 The acceleration of a particle is defined by the relation $a = -kx^{-2}$. The particle starts with no initial velocity at $x = 900$ mm, and it is observed that its velocity is 10 m/s when $x = 300$ mm. Determine (*a*) the value of k, (*b*) the velocity of the particle when $x = 500$ mm.

11.12 The acceleration of a particle is defined by the relation $a = -k/x$. It has been experimentally determined that $v = 4$ m/s when $x = 250$ mm and that $v = 3$ m/s when $x = 500$ mm. Determine (*a*) the velocity of the particle when $x = 750$ mm, (*b*) the position of the particle at which its velocity is zero.

11.13 The acceleration of an oscillating particle is defined by the relation $a = -kx$. Find the value of k such that $v = 24$ in./s when $x = 0$ and $x = 6$ in. when $v = 0$.

11.14 The acceleration of a particle is defined by the relation $a = 90 - 6x^2$, where a is expressed in in./s^2 and x in inches. The particle starts with no initial velocity at the position $x = 0$. Determine (*a*) the velocity when $x = 5$ in., (*b*) the position where the velocity is again zero, (*c*) the position where the velocity is maximum.

11.15 The acceleration of a particle is defined by the relation $a = -16x(1 + kx^2)$, where a is expressed in m/s^2 and x in meters. Knowing that $v = 20$ m/s when $x = 0$, determine the velocity when $x = 4$ m, for (*a*) $k = 0$, (*b*) $k = 0.002$, (*c*) $k = -0.002$.

11.16 The acceleration of a particle is defined by the relation $a = -60x^{-1.5}$, where a is expressed in m/s^2 and x in meters. Knowing that the particle starts with no initial velocity at $x = 4$ m, determine the velocity of the particle when (*a*) $x = 2$ m, (*b*) $x = 1$ m, (*c*) $x = 100$ mm.

11.17 The acceleration of a particle is defined by the relation $a = -3v$, where a is expressed in in./s^2 and v in in./s. Knowing that at $t = 0$ the velocity is 60 in./s, determine (*a*) the distance the particle will travel before coming to rest, (*b*) the time required for the particle to come to rest, (*c*) the time required for the velocity of the particle to be reduced to 1 percent of its initial value.

11.18 The acceleration of a particle is defined by the relation $a = -kv^2$, where a is expressed in ft/s^2 and v in ft/s. The particle starts at $x = 0$ with a velocity of 20 ft/s and when $x = 100$ ft the velocity is found to be 15 ft/s. Determine the distance the particle will travel (*a*) before its velocity drops to 10 ft/s, (*b*) before it comes to rest.

11.19 Solve Prob. 11.18, assuming that the acceleration of the particle is defined by the relation $a = -kv^3$.

11.20 The magnitude in m/s^2 of the deceleration due to air resistance of the nose cone of a small experimental rocket is known to be $0.0005v^2$, where v is expressed in m/s. If the nose cone is projected vertically from the ground with an initial velocity of 100 m/s, determine the maximum height that it will reach. [*Hint.* The total acceleration is $-(g + 0.0005v^2)$, where $g = 9.81$ m/s^2.]

11.21 Determine the velocity of the nose cone of Prob. 11.20 when it returns to the ground. (*Hint.* The total acceleration is $g - 0.0005v^2$, where $g = 9.81$ m/s^2.)

11.22 The acceleration of a particle is defined by the relation $a = -kv^{1.5}$. The particle starts at $t = 0$ and $x = 0$ with an initial velocity v_0. (*a*) Show that the velocity and position coordinate at any time t are related by the equation $x/t = \sqrt{v_0 v}$. (*b*) Knowing that for $v_0 = 36$ m/s the particle comes to rest after traveling 3 m, determine the velocity of the particle and the time when $x = 2$ m.

11.23 The velocity of a particle is defined by the relation $v = 40 - 0.2x$, where v is expressed in m/s and x in meters. Knowing that $x = 0$ at $t = 0$, determine (*a*) the distance traveled before the particle comes to rest, (*b*) the acceleration at $t = 0$, (*c*) the time when $x = 50$ m.

11.24 A projectile enters a resisting medium at $x = 0$ with an initial velocity $v_0 = 900$ ft/s and travels 4 in. before coming to rest. Assuming that the velocity of the projectile was defined by the relation $v = v_0 - kx$, where v is expressed in ft/s and x in feet, determine (*a*) the initial acceleration of the projectile, (*b*) the time required for the projectile to penetrate 3.9 in. into the resisting medium.

11.25 The acceleration due to gravity at an altitude y above the surface of the earth may be expressed as

$$a = \frac{-9.81}{\left(1 + \dfrac{y}{6.37 \times 10^6}\right)^2}$$

where a is measured in m/s^2 and y in meters. Using this expression, compute the height reached by a bullet fired vertically upward from the surface of the earth with the following initial velocities: (*a*) 300 m/s, (*b*) 3000 m/s, (*c*) 11.18 km/s.

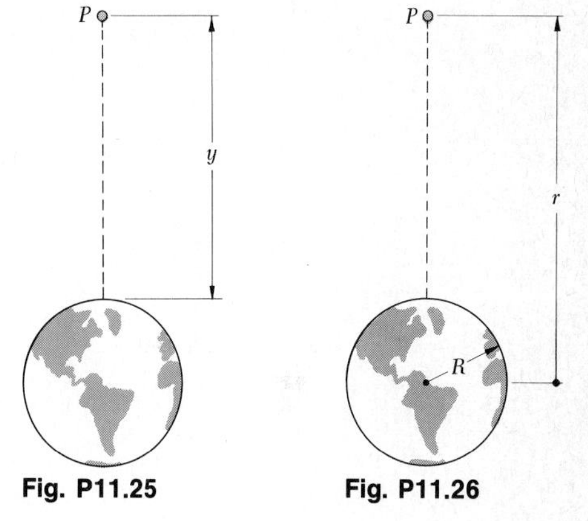

Fig. P11.25 **Fig. P11.26**

11.26 The acceleration due to gravity of a particle falling toward the earth is $a = -gR^2/r^2$, where r is the distance from the *center* of the earth to the particle, R is the radius of the earth, and g is the acceleration due to gravity at the surface of the earth. Derive an expression for the *escape velocity*, i.e., for the minimum velocity with which a particle should be projected vertically upward from the surface of the earth if it is not to return to the earth. (*Hint.* $v = 0$ for $r = \infty$.)

11.27 The acceleration of a particle is $a = k \sin(\pi t/T)$. Knowing that both the velocity and the position coordinate of the particle are zero when $t = 0$, determine (*a*) the equations of motion, (*b*) the maximum velocity, (*c*) the position at $t = 2T$, (*d*) the average velocity during the interval $t = 0$ to $t = 2T$.

11.28 The position of an oscillating particle is defined by the relation $x = A \sin(pt + \phi)$. Denoting the velocity and position coordinate when $t = 0$ by v_0 and x_0, respectively, show (*a*) that $\tan\phi = x_0 p/v_0$, (*b*) that the maximum value of the position coordinate is

$$A = \sqrt{x_0^2 + \left(\frac{v_0}{p}\right)^2}$$

11.4. Uniform Rectilinear Motion. Uniform rectilinear motion is a type of straight-line motion which is frequently encountered in practical applications. In this motion, the acceleration a of the particle is zero for every value of t. The velocity v is therefore constant, and Eq. (11.1) becomes

$$\frac{dx}{dt} = v = \text{constant}$$

The position coordinate x is obtained by integrating this equation. Denoting by x_0 the initial value of x, we write

$$\int_{x_0}^{x} dx = v \int_{0}^{t} dt$$
$$x - x_0 = vt$$

$$x = x_0 + vt \tag{11.5}$$

This equation may be used *only if the velocity of the particle is known to be constant.*

11.5. Uniformly Accelerated Rectilinear Motion. Uniformly accelerated rectilinear motion is another common type of motion. In this motion, the acceleration a of the particle is constant, and Eq. (11.2) becomes

$$\frac{dv}{dt} = a = \text{constant}$$

The velocity v of the particle is obtained by integrating this equation:

$$\int_{v_0}^{v} dv = a \int_{0}^{t} dt$$
$$v - v_0 = at$$

$$v = v_0 + at \tag{11.6}$$

where v_0 is the initial velocity. Substituting for v into (11.1), we write

$$\frac{dx}{dt} = v_0 + at$$

Denoting by x_0 the initial value of x and integrating, we have

$$\int_{x_0}^{x} dx = \int_{0}^{t} (v_0 + at)\, dt$$
$$x - x_0 = v_0 t + \tfrac{1}{2}at^2$$

$$x = x_0 + v_0 t + \tfrac{1}{2}at^2 \tag{11.7}$$

We may also use Eq. (11.4) and write

$$v\frac{dv}{dx} = a = \text{constant}$$
$$v\, dv = a\, dx$$

Integrating both sides, we obtain

$$\int_{v_0}^{v} v\,dv = a\int_{x_0}^{x} dx$$

$$\tfrac{1}{2}(v^2 - v_0^2) = a(x - x_0)$$

$$v^2 = v_0^2 + 2a(x - x_0) \tag{11.8}$$

The three equations we have derived provide useful relations among position coordinate, velocity, and time in the case of a uniformly accelerated motion, as soon as appropriate values have been substituted for a, v_0, and x_0. The origin O of the x axis should first be defined and a positive direction chosen along the axis; this direction will be used to determine the signs of a, v_0, and x_0. Equation (11.6) relates v and t and should be used when the value of v corresponding to a given value of t is desired, or inversely. Equation (11.7) relates x and t; Eq. (11.8) relates v and x. An important application of uniformly accelerated motion is the motion of a *freely falling body*. The acceleration of a freely falling body (usually denoted by g) is equal to 9.81 m/s^2, or 32.2 ft/s^2.

It is important to keep in mind that the three equations above may be used *only when the acceleration of the particle is known to be constant.* If the acceleration of the particle is variable, its motion should be determined from the fundamental equations (11.1) to (11.4), according to the methods outlined in Sec. 11.3.

11.6. Motion of Several Particles. When several particles move independently along the same line, independent equations of motion may be written for each particle. Whenever possible, time should be recorded from the same initial instant for all particles, and displacements should be measured from the same origin and in the same direction. In other words, a single clock and a single measuring tape should be used.

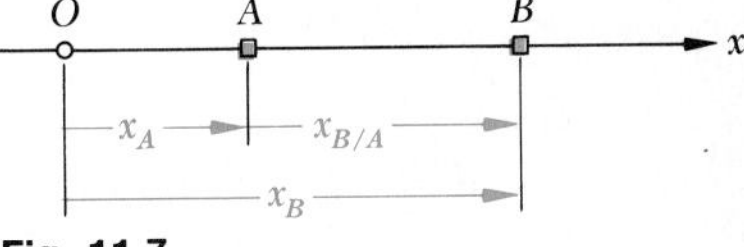

Fig. 11.7

Relative Motion of Two Particles. Consider two particles A and B moving along the same straight line (Fig. 11.7). If the position coordinates x_A and x_B are measured from the same origin, the difference $x_B - x_A$ defines the *relative position coordinate of B with respect to* A and is denoted by $x_{B/A}$. We write

$$x_{B/A} = x_B - x_A \quad \text{or} \quad x_B = x_A + x_{B/A} \tag{11.9}$$

A positive sign for $x_{B/A}$ means that B is to the right of A; a negative sign means that B is to the left of A, regardless of the position of A and B with respect to the origin.

The rate of change of $x_{B/A}$ is known as the *relative velocity of B with respect to* A and is denoted by $v_{B/A}$. Differentiating (11.9), we write

$$v_{B/A} = v_B - v_A \quad \text{or} \quad v_B = v_A + v_{B/A} \tag{11.10}$$

A positive sign for $v_{B/A}$ means that B is *observed from* A to move in the positive direction; a negative sign, that it is observed to move in the negative direction.

The rate of change of $v_{B/A}$ is known as the *relative acceleration of B with respect to A* and is denoted by $a_{B/A}$. Differentiating (11.10), we obtain†

$$a_{B/A} = a_B - a_A \qquad \text{or} \qquad a_B = a_A + a_{B/A} \tag{11.11}$$

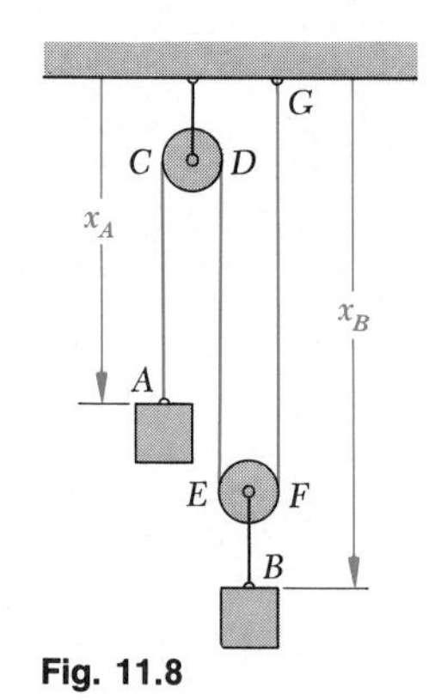

Fig. 11.8

Dependent Motions. Sometimes, the position of a particle will depend upon the position of another or of several other particles. The motions are then said to be dependent. For example, the position of block B in Fig. 11.8 depends upon the position of block A. Since the rope $ACDEFG$ is of constant length, and since the lengths of the portions of rope CD and EF wrapped around the pulleys remain constant, it follows that the sum of the lengths of the segments AC, DE, and FG is constant. Observing that the length of the segment AC differs from x_A only by a constant, and that, similarly, the lengths of the segments DE and FG differ from x_B only by a constant, we write

$$x_A + 2x_B = \text{constant}$$

Since only one of the two coordinates x_A and x_B may be chosen arbitrarily, we say that the system shown in Fig. 11.8 has *one degree of freedom.* From the relation between the position coordinates x_A and x_B, it follows that if x_A is given an increment Δx_A, that is, if block A is lowered by an amount Δx_A, the coordinate x_B will receive an increment $\Delta x_B = -\frac{1}{2}\Delta x_A$; that is, block B will rise by half the same amount; this may easily be checked directly from Fig. 11.8.

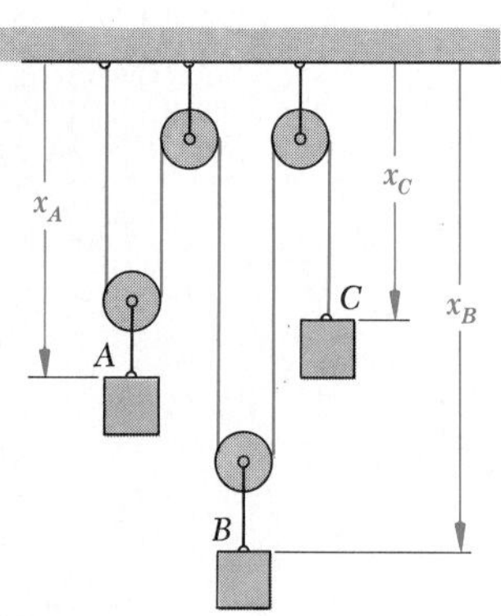

Fig. 11.9

In the case of the three blocks of Fig. 11.9, we may again observe that the length of the rope which passes over the pulleys is constant, and thus that the following relation must be satisfied by the position coordinates of the three blocks:

$$2x_A + 2x_B + x_C = \text{constant}$$

Since two of the coordinates may be chosen arbitrarily, we say that the system shown in Fig. 11.9 has *two degrees of freedom.*

When the relation existing between the position coordinates of several particles is *linear,* a similar relation holds between the velocities and between the accelerations of the particles. In the case of the blocks of Fig. 11.9, for instance, we differentiate twice the equation obtained and write

$$2\frac{dx_A}{dt} + 2\frac{dx_B}{dt} + \frac{dx_C}{dt} = 0 \qquad \text{or} \qquad 2v_A + 2v_B + v_C = 0$$

$$2\frac{dv_A}{dt} + 2\frac{dv_B}{dt} + \frac{dv_C}{dt} = 0 \qquad \text{or} \qquad 2a_A + 2a_B + a_C = 0$$

†Note that the product of the subscripts A and B/A used in the right-hand member of Eqs. (11.9), (11.10), and (11.11) is equal to the subscript B used in their left-hand member.

SAMPLE PROBLEM 11.4

A ball is thrown vertically upward from the 12-m level in an elevator shaft, with an initial velocity of 18 m/s. At the same instant an open-platform elevator passes the 5-m level, moving upward with a constant velocity of 2 m/s. Determine (*a*) when and where the ball will hit the elevator, (*b*) the relative velocity of the ball with respect to the elevator when the ball hits the elevator.

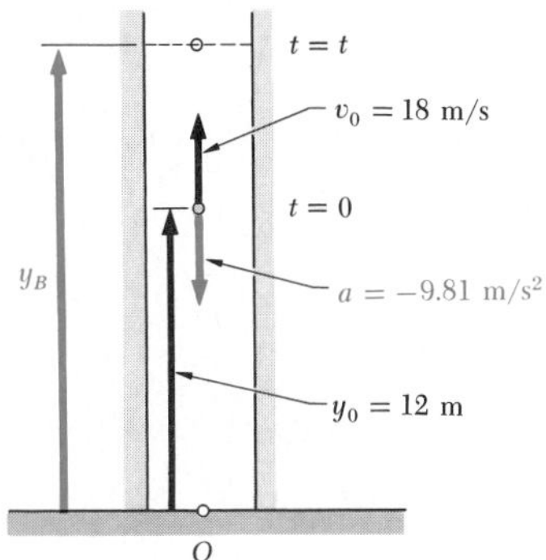

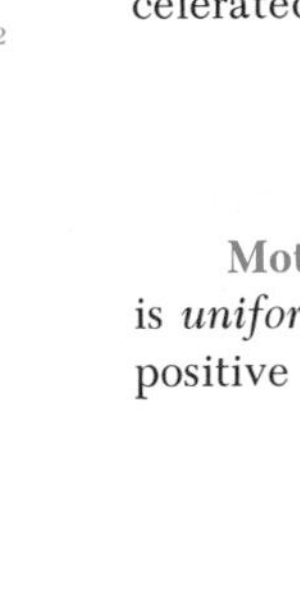

Motion of Ball. Since the ball has a constant acceleration, its motion is *uniformly accelerated.* Placing the origin O of the y axis at ground level and choosing its positive direction upward, we find that the initial position is $y_0 = +12$ m, the initial velocity is $v_0 = +18$ m/s, and the acceleration is $a = -9.81$ m/s². Substituting these values in the equations for uniformly accelerated motion, we write

$$v_B = v_0 + at \qquad v_B = 18 - 9.81t \tag{1}$$

$$y_B = y_0 + v_0 t + \tfrac{1}{2}at^2 \qquad y_B = 12 + 18t - 4.905t^2 \tag{2}$$

Motion of Elevator. Since the elevator has a constant velocity, its motion is *uniform.* Again placing the origin O at the ground level and choosing the positive direction upward, we note that $y_0 = +5$ m and write

$$v_E = +2 \text{ m/s} \tag{3}$$

$$y_E = y_0 + v_E t \qquad y_E = 5 + 2t \tag{4}$$

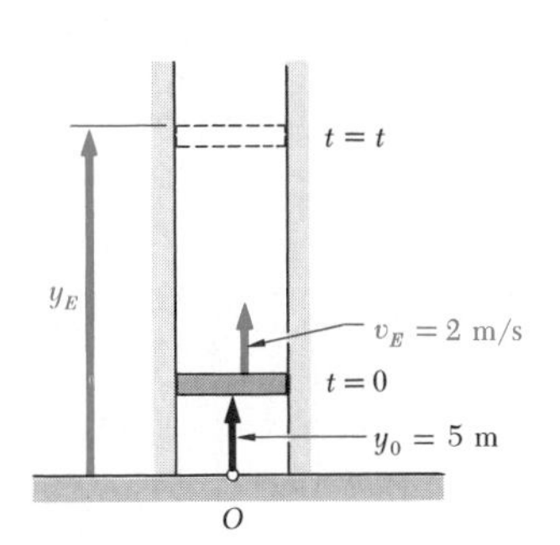

Ball Hits Elevator. We first note that the same time t and the same origin O were used in writing the equations of motion of both the ball and the elevator. We see from the figure that when the ball hits the elevator,

$$y_E = y_B \tag{5}$$

Substituting for y_E and y_B from (2) and (4) into (5), we have

$$5 + 2t = 12 + 18t - 4.905t^2$$

$$t = -0.39 \text{ s} \qquad \text{and} \qquad t = 3.65 \text{ s} \blacktriangleleft$$

Only the root $t = 3.65$ s corresponds to a time after the motion has begun. Substituting this value into (4), we have

$$y_E = 5 + 2(3.65) = 12.30 \text{ m}$$

Elevation from ground = 12.30 m ◀

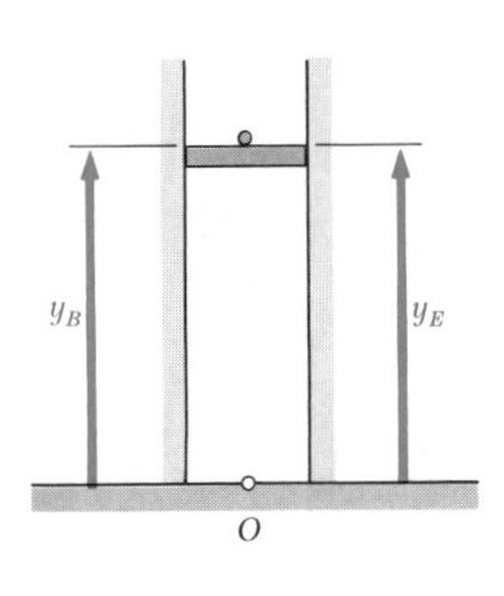

The relative velocity of the ball with respect to the elevator is

$$v_{B/E} = v_B - v_E = (18 - 9.81t) - 2 = 16 - 9.81t$$

When the ball hits the elevator at time $t = 3.65$ s, we have

$$v_{B/E} = 16 - 9.81(3.65) \qquad v_{B/E} = -19.81 \text{ m/s} \blacktriangleleft$$

The negative sign means that the ball is observed from the elevator to be moving in the negative sense (downward).

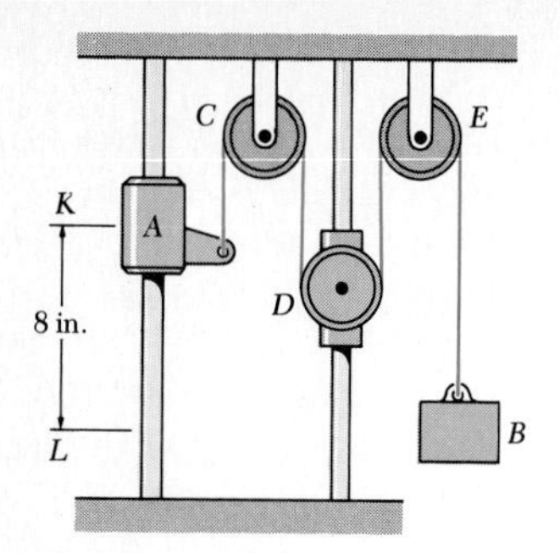

SAMPLE PROBLEM 11.5

Collar A and block B are connected by a cable passing over three pulleys C, D, and E as shown. Pulleys C and E are fixed, while D is attached to a collar which is pulled downward with a constant velocity of 3 in./s. At $t = 0$, collar A starts moving downward from position K with a constant acceleration and no initial velocity. Knowing that the velocity of collar A is 12 in./s as it passes through point L, determine the change in elevation, the velocity, and the acceleration of block B when collar A passes through L.

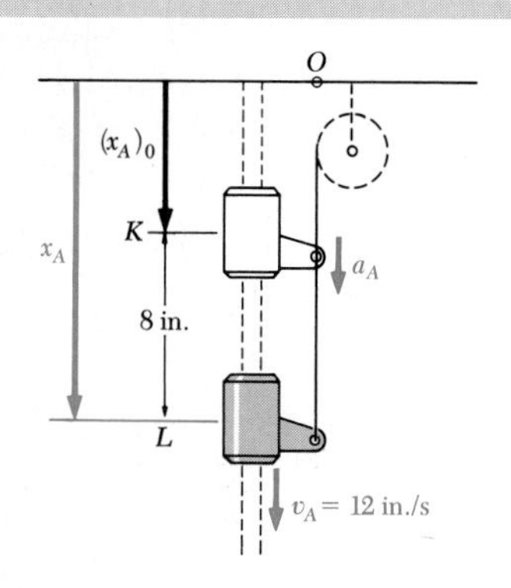

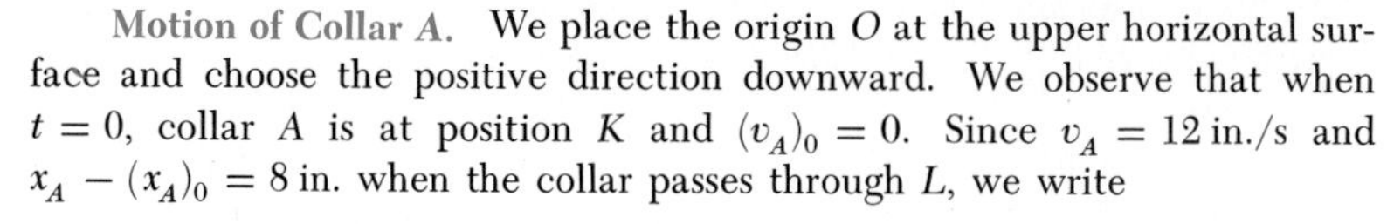

Motion of Collar A. We place the origin O at the upper horizontal surface and choose the positive direction downward. We observe that when $t = 0$, collar A is at position K and $(v_A)_0 = 0$. Since $v_A = 12$ in./s and $x_A - (x_A)_0 = 8$ in. when the collar passes through L, we write

$$v_A^2 = (v_A)_0^2 + 2a_A[x_A - (x_A)_0] \qquad (12)^2 = 0 + 2a_A(8)$$

$$a_A = 9 \text{ in./s}^2$$

The time at which collar A reaches point L is obtained by writing

$$v_A = (v_A)_0 + a_A t \qquad 12 = 0 + 9t \qquad t = 1.333 \text{ s}$$

Motion of Pulley D. Recalling that the positive direction is downward, we write

$$a_D = 0 \qquad v_D = 3 \text{ in./s} \qquad x_D = (x_D)_0 + v_D t = (x_D)_0 + 3t$$

When collar A reaches L, at $t = 1.333$ s, we have

$$x_D = (x_D)_0 + 3(1.333) = (x_D)_0 + 4$$

Thus,
$$x_D - (x_D)_0 = 4 \text{ in.}$$

Motion of Block B. We note that the total length of cable $ACDEB$ differs from the quantity $(x_A + 2x_D + x_B)$ only by a constant. Since the cable length is constant during the motion, this quantity must also remain constant. Thus considering the times $t = 0$ and $t = 1.333$ s, we write

$$x_A + 2x_D + x_B = (x_A)_0 + 2(x_D)_0 + (x_B)_0 \qquad (1)$$

$$[x_A - (x_A)_0] + 2[x_D - (x_D)_0] + [x_B - (x_B)_0] = 0 \qquad (2)$$

But we know that $x_A - (x_A)_0 = 8$ in. and $x_D - (x_D)_0 = 4$ in.; substituting these values in (2), we find

$$8 + 2(4) + [x_B - (x_B)_0] = 0 \qquad x_B - (x_B)_0 = -16 \text{ in.}$$

Thus: Change in elevation of $B = 16$ in. ↑ ◄

Differentiating (1) twice, we obtain equations relating the velocities and the accelerations of A, B, and D. Substituting for the velocities and accelerations of A and D at $t = 1.333$ s, we have

$$v_A + 2v_D + v_B = 0: \qquad 12 + 2(3) + v_B = 0$$

$$v_B = -18 \text{ in./s} \qquad v_B = 18 \text{ in./s} \uparrow ◄$$

$$a_A + 2a_D + a_B = 0: \qquad 9 + 2(0) + a_B = 0$$

$$a_B = -9 \text{ in./s}^2 \qquad a_B = 9 \text{ in./s}^2 \uparrow ◄$$

Problems

11.29 A ball is thrown vertically upward from a point on a tower located 25 m above the ground. Knowing that the ball strikes the ground 3 s after release, determine the speed with which the ball (*a*) was thrown upward, (*b*) strikes the ground.

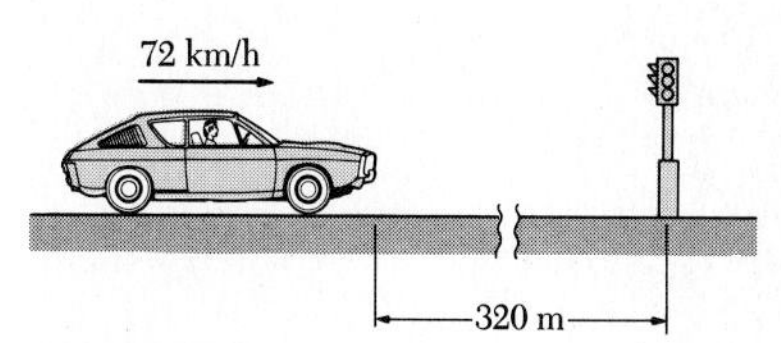

Fig. P11.30

11.30 A motorist is traveling at 72 km/h when she observes that a traffic light 320 m ahead of her turns red. The traffic light is timed to stay red for 22 s. If the motorist wishes to pass the light without stopping just as it turns green again, determine (*a*) the required uniform deceleration of the car, (*b*) the speed of the car as it passes the light.

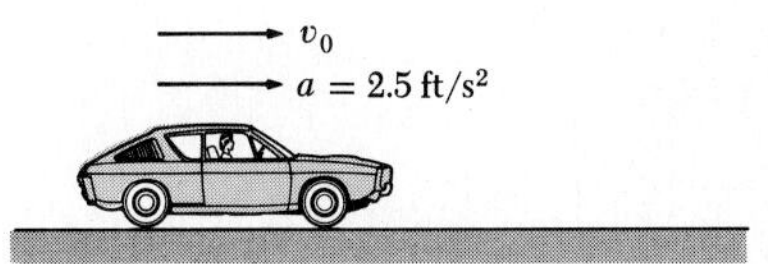

Fig. P11.31

11.31 An automobile travels 800 ft in 20 s while being accelerated at a constant rate of 2.5 ft/s^2. Determine (*a*) its initial velocity, (*b*) its final velocity, (*c*) the distance traveled during the first 10 s.

11.32 A stone is released from an elevator moving up at a speed of 12 ft/s and reaches the bottom of the shaft in 2.5 s. (*a*) How high was the elevator when the stone was released? (*b*) With what speed does the stone strike the bottom of the shaft?

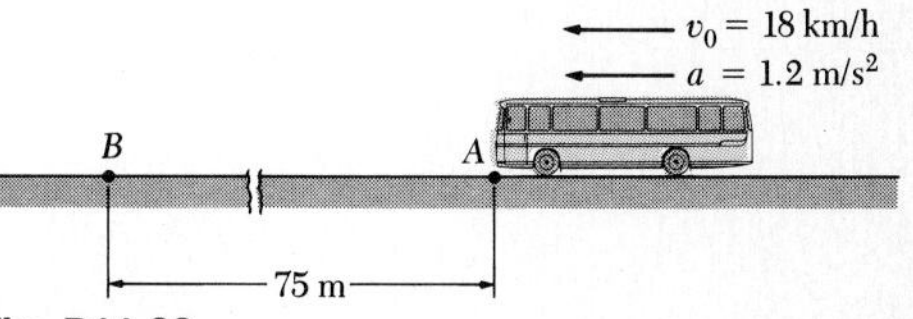

Fig. P11.33

11.33 A bus is accelerated at the rate of 1.2 m/s^2 as it travels from *A* to *B*. Knowing that the speed of the bus was $v_0 = 18$ km/h as it passed *A*, determine (*a*) the time required for the bus to reach *B*, (*b*) the corresponding speed as it passes *B*.

11.34 Solve Prob. 11.33, assuming that the velocity of the bus as it passed *A* was 36 km/h.

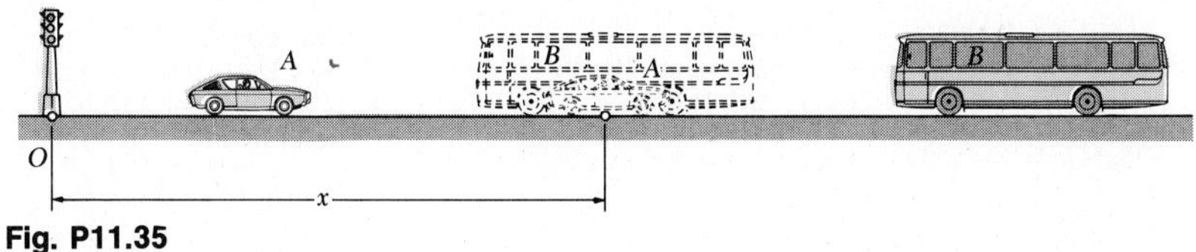

Fig. P11.35

11.35 Automobile *A* starts from *O* and accelerates at the constant rate of 0.8 m/s^2. A short time later it is passed by bus *B* which is traveling in the opposite direction at a constant speed of 5 m/s. Knowing that bus *B* passes point *O* 22 s after automobile *A* started from there, determine when and where the vehicles passed each other.

11.36 An open-platform elevator is moving down a mine shaft at a constant velocity v_e when the elevator platform hits and dislodges a stone. (*a*) Assuming that the stone starts falling with no initial velocity, show that the stone will hit the platform with a relative velocity of magnitude v_e. (*b*) If $v_e = 7.5$ m/s, determine when and where the stone will hit the elevator platform.

11.37 A freight elevator moving upward with a constant velocity of 6 ft/s passes a passenger elevator which is stopped. Four seconds later the passenger elevator starts upward with a constant acceleration of 2.4 ft/s^2. Determine (*a*) when and where the elevators will be at the same height, (*b*) the speed of the passenger elevator at that time.

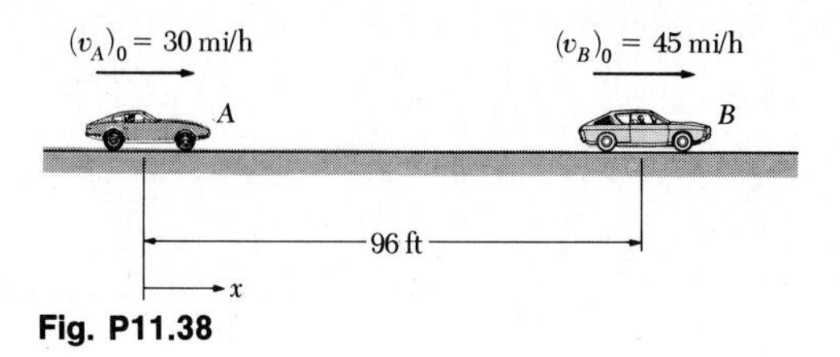

Fig. P11.38

11.38 Automobiles A and B are traveling in adjacent highway lanes and at $t = 0$ have the positions and speeds shown. Knowing that automobile A has a constant acceleration of 2 ft/s^2 and that B has a constant deceleration of 1.5 ft/s^2, determine (*a*) when and where A will overtake B, (*b*) the speed of each automobile at that time.

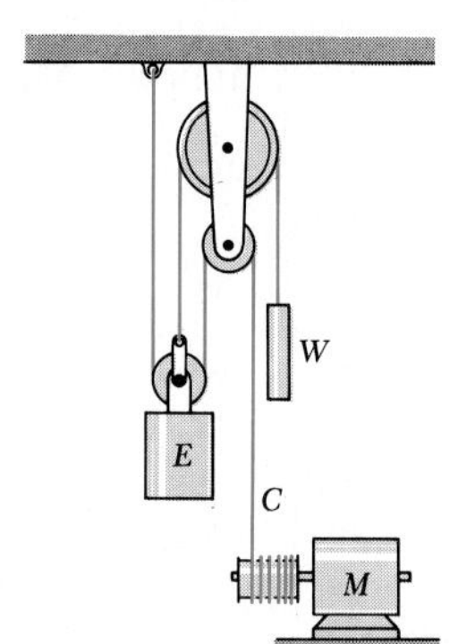

Fig. P11.39 and P11.40

11.39 The elevator shown in the figure moves upward at the constant velocity of 4 m/s. Determine (*a*) the velocity of the cable C, (*b*) the velocity of the counterweight W, (*c*) the relative velocity of the cable C with respect to the elevator, (*d*) the relative velocity of the counterweight W with respect to the elevator.

11.40 The elevator shown starts from rest and moves upward with a constant acceleration. If the counterweight W moves through 10 m in 4 s, determine (*a*) the acceleration of the elevator, (*b*) the acceleration of cable C, (*c*) the velocity of the elevator after 6 s.

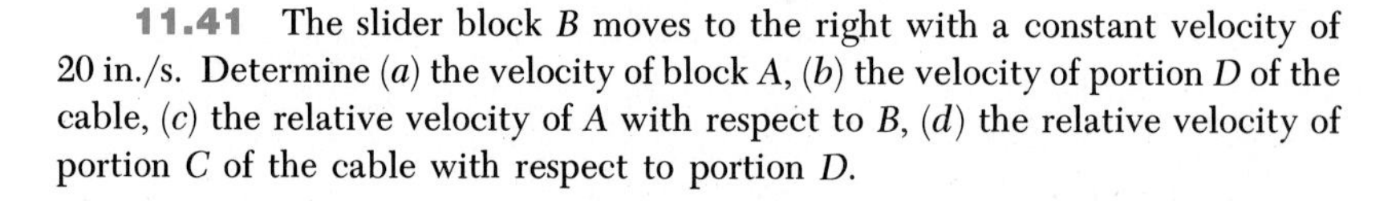
11.41 The slider block B moves to the right with a constant velocity of 20 in./s. Determine (*a*) the velocity of block A, (*b*) the velocity of portion D of the cable, (*c*) the relative velocity of A with respect to B, (*d*) the relative velocity of portion C of the cable with respect to portion D.

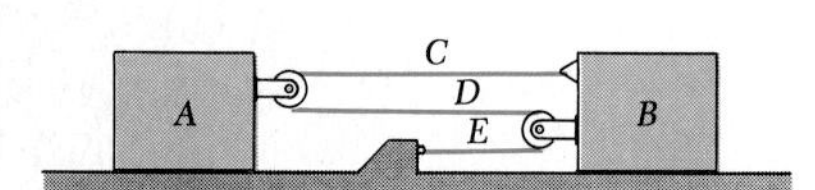

Fig. P11.41 and P11.42

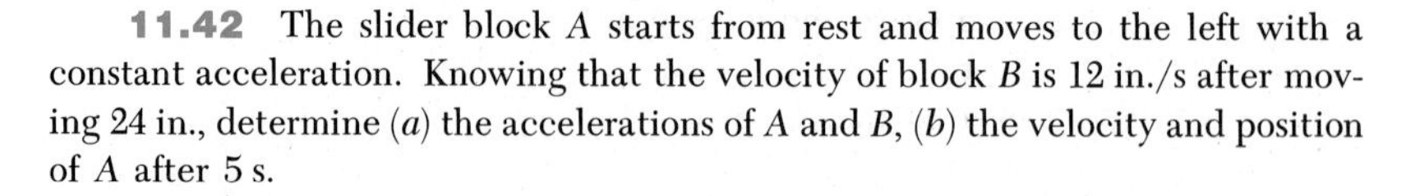
11.42 The slider block A starts from rest and moves to the left with a constant acceleration. Knowing that the velocity of block B is 12 in./s after moving 24 in., determine (*a*) the accelerations of A and B, (*b*) the velocity and position of A after 5 s.

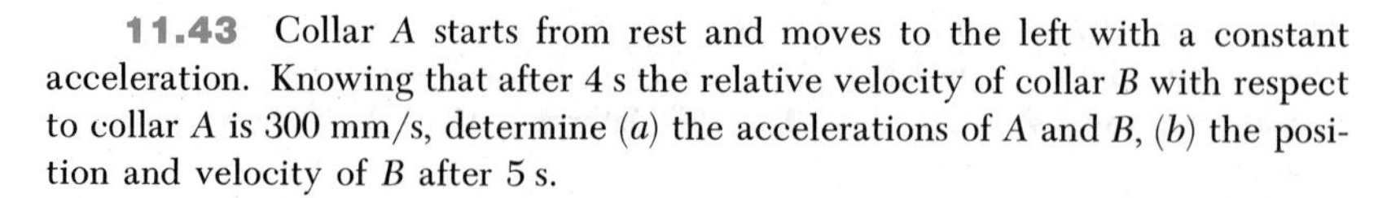
11.43 Collar A starts from rest and moves to the left with a constant acceleration. Knowing that after 4 s the relative velocity of collar B with respect to collar A is 300 mm/s, determine (*a*) the accelerations of A and B, (*b*) the position and velocity of B after 5 s.

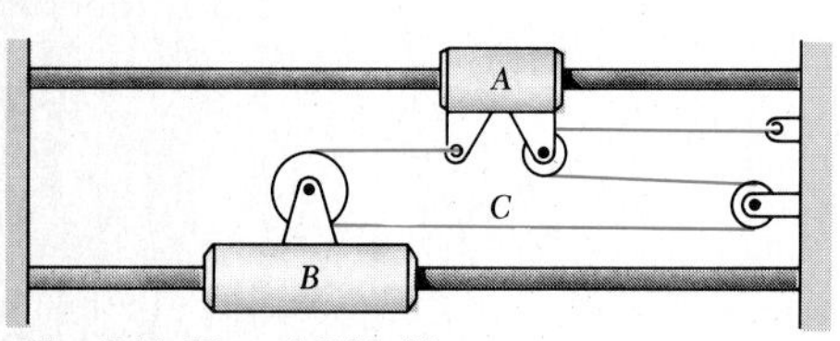

Fig. P11.43 and P11.44

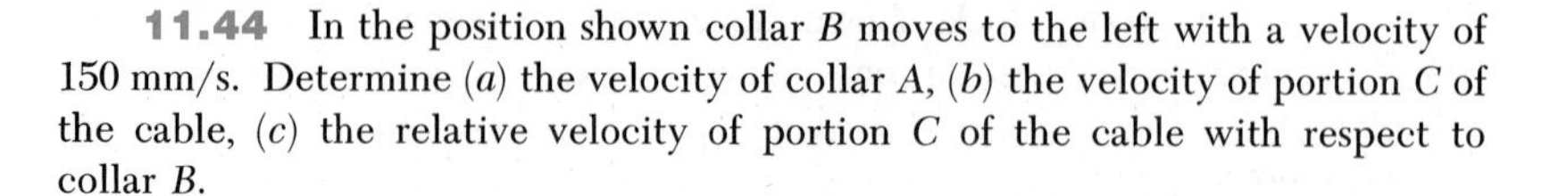
11.44 In the position shown collar B moves to the left with a velocity of 150 mm/s. Determine (*a*) the velocity of collar A, (*b*) the velocity of portion C of the cable, (*c*) the relative velocity of portion C of the cable with respect to collar B.

11.45 Under normal operating conditions, tape is transferred between the reels shown at a speed of 600 mm/s. At $t = 0$, portion A of the tape is moving to the right at a speed of 480 mm/s and has a constant acceleration. Knowing that portion B of the tape has a constant speed of 600 mm/s and that the speed of portion A reaches 600 mm/s at $t = 5$ s, determine (*a*) the acceleration and velocity of the compensator C at $t = 3$ s, (*b*) the distance through which C will have moved at $t = 5$ s.

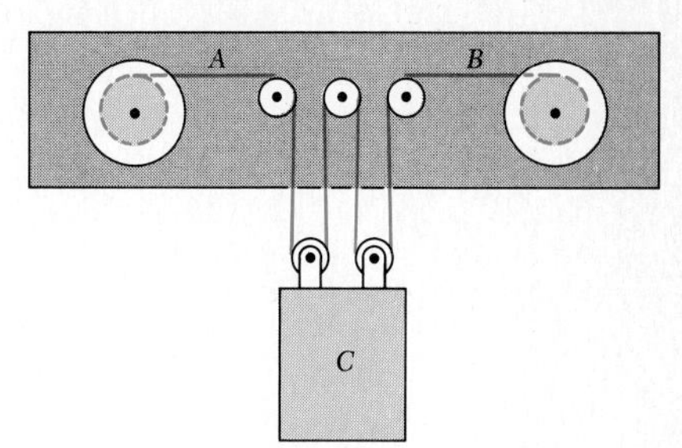

Fig. P11.45 and P11.46

11.46 Portions A and B of the tape shown start from rest at $t = 0$ and are each accelerated uniformly until a speed of 600 mm/s is reached; each portion of tape then moves at a constant speed of 600 mm/s. Knowing that portions A and B reached the speed of 600 mm/s in, respectively, 15 s and 12 s, determine (*a*) the acceleration and velocity of the compensator C at $t = 8$ s, (*b*) the distance through which C will have moved when both portions of tape reach their final speed.

11.47 Collar A starts from rest at $t = 0$ and moves upward with a constant acceleration of 2.5 in./s^2. Knowing that collar B moves downward with a constant velocity of 15 in./s, determine (*a*) the time at which the velocity of block C is zero, (*b*) the corresponding position of block C.

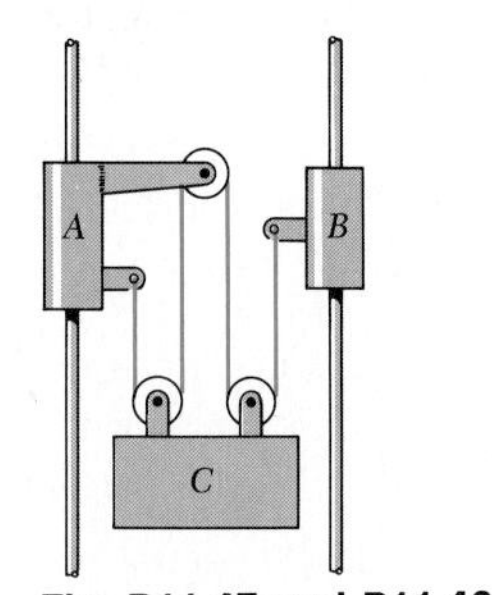

Fig. P11.47 and P11.48

11.48 Collars A and B start from rest and move with the following accelerations: $a_A = 3t$ in./s^2 upward and $a_B = 9$ in./s^2 downward. Determine (*a*) the time at which the velocity of block C is again zero, (*b*) the distance through which block C will have moved at that time.

11.49 (*a*) Choosing the positive sense downward for each block, express the velocity of A in terms of the velocities of B and C. (*b*) Knowing that both blocks A and C start from rest and move downward with the respective accelerations $a_A = 50$ mm/s^2 and $a_C = 110$ mm/s^2, determine the velocity and position of B after 3 s.

11.50 The three blocks shown move with constant velocities. Find the velocity of each block, knowing that the relative velocity of A with respect to C is 300 mm/s upward and that the relative velocity of B with respect to A is 200 mm/s downward.

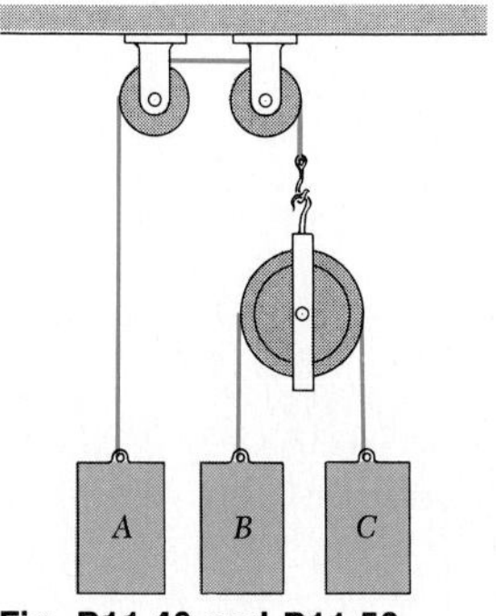

Fig. P11.49 and P11.50

* 11.7. Graphical Solution of Rectilinear-Motion Problems.

It was observed in Sec. 11.2 that the fundamental formulas

$$v = \frac{dx}{dt} \qquad \text{and} \qquad a = \frac{dv}{dt}$$

have a geometrical significance. The first formula expresses that the velocity at any instant is equal to the slope of the x–t curve at the same instant (Fig. 11.10). The second formula expresses that the acceleration is equal to

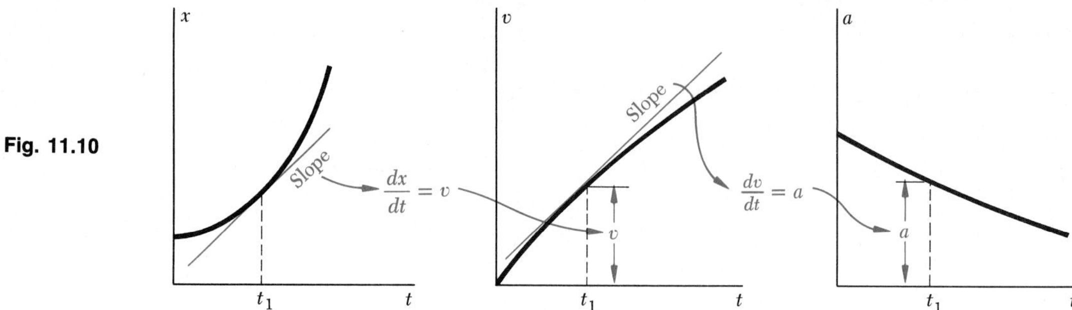

Fig. 11.10

the slope of the v–t curve. These two properties may be used to derive graphically the v–t and a–t curves of a motion when the x–t curve is known.

Integrating the two fundamental formulas from a time t_1 to a time t_2, we write

$$x_2 - x_1 = \int_{t_1}^{t_2} v\,dt \qquad \text{and} \qquad v_2 - v_1 = \int_{t_1}^{t_2} a\,dt \tag{11.12}$$

The first formula expresses that the area measured under the v–t curve from t_1 to t_2 is equal to the change in x during that time interval (Fig. 11.11). The second formula expresses similarly that the area measured under the a–t curve from t_1 to t_2 is equal to the change in v during that time interval. These two properties may be used to determine graphically the x–t curve of a motion when its v–t curve or its a–t curve is known (see Sample Prob. 11.6).

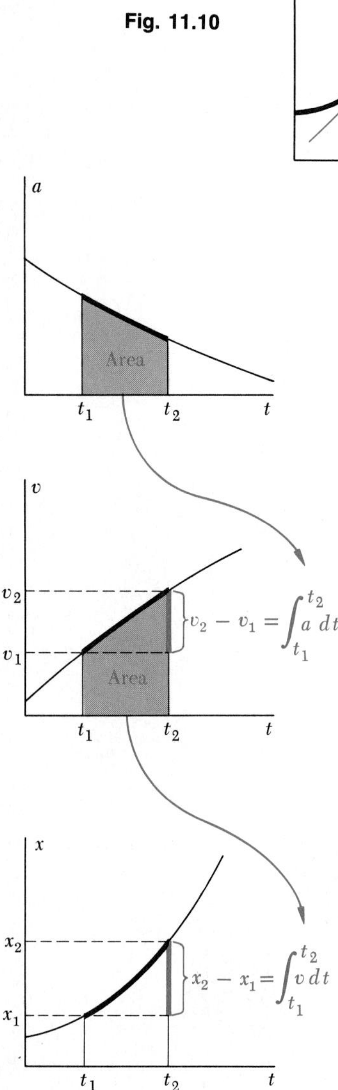

Fig. 11.11

Graphical solutions are particularly useful when the motion considered is defined from experimental data and when x, v, and a are not analytical functions of t. They may also be used to advantage when the motion consists of distinct parts and when its analysis requires writing a different equation for each of its parts. When using a graphical solution, however, one should be careful to note (1) that the area under the v–t curve measures the *change in* x, not x itself, and, similarly, that the area under the a–t curve measures the change in v; (2) that while an area above the t axis corresponds to an increase in x or v, an area located below the t axis measures a decrease in x or v.

It will be useful to remember in drawing motion curves that if the velocity is constant, it will be represented by a horizontal straight line; the position coordinate x will then be a linear function of t and will be represented by an oblique straight line. If the acceleration is constant and different from zero, it will be represented by a horizontal straight line; v will then be a linear function of t, represented by an oblique straight line; and x will be expressed as a second-degree polynomial in t, represented by a parabola. If the acceleration is a linear function of t, the velocity and the position coordinate will be equal, respectively, to second-degree and third-degree polynomials; a will then be represented by an oblique straight line, v by a parabola, and x by a cubic. In general, if the acceleration is a polynomial of degree n in t, the velocity will be a polynomial of degree $n + 1$ and the position coordinate a polynomial of degree $n + 2$; these polynomials are represented by motion curves of a corresponding degree.

*11.8. Other Graphical Methods.

An alternative graphical solution may be used to determine directly from the a–t curve the position of a particle at a given instant. Denoting, respectively, by x_0 and v_0 the values of x and v at $t = 0$ and by x_1 and v_1 their values at $t = t_1$, and observing that the area under the v–t curve may be divided into a rectangle of area $v_0 t_1$ and horizontal differential elements of area $(t_1 - t)dv$ (Fig. 11.12a), we write

$$x_1 - x_0 = \text{area under } v\text{–}t \text{ curve} = v_0 t_1 + \int_{v_0}^{v_1} (t_1 - t)\, dv$$

Substituting $dv = a\, dt$ in the integral, we obtain

$$x_1 - x_0 = v_0 t_1 + \int_0^{t_1} (t_1 - t) a\, dt$$

Referring to Fig. 11.12b, we note that the integral represents the first moment of the area under the a–t curve with respect to the line $t = t_1$ bounding the area on the right. This method of solution is known, therefore, as the *moment-area method.* If the abscissa $\bar{t}$ of the centroid C of the area is known, the position coordinate x_1 may be obtained by writing

$$x_1 = x_0 + v_0 t_1 + (\text{area under } a\text{–}t \text{ curve})(t_1 - \bar{t}) \tag{11.13}$$

If the area under the a–t curve is a composite area, the last term in (11.13) may be obtained by multiplying each component area by the distance from its centroid to the line $t = t_1$. Areas above the t axis should be considered as positive and areas below the t axis, as negative.

Another type of motion curve, the v–x curve, is sometimes used. If such a curve has been plotted (Fig. 11.13), the acceleration a may be obtained at any time by drawing the normal to the curve and *measuring the subnormal BC.* Indeed, observing that the angle between AC and AB is equal to the angle θ between the horizontal and the tangent at A (the slope of which is $\tan\theta = dv/dx$), we write

$$BC = AB \tan\theta = v\frac{dv}{dx}$$

and thus, recalling formula (11.4),

$$BC = a$$

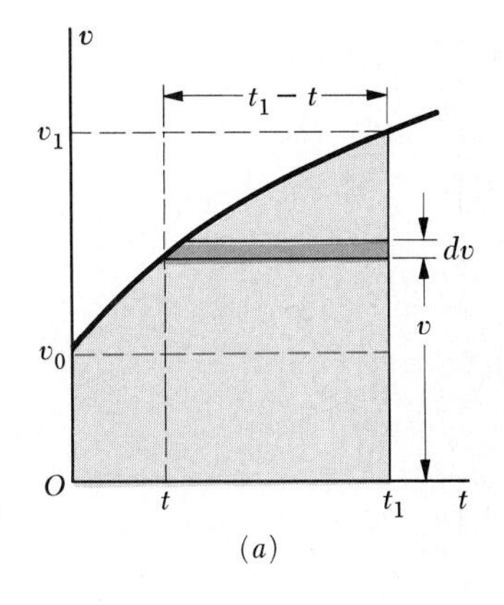

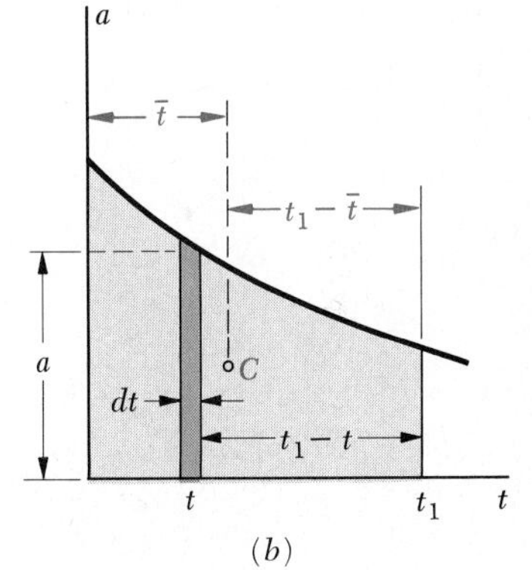

Fig. 11.12

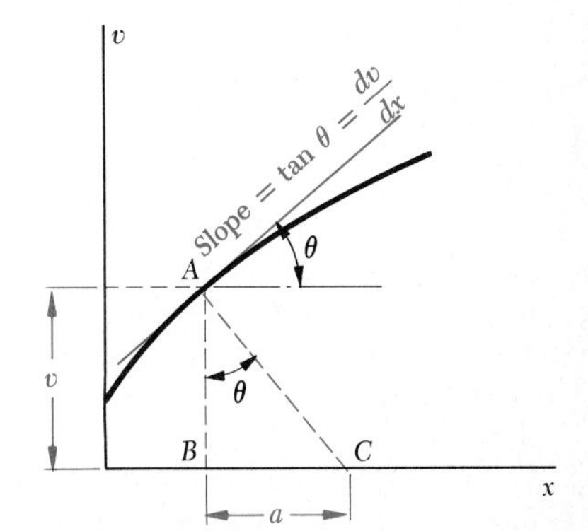

Fig. 11.13

SAMPLE PROBLEM 11.6

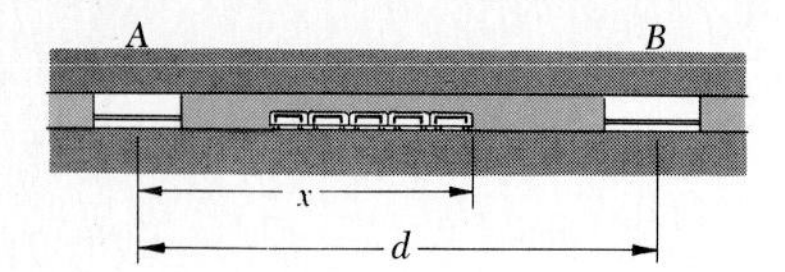

A subway train leaves station A; it gains speed at the rate of 4 ft/s^2 for 6 s and then at the rate of 6 ft/s^2 until it has reached the speed of 48 ft/s. The train maintains the same speed until it approaches station B; brakes are then applied, giving the train a constant deceleration and bringing it to a stop in 6 s. The total running time from A to B is 40 s. Draw the a–t, v–t, and x–t curves, and determine the distance between stations A and B.

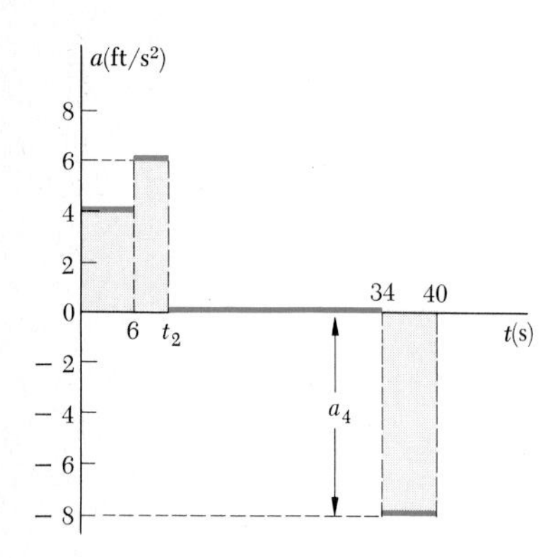

Acceleration-Time Curve. Since the acceleration is either constant or zero, the a–t curve is made of horizontal straight-line segments. The values of t_2 and a_4 are determined as follows:

$0 < t < 6$: Change in v = area under a–t curve

$$v_6 - 0 = (6 \text{ s})(4 \text{ ft/s}^2) = 24 \text{ ft/s}$$

$6 < t < t_2$: Since the velocity increases from 24 to 48 ft/s,

Change in v = area under a–t curve

$$48 - 24 = (t_2 - 6)(6 \text{ ft/s}^2) \qquad t_2 = 10 \text{ s}$$

$t_2 < t < 34$: Since the velocity is constant, the acceleration is zero.

$34 < t < 40$: Change in v = area under a–t curve

$$0 - 48 = (6 \text{ s})a_4 \qquad a_4 = -8 \text{ ft/s}^2$$

The acceleration being negative, the corresponding area is below the t axis; this area represents a decrease in velocity.

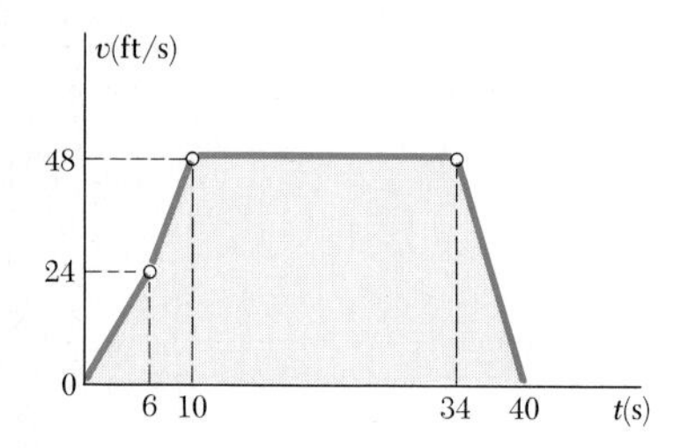

Velocity-Time Curve. Since the acceleration is either constant or zero, the v–t curve is made of segments of straight line connecting the points determined above.

Change in x = area under v–t curve

$0 < t < 6$: $\quad x_6 - 0 = \frac{1}{2}(6)(24) = 72 \text{ ft}$

$6 < t < 10$: $\quad x_{10} - x_6 = \frac{1}{2}(4)(24 + 48) = 144 \text{ ft}$

$10 < t < 34$: $\quad x_{34} - x_{10} = (24)(48) = 1152 \text{ ft}$

$34 < t < 40$: $\quad x_{40} - x_{34} = \frac{1}{2}(6)(48) = 144 \text{ ft}$

Adding the changes in x, we obtain the distance from A to B:

$$d = x_{40} - 0 = 1512 \text{ ft}$$

$d = 1512$ ft ◄

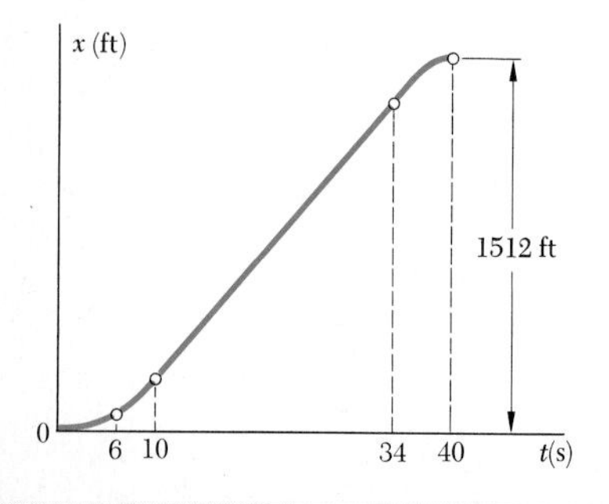

Position-Time Curve. The points determined above should be joined by three arcs of parabola and one segment of straight line. The construction of the x–t curve will be performed more easily and more accurately if we keep in mind that for any value of t the slope of the tangent to the x–t curve is equal to the value of v at that instant.

Problems

11.51 A particle moves in a straight line with the acceleration shown in the figure. Knowing that it starts from the origin with $v_0 = -14$ ft/s, plot the v–t and x–t curves for $0 < t < 15$ s and determine (*a*) the maximum value of the velocity of the particle, (*b*) the maximum value of its position coordinate.

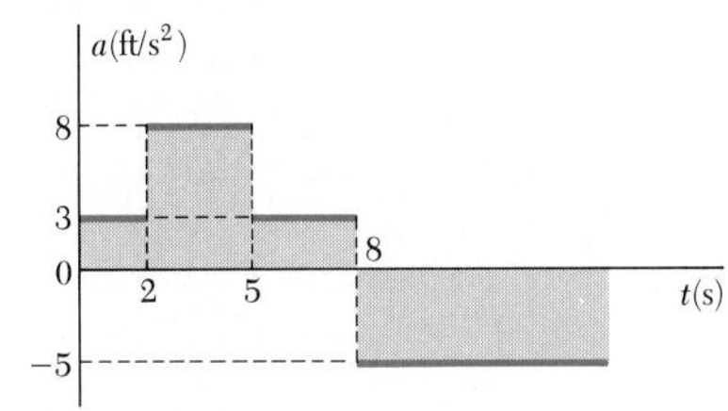

Fig. P11.51

11.52 For the particle and motion of Prob. 11.51, plot the v–t and x–t curves for $0 < t < 15$ s and determine the velocity of the particle, its position, and the total distance traveled after 10 s.

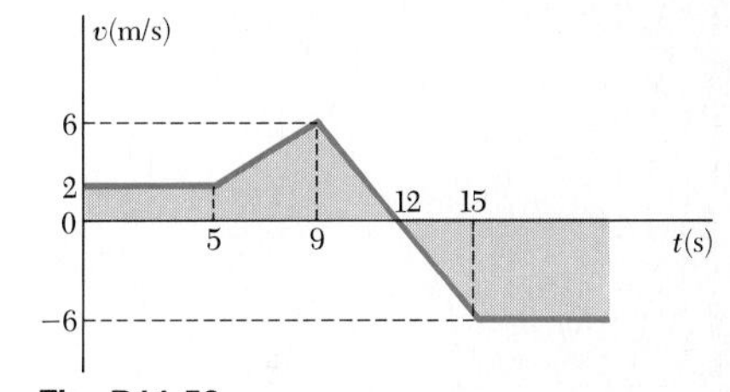

Fig. P11.53

11.53 A particle moves in a straight line with the velocity shown in the figure. Knowing that $x = -8$ m at $t = 0$, draw the a–t and x–t curves for $0 < t < 20$ s and determine (*a*) the maximum value of the position coordinate of the particle, (*b*) the values of t for which the particle is at a distance of 18 m from the origin.

11.54 For the particle and motion of Prob. 11.53, plot the a–t and x–t curves for $0 < t < 20$ s and determine (*a*) the total distance traveled by the particle during the period $t = 0$ to $t = 15$ s, (*b*) the two values of t for which the particle passes through the origin.

11.55 A bus starts from rest at point A and accelerates at the rate of 0.75 m/s^2 until it reaches a speed of 9 m/s. It then proceeds at 9 m/s until the brakes are applied; it comes to rest at point B, 27 m beyond the point where the brakes were applied. Assuming uniform deceleration and knowing that the distance between A and B is 180 m, determine the time required for the bus to travel from A to B.

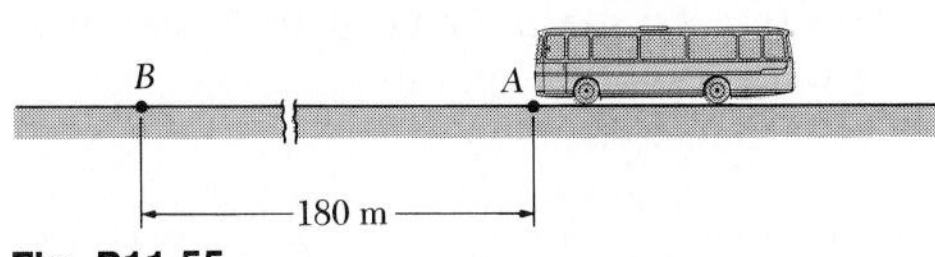

Fig. P11.55

11.56 A series of city traffic signals is timed so that an automobile traveling at a constant speed of 54 km/h will reach each signal just as it turns green. A motorist misses a signal and is stopped at signal *A*. Knowing that the next signal *B* is 300 m ahead and that the maximum acceleration of the automobile is 1.8 m/s^2, determine what the motorist should do to keep the maximum speed as small as possible, yet reach signal *B* just as it turns green. What is the maximum speed reached?

11.57 Firing a howitzer causes the barrel to recoil 42 in. before a braking mechanism brings it to rest. From a high-speed photographic record, it is found that the maximum value of the recoil velocity is 250 in./s and that this is reached 0.02 s after firing. Assuming that the recoil period consists of two phases during which the acceleration has, respectively, a constant positive value a_1 and a constant negative value a_2, determine (*a*) the values of a_1 and a_2, (*b*) the position of the barrel 0.02 s after firing, (*c*) the time at which the velocity of the barrel is zero.

11.58 During a finishing operation the bed of an industrial planer moves alternately 36 in. to the right and 36 in. to the left. The velocity of the bed is limited to a maximum value of 6 in./s to the right and 9 in./s to the left; the acceleration is successively equal to 3 in./s^2 to the right, zero, 3 in./s^2 to the left, zero, etc. Determine the time required for the bed to complete a full cycle, and draw the v–t and x–t curves.

11.59 A motorist is traveling at 72 km/h when she observes that a traffic signal 320 m ahead of her turns red. She knows that the signal is timed to stay red for 22 s. What should she do to pass the signal at 72 km/h just as it turns green again? Draw the v–t curve, selecting the solution which calls for the smallest possible deceleration and acceleration, and determine (*a*) the deceleration and acceleration in m/s^2, (*b*) the minimum speed reached in km/h.

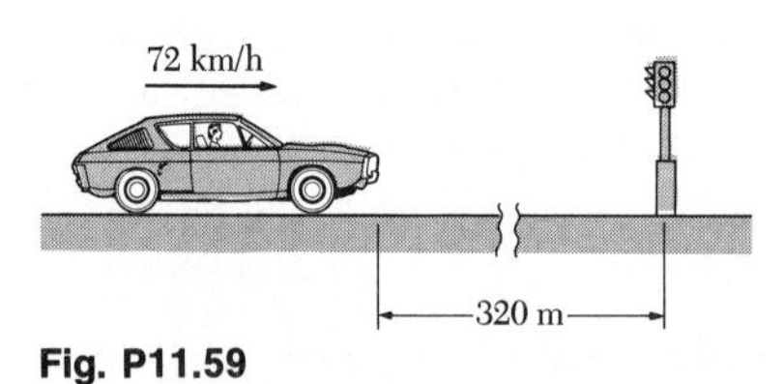

Fig. P11.59

11.60 Solve Prob. 11.59, knowing that the acceleration of the automobile cannot exceed 0.75 m/s^2.

11.61 An automobile at rest is passed by a truck traveling at a constant speed of 54 km/h. The automobile starts and accelerates for 10 s at a constant rate until it reaches a speed of 81 km/h. If the automobile then maintains a constant speed of 81 km/h, determine when and where it will overtake the truck, assuming that the automobile starts (*a*) just as the truck passes it, (*b*) 2 s after the truck has passed it.

11.62 A motorcycle and an automobile are both traveling at the constant speed of 40 mi/h; the motorcycle is 50 ft behind the automobile. The motorcyclist wants to pass the automobile, i.e., he wishes to place his motorcycle at B, 50 ft in front of the automobile, and then resume the speed of 40 mi/h. The maximum acceleration of the motorcycle is 6 ft/s^2 and the maximum deceleration obtained by applying the brakes is 18 ft/s^2. What is the shortest time in which the motorcyclist can complete the passing operation if he does not at any time exceed a speed of 55 mi/h? Draw the v–t curve.

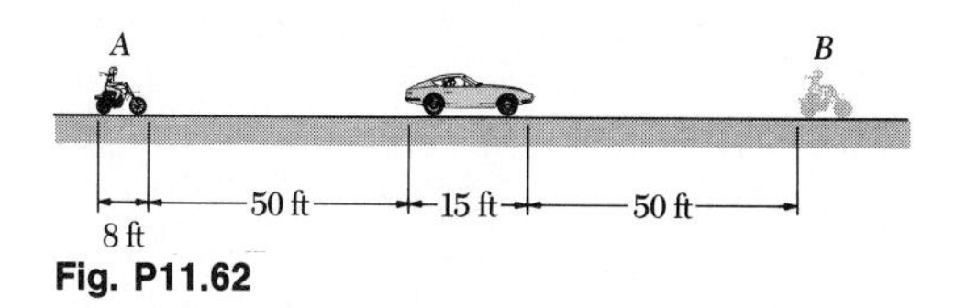

Fig. P11.62

11.63 Solve Prob. 11.62, assuming that the motorcyclist does not pay any attention to the speed limit while passing and concentrates on reaching position B and resuming a speed of 40 mi/h in the shortest possible time. What is the maximum speed reached? Draw the v–t curve.

11.64 A car and a truck are both traveling at a constant speed v_0; the car is 35 ft behind the truck. The truck driver suddenly applies his brakes, causing the truck to decelerate at the constant rate of 10 ft/s^2. Two seconds later the driver of the car applies her brakes and barely manages to avoid a rear-end collision. Determine the constant rate at which the car decelerated if (*a*) $v_0 = 60$ mi/h, (*b*) $v_0 = 30$ mi/h.

11.65 Car A is traveling at the constant speed v_A. It approaches car B, which is traveling in the same direction at the constant speed of 63 km/h. The driver of car B notices car A when it is still 50 m behind him and then accelerates at the constant rate of 0.8 m/s^2 to avoid being passed or struck by car A. Knowing that the closest that A comes to B is 10 m, determine the speed v_A of car A.

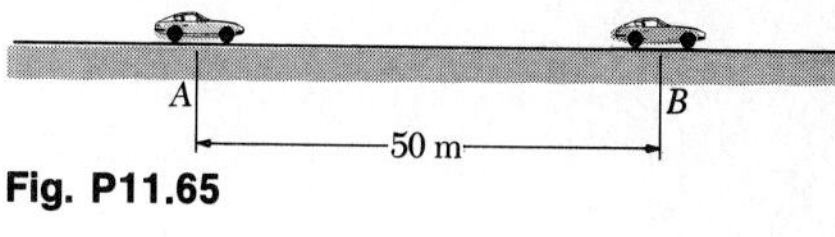

Fig. P11.65

11.66 A fighter plane flying horizontally in a straight line at 900 km/h is overtaking a bomber flying in the same straight line at 720 km/h. The pilot of the fighter plane fires an air-to-air missile at the bomber when his plane is 1150 m behind the bomber. The missile accelerates at a constant rate of 400 m/s^2 for 1 s and then travels at a constant speed. (*a*) How many seconds after firing will the missile reach the bomber? (*b*) If both planes continue at constant speeds, what will be the distance between the planes when the missile strikes the bomber?

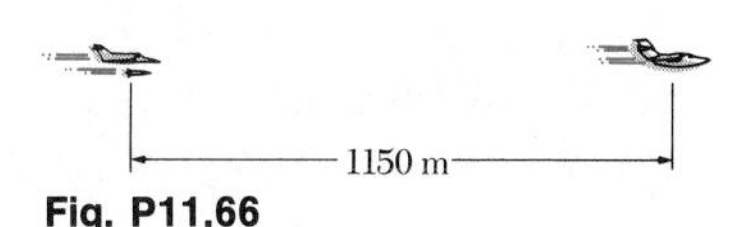

Fig. P11.66

11.67 The acceleration record shown was obtained for a small airplane traveling along a straight course. Knowing that $x = 0$ and $v = 50$ m/s when $t = 0$, determine (*a*) the velocity and position of the plane at $t = 20$ s, (*b*) its average velocity during the interval $6 \text{ s} < t < 14$ s.

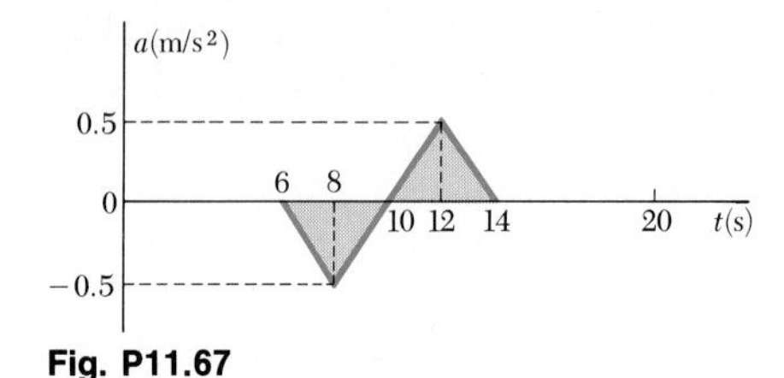

Fig. P11.67

11.68 A train starts at a station and accelerates uniformly at a rate of 1 ft/s² until it reaches a speed of 20 ft/s; it then proceeds at the constant speed of 20 ft/s. Determine the time and the distance traveled if its *average* velocity is (*a*) 12 ft/s, (*b*) 18 ft/s.

11.69 The rate of change of acceleration is known as the *jerk;* large or abrupt rates of change of acceleration cause discomfort to elevator passengers. If the jerk, or rate of change of the acceleration, of an elevator is limited to ± 1.5 ft/s² per second, determine (*a*) the shortest time required for an elevator, starting from rest, to rise 24 ft and stop, (*b*) the corresponding average velocity of the elevator.

11.70 In order to maintain passenger comfort, the acceleration of an elevator is limited to ± 1.2 m/s² and the jerk, or rate of change of acceleration, is limited to ± 0.4 m/s² per second. If the elevator starts from rest, determine (*a*) the shortest time required for it to attain a constant velocity of 6 m/s, (*b*) the distance traveled in that time, (*c*) the corresponding average velocity of the elevator.

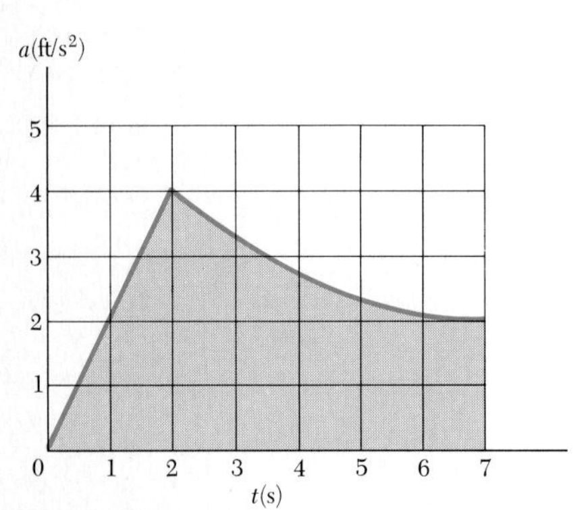

Fig. P11.71

11.71 The acceleration record shown was obtained for an automobile traveling on a straight highway. Knowing that the initial velocity of the automobile was 15 mi/h, determine the velocity and distance traveled when (*a*) $t = 3$ s, (*b*) $t = 6$ s.

11.72 A test sled is brought to rest in 5 s after the completion of an experiment. An accelerometer attached to the sled provides the acceleration record shown. Determine by approximate means (*a*) the initial velocity of the sled, (*b*) the distance that the sled travels during the interval $0 < t < 5$ s.

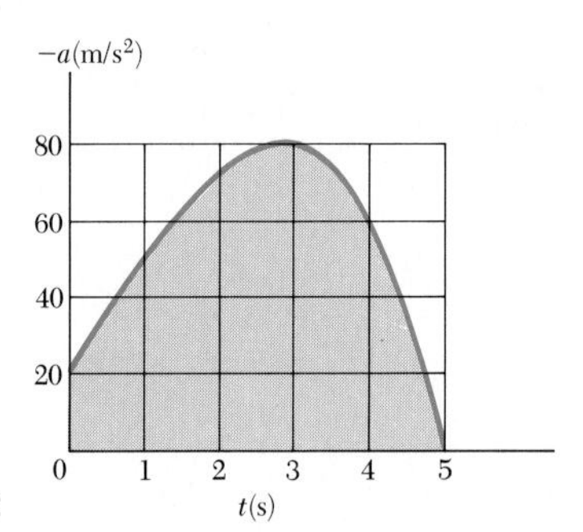

Fig. P11.72

11.73 The maximum possible deceleration of a passenger train under emergency conditions was determined experimentally; the results are shown (solid curve) in the figure. If the brakes are applied when the train is traveling at 108 km/h, determine by approximate means (*a*) the time required for the train to come to rest, (*b*) the distance traveled in that time.

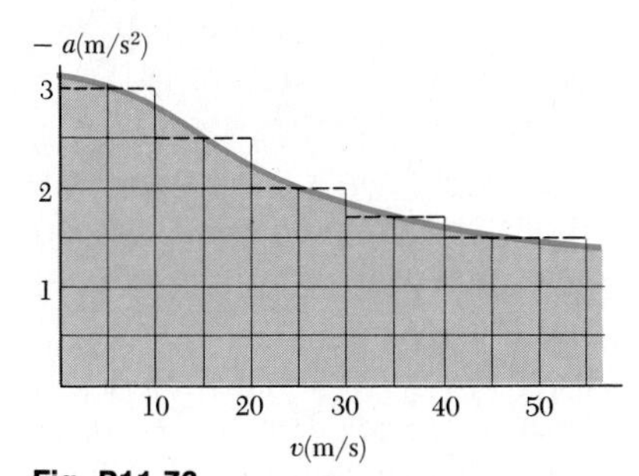

Fig. P11.73

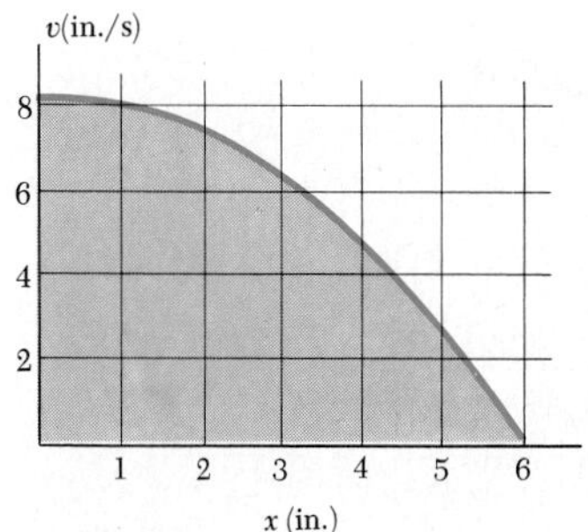

Fig. P11.74

11.74 The v–x curve shown was obtained experimentally during the motion of the bed of an industrial planer. Determine by approximate means the acceleration (*a*) when $x = 3$ in., (*b*) when $v = 4$ in./s.

11.75 Using the method of Sec. 11.8, derive the formula $x = x_0 + v_0 t + \frac{1}{2}at^2$ for the position coordinate of a particle in uniformly accelerated rectilinear motion.

11.76 The acceleration of an object subjected to the pressure wave of a large explosion is defined approximately by the curve shown. The object is initially at rest and is again at rest at time t_1. Using the method of Sec. 11.8, determine (*a*) the time t_1, (*b*) the distance through which the object is moved by the pressure wave.

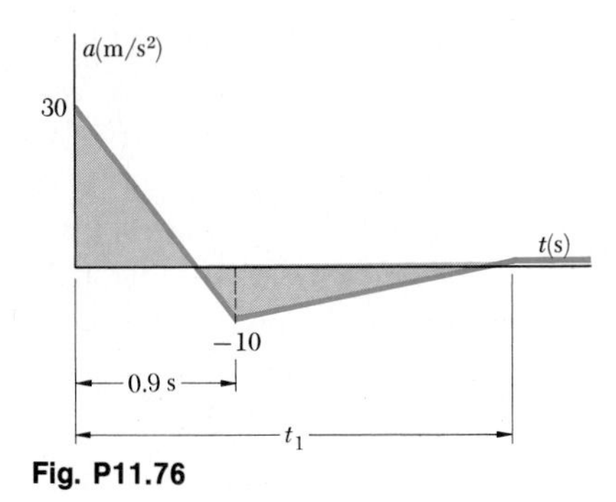

Fig. P11.76

11.77 For the particle of Prob. 11.53, draw the a–t curve and, using the method of Sec. 11.8, determine (*a*) the position of the particle when $t = 10$ s, (*b*) the maximum value of its position coordinate.

11.78 Using the method of Sec. 11.8, determine the position of the particle of Prob. 11.51 when $t = 8$ s.

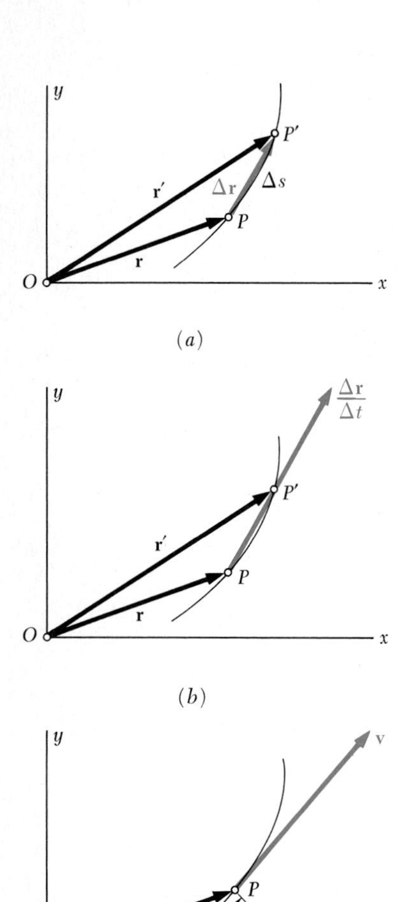

Fig. 11.14

11.9. Position Vector, Velocity, and Acceleration. When a particle moves along a curve other than a straight line, we say that the particle is in *curvilinear motion*. In this part of the chapter, we shall analyze the curvilinear motion of a particle in a given plane. To define the position P occupied by the particle in that plane at a given time t, we select a fixed reference system, such as the x and y axes shown in Fig. 11.14*a*, and draw the vector $\mathbf{r}$ joining the origin O and point P. Since the vector $\mathbf{r}$ is characterized by its magnitude r and its direction with respect to the reference axes, it completely defines the position of the particle with respect to those axes; the vector $\mathbf{r}$ is referred to as the *position vector* of the particle at time t.

Consider now the vector $\mathbf{r}'$ defining the position P' occupied by the same particle at a later time $t + \Delta t$. The vector $\Delta\mathbf{r}$ joining P and P' represents the change in the position vector during the time interval Δt since, as we may easily check from Fig. 11.14*a*, the vector $\mathbf{r}'$ is obtained by adding the vectors $\mathbf{r}$ and $\Delta\mathbf{r}$ according to the triangle rule.† We note that $\Delta\mathbf{r}$ represents a change in *direction* as well as a change in *magnitude* of the position vector $\mathbf{r}$. The *average velocity* of the particle over the time interval Δt is defined as the quotient of $\Delta\mathbf{r}$ and Δt. Since $\Delta\mathbf{r}$ is a vector and Δt is a scalar, the quotient $\Delta\mathbf{r}/\Delta t$ is a vector attached at P, of the same direction as $\Delta\mathbf{r}$, and of magnitude equal to the magnitude of $\Delta\mathbf{r}$ divided by Δt (Fig. 11.14*b*).

The *instantaneous velocity* of the particle at time t is obtained by choosing shorter and shorter time intervals Δt and, correspondingly, shorter and shorter vector increments $\Delta\mathbf{r}$. The instantaneous velocity is thus represented by the vector

$$\mathbf{v} = \lim_{\Delta t \to 0} \frac{\Delta\mathbf{r}}{\Delta t} \tag{11.14}$$

As Δt and $\Delta\mathbf{r}$ become shorter, the points P and P' get closer; the vector $\mathbf{v}$ obtained at the limit must therefore be tangent to the path of the particle (Fig. 11.14*c*). Its magnitude, denoted by v and called the *speed* of the particle, is obtained by substituting for the vector $\Delta\mathbf{r}$ in formula (11.14) its magnitude represented by the straight-line segment PP'. But the length of the segment PP' approaches the length Δs of the arc PP' as Δt decreases (Fig. 11.14*a*), and we may write

$$v = \lim_{\Delta t \to 0} \frac{PP'}{\Delta t} = \lim_{\Delta t \to 0} \frac{\Delta s}{\Delta t}$$

$$v = \frac{ds}{dt} \tag{11.15}$$

† Since the displacement from O to P' represented by $\mathbf{r}'$ is clearly equivalent to the two successive displacements represented by $\mathbf{r}$ and $\Delta\mathbf{r}$, we verify that displacements satisfy the triangle rule and, therefore, the parallelogram law of addition. Thus, *displacements are truly vectors*.

The speed v may thus be obtained by differentiating with respect to t the length s of the arc described by the particle.

It is recalled that in many problems of statics it was found convenient to resolve a force $\mathbf{F}$ into component forces $\mathbf{F}_x$ and $\mathbf{F}_y$. Similarly, we shall often find it convenient in kinematics to resolve the vector $\mathbf{v}$ into rectangular components $\mathbf{v}_x$ and $\mathbf{v}_y$. To obtain these components, we resolve the vector $\Delta\mathbf{r}$ into components $\overrightarrow{PP''}$ and $\overrightarrow{P''P'}$, respectively parallel to the x and y axes (Fig. 11.15a), and write

$$\Delta\mathbf{r} = \overrightarrow{PP''} + \overrightarrow{P''P'}$$

where the right-hand member represents a vector sum. Substituting in (11.14) the expression obtained for $\Delta\mathbf{r}$, we write

$$\mathbf{v} = \lim_{\Delta t \to 0} \frac{\overrightarrow{PP''}}{\Delta t} + \lim_{\Delta t \to 0} \frac{\overrightarrow{P''P'}}{\Delta t}$$

Since the first limit represents a vector parallel to the x axis, it must be equal to the component $\mathbf{v}_x$ of the velocity; similarly, the second limit must be equal to its component $\mathbf{v}_y$ (Fig. 11.15b). We write

$$\mathbf{v}_x = \lim_{\Delta t \to 0} \frac{\overrightarrow{PP''}}{\Delta t} \qquad \mathbf{v}_y = \lim_{\Delta t \to 0} \frac{\overrightarrow{P''P'}}{\Delta t}$$

The magnitudes v_x and v_y of the components of the velocity may be obtained by observing that the magnitudes of the vectors $\overrightarrow{PP''}$ and $\overrightarrow{P''P'}$ are respectively equal to the increments Δx and Δy of the coordinates x and y of the particle. We write

$$v_x = \lim_{\Delta t \to 0} \frac{\Delta x}{\Delta t} \qquad v_y = \lim_{\Delta t \to 0} \frac{\Delta y}{\Delta t}$$

$$v_x = \frac{dx}{dt} \qquad v_y = \frac{dy}{dt}$$

or, using dots to indicate time derivatives,

$$v_x = \dot{x} \qquad v_y = \dot{y} \tag{11.16}$$

The expressions obtained are known as the *scalar components* of the velocity. A positive value for v_x indicates that the vector component $\mathbf{v}_x$ is directed to the right, a negative value that it is directed to the left; similarly, a positive value for v_y indicates that $\mathbf{v}_y$ is directed upward, a negative value that it is directed downward. We verify from Fig. 11.15b that the magnitude and direction of the velocity $\mathbf{v}$ may be obtained from its scalar components v_x and v_y as follows:

$$v^2 = v_x^2 + v_y^2 \qquad \tan\alpha = \frac{v_y}{v_x} \tag{11.17}$$

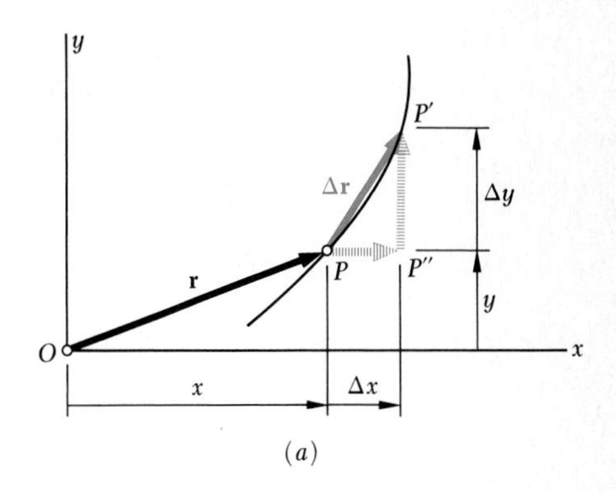

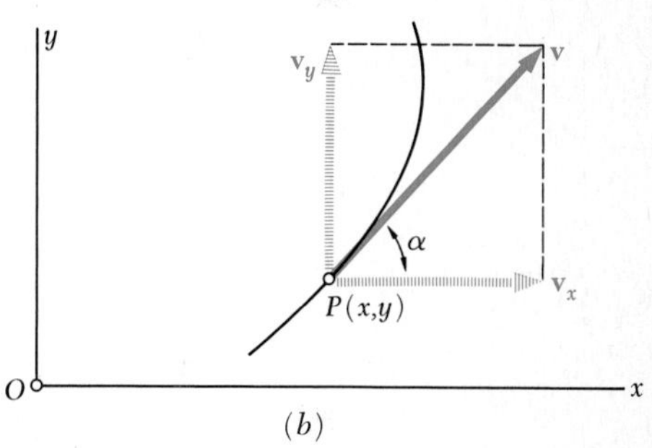

Fig. 11.15

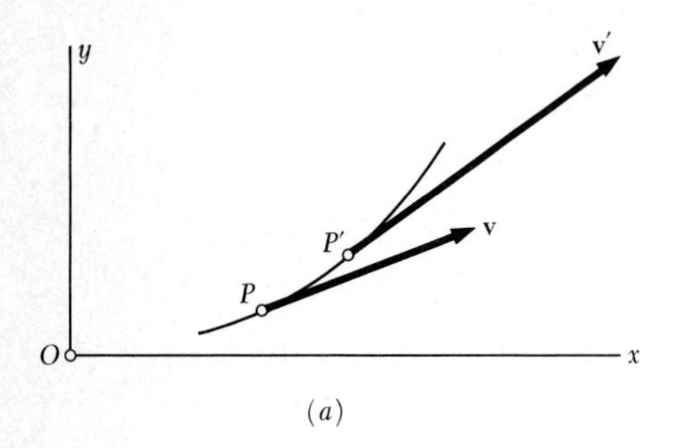

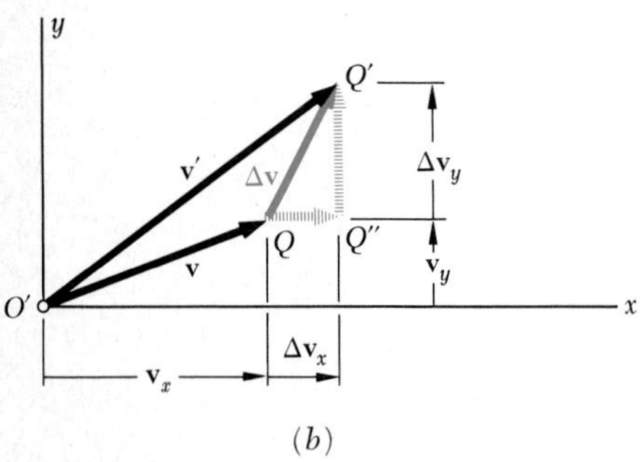

Fig. 11.16

Consider the velocity $\mathbf{v}$ of the particle at time t and also its velocity $\mathbf{v}'$ at a later time $t + \Delta t$ (Fig. 11.16*a*). Let us draw both vectors $\mathbf{v}$ and $\mathbf{v}'$ from the same origin O' (Fig. 11.16*b*). The vector $\Delta\mathbf{v}$ joining Q and Q' represents the change in the velocity of the particle during the time interval Δt, since the vector $\mathbf{v}'$ may be obtained by adding the vectors $\mathbf{v}$ and $\Delta\mathbf{v}$. We should note that $\Delta\mathbf{v}$ represents a change in the *direction* of the velocity as well as a change in *speed.* The *average acceleration* of the particle over the time interval Δt is defined as the quotient of $\Delta\mathbf{v}$ and Δt. Since $\Delta\mathbf{v}$ is a vector and Δt a scalar, the quotient $\Delta\mathbf{v}/\Delta t$ is a vector of the same direction as $\Delta\mathbf{v}$.

The *instantaneous acceleration* of the particle at time t is obtained by choosing smaller and smaller values for Δt and $\Delta\mathbf{v}$. The instantaneous acceleration is thus represented by the vector

$$\mathbf{a} = \lim_{\Delta t \to 0} \frac{\Delta\mathbf{v}}{\Delta t} \tag{11.18}$$

It should be noted that, in general, *the acceleration is not tangent to the path of the particle;* also, in general, the magnitude of the acceleration *does not* represent the rate of change of the speed of the particle.

To obtain the rectangular components $\mathbf{a}_x$ and $\mathbf{a}_y$ of the acceleration, we resolve the vector $\Delta\mathbf{v}$ into components $\overrightarrow{QQ''}$ and $\overrightarrow{Q''Q'}$, respectively parallel to the x and y axes (Fig. 11.16*b*), and write

$$\Delta\mathbf{v} = \overrightarrow{QQ''} + \overrightarrow{Q''Q'}$$

Substituting in (11.18) the expression obtained for $\Delta\mathbf{v}$, we write

$$\mathbf{a} = \lim_{\Delta t \to 0} \frac{\overrightarrow{QQ''}}{\Delta t} + \lim_{\Delta t \to 0} \frac{\overrightarrow{Q''Q'}}{\Delta t}$$

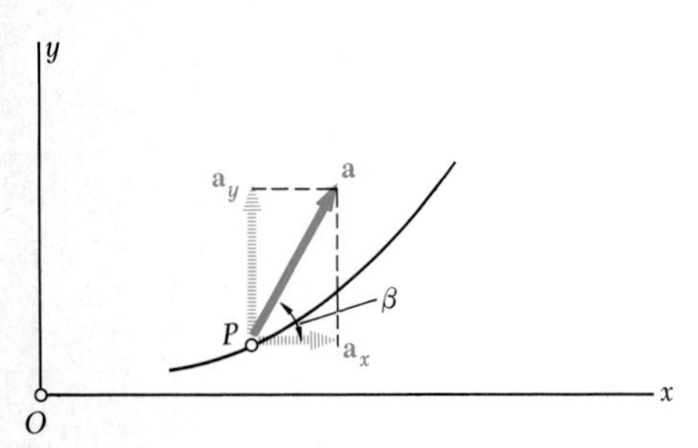

Fig. 11.17

Since the first limit represents a vector parallel to the x axis, it must be equal to the component $\mathbf{a}_x$ of the acceleration; similarly, the second limit must be equal to its component $\mathbf{a}_y$ (Fig. 11.17). We write

$$\mathbf{a}_x = \lim_{\Delta t \to 0} \frac{\overrightarrow{QQ''}}{\Delta t} \qquad \mathbf{a}_y = \lim_{\Delta t \to 0} \frac{\overrightarrow{Q''Q'}}{\Delta t}$$

The magnitudes a_x and a_y of the components of the acceleration, which are known as the scalar components of the acceleration, may be obtained by observing that the magnitudes of the vectors $\overrightarrow{QQ''}$ and $\overrightarrow{Q''Q'}$ are respectively equal to the increments Δv_x and Δv_y of the scalar components v_x and v_y of the velocity of the particle. We write

$$a_x = \lim_{\Delta t \to 0} \frac{\Delta v_x}{\Delta t} \qquad a_y = \lim_{\Delta t \to 0} \frac{\Delta v_y}{\Delta t}$$

$$a_x = \frac{dv_x}{dt} \qquad a_y = \frac{dv_y}{dt}$$

$$a_x = \dot{v}_x \qquad a_y = \dot{v}_y \tag{11.19}$$

or, substituting for v_x and v_y from (11.16),

$$a_x = \ddot{x} \qquad a_y = y \tag{11.20}$$

The magnitude and direction of the acceleration **a** may be obtained from its scalar components a_x and a_y (Fig. 11.17) by formulas similar to formulas (11.17).

11.10. Component Motions.

The use of rectangular components to describe the position, the velocity, and the acceleration of a particle is particularly effective when the component a_x of the acceleration is independent of the coordinate y and of the velocity component v_y and when the component a_y of the acceleration is independent of x and v_x. Equations (11.19) may then be integrated independently, and so may Eqs. (11.16). In other words, the motion of the particle in the x direction and its motion in the y direction may be considered separately.

In the case of the *motion of a projectile*, for example, it may be shown (Sec. 12.8), if the resistance of the air is neglected, that the components of the acceleration are

$$a_x = 0 \qquad a_y = -g$$

where g is 9.81 m/s^2 or 32.2 ft/s^2. Substituting into Eqs. (11.19), we write

$$\dot{v}_x = 0 \qquad \dot{v}_y = -g$$

$$v_x = (v_x)_0 \qquad v_y = (v_y)_0 + gt$$

where $(v_x)_0$ and $(v_y)_0$ are the components of the initial velocity. Substituting into Eqs. (11.16), and assuming that the projectile is fired from the origin O, we write

$$\dot{x} = (v_x)_0 \qquad \dot{y} = (v_y)_0 - gt$$

$$x = (v_x)_0 t \qquad y = (v_y)_0 t - \tfrac{1}{2}gt^2$$

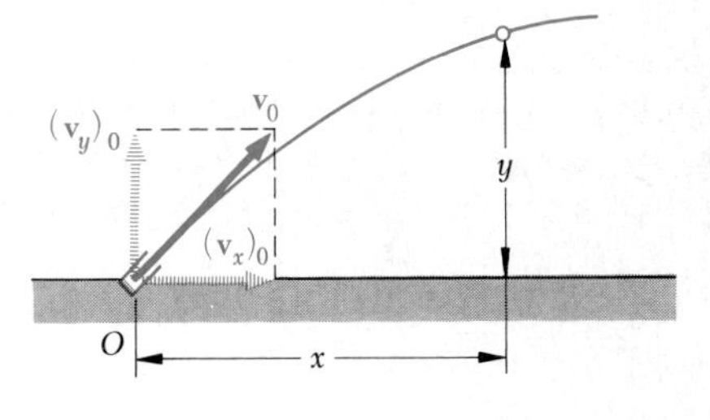

(*a*) Motion of a projectile

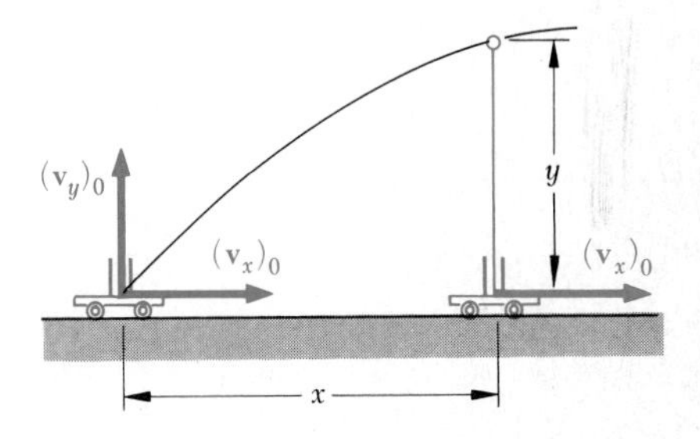

(*b*) Equivalent rectilinear motions

Fig. 11.18

These equations show that the motion of the projectile in the horizontal direction is uniform, while its motion in the vertical direction is uniformly accelerated. The motion of a projectile may thus be replaced by two independent rectilinear motions, which are easily visualized if we assume that the projectile is fired vertically with an initial velocity $(\mathbf{v}_y)_0$ from a platform moving with a constant horizontal velocity $(\mathbf{v}_x)_0$ (Fig. 11.18). The coordinate x of the projectile is equal at any instant to the distance traveled by the platform, while its coordinate y may be computed as if the projectile were moving along a vertical line.

It may be observed that the equations defining the coordinates x and y of a projectile at any instant are the parametric equations of a parabola. Thus, the trajectory of a projectile is *parabolic*. This result, however, ceases to be valid when the resistance of the air or the variation of g with altitude is taken into account.

11.11. Relative Motion. Consider two particles A and B moving in the same plane (Fig. 11.19); the vectors $\mathbf{r}_A$ and $\mathbf{r}_B$ define their position at any given instant with respect to a fixed system of axes xy centered at O. Consider now a system of axes $x'y'$ centered at A and parallel to the x and y axes. While the origin of these axes moves, their orientation remains the same; such a system of axes is said to be in *translation*. The vector $\mathbf{r}_{B/A}$ joining A and B defines the position of the particle B with respect to the moving reference system $x'y'$; it is therefore called the position vector of B relative to the reference system centered at A or, for short, the *position vector of B relative to A.*

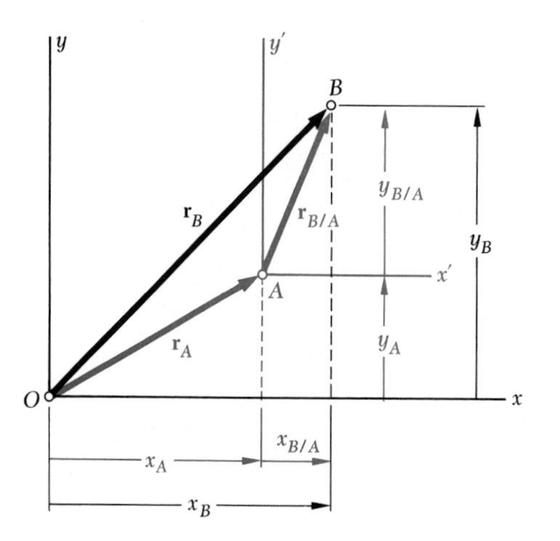

Fig. 11.19

We note from Fig. 11.19 that the position vector $\mathbf{r}_B$ of particle B may be obtained by adding by the triangle rule the position vector $\mathbf{r}_A$ of particle A and the position vector $\mathbf{r}_{B/A}$ of B relative to A; we write

$$\mathbf{r}_B = \mathbf{r}_A + \mathbf{r}_{B/A} \tag{11.21}$$

In terms of x and y components, this relation reads

$$x_B = x_A + x_{B/A} \qquad y_B = y_A + y_{B/A} \tag{11.22}$$

where x_B and y_B denote the coordinates of particle B with respect to the fixed x and y axes, while $x_{B/A}$ and $y_{B/A}$ represent the coordinates of B with respect to the moving x' and y' axes. Differentiating Eqs. (11.22) with respect to t and using dots to indicate time derivatives, we write

$$\dot{x}_B = \dot{x}_A + \dot{x}_{B/A} \qquad \dot{y}_B = \dot{y}_A + \dot{y}_{B/A} \tag{11.23}$$

The derivatives $\dot{x}_A$ and $\dot{y}_A$ represent the x and y components of the velocity $\mathbf{v}_A$ of particle A, while the derivatives $\dot{x}_B$ and $\dot{y}_B$ represent the components of the velocity $\mathbf{v}_B$ of particle B. The vector $\mathbf{v}_{B/A}$ of components $\dot{x}_{B/A}$ and $\dot{y}_{B/A}$ is known as the *relative velocity of B with respect to A* or, more accurately, as the relative velocity of B with respect to a reference system in translation with A. Writing Eqs. (11.23) in vector form, we have

$$\mathbf{v}_B = \mathbf{v}_A + \mathbf{v}_{B/A} \tag{11.24}$$

Differentiating Eqs. (11.23), we write

$$\ddot{x}_B = \ddot{x}_A + \ddot{x}_{B/A} \qquad \ddot{y}_B = \ddot{y}_A + \ddot{y}_{B/A} \tag{11.25}$$

where the first two derivatives in each equation represent the components of the accelerations $\mathbf{a}_B$ and $\mathbf{a}_A$. The vector $\mathbf{a}_{B/A}$ of components $\ddot{x}_{B/A}$ and $\ddot{y}_{B/A}$ is known as the *relative acceleration of B with respect to A* or, more accurately, as the relative acceleration of B with respect to a reference system in translation with A. Writing Eqs. (11.25) in vector form, we obtain

$$\mathbf{a}_B = \mathbf{a}_A + \mathbf{a}_{B/A} \tag{11.26}$$

The motion of B with respect to the fixed x and y axes is referred to as the *absolute motion of B*, while the motion of B with respect to the x' and y' axes attached at A is referred to as the *relative motion of B with respect to A*. The equations derived in this section show that the *absolute motion of B may be obtained by combining the motion of A and the relative motion of B with respect to A*. Equation (11.24), for example, expresses that the absolute velocity $\mathbf{v}_B$ of particle B may be obtained by adding vectorially the velocity of A and the velocity of B relative to A. Equation (11.26) expresses a similar property in terms of the accelerations.† We should keep in mind, however, that the x' and y' axes are in translation, i.e., that, while they move with A, they maintain the same orientation. As we shall see later (Sec. 15.11), different relations must be used when the moving reference axes rotate.

†Note that the product of the subscripts A and B/A used in the right-hand member of Eqs. (11.21) through (11.26) is equal to the subscript B used in their left-hand member.

SAMPLE PROBLEM 11.7

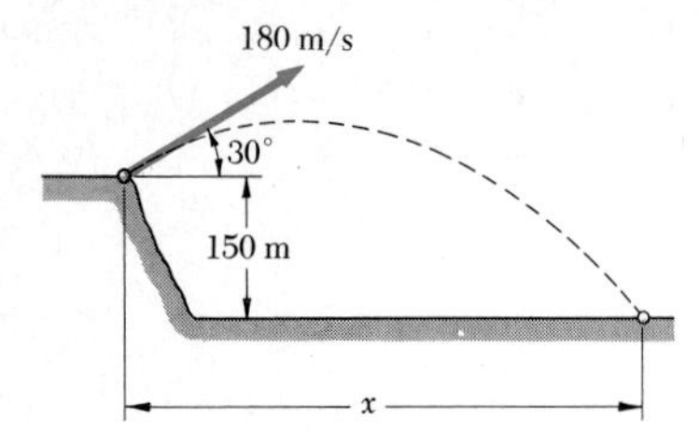

A projectile is fired from the edge of a 150-m cliff with an initial velocity of 180 m/s, at an angle of 30° with the horizontal. Neglecting air resistance, find (*a*) the horizontal distance from the gun to the point where the projectile strikes the ground, (*b*) the greatest elevation above the ground reached by the projectile.

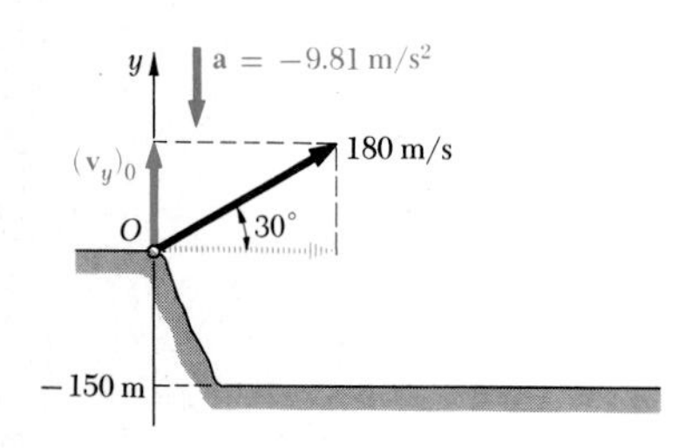

Solution. We shall consider separately the vertical and the horizontal motion.

Vertical Motion. Uniformly accelerated motion. Choosing the positive sense of the y axis upward and placing the origin O at the gun, we have

$$(v_y)_0 = (180 \text{ m/s}) \sin 30° = +90 \text{ m/s}$$
$$a = -9.81 \text{ m/s}^2$$

Substituting into the equations of uniformly accelerated motion, we have

$$v_y = (v_y)_0 + at \qquad v_y = 90 - 9.81t \qquad (1)$$
$$y = (v_y)_0 t + \tfrac{1}{2}at^2 \qquad y = 90t - 4.90t^2 \qquad (2)$$
$$v_y^2 = (v_y)_0^2 + 2ay \qquad v_y^2 = 8100 - 19.62y \qquad (3)$$

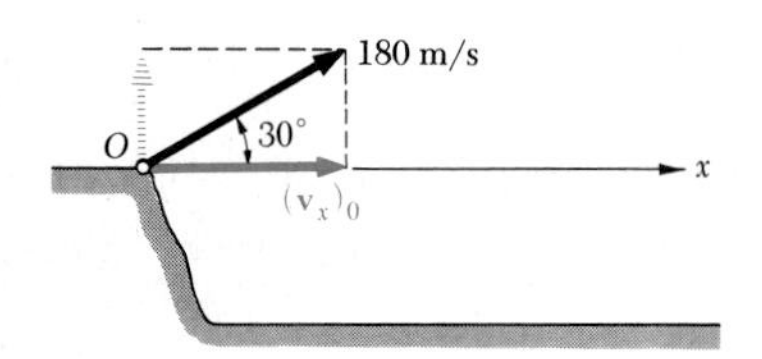

Horizontal Motion. Uniform motion. Choosing the positive sense of the x axis to the right, we have

$$(v_x)_0 = (180 \text{ m/s}) \cos 30° = +155.9 \text{ m/s}$$

Substituting into the equation of uniform motion, we obtain

$$x = (v_x)_0 t \qquad x = 155.9t \qquad (4)$$

***a.* Horizontal Distance.** When the projectile strikes the ground, we have

$$y = -150 \text{ m}$$

Carrying this value into Eq. (2) for the vertical motion, we write

$$-150 = 90t - 4.90t^2 \qquad t^2 - 18.37t - 30.6 = 0 \qquad t = 19.91 \text{ s}$$

Carrying $t = 19.91$ s into Eq. (4) for the horizontal motion, we obtain

$$x = 155.9(19.91) \qquad x = 3100 \text{ m} \blacktriangleleft$$

***b.* Greatest Elevation.** When the projectile reaches its greatest elevation, we have $v_y = 0$; carrying this value into Eq. (3) for the vertical motion, we write

$$0 = 8100 - 19.62y \qquad y = 413 \text{ m}$$

$$\text{Greatest elevation above ground} = 150 \text{ m} + 413 \text{ m}$$
$$= 563 \text{ m} \blacktriangleleft$$

SAMPLE PROBLEM 11.8

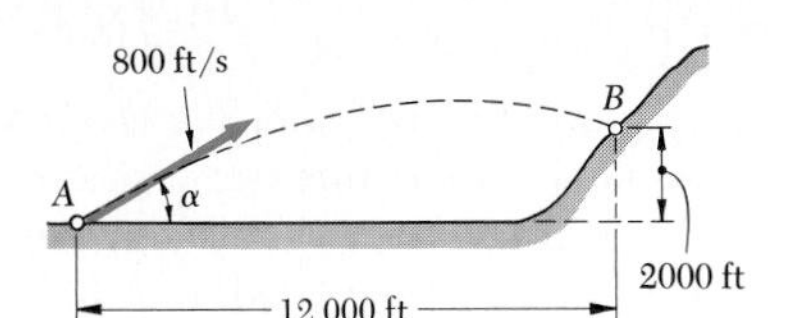

A projectile is fired with an initial velocity of 800 ft/s at a target B located 2000 ft above the gun A and at a horizontal distance of 12,000 ft. Neglecting air resistance, determine the value of the firing angle α.

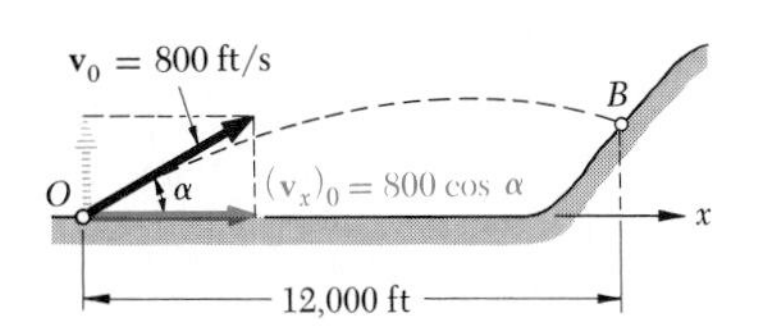

Solution. We shall consider separately the horizontal and the vertical motion.

Horizontal Motion. Placing the origin of coordinates at the gun, we have

$$(v_x)_0 = 800 \cos \alpha$$

Substituting into the equation of uniform horizontal motion, we obtain

$$x = (v_x)_0 t \qquad x = (800 \cos \alpha)t$$

The time required for the projectile to move through a horizontal distance of 12,000 ft is obtained by making x equal to 12,000 ft.

$$12{,}000 = (800 \cos \alpha)t$$

$$t = \frac{12{,}000}{800 \cos \alpha} = \frac{15}{\cos \alpha}$$

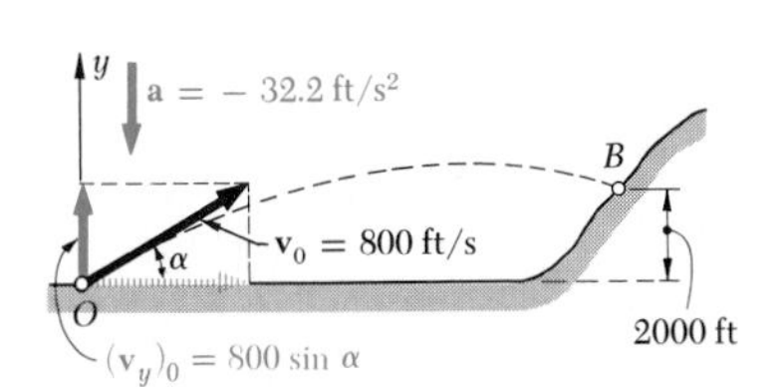

Vertical Motion

$$(v_y)_0 = 800 \sin \alpha \qquad a = -32.2 \text{ ft/s}^2$$

Substituting into the equation of uniformly accelerated vertical motion, we obtain

$$y = (v_y)_0 t + \tfrac{1}{2}at^2 \qquad y = (800 \sin \alpha)t - 16.1t^2$$

Projectile Hits Target. When $x = 12{,}000$ ft, we must have $y = 2000$ ft. Substituting for y and making t equal to the value found above, we write

$$2000 = 800 \sin \alpha \frac{15}{\cos \alpha} - 16.1\left(\frac{15}{\cos \alpha}\right)^2$$

Since $1/\cos^2 \alpha = \sec^2 \alpha = 1 + \tan^2 \alpha$, we have

$$2000 = 800(15) \tan \alpha - 16.1(15^2)(1 + \tan^2 \alpha)$$
$$3622 \tan^2 \alpha - 12{,}000 \tan \alpha + 5622 = 0$$

Solving this quadratic equation for $\tan \alpha$, we have

$$\tan \alpha = 0.565 \qquad \text{and} \qquad \tan \alpha = 2.75$$

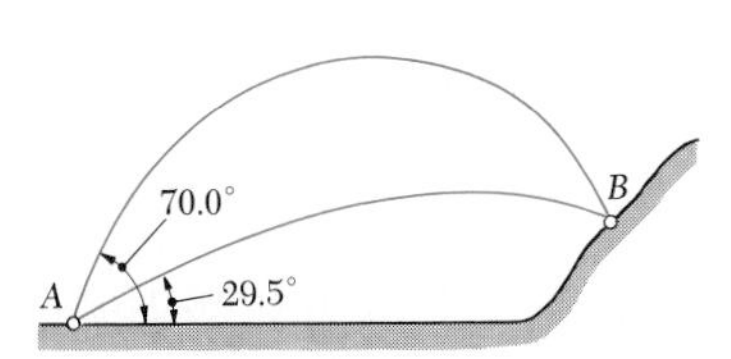

$$\alpha = 29.5° \qquad \text{and} \qquad \alpha = 70.0°$$ ◄

The target will be hit if either of these two firing angles is used (see figure).

SAMPLE PROBLEM 11.9

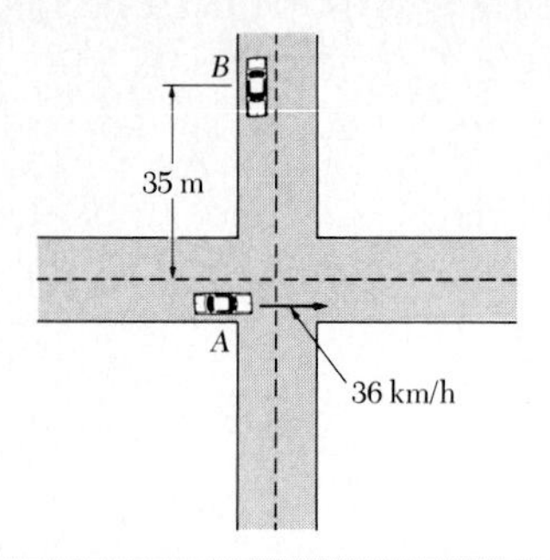

Automobile A is traveling east at the constant speed of 36 km/h. As automobile A crosses the intersection shown, automobile B starts from rest 35 m north of the intersection and moves south with a constant acceleration of 1.2 m/s². Determine the position, velocity, and acceleration of B relative to A five seconds after A crosses the intersection.

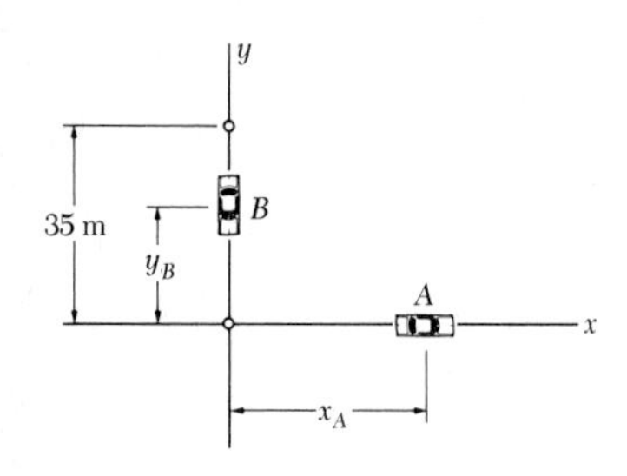

Solution. We choose x and y axes with origin at the intersection of the two streets and with positive senses directed respectively east and north.

Motion of Automobile A. First the speed is expressed in m/s:

$$v_A = \left(36\frac{\text{km}}{\text{h}}\right)\left(\frac{1000\text{ m}}{1\text{ km}}\right)\left(\frac{1\text{ h}}{3600\text{ s}}\right) = 10\text{ m/s}$$

Noting that the motion of A is uniform, we write, for any time t,

$$a_A = 0$$
$$v_A = +10\text{ m/s}$$
$$x_A = (x_A)_0 + v_A t = 0 + 10t$$

For $t = 5$ s, we have

$$a_A = 0 \qquad \mathbf{a}_A = 0$$
$$v_A = +10\text{ m/s} \qquad \mathbf{v}_A = 10\text{ m/s} \rightarrow$$
$$x_A = +(10\text{ m/s})(5\text{ s}) = +50\text{ m} \qquad \mathbf{r}_A = 50\text{ m} \rightarrow$$

Motion of Automobile B. We note that the motion of B is uniformly accelerated, and write

$$a_B = -1.2\text{ m/s}^2$$
$$v_B = (v_B)_0 + at = 0 - 1.2t$$
$$y_B = (y_B)_0 + (v_B)_0 t + \tfrac{1}{2}a_B t^2 = 35 + 0 - \tfrac{1}{2}(1.2)t^2$$

For $t = 5$ s, we have

$$a_B = -1.2\text{ m/s}^2 \qquad \mathbf{a}_B = 1.2\text{ m/s}^2 \downarrow$$
$$v_B = -(1.2\text{ m/s}^2)(5\text{ s}) = -6\text{ m/s} \qquad \mathbf{v}_B = 6\text{ m/s} \downarrow$$
$$y_B = 35 - \tfrac{1}{2}(1.2\text{ m/s}^2)(5\text{ s})^2 = +20\text{ m} \qquad \mathbf{r}_B = 20\text{ m} \uparrow$$

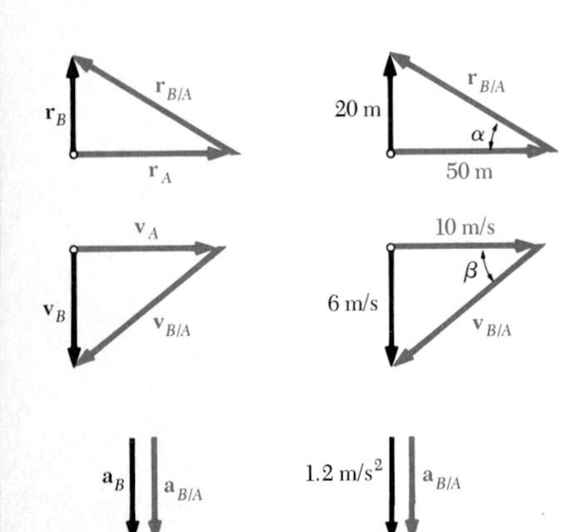

Motion of B Relative to A. We draw the triangle corresponding to the vector equation $\mathbf{r}_B = \mathbf{r}_A + \mathbf{r}_{B/A}$ and obtain the magnitude and direction of the position vector of B relative to A.

$$r_{B/A} = 53.9\text{ m} \qquad \alpha = 21.8° \qquad \mathbf{r}_{B/A} = 53.9\text{ m} \measuredangle 21.8° \blacktriangleleft$$

Proceeding in a similar fashion, we find the velocity and acceleration of B relative to A.

$$\mathbf{v}_B = \mathbf{v}_A + \mathbf{v}_{B/A}$$
$$v_{B/A} = 11.66\text{ m/s} \qquad \beta = 31.0° \qquad \mathbf{v}_{B/A} = 11.66\text{ m/s} \measuredangle 31.0° \blacktriangleleft$$
$$\mathbf{a}_B = \mathbf{a}_A + \mathbf{a}_{B/A} \qquad \mathbf{a}_{B/A} = 1.2\text{ m/s}^2 \downarrow \blacktriangleleft$$

Problems

Note. **Neglect air resistance in problems concerning projectiles.**

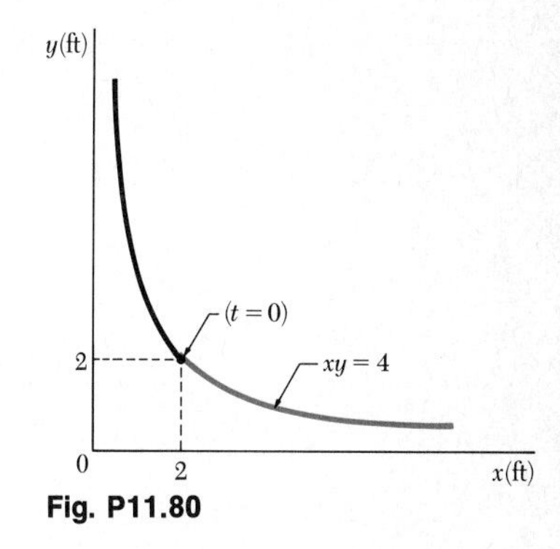

Fig. P11.80

11.79 The motion of a particle is defined by the equations $x = 1.5t^2 - 6t$ and $y = 6t^2 - 2t^3$, where x and y are expressed in inches and t in seconds. Determine the velocity and acceleration when (*a*) $t = 1$ s, (*b*) $t = 2$ s, (*c*) $t = 3$ s.

11.80 The motion of a particle is defined by the equations $x = 2(t + 1)^2$ and $y = 2(t + 1)^{-2}$, where x and y are expressed in feet and t in seconds. Show that the path of the particle is part of the rectangular hyperbola shown and determine the velocity and acceleration when (*a*) $t = 0$, (*b*) $t = \frac{1}{2}$ s.

11.81 The motion of a particle is defined by the equations $x = 2t^2 - 4t$ and $y = 2(t - 1)^2 - 4(t - 1)$, where x and y are expressed in meters and t in seconds. Determine (*a*) the magnitude of the smallest velocity reached by the particle, (*b*) the corresponding time, position, and direction of the velocity.

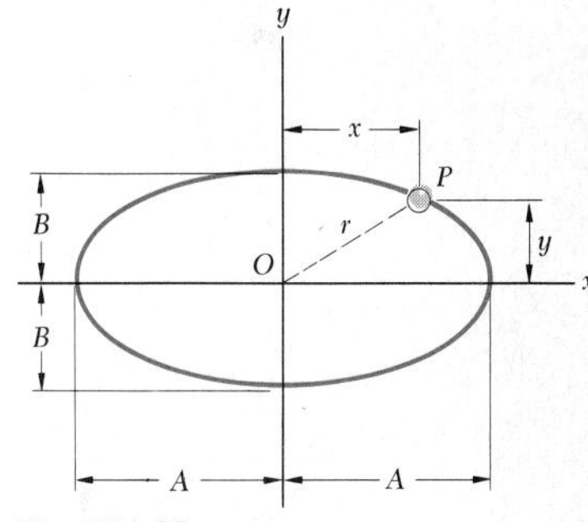

Fig. P11.83

11.82 The motion of a particle is defined by the equations $x = 120 \sin \frac{1}{2}\pi t$ and $y = 40t^2$, where x and y are expressed in millimeters and t in seconds. Determine the velocity and acceleration of the particle when (*a*) $t = 1$ s, (*b*) $t = 2$ s.

11.83 A particle moves in an elliptic path according to the equations $x = A \cos pt$ and $y = B \sin pt$. Show that the acceleration (*a*) is directed toward the origin, (*b*) is proportional to the distance from the origin to the particle.

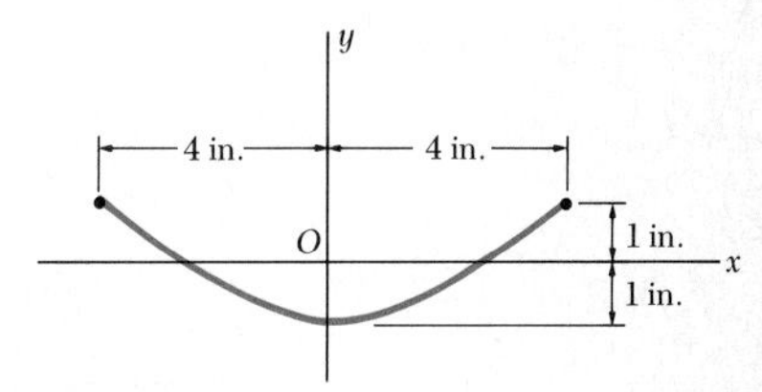

Fig. P11.84

11.84 The motion of a vibrating particle is defined by the equations $x = 4 \sin \pi t$ and $y = -\cos 2\pi t$, where x and y are expressed in inches and t in seconds. (*a*) Determine the velocity and acceleration when $t = 1$ s. (*b*) Show that the path of the particle is parabolic.

11.85 A particle moves in a circular path according to the equations $x = r \sin \omega t$ and $y = r \cos \omega t$. Show (*a*) that the magnitude v of the velocity is constant and equal to $r\omega$, (*b*) that the acceleration is of constant magnitude $r\omega^2$ (or v^2/r) and is directed toward the center of the circular path.

11.86 The motion of a vibrating particle is defined by the equations $x = A \sin pt$ and $y = B \sin (pt + \phi)$, where A, B, p, and ϕ are constants. If $A = B$, determine (*a*) the value of the constant ϕ for which the speed of the particle is constant, (*b*) the corresponding value of the speed, (*c*) the corresponding path of the particle.

11.87 A man standing at the 18-m level of a tower throws a stone in a horizontal direction. Knowing that the stone hits the ground 25 m from the bottom of the tower, determine (*a*) the initial velocity of the stone, (*b*) the distance at which a stone would hit the ground if it were thrown horizontally with the same velocity from the 22-m level of the tower.

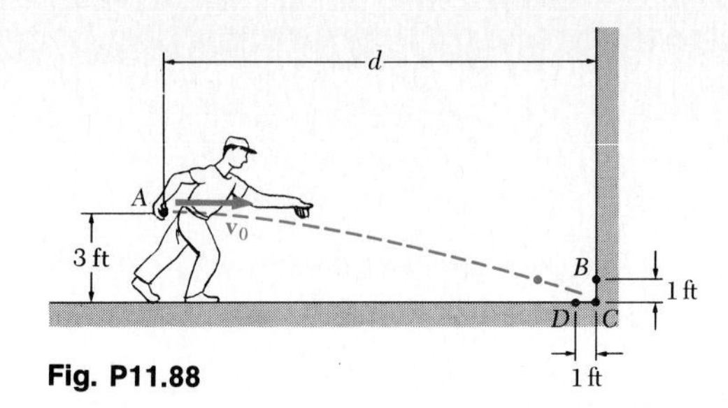

Fig. P11.88

11.88 A handball player throws a ball from A with a horizontal velocity $\mathbf{v}_0$. Knowing that $d = 20$ ft, determine (*a*) the value of v_0 for which the ball will strike the corner C, (*b*) the range of values of v_0 for which the ball will strike the corner region BCD.

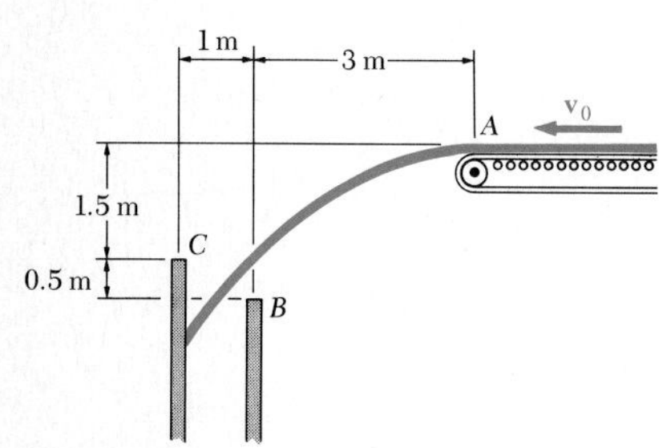

Fig. P11.90

11.89 Solve Prob. 11.88, assuming that the distance from point A to the wall is $d = 30$ ft.

11.90 Sand is discharged at A from a horizontal conveyor belt with an initial velocity $\mathbf{v}_0$. Determine the range of values of v_0 for which the sand will enter the vertical chute shown.

11.91 A projectile is fired with an initial velocity of 1000 ft/s at a target balloon located 1000 ft above the ground. Knowing that the projectile hits the balloon, determine (*a*) the horizontal distance d, (*b*) the time of flight of the projectile.

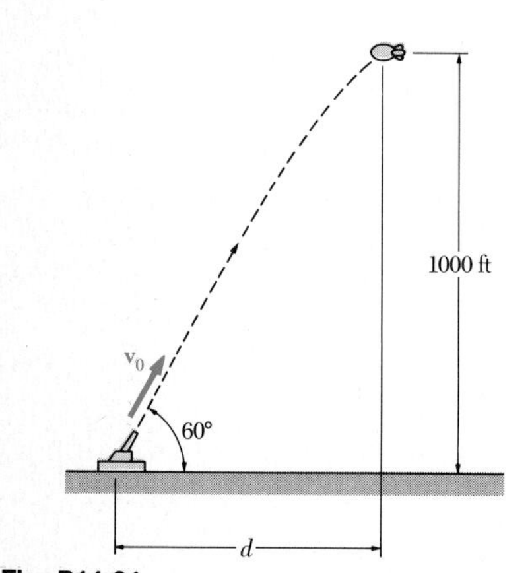

Fig. P11.91

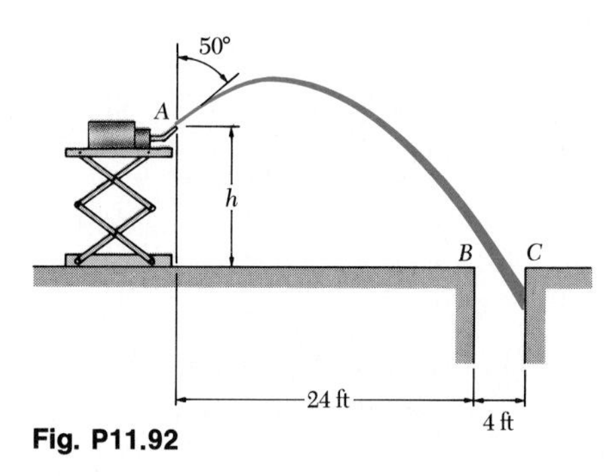

Fig. P11.92

11.92 A pump is located near the edge of the horizontal platform shown. The nozzle at A discharges water with an initial velocity of 25 ft/s at an angle of 50° with the vertical. Determine the range of values of the height h for which the water enters the opening BC.

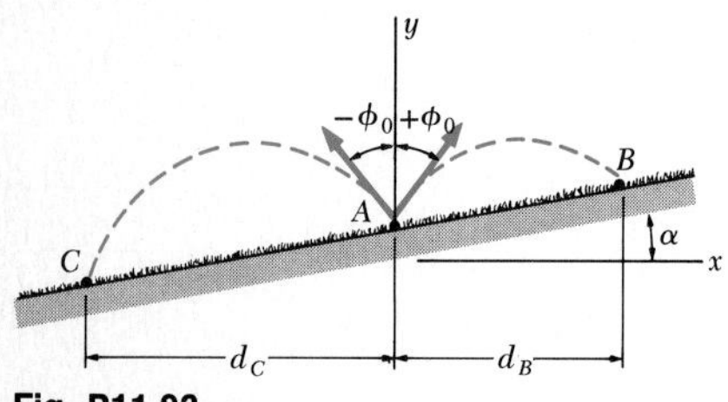

Fig. P11.93

11.93 An oscillating water sprinkler is operated at point A on an incline which forms an angle α with the horizontal. The sprinkler discharges water with an initial velocity $\mathbf{v}_0$ at an angle ϕ with the vertical which varies from $-\phi_0$ to $+\phi_0$. Knowing that $v_0 = 10$ m/s, $\phi_0 = 40°$, and $\alpha = 10°$, determine the horizontal distance between the sprinkler and points B and C which define the watered area.

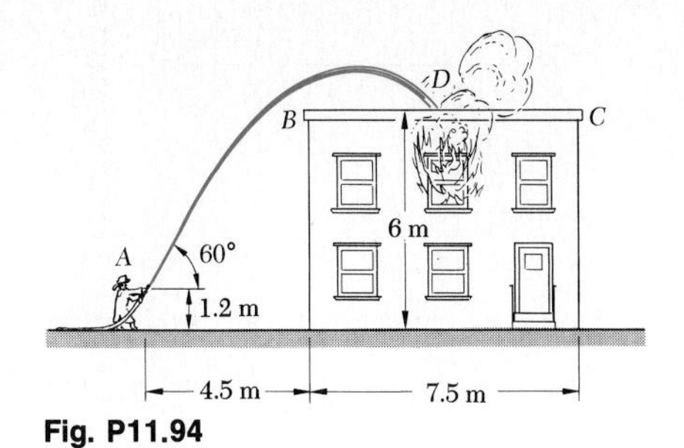

Fig. P11.94

11.94 A nozzle at A discharges water with an initial velocity of 12 m/s at an angle of 60° with the horizontal. Determine where the stream of water strikes the roof. Check that the stream will clear the edge of the roof.

11.95 In Prob. 11.94, determine the range of values of the initial velocity for which the water will fall on the roof.

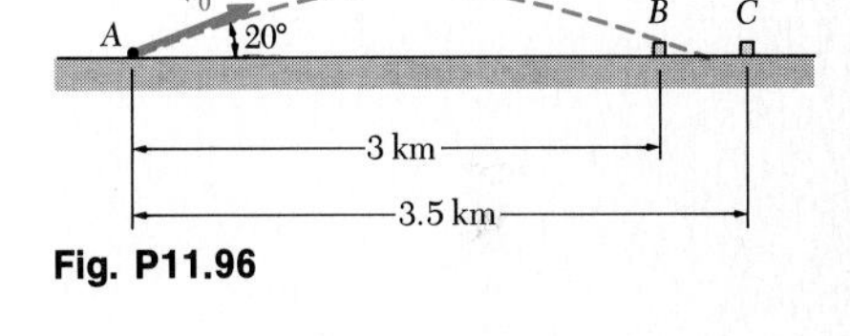

Fig. P11.96

11.96 A projectile is fired with an initial velocity $\mathbf{v}_0$ at an angle of 20° with the horizontal. Determine the required value of v_0 if the projectile is to hit (*a*) point B, (*b*) point C.

11.97 Sand is discharged at A from a conveyor belt and falls onto the top of a stockpile at B. Knowing that the conveyor belt forms an angle $\alpha = 20°$ with the horizontal, determine the speed v_0 of the belt.

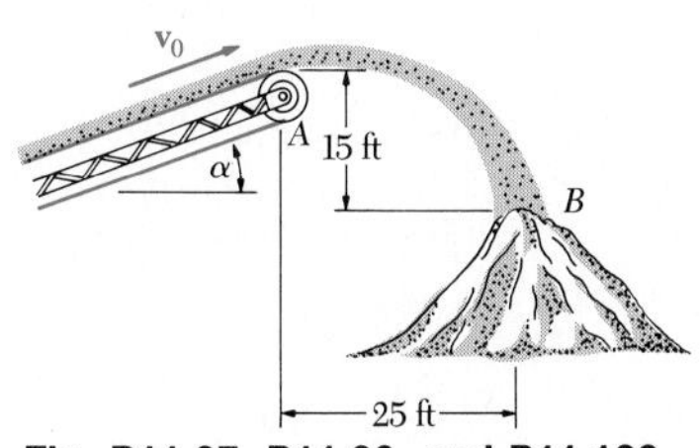

Fig. P11.97, P11.99, and P11.100

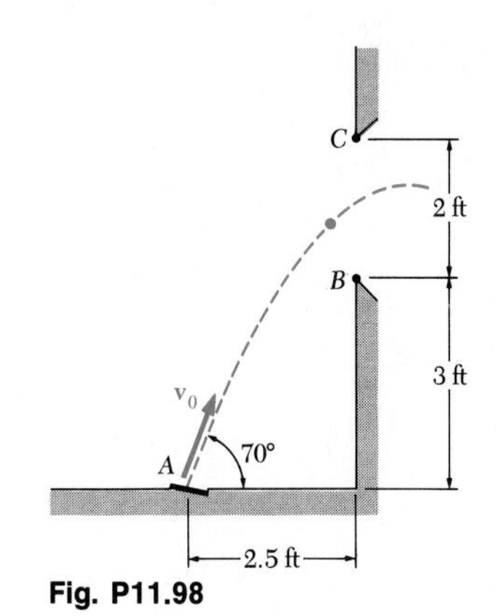

Fig. P11.98

11.98 A ball is dropped onto a pad at A and rebounds with a velocity $\mathbf{v}_0$ at an angle of 70° with the horizontal. Determine the range of values of v_0 for which the ball will enter the opening BC.

11.99 Knowing that the conveyor belt moves at the constant speed $v_0 = 24$ ft/s, determine the angle α for which the sand is deposited on the stockpile at B.

* **11.100** Knowing that the conveyor belt moves at the constant speed v_0, determine (*a*) the smallest value of v_0 for which sand can be deposited on the stockpile at B, (*b*) the corresponding value of α.

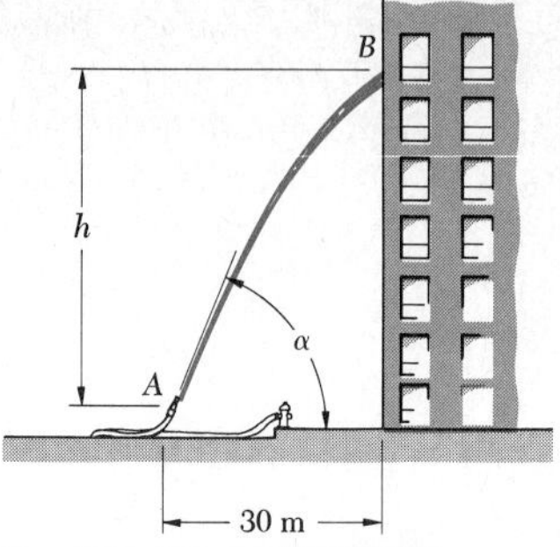

Fig. P11.101 and P11.102

***11.101** A fire nozzle discharges water with an initial velocity $\mathbf{v}_0$ of 24 m/s. Knowing that the nozzle is located 30 m from a building, determine (*a*) the maximum height h that can be reached by the water, (*b*) the corresponding angle α.

11.102 A fire nozzle located at A discharges water with an initial velocity $\mathbf{v}_0$ of 24 m/s. Knowing that the stream of water strikes the building at a height $h = 18$ m above the ground, determine the angle α.

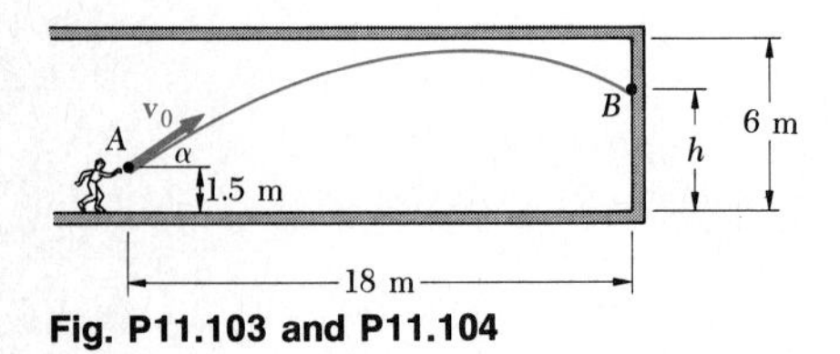

Fig. P11.103 and P11.104

11.103 A player throws a ball with an initial velocity $\mathbf{v}_0$ of 15 m/s from a point A located 1.5 m above the floor. Knowing that $h = 3$ m, determine the angle α for which the ball will strike the wall at point B.

11.104 A player throws a ball with an initial velocity $\mathbf{v}_0$ of 15 m/s from a point A located 1.5 m above the floor. Knowing that the ceiling of the gymnasium is 6 m high, determine the highest point B at which the ball can strike the wall 18 m away.

11.105 The velocities of boats A and C are as shown and the relative velocity of boat B with respect to A is $\mathbf{v}_{B/A} = 4$ m/s ∠ 50°. Determine (*a*) $v_{A/C}$, (*b*) $v_{C/B}$, (*c*) the change in position of B with respect to C during a 10-s interval. Also show that for any motion, $\mathbf{v}_{B/A} + \mathbf{v}_{C/B} + \mathbf{v}_{A/C} = 0$.

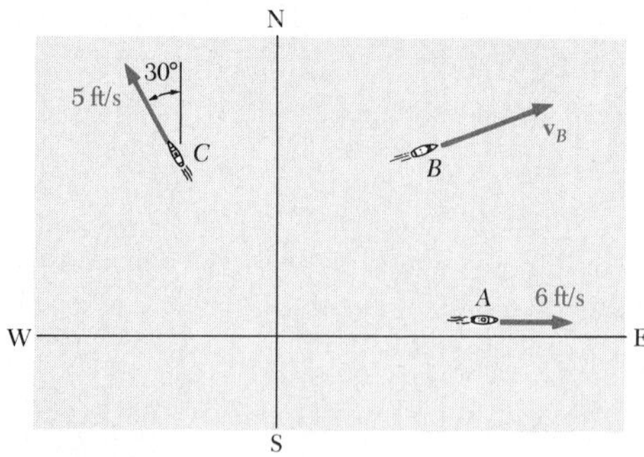

Fig. P11.105

11.106 Instruments in an airplane indicate that with respect to the air, the plane is moving north at a speed of 500 km/h. At the same time ground-based radar indicates that the plane is moving at a speed of 530 km/h in a direction 5° east of north. Determine the magnitude and direction of the velocity of the air.

11.107 Two airplanes A and B are flying at the same altitude; plane A is flying due east at a constant speed of 900 km/h, while plane B is flying southwest at a constant speed of 600 km/h. Determine the change in position of plane B relative to plane A, which takes place during a 2-min interval.

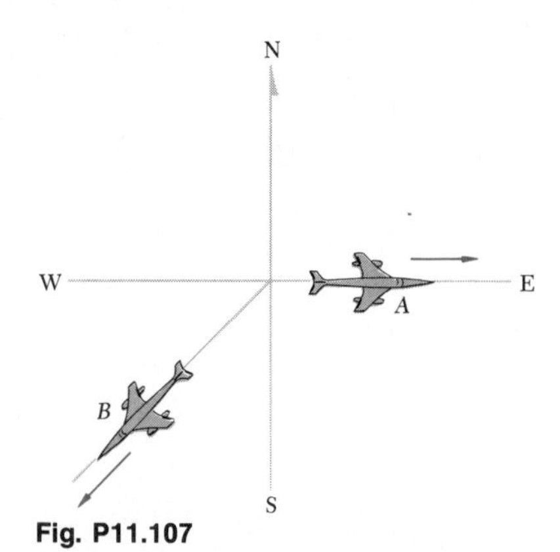

Fig. P11.107

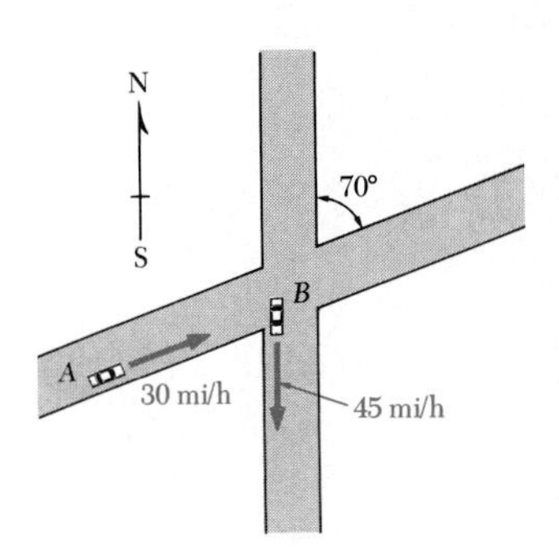

Fig. P11.108

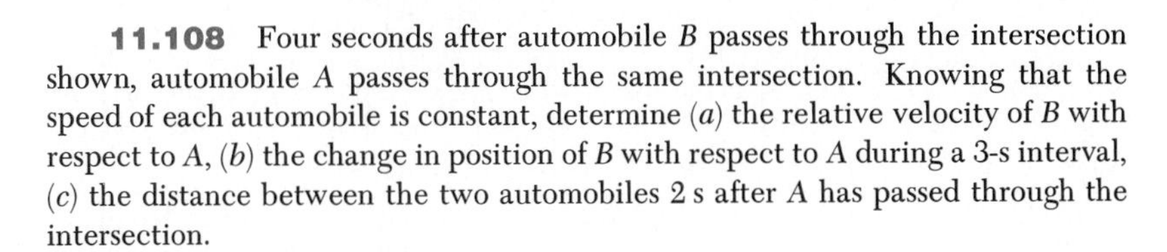

11.108 Four seconds after automobile B passes through the intersection shown, automobile A passes through the same intersection. Knowing that the speed of each automobile is constant, determine (*a*) the relative velocity of B with respect to A, (*b*) the change in position of B with respect to A during a 3-s interval, (*c*) the distance between the two automobiles 2 s after A has passed through the intersection.

11.109 Pin P moves with a constant speed of 90 mm/s in a counterclockwise sense along a circular slot which has been milled in the slider block A shown. Knowing that the block moves downward at a constant speed of 60 mm/s, determine the velocity of pin P when (*a*) $\theta = 30°$, (*b*) 120°.

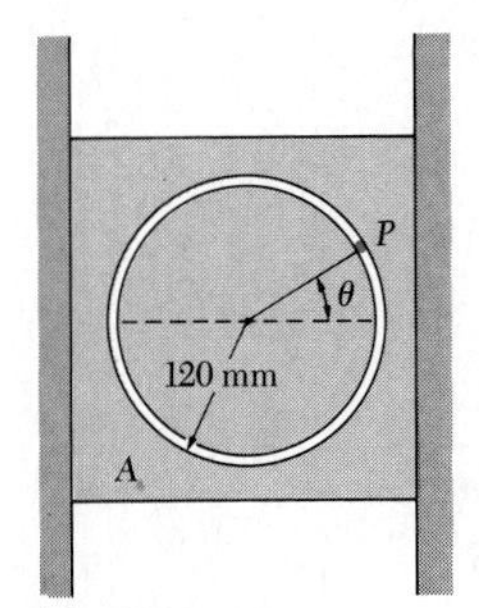

Fig. P11.109

11.110 At $t = 0$, wedge A starts moving to the right with a constant acceleration of 100 mm/s² and block B starts moving along the wedge toward the left with a constant acceleration of 150 mm/s² relative to the wedge. Determine (*a*) the acceleration of block B, (*b*) the velocity of block B when $t = 4$ s.

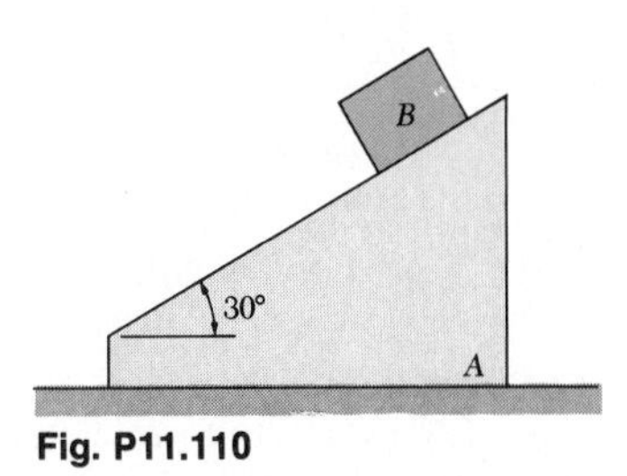

Fig. P11.110

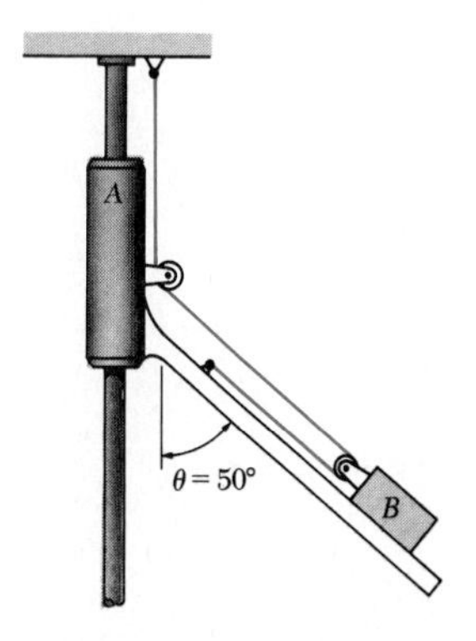

Fig. P11.111

11.111 Knowing that at the instant shown assembly A has a velocity of 16 in./s and an acceleration of 24 in./s² both directed downward, determine (*a*) the velocity of block B, (*b*) the acceleration of block B.

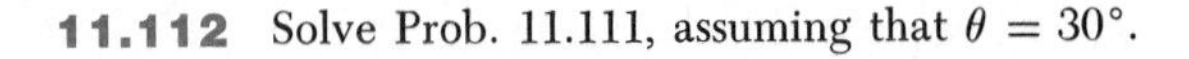

11.112 Solve Prob. 11.111, assuming that $\theta = 30°$.

11.113 Water is discharged at A with an initial velocity of 12 m/s and strikes a series of vanes at B. Knowing that the vanes move downward with a constant speed of 1.5 m/s, determine the velocity and acceleration of the water relative to the vane at B.

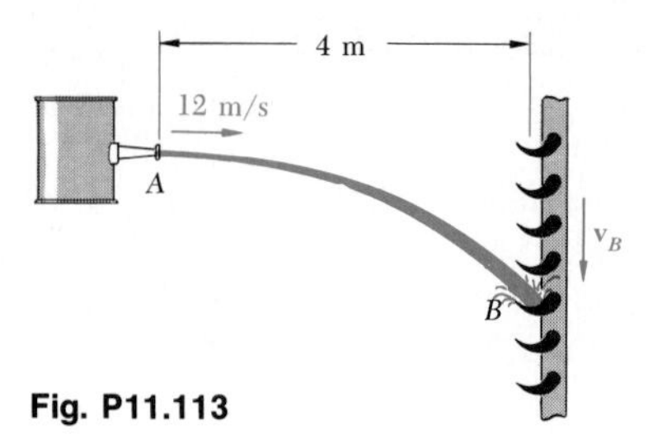

Fig. P11.113

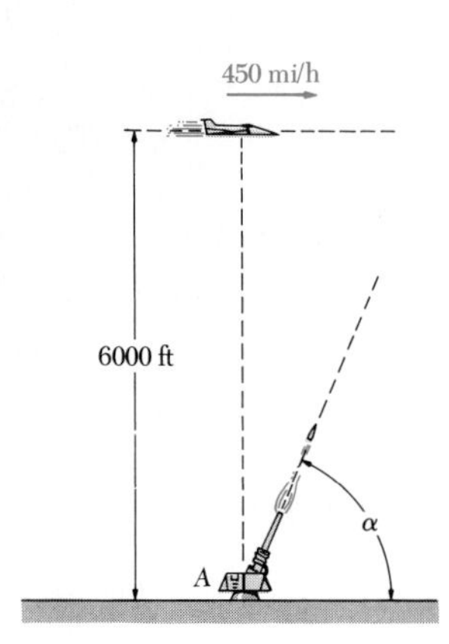

Fig. P11.114

11.114 An antiaircraft gun fires a shell as a plane passes directly over the position of the gun, at an altitude of 6000 ft. The muzzle velocity of the shell is 1500 ft/s. Knowing that the plane is flying horizontally at 450 mi/h, determine (*a*) the required firing angle if the shell is to hit the plane, (*b*) the velocity and acceleration of the shell relative to the plane at the time of impact.

11.115 As observed from a ship moving due east at 8 km/h, the wind appears to blow from the south. After the ship has changed course, and as it is moving due north at 8 km/h, the wind appears to blow from the southwest. Assuming that the wind velocity is constant during the period of observation, determine the magnitude and direction of the true wind velocity.

11.116 A small motorboat maintains a constant speed of 2 m/s relative to the water as it is maneuvering in a tidal current. When the boat is directed due east, it is observed from shore to move due south and when it is directed toward the northeast it is observed to move due west. Determine the speed and direction of the current.

11.117 During a rainstorm the paths of the raindrops appear to form an angle of 30° with the vertical and to be directed to the left when observed from a side window of a train moving at a speed of 10 mi/h. A short time later, after the speed of the train has increased to 15 mi/h, the angle between the vertical and the paths of the drops appears to be 45°. If the train were stopped, at what angle and with what velocity would the drops be observed to fall?

11.118 As the speed of the train of Prob. 11.117 increases, the angle between the vertical and the paths of the drops becomes equal to 60°. Determine the speed of the train at that time.

11.12. Tangential and Normal Components. We saw in Sec. 11.9 that the velocity of a particle is a vector tangent to the path of the particle but that, in general, the acceleration is not tangent to the path. It is sometimes convenient to resolve the acceleration into components directed, respectively, along the tangent and the normal to the path of the particle. The positive sense along the tangent is chosen to coincide with the sense of motion of the particle, and the positive sense along the normal is chosen toward the inside of the path (Fig. 11.20).

We consider again, as we did in Sec. 11.9, the velocity $\mathbf{v}$ of the particle at a time t and its velocity $\mathbf{v}'$ at a time $t + \Delta t$ (Fig. 11.20*a*). Drawing both vectors from the same origin O', and joining the points Q and Q', we obtain the small vector $\Delta\mathbf{v}$ representing the difference between the vectors $\mathbf{v}$ and $\mathbf{v}'$ (Fig. 11.20*b*). As we saw earlier, the average acceleration is defined as the quotient of $\Delta\mathbf{v}$ and Δt, and the instantaneous acceleration $\mathbf{a}$ as the limit of this quotient. To obtain the tangential component $\mathbf{a}_t$ and the normal component $\mathbf{a}_n$ of the acceleration, we measure the distance $O'Q'' = O'Q$ along the line $O'Q'$ and resolve the vector $\Delta\mathbf{v}$ into its components $\overrightarrow{QQ''}$ and $\overrightarrow{Q''Q'}$ (Fig. 11.20*b*); we write

$$\Delta\mathbf{v} = \overrightarrow{QQ''} + \overrightarrow{Q''Q'}$$

The component $\overrightarrow{QQ''}$ represents a change in the *direction* of the velocity, while the component $\overrightarrow{Q''Q'}$ represents a change in the *magnitude* of the velocity, i.e., a change in *speed*. Substituting in (11.18) the expression obtained for $\Delta\mathbf{v}$, we write

$$\mathbf{a} = \lim_{\Delta t\to 0}\frac{\Delta\mathbf{v}}{\Delta t} = \lim_{\Delta t\to 0}\frac{\overrightarrow{QQ''}}{\Delta t} + \lim_{\Delta t\to 0}\frac{\overrightarrow{Q''Q'}}{\Delta t}$$

As Δt approaches zero, the vector $\overrightarrow{Q''Q'}$ becomes parallel to the tangent to the path at P and the vector $\overrightarrow{QQ''}$ becomes parallel to the normal. We must have, therefore,

$$\mathbf{a}_t = \lim_{\Delta t\to 0}\frac{\overrightarrow{Q''Q'}}{\Delta t} \qquad \mathbf{a}_n = \lim_{\Delta t\to 0}\frac{\overrightarrow{QQ''}}{\Delta t} \tag{11.27}$$

Considering first the tangential component, we observe that the length of the vector $\overrightarrow{Q''Q'}$ measures the change in speed $v' - v = \Delta v$. The scalar component a_t of the acceleration is therefore

$$a_t = \lim_{\Delta t\to 0}\frac{\Delta v}{\Delta t}$$

$$a_t = \frac{dv}{dt} \tag{11.28}$$

Formula (11.28) expresses that the *tangential component* a_t *of the acceleration is equal to the rate of change of the speed of the particle.*

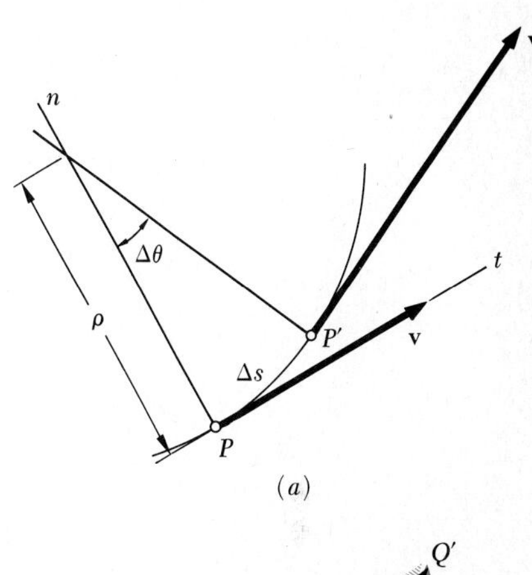

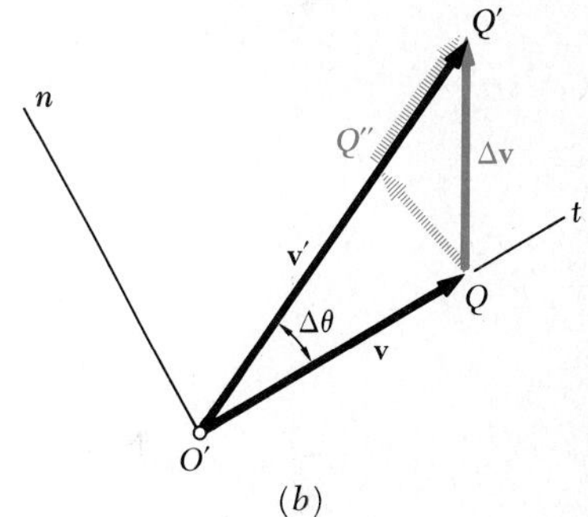

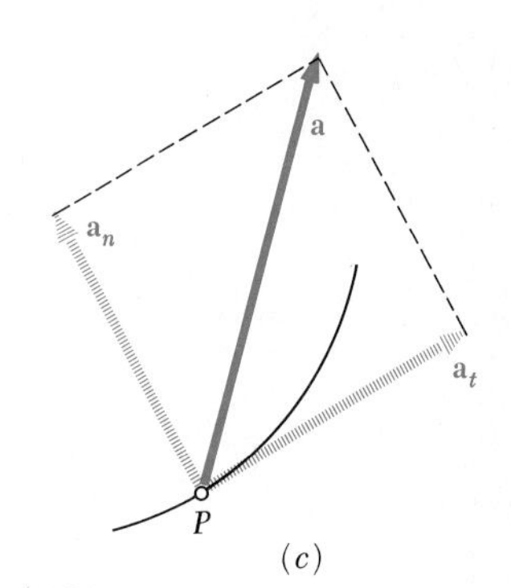

Fig. 11.20

Considering now the normal component of the acceleration, and denoting by $\Delta\theta$ the angle between the tangents to the path at P and P', we observe that, for small values of $\Delta\theta$, the length of the vector $\overrightarrow{QQ''}$ is equal to $v\,\Delta\theta$. Since the angle $\Delta\theta$ is equal to the angle between the normals at P and P' (Fig. 11.20*a*), we have $\Delta\theta = \Delta s/\rho$, where Δs is the length of the arc PP' and ρ is the radius of curvature of the path at P. It thus follows from the

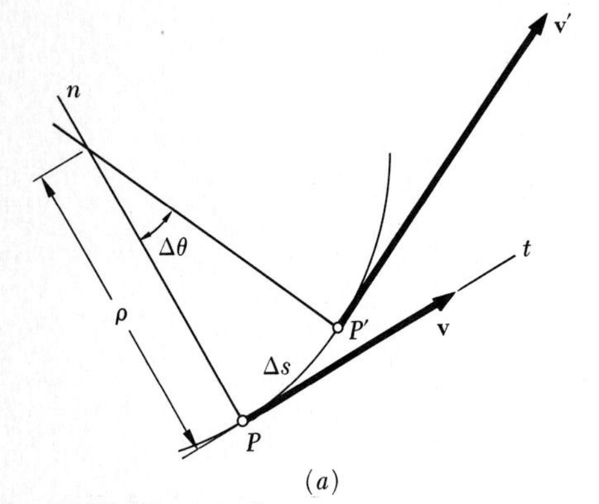

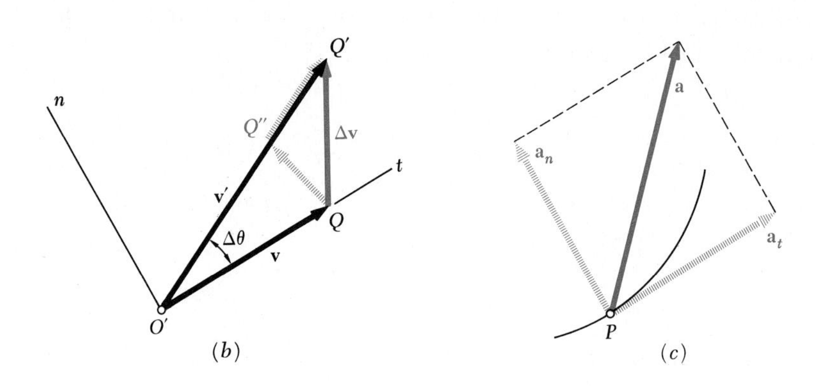

Fig. 11.20 (*Repeated*)

second of the relations (11.27) that the scalar component a_n of the acceleration is

$$a_n = \lim_{\Delta t \to 0} \frac{v\,\Delta\theta}{\Delta t} = \lim_{\Delta t \to 0} \frac{v}{\rho}\frac{\Delta s}{\Delta t} = \frac{v}{\rho}\frac{ds}{dt}$$

Recalling from (11.15) that $ds/dt = v$, we have

$$a_n = \frac{v^2}{\rho} \tag{11.29}$$

Formula (11.29) expresses that the *normal component* a_n of the acceleration at P is equal to the *square of the speed divided by the radius of curvature of the path at P.* It should be noted that the vector $\mathbf{a}_n$ is always directed *toward the center of curvature of the path* (Fig. 11.20*c*).

It appears from the above that the tangential component of the acceleration reflects a change in the speed of the particle, while its normal component reflects a change in the direction of motion of the particle. The acceleration of a particle will be zero only if both its components are zero. Thus, the acceleration of a particle moving with constant speed along a curve will not be zero, unless the particle happens to pass through a point of inflection of the curve (where the radius of curvature is infinite) or unless the curve is a straight line.

The fact that the normal component of the acceleration depends upon the radius of curvature of the path followed by the particle is taken into account in the design of structures or mechanisms as widely different as airplane wings, railroad tracks, and cams. In order to avoid sudden changes in the acceleration of the air particles flowing past a wing, wing profiles are designed without any sudden change in curvature. Similar care is taken in designing railroad curves, to avoid sudden changes in the acceleration of the cars (which would be hard on the equipment and unpleasant for the passengers). A straight section of track, for instance, is never directly followed by a circular section. Special transition sections are used, to help pass smoothly from the infinite radius of curvature of the straight section to the finite radius of the circular track. Likewise, in the design of high-speed cams, abrupt changes in acceleration are avoided by using transition curves which produce a continuous change in acceleration.

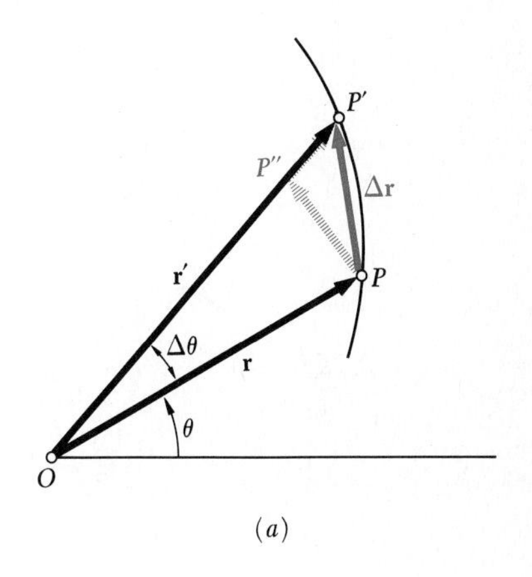

11.13. Radial and Transverse Components. In certain problems, the position of a particle is defined by its polar coordinates r and θ. It is then convenient to resolve the velocity and acceleration of the particle into components parallel and perpendicular, respectively, to the position vector $\mathbf{r}$. These components are called *radial* and *transverse components*.

Consider the vectors $\mathbf{r}$ and $\mathbf{r}'$ defining the position of the particle respectively at times t and $t + \Delta t$ (Fig. 11.21*a*). We recall that the velocity of the particle at time t is defined as the limit of the quotient of $\Delta\mathbf{r}$ and Δt, where $\Delta\mathbf{r}$ is the vector joining P and P'. To obtain the radial component $\mathbf{v}_r$ and the transverse component $\mathbf{v}_\theta$ of the velocity, we measure a distance $OP'' = OP$ along the line OP' and resolve the vector $\Delta\mathbf{r}$ into its components $\overrightarrow{PP''}$ and $\overrightarrow{P''P'}$. The velocity $\mathbf{v}$ may thus be expressed as

$$\mathbf{v} = \lim_{\Delta t\to 0} \frac{\Delta\mathbf{r}}{\Delta t} = \lim_{\Delta t\to 0} \frac{\overrightarrow{PP''}}{\Delta t} + \lim_{\Delta t\to 0} \frac{\overrightarrow{P''P'}}{\Delta t}$$

As Δt approaches zero, the vectors $\overrightarrow{P''P'}$ and $\overrightarrow{PP''}$ become parallel and perpendicular, respectively, to $\mathbf{r}$. We must have, therefore,

$$\mathbf{v}_r = \lim_{\Delta t\to 0} \frac{\overrightarrow{P''P'}}{\Delta t} \qquad \mathbf{v}_\theta = \lim_{\Delta t\to 0} \frac{\overrightarrow{PP''}}{\Delta t}$$

We observe that the length of the vector $\overrightarrow{P''P'}$ measures the change in magnitude $r' - r = \Delta r$ of the radius vector. (Note that the change in magnitude Δr should not be confused with the magnitude of the vector $\Delta\mathbf{r}$, which represents a change in direction as well as in magnitude.) On the other hand, the length of the vector $\overrightarrow{PP''}$ is equal to $r\,\Delta\theta$ for small values of $\Delta\theta$. The scalar components of the velocity are therefore

$$v_r = \lim_{\Delta t\to 0} \frac{\Delta r}{\Delta t} \qquad v_\theta = \lim_{\Delta t\to 0} \frac{r\,\Delta\theta}{\Delta t}$$

$$v_r = \frac{dr}{dt} \qquad v_\theta = r\frac{d\theta}{dt}$$

or, using dots to indicate time derivatives,

$$v_r = \dot{r} \qquad v_\theta = r\dot{\theta} \tag{11.30}$$

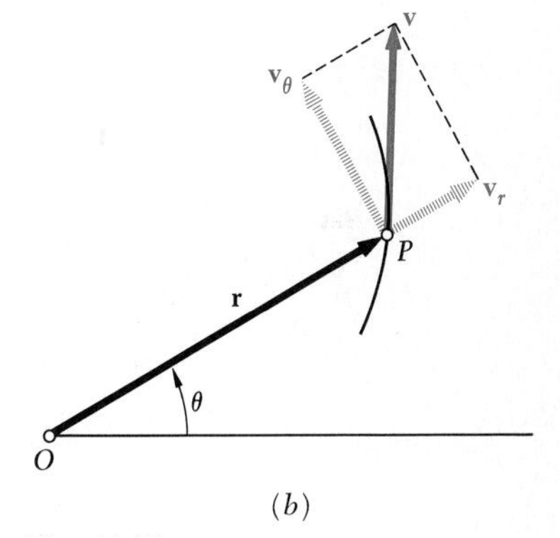

Fig. 11.21

Consider, now, the velocity $\mathbf{v}$ of P at a time t and its velocity $\mathbf{v}'$ at a time $t + \Delta t$ (Fig. 11.22*a*). Drawing the radial and transverse components of $\mathbf{v}$ and $\mathbf{v}'$ from the same origin O', we represent the change in velocity by the two vectors $\overrightarrow{RR'}$ and $\overrightarrow{TT'}$ (Fig. 11.22*b*). The vector $\overrightarrow{RR'}$ represents a change in magnitude and direction for the vector $\mathbf{v}_r$; similarly, the vector

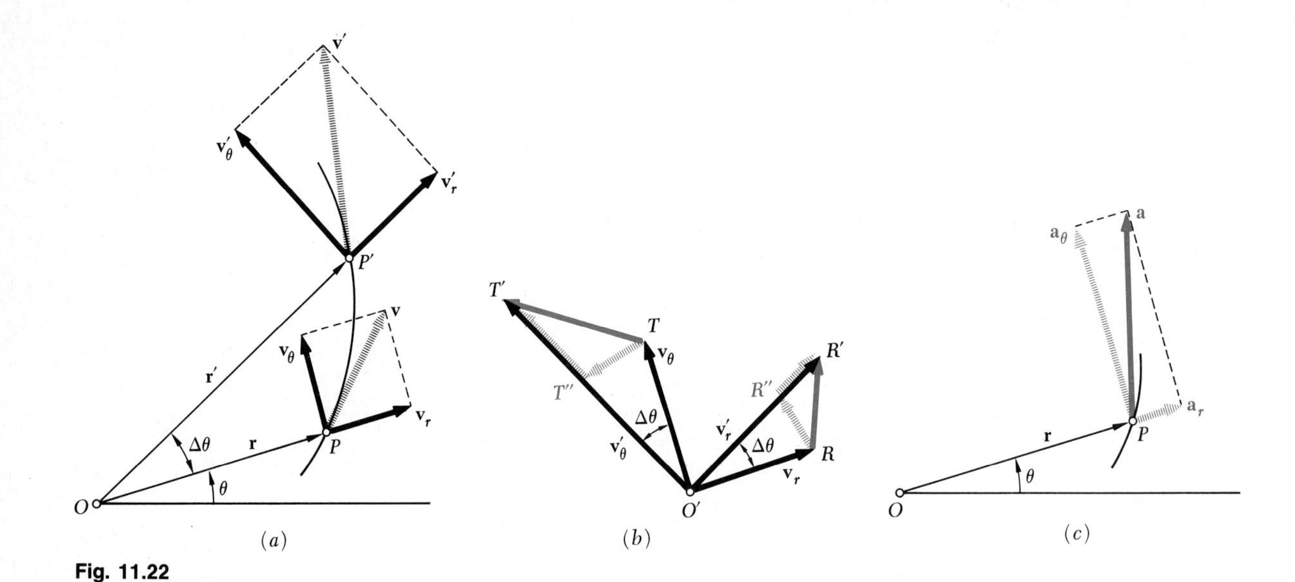

Fig. 11.22

$\overrightarrow{TT'}$ represents a change in magnitude and direction for the vector $\mathbf{v}_\theta$. The sum of the vectors $\overrightarrow{RR'}$ and $\overrightarrow{TT'}$ represents the total change in velocity $\Delta\mathbf{v}$. As we saw earlier, the acceleration of the particle is defined as the limit of the quotient of $\Delta\mathbf{v}$ and Δt. To obtain the radial component $\mathbf{a}_r$ and the transverse component $\mathbf{a}_\theta$ of the acceleration, we resolve the vector $\overrightarrow{RR'}$ into its components $\overrightarrow{RR''}$ and $\overrightarrow{R''R'}$ (with $O'R'' = O'R$) and the vector $\overrightarrow{TT'}$ into its components $\overrightarrow{TT''}$ and $\overrightarrow{T''T'}$ (with $O'T'' = O'T$). The acceleration $\mathbf{a}$ of the particle may then be expressed as

$$\mathbf{a} = \lim_{\Delta t \to 0} \frac{\Delta\mathbf{v}}{\Delta t} = \lim_{\Delta t \to 0} \left(\frac{\overrightarrow{RR'}}{\Delta t} + \frac{\overrightarrow{TT'}}{\Delta t} \right)$$

$$= \lim_{\Delta t \to 0} \left(\frac{\overrightarrow{RR''}}{\Delta t} + \frac{\overrightarrow{R''R'}}{\Delta t} + \frac{\overrightarrow{TT''}}{\Delta t} + \frac{\overrightarrow{T''T'}}{\Delta t} \right)$$

As Δt approaches zero, the vectors $\overrightarrow{R''R'}$ and $\overrightarrow{TT''}$ become parallel to $\mathbf{r}$, while the vectors $\overrightarrow{RR''}$ and $\overrightarrow{T''T'}$ become perpendicular to $\mathbf{r}$. We must have, therefore,

$$\mathbf{a}_r = \lim_{\Delta t \to 0}\left(\frac{\overrightarrow{R''R'}}{\Delta t} + \frac{\overrightarrow{TT''}}{\Delta t}\right) \qquad \mathbf{a}_\theta = \lim_{\Delta t \to 0}\left(\frac{\overrightarrow{T''T'}}{\Delta t} + \frac{\overrightarrow{RR''}}{\Delta t}\right)$$

From Fig. 11.22*b* we observe that, for small values of Δt, the lengths of the various vectors involved are

$$R''R' = \Delta(v_r) \qquad T''T' = \Delta(v_\theta)$$

$$RR'' = v_r\,\Delta\theta \qquad TT'' = v_\theta\,\Delta\theta$$

where $\Delta(v_r)$ and $\Delta(v_\theta)$ represent the *change in magnitude* of the vectors $\mathbf{v}_r$ and $\mathbf{v}_\theta$, respectively. Noting also that the vector $\overrightarrow{TT''}$ has a sense opposite to that of $\mathbf{r}$, we obtain the following expressions for the scalar components of the acceleration:

$$a_r = \lim_{\Delta t \to 0}\left[\frac{\Delta(v_r)}{\Delta t} - \frac{v_\theta\,\Delta\theta}{\Delta t}\right] \qquad a_\theta = \lim_{\Delta t \to 0}\left[\frac{\Delta(v_\theta)}{\Delta t} + \frac{v_r\,\Delta\theta}{\Delta t}\right]$$

$$a_r = \frac{d(v_r)}{dt} - v_\theta\frac{d\theta}{dt} \qquad a_\theta = \frac{d(v_\theta)}{dt} + v_r\frac{d\theta}{dt}$$

or, using dots to indicate time derivatives,

$$a_r = \dot{v}_r - v_\theta\dot{\theta} \qquad a_\theta = \dot{v}_\theta + v_r\dot{\theta} \qquad (11.31)$$

We should note that, in general, a_r *is not equal to* $\dot{v}_r$ and a_θ *is not equal to* $\dot{v}_\theta$. Each of the relations (11.31) contains an extra term which represents the *change in direction* of the vectors $\mathbf{v}_r$ and $\mathbf{v}_\theta$. Differentiating the relations (11.30), we obtain

$$\dot{v}_r = \ddot{r} \qquad \dot{v}_\theta = \frac{d}{dt}(r\dot{\theta}) = r\ddot{\theta} + \dot{r}\dot{\theta} \qquad (11.32)$$

Substituting from (11.30) and (11.32) into (11.31), we write

$$a_r = \ddot{r} - r\dot{\theta}^2 \qquad a_\theta = r\ddot{\theta} + 2\dot{r}\dot{\theta} \qquad (11.33)$$

A positive value for a_r indicates that the vector $\mathbf{a}_r$ has the same sense as $\mathbf{r}$; a positive value for a_θ indicates that the vector $\mathbf{a}_\theta$ points toward increasing values of θ.

In the case of a particle moving along a circle of center O, we have $r =$ constant, $\dot{r} = \ddot{r} = 0$, and the formulas (11.33) reduce to

$$a_r = -r\dot{\theta}^2 \qquad a_\theta = r\ddot{\theta} \qquad (11.34)$$

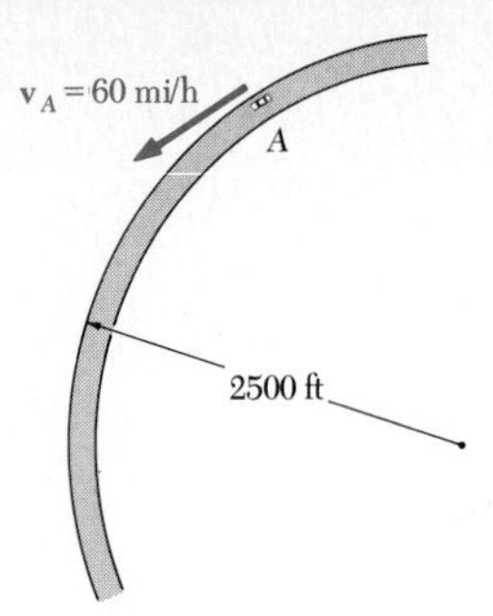

SAMPLE PROBLEM 11.10

A motorist is traveling on a curved section of highway of radius 2500 ft at the speed of 60 mi/h. The motorist suddenly applies the brakes, causing the automobile to slow down at a constant rate. Knowing that after 8 s the speed has been reduced to 45 mi/h, determine the acceleration of the automobile immediately after the brakes have been applied.

Tangential Component of Acceleration. First the speeds are expressed in ft/s.

$$60 \text{ mi/h} = \left(60\frac{\text{mi}}{\text{h}}\right)\left(\frac{5280 \text{ ft}}{1 \text{ mi}}\right)\left(\frac{1 \text{ h}}{3600 \text{ s}}\right) = 88 \text{ ft/s}$$

$$45 \text{ mi/h} = 66 \text{ ft/s}$$

Since the automobile slows down at a constant rate, we have

$$a_t = \text{average } a_t = \frac{\Delta v}{\Delta t} = \frac{66 \text{ ft/s} - 88 \text{ ft/s}}{8 \text{ s}} = -2.75 \text{ ft/s}^2$$

Normal Component of Acceleration. Immediately after the brakes have been applied, the speed is still 88 ft/s, and we have

$$a_n = \frac{v^2}{\rho} = \frac{(88 \text{ ft/s})^2}{2500 \text{ ft}} = 3.10 \text{ ft/s}^2$$

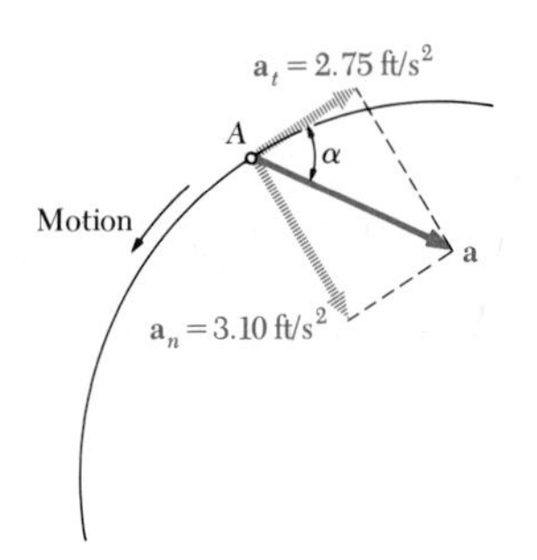

Magnitude and Direction of Acceleration. The magnitude and direction of the resultant **a** of the components $\mathbf{a}_n$ and $\mathbf{a}_t$ are

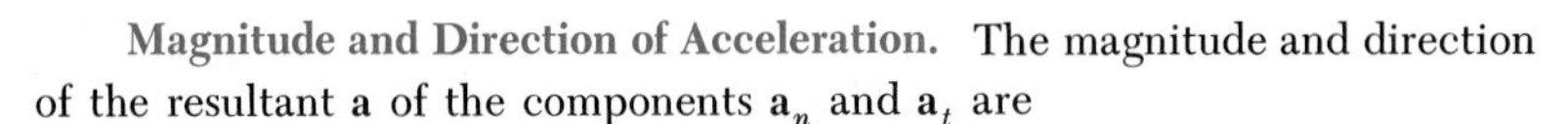

$$\tan \alpha = \frac{a_n}{a_t} = \frac{3.10 \text{ ft/s}^2}{2.75 \text{ ft/s}^2} \qquad \alpha = 48.4° \blacktriangleleft$$

$$a = \frac{a_n}{\sin \alpha} = \frac{3.10 \text{ ft/s}^2}{\sin 48.4°} \qquad a = 4.14 \text{ ft/s}^2 \blacktriangleleft$$

SAMPLE PROBLEM 11.11

Determine the minimum radius of curvature of the trajectory described by the projectile considered in Sample Prob. 11.7.

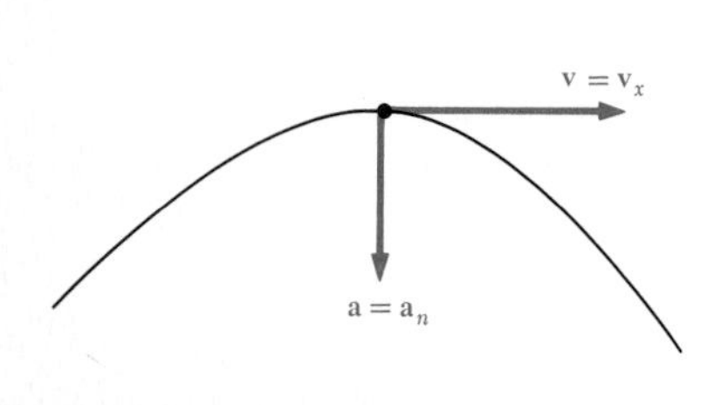

Solution. Since $a_n = v^2/\rho$, we have $\rho = v^2/a_n$. The radius will be small when v is small or when a_n is large. The speed v is minimum at the top of the trajectory since $v_y = 0$ at that point; a_n is maximum at that same point, since the direction of the vertical coincides with the direction of the normal. Therefore, the minimum radius of curvature occurs at the top of the trajectory. At this point, we have

$$v = v_x = 155.9 \text{ m/s} \qquad a_n = a = 9.81 \text{ m/s}^2$$

$$\rho = \frac{v^2}{a_n} = \frac{(155.9 \text{ m/s})^2}{9.81 \text{ m/s}^2} \qquad \rho = 2480 \text{ m} \blacktriangleleft$$

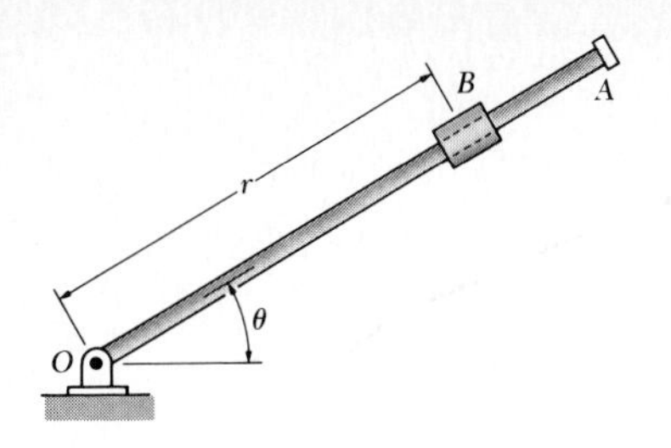

SAMPLE PROBLEM 11.12

The rotation of the 0.9-m arm OA about O is defined by the relation $\theta = 0.15t^2$, where θ is expressed in radians and t in seconds. Collar B slides along the arm in such a way that its distance from O is $r = 0.9 - 0.12t^2$, where r is expressed in meters and t in seconds. After the arm OA has rotated through 30°, determine (*a*) the total velocity of the collar, (*b*) the total acceleration of the collar, (*b*) the relative acceleration of the collar with respect to the arm.

Solution. We first find the time t at which $\theta = 30°$. Substituting $\theta = 30° = 0.524$ rad into the expression for θ, we obtain

$$\theta = 0.15t^2 \qquad 0.524 = 0.15t^2 \qquad t = 1.869 \text{ s}$$

Equations of Motion. Substituting $t = 1.869$ s in the expressions for r, θ, and their first and second derivatives, we have

$$\begin{aligned} r &= 0.9 - 0.12t^2 = 0.481 \text{ m} & \theta &= 0.15t^2 = 0.524 \text{ rad} \\ \dot{r} &= -0.24t = -0.449 \text{ m/s} & \dot{\theta} &= 0.30t = 0.561 \text{ rad/s} \\ \ddot{r} &= -0.24 = -0.240 \text{ m/s}^2 & \ddot{\theta} &= 0.30 = 0.300 \text{ rad/s}^2 \end{aligned}$$

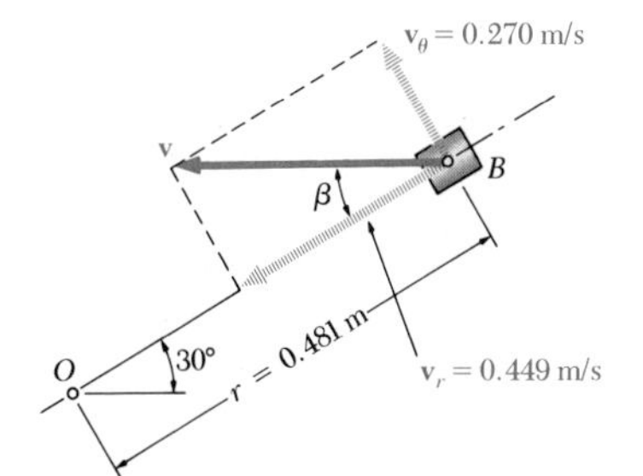

a. Velocity of B. Using Eqs. (11.30), we obtain the values of v_r and v_θ when $t = 1.869$ s:

$$\begin{aligned} v_r &= \dot{r} = -0.449 \text{ m/s} \\ v_\theta &= r\dot{\theta} = 0.481(0.561) = 0.270 \text{ m/s} \end{aligned}$$

Solving the right triangle shown, we obtain the magnitude and direction of the velocity,

$$v = 0.524 \text{ m/s} \qquad \beta = 31.0° \quad ◄$$

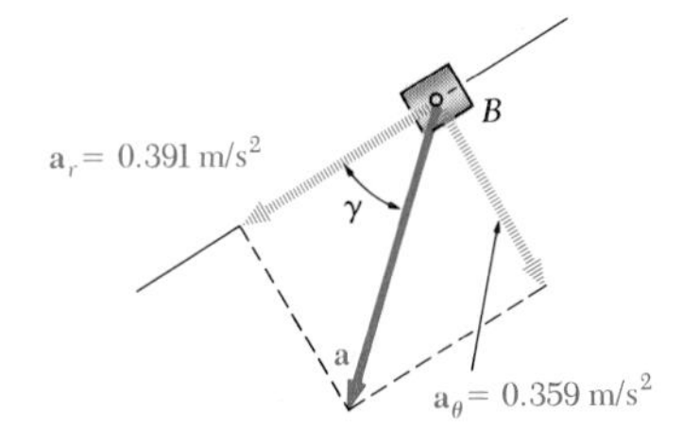

b. Acceleration of B. Using Eqs. (11.33), we obtain

$$\begin{aligned} a_r &= \ddot{r} - r\dot{\theta}^2 \\ &= -0.240 - 0.481(0.561)^2 = -0.391 \text{ m/s}^2 \\ a_\theta &= r\ddot{\theta} + 2\dot{r}\dot{\theta} \\ &= 0.481(0.300) + 2(-0.449)(0.561) = -0.359 \text{ m/s}^2 \end{aligned}$$

$$a = 0.531 \text{ m/s}^2 \qquad \gamma = 42.6° \quad ◄$$

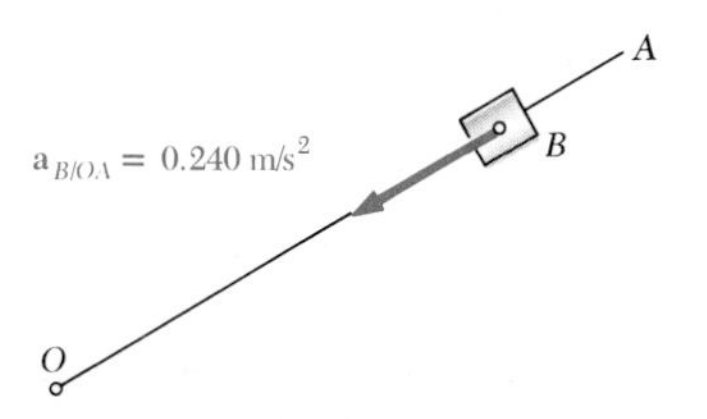

c. Acceleration of B with Respect to Arm OA. We note that the motion of the collar with respect to the arm is rectilinear and defined by the coordinate r. We write

$$a_{B/OA} = \ddot{r} = -0.240 \text{ m/s}^2$$

$$a_{B/OA} = 0.240 \text{ m/s}^2 \text{ toward } O. \quad ◄$$

Problems

11.119 What is the smallest radius which should be used for a highway curve if the normal component of the acceleration of a car traveling at 60 mi/h is not to exceed 2.5 ft/s^2?

11.120 A motorist drives along the circular exit ramp of a turnpike at the constant speed v_0. Knowing that the odometer indicates a distance of 0.6 km between point A where the automobile is going due south and B where it is going due north, determine the speed v_0 for which the normal component of the acceleration is $0.08g$.

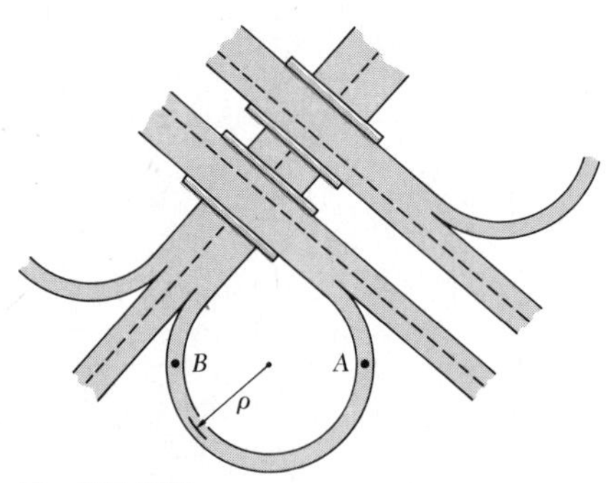

Fig. P11.120

11.121 During the test of an airplane it is desired that the normal component of the acceleration be equal to $6g$. Determine the radius of the horizontal circle along which the pilot should fly if the speed of the airplane is 1800 km/h.

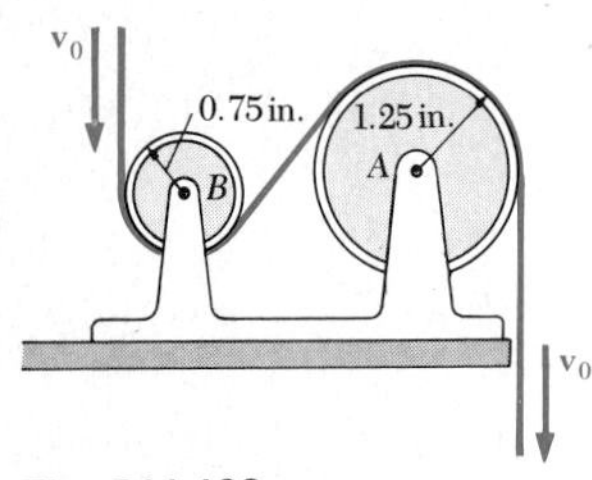

Fig. P11.122

11.122 A computer tape moves over two drums at a constant speed v_0. Knowing that the normal component of the acceleration of the portion of tape in contact with drum B is 400 ft/s^2, determine (a) the speed v_0, (b) the normal component of the acceleration of the portion of tape in contact with drum A.

11.123 A motorist is traveling on a curved portion of highway of radius 400 m at a speed of 90 km/h. The brakes are suddenly applied, causing the speed to decrease at a constant rate of 1.2 m/s^2. Determine the magnitude of the total acceleration of the automobile (a) immediately after the brakes have been applied, (b) 5 s later.

11.124 A bus starts from rest on a curve of 250-m radius and accelerates at the constant rate $a_t = 0.6$ m/s^2. Determine the distance and time that the bus will travel before the magnitude of its total acceleration is 0.75 m/s^2.

11.125 A motorist decreases the speed of an automobile at a constant rate from 45 to 30 mi/h over a distance of 750 ft along a curve of 1500-ft radius. Determine the magnitude of the total acceleration of the automobile after it has traveled 500 ft along the curve.

11.126 As a train enters a curve of radius 4000 ft at a speed of 60 mi/h, the brakes are applied sufficiently to cause the magnitude of the *total* acceleration of the train to be 2.5 ft/s². After 10 s the brakes are more fully applied so that the magnitude of the total acceleration is again 2.5 ft/s². If this second brake setting is then maintained, what is the total time required to bring the train to a stop?

11.127 Automobile A is traveling along a straight highway, while B is moving along a circular exit ramp of 80-m radius. The speed of A is being increased at the rate of 2 m/s² and the speed of B is being decreased at the rate of 1.2 m/s². For the position shown, determine (*a*) the velocity of A relative to B, (*b*) the acceleration of A relative to B.

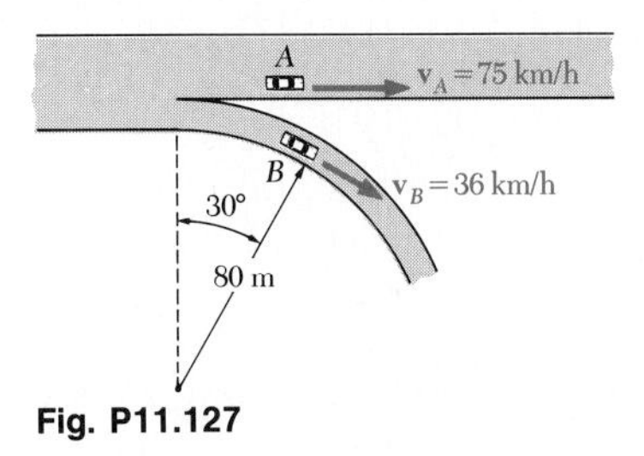

Fig. P11.127

11.128 Solve Prob. 11.127, assuming that the velocity of B is 18 km/h and is being decreased at the rate of 1.2 m/s².

11.129 A nozzle discharges a stream of water in the direction shown with an initial velocity of 25 ft/s. Determine the radius of curvature of the stream (*a*) as it leaves the nozzle, (*b*) at the maximum height of the stream.

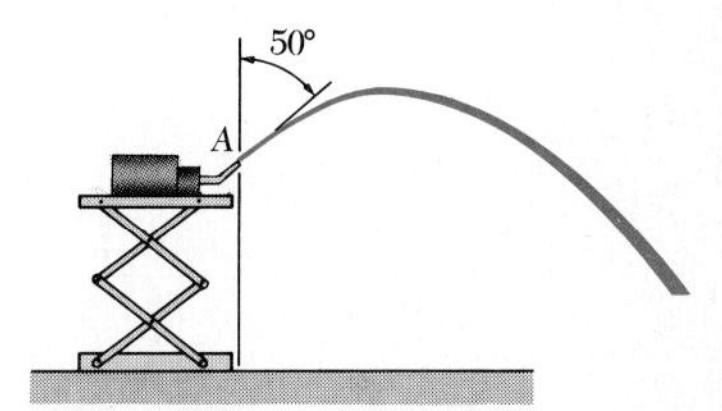

Fig. P11.129

11.130 For each of the two firing angles obtained in Sample Prob. 11.8, determine the radius of curvature of the trajectory described by the projectile as it leaves the gun.

11.131 Determine the radius of curvature of the trajectory described by the projectile of Sample Prob. 11.7 (*a*) as the projectile leaves the gun, (*b*) at the maximum elevation of the projectile.

11.132 From measurements of a photograph it has been found that as the stream of water shown left the nozzle at A, it had a radius of curvature of 25 m. Determine (*a*) the initial velocity $\mathbf{v}_A$ of the stream, (*b*) the radius of curvature of the stream at its maximum height.

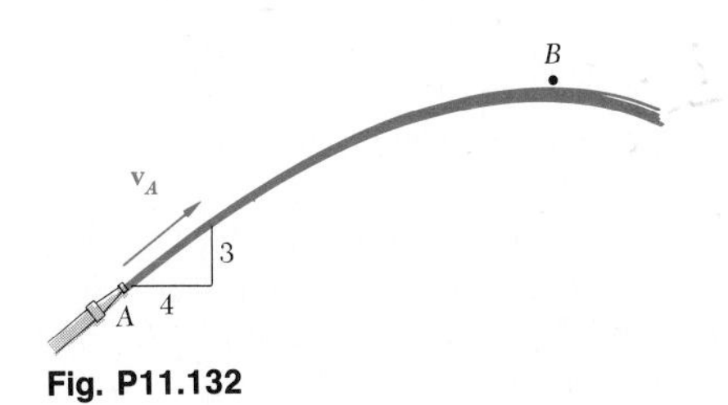

Fig. P11.132

11.133 A satellite will travel indefinitely in a circular orbit around the earth if the normal component of its acceleration is equal to $g(R/r)^2$, where $g = 32.2\ \text{ft/s}^2$, R = radius of the earth = 3960 mi, and r = distance from the center of the earth to the satellite. Determine the height above the surface of the earth at which a satellite will travel indefinitely at a speed of 16,500 mi/h.

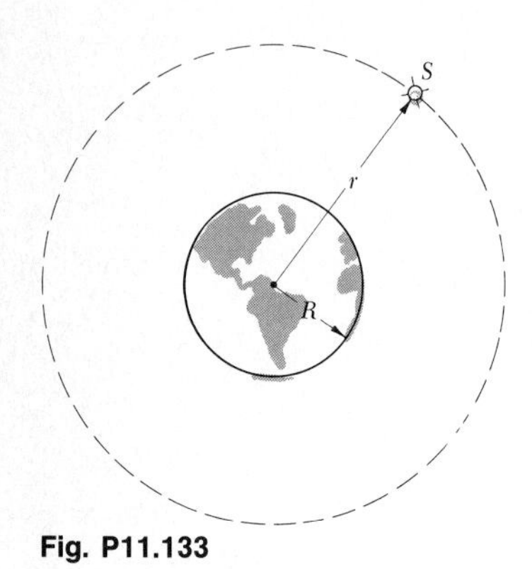

Fig. P11.133

11.134 Determine the speed of a space shuttle traveling in a circular orbit 140 mi above the surface of the earth. (See information given in Prob. 11.133.)

11.135 Knowing that the radius of the earth is 6370 km and that $g = 9.81\ \text{m/s}^2$, determine the height above the surface of the earth at which a satellite will travel indefinitely at a speed of 25 000 km/h. (See information given in Prob. 11.133.)

11.136 Determine the speed of a communication satellite in geosynchronous orbit at a height of 35 770 km above the surface of the earth. (See information given in Probs. 11.133 and 11.135.)

11.137 Show that the speed of an earth satellite traveling in a circular orbit is inversely proportional to the square root of the radius of its orbit and determine the minimum time required for a satellite to circle the earth. (See information given in Probs. 11.133 and 11.135.)

11.138 Assuming that the orbit of the moon is a circle of radius 384×10^3 km, determine the speed of the moon relative to the earth. (See information given in Probs. 11.133 and 11.135.)

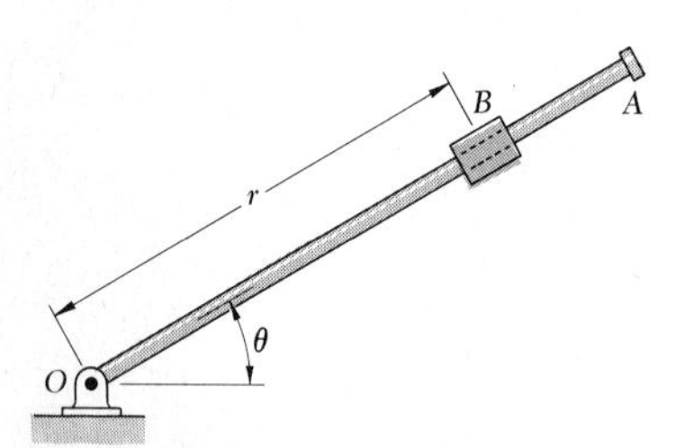

Fig. P11.139 and P11.140

11.139 The rotation of rod OA about O is defined by the relation $\theta = t^3 - 4t$, where θ is expressed in radians and t in seconds. Collar B slides along the rod in such a way that its distance from O is $r = 25t^3 - 50t^2$, where r is expressed in millimeters and t in seconds. When $t = 1$ s, determine (*a*) the velocity of the collar, (*b*) the total acceleration of the collar, (*c*) the acceleration of the collar relative to the rod.

11.140 The rotation of rod OA about O is defined by the relation $\theta = \frac{1}{2}\pi(4t - 3t^2)$, where θ is expressed in radians and t in seconds. Collar B slides along the rod in such a way that its distance from O is $r = 1.25t^2 - 0.9t^3$, where r is expressed in meters and t in seconds. When $t = 1$ s, determine (*a*) the velocity of the collar, (*b*) the total acceleration of the collar, (*c*) the acceleration of the collar relative to the rod.

11.141 The two-dimensional motion of a particle is defined by the relations $r = 2b \cos \omega t$ and $\theta = \omega t$, where b and ω are constants. Determine (*a*) the velocity and acceleration of the particle at any instant, (*b*) the radius of curvature of its path. What conclusion can you draw regarding the path of the particle?

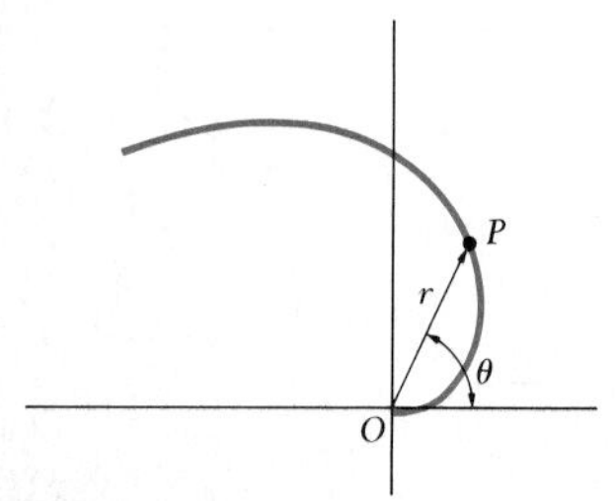

Fig. P11.142

11.142 The path of a particle P is an Archimedean spiral. The motion of the particle is defined by the relations $r = 10t$ and $\theta = 2\pi t$, where r is expressed in inches, t in seconds, and θ in radians. Determine the velocity and acceleration of the particle when (*a*) $t = 0$, (*b*) $t = 0.25$ s.

*11.143 The motion of particle P, along the elliptic path shown, is defined by the relations $r = 35/(1 - 0.75 \cos \pi t)$ and $\theta = \pi t$, where r is expressed in millimeters, t in seconds, and θ in radians. Determine the velocity and acceleration of the particle when (a) $t = 0$, (b) $t = 0.5$ s.

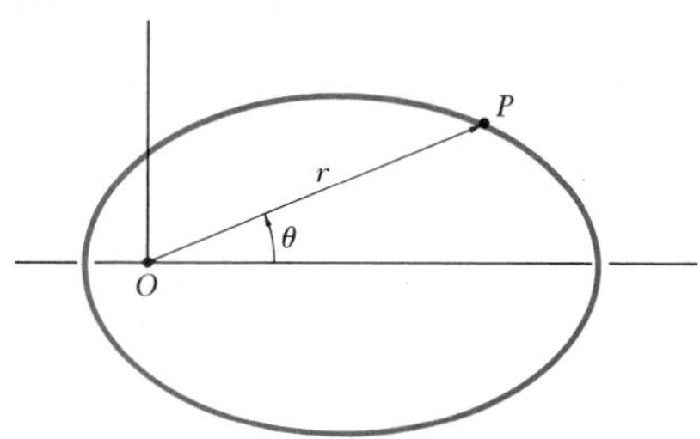

Fig. P11.143

*11.144 Solve Prob. 11.143, when (a) $t = 1$ s, (b) $t = 1.5$ s.

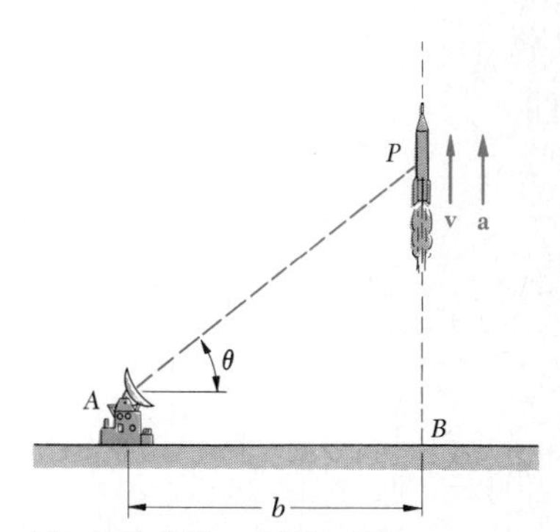

Fig. P11.145 and P11.149

11.145 A rocket is fired vertically from a launching pad at B. Its flight is tracked by radar from point A. Determine the velocity of the rocket in terms of b, θ, and $\dot{\theta}$.

11.146 The flight path of airplane B is a horizontal straight line and passes directly over a radar tracking station at A. Knowing that the airplane moves to the left with the constant velocity $\mathbf{v}_0$, determine $d\theta/dt$ in terms of v_0, h, and θ.

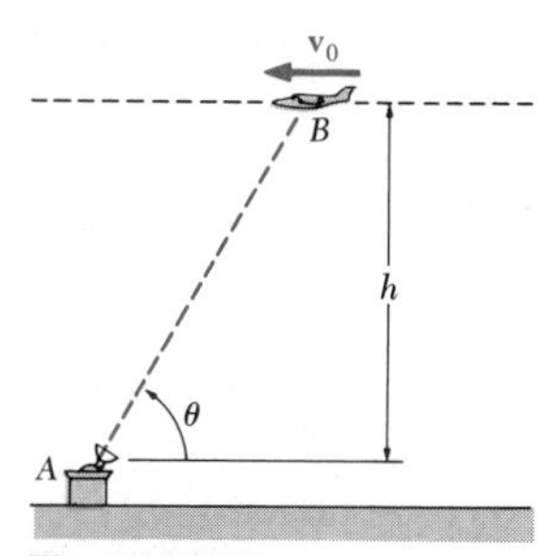

Fig. P11.146

11.147 Determine the acceleration of the rocket of Prob. 11.145 in terms of b, θ, $\dot{\theta}$, and $\ddot{\theta}$.

11.148 In Prob. 11.146, determine $d^2\theta/dt^2$ in terms of v_0, h, and θ.

11.149 A test rocket is fired vertically from a launching pad at B. When the rocket is at P the angle of elevation is $\theta = 42.0°$, and 0.5 s later it is $\theta = 43.2°$. Knowing that $b = 3$ km, determine approximately the speed of the rocket during the 0.5-s interval.

11.150 An airplane passes over a radar tracking station at A and continues to fly due east. When the airplane is at P, the distance and angle of elevation of the plane are, respectively, $r = 11{,}200$ ft and $\theta = 26.5°$. Two seconds later the radar station sights the plane at $r = 12{,}300$ ft and $\theta = 23.3°$. Determine approximately the speed and the angle of dive α of the plane during the 2-s interval.

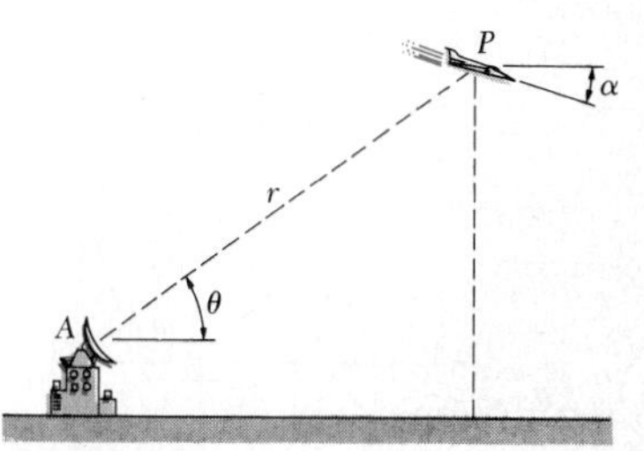

Fig. P11.150

11.151 and 11.152 A particle moves along the spiral shown; determine the magnitude of the velocity of the particle in terms of b, θ, and $\dot{\theta}$.

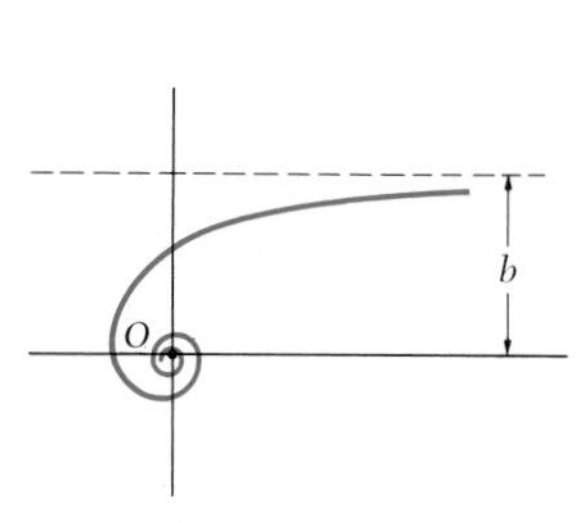

Hyperbolic spiral $r\theta = b$
Fig. P11.151 and P11.153

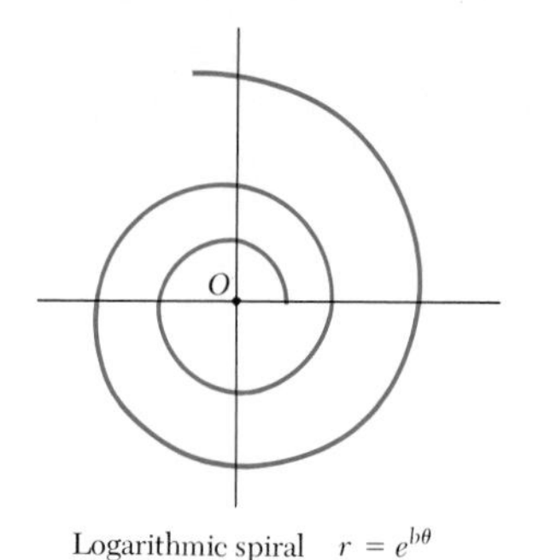

Logarithmic spiral $r = e^{b\theta}$
Fig. P11.152 and P11.154

11.153 and 11.154 A particle moves along the spiral shown. Knowing that $\dot{\theta}$ is constant and denoting this constant by ω, determine the magnitude of the acceleration of the particle in terms of b, θ, and ω.

11.155 As rod OA rotates, pin P moves along the parabola BCD. Knowing that the equation of the parabola is $r = 2b/(1 + \cos\theta)$ and that $\theta = kt$, determine the velocity and acceleration of P when (*a*) $\theta = 0$, (*b*) $\theta = 90°$.

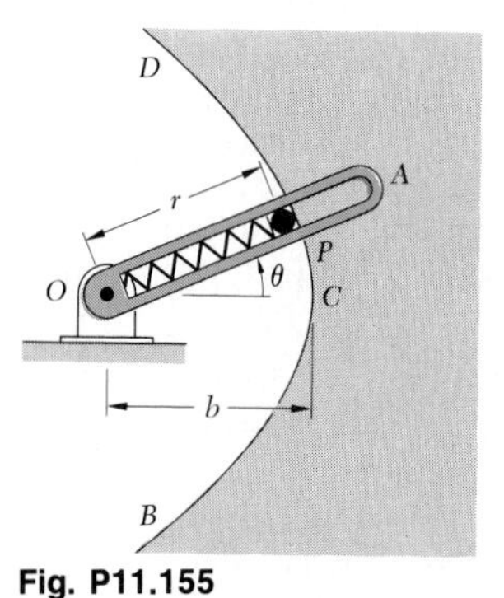

Fig. P11.155

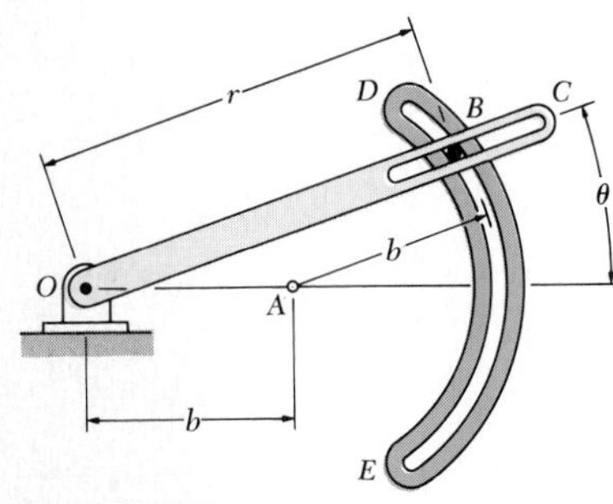

Fig. P11.156

11.156 The pin at B is free to slide along the circular slot DE and along the rotating rod OC. Assuming that the rod OC rotates at a constant rate $\dot{\theta}$, (*a*) show that the acceleration of pin B is of constant magnitude, (*b*) determine the direction of the acceleration of pin B.

11.14. Review and Summary. In the first half of the chapter, we analyzed the *rectilinear motion of a particle,* i.e., the motion of a particle along a straight line. To define the position P of the particle on that line, we chose a fixed origin O and a positive direction (Fig. 11.23). The distance x from O to P, with the appropriate sign, completely defines the position of the particle on the line and is called the *position coordinate* of the particle [Sec. 11.2].

Position coordinate of a particle in rectilinear motion

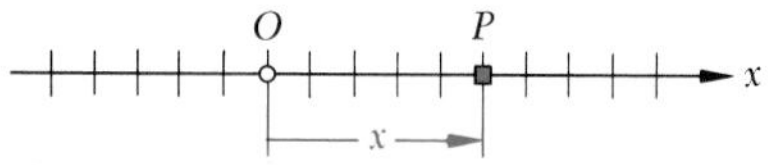

O P x x

Fig. 11.23

Velocity and acceleration in rectilinear motion

The *velocity* v of the particle was shown to be equal to the time derivative of the position coordinate x,

$$v = \frac{dx}{dt} \tag{11.1}$$

and the *acceleration* a was obtained by differentiating v with respect to t,

$$a = \frac{dv}{dt} \tag{11.2}$$

or

$$a = \frac{d^2x}{dt^2} \tag{11.3}$$

We also noted that a could be expressed as

$$a = v\frac{dv}{dx} \tag{11.4}$$

We observed that the velocity v and the acceleration a were represented by algebraic numbers which may be positive or negative. A positive value for v indicates that the particle moves in the positive direction, and a negative value that it moves in the negative direction. A positive value for a, however, may mean that the particle is truly accelerated (i.e., moves faster) in the positive direction, or that it is decelerated (i.e., moves more slowly) in the negative direction. A negative value for a is subject to a similar interpretation [Sample Prob. 11.1].

Determination of the velocity and acceleration by integration

In most problems, the conditions of motion of a particle are defined by the type of acceleration that the particle possesses and by the initial conditions [Sec. 11.3]. The velocity and position of the particle may then be obtained by integrating two of the equations (11.1) to (11.4). Which of these equations should be selected depends upon the type of acceleration involved [Sample Probs. 11.2 and 11.3].

Uniform rectilinear motion

Two types of motion are frequently encountered: the *uniform rectilinear motion* [Sec. 11.4], in which the velocity v of the particle is constant and

$$x = x_0 + vt \tag{11.5}$$

Uniformly accelerated rectilinear motion

and the *uniformly accelerated rectilinear motion* [Sec. 11.5], in which the acceleration a of the particle is constant and we have

$$v = v_0 + at \tag{11.6}$$

$$x = x_0 + v_0 t + \tfrac{1}{2}at^2 \tag{11.7}$$

$$v^2 = v_0^2 + 2a(x - x_0) \tag{11.8}$$

Relative motion of two particles

When two particles A and B move along the same straight line, we may wish to consider the *relative motion* of B with respect to A [Sec. 11.6].

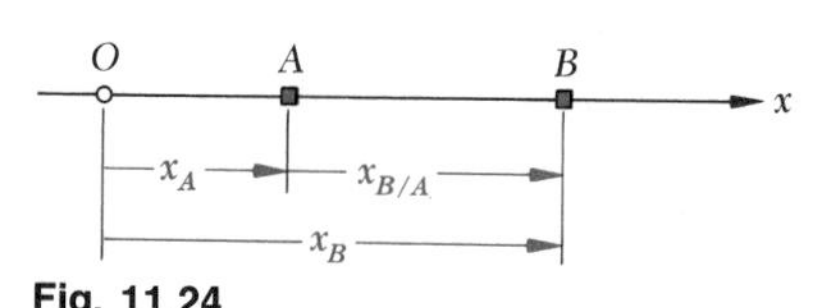

Fig. 11.24

Denoting by $x_{B/A}$ the *relative position coordinate* of B with respect to A (Fig. 11.24), we had

$$x_B = x_A + x_{B/A} \tag{11.9}$$

Differentiating Eq. (11.9) twice with respect to t, we obtained successively

$$v_B = v_A + v_{B/A} \tag{11.10}$$

$$a_B = a_A + a_{B/A} \tag{11.11}$$

where $v_{B/A}$ and $a_{B/A}$ represent, respectively, the *relative velocity* and the *relative acceleration* of B with respect to A.

Blocks connected by inextensible cords

When several blocks are *connected by inextensible cords*, it is possible to write a *linear relation* between their position coordinates. Similar relations may then be written between their velocities and between their accelerations and used to analyze their motion [Sample Prob. 11.5].

Graphical solutions

It is sometimes convenient to use a *graphical solution* for problems involving the rectilinear motion of a particle [Secs. 11.7 and 11.8]. The graphical solution most commonly used involves the x-t, v-t, and a-t curves [Sec. 11.7; Sample Prob. 11.6]. It was shown that, at any given time t,

$$v = \text{slope of } x\text{-}t \text{ curve}$$
$$a = \text{slope of } v\text{-}t \text{ curve}$$

while, over any given time interval from t_1 to t_2,

$$v_2 - v_1 = \text{area under } a\text{-}t \text{ curve}$$
$$x_2 - x_1 = \text{area under } v\text{-}t \text{ curve}$$

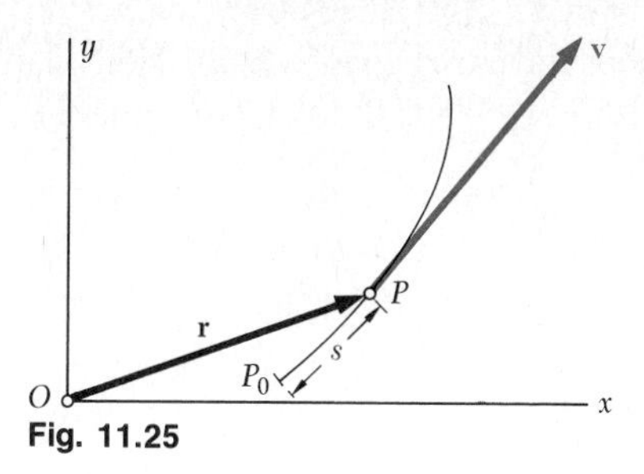

Fig. 11.25

Position vector and velocity in curvilinear motion

In the second half of the chapter, we analyzed the *curvilinear motion of a particle,* i.e., the motion of a particle along a curved path. Our analysis was limited to the motion of a particle in a plane. The position P of the particle at a given time [Sec. 11.9] was defined by the *position vector* $\mathbf{r}$ joining the origin O of the coordinates and point P (Fig. 11.25). The *velocity* $\mathbf{v}$ of the particle was defined by the relation

$$\mathbf{v} = \lim_{\Delta t \to 0} \frac{\Delta \mathbf{r}}{\Delta r} \tag{11.14}$$

and was found to be a *vector tangent to the path of the particle* and of magnitude v (called the *speed* of the particle) equal to the time derivative of the length s of the arc described by the particle:

$$v = \frac{ds}{dt} \tag{11.15}$$

Using rectangular components and denoting by x and y the coordinates of P, we also found that the components v_x and v_y of the velocity are equal to the time derivatives of x and y, respectively:

$$v_x = \dot{x} \qquad v_y = \dot{y} \tag{11.16}$$

Acceleration in curvilinear motion

The *acceleration* $\mathbf{a}$ of the particle was defined by the relation

$$\mathbf{a} = \lim_{\Delta t \to 0} \frac{\Delta \mathbf{v}}{\Delta t} \tag{11.18}$$

and we noted that, in general, *the acceleration is not tangent to the path of the particle.* Using rectangular components, we found that

$$a_x = \dot{v}_x \qquad a_y = \dot{v}_y \tag{11.19}$$

or, recalling Eq. (11.16),

$$a_x = \ddot{x} \qquad a_y = \ddot{y} \tag{11.20}$$

Rectangular components
Component motions

When the component a_x of the acceleration is independent of the coordinate y and of the velocity component v_y and when the component a_y of the acceleration is independent of x and v_x, Eqs. (11.20) may be integrated independently. The analysis of the given curvilinear motion may thus be reduced to the analysis of two independent rectilinear *component motions,* one in the x direction and the other in the y direction [Sec. 11.10].

Relative motion of two particles

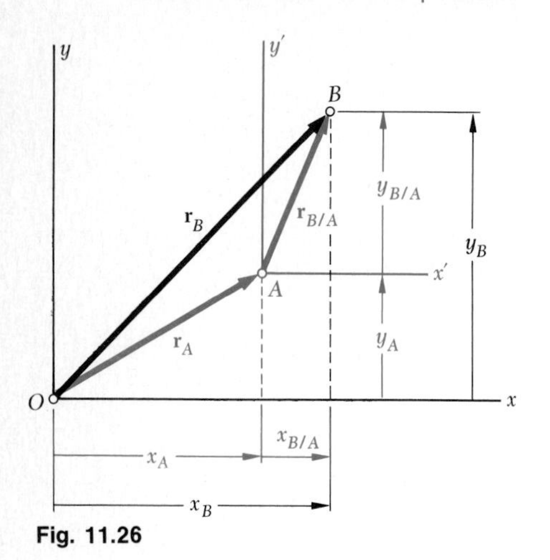

Fig. 11.26

Tangential and normal components

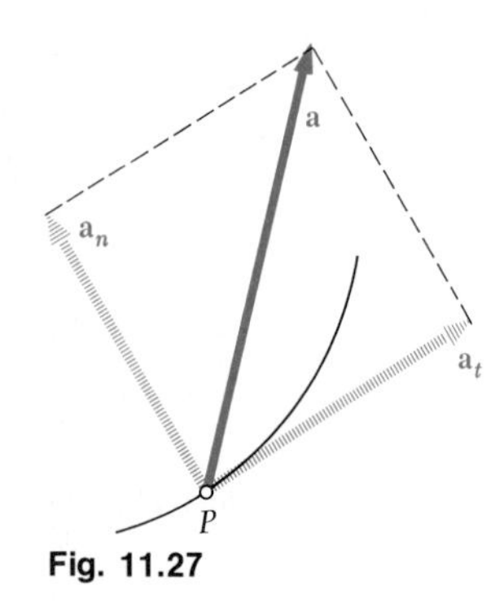

Fig. 11.27

Radial and transverse components

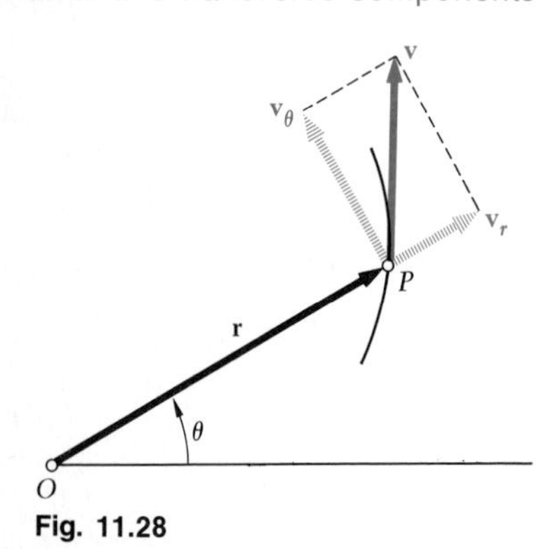

Fig. 11.28

This approach is particularly effective in the study of the *motion of projectiles* [Sample Probs. 11.7 and 11.8].

When two particles A and B move in the same plane, we may wish to consider the *relative motion* of B with respect to A, or more precisely, with respect to a system of axes attached to A and *in translation with* A [Sec. 11.11]. Denoting by $\mathbf{r}_{B/A}$ the *relative position vector* of B with respect to A (Fig. 11.26), we had

$$\mathbf{r}_B = \mathbf{r}_A + \mathbf{r}_{B/A} \tag{11.21}$$

Denoting by $\mathbf{v}_{B/A}$ and $\mathbf{a}_{B/A}$, respectively, the *relative velocity* and the *relative acceleration* of B with respect to A, we also showed that

$$\mathbf{v}_B = \mathbf{v}_A + \mathbf{v}_{B/A} \tag{11.24}$$

and

$$\mathbf{a}_B = \mathbf{a}_A + \mathbf{a}_{B/A} \tag{11.26}$$

It is sometimes convenient to resolve the velocity and acceleration of a particle into components other than the rectangular x and y components. Considering first *tangential and normal components* directed, respectively, along the tangent t to the path of the particle and along the normal n drawn toward the center of curvature of the path [Sec. 11.12], we observed that while the velocity $\mathbf{v}$ is directed along the tangent t, the acceleration $\mathbf{a}$ should be resolved into two components $\mathbf{a}_t$ and $\mathbf{a}_n$ (Fig. 11.27) of magnitude

$$a_t = \frac{dv}{dt} \tag{11.28}$$

and

$$a_n = \frac{v^2}{\rho} \tag{11.29}$$

where ρ denotes the radius of curvature of the path at point P [Sample Probs. 11.10 and 11.11].

When the position of a particle is defined by its polar coordinates r and θ, it is usually convenient to use *radial and transverse components* directed, respectively, along the position vector $\mathbf{r}$ of the particle and in the direction obtained by rotating $\mathbf{r}$ through 90° counterclockwise [Sec. 11.13]. Resolving the velocity $\mathbf{v}$ of the particle into radial and transverse components $\mathbf{v}_r$ and $\mathbf{v}_\theta$ (Fig 11.28), we found that

$$v_r = \dot{r} \qquad v_\theta = r\dot{\theta} \tag{11.30}$$

Resolving the acceleration $\mathbf{a}$ of the particle into radial and transverse components $\mathbf{a}_r$ and $\mathbf{a}_\theta$, we had

$$a_r = \ddot{r} - r\dot{\theta}^2 \qquad a_\theta = r\ddot{\theta} + 2\dot{r}\dot{\theta} \tag{11.33}$$

and noted that, in general, a_r is *not* equal to $\dot{v}_r$ (or $\ddot{r}$) and that a_θ is *not* equal to $\dot{v}_\theta$ [Sample Prob. 11.12].

Review Problems

11.157 The speed of a racing car is increased at a constant rate from 72 km/h to 108 km/h over a distance of 120 m along a curve of 200-m radius. Determine the magnitude of the total acceleration of the car after it has traveled 80 m along the curve.

11.158 The acceleration of a collar moving in a straight line is defined by the relation $a = 50 \sin \frac{1}{2}\pi t$, where a is expressed in mm/s^2 and t in seconds. Knowing that $x = 0$ and $v = 0$ when $t = 0$, determine (*a*) the maximum velocity of the collar, (*b*) its position at $t = 4$ s, (*c*) its average velocity over the interval $0 < t < 4$ s. Sketch the a–t, v–t, and x–t curves for the motion.

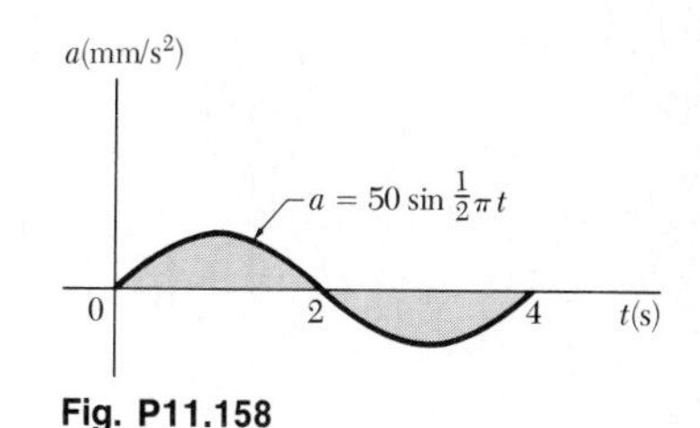

Fig. P11.158

11.159 The acceleration of a particle moving along the x axis is directed toward the origin O and is inversely proportional to the cube of the distance of the particle from O. The particle starts from rest at $x = 3$ ft, and it is observed that its velocity is 5 ft/s when $x = 2$ ft. Determine the velocity of the particle when (*a*) $x = 1$ ft, (*b*) $x = 0.5$ ft.

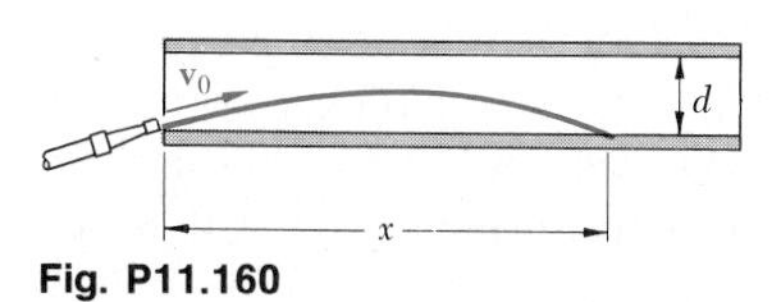

Fig. P11.160

11.160 A nozzle discharges a stream of water with an initial velocity $\mathbf{v}_0$ of 75 ft/s into the end of a horizontal pipe of inside diameter $d = 6$ ft. Determine the largest distance x that the stream can reach.

11.161 A ball is projected from point A with a velocity $\mathbf{v}_0$ which is perpendicular to the incline shown. Knowing that the ball strikes the incline at B, determine the range R in terms of v_0 and β.

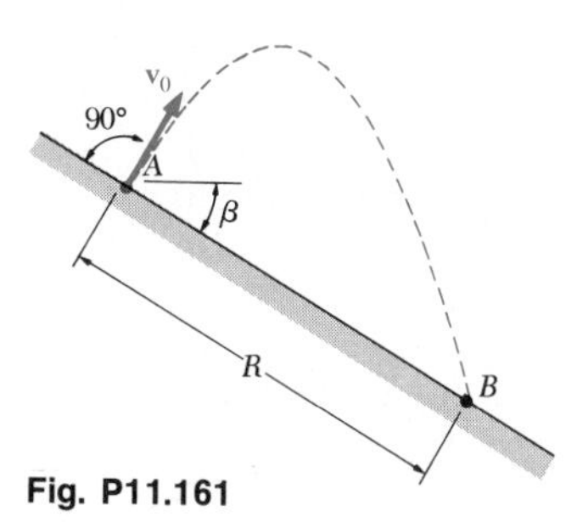

Fig. P11.161

11.162 In Prob. 11.161, determine the range R when $v_0 = 10$ m/s and $\beta = 30°$.

11.163 Airplane A is flying due east at 700 km/h, while airplane B is flying at 500 km/h at the same altitude and in a direction to the west of south. Knowing that the speed of B with respect to A is 1125 km/h, determine the direction of the flight path of B.

11.164 A stone is dropped from ground level into a mine shaft as an ore bucket passes a point 120 m below ground level. Determine where and when the stone will strike the bucket if the speed of the bucket is 8 m/s (*a*) upward, (*b*) downward.

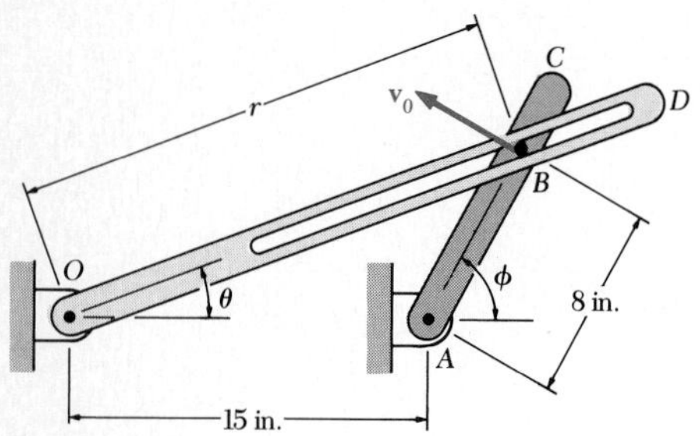

Fig. P11.165

11.165 Pin B is attached to the rotating arm AC and moves at a constant speed $v_0 = 115$ in./s. Knowing that pin B slides freely in a slot cut in arm OD, determine the rate $\dot{\theta}$ at which arm OD rotates and the radial component v_r of the velocity of pin B (*a*) when $\phi = 0$, (*b*) when $\phi = 90°$.

11.166 A police officer on a motorcycle is escorting a motorcade which is traveling at 30 mi/h. The police officer suddenly decides to take a new position in the motorcade, 200 ft ahead. Assuming that he accelerates and decelerates at the rate of 11 ft/s^2 and that he does not exceed at any time a speed of 45 mi/h, draw the a–t and v–t curves for his motion and determine (*a*) the shortest time in which he can occupy his new position in the motorcade, (*b*) the distance he will travel in that time.

11.167 In the position shown, collar A moves downward with a velocity of 80 mm/s and the relative velocity of block C with respect to collar B is 40 mm/s downward. Determine (*a*) the velocity of collar B, (*b*) the velocity of block C.

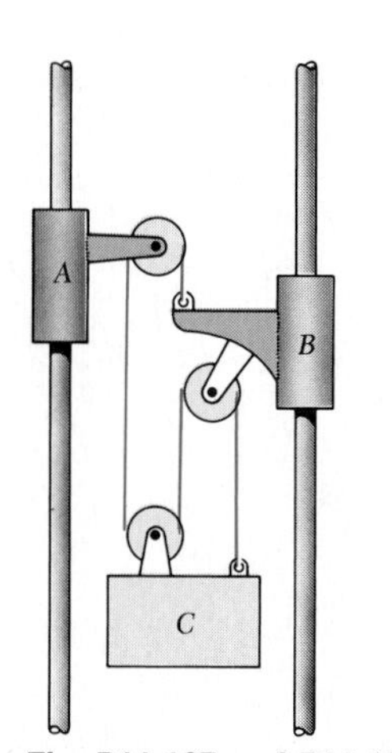

Fig. P11.167 and P11.168

11.168 Collars A and B start from rest and move with the following upward accelerations: $a_A = 160$ mm/s^2 and $a_B = 100$ mm/s^2. Determine the relative velocity of block C with respect to collar A after 4 s.

11.C1 Several test firings are to be made of the experimental rocket of Probs. 11.20 and 11.21. Write a computer program and calculate, for values of the initial velocity from 0 to 400 m/s at 25-m/s intervals, the maximum height reached by the nose cone of the rocket and its velocity as it returns to the ground.

11.C2 A nozzle at A discharges water with an initial velocity $\mathbf{v}_0$ at an angle of 60° with the horizontal. Write a computer program and use it to determine where the stream of water strikes the ground or the building for values of v_0 from 4 m/s to 16 m/s at 1-m/s intervals.

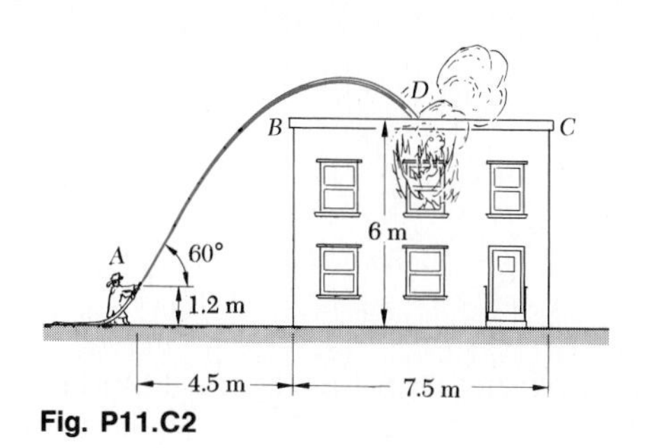

Fig. P11.C2

11.C3 A gun located on a plain fires a projectile with a velocity $\mathbf{v}_0$ at an angle α with the horizontal. After being fired, the projectile is subjected to the acceleration of gravity, $g = -32.2$ ft/s^2, directed vertically downward and to a deceleration due to air resistance, $a_t = -kv^2$, in a direction opposite to that of its velocity $\mathbf{v}$. Express the horizontal and vertical components a_x and a_y of the acceleration of the projectile in terms of g, k, and the components v_x and v_y of its velocity. Considering successive time intervals Δt and assuming the acceleration to remain constant over each time interval, use the expressions obtained for a_x and a_y to write a computer program to calculate at any instant the position of the projectile on its trajectory. Knowing that $v_0 = 1200$ ft/s, $\alpha = 40°$, and using time intervals $\Delta t = 0.5$ s, determine the range of the projectile and print the coordinates of the projectile along its trajectory at 5-s intervals, assuming (*a*) $k = 0$, (*b*) $k = 2 \times 10^{-6}$ ft^{-1}, (*c*) $k = 10 \times 10^{-6}$ ft^{-1}, (*d*) $k = 20 \times 10^{-6}$ ft^{-1}.

11.C4 As a train enters a curve of radius 4000 ft at a speed of 60 mi/h, the brakes are applied sufficiently to cause the magnitude of the total acceleration of the train to be 2.5 ft/s^2. After 1 s the brakes are more fully applied so that the magnitude of the total acceleration is again 2.5 ft/s^2. This brake setting is maintained for 1 s when the brakes are again more fully applied so that the total acceleration is again 2.5 ft/s^2. If this procedure is repeated each second, determine the total time required to bring the train to a stop.

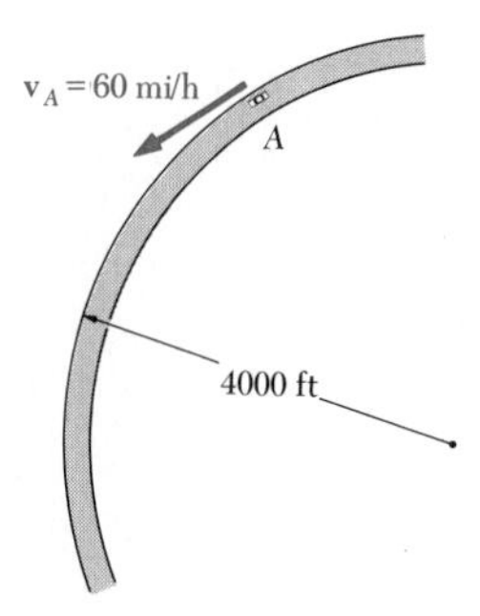

Fig. P11.C4

CHAPTER TWELVE

Kinetics of Particles: Newton's Second Law

12.1. Introduction. Newton's first and third laws of motion were used extensively in statics to study bodies at rest and the forces acting upon them. These two laws are also used in dynamics; in fact, they are sufficient for the study of the motion of bodies which have no acceleration. However, when bodies are accelerated, i.e., when the magnitude or the direction of their velocity changes, it is necessary to use Newton's second law of motion to relate the motion of the body with the forces acting on it.

In this chapter we shall discuss Newton's second law and apply it to the analysis of the motion of particles. As we state in Sec. 12.2, if the resultant of the forces acting on a particle is not zero, the particle will have an acceleration proportional to the magnitude of the resultant and in the direction of this resultant force. Moreover, the ratio of the magnitudes of the resultant force and of the acceleration may be used to define the *mass* of the particle.

Section 12.3 stresses the need for consistent units in the solution of dynamics problems and provides a review of the two systems used in this text, namely, the International System of Units (SI units) and the U.S. customary units.

In Sec. 12.4, Newton's second law will be used to derive the equations of motion for a single particle in terms of rectangular components. An alternative method of analysis, the method of *dynamic equilibrium*, will also be presented. In order to provide a basis for our study of the kinetics of rigid bodies in Chap. 16, we shall consider a *system of particles* in Sec. 12.5 and apply Newton's second law to each of the particles of the system. Defining the *effective force* of a particle as the product $m\mathbf{a}$ of its mass m and its acceleration $\mathbf{a}$, we shall show that *the external forces* acting on the various particles of the system *are equipollent to the effective forces*, i.e., the sums of the x, y, and z components of the external forces and of the effective forces are respectively equal, and the sums of their moments about each of the coordinate axes are also respectively equal. In Sec. 12.6, we shall define the *mass center* of a system of particles and describe the motion of that point.

In Sec. 12.7 and in the Sample Problems which follow, Newton's second law will be applied to the solution of problems involving one or several particles moving in a straight line, while in Sec. 12.8, we shall consider the curvilinear motion of a particle, using tangential and normal components, or radial and transverse components, of the forces and accelerations involved. We recall that an actual body—possibly as large as a car or a space vehicle—may be considered as a particle for the purpose of analyzing its motion, as long as the effect of the rotation of the body about its mass center may be ignored.

Section 12.9 deals with the motion of a particle under a *central force*, i.e., under a force directed toward or away from a fixed point O. An important application of this type of motion is provided by the orbital motion of bodies under gravitational attraction (Sec. 12.10).

Sections 12.11 through 12.13 are optional. They present a more extensive discussion of orbital motion and contain a number of problems related to space mechanics.

12.2. Newton's Second Law of Motion. Newton's second law may be stated as follows:

If the resultant force acting on a particle is not zero, the particle will have an acceleration proportional to the magnitude of the resultant and in the direction of this resultant force.

Newton's second law of motion may best be understood if we imagine the following experiment: A particle is subjected to a force $\mathbf{F}_1$ of constant direction and constant magnitude F_1. Under the action of that force, the particle will be observed to move in a straight line and *in the direction of the force* (Fig. 12.1a). By determining the position of the particle at various instants, we find that its acceleration has a constant magnitude a_1. If the experiment is repeated with forces $\mathbf{F}_2$, $\mathbf{F}_3$, etc., of different magnitude or direction (Fig. 12.1b and c), we find each time that the particle moves in the direction of the force acting on it and that the magnitudes a_1, a_2, a_3, etc., of the accelerations are proportional to the magnitudes F_1, F_2, F_3, etc., of the corresponding forces,

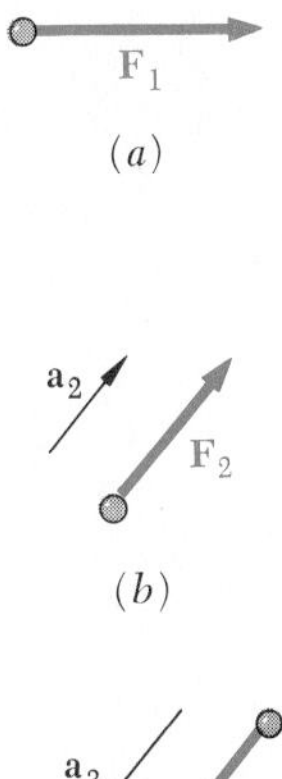

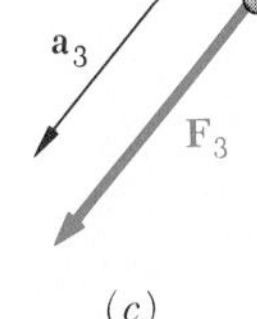

Fig. 12.1

$$\frac{F_1}{a_1} = \frac{F_2}{a_2} = \frac{F_3}{a_3} = \cdots = \text{constant}$$

The constant value obtained for the ratio of the magnitudes of the forces and accelerations is a characteristic of the particle under consideration. It is called the *mass* of the particle and is denoted by m. When a particle of mass m is acted upon by a force $\mathbf{F}$, the force $\mathbf{F}$ and the acceleration $\mathbf{a}$ of the particle must therefore satisfy the relation

$$\mathbf{F} = m\mathbf{a} \tag{12.1}$$

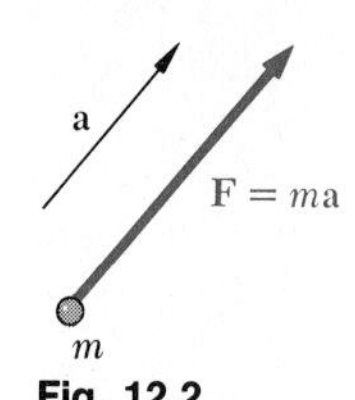

Fig. 12.2

This relation provides a complete formulation of Newton's second law; it expresses not only that the magnitudes of **F** and **a** are proportional but also (since m is a positive scalar) that the vectors **F** and **a** have the same direction (Fig. 12.2). We should note that Eq. (12.1) still holds when **F** is not constant but varies with t in magnitude or direction. The magnitudes of **F** and **a** remain proportional, and the two vectors have the same direction at any given instant. However, they will not, in general, be tangent to the path of the particle.

When a particle is subjected simultaneously to several forces, Eq. (12.1) should be replaced by

$$\Sigma \mathbf{F} = m\mathbf{a} \tag{12.2}$$

where $\Sigma \mathbf{F}$ represents the sum, or resultant, of all the forces acting on the particle.

It should be noted that the system of axes with respect to which the acceleration **a** is determined is not arbitrary. These axes must have a constant orientation with respect to the stars, and their origin must either be attached to the sun† or move with a constant velocity with respect to the sun. Such a system of axes is called a *newtonian frame of reference.*‡ A system of axes attached to the earth does *not* constitute a newtonian frame of reference, since the earth rotates with respect to the stars and is accelerated with respect to the sun. However, in most engineering applications, the acceleration **a** may be determined with respect to axes attached to the earth and Eqs. (12.1) and (12.2) used without any appreciable error. On the other hand, these equations do not hold if **a** represents a relative acceleration measured with respect to moving axes, such as axes attached to an accelerated car or to a rotating piece of machinery.

We may observe that if the resultant $\Sigma \mathbf{F}$ of the forces acting on the particle is zero, it follows from Eq. (12.2) that the acceleration **a** of the particle is also zero. If the particle is initially at rest ($\mathbf{v}_0 = 0$) with respect to the newtonian frame of reference used, it will thus remain at rest ($\mathbf{v} = 0$). If originally moving with a velocity $\mathbf{v}_0$, the particle will maintain a constant velocity $\mathbf{v} = \mathbf{v}_0$; that is, it will move with the constant speed v_0 in a straight line. This, we recall, is the statement of Newton's first law (Sec. 2.10). Thus, Newton's first law is a particular case of Newton's second law and may be omitted from the fundamental principles of mechanics.

12.3. Systems of Units. In using the fundamental equation $\mathbf{F} = m\mathbf{a}$, the units of force, mass, length, and time cannot be chosen arbitrarily. If they are, the magnitude of the force **F** required to give an acceleration **a** to the mass m will *not* be numerically equal to the product ma; it will be only proportional to this product. Thus, we may choose three of the four units arbitrarily but must choose the fourth unit so that the equation $\mathbf{F} = m\mathbf{a}$ is satisfied. The units are then said to form a system of consistent kinetic units.

† More accurately, to the mass center of the solar system.

‡ Since the stars are not actually fixed, a more rigorous definition of a newtonian frame of reference (also called *inertial system*) is *one with respect to which Eq.* (12.2) *holds.*

Two systems of consistent kinetic units are currently used by American engineers, the International System of Units (SI units†), and the U.S. customary units. Since both systems have been discussed in detail in Sec. 1.3, we shall describe them only briefly in this section.

International System of Units (SI Units). In this system, the base units are the units of length, mass, and time, and are called, respectively, the *meter* (m), the *kilogram* (kg), and the *second* (s). All three are arbitrarily defined (Sec. 1.3). The unit of force is a derived unit. It is called the *newton* (N) and is defined as the force which gives an acceleration of 1 m/s² to a mass of 1 kg (Fig. 12.3). From Eq. (12.1) we write

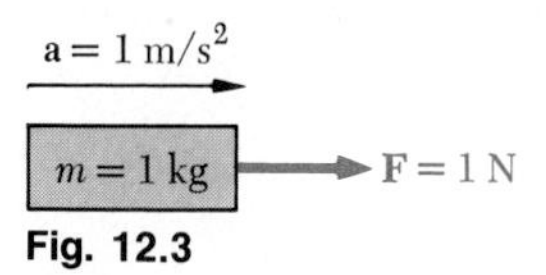

Fig. 12.3

$$1 \text{ N} = (1 \text{ kg})(1 \text{ m/s}^2) = 1 \text{ kg}\cdot\text{m/s}^2$$

The SI units are said to form an *absolute* system of units. This means that the three base units chosen are independent of the location where measurements are made. The meter, the kilogram, and the second may be used anywhere on the earth; they may even be used on another planet. They will always have the same significance.

The *weight* **W** of a body, or *force of gravity* exerted on that body, should, like any other force, be expressed in newtons. Since a body subjected to its own weight acquires an acceleration equal to the acceleration of gravity g, it follows from Newton's second law that the magnitude W of the weight of a body of mass m is

$$W = mg \tag{12.3}$$

Recalling that $g = 9.81 \text{ m/s}^2$, we find that the weight of a body of mass 1 kg (Fig. 12.4) is

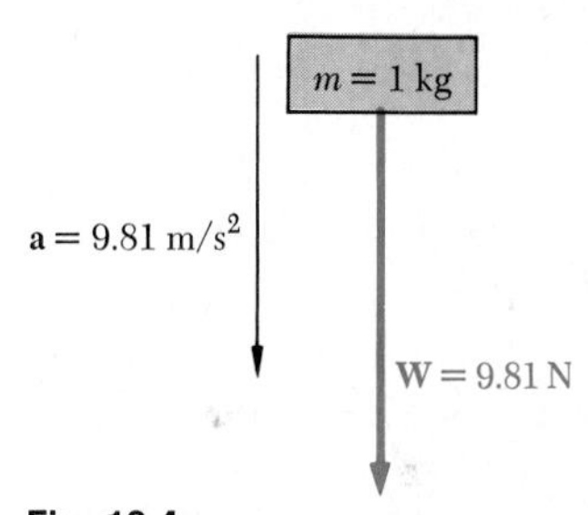

Fig. 12.4

$$W = (1 \text{ kg})(9.81 \text{ m/s}^2) = 9.81 \text{ N}$$

Multiples and submultiples of the units of length, mass, and force are frequently used in engineering practice. They are, respectively, the *kilometer* (km) and the *millimeter* (mm); the *megagram*‡ (Mg) and the *gram* (g); and the *kilonewton* (kN). By definition,

$$1 \text{ km} = 1000 \text{ m} \qquad 1 \text{ mm} = 0.001 \text{ m}$$
$$1 \text{ Mg} = 1000 \text{ kg} \qquad 1 \text{ g} = 0.001 \text{ kg}$$
$$1 \text{ kN} = 1000 \text{ N}$$

The conversion of these units to meters, kilograms, and newtons, respectively, can be effected by simply moving the decimal point three places to the right or to the left.

Units other than the units of mass, length, and time may all be expressed in terms of these three base units. For example, the unit of linear momentum may be obtained by recalling the definition of linear momentum and writing

$$mv = (\text{kg})(\text{m/s}) = \text{kg}\cdot\text{m/s}$$

† SI stands for *Système International d'Unités* (French).

‡ Also known as a *metric ton*.

U.S. Customary Units. Most practicing American engineers still commonly use a system in which the base units are the units of length, force, and time. These units are, respectively, the *foot* (ft), the *pound* (lb), and the *second* (s). The second is the same as the corresponding SI unit. The foot is defined as 0.3048 m. The pound is defined as the *weight* of a platinum standard, called the *standard pound* and kept at the National Bureau of Standards in Washington, the mass of which is 0.453 592 43 kg. Since the weight of a body depends upon the gravitational attraction of the earth, which varies with location, it is specified that the standard pound should be placed at sea level and at the latitude of 45° to properly define a force of 1 lb. Clearly the U.S. customary units do not form an absolute system of units. Because of their dependence upon the gravitational attraction of the earth, they are said to form a *gravitational* system of units.

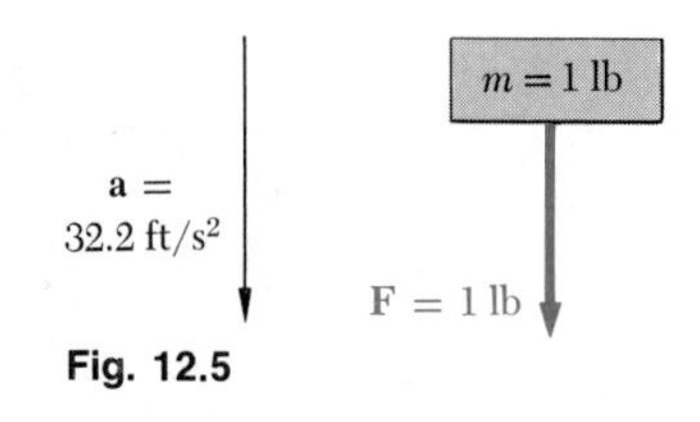

Fig. 12.5

While the standard pound also serves as the unit of mass in commercial transactions in the United States, it cannot be so used in engineering computations since such a unit would not be consistent with the base units defined in the preceding paragraph. Indeed, when acted upon by a force of 1 lb, that is, when subjected to its own weight, the standard pound receives the acceleration of gravity, $g = 32.2 \text{ ft/s}^2$ (Fig. 12.5), not the unit acceleration required by Eq. (12.1). The unit of mass consistent with the foot, the pound, and the second is the mass which receives an acceleration of 1 ft/s² when a force of 1 lb is applied to it (Fig. 12.6). This unit, sometimes called a *slug*, can be derived from the equation $F = ma$ after substituting 1 lb and 1 ft/s² for F and a, respectively. We write

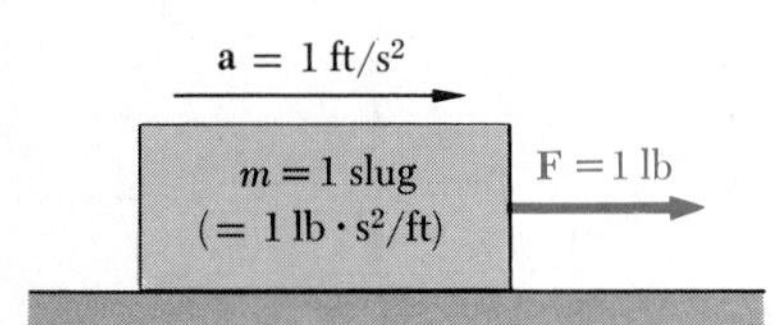

Fig. 12.6

$$F = ma \qquad 1 \text{ lb} = (1 \text{ slug})(1 \text{ ft/s}^2)$$

and obtain

$$1 \text{ slug} = \frac{1 \text{ lb}}{1 \text{ ft/s}^2} = 1 \text{ lb}\cdot\text{s}^2/\text{ft}$$

Comparing Figs. 12.5 and 12.6, we conclude that the slug is a mass 32.2 times larger than the mass of the standard pound.

The fact that bodies are characterized in the U.S. customary system of units by their weight in pounds rather than by their mass in slugs was a convenience in the study of statics, where we were dealing constantly with weights and other forces and only seldom with masses. However, in the study of kinetics, where forces, masses, and accelerations are involved, we repeatedly shall have to express in slugs the mass m of a body, the weight W of which has been given in pounds. Recalling Eq. (12.3), we shall write

$$m = \frac{W}{g} \tag{12.4}$$

where g is the acceleration of gravity ($g = 32.2 \text{ ft/s}^2$).

Units other than the units of force, length, and time may all be expressed in terms of these three base units. For example, the unit of linear momentum may be obtained by recalling the definition of linear momentum and writing

$$mv = (\text{lb}\cdot\text{s}^2/\text{ft})(\text{ft/s}) = \text{lb}\cdot\text{s}$$

The conversion from U.S. customary units to SI units, and vice versa, has been discussed in Sec. 1.4. We shall recall the conversion factors obtained, respectively, for the units of length, force, and mass:

Length:	$1 \text{ ft} = 0.3048 \text{ m}$
Force:	$1 \text{ lb} = 4.448 \text{ N}$
Mass:	$1 \text{ slug} = 1 \text{ lb}\cdot\text{s}^2/\text{ft} = 14.59 \text{ kg}$

Although it cannot be used as a consistent unit of mass, we also recall that the mass of the standard pound is, by definition,

$$1 \text{ pound-mass} = 0.4536 \text{ kg}$$

This constant may be used to determine the *mass* in SI units (kilograms) of a body which has been characterized by its *weight* in U.S. customary units (pounds).

12.4. Equations of Motion. Dynamic Equilibrium. Consider a particle of mass m acted upon by several forces. We recall from Sec. 12.2 that Newton's second law may be expressed by writing the equation

$$\Sigma \mathbf{F} = m\mathbf{a} \tag{12.2}$$

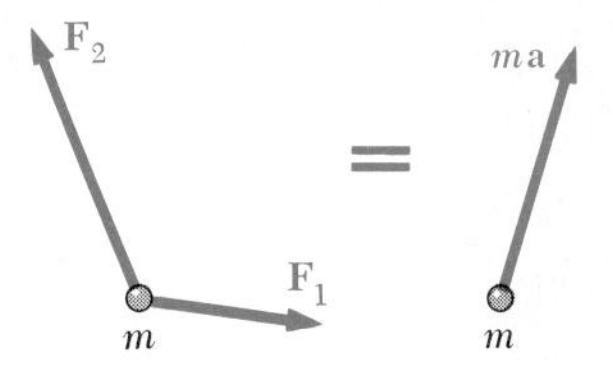

Fig. 12.7

which relates the forces acting on the particle and the vector $m\mathbf{a}$ (Fig. 12.7). In order to solve problems involving the motion of a particle, however, we shall replace Eq. (12.2) by equations involving scalar quantities. Using rectangular components, we write

$$\Sigma F_x = ma_x \qquad \Sigma F_y = ma_y \qquad \Sigma F_z = ma_z \tag{12.5}$$

Newton's second law may also be expressed by considering a vector of magnitude ma, but of *sense opposite* to that of the acceleration. This vector is denoted by $(m\mathbf{a})_{\text{rev}}$, where the subscript indicates that the sense of the acceleration has been reversed, and is called an *inertia vector*. If the inertia vector $(m\mathbf{a})_{\text{rev}}$ is added to the forces acting on the particle, *we obtain a system of vectors equivalent to zero* (Fig. 12.8). The particle may thus be considered to be in equilibrium under the given forces and the inertia vector. The particle is said to be in *dynamic equilibrium*, and the problem under consideration may be solved by using the methods developed earlier in statics. We may, for instance, write that the sums of the components of the vectors shown in Fig. 12.8, *including the inertia vector*, are zero,

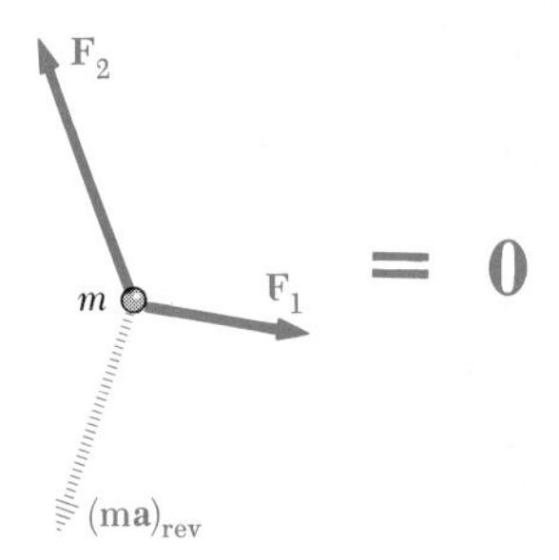

Fig. 12.8

$$\Sigma F_x = 0 \qquad \Sigma F_y = 0 \qquad \Sigma F_z = 0 \qquad \textbf{\textit{including inertia vector}} \tag{12.6}$$

or if the forces are coplanar, we may draw all vectors tip to tail, including again the inertia vector, to form a closed vector polygon.

Because they measure the resistance that particles offer when we try to set them in motion or when we try to change the conditions of their motion, inertia vectors are often called *inertia forces*. The inertia forces, however, are not forces like the forces found in statics, which are either contact forces or gravitational forces (weights). Many people, therefore, object to the use of the word force when referring to the vector $(m\mathbf{a})_{\text{rev}}$ or even avoid altogether the concept of dynamic equilibrium. Others point out that inertia forces and actual forces such as gravitational forces affect our senses in the same way and cannot be distinguished by physical measurements. A man riding in an elevator which is accelerated upward will have the feeling that his weight has suddenly increased; and no measurement made within the elevator could establish whether the elevator is truly accelerated or whether the force of attraction exerted by the earth has suddenly increased.

Sample Problems have been solved in this text by the direct application of Newton's second law as illustrated in Fig. 12.7, rather than by the method of dynamic equilibrium.

12.5. Systems of Particles. When a problem involves the motion of several particles, the equations of motion (12.5) may be written for each particle considered separately. However, in the case of a system involving a large number of particles, such as a rigid body, it will generally be more convenient to consider the system as a whole. In that case, we should note that the forces acting on a given particle of the system consist (1) of *external forces* exerted by bodies outside the system (such as the weight of the particle, which is exerted by the earth) and (2) of *internal forces* exerted by the other particles of the system. Figure 12.9 shows the particle P_1 of a system, the resultant $\mathbf{F}_1$ of the external forces exerted on P_1, and the internal forces $\mathbf{f}_{12}$, $\mathbf{f}_{13}$, etc., exerted by the other particles P_2, P_3, etc., of the system. According to Newton's second law, the resultant of the external and internal forces acting on P_1 is equal to the vector $m_1\mathbf{a}_1$ obtained by multiplying the acceleration $\mathbf{a}_1$ of the particle by its mass m_1. This vector is called the *effective force* of the particle P_1 since it represents the single force which should be applied to P_1 to produce the same acceleration $\mathbf{a}_1$.

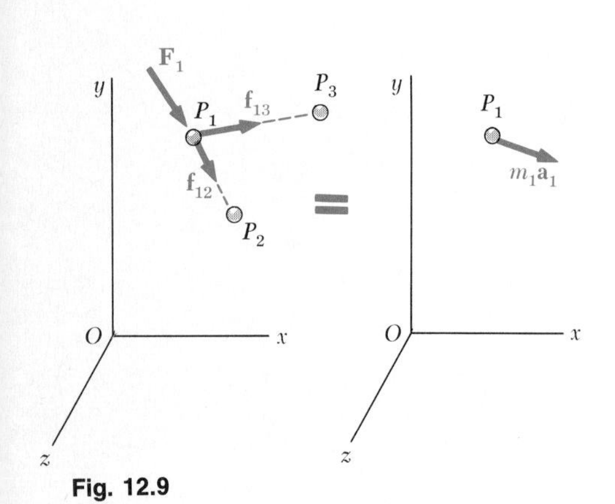

Fig. 12.9

We now consider simultaneously all the particles of the system (Fig. 12.10). Since the relation we have established holds for any particle, it follows that the system of all the external and internal forces acting on the various particles is equivalent to the system of the effective forces of the particles (i.e., one system may be replaced by the other). Thus the sum of the components in the x, y, or z direction of the external and internal forces must be equal to the sum of the components of the effective forces in the same direction. But the internal forces occur in pairs of *equal and opposite forces* (such as $\mathbf{f}_{12}$ and $\mathbf{f}_{21}$) and the sum of their components cancels out. We can therefore write the following relations between the components of the external and effective forces:

$$\Sigma(F_x)_{\text{ext}} = \Sigma ma_x \qquad \Sigma(F_y)_{\text{ext}} = \Sigma ma_y \qquad \Sigma(F_z)_{\text{ext}} = \Sigma ma_z \quad (12.7)$$

Consider now the moments about the x, y, or z axis of all the forces shown in Fig. 12.10. Since the internal forces occur in pairs of equal and

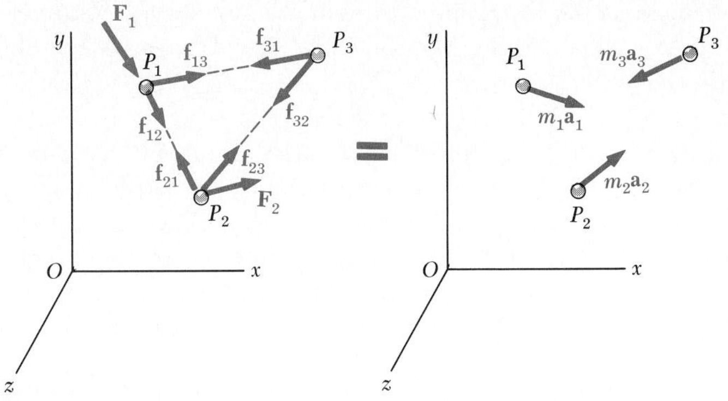

Fig. 12.10

opposite forces *directed along the same line of action,* the sum of their moments cancels out. Relations similar to Eqs. (12.7) therefore can be written between the moments of the external and effective forces.

Because the sums of their x, y, and z components and the sums of their moments about the x, y, and z axes are respectively equal, the system of the external forces and the system of the effective forces are said to be *equipollent*. This, however, does not mean that the two systems are *equivalent.* We note from Fig. 12.11 that the external force and the effective force acting on a given particle are *not* equal. The two systems of vectors, therefore, do not have the same effect on the individual particles; the two systems are *not* equivalent.

The distinction between equivalent and equipollent systems of vectors should be made in all cases involving a system of particles or a system of rigid bodies (as opposed to a single particle or a single rigid body). In all such cases we shall use a gray equals sign, as in Fig. 12.11, to indicate that the two systems of vectors are equipollent, i.e., have the same sums of

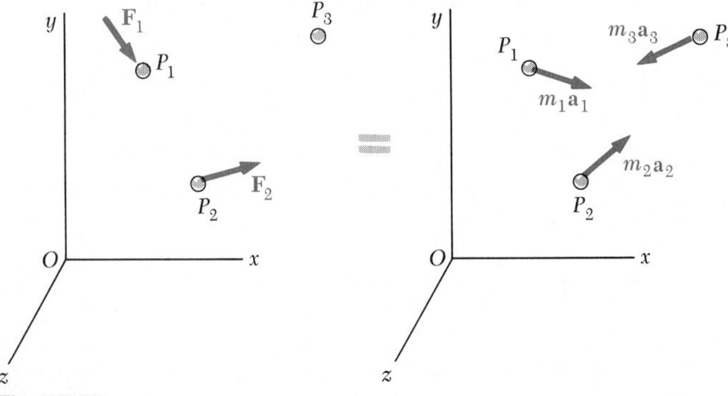

Fig. 12.11

components and the same sums of moments. We shall continue to use colored equals signs, as in Fig. 12.10, to indicate that two systems of vectors are truly equivalent.†

12.6. Motion of the Mass Center of a System of Particles. Equations (12.7) may be written in a modified form if the *mass center* of the system of particles is considered. The mass center of the system is the point G of coordinates $\bar{x}$, $\bar{y}$, $\bar{z}$ defined by the equations

$$(\Sigma m)\bar{x} = \Sigma mx \qquad (\Sigma m)\bar{y} = \Sigma my \qquad (\Sigma m)\bar{z} = \Sigma mz \tag{12.8}$$

Since $W = mg$, we note that G is also the center of gravity of the system. However, in order to avoid any confusion, we shall call G the *mass center* of the system of particles when discussing properties of the system associated with the *mass* of the particles, and we shall refer to it as the *center of gravity* of the system when considering properties associated with the *weight* of the particles. Particles located outside the gravitational field of the earth, for example, have a mass but no weight. We may then properly refer to their mass center, but obviously not to their center of gravity.‡

We differentiate twice with respect to t the relations (12.8):

$$(\Sigma m)\ddot{\bar{x}} = \Sigma m\ddot{x} \qquad (\Sigma m)\ddot{\bar{y}} = \Sigma m\ddot{y} \qquad (\Sigma m)\ddot{\bar{z}} = \Sigma m\ddot{z}$$

Recalling that the second derivatives of the coordinates of a point represent the components of the acceleration of that point, we have

$$(\Sigma m)\bar{a}_x = \Sigma ma_x \qquad (\Sigma m)\bar{a}_y = \Sigma ma_y \qquad (\Sigma m)\bar{a}_z = \Sigma ma_z \tag{12.9}$$

where $\bar{a}_x$, $\bar{a}_y$, $\bar{a}_z$ represent the components of the acceleration $\bar{\mathbf{a}}$ of the mass center G of the system. Combining the relations (12.7) and (12.9), we write the equations

$$\Sigma(F_x)_{\text{ext}} = (\Sigma m)\bar{a}_x \qquad \Sigma(F_y)_{\text{ext}} = (\Sigma m)\bar{a}_y \qquad \Sigma(F_z)_{\text{ext}} = (\Sigma m)\bar{a}_z \tag{12.10}$$

which define the motion of the mass center of the system. We note that Eqs. (12.10) are identical with the equations we would obtain for a particle of mass Σm, acted upon by all the external forces. We state therefore: *The mass center of a system of particles moves as if the entire mass of the system and all the external forces were concentrated at that point.*

† The relation established in this section between the external and the effective forces of a system of particles is often referred to as *d'Alembert's principle,* after the French mathematician Jean le Rond d'Alembert (1717–1783). However, d'Alembert's original statement refers to the motion of a system of connected bodies, with $\mathbf{f}_{12}$, $\mathbf{f}_{21}$, etc., representing constraint forces which, if applied by themselves, will not cause the system to move. Since this is in general not the case for the internal forces acting on a system of free particles, we shall postpone the consideration of d'Alembert's principle until the study of the motion of rigid bodies (Chap. 16).

‡ It may also be pointed out that the mass center and the center of gravity of a system of particles do not exactly coincide, since the weights of the particles are directed toward the center of the earth and thus do not truly form a system of parallel forces.

This principle is best illustrated by the motion of an exploding shell. We know that, if the resistance of the air is neglected, a shell may be assumed to travel along a parabolic path. After the shell has exploded, the mass center G of the fragments of shell will continue to travel along the same path. Indeed, point G must move as if the mass and the weight of all fragments were concentrated at G; it must move, therefore, as if the shell had not exploded.

The principle we have established shows that we may replace a rigid body by a particle of the same mass if we are interested in the motion of the mass center of the body and not in the motion of the body about its mass center. We should keep in mind, however, that Eqs. (12.9) and (12.10) relate only the components of the forces and vectors involved. To study the motion of a rigid body about its mass center, it will be necessary to also consider the relations between the moments of the external and effective forces (Chap. 16).

12.7. Rectilinear Motion of a Particle. Consider a particle of mass m moving in a straight line under the action of coplanar forces $\mathbf{F}_1$, $\mathbf{F}_2$, $\mathbf{F}_3$, etc. The conditions of motion of the particle are expressed by Eq. (12.2). Since the particle moves in a straight line, its acceleration $\mathbf{a}$ must be directed along that line. Choosing the x axis in the same direction and the y

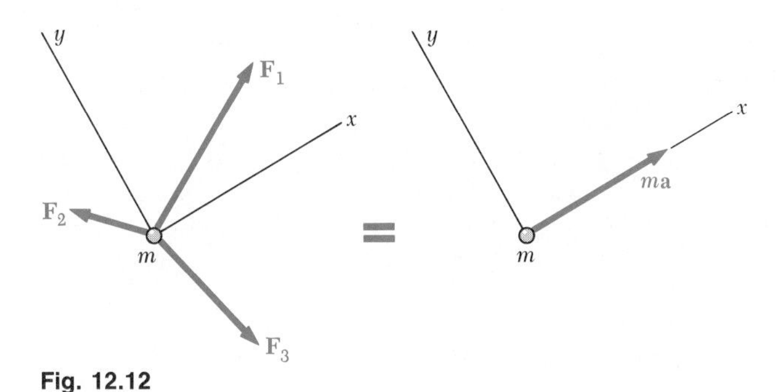

Fig. 12.12

axis in the plane of the forces (Fig. 12.12), we have $a_x = a$, $a_y = 0$ and write

$$\Sigma F_x = ma \qquad \Sigma F_y = 0 \tag{12.11}$$

The equations obtained may be solved for two unknowns.

When a problem involves two or more bodies, equations of motion should be written for each of the bodies (see Sample Probs. 12.3 and 12.4). We also recall from Sec. 12.2 that all accelerations should be measured with respect to a newtonian frame of reference. In most engineering applications accelerations may be determined with respect to axes attached to the earth, but relative accelerations measured with respect to moving axes, such as axes attached to an accelerated body, cannot be used in the equations of motion.

SAMPLE PROBLEM 12.1

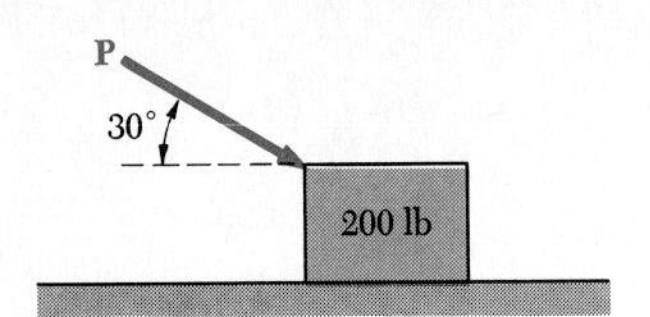

A 200-lb block rests on a horizontal plane. Find the magnitude of the force **P** required to give the block an acceleration of 10 ft/s² to the right. The coefficient of kinetic friction between the block and the plane is $\mu_k = 0.25$.

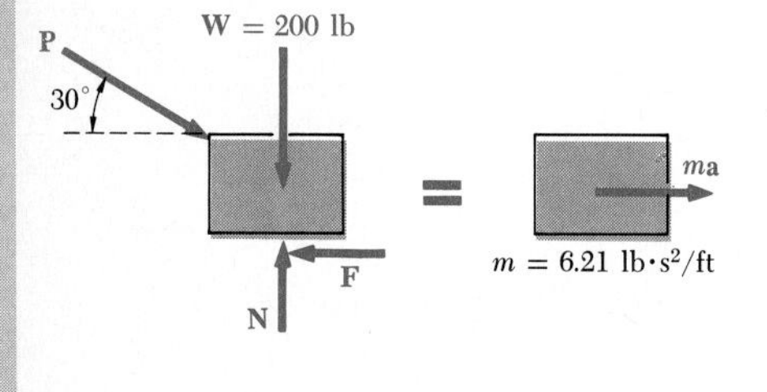

Solution. The mass of the block is

$$m = \frac{W}{g} = \frac{200 \text{ lb}}{32.2 \text{ ft/s}^2} = 6.21 \text{ lb} \cdot \text{s}^2/\text{ft}$$

We note that $F = \mu_k N = 0.25N$ and that $a = 10 \text{ ft/s}^2$. Expressing that the forces acting on the block are equivalent to the vector $m\mathbf{a}$, we write

$$\overset{+}{\rightarrow} \Sigma F_x = ma: \qquad P \cos 30° - 0.25N = (6.21 \text{ lb} \cdot \text{s}^2/\text{ft})(10 \text{ ft/s}^2)$$

$$P \cos 30° - 0.25N = 62.1 \text{ lb} \qquad (1)$$

$$+\uparrow \Sigma F_y = 0: \qquad N - P \sin 30° - 200 \text{ lb} = 0 \qquad (2)$$

Solving (2) for N and carrying the result into (1), we obtain

$$N = P \sin 30° + 200 \text{ lb}$$

$$P \cos 30° - 0.25(P \sin 30° + 200 \text{ lb}) = 62.1 \text{ lb} \qquad P = 151 \text{ lb}$$ ◄

SAMPLE PROBLEM 12.2

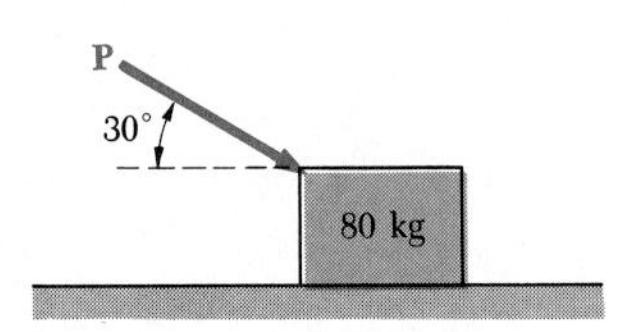

An 80-kg block rests on a horizontal plane. Find the magnitude of the force **P** required to give the block an acceleration of 2.5 m/s² to the right. The coefficient of kinetic friction between the block and the plane is $\mu_k = 0.25$.

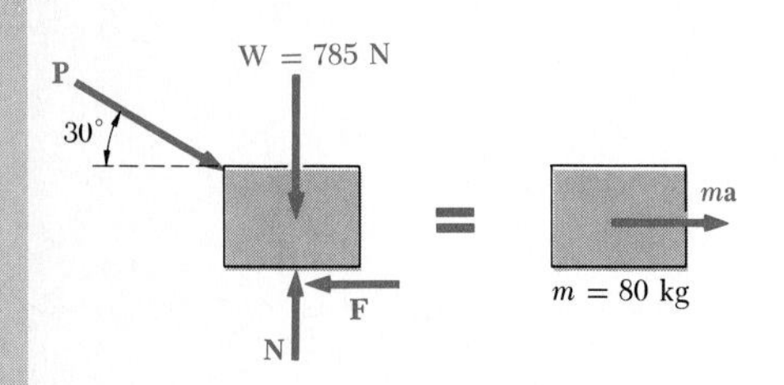

Solution. The weight of the block is

$$W = mg = (80 \text{ kg})(9.81 \text{ m/s}^2) = 785 \text{ N}$$

We note that $F = \mu_k N = 0.25N$ and that $a = 2.5 \text{ m/s}^2$. Expressing that the forces acting on the block are equivalent to the vector $m\mathbf{a}$, we write

$$\overset{+}{\rightarrow} \Sigma F_x = ma: \qquad P \cos 30° - 0.25N = (80 \text{ kg})(2.5 \text{ m/s}^2)$$

$$P \cos 30° - 0.25N = 200 \text{ N} \qquad (1)$$

$$+\uparrow \Sigma F_y = 0: \qquad N - P \sin 30° - 785 \text{ N} = 0 \qquad (2)$$

Solving (2) for N and carrying the result into (1), we obtain

$$N = P \sin 30° + 785 \text{ N}$$

$$P \cos 30° - 0.25(P \sin 30° + 785 \text{ N}) = 200 \text{ N} \qquad P = 535 \text{ N}$$ ◄

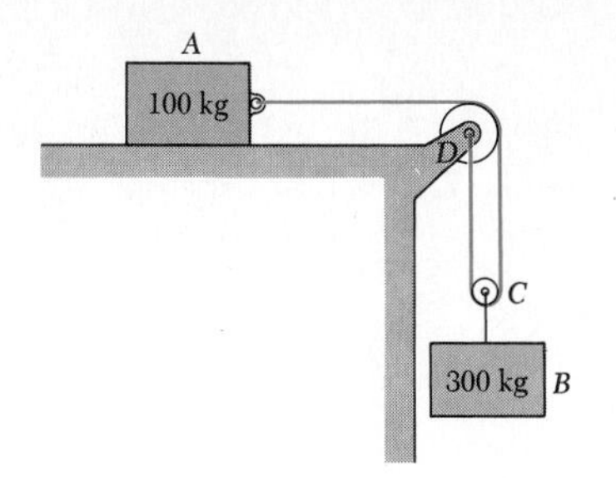

SAMPLE PROBLEM 12.3

The two blocks shown start from rest. The horizontal plane and the pulley are frictionless, and the pulley is assumed to be of negligible mass. Determine the acceleration of each block and the tension in each cord.

Kinematics. We note that if block A moves through x_A to the right, block B moves down through

$$x_B = \tfrac{1}{2}x_A$$

Differentiating twice with respect to t, we have

$$a_B = \tfrac{1}{2}a_A \qquad (1)$$

Kinetics. We shall apply Newton's second law successively to block A, block B, and pulley C.

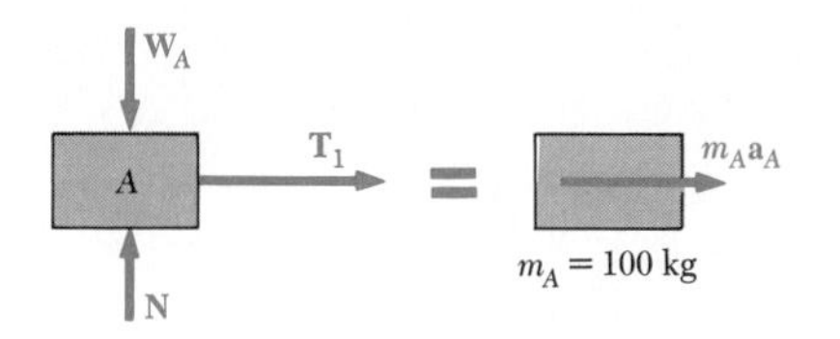

Block A. Denoting by T_1 the tension in cord ACD, we write

$$\xrightarrow{+} \Sigma F_x = m_A a_A: \qquad T_1 = 100a_A \qquad (2)$$

Block B. Observing that the weight of block B is

$$W_B = m_B g = (300 \text{ kg})(9.81 \text{ m/s}^2) = 2940 \text{ N}$$

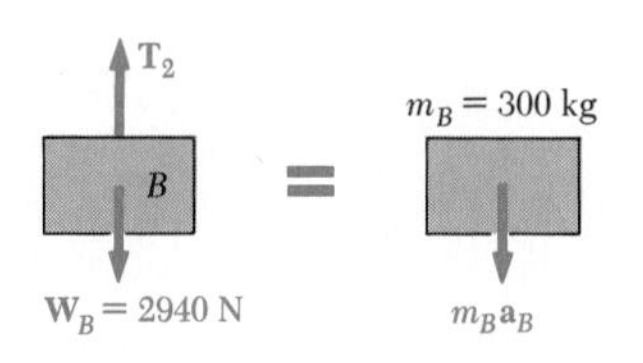

and denoting by T_2 the tension in cord BC, we write

$$+\downarrow \Sigma F_y = m_B a_B: \qquad 2940 - T_2 = 300a_B$$

or, substituting for a_B from (1),

$$2940 - T_2 = 300(\tfrac{1}{2}a_A)$$
$$T_2 = 2940 - 150a_A \qquad (3)$$

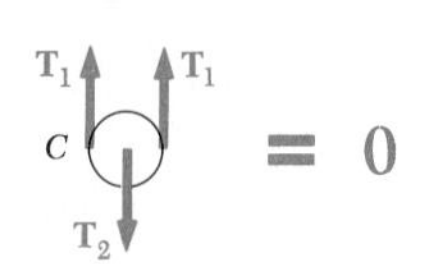

Pulley C. Since m_C is assumed to be zero, we have

$$+\downarrow \Sigma F_y = m_C a_C = 0: \qquad T_2 - 2T_1 = 0 \qquad (4)$$

Substituting for T_1 and T_2 from (2) and (3), respectively, into (4) we write

$$2940 - 150a_A - 2(100a_A) = 0$$
$$2940 - 350a_A = 0 \qquad a_A = 8.40 \text{ m/s}^2 \blacktriangleleft$$

Substituting the value obtained for a_A into (1) and (2), we have

$$a_B = \tfrac{1}{2}a_A = \tfrac{1}{2}(8.40 \text{ m/s}^2) \qquad a_B = 4.20 \text{ m/s}^2 \blacktriangleleft$$
$$T_1 = 100a_A = (100 \text{ kg})(8.40 \text{ m/s}^2) \qquad T_1 = 840 \text{ N} \blacktriangleleft$$

Recalling (4), we write

$$T_2 = 2T_1 \qquad T_2 = 2(840 \text{ N}) \qquad T_2 = 1680 \text{ N} \blacktriangleleft$$

We note that the value obtained for T_2 is *not* equal to the weight of block B.

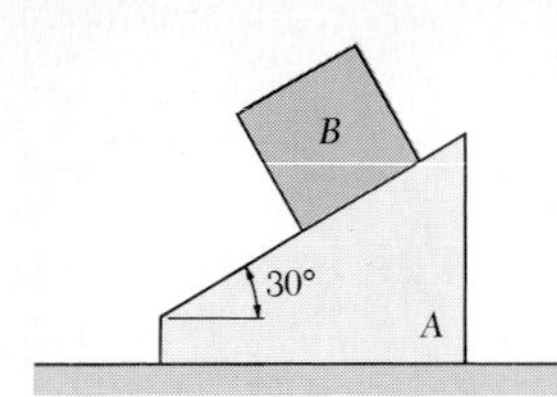

SAMPLE PROBLEM 12.4

The 12-lb block B starts from rest and slides on the 30-lb wedge A, which is supported by a horizontal surface. Neglecting friction, determine (*a*) the acceleration of the wedge, (*b*) the acceleration of the block relative to the wedge.

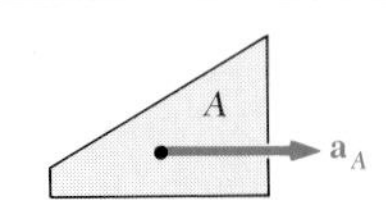

Kinematics. We shall first examine the accelerations of the wedge and of the block.

Wedge A. Since the wedge is constrained to move on the horizontal surface, its acceleration $\mathbf{a}_A$ is horizontal. We shall assume that it is directed to the right.

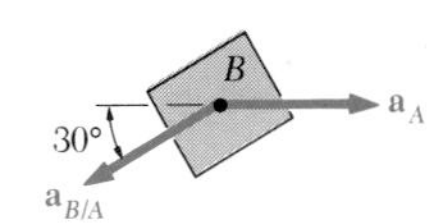

Block B. The acceleration $\mathbf{a}_B$ of block B may be expressed as the sum of the acceleration of A and of the acceleration of B relative to A. We have

$$\mathbf{a}_B = \mathbf{a}_A + \mathbf{a}_{B/A}$$

where $\mathbf{a}_{B/A}$ is directed along the sloping surface of the wedge.

Kinetics. We shall draw the free-body diagrams of the wedge and of the block and apply Newton's second law.

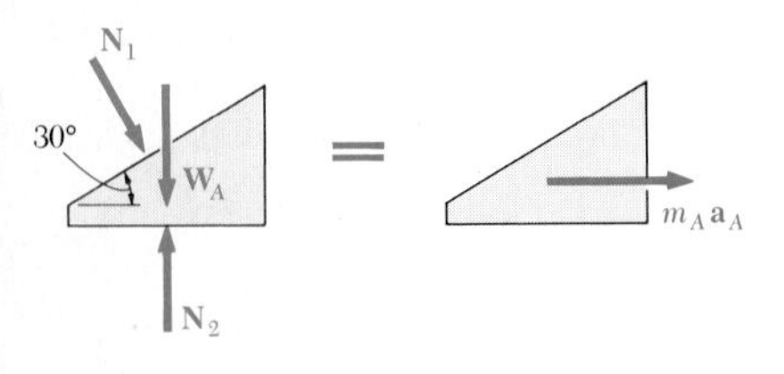

Wedge A. We denote respectively by $\mathbf{N}_1$ and $\mathbf{N}_2$ the forces exerted by the block and the horizontal surface on wedge A.

$$\overset{+}{\rightarrow} \Sigma F_x = m_A a_A: \qquad N_1 \sin 30° = m_A a_A$$

$$0.5N_1 = (W_A/g)a_A \tag{1}$$

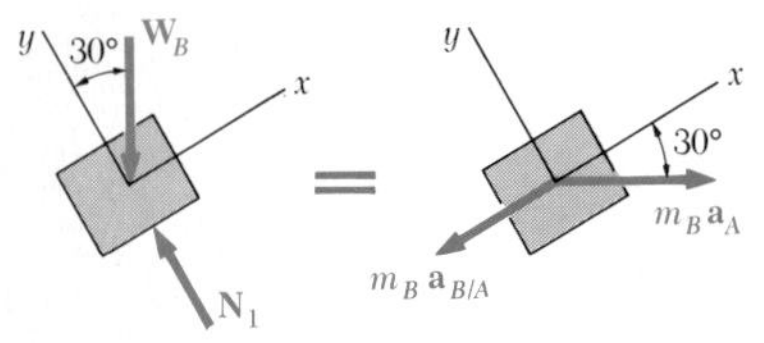

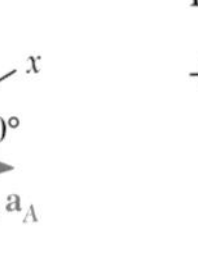

Block B. Using the coordinate axes shown and resolving $\mathbf{a}_B$ into its components $\mathbf{a}_A$ and $\mathbf{a}_{B/A}$, we write

$$+\nearrow \Sigma F_x = m_B a_x: \qquad -W_B \sin 30° = m_B a_A \cos 30° - m_B a_{B/A}$$

$$-W_B \sin 30° = (W_B/g)(a_A \cos 30° - a_{B/A})$$

$$a_{B/A} = a_A \cos 30° + g \sin 30° \tag{2}$$

$$+\nwarrow \Sigma F_y = m_B a_y: \qquad N_1 - W_B \cos 30° = -m_B a_A \sin 30°$$

$$N_1 - W_B \cos 30° = -(W_B/g)a_A \sin 30° \tag{3}$$

a. Acceleration of Wedge A. Substituting for N_1 from Eq. (1) into Eq. (3), we have

$$2(W_A/g)a_A - W_B \cos 30° = -(W_B/g)a_A \sin 30°$$

Solving for a_A and substituting the numerical data, we write

$$a_A = \frac{W_B \cos 30°}{2W_A + W_B \sin 30°} g = \frac{(12 \text{ lb}) \cos 30°}{2(30 \text{ lb}) + (12 \text{ lb}) \sin 30°}(32.2 \text{ ft/s}^2)$$

$$a_A = +5.07 \text{ ft/s}^2 \qquad \mathbf{a}_A = 5.07 \text{ ft/s}^2 \rightarrow \blacktriangleleft$$

b. Acceleration of Block B Relative to A. Substituting the value obtained for a_A into Eq. (2), we have

$$a_{B/A} = (5.07 \text{ ft/s}^2) \cos 30° + (32.2 \text{ ft/s}^2) \sin 30°$$

$$a_{B/A} = +20.5 \text{ ft/s}^2 \qquad \mathbf{a}_{B/A} = 20.5 \text{ ft/s}^2 \measuredangle 30° \blacktriangleleft$$

Problems

12.1 The acceleration due to gravity on the moon is 1.62 m/s^2. Determine (*a*) the weight in newtons, (*b*) the mass in kilograms, on the moon, of a gold bar, the mass of which has been officially designated as 2 kg.

12.2 The value of g at any latitude ϕ may be obtained from the formula

$$g = 32.09(1 + 0.0053 \sin^2 \phi) \text{ ft/s}^2$$

which takes into account the effect of the rotation of the earth, as well as the fact that the earth is not truly spherical. Determine to four significant figures (*a*) the weight in pounds, (*b*) the mass in pounds, (*c*) the mass in $\text{lb}\cdot\text{s}^2/\text{ft}$, at the latitudes of $0°$, $45°$, and $90°$, of a silver bar, the mass of which has been officially designated as 5 lb.

12.3 Two boxes are weighed on the scales shown: scale *a* is a lever scale; scale *b* is a spring scale. The scales are attached to the roof of an elevator. When the elevator is at rest, each scale indicates a mass of 12 kg. If the spring scale indicates a mass of 10 kg, determine (*a*) the acceleration of the elevator, (*b*) the mass indicated by the lever scale.

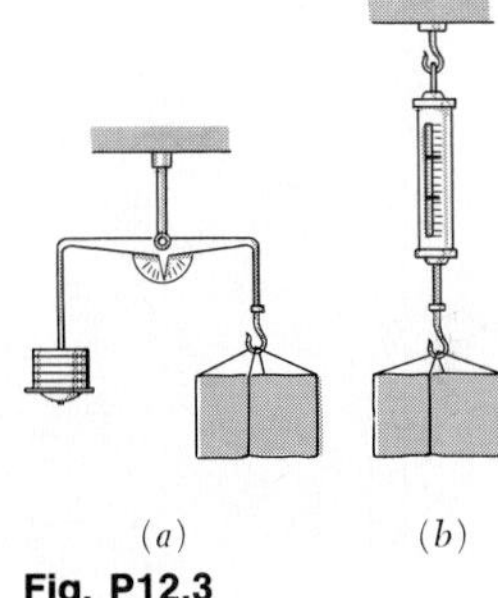

Fig. P12.3

12.4 A 400-kg satellite has been placed in a circular orbit 1500 km above the surface of the earth. The acceleration of gravity at this elevation is 6.43 m/s^2. Determine the linear momentum of the satellite, knowing that its orbital speed is 25.6×10^3 km/h.

12.5 Determine the maximum theoretical speed that may be achieved over a distance of 50 m by a car starting from rest, knowing that the coefficient of static friction is 0.80 between the tires and the pavement. Assume four-wheel drive.

12.6 Solve Prob. 12.5, knowing that 60 percent of the weight of the car is distributed over its front wheels and 40 percent over its rear wheels, and assuming (*a*) front-wheel drive, (*b*) rear-wheel drive.

12.7 A motorist traveling at a speed of 90 km/h suddenly applies the brakes and comes to a stop after skidding 50 m. Determine (*a*) the time required for the car to stop, (*b*) the coefficient of friction between the tires and the pavement.

12.8 A car has been traveling up a long 2 percent grade at a constant speed of 55 mi/h. If the driver does not change the setting of the throttle or shift gears as the car reaches the top of the hill, what will be the acceleration of the car as it starts moving down the 3 percent grade?

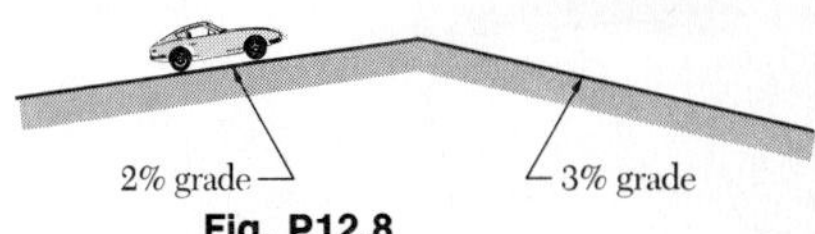

Fig. P12.8

12.9 A 3000-lb automobile is driven down a 5° incline at a speed of 50 mi/h when the brakes are applied, causing a total braking force of 1200 lb to be applied to the automobile. Determine the distance traveled by the automobile before it comes to a stop.

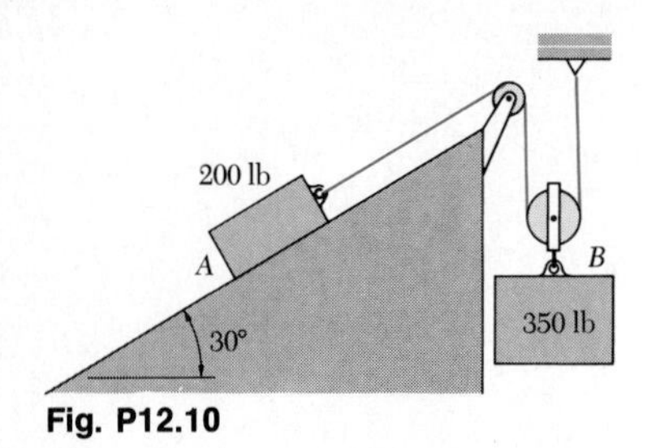

Fig. P12.10

12.10 The two blocks shown are originally at rest. Neglecting the masses of the pulleys and the effect of friction in the pulleys and between block A and the incline, determine (a) the acceleration of each block, (b) the tension in the cable.

12.11 Solve Prob. 12.10, assuming that the coefficients of friction between block A and the incline are $\mu_s = 0.25$ and $\mu_k = 0.20$.

12.12 A 1050-kg trailer is hitched to a 1200-kg car. The car and trailer are traveling at 90 km/h when the driver applies the brakes on both the car and the trailer. Knowing that the braking forces exerted on the car and the trailer are 4500 N and 3600 N, respectively, determine (a) the deceleration of the car and trailer, (b) the horizontal component of the force exerted by the trailer hitch on the car.

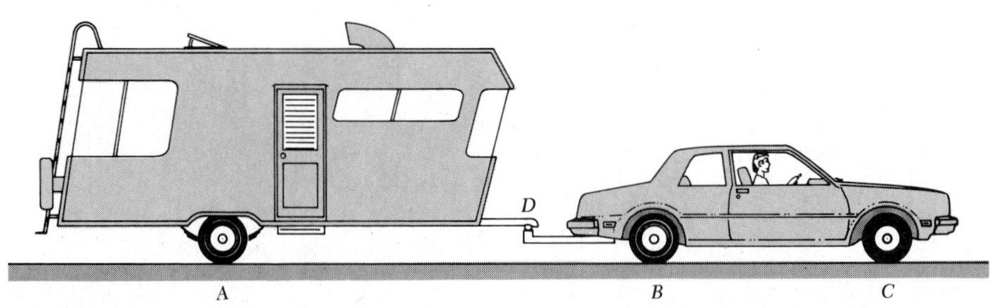

Fig. P12.12

12.13 Solve Prob. 12.12, assuming that the trailer brakes are inoperative.

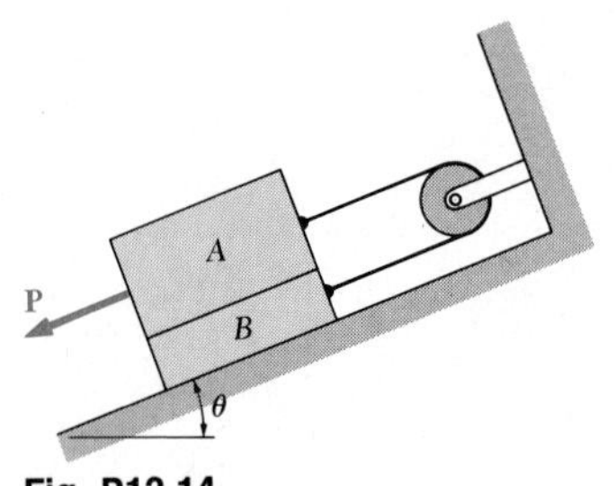

Fig. P12.14

12.14 Block A has a mass of 30 kg and block B a mass of 15 kg. The coefficients of friction between all surfaces of contact are $\mu_s = 0.15$ and $\mu_k = 0.10$. Knowing that $\theta = 30°$ and that the magnitude of the force $\mathbf{P}$ applied to block A is 250 N, determine (a) the acceleration of block A, (b) the tension in the cord.

12.15 Solve Prob. 12.14, assuming that the 250-N force $\mathbf{P}$ is applied to block B instead of block A.

12.16 Knowing that the system shown starts from rest, find the velocity at $t = 1.2$ s of (a) collar A, (b) collar B. Neglect the masses of the pulleys and the effect of friction.

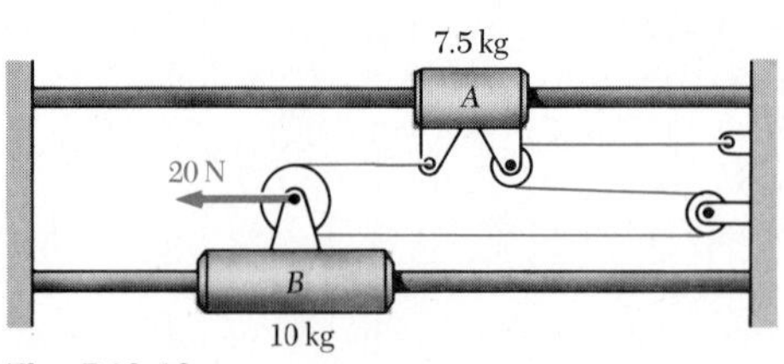

Fig. P12.16

12.17 Two packages are placed on a conveyor belt which is at rest. The coefficient of kinetic friction is 0.20 between the belt and package *A*, and 0.10 between the belt and package *B*. If the belt is suddenly started to the right and slipping occurs between the belt and the packages, determine (*a*) the acceleration of the packages, (*b*) the force exerted by package *A* on package *B*.

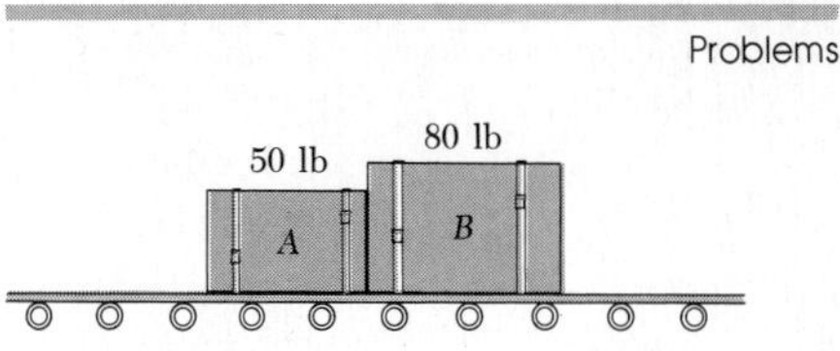

Fig. P12.17

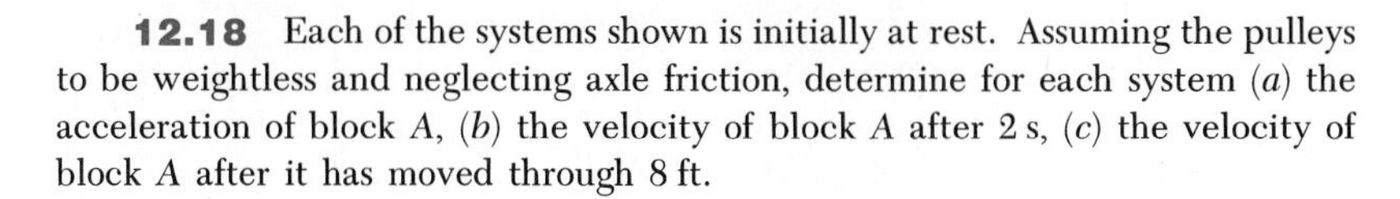

12.18 Each of the systems shown is initially at rest. Assuming the pulleys to be weightless and neglecting axle friction, determine for each system (*a*) the acceleration of block *A*, (*b*) the velocity of block *A* after 2 s, (*c*) the velocity of block *A* after it has moved through 8 ft.

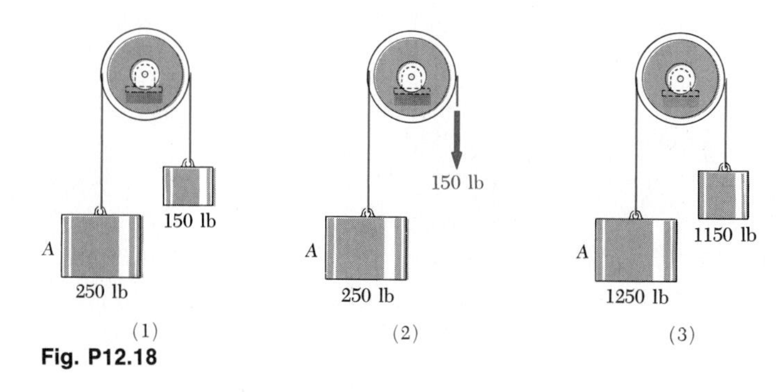

Fig. P12.18

12.19 The coefficients of friction between the load and the flat-bed trailer shown are $\mu_s = 0.50$ and $\mu_k = 0.40$. Knowing that the speed of the rig is 45 mi/h, determine the shortest distance in which the rig can be brought to a stop if the load is not to shift.

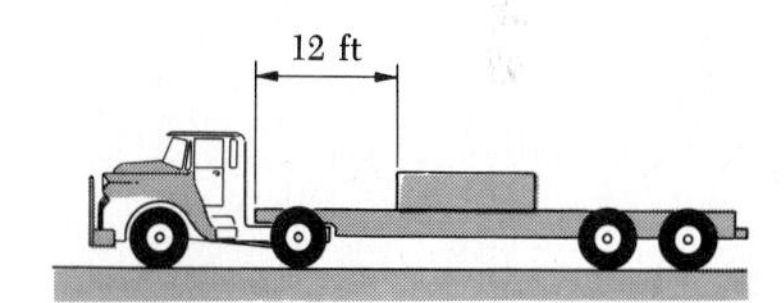

Fig. P12.19

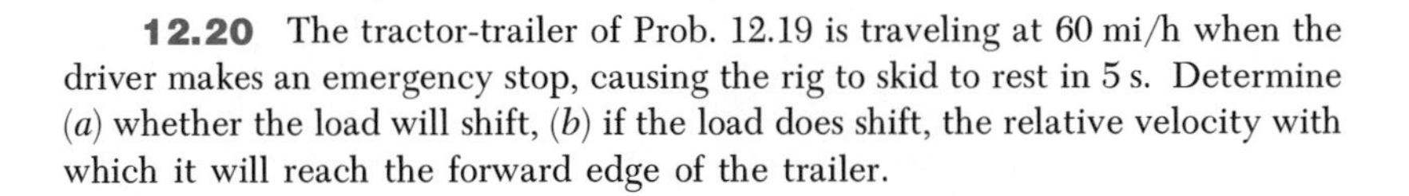

12.20 The tractor-trailer of Prob. 12.19 is traveling at 60 mi/h when the driver makes an emergency stop, causing the rig to skid to rest in 5 s. Determine (*a*) whether the load will shift, (*b*) if the load does shift, the relative velocity with which it will reach the forward edge of the trailer.

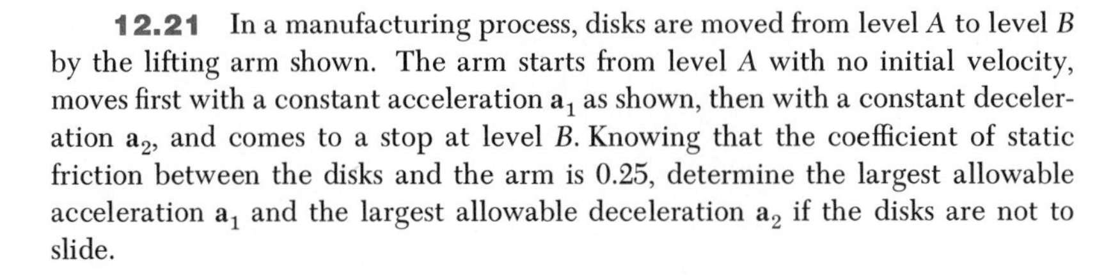

12.21 In a manufacturing process, disks are moved from level *A* to level *B* by the lifting arm shown. The arm starts from level *A* with no initial velocity, moves first with a constant acceleration $\mathbf{a}_1$ as shown, then with a constant deceleration $\mathbf{a}_2$, and comes to a stop at level *B*. Knowing that the coefficient of static friction between the disks and the arm is 0.25, determine the largest allowable acceleration $\mathbf{a}_1$ and the largest allowable deceleration $\mathbf{a}_2$ if the disks are not to slide.

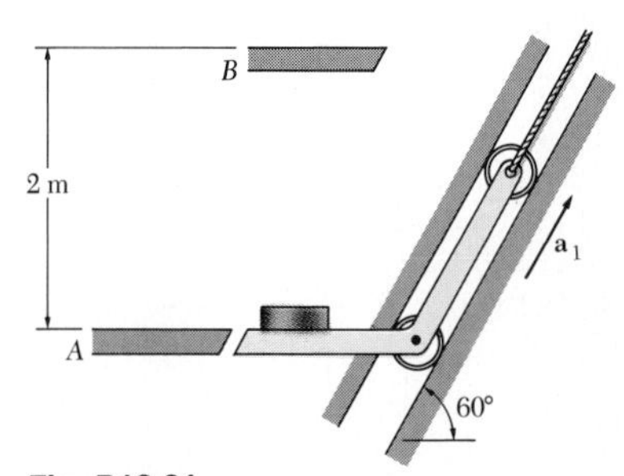

Fig. P12.21

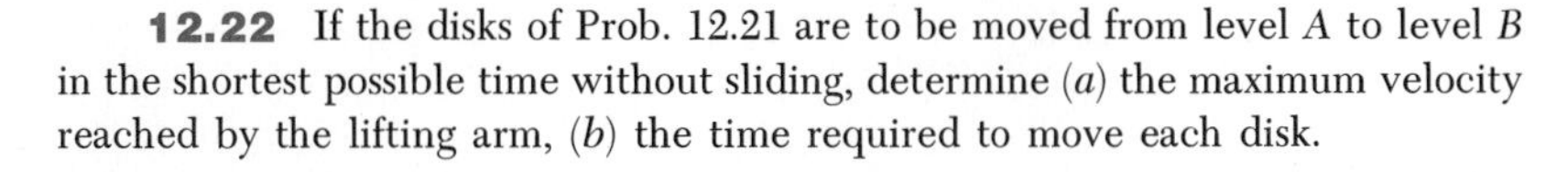

12.22 If the disks of Prob. 12.21 are to be moved from level *A* to level *B* in the shortest possible time without sliding, determine (*a*) the maximum velocity reached by the lifting arm, (*b*) the time required to move each disk.

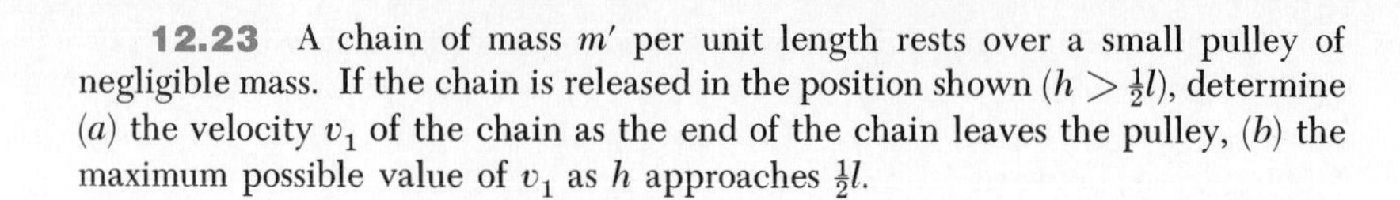

12.23 A chain of mass m' per unit length rests over a small pulley of negligible mass. If the chain is released in the position shown ($h > \frac{1}{2}l$), determine (a) the velocity v_1 of the chain as the end of the chain leaves the pulley, (b) the maximum possible value of v_1 as h approaches $\frac{1}{2}l$.

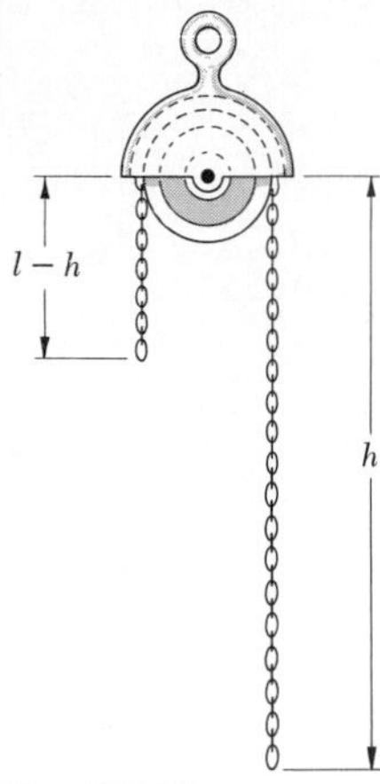

Fig. P12.23

12.24 An airplane has a mass of 30 Mg and its engines develop a total thrust of 50 kN during take-off. If the drag **D** exerted on the plane has a magnitude $D = 2.50\, v^2$, where v is expressed in meters per second and D in newtons, and if the plane becomes airborne at a speed of 270 km/h, determine the length of runway required for the plane to take off.

12.25 For the airplane of Prob. 12.24, determine the time required to take off.

12.26 A constant force **P** is applied to a piston and rod of total mass m in order to make them move in a cylinder filled with oil. As the piston moves, the oil is forced through orifices in the piston and exerts on the piston an additional force of magnitude kv, proportional to the speed v of the piston and in a direction opposite to its motion. Express the distance x traveled by the piston as a function of the time t, assuming that the piston starts from rest at $t = 0$ and $x = 0$.

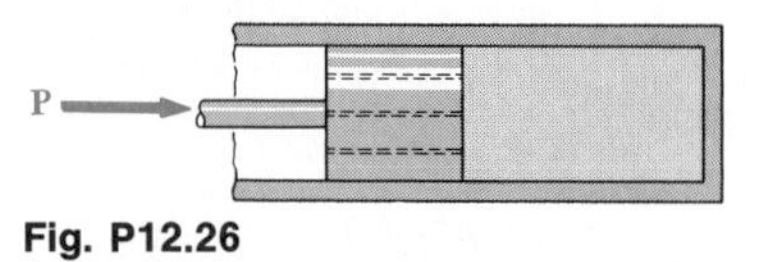

Fig. P12.26

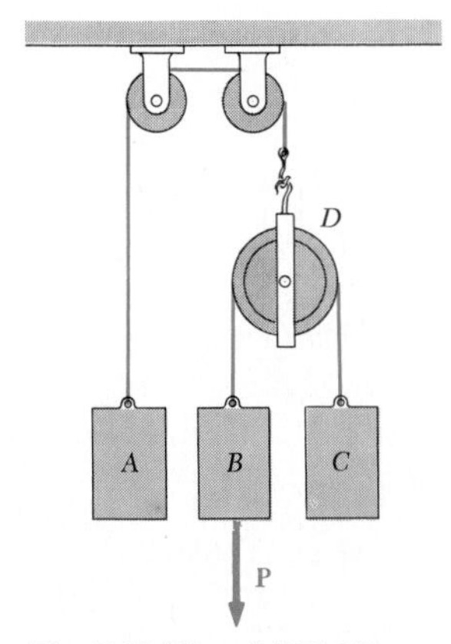

Fig. P12.27 and P12.28

12.27 Block A weighs 20 lb, and blocks B and C weigh 10 lb each. Knowing that $P = 2.5$ lb and that the blocks are initially at rest, determine at $t = 3$ s the velocity (a) of B relative to A, (b) of C relative to A. Neglect the weights of the pulleys and the effect of friction.

12.28 Block A weighs 20 lb, and blocks B and C weigh 10 lb each. Knowing that the blocks are initially at rest and that B moves through 8 ft in 2 s, determine (a) the magnitude of the force **P**, (b) the tension in the cord AD. Neglect the weights of the pulleys and the effect of friction.

12.29 An 8-kg block B rests as shown on a 12-kg bracket A. The coefficients of friction are $\mu_s = 0.40$ and $\mu_k = 0.30$ between block B and bracket A, and there is no friction in the pulley or between the bracket and the horizontal surface. If $P = 30$ N, determine the velocity of B relative to A after B has moved 400 mm with respect to A.

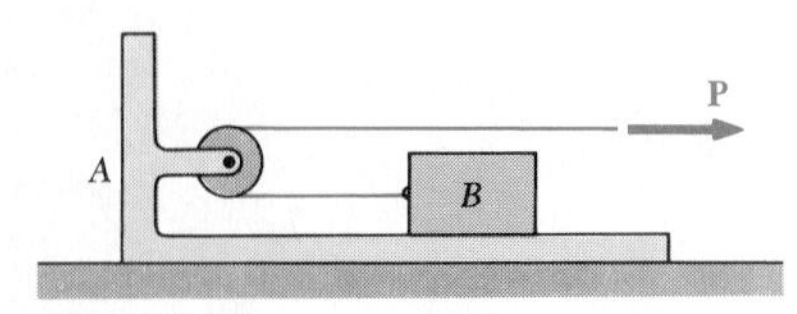

Fig. P12.29

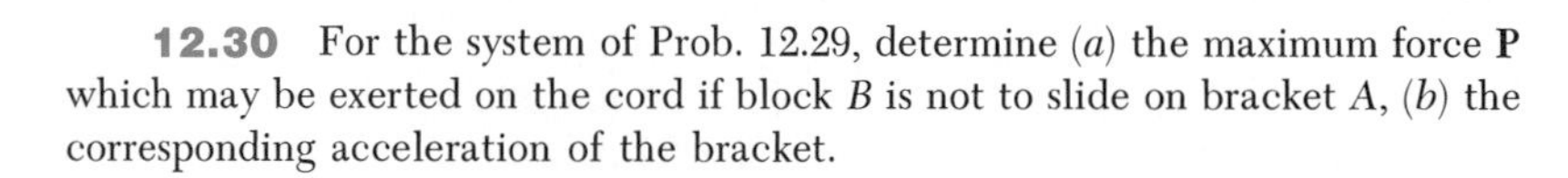

12.30 For the system of Prob. 12.29, determine (a) the maximum force **P** which may be exerted on the cord if block B is not to slide on bracket A, (b) the corresponding acceleration of the bracket.

12.31 A 12-lb block B rests as shown on the upper surface of a 30-lb wedge A. Neglecting friction, determine (*a*) the acceleration of A, (*b*) the acceleration of B relative to A, immediately after the system is released from rest.

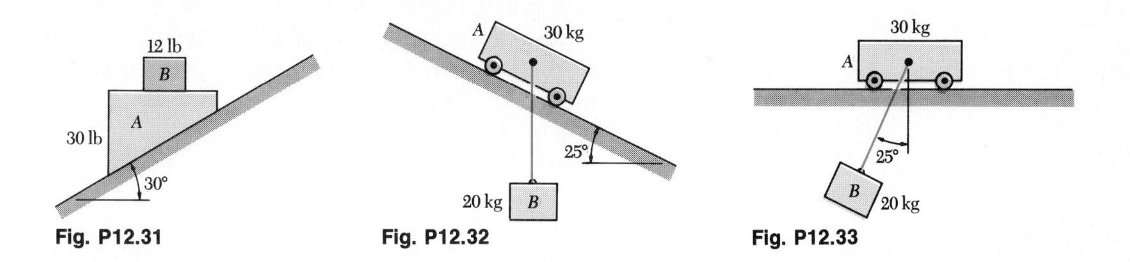

Fig. P12.31

Fig. P12.32

Fig. P12.33

12.32 and 12.33 A 20-kg block B is suspended from a 2-m cord attached to a 30-kg cart A. Neglecting friction, determine (*a*) the acceleration of the cart, (*b*) the tension in the cord, immediately after the system is released from rest in the position shown.

12.34 A 40-lb sliding panel is supported by rollers at B and C. A 25-lb counterweight A is attached to a cable as shown and, in cases *a* and *c*, is initially in contact with a vertical edge of the panel. Neglecting friction, determine in each case shown the acceleration of the panel and the tension in the cord immediately after the system is released from rest.

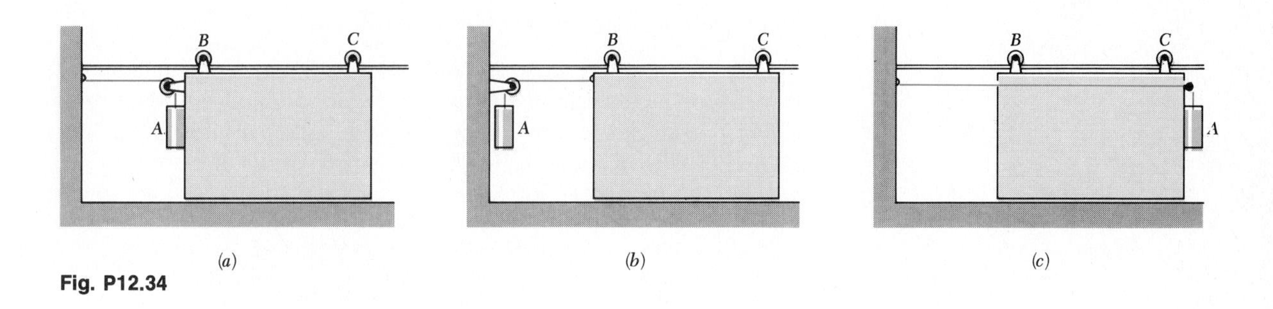

(*a*) (*b*) (*c*)

Fig. P12.34

12.8. Curvilinear Motion of a Particle. We saw in Chap. 11 that the acceleration of a particle moving in a plane along a curved path may be represented by its rectangular components $\mathbf{a}_x$ and $\mathbf{a}_y$, by its tangential and normal components $\mathbf{a}_t$ and $\mathbf{a}_n$, or by its radial and transverse components $\mathbf{a}_r$ and $\mathbf{a}_\theta$. We may therefore use any one of these methods of representation to express Newton's second law of motion.

Rectangular Components. Using the rectangular components $\mathbf{a}_x$ and $\mathbf{a}_y$ of the acceleration, and resolving the forces acting on the particle into their components $\mathbf{F}_x$ and $\mathbf{F}_y$, we write

$$\Sigma F_x = ma_x \qquad \Sigma F_y = ma_y \tag{12.12}$$

These equations may be solved for two unknowns.

Consider, as an example, the *motion of a projectile.* If the resistance of the air is neglected, the only force acting on the projectile after it has been fired is its weight **W**. Substituting $\Sigma F_x = 0$ and $\Sigma F_y = -W$ into Eqs. (12.12), we obtain

$$a_x = 0 \qquad a_y = -\frac{W}{m} = -g$$

where g is 9.81 m/s^2 or 32.2 ft/s^2. The equations obtained may be integrated independently, as was shown in Sec. 11.10 to obtain the velocity and displacement of the projectile at any instant.

Tangential and Normal Components. Using the tangential and normal components $\mathbf{a}_t$ and $\mathbf{a}_n$ of the acceleration, and resolving the forces acting on the particle into components $\mathbf{F}_t$ along the tangent to the path (in the direction of motion) and components $\mathbf{F}_n$ along the normal (toward the inside of the path), we write (Fig. 12.13)

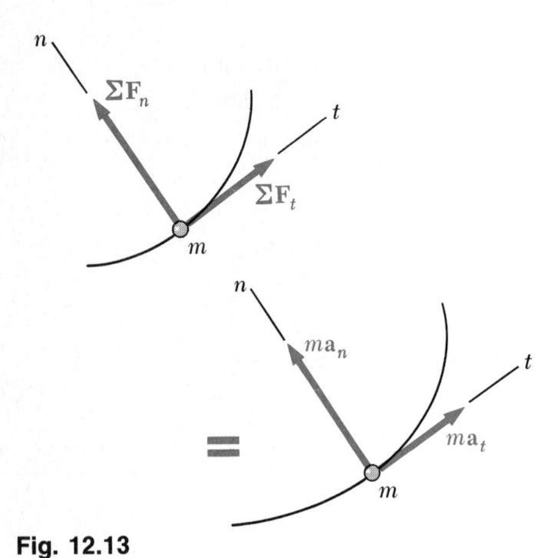

Fig. 12.13

$$\Sigma F_t = ma_t \qquad \Sigma F_n = ma_n \tag{12.13}$$

Substituting for a_t and a_n from (11.28) and (11.29), we have

$$\Sigma F_t = m\frac{dv}{dt} \qquad \Sigma F_n = m\frac{v^2}{\rho} \tag{12.14}$$

The equations obtained may be solved for two unknowns.

Radial and Transverse Components. Using the radial and transverse components $\mathbf{a}_r$ and $\mathbf{a}_\theta$ of the acceleration, and resolving the forces acting on the particle into components $\mathbf{F}_r$ along the line OP joining the origin O to the particle P, and components $\mathbf{F}_\theta$ in the direction perpendicular to OP (Fig. 12.14), we write

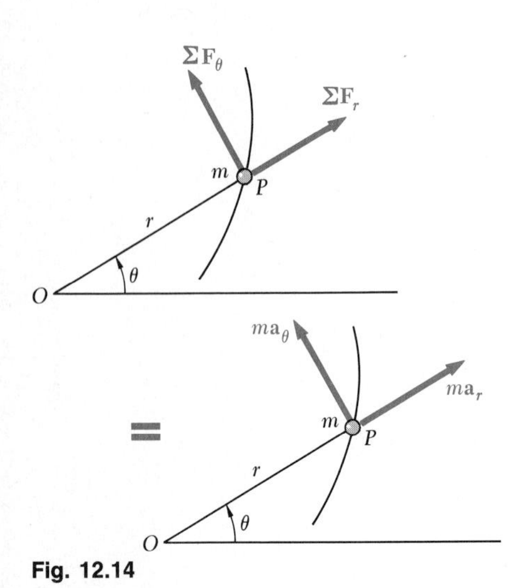

Fig. 12.14

$$\Sigma F_r = ma_r \qquad \Sigma F_\theta = ma_\theta \tag{12.15}$$

Substituting for a_r and a_θ from (11.33), we have

$$\Sigma F_r = m(\ddot{r} - r\dot{\theta}^2) \tag{12.16}$$
$$\Sigma F_\theta = m(r\ddot{\theta} + 2\dot{r}\dot{\theta}) \tag{12.17}$$

The equations obtained may be solved for two unknowns.

Dynamic Equilibrium. As an alternative method of solution, we may add the inertia vector $(m\mathbf{a})_{\text{rev}}$ to the forces acting on the particle and express that the system obtained is balanced. In practice, it is found convenient to resolve the inertia vector into two components. Resolving, for example, the inertia vector into its tangential and normal components, we obtain the system of vectors shown in Fig. 12.15. The tangential component of the inertia vector provides a measure of the resistance the particle offers to a change in speed, while its normal component (also called *centrifugal force*) represents the tendency of the particle to leave its curved path. We should note that either of these two components may be zero under special conditions: (1) if the particle starts from rest, its initial velocity is zero and the normal component of the inertia vector is zero at $t = 0$; (2) if the particle moves at constant speed along its path, the tangential component of the inertia vector is zero and only its normal component needs to be considered.

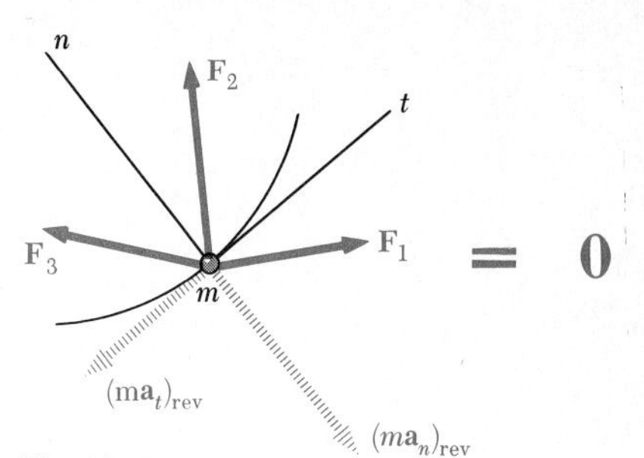

Fig. 12.15

12.9. Motion under a Central Force. When the only force acting on a particle P is a force $\mathbf{F}$ directed toward or away from a fixed point O, the particle is said to be moving *under a central force*, and the point O is referred to as the *center of force* (Fig. 12.16). Selecting O as the origin of coordinates and using radial and transverse components, we note that the transverse component $\mathbf{F}_\theta$ of the force $\mathbf{F}$ is zero. Thus, Eq. (12.17) yields

$$r\ddot{\theta} + 2\dot{r}\dot{\theta} = 0 \qquad \text{or} \qquad \frac{1}{r}\frac{d}{dt}(r^2\dot{\theta}) = 0$$

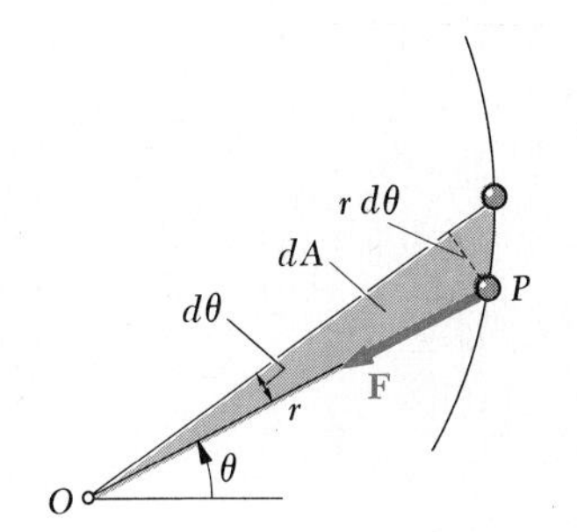

Fig. 12.16

It follows that, for any finite value of r, we must have

$$r^2\dot{\theta} = h \tag{12.18}$$

where h is a constant. Equation (12.18) may be given an interesting geometric interpretation. Observing from Fig. 12.16 that the radius vector OP sweeps an infinitesimal area $dA = \frac{1}{2}r^2\,d\theta$ as it rotates through an angle $d\theta$, and defining the *areal velocity* of the particle as the quotient dA/dt, we note that the left-hand member of Eq. (12.18) represents twice the areal velocity of the particle. We thus conclude that, *when a particle moves under a central force, its areal velocity is constant.*†

12.10. Newton's Law of Gravitation. The gravitational force exerted by the sun on a planet, or by the earth on an orbiting satellite, is an important example of a central force. In this section we shall learn how to determine the magnitude of a gravitational force.

In his *law of universal gravitation*, Newton states that two particles at a distance r from each other and, respectively, of mass M and m attract each other with equal and opposite forces $\mathbf{F}$ and $\mathbf{F}'$ directed along the line joining the particles (Fig. 12.17). The common magnitude F of the two forces is

$$F = G\frac{Mm}{r^2} \tag{12.19}$$

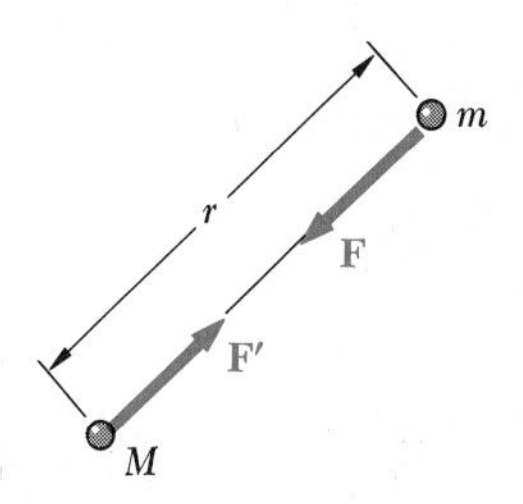

Fig. 12.17

† It will be seen in Sec. 14.11 that the left-hand member of Eq. (12.18) also represents the *angular momentum per unit mass* of the particle P. Thus, Eq. (12.18) may be considered as a statement of the *conservation of angular momentum* of a particle moving under a central force.

where G is a universal constant, called the *constant of gravitation*. Experiments show that the value of G is $(66.73 \pm 0.03) \times 10^{-12}\ \mathrm{m^3/kg \cdot s^2}$ in SI units, or approximately $34.4 \times 10^{-9}\ \mathrm{ft^4/lb \cdot s^4}$ in U.S. customary units. While gravitational forces exist between any pair of bodies, their effect is appreciable only when one of the bodies has a very large mass. The effect of gravitational forces is apparent in the case of the motion of a planet about the sun, of satellites orbiting about the earth, or of bodies falling on the surface of the earth.

Since the force exerted by the earth on a body of mass m located on or near its surface is defined as the weight $\mathbf{W}$ of the body, we may substitute the magnitude $W = mg$ of the weight for F, and the radius R of the earth for r, in Eq. (12.19). We obtain

$$W = mg = \frac{GM}{R^2}m \qquad \text{or} \qquad g = \frac{GM}{R^2} \tag{12.20}$$

where M is the mass of the earth. Since the earth is not truly spherical, the distance R from the center of the earth depends upon the point selected on its surface, and the values of W and g will thus vary with the altitude and latitude of the point considered. Another reason for the variation of W and g with the latitude is that a system of axes attached to the earth does not constitute a newtonian frame of reference (see Sec. 12.2). A more accurate definition of the weight of a body should therefore include a component representing the centrifugal force due to the rotation of the earth. Values of g at sea level vary from $9.781\ \mathrm{m/s^2}$, or $32.09\ \mathrm{ft/s^2}$, at the equator to $9.833\ \mathrm{m/s^2}$, or $32.26\ \mathrm{ft/s^2}$, at the poles.†

The force exerted by the earth on a body of mass m located in space at a distance r from its center may be found from Eq. (12.19). The computations will be somewhat simplified if we note that according to Eq. (12.20), the product of the constant of gravitation G and of the mass M of the earth may be expressed as

$$GM = gR^2 \tag{12.21}$$

where g and the radius R of the earth will be given their average values $g = 9.81\ \mathrm{m/s^2}$ and $R = 6.37 \times 10^6\ \mathrm{m}$ in SI units,‡ or $g = 32.2\ \mathrm{ft/s^2}$ and $R = (3960\ \mathrm{mi})(5280\ \mathrm{ft/mi})$ in U.S. customary units.

The discovery of the law of universal gravitation has often been attributed to the fact that, after observing an apple falling from a tree, Newton had reflected that the earth must attract an apple and the moon in much the same way. While it is doubtful that this incident actually took place, it may be said that Newton would not have formulated his law if he had not first perceived that the acceleration of a falling body must have the same cause as the acceleration which keeps the moon in its orbit. This basic concept of continuity of the gravitational attraction is more easily understood now, when the gap between the apple and the moon is being filled with long-range ballistic missiles and artificial earth satellites.

†A formula expressing g in terms of the latitude ϕ was given in Prob. 12.2.

‡The value of R is easily found if one recalls that the circumference of the earth is $2\pi R = 40 \times 10^6\ \mathrm{m}$.

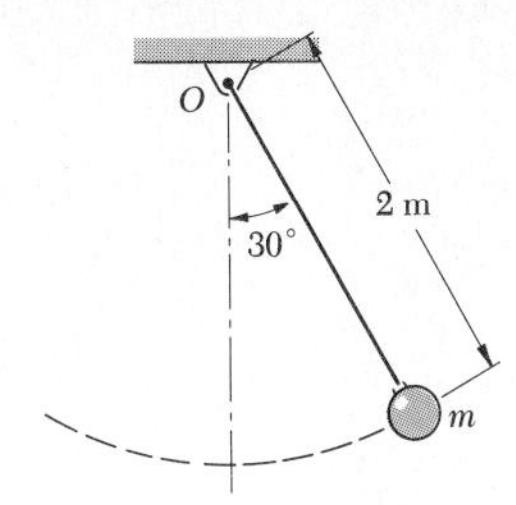

SAMPLE PROBLEM 12.5

The bob of a 2-m pendulum describes an arc of circle in a vertical plane. If the tension in the cord is 2.5 times the weight of the bob for the position shown, find the velocity and acceleration of the bob in that position.

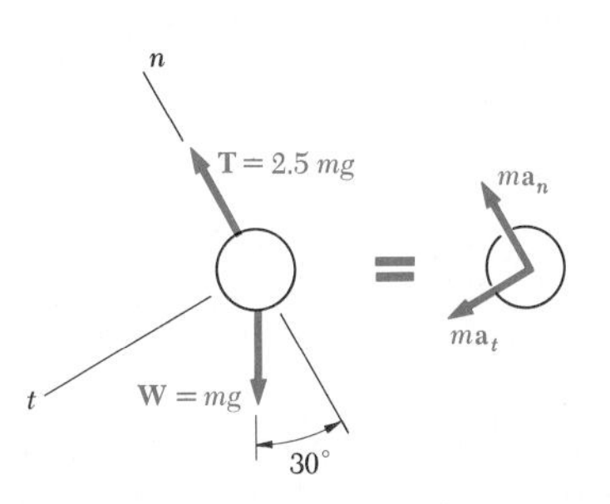

Solution. The weight of the bob is $W = mg$; the tension in the cord is thus 2.5 mg. Recalling that $\mathbf{a}_n$ is directed toward O and assuming $\mathbf{a}_t$ as shown, we apply Newton's second law and obtain

$+\swarrow \Sigma F_t = ma_t$: $\quad mg \sin 30° = ma_t$

$$a_t = g \sin 30° = +4.90 \text{ m/s}^2 \qquad \mathbf{a}_t = 4.90 \text{ m/s}^2 \swarrow \blacktriangleleft$$

$+\nwarrow \Sigma F_n = ma_n$: $\quad 2.5\, mg - mg \cos 30° = ma_n$

$$a_n = 1.634\, g = +16.03 \text{ m/s}^2 \qquad \mathbf{a}_n = 16.03 \text{ m/s}^2 \nwarrow \blacktriangleleft$$

Since $a_n = v^2/\rho$, we have $v^2 = \rho a_n = (2 \text{ m})(16.03 \text{ m/s}^2)$

$$v = \pm 5.66 \text{ m/s} \qquad \mathbf{v} = 5.66 \text{ m/s} \nearrow \text{(up or down)} \blacktriangleleft$$

SAMPLE PROBLEM 12.6

Determine the rated speed of a highway curve of radius $\rho = 400$ ft banked through an angle $\theta = 18°$. The rated speed of a banked curved road is the speed at which a car should travel if no lateral friction force is to be exerted on its wheels.

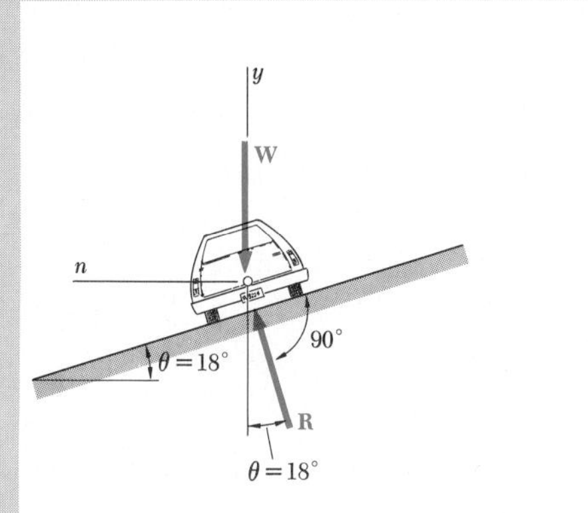

Solution. The car travels in a *horizontal* circular path of radius ρ. The normal component $\mathbf{a}_n$ of the acceleration is directed toward the center of the path; its magnitude is $a_n = v^2/\rho$, where v is the speed of the car in ft/s. The mass m of the car is W/g, where W is the weight of the car. Since no lateral friction force is to be exerted on the car, the reaction $\mathbf{R}$ of the road is shown perpendicular to the roadway. Applying Newton's second law, we write

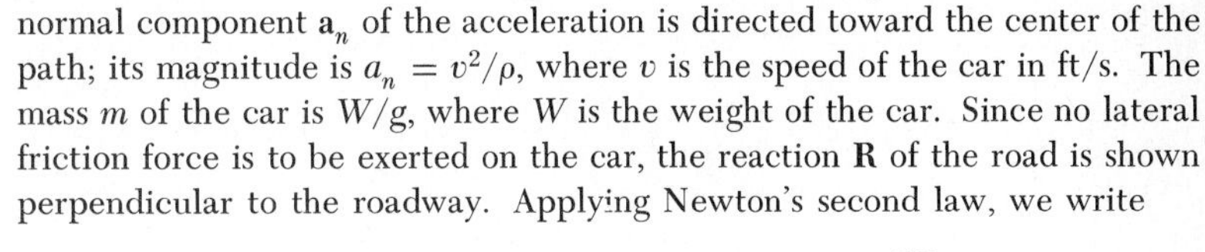

$+\uparrow \Sigma F_y = 0$: $\quad R \cos\theta - W = 0 \qquad R = \dfrac{W}{\cos\theta} \qquad (1)$

$\overset{+}{\leftarrow} \Sigma F_n = ma_n$: $\quad R \sin\theta = \dfrac{W}{g} a_n \qquad (2)$

Substituting for R from (1) into (2), and recalling that $a_n = v^2/\rho$,

$$\frac{W}{\cos\theta} \sin\theta = \frac{W}{g}\frac{v^2}{\rho} \qquad v^2 = g\rho \tan\theta$$

Substituting the given data, $\rho = 400$ ft and $\theta = 18°$, into this equation, we obtain

$$v^2 = (32.2 \text{ ft/s}^2)(400 \text{ ft}) \tan 18°$$
$$v = 64.7 \text{ ft/s} \qquad v = 44.1 \text{ mi/h} \blacktriangleleft$$

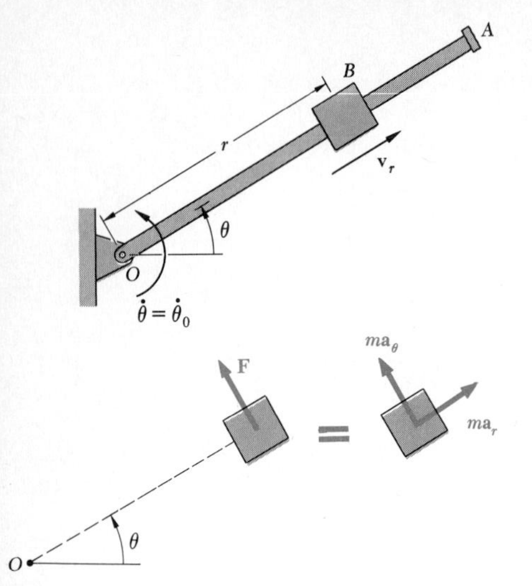

SAMPLE PROBLEM 12.7

A block B of mass m may slide freely on a frictionless arm OA which rotates in a horizontal plane at a constant rate $\dot\theta_0$. Knowing that B is released at a distance r_0 from O, express as a function of r, (*a*) the component v_r of the velocity of B along OA, (*b*) the magnitude of the horizontal force **F** exerted on B by the arm OA.

Solution. Since all other forces are perpendicular to the plane of the figure, the only force shown acting on B is the force **F** perpendicular to OA.

Equations of Motion. Using radial and transverse components,

$$+\nearrow \Sigma F_r = ma_r: \qquad 0 = m(\ddot r - r\dot\theta^2) \qquad (1)$$

$$+\nwarrow \Sigma F_\theta = ma_\theta: \qquad F = m(r\ddot\theta + 2\dot r\dot\theta) \qquad (2)$$

a. Component* $\mathbf{v}_r$ *of Velocity. Since $v_r = \dot r$, we have

$$\ddot r = \dot v_r = \frac{dv_r}{dt} = \frac{dv_r}{dr}\frac{dr}{dt} = v_r\frac{dv_r}{dr}$$

Substituting for $\ddot r$ into (1), recalling that $\dot\theta = \dot\theta_0$, and separating the variables,

$$v_r\,dv_r = \dot\theta_0^2 r\,dr$$

Multiplying by 2, and integrating form 0 to v_r and from r_0 to r,

$$v_r^2 = \dot\theta_0^2(r^2 - r_0^2) \qquad v_r = \dot\theta_0(r^2 - r_0^2)^{1/2} \blacktriangleleft$$

***b. Horizontal Force* F.** Making $\dot\theta = \dot\theta_0$, $\ddot\theta = 0$, $\dot r = v_r$ in Eq. (2), and substituting for v_r the expression obtained in part *a*,

$$F = 2m\dot\theta_0(r^2 - r_0^2)^{1/2}\dot\theta_0 \qquad F = 2m\dot\theta_0^2(r^2 - r_0^2)^{1/2} \blacktriangleleft$$

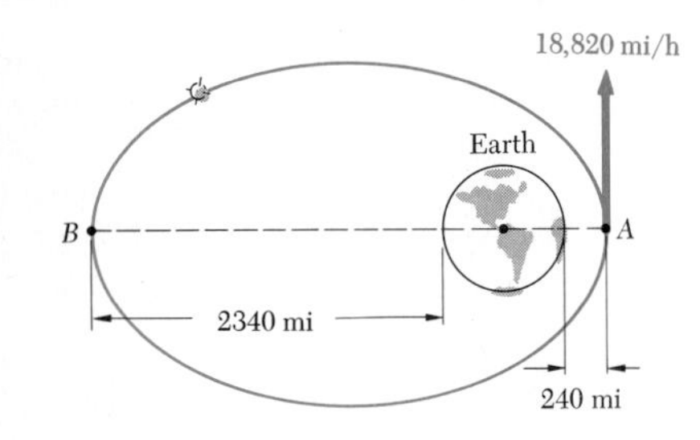

SAMPLE PROBLEM 12.8

A satellite is launched in a direction parallel to the surface of the earth with a velocity of 18,820 mi/h from an altitude of 240 mi. Determine the velocity of the satellite as it reaches its maximum altitude of 2340 mi. It is recalled that the radius of the earth is 3960 mi.

Solution. Since the satellite is moving under a central force directed toward the center O of the earth, we may use Eq. (12.18) and write

$$r_A^2\dot\theta_A = r_B^2\dot\theta_B = h \qquad (1)$$

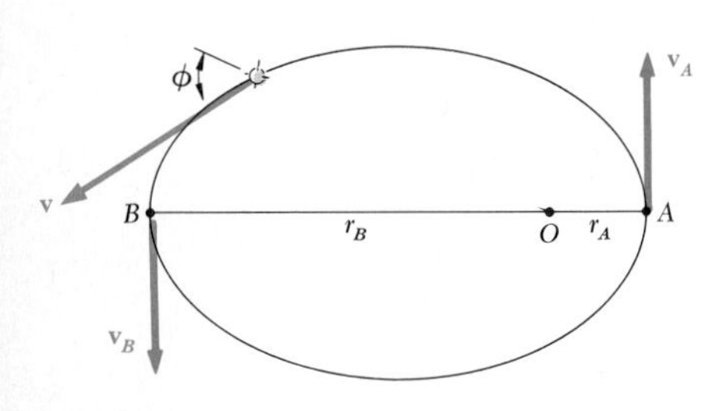

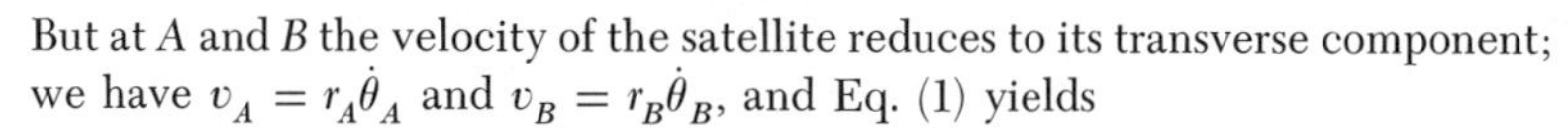

But at A and B the velocity of the satellite reduces to its transverse component; we have $v_A = r_A\dot\theta_A$ and $v_B = r_B\dot\theta_B$, and Eq. (1) yields

$$r_Av_A = r_Bv_B$$

$$v_B = v_A\frac{r_A}{r_B} = (18{,}820\text{ mi/h})\frac{3960\text{ mi} + 240\text{ mi}}{3960\text{ mi} + 2340\text{ mi}}$$

$$v_B = 12{,}550\text{ mi/h} \blacktriangleleft$$

Problems

12.35 A 5-lb ball revolves in a horizontal circle as shown. Knowing that $L = 4$ ft and that the maximum allowable tension in the cord is 12 lb, determine (*a*) the maximum allowable speed, (*b*) the corresponding value of the angle θ.

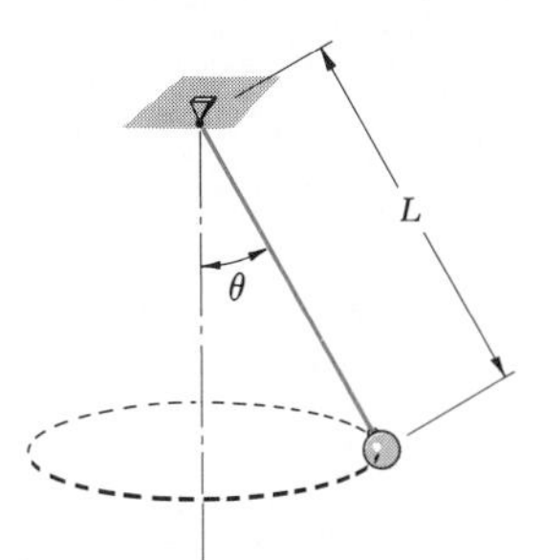

Fig. P12.35 and P12.36

12.36 A 3-kg ball revolves in a horizontal circle as shown at a constant speed of 1.2 m/s. Knowing that $L = 800$ mm, determine (*a*) the angle θ that the cord forms with the vertical, (*b*) the tension in the cord.

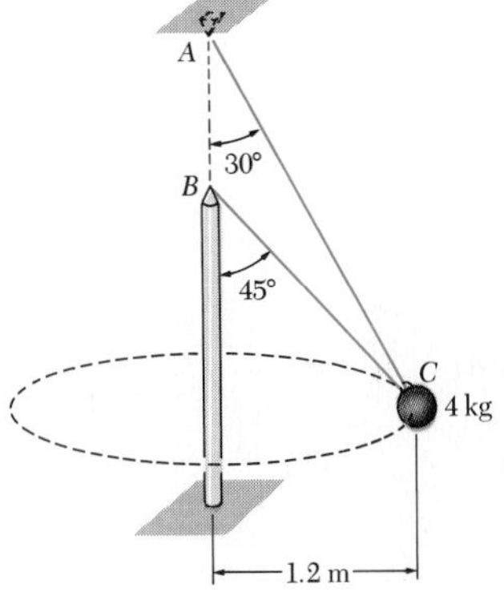

Fig. P12.37, P12.39, P12.40, and P12.42

12.37 and 12.38 A single wire *ACB* passes through a ring at *C* attached to a sphere which revolves at the constant speed v in the horizontal circle shown. Knowing that the tension is the same in both portions of the wire, determine the speed v.

12.39 Two wires *AC* and *BC* are tied at *C* to a sphere which revolves at the constant speed v in the horizontal circle shown. Determine the range of values of v for which both wires remain taut.

12.40 Two wires *AC* and *BC* are tied at *C* to a sphere which revolves at the constant speed v in the horizontal circle shown. Determine the range of the allowable values of v if the tension in either of the wires is not to exceed 35 N.

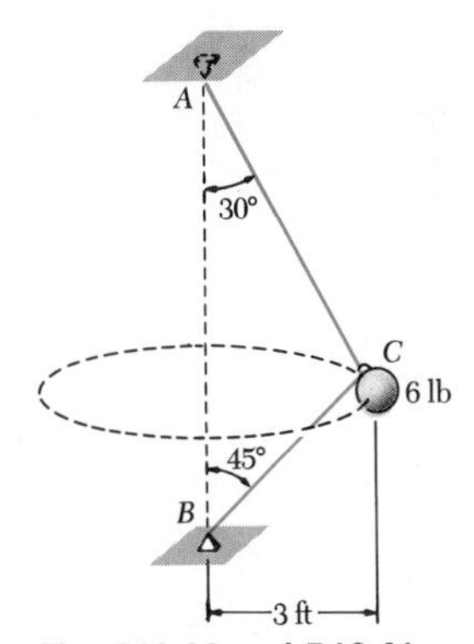

Fig. P12.38 and P12.41

* **12.41** Two wires *AC* and *BC* are tied at *C* to a sphere which revolves at the constant speed v in the horizontal circle shown. Determine the range of the allowable values of v if both wires are to remain taut and if the tension in either of the wires is not to exceed 12 lb.

* **12.42** Two wires *AC* and *BC* are tied at *C* to a sphere which revolves at the constant speed v in the horizontal circle shown. Determine the range of the allowable values of v if both wires are to remain taut and if the tension in either of the wires is not to exceed 50 N.

12.43 A small sphere of weight W is held as shown by two wires *AB* and *CD*. If wire *AB* is cut, determine the tension in the other wire (*a*) before *AB* is cut, (*b*) immediately after *AB* has been cut.

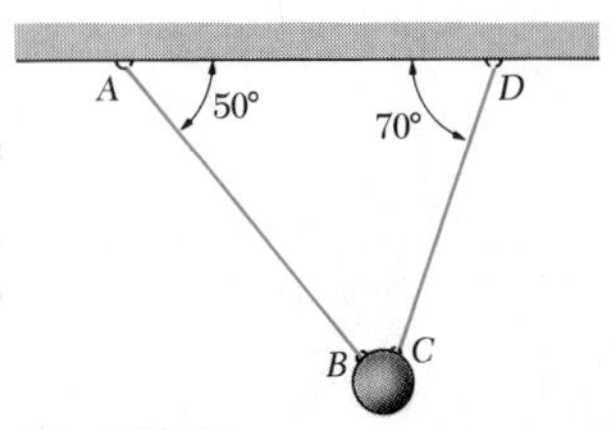

Fig. P12.43

12.44 Solve Prob. 12.43, assuming that wire *CD* is cut instead of wire *AB*.

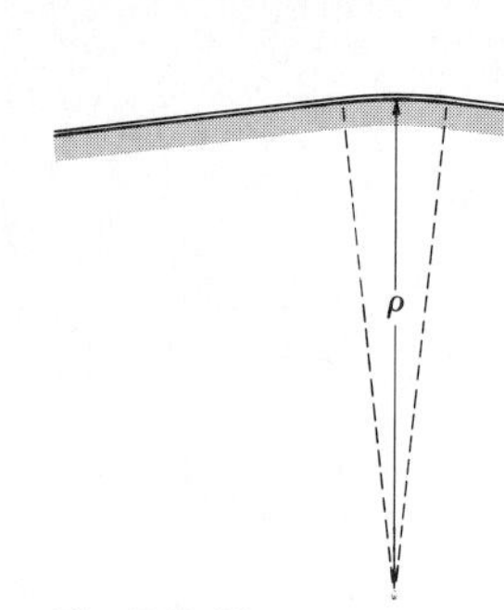

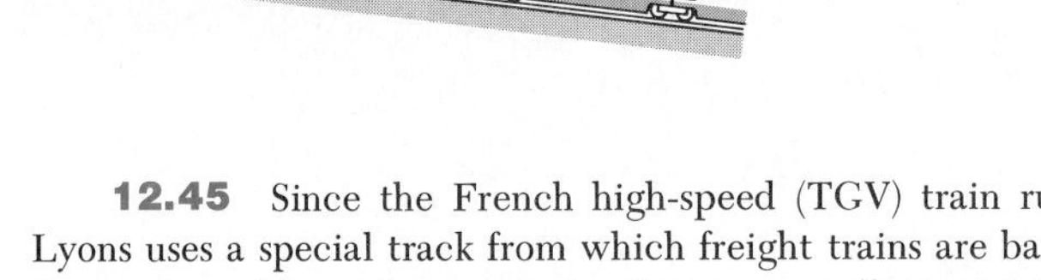

Fig. P12.45

12.45 Since the French high-speed (TGV) train running from Paris to Lyons uses a special track from which freight trains are barred, the designers of the track could use steeper grades than are usually permissible for railroads. On the other hand, because of the high speeds involved, they had to avoid sudden changes in grade and set a lower limit for the radius of curvature ρ of the vertical profile of the track. (*a*) Knowing that the train's maximum speed in test runs was 382 km/h, determine the smallest allowable value of ρ if the train is not to leave the track at that speed. (*b*) Assuming the value of ρ found in part *a*, determine the force exerted by his seat on an 80-kg passenger at the top and at the bottom of a hill when the train is traveling at 270 km/h.

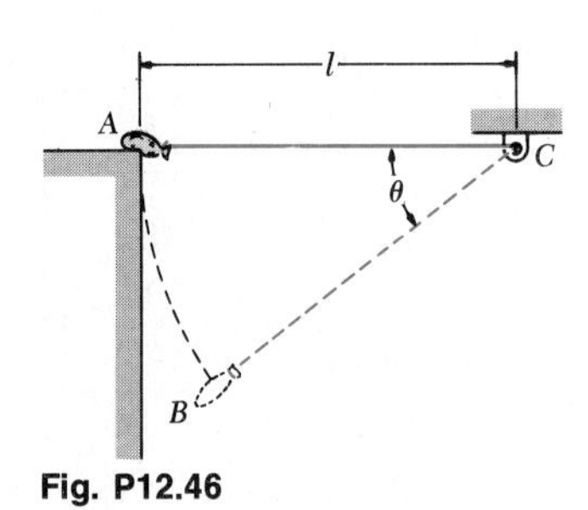

Fig. P12.46

12.46 A bag is gently pushed off the top of a wall at *A* and swings in a vertical plane at the end of a rope of length l. (*a*) For any position *B* of the bag, determine the tangential component a_t of its acceleration and obtain its velocity v by integration. (*b*) Determine the value of θ for which the rope will break, knowing that it can withstand a maximum tension equal to twice the weight of the bag.

12.47 The roller-coaster track shown is contained in a vertical plane. The portion of track between *A* and *B* is straight and horizontal, while the portions to the left of *A* and to the right of *B* have radii of curvature as indicated. A car is traveling at a speed of 45 mi/h when the brakes are suddenly applied, causing the wheels of the car to slide on the track ($\mu_k = 0.25$). Determine the initial deceleration of the car if the brakes are applied as the car (*a*) has almost reached *A*, (*b*) is traveling between *A* and *B*, (*c*) has just passed *B*.

Fig. P12.47

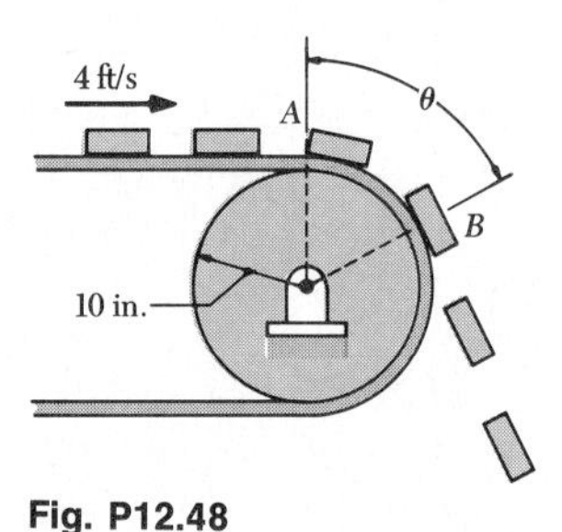

Fig. P12.48

12.48 A series of small packages, each weighing $\frac{3}{4}$ lb, are discharged from a conveyor belt as shown. Knowing that the coefficient of static friction between each package and the conveyor belt is 0.40, determine (*a*) the force exerted by the belt on a package just after it has passed point *A*, (*b*) the angle θ defining the point *B* where the packages first *slip* relative to the belt.

12.49 A 200-g block B fits inside a small cavity cut in arm OA, which rotates in the vertical plane at a constant rate such that $v = 2$ m/s. Knowing that the spring exerts on block B a force of magnitude $P = 0.8$ N and neglecting the effect of friction, determine the range of values of θ for which block B is in contact with the face of the cavity closest to the axis of rotation O.

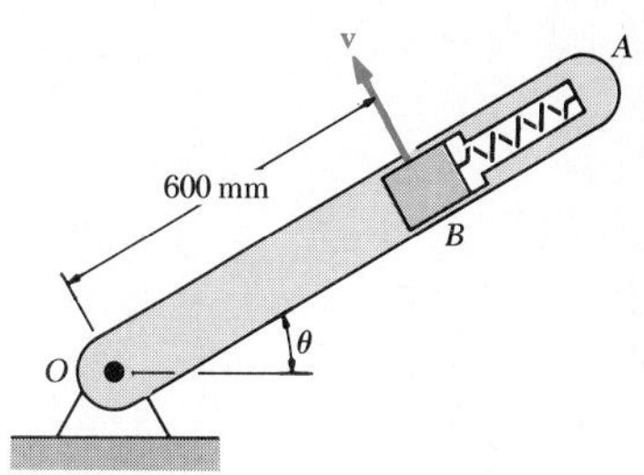

Fig. P12.49

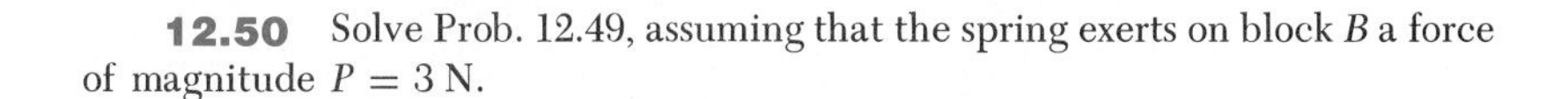
12.50 Solve Prob. 12.49, assuming that the spring exerts on block B a force of magnitude $P = 3$ N.

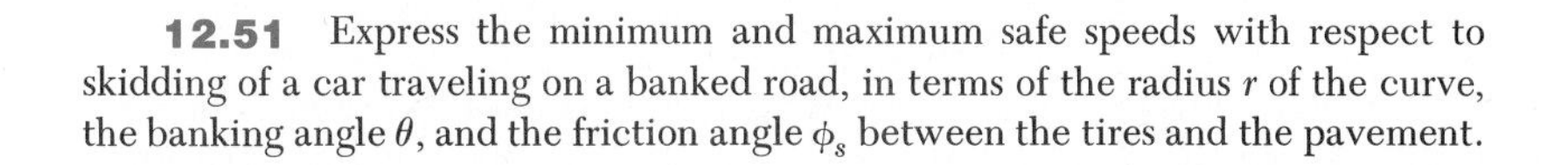
12.51 Express the minimum and maximum safe speeds with respect to skidding of a car traveling on a banked road, in terms of the radius r of the curve, the banking angle θ, and the friction angle ϕ_s between the tires and the pavement.

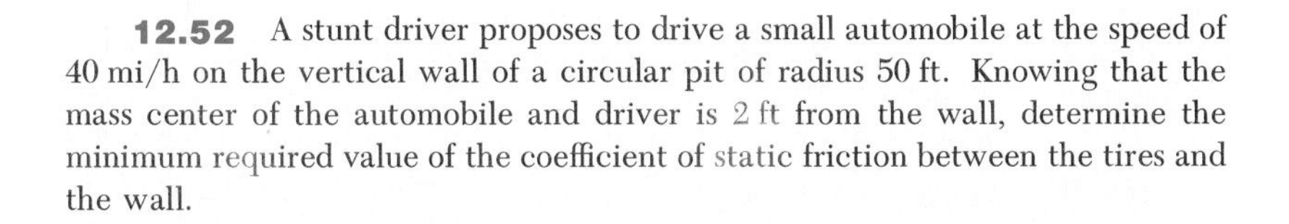
12.52 A stunt driver proposes to drive a small automobile at the speed of 40 mi/h on the vertical wall of a circular pit of radius 50 ft. Knowing that the mass center of the automobile and driver is 2 ft from the wall, determine the minimum required value of the coefficient of static friction between the tires and the wall.

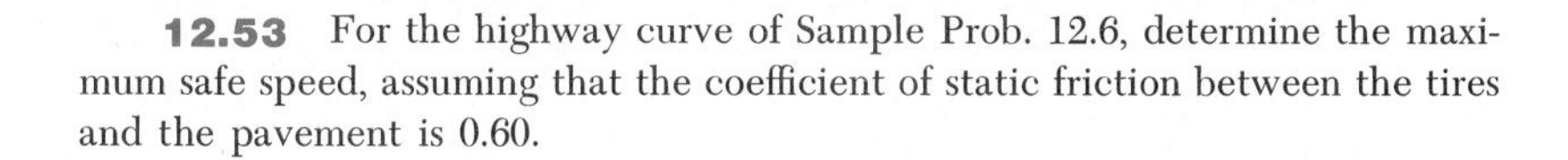
12.53 For the highway curve of Sample Prob. 12.6, determine the maximum safe speed, assuming that the coefficient of static friction between the tires and the pavement is 0.60.

12.54 A curve in a speed track has a radius of 150 m and a rated speed of 135 km/h. (See Sample Prob. 12.6 for the definition of rated speed.) Knowing that a racing car starts skidding on the curve when traveling at the speed of 300 km/h, determine (a) the banking angle θ, (b) the coefficient of static friction between the tires and the track under the prevailing conditions, (c) the minimum speed at which the same car could negotiate that curve.

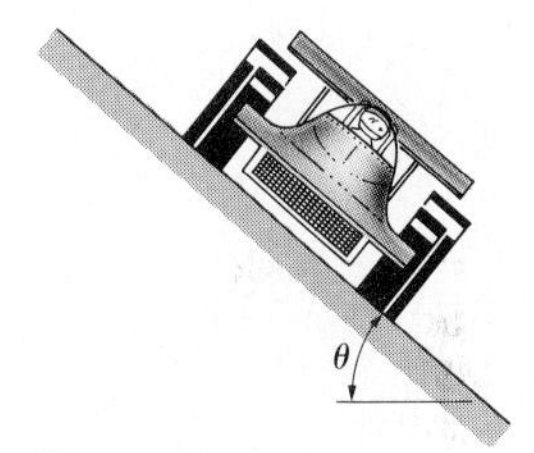

Fig. P12.54

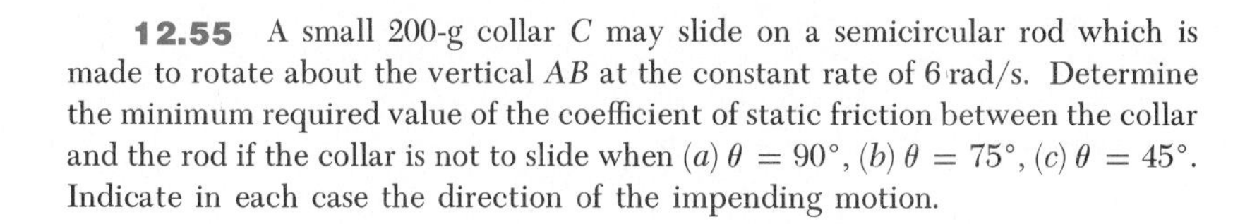
12.55 A small 200-g collar C may slide on a semicircular rod which is made to rotate about the vertical AB at the constant rate of 6 rad/s. Determine the minimum required value of the coefficient of static friction between the collar and the rod if the collar is not to slide when (a) $\theta = 90°$, (b) $\theta = 75°$, (c) $\theta = 45°$. Indicate in each case the direction of the impending motion.

12.56 For the collar and rod of Prob. 12.55, and assuming that the coefficients of friction are $\mu_s = 0.25$ and $\mu_k = 0.20$, determine the magnitude and direction of the friction force exerted on the collar immediately after it has been released on the rotating rod in the position corresponding to (a) $\theta = 70°$, (b) $\theta = 30°$. Indicate also in each case whether the collar will slide on the rod.

***12.57** For the collar and rod of Prob. 12.55, determine (a) the three values of θ for which the collar will not slide on the rod, assuming no friction between the collar and the rod, (b) the range of values of θ for which the collar will not slide on the rod, assuming a coefficient of static friction of 0.25. (*Hint.* In part b solve by trial the equation obtained for θ.)

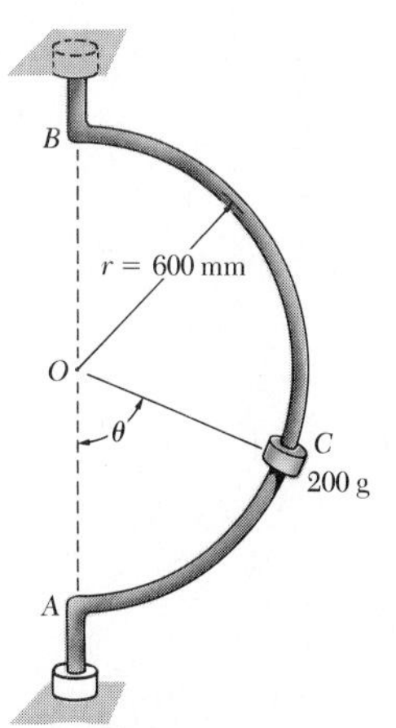

Fig. P12.55

12.58 A small block B is supported by a turntable which starting from rest is rotated in such a way that the block undergoes a constant tangential acceleration $a_t = 6\ \text{ft/s}^2$. Knowing that the coefficient of static friction between the block and the turntable is 0.60, determine (*a*) how long it will take for the block to start slipping on the turntable, (*b*) the speed v of the block at that instant.

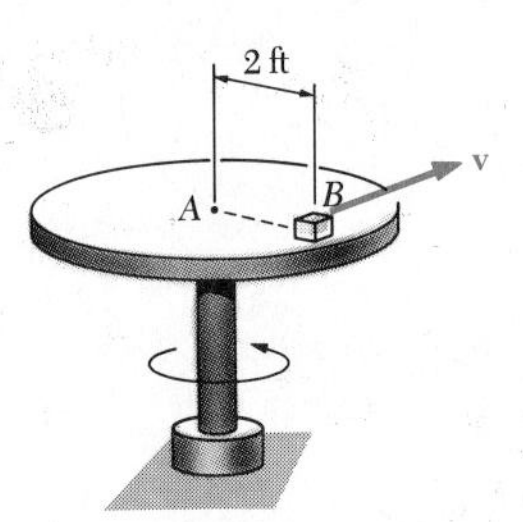

Fig. P12.58 and P12.59

12.59 A small block B is supported by a turntable which starting from rest is rotated in such a way that the block undergoes a constant tangential acceleration. Knowing that the coefficient of static friction between the block and the turntable is 0.50, determine the smallest interval of time in which the block can reach the speed $v = 5\ \text{ft/s}$ without slipping.

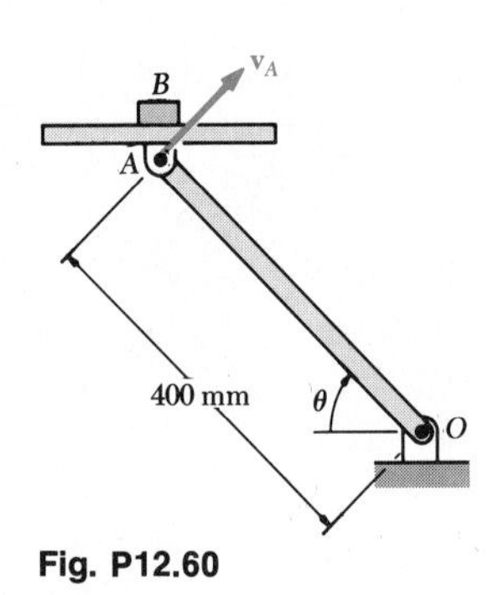

Fig. P12.60

12.60 A small block B is supported by a platform connected at A to rod OA. Point A describes a circle in a vertical plane at the constant speed v_A, while the platform is constrained to remain horizontal throughout its motion by the use of a special linkage (not shown in the figure). The coefficients of friction between the block and the platform are $\mu_s = 0.40$ and $\mu_k = 0.30$. Determine (*a*) the maximum allowable speed v_A if the block is not to slide on the platform, (*b*) the values of θ for which sliding is impending.

***12.61** For the system of Prob. 12.60, it is observed that block B slides intermittently on the platform when point A moves in a vertical circle at the constant speed $v_A = 1.22\ \text{m/s}$. Determine the values of θ for which the block starts sliding.

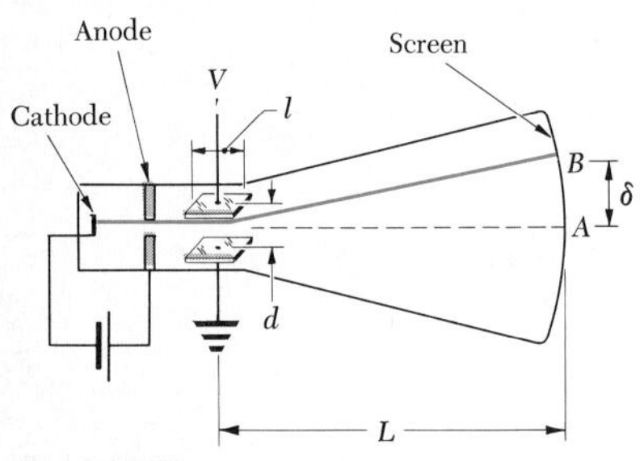

Fig. P12.62

12.62 In the cathode-ray tube shown, electrons emitted by the cathode and attracted by the anode pass through a small hole in the anode and keep traveling in a straight line with a speed v_0 until they strike the screen at A. However, if a difference of potential V is established between the two parallel plates, each electron will be subjected to a force $\mathbf{F}$ perpendicular to the plates while it travels between the plates and will strike the screen at point B at a distance δ from A. The magnitude of the force $\mathbf{F}$ is $F = eV/d$, where $-e$ is the charge of the electron and d is the distance between the plates. Derive an expression for the deflection δ in terms of V, v_0, the charge $-e$ of the electron, its mass m, and the dimensions d, l, and L.

12.63 In Prob. 12.62, determine the smallest allowable value of the ratio d/l in terms of e, m, v_0, and V if the electrons are not to strike the positive plate.

12.64 A manufacturer wishes to design a new cathode-ray tube which will be only half as long as his current model. If the size of the screen is to remain the same, how should the length l of the plates be modified if all the other characteristics of the circuit are to remain unchanged? (See Prob. 12.62 for description of cathode-ray tube.)

12.65 The two-dimensional motion of particle B is defined by the relations $r = 25\,t^3 - 50\,t^2$ and $\theta = t^3 - 4\,t$, where r is expressed in millimeters, t in seconds, and θ in radians. If the particle has a mass of 2 kg and moves in a horizontal plane, determine the radial and transverse components of the force acting on the particle when (*a*) $t = 0$, (*b*) $t = 1$ s.

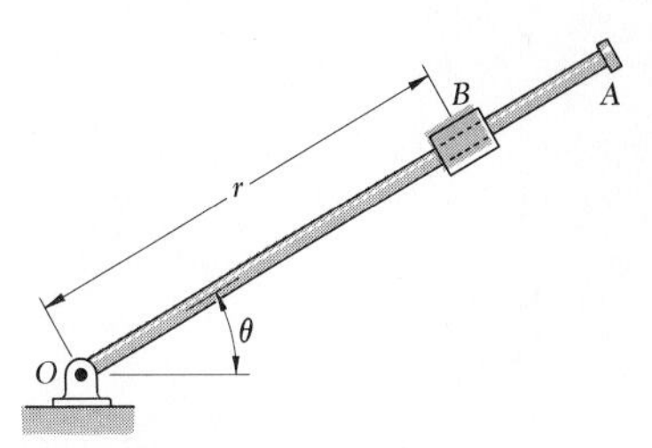

Fig. P12.65

12.66 For the motion defined in Prob. 12.65, determine the radial and transverse components of the force acting on the 2-kg particle as it passes again through the origin at $t = 2$ s.

12.67 For the motion defined in Sample Prob. 11.12, determine the force exerted by the arm OA on collar B and the radial force $\mathbf{Q}$ which must be applied to the collar (*a*) when $\theta = 0°$, (*b*) when $\theta = 30°$. Assume that collar B has a mass of 2 kg, and that the arm rotates in a vertical plane.

12.68 Pin B weighs 4 oz and is free to slide along the rotating arm OC and along the circular slot DE of radius $b = 20$ in. Neglecting friction and assuming that rod OC is made to rotate at the constant rate $\dot{\theta}_0 = 15$ rad/s in a horizontal plane, determine for any given value of θ (*a*) the radial and transverse components of the resultant force exerted on pin B, (*b*) the forces $\mathbf{P}$ and $\mathbf{Q}$ exerted on pin B, respectively, by rod OC and the wall of slot DE.

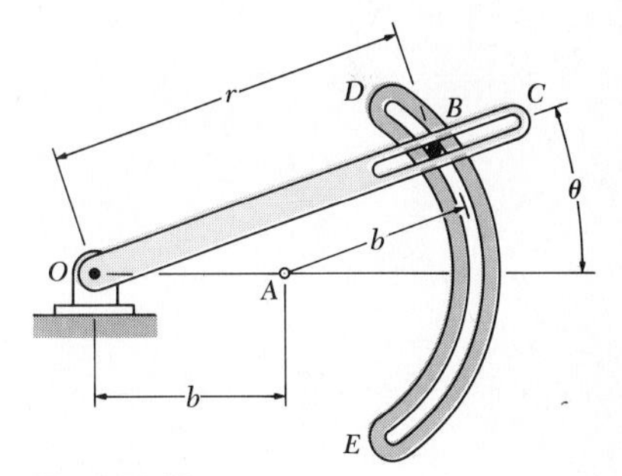

Fig. P12.68

12.69 Solve Prob. 12.68 for the position $\theta = 20°$, assuming that at that instant $\dot{\theta} = 15$ rad/s and $\ddot{\theta} = 250$ rad/s^2.

12.70 Slider C has a mass of 200 g and may move in a slot cut in arm AB, which rotates at the constant rate $\dot{\theta}_0 = 12$ rad/s in a horizontal plane. The slider is attached to a spring of constant $k = 36$ N/m, which is unstretched when $r = 0$. Knowing that the slider passes through the position $r = 400$ mm with a radial velocity $v_r = +1.8$ m/s, determine at that instant (*a*) the radial and transverse components of its acceleration, (*b*) its acceleration relative to arm AB, (*c*) the horizontal force exerted on the slider by arm AB.

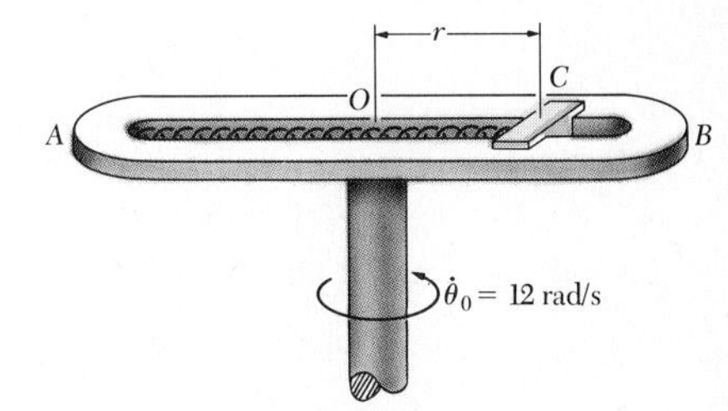

Fig. P12.70 and P12.71

* **12.71** Slider C has a mass of 200 g and may move in a slot cut in arm AB, which rotates at the constant rate $\dot{\theta}_0 = 12$ rad/s in a horizontal plane. The slider is attached to a spring of constant $k = 36$ N/m, which is unstretched when $r = 0$. Knowing that the slider is released with no radial velocity in the position $r = 500$ mm and neglecting friction, determine for the position $r = 300$ mm (*a*) the radial and transverse components of the velocity of the slider, (*b*) the radial and transverse components of its acceleration, (*c*) the horizontal force exerted on the slider by arm AB.

* **12.72** Solve Prob. 12.71, assuming that the spring is unstretched when slider C is located 45 mm to the left of the midpoint O of arm AB ($r = -45$ mm).

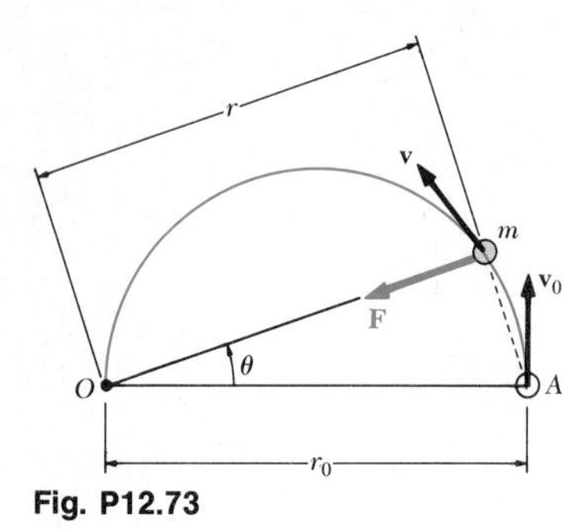

Fig. P12.73

12.73 A particle of mass m is projected from point A with an initial velocity $\mathbf{v}_0$ perpendicular to the line OA and moves under a central force $\mathbf{F}$ along a semicircular path of diameter OA. Observing that $r = r_0 \cos \theta$ and using Eq. (12.18), show that the speed v of the particle is inversely proportional to the square of the distance r from the particle to the center of force O.

12.74 A particle of mass m is projected from point A with an initial velocity $\mathbf{v}_0$ perpendicular to OA and moves under a central force $\mathbf{F}$ directed away from the center of force O. Knowing that the particle follows a path defined by the equation $r = r_0/\cos 2\theta$, and using Eq. (12.18), express the radial and transverse components of the velocity $\mathbf{v}$ of the particle as functions of θ.

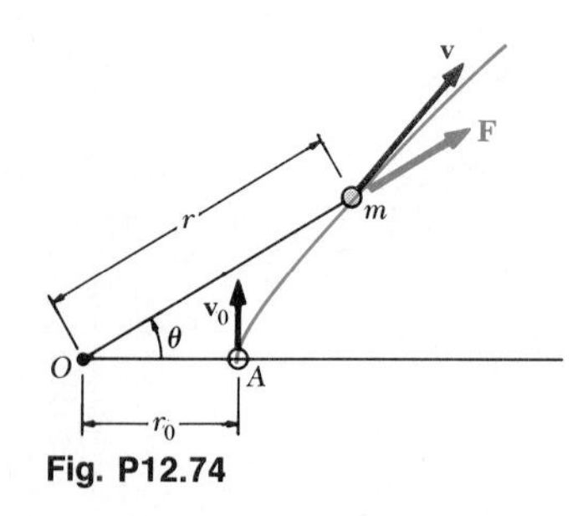

Fig. P12.74

12.75 For the particle of Prob. 12.73, (a) show that the central force $\mathbf{F}$ is inversely proportional to the fifth power of the distance r from the particle to the center of force O, (b) determine the magnitude of $\mathbf{F}$ for $\theta = 0$.

12.76 For the particle of Prob. 12.74, (a) show that the central force $\mathbf{F}$ is inversely proportional to the cube of the distance r from the particle to the center of force O, (b) determine the magnitude of $\mathbf{F}$ for $\theta = 0$.

12.77 Denoting by ρ the mean density of a planet, show that the minimum time required by a satellite to complete one full revolution about the planet is $(3\pi/G\rho)^{1/2}$, where G is the constant of gravitation.

12.78 Show that the radius r of the moon's orbit may be determined from the radius R of the earth, the acceleration of gravity g at the surface of the earth, and the time τ required by the moon to complete one full revolution about the earth. Compute r knowing that $\tau = 27.3$ days, giving the answer in both SI and U.S. customary units.

12.79 Communication satellites have been placed in a geosynchronous orbit, i.e., in a circular orbit such that they complete one full revolution about the earth in one sidereal day (23 h 56 min), and thus appear stationary with respect to the ground. Determine (a) the altitude of the satellites above the surface of the earth, (b) the velocity with which they describe their orbit. Give the answers in both SI and U.S. customary units.

12.80 The periodic times of two of the planet Jupiter's satellites, Io and Callisto, have been observed to be, respectively, 1 day 18 h 28 min and 16 days 16 h 32 min. Knowing that the radius of Callisto's orbit is 1.884×10^6 km, determine (a) the mass of the planet Jupiter, (b) the radius of Io's orbit. (The periodic time of a satellite is the time it requires to complete one full revolution about the planet.)

12.81 An Apollo spacecraft describes a circular orbit with a 1500-mi radius around the moon with a velocity of 3190 mi/h. In order to transfer it to a smaller circular orbit with a 1200-mi radius, the spacecraft is first placed on an elliptic path AB by reducing its velocity to 3000 mi/h as it passes through A. Determine (a) the velocity of the spacecraft as it approaches B on the elliptic path, (b) the value to which its velocity must be reduced at B to insert it into the smaller circular orbit.

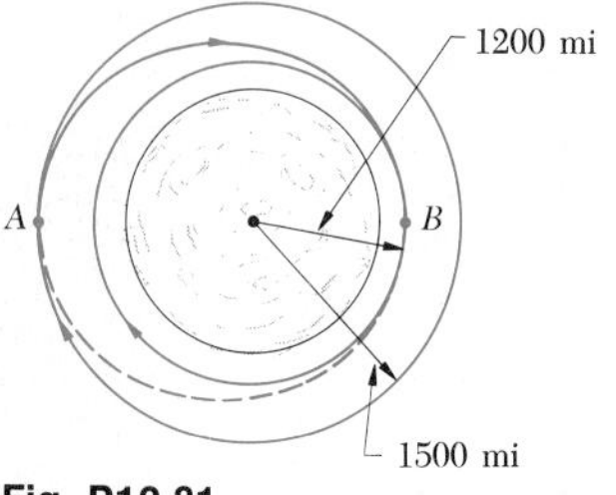

Fig. P12.81

12.82 A space probe is to be placed in a circular orbit of 8000-km radius about the planet Venus in a specified plane. As the probe reaches A, the point of its original trajectory closest to Venus, it is inserted in a first elliptic transfer orbit by reducing its speed by Δv_A. This orbit brings it to point B with a much reduced velocity. There the probe is inserted in a second transfer orbit located in the specified plane by changing the direction of its velocity and further reducing its speed by Δv_B. Finally, as the probe reaches point C, it is inserted in the desired circular orbit by reducing its speed by Δv_C. Knowing that the mass of Venus is 0.82 times the mass of the earth, that $r_A = 12 \times 10^3$ km and $r_B = 96 \times 10^3$ km, that the probe reaches A with a velocity of 7400 m/s and B with a velocity of 869 m/s, and that its speed is further reduced by 146 m/s at B, determine by how much the velocity of the probe should be reduced (*a*) at A, (*b*) at C.

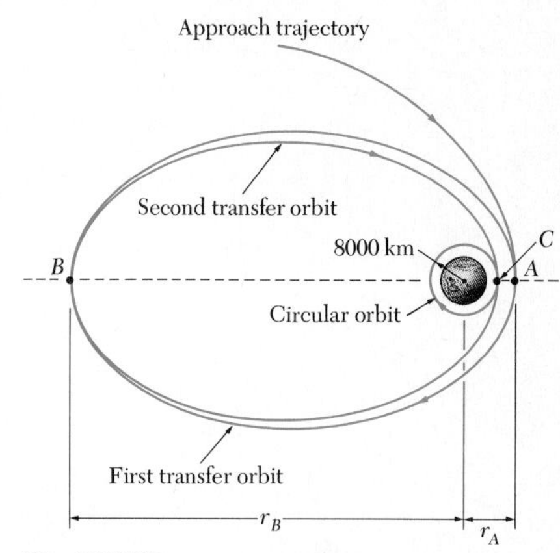

Fig. P12.82

12.83 For the space probe of Prob. 12.82, it is known that $r_A = 10 \times 10^3$ km and $r_B = 100 \times 10^3$ km, that the probe reaches A with a velocity of 8100 m/s, and that its speed is reduced by 400 m/s at A and by 2300 m/s at C. Determine (*a*) the velocity of the probe as it reaches B, (*b*) by how much its velocity should be reduced at B.

12.84 A space tug is used to place communication satellites into a geosynchronous orbit (see Prob. 12.79) at an altitude of 22,230 mi above the surface of the earth. The tug initially describes a circular orbit at an altitude of 220 mi and is inserted in an elliptic transfer orbit by firing its engine as it passes through A, thus increasing its velocity by 7910 ft/s. By how much should its velocity be increased as it reaches B to insert it in the geosynchronous orbit?

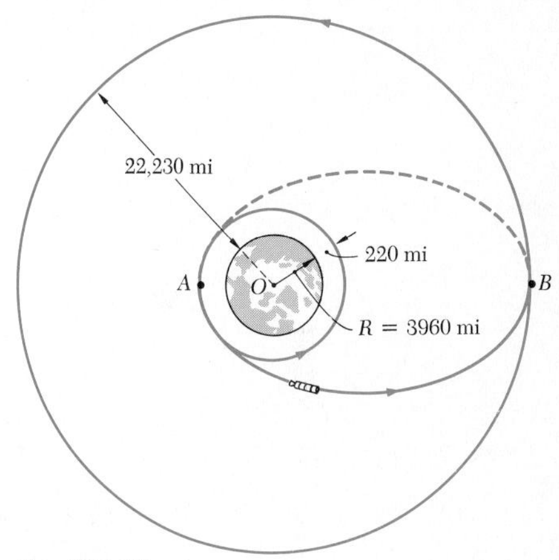

Fig. P12.84

12.85 A 5-lb ball is mounted on a horizontal rod which is free to rotate about a vertical shaft. In the position shown, the speed of the ball is $v_1 = 24$ in./s and the ball is held by a cord attached to the shaft. The cord is suddenly cut and the ball moves to position B as the rod rotates. Neglecting the mass of the rod, determine (*a*) the radial and transverse components of the acceleration of the ball immediately after the cord has been cut, (*b*) the acceleration of the ball relative to the rod at the same instant, (*c*) the speed of the ball after it has reached the stop B.

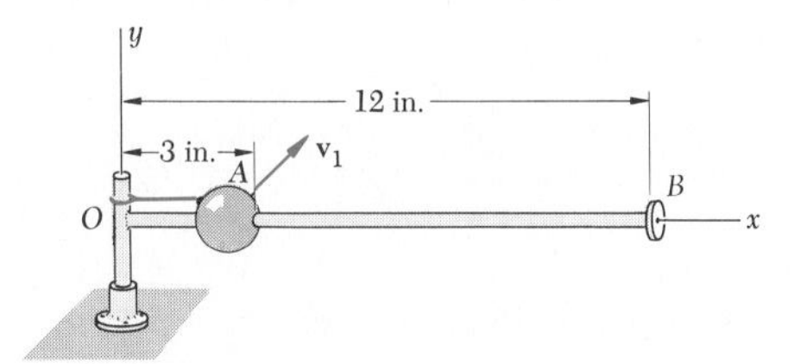

Fig. P12.85

12.86 A 4-oz ball slides on a smooth horizontal table at the end of a string which passes through a small hole in the table at O. When the length of string above the table is $r_1 = 18$ in., the speed of the ball is $v_1 = 5$ ft/s. Knowing that the breaking strength of the string is 4.00 lb, determine (*a*) the smallest distance r_2 which can be achieved by slowly drawing the string through the hole, (*b*) the corresponding speed v_2.

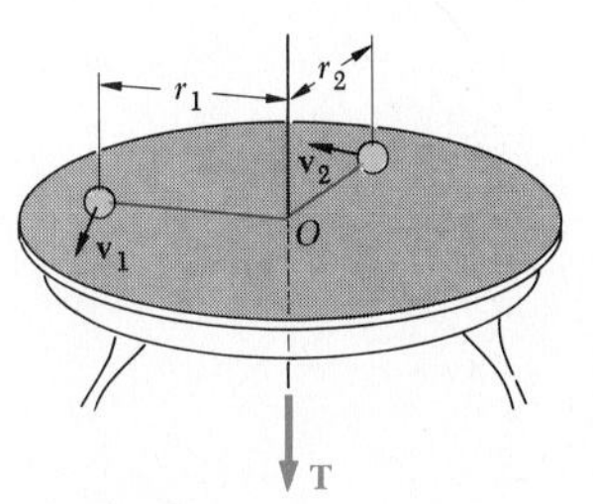

Fig. P12.86

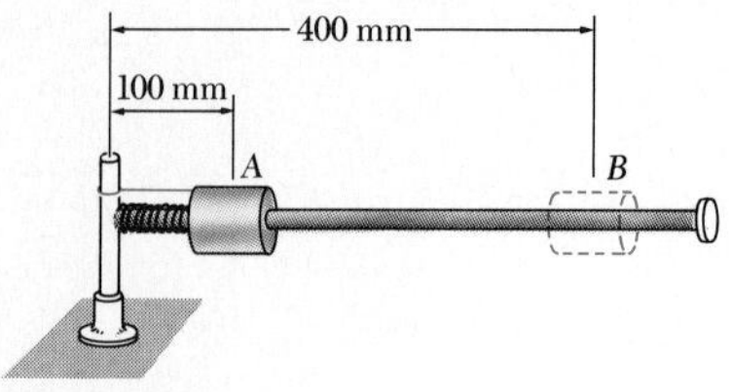

Fig. P12.87

12.87 A 250-g collar may slide on a horizontal rod which is free to rotate about a vertical shaft. The collar is initially held at A by a cord attached to the shaft and compresses a spring of constant 6 N/m, which is undeformed when the collar is located 500 mm from the shaft. As the rod rotates at the rate $\dot{\theta}_0 =$ 16 rad/s, the cord is cut and the collar moves out along the rod. Neglecting friction and the mass of the rod, determine for the position B of the collar (a) the transverse component of the velocity of the collar. (b) the radial and transverse components of its acceleration, (c) the acceleration of the collar relative to the rod.

* **12.88** In Prob. 12.87, determine for the position B of the collar, (a) the radial component of the velocity of the collar, (b) the value of $\ddot{\theta}$.

* 12.11. Trajectory of a Particle under a Central Force.

Consider a particle P moving under a central force $\mathbf{F}$. We propose to obtain the differential equation which defines its trajectory.

Assuming that the force $\mathbf{F}$ is directed toward the center of force O, we note that ΣF_r and ΣF_θ reduce, respectively, to $-F$ and zero in Eqs. (12.16) and (12.17). We therefore write

$$m(\ddot{r} - r\dot{\theta}^2) = -F \tag{12.22}$$

$$m(r\ddot{\theta} + 2\dot{r}\dot{\theta}) = 0 \tag{12.23}$$

But, as we saw in Sec. 12.9, the last equation yields

$$r^2\dot{\theta} = h \qquad \text{or} \qquad r^2\frac{d\theta}{dt} = h \tag{12.24}$$

Equation (12.24) may be used to eliminate the independent variable t from Eq. (12.22). Solving Eq. (12.24) for $\dot{\theta}$ or $d\theta/dt$, we have

$$\dot{\theta} = \frac{d\theta}{dt} = \frac{h}{r^2} \tag{12.25}$$

from which it follows that

$$\dot{r} = \frac{dr}{dt} = \frac{dr}{d\theta}\frac{d\theta}{dt} = \frac{h}{r^2}\frac{dr}{d\theta} = -h\frac{d}{d\theta}\left(\frac{1}{r}\right) \tag{12.26}$$

$$\ddot{r} = \frac{d\dot{r}}{dt} = \frac{d\dot{r}}{d\theta}\frac{d\theta}{dt} = \frac{h}{r^2}\frac{d\dot{r}}{d\theta}$$

or, substituting for $\dot{r}$ from (12.26),

$$\ddot{r} = \frac{h}{r^2}\frac{d}{d\theta}\left[-h\frac{d}{d\theta}\left(\frac{1}{r}\right)\right]$$

$$\ddot{r} = -\frac{h^2}{r^2}\frac{d^2}{d\theta^2}\left(\frac{1}{r}\right) \tag{12.27}$$

Substituting for $\dot{\theta}$ and $\ddot{r}$ from (12.25) and (12.27), respectively, into Eq. (12.22), and introducing the function $u = 1/r$, we obtain after reductions

$$\frac{d^2u}{d\theta^2} + u = \frac{F}{mh^2u^2} \tag{12.28}$$

In deriving Eq. (12.28), the force $\mathbf{F}$ was assumed directed toward O. The magnitude F should therefore be positive if $\mathbf{F}$ is actually directed toward O (attractive force) and negative if $\mathbf{F}$ is directed away from O (repulsive force). If F is a known function of r and thus of u, Eq. (12.28) is a differential equation in u and θ. This differential equation defines the trajectory followed by the particle under the central force $\mathbf{F}$. The equation of the trajectory will be obtained by solving the differential equation (12.28) for u as a function of θ and determining the constants of integration from the initial conditions.

***12.12. Application to Space Mechanics.** After the last stage of their launching rockets has burned out, earth satellites and other space vehicles are subjected to only the gravitational pull of the earth. Their motion may therefore be determined from Eqs. (12.24) and (12.28), which govern the motion of a particle under a central force, after F has been replaced by the expression obtained for the force of gravitational attraction.† Setting in Eq. (12.28)

$$F = \frac{GMm}{r^2} = GMmu^2$$

where M = mass of earth
m = mass of space vehicle
r = distance from center of earth to vehicle
$u = 1/r$

we obtain the differential equation

$$\frac{d^2u}{d\theta^2} + u = \frac{GM}{h^2} \tag{12.29}$$

where the right-hand member is observed to be a constant.

The solution of the differential equation (12.29) is obtained by adding the particular solution $u = GM/h^2$ to the general solution $u = C\cos(\theta - \theta_0)$ of the corresponding homogeneous equation (i.e., the equation obtained by setting the right-hand member equal to zero). Choosing

† It is assumed that the space vehicles considered here are attracted by only the earth and that their mass is negligible compared with the mass of the earth. If a vehicle moves very far from the earth, its path may be affected by the attraction of the sun, the moon, or another planet.

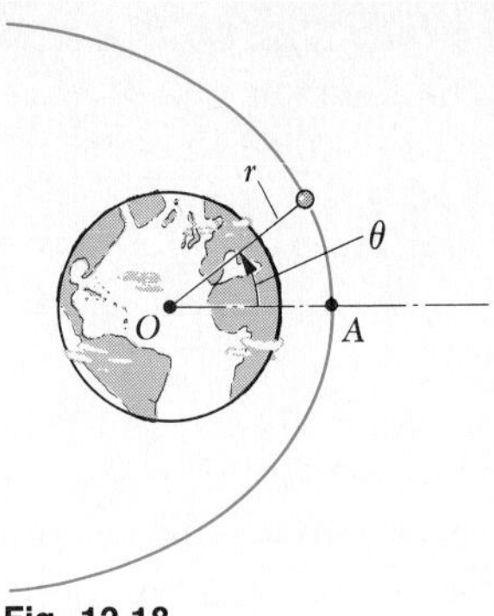

Fig. 12.18

the polar axis so that $\theta_0 = 0$, we write

$$\frac{1}{r} = u = \frac{GM}{h^2} + C\cos\theta \tag{12.30}$$

Equation (12.30) is the equation of a *conic section* (ellipse, parabola, or hyperbola) in the polar coordinates r and θ. The origin O of the coordinates, which is located at the center of the earth, is a *focus* of this conic section, and the polar axis is one of its axes of symmetry (Fig. 12.18).

The ratio of the constants C and GM/h^2 defines the *eccentricity* ε of the conic section; setting

$$\varepsilon = \frac{C}{GM/h^2} = \frac{Ch^2}{GM} \tag{12.31}$$

we may write Eq. (12.30) in the form

$$\frac{1}{r} = \frac{GM}{h^2}(1 + \varepsilon\cos\theta) \tag{12.30$'$}$$

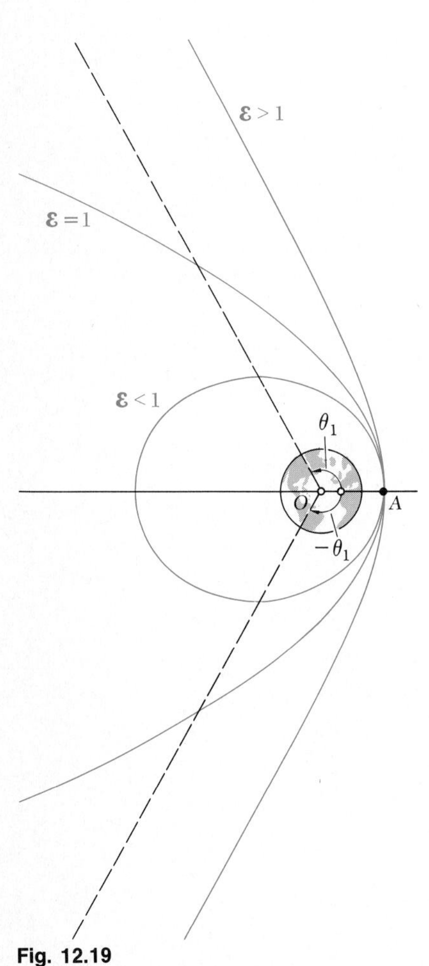

Fig. 12.19

Three cases may be distinguished:

1. $\varepsilon > 1$, or $C > GM/h^2$: There are two values θ_1 and $-\theta_1$ of the polar angle, defined by $\cos\theta_1 = -GM/Ch^2$, for which the right-hand member of Eq. (12.30) becomes zero. For both these values, the radius vector r becomes infinite; the conic section is a *hyperbola* (Fig. 12.19).
2. $\varepsilon = 1$, or $C = GM/h^2$: The radius vector becomes infinite for $\theta = 180°$; the conic section is a *parabola*.
3. $\varepsilon < 1$, or $C < GM/h^2$: The radius vector remains finite for every value of θ; the conic section is an *ellipse*. In the particular case when $\varepsilon = C = 0$, the length of the radius vector is constant; the conic section is a circle.

We shall see now how the constants C and GM/h^2 which characterize the trajectory of a space vehicle may be determined from the position and the velocity of the space vehicle at the beginning of its free flight. We shall assume, as is generally the case, that the powered phase of its flight has been programmed in such a way that as the last stage of the launching rocket burns out, the vehicle has a velocity parallel to the surface of the earth (Fig. 12.20). In other words, we shall assume that the space vehicle begins its free flight at the vertex A of its trajectory.†

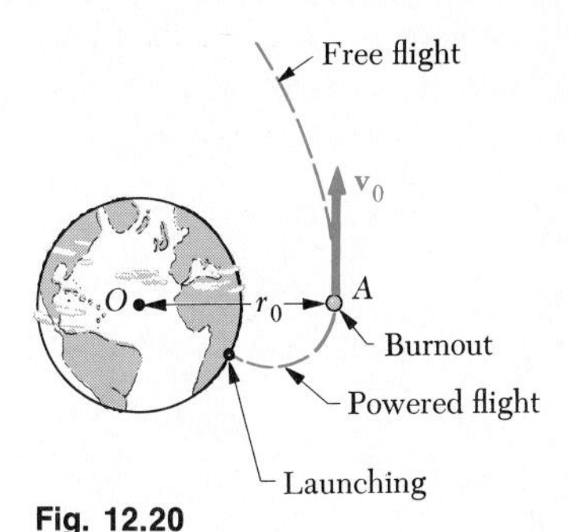

Fig. 12.20

Denoting, respectively, by r_0 and v_0 the radius vector and speed of the vehicle at the beginning of its free flight, we observe, since the velocity reduces to its transverse component, that $v_0 = r_0\dot{\theta}_0$. Recalling Eq. (12.24), we express the constant h as

$$h = r_0^2\dot{\theta}_0 = r_0 v_0 \tag{12.32}$$

The value obtained for h may be used to determine the constant GM/h^2. We also note that the computation of this constant will be simplified if we use the relation indicated in Sec. 12.10:

$$GM = gR^2 \tag{12.21}$$

where R is the radius of the earth ($R = 6.37 \times 10^6$ m or 3960 mi) and g the acceleration of gravity at the surface of the earth.

The constant C will be determined by setting $\theta = 0$, $r = r_0$ in Eq. (12.30); we obtain

$$C = \frac{1}{r_0} - \frac{GM}{h^2} \tag{12.33}$$

Substituting for h from (12.32), we may then easily express C in terms of r_0 and v_0.

Let us now determine the initial conditions corresponding to each of the three fundamental trajectories indicated above. Considering first the parabolic trajectory, we set C equal to GM/h^2 in Eq. (12.33) and eliminate h between Eqs. (12.32) and (12.33). Solving for v_0, we obtain

$$v_0 = \sqrt{\frac{2GM}{r_0}}$$

We may easily check that a larger value of the initial velocity corresponds to a hyperbolic trajectory and a smaller value, to an elliptic orbit. Since the value of v_0 obtained for the parabolic trajectory is the smallest value for which the space vehicle does not return to its starting point, it is called the *escape velocity*. We write therefore

$$v_{esc} = \sqrt{\frac{2GM}{r_0}} \qquad \text{or} \qquad v_{esc} = \sqrt{\frac{2gR^2}{r_0}} \tag{12.34}$$

if we make use of Eq. (12.21). We note that the trajectory will be (1) hyperbolic if $v_0 > v_{esc}$; (2) parabolic if $v_0 = v_{esc}$; (3) elliptic if $v_0 < v_{esc}$.

† Problems involving oblique launchings will be considered in Sec. 14.12.

Among the various possible elliptic orbits, one is of special interest, the *circular orbit,* which is obtained when $C = 0$. The value of the initial velocity corresponding to a circular orbit is easily found to be

$$v_{circ} = \sqrt{\frac{GM}{r_0}} \qquad \text{or} \qquad v_{circ} = \sqrt{\frac{gR^2}{r_0}} \tag{12.35}$$

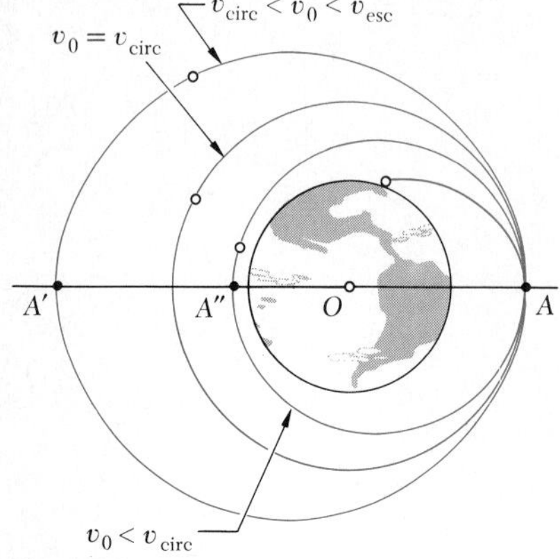

Fig. 12.21

if Eq. (12.21) is taken into account. We may note from Fig. 12.21 that for values of v_0 comprised between v_{circ} and v_{esc}, point A where free flight begins is the point of the orbit closest to the earth; this point is called the *perigee,* while point A', which is farthest away from the earth, is known as the *apogee.* For values of v_0 smaller than v_{circ}, point A becomes the apogee, while point A'', on the other side of the orbit, becomes the perigee. For values of v_0 much smaller than v_{circ}, the trajectory of the space vehicle intersects the surface of the earth; in such a case, the vehicle does not go into orbit.

Ballistic missiles, which are designed to hit the surface of the earth, also travel along elliptic trajectories. In fact, we should now realize that any object projected in vacuum with an initial velocity v_0 smaller than v_{esc} will move along an elliptic path. It is only when the distances involved are small that the gravitational field of the earth may be assumed uniform, and that the elliptic path may be approximated by a parabolic path, as was done earlier (Sec. 11.10) in the case of conventional projectiles.

Periodic Time. An important characteristic of the motion of an earth satellite is the time required by the satellite to describe its orbit. This time is known as the *periodic time* of the satellite and is denoted by τ. We first observe, in view of the definition of the areal velocity (Sec. 12.9), that τ may be obtained by dividing the area inside the orbit by the areal velocity. Since the area of an ellipse is equal to πab, where a and b denote, respectively, the semimajor and semiminor axes, and since the areal velocity is equal to $h/2$, we write

$$\tau = \frac{2\pi ab}{h} \tag{12.36}$$

While h may be readily determined from r_0 and v_0 in the case of a satellite launched in a direction parallel to the surface of the earth, the semiaxes a and b are not directly related to the initial conditions. Since, on the other hand, the values r_0 and r_1 of r corresponding to the perigee and apogee of the orbit may easily be determined from Eq. (12.30), we shall express the semiaxes a and b in terms of r_0 and r_1.

Consider the elliptic orbit shown in Fig. 12.22. The earth's center is located at O and coincides with one of the two foci of the ellipse, while the points A and A' represent, respectively, the perigee and apogee of the orbit. We easily check that

$$r_0 + r_1 = 2a$$

and thus

$$a = \tfrac{1}{2}(r_0 + r_1) \tag{12.37}$$

Recalling that the sum of the distances from each of the foci to any point of the ellipse is constant, we write

$$O'B + BO = O'A + OA = 2a \qquad \text{or} \qquad BO = a$$

On the other hand, we have $CO = a - r_0$. We may therefore write

$$b^2 = (BC)^2 = (BO)^2 - (CO)^2 = a^2 - (a - r_0)^2$$
$$b^2 = r_0(2a - r_0) = r_0 r_1$$

and thus

$$b = \sqrt{r_0 r_1} \tag{12.38}$$

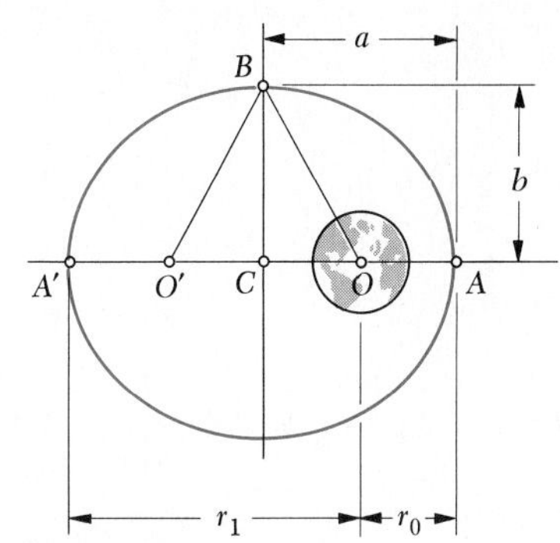

Fig. 12.22

Formulas (12.37) and (12.38) indicate that the semimajor and semiminor axes of the orbit are equal, respectively, to the arithmetic and geometric means of the maximum and minimum values of the radius vector. Once r_0 and r_1 have been determined, the lengths of the semiaxes may thus be easily computed and substituted for a and b in formula (12.36).

*** 12.13. Kepler's Laws of Planetary Motion.** The equations governing the motion of an earth satellite may be used to describe the motion of the moon around the earth. In that case, however, the mass of the moon is not negligible compared with the mass of the earth, and the results obtained are not entirely accurate.

The theory developed in the preceding sections may also be applied to the study of the motion of the planets around the sun. While another error is introduced by neglecting the forces exerted by the planets on one another, the approximation obtained is excellent. Indeed, the properties expressed by Eq. (12.30), where M now represents the mass of the sun, and by Eq. (12.24) had been discovered by the German astronomer Johann Kepler (1571–1630) from astronomical observations of the motion of the planets, even before Newton had formulated his fundamental theory.

Kepler's three *laws of planetary motion* may be stated as follows:

1. Each planet describes an ellipse, with the sun located at one of its foci.
2. The radius vector drawn from the sun to a planet sweeps equal areas in equal times.
3. The squares of the periodic times of the planets are proportional to the cubes of the semimajor axes of their orbits.

The first law states a particular case of the result established in Sec. 12.12, while the second law expresses that the areal velocity of each planet is constant (see Sec. 12.9). Kepler's third law may also be derived from the results obtained in Sec. 12.12.†

†See Prob. 12.115.

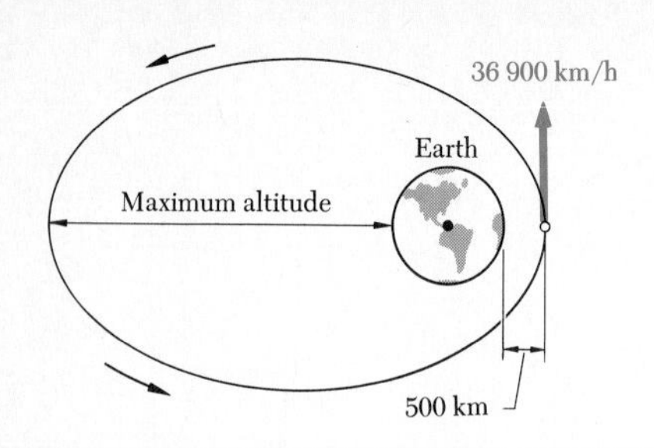

SAMPLE PROBLEM 12.9

A satellite is launched in a direction parallel to the surface of the earth with a velocity of 36 900 km/h from an altitude of 500 km. Determine (*a*) the maximum altitude reached by the satellite, (*b*) the periodic time of the satellite.

***a.* Maximum Altitude.** After launching, the satellite is subjected only to the gravitational attraction of the earth; its motion is thus governed by Eq. (12.30),

$$\frac{1}{r} = \frac{GM}{h^2} + C\cos\theta \qquad (1)$$

Since the radial component of the velocity is zero at the point of launching A, we have $h = r_0 v_0$. Recalling that the radius of the earth is $R = 6370$ km, we compute

$$r_0 = 6370 \text{ km} + 500 \text{ km} = 6870 \text{ km} = 6.87 \times 10^6 \text{ m}$$

$$v_0 = 36\,900 \text{ km/h} = \frac{36.9 \times 10^6 \text{ m}}{3.6 \times 10^3 \text{ s}} = 10.25 \times 10^3 \text{ m/s}$$

$$h = r_0 v_0 = (6.87 \times 10^6 \text{ m})(10.25 \times 10^3 \text{ m/s}) = 70.4 \times 10^9 \text{ m}^2\text{/s}$$

$$h^2 = 4.96 \times 10^{21} \text{ m}^4\text{/s}^2$$

Since $GM = gR^2$, where R is the radius of the earth, we have

$$GM = gR^2 = (9.81 \text{ m/s}^2)(6.37 \times 10^6 \text{ m})^2 = 398 \times 10^{12} \text{ m}^3\text{/s}^2$$

$$\frac{GM}{h^2} = \frac{398 \times 10^{12} \text{ m}^3\text{/s}^2}{4.96 \times 10^{21} \text{ m}^4\text{/s}^2} = 80.3 \times 10^{-9} \text{ m}^{-1}$$

Substituting this value into (1), we obtain

$$\frac{1}{r} = 80.3 \times 10^{-9} \text{ m}^{-1} + C\cos\theta \qquad (2)$$

Noting that at point A we have $\theta = 0$ and $r = r_0 = 6.87 \times 10^6$ m, we compute the constant C:

$$\frac{1}{6.87 \times 10^6 \text{ m}} = 80.3 \times 10^{-9} \text{ m}^{-1} + C\cos 0° \qquad C = 65.3 \times 10^{-9} \text{ m}^{-1}$$

At A', the point on the orbit farthest from the earth, we have $\theta = 180°$. Using (2), we compute the corresponding distance r_1:

$$\frac{1}{r_1} = 80.3 \times 10^{-9} \text{ m}^{-1} + (65.3 \times 10^{-9} \text{ m}^{-1}) \cos 180°$$

$$r_1 = 66.7 \times 10^6 \text{ m} = 66\,700 \text{ km}$$

Maximum altitude = 66 700 km − 6370 km = 60 300 km ◀

***b.* Periodic Time.** Since A and A' are the perigee and apogee, respectively, of the elliptic orbit, we use Eqs. (12.37) and (12.38) and compute the semimajor and semiminor axes of the orbit:

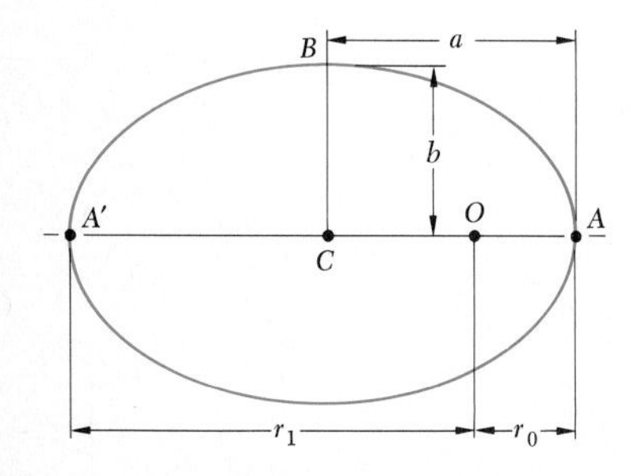

$$a = \tfrac{1}{2}(r_0 + r_1) = \tfrac{1}{2}(6.87 + 66.7)(10^6) \text{ m} = 36.8 \times 10^6 \text{ m}$$

$$b = \sqrt{r_0 r_1} = \sqrt{(6.87)(66.7)} \times 10^6 \text{ m} = 21.4 \times 10^6 \text{ m}$$

$$\tau = \frac{2\pi ab}{h} = \frac{2\pi(36.8 \times 10^6 \text{ m})(21.4 \times 10^6 \text{ m})}{70.4 \times 10^9 \text{ m}^2\text{/s}}$$

$$\tau = 70.3 \times 10^3 \text{ s} = 1171 \text{ min} = 19 \text{ h } 31 \text{ min} \quad ◀$$

Problems

12.89 For the particle of Prob. 12.73, and using Eq. (12.28), show that the central force **F** is inversely proportional to the fifth power of the distance r from the particle to the center of force O.

12.90 For the particle of Prob. 12.74, and using Eq. (12.28), show that the central force **F** is inversely proportional to the cube of the distance r from the particle to the center of force O.

12.91 A particle of mass m describes the hyperbolic spiral $r = b/\theta$ under a central force **F** directed toward the center of force O. Using Eq. (12.28), show that **F** is inversely proportional to the cube of the distance r from the particle to O.

12.92 A particle of mass m describes the logarithmic spiral $r = r_0 e^{b\theta}$ under a central force **F** directed toward the center of force O. Using Eq. (12.28), show that **F** is inversely proportional to the cube of the distance r from the particle to O.

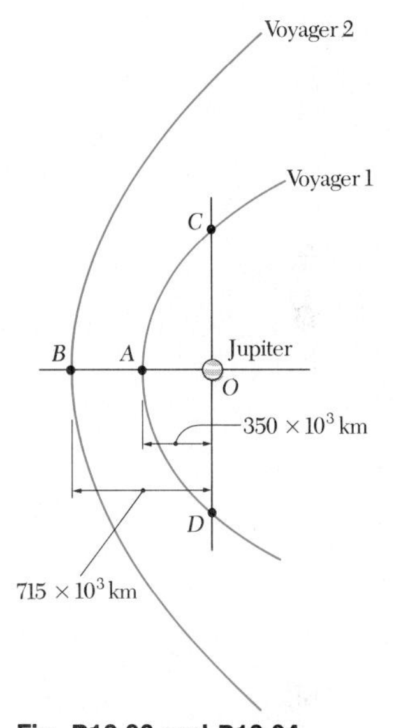

Fig. P12.93 and P12.94

12.93 It was observed that as the spacecraft Voyager 1 reached the point on its trajectory closest to the planet Jupiter, it was at a distance of 350×10^3 km from the center of the planet and had a velocity of 26.9 km/s. Determine the mass of Jupiter, assuming that the trajectory of the spacecraft was parabolic.

12.94 It was observed that as the spacecraft Voyager 2 reached the point on its trajectory closest to the planet Jupiter, it was at a distance of 715×10^3 km from the center of the planet. Assuming the trajectory of the spacecraft to be parabolic and using the data given in Prob. 12.93 for Voyager 1, determine the maximum velocity of Voyager 2 on its approach to Jupiter.

12.95 Solve Prob. 12.94, assuming that the eccentricity of the trajectory of the spacecraft Voyager 2 was $\varepsilon = 1.20$.

12.96 It was observed that as the spacecraft Voyager 1 reached the point of its trajectory closest to the planet Saturn, it was at a distance of 115×10^3 mi from the center of the planet and had a velocity of 68.8×10^3 ft/s. Knowing that Tethys, one of Saturn's satellites, describes a circular orbit of radius 183×10^3 mi at a speed of 37.2×10^3 ft/s, determine the eccentricity of the trajectory of Voyager 1 on its approach to Saturn.

12.97 A satellite describes an elliptic orbit about a planet of mass M. Denoting by r_0 and r_1, respectively, the minimum and maximum values of the distance r from the satellite to the center of the planet, derive the relation

$$\frac{1}{r_0} + \frac{1}{r_1} = \frac{2GM}{h^2}$$

where h is the angular momentum per unit mass of the satellite.

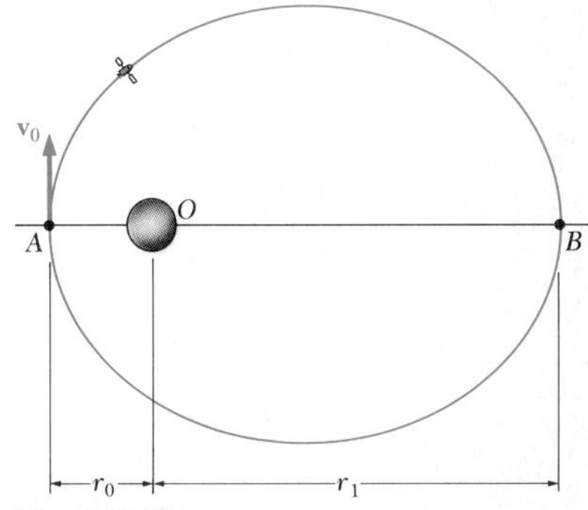

Fig. P12.97

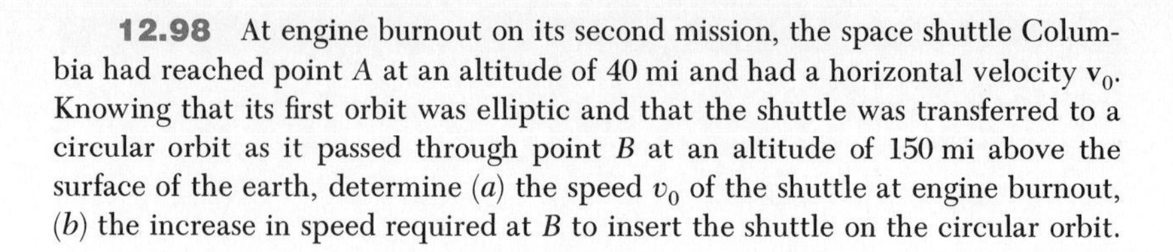

12.98 At engine burnout on its second mission, the space shuttle Columbia had reached point A at an altitude of 40 mi and had a horizontal velocity $\mathbf{v}_0$. Knowing that its first orbit was elliptic and that the shuttle was transferred to a circular orbit as it passed through point B at an altitude of 150 mi above the surface of the earth, determine (a) the speed v_0 of the shuttle at engine burnout, (b) the increase in speed required at B to insert the shuttle on the circular orbit.

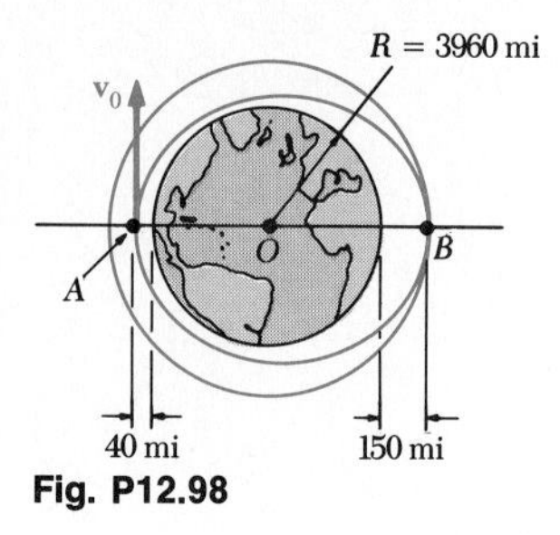

Fig. P12.98

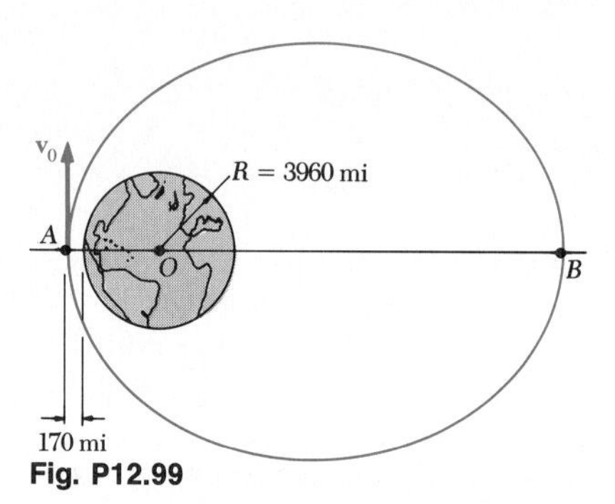

Fig. P12.99

12.99 At engine burnout an Explorer satellite was 170 mi above the surface of the earth and had a horizontal velocity $\mathbf{v}_0$ of magnitude 32.6×10^3 ft/s. Determine (a) the highest altitude reached by the satellite, (b) its velocity at its apogee B.

12.100 A space probe is describing a circular orbit about a planet of radius R. The altitude of the probe above the surface of the planet is αR and its velocity is v_0. In order to place the probe on an elliptic orbit which will bring it closer to the planet, its velocity is reduced from v_0 to βv_0, where $\beta < 1$, by firing its engine for a short interval of time. Determine the smallest permissible value of β if the probe is not to crash on the surface of the planet.

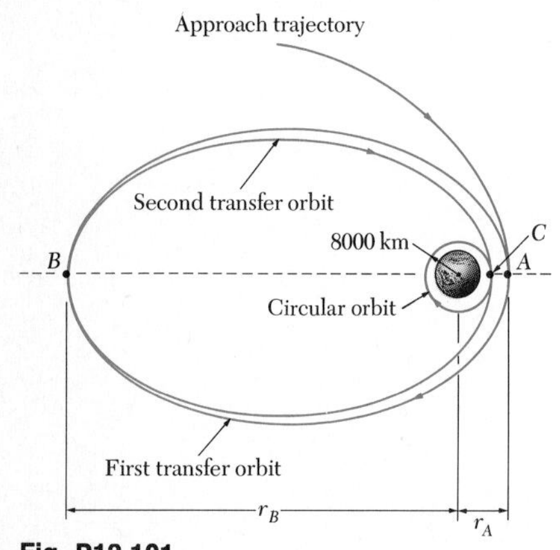

Fig. P12.101

12.101 A space probe is to be placed in a circular orbit of 8000-km radius about the planet Venus in a specified plane. As the probe reaches A, the point of its original trajectory closest to Venus, it is inserted in a first elliptic transfer orbit by reducing its speed by Δv_A. This orbit brings it to point B with a much reduced velocity. There the probe is inserted in a second transfer orbit located in the specified plane by changing the direction of its velocity and further reducing its speed by Δv_B. Finally, as the probe reaches point C, it is inserted in the desired circular orbit by reducing its speed by Δv_C. Knowing that the mass of Venus is 0.82 times the mass of the earth, that $r_A = 12.5 \times 10^3$ km and $r_B = 100 \times 10^3$ km, and that the probe approaches A on a parabolic trajectory, determine by how much the velocity of the probe should be reduced (a) at A, (b) at B, (c) at C.

12.102 For the space probe of Prob. 12.101, it is known that $r_A = 12.5 \times 10^3$ km and that the velocity of the probe is reduced to 6890 m/s as it passes through A. Determine (a) the distance from the center of Venus to point B, (b) the amounts by which the velocity of the probe should be reduced at B and C, respectively.

12.103 Determine the time needed for the space probe of Prob. 12.101 to travel from A to B on its first transfer orbit.

12.104 Determine the time needed for the space probe of Prob. 12.101 to travel from B to C on its second transfer orbit.

12.105 Determine the periodic time of the Explorer satellite of Prob. 12.99.

12.106 For the shuttle Columbia of Prob. 12.98, determine (*a*) the periodic time of the shuttle on its final circular orbit, (*b*) the time needed for the shuttle to travel from A to B on its original elliptic orbit.

12.107 Determine the time needed for the spacecraft Voyager 1 of Prob. 12.93 to travel from C to D on its parabolic trajectory.

12.108 Halley's comet travels in an elongated elliptic orbit for which the minimum distance from the sun is approximately $\frac{1}{2}r_E$, where $r_E = 92.9 \times 10^6$ mi is the mean distance from the sun to the earth. Knowing that the periodic time of Halley's comet is about 76 years, determine the maximum distance from the sun reached by the comet.

12.109 A space probe is describing a circular orbit of radius nR with a velocity v_0 about a planet of radius R and center O. As the probe passes through point A, its velocity is reduced from v_0 to βv_0, where $\beta < 1$, to place the probe on a crash trajectory. Show that the angle AOB, where B denotes the point of impact of the probe on the planet, depends only upon the values of n and β.

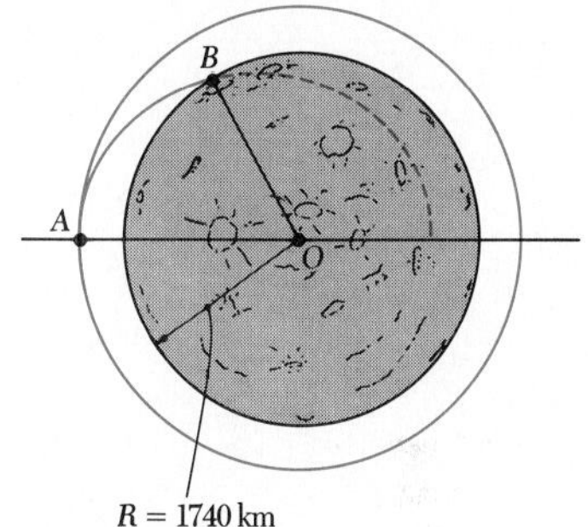

Fig. P12.110

12.110 Upon completion of their moon-exploration mission, the two astronauts forming the crew of an Apollo lunar excursion module (LEM) would rejoin the command module which had remained in a circular orbit around the moon. Before their return to earth, the astronauts would position their craft so that the LEM faced to the rear. As the command module passed through A, the LEM would be cast adrift and crash on the moon's surface at point B. Knowing that the command module was orbiting the moon at an altitude of 150 km and that the angle AOB was 60°, determine the velocity of the LEM relative to the command module as it was cast adrift. (*Hint.* Point A is the apogee of the elliptic crash trajectory. It is also recalled that the mass of the moon is 0.01230 times the mass of the earth.)

12.111 In Prob. 12.110, determine the angle AOB defining the point of impact of the LEM on the surface of the moon, assuming that the command module was orbiting the moon at an altitude of 120 km and that the LEM was cast adrift with a velocity of 40 m/s relative to the command module.

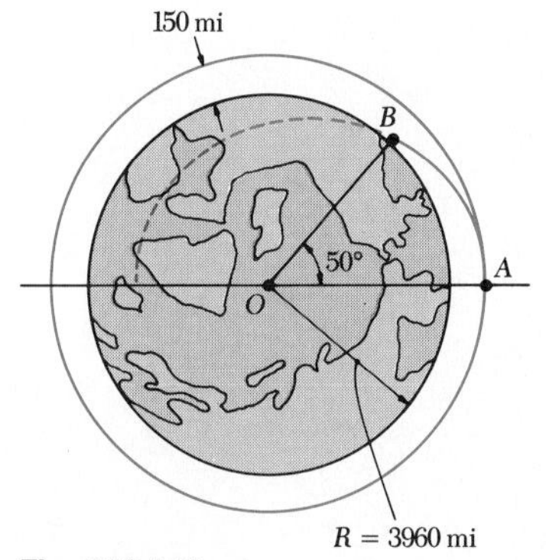

Fig. P12.112

12.112 A space shuttle is describing a circular orbit at an altitude of 150 mi above the surface of the earth. As it passes through A it fires its engine for a short interval of time to reduce its speed by 4 percent and begin its descent toward the earth. Determine the altitude of the shuttle at point B, knowing that the angle AOB is equal to 50°. (*Hint.* Point A is the apogee of the elliptic descent trajectory.)

12.113 A satellite describes an elliptic orbit about a planet. Denoting by r_0 and r_1 the distances corresponding, respectively, to the perigee and apogee of the orbit, show that the curvature of the orbit at each of these two points may be expressed as

$$\frac{1}{\rho} = \frac{1}{2}\left(\frac{1}{r_0} + \frac{1}{r_1}\right)$$

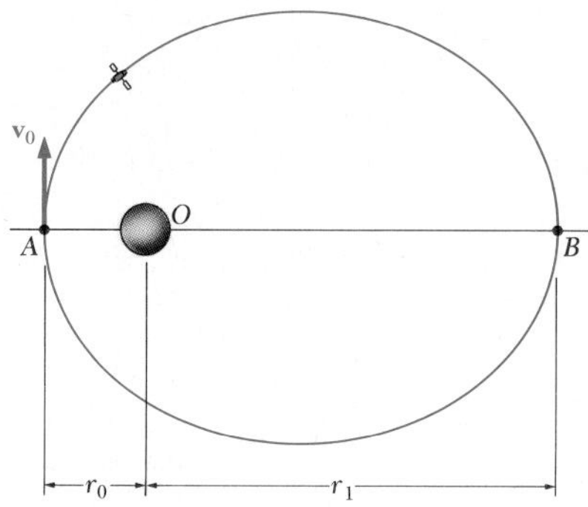

Fig. P12.113 and P12.114

12.114 (*a*) Express the eccentricity ε of the elliptic orbit described by a satellite about a planet in terms of the distances r_0 and r_1 corresponding, respectively, to the perigee and apogee of the orbit. (*b*) Use the result obtained in part *a* to determine the eccentricities of the two transfer orbits described in Prob. 12.101.

12.115 Derive Kepler's third law of planetary motion from Eqs. (12.30) and (12.36).

12.116 Show that for a satellite describing an elliptic orbit of semimajor axis a and eccentricity ε about a planet of mass M, the constant h used in Eq. (12.30) may be expressed as

$$h = \sqrt{GMa(1 - \varepsilon^2)}$$

12.14. Review and Summary. This chapter was devoted to Newton's second law and its application to the analysis of the motion of particles.

Newton's second law

Denoting by m the mass of a particle, by $\Sigma\mathbf{F}$ the sum, or resultant, of the forces acting on the particle, and by $\mathbf{a}$ the acceleration of the particle relative to a *newtonian frame of reference* [Sec. 12.2], we wrote

$$\Sigma\mathbf{F} = m\mathbf{a} \tag{12.2}$$

Consistent systems of units

Equation (12.2) holds only if a consistent system of units is used. With SI units, the forces should be expressed in newtons, the masses in kilograms, and the accelerations in m/s^2; with U.S. customary units, the forces should be expressed in pounds, the masses in $lb \cdot s^2/ft$ (also referred to as *slugs*), and the accelerations in ft/s^2 [Sec. 12.3].

To solve a problem involving the motion of a particle, Eq. (12.2) should be replaced by equations involving scalar quantities, such as

$$\Sigma F_x = ma_x \qquad \Sigma F_y = ma_y \qquad \Sigma F_z = ma_z \tag{12.5}$$

Equations of motion for a particle

It was also indicated [Sec. 12.4] that these equations may be replaced by equations similar to the equilibrium equations used in statics if a vector $(m\mathbf{a})_{rev}$ of magnitude ma but of sense opposite to that of the acceleration is added to the forces applied to the particle; the particle is then said to be in *dynamic equilibrium.*

Motion of a system of particles. External forces and effective forces

As a preparation to the study of the motion of rigid bodies in Chap. 16, we considered the motion of a *system of particles* [Sec. 12.5]. Distinguishing between the *external forces* exerted on the various particles of the system by bodies outside the system and the *internal forces* exerted by the particles on each other, and defining the *effective force* of a particle as the product $m\mathbf{a}$ of its mass m and acceleration $\mathbf{a}$, we concluded that *the external forces* applied to the system *are equipollent to the effective forces* of the various particles of the system. In other words, the sums of the x, y, and z components of the external forces and of the effective forces are respectively equal, and so are the sums of their moments about each of the coordinate axes.

Motion of the mass center of a system of particles

The *mass center* of a system of particles was defined [Sec. 12.6] as the point G of coordinates $\bar{x}$, $\bar{y}$, $\bar{z}$ satisfying the equations

$$(\Sigma m)\bar{x} = \Sigma mx \qquad (\Sigma m)\bar{y} = \Sigma my \qquad (\Sigma m)\bar{z} = \Sigma mz \tag{12.8}$$

where Σm represents the total mass of the system. It was shown that the components of the acceleration $\bar{\mathbf{a}}$ of G satisfy the equations

$$\Sigma(F_x)_{ext} = (\Sigma m)\,\bar{a}_x \quad \Sigma(F_y)_{ext} = (\Sigma m)\,\bar{a}_y \quad \Sigma(F_z)_{ext} = (\Sigma m)\,\bar{a}_z \tag{12.10}$$

Thus, *the mass center of a system of particles moves as if the entire mass of the system and all the external forces were concentrated at that point.*

Rectilinear motion of a particle

In Sec. 12.7, Newton's second law was applied to the analysis of the *rectilinear motion* of particles. In the Sample Problems which followed, we considered successively the motion of a single block [Sample Probs. 12.1 and 12.2], of two connected blocks [Sample Prob. 12.3], and of a system with two degrees of freedom [Sample Prob. 12.4].

Curvilinear motion of a particle

The second part of the chapter was devoted to the analysis of the *curvilinear motion* of a particle, and Newton's second law was expressed in terms of tangential and normal components, or radial and transverse components of the forces and accelerations involved [Sec. 12.8]. Using *tangential and normal components,* we had

$$\Sigma F_t = m\frac{dv}{dt} \qquad \Sigma F_n = m\frac{v^2}{\rho} \tag{12.14}$$

where v is the speed of the particle and ρ the radius of curvature of the path at the point considered [Sample Probs. 12.5 and 12.6]. Using *radial and transverse components*, we had

$$\Sigma F_r = m(\ddot{r} - r\dot{\theta}^2) \tag{12.16}$$
$$\Sigma F_\theta = m(r\ddot{\theta} + 2\dot{r}\dot{\theta}) \tag{12.17}$$

where r and θ are the polar coordinates of the particle [Sample Prob. 12.7].

Motion under a central force

In the particular case of the motion of a particle under a *central force* **F** [Sec. 12.9], Eq. (12.17) yields

$$r^2\dot{\theta} = h \tag{12.18}$$

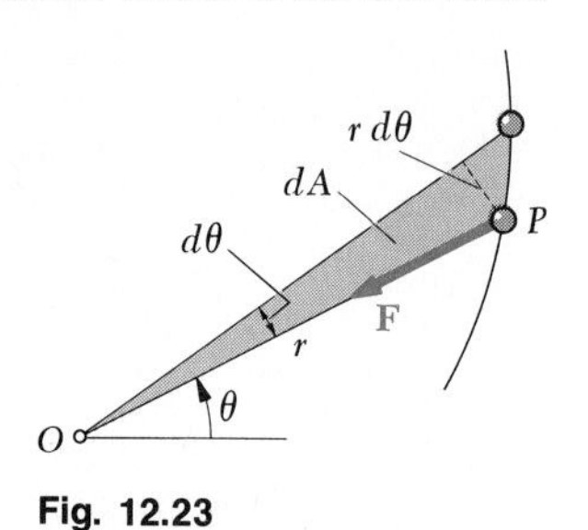

Fig. 12.23

where h is a constant [Sample Prob. 12.8]. It follows from Eq. (12.18) that the area swept per unit time by the radius OP (Fig. 12.23), or *areal velocity*, is constant. Since, as we shall see in Sec. 14.11, the left-hand member of Eq. (12.18) represents the angular momentum per unit mass of the particle, this equation also expresses that the angular momentum of the particle is conserved.

Newton's law of universal gravitation

An important application of the motion under a central force is provided by the orbital motion of bodies under gravitational attraction [Sec. 12.10]. According to *Newton's law of universal gravitation*, two particles at a distance r from each other and, respectively, of mass M and m attract each other with equal and opposite forces **F** and **F**′ directed along the line joining the particles (Fig. 12.24). The common magnitude F of the two forces is

$$F = G\frac{Mm}{r^2} \tag{12.19}$$

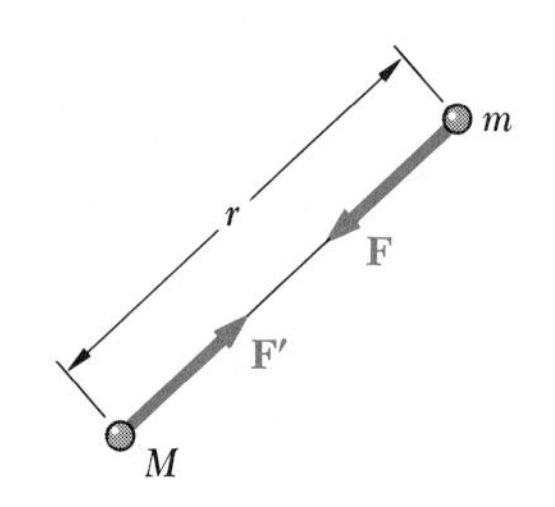

Fig. 12.24

where G is the *constant of gravitation*. In the case of a body of mass m subjected to the gravitational attraction of the earth, the product GM, where M is the mass of the earth, may be expressed as

$$GM = gR^2 \tag{12.21}$$

where $g = 9.81 \text{ m/s}^2 = 32.2 \text{ ft/s}^2$, and where R represents the radius of the earth.

Orbital motion

It was shown in Sec. 12.11 that a particle moving under a central force describes a trajectory defined by the differential equation

$$\frac{d^2u}{d\theta^2} + u = \frac{F}{mh^2u^2} \tag{12.28}$$

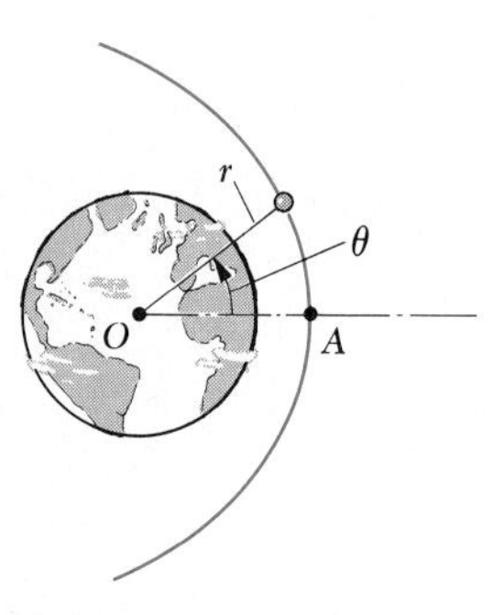

Fig. 12.25

where $F > 0$ corresponds to an attractive force and where $u = 1/r$. In the case of a particle moving under a force of gravitational attraction [Sec. 12.12], we substituted for F the expression given in Eq. (12.19). Measuring θ from the axis OA joining the focus O to the point A of the trajectory closest to O (Fig. 12.25), we found that the solution to Eq. (12.28) was

$$\frac{1}{r} = u = \frac{GM}{h^2} + C\cos\theta \tag{12.30}$$

This is the equation of a conic of eccentricity $\varepsilon = Ch^2/GM$. The conic is an *ellipse* if $\varepsilon < 1$, a *parabola* if $\varepsilon = 1$, and a *hyperbola* if $\varepsilon > 1$. The constants C and h may be determined from the initial conditions; if the particle is projected from point A ($\theta = 0$, $r = r_0$) with an initial velocity $\mathbf{v}_0$ perpendicular to OA, we have $h = r_0 v_0$ [Sample Prob. 12.9].

Escape velocity

It was also shown that the values of the initial velocity corresponding, respectively, to a parabolic and a circular trajectory were

$$v_{esc} = \sqrt{\frac{2GM}{r_0}} \tag{12.34}$$

$$v_{circ} = \sqrt{\frac{GM}{r_0}} \tag{12.35}$$

and that the first of these values, called the *escape velocity*, is the smallest value of v_0 for which the particle will not return to its starting point.

Periodic time

The *periodic time* τ of a planet or satellite was defined as the time required by that body to describe its orbit. It was shown that

$$\tau = \frac{2\pi ab}{h} \tag{12.36}$$

where $h = r_0 v_0$ and where a and b represent the semimajor and semiminor axes of the orbit. It was further shown that these semiaxes are respectively equal to the arithmetic and geometric means of the maximum and minimum values of the radius vector r.

Kepler's laws

The last section of the chapter [Sec. 12.13] presented *Kepler's laws of planetary motion* and showed that these empirical laws, obtained from early astronomical observations, confirm Newton's laws of motion as well as his law of gravitation.

Review Problems

12.117 A ship of total mass m is anchored in the middle of a river which is flowing with a constant velocity $\mathbf{v}_0$. The horizontal component of the force exerted on the ship by the anchor chain is $\mathbf{T}_0$. If the anchor chain suddenly breaks, determine the time required for the ship to attain a velocity equal to $\frac{1}{2}\mathbf{v}_0$. Assume that the frictional resistance of the water is proportional to the velocity of the ship relative to the water.

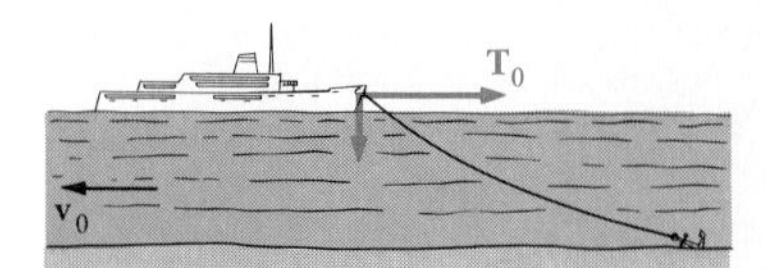

Fig. P12.117

12.118 A ball is attached to a cord and swung through a full circle in a vertical plane. Knowing that the radius of the circle is 1.25 m, determine the smallest velocity that the ball should have at the top of the circle if the cord is to remain taut.

12.119 The coefficients of friction between the 45-lb crate and the 30-lb cart are $\mu_s = 0.25$ and $\mu_k = 0.20$. If a force **P** of magnitude 20 lb is applied to the cart, determine the acceleration (*a*) of the cart, (*b*) of the crate, (*c*) of the crate with respect to the cart.

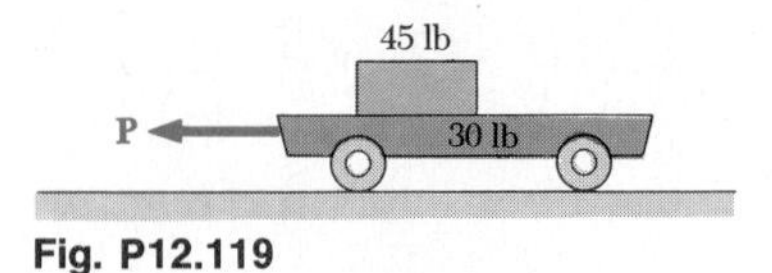

Fig. P12.119

12.120 For the cart and crate of Prob. 12.119, determine the acceleration of the cart as a function of the magnitude of the force **P**.

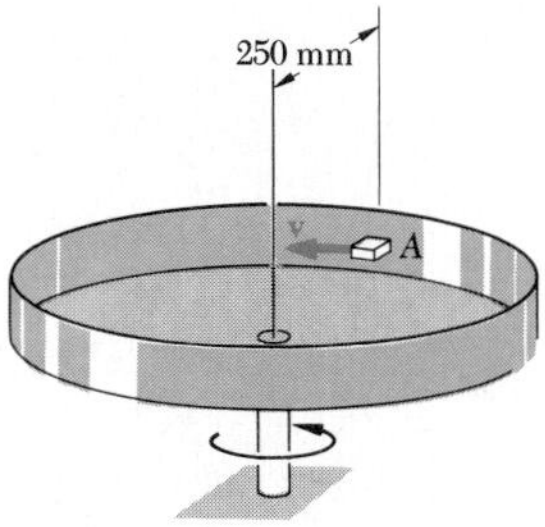

12.121 The assembly shown rotates about a vertical axis at a constant rate. Knowing that the coefficient of static friction between the small block *A* and the cylindrical wall is 0.30, determine the lowest speed *v* for which the block will remain in contact with the wall.

250 mm
Fig. P12.121

12.122 Knowing that blocks *B* and *C* strike the ground simultaneously and exactly 1 s after the system is released from rest, determine m_B and m_C in terms of m_A.

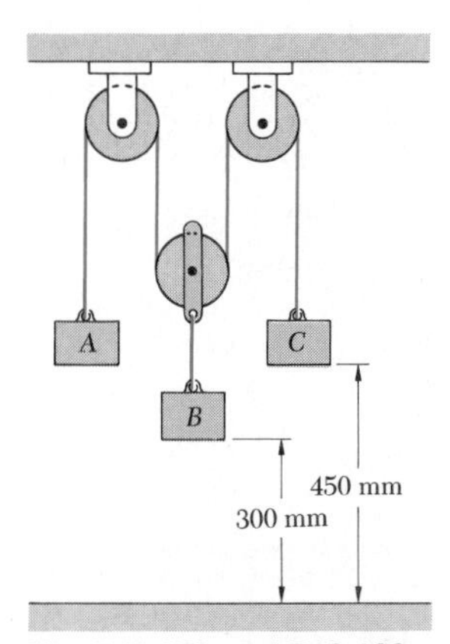

Fig. P12.122 and P12.123

12.123 Determine the acceleration of each block when $m_A = 5$ kg, $m_B = 15$ kg, and $m_C = 10$ kg. Which block strikes the ground first?

12.124 Two solid steel spheres, each of radius 100 mm, are placed so that their surfaces are in contact. (*a*) Determine the force of gravitational attraction between the spheres, knowing that the density of steel is 7850 kg/m^3. (*b*) If the spheres are moved 2 mm apart and released with zero velocity, determine the approximate time required for their gravitational attraction to bring them back into contact. (*Hint.* Assume that the gravitational forces remain constant.)

12.125 A bucket is attached to a rope of length $L = 3$ ft and is made to revolve in a horizontal circle. Drops of water leaking from the bucket fall and strike the floor along the perimeter of a circle of radius a. Determine the radius a when $\theta = 30°$.

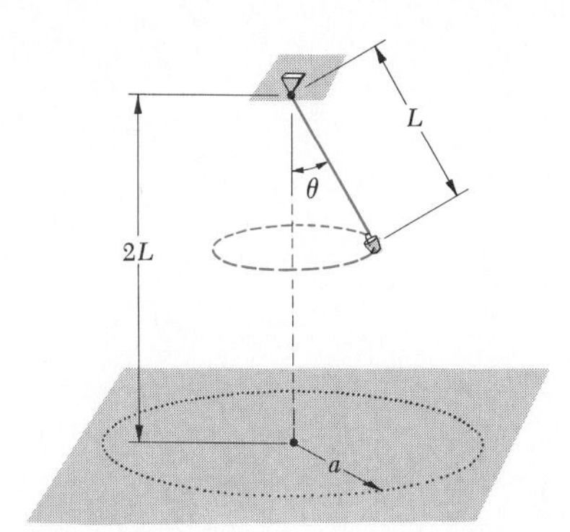

Fig. P12.125

12.126 A spacecraft is describing an elliptic orbit of minimum altitude $h_A = 1040$ mi and maximum altitude $h_B = 6040$ mi above the surface of the earth. (*a*) Determine the speed of the spacecraft at A. (*b*) If its engine is fired as the spacecraft passes through point A and its speed is increased by 10 percent, determine the maximum altitude reached by the spacecraft on its new orbit.

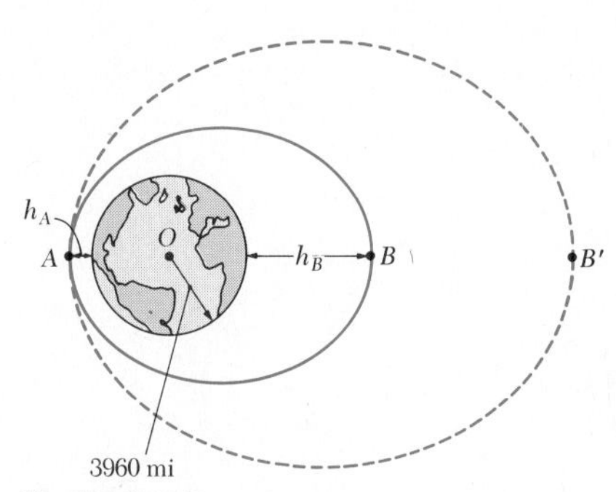

Fig. P12.126

12.127 Determine the approximate time required for an object to fall to the surface of the earth after being released with no velocity from a distance equal to the radius of the orbit of the moon, namely, 384×10^3 km. (*Hint.* Assume that the object is given a very small initial velocity in a transverse direction, say, $v_\theta = 1$ m/s, and determine the periodic time τ of the object on the resulting orbit. An examination of the orbit will show that the time of fall must be approximately equal to $\frac{1}{2}\tau$.)

12.128 Neglecting the effect of friction, determine (*a*) the acceleration of each block, (*b*) the tension in the cable.

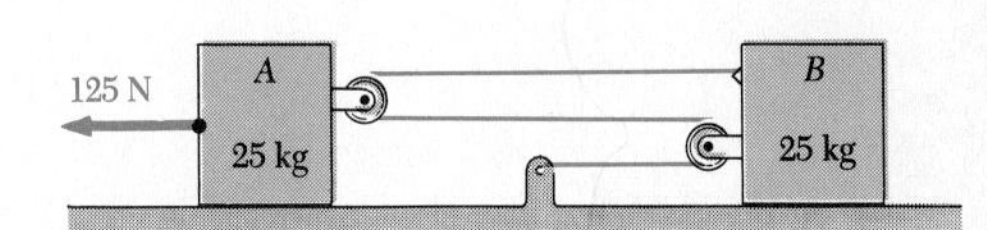

Fig. P12.128

The following problems are designed to be solved with a computer.

12.C1 A 3-kg ball revolves at the constant speed v in a horizontal circle as shown. The ball is attached to an elastic cord of undeformed length $L = 0.8$ m and spring constant k. Write a computer program and use it to calculate the speed v of the ball and the corresponding length $L + \Delta L$ of the cord for values of θ from 15 to 75° at 15° intervals when (*a*) $k = 20$ N/m, (*b*) $k = 200$ N/m, (*c*) $k = 2000$ N/m.

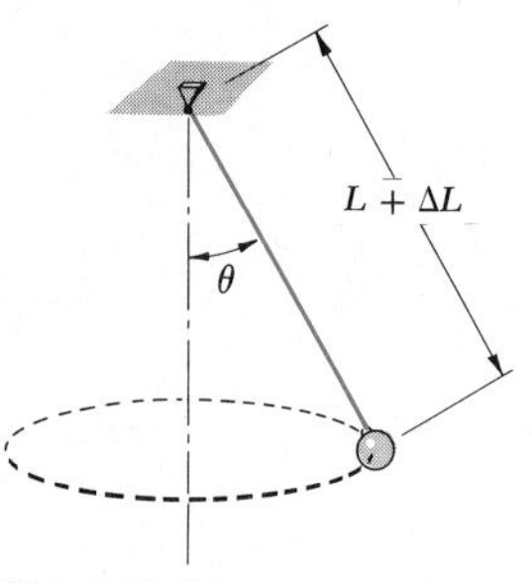

Fig. P12.C1

12.C2 The 12-lb block B is released from rest on the 30-lb wedge A, which is supported by a horizontal surface. Write a computer program which can be used to calculate the acceleration of the wedge and the acceleration of the block relative to the wedge. Denoting by μ the coefficient of friction at all surfaces, use this program to determine the accelerations for values of μ at 0.01 intervals from $\mu = 0$ to the value of μ for which the wedge does not move and then for values of μ at 0.1 intervals to the value of μ for which no motion occurs.

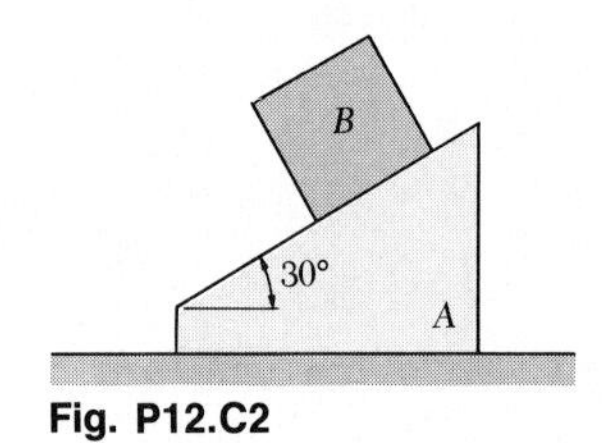

Fig. P12.C2

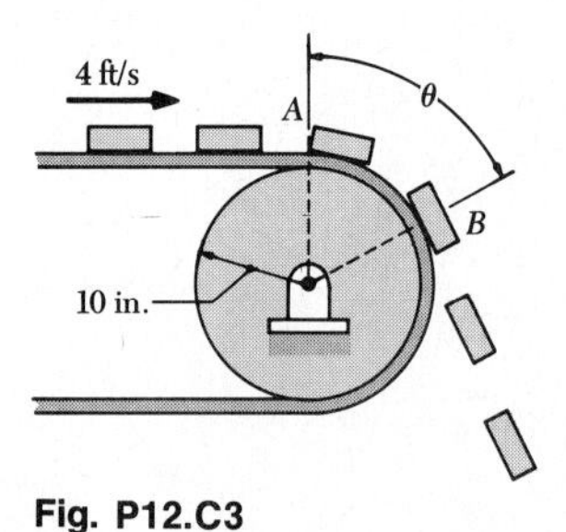

Fig. P12.C3

12.C3 A series of small packages are discharged from a conveyor belt as shown. The coefficients of friction are $\mu_s = 0.40$ and $\mu_k = 0.35$ between the belt and the packages and it is known that the packages first slip with respect to the belt when $\theta = 9.01°$ (see answer to Prob. 12.48). Write a computer program and use it to determine by numerical integration at intervals of time $\Delta t = 0.001$ s the angle θ defining the point where the packages will leave the belt.

12.C4 A space shuttle is describing a circular orbit with a velocity v_{circ} at an altitude of 150 mi above the surface of the earth. As it passes through A it fires its engine for a short interval of time to reduce its speed to βv_{circ} and begin its descent toward the earth. Knowing that it will enter the upper limit of the atmosphere at an altitude of 40 mi at point B, write a computer program and use it to calculate the angle AOB for values of β from 0 to 0.9 at 0.1 intervals and from 0.9 to 1.0 at 0.01 intervals.

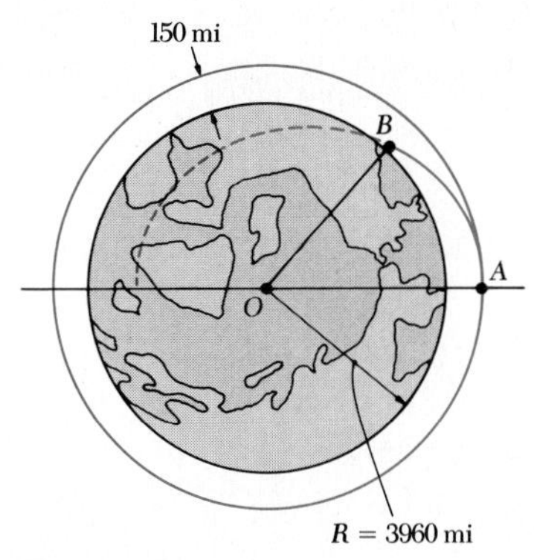

Fig. P12.C4

CHAPTER THIRTEEN
Kinetics of Particles: Work and Energy

13.1. Introduction. In the preceding chapter, most problems dealing with the motion of particles were solved through the use of the fundamental equation of motion $\mathbf{F} = m\mathbf{a}$. Given a particle acted upon by a force $\mathbf{F}$, we could solve this equation for the acceleration $\mathbf{a}$; then, by applying the principles of kinematics, we could determine from $\mathbf{a}$ the velocity and position of the particle at any time.

If the equation $\mathbf{F} = m\mathbf{a}$ and the principles of kinematics are combined, two additional methods of analysis may be obtained, the *method of work and energy* and the *method of impulse and momentum*. The advantage of these methods lies in the fact that they make the determination of the acceleration unnecessary. Indeed, the method of work and energy relates directly force, mass, velocity, and displacement, while the method of impulse and momentum relates force, mass, velocity, and time.

The method of work and energy will be considered in this chapter. In Secs. 13.2 through 13.5, we shall discuss the *work of a force* and the *kinetic energy of a particle* and apply the principle of work and energy to the solution of engineering problems involving the motion of one or several connected bodies.

Sections 13.6 and 13.7 are devoted to the concept of *potential energy* of a conservative force and to the application of the principle of conservation of energy to various problems of practical interest. Finally, the concepts of *power* and *efficiency* of a machine will be introduced in Sec. 13.8.

13.2. Work of a Force. We shall first define the terms *displacement* and *work* as they are used in mechanics.† Consider a particle which moves from a point A to a neighboring point A' (Fig. 13.1*a*). The small vector which joins A to A' is called the *displacement* of the particle and is denoted by the differential $d\mathbf{r}$, where $\mathbf{r}$ represents the position vector of A. Now, let us assume that a force $\mathbf{F}$, of magnitude F and forming an angle α with $d\mathbf{r}$, is acting on the particle. The *work of the force* $\mathbf{F}$ *during the displacement* $d\mathbf{r}$, of magnitude ds, is defined as the scalar quantity

$$dU = F\,ds\cos\alpha \tag{13.1}$$

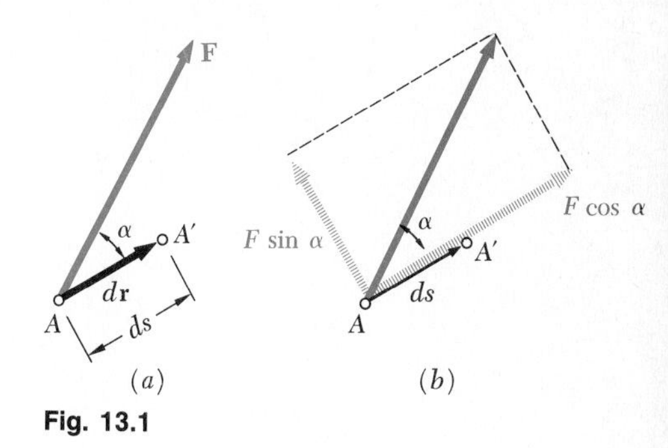

Fig. 13.1

Being a *scalar quantity*, work has a magnitude and a sign, but no direction. We also note that work should be expressed in units obtained by multiplying units of length by units of force. Thus, if U.S. customary units are used, work should be expressed in ft · lb or in · lb. If SI units are used, work should be expressed in N · m. The unit of work N · m is called a *joule* (J).‡ Recalling the conversion factors indicated in Sec. 12.2, we write

$$1\text{ ft}\cdot\text{lb} = (1\text{ ft})(1\text{ lb}) = (0.3048\text{ m})(4.448\text{ N}) = 1.356\text{ J}$$

It appears from (13.1) that the work dU may be considered as the product of the magnitude ds of the displacement and of the component $F\cos\alpha$ of the force $\mathbf{F}$ in the direction of $d\mathbf{r}$ (Fig. 13.1*b*). If this component and the displacement $d\mathbf{r}$ have the same sense, the work is positive; if they have opposite senses, the work is negative. Three particular cases are of special interest. If the force $\mathbf{F}$ has the same direction as $d\mathbf{r}$, the work dU reduces to $F\,ds$. If $\mathbf{F}$ has a direction opposite to that of $d\mathbf{r}$, the work is $dU = -F\,ds$. Finally, if $\mathbf{F}$ is perpendicular to $d\mathbf{r}$, the work dU is zero.

The work of $\mathbf{F}$ during a *finite* displacement of the particle from A_1 to A_2 (Fig. 13.2*a*) is obtained by integrating Eq. (13.1) from s_1 to s_2. This work, denoted by $U_{1\to2}$, is

$$U_{1\to2} = \int_{s_1}^{s_2} (F\cos\alpha)\,ds \tag{13.2}$$

where the variable of integration s measures the distance traveled by the particle along the path. The work $U_{1\to2}$ is thus represented by the area under the curve obtained by plotting $F\cos\alpha$ against s (Fig. 13.2*b*).

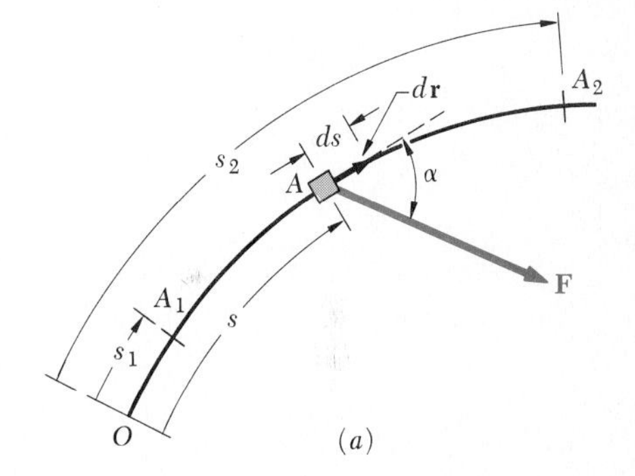

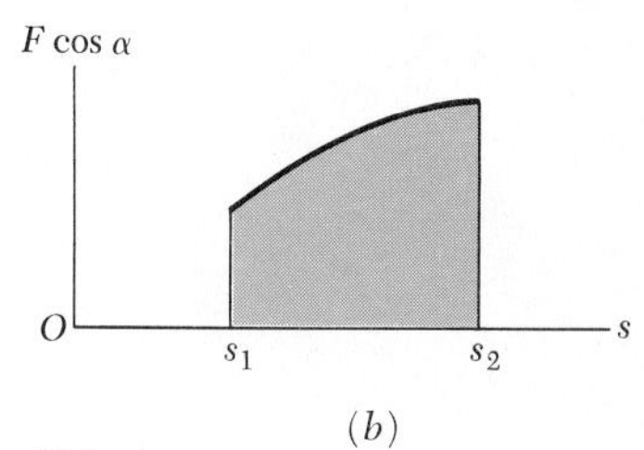

Fig. 13.2

† The definition of work was given in Sec. 10.2, and the basic properties of the work of a force were outlined in Secs. 10.2 and 10.6. For convenience, we repeat here the portions of this material which relate to the kinetics of particles.

‡ The joule (J) is the SI unit of *energy*, whether in mechanical form (work, potential energy, kinetic energy) or in chemical, electrical, or thermal form. We should note that even though N · m = J, the moment of a force must be expressed in N · m and not in joules, since the moment of a force is not a form of energy.

Work of a Constant Force in Rectilinear Motion. When a particle moving in a straight line is acted upon by a force **F** of constant magnitude and of constant direction (Fig. 13.3), formula (13.2) yields

$$U_{1\rightarrow 2} = (F \cos \alpha)\, \Delta x \qquad (13.3)$$

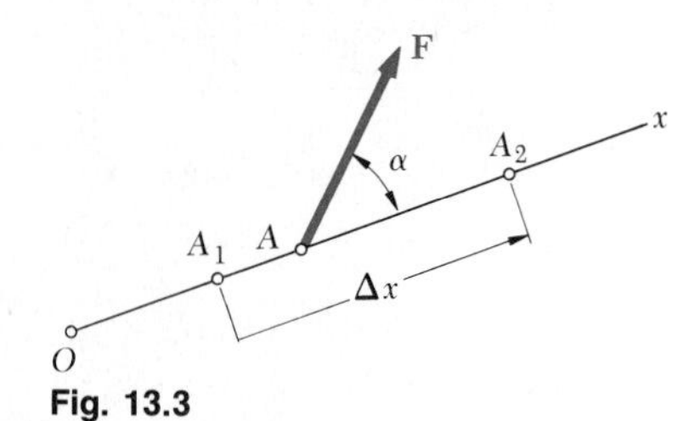

Fig. 13.3

where α = angle the force forms with direction of motion
Δx = displacement from A_1 to A_2

Work of the Force of Gravity. The work dU of a force **F** during a displacement $d\mathbf{r}$ may also be considered as the product of the magnitude F of the force **F** and of the component $ds \cos \alpha$ of the displacement $d\mathbf{r}$ along **F**. This view is particularly useful in the computation of the work of the weight **W** of a body, i.e., of the force of gravity exerted on that body. The work dU of **W** is equal to the product of W and of the vertical displacement

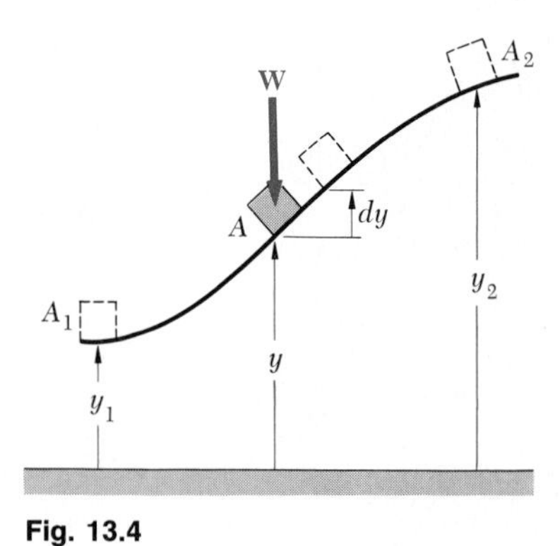

Fig. 13.4

of the center of gravity G of the body. With the y axis chosen upward, the work of **W** during a finite displacement (Fig. 13.4) is obtained by writing

$$dU = -W\, dy$$

$$U_{1\rightarrow 2} = -\int_{y_1}^{y_2} W\, dy = Wy_1 - Wy_2 \qquad (13.4)$$

or

$$U_{1\rightarrow 2} = -W(y_2 - y_1) = -W\, \Delta y \qquad (13.4')$$

where Δy is the vertical displacement from A_1 to A_2. The work of the weight **W** is thus equal to *the product of W and of the vertical displacement of the center of gravity of the body.* The work is *positive* when $\Delta y < 0$, that is, *when the body moves down.*

Work of the Force Exerted by a Spring. Consider a body A attached to a fixed point B by a spring; it is assumed that the spring is undeformed when the body is at A_0 (Fig. 13.5*a*). Experimental evidence shows that the magnitude of the force $\mathbf{F}$ exerted by the spring on body A is proportional to the deflection x of the spring measured from the position A_0. We have

$$F = kx \tag{13.5}$$

where k is the *spring constant,* expressed in N/m or kN/m if SI units are used and in lb/ft or lb/in. if U.S. customary units are used.†

The work of the force $\mathbf{F}$ exerted by the spring during a finite displacement of the body from $A_1(x = x_1)$ to $A_2(x = x_2)$ is obtained by writing

$$dU = -F\,dx = -kx\,dx$$

$$U_{1\to2} = -\int_{x_1}^{x_2} kx\,dx = \tfrac{1}{2}kx_1^2 - \tfrac{1}{2}kx_2^2 \tag{13.6}$$

Care should be taken to express k and x in consistent units. For example, if U.S. customary units are used, k should be expressed in lb/ft and x in feet, or k in lb/in. and x in inches; in the first case, the work is obtained in ft·lb, in the second case, in in·lb. We note that the work of the force $\mathbf{F}$ exerted by the spring on the body is *positive* when $x_2 < x_1$, that is, *when the spring is returning to its undeformed position.*

Since Eq. (13.5) is the equation of a straight line of slope k passing through the origin, the work $U_{1\to2}$ of $\mathbf{F}$ during the displacement from A_1 to A_2 may be obtained by evaluating the area of the trapezoid shown in Fig. 13.5*b*. This is done by computing F_1 and F_2 and multiplying the base Δx of the trapezoid by its mean height $\frac{1}{2}(F_1 + F_2)$. Since the work of the force $\mathbf{F}$ exerted by the spring is positive for a negative value of Δx, we write

$$U_{1\to2} = -\tfrac{1}{2}(F_1 + F_2)\,\Delta x \tag{13.6'}$$

Formula (13.6′) is usually more convenient to use than (13.6) and affords fewer chances of confusing the units involved.

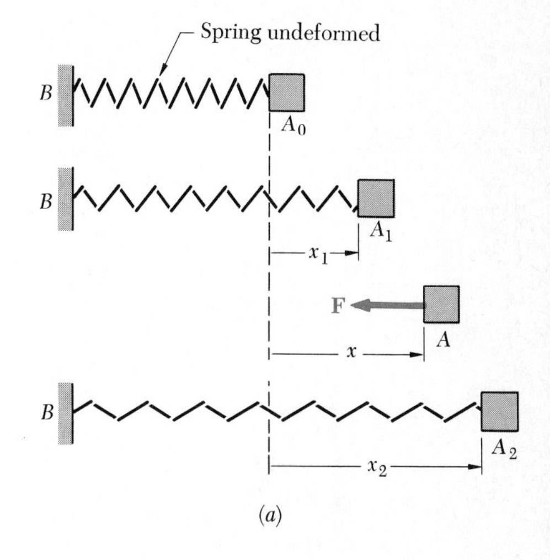

(*a*)

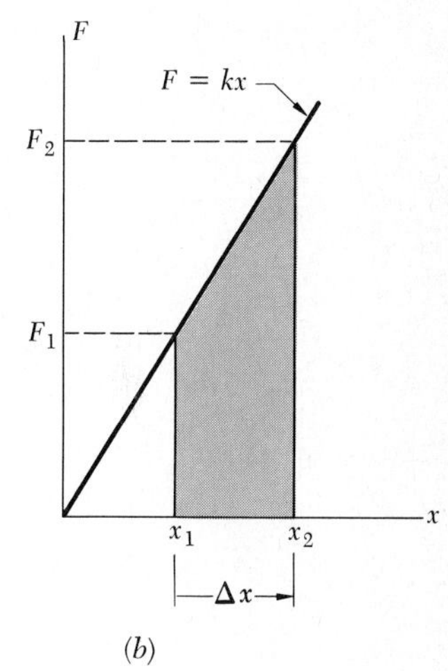

(*b*)

Fig. 13.5

Work of a Gravitational Force. We saw in Sec. 12.10 that two particles at distance r from each other and, respectively, of mass M and m, attract each other with equal and opposite forces $\mathbf{F}$ and $\mathbf{F}'$ directed along the line joining the particles, and of magnitude

$$F = G\frac{Mm}{r^2}$$

†The relation $F = kx$ is correct under static conditions only. Under dynamic conditions, formula (13.5) should be modified to take the inertia of the spring into account. However, the error introduced by using the relation $F = kx$ in the solution of kinetics problems is small if the mass of the spring is small compared with the other masses in motion.

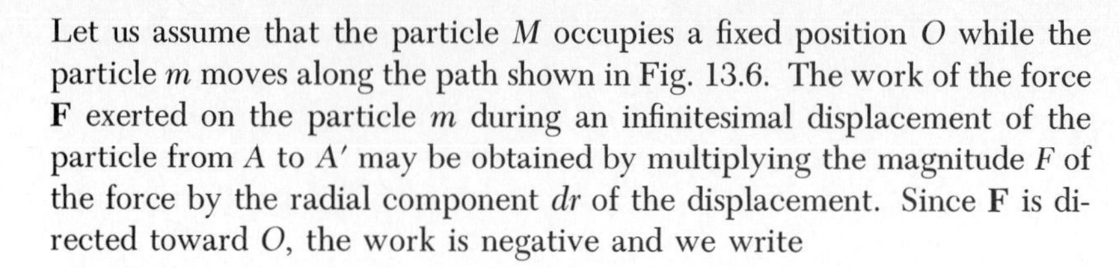

Let us assume that the particle M occupies a fixed position O while the particle m moves along the path shown in Fig. 13.6. The work of the force **F** exerted on the particle m during an infinitesimal displacement of the particle from A to A' may be obtained by multiplying the magnitude F of the force by the radial component dr of the displacement. Since **F** is directed toward O, the work is negative and we write

$$dU = -F\,dr = -G\frac{Mm}{r^2}\,dr$$

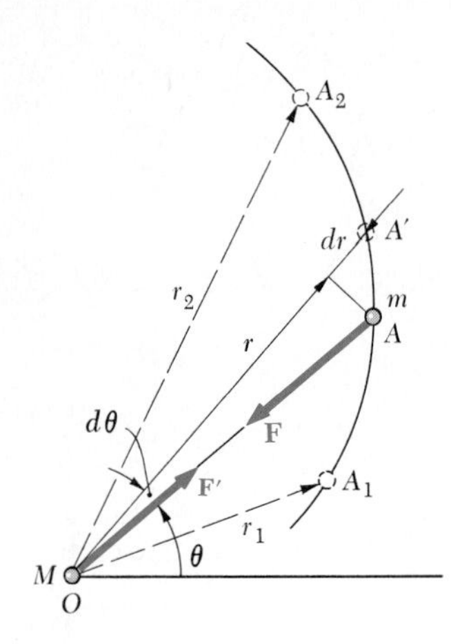

Fig. 13.6

The work of the gravitational force **F** during a finite displacement from $A_1(r = r_1)$ to $A_2(r = r_2)$ is therefore

$$U_{1\rightarrow 2} = -\int_{r_1}^{r_2} \frac{GMm}{r^2}\,dr = \frac{GMm}{r_2} - \frac{GMm}{r_1} \tag{13.7}$$

The formula obtained may be used to determine the work of the force exerted by the earth on a body of mass m at a distance r from the center of the earth, when r is larger than the radius R of the earth. The letter M represents then the mass of the earth; recalling the first of the relations (12.20), we may thus replace the product GMm in Eq. (13.7) by WR^2, where R is the radius of the earth ($R = 6.37 \times 10^6$ m or 3960 mi) and W the value of the weight of the body at the surface of the earth.

A number of forces frequently encountered in problems of kinetics *do no work*. They are forces applied to fixed points ($ds = 0$) or acting in a direction perpendicular to the displacement ($\cos\alpha = 0$). Among the forces which do no work are the following: the reaction at a frictionless pin when the body supported rotates about the pin, the reaction at a frictionless surface when the body in contact moves along the surface, the reaction at a roller moving along its track, and the weight of a body when its center of gravity moves horizontally.

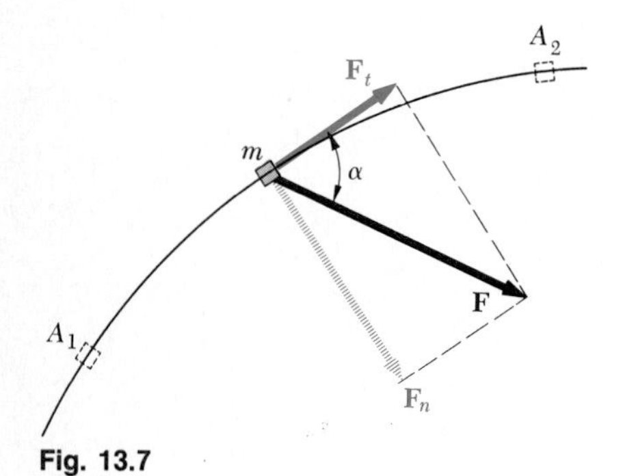

Fig. 13.7

13.3. Kinetic Energy of a Particle. Principle of Work and Energy. Consider a particle of mass m acted upon by a force **F** and moving along a path which is either rectilinear or curved (Fig. 13.7). Expressing Newton's second law in terms of the tangential components of the force and of the acceleration (see Sec. 12.8), we write

$$F_t = ma_t \qquad \text{or} \qquad F\cos\alpha = m\frac{dv}{dt}$$

where v is the speed of the particle. Recalling from Sec. 11.9 that $v = ds/dt$, we obtain

$$F\cos\alpha = m\frac{dv}{ds}\frac{ds}{dt} = mv\frac{dv}{ds}$$

$$(F\cos\alpha)\,ds = mv\,dv$$

Integrating from A_1, where $s = s_1$ and $v = v_1$, to A_2, where $s = s_2$ and $v = v_2$, we write

$$\int_{s_1}^{s_2} (F \cos \alpha)\, ds = m \int_{v_1}^{v_2} v\, dv = \tfrac{1}{2}mv_2^2 - \tfrac{1}{2}mv_1^2 \qquad (13.8)$$

The left-hand member of Eq. (13.8) represents the work $U_{1\rightarrow 2}$ of the force **F** exerted on the particle during the displacement from A_1 to A_2; as indicated in Sec. 13.2, the work $U_{1\rightarrow 2}$ is a scalar quantity. The expression $\frac{1}{2}mv^2$ is also a scalar quantity; it is defined as the kinetic energy of the particle and is denoted by T. We write

$$T = \tfrac{1}{2}mv^2 \qquad (13.9)$$

Substituting into (13.8), we have

$$U_{1\rightarrow 2} = T_2 - T_1 \qquad (13.10)$$

which expresses that, when a particle moves from A_1 to A_2 under the action of a force **F**, *the work of the force* **F** *is equal to the change in kinetic energy of the particle.* This is known as the *principle of work and energy.* Rearranging the terms in (13.10), we write

$$T_1 + U_{1\rightarrow 2} = T_2 \qquad (13.11)$$

Thus, *the kinetic energy of the particle at A_2 may be obtained by adding to its kinetic energy at A_1 the work done during the displacement from A_1 to A_2 by the force* **F** *exerted on the particle.* As Newton's second law from which it is derived, the principle of work and energy applies only with respect to a newtonian frame of reference (Sec. 12.2). The speed v used to determine the kinetic energy T should therefore be measured with respect to a newtonian frame of reference.

Since both work and kinetic energy are scalar quantities, their sum may be computed as an ordinary algebraic sum, the work $U_{1\rightarrow 2}$ being considered as positive or negative according to the direction of **F**. When several forces act on the particle, the expression $U_{1\rightarrow 2}$ represents the total work of the forces acting on the particle; it is obtained by adding algebraically the work of the various forces.

As noted above, the kinetic energy of a particle is a scalar quantity. It further appears from the definition $T = \frac{1}{2}mv^2$ that the kinetic energy is always positive, regardless of the direction of motion of the particle. Considering the particular case when $v_1 = 0$, $v_2 = v$, and substituting $T_1 = 0$, $T_2 = T$ into (13.10), we observe that the work done by the forces acting on the particle is equal to T. Thus, the kinetic energy of a particle moving with a speed v represents the work which must be done to bring the particle from rest to the speed v. Substituting $T_1 = T$ and $T_2 = 0$ into (13.10), we also note that when a particle moving with a speed v is brought to rest, the work done by the forces acting on the particle is $-T$. Assuming that no energy is dissipated into heat, we conclude that the work done by the forces exerted *by the particle* on the bodies which cause it to come to rest is equal to T. Thus, the kinetic energy of a particle also represents *the capacity to do work associated with the speed of the particle.*

The kinetic energy is measured in the same units as work, i.e., in joules if SI units are used, and in ft·lb if U.S. customary units are used. We check that, in SI units,

$$T = \tfrac{1}{2}mv^2 = \text{kg}(\text{m/s})^2 = (\text{kg}\cdot\text{m/s}^2)\text{m} = \text{N}\cdot\text{m} = \text{J}$$

while, in customary units,

$$T = \tfrac{1}{2}mv^2 = (\text{lb}\cdot\text{s}^2/\text{ft})(\text{ft/s})^2 = \text{lb}\cdot\text{ft}$$

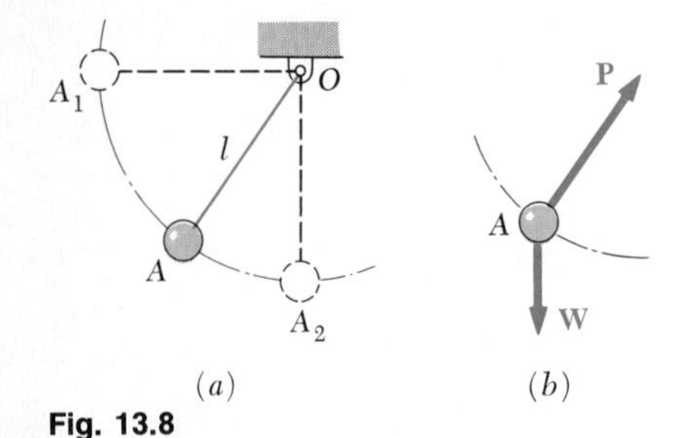

Fig. 13.8

13.4. Applications of the Principle of Work and Energy.

The application of the principle of work and energy greatly simplifies the solution of many problems involving forces, displacements, and velocities. Consider, for example, the pendulum OA consisting of a bob A of weight W attached to a cord of length l (Fig. 13.8*a*). The pendulum is released with no initial velocity from a horizontal position OA_1 and allowed to swing in a vertical plane. We wish to determine the speed of the bob as it passes through A_2, directly under O.

We first determine the work done during the displacement from A_1 to A_2 by the forces acting on the bob. We draw a free-body diagram of the bob, showing all the *actual* forces acting on it, i.e., the weight **W** and the force **P** exerted by the cord (Fig. 13.8*b*). (An inertia vector is not an actual force and *should not* be included in the free-body diagram.) We note that the force **P** does no work, since it is normal to the path; the only force which does work is thus the weight **W**. The work of **W** is obtained by multiplying its magnitude W by the vertical displacement l (Sec. 13.2); since the displacement is downward, the work is positive. We therefore write $U_{1\to2} = Wl$.

Considering, now, the kinetic energy of the bob, we find $T_1 = 0$ at A_1 and $T_2 = \frac{1}{2}(W/g)v_2^2$ at A_2. We may now apply the principle of work and energy; recalling formula (13.11), we write

$$T_1 + U_{1\to2} = T_2 \qquad 0 + Wl = \frac{1}{2}\frac{W}{g}v_2^2$$

Solving for v_2, we find $v_2 = \sqrt{2gl}$. We note that the speed obtained is that of a body falling freely from a height l.

We note the following advantages of the method of work and energy:

1. In order to find the speed at A_2, there is no need to determine the acceleration in an intermediate position A and to integrate the expression obtained from A_1 to A_2.
2. All quantities involved are scalars and may be added directly, without using x and y components.
3. Forces which do no work are eliminated from the solution of the problem.

What is an advantage in one problem, however, may become a disadvantage in another. It is evident, for instance, that the method of work and energy cannot be used to directly determine an acceleration. We also note that it should be supplemented by the direct application of Newton's second law in order to determine a force which is normal to the path of the particle, since such a force does no work. Suppose, for example, that we wish to determine the tension in the cord of the pendulum of Fig. 13.8a as the bob passes through A_2. We draw a free-body diagram of the bob in that position (Fig. 13.9) and express Newton's second law in terms of tangential and normal components. The equations $\Sigma F_t = ma_t$ and $\Sigma F_n = ma_n$ yield, respectively, $a_t = 0$ and

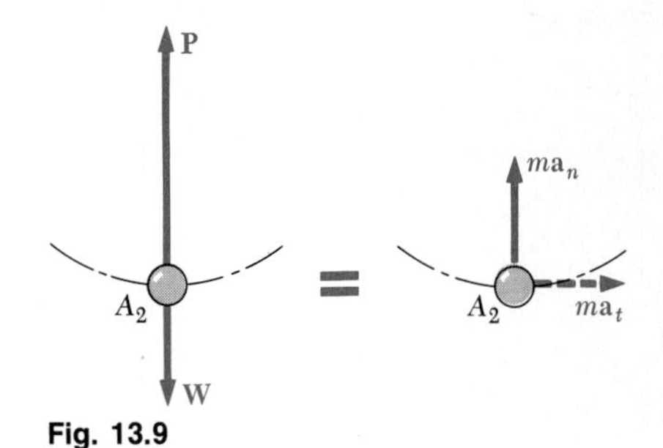

Fig. 13.9

$$P - W = ma_n = \frac{W}{g}\frac{v_2^2}{l}$$

But the speed at A_2 was determined earlier by the method of work and energy. Substituting $v_2^2 = 2gl$ and solving for P, we write

$$P = W + \frac{W}{g}\frac{2gl}{l} = 3W$$

Since friction forces have a direction opposite to that of the displacement of the body on which they act, *the work of friction forces is always negative*. This work represents energy dissipated into heat and always results in a decrease in the kinetic energy of the body involved (see Sample Prob. 13.3).

13.5. Systems of Particles. When a problem involves several particles, each particle may be considered separately and the principle of work and energy may be applied to each particle. Adding the kinetic energies of all the particles, and considering the work of all the forces involved, we may also write the equation of work and energy for the entire system. We have

$$T_1 + U_{1\to2} = T_2 \tag{13.11}$$

where T represents the arithmetic sum of the kinetic energies of the particles forming the system (all terms are positive) and $U_{1\to2}$ the work of all the forces acting on the various particles, whether these forces are *internal* or *external* from the point of view of the system as a whole.

The method of work and energy is particularly useful in solving problems involving a system of bodies connected by *inextensible cords or links*. In this case, the internal forces occur by pairs of equal and opposite forces, and the points of application of the forces in each pair *move through equal distances*. As a result, the work of the internal forces is zero and $U_{1\to2}$ reduces to the work of the *external forces only* (see Sample Prob. 13.2).

SAMPLE PROBLEM 13.1

An automobile weighing 4000 lb is driven down a 5° incline at a speed of 60 mi/h when the brakes are applied, causing a constant total braking force (applied by the road on the tires) of 1500 lb. Determine the distance traveled by the automobile as it comes to a stop.

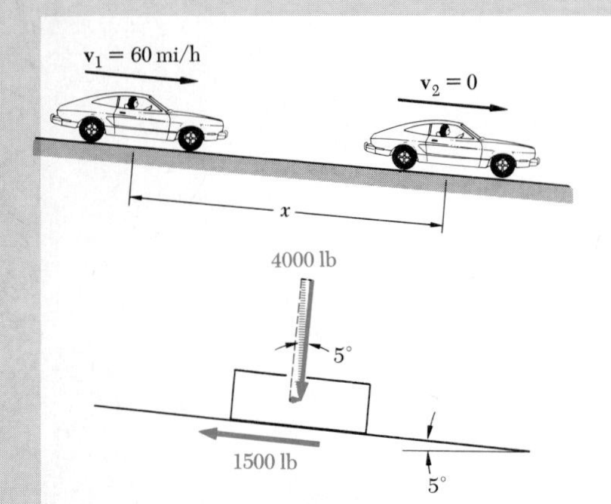

Solution. *Kinetic Energy*

Position 1: $v_1 = \left(60 \frac{\text{mi}}{\text{h}}\right)\left(\frac{5280 \text{ ft}}{1 \text{ mi}}\right)\left(\frac{1 \text{ h}}{3600 \text{ s}}\right) = 88 \text{ ft/s}$

$$T_1 = \tfrac{1}{2}mv_1^2 = \tfrac{1}{2}(4000/32.2)(88)^2 = 481{,}000 \text{ ft}\cdot\text{lb}$$

Position 2: $v_2 = 0 \qquad T_2 = 0$

Work $\quad U_{1\to2} = -1500x + (4000 \sin 5°)x = -1151x$

Principle of Work and Energy

$$T_1 + U_{1\to2} = T_2$$
$$481{,}000 - 1151x = 0 \qquad x = 418 \text{ ft}$$

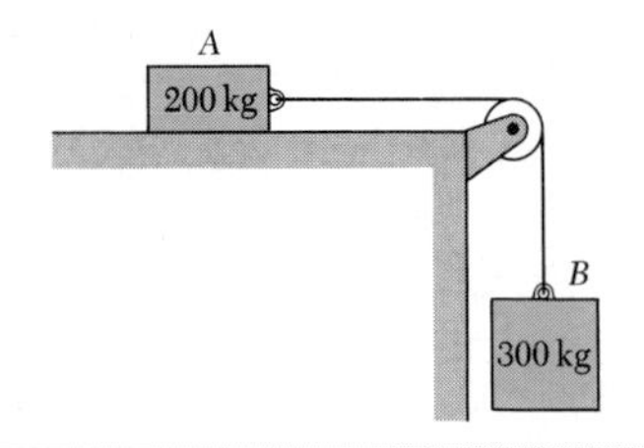

SAMPLE PROBLEM 13.2

Two blocks are joined by an inextensible cable as shown. If the system is released from rest, determine the velocity of block A after it has moved 2 m. Assume that the coefficient of kinetic friction between block A and the plane is $\mu_k = 0.25$ and that the pulley is weightless and frictionless.

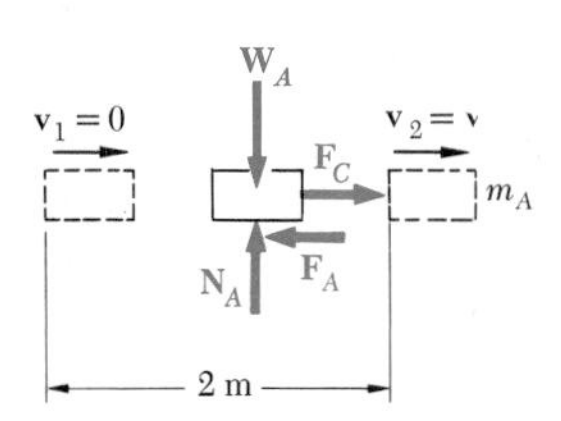

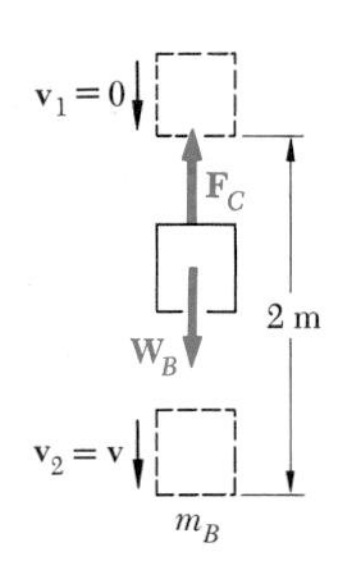

Solution. *Work and Energy for Block A.* We denote by $\mathbf{F}_A$ the friction force, by $\mathbf{F}_C$ the force exerted by the cable, and write

$$m_A = 200 \text{ kg} \qquad W_A = (200 \text{ kg})(9.81 \text{ m/s}^2) = 1962 \text{ N}$$
$$F_A = \mu_k N_A = \mu_k W_A = 0.25(1962 \text{ N}) = 490 \text{ N}$$

$T_1 + U_{1\to2} = T_2$: $\quad 0 + F_C(2 \text{ m}) - F_A(2 \text{ m}) = \tfrac{1}{2}m_A v^2$

$$F_C(2 \text{ m}) - (490 \text{ N})(2 \text{ m}) = \tfrac{1}{2}(200 \text{ kg})v^2 \qquad (1)$$

Work and Energy for Block B. We write

$$m_B = 300 \text{ kg} \qquad W_B = (300 \text{ kg})(9.81 \text{ m/s}^2) = 2940 \text{ N}$$

$T_1 + U_{1\to2} = T_2$: $\quad 0 + W_B(2 \text{ m}) - F_C(2 \text{ m}) = \tfrac{1}{2}m_B v^2$

$$(2940 \text{ N})(2 \text{ m}) - F_C(2 \text{ m}) = \tfrac{1}{2}(300 \text{ kg})v^2 \qquad (2)$$

Adding the left-hand and right-hand members of (1) and (2), we observe that the work of the forces exerted by the cable on A and B cancels out:

$$(2940 \text{ N})(2 \text{ m}) - (490 \text{ N})(2 \text{ m}) = \tfrac{1}{2}(200 \text{ kg} + 300 \text{ kg})v^2$$
$$4900 \text{ J} = \tfrac{1}{2}(500 \text{ kg})v^2 \qquad v = 4.43 \text{ m/s}$$

SAMPLE PROBLEM 13.3

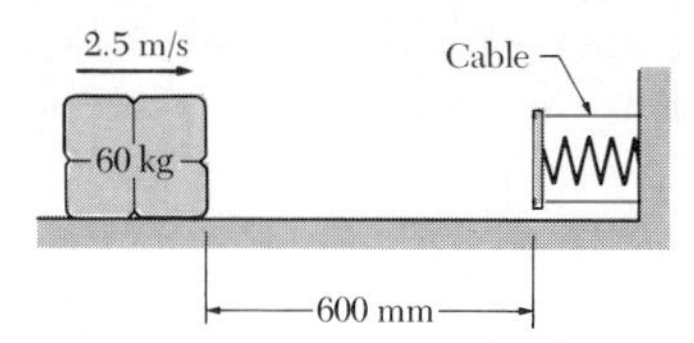

A spring is used to stop a 60-kg package which is sliding on a horizontal surface. The spring has a constant $k = 20$ kN/m and is held by cables so that it is initially compressed 120 mm. Knowing that the package has a velocity of 2.5 m/s in the position shown and that the maximum additional deflection of the spring is 40 mm, determine (*a*) the coefficient of kinetic friction between the package and the surface, (*b*) the velocity of the package as it passes again through the position shown.

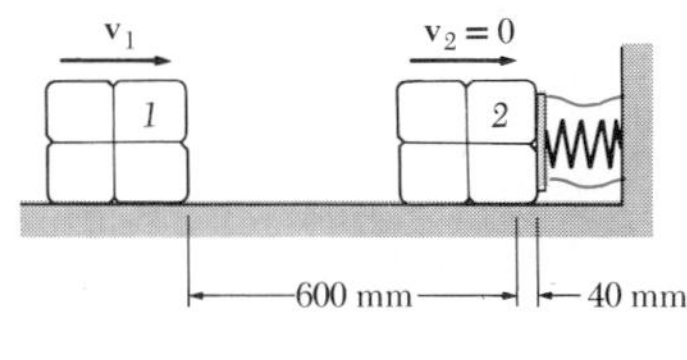

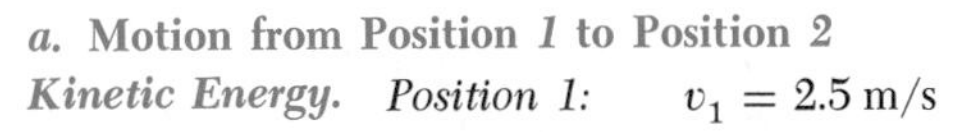

***a.* Motion from Position *1* to Position 2**

Kinetic Energy. *Position 1:* $v_1 = 2.5$ m/s

$$T_1 = \tfrac{1}{2}mv_1^2 = \tfrac{1}{2}(60 \text{ kg})(2.5 \text{ m/s})^2 = 187.5 \text{ N}\cdot\text{m} = 187.5 \text{ J}$$

Position 2 (maximum spring deflection): $v_2 = 0 \qquad T_2 = 0$

Work

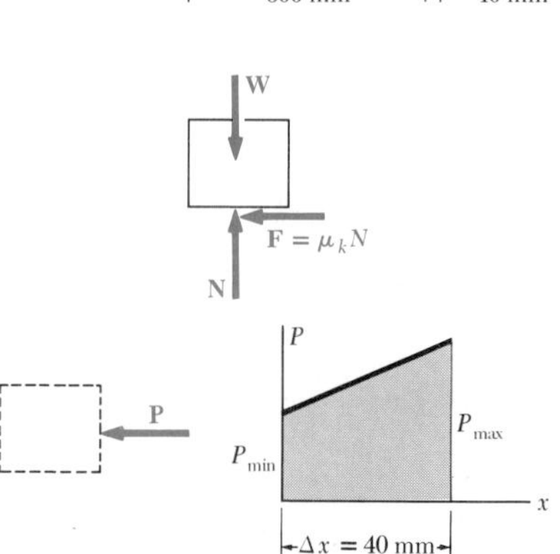

Friction Force **F**. We have

$$F = \mu_k N = \mu_k W = \mu_k mg = \mu_k(60 \text{ kg})(9.81 \text{ m/s}^2) = 588.6\,\mu_k$$

The work of **F** is negative and equal to

$$(U_{1\to2})_f = -Fx = -(588.6\,\mu_k)(0.600 \text{ m} + 0.040 \text{ m}) = -377\,\mu_k$$

Spring Force **P**. The variable force **P** exerted by the spring does an amount of negative work equal to the area under the force-deflection curve of the spring force. We have

$$P_{min} = kx_0 = (20 \text{ kN/m})(120 \text{ m}) = (20\,000 \text{ N/m})(0.120 \text{ m}) = 2400 \text{ N}$$
$$P_{max} = P_{min} + k\,\Delta x = 2400 \text{ N} + (20 \text{ kN/m})(40 \text{ mm}) = 3200 \text{ N}$$
$$(U_{1\to2})_e = -\tfrac{1}{2}(P_{min} + P_{max})\,\Delta x = -\tfrac{1}{2}(2400 \text{ N} + 3200 \text{ N})(0.040 \text{ m}) = -112.0 \text{ J}$$

The total work is thus

$$U_{1\to2} = (U_{1\to2})_f + (U_{1\to2})_e = -377\,\mu_k - 112.0 \text{ J}$$

Principle of Work and Energy

$T_1 + U_{1\to2} = T_2$: $\quad 187.5 \text{ J} - 377\,\mu_k - 112.0 \text{ J} = 0 \qquad \mu_k = 0.20$ ◄

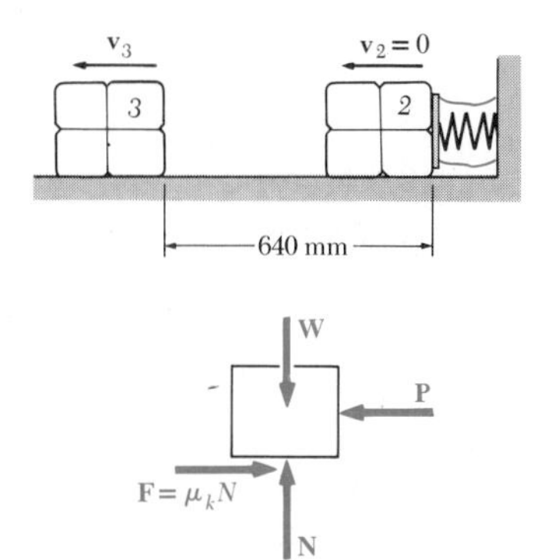

***b.* Motion from Position 2 to Position 3**

Kinetic Energy. *Position 2:* $v_2 = 0 \qquad T_2 = 0$

Position 3: $\quad T_3 = \tfrac{1}{2}mv_3^2 = \tfrac{1}{2}(60 \text{ kg})v_3^2$

Work. Since the distances involved are the same, the numerical values of the work of the friction force **F** and of the spring force **P** are the same as above. However, while the work of **F** is still negative, the work of **P** is now positive.

$$U_{2\to3} = -377\mu_k + 112.0 \text{ J} = -75.5 \text{ J} + 112.0 \text{ J} = +36.5 \text{ J}$$

Principle of Work and Energy

$T_2 + U_{2\to3} = T_3$: $\quad 0 + 36.5 \text{ J} = \tfrac{1}{2}(60 \text{ kg})v_3^2$

$v_3 = 1.103$ m/s $\qquad \mathbf{v}_3 = 1.103$ m/s ← ◄

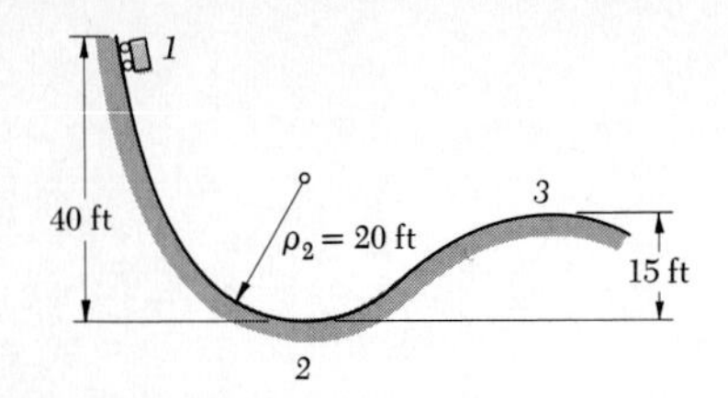

SAMPLE PROBLEM 13.4

A 2000-lb car starts from rest at point *1* and moves without friction down the track shown. (*a*) Determine the force exerted by the track on the car at point 2, where the radius of curvature of the track is 20 ft. (*b*) Determine the minimum safe value of the radius of curvature at point 3.

***a.* Force Exerted by the Track at Point 2.** The principle of work and energy is used to determine the velocity of the car as it passes through point 2.

Kinetic Energy: $\quad T_1 = 0 \qquad T_2 = \frac{1}{2}mv_2^2 = \frac{1}{2}\frac{W}{g}v_2^2$

Work. The only force which does work is the weight **W**. Since the vertical displacement from point *1* to point *2* is 40 ft downward, the work of the weight is

$$U_{1\to2} = +W(40\text{ ft})$$

Principle of Work and Energy

$$T_1 + U_{1\to2} = T_2 \qquad 0 + W(40\text{ ft}) = \frac{1}{2}\frac{W}{g}v_2^2$$

$$v_2^2 = 80g = 80(32.2) \qquad v_2 = 50.8\text{ ft/s}$$

Newton's Second Law at Point 2. The acceleration $\mathbf{a}_n$ of the car at point 2 has a magnitude $a_n = v_2^2/\rho$ and is directed upward. Since the external forces acting on the car are **W** and **N**, we write

$$+\uparrow\Sigma F_n = ma_n: \qquad -W + N = ma_n = \frac{W}{g}\frac{v_2^2}{\rho} = \frac{W}{g}\frac{80g}{20}$$

W   = ma_n N

$$N = 5W \qquad \mathbf{N} = 10{,}000\text{ lb}\uparrow$$ ◄

b.* Minimum Value of ρ at Point 3.** ***Principle of Work and Energy. Applying the principle of work and energy between point *1* and point *3*, we obtain

$$T_1 + U_{1\to3} = T_3 \qquad 0 + W(25\text{ ft}) = \frac{1}{2}\frac{W}{g}v_3^2$$

$$v_3^2 = 50g = 50(32.2) \qquad v_3 = 40.1\text{ ft/s}$$

Newton's Second Law at Point 3. The minimum safe value of ρ occurs when $\mathbf{N} = 0$. In this case, the acceleration $\mathbf{a}_n$, of magnitude $a_n = v_3^2/\rho$, is directed downward, and we write

$$+\downarrow\Sigma F_n = ma_n: \qquad W = \frac{W}{g}\frac{v_3^2}{\rho} = \frac{W}{g}\frac{50g}{\rho}$$

W   = ma_n N = 0

$$\rho = 50\text{ ft}$$ ◄

Problems

13.1 A 400-kg satellite was placed in a circular orbit 1500 km above the surface of the earth. At this elevation the acceleration of gravity is 6.43 m/s^2. Determine the kinetic energy of the satellite, knowing that its orbital speed is 25.6×10^3 km/h.

13.2 A stone which weighs 4 lb is dropped from a height h and strikes the ground with a velocity of 60 ft/s. (*a*) Find the kinetic energy of the stone as it strikes the ground and the height h from which it was dropped. (*b*) Solve part *a*, assuming that the same stone is dropped on the moon. (Acceleration of gravity on the moon = 5.31 ft/s^2.)

13.3 Determine the maximum theoretical speed that may be achieved over a distance of 50 m by a car starting from rest, knowing that the coefficient of static friction is 0.80 between the tires and the pavement and that 60 percent of the weight of the car is distributed over its front wheels and 40 percent over its rear wheels. Assume (*a*) front-wheel drive, (*b*) rear-wheel drive.

13.4 Solve Prob. 13.3, assuming four-wheel drive.

13.5 In an iron-ore mixing operation, a bucket full of ore is suspended from a traveling crane which is moving with a speed $v = 3.75$ m/s along a stationary bridge. If the crane suddenly stops, determine the additional horizontal distance through which the bucket will move.

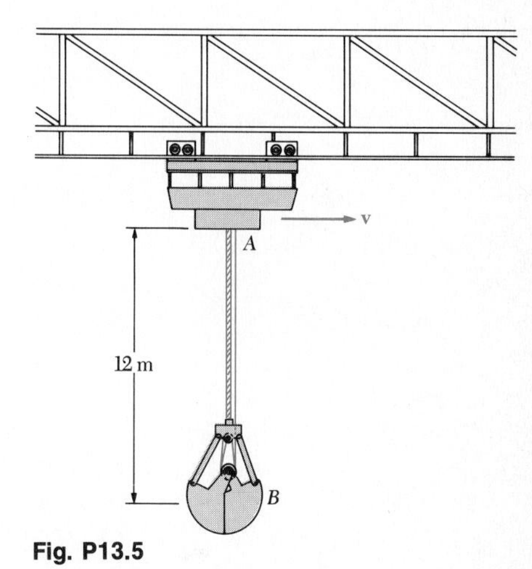

Fig. P13.5

13.6 Determine the speed v with which the crane of Prob. 13.5 was moving if the cables supporting the bucket B swing through an angle of 18° after the crane is brought to a sudden stop.

13.7 A 24-lb package is placed with no initial velocity at the top of a chute. Knowing that the coefficient of kinetic friction between the package and the chute is 0.25, determine (*a*) how far the package will slide on the horizontal portion of the chute, (*b*) the maximum velocity reached by the package, (*c*) the amount of energy dissipated due to friction between A and B.

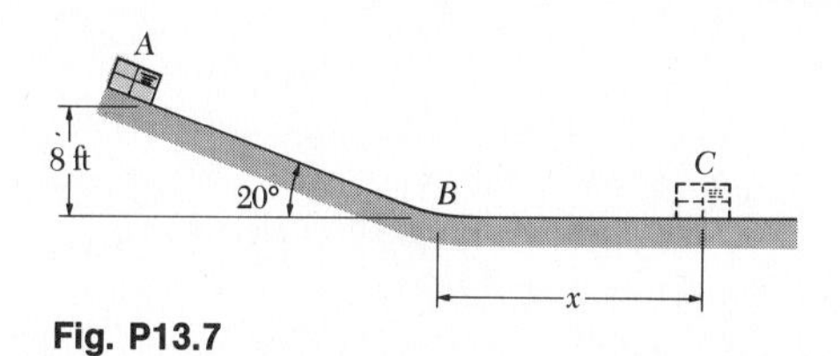

Fig. P13.7

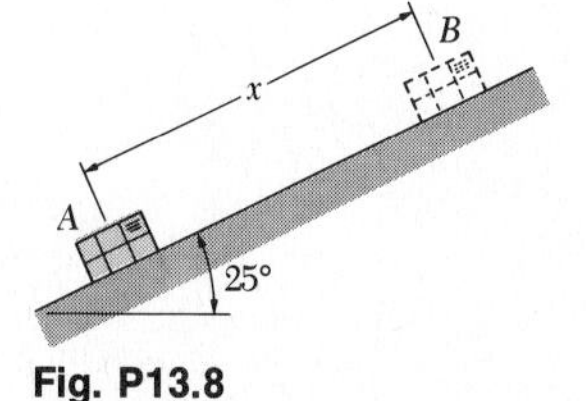

Fig. P13.8

13.8 A 20-lb package is projected up a 25° incline with an initial velocity of 24 ft/s. Knowing that the coefficient of kinetic friction between the package and the incline is 0.20, determine (*a*) the maximum distance x that the package will move up the incline, (*b*) the velocity of the package as it returns to its original position, (*c*) the total amount of energy dissipated due to friction.

13.9 A 1050-kg trailer is hitched to a 1200-kg car. The car and trailer are traveling at 90 km/h when the driver applies the brakes on both the car and the trailer. Knowing that the braking forces exerted on the car and the trailer are 4500 N and 3600 N, respectively, determine (*a*) the distance traveled by the car and trailer before they come to a stop, (*b*) the horizontal component of the force exerted by the trailer hitch on the car.

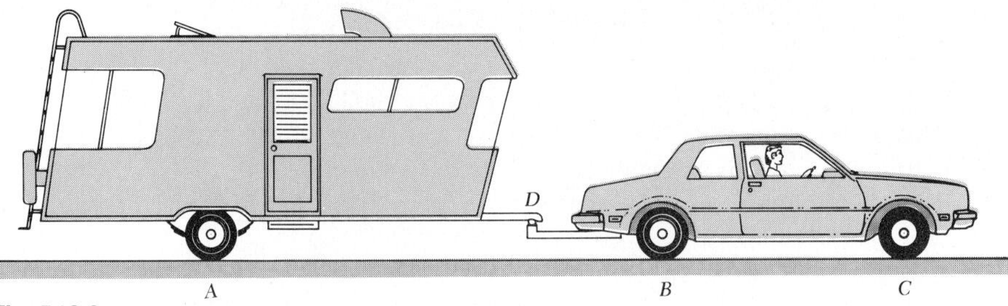

Fig. P13.9

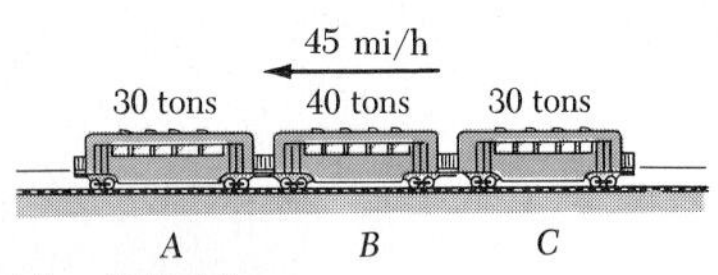

Fig. P13.10

13.10 The subway train shown is traveling at a speed of 45 mi/h when the brakes are fully applied on the wheels of cars *A* and *B*, causing them to slide on the track, but are not applied on the wheels of car *C*. Knowing that the coefficient of kinetic friction is 0.30 between the wheels and the track, determine (*a*) the distance required to bring the train to a stop, (*b*) the force in each coupling.

13.11 Solve Prob. 13.10, assuming that the brakes are applied only on the wheels of car *A*.

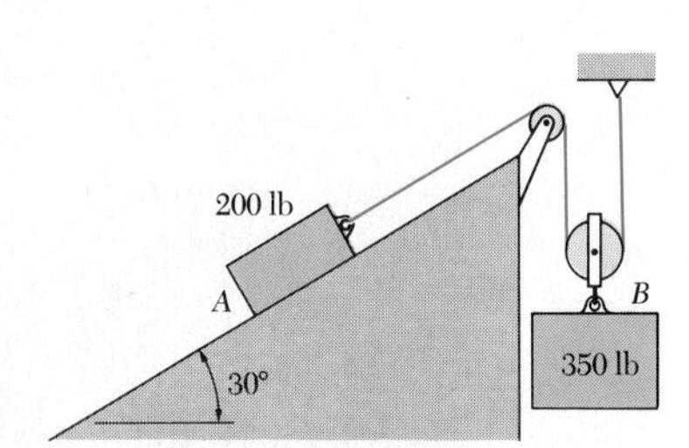

Fig. P13.13 and P13.15

13.12 Solve Prob. 13.9, assuming that the trailer brakes are inoperative.

13.13 The two blocks shown are originally at rest. Neglecting the masses of the pulleys and the effect of friction in the pulleys and between block *A* and the incline, determine (*a*) the velocity of block *A* after it has moved through 5 ft, (*b*) the tension in the cable.

13.14 Solve Prob. 12.18*c*, using the method of work and energy.

13.15 The two blocks shown start from rest and it is observed that the velocity of block *A* is 6.75 ft/s after it has moved through 5 ft. Neglecting the masses of the pulleys and the friction in the pulleys, determine (*a*) the amount of energy dissipated due to friction between block *A* and the incline, (*b*) the coefficient of kinetic friction between block *A* and the incline.

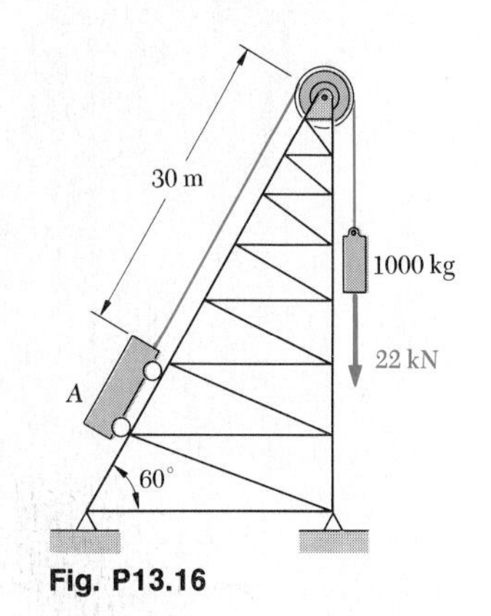

Fig. P13.16

13.16 The total mass of loading car *A* and its load is 3500 kg. The car is connected to a 1000-kg counterweight and is at rest when a constant 22-kN force is applied as shown. (*a*) If the force acts through the entire motion, what is the speed of the car after it has traveled 30 m? (*b*) If after the car has moved through a distance x the 22-kN force is removed, the car will coast to rest. After what distance x should the force be removed if the car is to come to rest after a total movement of 30 m?

13.17 Knowing that the system shown starts from rest, determine (*a*) the velocity of collar *A* after it has moved through 320 mm, (*b*) the corresponding velocity of collar *B*, (*c*) the tension in the cable. Neglect the masses of the pulleys and the effect of friction.

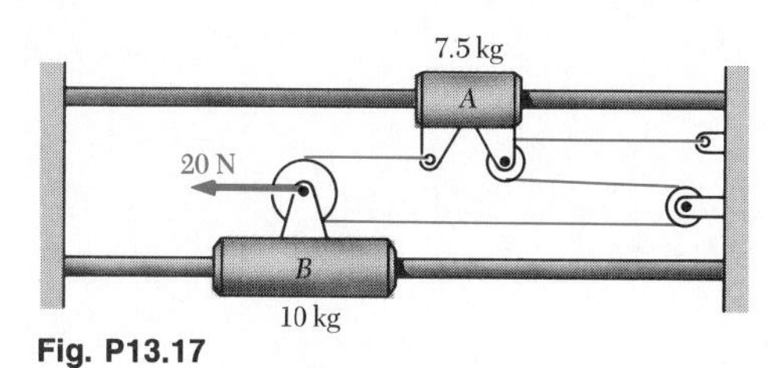

Fig. P13.17

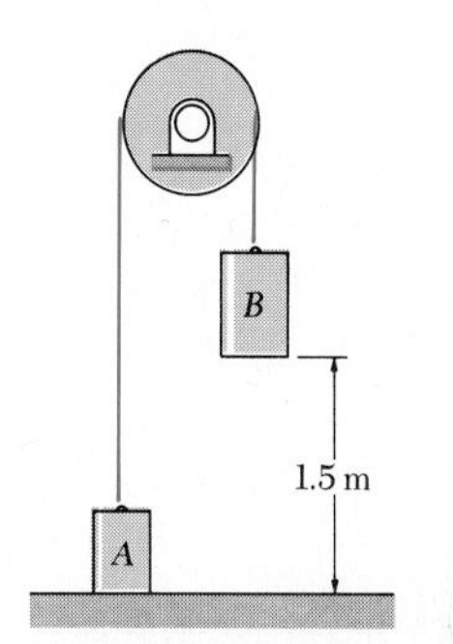

Fig. P13.18 and P13.19

13.18 Two blocks *A* and *B*, of mass 8 kg and 12 kg, respectively, hang from a cable which passes over a pulley of negligible mass. Knowing that the blocks are released from rest and that the energy dissipated by axle friction in the pulley is 10 J, determine (*a*) the velocity of block *B* as it strikes the ground, (*b*) the force exerted by the cable on each of the two blocks during the motion.

13.19 Two blocks *A* and *B*, of mass 12 kg and 15 kg, respectively, hang from a cable which passes over a pulley of negligible mass. The blocks are released from rest in the positions shown and block *B* is observed to strike the ground with a velocity of 1.6 m/s. Determine (*a*) the energy dissipated due to axle friction in the pulley, (*b*) the force exerted by the cable on each of the two blocks during the motion.

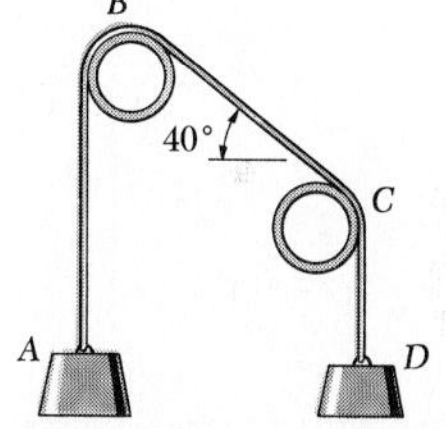

Fig. P13.20 and P13.21

13.20 Two blocks *A* and *D*, weighing, respectively, 125 lb and 300 lb, are attached to a rope which passes over two fixed pipes *B* and *C* as shown. It is observed that when the system is released from rest, block *A* acquires a velocity of 8 ft/s after moving 5 ft up. Determine (*a*) the force exerted by the rope on each of the two blocks during the motion, (*b*) the coefficient of kinetic friction between the rope and the pipes, (*c*) the energy dissipated due to friction.

13.21 Two blocks *A* and *D* are attached to a rope which passes over two fixed pipes *B* and *C* as shown. The coefficients of friction between the rope and the pipes are $\mu_s = 0.25$ and $\mu_k = 0.20$. Knowing that the masses of blocks *A* and *D* are, respectively, 50 kg and 125 kg and that the system is released from rest, determine (*a*) the velocity of *A* after it has moved 1.2 m up, (*b*) the force exerted by the rope on each of the two blocks during the motion, (*c*) the energy dissipated due to friction.

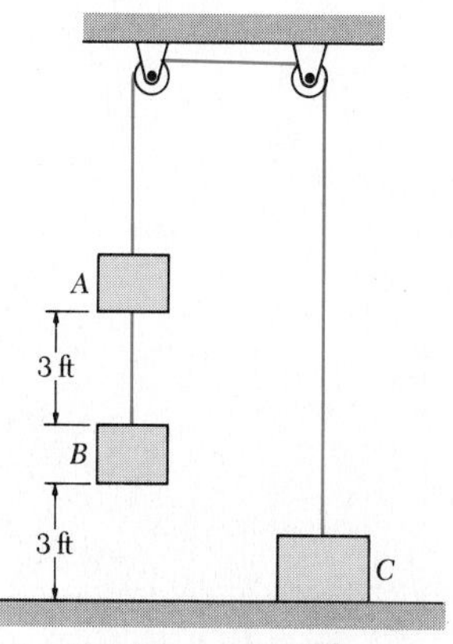

Fig. P13.22 and P13.23

13.22 Blocks *A* and *B* weigh 10 lb each, and block *C* weighs 12 lb. Knowing that the blocks are released from rest in the positions shown and neglecting friction, determine (*a*) the velocity of block *B* just before it strikes the ground, (*b*) the velocity of block *A* just before it strikes block *B*.

13.23 Blocks *A* and *B* weigh 10 lb each. Neglecting friction, determine the weight of block *C* so that when released from rest in the position shown, the system will come to rest again with block *A* just touching block *B*.

13.24 Four packages, each having a mass of 20 kg, are placed as shown on a conveyor belt which is disengaged from its drive motor. Package *1* is just to the right of the horizontal portion of the belt. If the system is released from rest, determine the velocities of packages *1* and *2* as they fall off the belt at point *A*. Assume that the mass of the belt and rollers is small compared with the mass of the packages.

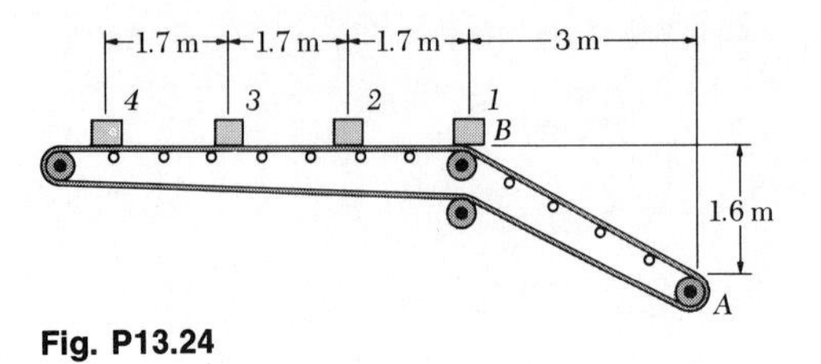

Fig. P13.24

13.25 Knowing that package *2* in Prob. 13.24 has a velocity of 4.72 m/s as it falls off the belt, determine the velocities of packages *3* and *4* as they fall off the belt at point *A*.

13.26 Two blocks *A* and *B*, of mass 5 kg and 6 kg, respectively, are connected by a cord which passes over pulleys as shown. A collar *C*, of mass 4 kg, is placed on block *A* and the system is released from rest. After the blocks have moved through 0.9 m, collar *C* is removed and the blocks continue to move. Determine the velocity of block *A* just before it strikes the ground.

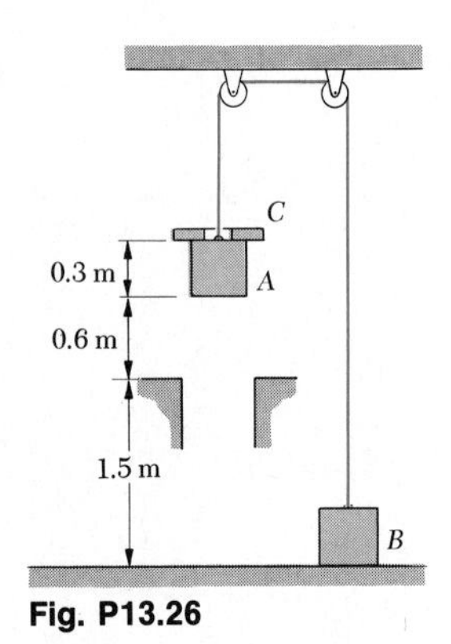

Fig. P13.26

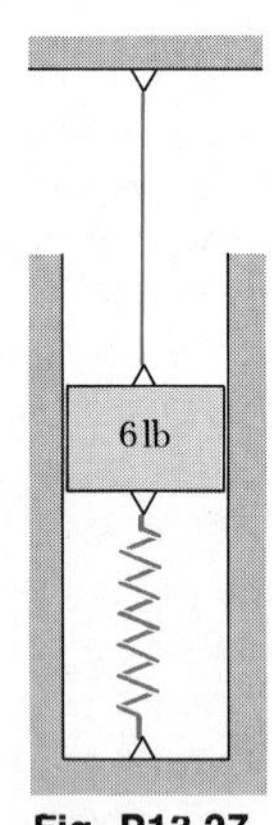

Fig. P13.27

13.27 A 6-lb block is attached to a cable and to a spring as shown. The constant of the spring is $k = 5$ lb/in. and the tension in the cable is 4 lb. If the cable is cut, determine (*a*) the maximum displacement of the block, (*b*) the maximum velocity of the block.

13.28 Solve Prob. 13.27, assuming that the initial tension in the cable is 8 lb.

13.29 An 8-kg plunger is released from rest in the position shown and is stopped by two nested springs; the constant of the outer spring is $k_1 = 4$ kN/m and the constant of the inner spring is $k_2 = 12$ kN/m. If the maximum deflection of the outer spring is observed to be 125 mm, determine the height h from which the plunger was released.

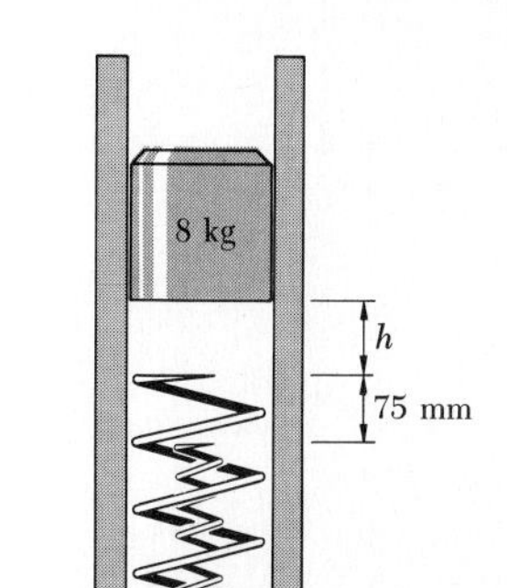

Fig. P13.29 and P13.30

13.30 An 8-kg plunger is released from rest in the position shown and is stopped by two nested springs; the constant of the outer spring is $k_1 = 4$ kN/m and the constant of the inner spring is $k_2 = 12$ kN/m. If the plunger is released from the height $h = 600$ mm, determine the maximum deflection of the outer spring.

13.31 It is shown in mechanics of materials that when an elastic beam AB supports a block of weight W at a given point D, the deflection y_{st} of point D (called the static deflection) is proportional to W. Show that if the same block is dropped from a height h onto the beam and hits it at D, the maximum deflection y_m of point D in the ensuing motion may be expressed as

$$y_m = y_{st}\left(1 + \sqrt{1 + \frac{2h}{y_{st}}}\right)$$

Note that this formula is approximate, since it is based on the assumption that the block does not bounce off the beam and that no energy is dissipated in the impact.

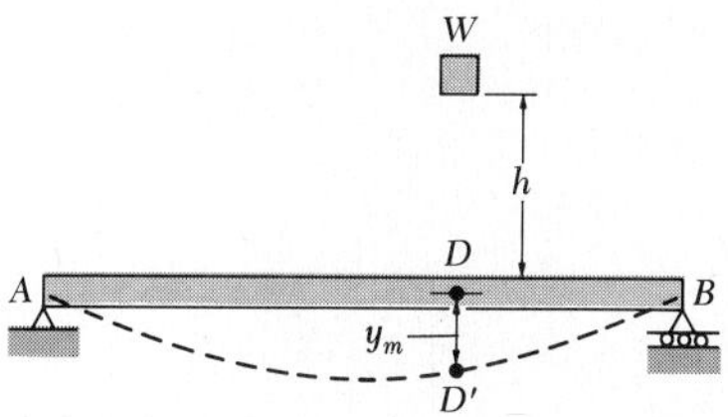

Fig. P13.31

13.32 Two types of energy-absorbing fenders designed to be used on a pier are statically loaded. The force-deflection curve for each type of fender is given in the graph. Determine the maximum deflection of each fender when a 100-ton ship moving at 1 mi/h strikes the fender and is brought to rest.

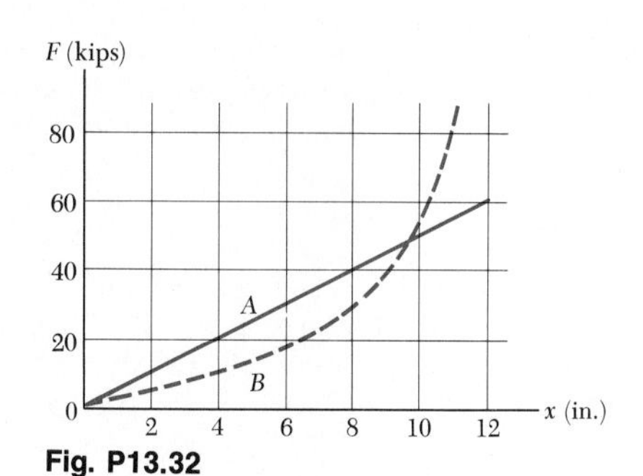

Fig. P13.32

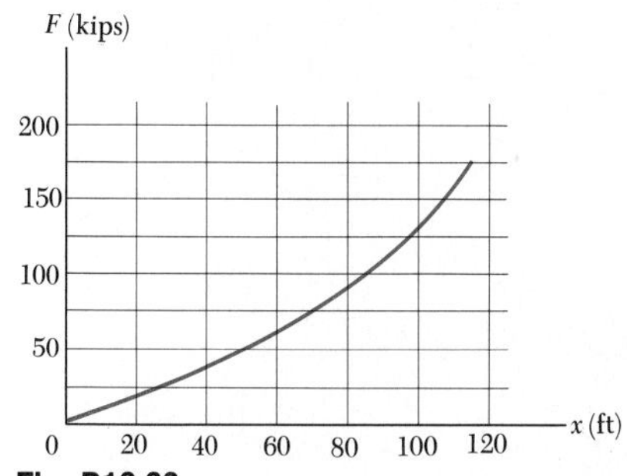

Fig. P13.33

13.33 A 12,000-lb airplane lands on an aircraft carrier and is caught by an arresting cable which is characterized by the force-deflection diagram shown. Knowing that the landing speed of the plane is 100 mi/h, determine (*a*) the distance required for the plane to come to rest, (*b*) the maximum rate of deceleration of the plane.

13.34 Nonlinear springs are classified as hard or soft, depending upon the curvature of their force-deflection curves (see figure). If a delicate instrument having a mass of 5 kg is placed on a spring of length l so that its base is just touching the undeformed spring and then inadvertently released from that position, determine the maximum deflection x_m of the spring and the maximum force F_m exerted by the spring, assuming (*a*) a linear spring of constant $k = 3$ kN/m, (*b*) a hard, nonlinear spring, for which $F = (3 \text{ kN/m})x(1 + 160x^2)$.

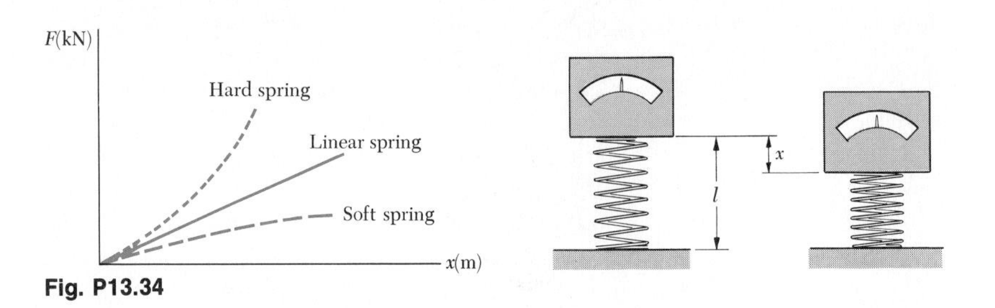

Fig. P13.34

13.35 Solve Prob. 13.34, assuming that in part *b*, the hard spring has been replaced by a soft, nonlinear spring, for which $F = (3 \text{ kN/m})x(1 - 160x^2)$.

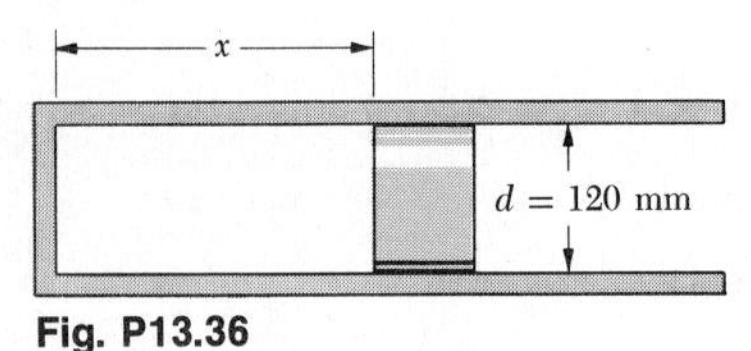

Fig. P13.36

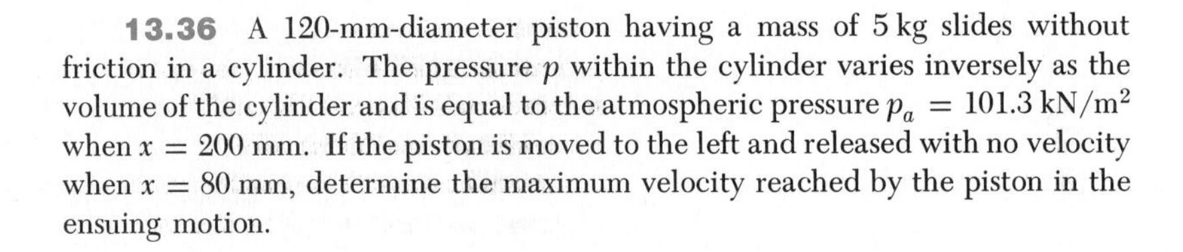

13.36 A 120-mm-diameter piston having a mass of 5 kg slides without friction in a cylinder. The pressure p within the cylinder varies inversely as the volume of the cylinder and is equal to the atmospheric pressure $p_a = 101.3 \text{ kN/m}^2$ when $x = 200$ mm. If the piston is moved to the left and released with no velocity when $x = 80$ mm, determine the maximum velocity reached by the piston in the ensuing motion.

13.37 In Prob. 13.36, determine the maximum value of the coordinate x after the piston has been released with no velocity in the position $x = 80$ mm.

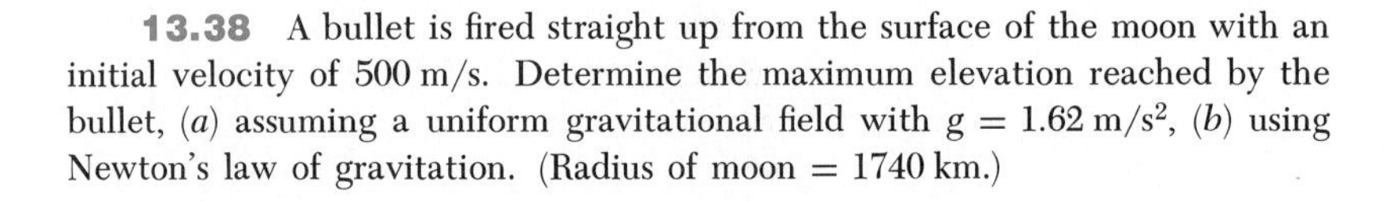

13.38 A bullet is fired straight up from the surface of the moon with an initial velocity of 500 m/s. Determine the maximum elevation reached by the bullet, (*a*) assuming a uniform gravitational field with $g = 1.62 \text{ m/s}^2$, (*b*) using Newton's law of gravitation. (Radius of moon = 1740 km.)

13.39 Determine the maximum velocity with which an object dropped from a very large distance will hit the surface (*a*) of the earth, (*b*) of the moon. Neglect the effect of the earth's atmosphere. (The radius of the moon is 1740 km and its mass is 0.01230 times the mass of the earth.)

13.40 A rocket is fired vertically from the ground. Knowing that at burnout the rocket is 50 mi above the ground and has a velocity of 15,000 ft/s, determine the highest altitude it will reach.

13.41 A rocket is fired vertically from the ground. What should be its velocity $\mathbf{v}_B$ at burnout, 50 mi above the ground, if it is to reach an altitude of 800 mi?

13.42 Sphere C and block A are both moving to the left with a velocity $\mathbf{v}_0$ when the block is suddenly stopped by the wall. Determine the smallest velocity $\mathbf{v}_0$ for which the sphere C will swing in a full circle about the pivot B (*a*) if BC is a slender rod of negligible mass, (*b*) if BC is a cord.

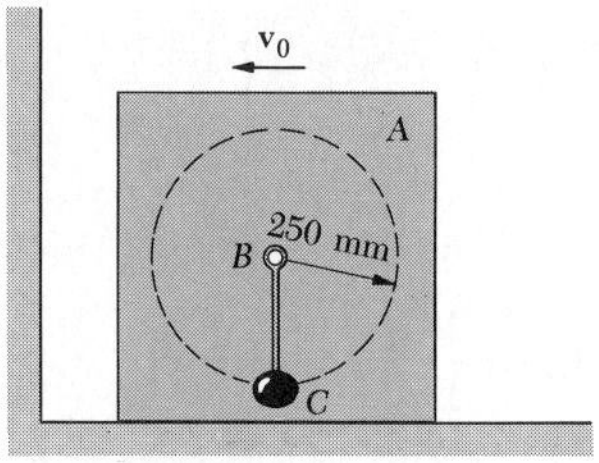

Fig. P13.42

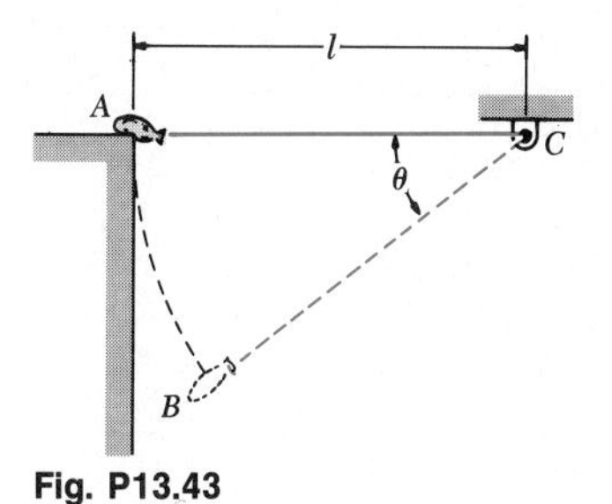

Fig. P13.43

13.43 A bag is gently pushed off the top of a wall at A and swings in a vertical plane at the end of a rope of length l. Determine the angle θ for which the rope will break, knowing that it can withstand a maximum tension 50 percent larger than the weight of the bag.

13.44 A roller coaster starts from rest at A, rolls down the track to B, describes a circular loop of 40-ft diameter, and moves up and down past point E. Knowing that $h = 60$ ft and assuming no energy loss due to friction, determine (*a*) the force exerted by his seat on a 160-lb rider at B and D, (*b*) the minimum value of the radius of curvature at E if the roller coaster is not to leave the track at that point.

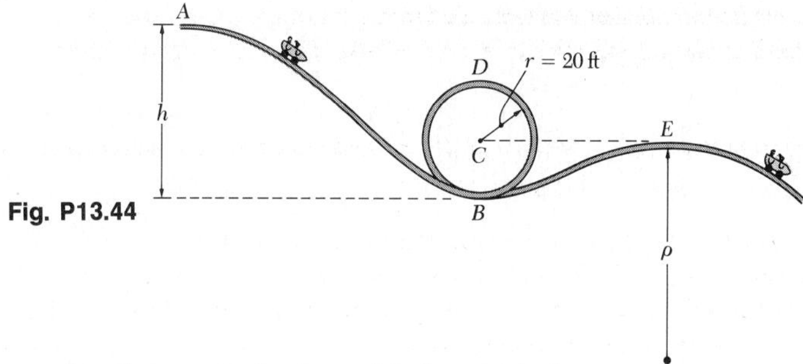

Fig. P13.44

13.45 In Prob. 13.44, determine the range of values of h for which the roller coaster will not leave the track at D or E, knowing that the radius of curvature at E is $\rho = 75$ ft. Assume no energy loss due to friction.

13.46 A small block slides with velocity $\mathbf{v}_0$ on the horizontal surface AB. Neglecting friction and knowing that $v_0 = 0.5\sqrt{gr}$, express in terms of r (*a*) the elevation h of point C where the block will leave the cylindrical surface BD, (*b*) the distance d from D to point E where it will hit the ground.

13.47 For the block of Prob. 13.46, and knowing that $r = 600$ mm, determine (*a*) the smallest value of v_0 for which the block leaves the surface ABD at point B and the corresponding value of d, (*b*) the smallest possible values of h and d (when v_0 approaches zero).

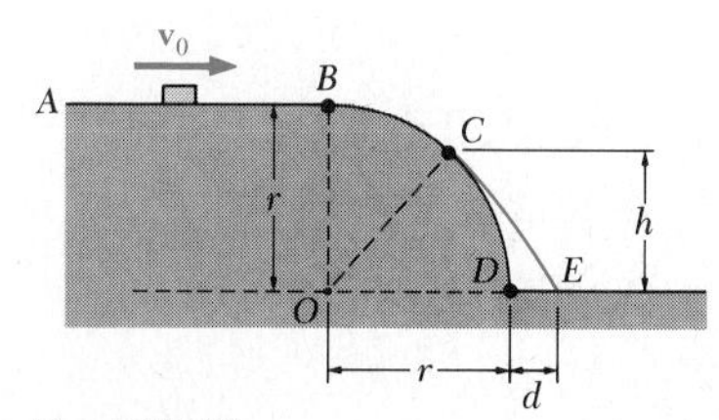

Fig. P13.46

13.6. Potential Energy.† Let us consider again a body of weight **W** which moves along a curved path from a point A_1 of elevation y_1 to a point A_2 of elevation y_2 (Fig. 13.4). We recall from Sec. 13.2 that the work of the force of gravity **W** during this displacement is

$$U_{1\to 2} = Wy_1 - Wy_2 \qquad (13.4)$$

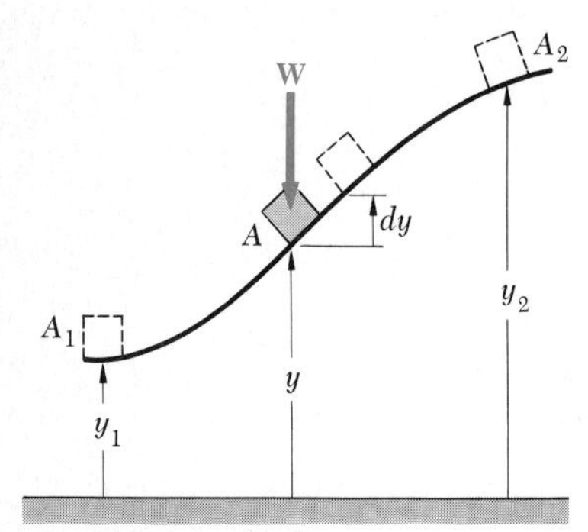

Fig. 13.4 (*repeated*)

The work of **W** may thus be obtained by subtracting the value of the function Wy corresponding to the second position of the body from its value corresponding to the first position. The work of **W** is independent of the actual path followed; it depends only upon the initial and final values of the function Wy. This function is called the *potential energy* of the body with respect to the *force of gravity* **W** and is denoted by V_g. We write

$$U_{1\to 2} = (V_g)_1 - (V_g)_2 \qquad \text{with } V_g = Wy \qquad (13.12)$$

We note that if $(V_g)_2 > (V_g)_1$, that is, *if the potential energy increases* during the displacement (as in the case considered here), *the work* $U_{1\to 2}$ *is negative.* If, on the other hand, the work of **W** is positive, the potential energy decreases. Therefore, the potential energy V_g of the body provides a measure of the work which may be done by its weight **W**. Since only the *change* in potential energy, and not the actual value of V_g, is involved in formula (13.12), an arbitrary constant may be added to the expression obtained for V_g. In other words, the level, or datum, from which the elevation y is measured may be chosen arbitrarily. Note that potential energy is expressed in the same units as work, i.e., in joules if SI units are used, and in ft·lb or in·lb if U.S. customary units are used.

It should be noted that the expression just obtained for the potential energy of a body with respect to gravity is valid only as long as the weight **W** of the body may be assumed to remain constant, i.e., as long as the displacements of the body are small compared with the radius of the earth. In the case of a space vehicle, however, we should take into consideration the variation of the force of gravity with the distance r from the center of the earth. Using the expression obtained in Sec. 13.2 for the work of a gravitational force, we write (Fig. 13.6)

$$U_{1\to 2} = \frac{GMm}{r_2} - \frac{GMm}{r_1} \qquad (13.7)$$

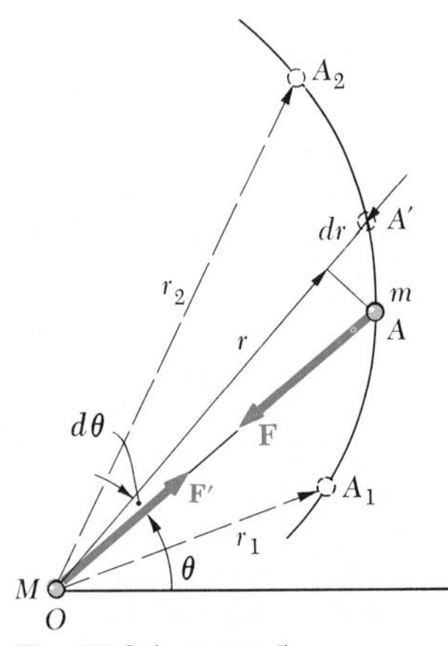

Fig. 13.6 (*repeated*)

The work of the force of gravity may therefore be obtained by subtracting the value of the function $-GMm/r$ corresponding to the second position of the body from its value corresponding to the first position. Thus, the expression which should be used for the potential energy V_g when the variation in the force of gravity cannot be neglected is

$$V_g = -\frac{GMm}{r} \qquad (13.13)$$

†Some of the material in this section has already been considered in Sec. 10.7.

Taking the first of the relations (12.20) into account, we write V_g in the alternative form

$$V_g = -\frac{WR^2}{r} \tag{13.13'}$$

where R is the radius of the earth and W the value of the weight of the body at the surface of the earth. When either of the relations (13.13) or (13.13′) is used to express V_g, the distance r should, of course, be measured from the center of the earth.† Note that V_g is always negative and that it approaches zero for very large values of r.

Consider, now, a body attached to a spring and moving from a position A_1, corresponding to a deflection x_1 of the spring, to a position A_2, corresponding to a deflection x_2 (Fig. 13.5). We recall from Sec. 13.2 that the work of the force **F** exerted by the spring on the body is

$$U_{1\rightarrow 2} = \tfrac{1}{2}kx_1^2 - \tfrac{1}{2}kx_2^2 \tag{13.6}$$

The work of the elastic force is thus obtained by subtracting the value of the function $\frac{1}{2}kx^2$ corresponding to the second position of the body from its value corresponding to the first position. This function is denoted by V_e and is called the *potential energy* of the body with respect to the *elastic force* **F**. We write

$$U_{1\rightarrow 2} = (V_e)_1 - (V_e)_2 \qquad \text{with } V_e = \tfrac{1}{2}kx^2 \tag{13.14}$$

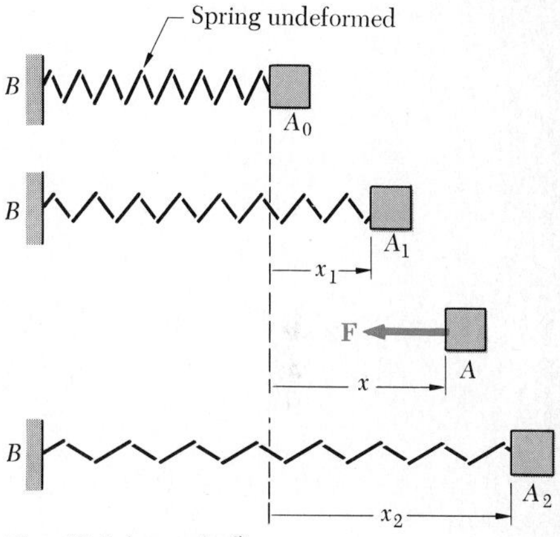

Fig. 13.5 ***(repeated)***

and observe that during the displacement considered, the work of the force **F** exerted by the spring on the body is negative and the potential energy V_e increases. We should note that the expression obtained for V_e is valid only if the deflection of the spring is measured from its undeformed position. On the other hand, formula (13.14) may be used even when the spring is rotated about its fixed end (Fig. 13.10*a*). The work of the elastic force depends only upon the initial and final deflections of the spring (Fig. 13.10*b*).

† The expressions given for V_g in (13.13) and (13.13′) are valid only when $r \geqq R$, that is, when the body considered is above the surface of the earth.

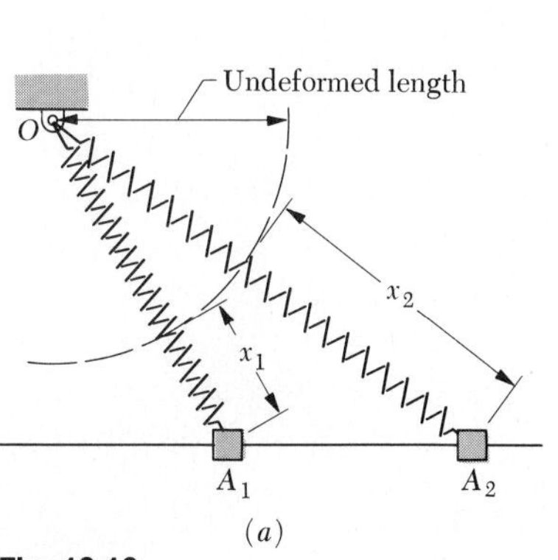

F
$F = kx$
$-U_{1\rightarrow 2}$
$(V_e)_1 = \frac{1}{2}kx_1^2$
$(V_e)_2 = \frac{1}{2}kx_2^2$
x
x_1
x_2
(b)

Fig. 13.10

The concept of potential energy may be used when forces other than gravity forces and elastic forces are involved. It remains valid as long as the elementary work dU of the force considered is an *exact differential*. It is then possible to find a function V, called potential energy, such that

$$dU = -dV$$

Integrating over a finite displacement, we obtain the general relationship

$$U_{1\rightarrow 2} = V_1 - V_2 \tag{13.15}$$

which expresses that *the work of the force is independent of the path followed and is equal to minus the change in potential energy.* A force which satisfies Eq. (13.15) is said to be a *conservative force.*

13.7. Conservation of Energy. We saw in the preceding section that the work of a conservative force, such as the weight of a particle or the force exerted by a spring, may be expressed as a change in potential energy. When a particle, or a system of particles, moves under the action of conservative forces, the principle of work and energy stated in Sec. 13.3 may be expressed in a modified form. Substituting for $U_{1\rightarrow 2}$ from (13.15) into (13.10), we write

$$V_1 - V_2 = T_2 - T_1$$

$$T_1 + V_1 = T_2 + V_2 \tag{13.16}$$

Formula (13.16) indicates that, when a system of particles moves under the action of conservative forces, *the sum of the kinetic energy and of the potential energy of the system remains constant.*† The sum $T + V$ is called the *total mechanical energy* of the system and is denoted by E.

Consider, for example, the pendulum analyzed in Sec. 13.4, which is released with no velocity from A_1 and allowed to swing in a vertical plane (Fig. 13.11). Measuring the potential energy from the level of A_2, we have, at A_1,

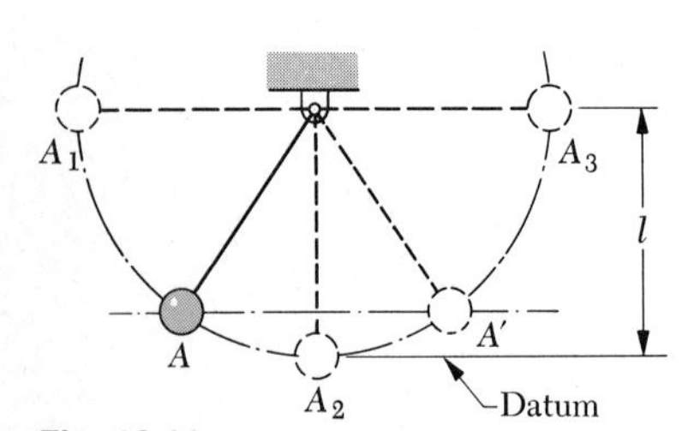

Fig. 13.11

$$T_1 = 0 \qquad V_1 = Wl \qquad T_1 + V_1 = Wl$$

Recalling that at A_2, the speed of the pendulum is $v_2 = \sqrt{2gl}$, we have

$$T_2 = \tfrac{1}{2}mv_2^2 = \frac{1}{2}\frac{W}{g}(2gl) = Wl \qquad V_2 = 0$$
$$T_2 + V_2 = Wl$$

† When the particles of a system move with respect to each other under the action of internal forces, the potential energy of the system must include the potential energy corresponding to the internal forces.

We thus check that the total mechanical energy $E = T + V$ of the pendulum is the same at A_1 and A_2. While the energy is entirely potential at A_1, it becomes entirely kinetic at A_2 and, as the pendulum keeps swinging to the right, the kinetic energy is transformed back into potential energy. At A_3, we shall have $T_3 = 0$ and $V_3 = Wl$.

Since the total mechanical energy of the pendulum remains constant and since its potential energy depends only upon its elevation, the kinetic energy of the pendulum will have the same value at any two points located on the same level. Thus, the speed of the pendulum is the same at A and at A' (Fig. 13.11). This result may be extended to the case of a particle moving along any given path, regardless of the shape of the path, as long as the only forces acting on the particle are its weight and the normal reaction of the path. The particle of Fig. 13.12, for example, which slides in a vertical plane along a frictionless track, will have the same speed at A, A', and A''.

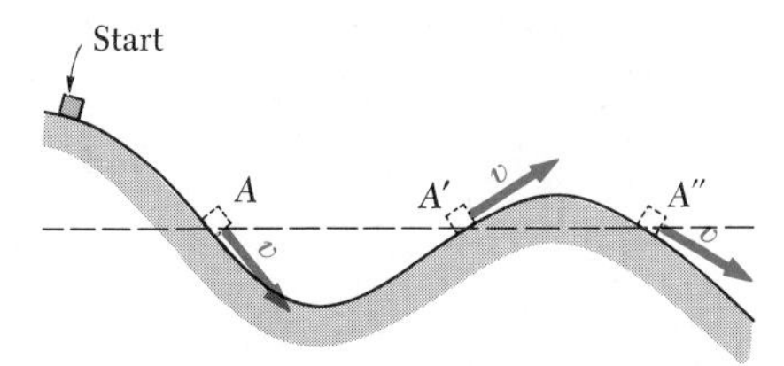

Fig. 13.12

While the weight of a particle and the force exerted by a spring are conservative forces, *friction forces are nonconservative forces*. In other words, *the work of a friction force cannot be expressed as a change in potential energy*. The work of a friction force depends upon the path followed by its point of application; and while the work $U_{1\rightarrow2}$ defined by (13.15) is positive or negative according to the sense of motion, *the work of a friction force*, as we noted in Sec. 13.4, *is always negative*. It follows that when a mechanical system involves friction, its total mechanical energy does not remain constant but decreases. The energy of the system, however, is not lost; it is transformed into heat, and the sum of the *mechanical energy* and of the *thermal energy* of the system remains constant.

Other forms of energy may also be involved in a system. For instance, a generator converts mechanical energy into *electric energy;* a gasoline engine converts *chemical energy* into mechanical energy; a nuclear reactor converts *mass* into thermal energy. If all forms of energy are considered, the energy of any system may be considered as constant and the principle of conservation of energy remains valid under all conditions.

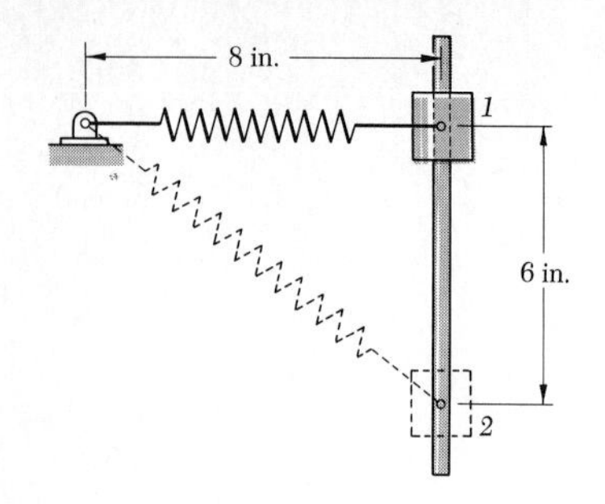

SAMPLE PROBLEM 13.5

A 20-lb collar slides without friction along a vertical rod as shown. The spring attached to the collar has an undeformed length of 4 in. and a constant of 3 lb/in. If the collar is released from rest in position *1*, determine its velocity after it has moved 6 in. to position *2*.

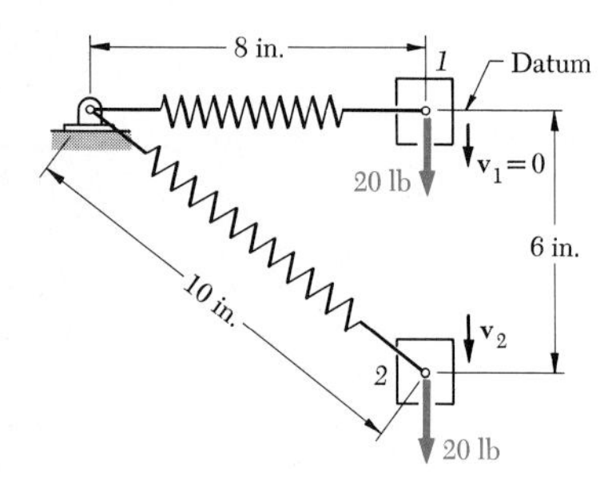

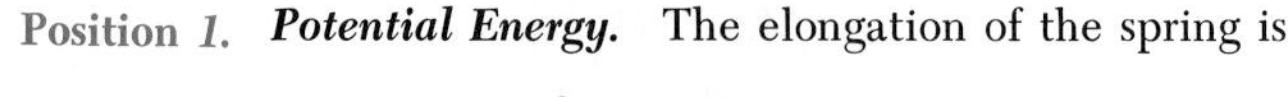

Position *1*. ***Potential Energy.*** The elongation of the spring is

$$x_1 = 8 \text{ in.} - 4 \text{ in.} = 4 \text{ in.}$$

and we have

$$V_e = \tfrac{1}{2}kx_1^2 = \tfrac{1}{2}(3 \text{ lb/in.})(4 \text{ in.})^2 = 24 \text{ in}\cdot\text{lb}$$

Choosing the datum as shown, we have $V_g = 0$. Therefore,

$$V_1 = V_e + V_g = 24 \text{ in}\cdot\text{lb} = 2 \text{ ft}\cdot\text{lb}$$

Kinetic Energy. Since the velocity in position *1* is zero, $T_1 = 0$.

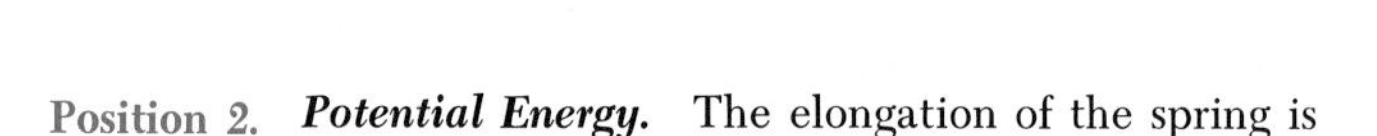

Position *2*. ***Potential Energy.*** The elongation of the spring is

$$x_2 = 10 \text{ in.} - 4 \text{ in.} = 6 \text{ in.}$$

and we have

$$V_e = \tfrac{1}{2}kx_2^2 = \tfrac{1}{2}(3 \text{ lb/in.})(6 \text{ in.})^2 = 54 \text{ in}\cdot\text{lb}$$
$$V_g = Wy = (20 \text{ lb})(-6 \text{ in.}) = -120 \text{ in}\cdot\text{lb}$$

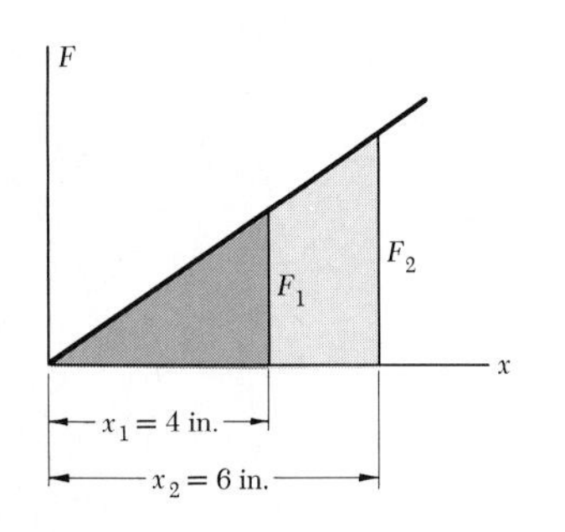

Therefore,

$$V_2 = V_e + V_g = 54 - 120 = -66 \text{ in}\cdot\text{lb} = -5.5 \text{ ft}\cdot\text{lb}$$

Kinetic Energy

$$T_2 = \tfrac{1}{2}mv_2^2 = \frac{1}{2}\frac{20}{32.2}v_2^2 = 0.311v_2^2$$

Conservation of Energy. Applying the principle of conservation of energy between positions *1* and *2*, we write

$$T_1 + V_1 = T_2 + V_2$$
$$0 + 2 \text{ ft}\cdot\text{lb} = 0.311v_2^2 - 5.5 \text{ ft}\cdot\text{lb}$$
$$v_2 = \pm 4.91 \text{ ft/s}$$

$\mathbf{v}_2 = 4.91 \text{ ft/s}\downarrow$ ◄

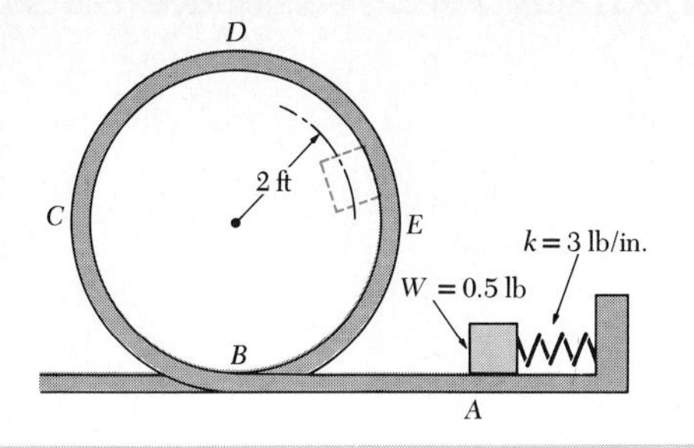

SAMPLE PROBLEM 13.6

The 0.5-lb pellet is pushed against the spring at A and released from rest. Neglecting friction, determine the smallest deflection of the spring for which the pellet will travel around the loop $ABCDE$ and remain at all times in contact with the loop.

Required Speed at Point D. As the pellet passes through the highest point D, its potential energy with respect to gravity is maximum; thus, at the same point its kinetic energy and its speed are minimum. Since the pellet must remain in contact with the loop, the force **N** exerted on the pellet by the loop must be equal to, or greater than, zero. Setting $\mathbf{N} = 0$, we compute the smallest possible speed v_D.

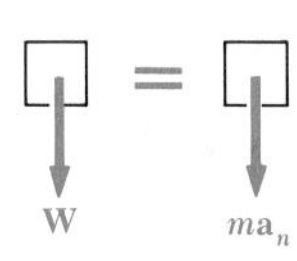

$$+\downarrow \Sigma F_n = ma_n: \qquad W = ma_n \qquad mg = ma_n \qquad a_n = g$$

$$a_n = \frac{v_D^2}{r}: \qquad v_D^2 = ra_n = rg = (2\text{ ft})(32.2\text{ ft/s}^2) = 64.4\text{ ft}^2/\text{s}^2$$

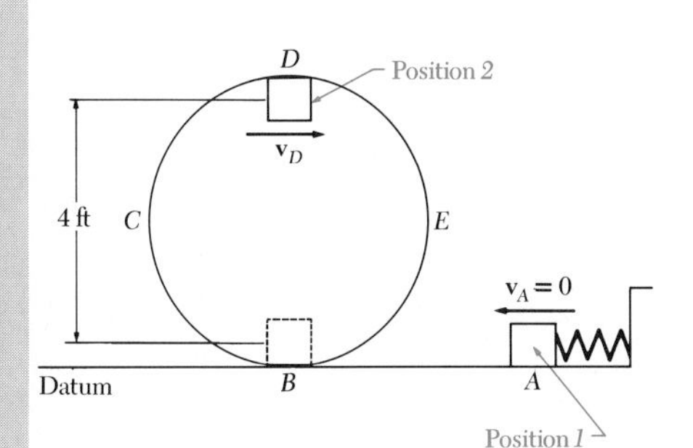

Position 1. ***Potential Energy.*** Denoting by x the deflection of the spring and noting that $k = 3\text{ lb/in.} = 36\text{ lb/ft}$, we write

$$V_e = \tfrac{1}{2}kx^2 = \tfrac{1}{2}(36\text{ lb/ft})x^2 = 18x^2$$

Choosing the datum at A, we have $V_g = 0$; therefore

$$V_1 = V_e + V_g = 18x^2$$

Kinetic Energy. Since the pellet is released from rest, $v_A = 0$ and we have $T_1 = 0$.

Position 2. ***Potential Energy.*** The spring is now undeformed; thus $V_e = 0$. Since the pellet is 4 ft above the datum, we have

$$V_g = Wy = (0.5\text{ lb})(4\text{ ft}) = 2\text{ ft}\cdot\text{lb}$$
$$V_2 = V_e + V_g = 2\text{ ft}\cdot\text{lb}$$

Kinetic Energy. Using the value of v_D^2 obtained above, we write

$$T_2 = \tfrac{1}{2}mv_D^2 = \frac{1}{2}\frac{0.5\text{ lb}}{32.2\text{ ft/s}^2}(64.4\text{ ft}^2/\text{s}^2) = 0.5\text{ ft}\cdot\text{lb}$$

Conservation of Energy. Applying the principle of conservation of energy between positions 1 and 2, we write

$$T_1 + V_1 = T_2 + V_2$$
$$0 + 18x^2 = 0.5\text{ ft}\cdot\text{lb} + 2\text{ ft}\cdot\text{lb}$$
$$x = 0.3727\text{ ft} \qquad\qquad x = 4.47\text{ in.} \blacktriangleleft$$

Problems

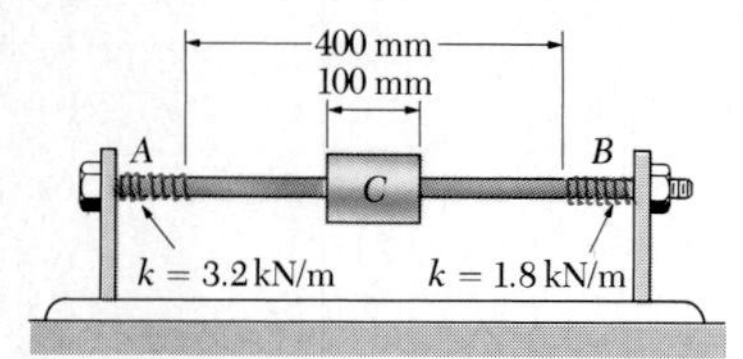

Fig. P13.48

13.48 A collar C of mass m slides without friction on a horizontal rod between springs A and B. If the collar is pushed to the left until spring A is compressed 60 mm and released, determine the distance through which the collar will travel and the maximum velocity it will reach (*a*) if $m = 0.5$ kg, (*b*) if $m = 2$ kg.

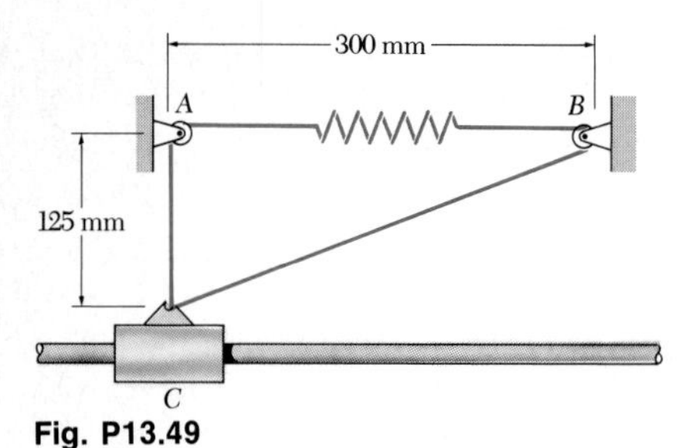

Fig. P13.49

13.49 A 0.8-kg collar C may slide without friction along a horizontal rod. It is attached to a spring of constant $k = 300$ N/m and 500-mm undeformed length. Knowing that the collar is released from rest in the position shown, determine the maximum velocity it will reach in the ensuing motion.

13.50 A 2-lb collar is attached to a spring and slides without friction along a circular rod which lies in a *horizontal* plane. The spring has a constant $k = 3$ lb/in. and is undeformed when the collar is at B. If the collar is released from rest at C, determine the speed of the collar as it passes through point B.

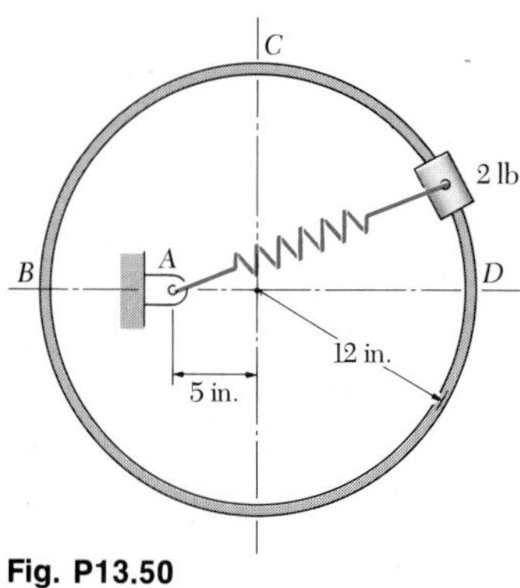

Fig. P13.50

13.51 It is possible for the collar of Prob. 13.50 to have a continuous, although nonuniform, motion along the rod. If the speed of the collar at B is to be twice the speed of the collar at D, determine (*a*) the required speed at D, (*b*) the corresponding speed at C.

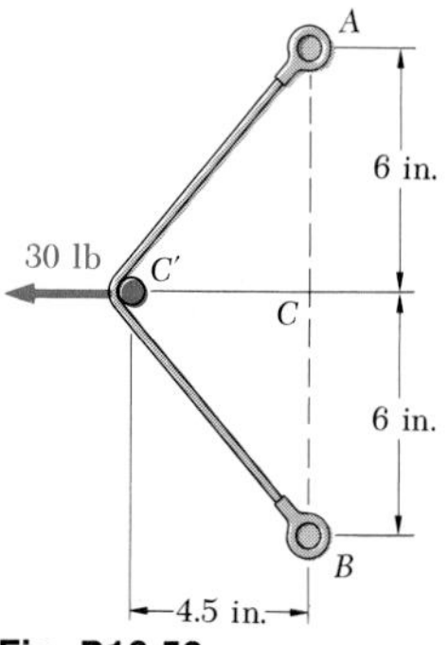

Fig. P13.52

13.52 An elastic cord is stretched between two points A and B, located 12 in. apart in the same horizontal plane. When stretched directly between A and B, the tension in the cord is 5 lb. The cord is then stretched as shown until its midpoint C has moved through 4.5 in. to C'; a force of 30 lb is required to hold the cord at C'. A 2-oz pellet is placed at C', and the cord is released. Determine the speed of the pellet as it passes through C.

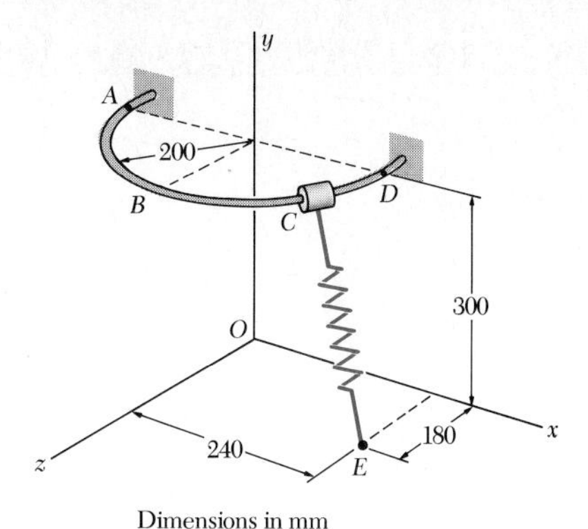

Fig. P13.53

13.53 A 600-g collar C may slide along a horizontal, semicircular rod ABD. The spring CE has an undeformed length of 250 mm and a spring constant of 135 N/m. Knowing that the collar is released from rest at A and neglecting friction, determine the speed of the collar (*a*) at B, (*b*) at D.

13.54 In Prob. 13.53, determine (*a*) the maximum speed of the collar, (*b*) the coordinates of the point where the collar reaches its maximum speed.

13.55 A 30-Mg railroad car starts from rest and coasts down a 1 percent incline for a distance of 20 m. It is stopped by a bumper having a spring constant of 2400 kN/m. (*a*) What is the speed of the car at the bottom of the incline? (*b*) How many millimeters will the spring be compressed?

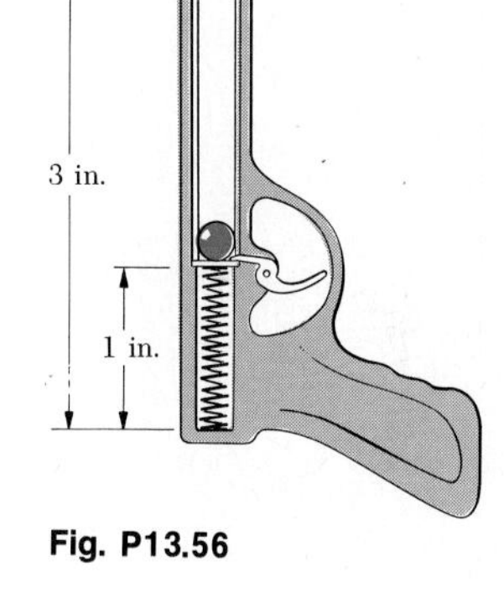

Fig. P13.56

13.56 A toy spring gun is used to shoot 1-oz bullets vertically upward. The undeformed length of the spring is 6 in.; it is compressed to a length of 1 in. when the gun is ready to be shot and expands to a length of 3 in. as the bullet leaves the gun. A force of 10 lb is required to maintain the spring in firing position when the length of the spring is 1 in. Determine (*a*) the velocity of the bullet as it leaves the gun, (*b*) the maximum height reached by the bullet.

13.57 A spring is used to stop a 150-lb package which is moving down a 20° incline. The spring has a constant $k = 150$ lb/in. and is held by cables so that it is initially compressed 4 in. Knowing that the velocity of the package is 10 ft/s when it is 30 ft from the spring and neglecting friction, determine the maximum additional deformation of the spring in bringing the package to rest.

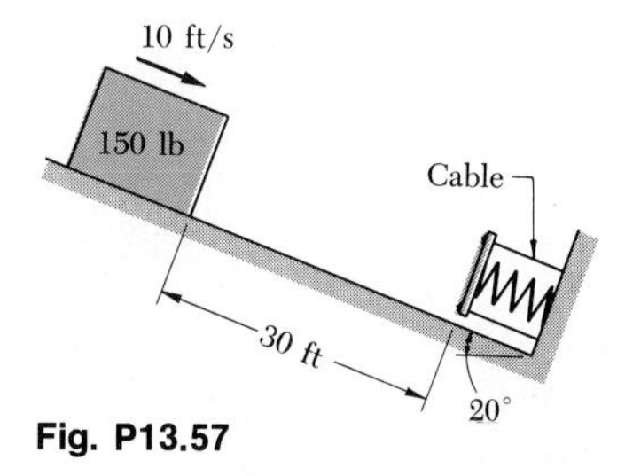

Fig. P13.57

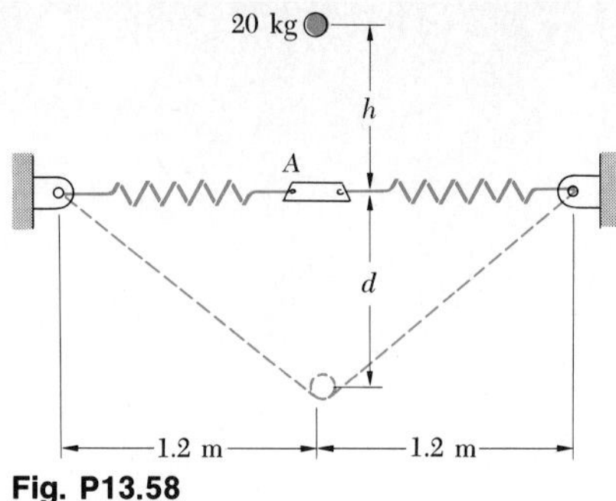

Fig. P13.58

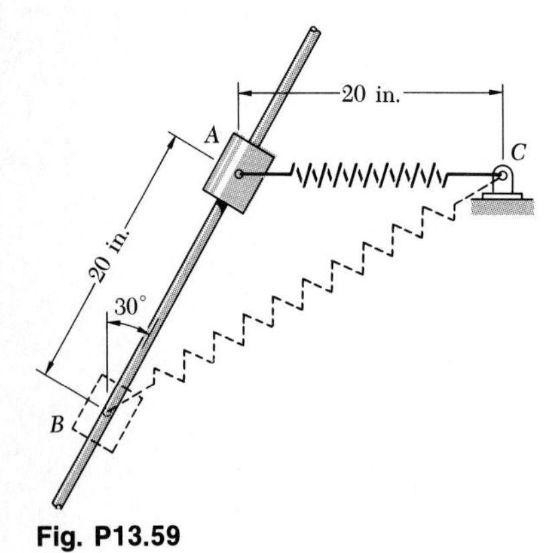

Fig. P13.59

13.58 Two springs are attached to a piece of cloth A of negligible mass, as shown. The initial tension in each spring is 500 N, and the spring constant of each spring is $k = 2$ kN/m. A 20-kg ball is released from a height h above A; the ball hits the cloth, causing it to move through a maximum distance $d = 0.9$ m. Determine the height h.

13.59 A 12-lb collar slides without friction along a rod which forms an angle of 30° with the vertical. The spring is unstretched when the collar is at A. If the collar is released from rest at A, determine the value of the spring constant k for which the collar has zero velocity at B.

13.60 A 2-kg collar is attached to a spring and slides without friction in a vertical plane along the curved rod ABC. The spring is undeformed when the collar is at C and its constant is 600 N/m. If the collar is released at A with no initial velocity, determine its velocity (*a*) as it passes through B, (*b*) as it reaches C.

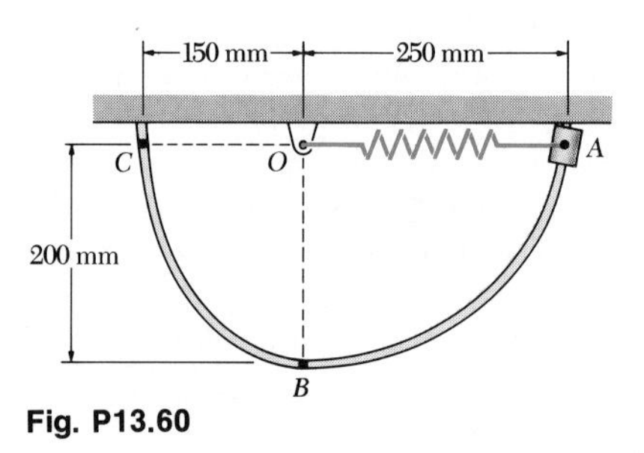

Fig. P13.60

13.61 Solve Prob. 13.60, assuming that the undeformed length of the spring is 120 mm and that its constant is 800 N/m.

13.62 Knowing that the rod of Prob. 13.60 has a radius of curvature of 200 mm at point B, determine the force exerted by the rod on the collar as it passes through point B.

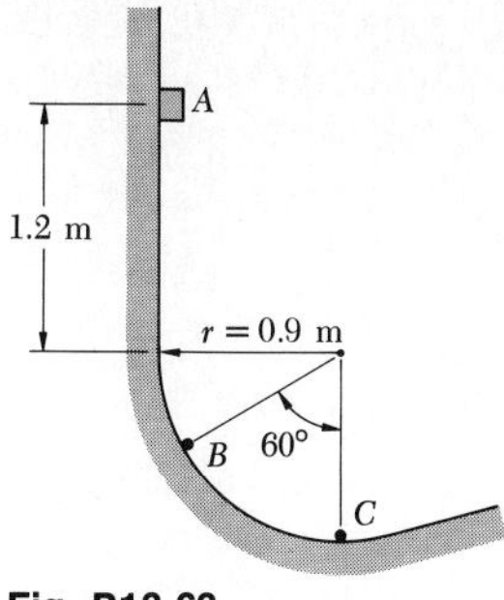

Fig. P13.63

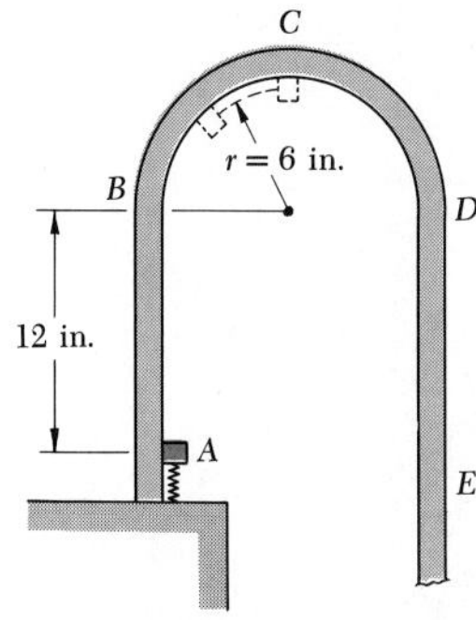

Fig. P13.64

13.63 A 200-g pellet is released from rest at A and slides without friction along the surface shown. Determine the force exerted by the surface on the pellet as it passes through (*a*) point B, (*b*) point C.

13.64 A $\frac{1}{2}$-lb pellet is released from rest at A when the spring is compressed 3 in. and travels around the loop $ABCDE$. Determine the smallest value of the spring constant for which the pellet will travel around the loop and will at all times remain in contact with the loop.

13.65 In Prob. 13.53, determine the force exerted by the rod on collar C as the collar passes through point B.

13.66 The pendulum shown is released from rest at A and swings through 90° before the cord touches the fixed peg B. Determine the smallest value of a for which the pendulum bob will describe a circle about the peg.

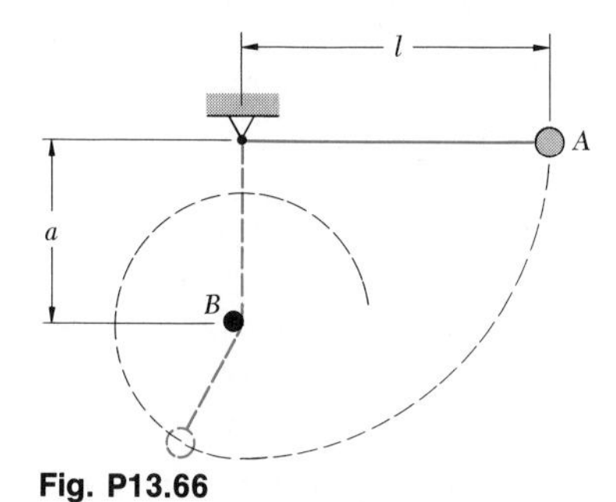

Fig. P13.66

13.67 (*a*) Determine the kinetic energy per unit mass which a missile must have after being fired from the surface of the earth if it is to reach an infinite distance from the earth. (*b*) What is the initial velocity of the missile (called *escape velocity*)? Give your answers in SI units and show that the answer to part *b* is independent of the firing angle.

13.68 A satellite of mass m describes a circular orbit of radius r around the earth. Express (*a*) its potential energy, (*b*) its kinetic energy, (*c*) its total energy, as a function of r. Denote by R the radius of the earth, by g the acceleration of gravity at the surface of the earth, and assume that the potential energy of the satellite is zero on its launching pad.

13.69 (*a*) Show by setting $r = R + y$ in formula (13.13′) and expanding in a power series in y/R that the expression obtained in (13.12) for the potential energy V_g due to gravity is a first-order approximation for the expression given in (13.13′). (*b*) Using the same expansion, derive a second-order approximation for V_g.

13.70 A lunar excursion module (LEM) was used in the Apollo moon-landing missions to save fuel by making it unnecessary to launch the entire Apollo spacecraft from the moon's surface on its return trip to the earth. Check the effectiveness of this approach by computing the energy per kilogram required for a spacecraft to escape the gravitational field of the moon if the spacecraft starts (*a*) from the moon's surface, (*b*) from a circular orbit 100 km above the moon's surface. Neglect the effect of the earth's gravitational field. (The radius of the moon is 1740 km and its mass is 0.01230 times the mass of the earth.)

13.71 During its fifth mission, the space shuttle Columbia ejected two communication satellites while describing a circular orbit, 185 mi above the surface of the earth. Knowing that one of these satellites weighed 8000 lb, determine (*a*) the additional energy required to place the satellite in a geosynchronous orbit (see Prob. 12.79) at an altitude of 22,230 mi above the surface of the earth, (*b*) the energy required to place it in the same orbit by launching it from the surface of the earth.

13.72 Show that the ratio of the potential and kinetic energies of an electron as it enters the plates of the cathode-ray tube of Prob. 12.62 is equal to $d\delta/lL$. (Place the datum at the surface of the positive plate.)

13.8. Power and Efficiency. *Power* is defined as the time rate at which work is done. In the selection of a motor or engine, power is a much more important criterion than is the actual amount of work to be performed. A small motor or a large power plant may both be used to do a given amount of work; but the small motor may require a month to do the work done by the power plant in a matter of minutes. If ΔU is the work done during the time interval Δt, then the average power during this time interval is

$$\text{Average power} = \frac{\Delta U}{\Delta t}$$

Letting Δt approach zero, we obtain at the limit

$$\text{Power} = \frac{dU}{dt} \tag{13.17}$$

Since $dU = (F \cos \alpha)\, ds$ and $v = ds/dt$, the power may also be expressed as follows:

$$\text{Power} = \frac{dU}{dt} = \frac{(F \cos \alpha)\, ds}{dt}$$

$$\text{Power} = (F \cos \alpha) v \qquad (13.18)$$

where v = magnitude of velocity of point of application of force $\mathbf{F}$
α = angle between $\mathbf{F}$ and velocity $\mathbf{v}$

Since power was defined as the time rate at which work is done, it should be expressed in units obtained by dividing units of work by the unit of time. Thus, if SI units are used, power should be expressed in J/s; this unit is called a *watt* (W). We have

$$1 \text{ W} = 1 \text{ J/s} = 1 \text{ N} \cdot \text{m/s}$$

If U.S. customary units are used, power should be expressed in ft·lb/s or in *horsepower* (hp), with the latter defined as

$$1 \text{ hp} = 550 \text{ ft} \cdot \text{lb/s}$$

Recalling from Sec. 13.2 that 1 ft·lb = 1.356 J, we verify that

$$1 \text{ ft} \cdot \text{lb/s} = 1.356 \text{ J/s} = 1.356 \text{ W}$$
$$1 \text{ hp} = 550(1.356 \text{ W}) = 746 \text{ W} = 0.746 \text{ kW}$$

The *mechanical efficiency* of a machine was defined in Sec. 10.5 as the ratio of the output work to the input work:

$$\eta = \frac{\text{output work}}{\text{input work}} \qquad (13.19)$$

This definition is based on the assumption that work is done at a constant rate. The ratio of the output to the input work is therefore equal to the ratio of the rates at which output and input work are done, and we have

$$\eta = \frac{\text{power output}}{\text{power input}} \qquad (13.20)$$

Because of energy losses due to friction, the output work is always smaller than the input work, and, consequently, the power output is always smaller than the power input. The mechanical efficiency of a machine is therefore always less than 1.

When a machine is used to transform mechanical energy into electric energy, or thermal energy into mechanical energy, its *overall efficiency* may be obtained from formula (13.20). The overall efficiency of a machine is always less than 1; it provides a measure of all the various energy losses involved (losses of electric or thermal energy as well as frictional losses). We should note that it is necessary, before using formula (13.20), to express the power output and the power input in the same units.

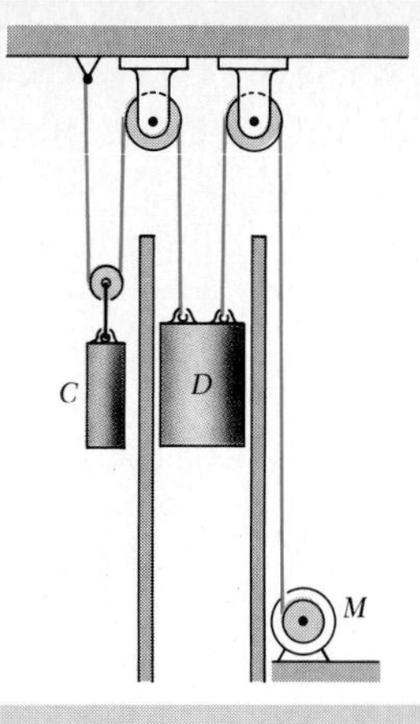

SAMPLE PROBLEM 13.7

The dumb-waiter D and its load have a combined weight of 600 lb, while the counterweight C weighs 800 lb. Determine the power delivered by the electric motor M when the dumb-waiter (*a*) is moving up at a constant speed of 8 ft/s, (*b*) has an instantaneous velocity of 8 ft/s and an acceleration of 2.5 ft/s², both directed upward.

Solution. Since the force $\mathbf{F}$ exerted by the motor cable has the same direction as the velocity $\mathbf{v}_D$ of the dumb-waiter, the power is equal to Fv_D, where $v_D = 8$ ft/s. To obtain the power, we must first determine $\mathbf{F}$ in each of the two given situations.

a. Uniform Motion. We have $\mathbf{a}_C = \mathbf{a}_D = 0$; both bodies are in equilibrium.

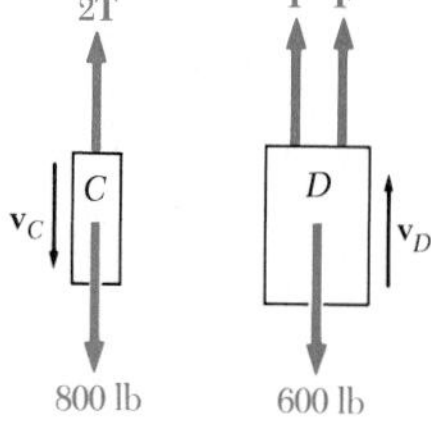
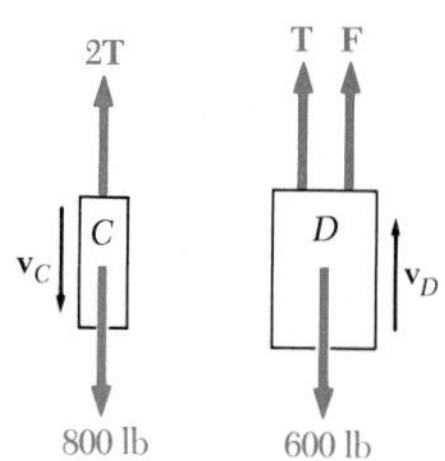

Free Body C: $\quad +\uparrow\Sigma F_y = 0: \quad 2T - 800 \text{ lb} = 0 \quad T = 400 \text{ lb}$

Free Body D: $\quad +\uparrow\Sigma F_y = 0: \quad F + T - 600 \text{ lb} = 0$

$$F = 600 \text{ lb} - T = 600 \text{ lb} - 400 \text{ lb} = 200 \text{ lb}$$

$$Fv_D = (200 \text{ lb})(8 \text{ ft/s}) = 1600 \text{ ft} \cdot \text{lb/s}$$

$$\text{Power} = (1600 \text{ ft} \cdot \text{lb/s})\frac{1 \text{ hp}}{550 \text{ ft} \cdot \text{lb/s}} = 2.91 \text{ hp} \quad ◄$$

b. Accelerated Motion. We have

$$\mathbf{a}_D = 2.5 \text{ ft/s}^2\uparrow \qquad \mathbf{a}_C = -\tfrac{1}{2}\mathbf{a}_D = 1.25 \text{ ft/s}^2\downarrow$$

The equations of motion are

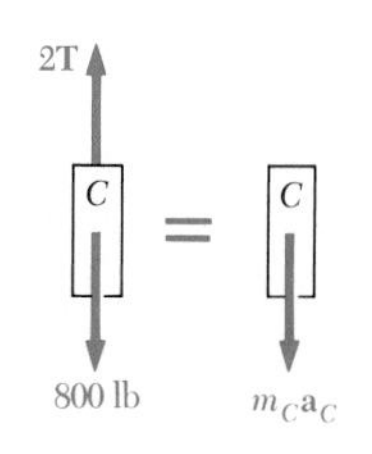

Free Body C: $\quad +\downarrow\Sigma F_y = m_C a_C: \quad 800 - 2T = \dfrac{800}{32.2}(1.25) \quad T = 384.5 \text{ lb}$

Free Body D: $\quad +\uparrow\Sigma F_y = m_D a_D: \quad F + T - 600 = \dfrac{600}{32.2}(2.5)$

$$F + 384.5 - 600 = 46.6 \qquad F = 262.1 \text{ lb}$$

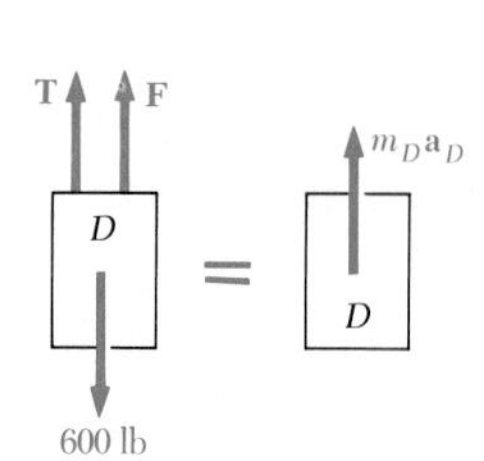

$$Fv_D = (262.1 \text{ lb})(8 \text{ ft/s}) = 2097 \text{ ft} \cdot \text{lb/s}$$

$$\text{Power} = (2097 \text{ ft} \cdot \text{lb/s})\frac{1 \text{ hp}}{550 \text{ ft} \cdot \text{lb/s}} = 3.81 \text{ hp} \quad ◄$$

Problems

13.73 A utility hoist can lift a 3000-kg crate at the rate of 25 m/min. Knowing that the hoist is run by a 15-kW motor, determine the overall efficiency of the hoist.

13.74 A 15-g bullet leaves a fixed rifle barrel 2 ms after being fired. Knowing that the muzzle velocity is 800 m/s and neglecting friction, determine the average power developed by the rifle.

13.75 A 75-kg man runs up a 6-m-high flight of stairs in 8 s. (*a*) What is the average power developed by the man? (*b*) If a 60-kg woman can develop 80 percent as much power, how long will it take her to run up a 4-m-high flight of stairs?

13.76 Crushed stone is being moved from a quarry at A to a loading dock at B at the rate of 500 tons/h. An electric generator is attached to the system in order to maintain a constant belt speed. Knowing that the efficiency of the belt-generator system is 0.70, determine the average power in kilowatts developed by the generator if the belt speed is (*a*) 6 ft/s, (*b*) 10 ft/s.

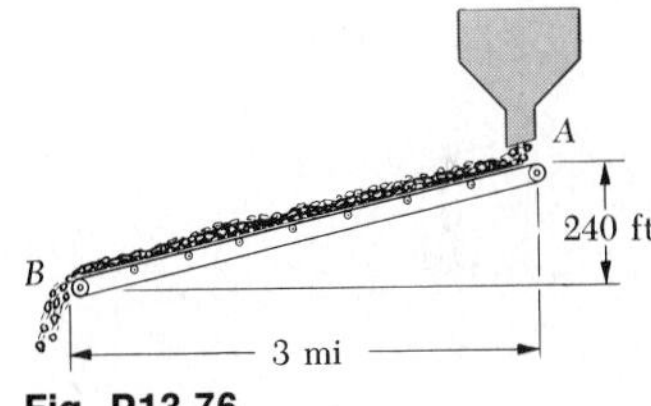

Fig. P13.76

13.77 The escalator shown is designed to transport 6000 persons per hour at a constant speed of 1.5 ft/s. Assuming an average weight of 150 lb per person, determine (*a*) the average power required, (*b*) the required capacity of the motor if the mechanical efficiency is 85 percent and if a 300 percent overload is to be allowed.

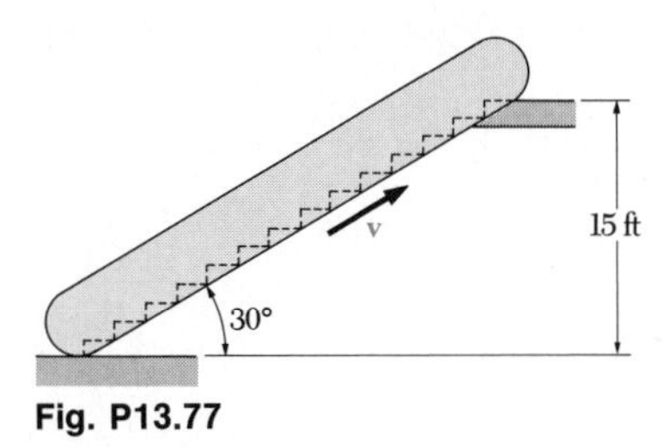

Fig. P13.77

13.78 A chair-lift is designed to transport 900 skiers per hour from the base A to the summit B. The average mass of a skier is 75 kg, and the average speed of the lift is 80 m/min. Determine (*a*) the average power required, (*b*) the required capacity of the motor if the mechanical efficiency is 80 percent and if a 250 percent overload is to be allowed.

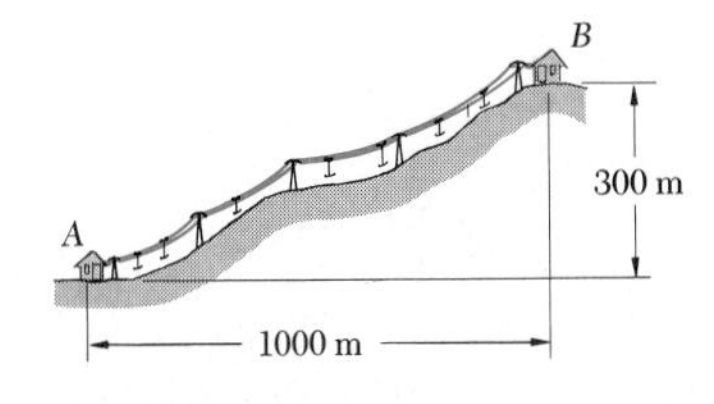

Fig. P13.78

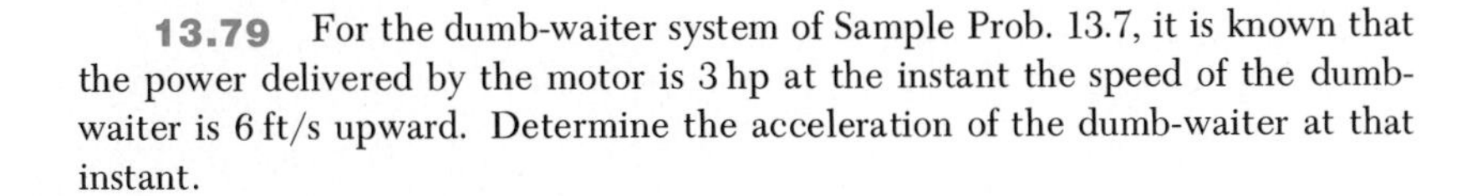

13.79 For the dumb-waiter system of Sample Prob. 13.7, it is known that the power delivered by the motor is 3 hp at the instant the speed of the dumb-waiter is 6 ft/s upward. Determine the acceleration of the dumb-waiter at that instant.

13.80 A train of total mass equal to 500 Mg starts from rest and accelerates uniformly to a speed of 90 km/h in 50 s. After reaching this speed, the train travels with a constant velocity. The track is horizontal and axle friction and rolling resistance result in a total force of 15 kN in a direction opposite to the direction of motion. Determine the power required as a function of time.

13.81 Solve Prob. 13.80, assuming that during the entire motion, the train is traveling up a 2.5 percent grade.

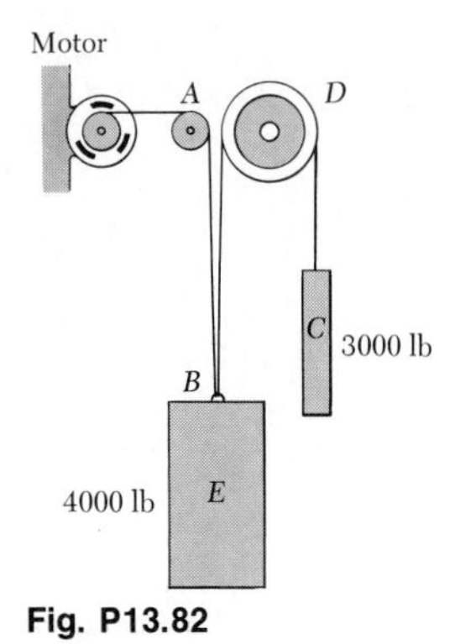

Fig. P13.82

13.82 Elevator E weighs 4000 lb when fully loaded and is connected to a 3000-lb counterweight C. Determine the power delivered by the electric motor when the elevator (*a*) is moving up at a constant speed of 20 ft/s, (*b*) has an instantaneous velocity of 20 ft/s and an acceleration of 3 ft/s², both directed upward.

13.83 The fluid transmission of a truck of mass m permits the engine to deliver an essentially constant power P to the driving wheels. Determine the time elapsed and the distance traveled as the speed is increased from v_0 to v_1.

13.84 The fluid transmission of a 12-Mg truck permits the engine to deliver an essentially constant power of 40 kW to the driving wheels. Determine the time required and the distance traveled as the speed of the truck is increased (*a*) from 24 km/h to 48 km/h, (*b*) from 48 km/h to 72 km/h.

13.85 The frictional resistance of a ship is known to vary directly as the 1.75 power of the speed v of the ship. A single tugboat at full power can tow the ship at a constant speed of 5.4 km/h by exerting a constant force of 250 kN. Determine (*a*) the power developed by the tugboat, (*b*) the maximum speed at which two tugboats, capable of delivering the same power, can tow the ship.

13.86 Determine the speed at which the single tugboat of Prob. 13.85 will tow the ship if the tugboat is developing half of its maximum power.

13.9. Review and Summary.

This chapter was devoted to the method of work and energy for the analysis of the motion of particles. We first defined the *work of a force* **F** *during a small displacement* **dr** [Sec. 13.2] as the scalar quantity

Work of a force

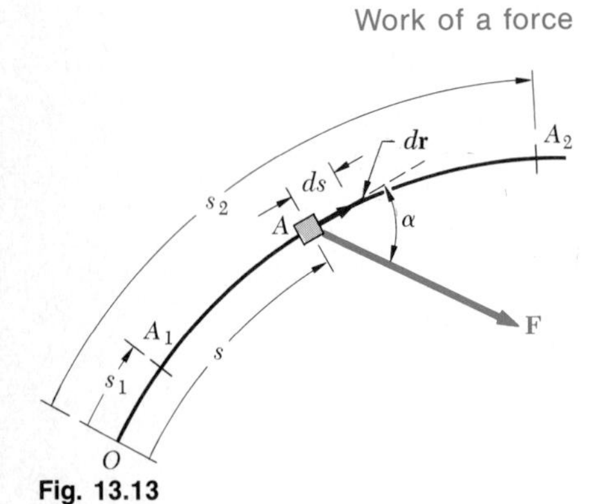

Fig. 13.13

$$dU = F\,ds \cos\alpha \tag{13.1}$$

where ds is the magnitude of the displacement $d\mathbf{r}$ and α the angle between **F** and $d\mathbf{r}$ (Fig. 13.13). The work of **F** during a *finite displacement* from A_1 to A_2 was denoted by $U_{1\to2}$ and obtained by integrating Eq. (13.1) from s_1 to s_2:

$$U_{1\to2} = \int_{s_1}^{s_2} (F\cos\alpha)\,ds \tag{13.2}$$

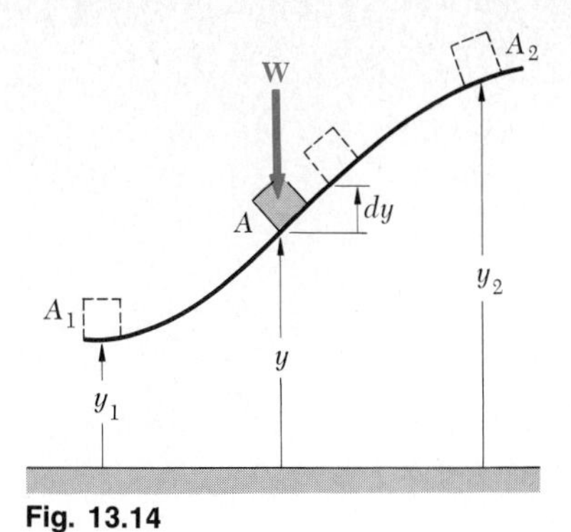

Fig. 13.14

Work of a weight

The *work of the weight* **W** *of a body* as its center of gravity moves from the elevation y_1 to y_2 (Fig. 13.14) may be obtained by making $F = W$ and $\alpha = 180°$ in Eq. (13.2):

$$U_{1\to2} = -\int_{y_1}^{y_2} W\,dy = Wy_1 - Wy_2 \tag{13.4}$$

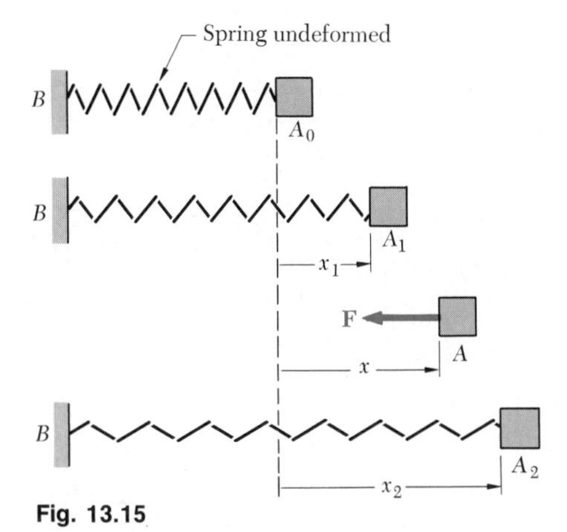

Fig. 13.15

Work of the force exerted by a spring

The *work of the force* **F** *exerted by a spring* on a body A as the spring is stretched from x_1 to x_2 (Fig. 13.15) may be obtained by making $F = kx$, where k is the constant of the spring, and $\alpha = 180°$ in Eq. (13.2):

$$U_{1\to2} = -\int_{x_1}^{x_2} kx\,dx = \tfrac{1}{2}kx_1^2 - \tfrac{1}{2}kx_2^2 \tag{13.6}$$

The work of **F** is therefore positive *when the spring is returning to its undeformed position.*

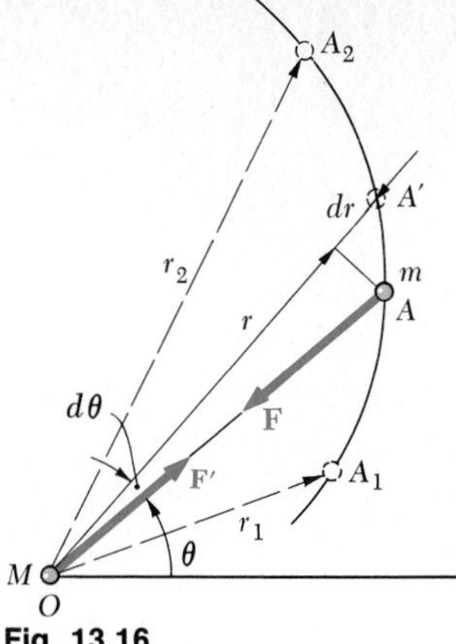

Fig. 13.16

Work of the gravitational force

The *work of the gravitational force* $\mathbf{F}$ exerted by a particle of mass M located at O on a particle of mass m as the latter moves from A_1 to A_2 (Fig. 13.16) was obtained by recalling from Sec. 12.10 the expression for the magnitude of $\mathbf{F}$ and writing

$$U_{1\rightarrow 2} = -\int_{r_1}^{r_2} \frac{GMm}{r^2}\, dr = \frac{GMm}{r_2} - \frac{GMm}{r_1} \qquad (13.7)$$

Kinetic energy of a particle

The *kinetic energy of a particle* of mass m moving with a velocity $\mathbf{v}$ [Sec. 13.3] was defined as the scalar quantity

$$T = \tfrac{1}{2}mv^2 \qquad (13.9)$$

Principle of work and energy

From Newton's second law we derived the *principle of work and energy,* which states that *the kinetic energy of a particle at A_2 may be obtained by adding to its kinetic energy at A_1 the work done during the displacement from A_1 to A_2 by the force* $\mathbf{F}$ *exerted on the particle:*

$$T_1 + U_{1\rightarrow 2} = T_2 \qquad (13.11)$$

Method of work and energy

The method of work and energy simplifies the solution of many problems dealing with forces, displacements, and velocities, since it does not require the determination of accelerations [Sec. 13.4]. We also note that it involves only scalar quantities and that forces which do no work need not be considered [Sample Probs. 13.1 and 13.3]. However, this method should be supplemented by the direct application of Newton's second law to determine a force normal to the path of the particle [Sample Prob. 13.4].

System of particles

The method of work and energy may also be applied to a system of particles [Sec. 13.5]. The work of all forces—internal and external—should then be considered, although the work of some internal forces may cancel out, as in the case of bodies connected by inextensible cords [Sample Prob. 13.2].

When the work of a force **F** is independent of the path followed [Sec. 13.6], the force **F** is said to be a *conservative force*, and its work is equal to *minus the change in the potential energy* V associated with **F**:

Conservative force. Potential energy

$$U_{1\rightarrow 2} = V_1 - V_2 \qquad (13.15)$$

The following expressions were obtained for the potential energy associated with each of the forces considered earlier:

Force of gravity (weight): $$V_g = Wy \qquad (13.12)$$

Gravitational force: $$V_g = -\frac{GMm}{r} \qquad (13.13)$$

Elastic force exerted by a spring: $$V_e = \tfrac{1}{2}kx^2 \qquad (13.14)$$

Principle of conservation of energy

Substituting for $U_{1\rightarrow 2}$ from Eq. (13.15) into Eq. (13.11) and rearranging the terms [Sec. 13.7], we obtained

$$T_1 + V_1 = T_2 + V_2 \qquad (13.16)$$

This is the *principle of conservation of energy*, which states that when a particle or a system of particles moves under the action of conservative forces, *the sum of its kinetic and potential energies remains constant.* The application of this principle facilitates the solution of problems involving only conservative forces [Sample Probs. 13.5 and 13.6].

Power and mechanical efficiency

The power developed by a machine and its mechanical efficiency were discussed in the last section of the chapter [Sec. 13.8; Sample Prob. 13.7]. *Power* was defined as the time rate dU/dt at which work is done and it was shown that

$$\text{Power} = \frac{dU}{dt} = (F \cos \alpha)v \qquad (13.18)$$

where **F** is the force exerted on the particle under consideration, **v** the velocity of the particle, and α the angle between **F** and **v**. The *mechanical efficiency*, denoted by η, was defined as the ratio of the output work to the input work, and it was shown that η could also be expressed as

$$\eta = \frac{\text{power output}}{\text{power input}} \qquad (13.20)$$

Review Problems

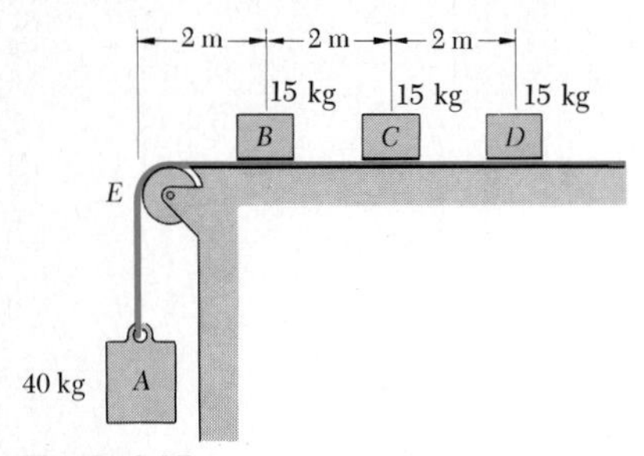

Fig. P13.87

13.87 Three 15-kg packages rest on a belt which passes over a pulley and is attached to a 40-kg block. Knowing that the coefficients of friction between the belt and the horizontal surface and also between the belt and the packages are $\mu_s = 0.50$ and $\mu_k = 0.40$, determine the speed of package B as it falls off the belt at E.

13.88 In Prob. 13.87, determine the speed of package C as it falls off the belt at E.

13.89 An elevator travels upward at a constant speed of 2 m/s. A child riding the elevator throws a 0.6-kg stone upward with a speed of 5 m/s *relative* to the elevator. Determine (*a*) the work done by the child in throwing the stone, (*b*) the difference in the values of the kinetic energy of the stone before and after it was thrown. (*c*) Why are the values obtained in parts *a* and *b* not the same?

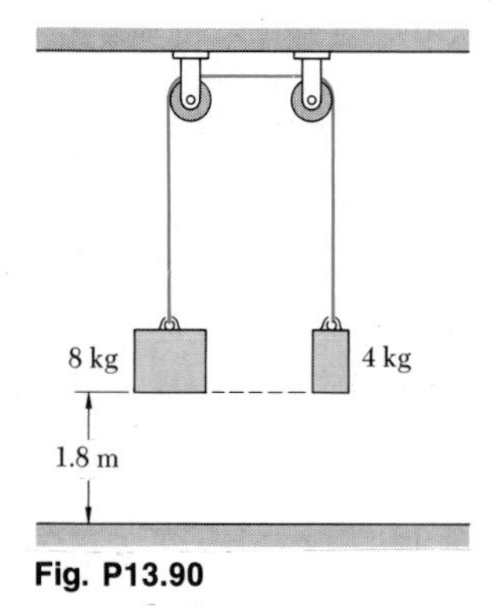

Fig. P13.90

13.90 Two cylinders are suspended from an inextensible cable as shown. If the system is released from rest, determine the maximum velocity attained by the 4-kg cylinder.

13.91 In Prob. 13.90, determine the maximum height above the floor to which the 4-kg cylinder will rise.

13.92 A 36,000-lb airplane lands on an aircraft carrier and is caught by an arresting cable. The cable is inextensible and is paid out at A and B from mechanisms located below deck and consisting of pistons moving in long oil-filled cylinders. Knowing that the piston-cylinder system maintains a constant tension of 90 kips in the cable during the entire landing, determine the landing speed of the airplane if it travels a distance $d = 100$ ft after being caught by the cable.

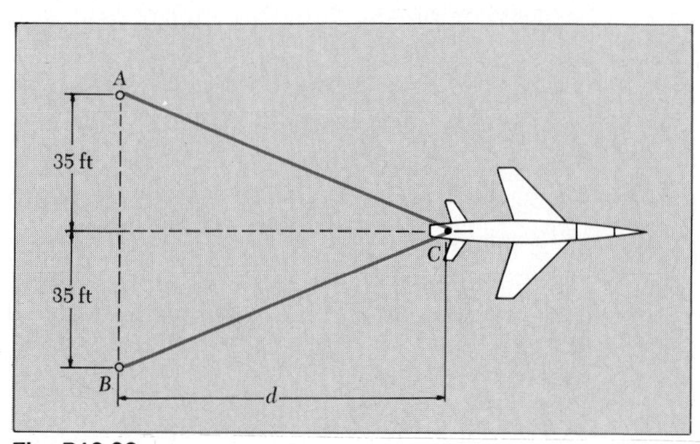

Fig. P13.92

13.93 For the airplane and aircraft carrier of Prob. 13.92, determine the constant value of the tension maintained in the arresting cable, knowing that the airplane lands with a speed of 120 mi/h and travels a distance $d = 120$ ft after being caught by the cable.

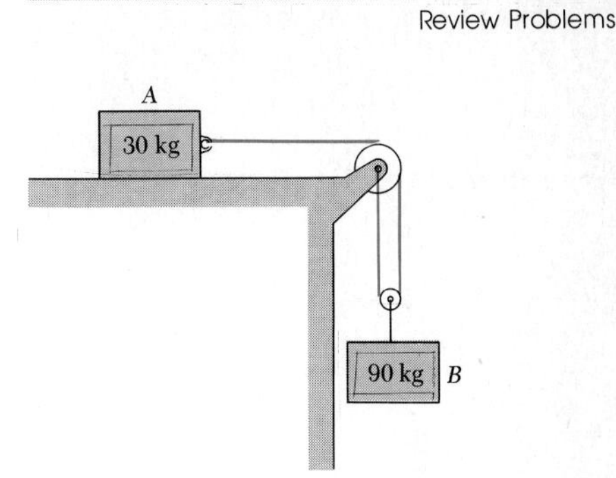

Fig. P13.94

13.94 Two blocks are joined by an inextensible cable as shown. If the system is released from rest, determine the velocity of block A after it has moved 1.75 m. Assume that μ_k equals 0.25 between block A and the plane and neglect the mass and friction of the pulleys.

13.95 A 1500-kg automobile travels 200 m while being accelerated at a uniform rate from 50 to 75 km/h. During the entire motion, the automobile is traveling on a horizontal road, and the rolling resistance is equal to 2 percent of the weight of the automobile. Determine (*a*) the maximum power required, (*b*) the power required to maintain a constant speed of 75 km/h.

13.96 The tension in the spring is zero when the arm ABC is horizontal ($\phi = 0$). If the 100-lb block is released when $\phi = 0$, determine the speed of the block (*a*) when $\phi = 90°$, (*b*) when $\phi = 180°$.

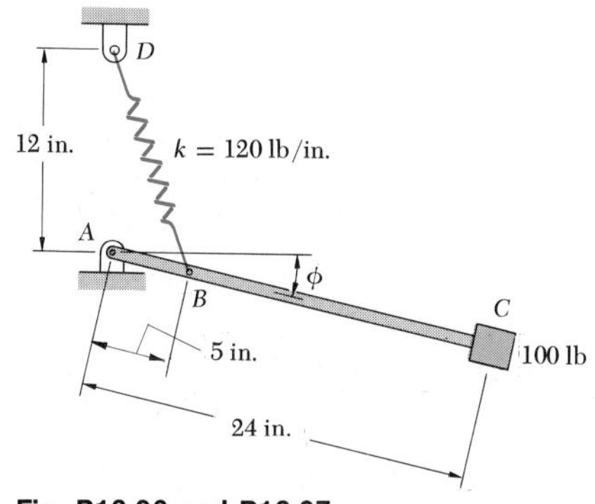

Fig. P13.96 and P13.97

13.97 The 100-lb block is released from rest when $\phi = 0$. If the speed of the block when $\phi = 90°$ is to be 8 ft/s, determine the required value of the initial tension in the spring.

13.98 A 6-oz collar may slide without friction on a rod in a vertical plane. The collar is released from rest when the spring is compressed 1.5 in. As the collar passes through point B, determine (*a*) the speed of the collar, (*b*) the force exerted by the rod on the collar.

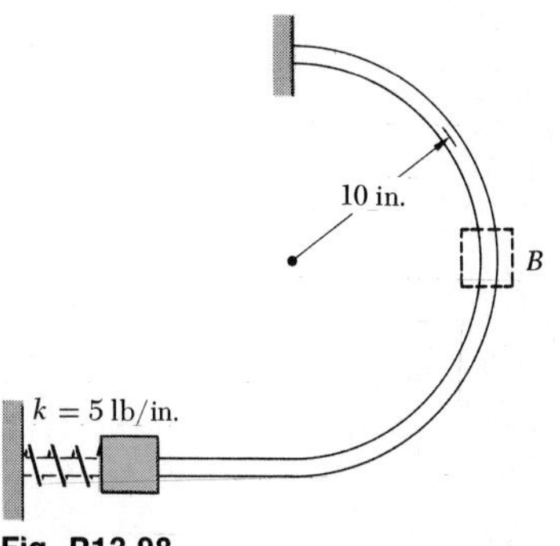

Fig. P13.98

The following problems are designed to be solved with a computer.

13.C1 A 10-lb bag is gently pushed off the top of a wall at A and swings in a vertical plane at the end of an 8-ft rope which can withstand a maximum tension F_m. Write a computer program which, for a given value of F_m, can be used to calculate (*a*) the difference in elevation h between point A and point B where the rope will break, (*b*) the distance d from the vertical wall to the point where the bag strikes the floor. Use this program to calculate h and d for values of F_m from 10 to 30 lb at 2-lb intervals.

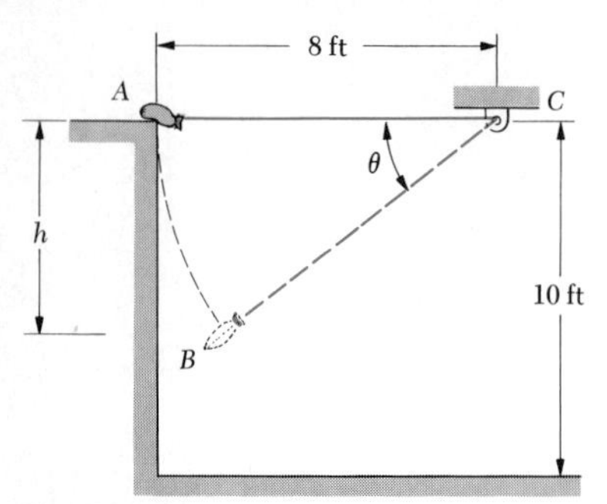

Fig. P13.C1

13.C2 A 12-lb collar slides without friction along a rod which forms an angle of 30° with the vertical. The spring is of constant k and is unstretched when the collar is at A. Knowing that the collar is released from rest at A, write a computer program and use it to calculate the velocity of the collar at point B for values of k from 0.1 to 2.0 lb/in. at 0.1-lb/in. intervals.

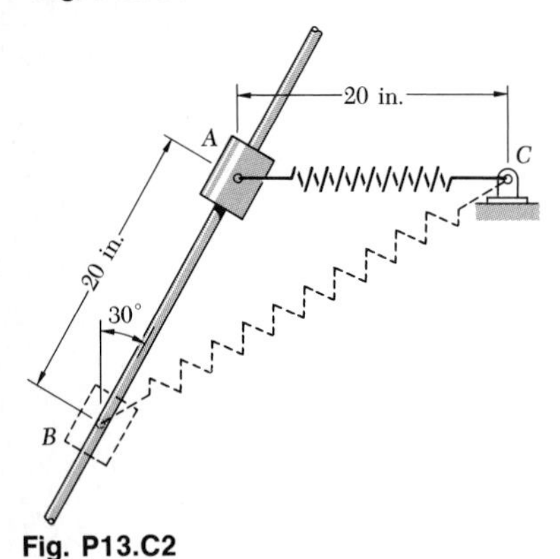

Fig. P13.C2

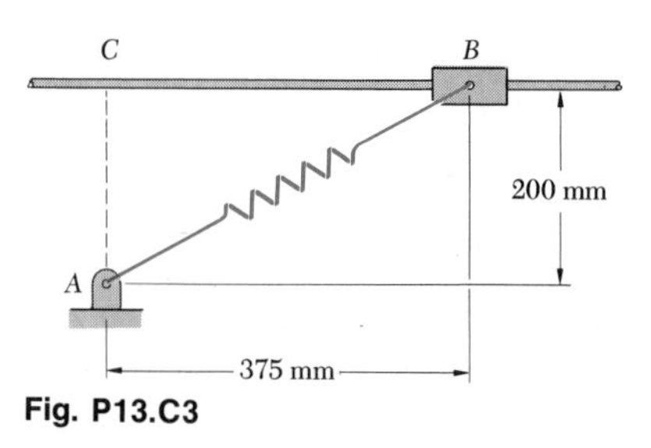

Fig. P13.C3

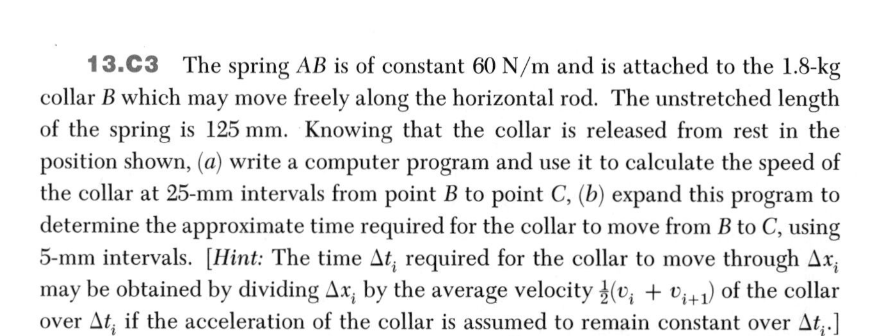
13.C3 The spring AB is of constant 60 N/m and is attached to the 1.8-kg collar B which may move freely along the horizontal rod. The unstretched length of the spring is 125 mm. Knowing that the collar is released from rest in the position shown, (*a*) write a computer program and use it to calculate the speed of the collar at 25-mm intervals from point B to point C, (*b*) expand this program to determine the approximate time required for the collar to move from B to C, using 5-mm intervals. [*Hint:* The time Δt_i required for the collar to move through Δx_i may be obtained by dividing Δx_i by the average velocity $\frac{1}{2}(v_i + v_{i+1})$ of the collar over Δt_i if the acceleration of the collar is assumed to remain constant over Δt_i.]

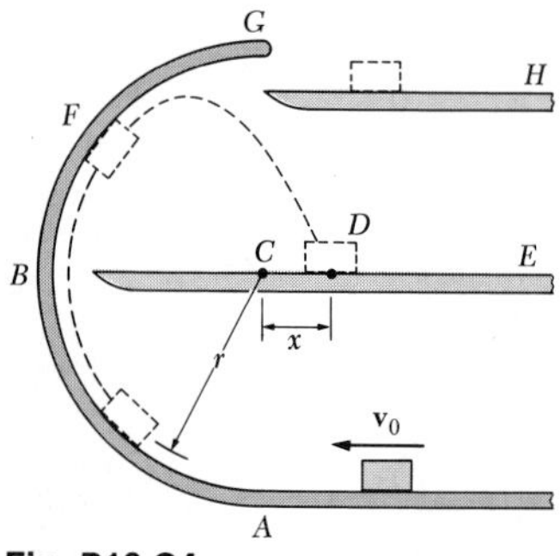

Fig. P13.C4

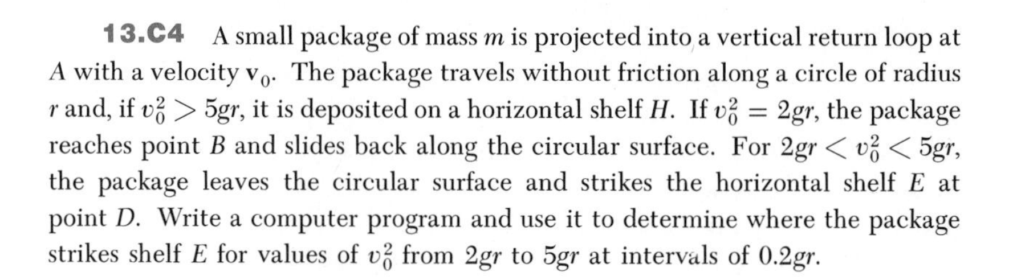
13.C4 A small package of mass m is projected into a vertical return loop at A with a velocity $\mathbf{v}_0$. The package travels without friction along a circle of radius r and, if $v_0^2 > 5gr$, it is deposited on a horizontal shelf H. If $v_0^2 = 2gr$, the package reaches point B and slides back along the circular surface. For $2gr < v_0^2 < 5gr$, the package leaves the circular surface and strikes the horizontal shelf E at point D. Write a computer program and use it to determine where the package strikes shelf E for values of v_0^2 from $2gr$ to $5gr$ at intervals of $0.2gr$.

CHAPTER FOURTEEN

Kinetics of Particles: Impulse and Momentum

14.1. Introduction. In Chap. 12, most problems dealing with the motion of particles were solved by the direct application of Newton's second law. In Chap. 13, the method of work and energy and the principle of conservation of energy were developed from Newton's second law and used to solve a number of problems involving the motion of a particle or of several connected bodies. In this chapter, we shall consider a third basic method for the solution of problems dealing with the motion of particles. This method is based on the *principle of impulse and momentum,* which is also derived from Newton's second law. It is of particular interest in problems involving force, mass, velocity, and time.

After defining the *linear momentum* of a particle and the *impulse* of a force, we shall derive the principle of impulse and momentum for a single particle (Sec. 14.2). This principle is particularly effective in the study of the *impulsive motion* of a particle, where very large forces are applied for a very short time interval (Sec. 14.3).

In Sec. 14.4, we shall extend the principle of impulse and momentum to a *system of particles,* and in Sec. 14.5, we shall find that the total momentum of a system of particles is conserved when the sum of the impulses of the external forces acting on the system is zero.

In Secs. 14.6 through 14.8, we shall consider the *central impact* of two bodies. It will be shown that a certain relation exists between the relative velocities of the two colliding bodies before and after impact. This relation may be used together with the fact that the total momentum of the two bodies is conserved to solve a number of problems of practical interest.

In Sec. 14.9, we shall learn to select from the three fundamental methods presented in Chaps. 12, 13, and 14 the method best suited for the solution of a given problem and we shall see how the method of work and energy and the method of impulse and momentum may be combined to solve problems involving a short impact phase during which impulsive forces must be taken into consideration. We shall also see that in some cases the two methods may be applied jointly to obtain equations which may be solved simultaneously for two or more unknowns.

The concept of *angular momentum* of a particle will be introduced in Sec. 14.10 and it will be shown in Sec. 14.11 that, when the only force acting on a particle is a *central force*, i.e., a force directed toward or away from a fixed point O, the angular momentum of the particle about O is conserved. The gravitational force exerted by the earth on a space vehicle being both central and conservative, the principles of conservation of angular momentum and of conservation of energy may be used jointly to study the motion of such a vehicle. This approach is particularly effective in the case of an oblique launching (Sec. 14.12).

In Sec. 14.13, the concepts of linear and angular momentum and of linear and angular impulse will be used to formulate the principle of impulse and momentum in a more general form applicable to any system of particles. The results obtained will be fundamental to our study in Chap. 18 of the motion of rigid bodies.

The last portion of this chapter is devoted to the study of variable systems of particles. In Sec. 14.15, we shall consider steady streams of particles, such as a stream of water diverted by a fixed vane or the flow of air through a jet engine, and we shall learn to determine the force exerted by the stream on the vane and the thrust developed by the engine. Finally, in Sec. 14.16, we shall analyze systems which gain mass by continually absorbing particles or lose mass by continually expelling particles. Among the various practical applications of this analysis will be the determination of the thrust developed by a rocket engine.

14.2. Principle of Impulse and Momentum. Consider a particle of mass m acted upon by a force $\mathbf{F}$. Expressing Newton's second law $\mathbf{F} = m\mathbf{a}$ in terms of the x and y components of the force and of the acceleration, we write

$$F_x = ma_x \qquad F_y = ma_y$$

$$F_x = m\frac{dv_x}{dt} \qquad F_y = m\frac{dv_y}{dt}$$

Since the mass m of the particle is constant, we have

$$F_x = \frac{d}{dt}(mv_x) \qquad F_y = \frac{d}{dt}(mv_y) \tag{14.1}$$

or, in vector form,

$$\mathbf{F} = \frac{d}{dt}(m\mathbf{v}) \tag{14.2}$$

The vector $m\mathbf{v}$ is called the *linear momentum*, or simply the *momentum*, of the particle. It has the same direction as the velocity of the particle and its magnitude is expressed in N · s or in lb · s. We check that, with SI units,

$$mv = (\text{kg})(\text{m/s}) = (\text{kg} \cdot \text{m/s}^2)\text{s} = \text{N} \cdot \text{s}$$

while, with U.S. customary units,

$$mv = (\text{lb} \cdot \text{s}^2/\text{ft})(\text{ft/s}) = \text{lb} \cdot \text{s}$$

Equation (14.2) expresses that *the force* **F** *acting on the particle is equal to the rate of change of the momentum of the particle.* It is in this form that the second law of motion was originally stated by Newton.

Multiplying both sides of Eqs. (14.1) by dt and integrating from a time t_1 to a time t_2, we write

$$F_x\,dt = d(mv_x) \qquad F_y\,dt = d(mv_y)$$

$$\int_{t_1}^{t_2} F_x\,dt = (mv_x)_2 - (mv_x)_1 \qquad \int_{t_1}^{t_2} F_y\,dt = (mv_y)_2 - (mv_y)_1$$

$$(mv_x)_1 + \int_{t_1}^{t_2} F_x\,dt = (mv_x)_2 \qquad (mv_y)_1 + \int_{t_1}^{t_2} F_y\,dt = (mv_y)_2 \tag{14.3}$$

or, in vector form,

$$m\mathbf{v}_1 + \int_{t_1}^{t_2} \mathbf{F}\,dt = m\mathbf{v}_2 \tag{14.4}$$

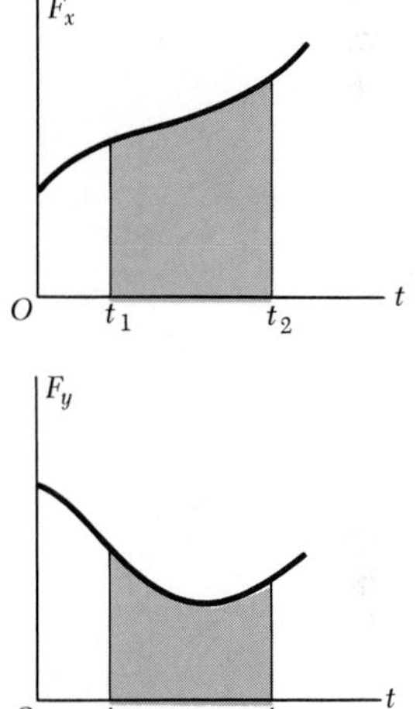

Fig. 14.1

The integral in Eq. (14.4) is a vector known as the *linear impulse,* or simply the *impulse,* of the force **F** during the interval of time considered. From Eqs. (14.3), we note that the components of the impulse of the force **F** are equal, respectively, to the areas under the curves obtained by plotting the components F_x and F_y against t (Fig. 14.1). In the case of a force **F** of constant magnitude and direction, the impulse is represented by the vector $\mathbf{F}(t_2 - t_1)$, which has the same direction as **F**. We easily check that the magnitude of the impulse of a force is expressed in N · s or in lb · s.

Equation (14.4) expresses that, when a particle is acted upon by a force **F** during a given time interval, *the final momentum* $m\mathbf{v}_2$ *of the particle may be obtained by adding vectorially its initial momentum* $m\mathbf{v}_1$ *and the impulse of the force* **F** *during the time interval considered* (Fig. 14.2). We write

$$m\mathbf{v}_1 + \mathbf{Imp}_{1\to 2} = m\mathbf{v}_2 \tag{14.5}$$

We note that, while kinetic energy and work are scalar quantities, momentum and impulse are vector quantities. To obtain an analytic solution, it is thus necessary to replace Eq. (14.5) by the two corresponding component equations (14.3).

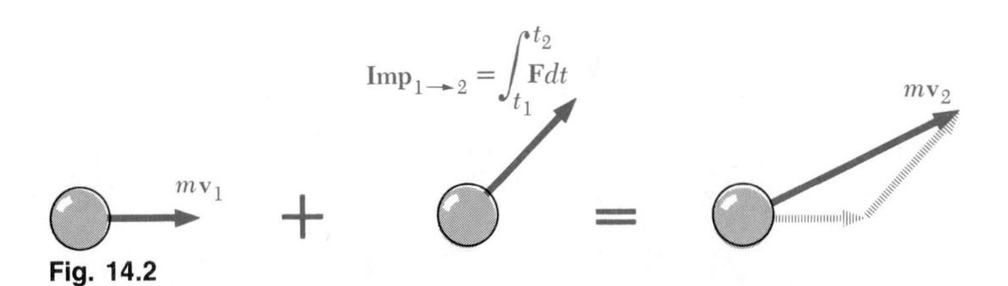

Fig. 14.2

When several forces act on a particle, the impulse of each of the forces must be considered. We have

$$m\mathbf{v}_1 + \Sigma\, \mathbf{Imp}_{1\to2} = m\mathbf{v}_2 \tag{14.6}$$

Again, the equation obtained represents a relation between vector quantities; in the actual solution of a problem, it should be replaced by the two corresponding component equations.

14.3. Impulsive Motion. In some problems, a very large force may act during a very short time interval on a particle and produce a definite change in momentum. Such a force is called an *impulsive force* and the resulting motion an *impulsive motion*. For example, when a baseball is struck, the contact between bat and ball takes place during a very short time interval Δt. But the average value of the force $\mathbf{F}$ exerted by the bat on the ball is very large, and the resulting impulse $\mathbf{F}\,\Delta t$ is large enough to change the sense of motion of the ball (Fig. 14.3).

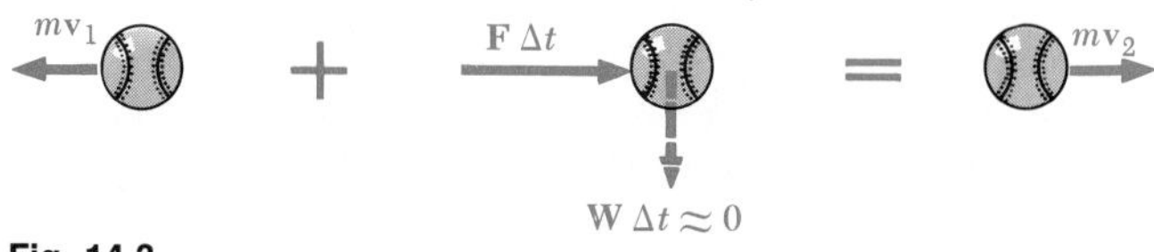

Fig. 14.3

When impulsive forces act on a particle, Eq. (14.6) becomes

$$m\mathbf{v}_1 + \Sigma\mathbf{F}\,\Delta t = m\mathbf{v}_2 \tag{14.7}$$

Any force which is not an impulsive force may be neglected, since the corresponding impulse $\mathbf{F}\,\Delta t$ is very small. *Nonimpulsive forces* include the weight of the body, the force exerted by a spring, or any other force which is *known* to be small compared with an impulsive force. Unknown reactions may or may not be impulsive; their impulse should therefore be included in Eq. (14.7) as long as it has not been proved negligible. The impulse of the weight of the baseball considered above, for example, may be neglected. If the motion of the bat is analyzed, the impulse of the weight of the bat may also be neglected. The impulses of the reactions of the player's hands on the bat, however, should be included; these impulses will not be negligible if the ball is incorrectly hit.

We note that the method of impulse and momentum is particularly effective in the analysis of the impulsive motion of a particle, since it involves only the initial and final velocities of the particle and the impulses of the forces exerted on the particle. The direct application of Newton's second law, on the other hand, would require the determination of the forces as functions of the time and the integration of the equations of motion over the time interval Δt.

14.4. Systems of Particles. When a problem involves the motion of several particles, each particle may be considered separately and Eq. (14.6) may be written for each particle. We may also add vectorially the momenta of all the particles and the impulses of all the forces involved. We write then

$$\Sigma m\mathbf{v}_1 + \Sigma\ \mathbf{Imp}_{1\to2} = \Sigma m\mathbf{v}_2$$

But since the internal forces occur in pairs of equal and opposite forces having the same line of action, and since the time interval from t_1 to t_2 is common to all the forces involved, the impulses of the internal forces cancel out, and only the impulses of the external forces need be considered.† We have

$$\Sigma m\mathbf{v}_1 + \Sigma\ \mathbf{Ext\ Imp}_{1\to2} = \Sigma m\mathbf{v}_2 \tag{14.8}$$

Formula (14.8) may be written in a modified form if the mass center of the system of particles is considered. From Sec. 12.6 we recall that the mass center of the system is the point G of coordinates $\bar{x}$ and $\bar{y}$ defined by the equations

$$(\Sigma m)\bar{x} = \Sigma mx \qquad (\Sigma m)\bar{y} = \Sigma my \tag{14.9}$$

Differentiating (14.9) with respect to t, we have

$$(\Sigma m)\bar{v}_x = \Sigma mv_x \qquad (\Sigma m)\bar{v}_y = \Sigma mv_y \tag{14.10}$$

Writing (14.8) in terms of x and y components, and substituting for Σmv_x and Σmv_y from (14.10), we obtain

$$\begin{aligned} (\Sigma m)(\bar{v}_x)_1 + \Sigma \int_{t_1}^{t_2} (F_x)_{\text{ext}}\,dt &= (\Sigma m)(\bar{v}_x)_2 \\ (\Sigma m)(\bar{v}_y)_1 + \Sigma \int_{t_1}^{t_2} (F_y)_{\text{ext}}\,dt &= (\Sigma m)(\bar{v}_y)_2 \end{aligned} \tag{14.11}$$

We note that Eqs. (14.11) are identical with the equations we would obtain for a particle of mass Σm acted upon by all the external forces. We thus check that the mass center of a system of particles moves as if the entire mass of the system and all the external forces were concentrated at that point.

The equations derived in this section relate only the components of the impulses and momenta of a system of particles. By considering the *moments* as well as the *components* of the vectors involved, we shall obtain in Sec. 14.13 a more meaningful formulation of the principle of impulse and momentum for a system of particles.

† We should note the difference between this statement and the corresponding statement made in Sec. 13.5 regarding the work of the forces of action and reaction between several particles. While the sum of the impulses of these forces is always zero, the sum of their work is zero only under special circumstances, e.g., when the various bodies involved are connected by inextensible cords or links and are thus constrained to move through equal distances.

14.5. Conservation of Momentum. We shall consider now the motion of a system of particles when the sum of the impulses of the external forces is zero. Equation (14.8) reduces then to

$$\Sigma m\mathbf{v}_1 = \Sigma m\mathbf{v}_2 \tag{14.12}$$

Thus, *when the sum of the impulses of the external forces acting on a system of particles is zero, the total momentum of the system remains constant.*

Introducing again the mass center G of the system, and using formulas (14.10), Eq. (14.12) reduces to

$$\overline{\mathbf{v}}_1 = \overline{\mathbf{v}}_2 \tag{14.13}$$

Thus, when the sum of the impulses of the external forces acting on a system of particles is zero, *the mass center of the system moves with a constant velocity.*

Two distinct cases of conservation of momentum are frequently encountered:

1. *The external forces acting on the system during the interval of time considered are balanced.* No matter how long the time interval is, we have Σ **Ext Imp**$_{1\to2} = 0$ and formula (14.12) applies. Consider, for example, two boats, of mass m_A and m_B, initially at rest, which

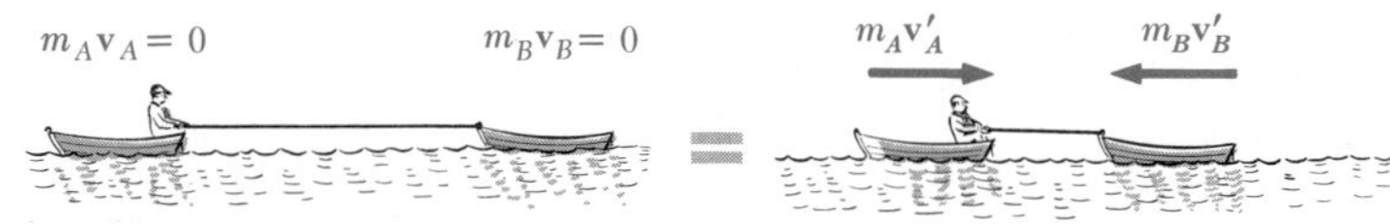

Fig. 14.4

are being pulled together (Fig. 14.4). If the resistance of the water is neglected, the only external forces acting on the boats are their weights and the buoyant forces exerted on them. Since these forces are balanced, we write

$$\Sigma m\mathbf{v}_1 = \Sigma m\mathbf{v}_2$$
$$0 = m_A\mathbf{v}'_A + m_B\mathbf{v}'_B$$

where $\mathbf{v}'_A$ and $\mathbf{v}'_B$ represent the velocities of the boats after a finite interval of time. The equation obtained indicates that the boats move in opposite directions (toward each other) with velocities inversely proportional to their masses. We also note that the mass center of the two boats, which was initially at rest, remains in the same position.

2. *The interval of time considered is very short, and all the external forces are nonimpulsive.* Again we have Σ **Ext Imp**$_{1\to2} = 0$, and formula (14.12) applies. Consider, for example, a bullet of mass m_A fired with a velocity $\mathbf{v}_A$ into a wooden sphere of mass m_B suspended

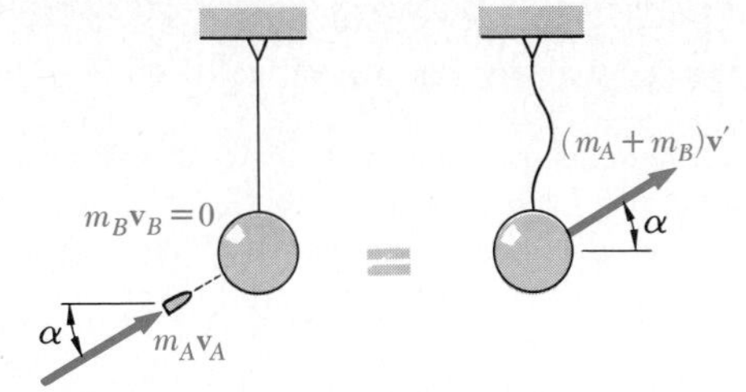

Fig. 14.5

from an inextensible wire and initially at rest (Fig. 14.5). The bullet penetrates the sphere and imparts to it a velocity $\mathbf{v}'$, which we propose to determine. After the sphere has been hit, the tension in the wire becomes zero and the combined weight of the sphere and bullet is unbalanced. But the impulse of the weight may be neglected since the time interval is very short. We write, therefore,

$$\Sigma m\mathbf{v}_1 = \Sigma m\mathbf{v}_2$$
$$m_A\mathbf{v}_A + 0 = (m_A + m_B)\,\mathbf{v}'$$

The equation obtained may be solved for $\mathbf{v}'$; we note that $\mathbf{v}'$ will have the same direction as $\mathbf{v}_A$.

Let us now consider the case when the bullet is fired downward into the sphere (Fig. 14.6). Since the wire is inextensible, it will prevent any downward motion of the sphere and exert on it a reaction **P**. This reaction is unknown and should therefore be assumed impulsive unless proved otherwise. Writing the general formula (14.8) in terms of horizontal and vertical components, and observing again that the impulse of the weight is negligible, we have

$$\Sigma m\mathbf{v}_1 + \Sigma\ \textbf{Ext Imp}_{1\rightarrow 2} = \Sigma m\mathbf{v}_2$$

$\xrightarrow{+}$ x components: $\quad m_A v_A \cos\alpha + 0 = (m_A + m_B)v'$

$+\uparrow y$ components: $\quad -m_A v_A \sin\alpha + P\,\Delta t = 0$

The first equation expresses that *the x component of the momentum is conserved;* it may be used to determine v'. The second equation indicates that the *y component of the linear momentum is not conserved;* this equation may be used to determine the magnitude $P\,\Delta t$ of the impulse of the force exerted by the wire. Thus, the momentum of the system considered is not conserved, and Eq. (14.12) does not hold, except in the x direction.

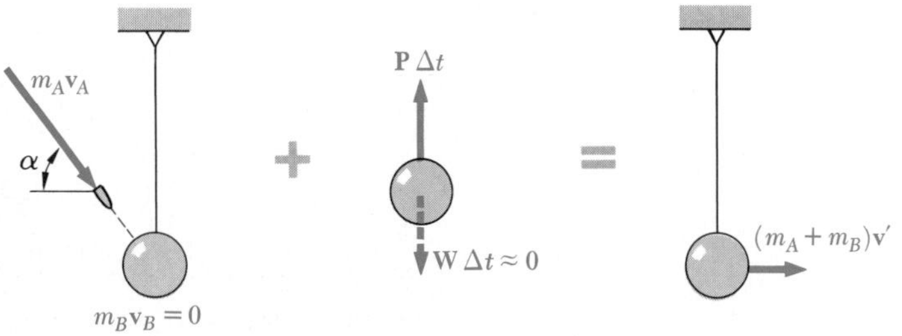

Fig. 14.6

SAMPLE PROBLEM 14.1

An automobile weighing 4000 lb is driven down a 5° incline at a speed of 60 mi/h when the brakes are applied, causing a constant total braking force (applied by the road on the tires) of 1500 lb. Determine the time required for the automobile to come to a stop.

Solution. We apply the principle of impulse and momentum. Since each force is constant in magnitude and direction, each corresponding impulse is equal to the product of the force and of the time interval t.

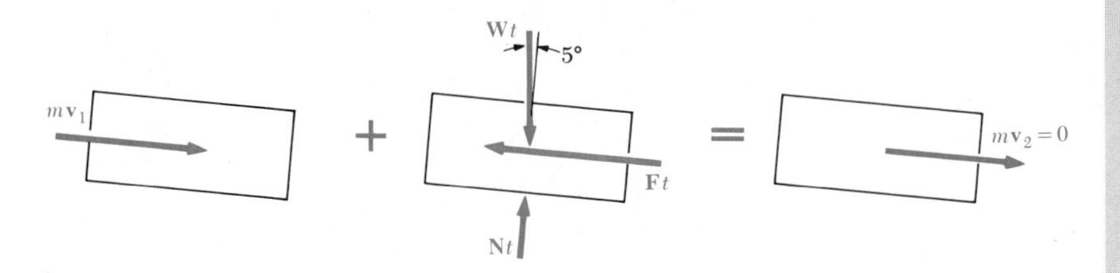

$$m\mathbf{v}_1 + \Sigma\,\mathbf{Imp}_{1\to 2} = m\mathbf{v}_2$$

$+\searrow x$ components: $mv_1 + (W \sin 5°)t - Ft = 0$

$(4000/32.2)(88 \text{ ft/s}) + (4000 \sin 5°)t - 1500t = 0$ $\qquad t = 9.49 \text{ s}$ ◀

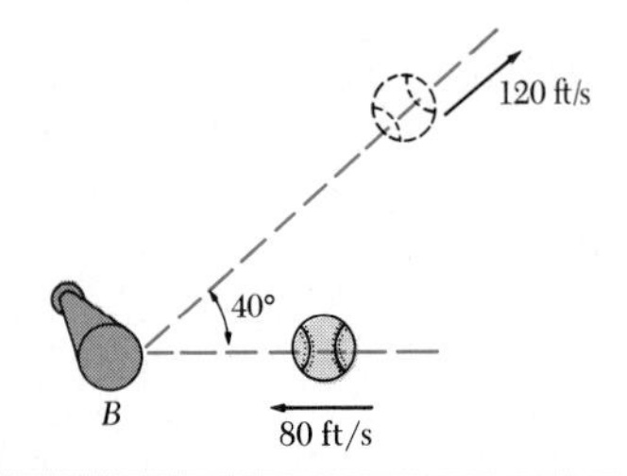

SAMPLE PROBLEM 14.2

A 4-oz baseball is pitched with a velocity of 80 ft/s toward a batter. After the ball is hit by the bat B, it has a velocity of 120 ft/s in the direction shown. If the bat and ball are in contact 0.015 s, determine the average impulsive force exerted on the ball during the impact.

Solution. We apply the principle of impulse and momentum to the ball. Since the weight of the ball is a nonimpulsive force, we shall neglect it.

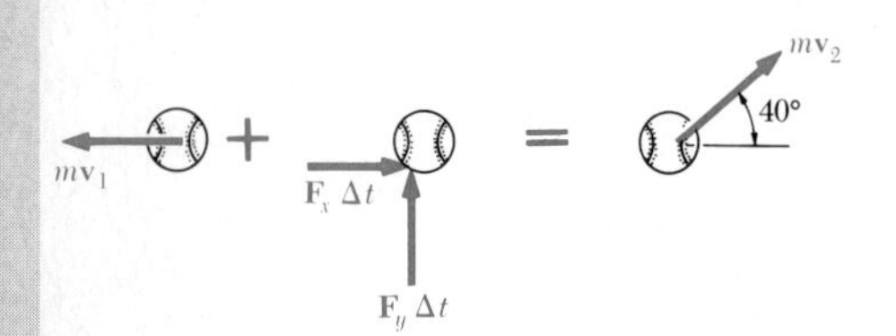

$$m\mathbf{v}_1 + \Sigma\,\mathbf{Imp}_{1\to 2} = m\mathbf{v}_2$$

$\xrightarrow{+}$ x components: $\qquad -mv_1 + F_x\,\Delta t = mv_2 \cos 40°$

$$-\frac{\frac{4}{16}}{32.2}(80 \text{ ft/s}) + F_x(0.015 \text{ s}) = \frac{\frac{4}{16}}{32.2}(120 \text{ ft/s}) \cos 40°$$

$$F_x = +89.0 \text{ lb}$$

$+\uparrow y$ components: $\qquad 0 + F_y\,\Delta t = mv_2 \sin 40°$

$$F_y(0.015 \text{ s}) = \frac{\frac{4}{16}}{32.2}(120 \text{ ft/s}) \sin 40°$$

$$F_y = +39.9 \text{ lb}$$

From its components F_x and F_y we determine the magnitude and direction of the force $\mathbf{F}$:

$\mathbf{F} = 97.5 \text{ lb} \measuredangle 24.2°$ ◀

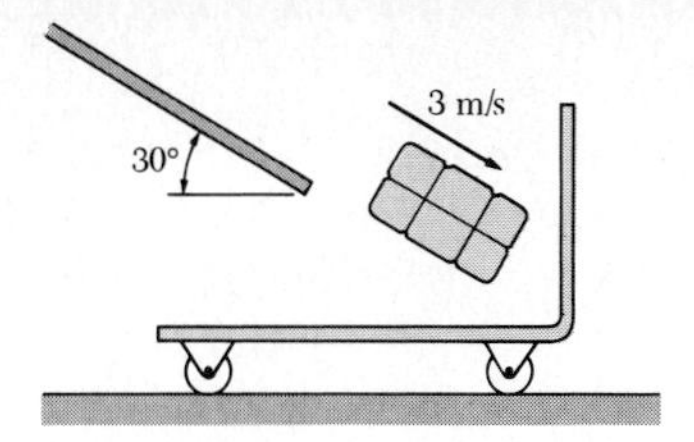

SAMPLE PROBLEM 14.3

A 10-kg package drops from a chute into a 25-kg cart with a velocity of 3 m/s. Knowing that the cart is initially at rest and may roll freely, determine (*a*) the final velocity of the cart, (*b*) the impulse exerted by the cart on the package, (*c*) the fraction of the initial energy lost in the impact.

Solution. We first apply the principle of impulse and momentum to the package-cart system to determine the velocity $\mathbf{v}_2$ of the cart and package. We then apply the same principle to the package alone to determine the impulse $\mathbf{F}\,\Delta t$ exerted on it.

a. Impulse-Momentum Principle: Package and Cart

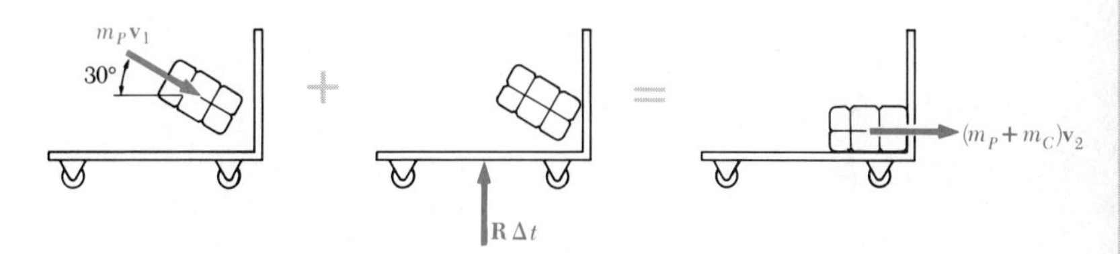

$$m_P\mathbf{v}_1 + \Sigma\,\mathbf{Imp}_{1\to 2} = (m_P + m_C)\mathbf{v}_2$$

$\xrightarrow{+}$ x components: $\quad m_P v_1 \cos 30° + 0 = (m_P + m_C)v_2$

$$(10\text{ kg})(3\text{ m/s})\cos 30° = (10\text{ kg} + 25\text{ kg})v_2$$

$$\mathbf{v}_2 = 0.742\text{ m/s} \rightarrow \blacktriangleleft$$

We note that the equation used expresses conservation of momentum in the x direction.

b. Impulse-Momentum Principle: Package

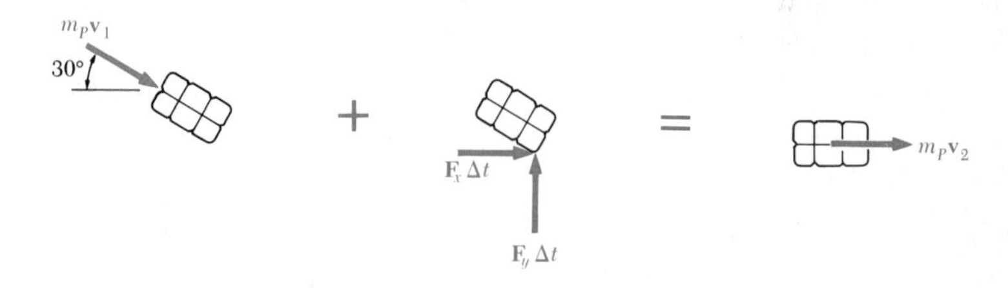

$$m_P\mathbf{v}_1 + \Sigma\,\mathbf{Imp}_{1\to 2} = m_P\mathbf{v}_2$$

$\xrightarrow{+}$ x components: $\quad m_P v_1 \cos 30° + F_x\,\Delta t = m_P v_2$

$$(10\text{ kg})(3\text{ m/s})\cos 30° + F_x\,\Delta t = (10\text{ kg})(0.742\text{ m/s})$$

$$F_x\,\Delta t = -18.56\text{ N}\cdot\text{s}$$

$+\uparrow y$ components: $\quad -m_P v_1 \sin 30° + F_y\,\Delta t = 0$

$$-(10\text{ kg})(3\text{ m/s})\sin 30° + F_y\,\Delta t = 0$$

$$F_y\,\Delta t = +15\text{ N}\cdot\text{s}$$

The impulse exerted on the package is $\quad \mathbf{F}\,\Delta t = 23.9\text{ N}\cdot\text{s} \measuredangle 38.9° \blacktriangleleft$

c. Fraction of Energy Lost. The initial and final energies are

$$T_1 = \tfrac{1}{2}m_P v_1^2 = \tfrac{1}{2}(10\text{ kg})(3\text{ m/s})^2 = 45\text{ J}$$

$$T_2 = \tfrac{1}{2}(m_P + m_C)v_2^2 = \tfrac{1}{2}(10\text{ kg} + 25\text{ kg})(0.742\text{ m/s})^2 = 9.63\text{ J}$$

The fraction of energy lost is $\quad \dfrac{T_1 - T_2}{T_1} = \dfrac{45\text{ J} - 9.63\text{ J}}{45\text{ J}} = 0.786 \blacktriangleleft$

Problems

14.1 A 50,000-ton ocean liner has an initial velocity of 3 mi/h. Neglecting the frictional resistance of the water, determine the time required to bring the liner to rest by using a single tugboat which exerts a constant force of 45 kips.

14.2 A 1200-kg automobile is moving at a speed of 90 km/h when the brakes are fully applied, causing all four wheels to skid. Determine the time required to stop the automobile (*a*) on dry pavement ($\mu_k = 0.75$), (*b*) on an icy road ($\mu_k = 0.10$).

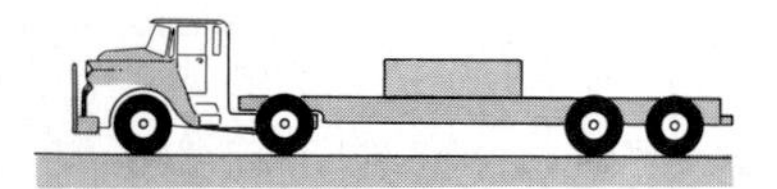

Fig. P14.3

14.3 The coefficients of friction between the load and the flatbed trailer shown are $\mu_s = 0.50$ and $\mu_k = 0.40$. Knowing that the speed of the rig is 72 km/h, determine the shortest time in which the rig can be brought to a stop if the load is not to shift.

14.4 A 15,000-lb plane lands on the deck of an aircraft carrier at a speed of 125 mi/h relative to the carrier and is brought to a stop in 2.80 s. Determine the average horizontal force exerted by the carrier on the plane (*a*) if the carrier is at rest, (*b*) if the carrier is moving at a speed of 18 knots in the same direction as the airplane. (1 knot = 1.152 mi/h.)

14.5 A gun of mass 40 Mg is designed to fire a 200-kg shell with an initial velocity of 600 m/s. Determine the average force required to hold the gun motionless if the shell leaves the gun 25 ms after being fired.

14.6 Using the principle of impulse and momentum, solve Prob. 12.16.

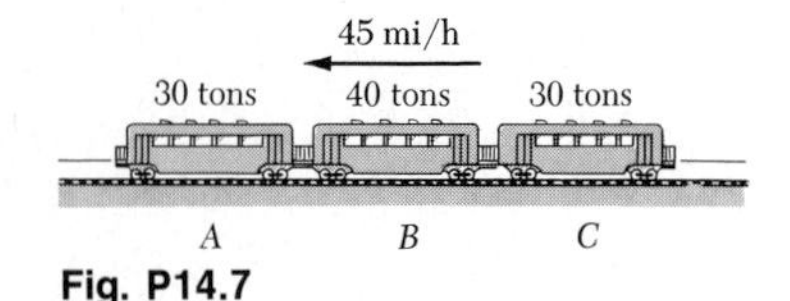

Fig. P14.7

14.7 The subway train shown is traveling at a speed of 45 mi/h when the brakes are fully applied on the wheels of cars *A* and *B*, causing them to slide on the track, but are not applied on the wheels of car *C*. Knowing that the coefficient of kinetic friction is 0.30 between the wheels and the track, determine (*a*) the time required to bring the train to a stop, (*b*) the force in each coupling.

14.8 Solve Prob. 14.7, assuming that the brakes are applied only on the wheels of car *A*.

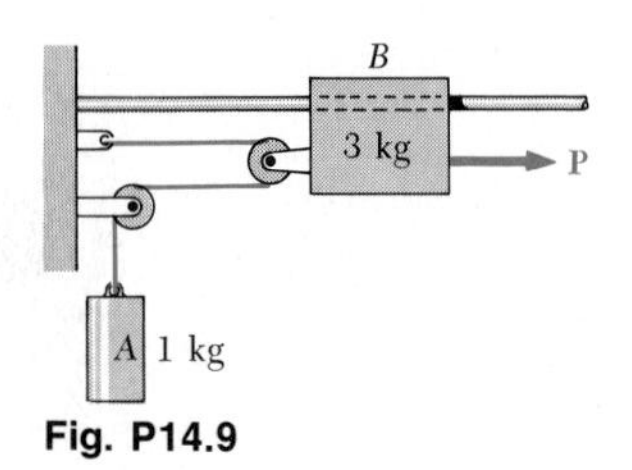

Fig. P14.9

14.9 The system shown is initially at rest. Neglecting friction, determine (*a*) the force **P** required if the velocity of collar *B* is to be 5 m/s after 2 s, (*b*) the corresponding tension in the cable.

14.10 Using the principle of impulse and momentum, solve Prob. 12.18*b*.

14.11 Two packages are placed on an incline as shown. The coefficients of friction are $\mu_s = 0.25$ and $\mu_k = 0.20$ between the incline and package A, and $\mu_s = 0.15$ and $\mu_k = 0.12$ between the incline and package B. Knowing that the packages are in contact when released, determine (*a*) the velocity of each package after 3 s, (*b*) the force exerted by package A on package B.

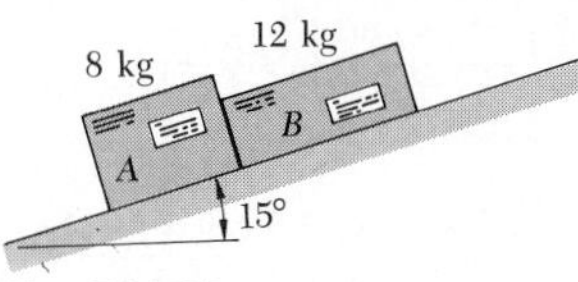

Fig. P14.11

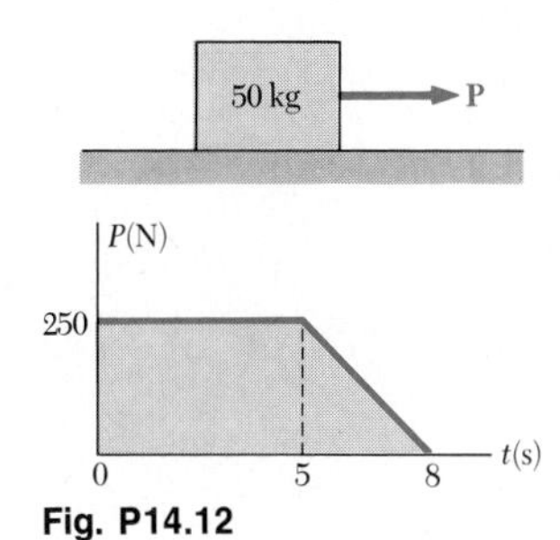

Fig. P14.12

14.12 A 50-kg block initially at rest is acted upon by a force **P** which varies as shown. Knowing that the coefficient of kinetic friction between the block and the horizontal surface is 0.20, determine the velocity of the block (*a*) at $t = 5$ s, (*b*) at $t = 8$ s.

14.13 In Prob. 14.12, determine (*a*) the maximum velocity reached by the block and the corresponding time, (*b*) the time at which the block comes to rest.

14.14 The pressure wave produced by an explosion exerts on a 50-lb block a force **P** whose variation with time may be approximated as shown. Knowing that the block was initially at rest and neglecting the effect of friction, determine (*a*) the maximum velocity reached by the block, (*b*) the velocity of the block at $t = 0.75$ s.

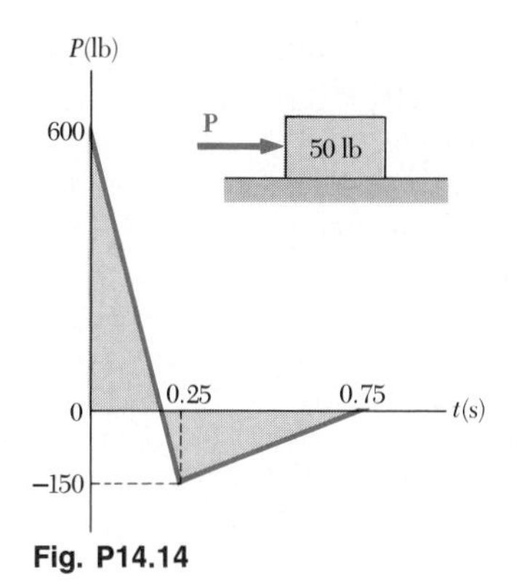

Fig. P14.14

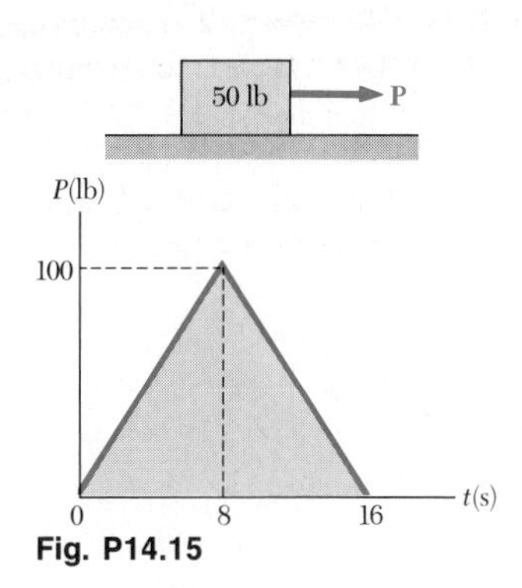

Fig. P14.15

14.15 A 50-lb block initially at rest is acted upon by a force **P** which varies as shown. Knowing that the coefficients of friction between the block and the horizontal surface are $\mu_s = 0.50$ and $\mu_k = 0.40$, determine the time at which the block will (*a*) start moving, (*b*) stop moving.

14.16 A 30-g bullet is fired with a velocity of 640 m/s into a wooden block which rests against a solid vertical wall. Knowing that the bullet is brought to rest in 0.8 ms, determine the average impulsive force exerted by the bullet on the block.

14.17 A 2500-lb car moving with a velocity of 3 mi/h hits a garage wall and is brought to rest in 0.05 s. Determine the average impulsive force exerted by the wall on the car bumper.

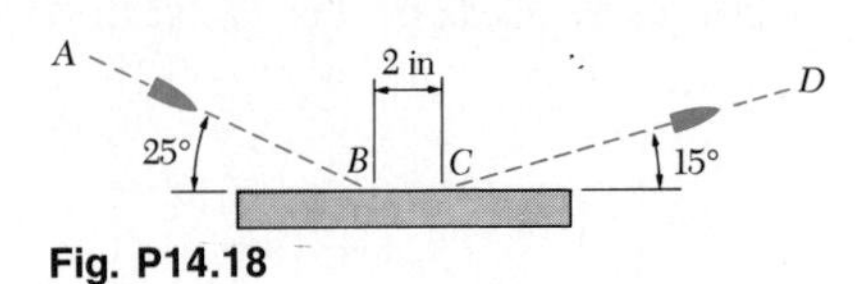

Fig. P14.18

14.18 A 1-oz steel-jacketed bullet is fired with a velocity of 2200 ft/s toward a steel plate and ricochets along the path CD with a velocity of 1800 ft/s. Knowing that the bullet leaves a 2-in. scratch on the surface of the plate and assuming that it has an average speed of 2000 ft/s while in contact with the plate, determine the magnitude and direction of the impulsive force exerted by the plate on the bullet.

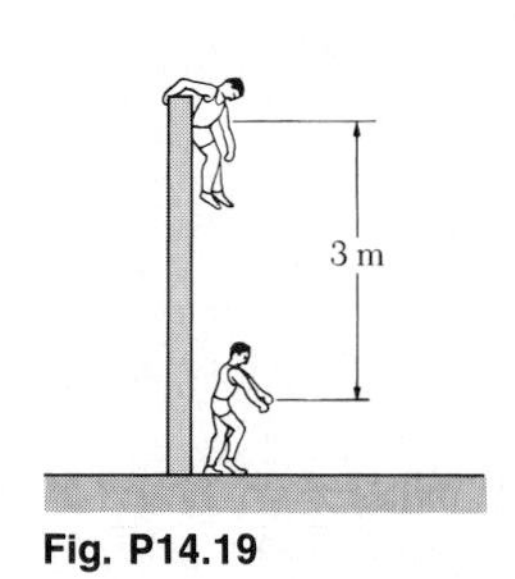

Fig. P14.19

14.19 After scaling a wall, a man lets himself drop 3 m to the ground. If his body comes to a complete stop 0.100 s after his feet first touch the ground, determine the vertical component of the average impulsive force exerted by the ground on his feet.

14.20 A 50-g bullet is fired with a horizontal velocity of 500 m/s into a 4-kg wooden block which is at rest on a frictionless, horizontal surface. Determine (a) the final velocity of the block, (b) the ratio of the final kinetic energy of the block and bullet to the initial kinetic energy of the bullet.

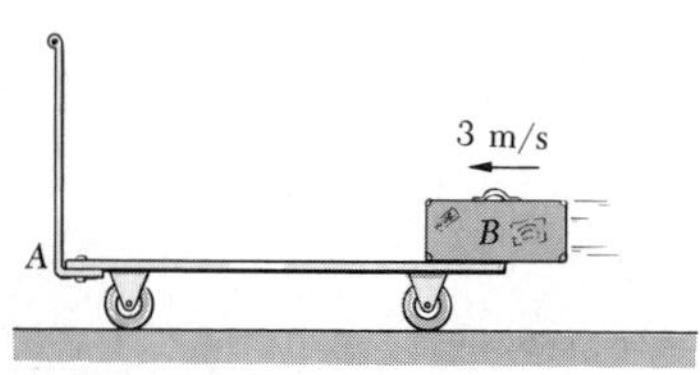

Fig. P14.21

14.21 An airline employee tosses a 15-kg suitcase with a horizontal velocity of 3 m/s onto a 35-kg baggage carrier. Knowing that the carrier is initially at rest and can roll freely, determine (a) the velocity of the carrier after the suitcase has slid to a relative stop on the carrier, (b) the ratio of the final kinetic energy of the carrier and suitcase to the initial kinetic energy of the suitcase.

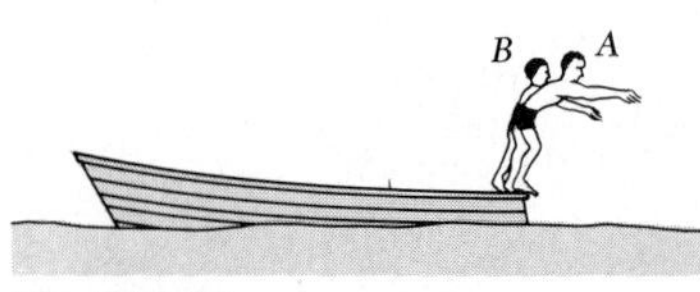

Fig. P14.22

14.22 Two swimmers A and B, of mass 75 kg and 50 kg, respectively, dive off the end of a 250-kg boat. Each swimmer dives so that his relative horizontal velocity with respect to the boat is 4 m/s. If the boat is initially at rest, determine its final velocity, assuming that (a) the two swimmers dive simultaneously, (b) swimmer A dives first, (c) swimmer B dives first.

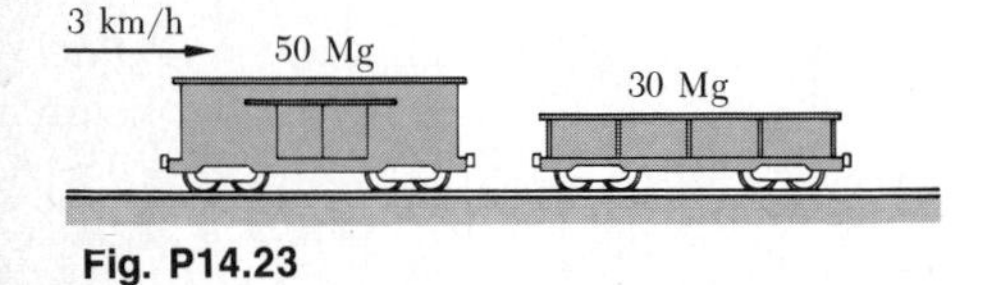

Fig. P14.23

14.23 A 50-Mg railroad car moving with a velocity of 3 km/h is to be coupled to a 30-Mg car which is at rest. Determine (a) the final velocity of the coupled cars, (b) the average impulsive force acting on each car if the coupling is completed in 0.4 s.

14.24 Car A was traveling due north through an intersection when it was hit broadside by car B which was traveling due east. While both drivers admitted having ignored the four-way stop signs at the intersection, each claimed that he was traveling at the 35-mi/h speed limit and that the other was traveling much faster. Knowing that car A weighs 2000 lb, car B 3600 lb, and that inspection of the scene of the accident showed that as a result of the impact the two cars got stuck together and skidded in a direction 40° north of east, determine (*a*) which of the two cars was actually traveling at 35 mi/h, (*b*) how fast the other car was moving.

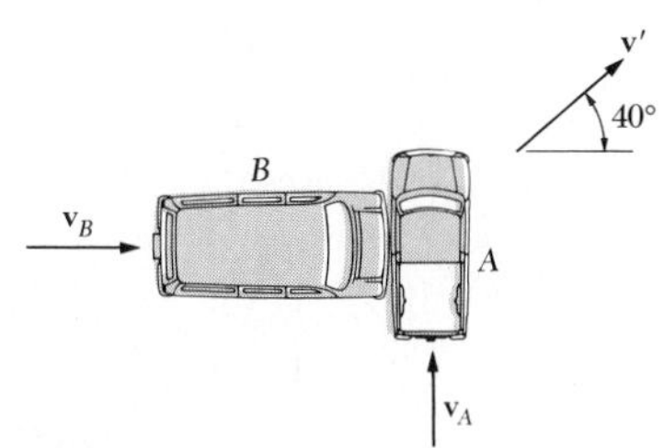

Fig. P14.24

14.25 An old 4000-lb gun fires a 20-lb shell with an initial velocity of 2000 ft/s at an angle of 30°. The gun rests on a horizontal surface and is free to move horizontally. Assuming that the barrel of the gun is rigidly attached to the frame (no recoil mechanism) and that the shell leaves the barrel 6 ms after firing, determine (*a*) the recoil velocity of the gun, (*b*) the resultant **R** of the vertical impulsive forces exerted by the ground on the gun.

Fig. P14.25

14.26 A small rivet connecting two pieces of sheet metal is being clinched by hammering. Determine the impulse exerted on the rivet and the energy absorbed by the rivet under each blow, knowing that the head of the hammer has a mass of 800 g and that it strikes the rivet with a velocity of 6 m/s. Assume that the hammer does not rebound and that the anvil is supported by springs and (*a*) has an infinite mass (rigid support), (*b*) has a mass of 4 kg.

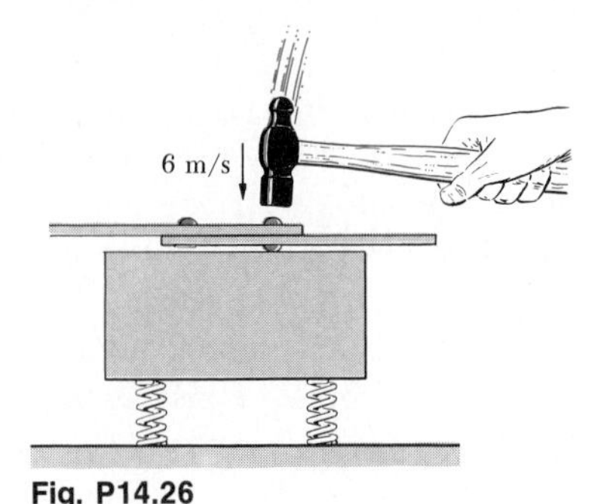

Fig. P14.26

14.27 In order to test the resistance of a chain to impact, the chain is suspended from an 80-kg block supported by two columns. A rod attached to the last link of the chain is then hit by a 20-kg cylinder dropped from a 1.2-m height. Determine the initial impulse exerted on the chain and the energy absorbed by the chain, assuming that the cylinder does not rebound and that the columns supporting the 80-kg block (*a*) are perfectly rigid, (*b*) are equivalent to two perfectly elastic springs.

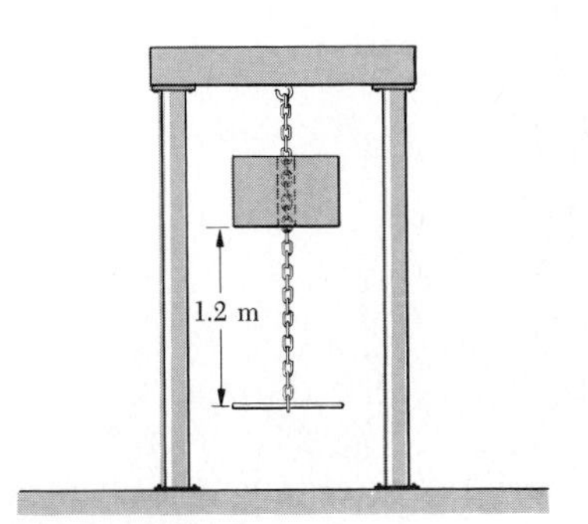

Fig. P14.27

14.6. Impact. A collision between two bodies which occurs in a very small interval of time, and during which the two bodies exert on each other relatively large forces, is called an *impact*. The common normal to the surfaces in contact during the impact is called the *line of impact*. If the mass centers of the two colliding bodies are located on this line, the impact is a *central impact*. Otherwise, the impact is said to be *eccentric*. We shall limit our present study to that of the central impact of two particles and postpone until later the analysis of the eccentric impact of two rigid bodies (Sec. 18.8).

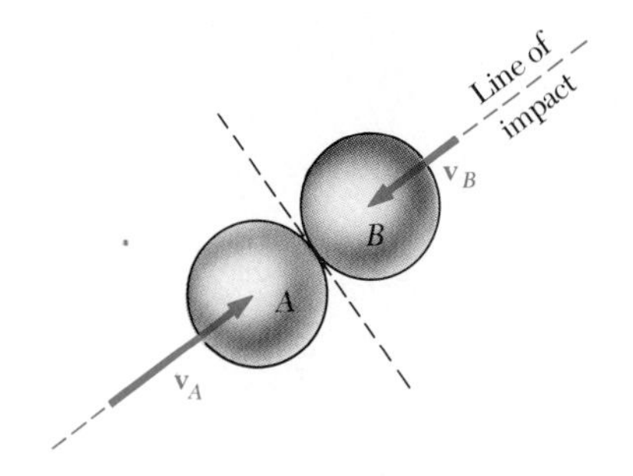

(*a*) Direct central impact

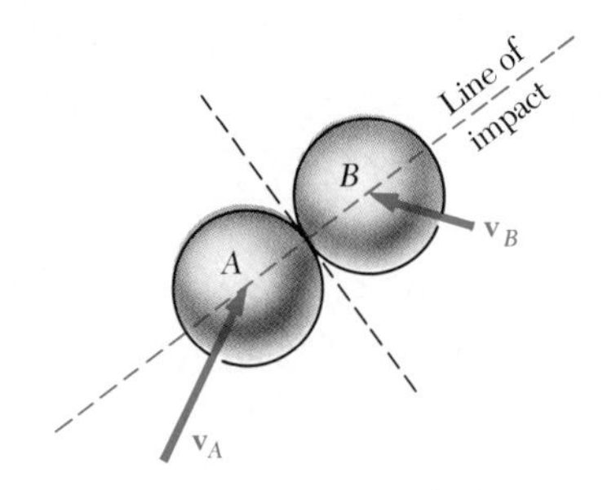

(*b*) Oblique central impact

Fig. 14.7

If the velocities of the two particles are directed along the line of impact, the impact is said to be a *direct impact* (Fig. 14.7*a*). If, on the other hand, either or both particles move along a line other than the line of impact, the impact is said to be an *oblique impact* (Fig. 14.7*b*).

14.7. Direct Central Impact. Consider two particles A and B, of mass m_A and m_B, which are moving in the same straight line and to the right with known velocities $\mathbf{v}_A$ and $\mathbf{v}_B$ (Fig. 14.8*a*). If $\mathbf{v}_A$ is larger than $\mathbf{v}_B$, particle A will eventually strike particle B. Under the impact, the two particles will *deform* and, at the end of the period of deformation, they will have the same velocity $\mathbf{u}$ (Fig. 14.8*b*). A period of *restitution* will then take place, at the end of which, depending upon the magnitude of the impact forces and upon the materials involved, the two particles either will have regained their original shape or will stay permanently deformed. Our purpose here is to determine the velocities $\mathbf{v}'_A$ and $\mathbf{v}'_B$ of the particles at the end of the period of restitution (Fig. 14.8*c*).

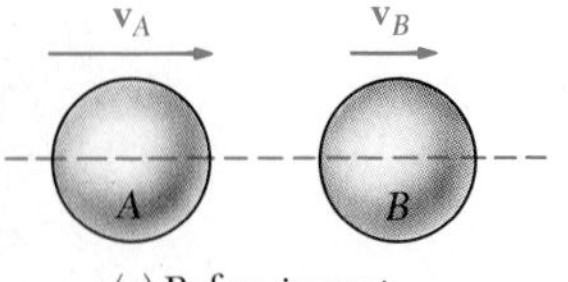

(*a*) Before impact

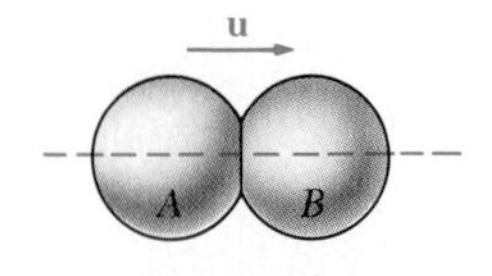

(*b*) At maximum deformation

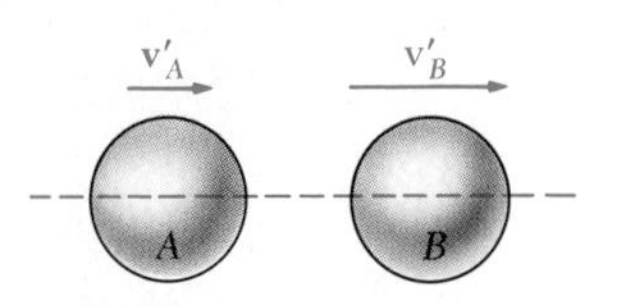

(*c*) After impact

Fig. 14.8

Considering first the two particles as a single system, we note that there is no impulsive, external force. Thus, *the total momentum of the system is conserved*, and we write

$$m_A\mathbf{v}_A + m_B\mathbf{v}_B = m_A\mathbf{v}'_A + m_B\mathbf{v}'_B \tag{14.14}$$

Since all the velocities considered are directed along the same axis, we may replace the equation obtained by the following relation involving only scalar components:

$$m_Av_A + m_Bv_B = m_Av'_A + m_Bv'_B \tag{14.15}$$

A positive value for any of the scalar quantities v_A, v_B, v'_A, or v'_B means that the corresponding vector is directed to the right; a negative value indicates that the corresponding vector is directed to the left.

To obtain the velocities $\mathbf{v}'_A$ and $\mathbf{v}'_B$, it is necessary to establish a second relation between the scalars v'_A and v'_B. For this purpose, we shall consider now the motion of particle A during the period of deformation and apply the principle of impulse and momentum. Since the only impulsive force acting on A during this period is the force $\mathbf{P}$ exerted by B (Fig. 14.9*a*), we write, using again scalar components,

$$m_A v_A - \int P\,dt = m_A u \qquad (14.16)$$

where the integral extends over the period of deformation. Considering now the motion of A during the period of restitution, and denoting by $\mathbf{R}$ the force exerted by B on A during this period (Fig. 14.9*b*), we write

$$m_A u - \int R\,dt = m_A v'_A \qquad (14.17)$$

where the integral extends over the period of restitution.

(*a*) Period of deformation

(*b*) Period of restitution

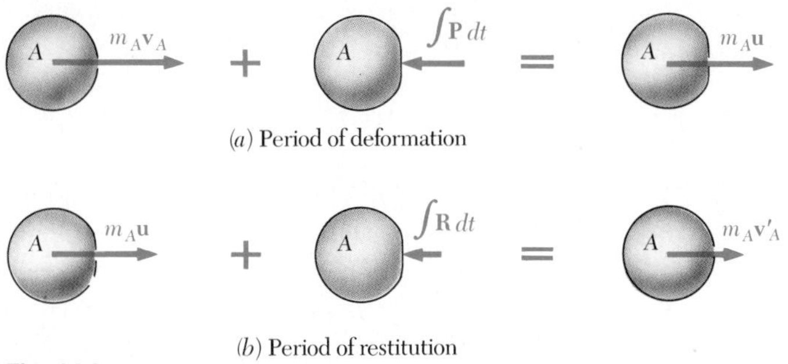

Fig. 14.9

In general, the force $\mathbf{R}$ exerted on A during the period of restitution differs from the force $\mathbf{P}$ exerted during the period of deformation, and the magnitude $\int R\,dt$ of its impulse is smaller than the magnitude $\int P\,dt$ of the impulse of $\mathbf{P}$. The ratio of the magnitudes of the impulses corresponding, respectively, to the period of restitution and to the period of deformation is called the *coefficient of restitution* and is denoted by e. We write

$$e = \frac{\int R\,dt}{\int P\,dt} \qquad (14.18)$$

The value of the coefficient e is always between 0 and 1 and depends to a large extent on the two materials involved. However, it also varies considerably with the impact velocity and the shape and size of the two colliding bodies.

Solving Eqs. (14.16) and (14.17) for the two impulses and substituting into (14.18), we write

$$e = \frac{u - v'_A}{v_A - u} \qquad (14.19)$$

A similar analysis of particle B leads to the relation

$$e = \frac{v'_B - u}{u - v_B} \tag{14.20}$$

Since the quotients in (14.19) and (14.20) are equal, they are also equal to the quotient obtained by adding, respectively, their numerators and their denominators. We have, therefore,

$$e = \frac{(u - v'_A) + (v'_B - u)}{(v_A - u) + (u - v_B)} = \frac{v'_B - v'_A}{v_A - v_B}$$

and

$$v'_B - v'_A = e(v_A - v_B) \tag{14.21}$$

Since $v'_B - v'_A$ represents the relative velocity of the two particles after impact and $v_A - v_B$ their relative velocity before impact, formula (14.21) expresses that *the relative velocity of the two particles after impact may be obtained by multiplying their relative velocity before impact by the coefficient of restitution.* This property is used to determine experimentally the value of the coefficient of restitution of two given materials.

The velocities of the two particles after impact may now be obtained by solving Eqs. (14.15) and (14.21) simultaneously for v'_A and v'_B. It is recalled that the derivation of Eqs. (14.15) and (14.21) was based on the assumption that particle B is located to the right of A, and that both particles are initially moving to the right. If particle B is initially moving to the left, the scalar v_B should be considered negative. The same sign convention holds for the velocities after impact: a positive sign for v'_A will indicate that particle A moves to the right after impact and a negative sign, that it moves to the left.

Two particular cases of impact are of special interest:

1. $e = 0$, *perfectly plastic impact.* When $e = 0$, Eq. (14.21) yields $v'_B = v'_A$. There is no period of restitution, and both particles stay together after impact. Substituting $v'_B = v'_A = v'$ into Eq. (14.15), which expresses that the total momentum of the particles is conserved, we write

$$m_A v_A + m_B v_B = (m_A + m_B)v' \tag{14.22}$$

   This equation may be solved for the common velocity v' of the two particles after impact.
2. $e = 1$, *perfectly elastic impact.* When $e = 1$, Eq. (14.21) reduces to

$$v'_B - v'_A = v_A - v_B \tag{14.23}$$

   which expresses that the relative velocities before and after impact are equal. The impulses received by each particle during the period of deformation and during the period of restitution are equal.

The particles move away from each other after impact with the same velocity with which they approached each other before impact. The velocities v'_A and v'_B may be obtained by solving Eqs. (14.15) and (14.23) simultaneously.

It is worth noting that *in the case of a perfectly elastic impact, the energy of the system,* as well as its momentum, *is conserved.* Equations (14.15) and (14.23) may be written as follows:

$$m_A(v_A - v'_A) = m_B(v'_B - v_B) \qquad (14.15')$$

$$v_A + v'_A = v_B + v'_B \qquad (14.23')$$

Multiplying (14.15′) and (14.23′) member by member, we have

$$m_A(v_A - v'_A)(v_A + v'_A) = m_B(v'_B - v_B)(v'_B + v_B)$$

$$m_A v_A^2 - m_A(v'_A)^2 = m_B(v'_B)^2 - m_B v_B^2$$

Rearranging the terms in the equation obtained, and multiplying by $\frac{1}{2}$, we write

$$\tfrac{1}{2}m_A v_A^2 + \tfrac{1}{2}m_B v_B^2 = \tfrac{1}{2}m_A(v'_A)^2 + \tfrac{1}{2}m_B(v'_B)^2 \qquad (14.24)$$

which expresses that the kinetic energy of the system is conserved. It should be noted, however, that *in the general case of impact,* i.e., when e is not equal to 1, *the energy of the system is not conserved.* This may be shown in any given case by comparing the kinetic energies before and after impact. The lost kinetic energy is in part transformed into heat and in part spent in generating elastic waves within the two colliding bodies.

14.8. Oblique Central Impact. Let us now consider the case when the velocities of the two colliding particles are *not* directed along the line of impact (Fig. 14.10). As indicated in Sec. 14.6, the impact is said to

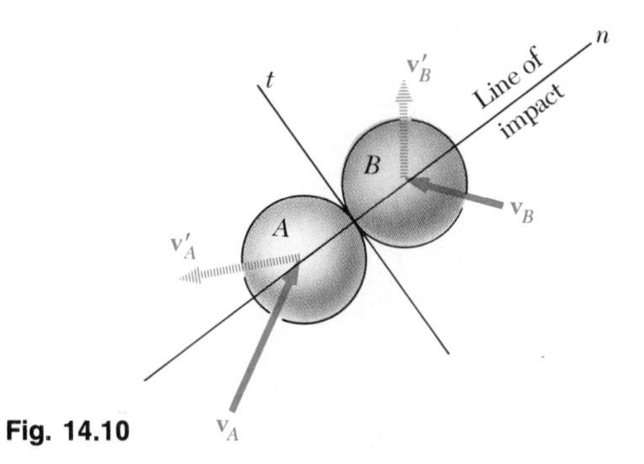

Fig. 14.10

be *oblique.* Since the velocities $\mathbf{v}'_A$ and $\mathbf{v}'_B$ of the particles after impact are unknown in direction as well as in magnitude, their determination will require the use of four independent equations.

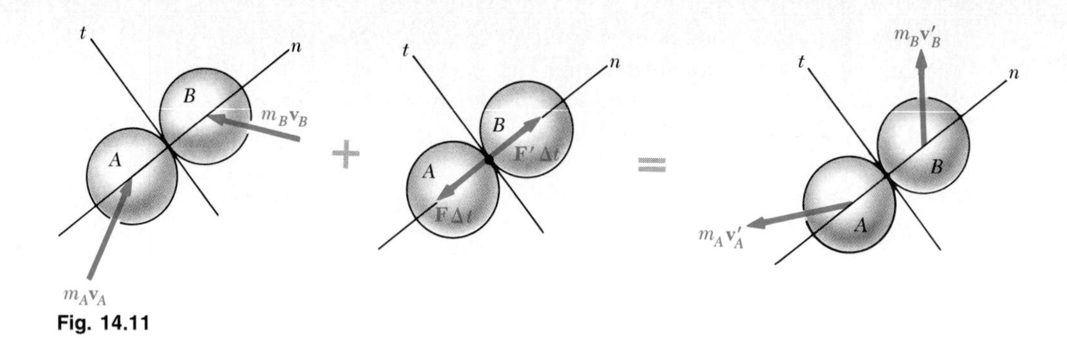

Fig. 14.11

We choose as coordinate axes the n axis along the line of impact, i.e., along the common normal to the surfaces in contact, and the t axis along their common tangent. Assuming that the particles are perfectly *smooth and frictionless*, we observe that the only impulses exerted on the particles during the impact are due to internal forces directed along the line of impact, i.e., along the n axis (Fig. 14.11). It follows that

1. The component along the t axis of the momentum of each particle, considered separately, is conserved; hence the t component of the velocity of each particle remains unchanged. We write

$$(v_A)_t = (v'_A)_t \qquad (v_B)_t = (v'_B)_t \tag{14.25}$$

2. The component along the n axis of the total momentum of the two particles is conserved. We write

$$m_A(v_A)_n + m_B(v_B)_n = m_A(v'_A)_n + m_B(v'_B)_n \tag{14.26}$$

3. The component along the n axis of the relative velocity of the two particles after impact is obtained by multiplying the n component of their relative velocity before impact by the coefficient of restitution. Indeed, a derivation similar to that given in Sec. 14.7 for direct central impact yields

$$(v'_B)_n - (v'_A)_n = e[(v_A)_n - (v_B)_n] \tag{14.27}$$

We have thus obtained four independent equations which may be solved for the components of the velocities of A and B after impact. This method of solution is illustrated in Sample Prob. 14.6.

Our analysis of the oblique central impact of two particles has been based so far on the assumption that both particles moved freely before and after the impact. We shall now examine the case when one or both of the colliding particles is constrained in its motion. Consider, for instance, the collision between block A, which is constrained to move on a horizontal surface, and ball B, which is free to move in the plane of the figure (Fig. 14.12). Assuming no friction between the block and the ball, or between

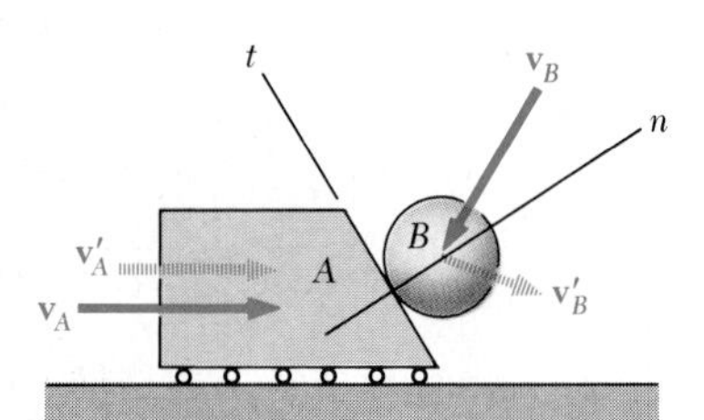

Fig. 14.12

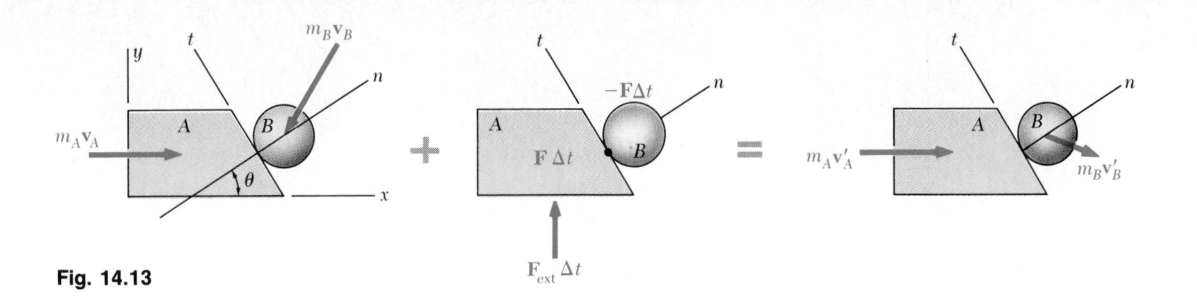

Fig. 14.13

the block and the horizontal surface, we note that the impulses exerted on the system consist of the impulses of the internal forces **F** and **F**′ directed along the line of impact, i.e., along the n axis, and of the impulse of the external force $\mathbf{F}_{ext}$ exerted by the horizontal surface on block A and directed along the vertical (Fig. 14.13).

The velocities of block A and ball B immediately after the impact are represented by three unknowns, namely, the magnitude of the velocity $\mathbf{v}'_A$ of block A, which is known to be horizontal, and the magnitude and direction of the velocity $\mathbf{v}'_B$ of ball B. We shall therefore write three equations by expressing that

1. The component along the t axis of the momentum of ball B is conserved; hence the t component of the velocity of ball B remains unchanged. We write

$$(v_B)_t = (v'_B)_t \qquad (14.28)$$

2. Since the impulses $\mathbf{F}\,\Delta t$ and $\mathbf{F}'\,\Delta t$ are equal and opposite, and since the impulse $\mathbf{F}_{ext}\,\Delta t$ is vertical, the component along the horizontal x axis of the total momentum of the system is conserved. We write

$$m_A v_A + m_B(v_B)_x = m_A v'_A + m_B(v'_B)_x \qquad (14.29)$$

3. The component along the n axis of the relative velocity of block A and ball B after impact is obtained by multiplying the n component of their relative velocity before impact by the coefficient of restitution. We write again

$$(v'_B)_n - (v'_A)_n = e[(v_A)_n - (v_B)_n] \qquad (14.27)$$

We should note, however, that in the case considered here, the validity of Eq. (14.27) cannot be established through a mere extension of the derivation given in Sec. 14.7 for the direct central impact of two particles moving in a straight line. Indeed, these particles were not subjected to any external impulse, while block A in the present analysis is subjected to the impulse exerted by the horizontal surface. To prove that Eq. (14.27) is still

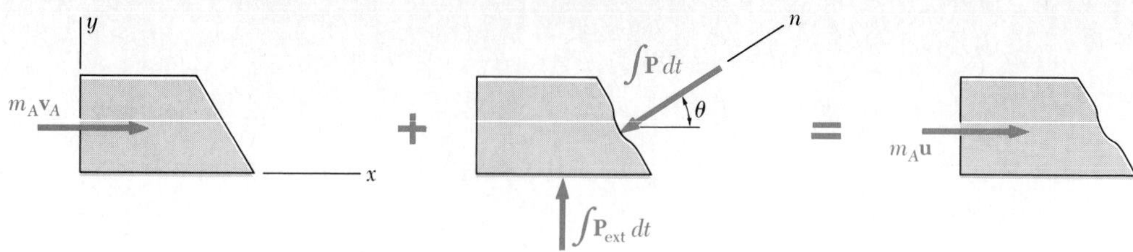

Fig. 14.14

valid, we shall first apply the principle of impulse and momentum to block A over the period of deformation (Fig. 14.14). Considering only the horizontal components, we write

$$m_A v_A - (\int P\,dt)\cos\theta = m_A u \tag{14.30}$$

where the integral extends over the period of deformation and where **u** represents the velocity of block A at the end of that period. Considering now the period of restitution, we write in a similar way

$$m_A u - (\int R\,dt)\cos\theta = m_A v'_A \tag{14.31}$$

where the integral extends over the period of restitution.

Recalling from Sec. 14.7 the definition of the coefficient of restitution, we write

$$e = \frac{\int R\,dt}{\int P\,dt} \tag{14.18}$$

Solving Eqs. (14.30) and (14.31) for the integrals $\int P\,dt$ and $\int R\,dt$, and substituting into Eq. (14.18), we have, after reductions,

$$e = \frac{u - v'_A}{v_A - u}$$

or, multiplying all velocities by $\cos\theta$ to obtain their projections on the line of impact,

$$e = \frac{u_n - (v'_A)_n}{(v_A)_n - u_n} \tag{14.32}$$

We note that Eq. (14.32) is identical to Eq. (14.19) of Sec. 14.7, except for the subscripts n which are used here to indicate that we are considering velocity components along the line of impact. Since the motion of ball B is unconstrained, the proof of Eq. (14.27) may be completed in the same manner as the derivation of Eq. (14.21) of Sec. 14.7. Thus, we conclude that the relation (14.27) between the components along the line of impact of the relative velocities of two colliding particles remains valid when one of the particles is constrained in its motion. The validity of this relation may easily be extended to the case when both particles are constrained in their motion.

SAMPLE PROBLEM 14.4

A 20-Mg railroad car moving at a speed of 0.5 m/s to the right collides with a 35-Mg car which is at rest. If after the collision the 35-Mg car is observed to move to the right at a speed of 0.3 m/s, determine the coefficient of restitution between the two cars.

Solution. We consider the system consisting of the two cars and express that the total momentum is conserved.

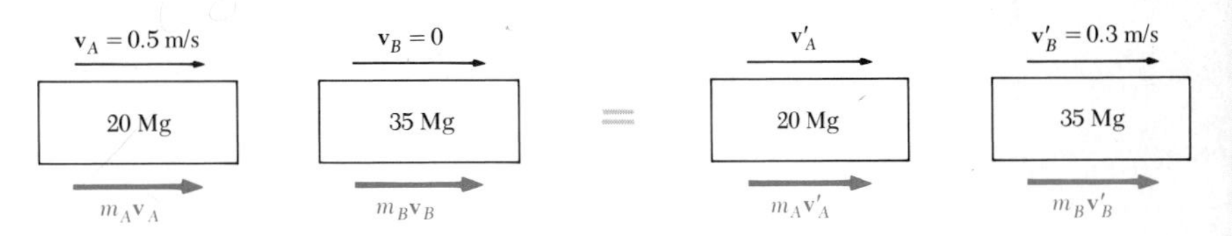

$$m_A\mathbf{v}_A + m_B\mathbf{v}_B = m_A\mathbf{v}'_A + m_B\mathbf{v}'_B$$
$$(20\text{ Mg})(+0.5\text{ m/s}) + (35\text{ Mg})(0) = (20\text{ Mg})v'_A + (35\text{ Mg})(+0.3\text{ m/s})$$
$$v'_A = -0.025\text{ m/s} \qquad \mathbf{v}'_A = 0.025\text{ m/s} \leftarrow$$

The coefficient of restitution is obtained by writing

$$e = \frac{v'_B - v'_A}{v_A - v_B} = \frac{+0.3 - (-0.025)}{+0.5 - 0} = \frac{0.325}{0.5} \qquad e = 0.65 \blacktriangleleft$$

SAMPLE PROBLEM 14.5

A ball is thrown against a frictionless, vertical wall. Immediately before the ball strikes the wall, its velocity has a magnitude v and forms an angle of 30° with the horizontal. Knowing that $e = 0.90$, determine the magnitude and direction of the velocity of the ball as it rebounds from the wall.

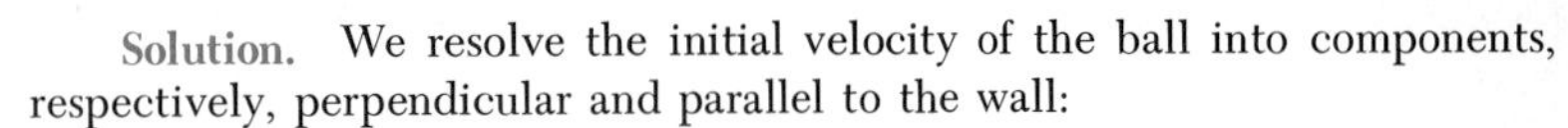

Solution. We resolve the initial velocity of the ball into components, respectively, perpendicular and parallel to the wall:

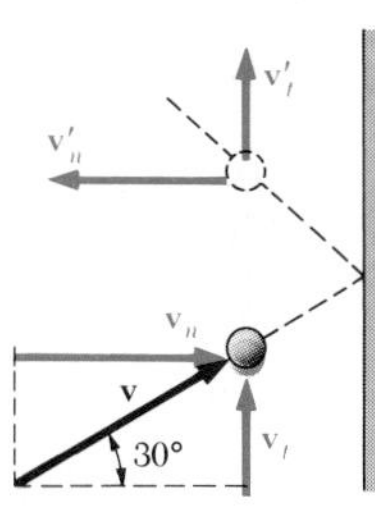

$$v_n = v\cos 30° = 0.866v \qquad v_t = v\sin 30° = 0.500v$$

Motion Parallel to the Wall. Since the wall is frictionless, the impulse it exerts on the ball is perpendicular to the wall. Thus, the component parallel to the wall of the momentum of the ball is conserved and we have

$$\mathbf{v}'_t = \mathbf{v}_t = 0.500v \uparrow$$

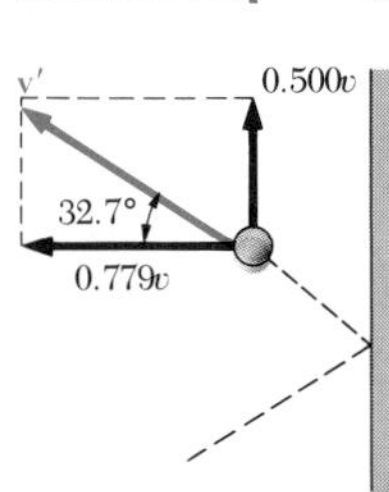

Motion Perpendicular to the Wall. Since the mass of the wall (and earth) is essentially infinite, expressing that the total momentum of the ball and wall is conserved would yield no useful information. Using the relation (14.27) between relative velocities, we write

$$0 - v'_n = e(v_n - 0)$$
$$v'_n = -0.90(0.866v) = -0.779v \qquad \mathbf{v}'_n = 0.779v \leftarrow$$

Resultant Motion. Adding vectorially the components $\mathbf{v}'_n$ and $\mathbf{v}'_t$,

$$\mathbf{v}' = 0.926v \measuredangle 32.7° \blacktriangleleft$$

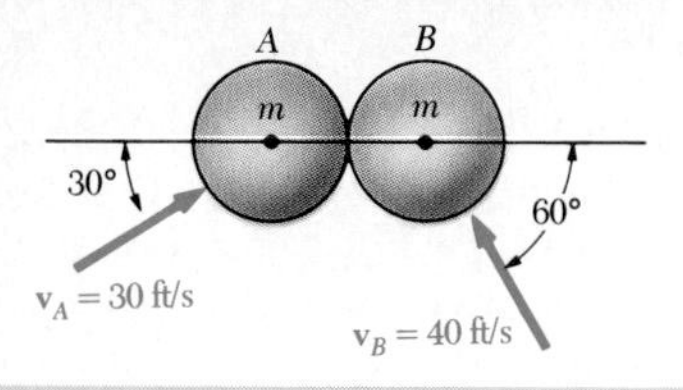

SAMPLE PROBLEM 14.6

The magnitude and direction of the velocities of two identical frictionless balls before they strike each other are as shown. Assuming $e = 0.90$, determine the magnitude and direction of the velocity of each ball after the impact.

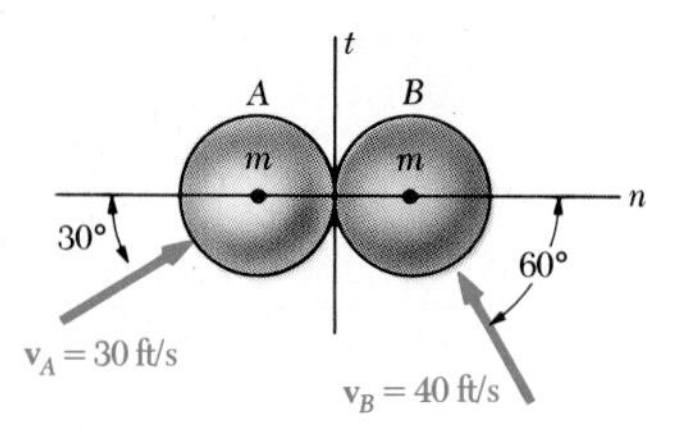

Solution. The impulsive forces acting between the balls during the impact are directed along a line joining the centers of the balls called the *line of impact.* Resolving the velocities into components directed, respectively, along the line of impact and along the common tangent to the surfaces in contact, we write

$$(v_A)_n = v_A \cos 30° = +26.0 \text{ ft/s}$$
$$(v_A)_t = v_A \sin 30° = +15.0 \text{ ft/s}$$
$$(v_B)_n = -v_B \cos 60° = -20.0 \text{ ft/s}$$
$$(v_B)_t = v_B \sin 60° = +34.6 \text{ ft/s}$$

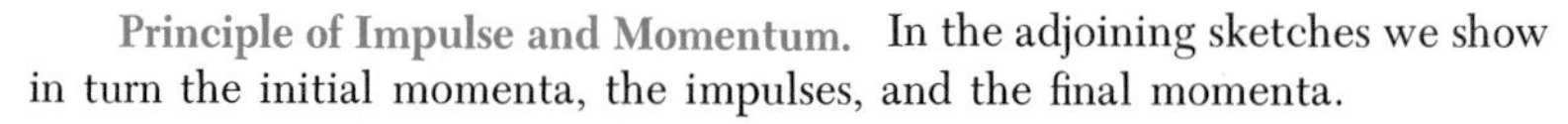

Principle of Impulse and Momentum. In the adjoining sketches we show in turn the initial momenta, the impulses, and the final momenta.

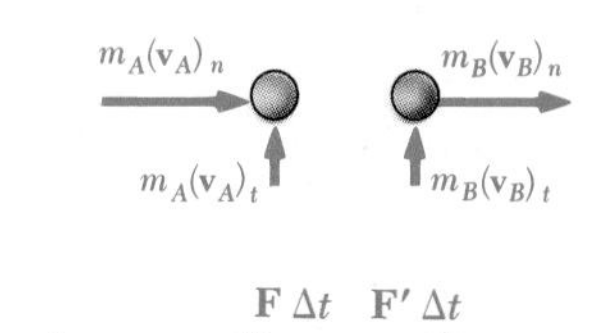

Motion along the Common Tangent. Considering only the t components, we apply the principle of impulse and momentum to each ball *separately*. Since the impulsive forces are directed along the line of impact, the t component of the momentum, and hence the t component of the velocity of each ball, is unchanged. We have

$$(\mathbf{v}'_A)_t = 15.0 \text{ ft/s} \uparrow \qquad (\mathbf{v}'_B)_t = 34.6 \text{ ft/s} \uparrow$$

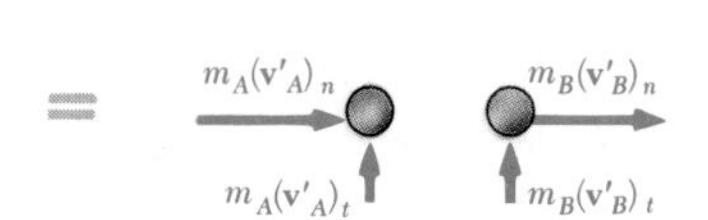

Motion along the Line of Impact. In the n direction, we consider the two balls as a single system and note that by Newton's third law, the internal impulses $\mathbf{F}\,\Delta t$ and $\mathbf{F}'\,\Delta t$ cancel. We thus write that the total momentum of the balls is conserved:

$$m_A(v_A)_n + m_B(v_B)_n = m_A(v'_A)_n + m_B(v'_B)_n$$
$$m(26.0) + m(-20.0) = m(v'_A)_n + m(v'_B)_n$$
$$(v'_A)_n + (v'_B)_n = 6.0 \quad (1)$$

Using the relation (14.27) between relative velocities, we write

$$(v'_B)_n - (v'_A)_n = e[(v_A)_n - (v_B)_n]$$
$$(v'_B)_n - (v'_A)_n = (0.90)[26.0 - (-20.0)]$$
$$(v'_B)_n - (v'_A)_n = 41.4 \quad (2)$$

Solving Eqs. (1) and (2) simultaneously, we obtain

$$(v'_A)_n = -17.7 \qquad (v'_B)_n = +23.7$$
$$(\mathbf{v}'_A)_n = 17.7 \text{ ft/s} \leftarrow \qquad (\mathbf{v}'_B)_n = 23.7 \text{ ft/s} \rightarrow$$

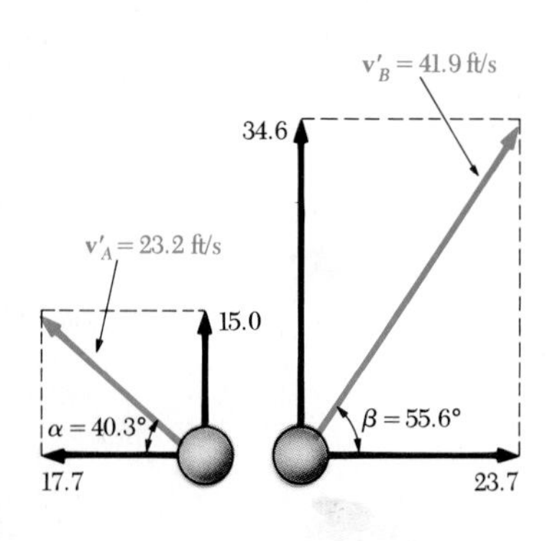

Resultant Motion. Adding vectorially the velocity components of each ball, we obtain

$$\mathbf{v}'_A = 23.2 \text{ ft/s} ⦫ 40.3° \qquad \mathbf{v}'_B = 41.9 \text{ ft/s} ⦪ 55.6° \blacktriangleleft$$

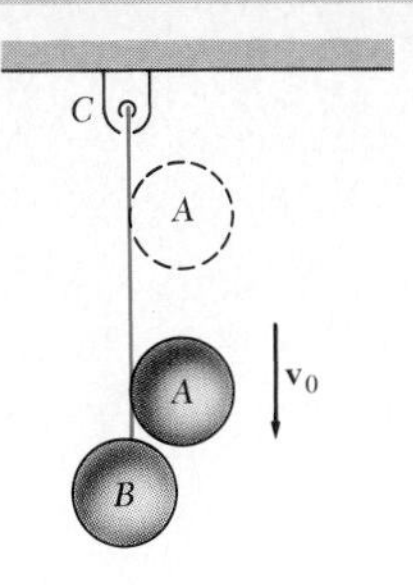

SAMPLE PROBLEM 14.7

Ball B is hanging from an inextensible cord BC. An identical ball A is released from rest when it is just touching the cord and acquires a velocity $\mathbf{v}_0$ before striking ball B. Assuming perfectly elastic impact ($e = 1$) and no friction, determine the velocity of each ball immediately after impact.

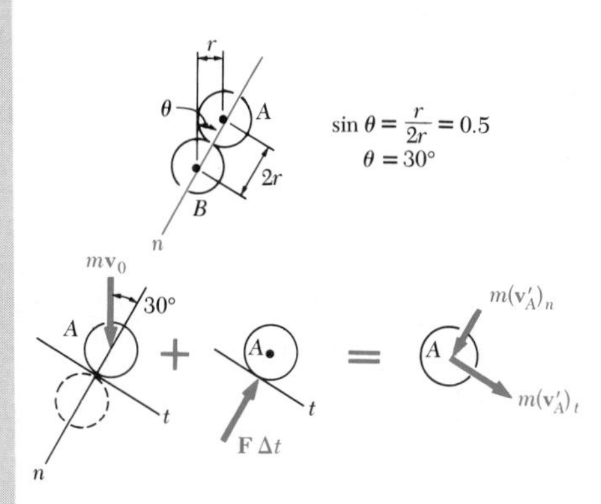

Solution. Since ball B is constrained to move in a circle of center C, its velocity $\mathbf{v}'_B$ after impact must be horizontal. Thus the problem involves three unknowns: the magnitude v'_B of the velocity of B, and the magnitude and direction of the velocity $\mathbf{v}'_A$ of A after impact.

Impulse-Momentum Principle: Ball A

$$m\mathbf{v}_A + \mathbf{F}\,\Delta t = m\mathbf{v}'_A$$

$+\searrow t$ components:
$$mv_0 \sin 30° + 0 = m(v'_A)_t$$
$$(v'_A)_t = 0.5v_0 \qquad (1)$$

We note that the equation used expresses conservation of the momentum of ball A along the common tangent to balls A and B.

Impulse-Momentum Principle: Balls A and B

$$m\mathbf{v}_A + \mathbf{T}\,\Delta t = m\mathbf{v}'_A + m\mathbf{v}'_B$$

$\overset{+}{\rightarrow} x$ components:
$$0 = m(v'_A)_t \cos 30° - m(v'_A)_n \sin 30° - mv'_B$$

We note that the equation obtained expresses conservation of the total momentum in the x direction. Substituting for $(v'_A)_t$ from Eq. (1) and rearranging terms, we write

$$0.5(v'_A)_n + v'_B = 0.433v_0 \qquad (2)$$

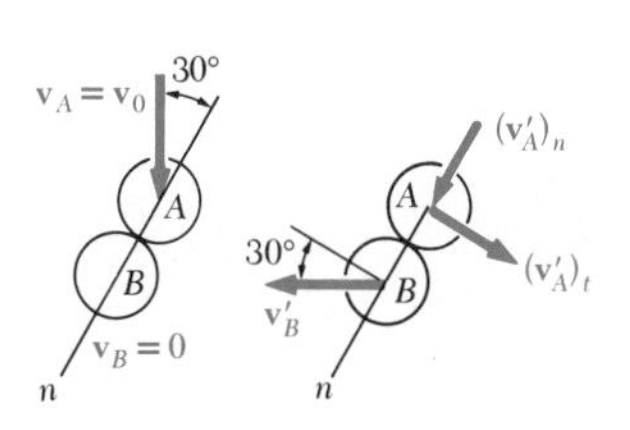

Relative Velocities along the Line of Impact. Since $e = 1$, Eq. (14.27) yields

$$(v'_B)_n - (v'_A)_n = (v_A)_n - (v_B)_n$$
$$v'_B \sin 30° - (v'_A)_n = v_0 \cos 30° - 0$$
$$0.5v'_B - (v'_A)_n = 0.866v_0 \qquad (3)$$

Solving Eqs. (2) and (3) simultaneously, we obtain

$$(v'_A)_n = -0.520v_0 \qquad v'_B = 0.693v_0$$
$$\mathbf{v}'_B = 0.693v_0 \leftarrow \blacktriangleleft$$

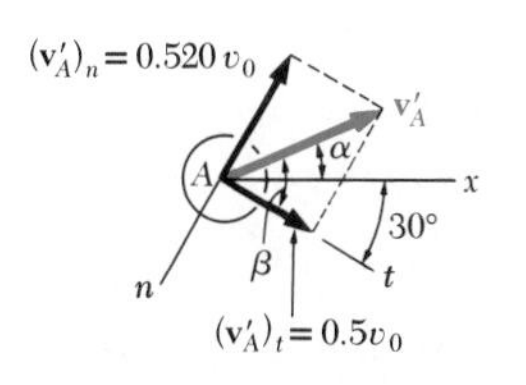

Recalling Eq. (1) we draw the adjoining sketch and obtain by trigonometry

$$v'_A = 0.721v_0 \qquad \beta = 46.1° \qquad \alpha = 46.1° - 30° = 16.1°$$
$$\mathbf{v}'_A = 0.721v_0 \measuredangle 16.1° \blacktriangleleft$$

Problems

14.28 The velocities of two collars before impact are as shown. If after impact the velocity of collar A is observed to be 5.4 m/s to the left, determine (a) the velocity of collar B after impact, (b) the coefficient of restitution between the two collars.

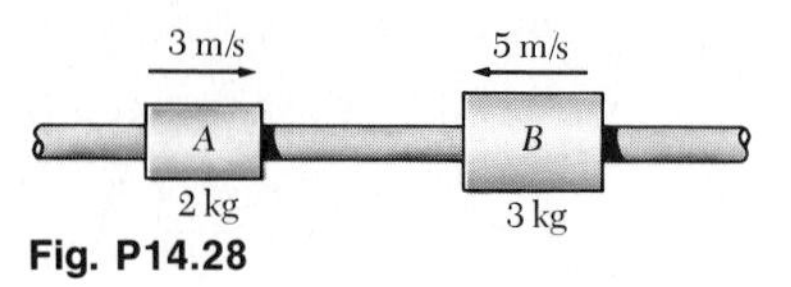

Fig. P14.28

14.29 Solve Prob. 14.28, assuming that the velocity of collar A after impact is observed to be 3 m/s to the left.

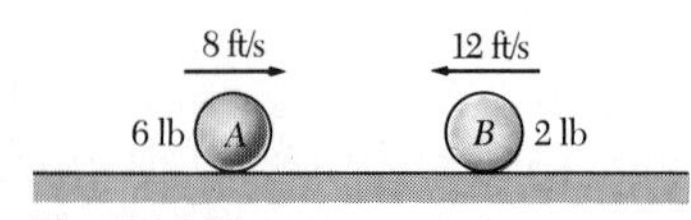

Fig. P14.30

14.30 Two small spheres A and B are made of different materials and have the weights indicated. They are moving on a frictionless, horizontal surface with the velocities shown when they hit each other. Knowing that the coefficient of restitution between the spheres is $e = 0.80$, determine (a) the velocity of each sphere after impact, (b) the energy loss due to the impact.

14.31 Solve Prob. 14.30, assuming that sphere A is moving to the left at the speed of 8 ft/s.

14.32 Two identical cars B and C are at rest on a loading dock with their brakes released. Car A of the same model, which has been pushed by dock workers, hits car B with a velocity of 1.6 m/s, causing a series of collisions among the three cars. Assuming a coefficient of restitution $e = 0.70$ between the bumpers, determine the velocity of each car after *all* collisions have taken place.

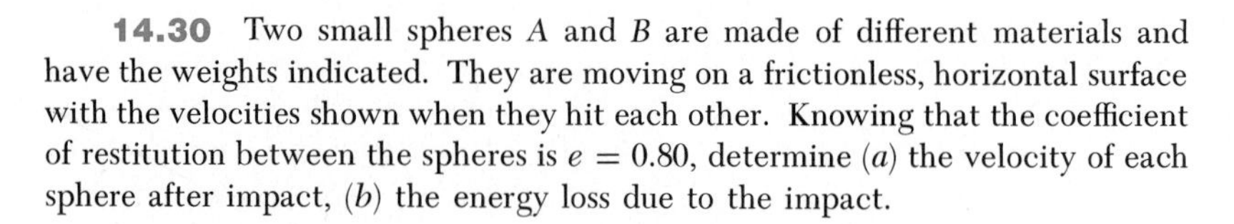

Fig. P14.32

14.33 In Prob. 14.32, it is observed that the final velocity of car C is 0.9 m/s to the right. Determine (a) the coefficient of restitution e between the bumpers, (b) the velocities of cars A and B after *all* collisions have taken place.

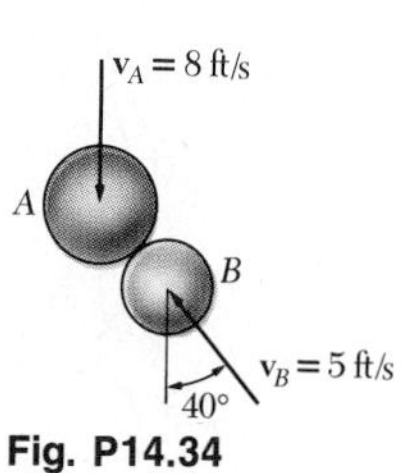

Fig. P14.34

14.34 A 2.5-lb ball A is falling vertically with a velocity of magnitude $v_A = 8$ ft/s when it is hit as shown by a 1.5-lb ball B which has a velocity of magnitude $v_B = 5$ ft/s. Knowing that the coefficient of restitution between the two balls is $e = 0.75$ and assuming no friction, determine the velocity of each ball immediately after impact.

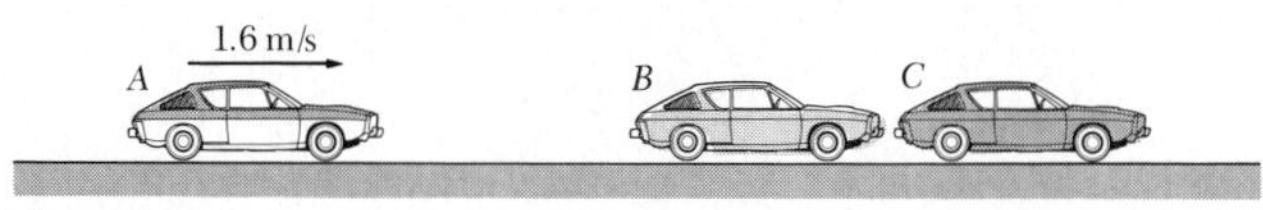

14.35 Two identical billiard balls may move freely on a horizontal table. Ball A has a velocity $\mathbf{v}_0$ as shown and hits ball B, which is at rest, at a point C defined by $\theta = 30°$. Knowing that the coefficient of restitution between the two balls is $e = 0.90$ and assuming no friction, determine the velocity of each ball after impact.

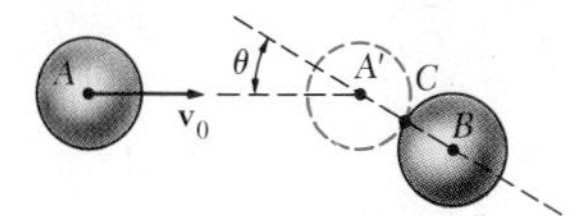

Fig. P14.35 and P14.36

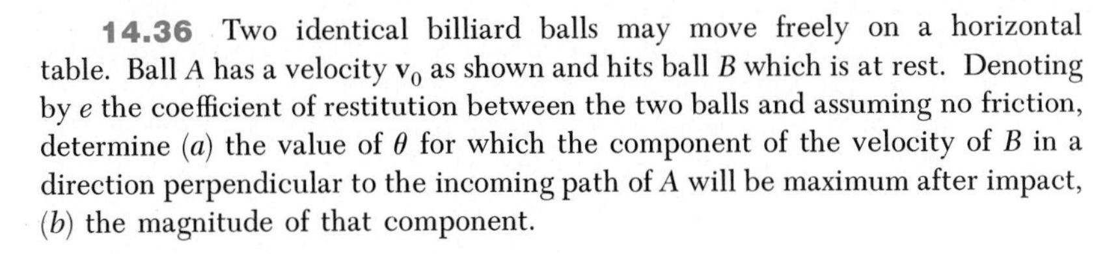

14.36 Two identical billiard balls may move freely on a horizontal table. Ball A has a velocity $\mathbf{v}_0$ as shown and hits ball B which is at rest. Denoting by e the coefficient of restitution between the two balls and assuming no friction, determine (a) the value of θ for which the component of the velocity of B in a direction perpendicular to the incoming path of A will be maximum after impact, (b) the magnitude of that component.

14.37 A billiard player wishes to have ball A hit ball B obliquely and then ball C squarely. Assuming perfectly elastic impact $(e = 1)$ and denoting by r the radius of the balls and by d the distance between the centers of B and C, determine (a) the angle θ defining point D where ball B should be hit, (b) the ranges of values of angles ABC and ACB for which this play is possible.

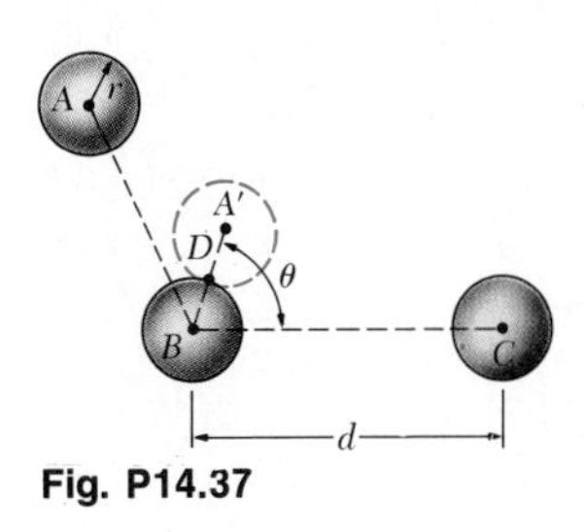

Fig. P14.37

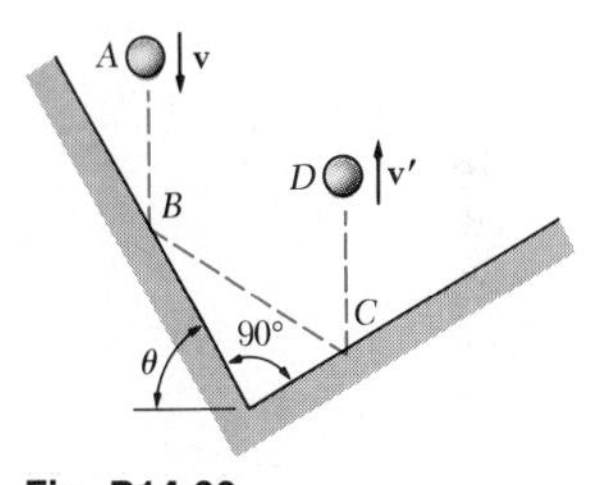

Fig. P14.38

14.38 A ball falling vertically hits a 90° corner at B with a velocity $\mathbf{v}$. Show that after hitting the corner again at C, it will rebound vertically with a velocity $\mathbf{v}'$ of magnitude ev, where e is the coefficient of restitution between the ball and the corner. Neglect friction and assume that the short intermediate path BC is a straight line.

14.39 In Prob. 14.38, determine the required value of θ if the path BC is to be horizontal.

14.40 A small ball A is dropped from a height h onto a rigid, frictionless plate at B and bounces to point C at the same elevation as B. Knowing that $\theta = 20°$ and that the coefficient of restitution between the ball and the plate is $e = 0.40$, determine the distance d.

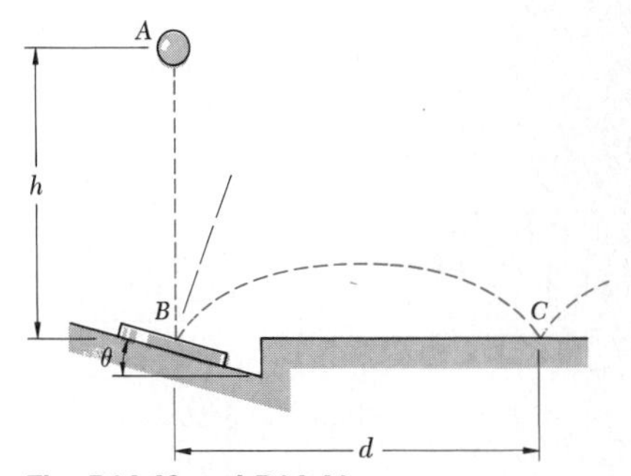

Fig. P14.40 and P14.41

14.41 A small ball A is dropped from a height h onto a rigid, frictionless plate at B and bounces to point C at the same elevation as B. Determine the value of θ for which the distance d is maximum and the corresponding value of d, assuming that the coefficient of restitution between the ball and the plate is (a) $e = 1$, (b) $e = 0.30$.

14.42 A ball moving with the horizontal velocity $\mathbf{v}_0$ drops from A through the vertical distance $h_0 = 10$ in. to a frictionless floor. Knowing that the ball hits the floor at a distance $d_0 = 4$ in. from B and that the coefficient of restitution between the ball and the floor is $e = 0.80$, determine (*a*) the height h_1 and length d_1 of the first bounce, (*b*) the height h_2 and length d_2 of the second bounce.

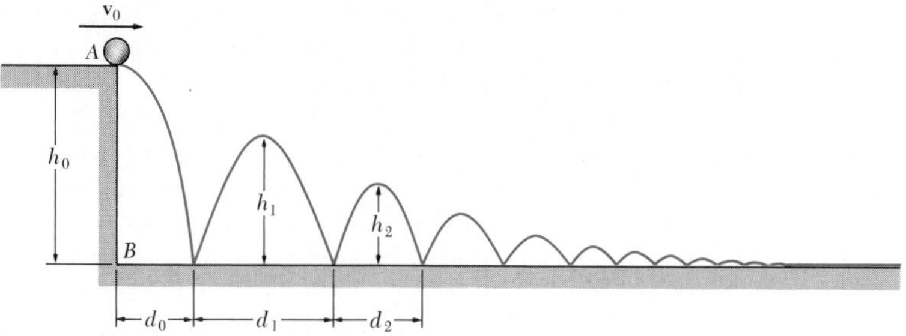

Fig. P14.42, P14.43, and P14.44

14.43 A ball moving with the horizontal velocity $\mathbf{v}_0$ drops from A through the vertical distance $h_0 = 16$ in. to a frictionless floor. Knowing that the ball hits the floor at a distance $d_0 = 4$ in. from B and that the height of its first bounce is $h_1 = 9$ in., determine (*a*) the coefficient of restitution between the ball and the floor, (*b*) the length d_1 of the first bounce.

14.44 A ball moving with a horizontal velocity of magnitude $v_0 = 0.2$ m/s drops from A to a frictionless floor. Knowing that the ball hits the floor at a distance $d_0 = 60$ mm from B and that the length of its first bounce is $d_1 =$ 96 mm, determine (*a*) the coefficient of restitution between the ball and the floor, (*b*) the height h_1 of the first bounce.

14.45 In Prob. 14.44, determine how long the ball will keep bouncing after first hitting the floor.

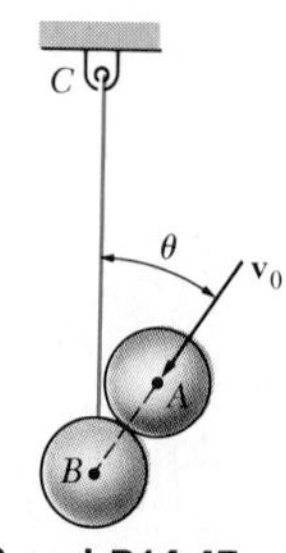

Fig. P14.46 and P14.47

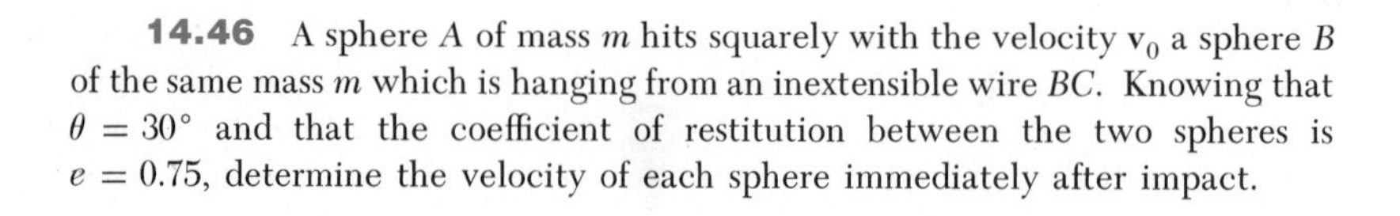

14.46 A sphere A of mass m hits squarely with the velocity $\mathbf{v}_0$ a sphere B of the same mass m which is hanging from an inextensible wire BC. Knowing that $\theta = 30°$ and that the coefficient of restitution between the two spheres is $e = 0.75$, determine the velocity of each sphere immediately after impact.

14.47 A sphere A of mass m hits squarely with the velocity $\mathbf{v}_0$ a sphere B of the same mass m which is hanging from an inextensible wire BC. Denoting by e the coefficient of restitution between the two spheres, determine (*a*) the required value of the angle θ if the velocity of sphere A is to be zero immediately after impact, (*b*) the corresponding velocity of sphere B.

14.48 A sphere A of mass $m_A = 2$ kg is released from rest in the position shown and strikes the frictionless, inclined surface of a wedge B of mass $m_B = 6$ kg with a velocity of magnitude $v_0 = 3$ m/s. The wedge, which is supported by rollers and may move freely in the horizontal direction, is initially at rest. Knowing that $\theta = 30°$ and that the coefficient of restitution between the sphere and the wedge is $e = 0.80$, determine the velocities of the sphere and of the wedge immediately after impact.

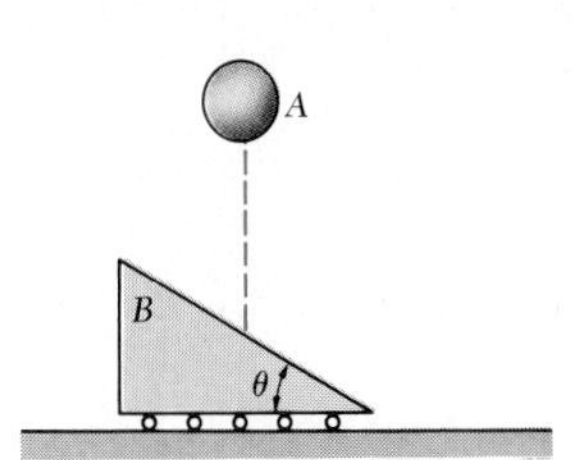

Fig. P14.48 and P14.49

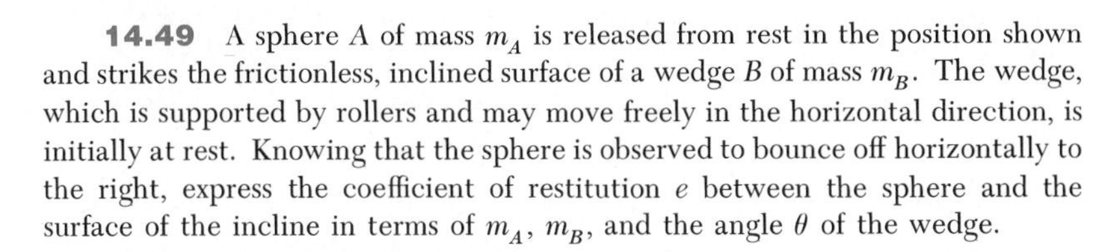

14.49 A sphere A of mass m_A is released from rest in the position shown and strikes the frictionless, inclined surface of a wedge B of mass m_B. The wedge, which is supported by rollers and may move freely in the horizontal direction, is initially at rest. Knowing that the sphere is observed to bounce off horizontally to the right, express the coefficient of restitution e between the sphere and the surface of the incline in terms of m_A, m_B, and the angle θ of the wedge.

14.50 In Prob. 14.49, knowing that $m_A = 1.5$ kg, $m_B = 6$ kg, and $e = 0.75$, determine the angle θ of the wedge for which the sphere bounces off horizontally to the right, assuming that the wedge (*a*) is free to move in the horizontal direction, (*b*) is rigidly attached to the ground.

14.9. Problems Involving Energy and Momentum. We have now at our disposal three different methods for the solution of kinetics problems: the direct application of Newton's second law, $\Sigma\mathbf{F} = m\mathbf{a}$, the method of work and energy, and the method of impulse and momentum. To derive maximum benefit from these three methods, we should be able to choose the method best suited for the solution of a given problem. We should also be prepared to use different methods for solving the various parts of a problem when such a procedure seems advisable.

We have already seen that the method of work and energy is in many cases more expeditious than the direct application of Newton's second law. As indicated in Sec. 13.4, however, the method of work and energy has limitations, and it must sometimes be supplemented by the use of $\Sigma\mathbf{F} = m\mathbf{a}$. This is the case, for example, when we wish to determine an acceleration or a normal force.

There is generally no great advantage in using the method of impulse and momentum for the solution of problems involving no impulsive forces. It will usually be found that the equation $\Sigma\mathbf{F} = m\mathbf{a}$ yields a solution just as fast and that the method of work and energy, if it applies, is more rapid and more convenient. However, the method of impulse and momentum is the only practicable method in problems of impact. A solution based on the direct application of $\Sigma\mathbf{F} = m\mathbf{a}$ would be unwieldy, and the method of work and energy cannot be used since impact (unless perfectly elastic) involves a loss of mechanical energy.

The solution of many problems involving a short impact phase may be divided into several parts. While the part corresponding to the impact phase calls for the use of the method of impulse and momentum and of the relation between relative velocities, the other parts may usually be solved by the method of work and energy. The use of the equation $\Sigma \mathbf{F} = m\mathbf{a}$ will be necessary, however, if the problem involves the determination of a normal force.

Consider, for example, a pendulum A, of mass m_A and length l, which is released with no velocity from a position A_1 (Fig. 14.15*a*). The pendulum swings freely in a vertical plane and hits a second pendulum B, of mass m_B and same length l, which is initially at rest. After the impact (with coefficient of restitution e), pendulum B swings through an angle θ that we wish to determine.

The solution of the problem may be divided into three parts:

1. *Pendulum A swings from A_1 to A_2.* The principle of conservation of energy may be used to determine the velocity $(\mathbf{v}_A)_2$ of the pendulum at A_2 (Fig. 14.15*b*).
2. *Pendulum A hits pendulum B.* Using the fact that the total momentum of the two pendulums is conserved and the relation between their relative velocities, we determine the velocities $(\mathbf{v}_A)_3$ and $(\mathbf{v}_B)_3$ of the two pendulums after impact (Fig. 14.15*c*).
3. *Pendulum B swings from B_3 to B_4.* Applying the principle of conservation of energy to pendulum B, we determine the maximum elevation y_4 reached by that pendulum (Fig. 14.15*d*). The angle θ may then be determined by trigonometry.

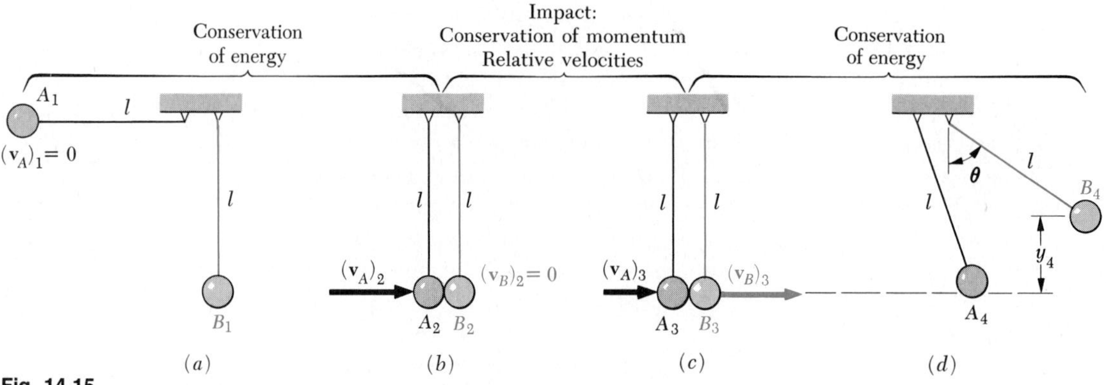

Fig. 14.15

We note that the method of solution just described should be supplemented by the use of $\Sigma \mathbf{F} = m\mathbf{a}$ if the tensions in the cords holding the pendulums are to be determined.

We observed in Sec. 14.7 that, in the case of the perfectly elastic impact of two particles, the energy of the system, as well as its momentum, is conserved. There are many other situations involving a system of two or more particles where the principle of conservation of energy and the principle of impulse and momentum may be used together to obtain two or more equations which may be solved simultaneously (see Sample Prob. 14.9).

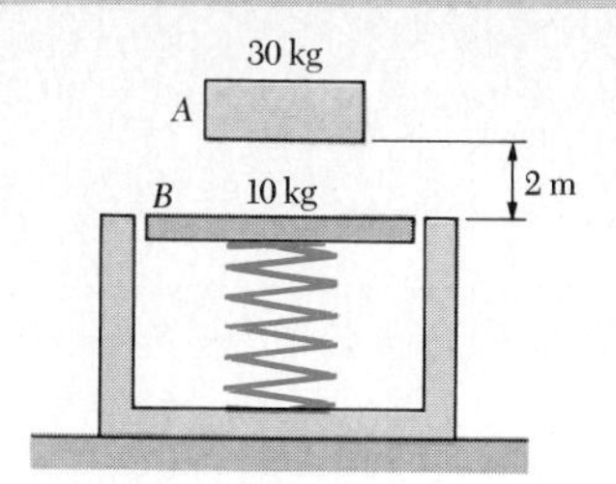

SAMPLE PROBLEM 14.8

A 30-kg block is dropped from a height of 2 m onto the 10-kg pan of a spring scale. Assuming the impact to be perfectly plastic, determine the maximum deflection of the pan. The constant of the spring is $k = 20$ kN/m.

Solution. The impact between the block and the pan *must* be treated separately; therefore we divide the solution into three parts.

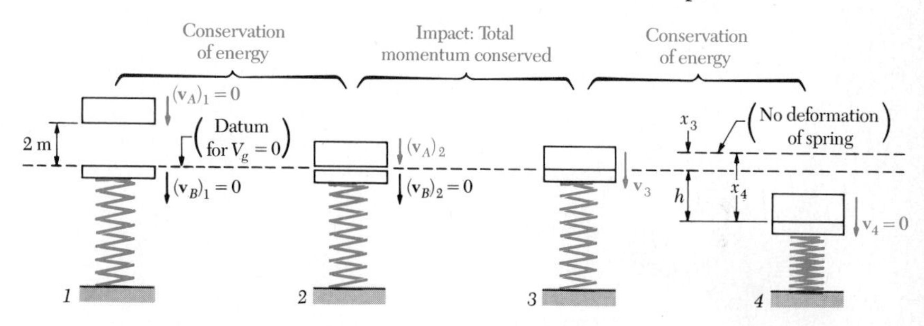

Conservation of Energy. Block: $W_A = (30 \text{ kg})(9.81 \text{ m/s}^2) = 294 \text{ N}$

$$T_1 = \tfrac{1}{2}m_A(v_A)_1^2 = 0 \qquad V_1 = W_A y = (294 \text{ N})(2 \text{ m}) = 588 \text{ J}$$

$$T_2 = \tfrac{1}{2}m_A(v_A)_2^2 = \tfrac{1}{2}(30 \text{ kg})(v_A)_2^2 \qquad V_2 = 0$$

$$T_1 + V_1 = T_2 + V_2: \qquad 0 + 588 \text{ J} = \tfrac{1}{2}(30 \text{ kg})(v_A)_2^2 + 0$$

$$(v_A)_2 = +6.26 \text{ m/s} \qquad (\mathbf{v}_A)_2 = 6.26 \text{ m/s} \downarrow$$

Impact: Conservation of Momentum. Since the impact is perfectly plastic, $e = 0$; the block and pan move together after the impact.

$$m_A(v_A)_2 + m_B(v_B)_2 = (m_A + m_B)v_3$$

$$(30 \text{ kg})(6.26 \text{ m/s}) + 0 = (30 \text{ kg} + 10 \text{ kg})v_3$$

$$v_3 = +4.70 \text{ m/s} \qquad \mathbf{v}_3 = 4.70 \text{ m/s} \downarrow$$

Conservation of Energy. Initially the spring supports the weight W_B of the pan; thus the initial deflection of the spring is

$$x_3 = \frac{W_B}{k} = \frac{(10 \text{ kg})(9.81 \text{ m/s}^2)}{20 \times 10^3 \text{ N/m}} = \frac{98.1 \text{ N}}{20 \times 10^3 \text{ N/m}} = 4.91 \times 10^{-3} \text{ m}$$

Denoting by x_4 the total maximum deflection of the spring, we write

$$T_3 = \tfrac{1}{2}(m_A + m_B)v_3^2 = \tfrac{1}{2}(30 \text{ kg} + 10 \text{ kg})(4.70 \text{ m/s})^2 = 442 \text{ J}$$

$$V_3 = V_g + V_e = 0 + \tfrac{1}{2}kx_3^2 = \tfrac{1}{2}(20 \times 10^3)(4.91 \times 10^{-3})^2 = 0.241 \text{ J}$$

$$T_4 = 0$$

$$V_4 = V_g + V_e = (W_A + W_B)(-h) + \tfrac{1}{2}kx_4^2 = -(392)h + \tfrac{1}{2}(20 \times 10^3)x_4^2$$

Noting that the displacement of the pan is $h = x_4 - x_3$, we write

$$T_3 + V_3 = T_4 + V_4:$$

$$442 + 0.241 = 0 - 392(x_4 - 4.91 \times 10^{-3}) + \tfrac{1}{2}(20 \times 10^3)x_4^2$$

$$x_4 = 0.230 \text{ m} \qquad h = x_4 - x_3 = 0.230 \text{ m} - 4.91 \times 10^{-3} \text{ m}$$

$$h = 0.225 \text{ m} \qquad h = 225 \text{ mm} \blacktriangleleft$$

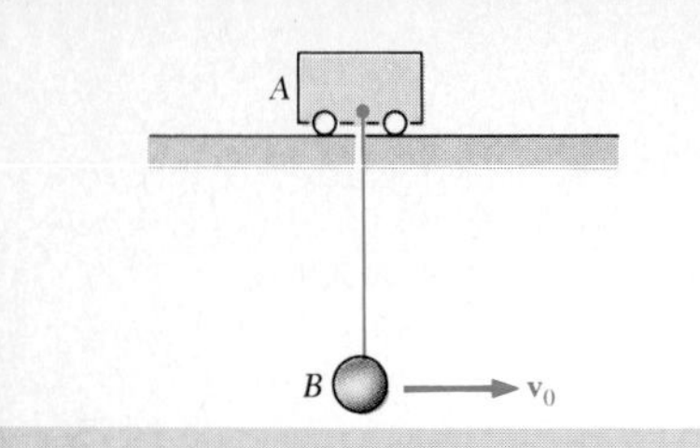

SAMPLE PROBLEM 14.9

Ball B, of mass m_B, is suspended from a cord of length l attached to a cart A, of mass m_A, which may roll freely on a frictionless horizontal track. If the ball is given an initial horizontal velocity $\mathbf{v}_0$ while the cart is at rest, determine (a) the velocity of B as it reaches its maximum elevation, (b) the maximum vertical distance h through which B will rise.

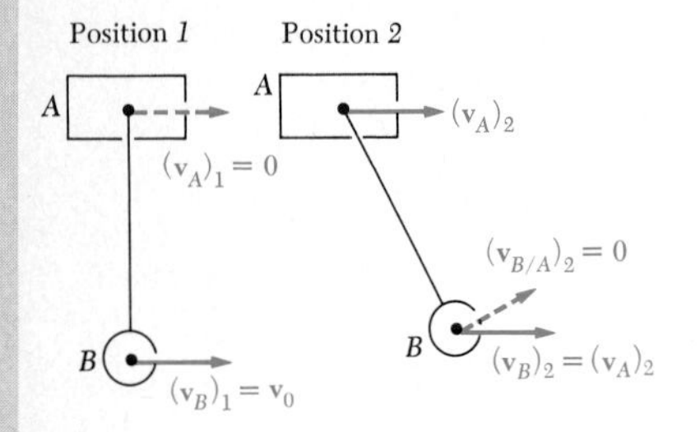

Solution. We shall apply the impulse-momentum principle and the principle of conservation of energy to the cart-ball system between its initial position *1* and the position 2 when B reaches its maximum elevation.

Velocities

Position *1*: $\qquad (\mathbf{v}_A)_1 = 0 \qquad (\mathbf{v}_B)_1 = \mathbf{v}_0 \qquad (1)$

Position 2: When ball B reaches its maximum elevation, its velocity $(\mathbf{v}_{B/A})_2$ relative to its support A is zero. Thus, at that instant, its absolute velocity is

$$(\mathbf{v}_B)_2 = (\mathbf{v}_A)_2 + (\mathbf{v}_{B/A})_2 = (\mathbf{v}_A)_2 \qquad (2)$$

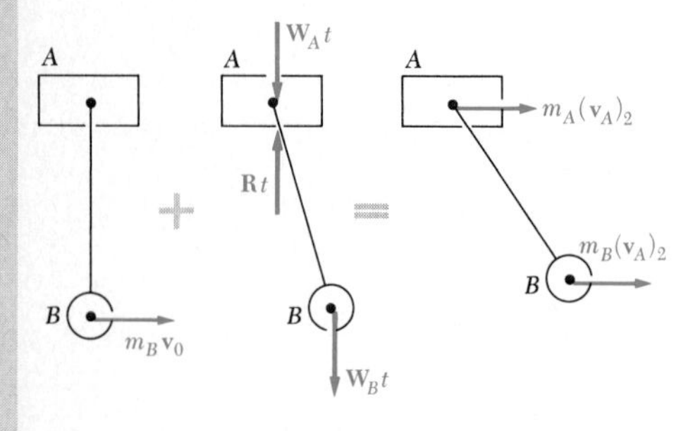

Impulse-Momentum Principle. Noting that the external impulses consist of $\mathbf{W}_A t$, $\mathbf{W}_B t$, and $\mathbf{R}t$, where $\mathbf{R}$ is the reaction of the track on the cart, and recalling (1) and (2), we draw the impulse-momentum diagram and write

$$\Sigma m\mathbf{v}_1 + \Sigma \text{ Ext Imp}_{1\to2} = \Sigma m\mathbf{v}_2$$

$\xrightarrow{+}$ x components: $\qquad m_B v_0 = (m_A + m_B)(v_A)_2$

which expresses that the total momentum of the system is conserved in the horizontal direction. Solving for $(v_A)_2$:

$$(v_A)_2 = \frac{m_B}{m_A + m_B} v_0 \qquad\qquad (\mathbf{v}_B)_2 = (\mathbf{v}_A)_2 = \frac{m_B}{m_A + m_B} v_0 \rightarrow \blacktriangleleft$$

Conservation of Energy

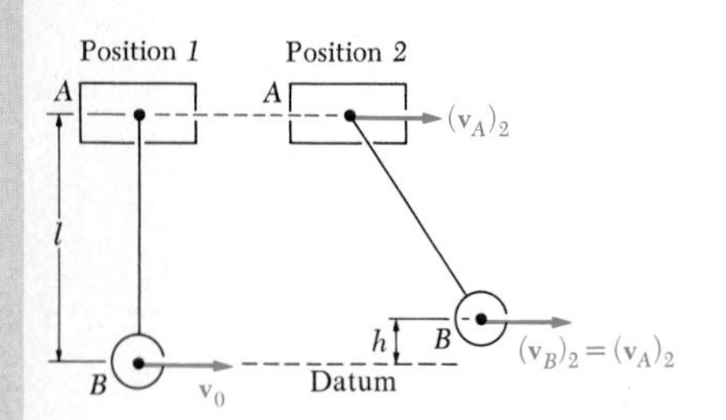

Position *1*. ***Potential Energy:*** $\qquad V_1 = m_A g l$

Kinetic Energy: $\qquad T_1 = \frac{1}{2} m_B v_0^2$

Position 2. ***Potential Energy:*** $\qquad V_2 = m_A g l + m_B g h$

Kinetic Energy: $\qquad T_2 = \frac{1}{2}(m_A + m_B)(v_A)_2^2$

$T_1 + V_1 = T_2 + V_2$: $\qquad \frac{1}{2} m_B v_0^2 + m_A g l = \frac{1}{2}(m_A + m_B)(v_A)_2^2 + m_A g l + m_B g h$

Solving for h, we have

$$h = \frac{v_0^2}{2g} - \frac{m_A + m_B}{m_B} \frac{(v_A)_2^2}{2g}$$

or, substituting for $(v_A)_2$ the expression found above,

$$h = \frac{v_0^2}{2g} - \frac{m_B}{m_A + m_B} \frac{v_0^2}{2g} \qquad\qquad h = \frac{m_A}{m_A + m_B} \frac{v_0^2}{2g} \blacktriangleleft$$

Remark. For $m_A \gg m_B$, the answers obtained reduce to $(\mathbf{v}_B)_2 = (\mathbf{v}_A)_2 = 0$ and $h = v_0^2/2g$; B oscillates as a simple pendulum with A fixed. For $m_A \ll m_B$, they reduce to $(\mathbf{v}_B)_2 = (\mathbf{v}_A)_2 = \mathbf{v}_0$ and $h = 0$; A and B move with the same constant velocity $\mathbf{v}_0$.

Problems

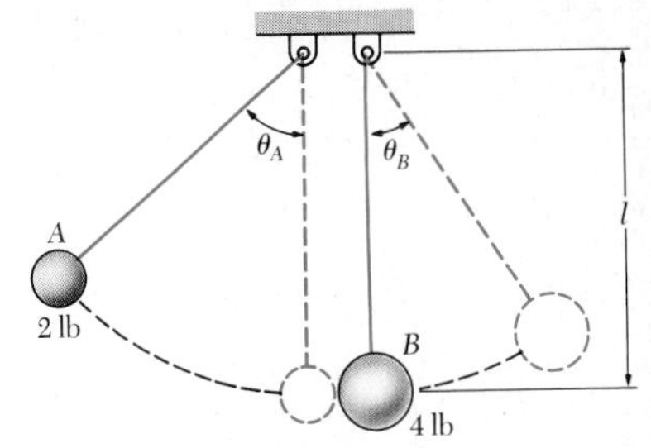

Fig. P14.51 and P14.52

14.51 A 2-lb sphere A is released from rest when $\theta_A = 50°$ and strikes a 4-lb sphere B which is at rest. Knowing that the coefficient of restitution is $e = 0.80$, determine the values of θ_A and θ_B corresponding to the highest positions to which the spheres will rise after impact.

14.52 A 2-lb sphere A is released from rest when $\theta_A = 60°$ and strikes a 4-lb sphere B which is at rest. Knowing that the velocity of sphere A is zero after impact, determine (*a*) the coefficient of restitution e, (*b*) the value of θ_B corresponding to the highest position to which sphere B will rise.

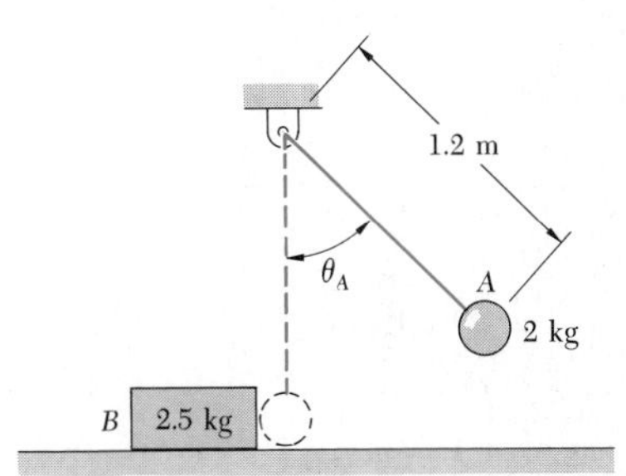

Fig. P14.53 and P14.54

14.53 A 2-kg sphere A is released from rest when $\theta_A = 60°$ and strikes a 2.5-kg block B which is at rest. It is observed that the velocity of the sphere is zero after impact and that the block moves through 1.5 m before coming to rest. Determine (*a*) the coefficient of restitution between the sphere and the block, (*b*) the coefficient of kinetic friction between the block and the floor.

14.54 A 2-kg sphere A is released from rest when $\theta_A = 90°$ and strikes a 2.5-kg block B which is at rest. Knowing that the coefficient of restitution between the sphere and the block is 0.75 and that the coefficient of kinetic friction between the block and the floor is 0.25, determine (*a*) how far block B will move, (*b*) the percentage of the initial energy lost in friction between the block and the floor.

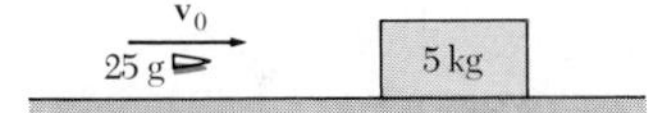

Fig. P14.55

14.55 A 25-g bullet is fired with a velocity of magnitude $v_0 = 550$ m/s into a 5-kg block of wood. Knowing that the coefficient of kinetic friction between the block and the floor is 0.30, determine (*a*) how far the block will move, (*b*) the percentage of the initial energy lost in friction between the block and the floor.

14.56 A ballistic pendulum consisting of a 30-kg block suspended as shown from two wires 1.8-m long is used to measure the muzzle velocity of a rifle. If the pendulum swings through a horizontal distance $d = 250$ mm when a 40-g bullet is fired into it, determine the muzzle velocity v_0 of the rifle.

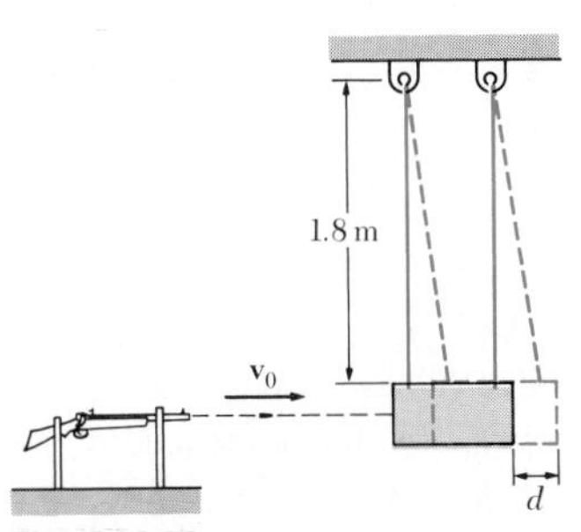

Fig. P14.56

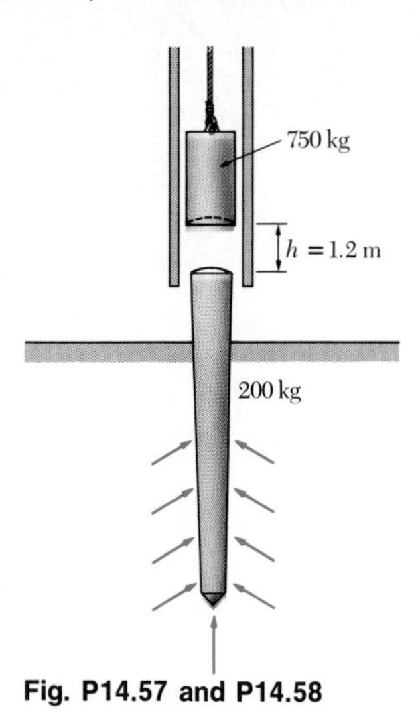

Fig. P14.57 and P14.58

14.57 It is desired to drive a 200-kg pile into the ground until the resistance to its penetration is 100 kN. Each blow of the 750-kg hammer is the result of a 1.2-m free fall onto the top of the pile. Determine how far the pile will be driven into the ground by a single blow when the 100-kN resistance is achieved. Assume that the impact is perfectly plastic.

14.58 The 750-kg hammer of a drop-hammer pile driver falls from a height of 1.2 m onto the top of a 200-kg pile. The pile is driven 100 mm into the ground. Assuming perfectly plastic impact, determine the average resistance of the ground to penetration.

14.59 A 2-oz ball is dropped from a height h_0 onto an 8-oz plate. The ball is observed to rebound to a height $h_1 = 22.5$ in. when the plate rests directly on hard ground and to a height $h_2 = 6.40$ in. when a foam-rubber mat is placed between the plate and the ground. Determine (*a*) the coefficient of restitution between the ball and the plate, (*b*) the height h_0 from which the ball was dropped.

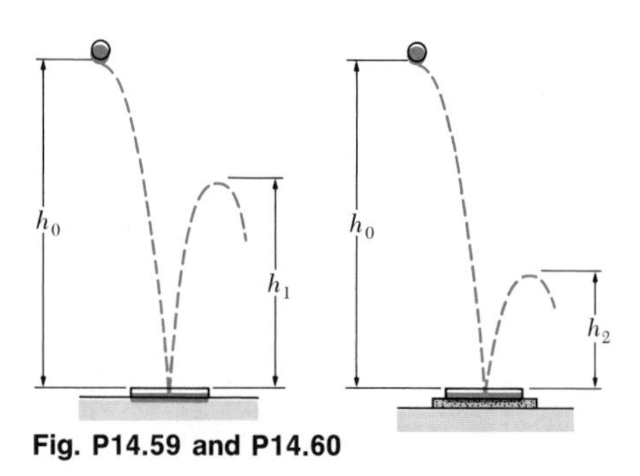

Fig. P14.59 and P14.60

14.60 A 1.5-oz ball is dropped from a height $h_0 = 25$ in. onto a small plate. The ball is observed to rebound to a height $h_1 = 16$ in. when the plate rests directly on hard ground and to a height $h_2 = 9$ in. when a foam-rubber mat is placed between the plate and the ground. Determine (*a*) the coefficient of restitution between the ball and the plate, (*b*) the weight of the plate.

14.61 A space vehicle which is drifting in space consists of two parts *A* and *B* connected by explosive bolts. Parts *A* and *B* compress four springs, each of which has a potential energy of 120 ft·lb. When the bolts are exploded the springs expand, causing parts *A* and *B* to move away from each other. Determine the relative velocity of *B* with respect to *A*, knowing that (*a*) each part weighs 2000 lb, (*b*) part *A* weighs 3000 lb and part *B* 1000 lb.

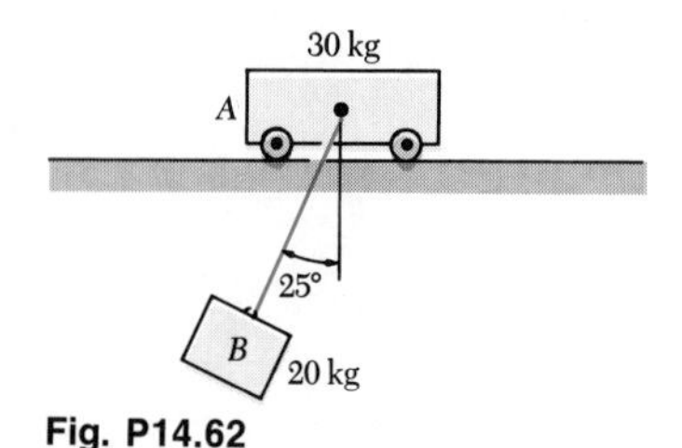

Fig. P14.62

14.62 A 20-kg block *B* is suspended from a 2-m cord attached to a 30-kg cart *A*, which may roll freely on a frictionless, horizontal track. If the system is released from rest in the position shown, determine the velocities of *A* and *B* as *B* passes directly under *A*.

14.63 Assuming that the cart A and the ball B of Sample Prob. 14.9 have the same mass m, describe the entire motion of the system after B has been given its initial velocity $\mathbf{v}_0$, specifying the velocities of A and B for the following successive values of the angle θ (assumed positive counterclockwise) that the cord will form with the vertical: (*a*) $\theta = \theta_{max}$, (*b*) $\theta = 0$, (*c*) $\theta = \theta_{min}$.

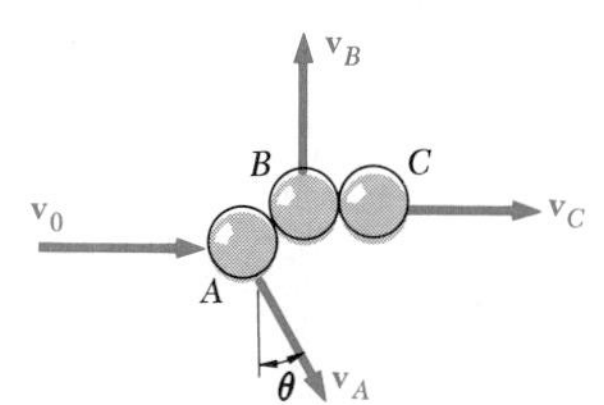

Fig. P14.64

14.64 In a game of billiards, ball A is moving with a 4-m/s velocity $\mathbf{v}_0$ when it strikes balls B and C which are at rest side by side, After the collision, the three balls are observed to move in the directions shown, with $\theta = 25°$. Assuming frictionless surfaces and perfectly elastic impacts (i.e., conservation of energy), determine the magnitudes of the velocities $\mathbf{v}_A$, $\mathbf{v}_B$, and $\mathbf{v}_C$.

***14.65** Three spheres, each of mass m, may slide freely on a frictionless, horizontal surface. Spheres A and B are attached to an inextensible, inelastic cord of length l and are at rest in the position shown when sphere B is struck squarely by sphere C which is moving to the right with a velocity $\mathbf{v}_0$. Knowing that the cord is taut when sphere B is struck by sphere C and assuming perfectly elastic impact between B and C, and thus conservation of energy for the entire system, determine the velocity of each sphere immediately after impact.

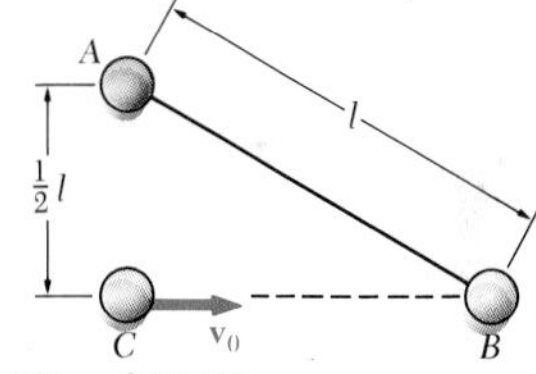

Fig. P14.65

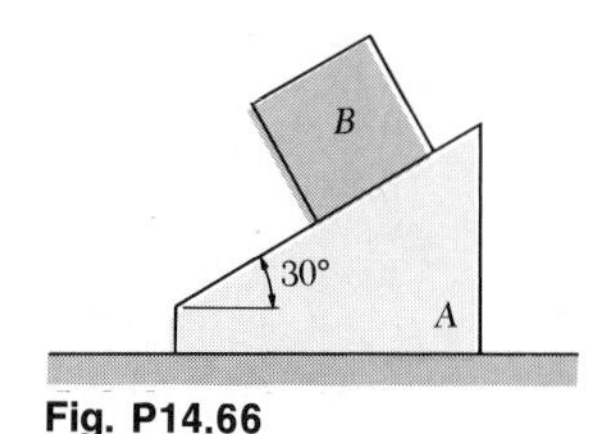

Fig. P14.66

***14.66** A 12-lb block B starts from rest and slides on the 30-lb wedge A, which is supported by a horizontal surface. Neglecting friction, determine (*a*) the velocity of B relative to A after it has slid 3 ft down the inclined surface of the wedge, (*b*) the corresponding velocity of A.

14.67 Ball B is hanging from an inextensible cord. An identical ball A is released from rest when it is just touching the cord and drops through the vertical distance $h_A = 150$ mm before striking ball B. Assuming perfectly elastic impact ($e = 1$) and no friction, determine the resulting maximum vertical displacement h_B of ball B.

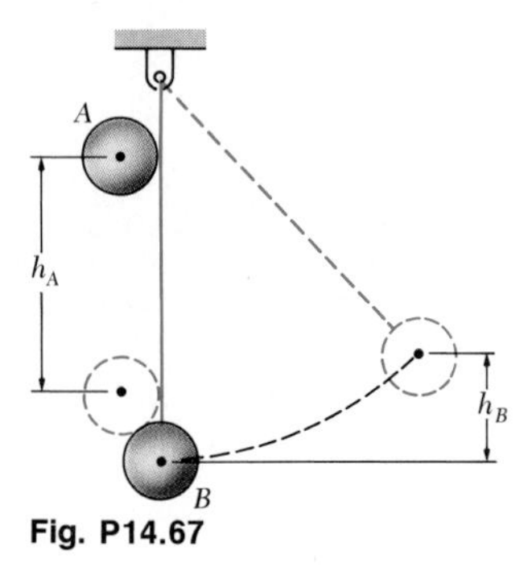

Fig. P14.67

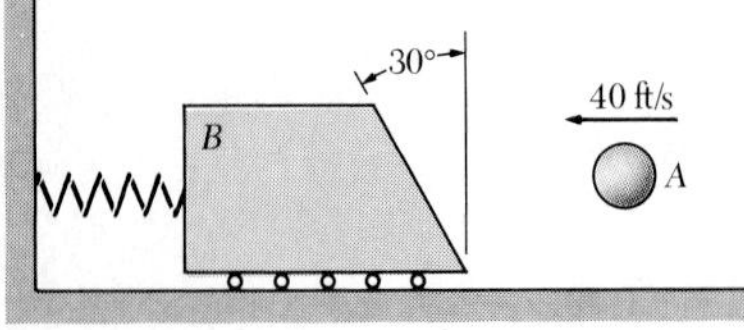

Fig. P14.68

14.68 A 2-lb sphere A is moving to the left with a velocity of 40 ft/s when it strikes the inclined surface of a 5-lb block B which is at rest. The block is supported by rollers and is attached to a spring of constant $k = 12$ lb/in. Knowing that the coefficient of restitution between the sphere and the block is $e = 0.75$ and neglecting friction, determine the maximum deflection of the spring.

14.69 A bumper is designed to protect a 2500-lb automobile from damage when it hits a rigid wall at speeds up to 9 mi/h. Assuming perfectly plastic impact, determine (*a*) the energy absorbed by the bumper during the impact, (*b*) the speed at which the automobile can hit another 2500-lb automobile without incurring any damage if the other automobile is similarly protected and is at rest with its brakes released.

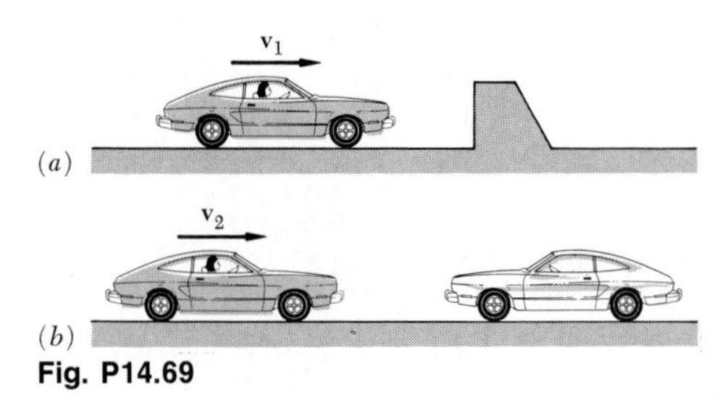

Fig. P14.69

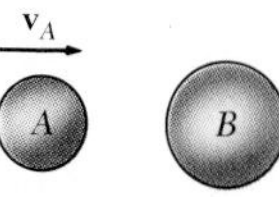

Fig. P14.70

* **14.70** A ball of mass m_A moving with a velocity $\mathbf{v}_A$ strikes squarely a second ball of mass m_B which is at rest. Denoting by e the coefficient of restitution between the two balls, show that the percentage of energy lost during the impact is

$$\frac{100(1 - e^2)}{1 + (m_A/m_B)}$$

14.10. Angular Momentum. Consider a particle of mass m moving in the xy plane. As we saw in Sec. 14.2, the momentum of the particle at a given instant is defined as the vector $m\mathbf{v}$ obtained by multiplying the velocity $\mathbf{v}$ of the particle by its mass m. The moment about O of the vector $m\mathbf{v}$ is called the *moment of momentum*, or the *angular momentum*, of the particle about O at that instant and is denoted by H_O. Resolving the vector $m\mathbf{v}$ into its x and y components (Fig. 14.16*a*), and recalling that counterclockwise is positive, we write

$$H_O = x(mv_y) - y(mv_x)$$

$$H_O = m(xv_y - yv_x) \tag{14.33}$$

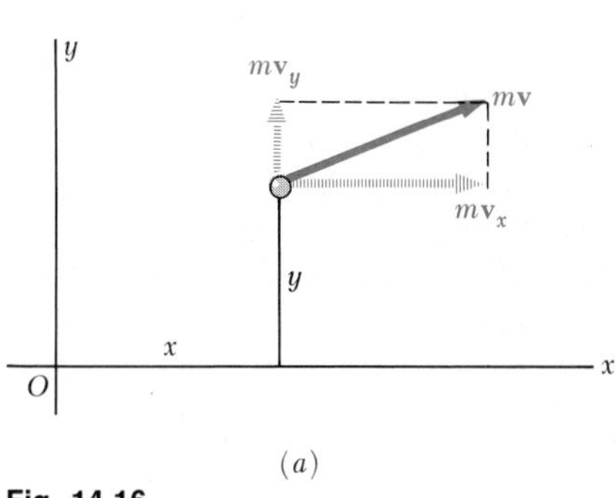

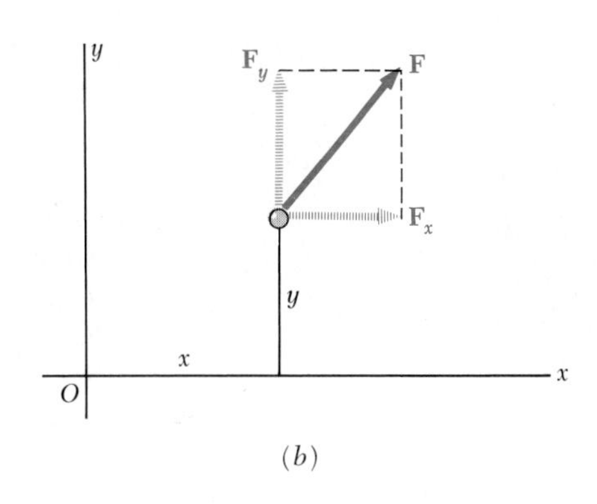

Fig. 14.16

We shall now compute the derivative of the angular momentum H_O with respect to t; we have

$$\dot{H}_O = m\frac{d}{dt}(xv_y - yv_x)$$
$$= m(\dot{x}v_y + x\dot{v}_y - \dot{y}v_x - y\dot{v}_x)$$

or, since $\dot{x} = v_x$ and $\dot{y} = v_y$,

$$\dot{H}_O = m(x\dot{v}_y - y\dot{v}_x) \tag{14.34}$$

But, according to Newton's second law, we have

$$m\dot{v}_x = ma_x = F_x \qquad m\dot{v}_y = ma_y = F_y$$

where F_x and F_y are the rectangular components of the force $\mathbf{F}$ acting on the particle. Equation (14.34) may therefore be written in the form

$$\dot{H}_O = xF_y - yF_x$$

or, since the right-hand member of the equation obtained represents the moment M_O of $\mathbf{F}$ about O (Fig. 14.16*b*),

$$M_O = \dot{H}_O \tag{14.35}$$

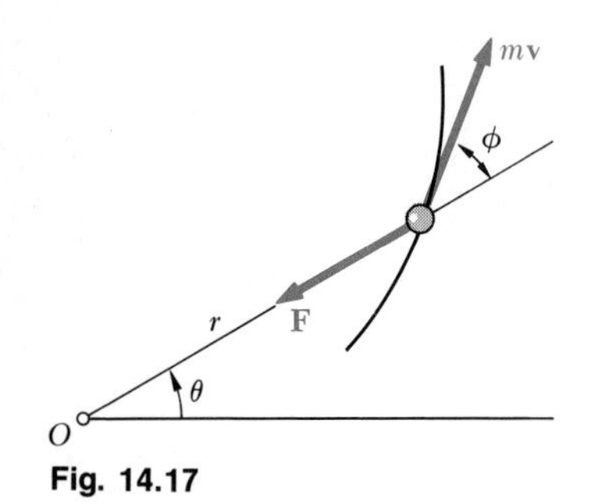

Fig. 14.17

14.11. Conservation of Angular Momentum. If the moment M_O about O of the force $\mathbf{F}$ acting on the particle of the preceding section is zero for every value of t, Eq. (14.35) yields

$$\dot{H}_O = 0$$

for any t; integrating with respect to t, we obtain

$$H_O = \text{constant} \tag{14.36}$$

Thus, if the moment about O of the force $\mathbf{F}$ acting on a particle is zero for every value of t, *the angular momentum of the particle about O is conserved.*

Clearly, the angular momentum of a particle about a fixed point O will be conserved if the resultant of the forces acting on the particle is zero. But the angular momentum of the particle about O will also be conserved when the particle is acted upon by an unbalanced force $\mathbf{F}$ *if the force* $\mathbf{F}$ *is a central force whose line of action passes through O* (Fig. 14.17).

The angular momentum H_O of a particle was expressed in Eq. (14.33) in terms of the rectangular coordinates of the particle and of the rectangular components of its momentum. In the case of a particle moving under a central force $\mathbf{F}$, it is found more convenient to use polar coordinates. Recalling that H_O is, by definition, the moment about O of the momentum $m\mathbf{v}$ of the particle, and denoting by ϕ the angle formed by the radius vector and the tangent to the path of the particle, we write Eq. (14.36) in the form

$$H_O = r(mv \sin \phi) = \text{constant} \tag{14.37}$$

Observing that $v \sin \phi$ represents the transverse component v_θ of the velocity, which is equal to $r\dot{\theta}$ (Sec. 11.13), we may also write

$$H_O = mr^2\dot{\theta} = \text{constant} \tag{14.38}$$

Recalling that, in Sec. 12.9, we found that a particle moving under a central force satisfies the equation

$$r^2\dot{\theta} = h \tag{12.18}$$

and comparing this equation with Eq. (14.38), we note that the constant h represents the *angular momentum per unit mass,* H_O/m, of the particle. Since h was shown in Sec. 12.9 to also represent twice the areal velocity of the particle, it follows that stating that the areal velocity of the particle is constant—as we did in Sec. 12.9—is equivalent to stating that its angular momentum is conserved.

14.12. Motion under a Conservative Central Force. Application to Space Mechanics. When a particle moves under a central force which is also a *conservative force,* both the principle of conservation of angular momentum and the principle of conservation of energy may be used to study its motion.

Consider, for example, a space vehicle moving under the earth's gravitational force. We shall assume that it begins its free flight at point P_0 at a

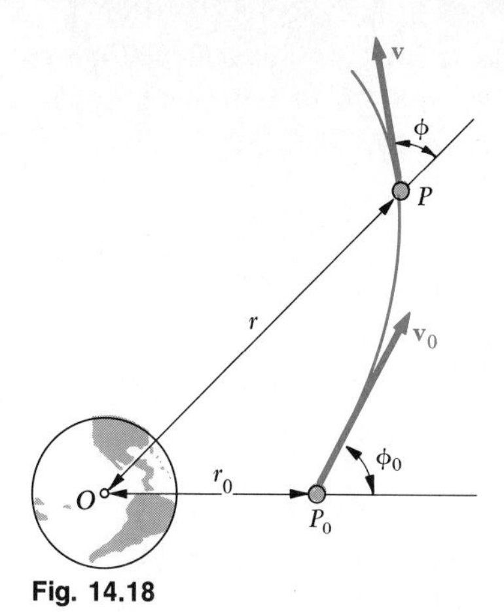

Fig. 14.18

distance r_0 from the center of the earth, with a velocity $\mathbf{v}_0$ forming an angle ϕ_0 with the radius vector OP_0 (Fig. 14.18). Let P be a point of the trajectory described by the vehicle; we denote by r the distance from O to P, by $\mathbf{v}$ the velocity of the vehicle at P, and by ϕ the angle formed by $\mathbf{v}$ and the radius vector OP. Applying the principle of conservation of angular momentum about O between P_0 and P (Sec. 14.11), we write

$$r_0 m v_0 \sin \phi_0 = r m v \sin \phi \tag{14.39}$$

Recalling the expression (13.13) obtained for the potential energy due to a gravitational force, we apply the principle of conservation of energy between P_0 and P and write

$$T_0 + V_0 = T + V$$

$$\tfrac{1}{2} m v_0^2 - \frac{GMm}{r_0} = \tfrac{1}{2} m v^2 - \frac{GMm}{r} \tag{14.40}$$

where M is the mass of the earth.

Equation (14.40) may be solved for the magnitude v of the velocity of the vehicle at P when the distance r from O to P is known; Eq. (14.39) may then be used to determine the angle ϕ that the velocity forms with the radius vector OP.

Equations (14.39) and (14.40) may also be used to determine the maximum and minimum values of r in the case of a satellite launched from P_0 in a direction forming an angle ϕ_0 with the vertical OP_0 (Fig. 14.19). The desired values of r are obtained by making $\phi = 90°$ in (14.39) and eliminating v between Eqs. (14.39) and (14.40).

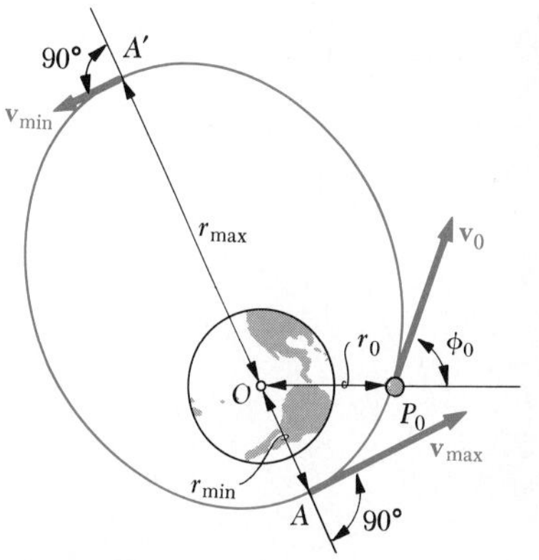

Fig. 14.19

It should be noted that the application of the principles of conservation of energy and of conservation of angular momentum leads to a more fundamental formulation of the problems of space mechanics than does the method indicated in Sec. 12.12. In all cases involving oblique launchings, it will also result in much simpler computations. And while the method of Sec. 12.12 must be used when the actual trajectory or the periodic time of a space vehicle is to be determined, the calculations will be simplified if the conservation principles are first used to compute the maximum and minimum values of the radius vector r.

14.13. Principle of Impulse and Momentum for a System of Particles. The moment of momentum, or angular momentum, about O of a system of particles is defined as the sum of the moments about O of the momenta of the various particles of the system; we write

$$H_O = \Sigma m(xv_y - yv_x) \tag{14.41}$$

Since Eq. (14.35) is satisfied for each particle, we may write that the sum of the moments of all the forces acting on the various particles of the system is equal to the rate of change of the angular momentum of the system. However, since the internal forces occur by pairs of equal and opposite forces having the same line of action, the sum of their moments is zero and we need consider only the sum of the moments of the external forces. We thus write

$$\Sigma(M_O)_{\text{ext}} = \dot{H}_O \tag{14.42}$$

where H_O is the angular momentum of the system defined by relation (14.41).

We saw in Sec. 14.2 that the linear impulse of a force **F**, of components F_x and F_y, during a given time interval may be defined by the integrals of F_x and F_y over that time interval. We shall now, in a similar way, define the *angular impulse* of a couple **M** over a time interval from t_1 to t_2 as the integral

$$\int_{t_1}^{t_2} M\,dt$$

of the moment M of the couple over the given time interval. This new concept, as we shall see presently, will enable us to express in a more general form the principle of impulse and momentum for a system of particles.

Consider a system of particles and the *external forces* acting on the various particles. Each external force **F** may be replaced by an equal force **F** attached at the origin O, plus a couple $\mathbf{M}_O$ of moment equal to the moment about O of the original force (Chap. 3). Now, we shall draw three separate sketches (Fig. 14.20). The first and third sketches will show the systems of vectors representing the momenta of the various particles at a time t_1 and a time t_2, respectively. The second sketch will show the resultant of the vectors representing the linear impulses of the external forces **F** during the given time interval, as well as the sum of the couples representing the angular impulses of the couples $\mathbf{M}_O$ during the same time interval. We recall from Sec. 14.4 that the sums of the x and y components of the vectors shown in parts a and b of Fig. 14.20 must be respectively equal to the sums of the x and y components of the vectors in part c of the same figure. Indeed, rewriting Eq. (14.8) in terms of x and y components, we have

$$\begin{aligned}\sum (mv_x)_1 + \sum \int_{t_1}^{t_2} (F_x)_{\text{ext}}\, dt &= \sum (mv_x)_2 \\ \sum (mv_y)_1 + \sum \int_{t_1}^{t_2} (F_y)_{\text{ext}}\, dt &= \sum (mv_y)_2\end{aligned} \tag{14.43}$$

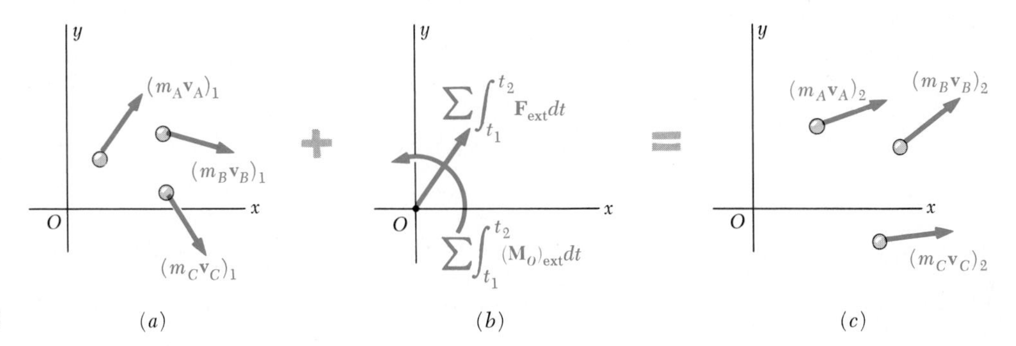

Fig. 14.20

Returning now to Eq. (14.42) and integrating both members over the time interval from t_1 to t_2, we write

$$(H_O)_1 + \sum \int_{t_1}^{t_2} (M_O)_{\text{ext}}\, dt = (H_O)_2 \tag{14.44}$$

The equation obtained expresses that the sum of the moments about O of the vectors shown in parts a and b of Fig. 14.20 must be equal to the sum of the moments about O of the vectors shown in part c of the same figure. Together, Eqs. (14.43) and (14.44) thus express that *the momenta of the particles at time t_1 and the impulses of the external forces from t_1 to t_2 form a system of vectors equipollent to the system of the momenta of the particles at time t_2.*† We write

$$\textbf{Syst Momenta}_1 + \textbf{Syst Ext Imp}_{1\to 2} = \textbf{Syst Momenta}_2 \tag{14.45}$$

This more general statement of the principle of impulse and momentum for a system of particles will prove particularly useful for the study of the motion of rigid bodies (Chap. 18).

When no external force acts on the particles of a system, the integrals in Eqs. (14.43) and (14.44) are zero, and part b of Fig. 14.20 may be omitted. The system of the momenta of the particles at time t_1 is equipollent to the system of the momenta at time t_2:

$$\textbf{Syst Momenta}_1 = \textbf{Syst Momenta}_2 \tag{14.46}$$

† Since we are dealing here with a system of particles, and not with a single rigid body, we cannot conclude that the two systems of vectors are actually equivalent. Hence we use gray plus and equals signs in Fig. 14.20 (cf. Sec. 12.5).

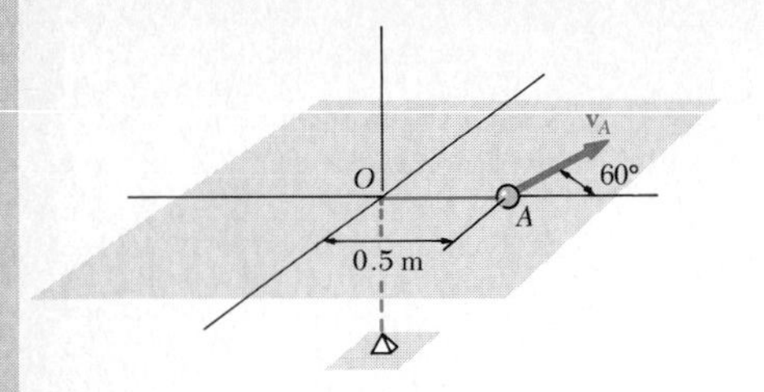

SAMPLE PROBLEM 14.10

A sphere of mass $m = 0.6$ kg is attached to an elastic cord of constant $k = 100$ N/m, which is undeformed when the sphere is located at the origin O. Knowing that the sphere may slide without friction on the horizontal surface and that in the position shown its velocity $\mathbf{v}_A$ has a magnitude of 20 m/s, determine (*a*) the maximum and minimum distances from the sphere to the origin O, (*b*) the corresponding values of its speed.

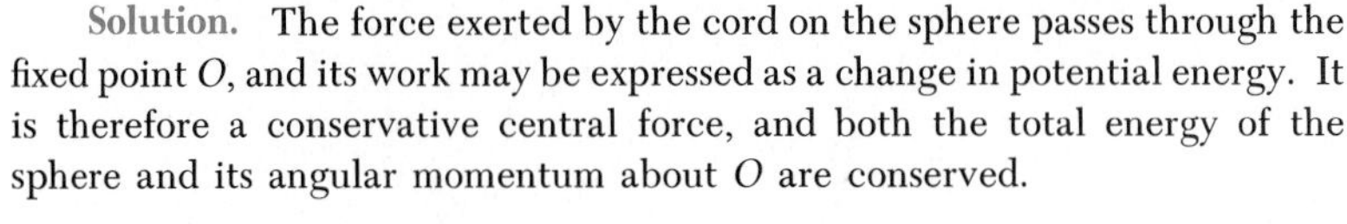

Solution. The force exerted by the cord on the sphere passes through the fixed point O, and its work may be expressed as a change in potential energy. It is therefore a conservative central force, and both the total energy of the sphere and its angular momentum about O are conserved.

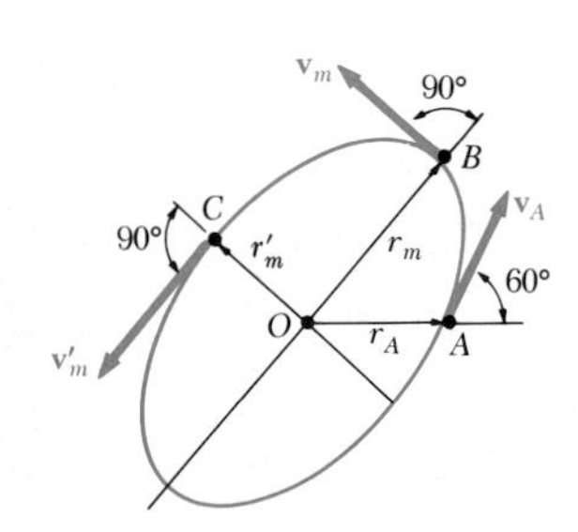

Conservation of Angular Momentum about O. At point B, where the distance from O is maximum, the velocity of the sphere is perpendicular to OB and the angular momentum is $r_m v_m$. A similar property holds at point C, where the distance from O is minimum. Expressing conservation of angular momentum between A and B, we write

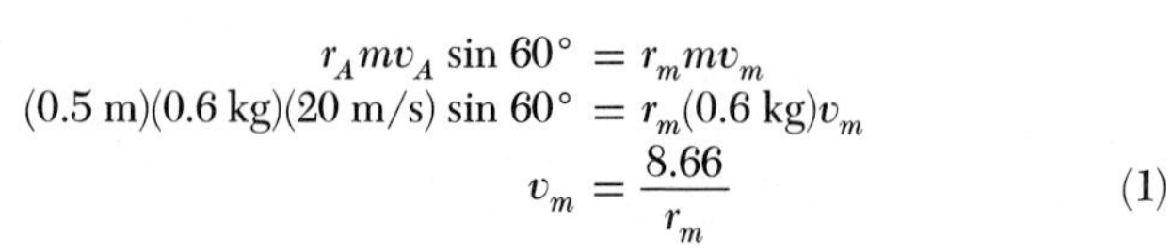

$$r_A m v_A \sin 60° = r_m m v_m$$
$$(0.5 \text{ m})(0.6 \text{ kg})(20 \text{ m/s}) \sin 60° = r_m(0.6 \text{ kg})v_m$$
$$v_m = \frac{8.66}{r_m} \qquad (1)$$

Conservation of Energy

At point A:
$$T_A = \tfrac{1}{2}mv_A^2 = \tfrac{1}{2}(0.6 \text{ kg})(20 \text{ m/s})^2 = 120 \text{ J}$$
$$V_A = \tfrac{1}{2}kr_A^2 = \tfrac{1}{2}(100 \text{ N/m})(0.5 \text{ m})^2 = 12.5 \text{ J}$$

At point B:
$$T_B = \tfrac{1}{2}mv_m^2 = \tfrac{1}{2}(0.6 \text{ kg})v_m^2 = 0.3v_m^2$$
$$V_B = \tfrac{1}{2}kr_m^2 = \tfrac{1}{2}(100 \text{ N/m})r_m^2 = 50r_m^2$$

Applying the principle of conservation of energy between points A and B, we write

$$T_A + V_A = T_B + V_B$$
$$120 + 12.5 = 0.3v_m^2 + 50r_m^2 \qquad (2)$$

***a.* Maximum and Minimum Values of Distance.** Substituting for v_m from Eq. (1) into Eq. (2) and solving for r_m^2, we obtain

$$r_m^2 = 2.468, \text{ or } 0.1824 \qquad r_m = 1.571 \text{ m}, \; r'_m = 0.427 \text{ m} \blacktriangleleft$$

***b.* Corresponding Values of Speed.** Substituting the values obtained for r_m and r'_m into Eq. (1), we have

$$v_m = \frac{8.66}{1.571} \qquad v_m = 5.51 \text{ m/s} \blacktriangleleft$$

$$v'_m = \frac{8.66}{0.427} \qquad v'_m = 20.3 \text{ m/s} \blacktriangleleft$$

Note. It may be shown that the path of the sphere is an ellipse of *center O.*

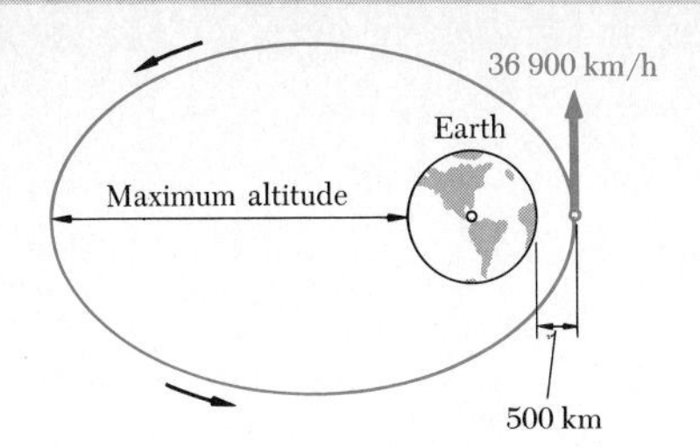

SAMPLE PROBLEM 14.11

A satellite is launched in a direction parallel to the surface of the earth with a velocity of 36 900 km/h from an altitude of 500 km. Determine (*a*) the maximum altitude reached by the satellite, (*b*) the maximum allowable error in the direction of launching if the satellite is to go into orbit and come no closer than 200 km to the surface of the earth.

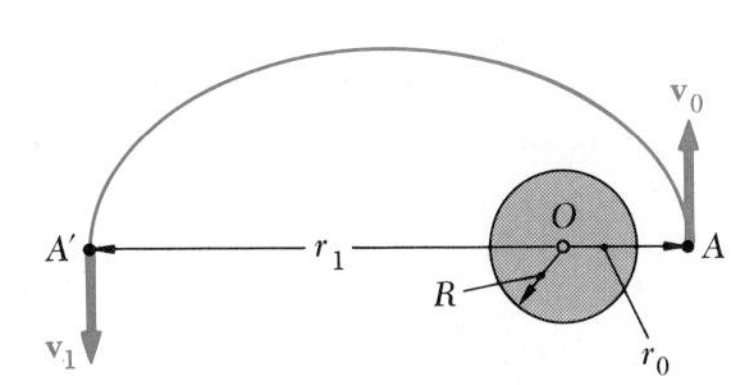

***a.* Maximum Altitude.** We denote by A' the point of the orbit farthest from the earth and by r_1 the corresponding distance from the center of the earth. Since the satellite is in free flight between A and A', we apply the principle of conservation of energy:

$$T_A + V_A = T_{A'} + V_{A'}$$

$$\tfrac{1}{2}mv_0^2 - \frac{GMm}{r_0} = \tfrac{1}{2}mv_1^2 - \frac{GMm}{r_1} \qquad (1)$$

Since the only force acting on the satellite is the force of gravity, which is a central force, the angular momentum of the satellite about O is conserved. Considering points A and A', we write

$$r_0 m v_0 = r_1 m v_1 \qquad v_1 = v_0 \frac{r_0}{r_1} \qquad (2)$$

Substituting this expression for v_1 into Eq. (1) and dividing each term by the mass m, we obtain after rearranging the terms,

$$\tfrac{1}{2}v_0^2\left(1 - \frac{r_0^2}{r_1^2}\right) = \frac{GM}{r_0}\left(1 - \frac{r_0}{r_1}\right) \qquad 1 + \frac{r_0}{r_1} = \frac{2GM}{r_0 v_0^2} \qquad (3)$$

Recalling that the radius of the earth is $R = 6370$ km, we compute

$$r_0 = 6370 \text{ km} + 500 \text{ km} = 6870 \text{ km} = 6.87 \times 10^6 \text{ m}$$
$$v_0 = 36\,900 \text{ km/h} = (36.9 \times 10^6 \text{ m})/(3.6 \times 10^3 \text{ s}) = 10.25 \times 10^3 \text{ m/s}$$
$$GM = gR^2 = (9.81 \text{ m/s}^2)(6.37 \times 10^6 \text{ m})^2 = 398 \times 10^{12} \text{ m}^3/\text{s}^2$$

Substituting these values into (3), we obtain $r_1 = 66.8 \times 10^6$ m.

Maximum altitude $= 66.8 \times 10^6 \text{ m} - 6.37 \times 10^6 \text{ m} = 60.4 \times 10^6 \text{ m}$
$= 60\,400$ km ◄

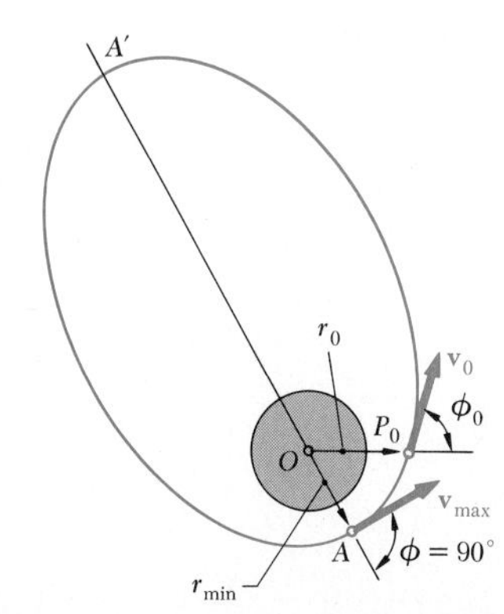

***b.* Allowable Error in Direction of Launching.** The satellite is launched from P_0 in a direction forming an angle ϕ_0 with the vertical OP_0. The value of ϕ_0 corresponding to $r_{min} = 6370 \text{ km} + 200 \text{ km} = 6570$ km is obtained by applying the principles of conservation of energy and of conservation of angular momentum between P_0 and A:

$$\tfrac{1}{2}mv_0^2 - \frac{GMm}{r_0} = \tfrac{1}{2}mv_{max}^2 - \frac{GMm}{r_{min}} \qquad (4)$$

$$r_0 m v_0 \sin \phi_0 = r_{min} m v_{max} \qquad (5)$$

Solving (5) for v_{max} and then substituting for v_{max} into (4), we may solve (4) for $\sin \phi_0$. Using the values of v_0 and GM computed in part *a* and noting that $r_0/r_{min} = 6870/6570 = 1.0457$, we find

$\sin \phi_0 = 0.9801 \qquad \phi_0 = 90° \pm 11.5° \qquad$ Allowable error $= \pm 11.5°$ ◄

Problems

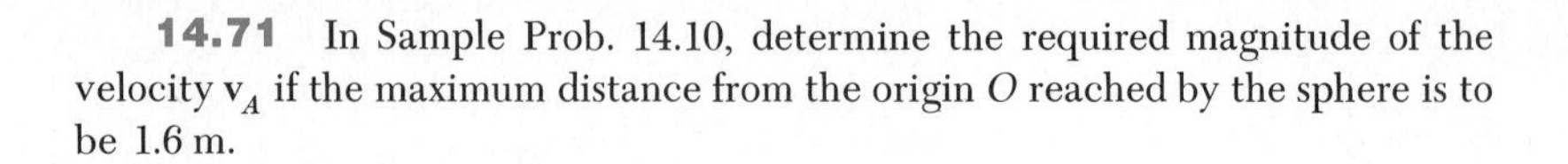
14.71 In Sample Prob. 14.10, determine the required magnitude of the velocity $\mathbf{v}_A$ if the maximum distance from the origin O reached by the sphere is to be 1.6 m.

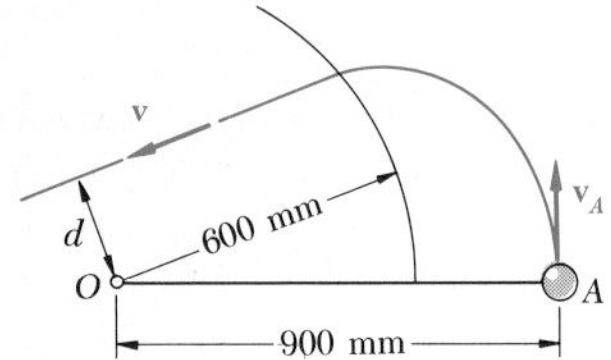

Fig. P14.72

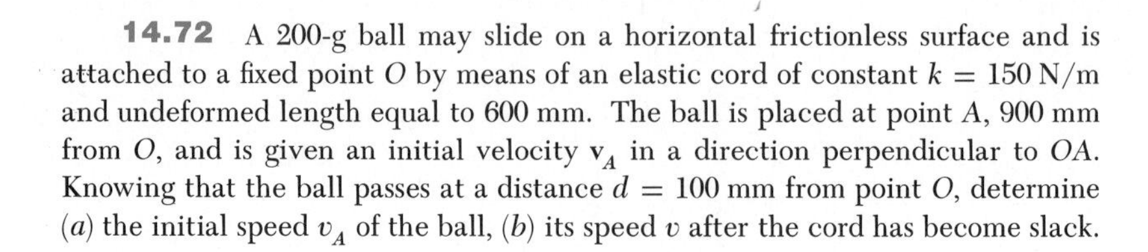
14.72 A 200-g ball may slide on a horizontal frictionless surface and is attached to a fixed point O by means of an elastic cord of constant $k = 150$ N/m and undeformed length equal to 600 mm. The ball is placed at point A, 900 mm from O, and is given an initial velocity $\mathbf{v}_A$ in a direction perpendicular to OA. Knowing that the ball passes at a distance $d = 100$ mm from point O, determine (*a*) the initial speed v_A of the ball, (*b*) its speed v after the cord has become slack.

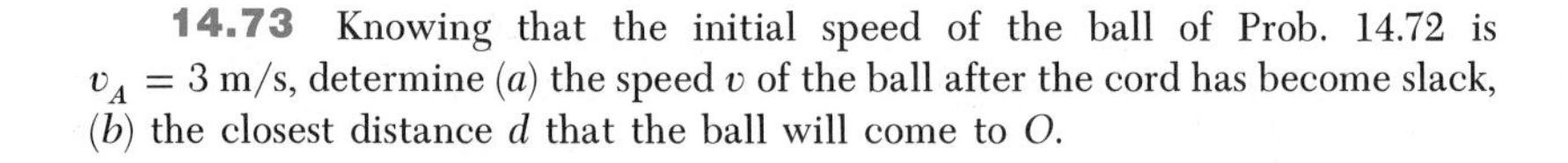
14.73 Knowing that the initial speed of the ball of Prob. 14.72 is $v_A = 3$ m/s, determine (*a*) the speed v of the ball after the cord has become slack, (*b*) the closest distance d that the ball will come to O.

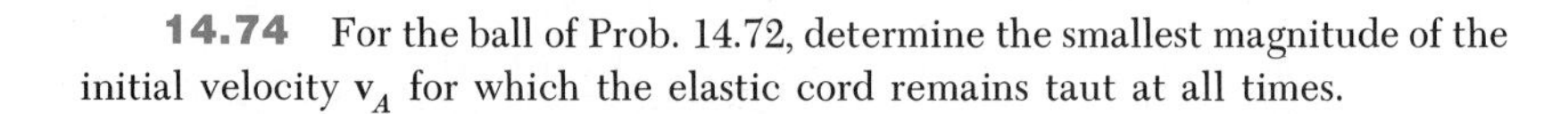
14.74 For the ball of Prob. 14.72, determine the smallest magnitude of the initial velocity $\mathbf{v}_A$ for which the elastic cord remains taut at all times.

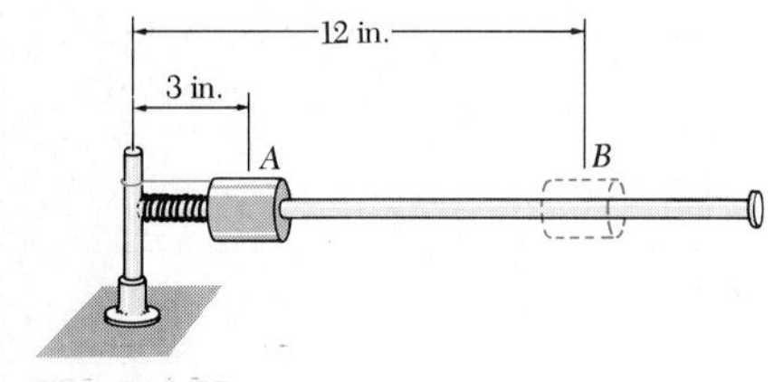

Fig. P14.75

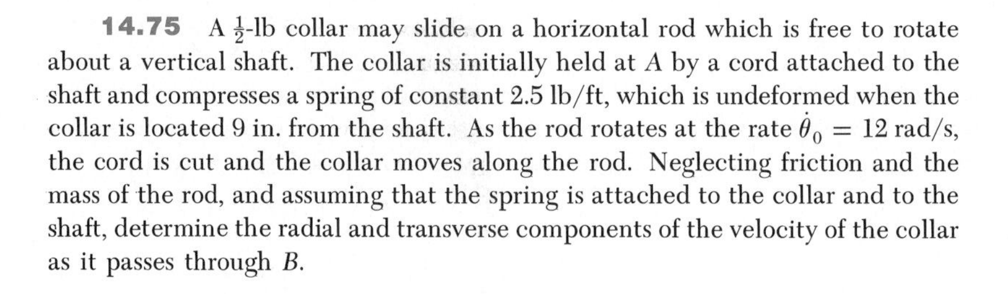
14.75 A $\frac{1}{2}$-lb collar may slide on a horizontal rod which is free to rotate about a vertical shaft. The collar is initially held at A by a cord attached to the shaft and compresses a spring of constant 2.5 lb/ft, which is undeformed when the collar is located 9 in. from the shaft. As the rod rotates at the rate $\dot{\theta}_0 = 12$ rad/s, the cord is cut and the collar moves along the rod. Neglecting friction and the mass of the rod, and assuming that the spring is attached to the collar and to the shaft, determine the radial and transverse components of the velocity of the collar as it passes through B.

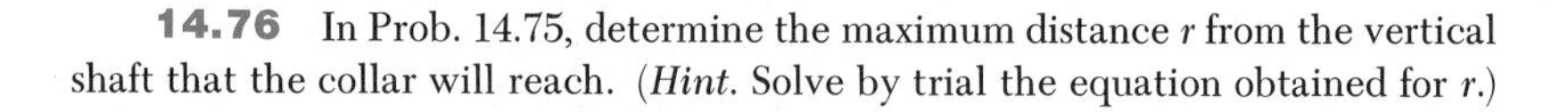
14.76 In Prob. 14.75, determine the maximum distance r from the vertical shaft that the collar will reach. (*Hint.* Solve by trial the equation obtained for r.)

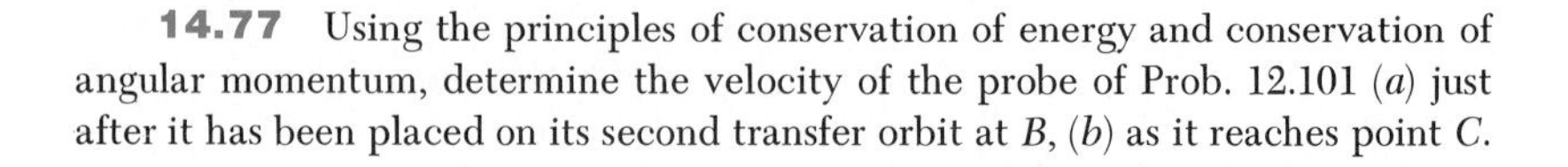
14.77 Using the principles of conservation of energy and conservation of angular momentum, determine the velocity of the probe of Prob. 12.101 (*a*) just after it has been placed on its second transfer orbit at B, (*b*) as it reaches point C.

14.78 through 14.80 Using the principles of conservation of energy and conservation of angular momentum, solve the following problems:

14.78 Prob. 12.98.

14.79 Prob. 12.99.

14.80 Prob. 12.102*a*.

14.81 After completing their moon-exploration mission, the two astronauts forming the crew of an Apollo lunar excursion module (LEM) prepared to rejoin the command module which was orbiting the moon at an altitude of 140 km. They fired the LEM's engine, brought it along a curved path to a point A, 8 km above the moon's surface, and shut off the engine. Knowing that the LEM was moving at that time in a direction parallel to the moon's surface and that it then coasted along an elliptic path to a rendezvous at B with the command module, determine (a) the speed of the LEM at engine shutoff, (b) the relative velocity with which the command module approached the LEM at B. (The radius of the moon is 1740 km and its mass is 0.01230 times the mass of the earth.)

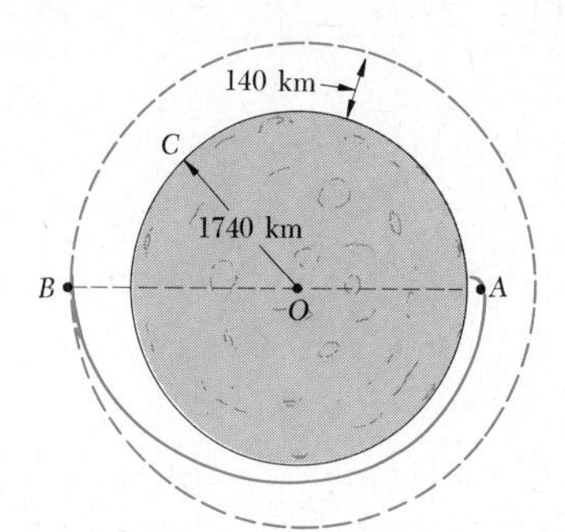

Fig. P14.81

14.82 While describing a circular orbit, 185 mi above the surface of the earth, a space shuttle ejects at point A an inertial upper stage (IUS) carrying a communication satellite to be placed in a geosynchronous orbit (see Prob. 12.79) at an altitude of 22,230 mi above the surface of the earth. Determine (a) the velocity of the IUS relative to the shuttle after its engine has been fired at A, (b) the increase in velocity required at B to place the satellite in its final orbit.

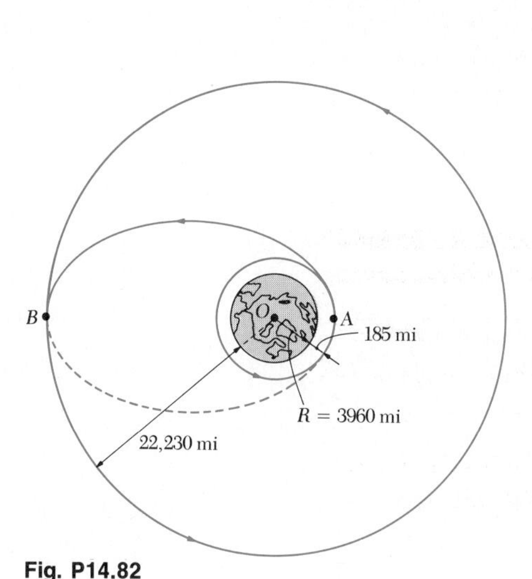

Fig. P14.82

14.83 Upon the LEM's return to the command module, the Apollo spacecraft of Prob. 14.81 was turned around so that the LEM faced to the rear. The LEM was then cast adrift with a velocity of 200 m/s relative to the command module. Determine the magnitude and direction (angle ϕ formed with the vertical OC) of the velocity $\mathbf{v}_C$ of the LEM just before it crashed at C on the moon's surface.

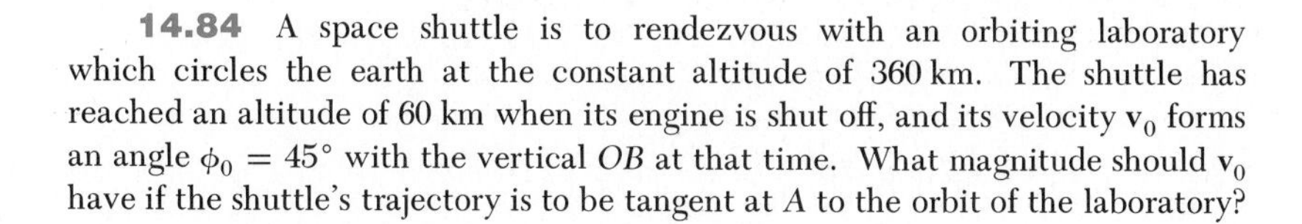

14.84 A space shuttle is to rendezvous with an orbiting laboratory which circles the earth at the constant altitude of 360 km. The shuttle has reached an altitude of 60 km when its engine is shut off, and its velocity $\mathbf{v}_0$ forms an angle $\phi_0 = 45°$ with the vertical OB at that time. What magnitude should $\mathbf{v}_0$ have if the shuttle's trajectory is to be tangent at A to the orbit of the laboratory?

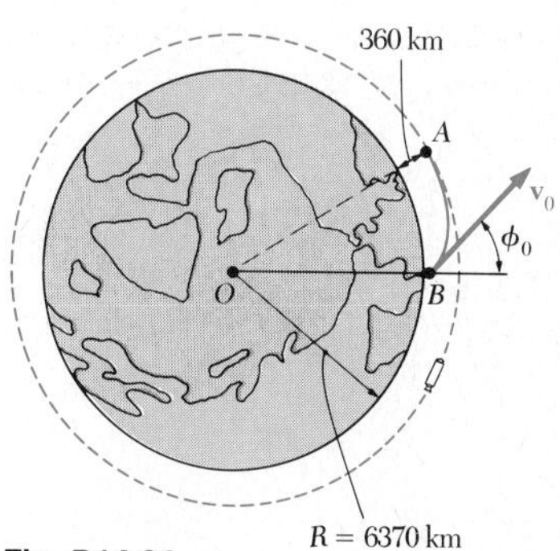

Fig. P14.84

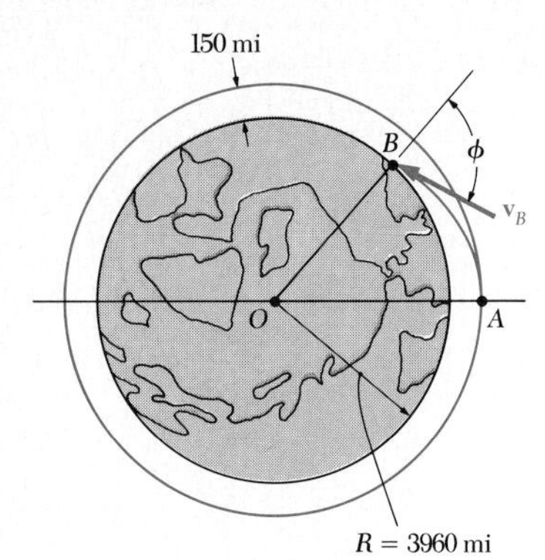

Fig. P14.85 and P14.86

14.85 A space shuttle is describing a circular orbit at an altitude of 150 mi above the surface of the earth. As it passes through A, it fires its engine for a short interval of time to reduce its speed by 4 percent and begin its descent toward the earth. Determine the magnitude and direction (angle ϕ formed with the vertical OB) of the velocity $\mathbf{v}_B$ of the shuttle as it reaches point B at an altitude of 30 mi.

14.86 A space shuttle is describing a circular orbit at an altitude of 150 mi above the surface of the earth. As it passes through A, it fires its engine to reduce its speed and begin its descent toward the earth. As it reaches point B at an altitude of 30 mi, its velocity $\mathbf{v}_B$ is observed to form an angle $\phi = 75°$ with the vertical. What is the magnitude of its velocity at that instant?

14.87 At engine burnout, an experimental rocket has reached an altitude of 400 km and has a velocity $\mathbf{v}_0$ of magnitude 7500 m/s forming an angle of 35° with the vertical. Determine the maximum altitude reached by the rocket.

14.88 At engine burnout, an experimental rocket has reached an altitude of 400 km and has a velocity $\mathbf{v}_0$ of magnitude 8000 m/s. What angle should $\mathbf{v}_0$ form with the vertical if the rocket is to reach a maximum altitude of 4000 km?

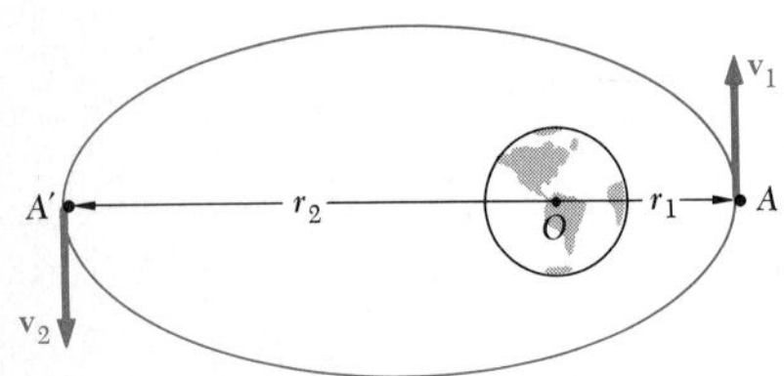

Fig. P14.89 and P14.90

14.89 Show that the values v_1 and v_2 of the speed of an earth satellite at the perigee A and the apogee A' of an elliptic orbit are defined by the relations

$$v_1^2 = \frac{2GM}{r_1 + r_2}\frac{r_2}{r_1} \qquad v_2^2 = \frac{2GM}{r_1 + r_2}\frac{r_1}{r_2}$$

where M is the mass of the earth, and r_1 and r_2 represent, respectively, the minimum and maximum distance of the orbit to the center of the earth.

14.90 Show that the total energy E of an earth satellite of mass m describing an elliptic orbit is $E = -GMm/(r_1 + r_2)$, where M is the mass of the earth, and r_1 and r_2 represent, respectively, the minimum and maximum distance of the orbit to the center of the earth. (It is recalled that the gravitational potential energy of a satellite was defined as being zero at an infinite distance from the earth.)

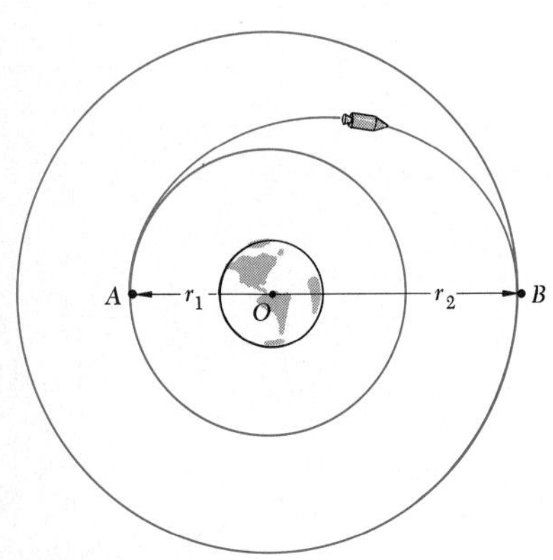

Fig. P14.91

14.91 A spacecraft of mass m describes a circular orbit of radius r_1 around the earth. (*a*) Show that the additional energy ΔE which must be imparted to the spacecraft to transfer it to a circular orbit of larger radius r_2 is

$$\Delta E = \frac{GMm(r_2 - r_1)}{2r_1 r_2}$$

where M is the mass of the earth. (*b*) Further show that if the transfer from one circular orbit to the other is executed by placing the spacecraft on a transitional semielliptic path AB, the amounts of energy ΔE_A and ΔE_B which must be imparted at A and B are, respectively, proportional to r_2 and r_1:

$$\Delta E_A = \frac{r_2}{r_1 + r_2}\Delta E \qquad \Delta E_B = \frac{r_1}{r_1 + r_2}\Delta E$$

14.92 A satellite is projected into space with a velocity $\mathbf{v}_0$ at a distance r_0 from the center of the earth by the last stage of its launching rocket. The velocity $\mathbf{v}_0$ was designed to send the satellite into a circular orbit of radius r_0. However, owing to a malfunction of control, the satellite is not projected horizontally but at an angle α with the horizontal and, as a result, is propelled into an elliptic orbit. Determine the maximum and minimum values of the distance from the center of the earth to the satellite.

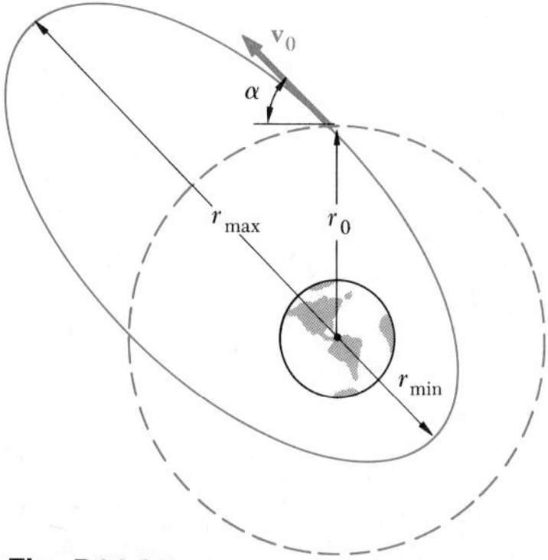

Fig. P14.92

14.93 A missile is fired from the ground with an initial velocity $\mathbf{v}_0$ forming an angle ϕ_0 with the vertical. If the missile is to reach a maximum altitude equal to αR, where R is the radius of the earth, (*a*) show that the required angle ϕ_0 is defined by the relation

$$\sin \phi_0 = (1 + \alpha)\sqrt{1 - \frac{\alpha}{1 + \alpha}\left(\frac{v_{esc}}{v_0}\right)^2}$$

where v_{esc} is the escape velocity, (*b*) determine the range of allowable values of v_0.

* **14.94** Using the answers obtained in Prob. 14.92, show that the intended circular orbit and the resulting elliptic orbit intersect at the ends of the minor axis of the elliptic orbit.

* **14.95** (*a*) Express in terms of r_{min} and v_{max} the angular momentum per unit mass, h, and the total energy per unit mass, E/m, of a space vehicle moving under the gravitational attraction of a planet of mass M (Fig. 14.19). (*b*) Eliminating v_{max} between the equations obtained, derive the formula

$$\frac{1}{r_{min}} = \frac{GM}{h^2}\left[1 + \sqrt{1 + \frac{2E}{m}\left(\frac{h}{GM}\right)^2}\right]$$

(*c*) Show that the eccentricity ε of the trajectory of the vehicle may be expressed as

$$\varepsilon = \sqrt{1 + \frac{2E}{m}\left(\frac{h}{GM}\right)^2}$$

(*d*) Further show that the trajectory of the vehicle is a hyperbola, an ellipse, or a parabola, depending on whether E is positive, negative, or zero.

***14.14. Variable Systems of Particles.** All the systems of particles considered so far consisted of well-defined particles. These systems did not gain or lose any particles during their motion. In a large number of engineering applications, however, it is necessary to consider *variable systems of particles,* i.e., systems which are continually gaining or losing particles, or doing both at the same time. Consider, for example, a hydraulic turbine. Its analysis involves the determination of the forces exerted by a stream of water on rotating blades, and we note that the particles of water in contact with the blades form an ever-changing system which continually acquires and loses particles. Rockets furnish another example of variable systems, since their propulsion depends upon the continual ejection of fuel particles.

We recall that all the kinetics principles established so far were derived for constant systems of particles, which neither gain nor lose particles. We must therefore find a way to reduce the analysis of a variable system of particles to that of an auxiliary constant system. The procedure to follow is indicated in Secs. 14.15 and 14.16 for two broad categories of applications.

***14.15. Steady Stream of Particles.** Consider a steady stream of particles, such as a stream of water diverted by a fixed vane or a flow of air through a duct or through a blower. In order to determine the resultant of the forces exerted on the particles in contact with the vane, duct, or blower, we isolate these particles and denote by S the system thus defined (Fig. 14.21). We observe that S is a variable system of particles, since it continually gains particles flowing in and loses an equal number of particles flowing out. Therefore, the kinetics principles that have been established so far cannot be directly applied to S.

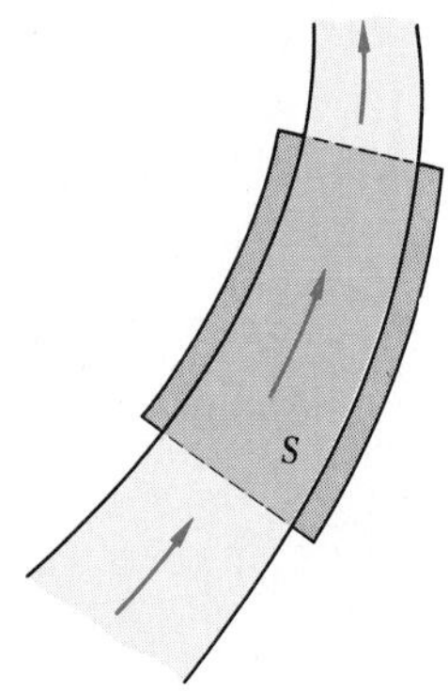

Fig. 14.21

However, we may easily define an auxiliary system of particles which does remain constant for a short interval of time Δt. Consider at time t the system S *plus* the particles which will enter S during the interval of time Δt (Fig. 14.22*a*). Next, consider at time $t + \Delta t$ the system S *plus* the particles which have left S during the interval Δt (Fig. 14.22*c*). Clearly, *the same particles are involved in both cases,* and we may apply to these particles the principle of impulse and momentum. Since the total mass m of the system S remains constant, the particles entering the system and those leaving the system in the time Δt must have the same mass Δm. Denoting by $\mathbf{v}_A$ and $\mathbf{v}_B$, respectively, the velocities of the particles entering S at A and leaving S at B, we represent the momentum of the particles entering S by $(\Delta m)\mathbf{v}_A$ (Fig. 14.22*a*) and the momentum of the particles leaving S by $(\Delta m)\mathbf{v}_B$ (Fig. 14.22*c*). We also represent the momenta $m_i\mathbf{v}_i$ of the particles forming S and the impulses of the forces exerted on S by the appropriate vectors, and indicate by gray plus and equals signs that the system of the momenta and impulses in parts a and b of Fig. 14.22 is equipollent to the system of the momenta in part c of the same figure.

Since the resultant $\Sigma m_i\mathbf{v}_i$ of the momenta of the particles of S is found on both sides of the equals sign, it may be omitted. We conclude that *the system formed by the momentum* $(\Delta m)\mathbf{v}_A$ *of the particles entering S in the*

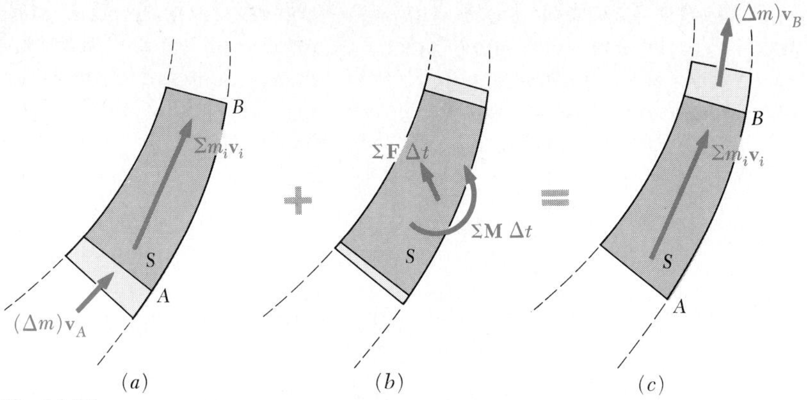

Fig. 14.22

time Δt *and the impulses of the forces exerted on S during that time is equipollent to the momentum* $(\Delta m)\mathbf{v}_B$ *of the particles leaving S in the same time* Δt. We may therefore write

$$(\Delta m)\mathbf{v}_A + \Sigma\mathbf{F}\,\Delta t = (\Delta m)\mathbf{v}_B \tag{14.47}$$

A similar equation may be obtained by taking the moments of the vectors involved (see Sample Prob. 14.12). Dividing all terms of Eq. (14.47) by Δt and letting Δt approach zero, we obtain at the limit

$$\Sigma\mathbf{F} = \frac{dm}{dt}(\mathbf{v}_B - \mathbf{v}_A) \tag{14.48}$$

where $\mathbf{v}_B - \mathbf{v}_A$ represents the difference between the *vectors* $\mathbf{v}_B$ and $\mathbf{v}_A$.

If SI units are used, dm/dt is expressed in kg/s and the velocities in m/s; we check that both members of Eq. (14.48) are expressed in the same units (newtons). If U.S. customary units are used, dm/dt must be expressed in slugs/s and the velocities in ft/s; we check again that both members of the equation are expressed in the same units (pounds).†

The principle we have established may be used to analyze a large number of engineering applications. Some of the most common are indicated below.

Fluid Stream Diverted by a Vane. If the vane is fixed, the method of analysis given above may be applied directly to find the force **F** exerted by the vane on the stream. We note that **F** is the only force which needs to be considered since the pressure in the stream is constant (atmospheric pressure). The force exerted by the stream on the vane will be equal and

†It is often convenient to express the mass rate of flow dm/dt as the product ρQ, where ρ is the density of the stream (mass per unit volume) and Q its volume rate of flow (volume per unit time). If SI units are used, ρ is expressed in kg/m^3 (for instance, $\rho = 1000$ kg/m^3 for water) and Q in m^3/s. However, if U.S. customary units are used, ρ will generally have to be computed from the corresponding specific weight γ (weight per unit volume), $\rho = \gamma/g$. Since γ is expressed in lb/ft^3 (for instance, $\gamma = 62.4$ lb/ft^3 for water), ρ is obtained in slugs/ft^3. The volume rate of flow Q is expressed in ft^3/s.

opposite to **F**. If the vane moves with a constant velocity, the stream is not steady. However, it will appear steady to an observer moving with the vane. We should therefore choose a system of axes moving with the vane. Since this system of axes is not accelerated, Eq. (14.47) may still be used, but $\mathbf{v}_A$ and $\mathbf{v}_B$ must be replaced by the *relative velocities* of the stream with respect to the vane (see Sample Prob. 14.13).

Fluid Flowing through a Pipe. The force exerted by the fluid on a pipe transition such as a bend or a contraction may be determined by considering the system of particles S in contact with the transition. Since, in general, the pressure in the flow will vary, we should also consider the forces exerted on S by the adjoining portions of the fluid.

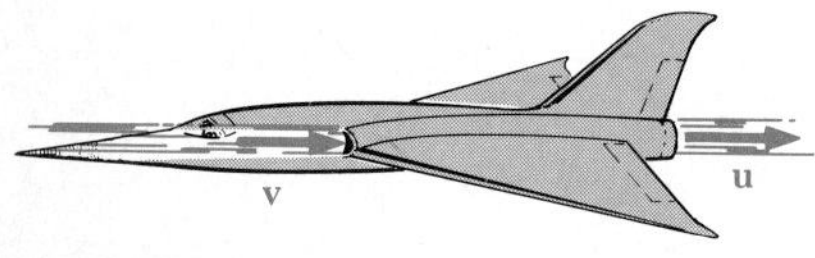

Fig. 14.23

Jet Engine. In a jet engine, air enters with no velocity through the front of the engine and leaves through the rear with a high velocity. The energy required to accelerate the air particles is obtained by burning fuel. While the exhaust gases contain burned fuel, the mass of the fuel is small compared with the mass of the air flowing through the engine and usually may be neglected. Thus, the analysis of a jet engine reduces to that of an air stream. This stream may be considered as a steady stream if all velocities are measured with respect to the airplane. The air stream shall be assumed, therefore, to enter the engine with a velocity **v** of magnitude equal to the speed of the airplane and to leave with a velocity **u** equal to the relative velocity of the exhaust gases (Fig. 14.23). Since the intake and exhaust pressures are nearly atmospheric, the only external force which needs to be considered is the force exerted by the engine on the air stream. This force is equal and opposite to the thrust.†

Fan. We consider the system of particles S shown in Fig. 14.24. The velocity $\mathbf{v}_A$ of the particles entering the system is assumed equal to zero,

†Note that if the airplane is accelerated, it cannot be used as a newtonian frame of reference. The same result will be obtained for the thrust, however, by using a reference frame at rest with respect to the atmosphere, since the air particles will then be observed to enter the engine with no velocity and to leave it with a velocity of magnitude $u - v$.

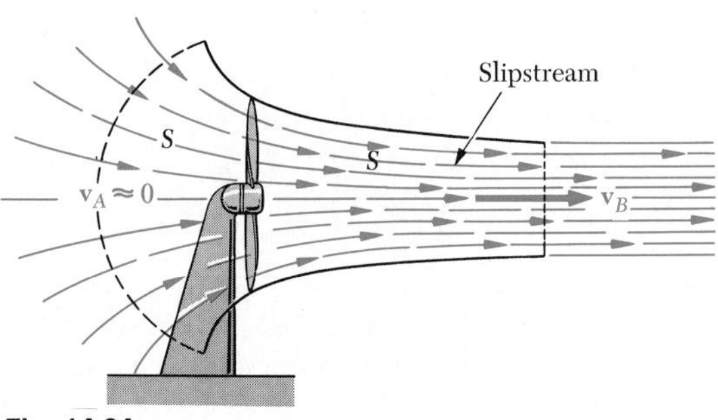

Fig. 14.24

and the velocity $\mathbf{v}_B$ of the particles leaving the system is the velocity of the *slipstream*. The rate of flow may be obtained by multiplying v_B by the cross-sectional area of the slipstream. Since the pressure all around S is atmospheric, the only external force acting on S is the thrust of the fan.

Airplane Propeller. In order to obtain a steady stream of air, velocities should be measured with respect to the airplane. Thus, the air particles will be assumed to enter the system with a velocity $\mathbf{v}$ of magnitude equal to the speed of the airplane and to leave with a velocity $\mathbf{u}$ equal to the relative velocity of the slipstream.

* 14.16. Systems Gaining or Losing Mass.

We shall now analyze a different type of variable system of particles, namely, a system which gains mass by continually absorbing particles or loses mass by continually expelling particles. Consider the system S shown in Fig. 14.25. Its mass, equal to m at the instant t, increases by Δm in the interval of time Δt.

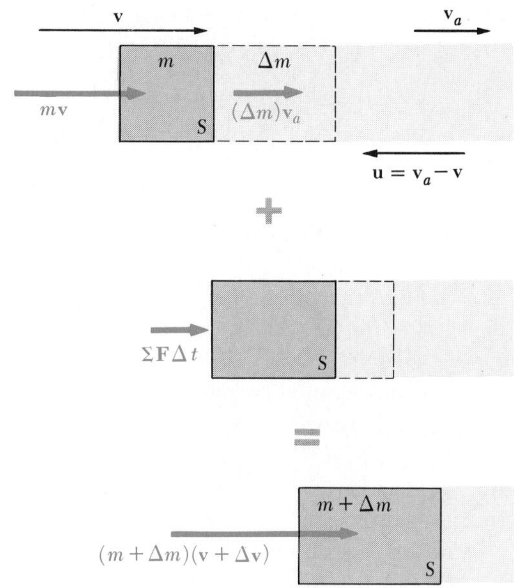

Fig. 14.25

In order to apply the principle of impulse and momentum to the analysis of this system, we must consider at time t the system S *plus* the particles of mass Δm which S absorbs during the time interval Δt. The velocity of S at time t is denoted by $\mathbf{v}$, and its velocity at time $t + \Delta t$ is denoted by $\mathbf{v} + \Delta\mathbf{v}$, while the absolute velocity of the particles which are absorbed is denoted by $\mathbf{v}_a$. Applying the principle of impulse and momentum, we write

$$m\mathbf{v} + (\Delta m)\mathbf{v}_a + \Sigma\mathbf{F}\,\Delta t = (m + \Delta m)(\mathbf{v} + \Delta\mathbf{v})$$

$$m\mathbf{v} + (\Delta m)\mathbf{v}_a + \Sigma\mathbf{F}\,\Delta t = (m + \Delta m)(\mathbf{v} + \Delta\mathbf{v}) \qquad (repeated)$$

Solving for the sum $\Sigma\mathbf{F}\,\Delta t$ of the impulses of the external forces acting on S (excluding the forces exerted by the particles being absorbed), we have

$$\Sigma\mathbf{F}\,\Delta t = m\,\Delta\mathbf{v} + \Delta m(\mathbf{v} - \mathbf{v}_a) + (\Delta m)(\Delta\mathbf{v}) \qquad (14.49)$$

Introducing the *relative velocity* $\mathbf{u}$ with respect to S of the particles which are absorbed, we write $\mathbf{u} = \mathbf{v}_a - \mathbf{v}$ and note, since $v_a < v$, that the relative velocity $\mathbf{u}$ is directed to the left, as shown in Fig. 14.25. Neglecting the last term in Eq. (14.49), which is of the second order, we write

$$\Sigma\mathbf{F}\,\Delta t = m\,\Delta\mathbf{v} - (\Delta m)\mathbf{u}$$

Dividing through by Δt and letting Δt approach zero, we have at the limit†

$$\Sigma\mathbf{F} = m\frac{d\mathbf{v}}{dt} - \frac{dm}{dt}\mathbf{u} \qquad (14.50)$$

Rearranging the terms and recalling that $d\mathbf{v}/dt = \mathbf{a}$, where $\mathbf{a}$ is the acceleration of the system S, we write

$$\Sigma\mathbf{F} + \frac{dm}{dt}\mathbf{u} = m\mathbf{a} \qquad (14.51)$$

which shows that the action on S of the particles being absorbed is equivalent to a thrust

$$\mathbf{P} = \frac{dm}{dt}\mathbf{u} \qquad (14.52)$$

which tends to slow down the motion of S, since the relative velocity $\mathbf{u}$ of the particles is directed to the left. If SI units are used, dm/dt is expressed in kg/s, the relative velocity u in m/s, and the corresponding thrust in newtons. If U.S. customary units are used, dm/dt must be expressed in slugs/s and u in ft/s; the corresponding thrust will then be expressed in pounds.‡

The equations obtained may also be used to determine the motion of a system S losing mass. In this case, the rate of change of mass is negative, and the action on S of the particles being expelled is equivalent to a thrust in the direction of $-\mathbf{u}$, that is, in the direction opposite to that in which the particles are being expelled. A *rocket* represents a typical case of a system continually losing mass (see Sample Prob. 14.14).

† When the absolute velocity $\mathbf{v}_a$ of the particles absorbed is zero, we have $\mathbf{u} = -\mathbf{v}$, and formula (14.50) becomes

$$\Sigma\mathbf{F} = \frac{d}{dt}(m\mathbf{v})$$

Comparing the formula obtained to Eq. (14.2) of Sec. 14.2, we observe that Newton's second law may be applied to a system gaining mass, *provided that the particles absorbed are initially at rest.* It may also be applied to a system losing mass, *provided that the velocity of the particles expelled is zero* with respect to the frame of reference selected.

‡ See footnote on page 631.

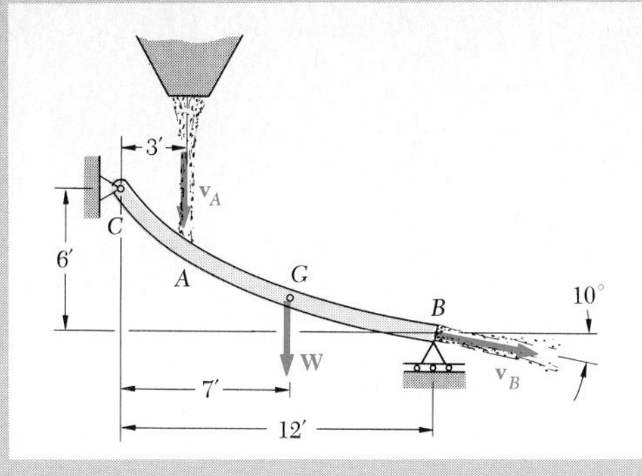

SAMPLE PROBLEM 14.12

Grain falls from a hopper onto a chute CB at the rate of 240 lb/s. It hits the chute at A with a velocity of 20 ft/s and leaves at B with a velocity of 15 ft/s, forming an angle of 10° with the horizontal. Knowing that the combined weight of the chute and of the grain it supports is a force $\mathbf{W}$ of magnitude 600 lb applied at G, determine the reaction at the roller-support B and the components of the reaction at the hinge C.

Solution. We apply the principle of impulse and momentum for the time interval Δt to the system consisting of the chute, the grain it supports, and the amount of grain which hits the chute in the interval Δt. Since the chute does not move, it has no momentum. We also note that the sum $\Sigma m_i \mathbf{v}_i$ of the momenta of the particles supported by the chute is the same at t and $t + \Delta t$ and thus may be omitted.

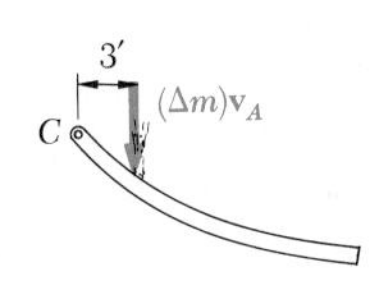

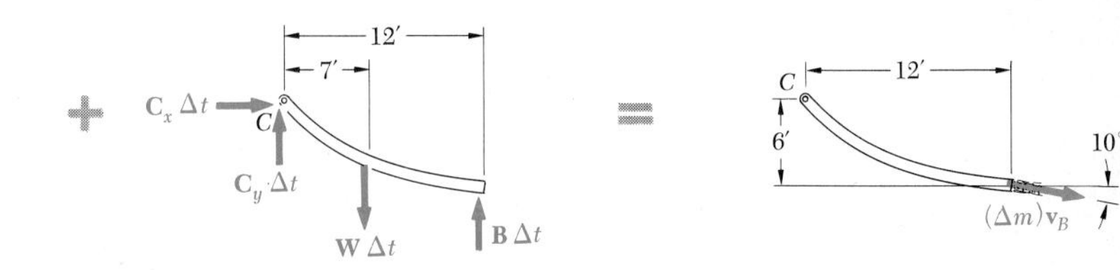

Since the system formed by the momentum $(\Delta m)\mathbf{v}_A$ and the impulses is equipollent to the momentum $(\Delta m)\mathbf{v}_B$, we write

$\xrightarrow{+}$ x components: $\quad C_x\,\Delta t = (\Delta m)v_B \cos 10°$ (1)

$+\uparrow y$ components: $\quad -(\Delta m)v_A + C_y\,\Delta t - W\,\Delta t + B\,\Delta t = -(\Delta m)v_B \sin 10°$ (2)

$+\circlearrowleft$ moments about C: $\quad -3(\Delta m)v_A - 7(W\,\Delta t) + 12(B\,\Delta t) = 6(\Delta m)v_B \cos 10° - 12(\Delta m)v_B \sin 10°$ (3)

Using the given data, $W = 600$ lb, $v_A = 20$ ft/s, $v_B = 15$ ft/s, $\Delta m/\Delta t = 240/32.2 = 7.45$ slugs/s, and solving Eq. (3) for B and Eq. (1) for C_x,

$$12B = 7(600) + 3(7.45)(20) + 6(7.45)(15)(\cos 10° - 2 \sin 10°)$$

$$12B = 5075 \qquad B = 423 \text{ lb} \qquad \mathbf{B} = 423 \text{ lb} \uparrow \blacktriangleleft$$

$$C_x = (7.45)(15) \cos 10° = 110.1 \text{ lb} \qquad \mathbf{C}_x = 110.1 \text{ lb} \rightarrow \blacktriangleleft$$

Substituting for B and solving Eq. (2) for C_y,

$$C_y = 600 - 423 + (7.45)(20 - 15 \sin 10°) = 307 \text{ lb}$$

$$\mathbf{C}_y = 307 \text{ lb} \uparrow \blacktriangleleft$$

SAMPLE PROBLEM 14.13

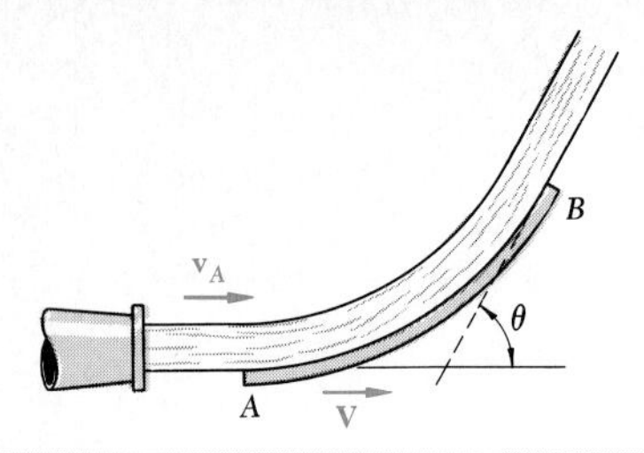

A nozzle discharges a stream of water of cross-sectional area A with a velocity $\mathbf{v}_A$. The stream is deflected by a *single* blade which moves to the right with a constant velocity $\mathbf{V}$. Assuming that the water moves along the blade at constant speed, determine (*a*) the components of the force $\mathbf{F}$ exerted by the blade on the stream, (*b*) the velocity $\mathbf{V}$ for which maximum power is developed.

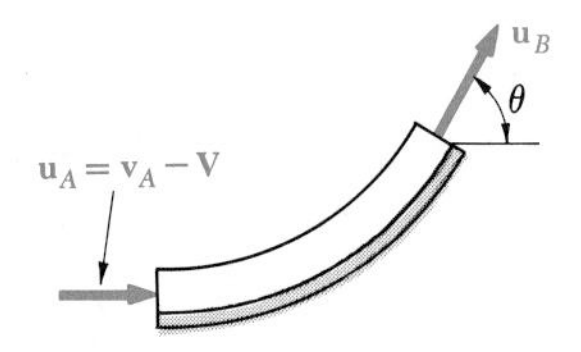

a. **Components of Force Exerted on Stream.** We choose a coordinate system which moves with the blade at a constant velocity $\mathbf{V}$. The particles of water strike the blade with a relative velocity $\mathbf{u}_A = \mathbf{v}_A - \mathbf{V}$ and leave the blade with a relative velocity $\mathbf{u}_B$. Since the particles move along the blade at a constant speed, the relative velocities $\mathbf{u}_A$ and $\mathbf{u}_B$ have the same magnitude u. Denoting the density of water by ρ, the mass of the particles striking the blade during the time interval Δt is $\Delta m = A\rho(v_A - V)\,\Delta t$; an equal mass of particles leaves the blade during Δt. We apply the principle of impulse and momentum to the system formed by the particles in contact with the blade and by those striking the blade in the time Δt.

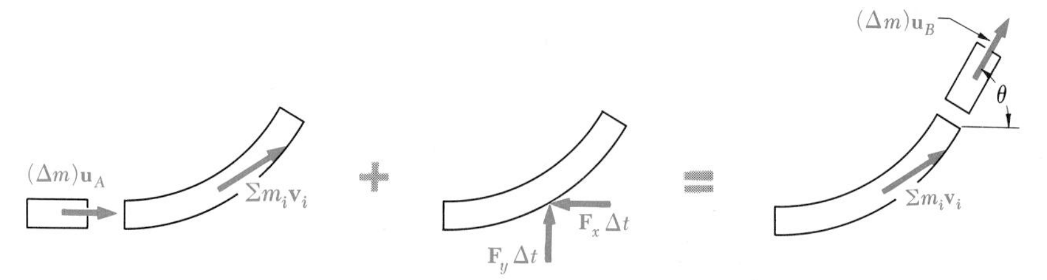

Recalling that $\mathbf{u}_A$ and $\mathbf{u}_B$ have the same magnitude u, and omitting the momentum $\Sigma m_i \mathbf{v}_i$ which appears on both sides, we write

$\xrightarrow{+}$ x components: $\quad (\Delta m)u - F_x\,\Delta t = (\Delta m)u\cos\theta$

$+\uparrow y$ components: $\quad +F_y\,\Delta t = (\Delta m)u\sin\theta$

Substituting $\Delta m = A\rho(v_A - V)\,\Delta t$ and $u = v_A - V$, we obtain

$$\mathbf{F}_x = A\rho(v_A - V)^2(1 - \cos\theta) \leftarrow \qquad \mathbf{F}_y = A\rho(v_A - V)^2 \sin\theta \uparrow \quad ◄$$

b. **Velocity of Blade for Maximum Power.** The power is obtained by multiplying the velocity V of the blade by the component F_x of the force exerted by the stream on the blade.

$$\text{Power} = F_x V = A\rho(v_A - V)^2(1 - \cos\theta)V$$

Differentiating the power with respect to V and setting the derivative equal to zero, we obtain

$$\frac{d(\text{power})}{dV} = A\rho(v_A^2 - 4v_A V + 3V^2)(1 - \cos\theta) = 0$$

$$V = v_A \qquad V = \tfrac{1}{3}v_A \qquad \text{For maximum power } \mathbf{V} = \tfrac{1}{3}v_A \rightarrow \quad ◄$$

Note. These results are valid only when a *single* blade deflects the stream. Different results are obtained when a series of blades deflects the stream, as in a Pelton-wheel turbine. (See Prob. 14.120.)

SAMPLE PROBLEM 14.14

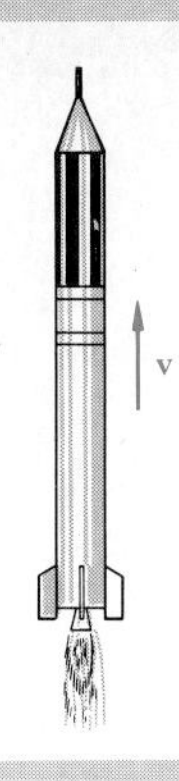

A rocket of initial mass m_0 (including shell and fuel) is fired vertically at time $t = 0$. The fuel is consumed at a constant rate $q = dm/dt$ and is expelled at a constant speed u relative to the rocket. Derive an expression for the velocity of the rocket at time t, neglecting the resistance of the air.

Solution. At time t, the mass of the rocket shell and remaining fuel is $m = m_0 - qt$, and the velocity is **v**. During the time interval Δt, a mass of fuel $\Delta m = q\,\Delta t$ is expelled with a speed u relative to the rocket. Denoting by $\mathbf{v}_e$ the absolute velocity of the expelled fuel, we apply the principle of impulse and momentum between time t and time $t + \Delta t$.

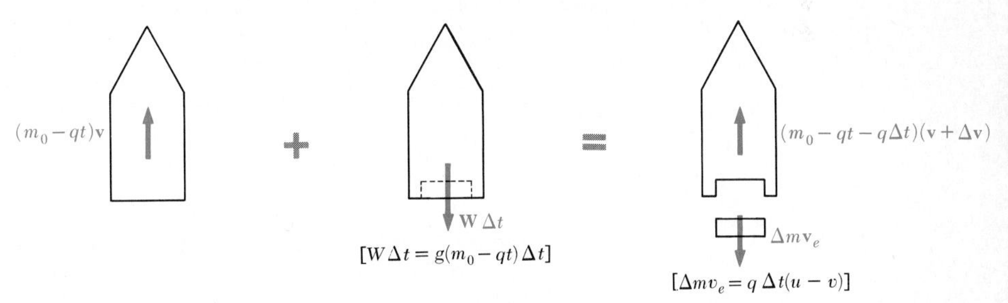

We write

$$(m_0 - qt)v - g(m_0 - qt)\,\Delta t = (m_0 - qt - q\,\Delta t)(v + \Delta v) - q\,\Delta t(u - v)$$

Dividing through by Δt, and letting Δt approach zero, we obtain

$$-g(m_0 - qt) = (m_0 - qt)\frac{dv}{dt} - qu$$

Separating variables and integrating from $t = 0$, $v = 0$ to $t = t$, $v = v$,

$$dv = \left(\frac{qu}{m_0 - qt} - g\right)dt \qquad \int_0^v dv = \int_0^t \left(\frac{qu}{m_0 - qt} - g\right)dt$$

$$v = [-u \ln (m_0 - qt) - gt]_0^t \qquad v = u \ln \frac{m_0}{m_0 - qt} - gt \blacktriangleleft$$

Remark. The mass remaining at time t_f, after all the fuel has been expended, is equal to the mass of the rocket shell $m_s = m_0 - qt_f$, and the maximum velocity attained by the rocket is $v_m = u \ln (m_0/m_s) - gt_f$. Assuming that the fuel is expelled in a relatively short period of time, the term gt_f is small and we have $v_m \approx u \ln (m_0/m_s)$. In order to escape the gravitational field of the earth, a rocket must reach a velocity of 11.18 km/s. Assuming $u = 2200$ m/s and $v_m = 11.18$ km/s, we obtain $m_0/m_s = 161$. Thus, to project each kilogram of the rocket shell into space, it is necessary to consume more than 161 kg of fuel if a propellant yielding $u = 2200$ m/s is used.

Problems

Note. In the following problems use $\rho = 1000 \text{ kg/m}^3$ for the density of water in SI units, and $\gamma = 62.4 \text{ lb/ft}^3$ for its specific weight in U.S. customary units. (See footnote on page 631.)

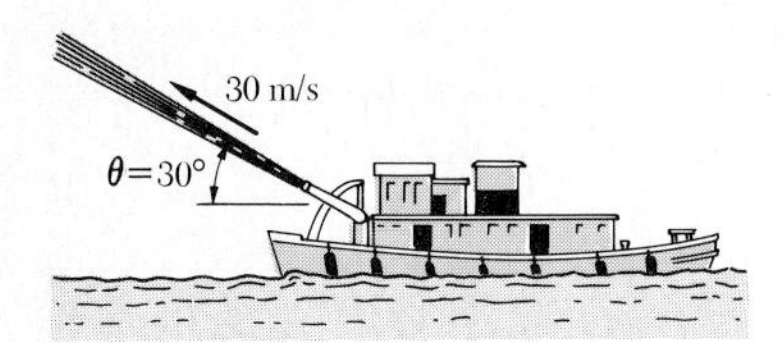

Fig. P14.96

14.96 A hose discharges water at the rate of $9 \text{ m}^3/\text{min}$ from the stern of a 20-Mg fireboat. If the velocity of the water stream is 30 m/s, determine the reaction on the boat.

14.97 Tree limbs and branches are being fed at D at the rate of 5 kg/s into a shredder which spews the resulting wood chips at C with a velocity of 25 m/s. Determine the horizontal component of the force exerted by the shredder on the truck hitch at A.

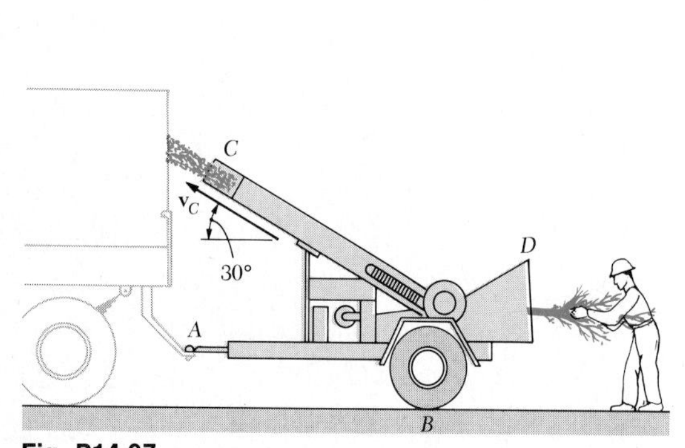

Fig. P14.97

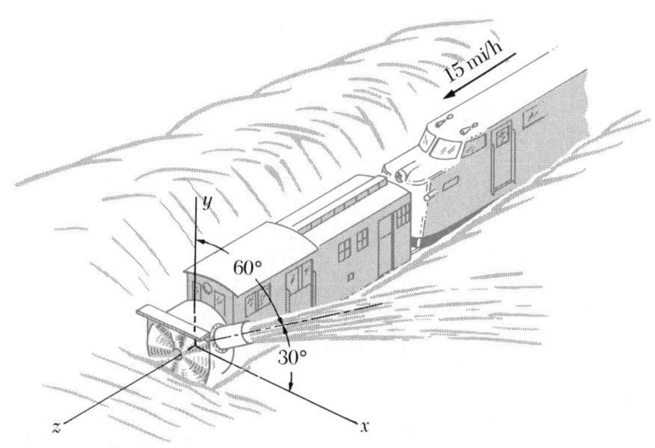

Fig. P14.98

14.98 A rotary power plow is used to remove snow from a level section of railroad track. The plow car is placed ahead of an engine which propels it at a constant speed of 15 mi/h. The plow clears 180 tons of snow per minute, projecting it in the direction shown with a velocity of 40 ft/s relative to the plow car. Neglecting rolling friction, determine (*a*) the magnitude of the force **P** exerted by the engine on the plow car, (*b*) the lateral force exerted on the plow car by the track.

14.99 For the rotary power plow of Prob. 14.98, determine the angle that the discharge duct should form with the z axis if at the speed considered, the force **P** exerted by the engine on the plow car is to be zero.

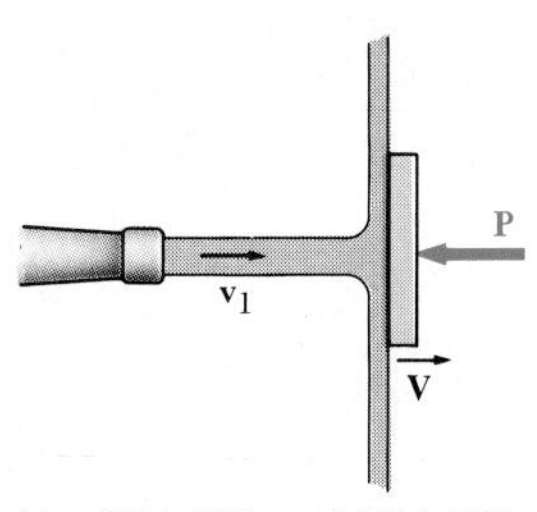

Fig. P14.100 and P14.101

14.100 A stream of water of cross-sectional area A and velocity $\mathbf{v}_1$ strikes a plate which is held motionless by a force **P**. Determine the magnitude of **P**, knowing that $A = 600 \text{ mm}^2$, $v_1 = 30$ m/s, and $V = 0$.

14.101 A stream of water of cross-sectional area A and velocity $\mathbf{v}_1$ strikes a plate which moves to the right with a velocity **V**. Determine the magnitude of **V**, knowing that $A = 400 \text{ mm}^2$, $v_1 = 35$ m/s, and $P = 360$ N.

14.102 Water flows in a continuous sheet from between two plates A and B with a velocity $\mathbf{v}$ of magnitude 120 ft/s. The stream is split into two parts by a smooth horizontal plate C. Knowing that the rates of flow in each of the two resulting streams are, respectively, $Q_1 = 40$ gal/min and $Q_2 = 160$ gal/min, determine (*a*) the angle θ, (*b*) the total force exerted by the streams on the plate ($1 \text{ ft}^3 = 7.48$ gal).

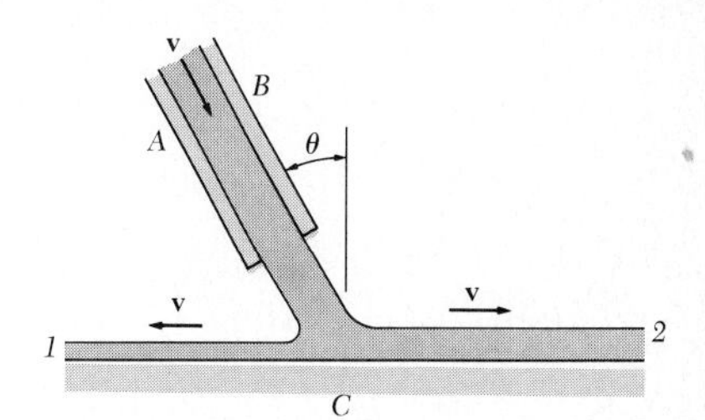

Fig. P14.102 and P14.103

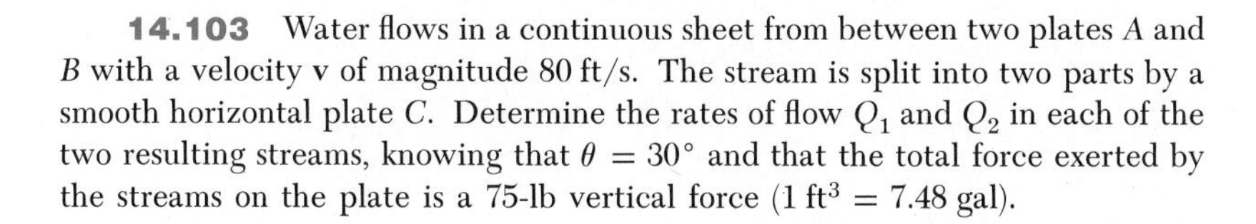
14.103 Water flows in a continuous sheet from between two plates A and B with a velocity $\mathbf{v}$ of magnitude 80 ft/s. The stream is split into two parts by a smooth horizontal plate C. Determine the rates of flow Q_1 and Q_2 in each of the two resulting streams, knowing that $\theta = 30°$ and that the total force exerted by the streams on the plate is a 75-lb vertical force ($1 \text{ ft}^3 = 7.48$ gal).

14.104 The nozzle shown discharges water at the rate of 300 gal/min. Knowing that at both A and B the stream of water moves with a velocity of magnitude 80 ft/s and neglecting the weight of the vane, determine the components of the reactions at C and D ($1 \text{ ft}^3 = 7.48$ gal).

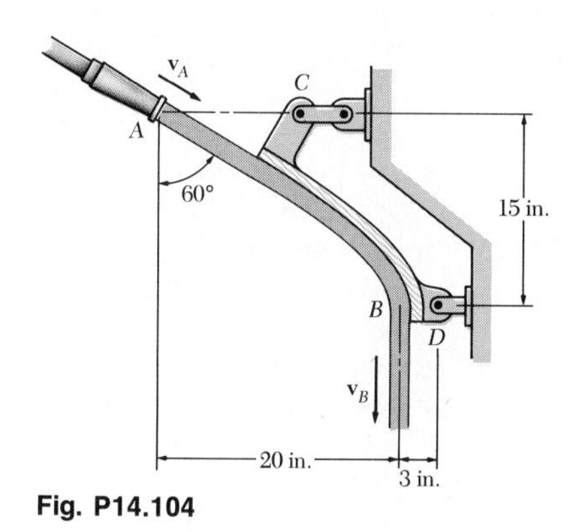

Fig. P14.104

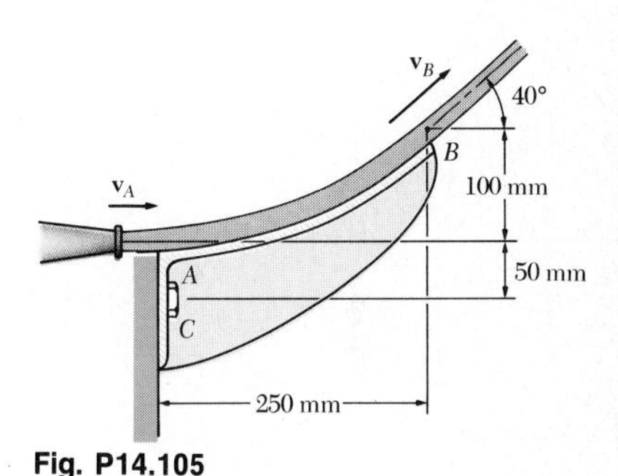

Fig. P14.105

14.105 The nozzle shown discharges water at the rate of $0.95 \text{ m}^3/\text{min}$. Knowing that at both A and B the water moves with a velocity of magnitude 40 m/s and neglecting the weight of the vane, determine the force-couple system which must be applied at C to hold the vane in place.

14.106 Knowing that the blade AB of Sample Prob. 14.13 is in the shape of an arc of circle, show that the resultant force $\mathbf{F}$ exerted by the blade on the stream is applied at the midpoint C of the arc AB. (*Hint.* First show that the line of action of $\mathbf{F}$ must pass through the center O of the circle.)

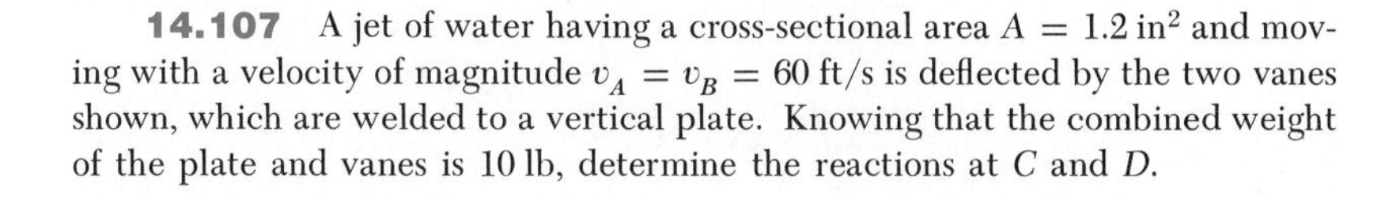
14.107 A jet of water having a cross-sectional area $A = 1.2 \text{ in}^2$ and moving with a velocity of magnitude $v_A = v_B = 60$ ft/s is deflected by the two vanes shown, which are welded to a vertical plate. Knowing that the combined weight of the plate and vanes is 10 lb, determine the reactions at C and D.

Fig. P14.107

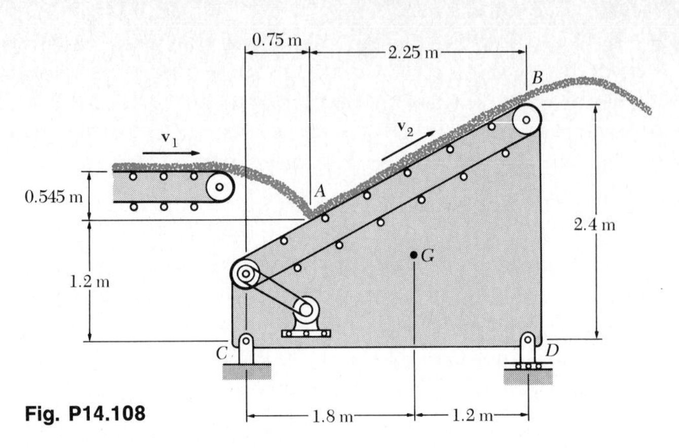

Fig. P14.108

14.108 Coal is being discharged from a first conveyor belt at the rate of 120 kg/s. It is received at A by a second belt which discharges it again at B. Knowing that $v_1 = 3$ m/s and $v_2 = 4.25$ m/s and that the second conveyor belt assembly and the coal it supports have a total mass of 472 kg, determine the reactions at C and D.

14.109 Solve Prob. 14.108, assuming that the velocity of the second conveyor belt is decreased to $v_2 = 2.55$ m/s and that the mass of coal it supports is increased by 48 kg.

14.110 The total drag due to air friction on a jet airplane traveling at 900 km/h is 30 kN. Knowing that the exhaust velocity is 650 m/s relative to the airplane, determine the mass of air which must pass through the engine per second to maintain the speed of 900 km/h in level flight.

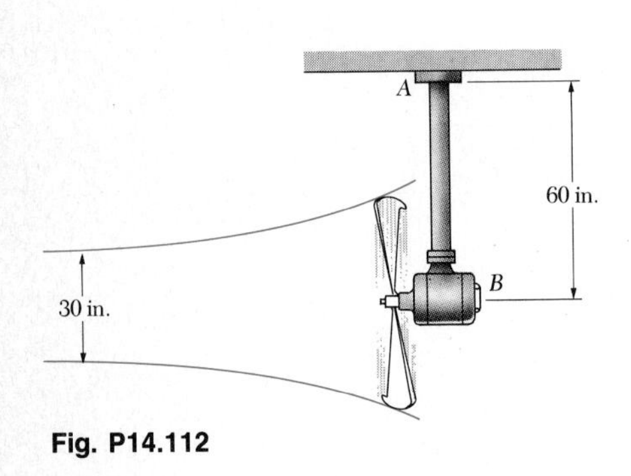

Fig. P14.112

14.111 While cruising in level flight at a speed of 600 mi/h, a jet airplane scoops in air at the rate of 170 lb/s and discharges it with a velocity of 2000 ft/s relative to the airplane. Determine the total drag due to air friction on the airplane.

14.112 For the ceiling-mounted fan shown, determine the maximum allowable air velocity in the slipstream if the bending moment in the supporting rod AB is not to exceed 80 lb·ft. Assume $\gamma = 0.076$ lb/ft^3 for air and neglect the approach velocity of the air.

14.113 The jet engine shown scoops in air at A at the rate of 75 kg/s and discharges it at B with a velocity of 800 m/s relative to the airplane. Determine the magnitude and line of action of the propulsive thrust developed by the engine when the speed of the airplane is (a) 500 km/h, (b) 1000 km/h.

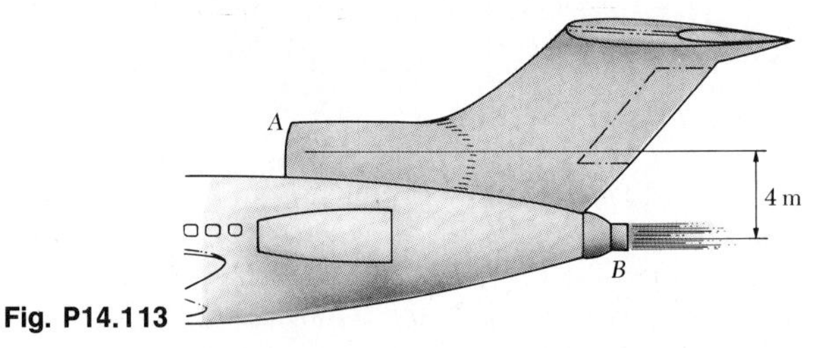

Fig. P14.113

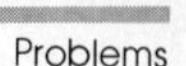

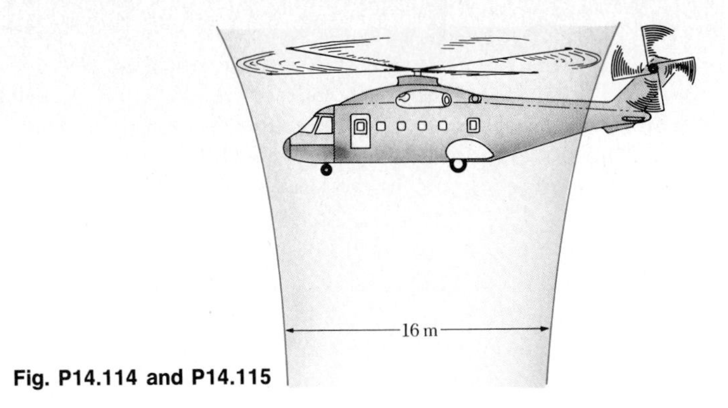

Fig. P14.114 and P14.115

14.114 The helicopter shown has a mass of 10 Mg when empty and can produce a maximum downward air speed of 25 m/s in its 16-m-diameter slipstream. Assuming $\rho = 1.21\ \text{kg/m}^3$ for air, determine the maximum combined payload and fuel load the helicopter can carry while hovering in midair.

14.115 The helicopter shown has a mass of 10 Mg. Assuming $\rho = 1.21\ \text{kg/m}^3$ for air, determine the downward air speed the helicopter produces in its 16-m-diameter slipstream while hovering in midair with a combined payload and fuel load of 4000 kg.

Fig. P14.116

14.116 A jet airliner is cruising at the speed of 990 km/h with each of its three engines discharging air with a velocity of 675 m/s relative to the plane. Determine the speed of the airliner after it has lost the use of (*a*) one of its engines, (*b*) two of its engines. Assume that the drag due to air friction is proportional to the square of the speed and that the remaining engines keep operating at the same rate.

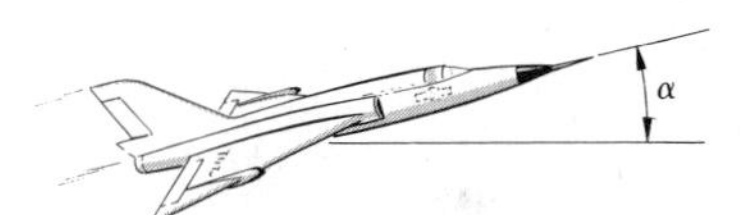

Fig. P14.117

14.117 A 30,000-lb jet airplane maintains a constant speed of 540 mi/h while climbing at an angle $\alpha = 10°$. The airplane scoops in air at the rate of 500 lb/s and discharges it with a velocity of 2250 ft/s relative to the airplane. If the pilot changes to a horizontal flight and the same engine conditions are maintained, determine (*a*) the initial acceleration of the plane, (*b*) the maximum horizontal speed attained. Assume that the drag due to air friction is proportional to the square of the speed.

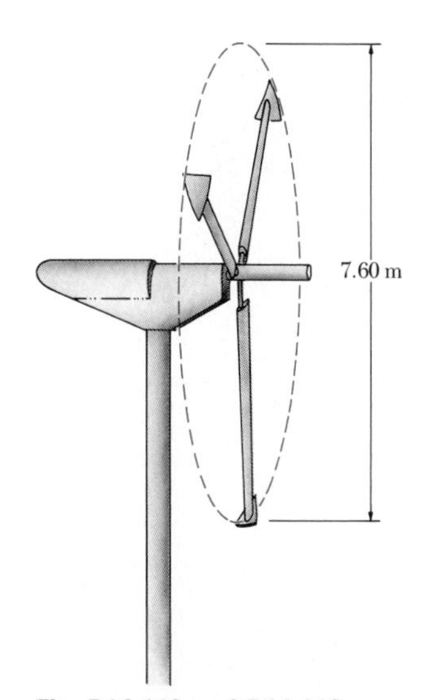

Fig. P14.118 and P14.119

14.118 The wind turbine-generator shown has an output-power rating of 15 kW for a wind speed of 43.2 km/h. For the given wind speed, determine (*a*) the kinetic energy of the air particles entering the 7.60-m-diameter circle per second, (*b*) the efficiency of this energy-conversion system. Assume $\rho = 1.21\ \text{kg/m}^3$ for air.

14.119 For a given wind speed, the wind turbine-generator shown produces 20 kW of electric power and, as an energy-conversion system, has an efficiency of 0.30. Determine (*a*) the kinetic energy of the air particles entering the 7.60-m-diameter circle per second, (*b*) the wind speed.

14.120 In a Pelton-wheel turbine, a stream of water is deflected by a series of blades so that the rate at which water is deflected by the blades is equal to the rate at which water issues from the nozzle ($\Delta m/\Delta t = A\rho v_A$). Using the same notation as in Sample Prob. 14.13, (*a*) determine the velocity **V** of the blades for which maximum power is developed, (*b*) derive an expression for the maximum power, (*c*) derive an expression for the mechanical efficiency.

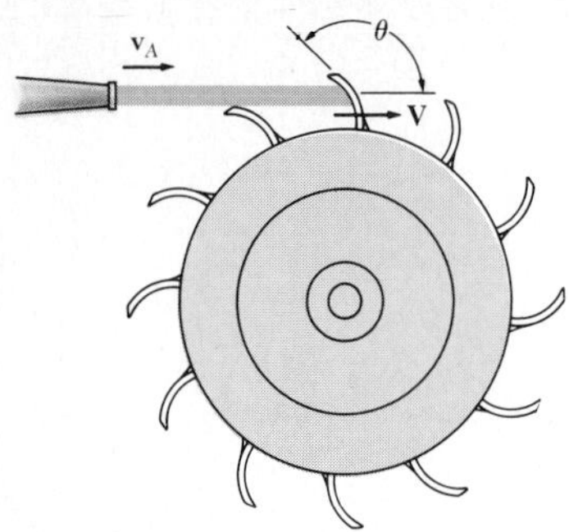

Fig. P14.120

14.121 While cruising in level flight at a speed of 600 mi/h, a jet airplane scoops in air at the rate of 200 lb/s and discharges it with a velocity of 2200 ft/s relative to the airplane. Determine (*a*) the power actually used to propel the airplane, (*b*) the total power developed by the engine, (*c*) the mechanical efficiency of the airplane.

14.122 A circular reentrant orifice (also called Borda's mouthpiece) of diameter D is placed at a depth h below the surface of a tank. Knowing that the speed of the issuing stream is $v = \sqrt{2gh}$ and assuming that the speed of approach v_1 is zero, show that the diameter of the stream is $d = D/\sqrt{2}$. (*Hint.* Consider the section of water indicated, and note that P is equal to the pressure at a depth h multiplied by the area of the orifice.)

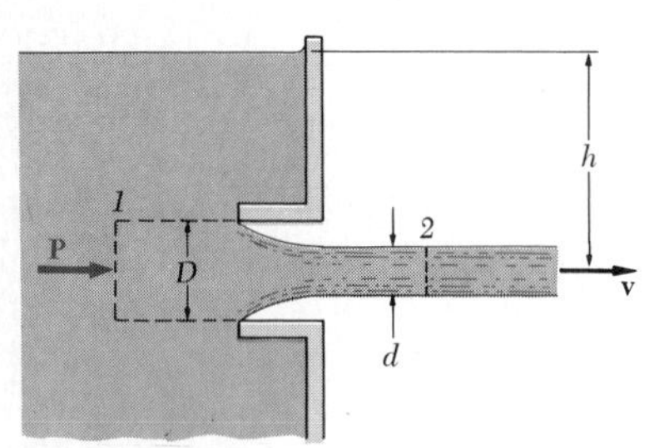

Fig. P14.122

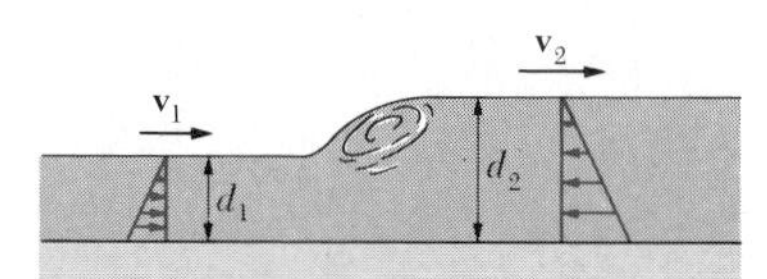

Fig. P14.123

* **14.123** The depth of water flowing in a rectangular channel of width b at a speed v_1 and depth d_1 increases to a depth d_2 at a *hydraulic jump*. Express the rate of flow Q in terms of b, d_1, d_2, and g.

* **14.124** Determine the rate of flow in the channel of Prob. 14.123, knowing that $d_1 = 0.8$ m, $d_2 = 1.2$ m, and the rectangular channel is 3 m wide.

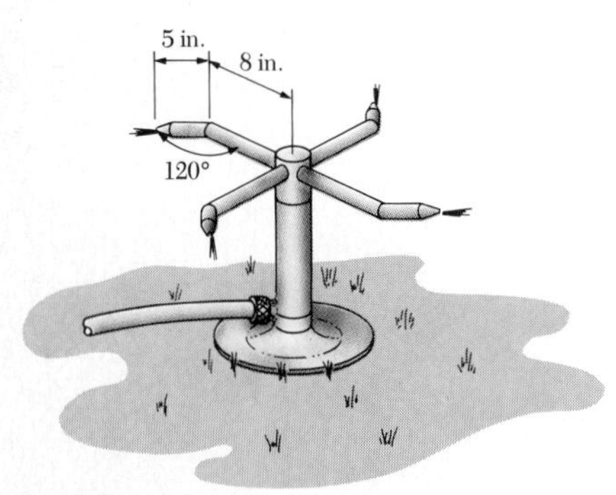

Fig. P14.125

14.125 A garden sprinkler has four rotating arms, each of which consists of two horizontal straight sections of pipe forming an angle of 120°. Each arm discharges water at the rate of 4 gal/min with a velocity of 50 ft/s relative to the arm. Knowing that the friction between the moving and the stationary parts of the sprinkler is equivalent to a couple of moment $M = 0.250$ lb·ft, determine the constant rate at which the sprinkler rotates (1 ft^3 = 7.48 gal).

14.126 A chain of length l and total mass m lies in a pile on the floor. If its end A is raised vertically at a constant speed v, determine (*a*) the force **P** applied to A at the time when half the chain is off the floor, (*b*) the reaction exerted by the floor at that time.

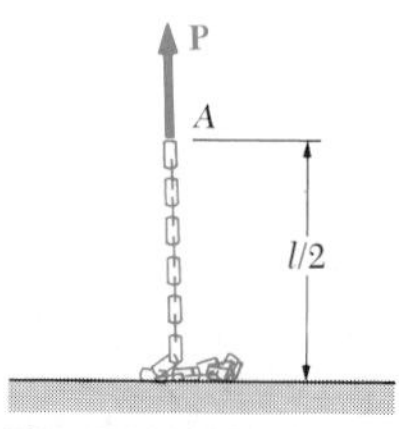

Fig. P14.126

14.127 Solve Prob. 14.126, assuming that the end A of the chain is being *lowered* to the floor at a constant speed v.

14.128 The ends of a chain lie in piles at A and C. When given an initial speed v, the chain keeps moving freely at that speed over the pulley at B. Neglecting friction, determine the required value of h.

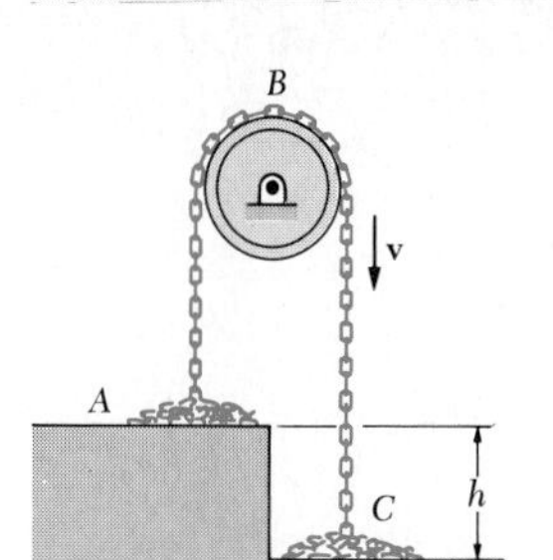

Fig. P14.128 and P14.129

14.129 The ends of a chain lie in piles at A and C. When released from rest at time $t = 0$, the chain moves over the pulley at B, which has a negligible mass. Denoting by L the length of chain connecting the two piles and neglecting friction, determine the speed v of the chain at time t.

14.130 For the chain of Prob. 14.129, it is known that $h = 300$ mm, and that the length of chain connecting the two piles is $L = 4$ m. Neglecting friction, determine at $t = 2$ s (a) the speed v of the chain, (b) the length of chain which has been transferred from pile A to pile C.

14.131 Determine the speed v of the chain of Prob. 14.129, after a length x of chain has been transferred from pile A to pile C.

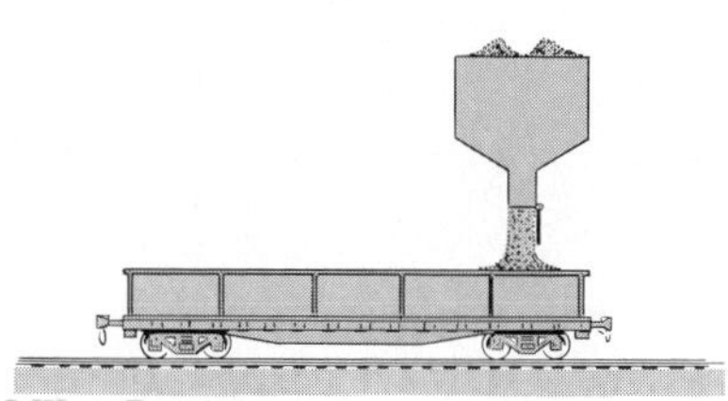
Fig. P14.132

14.132 A railroad car of length L and mass m_0 when empty is moving freely on a horizontal track while being loaded with sand from a stationary chute at the rate $q = dm/dt$. Knowing that the car was approaching the chute with a speed v_0, determine after the car has cleared the chute (a) the mass of the car and its load, (b) the speed of the car.

14.133 A possible method for reducing the speed of a training plane as it lands on an aircraft carrier consists in having the tail of the plane hook into the end of a heavy chain of length l which lies in a pile below deck. Denoting by m the mass of the plane and by v_0 its speed at touchdown, and assuming no other retarding force, determine (a) the required mass of the chain if the speed of the plane is to be reduced to βv_0, where $\beta < 1$, (b) the maximum value of the force exerted by the chain on the plane.

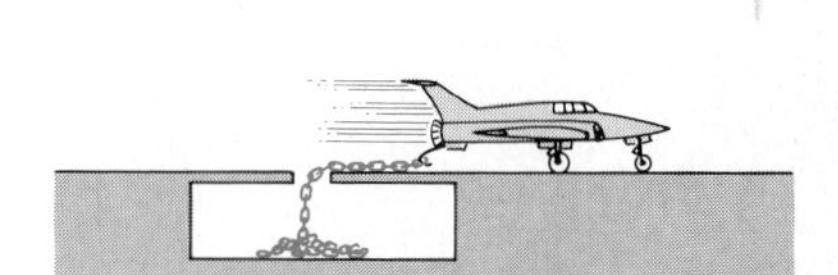
Fig. P14.133 and P14.134

14.134 As a 15,000-lb training plane lands on an aircraft carrier at a speed of 120 mi/h, its tail hooks into the end of a long chain which lies in a pile below deck. Knowing that the chain weighs 25 lb/ft and assuming no other retarding force, determine the maximum deceleration of the plane.

14.135 For the railroad car of Prob. 14.132, express in terms of t the magnitude of the horizontal force **P** which should be applied to the car to keep it moving, (a) at the constant speed v_0, (b) with a constant acceleration a, while being loaded. Assume that $t = 0$ when the loading operation begins.

14.136 The main propulsion system of a space shuttle consists of three identical rocket engines, each of which burns the hydrogen-oxygen propellant at the rate of 800 lb/s and ejects it with a relative velocity of 12,000 ft/s. Determine the total thrust provided by the three engines.

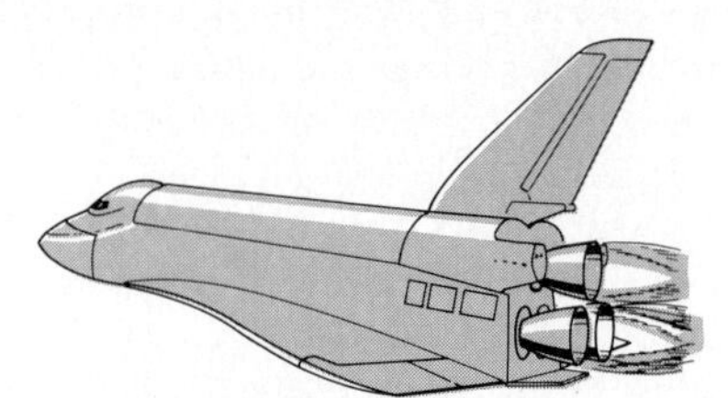

Fig. P14.136 and P14.137

14.137 The main propulsion system of a space shuttle consists of three identical rocket engines which provide a total thrust of 1250 kips. Determine the rate at which the hydrogen-oxygen propellant is burned by each of the three engines, knowing that it is ejected with a relative velocity of 12,000 ft/s.

14.138 A rocket has a mass of 1200 kg, including 1000 kg of fuel, which is consumed at the rate of 12.5 kg/s and ejected with a relative velocity of 4000 m/s. Knowing that the rocket is fired vertically from the ground, determine its acceleration (*a*) as it is fired, (*b*) as the last particle of fuel is being consumed.

14.139 The acceleration of a rocket is observed to be 30 m/s^2 at $t = 0$, as it is fired vertically from the ground, and 350 m/s^2 at $t = 80$ s. Knowing that the fuel is consumed at the rate of 10 kg/s, determine (*a*) the initial mass of the rocket, (*b*) the relative velocity with which the fuel is ejected.

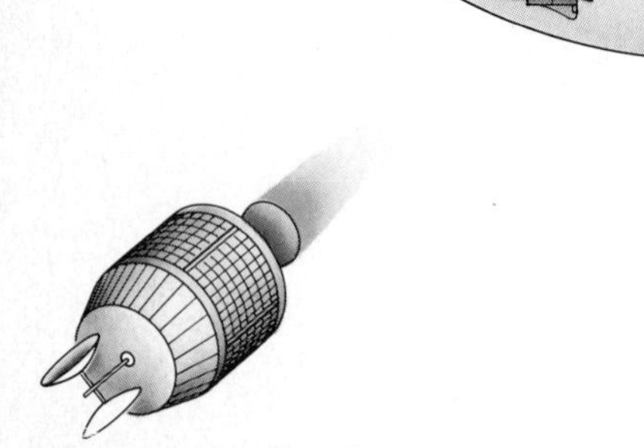

Fig. P14.140 and P14.141

14.140 A weather satellite of mass 4000 kg, including fuel, has been ejected from a space shuttle describing a low circular orbit around the earth. After the satellite has slowly drifted to a safe distance from the shuttle, its engine is fired to increase its velocity by 2430 m/s as a first step to its transfer to a geosynchronous orbit. Knowing that the fuel is ejected with a relative velocity of 3800 m/s, determine the mass of fuel consumed in this maneuver.

14.141 Determine the increase in velocity of the weather satellite of Prob. 14.140 after 1200 kg of fuel has been consumed.

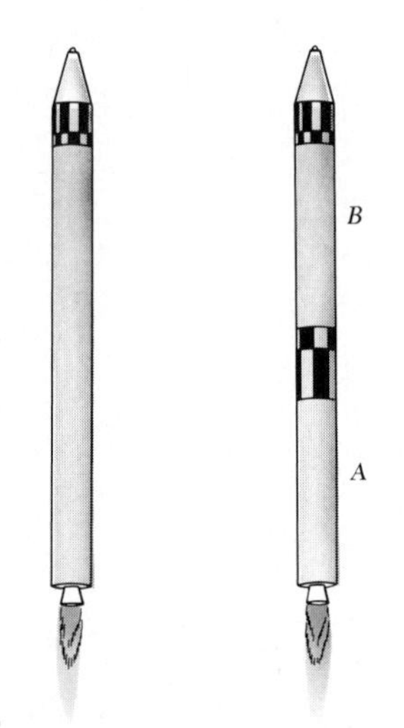

Fig. P14.142 **Fig. P14.143**

14.142 An 800-lb spacecraft is mounted on top of a rocket weighing 38,400 lb, including 36,000 lb of fuel. Knowing that the fuel is consumed at the rate of 400 lb/s and ejected with a relative velocity of 12,000 ft/s, determine the maximum speed imparted to the spacecraft when the rocket is fired vertically from the ground.

14.143 The rocket used to launch the 800-lb spacecraft of Prob. 14.142 is redesigned to include two stages *A* and *B*, each weighing 19,200 lb, including 18,000 lb of fuel. The fuel is again consumed at the rate of 400 lb/s and ejected with a relative velocity of 12,000 ft/s. Knowing that when stage *A* expels its last particle of fuel, its casing is released and jettisoned, determine (*a*) the speed of the rocket at that instant, (*b*) the maximum speed imparted to the spacecraft.

14.144 Determine the altitude reached by the spacecraft of Prob. 14.142 when all the fuel of its launching rocket has been consumed.

14.145 For the spacecraft and the two-stage launching rocket of Prob. 14.143, determine the altitude at which (*a*) stage *A* of the rocket is released, (*b*) the fuel of both stages has been consumed.

14.146 Knowing that the engine of the weather satellite of Prob. 14.140 was fired for 126 s, determine the distance separating the weather satellite from the space shuttle as the engine was turned off.

14.147 For the rocket of Prob. 14.138, determine (*a*) the altitude at which all the fuel has been consumed, (*b*) the velocity of the rocket at that time.

14.148 In a rocket, the kinetic energy imparted to the consumed and ejected fuel is wasted as far as propelling the rocket is concerned. The useful power is equal to the product of the force available to propel the rocket and the speed of the rocket. If v is the speed of the rocket and u is the relative speed of the expelled fuel, show that the mechanical efficiency of the rocket is $\eta = 2uv/(u^2 + v^2)$. Explain why $\eta = 1$ when $u = v$.

14.149 In a jet airplane, the kinetic energy imparted to the exhaust gases is wasted as far as propelling the airplane is concerned. The useful power is equal to the product of the force available to propel the airplane and the speed of the airplane. If v is the speed of the airplane and u is the relative speed of the expelled gases, show that the mechanical efficiency of the airplane is $\eta = 2v/(u + v)$. Explain why $\eta = 1$ when $u = v$.

14.17. Review and Summary. This chapter was devoted to the method of impulse and momentum and to its application to the solution of various types of problems involving the motion of particles.

Principle of impulse and momentum for a particle

The *linear momentum of a particle* was defined [Sec. 14.2] as the product $m\mathbf{v}$ of the mass m of the particle and its velocity $\mathbf{v}$. From Newton's second law, $\mathbf{F} = m\mathbf{a}$, we derived the relation

$$m\mathbf{v}_1 + \int_{t_1}^{t_2} \mathbf{F}\,dt = m\mathbf{v}_2 \tag{14.4}$$

where $m\mathbf{v}_1$ and $m\mathbf{v}_2$ represent the momentum of the particle at a time t_1 and a time t_2, respectively, and where the integral defines the *linear impulse of the force* $\mathbf{F}$ during the corresponding time interval. We wrote therefore

$$m\mathbf{v}_1 + \mathbf{Imp}_{1\to2} = m\mathbf{v}_2 \tag{14.5}$$

which expresses the principle of impulse and momentum for a particle. When the particle considered is subjected to several forces, the sum of the impulses of these forces should be used; we had

$$m\mathbf{v}_1 + \Sigma\,\mathbf{Imp}_{1\to2} = m\mathbf{v}_2 \tag{14.6}$$

Since Eqs. (14.5) and (14.6) involve *vector quantities,* it is necessary to consider separately their x and y components when applying them to the solution of a given problem [Sample Probs. 14.1 and 14.2].

Impulsive motion

The method of impulse and momentum is particularly effective in the study of the *impulsive motion* of a particle, when very large forces, called *impulsive forces*, are applied for a very short interval of time Δt, since this method involves the impulses $\mathbf{F}\,\Delta t$ of the forces, rather than the forces themselves. Neglecting the impulse of any nonimpulsive force, we wrote

$$m\mathbf{v}_1 + \Sigma\mathbf{F}\,\Delta t = m\mathbf{v}_2 \qquad (14.7)$$

Principle of impulse and momentum for a system of particles

When the principle of impulse and momentum is applied to a *system of particles* [Sec. 14.4], the impulses of the internal forces cancel out and we have

$$\Sigma m\mathbf{v}_1 + \Sigma\ \mathbf{Ext\ Imp}_{1\rightarrow 2} = \Sigma m\mathbf{v}_2 \qquad (14.8)$$

We note again that, since vector quantities are involved, their x and y components should be considered separately in the solution of a given problem [Sample Prob. 14.3].

In the particular case *when the sum of the impulses of the external forces is zero,* Eq. (14.8) reduces to $\Sigma m\mathbf{v}_1 = \Sigma m\mathbf{v}_2$, i.e., *the total momentum of the system is conserved,* and the velocity of the mass center G of the system remains unchanged: $\overline{\mathbf{v}}_1 = \overline{\mathbf{v}}_2$ [Sec. 14.5].

Direct central impact

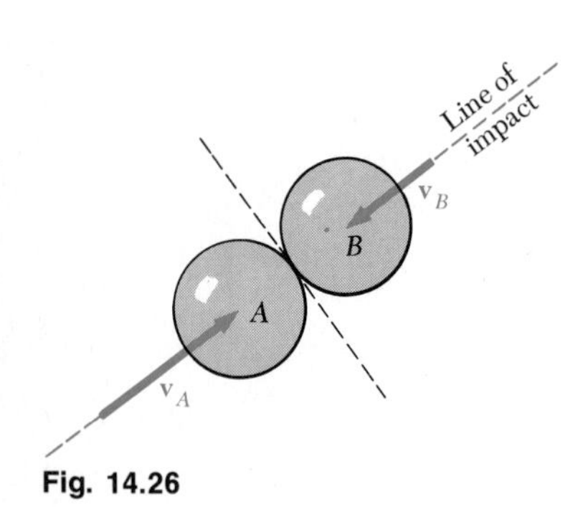

Fig. 14.26

In Secs. 14.6 through 14.8, we considered the *central impact* of two colliding bodies. In the case of a *direct central impact* [Sec. 14.7], the two colliding bodies A and B were moving along the *line of impact* with velocities $\mathbf{v}_A$ and $\mathbf{v}_B$, respectively (Fig. 14.26). Two equations could be used to determine their velocities $\mathbf{v}'_A$ and $\mathbf{v}'_B$ after the impact. The first expressed *conservation of momentum* for the two-body system,

$$m_A v_A + m_B v_B = m_A v'_A + m_B v'_B \qquad (14.15)$$

where a positive sign indicates that the corresponding velocity is directed to the right, while the second related the *relative velocities* of the two bodies before and after the impact,

$$v'_B - v'_A = e(v_A - v_B) \qquad (14.21)$$

The constant e is known as the *coefficient of restitution;* its value lies between 0 and 1 and depends in a large measure on the materials involved. When $e = 0$, the impact is said to be *perfectly plastic;* when $e = 1$, it is said to be *perfectly elastic* [Sample Prob. 14.4].

Oblique central impact

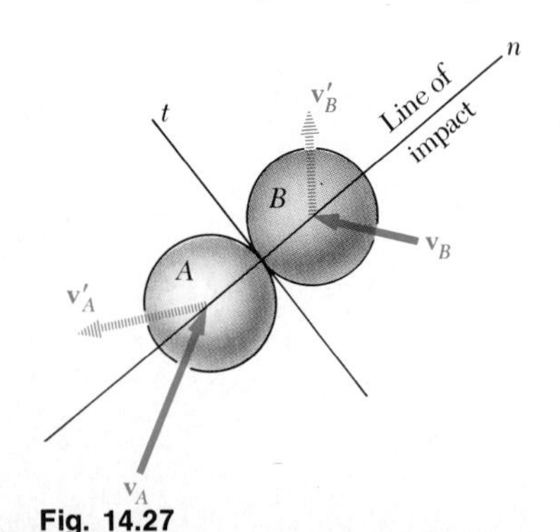

Fig. 14.27

In the case of an *oblique central impact* [Sec. 14.8], the velocities of the two colliding bodies before and after the impact were resolved into n components along the line of impact and t components along the common tangent to the surfaces in contact (Fig. 14.27). We observed that the t component of the velocity of each body remained unchanged, while the n components satisfied equations similar to Eqs. (14.15) and (14.21) [Sample Probs. 14.5 and 14.6]. While this method was developed for bodies moving freely before and after the impact, it was shown that it could be extended to the case when one or both of the colliding bodies is constrained in its motion [Sample Prob. 14.7].

In Sec. 14.9, we discussed the relative advantages of the three fundamental methods presented in this chapter and the preceding two, namely, Newton's second law, work and energy, and impulse and momentum. We noted that the method of work and energy and the method of impulse and momentum may be combined to solve problems involving a short impact phase during which impulsive forces must be taken into consideration [Sample Prob. 14.8]. We also saw that in some cases these two methods may be applied jointly to obtain equations which may be solved simultaneously for two or more unknowns [Sample Prob. 14.9].

Using the three fundamental methods of kinetic analysis

Angular momentum of a particle

The *angular momentum* H_O *of a particle* was defined as the *moment of the momentum* $m\mathbf{v}$ of the particle about the origin O of the coordinates [Sec. 14.10]. Resolving $m\mathbf{v}$ into x and y components (Fig. 14.28), we had

$$H_O = m(xv_y - yv_x) \tag{14.33}$$

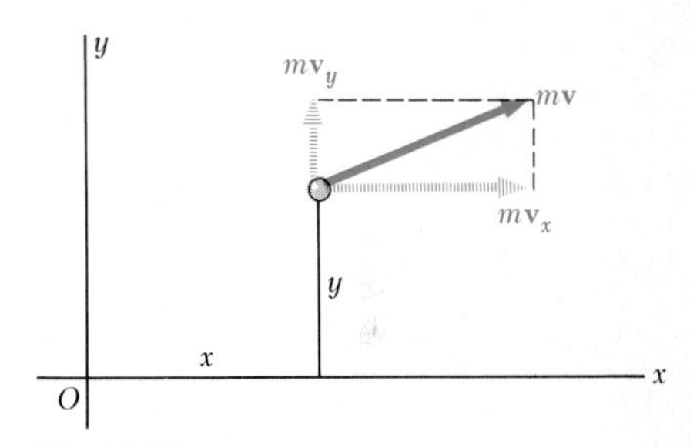

Fig. 14.28

Denoting by M_O the moment about O of the resultant $\mathbf{F}$ of the forces exerted on the particle, we derived from Newton's second law the equation

$$M_O = \dot{H}_O \tag{14.35}$$

Thus, *the moment* M_O *of* $\mathbf{F}$ *is equal to the rate of change of the angular momentum* H_O *of the particle.*

Motion under a central force

If the moment M_O *is zero* for every value of the time t, it follows from Eq. (14.35) that $\dot{H}_O$ is also zero for every value of t, and thus that *the angular momentum* H_O *of the particle is conserved* [Sec. 14.11]. This situation occurs, for example, when a particle moves under a *central force* $\mathbf{F}$. Using polar coordinates (Fig. 14.29), we wrote

$$H_O = r(mv \sin \phi) = \text{constant} \tag{14.37}$$

which reduces to the equation

$$r^2\dot{\theta} = h \tag{12.18}$$

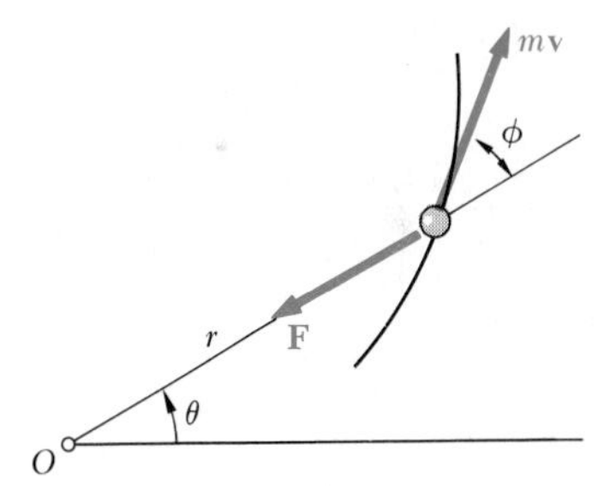

Fig. 14.29

which was previously derived in Sec. 12.9 and where the constant h is now seen to represent the angular momentum per unit mass, H_O/m, of the particle.

Motion under a gravitational force

The *gravitational force* exerted by the earth on a space vehicle being both *central and conservative*, the principles of conservation of angular momentum and of conservation of energy were used jointly to study the motion of such a vehicle [Sec. 14.12]. This approach was found particularly effective in the case of an *oblique launching*. Considering the initial position P_0 and an arbitrary position P of the vehicle (Fig. 14.30), we wrote

$$(H_O)_0 = H_O: \qquad r_0 m v_0 \sin \phi_0 = rmv \sin \phi \tag{14.39}$$

$$T_0 + V_0 = T + V: \qquad \tfrac{1}{2}mv_0^2 - \frac{GMm}{r_0} = \tfrac{1}{2}mv^2 - \frac{GMm}{r} \tag{14.40}$$

where m was the mass of the vehicle and M the mass of the earth.

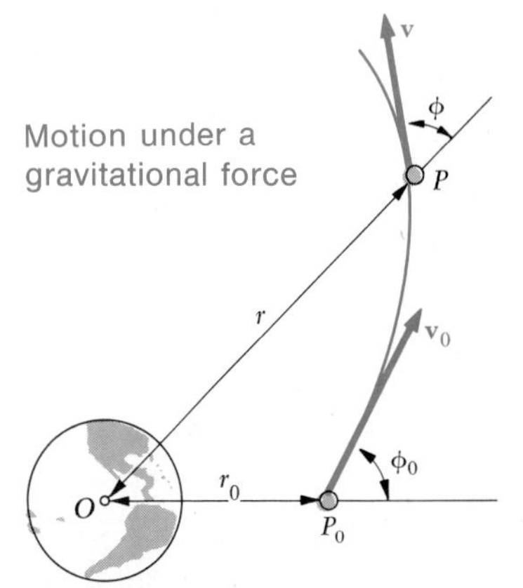

Fig. 14.30

General formulation of the principle of impulse and momentum

In Sec. 14.13, we formulated the principle of impulse and momentum for a *system of particles* in a more general form, expressing that *the momenta of the particles at time t_1 and the impulses of the external forces from t_1 to t_2 form a system of vectors equipollent to the system of the momenta of the particles at time t_2*. We wrote

$$\textbf{Syst Momenta}_1 + \textbf{Syst Ext Imp}_{1\to2} = \textbf{Syst Momenta}_2 \quad (14.45)$$

This statement, which implies that the *moments* of the vectors involved are equal, and not just their components, will be fundamental to our study of the motion of rigid bodies in Chap. 18.

Steady stream of particles

In the last part of the chapter, we considered *variable systems of particles*. First we considered a *steady stream of particles*, such as a stream of water diverted by a fixed vane or the flow of air through a jet engine [Sec. 14.15]. Applying the principle of impulse and momentum to a system S of particles during a time interval Δt, and including the particles which enter the system at A during that time interval and those (of the same mass Δm) which leave the system at B, we concluded that *the system formed by the momentum $(\Delta m)\mathbf{v}_A$ of the particles entering S in the time Δt and the impulse of the forces exerted on S during that time is equipollent to the momentum $(\Delta m)\mathbf{v}_B$ of the particles leaving S in the same time Δt* (Fig. 14.31).

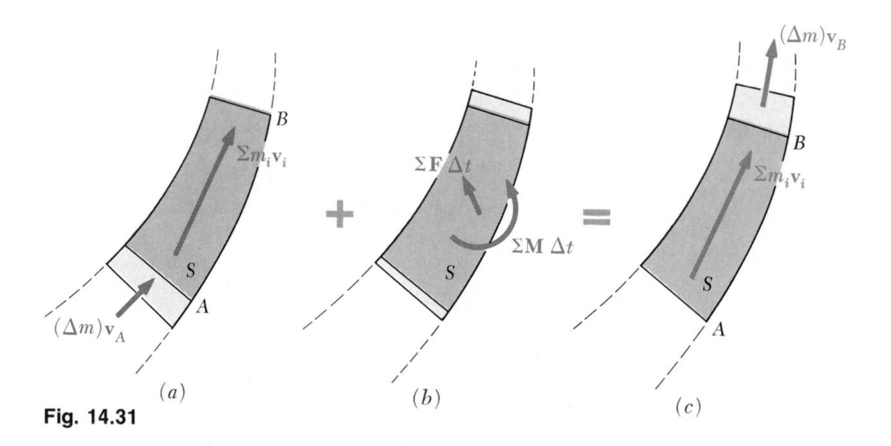

Fig. 14.31

Equating the x components, y components, and moments about a fixed point of the vectors involved, we could obtain as many as three equations, which could be solved for the desired unknowns [Sample Probs. 14.12 and 14.13]. From this result, we could also derive the following expression for the resultant $\Sigma\mathbf{F}$ of the forces exerted on S,

$$\Sigma\mathbf{F} = \frac{dm}{dt}(\mathbf{v}_B - \mathbf{v}_A) \quad (14.48)$$

where $\mathbf{v}_B - \mathbf{v}_A$ represents the difference between the *vectors* $\mathbf{v}_B$ and $\mathbf{v}_A$ and where dm/dt is the mass rate of flow of the stream (see footnote, page 631).

Systems gaining or losing mass

Considering next a system of particles gaining mass by continually absorbing particles or losing mass by continually expelling particles [Sec. 14.16], as in the case of a rocket, we applied the principle of impulse and momentum to the system during a time interval Δt, being careful to include the particles gained or lost during that time interval [Sample Prob. 14.14]. We also noted that the action on a system S of the particles being *absorbed* by S was equivalent to a thrust

$$\mathbf{P} = \frac{dm}{dt}\mathbf{u} \tag{14.52}$$

where dm/dt is the rate at which mass is being absorbed, and $\mathbf{u}$ the velocity of the particles *relative to* S. In the case of particles being *expelled* by S, the rate dm/dt is negative and the thrust $\mathbf{P}$ is exerted in a direction opposite to that in which the particles are being expelled.

Review Problems

14.150 A 15-Mg truck and a 45-Mg railroad flatcar are both at rest with their brakes released. An engine bumps the flatcar and causes the flatcar alone to start moving with a velocity of 0.8 m/s to the right. Assuming $e = 1$ between the truck and the ends of the flatcar and neglecting friction, determine the velocities of the truck and of the flatcar after (*a*) end *A* strikes the truck, (*b*) the truck strikes end *B*.

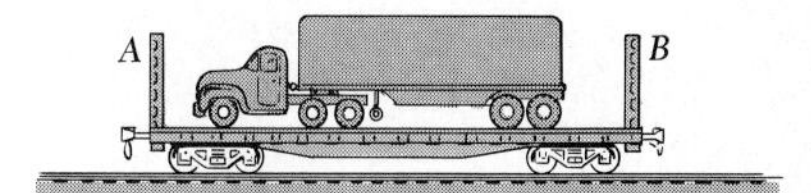

Fig. P14.150

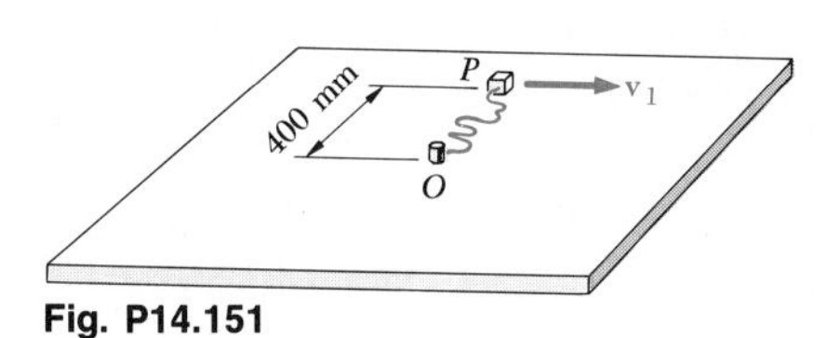

Fig. P14.151

14.151 A 500-g block *P* rests on a frictionless horizontal table at a distance of 400 mm from a fixed pin *O*. The block is attached to pin *O* by an elastic cord of constant $k = 100$ N/m and of undeformed length 900 mm. If the block is set in motion to the right as shown, determine (*a*) the speed v_1 for which the distance from *O* to the block *P* will reach a maximum value of 1.2 m, (*b*) the speed v_2 when $OP = 1.2$ m, (*c*) the radius of curvature of the path of the block when $OP = 1.2$ m.

14.152 A ball is dropped from a height *h* above the landing and bounces down a flight of stairs. Denoting by *e* the coefficient of restitution, determine the value of *h* for which the ball will bounce the same height above each step.

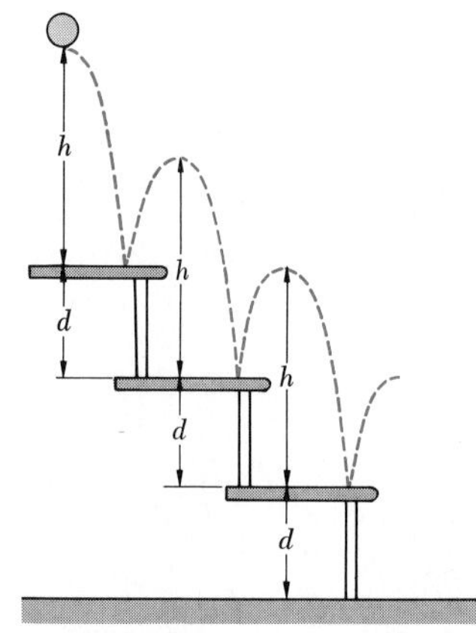

Fig. P14.152

14.153 Collar A is dropped 30 in. onto collar B, which is resting on a spring of constant $k = 4$ lb/in. Assuming perfectly plastic impact, determine (*a*) the maximum deflection of collar B, (*b*) the energy lost during the impact.

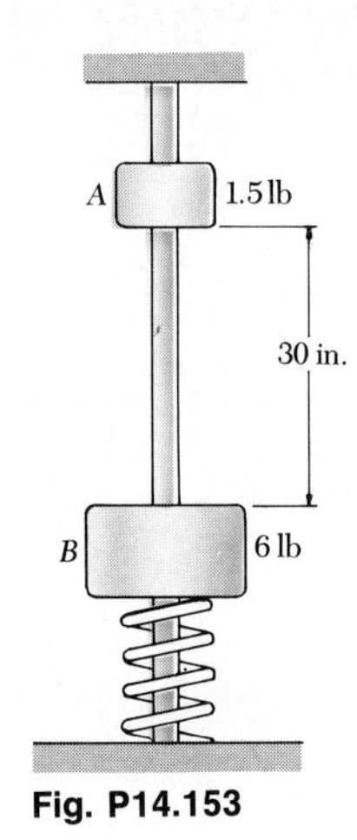

Fig. P14.153

14.154 A light train made of two cars travels at 45 mi/h. Car A weighs 20 tons and car B weighs 25 tons. When the brakes are applied, a constant braking force of 4000 lb is applied to each car. Determine (*a*) the time required for the train to stop after the brakes are applied, (*b*) the force in the coupling between the cars while the train is slowing down.

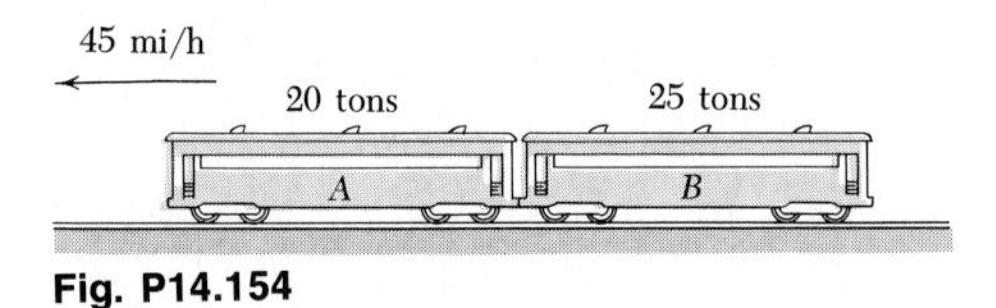

Fig. P14.154

14.155 A dime which is at rest on a *rough* surface is struck squarely by a half dollar moving to the right. After the impact, each coin slides and comes to rest; the dime slides 480 mm to the right, and the half dollar slides 95 mm to the right. Assuming that the coefficient of friction is the same for each coin, determine the value of the coefficient of restitution between the coins. (*Masses:* half-dollar, 12.50 g; quarter dollar, 6.25 g; dime, 2.50 g.)

14.156 A machine part is forged in a small drop forge. The hammer has a mass of 200 kg and is dropped from a height of 1.4 m. Determine the initial impulse exerted on the machine part and the energy absorbed by it, assuming that $e = 0$ and that the 500-kg anvil (*a*) is resting directly on hard ground, (*b*) is supported by springs.

14.157 A spacecraft is launched from the ground to rendezvous with a skylab orbiting the earth at a constant altitude of 1040 mi. The spacecraft has reached at burnout an altitude of 40 miles and a velocity $\mathbf{v}_0$ of magnitude 20,000 ft/s. What is the angle ϕ_0 that $\mathbf{v}_0$ should form with the vertical OB if the trajectory of the spacecraft is to be tangent at A to the orbit of the skylab?

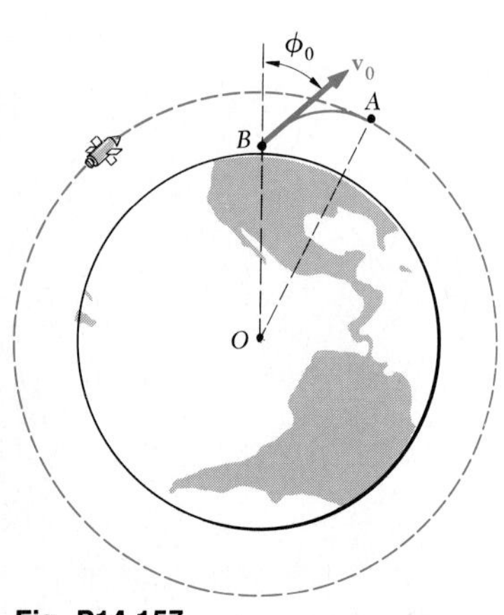

Fig. P14.157

14.158 Three identical freight cars have the velocities indicated. Assuming that car B is first hit by car A, determine the velocity of each car after all collisions have taken place, if (a) all three cars get automatically coupled, (b) cars A and B get automatically and tightly coupled, while cars B and C bounce off each other with a coefficient of restitution $e = 1$.

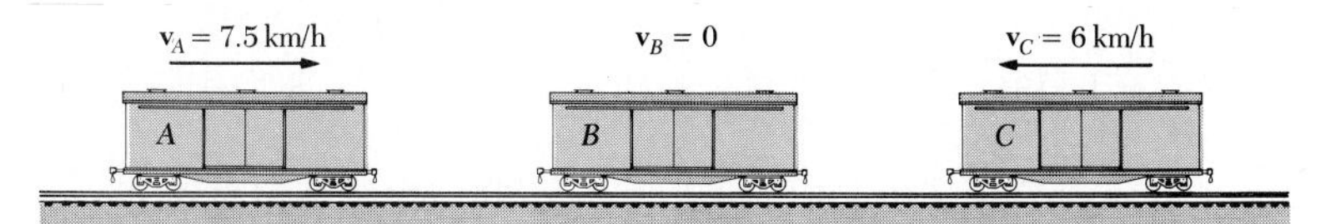

Fig. P14.158

14.159 Solve Prob. 14.158, assuming that car B is first hit by car C.

14.160 The slipstream of a fan has a diameter of 20 in. and a velocity of 35 ft/s relative to the fan. Assuming $\gamma = 0.076\ \text{lb/ft}^3$ for air and neglecting the velocity of approach of the air, determine the force required to hold the fan motionless.

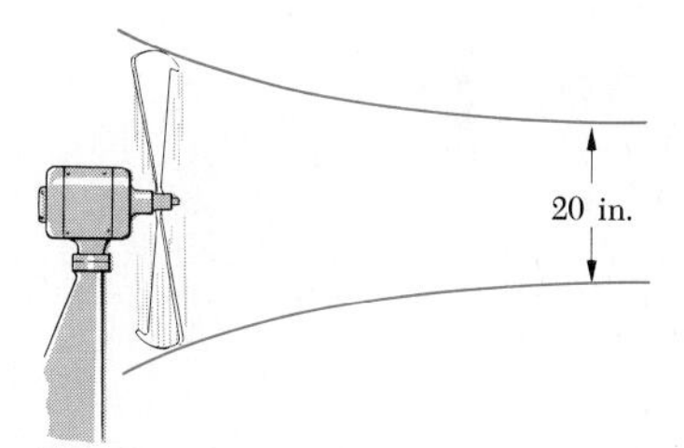

Fig. P14.160

14.161 A chain of length l and mass m falls through a small hole in a plate. Initially, when y is very small, the chain is at rest. In each case shown, determine (a) the acceleration of the first link A as a function of y, (b) the velocity of the chain as the last link passes through the hole. In case *1* assume that the individual links are at rest until they fall through the hole; in case 2 assume that at any instant all links have the same speed. Ignore the effect of friction.

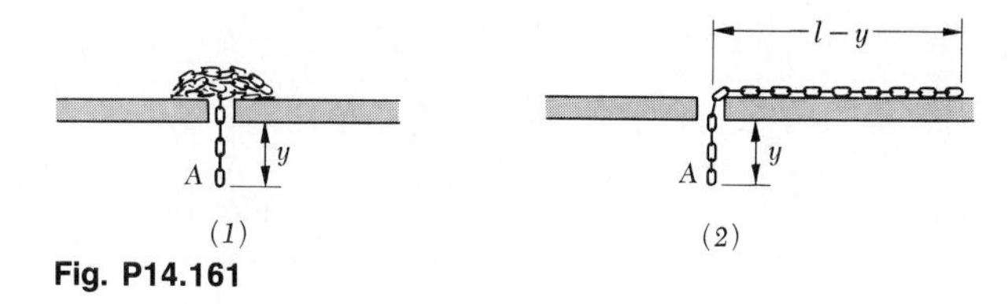

Fig. P14.161

The following problems are designed to be solved with a computer.

14.C1 A block of weight W initially at rest is acted upon by a force **P** which varies as shown. The coefficients of friction between the block and the horizontal surface are $\mu_s = 0.50$ and $\mu_k = 0.40$. Write a computer program and use it to calculate, for values of W from 100 to 225 lb at 5-lb intervals, (a) the time at which the block will start moving, (b) the maximum velocity reached by the block, (c) the time at which the block will stop moving.

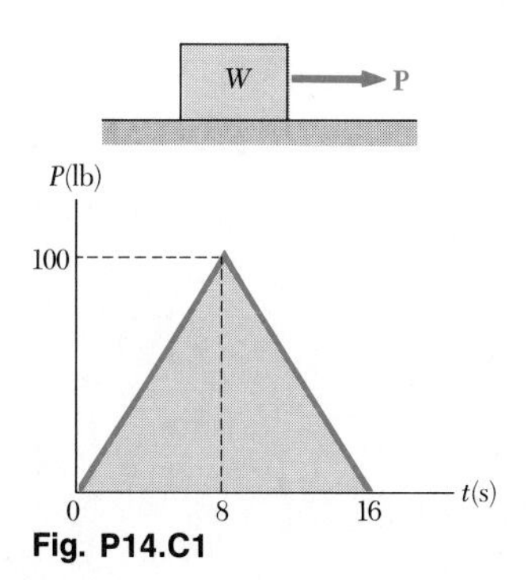

Fig. P14.C1

14.C2 At engine burnout a satellite has reached an altitude of 2400 km and has a velocity $\mathbf{v}_0$ of magnitude 8100 m/s forming an angle ϕ_0 with the vertical. Write a computer program and use it to determine the maximum and minimum heights reached by the satellite for values of ϕ_0 from 60 to 120° at 5° intervals. Assuming that if the satellite gets closer than 300 km from the earth's surface, it will soon burn up, indicate those values of ϕ_0 for which a permanent orbit is not achieved.

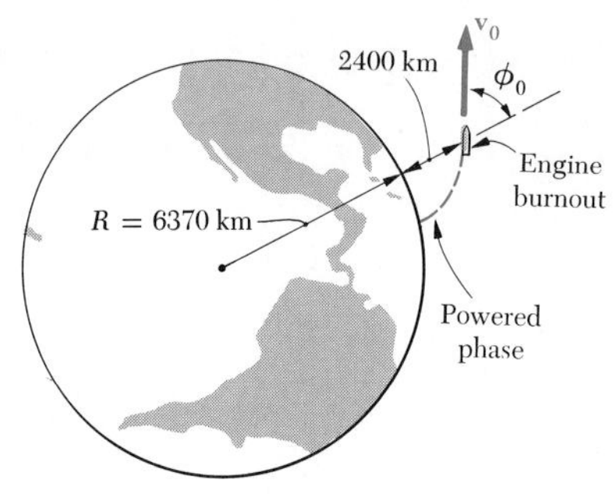

Fig. P14.C2

14.C3 In a game of billiards, a player wishes to hit ball B squarely with ball A at a distance nd, where d is the diameter of the balls. However, the player projects ball A with a velocity $\mathbf{v}_0$ forming a small angle θ_0 with line AB. Denoting by e the coefficient of restitution, write a computer program to calculate the magnitude of the velocity $\mathbf{v}'_A$ of ball A after the impact and the angle θ'_A that $\mathbf{v}'_A$ will form with line AB. Using this program and knowing that $n = 10$ and $v_0 = 4$ m/s, determine v'_A and θ'_A for values of θ_0 from 0 to 0.5° at 0.05° intervals, and from 0.5 to 5.5° at 0.5° intervals, assuming (*a*) $e = 0.95$, (*b*) $e = 0.90$, (*c*) $e = 0.70$.

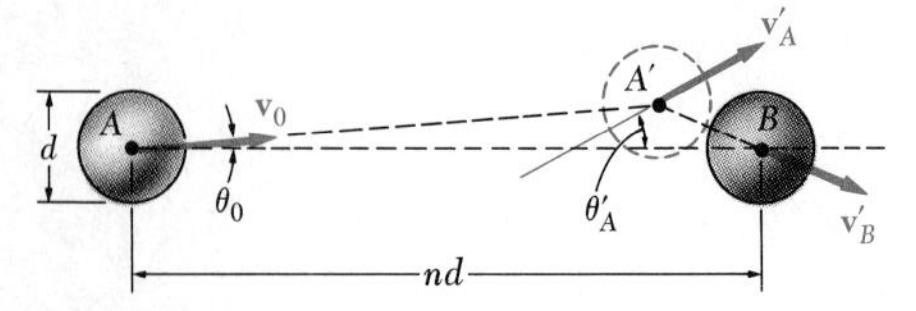

Fig. P14.C3

14.C4 A sphere A of mass $m_A = 2$ kg is released from rest in the position shown and strikes the frictionless, inclined surface of a wedge B of mass $m_B = 6$ kg with a velocity of magnitude $v_0 = 3$ m/s. The wedge, which is supported by rollers and may move freely in the horizontal direction, is initially at rest. Denoting by e the coefficient of restitution between the sphere and the wedge, write a computer program which can be used to calculate the velocities $\mathbf{v}'_A$ of the sphere and $\mathbf{v}'_B$ of the wedge immediately after impact. Use this program to determine $\mathbf{v}'_A$ and $\mathbf{v}'_B$ for values of θ from 0 to 90° at 10° intervals when (*a*) $e = 1$, (*b*) $e = 0.8$.

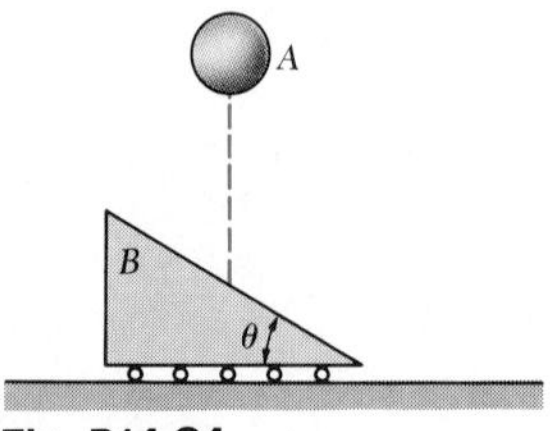

Fig. P14.C4

CHAPTER FIFTEEN

Kinematics of Rigid Bodies

15.1. Introduction. In this chapter, we shall study the kinematics of *rigid bodies*. We shall investigate the relations existing between the time, the positions, the velocities, and the accelerations of the various particles forming a rigid body. Our study will be limited to that of the *plane motion* of a rigid body, which is characterized by the fact that each particle of the body remains at a constant distance from a fixed reference plane. Thus all particles of the body move in parallel planes, and the motion of the body may be represented by the motion of a representative slab in the reference plane.

The various types of plane motion may be grouped as follows:

1. *Translation.* A motion is said to be a translation if any straight line drawn on the body keeps the same direction during the motion. It may also be observed that in a translation all the particles forming the body move along parallel paths. If these paths are straight lines, the motion is said to be a *rectilinear translation* (Fig. 15.1*a*); if the paths are curved lines, the motion is a *curvilinear translation* (Fig. 15.1*b*).

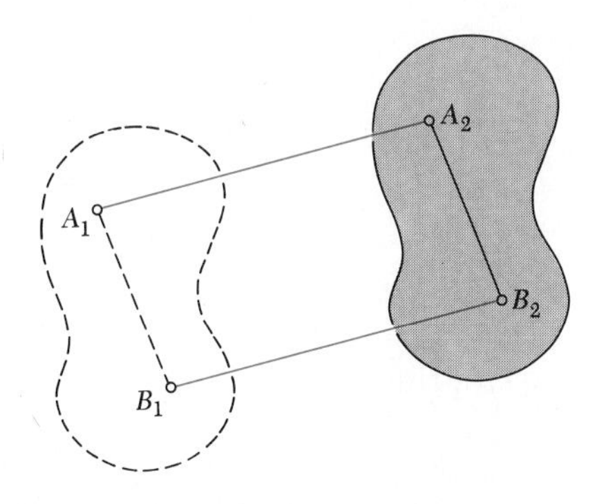

Fig. 15.1 (*a*) Rectilinear translation

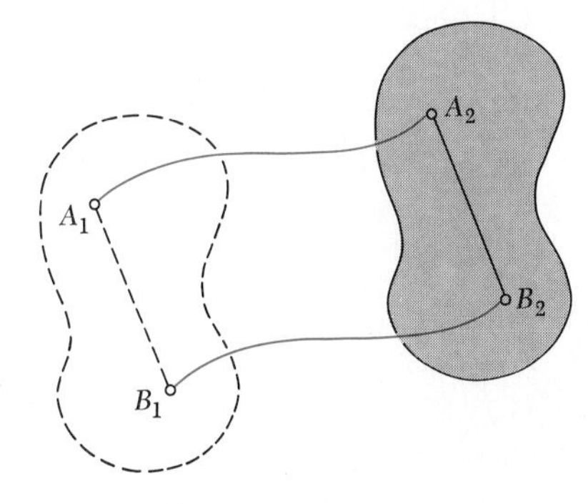

(*b*) Curvilinear translation

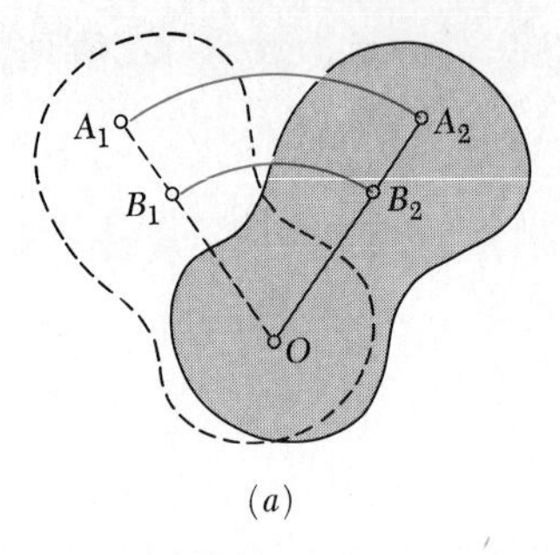

(a)

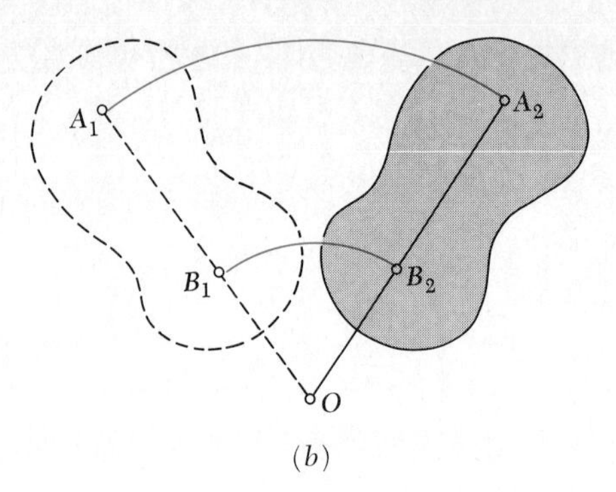

(b)

Fig. 15.2

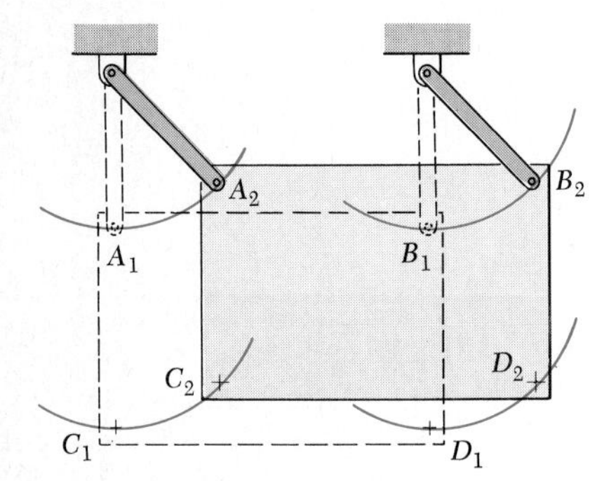

(a) Curvilinear translation

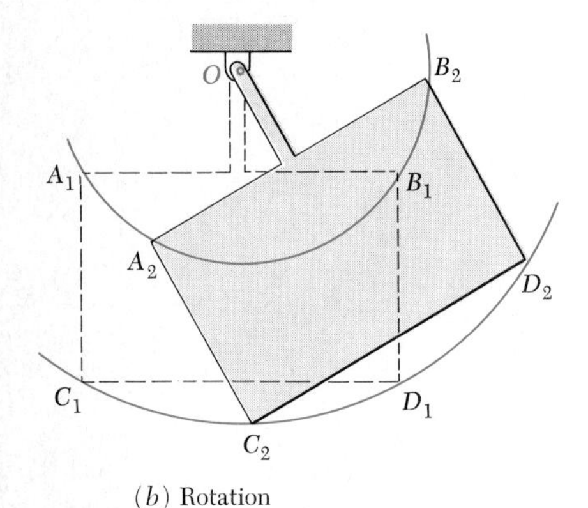

(b) Rotation

Fig. 15.3

2. *Rotation.* A motion is said to be a rotation when the particles of the representative slab move along concentric circles. If the common center O of the circles is located on the slab, the corresponding point remains fixed (Fig. 15.2*a*). In many cases of rotation, however, the common center O is located outside the slab, and none of the points of the slab remains fixed (Fig. 15.2*b*).

   Rotation should not be confused with certain types of curvilinear translation. For example, the plate shown in Fig. 15.3*a* is in curvilinear translation, with all its particles moving along *parallel* circles, while the plate shown in Fig. 15.3*b* is in rotation, with all its particles moving along *concentric* circles. In the first case, any given straight line drawn on the plate will maintain the same direction, while, in the second case, point O remains fixed.

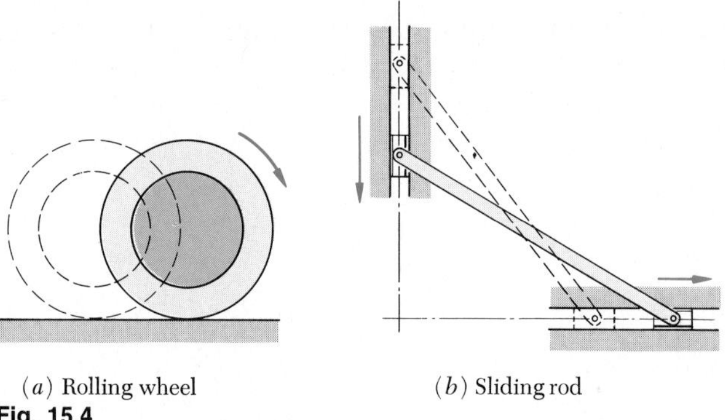
(a) Rolling wheel (b) Sliding rod

Fig. 15.4

3. *General Plane Motion.* Any plane motion which is neither a translation nor a rotation is referred to as a general plane motion. Two examples of general plane motion are given in Fig. 15.4.

After a brief discussion in Sec. 15.2 of the motion of translation, we shall consider in Sec. 15.3 the rotation of a slab about a fixed axis perpendicular to the plane of the slab. We shall define the *angular velocity* and *angular acceleration* of the slab and learn in Sec. 15.4 to express the velocity and acceleration of a given point of the slab in terms of its position and the angular velocity and angular acceleration of the slab.

The following sections are devoted to the study of the general plane motion of a slab and to its application to the analysis of mechanisms such as gears, connecting rods, and pin-connected linkages. Resolving the plane motion of a slab into a translation and a rotation (Secs. 15.5 and 15.6), we shall then express the velocity of a point B of the slab as the sum of the velocity of a reference point A and of the velocity of B relative to a frame of reference translating with A (i.e., moving with A but not rotating). The same approach is used later in Sec. 15.8 to express the acceleration of B in terms of the acceleration of A and of the acceleration of B relative to a frame translating with A.

An alternative method for the analysis of velocities in plane motion, based on the concept of *instantaneous center of rotation,* is given in Sec. 15.7; and still another method of analysis, based on the use of parametric expressions for the coordinates of a given point, is presented in Sec. 15.9.

The motion of a particle relative to a rotating frame of reference and the concept of *Coriolis acceleration* are discussed in Secs. 15.10 and 15.11, and the results obtained are applied to the analysis of the plane motion of mechanisms containing parts which slide on each other.

15.2. Translation. Consider a rigid slab in translation (either rectilinear or curvilinear translation), and denote by A and B the positions of two of its particles at time t. At time $t + \Delta t$, the two particles will occupy new positions A' and B' (Fig. 15.5*a*). Now, from the very definition of a translation, the line $A'B'$ must be parallel to AB. Thus, $ABB'A'$ is a parallelogram, and the two vectors $\Delta\mathbf{r}_A$ and $\Delta\mathbf{r}_B$ representing the displacements of the two particles during the time interval Δt must have the same magnitude and direction. We have, therefore, $\Delta\mathbf{r}_A/\Delta t = \Delta\mathbf{r}_B/\Delta t$ and, at the limit, as Δt approaches zero,

$$\mathbf{v}_A = \mathbf{v}_B \tag{15.1}$$

Since relation (15.1) holds at any time, the vectors $\Delta\mathbf{v}_A$ and $\Delta\mathbf{v}_B$ representing the change in velocity of the two particles during the time interval Δt must be equal. We have, therefore, $\Delta\mathbf{v}_A/\Delta t = \Delta\mathbf{v}_B/\Delta t$ and, at the limit, as Δt approaches zero,

$$\mathbf{a}_A = \mathbf{a}_B \tag{15.2}$$

Thus, *when a rigid slab is in translation, all the points of the slab have the same velocity and the same acceleration at any given instant* (Fig. 15.5*b* and *c*). In the case of curvilinear translation, the velocity and acceleration change in direction as well as in magnitude at every instant. In the case of rectilinear translation, all particles of the slab move along parallel straight lines, and their velocity and acceleration keep the same direction during the entire motion.

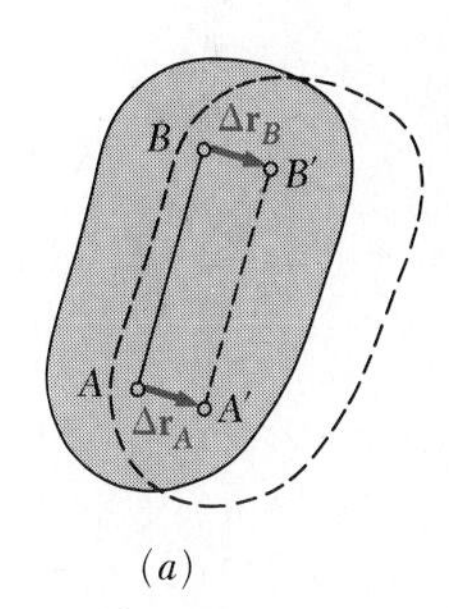

(*a*)

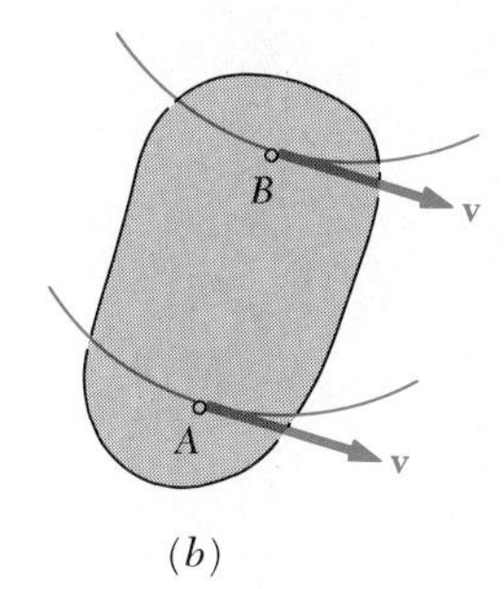

(*b*)

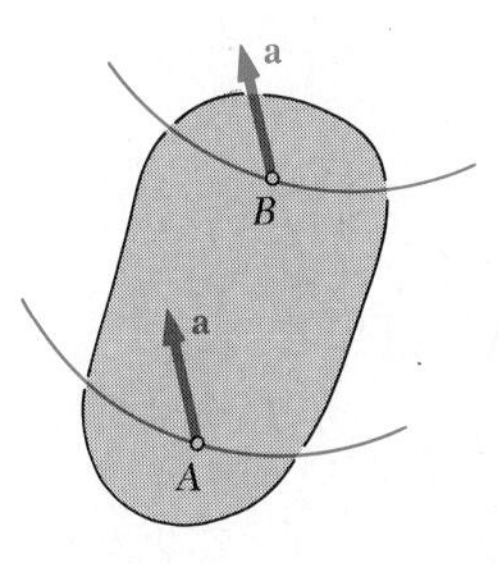

(*c*)

Fig. 15.5

15.3. Rotation. Consider a rigid slab which rotates about a fixed axis perpendicular to the plane of the slab and intersecting it at point O. Let P be a point of the slab (Fig. 15.6). The position of the slab will be entirely defined if the angle θ the line OP forms with a fixed direction OQ is given; the angle θ is known as the *angular coordinate* of the slab. The angular coordinate is defined as positive when counterclockwise and will be expressed in radians (rad) or, occasionally, in degrees (°) or revolutions (rev). We recall that

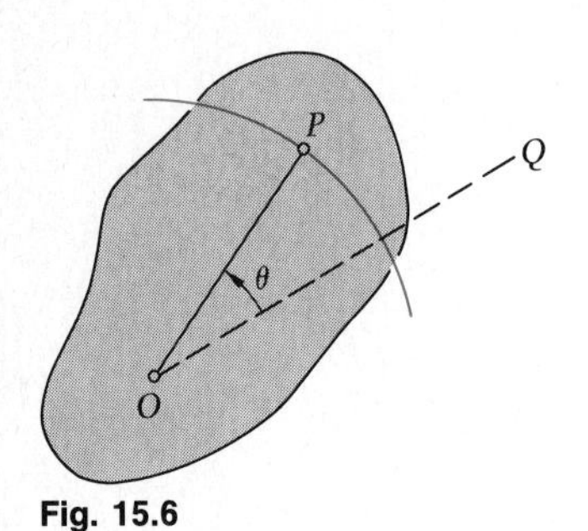

Fig. 15.6

$$1 \text{ rev} = 2\pi \text{ rad} = 360°$$

In Chap. 11, the first and second derivatives, respectively, of the coordinate x of a particle were used to define the velocity and acceleration of a particle moving in a straight line. Similarly, the first derivative of the angular coordinate θ of a rotating slab may be used to define the *angular velocity* ω of the slab,

$$\omega = \frac{d\theta}{dt} \tag{15.3}$$

and its second derivative may be used to define the *angular acceleration* α of the slab,

$$\alpha = \frac{d\omega}{dt} = \frac{d^2\theta}{dt^2} \tag{15.4}$$

An alternative form for the angular acceleration is obtained by solving (15.3) for dt and substituting into (15.4). We write $dt = d\theta/\omega$ and

$$\alpha = \omega\frac{d\omega}{d\theta} \tag{15.5}$$

The angular velocity ω is expressed in rad/s or in rpm (revolutions per minute). The angular acceleration α is generally expressed in rad/s^2.

Since the angular velocity and the angular acceleration of a slab have a *sense,* as well as a *magnitude,* they should be considered as *vector quantities.*† We may, for instance, represent the angular velocity of the slab of Fig. 15.6 by a vector $\boldsymbol{\omega}$ of magnitude ω directed along the axis of rotation of the slab. This vector will be made to point out of the plane of the figure if

† It may be shown, in the more general case of a rigid body rotating simultaneously about axes having different directions, that angular velocities and angular accelerations obey the parallelogram law of addition.

the angular velocity is counterclockwise, and into the plane of the figure if the angular velocity is clockwise. The angular acceleration may similarly be represented by a vector $\boldsymbol{\alpha}$ of magnitude α directed along the axis of rotation and pointing out of, or into, the plane of the figure, depending upon the sense of the angular acceleration. However, since in the cases considered in this chapter the axis of rotation will always be perpendicular to the plane of the figure, we need to specify only the sense and the magnitude of the angular velocity and of the angular acceleration. This may be done by using the scalar quantities ω and α, together with a plus or minus sign. As in the case of the angular coordinate θ, the plus sign corresponds to counterclockwise, and the minus sign to clockwise.

It should be observed that the values obtained for the angular velocity and the angular acceleration of a slab are independent of the choice of the reference point P. If a point P' were chosen instead of P (Fig. 15.7), the angular coordinate would be θ', which differs from θ only by the constant angle POP'. The successive derivatives of θ and θ' are thus equal, and the same values are obtained for ω and α.

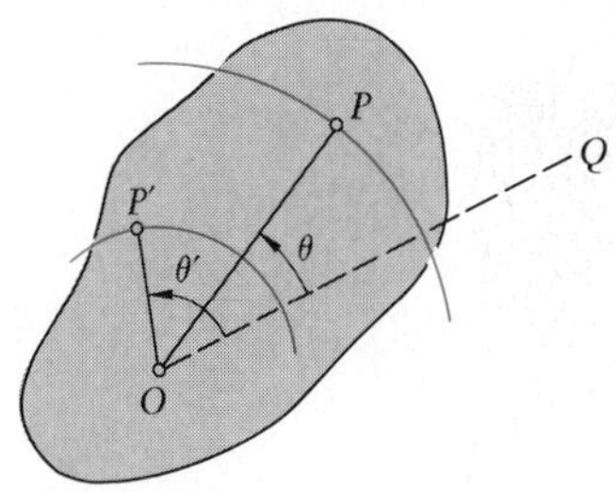

Fig. 15.7

The motion of a slab rotating about a fixed point O is said to be *known* when its coordinate θ may be expressed as a known function of t. In practice, however, the rotation of a slab is seldom defined by a relation between θ and t. More often, the conditions of motion will be specified by the type of angular acceleration that the slab possesses. For example, α may be given as a function of t, or as a function of θ, or as a function of ω. Equations (15.3) to (15.5) may then be used to determine the angular velocity ω and the angular coordinate θ, the procedure to follow being the same as that described in Sec. 11.3 for the rectilinear motion of a particle.

Two particular cases of rotation are frequently encountered:

1. *Uniform Rotation.* This case is characterized by the fact that the angular acceleration is zero. The angular velocity ω is thus constant, and the angular coordinate is given by the formula

$$\theta = \theta_0 + \omega t \tag{15.6}$$

2. *Uniformly Accelerated Rotation.* In this case, the angular acceleration α is constant. The following formulas relating angular velocity, angular coordinate, and time may then be derived in a manner similar to that described in Sec. 11.5. The similitude between the formulas derived here and those obtained for the rectilinear uniformly accelerated motion of a particle is easily noted.

$$\omega = \omega_0 + at \tag{15.7}$$

$$\theta = \theta_0 + \omega_0 t + \tfrac{1}{2}\alpha t^2 \tag{15.8}$$

$$\omega^2 = \omega_0^2 + 2\alpha(\theta - \theta_0) \tag{15.9}$$

It should be emphasized that formula (15.6) may be used only when $\alpha = 0$, and formulas (15.7) to (15.9) only when $\alpha =$ constant. In any other case, the general formulas (15.3) to (15.5) should be used.

15.4. Linear and Angular Velocity, Linear and Angular Acceleration in Rotation. Consider again a point P on a rotating slab (Fig. 15.8*a*). Point P describes a circle of center O when the slab describes a full rotation. The arc s that P describes while the slab rotates through an

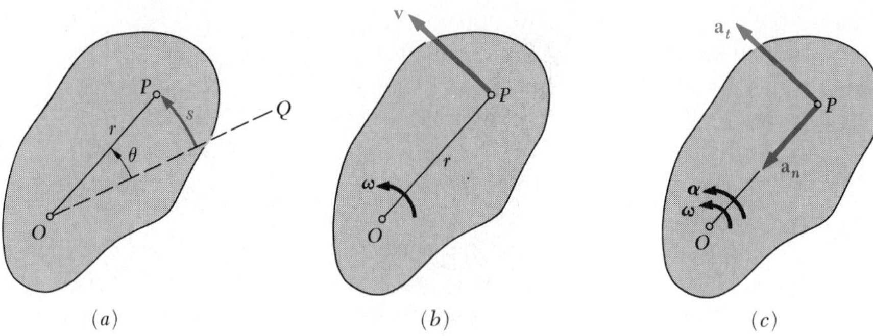

Fig. 15.8

angle θ depends not only upon θ but also upon the distance r from O to P. Expressing θ in radians and r and s in the same units of length, we write

$$s = r\theta \tag{15.10}$$

Differentiating, we obtain the following relation between the magnitude v of the *linear velocity* of P (Fig. 15.8*b*) and the *angular velocity* ω of the slab,

$$\frac{ds}{dt} = r\frac{d\theta}{dt} \qquad v = r\omega \tag{15.11}$$

While the value of the angular velocity ω is independent of the choice of P, the magnitude v of the linear velocity clearly depends upon the distance r. Recalling the definition of the tangential and normal components of the *linear acceleration* of a point P (Sec. 11.12), we write (Fig. 15.8*c*)

$$a_t = \frac{dv}{dt} = r\frac{d\omega}{dt} \qquad a_t = r\alpha \tag{15.12}$$

$$a_n = \frac{v^2}{r} = \frac{(r\omega)^2}{r} \qquad a_n = r\omega^2 \tag{15.13}$$

SAMPLE PROBLEM 15.1

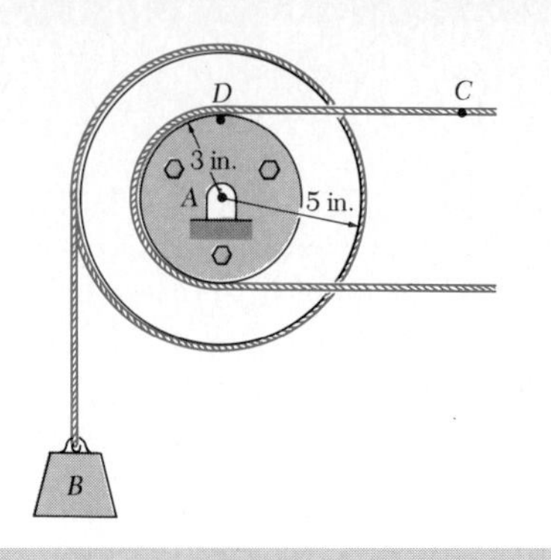

Load B is connected to a double pulley by one of the two inextensible cables shown. The motion of the pulley is controlled by cable C, which has a constant acceleration of 9 in./s^2 and an initial velocity of 12 in./s, both directed to the right. Determine (*a*) the number of revolutions executed by the pulley in 2 s, (*b*) the velocity and change in position of the load B after 2 s, and (*c*) the acceleration of point D on the rim of the inner pulley at $t = 0$.

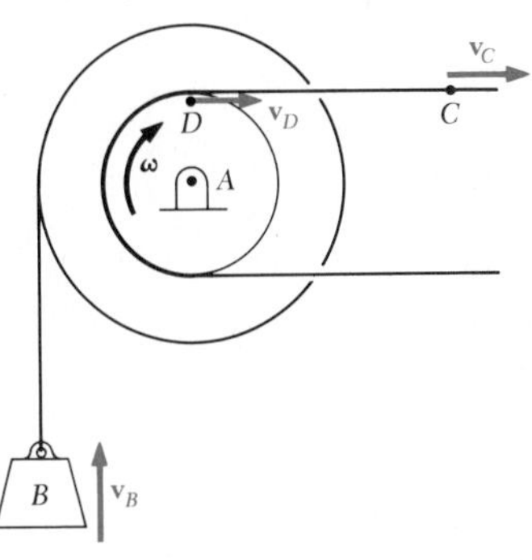

***a.* Motion of Pulley.** Since the cable is inextensible, the velocity of point D is equal to the velocity of point C and the tangential component of the acceleration of D is equal to the acceleration of C.

$$(\mathbf{v}_D)_0 = (\mathbf{v}_C)_0 = 12 \text{ in./s} \rightarrow \qquad (\mathbf{a}_D)_t = \mathbf{a}_C = 9 \text{ in./s}^2 \rightarrow$$

Noting that the distance from D to the center of the pulley is 3 in., we write

$$(v_D)_0 = r\omega_0 \qquad 12 \text{ in./s} = (3 \text{ in.})\omega_0 \qquad \boldsymbol{\omega}_0 = 4 \text{ rad/s} \downarrow$$

$$(a_D)_t = r\alpha \qquad 9 \text{ in./s}^2 = (3 \text{ in.})\alpha \qquad \boldsymbol{\alpha} = 3 \text{ rad/s}^2 \downarrow$$

Using the equations of uniformly accelerated motion, we obtain, for $t = 2$ s,

$$\omega = \omega_0 + \alpha t = 4 \text{ rad/s} + (3 \text{ rad/s}^2)(2 \text{ s}) = 10 \text{ rad/s}$$

$$\boldsymbol{\omega} = 10 \text{ rad/s} \downarrow$$

$$\theta = \omega_0 t + \tfrac{1}{2}\alpha t^2 = (4 \text{ rad/s})(2 \text{ s}) + \tfrac{1}{2}(3 \text{ rad/s}^2)(2 \text{ s})^2 = 14 \text{ rad}$$

$$\theta = 14 \text{ rad} \downarrow$$

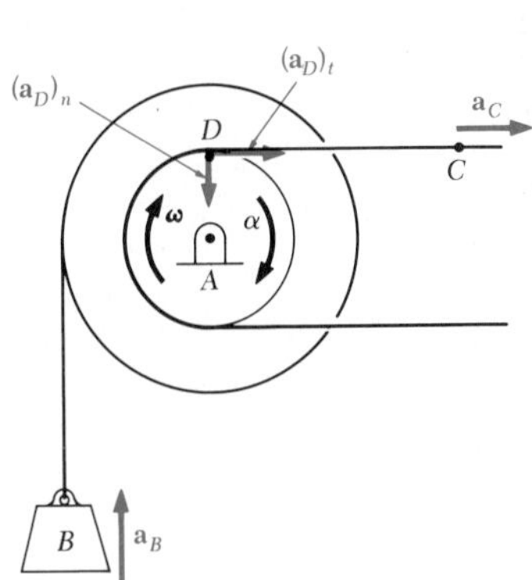

$$\text{Number of revolutions} = (14 \text{ rad})\left(\frac{1 \text{ rev}}{2\pi \text{ rad}}\right) = 2.23 \text{ rev} \blacktriangleleft$$

***b.* Motion of Load *B*.** Using the following relations between linear and angular motion, with $r = 5$ in., we write

$$v_B = r\omega = (5 \text{ in.})(10 \text{ rad/s}) = 50 \text{ in./s} \qquad \mathbf{v}_B = 50 \text{ in./s} \uparrow \blacktriangleleft$$

$$\Delta y_B = r\theta = (5 \text{ in.})(14 \text{ rad}) = 70 \text{ in.} \qquad \Delta y_B = 70 \text{ in. upward} \blacktriangleleft$$

***c.* Acceleration of Point *D* at *t* = 0.** The tangential component of the acceleration is

$$(\mathbf{a}_D)_t = \mathbf{a}_C = 9 \text{ in./s}^2 \rightarrow$$

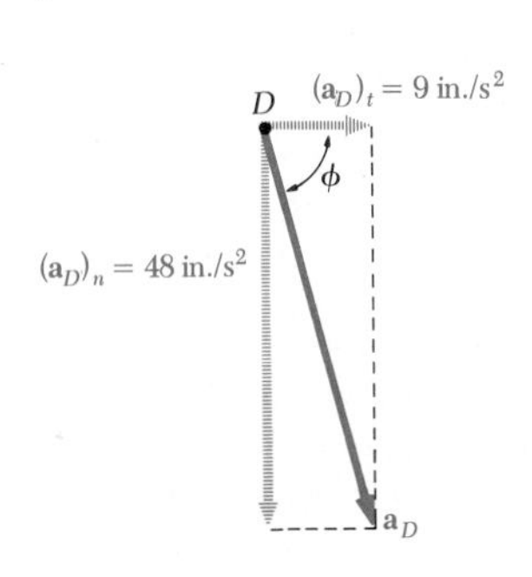

Since, at $t = 0$, $\omega_0 = 4$ rad/s, the normal component of the acceleration is

$$(a_D)_n = r_D\omega_0^2 = (3 \text{ in.})(4 \text{ rad/s})^2 = 48 \text{ in./s}^2 \qquad (\mathbf{a}_D)_n = 48 \text{ in./s}^2 \downarrow$$

The magnitude and direction of the total acceleration may be obtained by writing

$$\tan\phi = (48 \text{ in./s}^2)/(9 \text{ in./s}^2) \qquad \phi = 79.4°$$

$$a_D \sin 79.4° = 48 \text{ in./s}^2 \qquad a_D = 48.8 \text{ in./s}^2$$

$$\mathbf{a}_D = 48.8 \text{ in./s}^2 \searrow 79.4° \blacktriangleleft$$

Problems

15.1 The motion of a cam is defined by the relation $\theta = 0.3t(12 - t^2)$ where θ is expressed in radians and t in seconds. Determine the angular coordinate, the angular velocity, and the angular acceleration of the cam when (*a*) $t = 0$, (*b*) $t = 3$ s.

15.2 For the cam of Prob. 15.1 determine (*a*) the time at which the angular velocity is zero, (*b*) the corresponding angular coordinate and angular acceleration.

15.3 The motion of an oscillating crank is defined by the relation $\theta = \theta_0 \sin(2\pi t/T)$, where θ is expressed in radians and t in seconds. Knowing that $\theta_0 = 1.2$ rad and $T = 0.5$ s, determine the maximum angular velocity and the maximum angular acceleration of the crank.

15.4 The motion of a disk rotating in an oil bath is defined by the relation $\theta = \theta_0(1 - e^{-t/4})$, where θ is expressed in radians and t in seconds. Knowing that $\theta_0 = 0.80$ rad, determine the angular coordinate, velocity, and acceleration of the disk when (*a*) $t = 0$, (*b*) $t = 4$ s, (*c*) $t = \infty$.

15.5 The rotor of an electric motor has a speed of 1800 rpm when the power is cut off. The rotor is then observed to come to rest after executing 625 revolutions. Assuming uniformly accelerated motion, determine (*a*) the angular acceleration, (*b*) the time required for the rotor to come to rest.

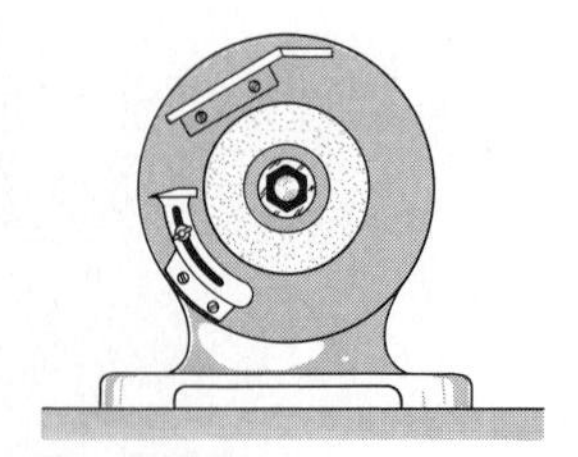

Fig. P15.6

15.6 A small grinding wheel is attached to the shaft of an electric motor which has a rated speed of 1800 rpm. When the power is turned on, the unit reaches its rated speed in 5 s, and when the power is turned off, the unit coasts to rest in 90 s. Assuming uniformly accelerated motion, determine the number of revolutions that the motor executes (*a*) in reaching its rated speed, (*b*) in coasting to rest.

15.7 During the starting phase of a computer, it is observed that a storage disk which was initially at rest executed 2.5 revolutions in 0.3 s. Assuming that the motion was uniformly accelerated, determine (*a*) the angular acceleration of the disk, (*b*) the final velocity of the disk.

15.8 The rotor of a steam turbine is rotating at a speed of 9300 rpm when the steam supply is suddenly cut off. It is observed that 6 min are required for the rotor to come to rest. Assuming uniformly accelerated motion, determine (*a*) the angular acceleration, (*b*) the total number of revolutions that the rotor executes before coming to rest.

15.9 The angular acceleration of the rotor of an electric motor is directly proportional to the time t. The rotor starts at $t = 0$ with no initial velocity; after 2.5 s, the rotor has completed 5 revolutions. Write the equations of motion for the rotor, and determine the angular velocity at $t = 2.5$ s.

15.10 The angular acceleration of a disk is defined by the relation $\alpha = -2.5\omega$, where α is expressed in rad/s^2 and ω in rad/s. Knowing that at $t = 0$ the angular velocity is 360 rpm, determine (*a*) the number of revolutions that the disk will execute before coming to rest, (*b*) the time required for the disk to come to rest, (*c*) the time required for the angular velocity of the disk to be reduced to 1 percent of its initial value.

15.11 The earth makes one complete revolution about the sun in 365.24 days. Assuming that the orbit of the earth is circular and has a radius of 93,000,000 mi, determine the velocity and acceleration of the earth.

15.12 The earth makes one complete revolution on its axis in 23.93 h. Knowing that the mean radius of the earth is 3960 mi, determine the linear velocity and acceleration of a point on the surface of the earth (*a*) at the equator, (*b*) at Philadelphia, latitude 40° north, (*c*) at the North Pole.

15.13 A series of small machine components being moved by a conveyor belt pass over the 180-mm-radius idler pulley shown. At the instant shown, the velocity of point A is 450 mm/s to the left and its acceleration is 315 mm/s^2 to the right. Determine (*a*) the angular velocity and angular acceleration of the idler pulley, (*b*) the total acceleration of the machine component at B.

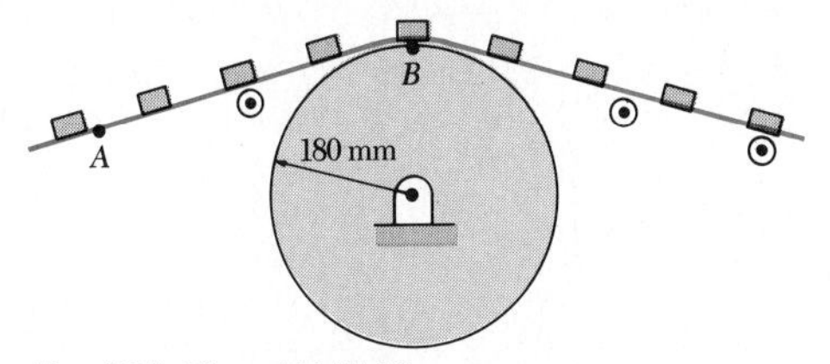

Fig. P15.13 and P15.14

15.14 A series of small machine components being moved by a conveyor belt pass over the 180-mm-radius idler pulley shown. At the instant shown, the angular velocity of the idler pulley is 3 rad/s clockwise. Determine the angular acceleration of the pulley for which the total acceleration of the machine component at B is 2 m/s^2.

15.15 It is known that the static-friction force between the small block B and the plate will be exceeded and that the block will start sliding on the plate when the total acceleration of the block reaches 3 m/s^2. If the plate starts from rest at $t = 0$ and is accelerated at the constant rate of 4 rad/s^2, determine the time t and the angular velocity of the plate when the block starts sliding, assuming $r = 200$ mm.

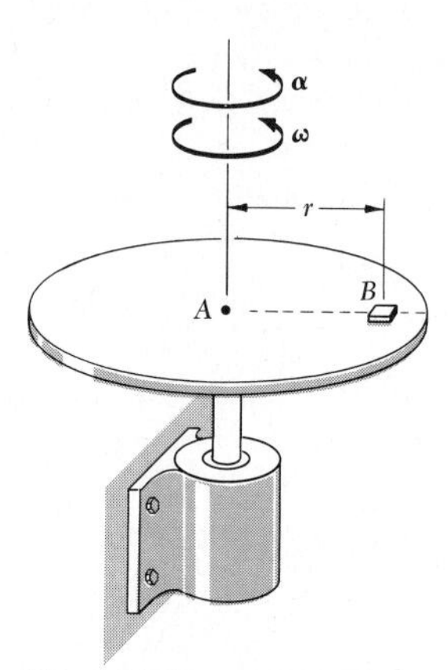

Fig. P15.15 and P15.16

15.16 A small block B rests on a horizontal plate which rotates about a fixed axis. The plate starts from rest at $t = 0$ and is accelerated at the constant rate of 0.5 rad/s^2. Knowing that $r = 200$ mm, determine the magnitude of the total acceleration of the block when (*a*) $t = 0$, (*b*) $t = 1$ s, (*c*) $t = 2$ s.

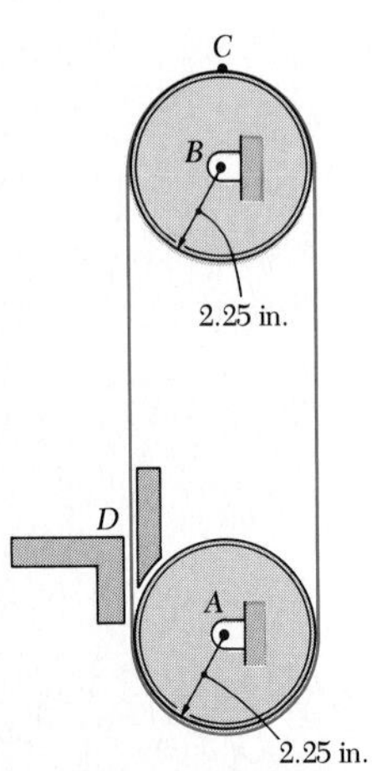

Fig. P15.17 and P15.18

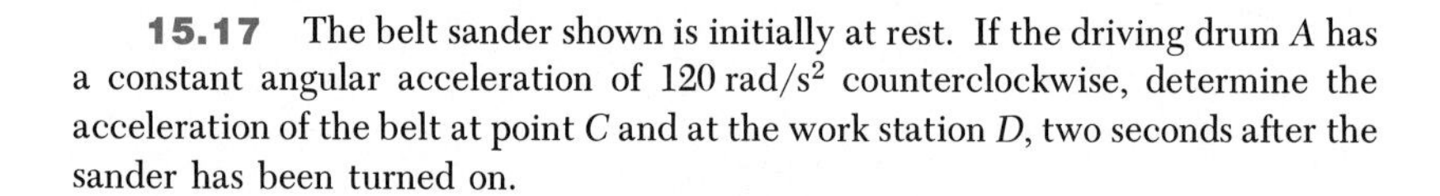
15.17 The belt sander shown is initially at rest. If the driving drum A has a constant angular acceleration of 120 rad/s^2 counterclockwise, determine the acceleration of the belt at point C and at the work station D, two seconds after the sander has been turned on.

15.18 Drum A of the belt sander shown rotates counterclockwise at a rated speed of 3450 rpm. When the power is turned off it is observed that the sander coasts from its rated speed to rest in 5 s. Assuming uniformly decelerated motion, determine the velocity and acceleration of point C of the belt (*a*) immediately before the power is turned off, (*b*) 4.5 s later.

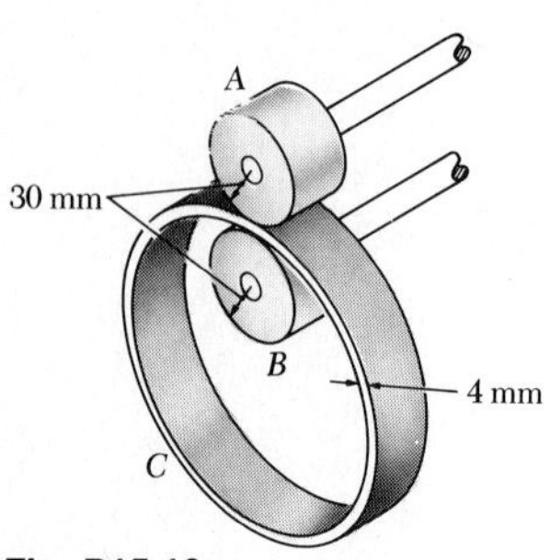

Fig. P15.19

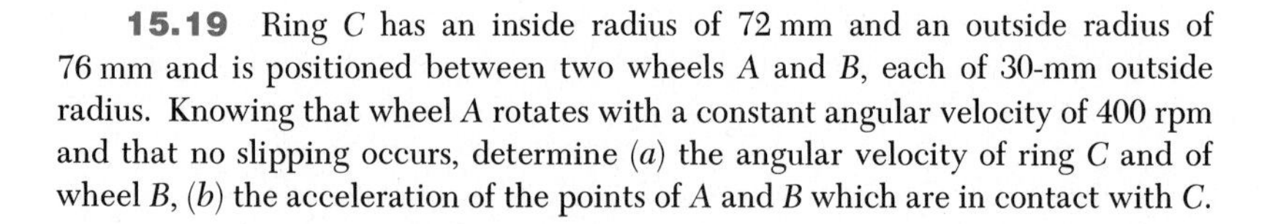
15.19 Ring C has an inside radius of 72 mm and an outside radius of 76 mm and is positioned between two wheels A and B, each of 30-mm outside radius. Knowing that wheel A rotates with a constant angular velocity of 400 rpm and that no slipping occurs, determine (*a*) the angular velocity of ring C and of wheel B, (*b*) the acceleration of the points of A and B which are in contact with C.

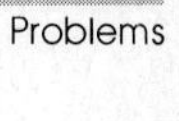

15.20 A computer tape moves over two drums. During a 3-s interval the speed of the tape is increased uniformly from $v_0 = 2$ ft/s to $v_1 = 5$ ft/s. Knowing that the tape does not slip on the drums, determine (*a*) the angular acceleration of drum *A*, (*b*) the number of revolutions executed by drum *A* during the 3-s interval.

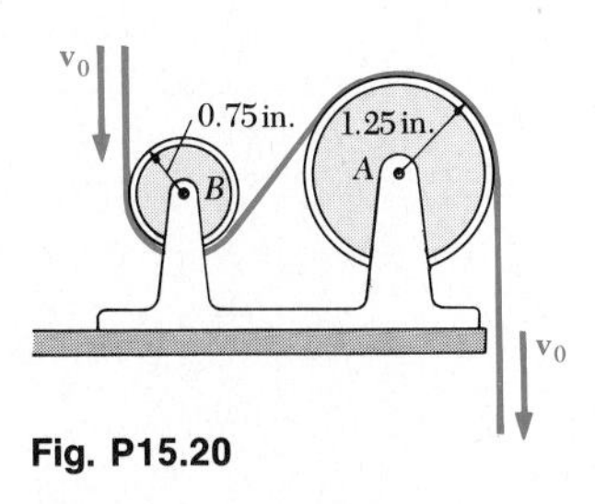

Fig. P15.20

15.21 Solve Prob. 15.20, considering drum *B* instead of drum *A*.

15.22 A mixing drum of 125-mm outside radius rests on two casters, each of 25-mm radius. The drum executes 12 rev during the time interval *t*, while its angular velocity is being increased uniformly from 25 to 45 rpm. Knowing that no slipping occurs between the drum and the casters, determine (*a*) the angular acceleration of the casters, (*b*) the time interval *t*.

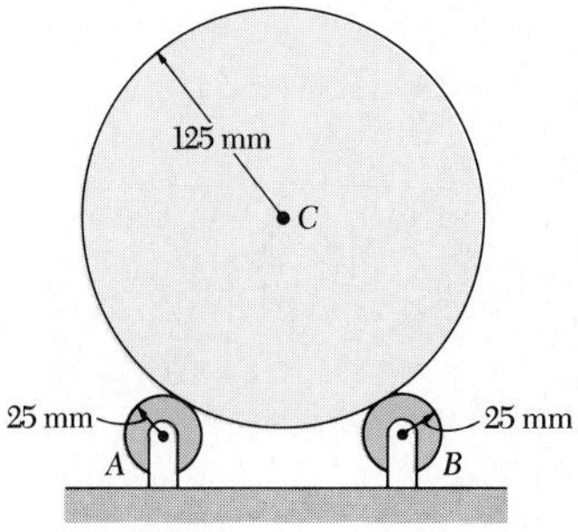

Fig. P15.22

15.23 A pulley and two loads are connected by inextensible cords as shown. Load *A* has a constant acceleration of 2.5 m/s^2 and an initial velocity of 3.75 m/s, both directed upward. Determine (*a*) the number of revolutions executed by the pulley in 3 s, (*b*) the velocity and position of load *B* after 3 s, (*c*) the acceleration of point *C* on the rim of the pulley at $t = 0$.

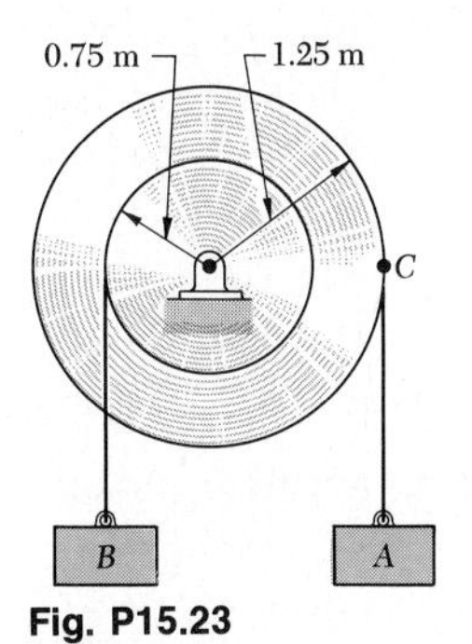

Fig. P15.23

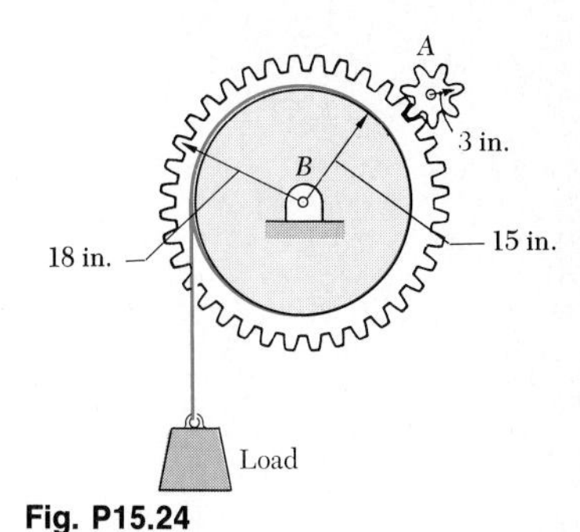

Fig. P15.24

15.24 A load is to be raised 20 ft by the hoisting system shown. Assuming gear *A* is initially at rest, accelerates uniformly to a speed of 120 rpm in 5 s, and then maintains a constant speed of 120 rpm, determine (*a*) the number of revolutions executed by gear *A* in raising the load, (*b*) the time required to raise the load.

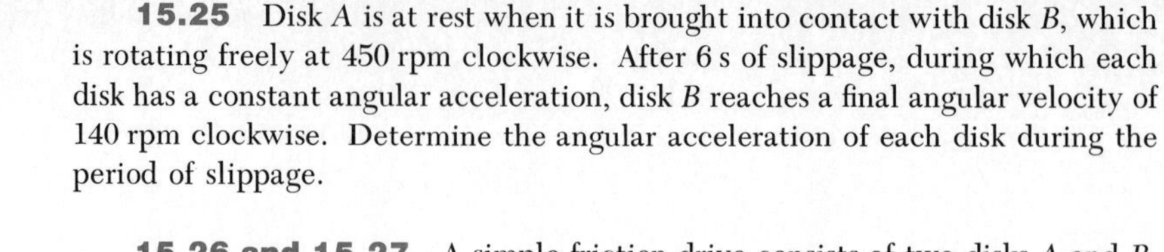

15.25 Disk A is at rest when it is brought into contact with disk B, which is rotating freely at 450 rpm clockwise. After 6 s of slippage, during which each disk has a constant angular acceleration, disk B reaches a final angular velocity of 140 rpm clockwise. Determine the angular acceleration of each disk during the period of slippage.

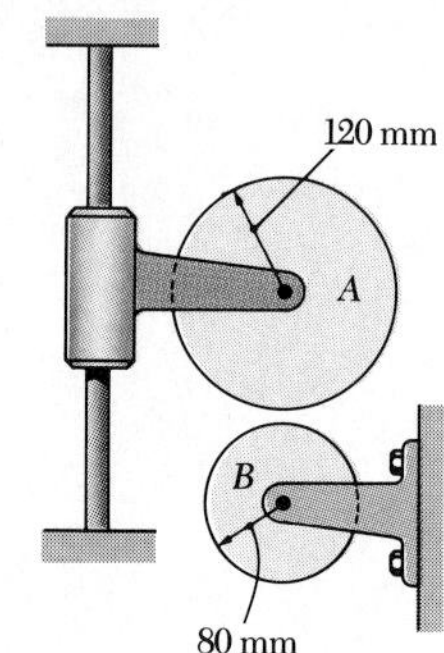

Fig. P15.25 and P15.26

15.26 and 15.27 A simple friction drive consists of two disks A and B. Initially, disk B has a clockwise angular velocity of 500 rpm, and disk A is at rest. It is known that disk B will coast to rest in 60 s. However, rather than waiting until both disks are at rest to bring them together, disk A is given a constant angular acceleration of 3 rad/s^2 counterclockwise. Determine (*a*) at what time the disks may be brought together if they are not to slip, (*b*) the angular velocity of each disk as contact is made.

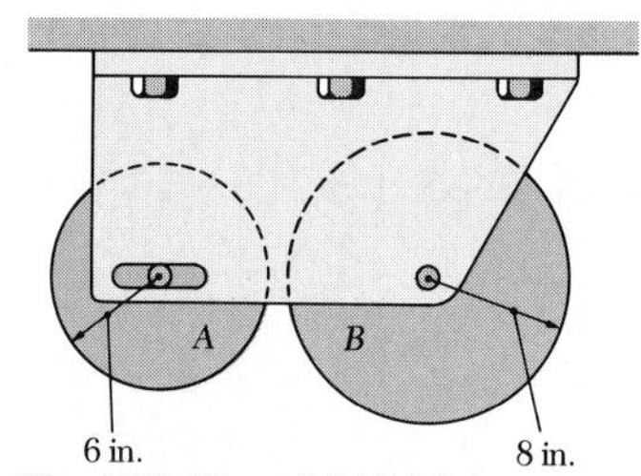

Fig. P15.27 and P15.28

15.28 Two friction wheels A and B are both rotating freely at 240 rpm counterclockwise when they are brought into contact. After 8 s of slippage, during which each wheel has a constant angular acceleration, wheel B reaches a final angular velocity of 60 rpm counterclockwise. Determine (*a*) the angular acceleration of each wheel during the period of slippage, (*b*) the time at which the angular velocity of wheel A is equal to zero.

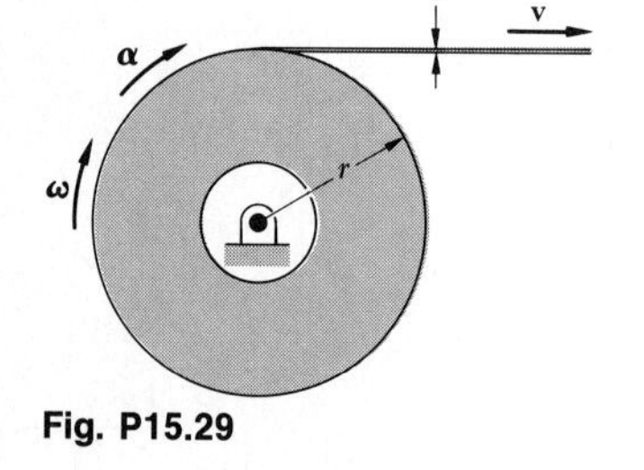

Fig. P15.29

* **15.29** In a continuous printing process, paper is drawn into the presses at a constant speed v. Denoting by r the radius of paper on the roll at any given time and by b the thickness of the paper, derive an expression for the angular acceleration of the paper roll.

15.5. General Plane Motion.

As indicated in Sec. 15.1, we understand by general plane motion a plane motion which is neither a translation nor a rotation. As we shall presently see, however, *a general plane motion may always be considered as the sum of a translation and a rotation.*

Consider, for example, a wheel rolling on a straight track (Fig. 15.9). Over a certain interval of time, two given points A and B will have moved, respectively, from A_1 to A_2 and from B_1 to B_2. The same result could be obtained through a translation which would bring A and B into A_2 and B_1' (the line AB remaining vertical), followed by a rotation about A bringing B into B_2. Although the original rolling motion differs from the combination of translation and rotation when these motions are taken in succession, the original motion may be completely duplicated by a combination of simultaneous translation and rotation.

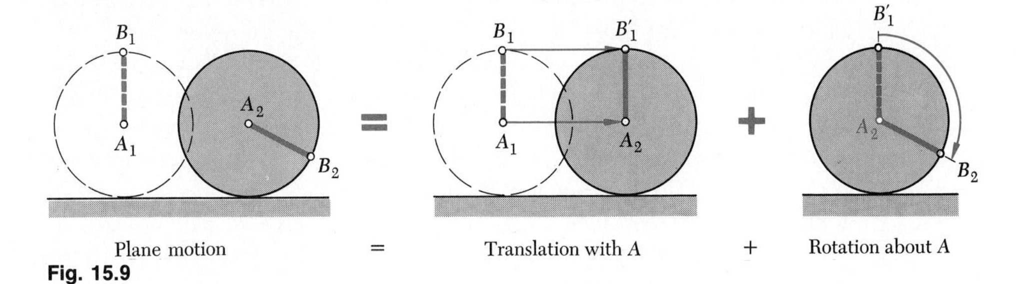

Fig. 15.9

Another example of plane motion is given in Fig. 15.10, which represents a rod whose extremities slide, respectively, along a horizontal and a vertical track. This motion may be replaced by a translation in a horizontal direction and a rotation about A (Fig. 15.10a) or by a translation in a vertical direction and a rotation about B (Fig. 15.10b).

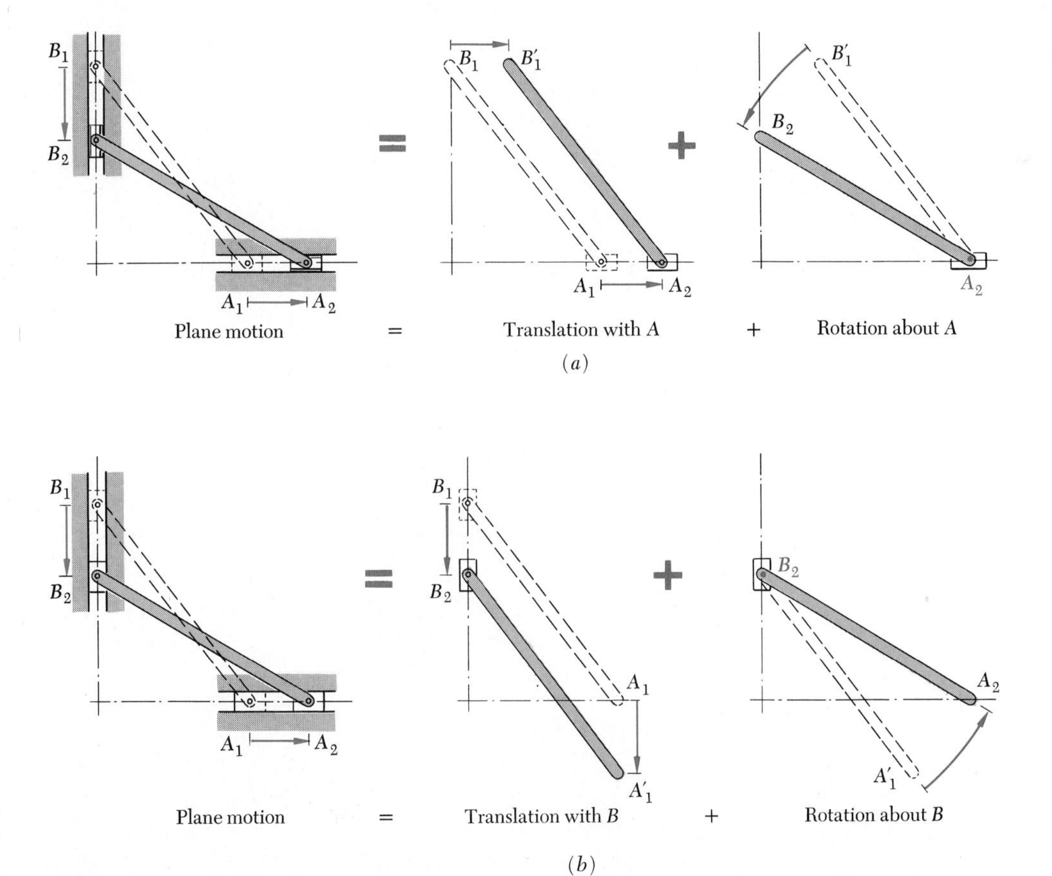

Fig. 15.10

In general, any plane motion of a rigid slab may be replaced by a translation in which all the particles of the slab move along paths parallel to the path actually followed by some arbitrary reference point A, and by a rotation about the reference point A. We saw in Sec. 11.11 that, when two particles A and B move in a plane, the absolute motion of B may be obtained by combining the motion of A and the relative motion of B with respect to A. We recall that by "absolute motion of B" we mean the motion of B with respect to fixed axes, while by "relative motion of B with respect to A" we mean the motion of B with respect to a system of axes whose origin moves with A but which does not rotate. Now, in the present case, the two particles belong to the same rigid body. Particle B must thus remain at a constant distance from A and, *to an observer moving with A, but not rotating, particle B will appear to describe an arc of circle centered at A.* Considering successively each particle of the slab, and resolving its motion into the motion of A and its relative motion about A, we thus conclude that any plane motion of a rigid slab is the sum of a translation with A and a rotation about A.

15.6. Absolute and Relative Velocity in Plane Motion.

We saw in the preceding section that any plane motion of a slab may be replaced by a translation defined by the motion of an arbitrary reference point A, and a simultaneous rotation about A. The absolute velocity $\mathbf{v}_B$ of a particle B of the slab is obtained from the relative-velocity formula derived in Sec. 11.11,

$$\mathbf{v}_B = \mathbf{v}_A + \mathbf{v}_{B/A} \qquad (15.14)$$

where the right-hand member represents a vector sum. The velocity $\mathbf{v}_A$

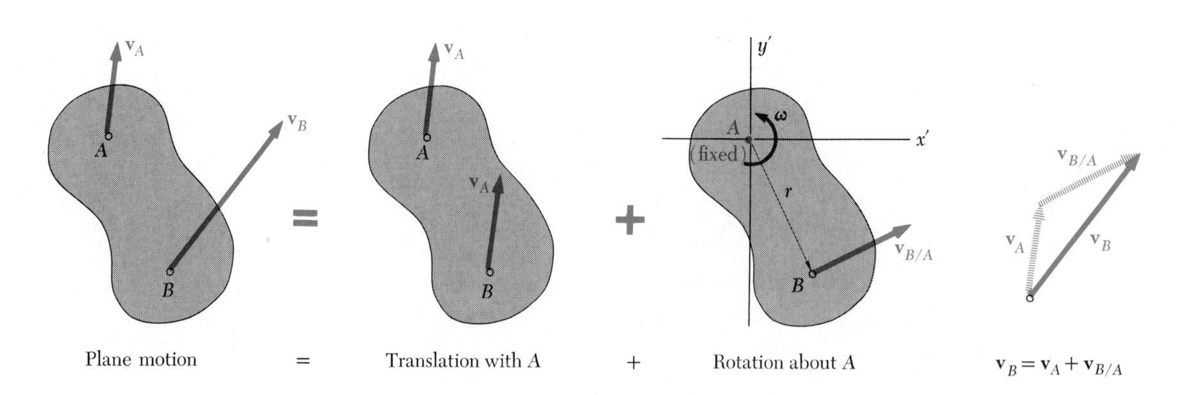

Fig. 15.11

corresponds to the translation of the slab with A, while the relative velocity $\mathbf{v}_{B/A}$ is associated with the rotation of the slab about A and is measured with respect to axes centered at A and of fixed orientation (Fig. 15.11). Denoting by r the distance from A to B, and by ω the magnitude of the angular velocity $\boldsymbol{\omega}$ of the slab with respect to axes of fixed orientation, we find that the magnitude of the relative velocity $\mathbf{v}_{B/A}$ is

$$v_{B/A} = r\omega \tag{15.15}$$

As an example, we shall consider again the rod AB of Fig. 15.10. Assuming that the velocity $\mathbf{v}_A$ of end A is known, we propose to find the velocity $\mathbf{v}_B$ of end B and the angular velocity $\boldsymbol{\omega}$ of the rod, in terms of the velocity $\mathbf{v}_A$, the length l, and the angle θ. Choosing A as a reference point, we express that the given motion is equivalent to a translation with A and a simultaneous rotation about A (Fig. 15.12). The absolute velocity of B must therefore be equal to the vector sum

$$\mathbf{v}_B = \mathbf{v}_A + \mathbf{v}_{B/A} \tag{15.14}$$

We note that while the direction of $\mathbf{v}_{B/A}$ is known, its magnitude $l\omega$ is unknown. However, this is compensated for by the fact that the direction of $\mathbf{v}_B$ is known. We may therefore complete the diagram of Fig. 15.12. Solving for the magnitudes v_B and ω, we write

$$v_B = v_A \tan\theta \qquad \omega = \frac{v_{B/A}}{l} = \frac{v_A}{l\cos\theta} \tag{15.16}$$

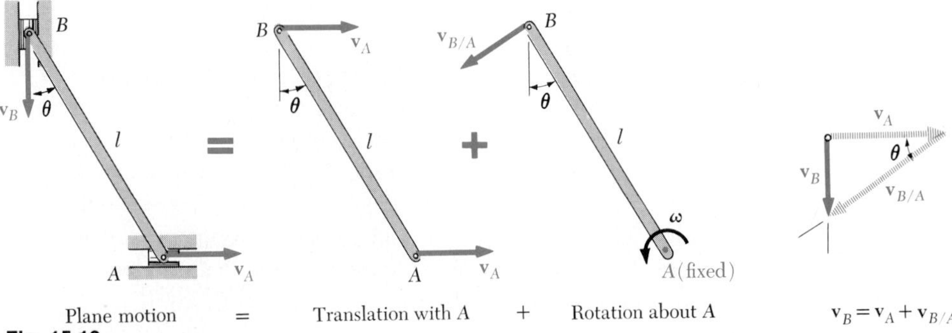

Fig. 15.12

The same result may be obtained by using B as a point of reference. Resolving the given motion into a translation with B and a simultaneous rotation about B (Fig. 15.13), we write the equation

$$\mathbf{v}_A = \mathbf{v}_B + \mathbf{v}_{A/B} \tag{15.17}$$

which is represented graphically in Fig. 15.13. We note that $\mathbf{v}_{A/B}$ and $\mathbf{v}_{B/A}$ have the same magnitude $l\omega$ but opposite sense. The sense of the relative velocity depends, therefore, upon the point of reference which has been selected and should be carefully ascertained from the appropriate diagram (Fig. 15.12 or 15.13).

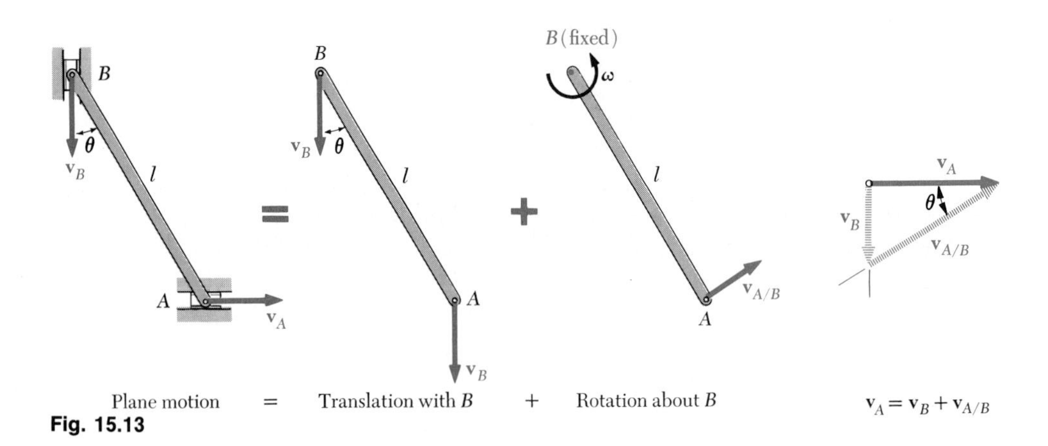

Fig. 15.13

Finally, we observe that the angular velocity ω of the rod in its rotation about B is the same as in its rotation about A. It is measured in both cases by the rate of change of the angle θ. This result is quite general; we should therefore bear in mind that *the angular velocity* ω *of a rigid body in plane motion is independent of the reference point.*

Most mechanisms consist not of one but of *several* moving parts. When the various parts of a mechanism are pin-connected, its analysis may be carried out by considering each part as a rigid body, while keeping in mind that the points where two parts are connected must have the same absolute velocity (see Sample Prob. 15.3). A similar analysis may be used when gears are involved, since the teeth in contact must also have the same absolute velocity. However, when a mechanism contains parts which slide on each other, the relative velocity of the parts in contact must be taken into account (see Secs. 15.10 and 15.11).

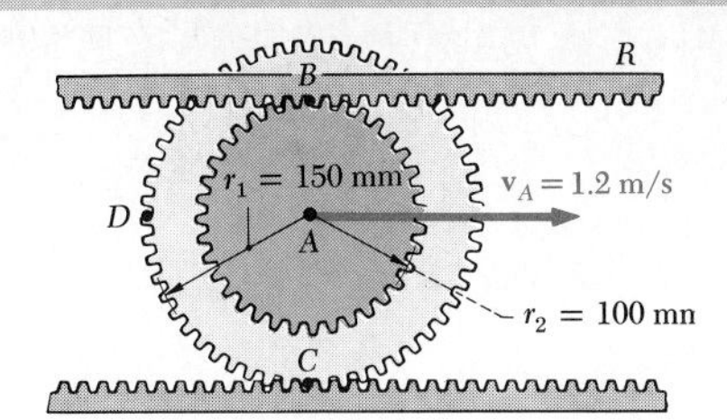

SAMPLE PROBLEM 15.2

The double gear shown rolls on the stationary lower rack; the velocity of its center A is 1.2 m/s directed to the right. Determine (*a*) the angular velocity of the gear, (*b*) the velocities of the upper rack R and of point D of the gear.

a. **Angular Velocity of the Gear.** Since the gear rolls on the lower rack, its center A moves through a distance equal to the outer circumference $2\pi r_1$ for each full revolution of the gear. Noting that 1 rev = 2π rad, we obtain the coordinate x_A in terms of the corresponding angular coordinate θ (in radians) of the gear by a proportion,

$$\frac{x_A}{2\pi r_1} = \frac{\theta}{2\pi} \qquad x_A = r_1\theta$$

Differentiating with respect to the time t and substituting the known values $v_A = 1.2$ m/s and $r_1 = 150 \text{ mm} = 0.150$ m, we obtain

$$v_A = r_1\omega \qquad 1.2 \text{ m/s} = (0.150 \text{ m})\omega \qquad \omega = 8 \text{ rad/s} \downarrow \quad \blacktriangleleft$$

b. **Velocities.** The rolling motion is resolved into two component motions: a translation with the center A and a rotation about the center A. In the translation, all points of the gear move with the same velocity $\mathbf{v}_A$. In the rotation, each point P of the gear moves about A with a relative velocity of magnitude $r\omega$, where r is the distance from A.

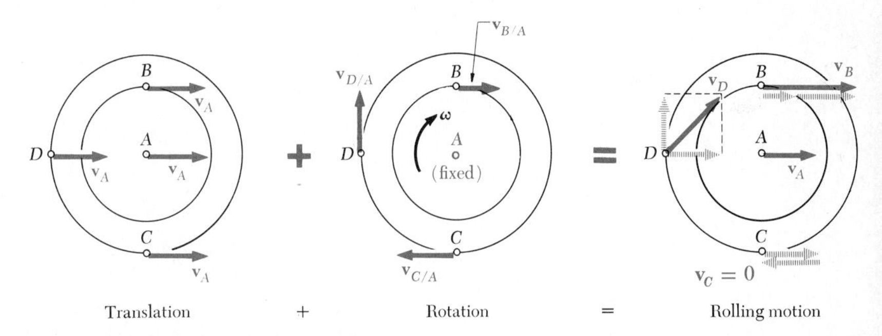

Velocity of Upper Rack. The velocity of the upper rack is equal to the velocity of point B; we write

$$\mathbf{v}_R = \mathbf{v}_B = \mathbf{v}_A + \mathbf{v}_{B/A}$$

where

$$\mathbf{v}_A = 1.2 \text{ m/s} \rightarrow$$

$$v_{B/A} = r_2\omega = (0.100 \text{ m})(8 \text{ rad/s}) = 0.8 \text{ m/s} \qquad \mathbf{v}_{B/A} = 0.8 \text{ m/s} \rightarrow$$

Since $\mathbf{v}_A$ and $\mathbf{v}_{B/A}$ are collinear, we add their magnitudes and obtain

$$\mathbf{v}_R = 2 \text{ m/s} \rightarrow \quad \blacktriangleleft$$

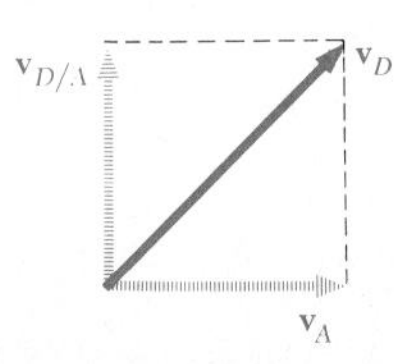

Velocity of Point D: $\mathbf{v}_D = \mathbf{v}_A + \mathbf{v}_{D/A}$

where

$$\mathbf{v}_A = 1.2 \text{ m/s} \rightarrow$$

$$v_{D/A} = r_1\omega = (0.150 \text{ m})(8 \text{ rad/s}) = 1.2 \text{ m/s} \qquad \mathbf{v}_{D/A} = 1.2 \text{ m/s} \uparrow$$

Adding these velocities vectorially, we obtain $\mathbf{v}_D = 1.697 \text{ m/s} \angle 45° \quad \blacktriangleleft$

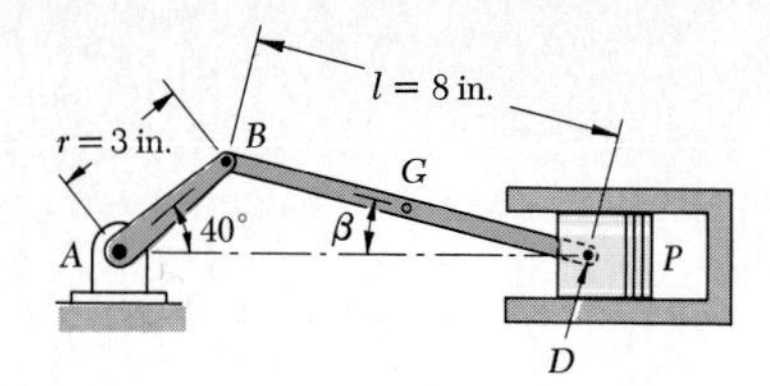

SAMPLE PROBLEM 15.3

In the engine system shown, the crank *AB* has a constant clockwise angular velocity of 2000 rpm. For the crank position indicated, determine (*a*) the angular velocity of the connecting rod *BD*, (*b*) the velocity of the piston *P*.

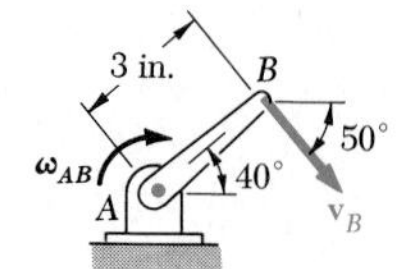

Motion of Crank *AB*. The crank *AB* rotates about point *A*. Expressing ω_{AB} in rad/s and writing $v_B = r\omega_{AB}$, we obtain

$$\omega_{AB} = \left(2000\frac{\text{rev}}{\text{min}}\right)\left(\frac{1\text{ min}}{60\text{ s}}\right)\left(\frac{2\pi\text{ rad}}{1\text{ rev}}\right) = 209.4\text{ rad/s}$$

$$v_B = (AB)\omega_{AB} = (3\text{ in.})(209.4\text{ rad/s}) = 628.3\text{ in./s}$$

$$\mathbf{v}_B = 628.3\text{ in./s} \searrow 50°$$

Motion of Connecting Rod *BD*. We consider this motion as a general plane motion. Using the law of sines, we compute the angle β between the connecting rod and the horizontal:

$$\frac{\sin 40°}{8\text{ in.}} = \frac{\sin\beta}{3\text{ in.}} \qquad \beta = 13.95°$$

The velocity $\mathbf{v}_D$ of the point *D* where the rod is attached to the piston must be horizontal, while the velocity of point *B* is equal to the velocity $\mathbf{v}_B$ obtained above. Resolving the motion of *BD* into a translation with *B* and a rotation about *B*, we obtain

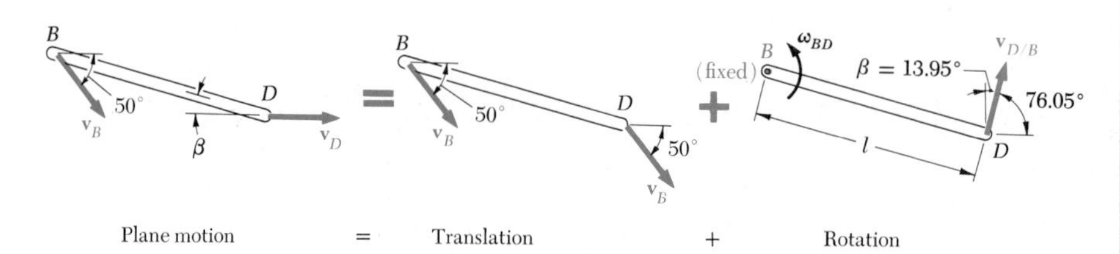

Expressing the relation between the velocities $\mathbf{v}_D$, $\mathbf{v}_B$, and $\mathbf{v}_{D/B}$, we write

$$\mathbf{v}_D = \mathbf{v}_B + \mathbf{v}_{D/B}$$

We draw the vector diagram corresponding to this equation. Recalling that $\beta = 13.95°$, we determine the angles of the triangle and write

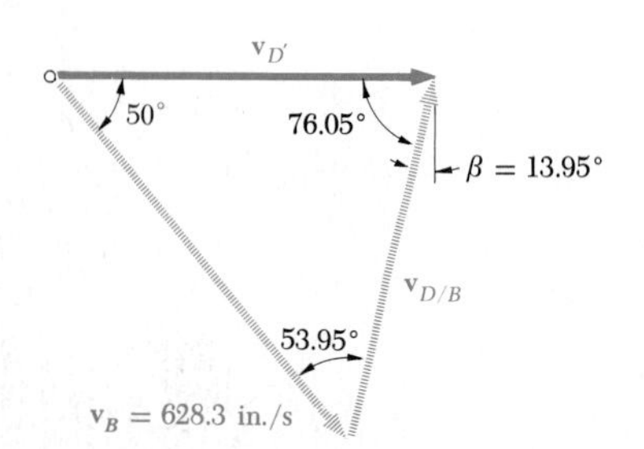

$$\frac{v_D}{\sin 53.95°} = \frac{v_{D/B}}{\sin 50°} = \frac{628.3\text{ in./s}}{\sin 76.05°}$$

$$v_{D/B} = 495.9\text{ in./s} \qquad \mathbf{v}_{D/B} = 495.9\text{ in./s} \measuredangle 76.05°$$

$$v_D = 523.4\text{ in./s} = 43.6\text{ ft/s} \qquad \mathbf{v}_D = 43.6\text{ ft/s} \rightarrow$$

$$\mathbf{v}_P = \mathbf{v}_D = 43.6\text{ ft/s} \rightarrow \blacktriangleleft$$

Since $v_{D/B} = l\omega_{BD}$, we have

$$495.9\text{ in./s} = (8\text{ in.})\omega_{BD} \qquad \omega_{BD} = 62.0\text{ rad/s} \circlearrowleft \blacktriangleleft$$

Problems

15.30 An automobile travels to the right at a constant speed of 45 mi/h. If the diameter of a wheel is 22 in., determine the velocities of points B, C, D, and E on the rim of the wheel.

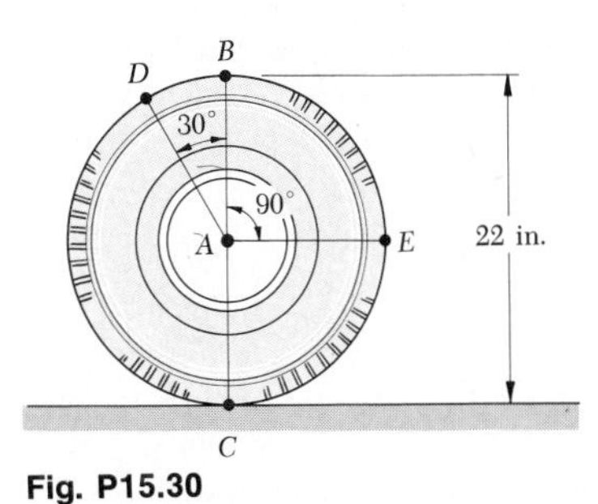

Fig. P15.30

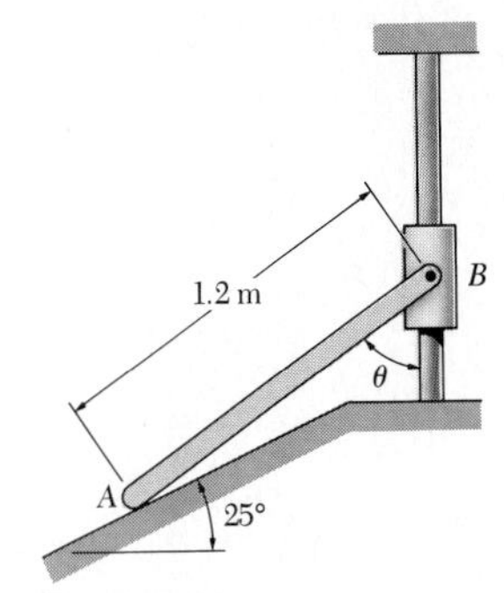

Fig. P15.31

15.31 Collar B moves upward with a constant velocity of 1.8 m/s. At the instant when $\theta = 50°$, determine (*a*) the angular velocity of rod AB, (*b*) the velocity of end A of the rod.

15.32 Solve Prob. 15.31, assuming that $\theta = 35°$.

15.33 Rod AB is 30 in. long and slides with its ends in contact with the floor and the inclined plane. End A moves with a constant velocity of 25 in./s to the right. At the instant when $\theta = 25°$, determine (*a*) the angular velocity of the rod, (*b*) the velocity of end B.

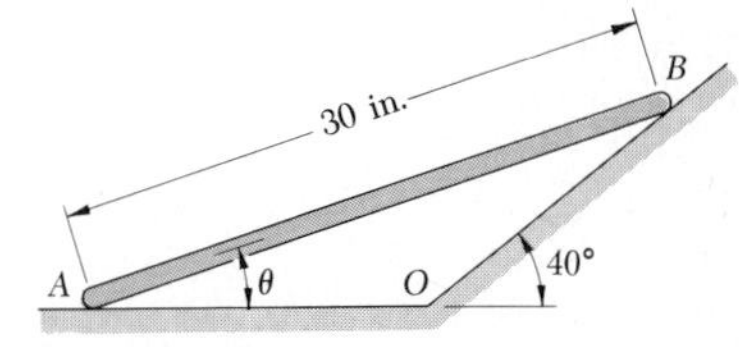

Fig. P15.33

15.34 Collar B moves with a constant velocity of 2.4 m/s to the right. At the instant when $\theta = 20°$, determine (*a*) the angular velocity of rod AB, (*b*) the velocity of collar A.

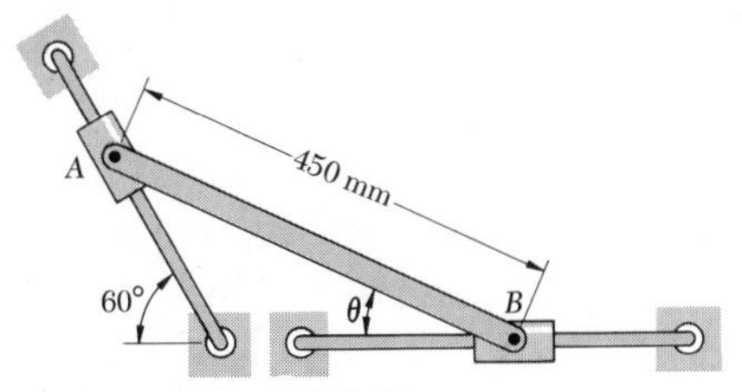

Fig. P15.34 and P15.35

15.35 Collar A moves with a constant velocity of 3 m/s upward along the inclined rod. At the instant when $\theta = 35°$, determine (*a*) the angular velocity of rod AB, (*b*) the velocity of collar B.

15.36 A 16-in. rod AB is guided by wheels at A and B which roll in the track shown. Knowing that $\theta = 60°$ and that B moves at a constant speed $v_B = 30$ in./s, determine (*a*) the angular velocity of the rod, (*b*) the velocity of A.

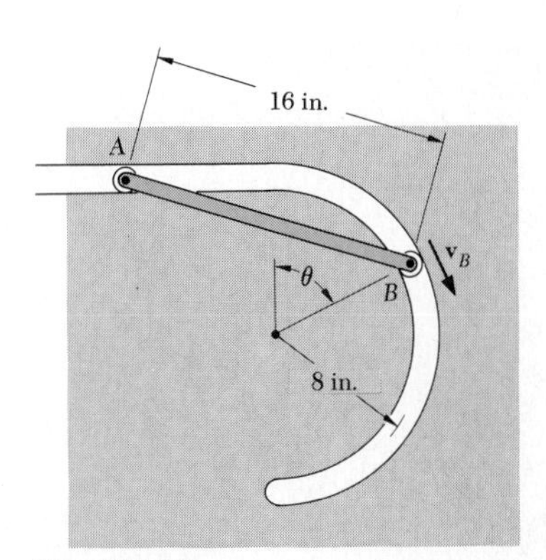

Fig. P15.36

15.37 Solve Prob. 15.36 assuming $\theta = 120°$.

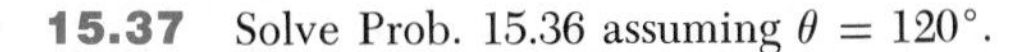

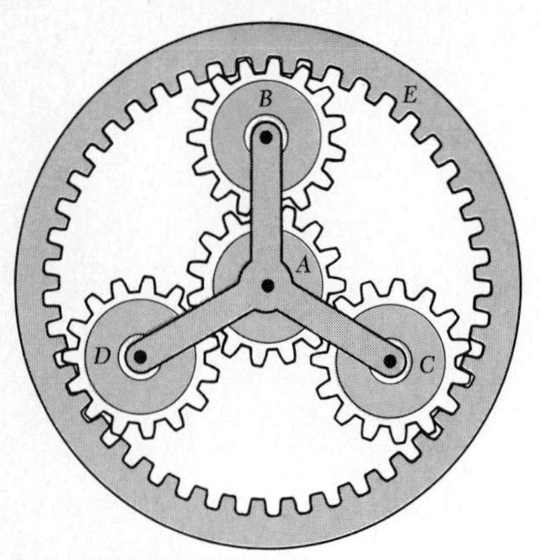

Fig. P15.38 and P15.39

15.38 In the planetary gear system shown, the radius of gears A, B, C, and D is a and the radius of the outer gear E is $3a$. Knowing that the angular velocity of gear A is $\boldsymbol{\omega}_A$ clockwise and that the outer gear E is stationary, determine (*a*) the angular velocity of each planetary gear, (*b*) the angular velocity of the spider connecting the planetary gears.

15.39 In the planetary gear system shown, the radius of gears A, B, C, and D is 45 mm and the radius of the outer gear E is 135 mm. Knowing that gear E has an angular velocity of 120 rpm clockwise and that the central gear A has an angular velocity of 150 rpm clockwise, determine (*a*) the angular velocity of each planetary gear, (*b*) the angular velocity of the spider connecting the planetary gears.

15.40 Gear A rotates clockwise with a constant angular velocity of 60 rpm. Determine the angular velocity of gear B if the angular velocity of arm AB is (*a*) 40 rpm counterclockwise, (*b*) 40 rpm clockwise.

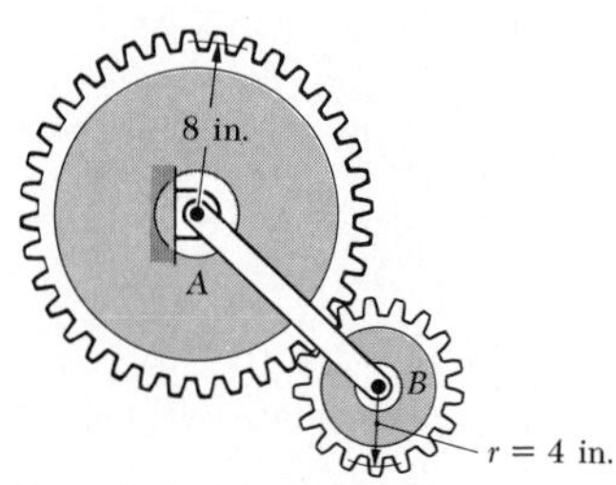

Fig. P15.40 and P15.41

15.41 Gear A rotates clockwise with a constant angular velocity of 150 rpm. Knowing that at the same time arm AB rotates clockwise with a constant angular velocity of 125 rpm, determine the angular velocity of gear B.

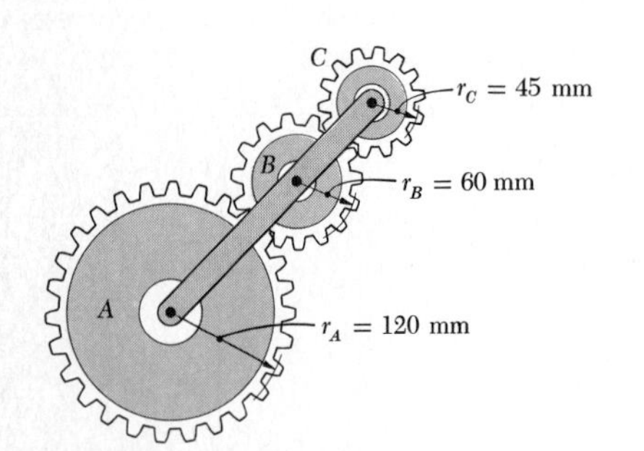

Fig. P15.42 and P15.43

15.42 Three gears A, B, and C are pinned at their centers to rod ABC. Knowing that gear A does not rotate, determine the angular velocity of gears B and C when the rod ABC rotates clockwise with a constant angular velocity of 48 rpm.

15.43 Three gears A, B, and C are pinned at their centers to rod ABC. Knowing that gear C does not rotate, determine the angular velocity of gears A and B when the rod ABC rotates clockwise with an angular velocity of 48 rpm.

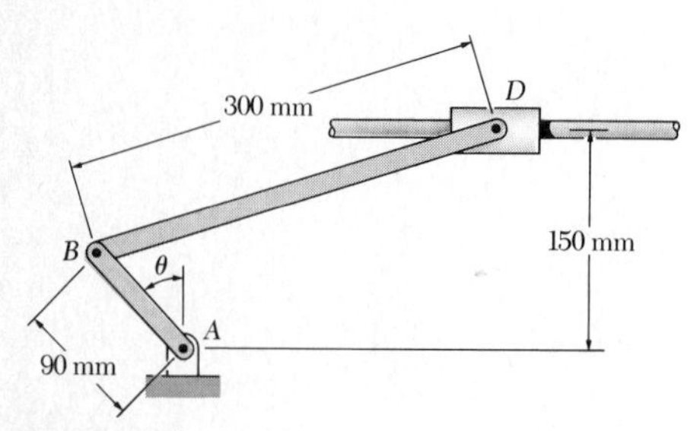

Fig. P15.44

15.44 Crank AB has a constant angular velocity of 200 rpm counterclockwise. Determine the angular velocity of rod BD and the velocity of collar D when (*a*) $\theta = 0$, (*b*) $\theta = 90°$, (*c*) $\theta = 180°$.

15.45 In the engine system shown, $l = 10$ in. and $b = 3$ in.; the crank AB rotates with a constant angular velocity of 750 rpm clockwise. Determine the velocity of the piston P and the angular velocity of the connecting rod for the position corresponding to (*a*) $\theta = 0$, (*b*) $\theta = 90°$, (*c*) $\theta = 180°$.

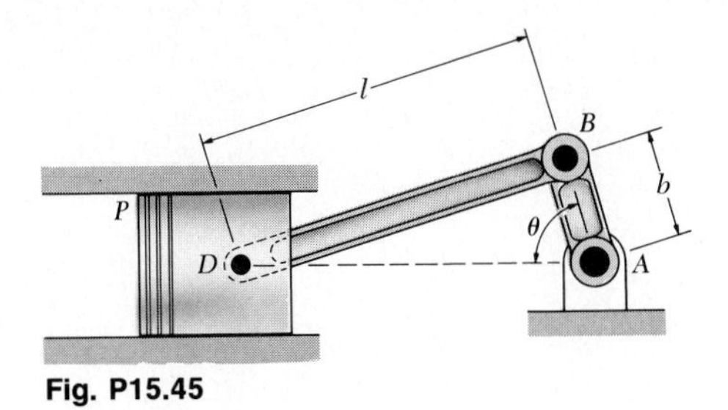

Fig. P15.45

15.46 Solve Prob. 15.45 for the position corresponding to $\theta = 30°$.

15.47 Solve Prob. 15.44 for the position corresponding to $\theta = 60°$.

15.48 through 15.51 In the position shown, bar AB has a constant angular velocity of 3 rad/s counterclockwise. Determine the angular velocity of bars BD and DE.

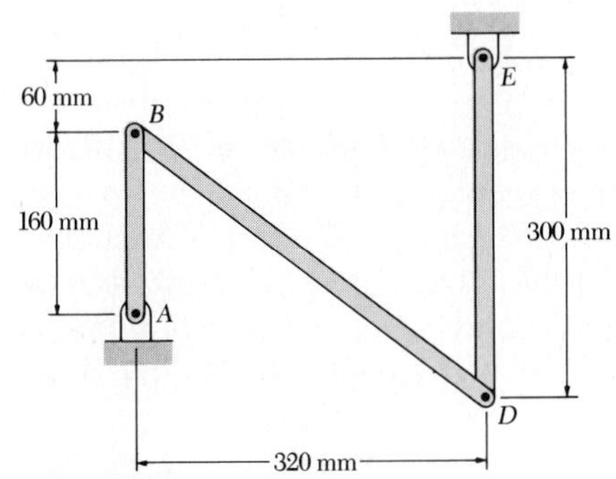

Fig. P15.48

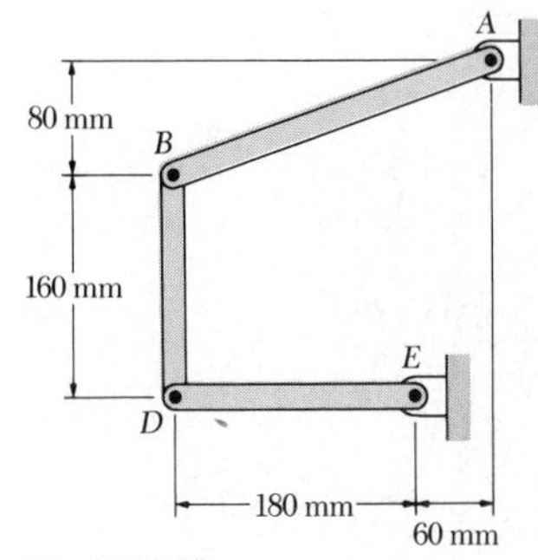

Fig. P15.49

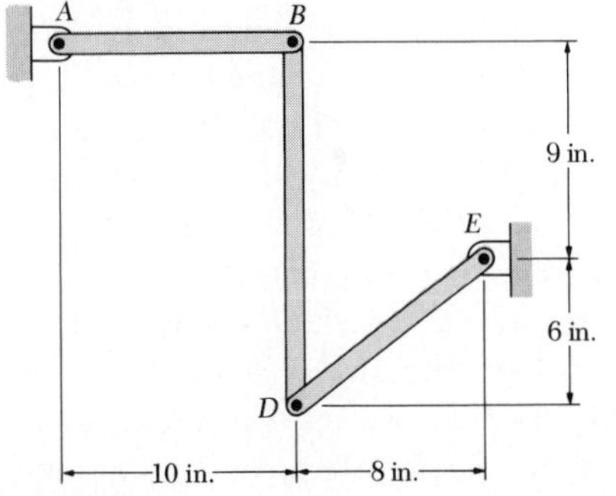

Fig. P15.50

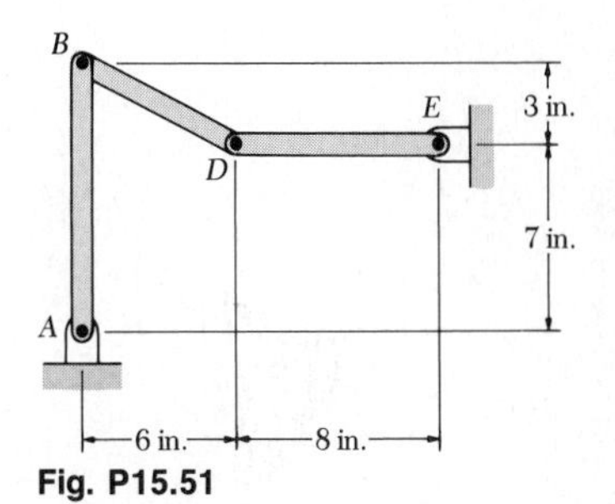

Fig. P15.51

15.52 The flanged wheel shown rolls to the right with a constant velocity of 1.5 m/s. Knowing that rod AB is 1.2 m long, determine the velocity of A and the angular velocity of the rod when (*a*) $\beta = 0$, (*b*) $\beta = 90°$.

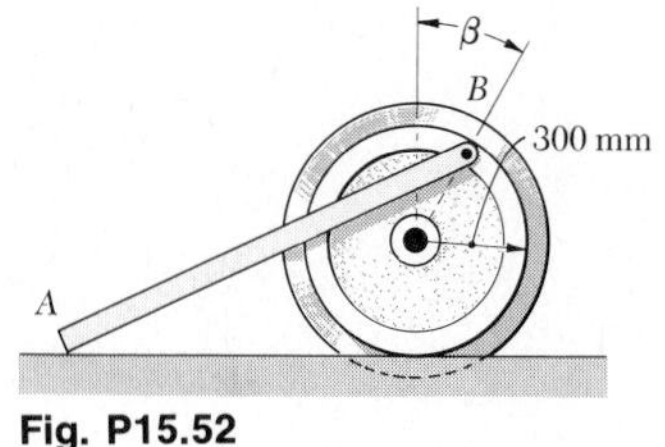

Fig. P15.52

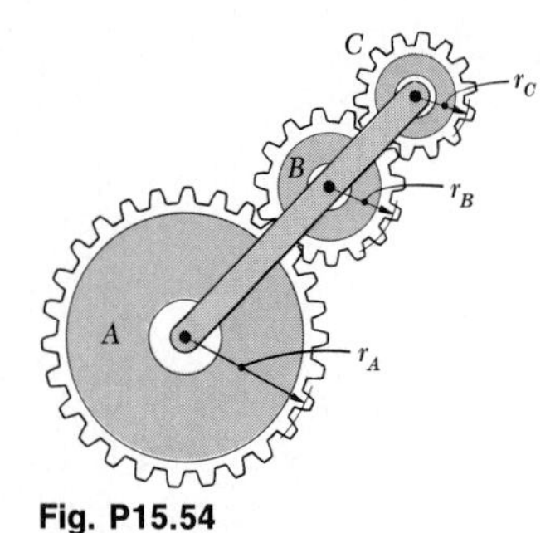

Fig. P15.54

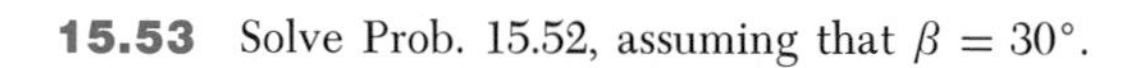

15.53 Solve Prob. 15.52, assuming that $\beta = 30°$.

***15.54** For the gearing system shown, derive an expression for the angular velocity ω_C of gear C and show that ω_C is independent of the radius of gear B. Assume that point A is fixed and denote the angular velocities of rod ABC and gear A by ω_{ABC} and ω_A, respectively.

15.7. Instantaneous Center of Rotation in Plane Motion. Consider the general plane motion of a slab. We shall show that at any given instant the velocities of the various particles of the slab are the same as if the slab were rotating about a certain axis perpendicular to the plane of the slab, called the *instantaneous axis of rotation*. This axis intersects the plane of the slab at a point C, called the *instantaneous center of rotation* of the slab.

To prove our statement, we first recall that the plane motion of a slab may always be replaced by a translation defined by the motion of an arbitrary reference point A, and by a rotation about A. As far as the velocities are concerned, the translation is characterized by the velocity $\mathbf{v}_A$ of the reference point A and the rotation is characterized by the angular velocity $\boldsymbol{\omega}$ of the slab (which is independent of the choice of A). Thus, the velocity $\mathbf{v}_A$ of point A and the angular velocity $\boldsymbol{\omega}$ of the slab define completely the velocities of all the other particles of the slab (Fig. 15.14*a*). Now let us assume that $\mathbf{v}_A$ and $\boldsymbol{\omega}$ are known and that they are both different from zero. (If $\mathbf{v}_A = 0$, point A is itself the instantaneous center of rotation, and if $\boldsymbol{\omega} = 0$, all the particles have the same velocity $\mathbf{v}_A$.) These velocities could be obtained by letting the slab rotate with the angular velocity $\boldsymbol{\omega}$ about a point C located on the perpendicular to $\mathbf{v}_A$ at a distance $r = v_A/\omega$ from A

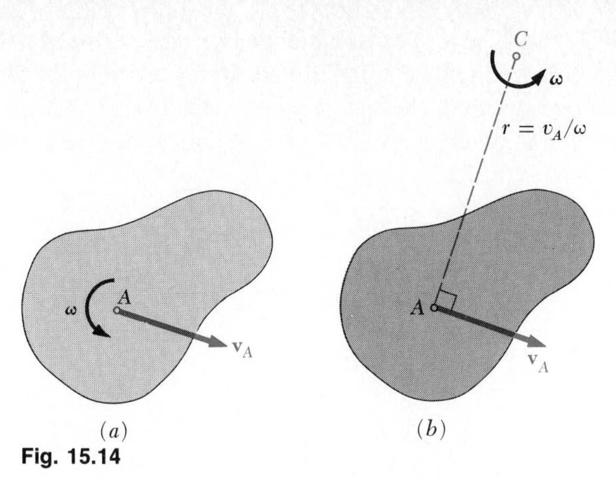

Fig. 15.14

as shown in Fig. 15.14*b*. We check that the velocity of A would be perpendicular to AC and that its magnitude would be $r\omega = (v_A/\omega)\omega = v_A$. Thus the velocities of all the other particles of the slab would be the same as originally defined. Therefore, *as far as the velocities are concerned, the slab seems to rotate about the instantaneous center C* at the instant considered.

The position of the instantaneous center may be defined in two other ways. If the directions of the velocities of two particles A and B of the slab are known and if they are different, the instantaneous center C is obtained by drawing the perpendicular to $\mathbf{v}_A$ through A and the perpendicular to $\mathbf{v}_B$ through B and determining the point at which these two lines intersect (Fig. 15.15*a*). If the velocities $\mathbf{v}_A$ and $\mathbf{v}_B$ of two particles A and B are perpendicular to the line AB and if their magnitudes are known, the instantaneous center may be found by intersecting the line AB with the line joining the extremities of the vectors $\mathbf{v}_A$ and $\mathbf{v}_B$ (Fig. 15.15*b*). Note that if $\mathbf{v}_A$ and $\mathbf{v}_B$ were parallel in Fig. 15.15*a* or if $\mathbf{v}_A$ and $\mathbf{v}_B$ had the same magnitude in Fig. 15.15*b*, the instantaneous center C would be at an infinite distance and $\boldsymbol{\omega}$ would be zero; all points of the slab would have the same velocity.

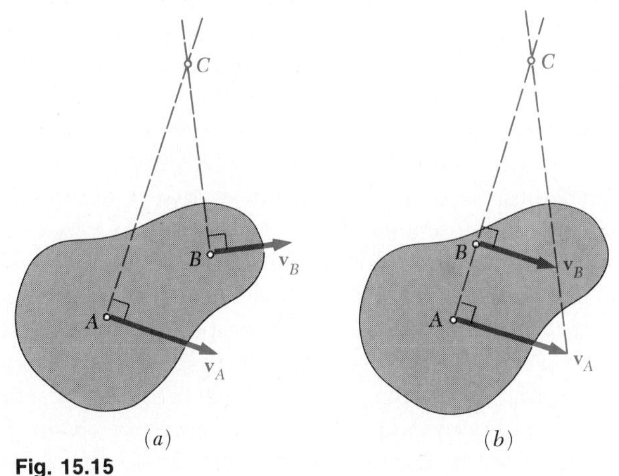

Fig. 15.15

To see how the concept of instantaneous center of rotation may be put to use, let us consider again the rod of Sec. 15.6. Drawing the perpendicular to $\mathbf{v}_A$ through A and the perpendicular to $\mathbf{v}_B$ through B (Fig. 15.16), we obtain the instantaneous center C. At the instant considered, the velocities

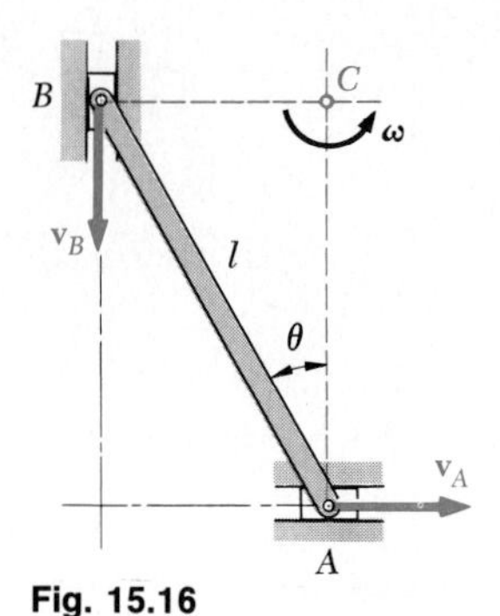

Fig. 15.16

of all the particles of the rod are thus the same as if the rod rotated about C. Now, if the magnitude v_A of the velocity of A is known, the magnitude ω of the angular velocity of the rod may be obtained by writing

$$\omega = \frac{v_A}{AC} = \frac{v_A}{l \cos \theta}$$

The magnitude of the velocity of B may then be obtained by writing

$$v_B = (BC)\omega = l \sin \theta \frac{v_A}{l \cos \theta} = v_A \tan \theta$$

Note that only *absolute* velocities are involved in the computation.

The instantaneous center of a slab in plane motion may be located either on the slab or outside the slab. If it is located on the slab, the particle C coinciding with the instantaneous center at a given instant t must have zero velocity at that instant. However, it should be noted that the instantaneous center of rotation is valid only at a given instant. Thus, the particle C of the slab which coincides with the instantaneous center at time t will generally not coincide with the instantaneous center at time $t + \Delta t$; while its velocity is zero at time t, it will probably be different from zero at time $t + \Delta t$. This means that, in general, the particle C *does not have zero acceleration* and, therefore, that the *accelerations* of the various particles of the slab *cannot* be determined as if the slab were rotating about C.

As the motion of the slab proceeds, the instantaneous center moves in space. But it was just pointed out that the position of the instantaneous center on the slab keeps changing. Thus, the instantaneous center describes one curve in space, called the *space centrode*, and another curve on the slab, called the *body centrode* (Fig. 15.17). It may be shown that at any instant, these two curves are tangent at C and that as the slab moves, the body centrode appears to *roll* on the space centrode.

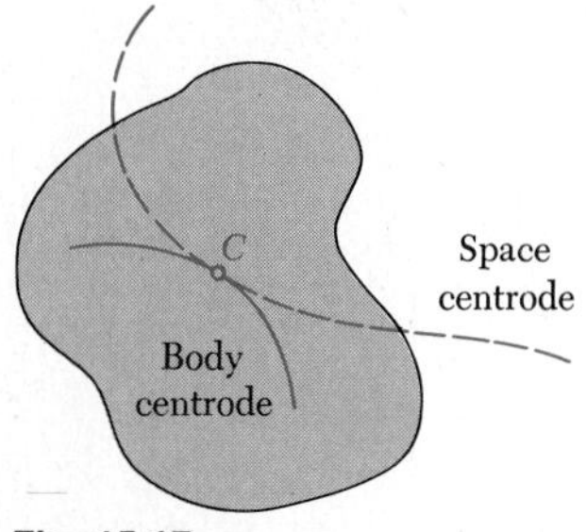

Fig. 15.17

SAMPLE PROBLEM 15.4

Solve Sample Prob. 15.2, using the method of the instantaneous center of rotation.

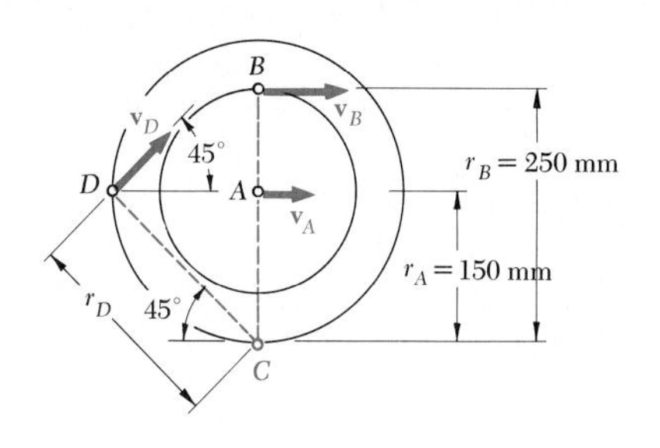

***a.* Angular Velocity of the Gear.** Since the gear rolls on the stationary lower rack, the point of contact C of the gear with the rack has no velocity; point C is therefore the instantaneous center of rotation. We write

$$v_A = r_A\omega \qquad 1.2 \text{ m/s} = (0.150 \text{ m})\omega \qquad \omega = 8 \text{ rad/s} \downarrow$$

***b.* Velocities.** All points of the gear seem to rotate about the instantaneous center as far as velocities are concerned.

Velocity of Upper Rack. Recalling that $v_R = v_B$, we write

$$v_R = v_B = r_B\omega \qquad v_R = (0.250 \text{ m})(8 \text{ rad/s}) = 2 \text{ m/s}$$

$$\mathbf{v}_R = 2 \text{ m/s} \rightarrow$$

Velocity of Point D. Since $r_D = (0.150 \text{ m})\sqrt{2} = 0.2121$ m, we write

$$v_D = r_D\omega \qquad v_D = (0.2121 \text{ m})(8 \text{ rad/s}) = 1.697 \text{ m/s}$$

$$\mathbf{v}_D = 1.697 \text{ m/s} \measuredangle 45°$$

SAMPLE PROBLEM 15.5

Solve Sample Prob. 15.3, using the method of the instantaneous center of rotation.

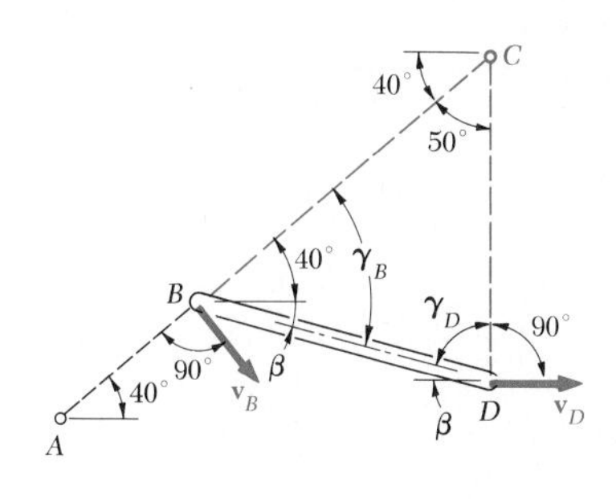

Motion of Crank *AB*. Referring to Sample Prob. 15.3, we obtain the velocity of point B; $\mathbf{v}_B = 628.3$ in./s $\measuredangle$ 50°.

Motion of the Connecting Rod *BD*. We first locate the instantaneous center C by drawing lines perpendicular to the absolute velocities $\mathbf{v}_B$ and $\mathbf{v}_D$. Recalling from Sample Prob. 15.3 that $\beta = 13.95°$ and that $BD = 8$ in., we solve the triangle BCD.

$$\gamma_B = 40° + \beta = 53.95° \qquad \gamma_D = 90° - \beta = 76.05°$$

$$\frac{BC}{\sin 76.05°} = \frac{CD}{\sin 53.95°} = \frac{8 \text{ in.}}{\sin 50°}$$

$$BC = 10.14 \text{ in.} \qquad CD = 8.44 \text{ in.}$$

Since the connecting rod BD seems to rotate about point C, we write

$$v_B = (BC)\omega_{BD}$$
$$628.3 \text{ in./s} = (10.14 \text{ in.})\omega_{BD}$$
$$\omega_{BD} = 62.0 \text{ rad/s} \uparrow$$
$$v_D = (CD)\omega_{BD} = (8.44 \text{ in.})(62.0 \text{ rad/s})$$
$$= 523 \text{ in./s} = 43.6 \text{ ft/s}$$
$$\mathbf{v}_P = \mathbf{v}_D = 43.6 \text{ ft/s} \rightarrow$$

Problems

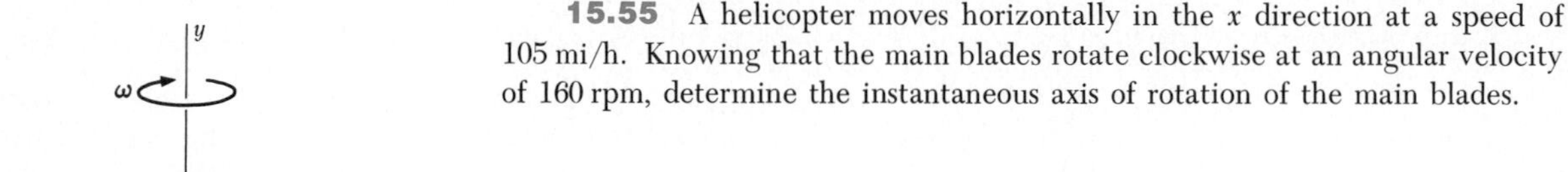

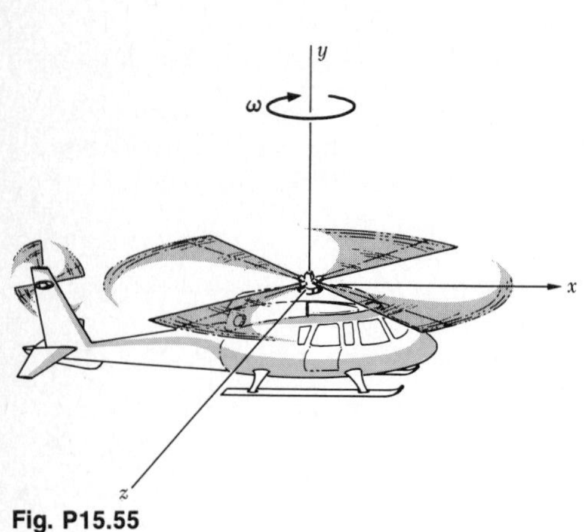

Fig. P15.55

15.55 A helicopter moves horizontally in the x direction at a speed of 105 mi/h. Knowing that the main blades rotate clockwise at an angular velocity of 160 rpm, determine the instantaneous axis of rotation of the main blades.

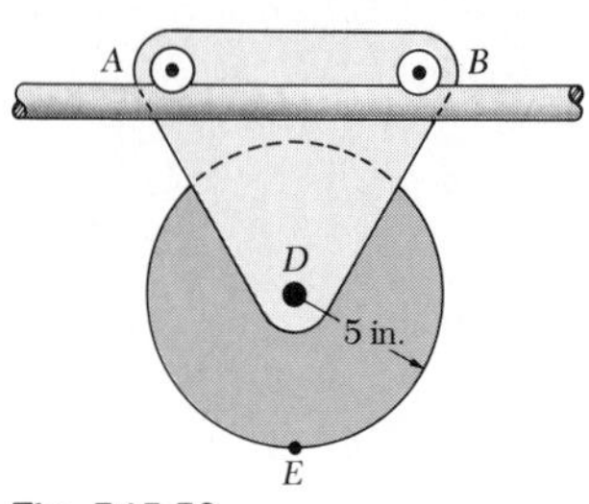

Fig. P15.56

15.56 The trolley shown moves to the left along a horizontal pipe support at a speed of 24 in./s. Knowing that the 5-in.-radius disk has an angular velocity of 8 rad/s counterclockwise, determine (a) the instantaneous center of rotation of the disk, (b) the velocity of point E.

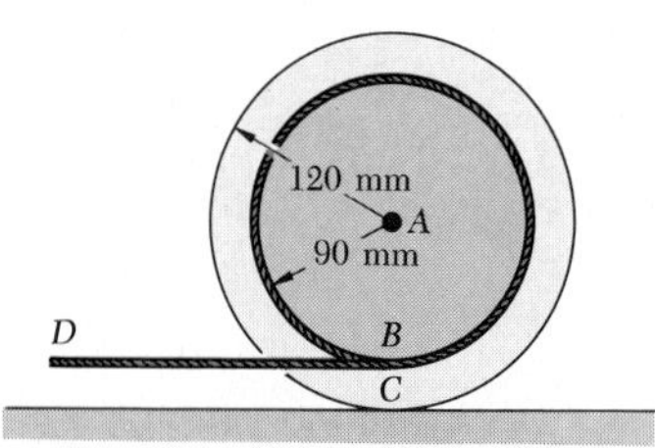

Fig. P15.57

15.57 A drum of radius 90 mm is mounted on a cylinder of radius 120 mm. A cord is wound around the drum, and its extremity D is pulled to the left at a constant velocity of 150 mm/s, causing the cylinder to roll without sliding. Determine (a) the angular velocity of the cylinder, (b) the velocity of the center of the cylinder, (c) the length of cord which is wound or unwound per second.

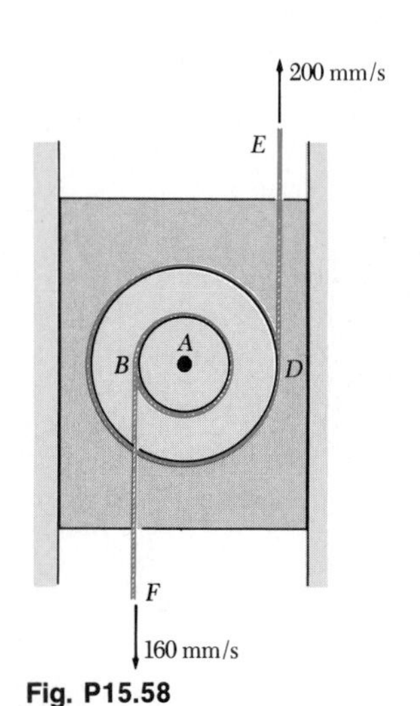

Fig. P15.58

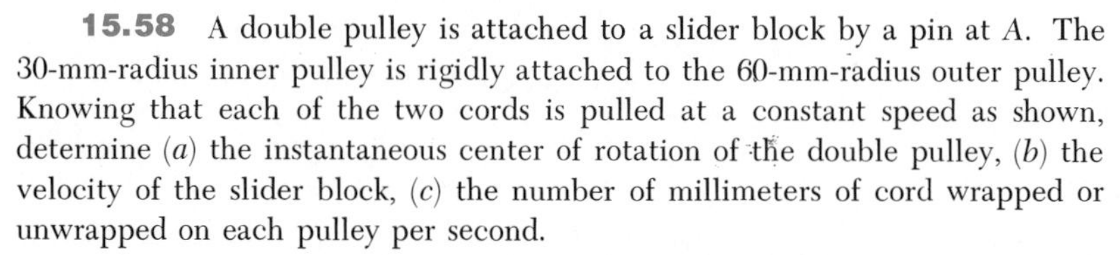

15.58 A double pulley is attached to a slider block by a pin at A. The 30-mm-radius inner pulley is rigidly attached to the 60-mm-radius outer pulley. Knowing that each of the two cords is pulled at a constant speed as shown, determine (a) the instantaneous center of rotation of the double pulley, (b) the velocity of the slider block, (c) the number of millimeters of cord wrapped or unwrapped on each pulley per second.

15.59 Knowing that at the instant shown the angular velocity of rod *BE* is 3 rad/s counterclockwise, determine (*a*) the angular velocity of rod *AD*, (*b*) the velocity of collar *D*, (*c*) the velocity of point *A*.

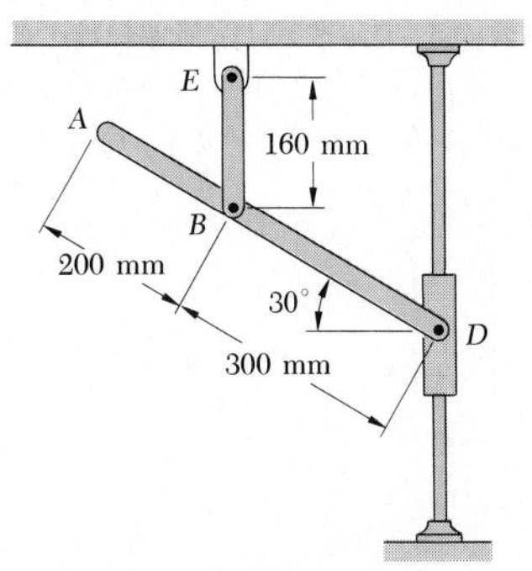

Fig. P15.59 and P15.60

15.60 Knowing that at the instant shown the velocity of collar *D* is 1.8 m/s upward, determine (*a*) the angular velocity of rod *AD*, (*b*) the velocity of point *B*, (*c*) the velocity of point *A*.

15.61 Knowing that at the instant shown the velocity of collar *D* is 48 in./s upward, determine (*a*) the instantaneous center of rotation of link *BD*, (*b*) the angular velocities of crank *AB* and link *BD*, (*c*) the velocity of the midpoint of link *BD*.

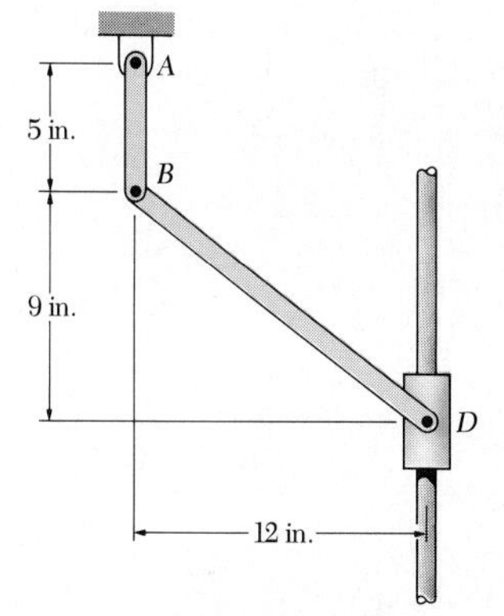

Fig. P15.61 and P15.62

15.62 Knowing that at the instant shown the angular velocity of crank *AB* is 2.7 rad/s clockwise, determine (*a*) the angular velocity of link *BD*, (*b*) the velocity of collar *D*, (*c*) the velocity of the midpoint of link *BD*.

15.63 The rod *BDE* is partially guided by a wheel at *D* which rolls in a vertical track. Knowing that at the instant shown the angular velocity of crank *AB* is 5 rad/s clockwise and that $\beta = 30°$, determine (*a*) the angular velocity of the rod, (*b*) the velocity of point *E*.

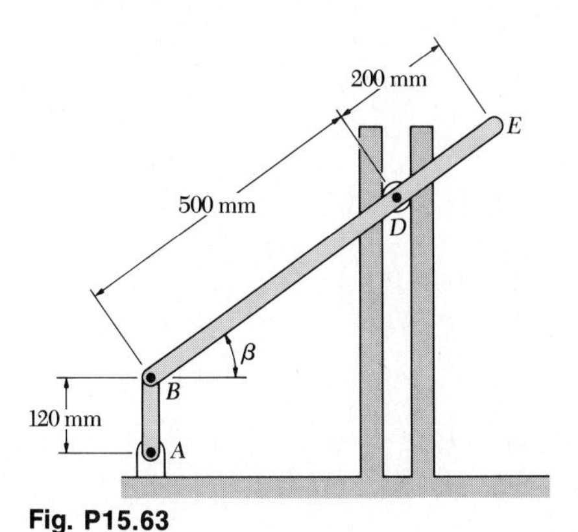

Fig. P15.63

15.64 Solve Prob. 15.63, assuming that $\beta = 40°$.

15.65 Two links AB and BD, each 500 mm long, are connected at B and guided by hydraulic cylinders attached at A and D. Knowing that D is stationary and that the velocity of A is 1.5 m/s to the right, determine at the instant shown (*a*) the angular velocity of each link, (*b*) the velocity of B.

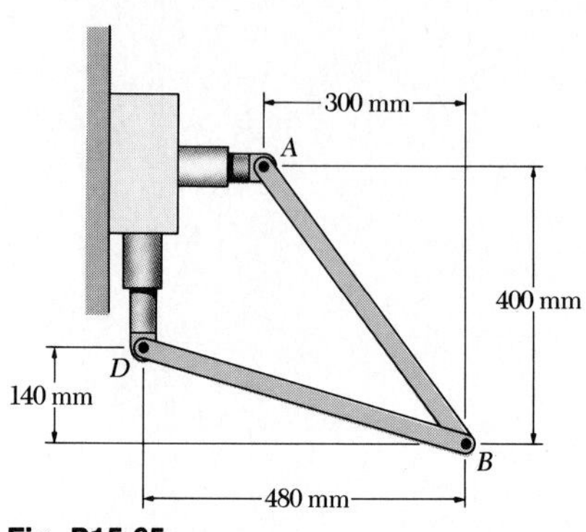

Fig. P15.65

15.66 Solve Prob. 15.65, assuming that A is stationary and that the velocity of D is 1.5 m/s downward.

15.67 Two slots have been cut in the plate FG and the plate has been placed so that the slots fit two fixed pins A and B. Knowing that at the instant shown the angular velocity of crank DE is 8 rad/s clockwise, determine (*a*) the velocity of point F, (*b*) the velocity of point G.

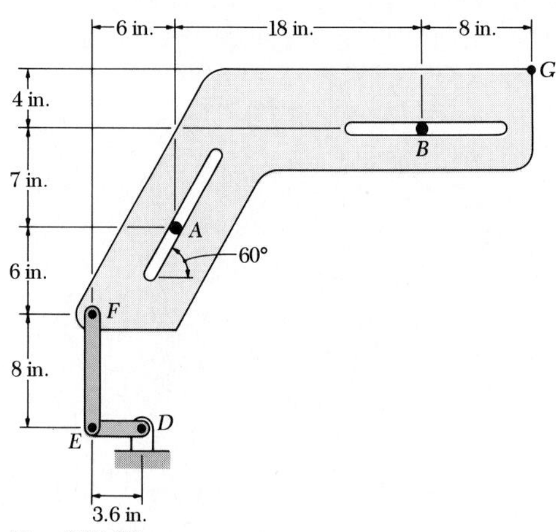

Fig. P15.67

15.68 In Prob. 15.67, determine the velocity of the point of the plate which is in contact with (*a*) pin A, (*b*) pin B.

15.69 Two links AB and BD, each of length 30 in., are connected to three collars. At the instant shown $\beta = 70°$ and the velocity of collar A is 90 in./s to the right. Determine (*a*) the corresponding value of γ, (*b*) the velocity of collar D.

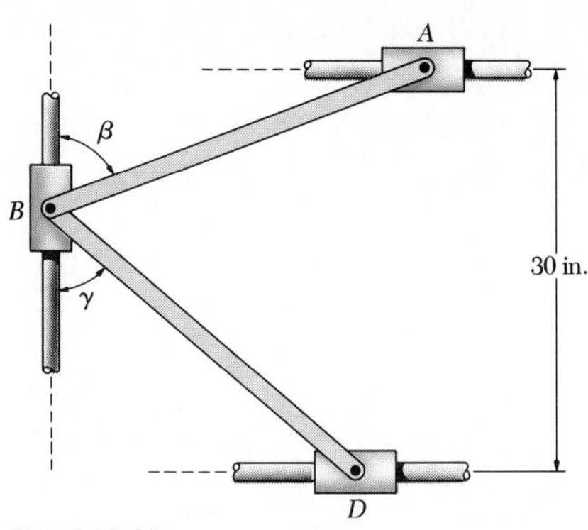

Fig. P15.69

15.70 Solve Prob. 15.69, assuming that $\beta = 55°$.

15.71 and 15.72 Two 500-mm rods are pin-connected at D as shown. Knowing that B moves to the right with a constant velocity of 480 mm/s, determine at the instant shown (*a*) the angular velocity of each rod, (*b*) the velocity of E.

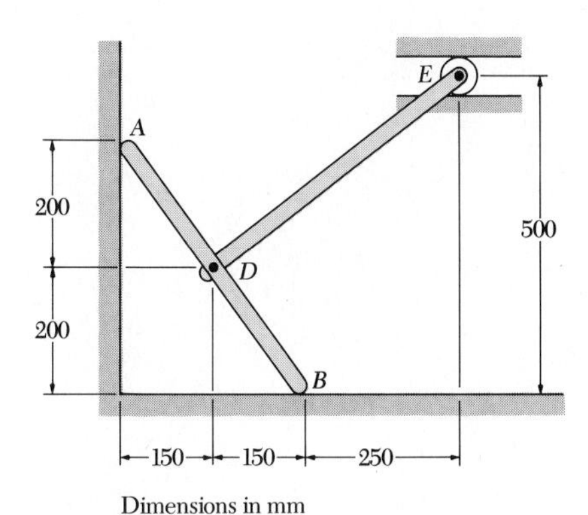

Fig. P15.71

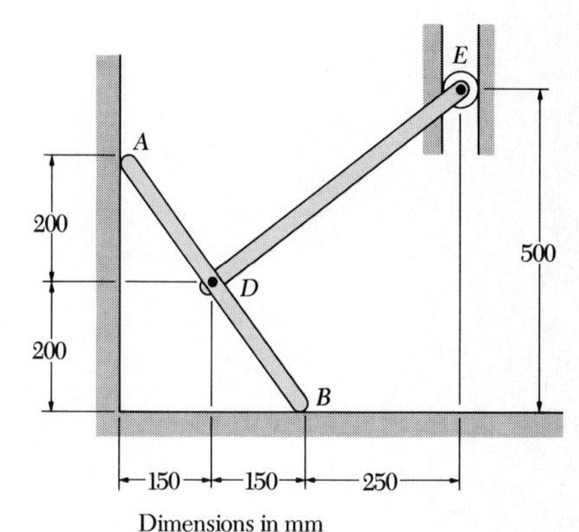

Fig. P15.72

15.73 Describe the space centrode and the body centrode of link AB of Prob. 15.69 as collar B moves upward. (*Note.* The body centrode need not lie on a physical portion of the link.)

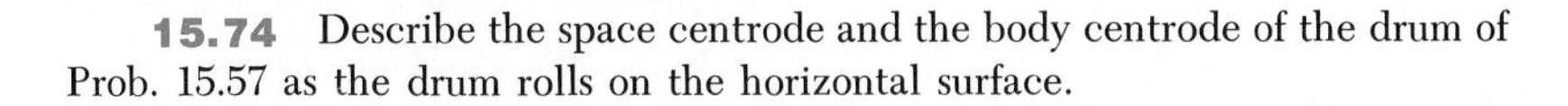

15.74 Describe the space centrode and the body centrode of the drum of Prob. 15.57 as the drum rolls on the horizontal surface.

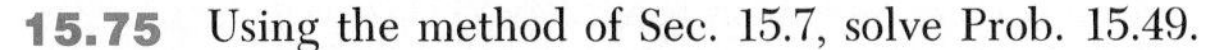

15.75 Using the method of Sec. 15.7, solve Prob. 15.49.
15.76 Using the method of Sec. 15.7, solve Prob. 15.50.
15.77 Using the method of Sec. 15.7, solve Prob. 15.51.
15.78 Using the method of Sec. 15.7, solve Prob. 15.52.

15.8. Absolute and Relative Acceleration in Plane Motion. We saw in Sec. 15.5 that any plane motion may be replaced by a translation defined by the motion of an arbitrary reference point A, and by a rotation about A. This property was used in Sec. 15.6 to determine the velocity of the various points of a moving slab. We shall now use the same property to determine the acceleration of the points of the slab.

We first recall that the absolute acceleration $\mathbf{a}_B$ of a particle of the slab may be obtained from the relative-acceleration formula derived in Sec. 11.11,

$$\mathbf{a}_B = \mathbf{a}_A + \mathbf{a}_{B/A} \tag{15.18}$$

where the right-hand member represents a vector sum. The acceleration $\mathbf{a}_A$ corresponds to the translation of the slab with A, while the relative acceleration $\mathbf{a}_{B/A}$ is associated with the rotation of the slab about A and is measured with respect to axes of fixed orientation centered at A (Fig. 15.18). Denoting by r the distance from A to B and, respectively, by $\boldsymbol{\omega}$ and $\boldsymbol{\alpha}$ the angular velocity and angular acceleration of the slab with respect to

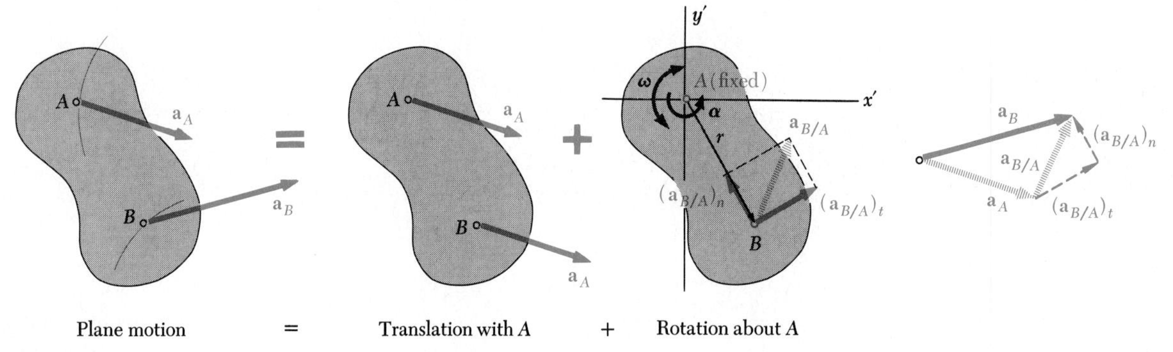

Fig. 15.18

axes of fixed orientation, we find that the relative acceleration $\mathbf{a}_{B/A}$ of B with respect to A consists of two components, a *normal component* $(\mathbf{a}_{B/A})_n$ of magnitude $r\omega^2$ directed toward A, and a *tangential component* $(\mathbf{a}_{B/A})_t$ of magnitude $r\alpha$ perpendicular to the line AB.

As an example, we shall consider again the rod AB whose extremities slide, respectively, along a horizontal and a vertical track (Fig. 15.19). Assuming that the velocity $\mathbf{v}_A$ and the acceleration $\mathbf{a}_A$ of A are known, we propose to determine the acceleration $\mathbf{a}_B$ of B and the angular acceleration $\boldsymbol{\alpha}$ of the rod. Choosing A as a reference point, we express that the given motion is equivalent to a translation with A and a rotation about A. The

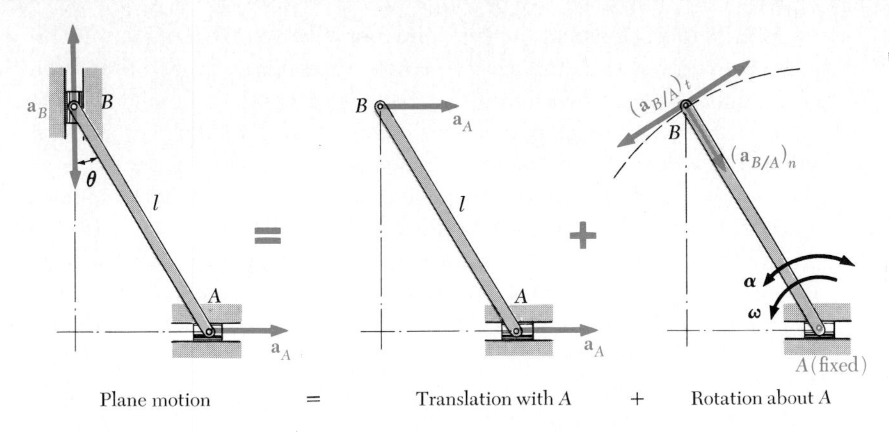

Fig. 15.19

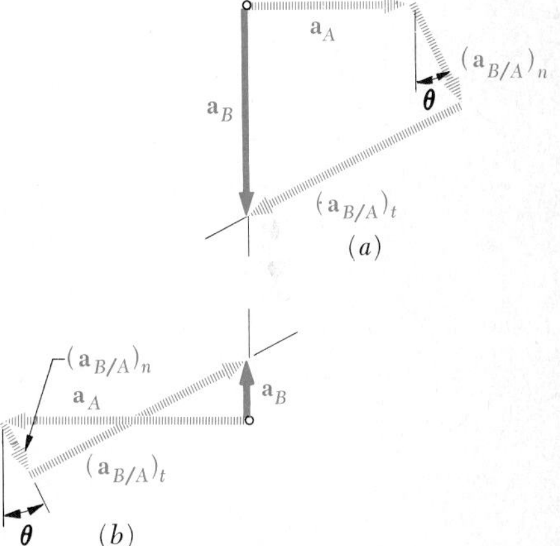

absolute acceleration of B must be equal to the sum

$$
\begin{aligned}
\mathbf{a}_B &= \mathbf{a}_A + \mathbf{a}_{B/A} \\
&= \mathbf{a}_A + (\mathbf{a}_{B/A})_n + (\mathbf{a}_{B/A})_t
\end{aligned}
\tag{15.19}
$$

where $(\mathbf{a}_{B/A})_n$ has the magnitude $l\omega^2$ and is *directed toward* A, while $(\mathbf{a}_{B/A})_t$ has the magnitude $l\alpha$ and is perpendicular to AB. There is no way of telling at the present time whether the tangential component $(\mathbf{a}_{B/A})_t$ is directed to the left or to the right, and students should not rely on their "intuition" in this matter. We shall therefore indicate both possible directions for this component in Fig. 15.19. Similarly, we indicate both possible senses for $\mathbf{a}_B$, since we do not know whether point B is accelerated upward or downward.

Equation (15.19) has been expressed geometrically in Fig. 15.20. Four different vector polygons may be obtained, depending upon the sense of $\mathbf{a}_A$ and the relative magnitude of a_A and $(a_{B/A})_n$. If we are to determine a_B and α from one of these diagrams, we must know not only a_A and θ but also ω. The angular velocity of the rod should therefore be separately determined by one of the methods indicated in Secs. 15.6 and 15.7. The values of a_B and α may then be obtained by considering successively the x and y components of the vectors shown in Fig. 15.20. In the case of polygon a, for example, we write

$$\xrightarrow{+} x \text{ components:} \qquad 0 = a_A + l\omega^2 \sin\theta - l\alpha\cos\theta$$

$$+\uparrow y \text{ components:} \qquad -a_B = -l\omega^2\cos\theta - l\alpha\sin\theta$$

and solve for a_B and α. The two unknowns may also be obtained by direct measurement on the vector polygon. In that case, care should be taken to draw first the known vectors $\mathbf{a}_A$ and $(\mathbf{a}_{B/A})_n$.

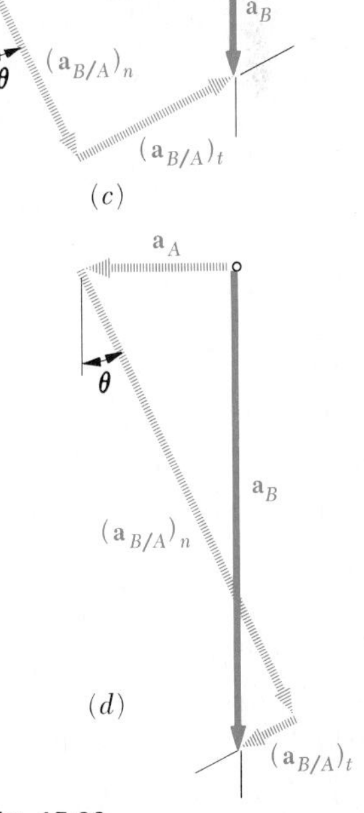

Fig. 15.20

It is quite evident that the determination of accelerations is considerably more involved than the determination of velocities. Yet in the example considered here, the extremities A and B of the rod were moving along straight tracks, and the diagrams drawn were relatively simple. If A and B had moved along curved tracks, the accelerations $\mathbf{a}_A$ and $\mathbf{a}_B$ should have been resolved into normal and tangential components and the solution of the problem would have involved six different vectors.

When a mechanism consists of several moving parts which are pin-connected, its analysis may be carried out by considering each part as a rigid body, while keeping in mind that the points where two parts are connected must have the same absolute acceleration (see Sample Prob. 15.7). In the case of meshed gears, the tangential components of the accelerations of the teeth in contact are equal, but their normal components are different.

***15.9. Analysis of Plane Motion in Terms of a Parameter.** In the case of certain mechanisms, it is possible to express the coordinates x and y of all the significant points of the mechanism by means of simple analytic expressions containing a single parameter. It may be advantageous in such a case to determine directly the absolute velocity and the absolute acceleration of the various points of the mechanism, since the components of the velocity and of the acceleration of a given point may be obtained by differentiating the coordinates x and y of that point.

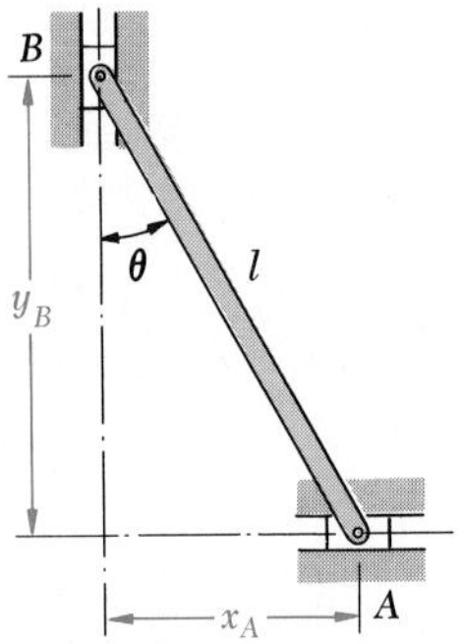

Let us consider again the rod AB whose extremities slide, respectively, in a horizontal and a vertical track (Fig. 15.21). The coordinates x_A and y_B of the extremities of the rod may be expressed in terms of the angle θ the rod forms with the vertical:

$$x_A = l \sin\theta \qquad y_B = l\cos\theta \tag{15.20}$$

Fig. 15.21

Differentiating Eqs. (15.20) twice with respect to t, we write

$$\begin{aligned} v_A &= \dot{x}_A = l\dot{\theta}\cos\theta \\ a_A &= \ddot{x}_A = -l\dot{\theta}^2\sin\theta + l\ddot{\theta}\cos\theta \end{aligned}$$

$$\begin{aligned} v_B &= \dot{y}_B = -l\dot{\theta}\sin\theta \\ a_B &= \ddot{y}_B = -l\dot{\theta}^2\cos\theta - l\ddot{\theta}\sin\theta \end{aligned}$$

Recalling that $\dot{\theta} = \omega$ and $\ddot{\theta} = \alpha$, we obtain

$$v_A = l\omega\cos\theta \qquad v_B = -l\omega\sin\theta \tag{15.21}$$

$$a_A = -l\omega^2\sin\theta + l\alpha\cos\theta \qquad a_B = -l\omega^2\cos\theta - l\alpha\sin\theta \tag{15.22}$$

We note that a positive sign for v_A or a_A indicates that the velocity $\mathbf{v}_A$ or the acceleration $\mathbf{a}_A$ is directed to the right; a positive sign for v_B or a_B indicates that $\mathbf{v}_B$ or $\mathbf{a}_B$ is directed upward. Equations (15.21) may be used, for example, to determine v_B and ω when v_A and θ are known. Substituting for ω in (15.22), we may then determine a_B and α if a_A is known.

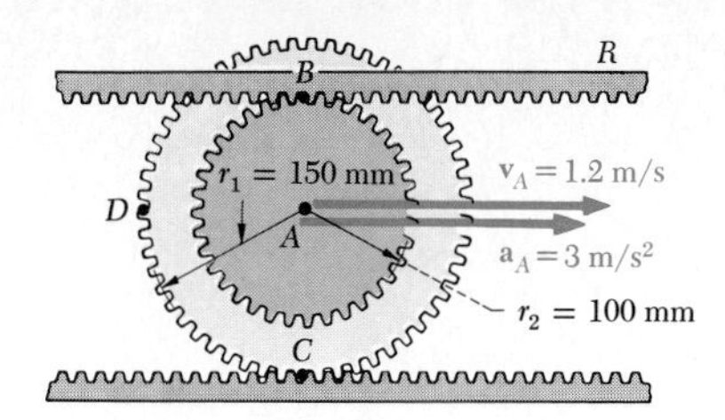

SAMPLE PROBLEM 15.6

The center of the double gear of Sample Prob. 15.2 has a velocity of 1.2 m/s to the right and an acceleration of 3 m/s² to the right. Recalling that the lower rack is stationary, determine (*a*) the angular acceleration of the gear, (*b*) the acceleration of points *B*, *C*, and *D* of the gear.

***a.* Angular Acceleration of the Gear.** In Sample Prob. 15.2, we found that $x_A = r_1\theta$ and $v_A = r_1\omega$. Differentiating the latter with respect to time, we obtain $a_A = r_1\alpha$.

$$v_A = r_1\omega \qquad 1.2 \text{ m/s} = (0.150 \text{ m})\omega \qquad \omega = 8 \text{ rad/s} \downarrow$$
$$a_A = r_1\alpha \qquad 3 \text{ m/s}^2 = (0.150 \text{ m})\alpha \qquad \alpha = 20 \text{ rad/s}^2 \downarrow \blacktriangleleft$$

***b.* Accelerations.** The rolling motion of the gear is resolved into a translation with *A* and a rotation about *A*.

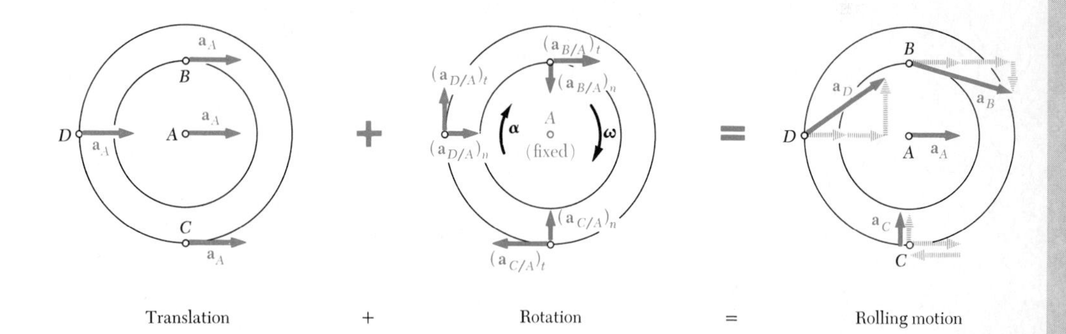

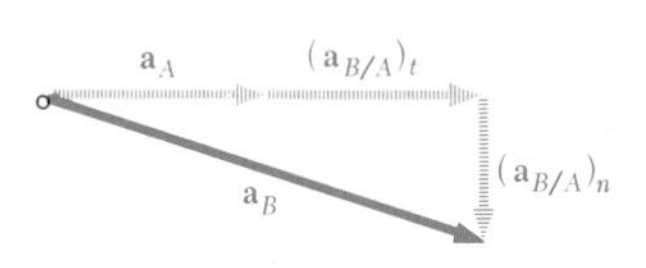

Acceleration of Point B. Adding vectorially the accelerations corresponding to the translation and to the rotation, we obtain

$$\mathbf{a}_B = \mathbf{a}_A + \mathbf{a}_{B/A} = \mathbf{a}_A + (\mathbf{a}_{B/A})_t + (\mathbf{a}_{B/A})_n$$

where

$$\mathbf{a}_A = 3 \text{ m/s}^2 \rightarrow$$

$$(a_{B/A})_t = r_2\alpha = (0.100 \text{ m})(20 \text{ rad/s}^2) \qquad (\mathbf{a}_{B/A})_t = 2 \text{ m/s}^2 \rightarrow$$
$$(a_{B/A})_n = r_2\omega^2 = (0.100 \text{ m})(8 \text{ rad/s})^2 \qquad (\mathbf{a}_{B/A})_n = 6.4 \text{ m/s}^2 \downarrow$$

We have $\quad \mathbf{a}_B = [3 \text{ m/s}^2 \rightarrow] + [2 \text{ m/s}^2 \rightarrow] + [6.4 \text{ m/s}^2 \downarrow]$

$$\mathbf{a}_B = 8.12 \text{ m/s}^2 \searrow 52.0° \blacktriangleleft$$

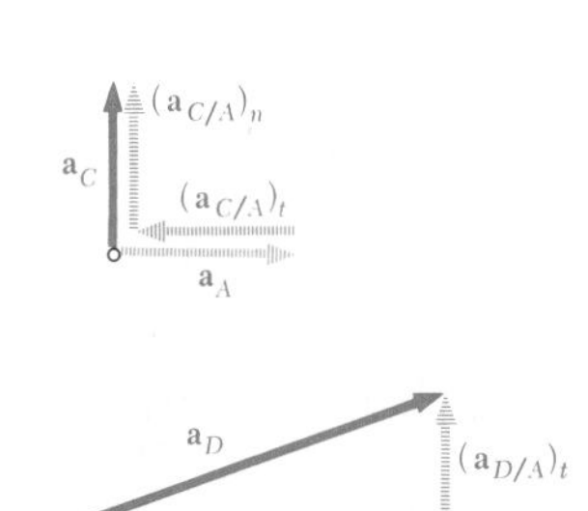

Acceleration of Point C

$$\mathbf{a}_C = \mathbf{a}_A + \mathbf{a}_{C/A} = \mathbf{a}_A + (\mathbf{a}_{C/A})_t + (\mathbf{a}_{C/A})_n$$
$$= [3 \text{ m/s}^2 \rightarrow] + [(0.150 \text{ m})(20 \text{ rad/s}^2) \leftarrow] + [(0.150 \text{ m})(8 \text{ rad/s})^2 \uparrow]$$
$$\mathbf{a}_C = 9.60 \text{ m/s}^2 \uparrow \blacktriangleleft$$

Acceleration of Point D

$$\mathbf{a}_D = \mathbf{a}_A + \mathbf{a}_{D/A} = \mathbf{a}_A + (\mathbf{a}_{D/A})_t + (\mathbf{a}_{D/A})_n$$
$$= [3 \text{ m/s}^2 \rightarrow] + [(0.150 \text{ m})(20 \text{ rad/s}^2) \uparrow] + [(0.150 \text{ m})(8 \text{ rad/s})^2 \rightarrow]$$
$$\mathbf{a}_D = 12.95 \text{ m/s}^2 \angle 13.4° \blacktriangleleft$$

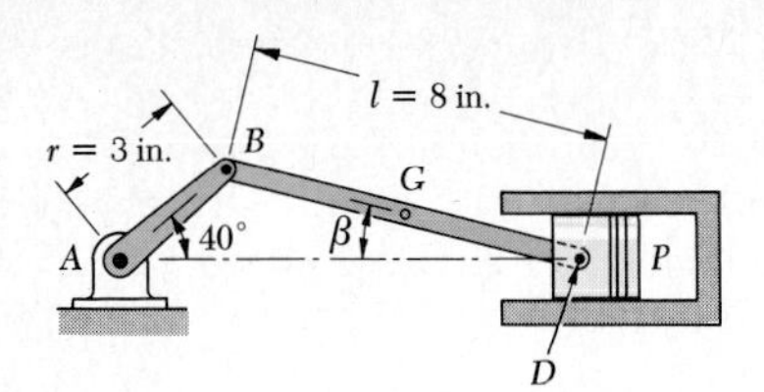

SAMPLE PROBLEM 15.7

Crank AB of the engine system of Sample Prob. 15.3 has a constant clockwise angular velocity of 2000 rpm. For the crank position shown, determine the angular acceleration of the connecting rod BD and the acceleration of point D.

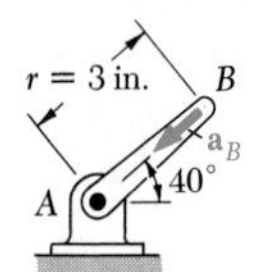

Motion of Crank AB. Since the crank rotates about A with constant $\omega_{AB} = 2000 \text{ rpm} = 209.4 \text{ rad/s}$, we have $\alpha_{AB} = 0$. The acceleration of B is therefore directed toward A and has a magnitude

$$a_B = r\omega_{AB}^2 = (\tfrac{3}{12}\text{ ft})(209.4 \text{ rad/s})^2 = 10{,}962 \text{ ft/s}^2$$
$$\mathbf{a}_B = 10{,}962 \text{ ft/s}^2 \measuredangle 40°$$

Motion of the Connecting Rod BD. The angular velocity $\boldsymbol{\omega}_{BD}$ and the value of β were obtained in Sample Prob. 15.3:

$$\boldsymbol{\omega}_{BD} = 62.0 \text{ rad/s} \circlearrowleft \qquad \beta = 13.95°$$

The motion of BD is resolved into a translation with B and a rotation about B. The relative acceleration $\mathbf{a}_{D/B}$ is resolved into normal and tangential components:

$$(a_{D/B})_n = (BD)\omega_{BD}^2 = (\tfrac{8}{12}\text{ ft})(62.0 \text{ rad/s})^2 = 2563 \text{ ft/s}^2$$
$$(\mathbf{a}_{D/B})_n = 2563 \text{ ft/s}^2 \measuredangle 13.95°$$
$$(a_{D/B})_t = (BD)\alpha_{BD} = (\tfrac{8}{12})\alpha_{BD} = 0.6667\alpha_{BD}$$
$$(\mathbf{a}_{D/B})_t = 0.6667\alpha_{BD} \measuredangle 76.05°$$

While $(\mathbf{a}_{D/B})_t$ must be perpendicular to BD, its sense is not known.

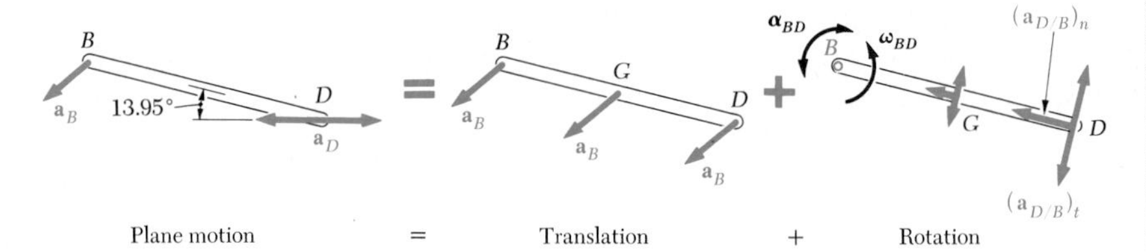

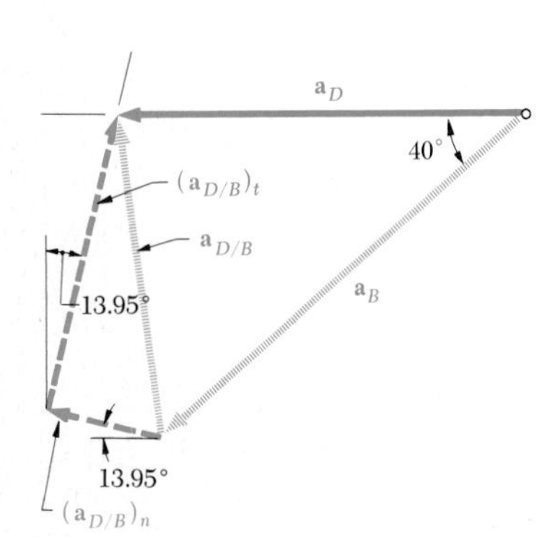

Noting that the acceleration $\mathbf{a}_D$ must be horizontal, we write

$$\mathbf{a}_D = \mathbf{a}_B + \mathbf{a}_{D/B} = \mathbf{a}_B + (\mathbf{a}_{D/B})_n + (\mathbf{a}_{D/B})_t$$
$$[a_D \leftrightarrow] = [10{,}962 \measuredangle 40°] + [2563 \measuredangle 13.95°] + [0.6667\alpha_{BD} \measuredangle 76.05°]$$

Equating x and y components, we obtain the following scalar equations:

$\xrightarrow{+}$ x components:
$$-a_D = -10{,}962 \cos 40° - 2563 \cos 13.95° + 0.6667\alpha_{BD} \sin 13.95°$$
$+\uparrow y$ components:
$$0 = -10{,}962 \sin 40° + 2563 \sin 13.95° + 0.6667\alpha_{BD} \cos 13.95°$$

Solving the equations simultaneously, we obtain $\alpha_{BD} = +9940 \text{ rad/s}^2$ and $a_D = +9290 \text{ ft/s}^2$. The positive signs indicate that the senses shown on the vector polygon are correct; we write

$$\boldsymbol{\alpha}_{BD} = 9940 \text{ rad/s}^2 \circlearrowleft \blacktriangleleft$$
$$\mathbf{a}_D = 9290 \text{ ft/s}^2 \leftarrow \blacktriangleleft$$

Problems

15.79 A 4-m steel beam is lowered by means of two cables unwinding at the same speed from overhead cranes. As the beam approaches the ground, the crane operators apply brakes to slow down the unwinding motion. At the instant considered the deceleration of the cable attached at A is 5 m/s², while that of the cable attached at B is 2 m/s². Determine (*a*) the angular acceleration of the beam, (*b*) the acceleration of point C.

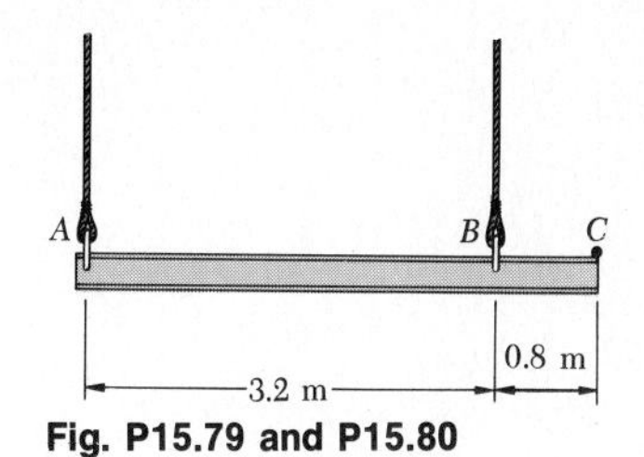

Fig. P15.79 and P15.80

15.80 The acceleration of point C is 0.4 m/s² downward and the angular acceleration of the beam is 0.75 rad/s² clockwise. Knowing that the angular velocity of the beam is zero at the instant considered, determine the acceleration of each cable.

15.81 A tape is wrapped around a 10-in.-diameter disk which is at rest on a horizontal table. A force **P** applied as shown produces the following accelerations: $\mathbf{a}_A = 30$ in./s² to the right, $\boldsymbol{\alpha} = 4$ rad/s² counterclockwise as viewed from above. Determine the acceleration (*a*) of point B, (*b*) of point C.

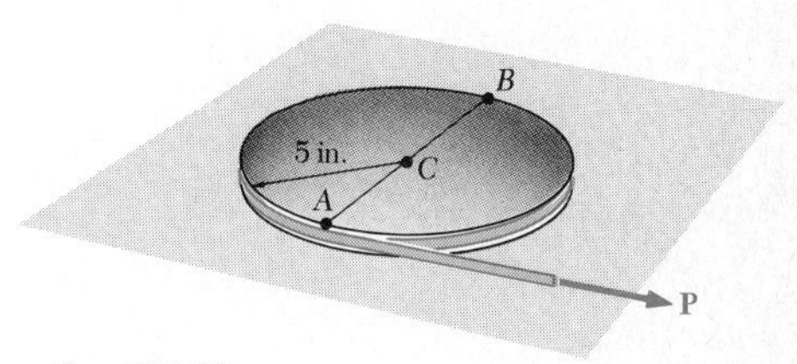

Fig. P15.81

15.82 In Prob. 15.81, determine the point of the plate which (*a*) has no acceleration, (*b*) has an acceleration of 22 in./s² to the right.

15.83 A drum of radius 90 mm is mounted on a cylinder of radius 120 mm. A cord is wound around the drum and is pulled in such a way that point D has a velocity of 90 mm/s and an acceleration of 450 mm/s², both directed to the left. Assuming that the cylinder rolls without slipping, determine the acceleration (*a*) of point A, (*b*) of point B, (*c*) of point C.

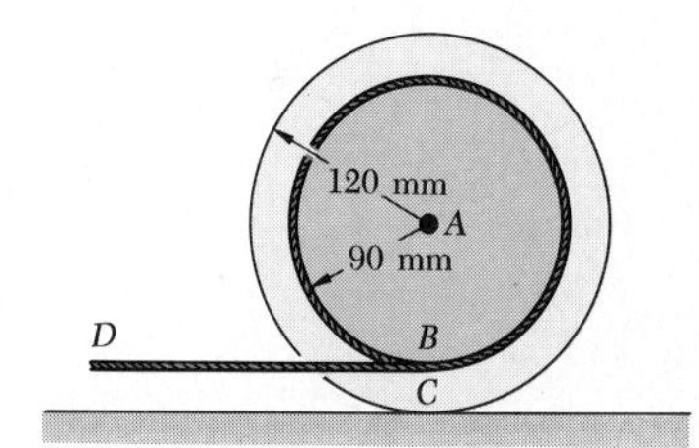

Fig. P15.83

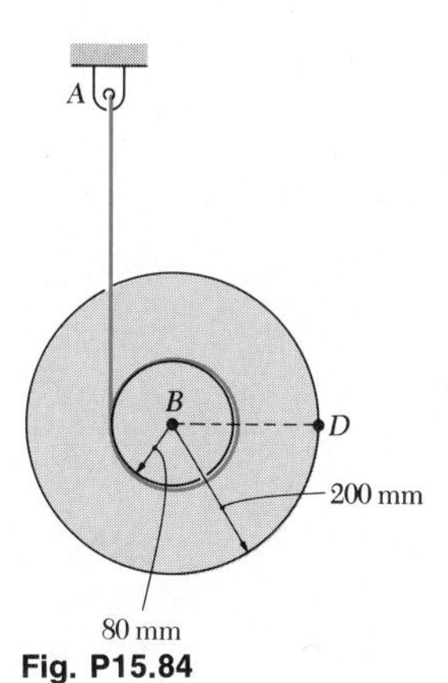

Fig. P15.84

15.84 At the instant shown the center B of the double pulley has a velocity of 0.6 m/s and an acceleration of 2.4 m/s², both directed downward. Knowing that the cord wrapped around the inner pulley is attached to a fixed support at A, determine the acceleration of point D.

15.85 An automobile travels to the right at a constant speed of 45 mi/h. Knowing that the diameter of a wheel is 22 in., determine the acceleration (*a*) of point *C*, (*b*) of point *D*.

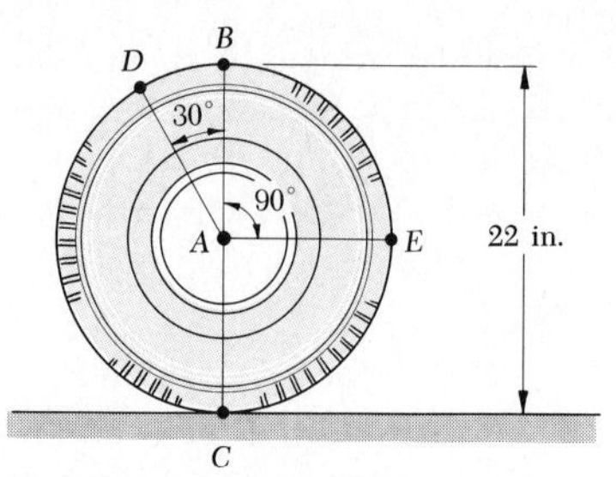

Fig. P15.85 and P15.86

15.86 An automobile is started from rest and has a constant acceleration of 7 ft/s^2 to the right. Knowing that the wheel shown rolls without sliding, determine the speed of the automobile at which the magnitude of the acceleration of point *B* of the wheel is 50 ft/s^2.

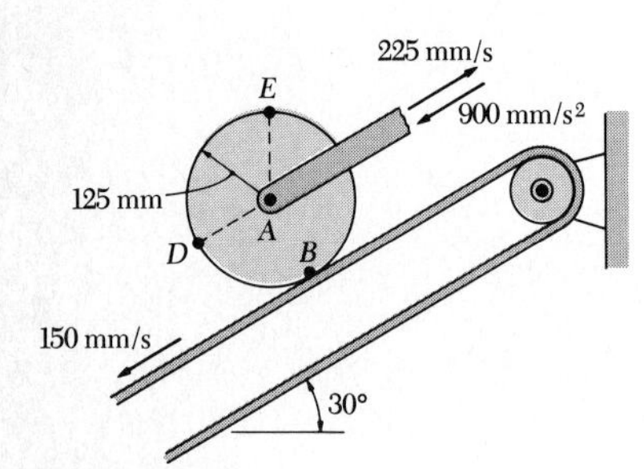

Fig. P15.87

15.87 The 125-mm-radius drum rolls without slipping on a portion of a belt which moves downward to the left with a constant velocity of 150 mm/s. Knowing that at a given instant the velocity and acceleration of the center *A* of the drum are as shown, determine the acceleration of point *D*.

15.88 In Prob. 15.87, determine the acceleration of point *E*.

15.89 Crank *AB* has a constant angular velocity of 200 rpm counterclockwise. Determine the acceleration of collar *D* when (*a*) $\theta = 0$, (*b*) $\theta = 90°$.

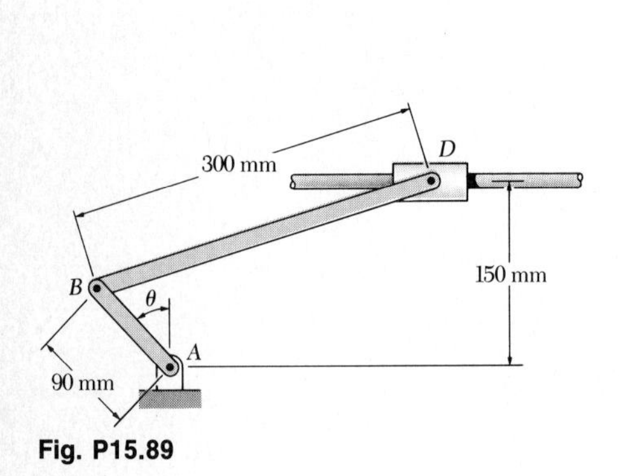

Fig. P15.89

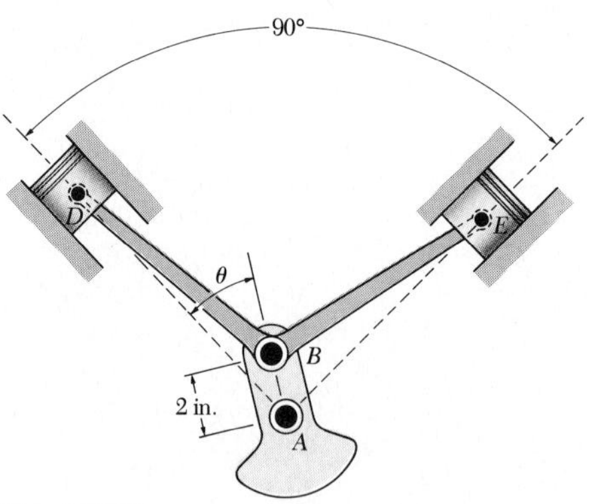

Fig. P15.90

15.90 In the two-cylinder air compressor shown the connecting rods *BD* and *BE* are each 8 in. long and the crank *AB* rotates about the fixed point *A* with a constant angular velocity of 1800 rpm clockwise. Determine the acceleration of each piston when $\theta = 0$.

15.91 In Prob. 15.90, determine the acceleration of piston *E* when $\theta = 45°$.

15.92 Arm AB rotates with a constant angular velocity of 90 rpm clockwise. Knowing that gear A does not rotate, determine the acceleration of the tooth of gear B which is in contact with gear A.

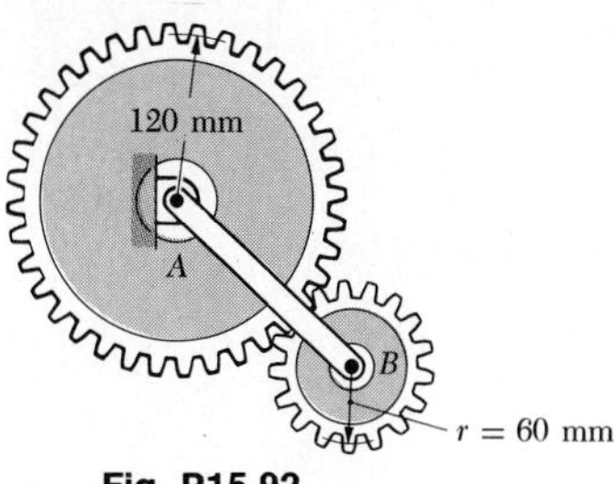

Fig. P15.92

15.93 At the instant shown, the disk rotates with a constant angular velocity ω_0 clockwise. Determine the angular velocities and the angular accelerations of the rods AB and BC.

15.94 At the instant shown, the disk rotates with a constant angular velocity ω_0 clockwise. Determine the angular velocities and the angular accelerations of the rods DE and EF.

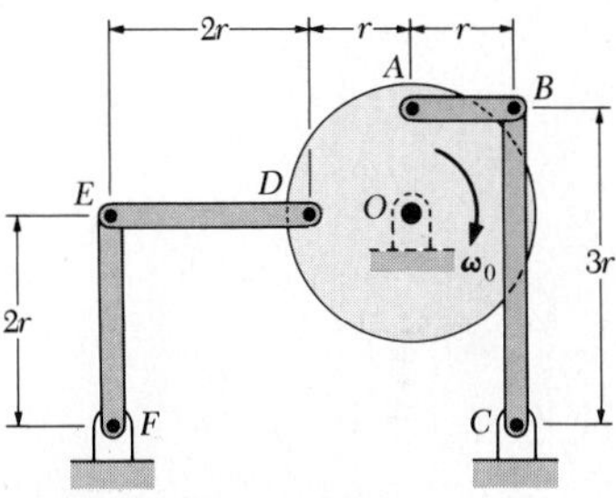

Fig. P15.93 and P15.94

15.95 and 15.96 For the linkage indicated, determine the angular acceleration (a) of bar BD, (b) of bar DE.

15.95 Linkage of Prob. 15.51.
15.96 Linkage of Prob. 15.48.

15.97 End A of rod AB moves to the right with a constant velocity of 2.5 m/s. For the position shown, determine (a) the angular acceleration of rod AB, (b) the acceleration of the midpoint G of rod AB.

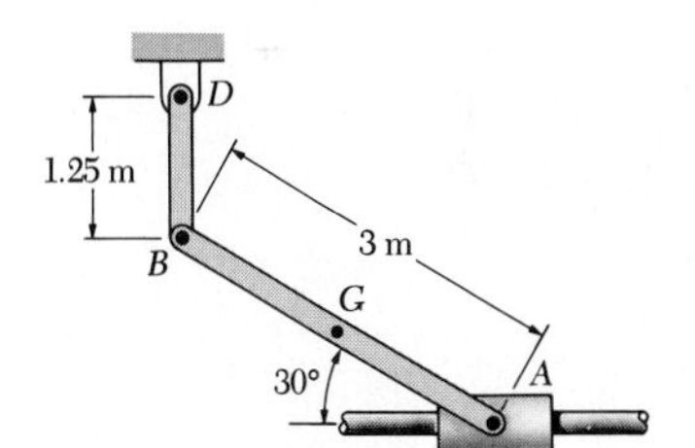

Fig. P15.97 and P15.98

15.98 In the position shown, end A of rod AB has a velocity of 2.5 m/s and an acceleration of 1.5 m/s^2, both directed to the right. Determine (a) the angular acceleration of rod AB, (b) the acceleration of the midpoint G of rod AB.

15.99 In the position shown, end A of rod AB has a velocity of 3 ft/s and an acceleration of 2.5 ft/s^2, both directed to the left. Determine (a) the angular acceleration of rod AB, (b) the acceleration of the midpoint G of rod AB.

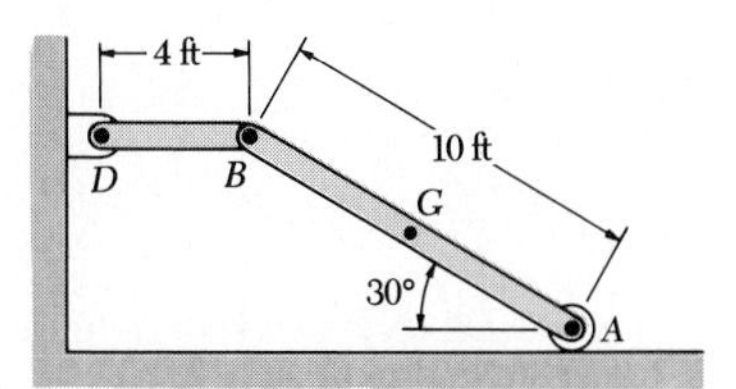

Fig. P15.99 and P15.100

15.100 End A of rod AB moves to the left with a constant velocity of 3 ft/s. For the position shown, determine (a) the angular acceleration of rod AB, (b) the acceleration of the midpoint G of rod AB.

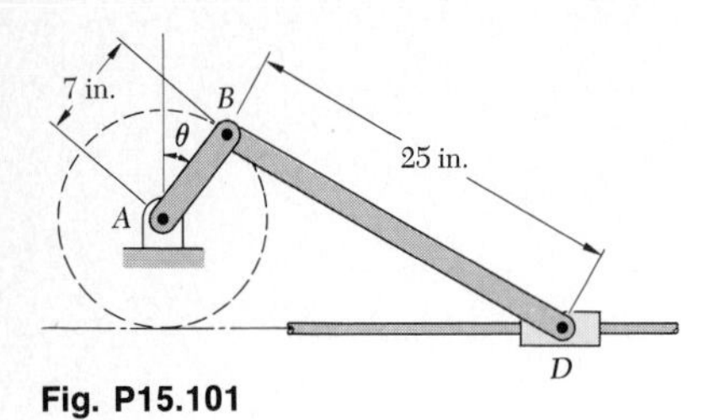

Fig. P15.101

15.101 Crank AB has a constant angular velocity of 8 rad/s clockwise. Determine the acceleration of the midpoint of rod BD when (a) $\theta = 0$, (b) $\theta = 90°$.

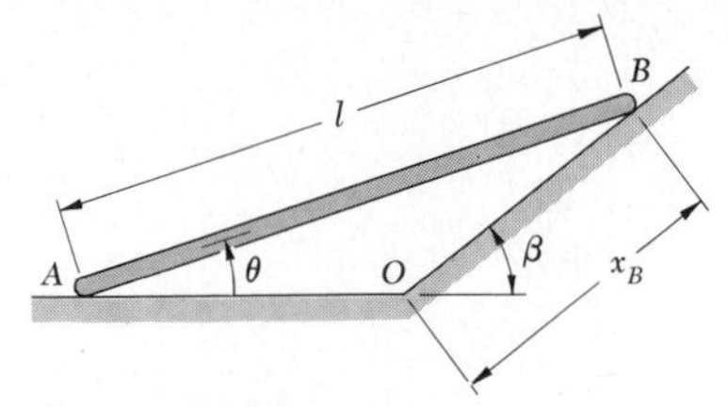

Fig. P15.102

* **15.102** Rod AB slides with its ends in contact with the floor and the inclined plane. Using the method of Sec. 15.9, derive an expression for the angular velocity of the rod in terms of v_B, θ, l, and β.

* **15.103** Derive an expression for the angular acceleration of rod AB of Prob. 15.102 in terms of v_B, θ, l, and β, knowing that the acceleration of point B is zero.

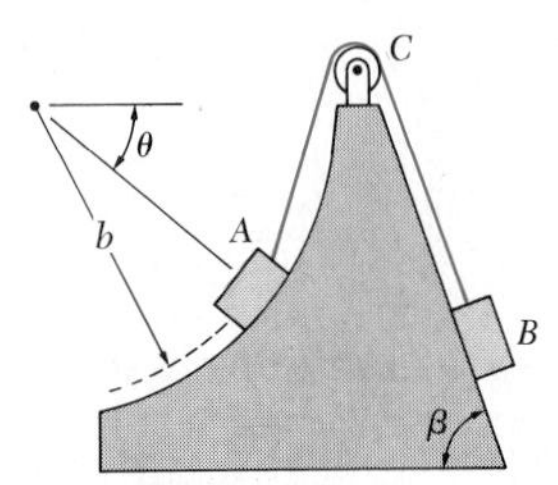

Fig. P15.104

* **15.104** Blocks A and B are connected by a cable AB which passes over a pulley at C. Knowing that block A moves downward along the curved surface with a constant speed v_0, determine (a) the velocity of block B, (b) the acceleration of block B.

* **15.105** The drive disk of the Scotch crosshead mechanism shown has an angular velocity $\boldsymbol{\omega}$ and an angular acceleration $\boldsymbol{\alpha}$, both directed clockwise. Using the method of Sec. 15.9, derive an expression for (a) the velocity of point A, (b) the acceleration of point A.

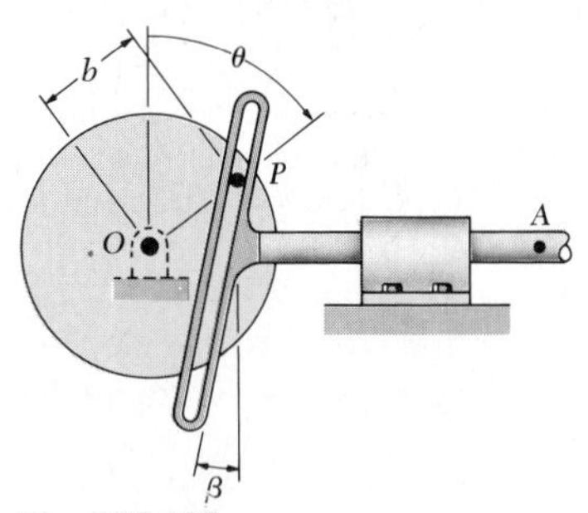

Fig. P15.105

* **15.106** Solve Prob. 15.105, assuming that $\beta = 0$.

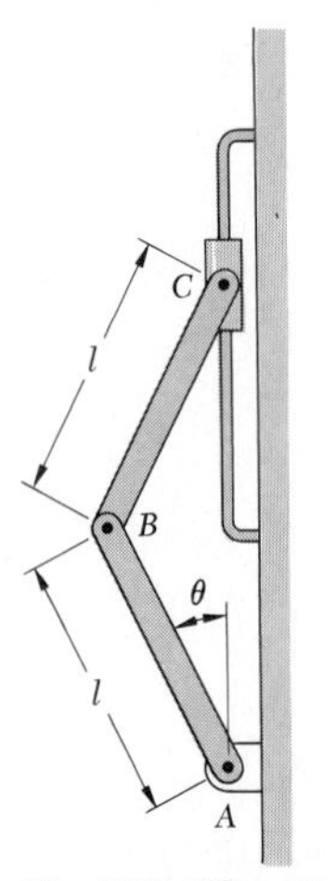

Fig. P15.107 and P15.108

* **15.107** Knowing that rod AB rotates with an angular velocity $\boldsymbol{\omega}$ and with an angular acceleration $\boldsymbol{\alpha}$, both counterclockwise, derive expressions for the velocity and acceleration of collar C.

* **15.108** Knowing that collar C moves upward with a constant velocity $\mathbf{v}_0$, derive expressions for the angular velocity and angular acceleration of rod AB in the position shown.

***15.109** The crank AB rotates with a constant clockwise angular velocity ω, and $\theta = 0$ at $t = 0$. Using the method of Sec. 15.9, derive an expression for the velocity of the piston P in terms of the time t.

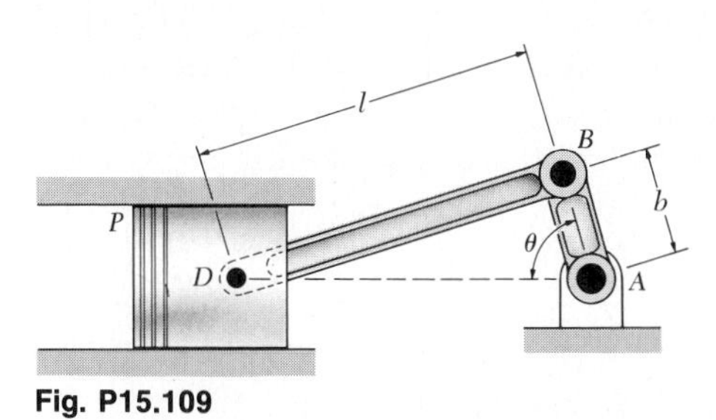

Fig. P15.109

***15.110** Pin C is attached to rod CD and slides in a slot cut in arm AB. Knowing that rod CD moves vertically upward with a constant velocity $\mathbf{v}_0$, derive an expression for (*a*) the angular velocity of arm AB, (*b*) the components of the velocity of point A.

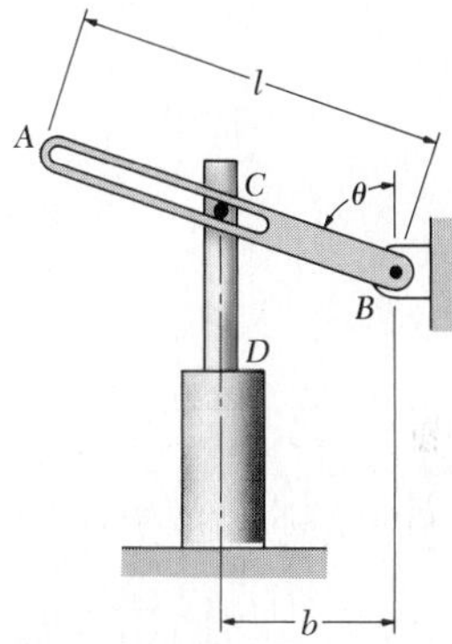

Fig. P15.110

***15.111** In Prob. 15.110, derive an expression for the angular acceleration of arm AB.

***15.112** A disk of radius r rolls to the right with a constant velocity $\mathbf{v}$. Denoting by P the point of the rim in contact with the ground at $t = 0$, derive expressions for the horizontal and vertical components of the velocity of P at any time t. (The curve described by point P is called a *cycloid*.)

***15.113** The position of a factory window is controlled by the rack and pinion shown. Knowing that the pinion C has a radius r and rotates counterclockwise at a constant rate ω, derive an expression for the angular velocity of the window.

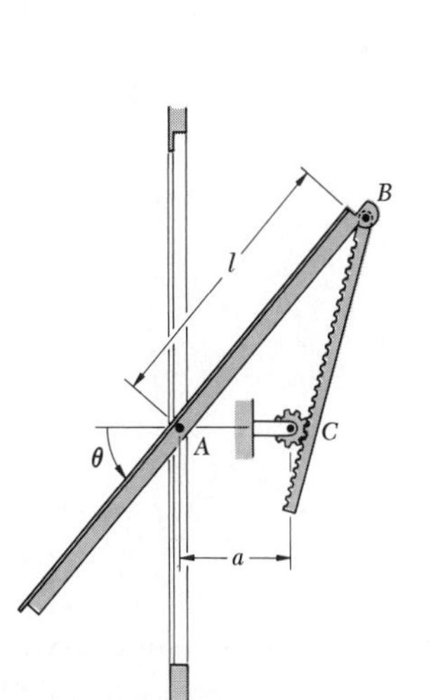

Fig. P15.113

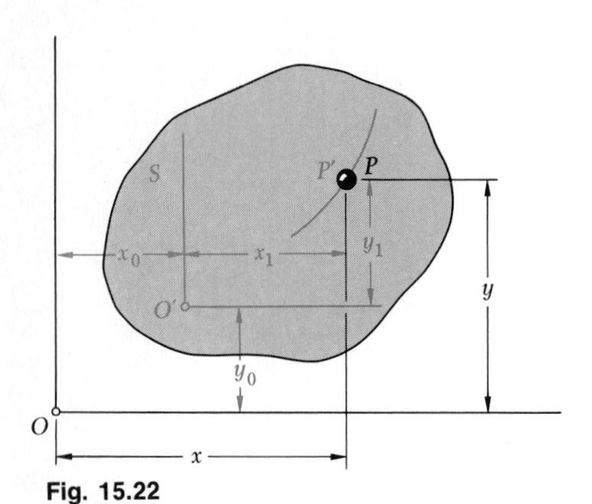

Fig. 15.22

***15.10. Particle Moving on a Slab in Translation.** Let us consider the motion of a particle P which describes a path on a slab S which is itself in translation. The motion of P may be analyzed, either in terms of its coordinates x and y with respect to a fixed set of axes, or in terms of its coordinates x_1 and y_1 with respect to a set of axes attached to the slab S and moving with it (Fig. 15.22). We propose to determine the relation existing between the absolute motion of P with respect to the fixed axes and its relative motion with respect to the axes moving with S.

Denoting by x_0 and y_0 the coordinates of O' with respect to the fixed axes, we write

$$x = x_0 + x_1 \qquad y = y_0 + y_1 \tag{15.23}$$

Differentiating with respect to t and using dots to indicate time derivatives, we have

$$\dot{x} = \dot{x}_0 + \dot{x}_1 \qquad \dot{y} = \dot{y}_0 + \dot{y}_1 \tag{15.24}$$

Now, $\dot{x}$ and $\dot{y}$ represent the components of the absolute velocity $\mathbf{v}_P$ of P, while $\dot{x}_1$ and $\dot{y}_1$ represent the components of the velocity $\mathbf{v}_{P/S}$ of P with respect to S. On the other hand, $\dot{x}_0$ and $\dot{y}_0$ represent the components of the velocity of O' or, since the slab S is in translation, the components of the velocity of any other point of S. We may, for example, consider that $\dot{x}_0$ and $\dot{y}_0$ represent the components of the velocity $\mathbf{v}_{P'}$ of the point P' *of the slab* S *which happens to coincide with the particle P at the instant considered.* We thus write

$$\mathbf{v}_P = \mathbf{v}_{P'} + \mathbf{v}_{P/S} \tag{15.25}$$

where the right-hand member represents a vector sum. We note that the velocity $\mathbf{v}_P$ reduces to $\mathbf{v}_{P/S}$ if the slab is stopped and if the particle is allowed to keep moving on S; it reduces to the velocity $\mathbf{v}_{P'}$ of the coinciding point P' if the particle P is immobilized on S while S is allowed to keep moving. Thus, formula (15.25) expresses that the absolute velocity of P may be obtained by adding vectorially these two partial velocities.

Differentiating Eqs. (15.24) with respect to t, we obtain

$$\ddot{x} = \ddot{x}_0 + \ddot{x}_1 \qquad \ddot{y} = \ddot{y}_0 + \ddot{y}_1 \tag{15.26}$$

or, in vector form,

$$\mathbf{a}_P = \mathbf{a}_{P'} + \mathbf{a}_{P/S} \tag{15.27}$$

where $\mathbf{a}_{P/S}$ = acceleration of P with respect to S
$\mathbf{a}_{P'}$ = acceleration of point P' of S coinciding with P at the instant considered

Formula (15.27) may be interpreted in the same way as formula (15.25).

Formulas (15.25) and (15.27) actually restate the results obtained in Sec. 11.11. Since S is in translation, the x_1 and y_1 axes, respectively, remain parallel to the fixed x and y axes, and the velocity $\mathbf{v}_{P/S}$ is equal to the velocity $\mathbf{v}_{P/P'}$ relative to point P' as it was defined in Sec. 11.11; similarly, the acceleration $\mathbf{a}_{P/S}$ is equal to the acceleration $\mathbf{a}_{P/P'}$ relative to P'.

*15.11. Particle Moving on a Rotating Slab. Coriolis Acceleration.

We shall now consider the motion of a particle P which describes a path on a slab S which is itself rotating about a fixed point O. The motion of P may be analyzed either in terms of its polar coordinates r and θ with respect to fixed axes or in terms of its coordinates r and θ_1 with respect to axes attached to the slab S and rotating with it (Fig. 15.23). Again we propose to determine the relation existing between the absolute motion of P and its relative motion with respect to S.

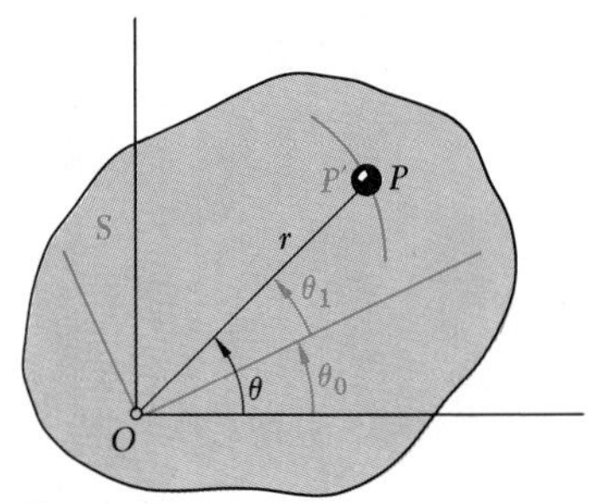

Fig. 15.23

We shall use formulas (11.30) to express the radial and transverse components of the absolute velocity $\mathbf{v}_P$ of P. Observing that $\theta = \theta_0 + \theta_1$, where θ_0 denotes the angular displacement of the slab at the instant considered, we write

$$(v_P)_r = \dot{r} \qquad (v_P)_\theta = r\dot{\theta} = r(\dot{\theta}_0 + \dot{\theta}_1) \tag{15.28}$$

Considering the particular case when P is immobilized on S and S is allowed to rotate, $\mathbf{v}_P$ reduces to the velocity $\mathbf{v}_{P'}$ of the point P' of the slab S which happens to coincide with P at the instant considered. Making r = constant and θ_1 = constant in (15.28), we obtain

$$(v_{P'})_r = 0 \qquad (v_{P'})_\theta = r\dot{\theta}_0 \tag{15.29}$$

Considering now the particular case when the slab is maintained fixed and P is allowed to move, $\mathbf{v}_P$ reduces to the relative velocity $\mathbf{v}_{P/S}$. Making θ_0 constant in (15.28), we obtain, therefore,

$$(v_{P/S})_r = \dot{r} \qquad (v_{P/S})_\theta = r\dot{\theta}_1 \tag{15.30}$$

Comparing formulas (15.28) to (15.30), we find that

$$(v_P)_r = (v_{P'})_r + (v_{P/S})_r \qquad (v_P)_\theta = (v_{P'})_\theta + (v_{P/S})_\theta$$

or, in vector form,

$$\mathbf{v}_P = \mathbf{v}_{P'} + \mathbf{v}_{P/S} \tag{15.31}$$

Formula (15.31) expresses that $\mathbf{v}_P$ may be obtained by adding vectorially the velocities corresponding to the two particular cases considered above.

Using formula (11.33), we now express the radial and transverse components of the absolute acceleration $\mathbf{a}_P$ of P.

$$\begin{aligned}(a_P)_r &= \ddot{r} - r\dot{\theta}^2 = \ddot{r} - r(\dot{\theta}_0 + \dot{\theta}_1)^2 \\ &= \ddot{r} - r(\dot{\theta}_0^2 + 2\dot{\theta}_0\dot{\theta}_1 + \dot{\theta}_1^2)\end{aligned} \tag{15.32}$$

$$\begin{aligned}(a_P)_\theta &= r\ddot{\theta} + 2\dot{r}\dot{\theta} \\ &= r(\ddot{\theta}_0 + \ddot{\theta}_1) + 2\dot{r}(\dot{\theta}_0 + \dot{\theta}_1)\end{aligned} \tag{15.33}$$

Repeating for convenience the equations just obtained,

$$(a_P)_r = \ddot{r} - r(\dot{\theta}_0^2 + 2\dot{\theta}_0\dot{\theta}_1 + \dot{\theta}_1^2) \qquad (15.32)$$

$$(a_P)_\theta = r(\ddot{\theta}_0 + \ddot{\theta}_1) + 2\dot{r}(\dot{\theta}_0 + \dot{\theta}_1) \qquad (15.33)$$

and considering the particular case when r = constant and θ_1 = constant (P immobilized on S), we write

$$(a_{P'})_r = -r\dot{\theta}_0^2 \qquad (a_{P'})_\theta = r\ddot{\theta}_0 \qquad (15.34)$$

Considering now the particular case when θ_0 = constant (slab fixed), we obtain

$$(a_{P/S})_r = \ddot{r} - r\dot{\theta}_1^2 \qquad (a_{P/S})_\theta = r\ddot{\theta}_1 + 2\dot{r}\dot{\theta}_1 \qquad (15.35)$$

Comparing formulas (15.32) to (15.35), we find that the absolute acceleration $\mathbf{a}_P$ *cannot* be obtained by adding the accelerations $\mathbf{a}_{P'}$ and $\mathbf{a}_{P/S}$ corresponding to the two particular cases considered above. We have instead

$$\mathbf{a}_P = \mathbf{a}_{P'} + \mathbf{a}_{P/S} + \mathbf{a}_c \qquad (15.36)$$

where $\mathbf{a}_c$ is a vector of components

$$(a_c)_r = -2r\dot{\theta}_0\dot{\theta}_1 \qquad (a_c)_\theta = 2\dot{r}\dot{\theta}_0$$

Noting that $\dot{\theta}_0$ represents the angular velocity ω of S, and recalling formulas (15.30), we have

$$(a_c)_r = -2\omega(v_{P/S})_\theta \qquad (a_c)_\theta = 2\omega(v_{P/S})_r \qquad (15.37)$$

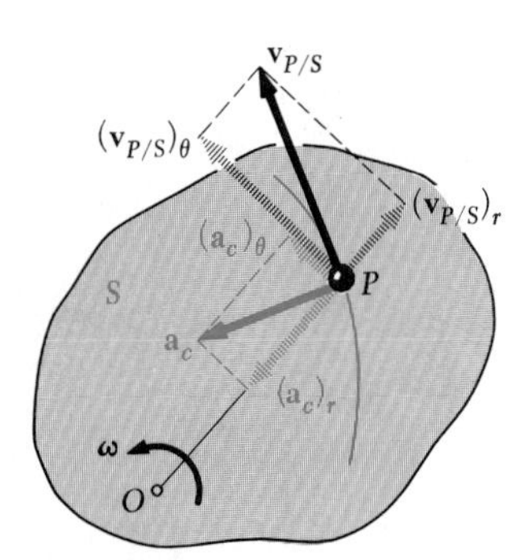

Fig. 15.24

The vector $\mathbf{a}_c$ is called the *complementary acceleration*, or *Coriolis acceleration*, after the French mathematician de Coriolis (1792–1843). The first of formulas (15.37) indicates that the vector $(\mathbf{a}_c)_r$ is obtained by multiplying the vector $(\mathbf{v}_{P/S})_\theta$ by 2ω and rotating it through 90° in the sense of rotation of the slab (Fig. 15.24); the second formula defines $(\mathbf{a}_c)_\theta$ from $(\mathbf{v}_{P/S})_r$ in a similar way. *The Coriolis acceleration* $\mathbf{a}_c$ *is thus a vector perpendicular to the relative velocity* $\mathbf{v}_{P/S}$, *and of magnitude equal to* $2\omega v_{P/S}$; *the sense of* $\mathbf{a}_c$ *is obtained by rotating the vector* $\mathbf{v}_{P/S}$ *through 90° in the sense of rotation of S.*

Comparing formulas (15.36) and (11.26), we note that the acceleration $\mathbf{a}_{P/S}$ of P relative to the slab S is *not* equal to the acceleration $\mathbf{a}_{P/P'}$ of P relative to the point P' of S; this is due to the fact that the first acceleration is defined with respect to *rotating axes*, while the second is defined with respect to *axes of fixed orientation*.

The following example will help in understanding the physical meaning of the Coriolis acceleration. Consider a collar P which is made to slide at a constant relative speed u along a rod OB rotating at a constant angular velocity $\boldsymbol{\omega}$ about O (Fig. 15.25*a*). According to formula (15.36), the absolute acceleration of P may be obtained by adding vectorially the acceleration $\mathbf{a}_A$ of the point A of the rod coinciding with P, the relative acceleration $\mathbf{a}_{P/OB}$ of P with respect to the rod, and the Coriolis acceleration $\mathbf{a}_c$. Since the angular velocity $\boldsymbol{\omega}$ of the rod is constant, $\mathbf{a}_A$ reduces to its normal component $(\mathbf{a}_A)_n$ of magnitude $r\omega^2$; and since u is constant, the relative acceleration $\mathbf{a}_{P/OB}$ is zero. According to the definition given above, the Coriolis acceleration is a vector perpendicular to OB, of magnitude $2\omega u$, and directed as shown in the figure. The acceleration of the collar P consists, therefore, of the two vectors shown in Fig. 15.25a. Note that the result obtained may be checked by applying the relation (11.44).

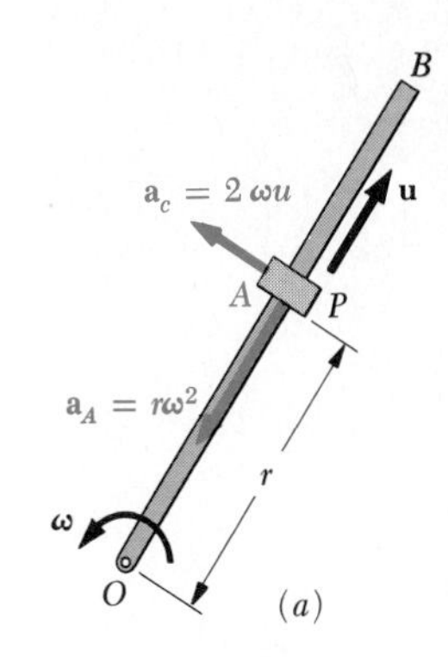

To understand better the significance of the Coriolis acceleration, we shall consider the absolute velocity of P at time t and at time $t + \Delta t$ (Fig. 15.25b). At time t, the velocity may be resolved into its components $\mathbf{u}$ and $\mathbf{v}_A$, and at time $t + \Delta t$ into its components $\mathbf{u}'$ and $\mathbf{v}_{A'}$. Drawing these components from the same origin (Fig. 15.25c), we note that the change in velocity during the time Δt may be represented by the sum of three vectors, $\overrightarrow{RR'}$, $\overrightarrow{TT''}$, and $\overrightarrow{T''T'}$. The vector $\overrightarrow{TT''}$ measures the change in direction of the velocity $\mathbf{v}_A$, and the quotient $\overrightarrow{TT''}/\Delta t$ represents the acceleration $\mathbf{a}_A$ when Δt approaches zero. We check that the direction of $\overrightarrow{TT''}$ is that of $\mathbf{a}_A$ when Δt approaches zero and that

$$\lim_{\Delta t \to 0} \frac{TT''}{\Delta t} = \lim_{\Delta t \to 0} v_A \frac{\Delta\theta}{\Delta t} = r\omega\omega = r\omega^2 = a_A$$

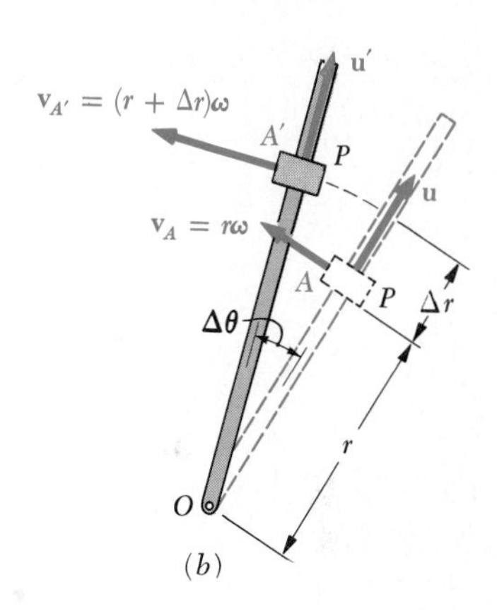

The vector $\overrightarrow{RR'}$ measures the change in direction of $\mathbf{u}$ due to the rotation of the rod; the vector $\overrightarrow{T''T'}$ measures the change in magnitude of $\mathbf{v}_A$ due to the motion of P on the rod. The vectors $\overrightarrow{RR'}$ and $\overrightarrow{T''T'}$ result from the *combined effect* of the relative motion of P and of the rotation of the rod; they would vanish if *either* of these two motions stopped. We may easily verify that the sum of these two vectors defines the Coriolis acceleration. Their direction is that of $\mathbf{a}_c$ when Δt approaches zero, and since $RR' = u\,\Delta\theta$ and $T''T' = v_{A'} - v_A = (r + \Delta r)\omega - r\omega = \omega\,\Delta r$, we check that

$$\lim_{\Delta t \to 0} \left(\frac{RR'}{\Delta t} + \frac{T''T'}{\Delta t} \right) = \lim_{\Delta t \to 0} \left(u \frac{\Delta\theta}{\Delta t} + \omega \frac{\Delta r}{\Delta t} \right) = u\omega + \omega u = 2\omega u = a_c$$

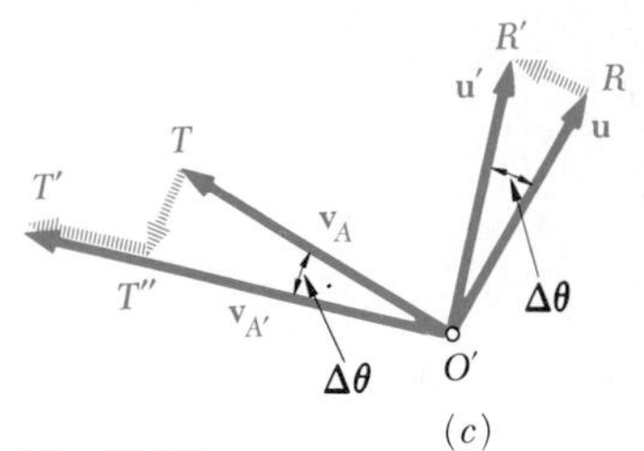

Fig. 15.25

Formulas (15.31) and (15.36) may be used to analyze the motion of mechanisms which contain parts sliding on each other. They make it possible, for example, to relate the absolute and relative motions of sliding pins and collars (see Sample Probs. 15.8 and 15.9). The concept of Coriolis acceleration is also very useful in the study of long-range projectiles and of other bodies whose motions are appreciably affected by the rotation of the earth. As was pointed out in Sec. 12.2, a system of axes attached to the earth does not truly constitute a newtonian frame of reference; such a system of axes should actually be considered as rotating. The formulas derived in this section will therefore facilitate the study of the motion of bodies with respect to axes attached to the earth.

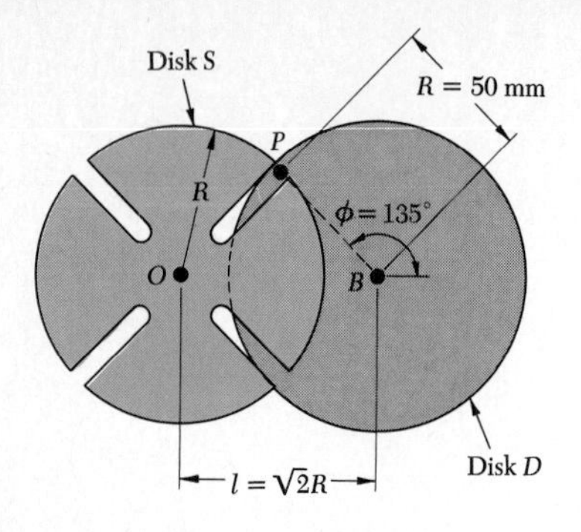

SAMPLE PROBLEM 15.8

The Geneva mechanism shown is used in many counting instruments and in other applications where an intermittent rotary motion is required. Disk D rotates with a constant counterclockwise angular velocity $\boldsymbol{\omega}_D$ of 10 rad/s. A pin P is attached to disk D and slides along one of several slots cut in disk S. It is desirable that the angular velocity of disk S be zero as the pin enters and leaves each slot; in the case of four slots, this will occur if the distance between the centers of the disks is $l = \sqrt{2}\,R$.

At the instant when $\phi = 150°$, determine (*a*) the angular velocity of disk S, (*b*) the velocity of pin P relative to disk S.

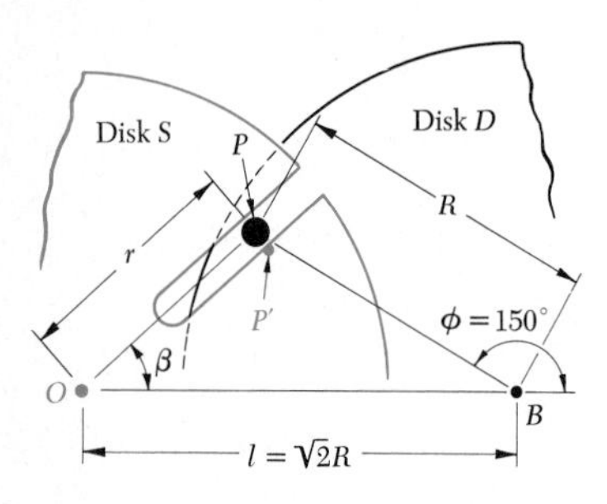

Solution. We solve triangle OPB, which corresponds to the position $\phi = 150°$. Using the law of cosines, we write

$$r^2 = R^2 + l^2 - 2Rl\cos 30° = 0.551R^2 \qquad r = 0.742R = 37.1 \text{ mm}$$

From the law of sines,

$$\frac{\sin\beta}{R} = \frac{\sin 30°}{r} \qquad \sin\beta = \frac{\sin 30°}{0.742} \qquad \beta = 42.4°$$

Since pin P is attached to disk D, and since disk D rotates about point B, the magnitude of the absolute velocity of P is

$$v_P = R\omega_D = (50 \text{ mm})(10 \text{ rad/s}) = 500 \text{ mm/s}$$
$$\mathbf{v}_P = 500 \text{ mm/s} \nearrow 60°$$

We consider now the motion of pin P along the slot in disk S. Denoting by P' the point of disk S which coincides with P at the instant considered, we write

$$\mathbf{v}_P = \mathbf{v}_{P'} + \mathbf{v}_{P/S}$$

Noting that $\mathbf{v}_{P'}$ is perpendicular to the radius OP and that $\mathbf{v}_{P/S}$ is directed along the slot, we draw the velocity triangle corresponding to the above equation. From the triangle, we compute

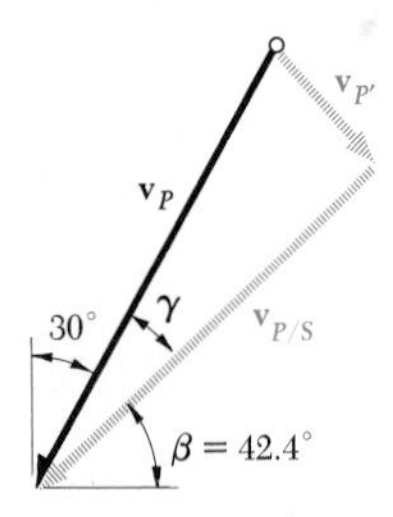

$$\gamma = 90° - 42.4° - 30° = 17.6°$$
$$v_{P'} = v_P \sin\gamma = (500 \text{ mm/s})\sin 17.6°$$
$$\mathbf{v}_{P'} = 151.2 \text{ mm/s} \searrow 42.4°$$
$$v_{P/S} = v_P \cos\gamma = (500 \text{ mm/s})\cos 17.6°$$
$$\mathbf{v}_{P/S} = 477 \text{ mm/s} \nearrow 42.4° \quad ◀$$

Since $\mathbf{v}_{P'}$ is perpendicular to the radius OP, we write

$$v_{P'} = r\omega_S \qquad 151.2 \text{ mm/s} = (37.1 \text{ mm})\omega_S$$
$$\boldsymbol{\omega}_S = 4.08 \text{ rad/s} \circlearrowright \quad ◀$$

SAMPLE PROBLEM 15.9

In the Geneva mechanism of Sample Prob. 15.8, disk D rotates with a constant counterclockwise angular velocity $\boldsymbol{\omega}_D$ of 10 rad/s. At the instant when $\phi = 150°$, determine the angular acceleration of disk S.

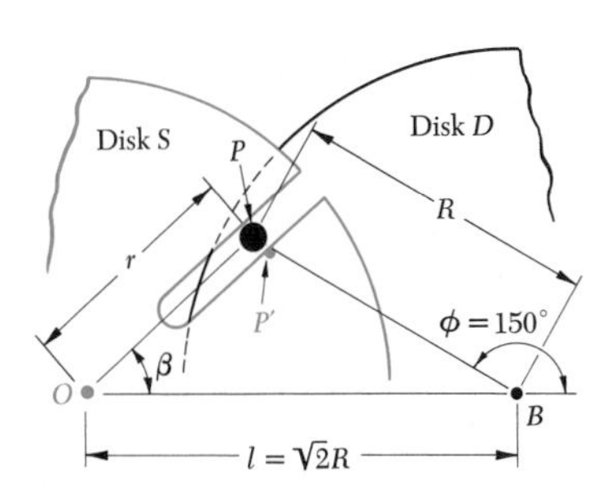

Solution. Referring to Sample Prob. 15.8, we obtain the angular velocity of disk S and the velocity of the pin relative to disk S:

$$\omega_S = 4.08 \text{ rad/s} \circlearrowright$$

$$\beta = 42.4° \qquad \mathbf{v}_{P/S} = 477 \text{ mm/s} \swarrow 42.4°$$

Since pin P moves with respect to the rotating disk S, we write

$$\mathbf{a}_P = \mathbf{a}_{P'} + \mathbf{a}_{P/S} + \mathbf{a}_c \tag{1}$$

Each term of this vector equation is investigated separately.

***Absolute Acceleration* $\mathbf{a}_P$.** Since disk D rotates with constant ω, the absolute acceleration $\mathbf{a}_P$ is directed toward B.

$$a_P = R\omega_D^2 = (50 \text{ mm})(10 \text{ rad/s})^2 = 5000 \text{ mm/s}^2$$

$$\mathbf{a}_P = 5000 \text{ mm/s}^2 \searrow 30°$$

***Acceleration* $\mathbf{a}_{P'}$ *of the Coinciding Point* P'.** The acceleration $\mathbf{a}_{P'}$ of the point P' of disk S which coincides with P at the instant considered is resolved into normal and tangential components. (We recall from Sample Prob. 15.8 that $r = 37.1$ mm.)

$$(a_{P'})_n = r\omega_S^2 = (37.1 \text{ mm})(4.08 \text{ rad/s})^2 = 618 \text{ mm/s}^2$$

$$(\mathbf{a}_{P'})_n = 618 \text{ mm/s}^2 \swarrow 42.4°$$

$$(a_{P'})_t = r\alpha_S = 37.1\alpha_S \qquad (\mathbf{a}_{P'})_t = 37.1\alpha_S \searrow 42.4°$$

***Relative Acceleration* $\mathbf{a}_{P/S}$.** Since the pin P moves in a straight slot cut in disk S, the relative acceleration $\mathbf{a}_{P/S}$ must be parallel to the slot; i.e., its direction must be $\swarrow 42.4°$.

***Coriolis Acceleration* $\mathbf{a}_c$.** Rotating the relative velocity $\mathbf{v}_{P/S}$ through 90° in the sense of $\boldsymbol{\omega}_S$, we obtain the direction of the Coriolis component of the acceleration.

$$a_c = 2\omega_S v_{P/S} = 2(4.08 \text{ rad/s})(477 \text{ mm/s}) = 3890 \text{ mm/s}^2$$

$$\mathbf{a}_c = 3890 \text{ mm/s}^2 \nwarrow 42.4°$$

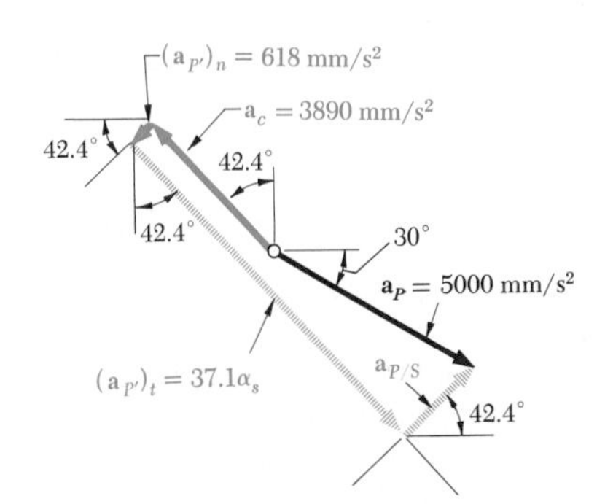

We rewrite Eq. (1) and substitute the accelerations found above:

$$\mathbf{a}_P = (\mathbf{a}_{P'})_n + (\mathbf{a}_{P'})_t + \mathbf{a}_{P/S} + \mathbf{a}_c$$

$$[5000 \searrow 30°] = [618 \swarrow 42.4°] + [37.1\alpha_S \searrow 42.4°]$$

$$+ [a_{P/S} \swarrow 42.4°] + [3890 \nwarrow 42.4°]$$

Equating components in a direction perpendicular to the slot:

$$5000 \cos 17.6° = 37.1\alpha_S - 3890$$

$$\alpha_S = 233 \text{ rad/s}^2 \circlearrowright \blacktriangleleft$$

Problems

15.114 and 15.115 Two rotating rods are connected by a slider block P. The rod attached at A rotates with a constant clockwise angular velocity $\boldsymbol{\omega}_A$. For the given data, determine for the position shown (*a*) the angular velocity of the rod attached at B, (*b*) the relative velocity of the slider block P with respect to the rod on which it slides.

15.114 $b = 8$ in., $\omega_A = 6$ rad/s.
15.115 $b = 250$ mm, $\omega_A = 8$ rad/s.

Fig. P15.114 and P15.116

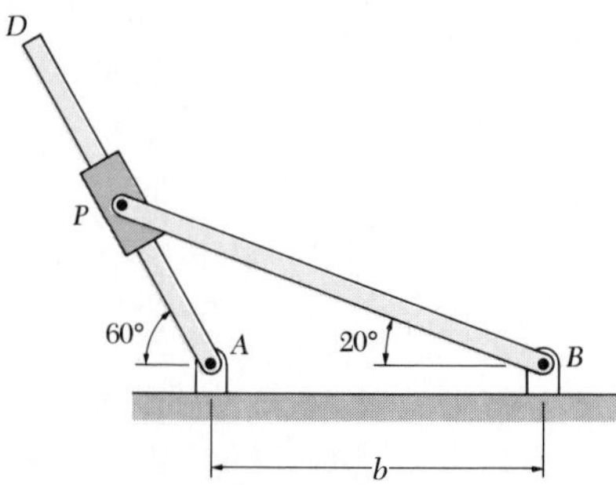

Fig. P15.115 and P15.117

15.116 and 15.117 Two rotating rods are connected by a slider block P. The velocity $\mathbf{v}_0$ of the slider block relative to the rod on which it slides is constant and is directed outward. For the given data, determine the angular velocity of each rod for the position shown.

15.116 $b = 250$ mm, $v_0 = 360$ mm/s.
15.117 $b = 8$ in., $v_0 = 12$ in./s.

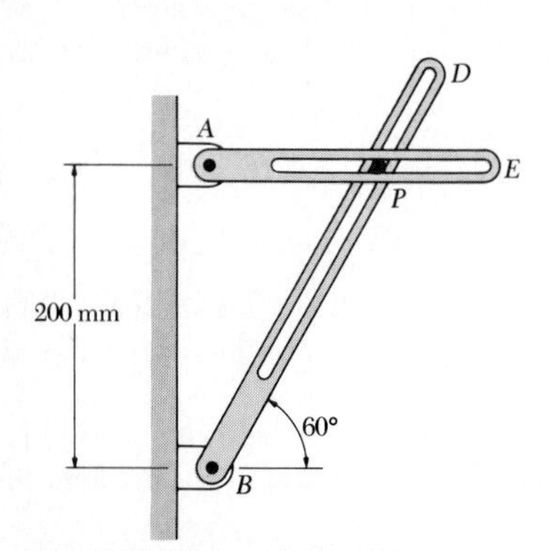

Fig. P15.118 and P15.119

15.118 The motion of pin P is guided by slots cut in rods AE and BD. Knowing that the rods rotate with the constant angular velocities $\boldsymbol{\omega}_A = 4$ rad/s ↻ and $\boldsymbol{\omega}_B = 3$ rad/s ↻, determine the velocity of pin P for the position shown.

15.119 The motion of pin P is guided by slots cut in rods AE and BD. Knowing that the rods rotate with the constant angular velocities $\boldsymbol{\omega}_A = 4$ rad/s ↺ and $\boldsymbol{\omega}_B = 3$ rad/s ↻, determine the velocity of pin P for the position shown.

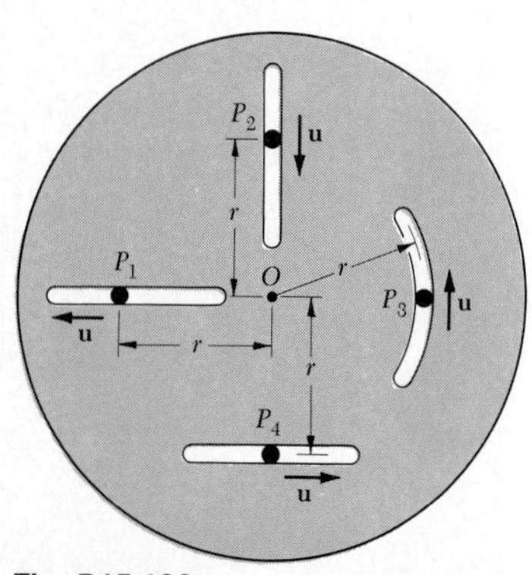

Fig. P15.120

15.120 Four pins slide in four separate slots cut in a circular plate as shown. When the plate is at rest, each pin has a velocity directed as shown and of the same constant magnitude u. If each pin maintains the same velocity in relation to the plate when the plate rotates about O with a constant *counterclockwise* angular velocity $\boldsymbol{\omega}$, determine the acceleration of each pin.

15.121 Solve Prob. 15.120, assuming that the plate rotates about O with a constant *clockwise* angular velocity $\boldsymbol{\omega}$.

15.122 At the instant shown the length of the boom is being decreased at the constant rate of 8 in./s and the boom is being lowered at the constant rate of 0.08 rad/s. Knowing that $\theta = 30°$, determine (*a*) the velocity of point *B*, (*b*) the acceleration of point *B*.

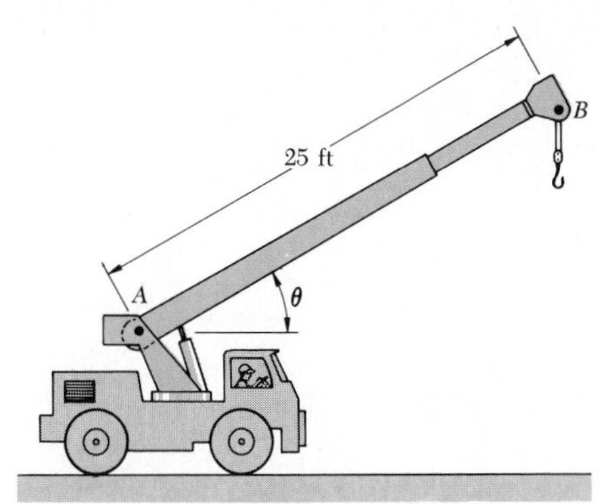

Fig. P15.122 and P15.123

15.123 At the instant shown the length of the boom is being increased at the constant rate of 8 in./s and the boom is being lowered at the constant rate of 0.08 rad/s. Knowing that $\theta = 30°$, determine (*a*) the velocity of point *B*, (*b*) the acceleration of point *B*.

15.124 Water flows at a constant rate through a straight pipe *OB* which rotates counterclockwise with a constant angular velocity of 150 rpm. If the velocity of the water relative to the pipe is 6 m/s, determine the total acceleration (*a*) of the particle of water P_1, (*b*) of the particle of water P_2.

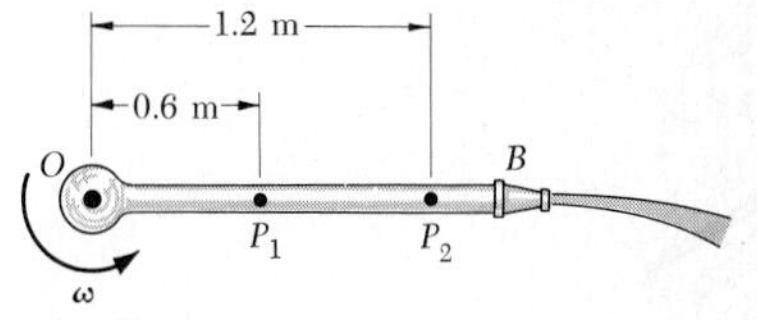

Fig. P15.124

15.125 In the automated welding setup shown, the position of the two welding tips *G* and *H* is controlled by the hydraulic cylinder *D* and the rod *BC*. The cylinder is bolted to the vertical plate which at the instant shown rotates counterclockwise about *A* with a constant angular velocity of 1.2 rad/s. Knowing that at the same instant the length *EF* of the welding asssembly is increasing at the constant rate of 300 mm/s, determine (*a*) the velocity of tip *H*, (*b*) the acceleration of tip *H*.

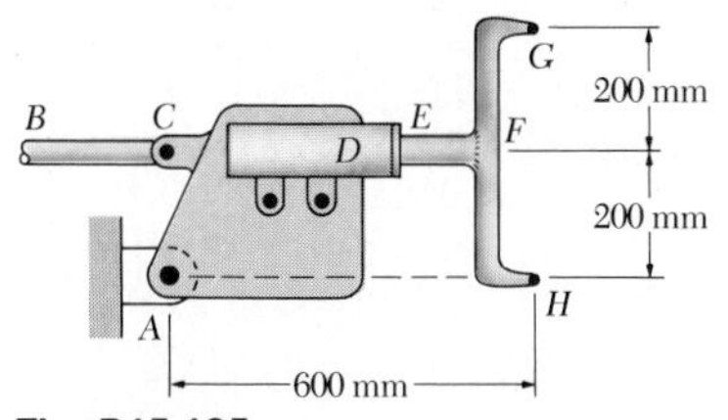

Fig. P15.125

15.126 In Prob. 15.125, determine (*a*) the velocity of tip *G*, (*b*) the acceleration of tip *G*.

15.127 The motion of blade *D* is controlled by the robot arm *ABC*. At the instant shown the arm is rotating clockwise at the constant rate $\omega = 1.5$ rad/s and the length of portion *BC* of the arm is being decreased at the constant rate of 180 mm/s. Determine (*a*) the velocity of *D*, (*b*) the acceleration of *D*.

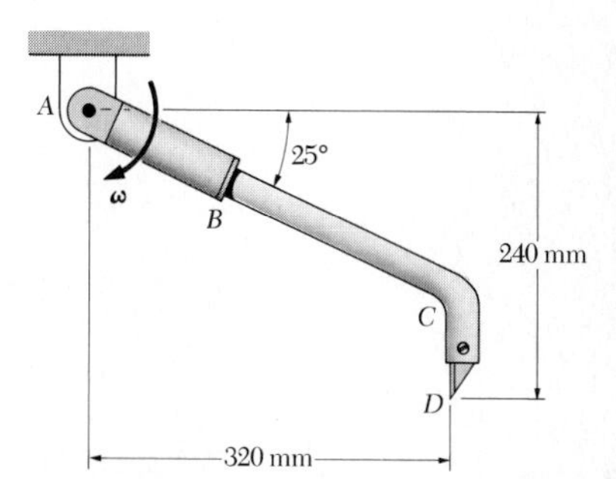

Fig. P15.127

15.128 The hydraulic cylinder CD is welded to an arm which rotates clockwise about A at the constant rate $\omega = 3$ rad/s. Knowing that in the position shown the rod BE is being moved to the right at the constant rate of 12 in./s with respect to the cylinder, determine (*a*) the velocity of point B, (*b*) the acceleration of point B.

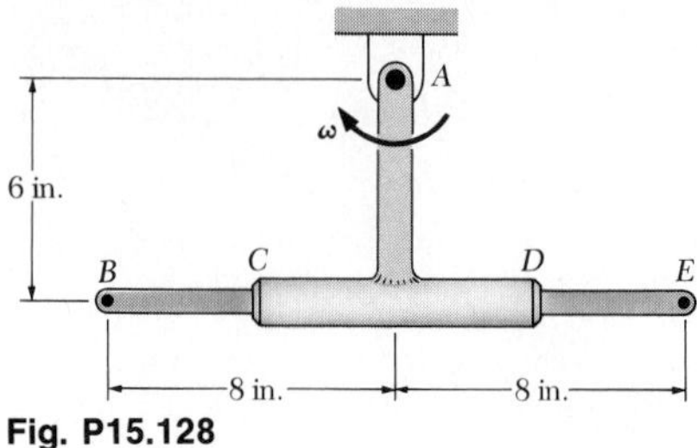

Fig. P15.128

15.129 In Prob. 15.128, determine (*a*) the velocity of point E, (*b*) the acceleration of point E.

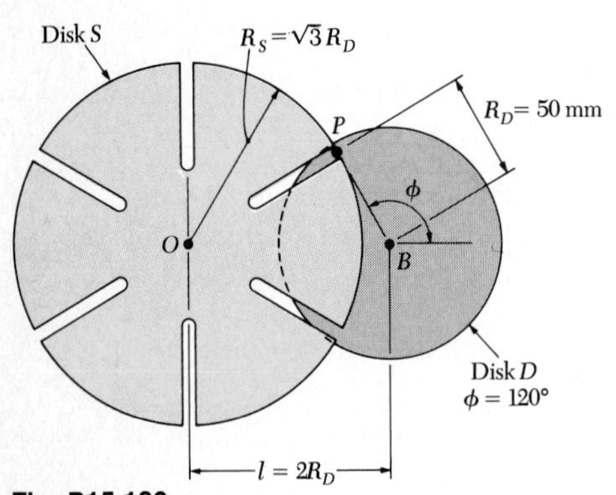

Fig. P15.130

15.130 The Geneva mechanism shown is used to provide an intermittent rotary motion of disk S. Disk D rotates with a constant counterclockwise angular velocity $\boldsymbol{\omega}_D$ of 10 rad/s. A pin P is attached to disk D and may slide in one of the six equally spaced slots cut in disk S. It is desirable that the angular velocity of disk S be zero as the pin enters and leaves each of the six slots; this will occur if the distance between the centers of the disks and the radii of the disks are related as shown. Determine the angular velocity and angular acceleration of disk S at the instant when $\phi = 150°$.

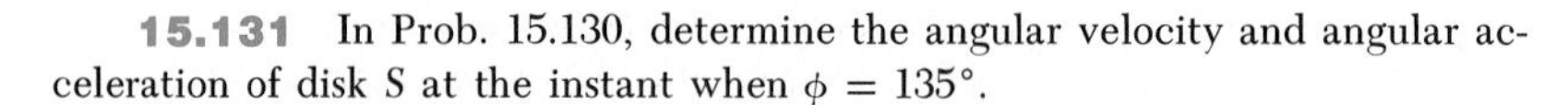

15.131 In Prob. 15.130, determine the angular velocity and angular acceleration of disk S at the instant when $\phi = 135°$.

15.132 In Prob. 15.114, determine the angular acceleration of the rod attached at B.

15.133 In Prob. 15.117, determine the angular acceleration of the rod attached at B.

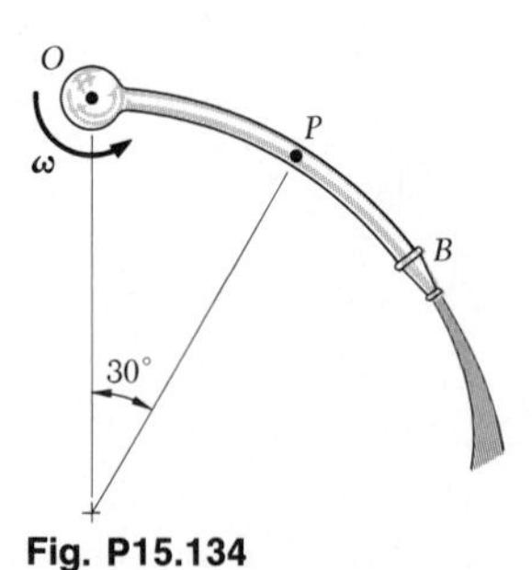

Fig. P15.134

15.134 Water flows through the curved pipe OB, which has a uniform radius of 15 in. and which rotates with a constant counterclockwise angular velocity of 120 rpm. If the velocity of the water relative to the pipe is 40 ft/s, determine the total acceleration of the particle of water P.

15.135 Solve Prob. 15.134, assuming that the curved pipe rotates with a constant clockwise angular velocity of 120 rpm.

15.136 Rod AD is bent in the shape of an arc of circle of radius $b = 120$ mm. The position of the rod is controlled by pin B which slides in a horizontal slot and also slides along the rod. Knowing that at the instant shown pin B moves to the right with a constant speed of 40 mm/s, determine (*a*) the angular velocity of the rod, (*b*) the angular acceleration of the rod.

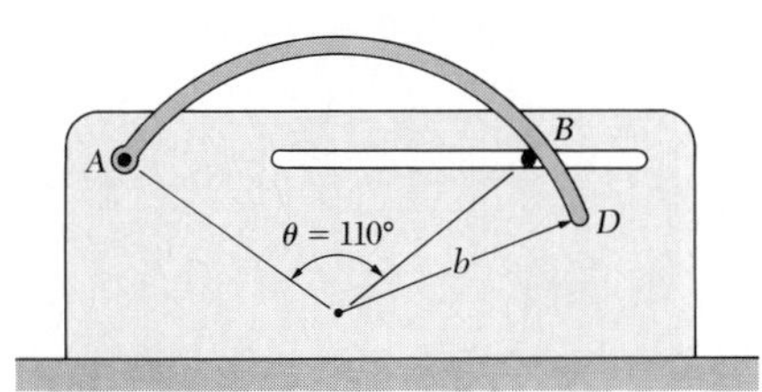

Fig. P15.136

15.137 Solve Prob. 15.136 when $\theta = 90°$.

* **15.138** In Prob. 15.118, determine the acceleration of pin P.

* **15.139** In Prob. 15.119, determine the acceleration of pin P.

* **15.140** Rod AB passes through a collar which is welded to link DE. Knowing that at the instant shown block A moves to the right with a constant speed of 75 in./s, determine (*a*) the angular velocity of rod AB, (*b*) the velocity relative to the collar of the point of the rod in contact with the collar, (*c*) the acceleration of the point of the rod in contact with the collar. (*Hint.* Rod AB and link DE have the same $\boldsymbol{\omega}$ and the same $\boldsymbol{\alpha}$.)

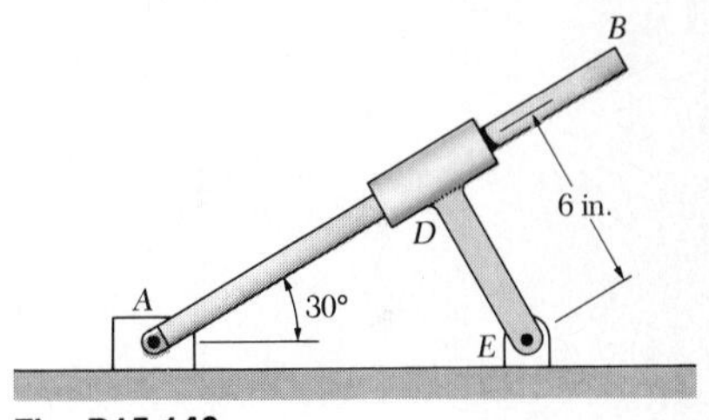

Fig. P15.140

* **15.141** Solve Prob. 15.140, assuming that block A moves to the left with a constant speed of 75 in./s.

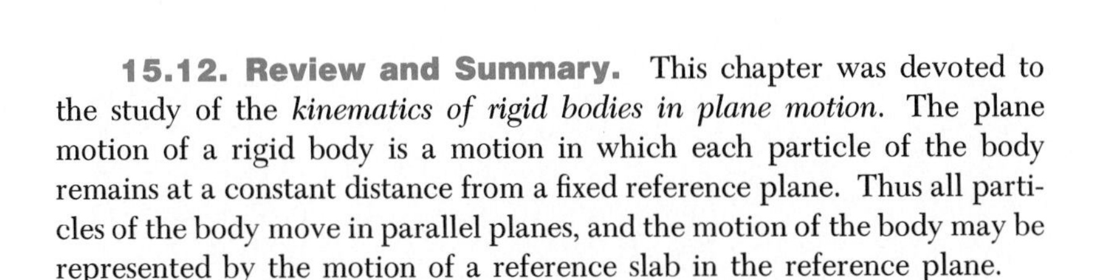

Rigid body in plane motion

15.12. Review and Summary. This chapter was devoted to the study of the *kinematics of rigid bodies in plane motion.* The plane motion of a rigid body is a motion in which each particle of the body remains at a constant distance from a fixed reference plane. Thus all particles of the body move in parallel planes, and the motion of the body may be represented by the motion of a reference slab in the reference plane.

Translation

We first considered the *translation* of a rigid slab [Sec. 15.2] and observed that in such a motion, *all the points of the slab have the same velocity and the same acceleration at any given instant.*

Rotation

Next we considered the *rotation* of a rigid slab about a fixed axis perpendicular to the plane of the slab and intersecting it at a point O [Sec. 15.3]. We noted that the position of the slab is completely defined by the angle θ that a given line OP drawn on the slab forms with a fixed direction OQ (Fig. 15.26). We also defined the *angular velocity* of the slab as

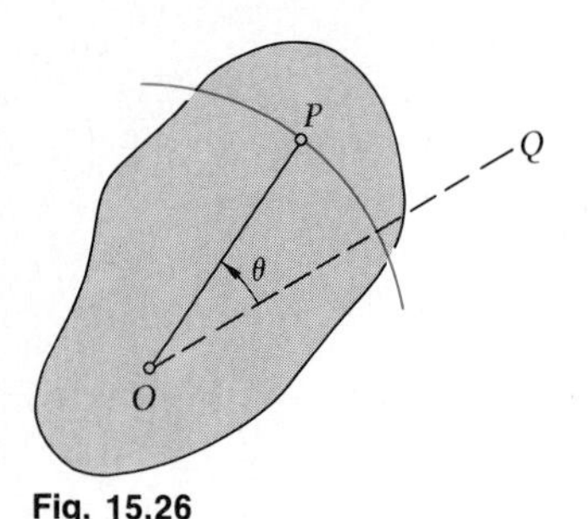

Fig. 15.26

$$\omega = \frac{d\theta}{dt} \tag{15.3}$$

and its *angular acceleration* as

$$\alpha = \frac{d\omega}{dt} = \frac{d^2\theta}{dt^2} \tag{15.4}$$

or

$$\alpha = \omega\frac{d\omega}{d\theta} \tag{15.5}$$

Two particular cases of rotation are frequently encountered: *uniform rotation* and *uniformly accelerated rotation.* Problems involving either of these motions may be solved by using equations similar to those used in Secs. 11.4 and 11.5 for the uniform rectilinear motion and the uniformly accelerated rectilinear motion of a particle, but where x, v, and a are replaced by θ, ω, and α, respectively [Sample Prob. 15.1].

Velocities and accelerations in rotation

We saw in Sec. 15.4 that when a rigid slab rotates through an angle θ about an axis through O, any given point P of the slab describes an arc of a circle centered at O (Fig. 15.27) and of length

$$s = r\theta \tag{15.10}$$

where θ is expressed in radians, and that the velocity of P is

$$v = r\omega \tag{15.11}$$

The acceleration of P may be resolved into tangential and normal components [Sample Prob. 15.1], respectively equal to

$$a_t = r\alpha \tag{15.12}$$
$$a_n = r\omega^2 \tag{15.13}$$

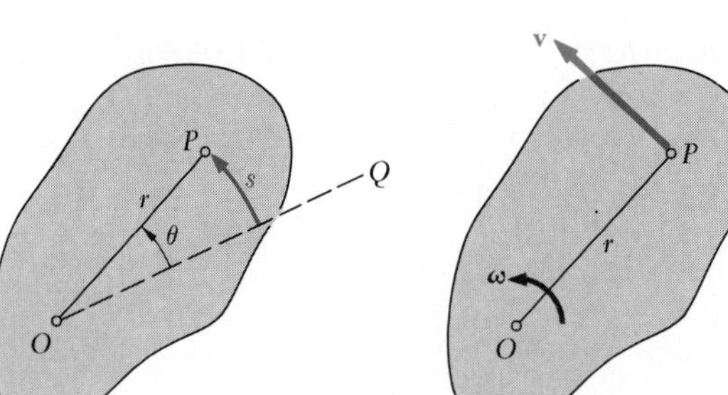

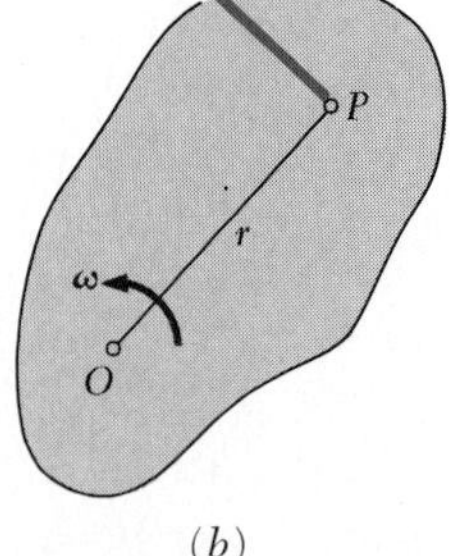

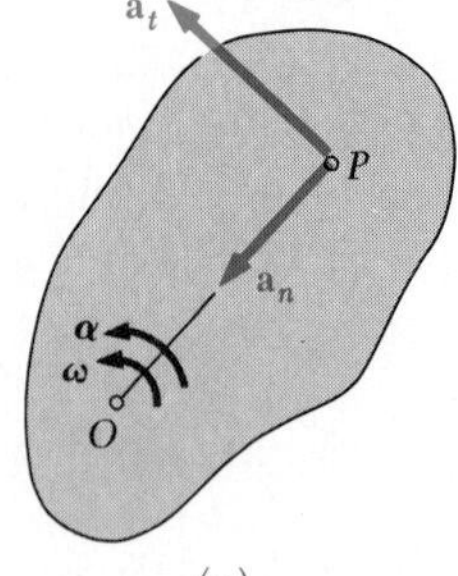

Fig. 15.27 (a) (b) (c)

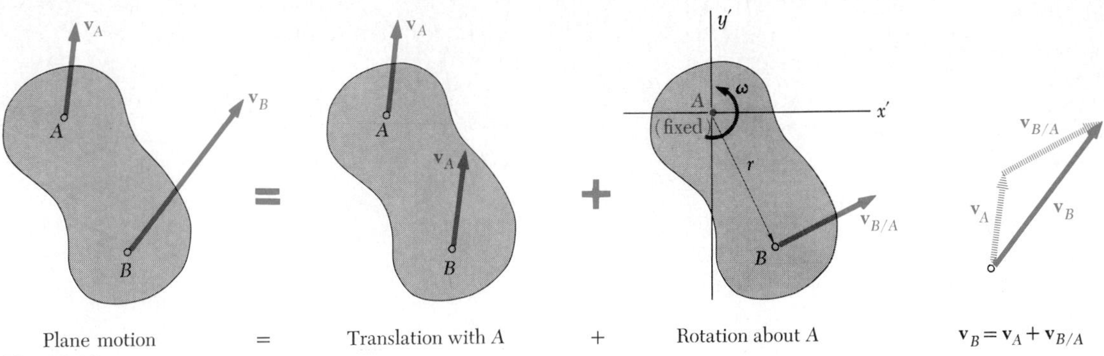

Fig. 15.28

Velocities in plane motion

The *most general plane motion* of a rigid slab may be considered as the *sum of a translation and a rotation* [Sec. 15.5]. For example, the slab shown in Fig. 15.28 may be assumed to translate with point A, while simultaneously rotating about A. It follows [Sec. 15.6] that the velocity of any point B of the slab may be expressed as

$$\mathbf{v}_B = \mathbf{v}_A + \mathbf{v}_{B/A} \tag{15.14}$$

where $\mathbf{v}_{B/A}$ is the relative velocity of B with respect to the $x'y'$ axes translating with A; the magnitude of $\mathbf{v}_{B/A}$ is

$$v_{B/A} = r\omega \tag{15.15}$$

The fundamental equation (15.14) relating the absolute velocities of points A and B and the relative velocity of B with respect to A was expressed in the form of a vector diagram and used to solve problems involving the motion of various types of mechanisms [Sample Probs. 15.2 and 15.3].

Instantaneous center of rotation

Another approach to the solution of problems involving the velocities of the points of a rigid slab in plane motion was presented in Sec. 15.7 and used in Sample Probs. 15.4 and 15.5. It is based on the determination of the *instantaneous center of rotation* C of the slab (Fig. 15.29).

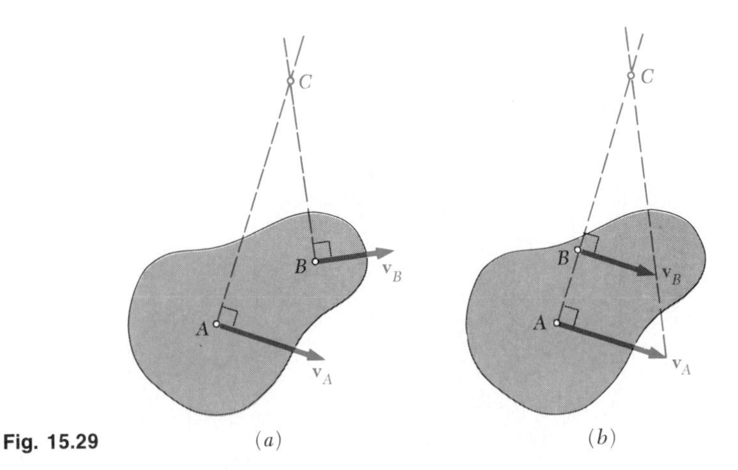

Fig. 15.29 (a) (b)

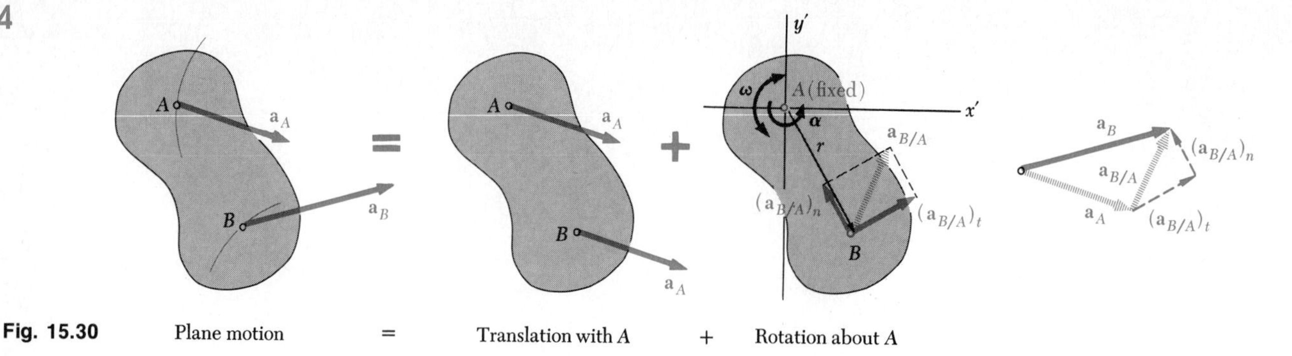

Fig. 15.30 Plane motion = Translation with A + Rotation about A

Accelerations in plane motion

The fact that any plane motion of a rigid slab may be considered as the sum of a translation of the slab with a reference point A and a rotation about A was used in Sec. 15.8 to relate the absolute accelerations of any two points A and B of the slab and the relative acceleration of B with respect to A. We had

$$\mathbf{a}_B = \mathbf{a}_A + \mathbf{a}_{B/A} \tag{15.18}$$

where $\mathbf{a}_{B/A}$ consisted of a *normal component* $(\mathbf{a}_{B/A})_n$ of magnitude $r\omega^2$ directed toward A, and a *tangential component* $(\mathbf{a}_{B/A})_t$ of magnitude $r\alpha$ perpendicular to the line AB (Fig. 15.30). The fundamental relation (15.18) was expressed in terms of a vector diagram and used to determine the accelerations of given points of various mechanisms [Sample Probs. 15.6 and 15.7]. It should be noted that the instantaneous center of rotation C considered in Sec. 15.7 cannot be used for the determination of accelerations, since point C, in general, does *not* have zero acceleration.

Coordinates expressed in terms of a parameter

In the case of certain mechanisms, it is possible to express the coordinates x and y of all significant points of the mechanism by means of simple analytic expressions containing a *single parameter*. The components of the absolute velocity and acceleration of a given point may then be obtained by differentiating twice with respect to the time t the coordinates x and y of that point [Sec. 15.9].

Particle moving on a slab in translation

The last part of the chapter was devoted to the kinematic analysis of a particle moving on a slab which is either in translation or in rotation about a fixed axis. Considering first the case of *a particle P moving on a slab S in translation* [Sec. 15.10] , we expressed the absolute velocity $\mathbf{v}_P$ of P as the sum of the velocity $\mathbf{v}_{P/S}$ of P with respect to axes attached to S, and of the velocity $\mathbf{v}_{P'}$ of *the point P′ of the slab coinciding with the particle P at the instant considered:*

$$\mathbf{v}_P = \mathbf{v}_{P'} + \mathbf{v}_{P/S} \tag{15.25}$$

We found that the absolute acceleration $\mathbf{a}_P$ of the particle could be similarly expressed as

$$\mathbf{a}_P = \mathbf{a}_{P'} + \mathbf{a}_{P/S} \tag{15.27}$$

where $\mathbf{a}_{P/S}$ = acceleration of P with respect to S
$\mathbf{a}_{P'}$ = acceleration of point P' of S coinciding with P

In the case of a particle P moving on a slab S rotating about a fixed axis [Sec. 15.11], we found that the absolute velocity of P could again be expressed as

Particle moving on a rotating slab

$$\mathbf{v}_P = \mathbf{v}_{P'} + \mathbf{v}_{P/S} \qquad (15.31)$$

where $\mathbf{v}_{P/S}$ = velocity of P with respect to S
$\mathbf{v}_{P'}$ = velocity of point P' of S coinciding with P

However, the expression for the acceleration of P was found to contain an additional term $\mathbf{a}_c$ called the *complementary acceleration* or *Coriolis acceleration*. We had

$$\mathbf{a}_P = \mathbf{a}_{P'} + \mathbf{a}_{P/S} + \mathbf{a}_c \qquad (15.36)$$

where $\mathbf{a}_{P/S}$ = acceleration of P with respect to S
$\mathbf{a}_{P'}$ = acceleration of point P' of S coinciding with P
$\mathbf{a}_c$ = Coriolis acceleration

Coriolis acceleration

The Coriolis acceleration $\mathbf{a}_c$ was found to have a magnitude $a_c = 2\omega v_{P/S}$, where ω is the angular velocity of the slab S, and *to point in the direction obtained by rotating the vector* $\mathbf{v}_{P/S}$ *through 90° in the sense of rotation of* S. Formulas (15.31) and (15.36) may be used to analyze the motion of mechanisms which contain parts sliding on each other [Sample Probs. 15.8 and 15.9].

Review Problems

15.142 The flanged wheel rolls without slipping on the horizontal rail. If at a given instant the velocity and acceleration of the center of the wheel are as shown, determine the acceleration (*a*) of point *B*, (*b*) of point *C*, (*c*) of point *D*.

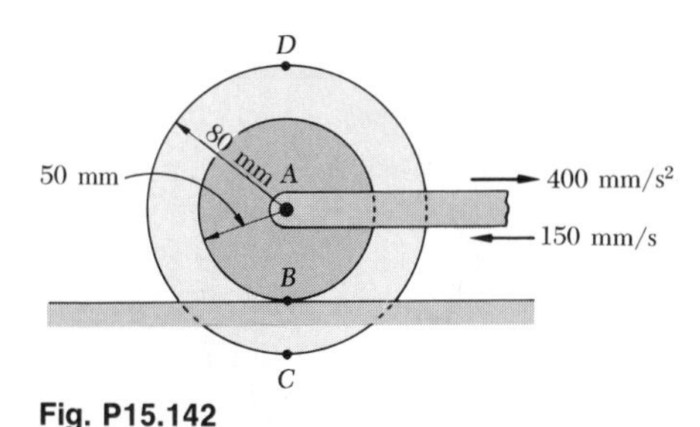

Fig. P15.142

15.143 Rod AC of length $2b$ is attached to a collar at A and passes through a pivoted collar at B. Knowing that the collar A moves upward with a constant velocity $\mathbf{v}_A$, derive an expression for (*a*) the angular velocity of rod AC, (*b*) the velocity of the point of the rod in contact with the pivoted collar B, (*c*) the angular acceleration of rod AC.

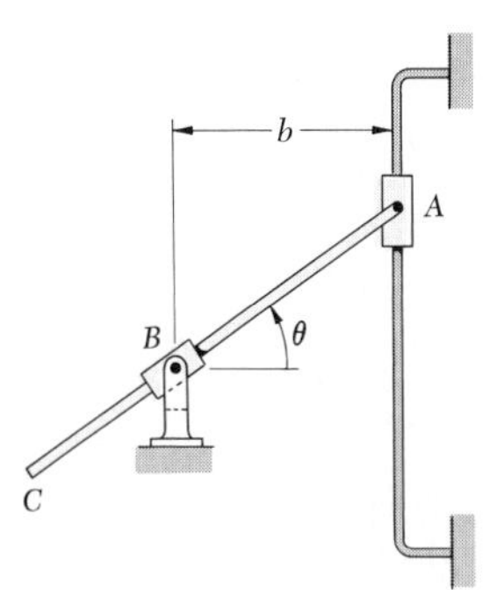

Fig. P15.143

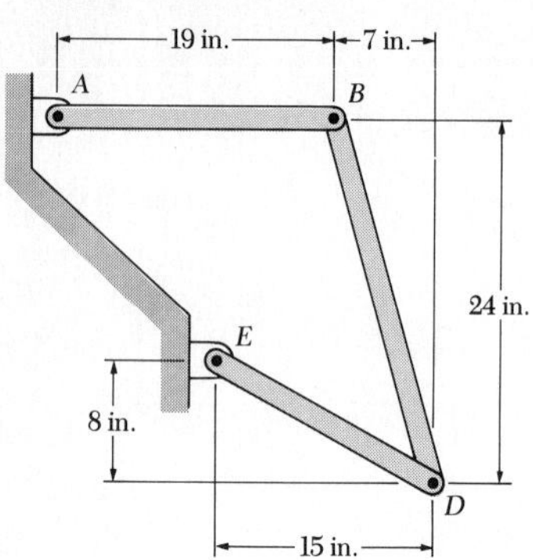

Fig. P15.144

15.144 In the position shown, bar AB has a constant angular velocity of 4 rad/s counterclockwise. Determine the angular velocity (*a*) of bar BD, (*b*) of bar DE.

15.145 For the linkage of Prob. 15.144, determine the angular acceleration (*a*) of bar BD, (*b*) of bar DE.

15.146 It takes 1.2 s for the turntable of a 33-rpm record player to reach full speed after being started. Assuming uniformly accelerated motion, determine (*a*) the angular acceleration of the turntable, (*b*) the normal and tangential components of the acceleration of a point on the rim of the 300-mm-diameter turntable just before the speed of 33 rpm is reached, (*c*) the total acceleration of the same point at that time.

15.147 The telescoping arm AB is used to raise a worker to the elevation of electric and telephone wires. At the instant shown the length of the arm is being increased at the constant rate of 0.25 m/s and the arm is being raised at the constant rate of 0.40 rad/s. Knowing that $\theta = 30°$, determine (*a*) the velocity of point B, (*b*) the acceleration of point B.

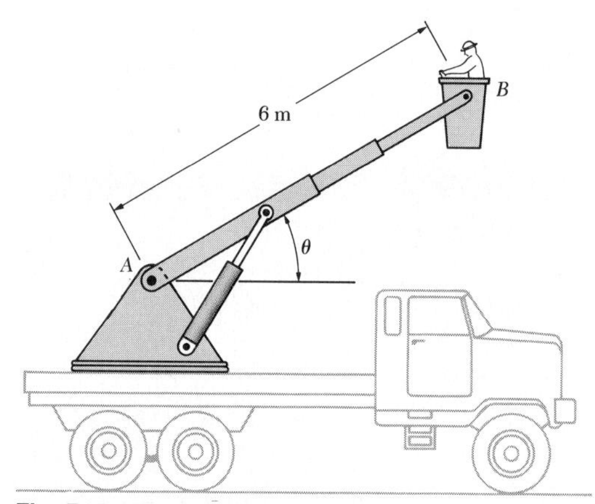

Fig. P15.147 and P15.148

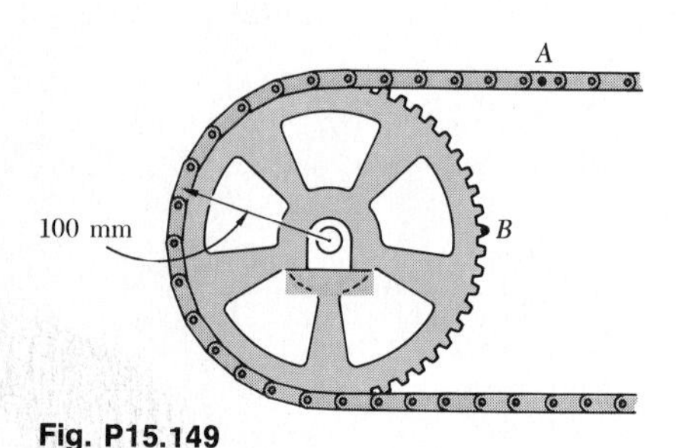

Fig. P15.149

15.148 At the instant shown the length of arm AB is being decreased at the constant rate of 0.35 m/s and the arm is being lowered at the constant rate of 0.25 rad/s. Knowing that $\theta = 20°$, determine (*a*) the velocity of point B, (*b*) the acceleration of point B.

15.149 The sprocket wheel and chain are initially at rest. If the acceleration of point A of the chain has a constant magnitude of 120 mm/s^2 and is directed to the left, determine the acceleration of sprocket B as the wheel is completing a full revolution.

15.150 Rod ABD is guided by wheels at A and B which roll in horizontal and vertical tracks as shown. Knowing that $\beta = 60°$ and that point A moves to the left with a constant velocity of 30 in./s, determine (*a*) the angular velocity of the rod, (*b*) the velocity of point D.

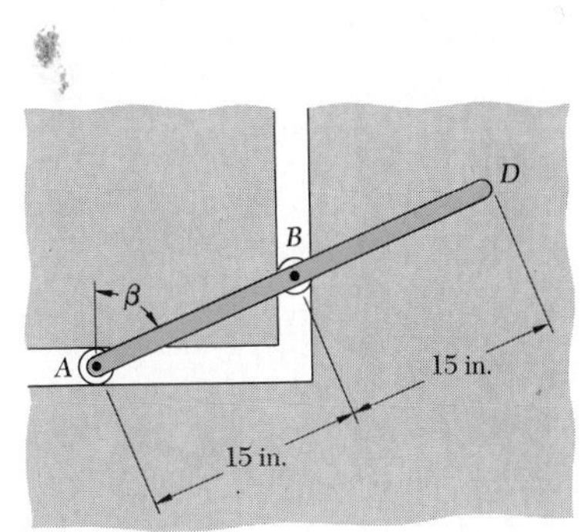

Fig. P15.150

15.151 In Prob. 15.150, determine (*a*) the angular acceleration of the rod, (*b*) the acceleration of point D.

15.152 Two collars C and D move along the vertical rod shown. Knowing that the velocity of collar C is 660 mm/s downward, determine (*a*) the velocity of collar D, (*b*) the angular velocity of member AB.

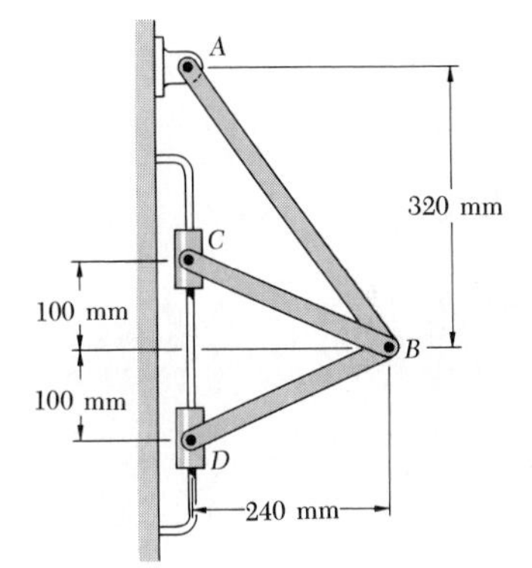

Fig. P15.152

15.153 An 18-in. rod rests on a frictionless horizontal table. A force **P** applied as shown produces the following accelerations: $\mathbf{a}_A = 30$ in./s² to the right, $\boldsymbol{\alpha} = 2.5$ rad/s² clockwise as viewed from above. Determine (*a*) the acceleration of point B, (*b*) the point of the rod which has no acceleration.

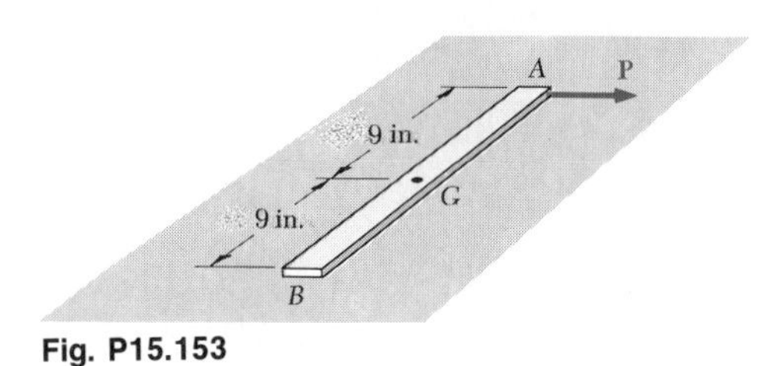

Fig. P15.153

The following problems are designed to be solved with a computer.

15.C1 A 400-mm rod AB is guided by wheels at A and B which roll in the track shown. Knowing that B moves at a constant speed $v_B = 1.2$ m/s, write a computer program and use it to calculate the angular velocity of the rod and the velocity of A for values of θ from 0 to 150° at 10° intervals.

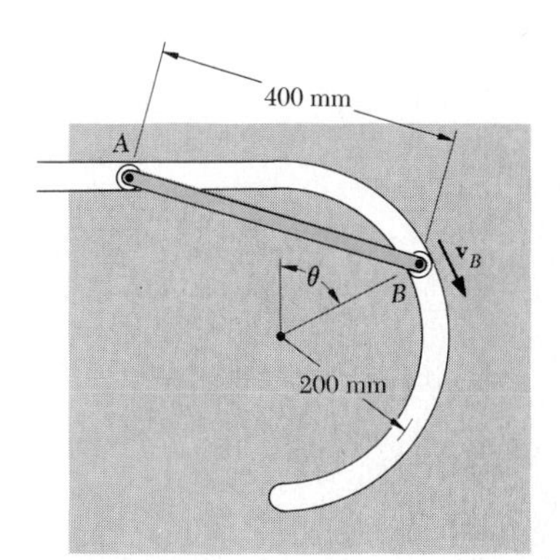

Fig. P15.C1

15.C2 Collar A slides upward at a constant speed $v_A = 3$ in./s. Write a computer program and use it to calculate the velocity of point B for values of θ from 0 to 70° at 5° intervals for (*a*) $b = 2$ in., (*b*) $b = 3$ in., (*c*) $b = 4$ in.

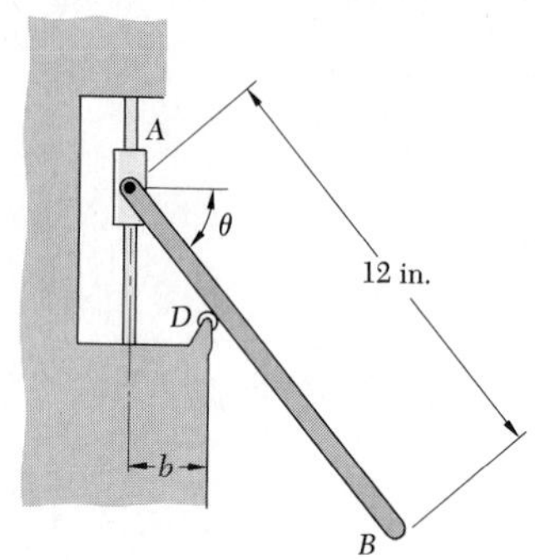

Fig. P15.C2

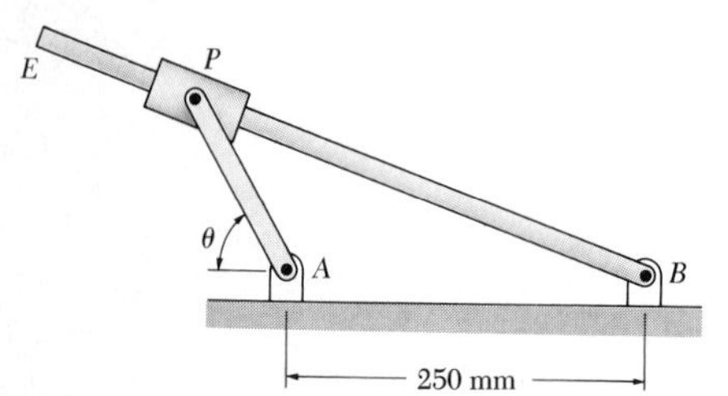

Fig. P15.C3

15.C3 Two rotating rods are connected by a slider block P. The rod AP is 125 mm long and rotates with a constant angular velocity $\omega_A = 10$ rad/s clockwise. Write a computer program and use it to calculate the angular velocity and angular acceleration of rod BE for values of θ from 0 to 180° at 10° intervals.

15.C4 In the engine system shown, $l = 8$ in. and $b = 3$ in. During a test of the system, crank AB is rotated with a constant angular velocity of 2000 rpm clockwise. Write a computer program and use it to calculate, for values of θ from 0 to 180° at 10° intervals, (*a*) the angular velocity and angular acceleration of the connecting rod BD, (*b*) the velocity and acceleration of the piston P.

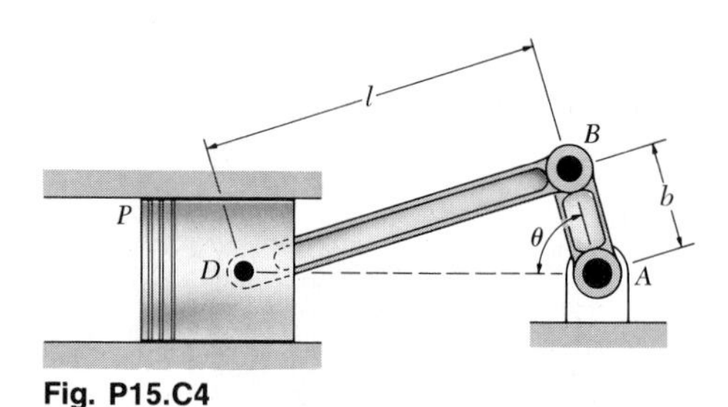

Fig. P15.C4

CHAPTER SIXTEEN

Kinetics of Rigid Bodies: Forces and Accelerations

16.1. Introduction. In this chapter and in Chaps. 17 and 18, we shall study the *kinetics of rigid bodies*, i.e., the relations existing between the forces acting on a rigid body, the shape and mass of the body, and the motion produced. In Chaps. 12 to 14, we studied similar relations, assuming then that the body could be considered as a particle, i.e., that its mass could be concentrated in one point and that all forces acted at that point. We shall now take the shape of the body into account, as well as the exact location of the points of application of the forces. Besides, we shall be concerned not only with the motion of the body as a whole but also with the rotation of the body about its mass center.

In this chapter, our study of the motion of rigid bodies will be based directly on the equation $\mathbf{F} = m\mathbf{a}$, while in the next two chapters we shall make use of the principles of work and energy and of impulse and momentum. Our study will be limited to that of the *plane motion* of rigid bodies. This motion was defined in Sec. 15.1 as a motion in which each particle of the body remains at a constant distance from a fixed reference plane. In the study of the kinetics of plane motion, the reference plane is chosen so that it contains the mass center of the body. Our study will be further limited to that of plane slabs and of bodies which are symmetrical with respect to the reference plane,† except for Sec. 16.6, where the motion of three-dimensional bodies which are not symmetrical with respect to that plane will be considered.

† Or, more generally, bodies which have a principal centroidal axis of inertia perpendicular to the reference plane.

Using the results obtained in Sec. 12.5 for a system of particles, we shall derive *d'Alembert's principle* for a rigid body in Sec. 16.2. This principle states that the external forces acting on a rigid body are equivalent to the effective forces of the various particles forming the body. Furthermore, it will be shown that in the case of a rigid slab—or, more generally, of a body symmetrical with respect to the reference plane—the effective forces reduce to a force-couple system consisting of a vector $m\bar{\mathbf{a}}$ attached at the mass center G of the body and a couple $\bar{I}\boldsymbol{\alpha}$. This result may be expressed in the form of a *free-body-diagram equation* which will be used to solve various problems involving the plane motion of rigid bodies and systems of rigid bodies.

In Secs. 16.3 and 16.4, we shall consider problems involving the translation, centroidal rotation, or unconstrained plane motion of rigid bodies, while in Sec. 16.5 we shall consider problems involving the noncentroidal rotation, rolling motion, or other partially constrained motion of rigid bodies.

Finally, we shall analyze in Sec. 16.6 the rotation about a fixed axis of three-dimensional rigid bodies which are *not* symmetrical with respect to the reference plane, and it will be shown why rotating systems should be both *statically* and *dynamically balanced.*

16.2. Plane Motion of a Rigid Body. D'Alembert's Principle. Consider a rigid slab of mass m moving under the action of several forces $\mathbf{F}_1$, $\mathbf{F}_2$, $\mathbf{F}_3$, etc., contained in the plane of the slab (Fig. 16.1*a*). We may regard the slab as being made of a large number of particles and use the results obtained in Sec. 12.5 for a system of particles. It was shown at that time that the system of the external forces acting on the particles is equipollent to the system of the effective forces of the particles. In other words, the sum of the components of the external forces in any given direction is equal to the sum of the components of the effective forces in that direction, and the sum of the moments of the external forces about any given axis is equal to the sum of the moments of the effective forces about that axis.

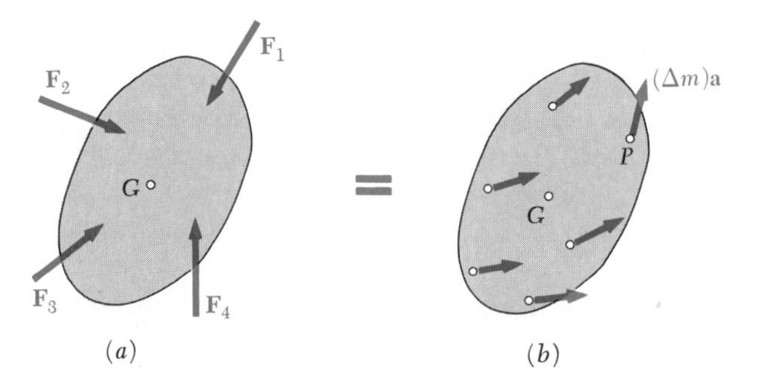

Fig. 16.1

It follows that the external forces $\mathbf{F}_1$, $\mathbf{F}_2$, $\mathbf{F}_3$, etc., acting on the slab of Fig. 16.1 are equipollent to the system of the effective forces of the particles forming the slab. But since we are now dealing with a rigid body, the two systems of forces are not only equipollent, they are also *equivalent;* i.e., they both have the same effect on the slab (Sec. 3.12). This is shown in Fig. 16.1, where a colored equals sign has been used to connect the system of the external forces and the system of the effective forces.

We may thus state that *the external forces acting on a rigid body are equivalent to the effective forces of the various particles forming the body.* This statement is referred to as *d'Alembert's principle,* after the French mathematician Jean le Rond d'Alembert (1717–1783), even though d'Alembert's original statement was written in a somewhat different form.

Again, because we are dealing with a rigid body, we may reduce the system of the effective forces to a force-couple system attached at a point of our choice (Sec. 3.11). By selecting the mass center G of the slab as the reference point, we shall be able to obtain a particularly simple result.

In order to determine the force-couple system at G which is equivalent to the system of the effective forces, we must evaluate the sums of the x and y components of the effective forces and the sum of their moments about G. We first write

$$\Sigma(F_x)_{\text{eff}} = \Sigma(\Delta m)a_x \qquad \Sigma(F_y)_{\text{eff}} = \Sigma(\Delta m)a_y \tag{16.1}$$

But, according to Eqs. (12.9) of Sec. 12.6, we have

$$\Sigma(\Delta m)a_x = (\Sigma\,\Delta m)\bar{a}_x \qquad \Sigma(\Delta m)a_y = (\Sigma\,\Delta m)\bar{a}_y \tag{16.2}$$

where $\bar{a}_x$ and $\bar{a}_y$ are the components of the acceleration $\bar{\mathbf{a}}$ of the mass center G. Noting that $\Sigma\,\Delta m$ represents the total mass m of the slab, and substituting from (16.2) into (16.1), we have

$$\Sigma(F_x)_{\text{eff}} = m\bar{a}_x \qquad \Sigma(F_y)_{\text{eff}} = m\bar{a}_y \tag{16.3}$$

Turning our attention to the determination of the sum of the moments about G of the effective forces, we recall from Sec. 15.8 that the acceleration $\mathbf{a}$ of any given particle P of the slab (Fig. 16.2) may be expressed as the sum of the acceleration $\bar{\mathbf{a}}$ of the mass center G and of the acceleration $\mathbf{a}'$ of P relative to a frame $Gx'y'$ attached to G and of fixed orientation:

$$\mathbf{a} = \bar{\mathbf{a}} + \mathbf{a}'$$

Resolving $\mathbf{a}'$ into normal and tangential components, we write

$$\mathbf{a} = \bar{\mathbf{a}} + \mathbf{a}'_n + \mathbf{a}'_t \tag{16.4}$$

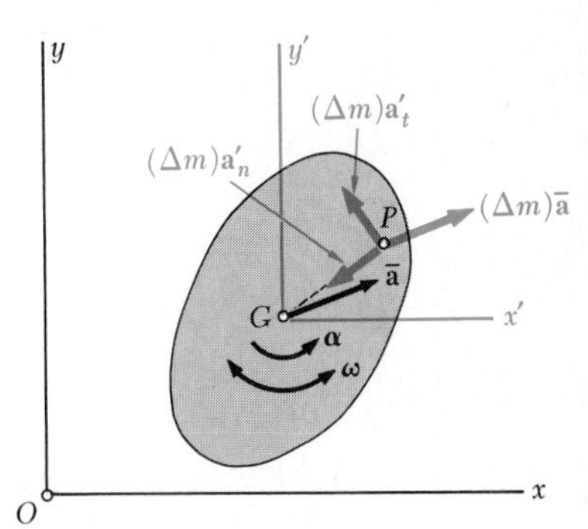

Fig. 16.2

Each effective force $(\Delta m)\mathbf{a}$ may thus be resolved into the three component effective forces shown in Fig. 16.2. The effective forces $(\Delta m)\bar{\mathbf{a}}$ obtained in this fashion are associated with a translation of the slab with G, while the effective forces $(\Delta m)\mathbf{a}'_n$ and $(\Delta m)\mathbf{a}'_t$ are associated with a rotation of the slab about G.

Considering first the effective forces $(\Delta m)\bar{\mathbf{a}}$ associated with the translation of the slab, and resolving them into rectangular components (Fig. 16.3), we find that the sum of their moments about G is

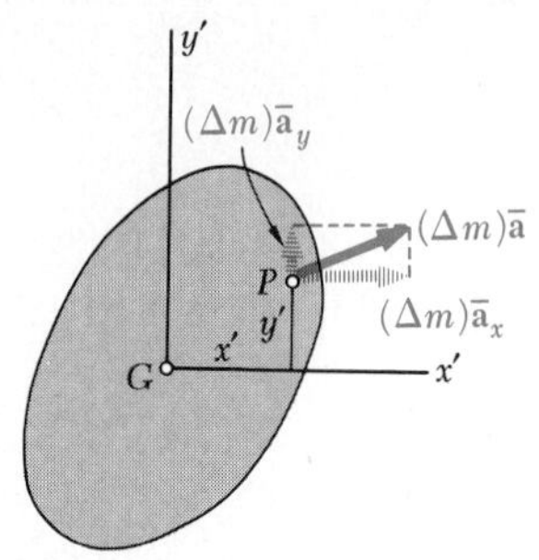

Fig. 16.3

$$+\circlearrowleft \Sigma(M_G)_{\text{eff, translation}} = \Sigma x'(\Delta m)\bar{a}_y - \Sigma y'(\Delta m)\bar{a}_x \\ = \bar{a}_y \Sigma x' \, \Delta m - \bar{a}_x \Sigma y' \, \Delta m \tag{16.5}$$

Now, according to the definition of the mass center of a system of particles (Sec. 12.6), we have

$$\Sigma x' \, \Delta m = m\bar{x}' \qquad \Sigma y' \, \Delta m = m\bar{y}' \tag{16.6}$$

where $\bar{x}'$ and $\bar{y}'$ are the coordinates of the mass center G and m the total mass. But $\bar{x}' = \bar{y}' = 0$, since the origin of the x' and y' coordinate axes was chosen at G, and Eq. (16.5) yields

$$\Sigma(M_G)_{\text{eff, translation}} = 0 \tag{16.7}$$

Considering now the effective forces $(\Delta m)\mathbf{a}'_n$ and $(\Delta m)\mathbf{a}'_t$ associated with the rotation of the slab about G (Fig. 16.4), we note that the moment of $(\Delta m)\mathbf{a}'_n$ about G is zero and that the moment of $(\Delta m)\mathbf{a}'_t$ about G is $r'(\Delta m)a'_t$. Recalling that $a'_t = r'\alpha$, and considering all the particles of the slab, we have

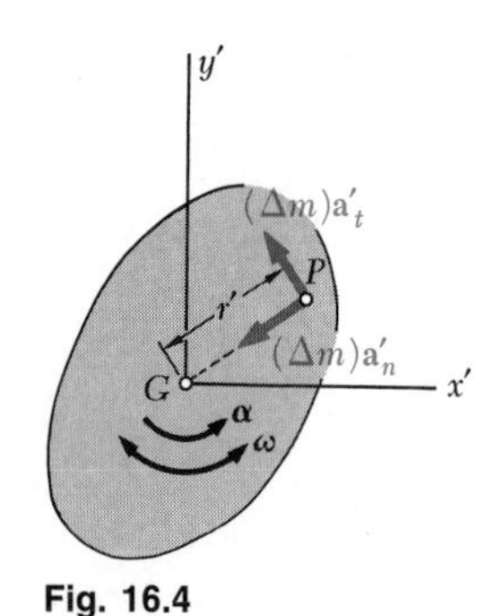

Fig. 16.4

$$+\circlearrowleft \Sigma(M_G)_{\text{eff, rotation}} = \Sigma r'(\Delta m)r'\alpha = \alpha \Sigma r'^2 \, \Delta m$$

Since the sum $\Sigma r'^2 \, \Delta m$ represents the moment of inertia $\bar{I}$ of the slab about an axis through G perpendicular to the slab, we write

$$\Sigma(M_G)_{\text{eff, rotation}} = \bar{I}\alpha \tag{16.8}$$

The sum of the moments about G of the effective forces of the slab is obtained by combining the information provided by Eqs. (16.7) and (16.8). We have

$$\Sigma(M_G)_{\text{eff}} = \bar{I}\alpha \tag{16.9}$$

Equations (16.3) and (16.9) indicate that the system of the effective forces shown in Fig. 16.1*b* may be replaced by an equivalent force-couple system consisting of a vector $m\bar{\mathbf{a}}$ attached at G and a couple of moment $\bar{I}\alpha$ and of the same sense as the angular acceleration $\boldsymbol{\alpha}$ (Fig. 16.5). Recalling from Sec. 15.3 that the angular acceleration of a rigid body rotating about a fixed axis may be represented by a vector $\boldsymbol{\alpha}$ of magnitude α directed along that axis, we note that the couple obtained may be represented by the vector $\bar{I}\boldsymbol{\alpha}$ of magnitude $\bar{I}\alpha$ directed along the axis of rotation.

In the case of the plane motion of a rigid slab,† d'Alembert's principle may thus be restated as follows: *The external forces acting on the body are equivalent to a force-couple system consisting of a vector $m\bar{\mathbf{a}}$ attached at the mass center G of the body and a couple $\bar{I}\boldsymbol{\alpha}$.*

† Or, more generally, in the case of the plane motion of a rigid body which has a principal centroidal axis of inertia perpendicular to the reference plane.

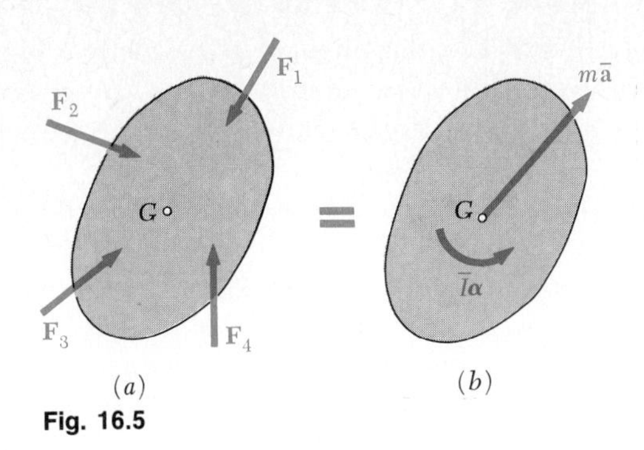

Fig. 16.5

The relation shown in Fig. 16.5 can be expressed algebraically by writing three equations relating respectively the x and y components and the moments about any given point A of the external and effective forces in Fig. 16.5. If the moments are computed about the mass center G of the rigid body, these equations of motion read

$$\Sigma F_x = m\bar{a}_x \qquad \Sigma F_y = m\bar{a}_y \qquad \Sigma M_G = \bar{I}\alpha \tag{16.10}$$

Translation. In the particular case of a body in translation, the angular acceleration of the body is identically equal to zero and its effective forces reduce to the vector $m\bar{\mathbf{a}}$ attached at G (Fig. 16.6). Thus, the resultant of the external forces acting on a rigid body in translation passes through the mass center of the body and is equal to $m\bar{\mathbf{a}}$.

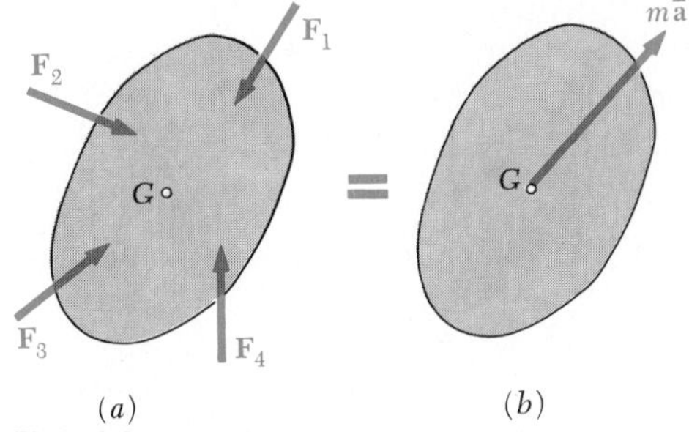

Fig. 16.6 Translation.

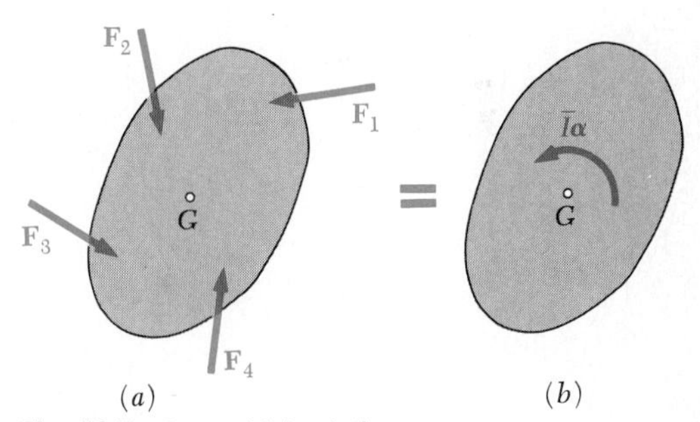

Fig. 16.7 Centroidal rotation.

Centroidal Rotation. When a slab, or, more generally, a body symmetrical with respect to the reference plane, rotates about a fixed axis perpendicular to the reference plane and passing through its mass center G, we say that the body is in *centroidal rotation*. Since the acceleration $\bar{\mathbf{a}}$ is identically equal to zero, the effective forces of the body reduce to the couple $\bar{I}\boldsymbol{\alpha}$ (Fig. 16.7). Thus, the external forces acting on a body in centroidal rotation are equivalent to a couple $\bar{I}\boldsymbol{\alpha}$.

General Plane Motion. Comparing Fig. 16.5 with Figs. 16.6 and 16.7, we observe that from the point of view of *kinetics*, the most general plane motion of a rigid body symmetrical with respect to the reference plane may be replaced by the sum of a translation and a centroidal rotation.

We should note that this statement is more restrictive than the similar statement made earlier from the point of view of *kinematics* (Sec. 15.5), since we now require that the mass center of the body be selected as the reference point.

Referring to Eqs. (16.10), we observe that the first two equations are identical with the equations of motion of a particle of mass m acted upon by the given forces $\mathbf{F}_1$, $\mathbf{F}_2$, $\mathbf{F}_3$, etc. We thus check that *the mass center G of a rigid body in plane motion moves as if the entire mass of the body were concentrated at that point, and as if all the external forces acted on it.* We recall that this result has already been obtained in Sec. 12.6 in the general case of a system of particles, the particles being not necessarily rigidly connected. We should note, however, that the system of the external forces does not, in general, reduce to a single vector $m\bar{\mathbf{a}}$ attached at G. Therefore, in the general case of the plane motion of a rigid body, *the resultant of the external forces acting on the body does not pass through the mass center of the body.*

Finally, we may observe that the last of Eqs. (16.10) would still be valid if the rigid body while subjected to the same applied forces were constrained to rotate about a fixed axis through G. Thus, *a rigid body in plane motion rotates about its mass center as if this point were fixed.*

16.3. Solution of Problems Involving the Plane Motion of a Rigid Body. We saw in Sec. 16.2 that when a rigid body is in plane motion, there exists a fundamental relation between the forces $\mathbf{F}_1$, $\mathbf{F}_2$, $\mathbf{F}_3$, etc., acting on the body, the acceleration $\bar{\mathbf{a}}$ of its mass center, and the angular acceleration $\boldsymbol{\alpha}$ of the body. This relation, which is represented in Fig. 16.8 in the form of a *free-body-diagram equation,* may be used to determine the acceleration $\bar{\mathbf{a}}$ and the angular acceleration $\boldsymbol{\alpha}$ produced by a given system of forces acting on a rigid body or, conversely, to determine the forces which produce a given motion of the rigid body.

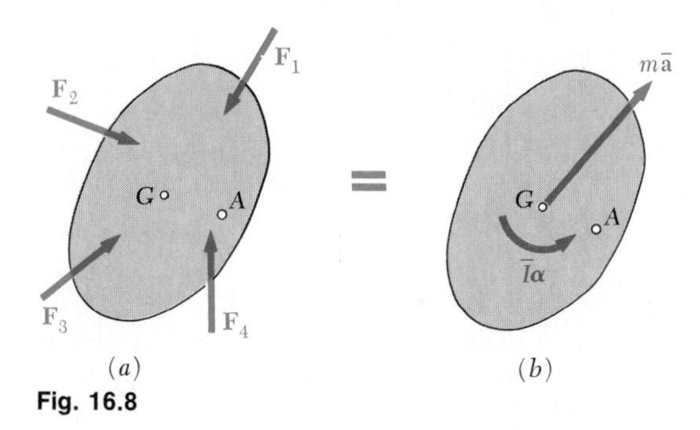

Fig. 16.8

While the three algebraic equations (16.10) may be used to solve problems of plane motion,† our experience in statics suggests that the solution

† We recall that the last of Eqs. (16.10) is valid only in the case of the plane motion of a rigid body symmetrical with respect to the reference plane. The plane motion of bodies which are not symmetrical with respect to the reference plane is discussed in Sec. 16.6.

of many problems involving rigid bodies could be simplified by an appropriate choice of the point about which the moments of the forces are computed. It is therefore preferable to remember the relation existing between the forces and the accelerations in the pictorial form shown in Fig. 16.8, and to derive from this fundamental relation the component or moment equations which fit best the solution of the problem under consideration. The equations most frequently used are

$$\Sigma F_x = \Sigma(F_x)_{\text{eff}} \qquad \Sigma F_y = \Sigma(F_y)_{\text{eff}} \qquad \Sigma M_A = \Sigma(M_A)_{\text{eff}} \tag{16.11}$$

We easily verify that the first two equations reduce to the corresponding equations in (16.10). The third equation, however, will generally be different from the corresponding equation in (16.10), since the selection of point A, about which the moments of the external and effective forces are to be computed, will be guided by our desire to simplify the solution of the problem as much as possible. In Sample Prob. 16.1, for example, the three unknown forces $\mathbf{F}_A$, $\mathbf{F}_B$, and $\mathbf{N}_A$ may be eliminated from this equation by equating moments about A. It should be noted, however, that, unless A happens to coincide with the mass center G, the determination of $\Sigma(M_A)_{\text{eff}}$ will involve the computation of the moments of the components of the vector $m\bar{\mathbf{a}}$, as well as the computation of the moment of the couple $\bar{I}\boldsymbol{\alpha}$.

The fundamental relation shown in Fig. 16.8 may be presented in an alternative form if we add to the external forces an inertia vector $(m\bar{\mathbf{a}})_{\text{rev}}$ of sense opposite to that of $\bar{\mathbf{a}}$, attached at G, and an inertia couple $(\bar{I}\boldsymbol{\alpha})_{\text{rev}}$ of moment equal in magnitude to $\bar{I}\alpha$ and of sense opposite to that of $\boldsymbol{\alpha}$ (Fig. 16.9). The system obtained is equivalent to zero, and the rigid body is said to be in dynamic equilibrium.

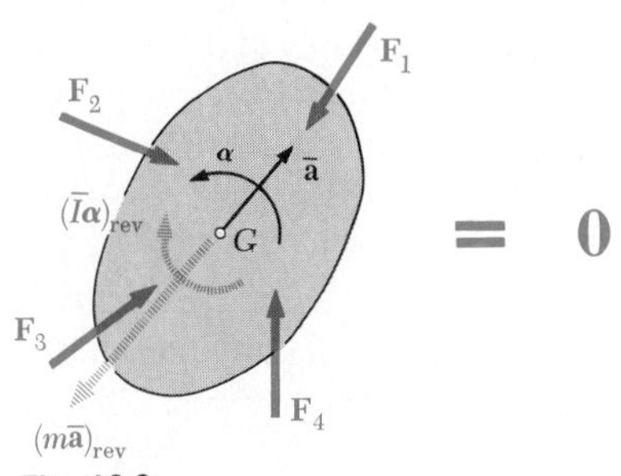

Fig. 16.9

Whether the principle of equivalence of external and effective forces is directly applied, as in Fig. 16.8, or whether the concept of dynamic equilibrium is introduced, as in Fig. 16.9, the use of free-body-diagram equations showing vectorially the relationship existing between the forces applied on the rigid body and the resulting linear and angular accelerations presents considerable advantages over the blind application of formulas (16.10). These advantages may be summarized as follows:

1. A much clearer understanding of the effect of the forces on the motion of the body will result from the use of a pictorial representation.
2. This approach makes it possible to divide the solution of a dynamics problem into two parts: In the first part, the analysis of the kinematic and kinetic characteristics of the problem leads to the free-body diagrams of Fig. 16.8 or 16.9; in the second part, the diagram obtained is used to analyze by the methods of Chap. 3 the various forces and vectors involved.
3. A unified approach is provided for the analysis of the plane motion of a rigid body, regardless of the particular type of motion involved. While the kinematics of the various motions considered may vary from one case to the other, the approach to the kinetics of the motion is consistently the same. In every case we shall draw a diagram showing the external forces, the vector $m\bar{\mathbf{a}}$ associated with the motion of G, and the couple $\bar{I}\boldsymbol{\alpha}$ associated with the rotation of the body about G.
4. The resolution of the plane motion of a rigid body into a translation and a centroidal rotation, which is used here, is a basic concept which may be applied effectively throughout the study of mechanics. We shall use it again in Chap. 17 with the method of work and energy and in Chap. 18 with the method of impulse and momentum.

16.4. Systems of Rigid Bodies. The method described in the preceding section may also be used in problems involving the plane motion of several connected rigid bodies. A diagram similar to Fig. 16.8 or Fig. 16.9 may be drawn for each part of the system. The equations of motion obtained from these diagrams are solved simultaneously.

In some cases, as in Sample Prob. 16.3, a single diagram may be drawn for the entire system. This diagram should include all the external forces, as well as the vectors $m\bar{\mathbf{a}}$ and the couples $\bar{I}\boldsymbol{\alpha}$ associated with the various parts of the system. However, internal forces, such as the forces exerted by connecting cables, may be omitted since they occur in pairs of equal and opposite forces and are thus equipollent to zero. The equations obtained by expressing that the system of the external forces is equipollent to the system of the effective forces may be solved for the remaining unknowns.†

This second approach may not be used in problems involving more than three unknowns, since only three equations of motion are available when a single diagram is used. We shall not elaborate upon this point, since the discussion involved would be completely similar to that given in Sec. 6.11 in the case of the equilibrium of a system of rigid bodies.

†Note that we cannot speak of *equivalent* systems since we are not dealing with a single rigid body.

SAMPLE PROBLEM 16.1

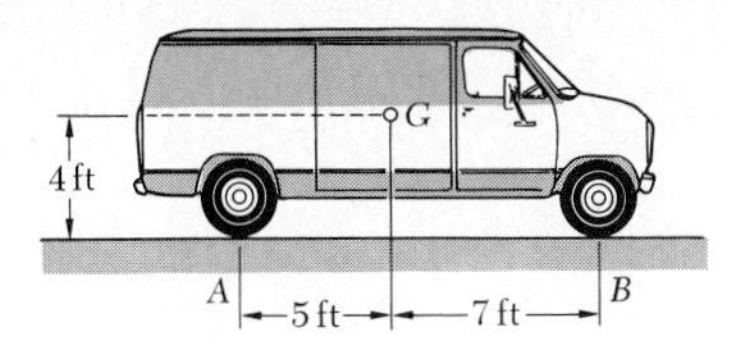

When the forward speed of the truck shown was 30 ft/s, the brakes were suddenly applied, causing all four wheels to stop rotating. It was observed that the truck skidded to rest in 20 ft. Determine the magnitude of the normal reaction and of the friction force at each wheel as the truck skidded to rest.

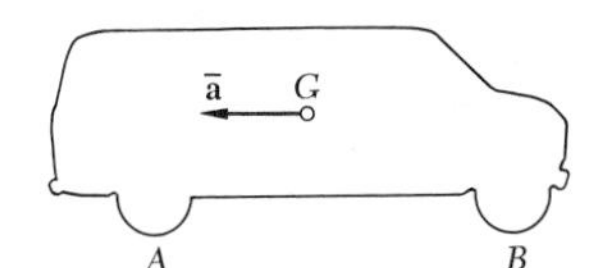

Kinematics of Motion. Choosing the positive sense to the right and using the equations of uniformly accelerated motion, we write

$$\bar{v}_0 = +30 \text{ ft/s} \qquad \bar{v}^2 = \bar{v}_0^2 + 2\bar{a}\bar{x} \qquad 0 = (30)^2 + 2\bar{a}(20)$$

$$\bar{a} = -22.5 \text{ ft/s}^2 \qquad \bar{\mathbf{a}} = 22.5 \text{ ft/s}^2 \leftarrow$$

Equations of Motion. The external forces consist of the weight $\mathbf{W}$ of the truck and of the normal reactions and friction forces at the wheels. (The vectors $\mathbf{N}_A$ and $\mathbf{F}_A$ represent the sum of the reactions at the rear wheels, while $\mathbf{N}_B$ and $\mathbf{F}_B$ represent the sum of the reactions at the front wheels.) Since the truck is in translation, the effective forces reduce to the vector $m\bar{\mathbf{a}}$ attached at G. Three equations of motion are obtained by expressing that the system of the external forces is equivalent to the system of the effective forces.

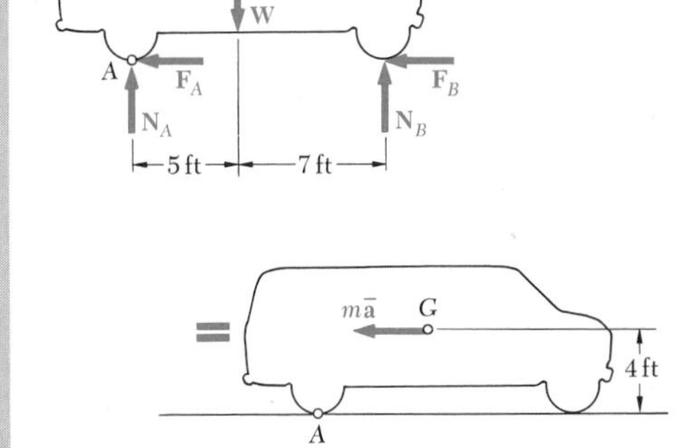

$$+\uparrow \Sigma F_y = \Sigma(F_y)_{\text{eff}}: \qquad N_A + N_B - W = 0$$

Since $F_A = \mu_k N_A$ and $F_B = \mu_k N_B$ where μ_k is the coefficient of kinetic friction, we find that

$$F_A + F_B = \mu_k(N_A + N_B) = \mu_k W$$

$$\overset{+}{\rightarrow} \Sigma F_x = \Sigma(F_x)_{\text{eff}}: \quad -(F_A + F_B) = -m\bar{a}$$

$$-\mu_k W = -\frac{W}{32.2 \text{ ft/s}^2}(22.5 \text{ ft/s}^2)$$

$$\mu_k = 0.699$$

$$+\circlearrowleft \Sigma M_A = \Sigma(M_A)_{\text{eff}}: \; -W(5 \text{ ft}) + N_B(12 \text{ ft}) = m\bar{a}(4 \text{ ft})$$

$$-W(5 \text{ ft}) + N_B(12 \text{ ft}) = \frac{W}{32.2 \text{ ft/s}^2}(22.5 \text{ ft/s}^2)(4 \text{ ft})$$

$$N_B = 0.650W$$

$$F_B = \mu_k N_B = (0.699)(0.650W) \qquad F_B = 0.454W$$

$$+\uparrow \Sigma F_y = \Sigma(F_y)_{\text{eff}}: \quad N_A + N_B - W = 0$$

$$N_A + 0.650W - W = 0$$

$$N_A = 0.350W$$

$$F_A = \mu_k N_A = (0.699)(0.350W) \qquad F_A = 0.245W$$

Reactions at Each Wheel. Recalling that the values computed above represent the sum of the reactions at the two front wheels or the two rear wheels, we obtain the magnitude of the reactions at each wheel by writing

$$N_{\text{front}} = \tfrac{1}{2}N_B = 0.325W \qquad N_{\text{rear}} = \tfrac{1}{2}N_A = 0.175W \blacktriangleleft$$

$$F_{\text{front}} = \tfrac{1}{2}F_B = 0.227W \qquad F_{\text{rear}} = \tfrac{1}{2}F_A = 0.122W \blacktriangleleft$$

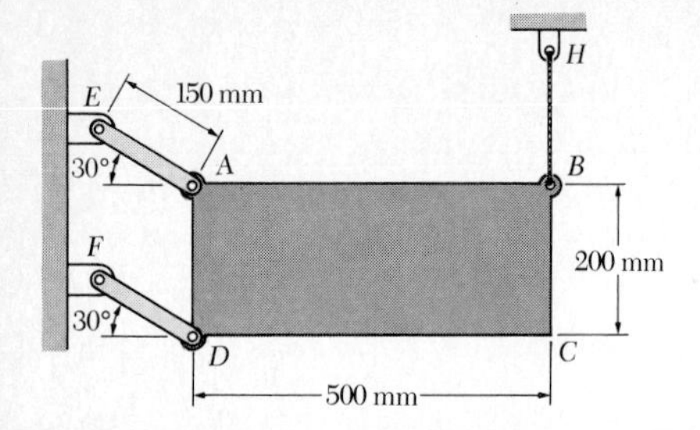

SAMPLE PROBLEM 16.2

The thin plate $ABCD$ has a mass of 8 kg and is held in the position shown by the wire BH and two links AE and DF. Neglecting the mass of the links, determine immediately after wire BH has been cut (*a*) the acceleration of the plate, (*b*) the force in each link.

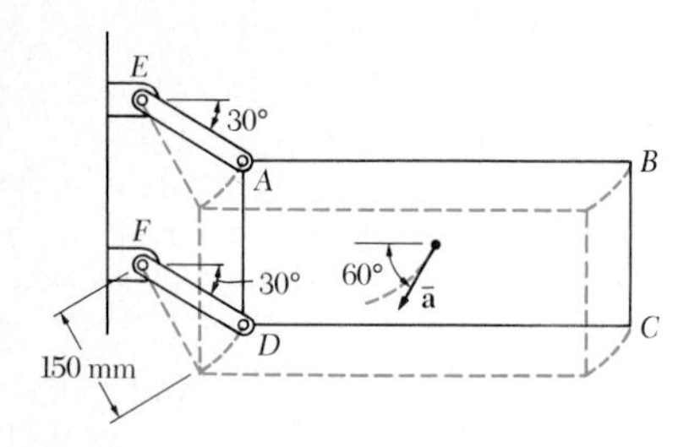

Kinematics of Motion. After wire BH has been cut, we observe that corners A and D move along parallel circles of radius 150 mm centered, respectively, at E and F. The motion of the plate is thus a curvilinear translation; the particles forming the plate move along parallel circles of radius 150 mm.

At the instant wire BH is cut, the velocity of the plate is zero. Thus the acceleration $\bar{\mathbf{a}}$ of the mass center G of the plate is tangent to the circular path which will be described by G.

Equations of Motion. The external forces consist of the weight $\mathbf{W}$ and the forces $\mathbf{F}_{AE}$ and $\mathbf{F}_{DF}$ exerted by the links. Since the plate is in translation, the effective forces reduce to the vector $m\bar{\mathbf{a}}$ attached at G and directed along the t axis. We shall now express that the system of the external forces is equivalent to the system of the effective forces.

a. Acceleration of the Plate.

$+\swarrow \Sigma F_t = \Sigma(F_t)_{\text{eff}}$:

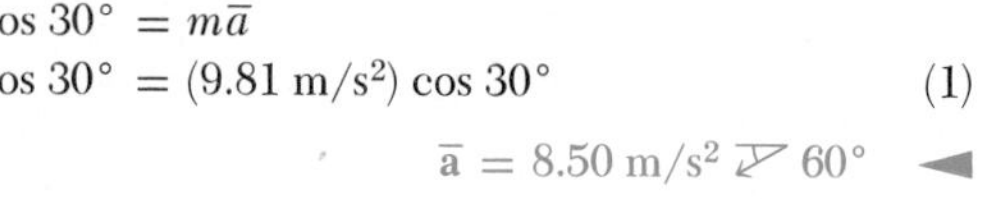

$$W \cos 30° = m\bar{a}$$
$$mg \cos 30° = m\bar{a}$$
$$\bar{a} = g \cos 30° = (9.81 \text{ m/s}^2) \cos 30° \qquad (1)$$
$$\bar{\mathbf{a}} = 8.50 \text{ m/s}^2 \nearrow 60° \blacktriangleleft$$

b. Forces in Links AE and DF.

$+\nwarrow \Sigma F_n = \Sigma(F_n)_{\text{eff}}$: $\quad F_{AE} + F_{DF} - W \sin 30° = 0 \qquad (2)$

$+\circlearrowright \Sigma M_G = \Sigma(M_G)_{\text{eff}}$:

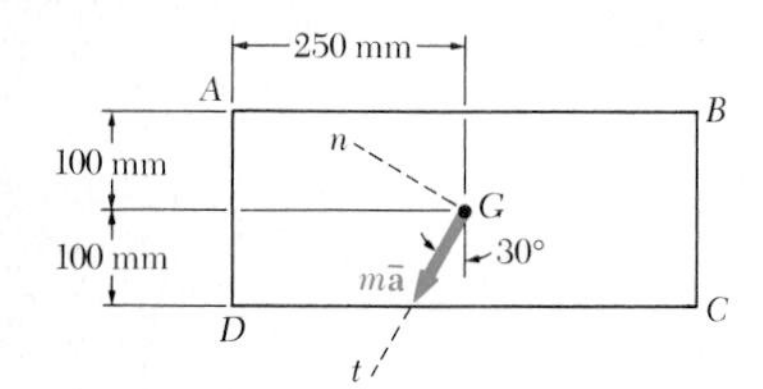

$$(F_{AE} \sin 30°)(250 \text{ mm}) - (F_{AE} \cos 30°)(100 \text{ mm})$$
$$+(F_{DF} \sin 30°)(250 \text{ mm}) + (F_{DF} \cos 30°)(100 \text{ mm}) = 0$$
$$38.4F_{AE} + 211.6F_{DF} = 0$$
$$F_{DF} = -0.1815F_{AE} \qquad (3)$$

Substituting for F_{DF} from (3) into (2), we write

$$F_{AE} - 0.1815F_{AE} - W \sin 30° = 0$$
$$F_{AE} = 0.6109W$$
$$F_{DF} = -0.1815(0.6109W) = -0.1109W$$

Noting that $W = mg = (8 \text{ kg})(9.81 \text{ m/s}^2) = 78.48$ N, we have

$$F_{AE} = 0.6109(78.48 \text{ N}) \qquad F_{AE} = 47.9 \text{ N } T \blacktriangleleft$$
$$F_{DF} = -0.1109(78.48 \text{ N}) \qquad F_{DF} = 8.70 \text{ N } C \blacktriangleleft$$

SAMPLE PROBLEM 16.3

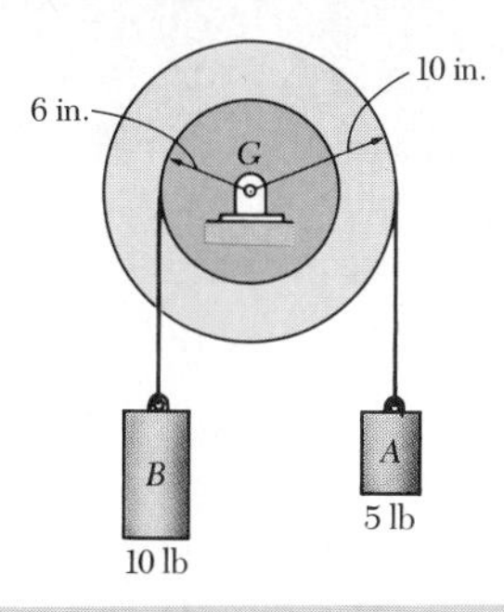

A pulley weighing 12 lb and having a radius of gyration of 8 in. is connected to two cylinders as shown. Assuming no axle friction, determine the angular acceleration of the pulley and the acceleration of each cylinder.

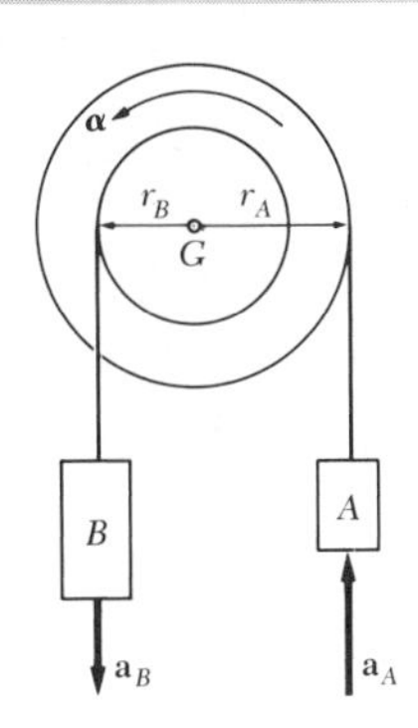

Sense of Motion. Although an arbitrary sense of motion may be assumed (since no friction forces are involved) and later checked by the sign of the answer, we may prefer first to determine the actual sense of rotation of the pulley. We first find the weight of cylinder B required to maintain the equilibrium of the pulley when it is acted upon by the 5-lb cylinder A. We write

$$+\circlearrowleft\Sigma M_G = 0: \qquad W_B(6\text{ in.}) - (5\text{ lb})(10\text{ in.}) = 0 \qquad W_B = 8.33\text{ lb}$$

Since cylinder B actually weighs 10 lb, the pulley will rotate counterclockwise.

Kinematics of Motion. Assuming $\boldsymbol{\alpha}$ counterclockwise and noting that $a_A = r_A\alpha$ and $a_B = r_B\alpha$, we obtain

$$\mathbf{a}_A = (\tfrac{10}{12}\text{ ft})\alpha\uparrow \qquad \mathbf{a}_B = (\tfrac{6}{12}\text{ ft})\alpha\downarrow$$

Equations of Motion. A single system consisting of the pulley and the two cylinders is considered. Forces external to this system consist of the weights of the pulley and the two cylinders and of the reaction at G. (The forces exerted by the cables on the pulley and on the cylinders are internal to the system considered and cancel out.) Since the motion of the pulley is a centroidal rotation and the motion of each cylinder is a translation, the effective forces reduce to the couple $\bar{I}\boldsymbol{\alpha}$ and the two vectors $m\mathbf{a}_A$ and $m\mathbf{a}_B$. The centroidal moment of inertia of the pulley is

$$\bar{I} = m\bar{k}^2 = \frac{W}{g}\bar{k}^2 = \frac{12\text{ lb}}{32.2\text{ ft/s}^2}(\tfrac{8}{12}\text{ ft})^2 = 0.1656\text{ lb}\cdot\text{ft}\cdot\text{s}^2$$

Since the system of the external forces is equipollent to the system of the effective forces, we write

$$+\circlearrowleft\Sigma M_G = \Sigma(M_G)_{\text{eff}}:$$

$$(10\text{ lb})(\tfrac{6}{12}\text{ ft}) - (5\text{ lb})(\tfrac{10}{12}\text{ ft}) = +\bar{I}\alpha + m_B a_B(\tfrac{6}{12}\text{ ft}) + m_A a_A(\tfrac{10}{12}\text{ ft})$$

$$(10)(\tfrac{6}{12}) - (5)(\tfrac{10}{12}) = 0.1656\alpha + \tfrac{10}{32.2}(\tfrac{6}{12}\alpha)(\tfrac{6}{12}) + \tfrac{5}{32.2}(\tfrac{10}{12}\alpha)(\tfrac{10}{12})$$

$$\alpha = +2.374\text{ rad/s}^2 \qquad \boldsymbol{\alpha} = 2.37\text{ rad/s}^2\circlearrowleft \blacktriangleleft$$

$$a_A = r_A\alpha = (\tfrac{10}{12}\text{ ft})(2.374\text{ rad/s}^2) \qquad \mathbf{a}_A = 1.978\text{ ft/s}^2\uparrow \blacktriangleleft$$

$$a_B = r_B\alpha = (\tfrac{6}{12}\text{ ft})(2.374\text{ rad/s}^2) \qquad \mathbf{a}_B = 1.187\text{ ft/s}^2\downarrow \blacktriangleleft$$

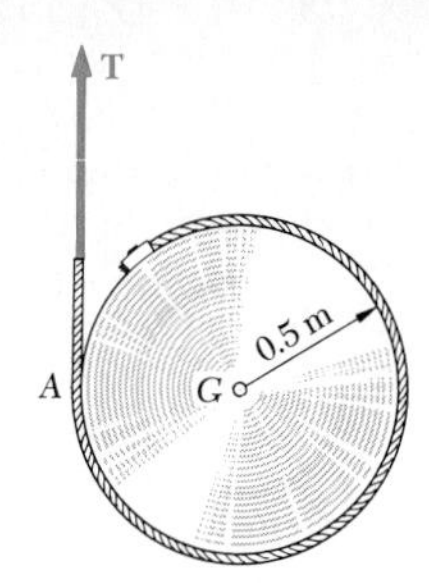

SAMPLE PROBLEM 16.4

A cord is wrapped around a homogeneous disk of radius $r = 0.5$ m and mass $m = 15$ kg. If the cord is pulled upward with a force **T** of magnitude 180 N, determine (*a*) the acceleration of the center of the disk, (*b*) the angular acceleration of the disk, (*c*) the acceleration of the cord.

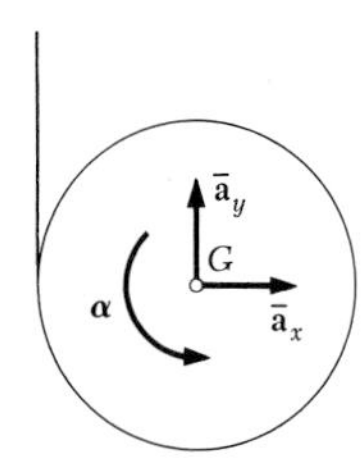

Equations of Motion. We assume that the components $\bar{\mathbf{a}}_x$ and $\bar{\mathbf{a}}_y$ of the acceleration of the center are directed, respectively, to the right and upward and that the angular acceleration of the disk is counterclockwise. The external forces acting on the disk consist of the weight **W** and the force **T** exerted by the cord. This system is equivalent to the system of the effective forces, which consists of a vector of components $m\bar{\mathbf{a}}_x$ and $m\bar{\mathbf{a}}_y$ attached at G and a couple $\bar{I}\boldsymbol{\alpha}$. We write

$$\overset{+}{\rightarrow} \Sigma F_x = \Sigma(F_x)_{\text{eff}}: \qquad 0 = m\bar{a}_x \qquad \bar{\mathbf{a}}_x = 0 \blacktriangleleft$$

$$+\uparrow \Sigma F_y = \Sigma(F_y)_{\text{eff}}: \qquad T - W = m\bar{a}_y$$

$$\bar{a}_y = \frac{T - W}{m}$$

Since $T = 180$ N, $m = 15$ kg, and $W = (15\text{ kg})(9.81\text{ m/s}^2) = 147.1$ N, we have

$$\bar{a}_y = \frac{180\text{ N} - 147.1\text{ N}}{15\text{ kg}} = +2.19\text{ m/s}^2 \qquad \bar{\mathbf{a}}_y = 2.19\text{ m/s}^2 \uparrow \blacktriangleleft$$

$$+\circlearrowleft \Sigma M_G = \Sigma(M_G)_{\text{eff}}: \qquad -Tr = \bar{I}\alpha$$

$$-Tr = (\tfrac{1}{2}mr^2)\alpha$$

$$\alpha = -\frac{2T}{mr} = -\frac{2(180\text{ N})}{(15\text{ kg})(0.5\text{ m})} = -48.0\text{ rad/s}^2$$

$$\boldsymbol{\alpha} = 48.0\text{ rad/s}^2 \circlearrowright \blacktriangleleft$$

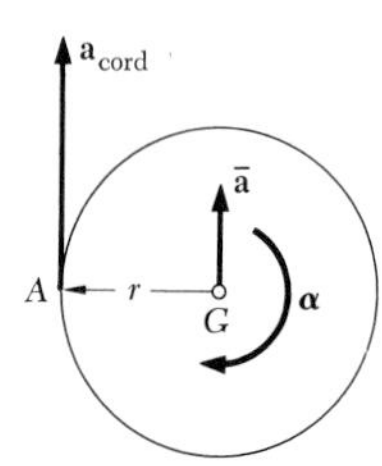

Acceleration of Cord. Since the acceleration of the cord is equal to the tangential component of the acceleration of point A on the disk, we write

$$\mathbf{a}_{\text{cord}} = (\mathbf{a}_A)_t = \bar{\mathbf{a}} + (\mathbf{a}_{A/G})_t$$
$$= [2.19\text{ m/s}^2 \uparrow] + [(0.5\text{ m})(48\text{ rad/s}^2) \uparrow]$$

$$\mathbf{a}_{\text{cord}} = 26.2\text{ m/s}^2 \uparrow \blacktriangleleft$$

SAMPLE PROBLEM 16.5

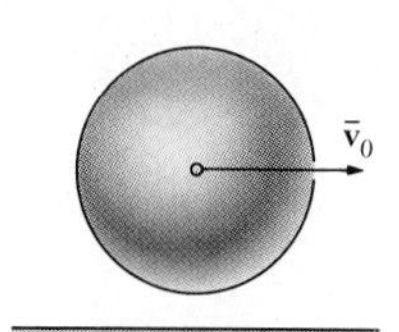

A uniform sphere of mass m and radius r is projected along a rough horizontal surface with a linear velocity $\bar{\mathbf{v}}_0$ and no angular velocity. Denoting by μ_k the coefficient of kinetic friction between the sphere and the floor, determine (*a*) the time t_1 at which the sphere will start rolling without sliding, (*b*) the linear velocity and angular velocity of the sphere at time t_1.

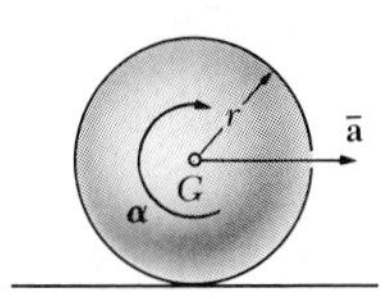

Equations of Motion. The positive sense is chosen to the right for $\bar{\mathbf{a}}$ and clockwise for $\boldsymbol{\alpha}$. The external forces acting on the sphere consist of the weight **W**, the normal reaction **N**, and the friction force **F**. Since the point of the sphere in contact with the surface is sliding to the right, the friction force **F** is directed to the left. While the sphere is sliding, the magnitude of the friction force is $F = \mu_k N$. The effective forces consist of the vector $m\bar{\mathbf{a}}$ attached at G and the couple $\bar{I}\boldsymbol{\alpha}$. Expressing that the system of the external forces is equivalent to the system of the effective forces, we write

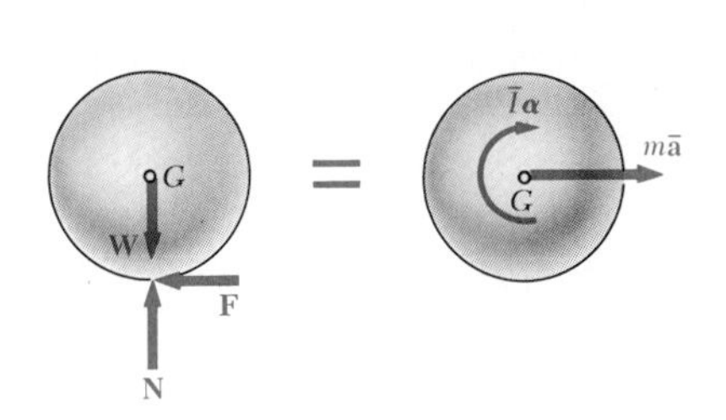

$$+\uparrow \Sigma F_y = \Sigma(F_y)_{\text{eff}}: \qquad N - W = 0$$

$$N = W = mg \qquad F = \mu_k N = \mu_k mg$$

$$\xrightarrow{+} \Sigma F_x = \Sigma(F_x)_{\text{eff}}: \qquad -F = m\bar{a} \qquad -\mu_k mg = m\bar{a} \qquad \bar{a} = -\mu_k g$$

$$+\circlearrowright \Sigma M_G = \Sigma(M_G)_{\text{eff}}: \qquad Fr = \bar{I}\alpha$$

Noting that $\bar{I} = \frac{2}{5}mr^2$ and substituting the value obtained for F, we write

$$(\mu_k mg)r = \tfrac{2}{5}mr^2\alpha \qquad \alpha = \frac{5}{2}\frac{\mu_k g}{r}$$

Kinematics of Motion. As long as the sphere both rotates and slides, its linear and angular motions are uniformly accelerated.

$$t = 0,\ \bar{v} = \bar{v}_0 \qquad \bar{v} = \bar{v}_0 + \bar{a}t = \bar{v}_0 - \mu_k g t \tag{1}$$

$$t = 0,\ \omega_0 = 0 \qquad \omega = \omega_0 + \alpha t = 0 + \left(\frac{5}{2}\frac{\mu_k g}{r}\right)t \tag{2}$$

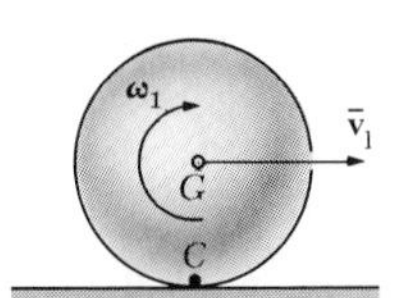

The sphere will start rolling without sliding when the velocity $\mathbf{v}_C$ of the point of contact C is zero. At that time, $t = t_1$, point C becomes the instantaneous center of rotation, and we have

$$\bar{v}_1 = r\omega_1 \tag{3}$$

Substituting in (3) the values obtained for $\bar{v}_1$ and ω_1 by making $t = t_1$ in (1) and (2), respectively, we write

$$\bar{v}_0 - \mu_k g t_1 = r\left(\frac{5}{2}\frac{\mu_k g}{r}t_1\right) \qquad t_1 = \frac{2}{7}\frac{\bar{v}_0}{\mu_k g} \blacktriangleleft$$

Substituting for t_1 into (2), we have

$$\omega_1 = \frac{5}{2}\frac{\mu_k g}{r}t_1 = \frac{5}{2}\frac{\mu_k g}{r}\left(\frac{2}{7}\frac{\bar{v}_0}{\mu_k g}\right) \qquad \omega_1 = \frac{5}{7}\frac{\bar{v}_0}{r} \qquad \boldsymbol{\omega}_1 = \frac{5}{7}\frac{\bar{v}_0}{r} \circlearrowright \blacktriangleleft$$

$$\bar{v}_1 = r\omega_1 = r\left(\frac{5}{7}\frac{\bar{v}_0}{r}\right) \qquad \bar{v}_1 = \tfrac{5}{7}\bar{v}_0 \qquad \bar{\mathbf{v}}_1 = \tfrac{5}{7}\bar{v}_0 \rightarrow \blacktriangleleft$$

Problems

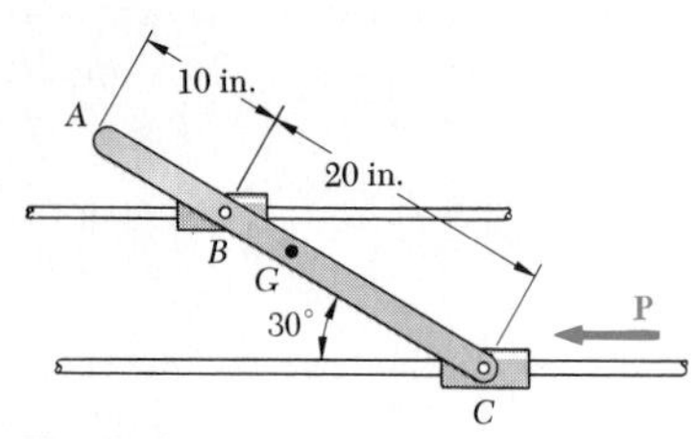

Fig. P16.1

16.1 A uniform rod *ABC* weighs 16 lb and is connected to two collars of negligible weight which slide on horizontal, frictionless rods located in the same vertical plane. If a force **P** of magnitude 4 lb is applied at *C*, determine (*a*) the acceleration of the rod, (*b*) the reactions at *B* and *C*.

16.2 In Prob. 16.1, determine (*a*) the required magnitude of **P** if the reaction at *B* is to be 8 lb upward, (*b*) the corresponding acceleration of the rod.

16.3 and 16.4 The motion of the 1.5-kg rod *AB* is guided by two small wheels which roll freely in horizontal slots. If a force **P** of magnitude 5 N is applied at *B*, determine (*a*) the acceleration of the rod, (*b*) the reactions at *A* and *B*.

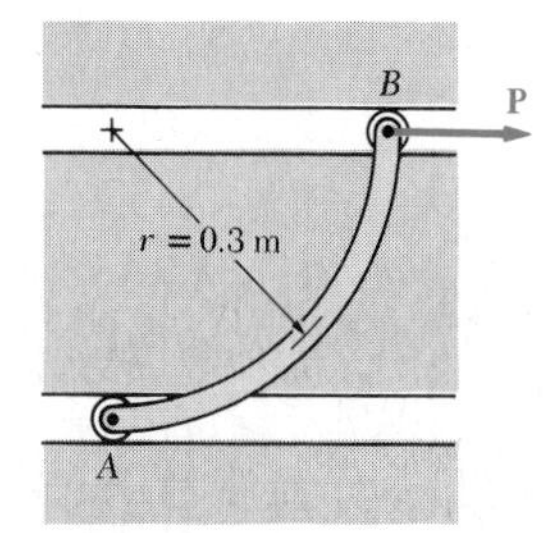

Fig. P16.3

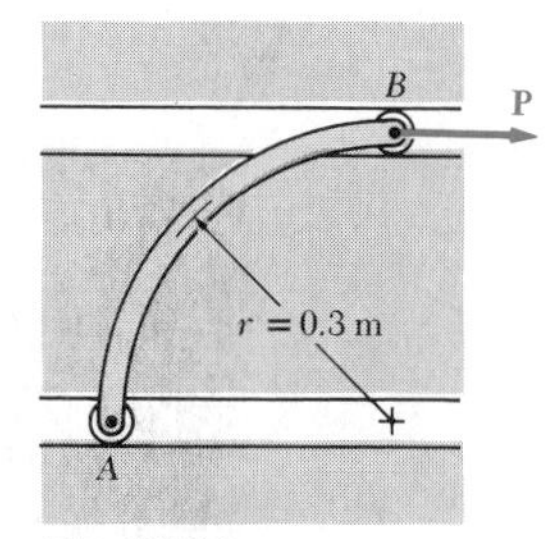

Fig. P16.4

16.5 A uniform rod *BC* of mass 3 kg is connected to a collar *A* by a 0.2-m cord *AB*. Neglecting the mass of the collar and cord, determine (*a*) the smallest constant acceleration $\mathbf{a}_A$ for which the cord and the rod lie in a straight line, (*b*) the corresponding tension in the cord.

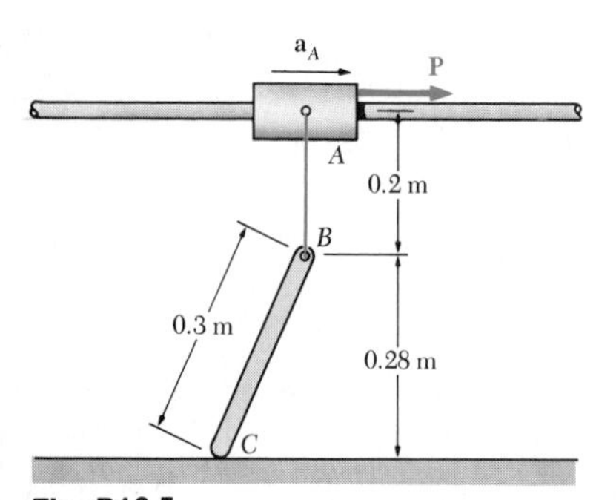

Fig. P16.5

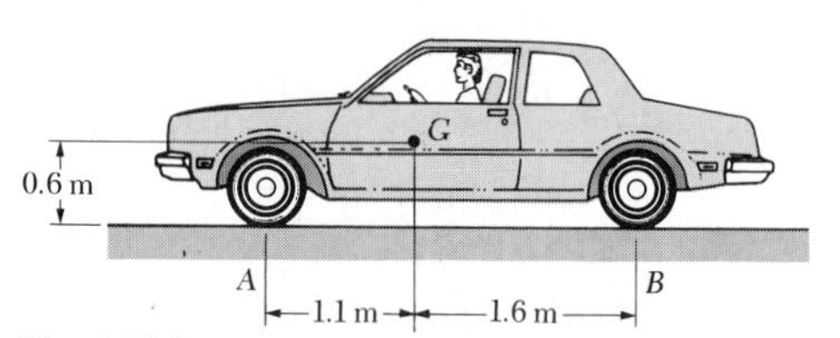

Fig. P16.6

16.6 Knowing that the coefficient of static friction between the tires and road is 0.80 for the car shown, determine the maximum possible acceleration on a level road, assuming (*a*) four-wheel drive, (*b*) conventional rear-wheel drive, (*c*) front-wheel drive.

16.7 Determine the distance through which the truck of Sample Prob. 16.1 will skid if (*a*) the rear-wheel brakes fail to operate, (*b*) the front-wheel brakes fail to operate.

16.8 A 20-kg rectangular panel is suspended from two skids A and B and is maintained in the position shown by a wire CD. Knowing that the coefficient of kinetic friction between each skid and the inclined track is $\mu_k = 0.15$, determine the normal component of the reaction at B immediately after wire CD has been cut.

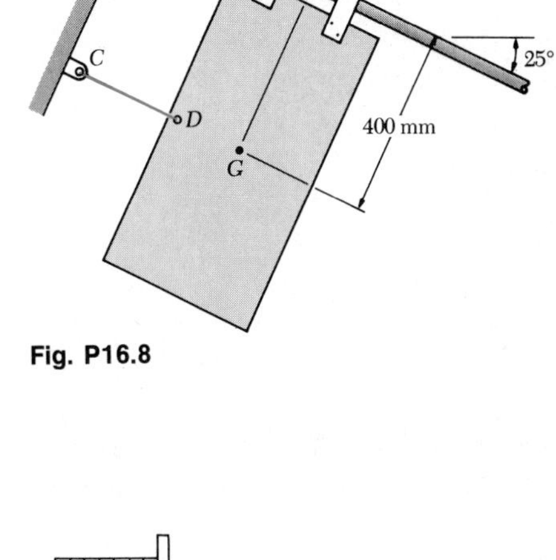

Fig. P16.8

16.9 In Prob. 16.8, determine (*a*) the maximum value of μ_k for which the skid at B remains in contact with the track after wire CD has been cut, (*b*) the corresponding acceleration of the panel.

16.10 Cylindrical cans are transported from one elevation to another by the moving horizontal arms shown. Assuming that $\mu_s = 0.25$ between the cans and the arms, determine (*a*) the magnitude of the upward acceleration **a** for which the cans slide on the horizontal arms, (*b*) the smallest ratio h/d for which the cans tip before they slide.

16.11 Solve Prob. 16.10, assuming that the acceleration **a** of the horizontal arms is directed downward.

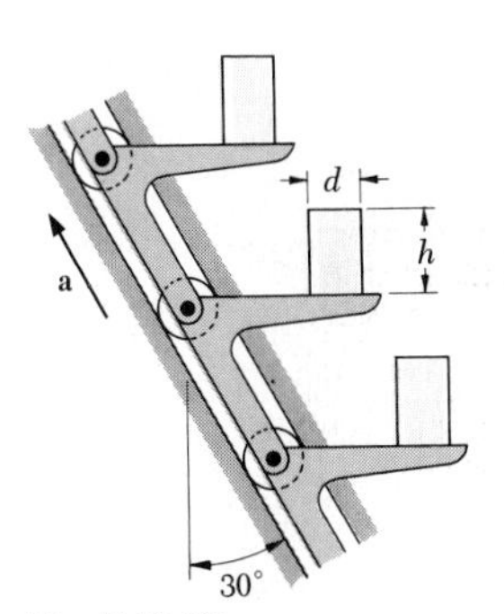

Fig. P16.10

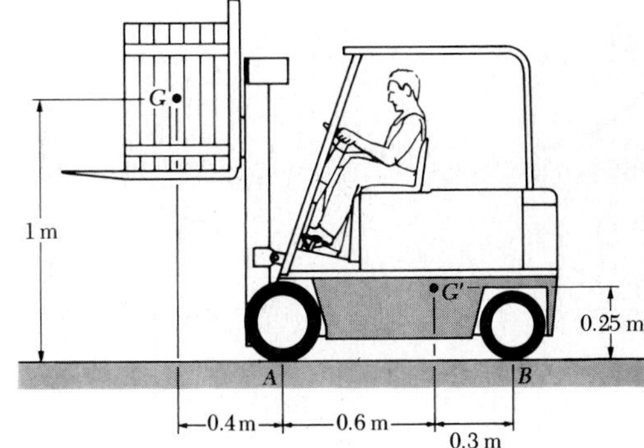

Fig. P16.12

16.12 A 2500-kg fork-lift truck carries the 1200-kg crate at the height shown. The truck is moving to the left when the brakes are applied, causing a deceleration of 3 m/s^2. Knowing that the coefficient of static friction between the crate and the fork lift is 0.60, determine the vertical component of the reaction at each wheel.

16.13 In Prob. 16.12, determine the maximum allowable deceleration of the truck if the crate is not to slide forward and if the truck is not to tip forward.

16.14 A 70-lb cabinet is mounted on casters which allow it to move freely ($\mu = 0$) on the floor. If a 35-lb force is applied as shown, determine (*a*) the acceleration of the cabinet, (*b*) the range of values of h for which the cabinet will not tip.

16.15 Solve Prob. 16.14, assuming that the casters are locked and slide along the rough floor ($\mu_k = 0.20$).

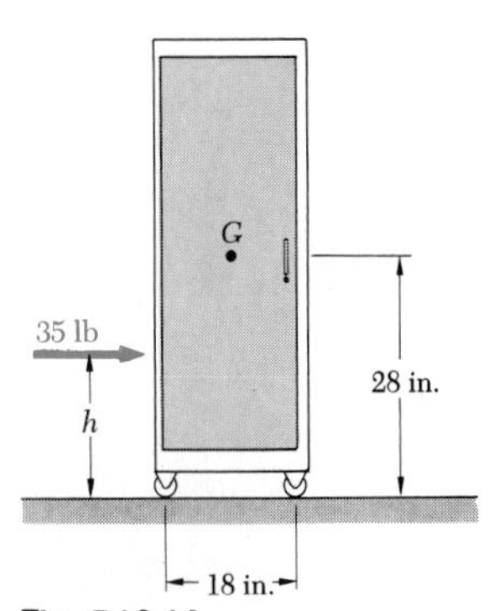

Fig. P16.14

16.16 A completely filled barrel and its contents have a combined mass of 125 kg. A 60-kg cylinder is connected to the barrel as shown. Knowing that the coefficients of friction between the barrel and the floor are $\mu_s = 0.35$ and $\mu_k = 0.30$, determine (*a*) the acceleration of the barrel, (*b*) the range of values of h for which the barrel will not tip.

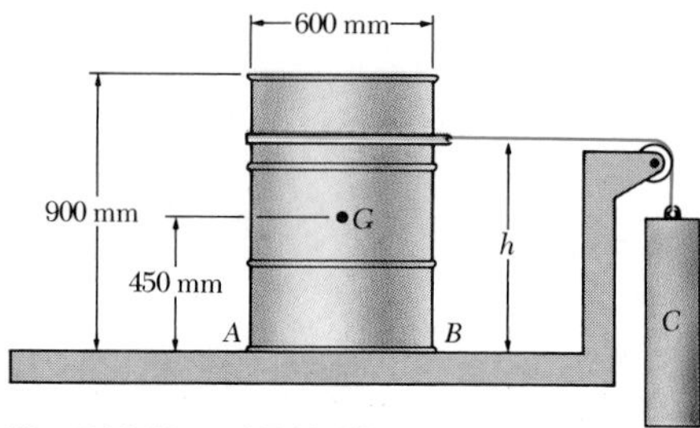

Fig. P16.16 and P16.17

16.17 A completely filled barrel and its contents have a combined mass of 160 kg. A cylinder C of mass m_C is connected to the barrel at a height $h = 700$ mm. Knowing that the coefficients of friction between the barrel and the floor are $\mu_s = 0.40$ and $\mu_k = 0.35$, determine the range of values of the mass of cylinder C for which the barrel will not tip.

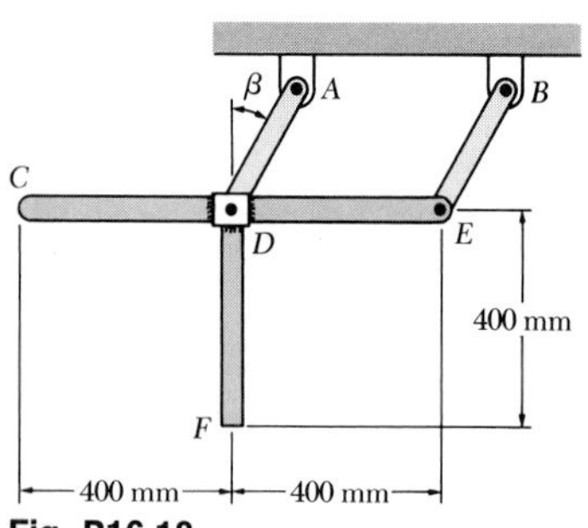

Fig. P16.18

16.18 Three uniform rods CD, DE, and DF, each of mass 1.8 kg, are welded together and are pin-connected to two links AD and BE. Neglecting the mass of the links, determine the force in each link immediately after the system has been released from rest with $\beta = 30°$.

16.19 Solve Prob. 16.18, knowing that the system has been released from rest with $\beta = 60°$.

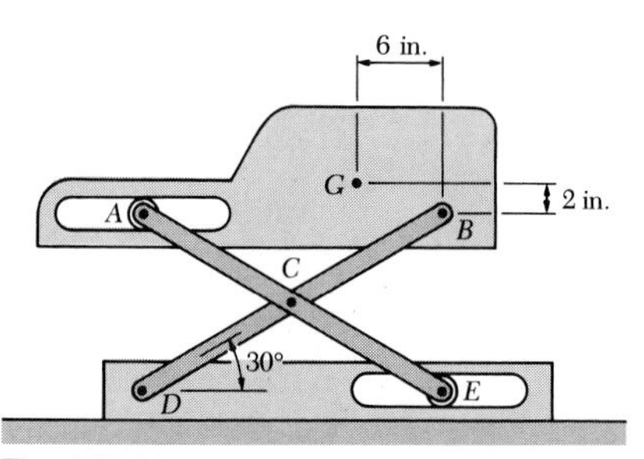

Fig. P16.20

16.20 Members ACE and DCB are each 24 in. long and are connected by a pin at C. The mass center of the 15-lb member AB is located at G. Determine (*a*) the acceleration of AB immediately after the system has been released from rest in the position shown, (*b*) the corresponding force exerted by roller A on member AB. Neglect the weight of members ACE and DCB.

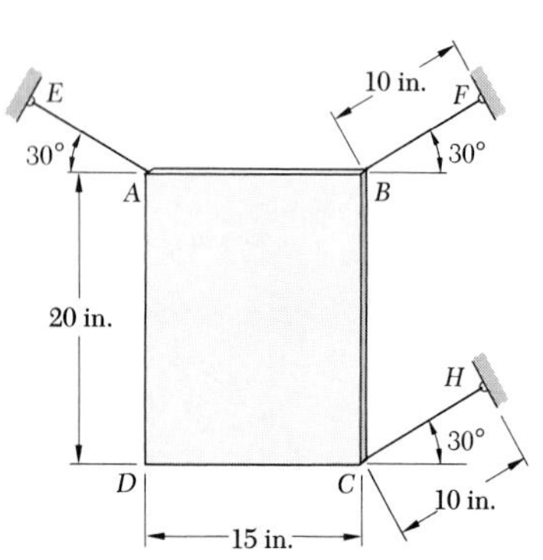

Fig. P16.21

16.21 The thin plate $ABCD$ weighs 18 lb and is held in position by the three inextensible wires AE, BF, and CH. Wire AE is then cut. Determine (*a*) the acceleration of the plate, (*b*) the tension in wires BF and CH immediately after wire AE has been cut.

16.22 A uniform semicircular plate of mass 5 kg is attached to two links *AB* and *DE*, each 250 mm long, and moves under its own weight. Neglecting the mass of the links and knowing that in the position shown the velocity of the plate is 1.8 m/s, determine the force in each link.

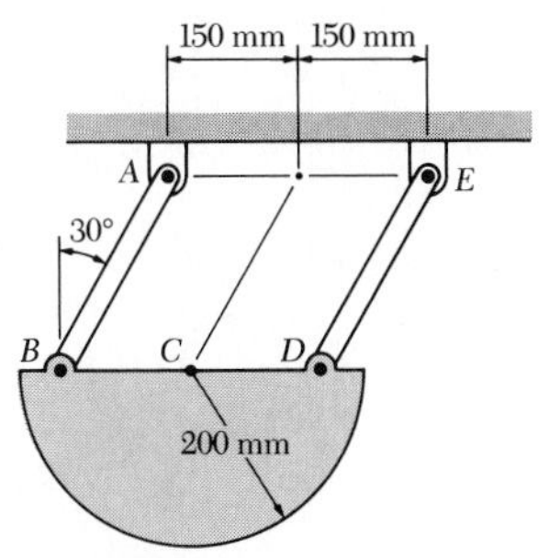

Fig. P16.22

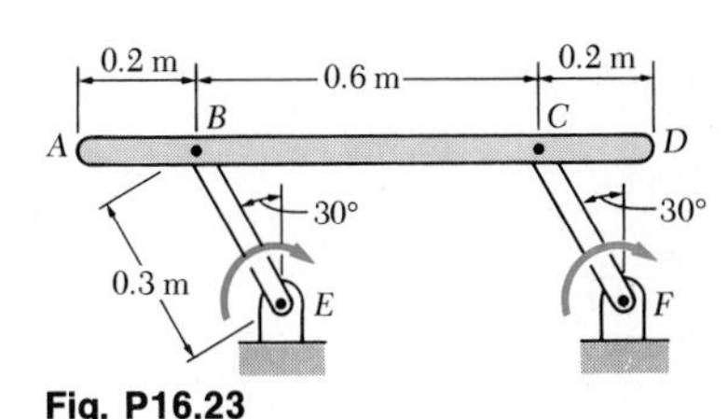

Fig. P16.23

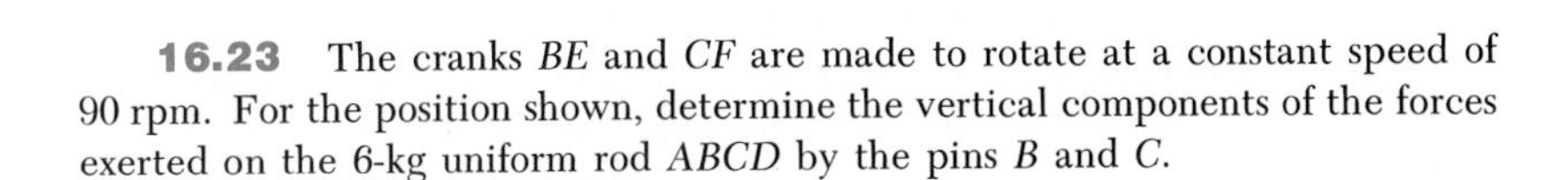

16.23 The cranks *BE* and *CF* are made to rotate at a constant speed of 90 rpm. For the position shown, determine the vertical components of the forces exerted on the 6-kg uniform rod *ABCD* by the pins *B* and *C*.

16.24 The T-shaped rod *ABCD* is guided by two pins which slide freely in parallel curved slots of radius 5 in. The rod weighs 4 lb, and its mass center is located at point *G*. Knowing that for the position shown the *horizontal* component of the velocity of *D* is 3 ft/s to the right and the *horizontal* component of the acceleration of *D* is 20 ft/s² to the right, determine the magnitude of the force **P**.

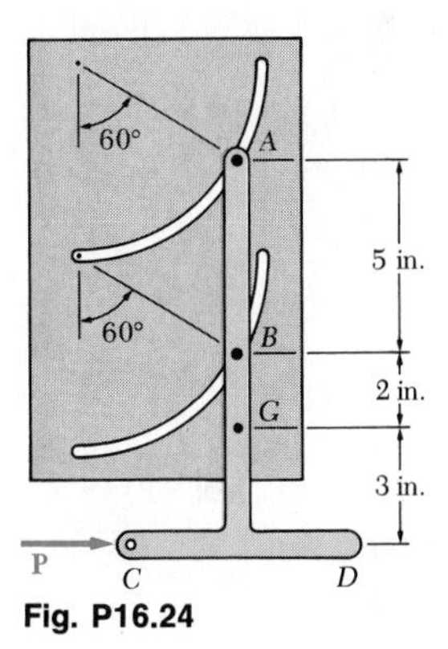

Fig. P16.24

16.25 Solve Prob. 16.24, knowing that for the position shown the *horizontal* component of the velocity of *D* is 3 ft/s to the right and the *horizontal* component of the acceleration of *D* is zero.

***16.26** A 16-lb block is placed on a 4-lb platform *BD* which is held in the position shown by three wires. Determine the accelerations of the block and of the platform immediately after wire *AB* has been cut. Assume that the block (*a*) is rigidly attached to the platform, (*b*) can slide without friction on the platform.

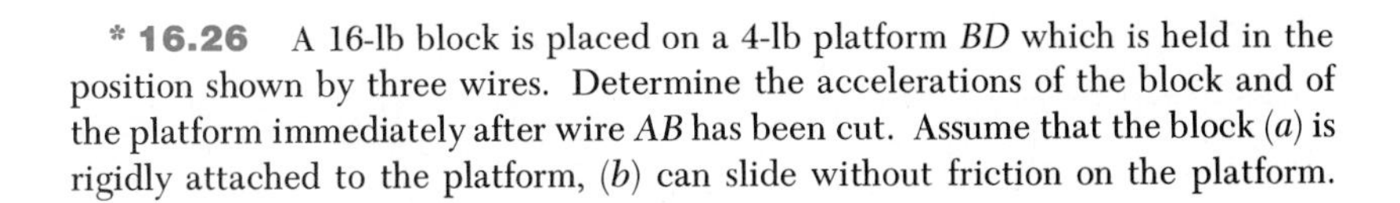

***16.27** The coefficients of friction between the 16-lb block and the 4-lb platform *BD* are $\mu_s = 0.50$ and $\mu_k = 0.40$. Determine the accelerations of the block and of the platform immediately after wire *AB* has been cut.

Fig. P16.26 and P16.27

***16.28** Draw the shear and bending-moment diagrams for the horizontal rod *CD* of Prob. 16.18.

***16.29** Draw the shear and bending-moment diagrams for the horizontal rod *ABCD* of Prob. 16.23.

*** 16.30** Two uniform rods AB and CD, each of mass 2.5 kg, are welded together and are attached to two links CE and DF. Neglecting the mass of the links, (a) determine the acceleration of the system immediately after it has been released in the position shown, (b) draw the shear and bending-moment diagrams for rod AB at that instant.

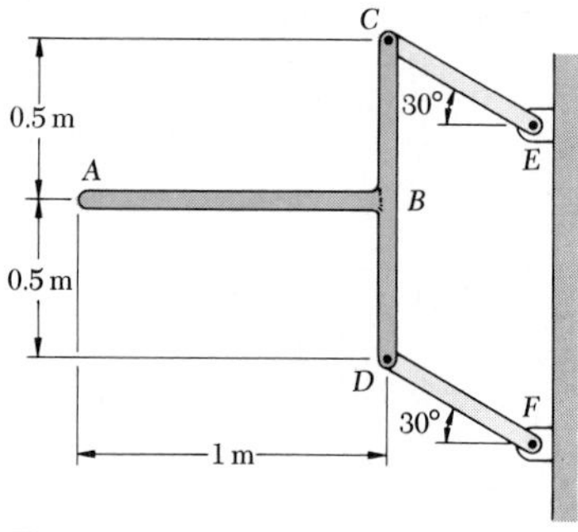

Fig. P16.30

*** 16.31** Draw the shear and bending-moment diagrams for rod CD of Prob. 16.30 immediately after the system has been released from the position shown.

16.32 An electric motor is rotating at 1800 rpm when the load and power are cut off. The rotor weighs 125 lb and has a radius of gyration of 6 in. If the kinetic friction of the rotor produces a couple of moment 9 lb·in., how many revolutions will the rotor execute before stopping?

16.33 A turbine-generator unit is shut off when its rotor is rotating at 3600 rpm; it is observed that the rotor coasts to rest in 8.4 min. Knowing that the 1600-kg rotor has a radius of gyration of 220 mm, determine the average value of the moment of the couple due to bearing friction.

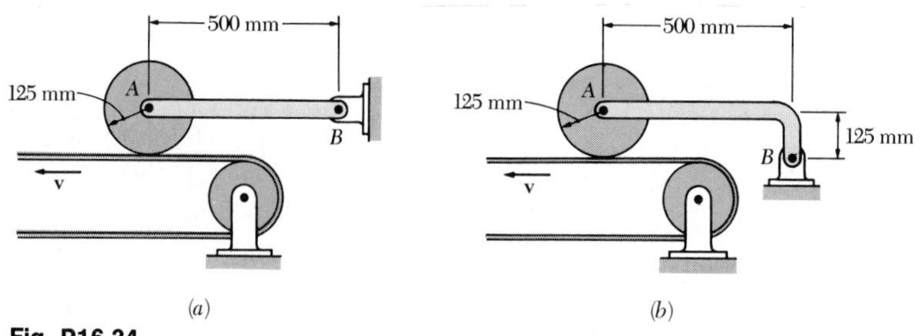

Fig. P16.34

16.34 The 4-kg disk is at rest when it is placed in contact with a conveyor belt moving at a constant speed. The link AB connecting the center of the disk to the support at B is of negligible weight. Knowing that the coefficient of kinetic friction between the disk and the belt is 0.40, determine for each of the arrangements shown the angular acceleration of the disk while slipping occurs.

16.35 Solve Prob. 16.34, assuming that the direction of motion of the conveyor belt is reversed.

16.36 Three disks of the same thickness and same material are attached to a shaft as shown. Disks A and B are each of mass 3 kg and radius $r = 250$ mm. A couple $\mathbf{M}$ of moment 35 N·m is applied to disk A when the system is at rest. Determine the radius nr of disk C if the angular acceleration of the system is to be 50 rad/s^2.

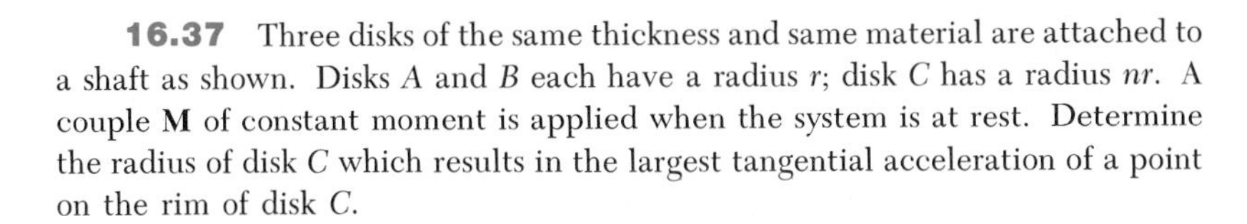

16.37 Three disks of the same thickness and same material are attached to a shaft as shown. Disks A and B each have a radius r; disk C has a radius nr. A couple $\mathbf{M}$ of constant moment is applied when the system is at rest. Determine the radius of disk C which results in the largest tangential acceleration of a point on the rim of disk C.

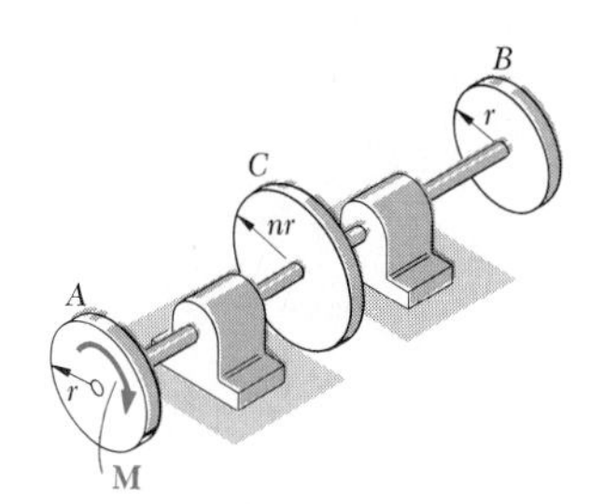

Fig. P16.36 and P16.37

16.38 The 10-in.-radius brake drum is attached to a larger flywheel which is not shown. The total mass moment of inertia of the flywheel and drum is 13.5 lb·ft·s^2. Knowing that the initial angular velocity is 180 rpm clockwise, determine the force which must be exerted by the hydraulic cylinder if the system is to stop in 50 revolutions.

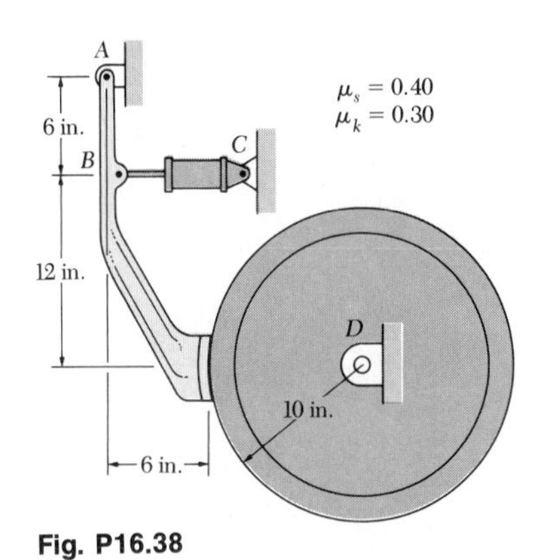

Fig. P16.38

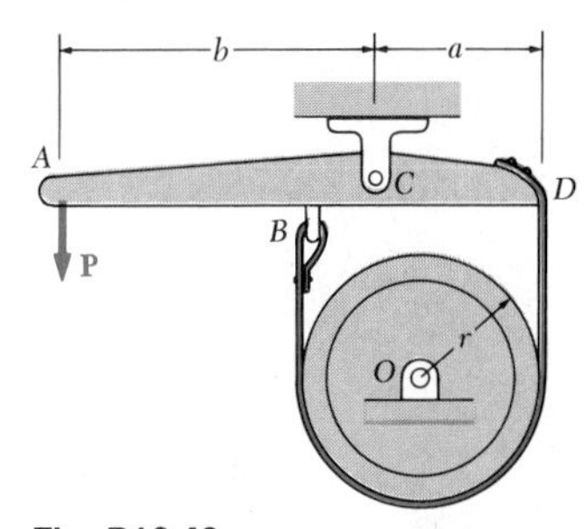

Fig. P16.40

16.39 Solve Prob. 16.38, assuming that the initial angular velocity of the flywheel is 180 rpm counterclockwise.

16.40 The 125-mm-radius brake drum is attached to a flywheel which is not shown. The drum and flywheel together have a mass of 325 kg and a radius of gyration of 725 mm. The coefficient of kinetic friction between the brake band and the drum is 0.40. Knowing that a force $\mathbf{P}$ of magnitude 50 N is applied at A when the angular velocity is 240 rpm counterclockwise, determine the time required to stop the flywheel when $a = 200$ mm and $b = 250$ mm.

16.41 Solve Prob. 16.40, assuming that the coefficient of kinetic friction is 0.30 between the brake band and the drum.

16.42 The double pulley shown has a total mass of 6 kg and a centroidal radius of gyration of 135 mm. Five collars, each of mass 1.2 kg, are attached to cords A and B as shown. When the system is at rest and in equilibrium, one collar is removed from cord A. Neglecting friction, determine (a) the angular acceleration of the pulley, (b) the velocity of cord A at $t = 2.5$ s.

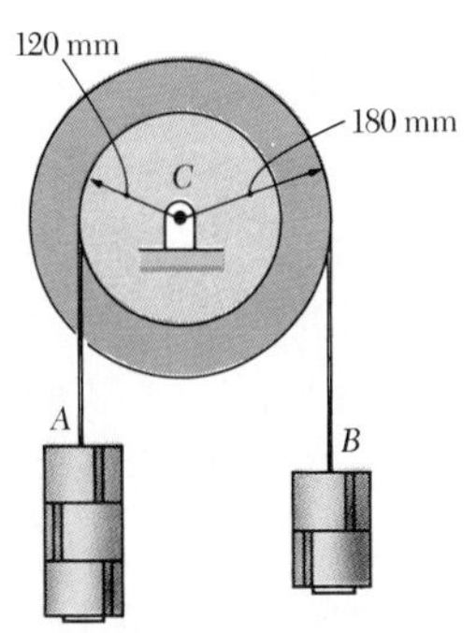

Fig. P16.42

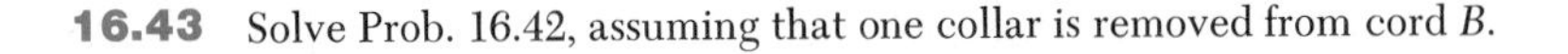

16.43 Solve Prob. 16.42, assuming that one collar is removed from cord B.

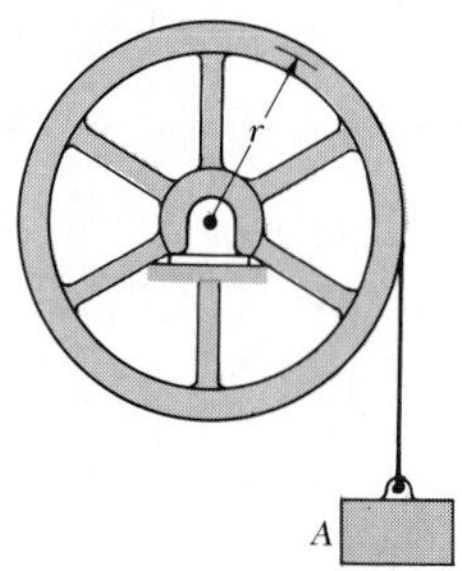

Fig. P16.44, P16.45, and P16.46

16.44 In order to determine the mass moment of inertia of a flywheel of radius $r = 2$ ft, a block A of weight 20 lb is attached to a cord wrapped around the rim of the flywheel. The block is released from rest and is observed to fall 10 ft in 4.60 s. To eliminate bearing friction from the computation, a second block of weight 40 lb is used and is observed to fall 10 ft in 3.10 s. Assuming that the moment of the couple due to bearing friction is constant, determine the mass moment of inertia of the flywheel.

16.45 The flywheel shown has a mass m_F and a radius of gyration $\bar{k}$. A block A of mass m is attached to a wire wrapped around the rim of radius r. Neglecting the effect of friction, derive a formula for the acceleration of the block in terms of m_F, m, r, and $\bar{k}$.

16.46 The flywheel shown has a mass of 100 kg and a radius of gyration of 300 mm. A block A of mass 15 kg is attached to a wire wrapped around the rim of radius $r = 400$ mm. The system is released from rest. Neglecting the effect of friction, determine (*a*) the acceleration of block A, (*b*) the speed of block A after it has moved 3 m.

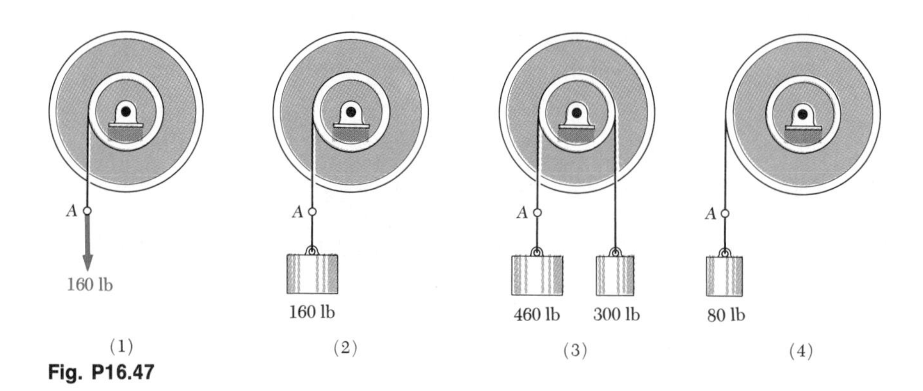

Fig. P16.47

16.47 Each of the double pulleys shown has a mass moment of inertia of 15 lb·ft·s^2 and is initially at rest. The outside radius is 18 in., and the inner radius is 9 in. Determine (*a*) the angular acceleration of each pulley, (*b*) the angular velocity of each pulley at $t = 3$ s, (*c*) the angular velocity of each pulley after point A on the cord has moved 10 ft.

Fig. P16.48

16.48 A coder C, used to record in digital form the rotation of a shaft S, is connected to the shaft by means of the gear train shown, which consists of four gears of the same thickness and of the same material. Two of the gears have a radius r and the other two a radius nr. Let I_R denote the ratio M/α of the moment M of the couple applied to the shaft S and of the resulting angular acceleration α of S. (I_R is sometimes called the "reflected moment of inertia" of the coder and gear train.) Determine I_R in terms of the gear ratio n, the moment of inertia I_0 of the first gear, and the moment of inertia I_C of the coder. Neglect the moments of inertia of the shafts.

16.49 Each of the gears A and B has a mass of 2 kg and a radius of gyration of 75 mm; gear C has a mass of 10 kg and a radius of gyration of 225 mm. If a couple $\mathbf{M}$ of constant moment 6 N·m is applied to gear C, determine (a) the angular acceleration of gear A, (b) the tangential force which gear C exerts on gear A.

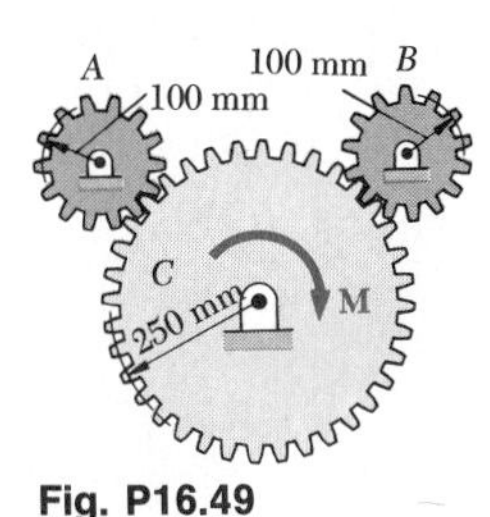

Fig. P16.49

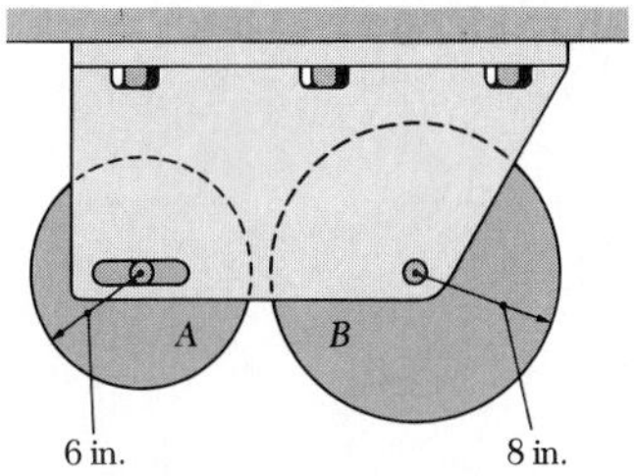

A
B
6 in.
8 in.

Fig. P16.50

16.50 Disk A weighs 5 lb and has an initial angular velocity of 900 rpm clockwise; disk B weighs 10 lb and is initially at rest. The disks are brought together by applying a horizontal force $\mathbf{P}$ of magnitude $P = 3$ lb to the axle of disk A. Knowing that $\mu_k = 0.25$ and neglecting bearing friction, determine (a) the angular acceleration of each disk, (b) the final angular velocity of each disk.

16.51 Solve Prob. 16.50, assuming that disk A was initially at rest and disk B had an angular velocity of 900 rpm clockwise.

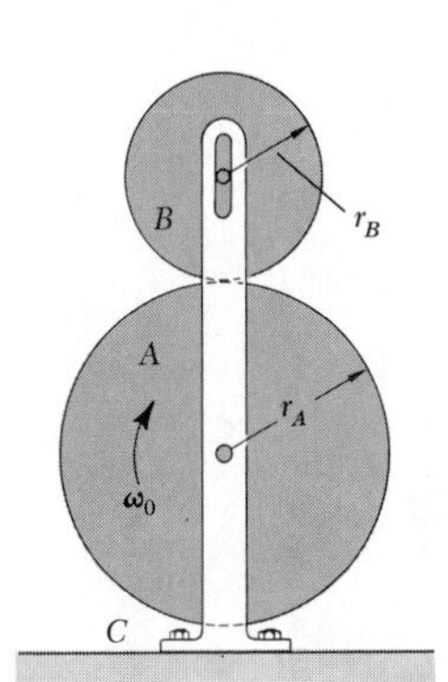

Fig. P16.52 and P16.53

16.52 Disk A has a mass $m_A = 3$ kg, a radius $r_A = 200$ mm, and an initial angular velocity $\omega_0 = 240$ rpm clockwise. Disk B has a mass $m_B = 1.2$ kg, a radius $r_B = 120$ mm, and is at rest when it is brought into contact with disk A. Knowing that $\mu_k = 0.30$ between the disks and neglecting bearing friction, determine (a) the angular acceleration of each disk, (b) the reaction at the support C.

16.53 Disk B is at rest when it is brought into contact with disk A, which has an initial angular velocity ω_0. Show that (a) the final angular velocities of the disks are independent of the coefficient of friction μ_k between the disks as long as $\mu_k \neq 0$, (b) the final angular velocity of disk A depends only upon ω_0 and the ratio of the masses m_A and m_B of the two disks.

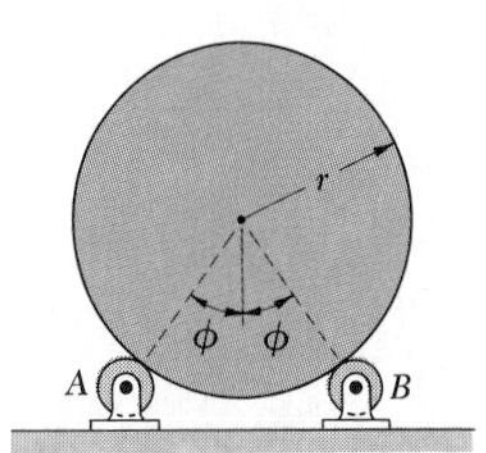

Fig. P16.54

16.54 A cylinder of radius r and mass m rests on two small casters A and B as shown. Initially, the cylinder is at rest and is set in motion by rotating caster B clockwise at high speed so that slipping occurs between the cylinder and caster B. Denoting by μ_k the coefficient of kinetic friction and neglecting the moment of inertia of the free caster A, derive an expression for the angular acceleration of the cylinder.

16.55 In Prob. 16.54, assume that no slipping can occur between caster B and the cylinder (such a case would exist if the cylinder and caster had gear teeth along their rims). Derive an expression for the maximum allowable counterclockwise acceleration $\boldsymbol{\alpha}$ of the cylinder if it is not to lose contact with the caster at A.

16.56 A 6-kg bar is held between four disks as shown. Each disk has a mass of 3 kg and a diameter of 200 mm. The disks may rotate freely, and the normal reaction exerted by each disk on the bar is sufficient to prevent slipping. If the bar is released from rest, determine (*a*) its acceleration immediately after release, (*b*) its velocity after it has dropped 0.75 m.

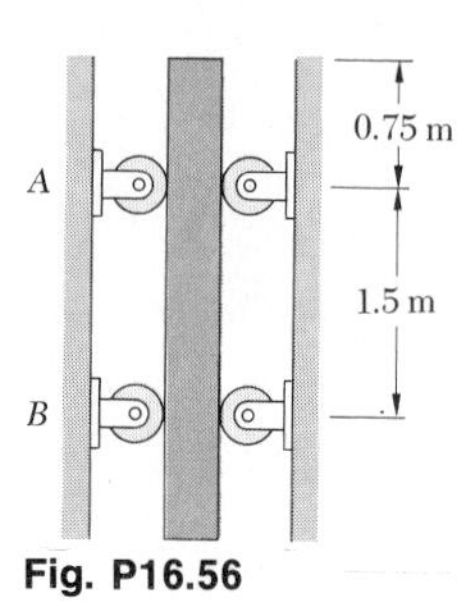

Fig. P16.56

16.57 Show that the system of the effective forces for a rigid slab in plane motion reduces to a single vector, and express the distance from the mass center G of the slab to the line of action of this vector in terms of the centroidal radius of gyration $\bar{k}$ of the slab, the magnitude $\bar{a}$ of the acceleration of G, and the angular acceleration α.

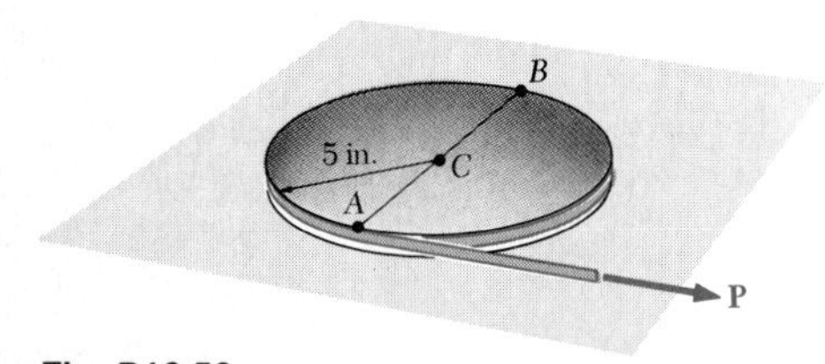

Fig. P16.58

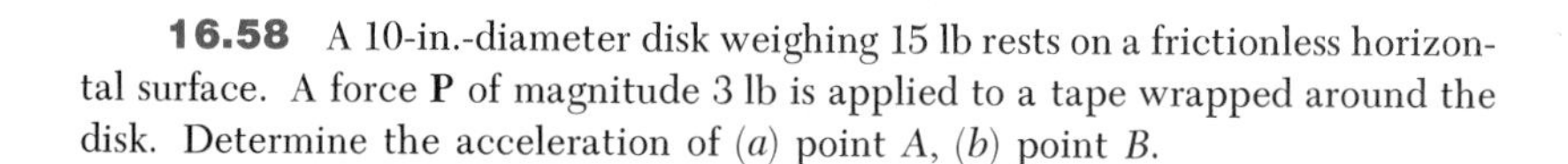

16.58 A 10-in.-diameter disk weighing 15 lb rests on a frictionless horizontal surface. A force **P** of magnitude 3 lb is applied to a tape wrapped around the disk. Determine the acceleration of (*a*) point A, (*b*) point B.

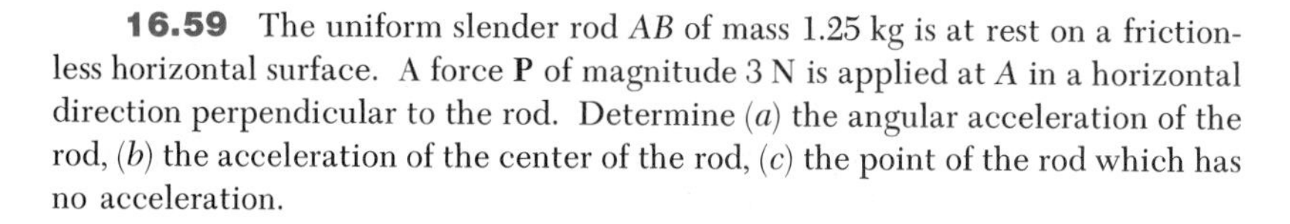

16.59 The uniform slender rod AB of mass 1.25 kg is at rest on a frictionless horizontal surface. A force **P** of magnitude 3 N is applied at A in a horizontal direction perpendicular to the rod. Determine (*a*) the angular acceleration of the rod, (*b*) the acceleration of the center of the rod, (*c*) the point of the rod which has no acceleration.

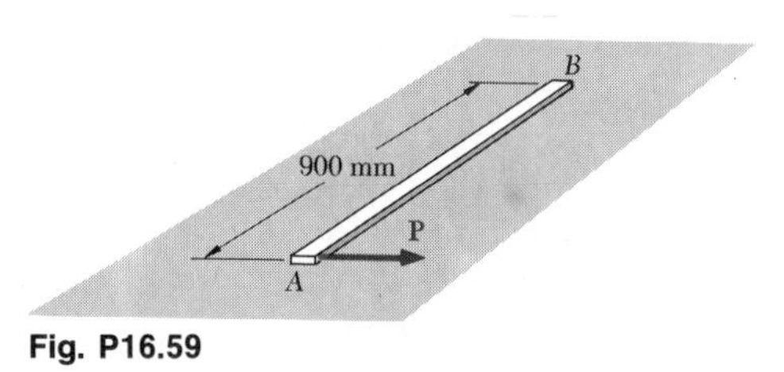

Fig. P16.59

16.60 (*a*) In Prob. 16.59, determine the point of the rod AB at which the force **P** should be applied if the acceleration of point B is to be zero. (*b*) Knowing that the magnitude of **P** is 3 N, determine the corresponding angular acceleration of the rod and the acceleration of the center of the rod.

16.61 Shortly after being fired, the experimental rocket shown weighs 25,000 lb and is moving upward with an acceleration of 45 ft/s^2. If at that instant the rocket engine A fails, while rocket engine B continues to operate, determine (*a*) the acceleration of the mass center of the rocket, (*b*) the angular acceleration of the rocket. Assume that the rocket is a uniform slender rod 48 ft long.

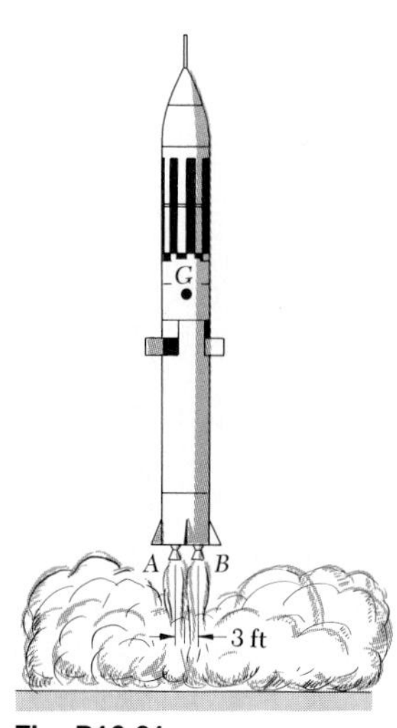

Fig. P16.61

16.62 A 4-m beam of mass 250 kg is lowered from a considerable height by means of two cables unwinding from overhead cranes. As the beam approaches the ground, the crane operators apply brakes to slow the unwinding motion. Determine the acceleration of each cable at that instant, knowing that $T_A = 1000$ N and $T_B = 1800$ N.

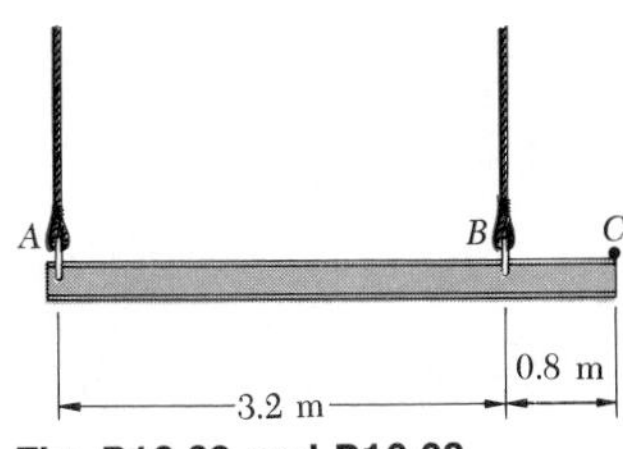

Fig. P16.62 and P16.63

16.63 A 4-m beam of mass 250 kg is lowered from a considerable height by means of two cables unwinding from overhead cranes. As the beam approaches the ground, the crane operators apply brakes to slow the unwinding motion. The deceleration of cable A is 2 m/s² and that of cable B is 0.5 m/s². Determine the tension in each cable.

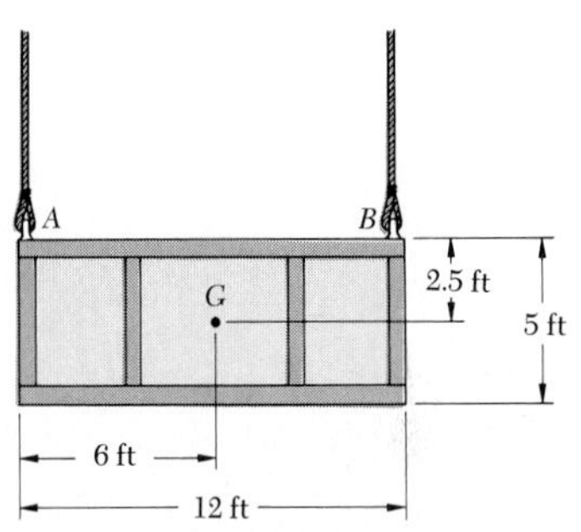

Fig. P16.64 and P16.65

16.64 The 800-lb crate shown is being lowered by means of two overhead cranes. Knowing that at the instant shown the deceleration of cable A is 21 ft/s² and that of cable B is 3 ft/s², determine the tension in each cable.

16.65 The 800-lb crate shown is being lowered by means of two overhead cranes. As the crate approaches the ground, the crane operators apply brakes to slow the motion. Determine the acceleration of each cable at that instant, knowing that $T_A = 650$ lb and $T_B = 550$ lb.

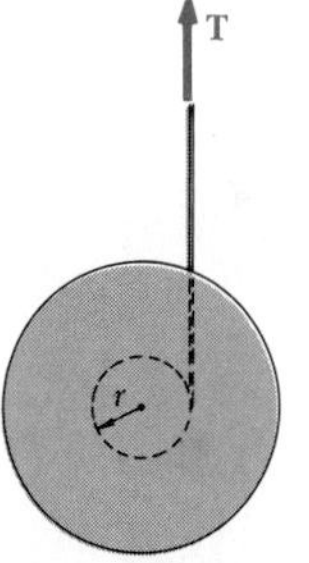

Fig. P16.66

16.66 By pulling on the cord of a yo-yo just fast enough, a person manages to make the yo-yo spin counterclockwise, while remaining at a constant height above the floor. Denoting the weight of the yo-yo by W, the radius of the inner drum on which the cord is wound by r, and the radius of gyration of the yo-yo by $\bar{k}$, determine (a) the tension in the cord, (b) the angular acceleration of the yo-yo.

16.67 For the disk and cord of Sample Prob. 16.4, determine the magnitude of the force **T** and the corresponding angular acceleration of the disk, knowing that the acceleration of the center of the disk is 0.8 m/s² (a) downward, (b) upward.

16.68 Structural steel beams are formed by passing them through successive pairs of rolls. The roll shown weighs 2800 lb, has a centroidal radius of gyration of 6 in., and is lifted by two cables looped around its shaft. Knowing that for each cable $T_A = 700$ lb and $T_B = 725$ lb, determine (a) the angular acceleration of the roll, (b) the acceleration of its mass center.

16.69 The steel roll shown weighs 2800 lb, has a centroidal radius of gyration of 6 in., and is being lowered by two cables looped around its shaft. Knowing that at the instant shown the acceleration of the roll is 5 in./s² downward and that for each cable $T_A = 680$ lb, determine (a) the corresponding value of the tension T_B, (b) the angular acceleration of the roll.

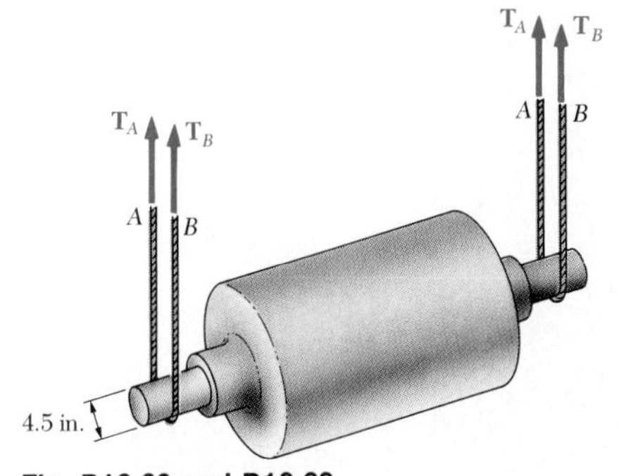

Fig. P16.68 and P16.69

16.70 through 16.73 A uniform slender bar AB of mass m is suspended from two springs as shown. If spring 2 breaks, determine at that instant (*a*) the angular acceleration of the bar, (*b*) the acceleration of point A, (*c*) the acceleration of point B.

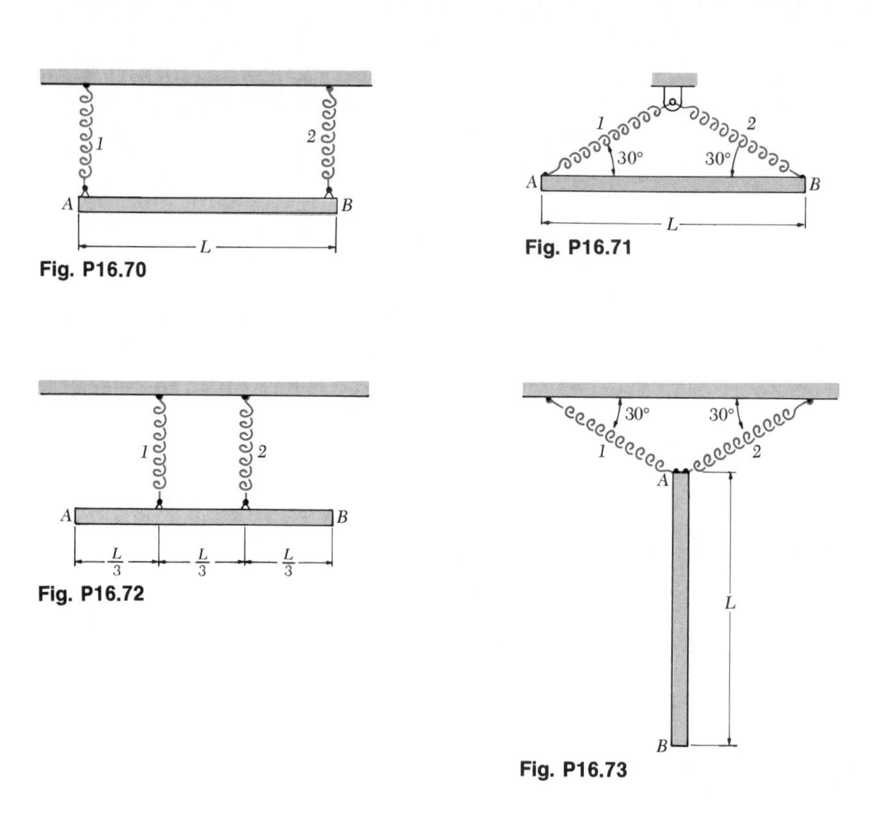

Fig. P16.70

Fig. P16.71

Fig. P16.72

Fig. P16.73

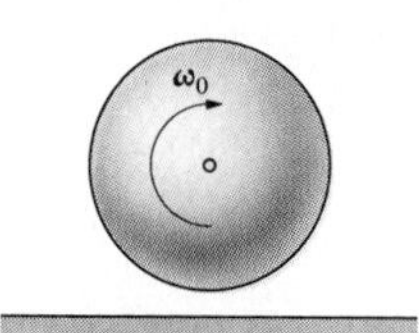

Fig. P16.74

16.74 A sphere of radius r and mass m is placed on a horizontal surface with no linear velocity but with a clockwise angular velocity $\boldsymbol{\omega}_0$. Denoting by μ_k the coefficient of kinetic friction between the sphere and the floor, determine (*a*) the time t_1 at which the sphere will start rolling without sliding, (*b*) the linear and angular velocities of the sphere at time t_1.

16.75 Solve Prob. 16.74, assuming that the sphere is replaced by a uniform disk of radius r and mass m.

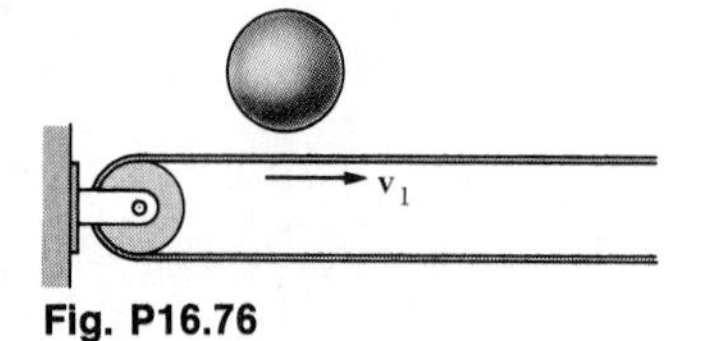

Fig. P16.76

16.76 A uniform sphere of radius r and mass m is placed with no initial velocity on a belt which moves to the right with a constant velocity $\mathbf{v}_1$. Denoting by μ_k the coefficient of kinetic friction between the sphere and the belt, determine (*a*) the time t_1 at which the sphere will start rolling without sliding, (*b*) the linear velocity and the angular velocity of the sphere at time t_1.

16.77 Solve Prob. 16.76, assuming that the sphere is replaced by a wheel of radius r, mass m, and centroidal radius of gyration $\bar{k}$.

16.5. Constrained Plane Motion. Most engineering applications deal with rigid bodies which are moving under given constraints. Cranks, for example, must rotate about a fixed axis, wheels must roll without sliding, connecting rods must describe certain prescribed motions. In all such cases, definite relations exist between the components of the acceleration $\bar{\mathbf{a}}$ of the mass center G of the body considered and its angular acceleration $\boldsymbol{\alpha}$; the corresponding motion is said to be a *constrained motion.*

The solution of a problem involving a constrained plane motion calls first for a *kinematic analysis* of the problem. Consider, for example, a slender rod AB of length l and mass m whose extremities are connected to blocks of negligible mass which slide along horizontal and vertical frictionless tracks. The rod is pulled by a force $\mathbf{P}$ applied at A (Fig. 16.10). We know from Sec. 15.8 that the acceleration $\bar{\mathbf{a}}$ of the mass center G of the rod may be determined at any given instant from the position of the rod, its angular velocity, and its angular acceleration at that instant. Suppose, for instance, that the values of θ, ω, and α are known at a given instant and that we wish to determine the corresponding value of the force $\mathbf{P}$, as well as the reactions at A and B. We should first *determine the components $\bar{a}_x$ and $\bar{a}_y$ of the acceleration of the mass center* G by the method of Sec. 15.8. We next apply d'Alembert's principle (Fig. 16.11), using the expressions obtained for $\bar{a}_x$ and $\bar{a}_y$. The unknown forces $\mathbf{P}$, $\mathbf{N}_A$, and $\mathbf{N}_B$ may then be determined by writing and solving the appropriate equations.

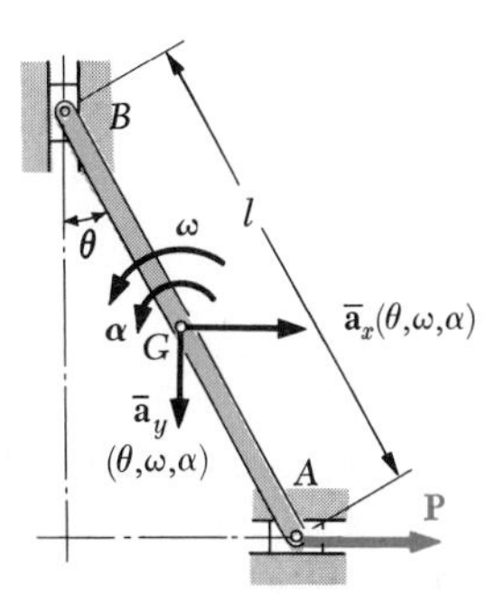

Fig. 16.10

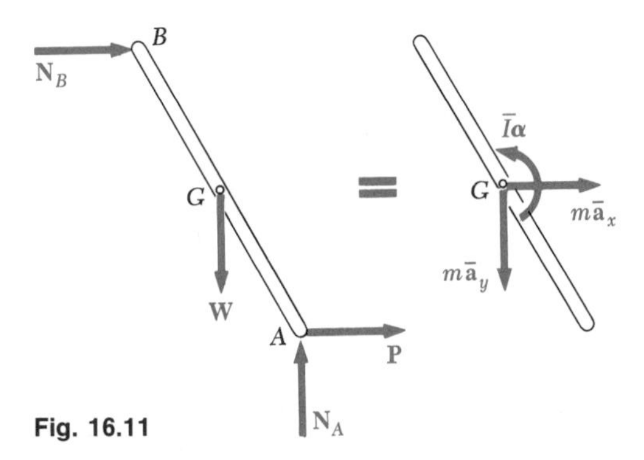

Fig. 16.11

Suppose now that the applied force $\mathbf{P}$, the angle θ, and the angular velocity ω of the rod are known at a given instant and that we wish to find the angular acceleration α of the rod and the components $\bar{a}_x$ and $\bar{a}_y$ of the acceleration of its mass center at that instant, as well as the reactions at A and B. The preliminary kinematic study of the problem will have for its object *to express the components $\bar{a}_x$ and $\bar{a}_y$ of the acceleration of G in terms of the angular acceleration α of the rod.* This will be done by first expressing the acceleration of a suitable reference point such as A in terms of the angular acceleration α. The components $\bar{a}_x$ and $\bar{a}_y$ of the acceleration of G may then be determined in terms of α, and the expressions obtained carried into Fig. 16.11. Three equations may then be derived in terms of α, N_A, and N_B, and solved for the three unknowns (see Sample Prob. 16.10). Note that the method of dynamic equilibrium may also be used to carry out the solution of the two types of problems we have considered (Fig. 16.12).

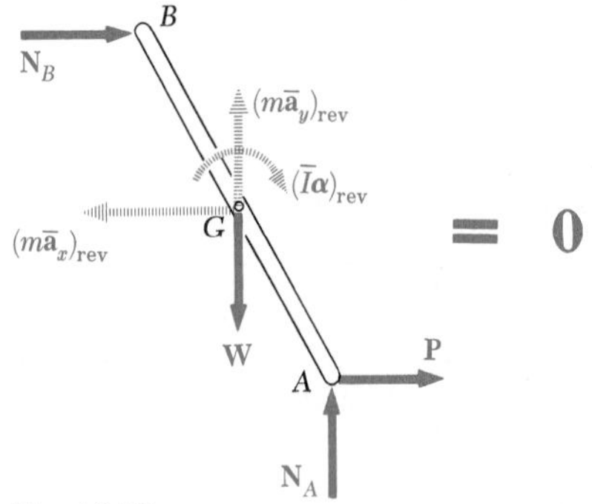

Fig. 16.12

When a mechanism consists of *several moving parts*, the approach just described may be used with each part of the mechanism. The procedure required to determine the various unknowns is then similar to the procedure followed in the case of the equilibrium of a system of connected rigid bodies (Sec. 6.11).

We have analyzed earlier two particular cases of constrained plane motion, the translation of a rigid body, in which the angular acceleration of the body is constrained to be zero, and the centroidal rotation, in which the acceleration $\bar{\mathbf{a}}$ of the mass center of the body is constrained to be zero. Two other particular cases of constrained plane motion are of special interest, the *noncentroidal rotation* of a rigid body and the *rolling motion* of a disk or wheel. These two cases should be analyzed by one of the general methods described above. However, in view of the range of their applications, they deserve a few special comments.

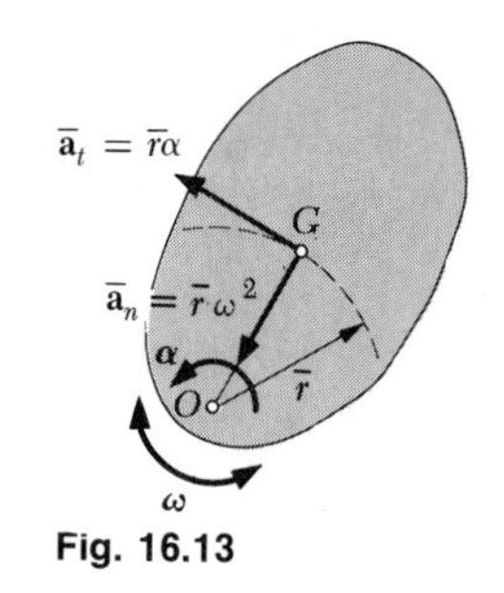

Fig. 16.13

Noncentroidal Rotation. This is the motion of a rigid body constrained to rotate about a fixed axis which does not pass through its mass center. Such a motion is called a *noncentroidal rotation*. The mass center G of the body moves along a circle of radius $\bar{r}$ centered at the point O, where the axis of rotation intersects the plane of reference (Fig. 16.13). Denoting, respectively, by $\boldsymbol{\omega}$ and $\boldsymbol{\alpha}$ the angular velocity and the angular acceleration of the line OG, we obtain the following expressions for the tangential and normal components of the acceleration of G:

$$\bar{a}_t = \bar{r}\alpha \qquad \bar{a}_n = \bar{r}\omega^2 \tag{16.12}$$

Since line OG belongs to the body, its angular velocity $\boldsymbol{\omega}$ and its angular acceleration $\boldsymbol{\alpha}$ also represent the angular velocity and the angular acceleration of the body in its motion relative to G. Equations (16.12) define, therefore, the kinematic relation existing between the motion of the mass center G and the motion of the body about G. They should be used to eliminate $\bar{a}_t$ and $\bar{a}_n$ from the equations obtained by applying d'Alembert's principle (Fig. 16.14) or the method of dynamic equilibrium (Fig. 16.15).

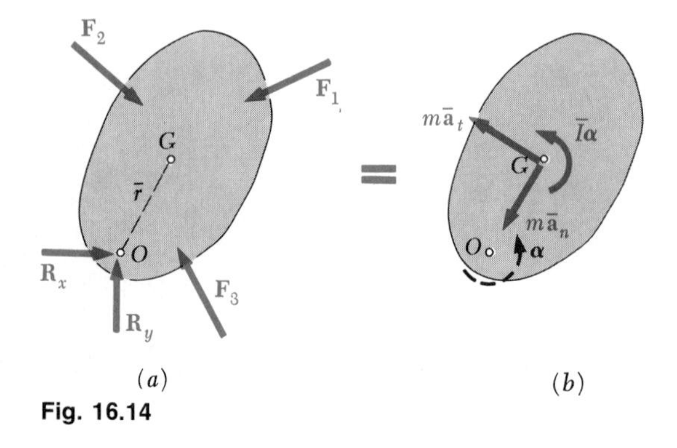

Fig. 16.14

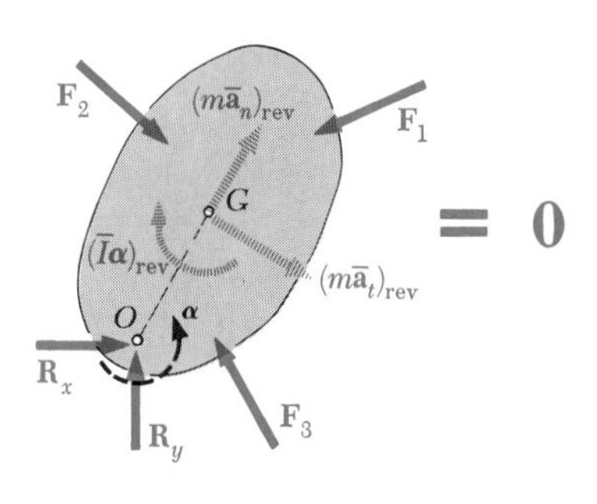

Fig. 16.15

An interesting relation may be obtained by equating the moments about the fixed point O of the forces and vectors shown, respectively, in parts a and b of Fig. 16.14. We write

$$+\circlearrowleft \Sigma M_O = \bar{I}\alpha + (m\bar{r}\alpha)\bar{r} = (\bar{I} + m\bar{r}^2)\alpha$$

But according to the parallel-axis theorem, we have $\bar{I} + m\bar{r}^2 = I_O$, where I_O denotes the moment of inertia of the rigid body about the fixed axis. We therefore write

$$\Sigma M_O = I_O\alpha \qquad (16.13)$$

Although formula (16.13) expresses an important relation between the sum of the moments of the external forces about the fixed point O and the product $I_O\alpha$, it should be clearly understood that this formula *does not mean* that the system of the external forces is equivalent to a couple of moment $I_O\alpha$. The system of the effective forces, and thus the system of the external forces, reduces to a couple only when O coincides with G—that is, *only when the rotation is centroidal* (Sec. 16.2). In the more general case of noncentroidal rotation, the system of the external forces does not reduce to a couple.

A particular case of noncentroidal rotation is of special interest—the case of *uniform rotation,* in which the angular velocity $\boldsymbol{\omega}$ is constant. Since $\boldsymbol{\alpha}$ is zero, the inertia couple in Fig. 16.15 vanishes and the inertia vector reduces to its normal component. This component (also called *centrifugal force*) represents the tendency of the rigid body to break away from the axis of rotation.

Rolling Motion. Another important case of plane motion is the motion of a disk or wheel rolling on a plane surface. If the disk is constrained to roll without sliding, the acceleration $\bar{\mathbf{a}}$ of its mass center G and its angular acceleration $\boldsymbol{\alpha}$ are not independent. Assuming that the disk is balanced, so that its mass center and its geometric center coincide, we first write that the distance $\bar{x}$ traveled by G during a rotation θ of the disk is $\bar{x} = r\theta$, where r is the radius of the disk. Differentiating this relation twice, we write

$$\bar{a} = r\alpha \qquad (16.14)$$

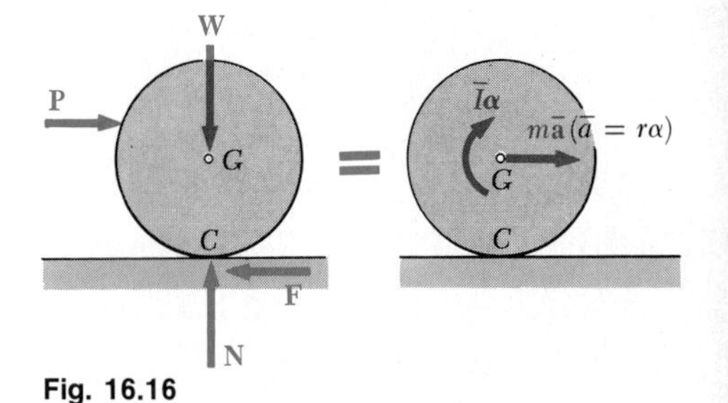

Fig. 16.16

Recalling that the system of the effective forces in plane motion reduces to a vector $m\bar{\mathbf{a}}$ and a couple $\bar{I}\boldsymbol{\alpha}$, we find that in the particular case of the rolling motion of a balanced disk, the effective forces reduce to a vector of magnitude $mr\alpha$ attached at G and to a couple of magnitude $\bar{I}\alpha$. We may thus express that the external forces are equivalent to the vector and couple shown in Fig. 16.16.

When a disk *rolls without sliding*, there is no relative motion between the point of the disk in contact with the ground and the ground itself. As far as the computation of the friction force **F** is concerned, a rolling disk may thus be compared with a block at rest on a surface. The magnitude F of the friction force may have any value, as long as this value does not exceed the maximum value $F_m = \mu_s N$, where μ_s is the coefficient of static friction and N the magnitude of the normal force. In the case of a rolling disk, the magnitude F of the friction force should therefore be determined independently of N by solving the equation obtained from Fig. 16.16.

When *sliding is impending*, the friction force reaches its maximum value $F_m = \mu_s N$ and may be obtained from N.

When the disk *rotates and slides* at the same time, a relative motion exists between the point of the disk which is in contact with the ground and the ground itself, and the force of friction has the magnitude $F_k = \mu_k N$, where μ_k is the coefficient of kinetic friction. In this case, however, the motion of the mass center G of the disk and the rotation of the disk about G are independent, and $\bar{a}$ is not equal to $r\alpha$.

These three different cases may be summarized as follows:

Rolling, no sliding:	$F \leq \mu_s N$	$\bar{a} = r\alpha$
Rolling, sliding impending:	$F = \mu_s N$	$\bar{a} = r\alpha$
Rotating and sliding:	$F = \mu_k N$	$\bar{a}$ *and* α *independent*

When it is not known whether or not a disk slides, it should first be assumed that the disk rolls without sliding. If F is found smaller than, or equal to, $\mu_s N$, the assumption is proved correct. If F is found larger than $\mu_s N$, the assumption is incorrect and the problem should be started again, assuming rotating and sliding.

When a disk is *unbalanced*, i.e., when its mass center G does not coincide with its geometric center O, the relation (16.14) does not hold between $\bar{a}$ and α. However, a similar relation holds between the magnitude a_O of the acceleration of the geometric center and the angular acceleration α of an unbalanced disk which rolls without sliding. We have

$$a_O = r\alpha \tag{16.15}$$

To determine $\bar{a}$ in terms of the angular acceleration α and the angular velocity ω of the disk, we may use the relative-acceleration formula

$$\begin{aligned}\bar{\mathbf{a}} = \mathbf{a}_G &= \mathbf{a}_O + \mathbf{a}_{G/O} \\ &= \mathbf{a}_O + (\mathbf{a}_{G/O})_t + (\mathbf{a}_{G/O})_n\end{aligned} \tag{16.16}$$

where the three component accelerations obtained have the directions indicated in Fig. 16.17 and the magnitudes $a_O = r\alpha$, $(a_{G/O})_t = (OG)\alpha$, and $(a_{G/O})_n = (OG)\omega^2$.

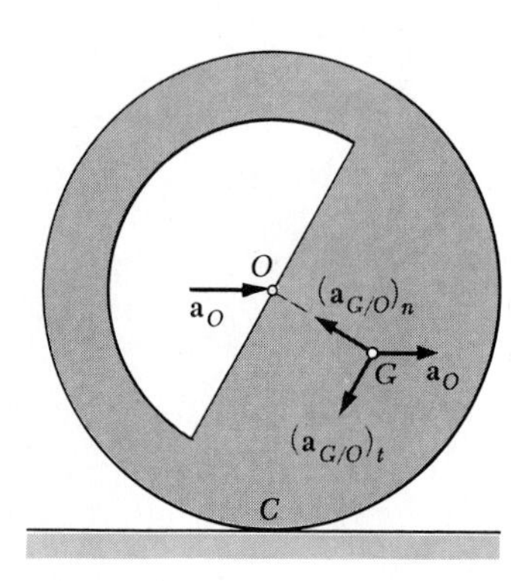

Fig. 16.17

SAMPLE PROBLEM 16.6

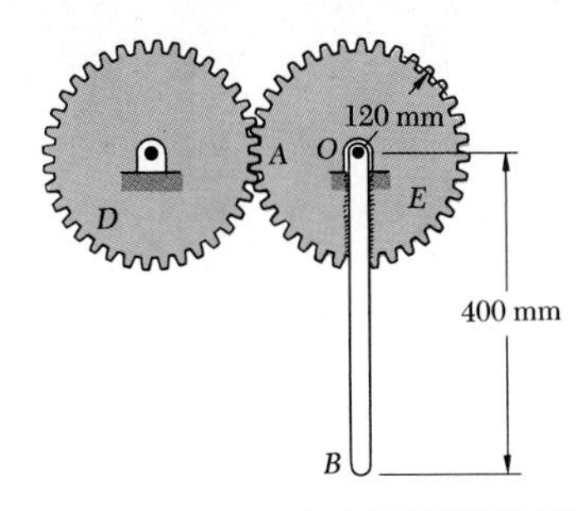

The portion *AOB* of a mechanism consists of a 400-mm steel rod *OB* welded to a gear *E* of radius 120 mm which may rotate about a horizontal shaft *O*. It is actuated by a gear *D* and, at the instant shown, has a clockwise angular velocity of 8 rad/s and a counterclockwise angular acceleration of 40 rad/s^2. Knowing that rod *OB* has a mass of 3 kg and gear *E* a mass of 4 kg and a radius of gyration of 85 mm, determine (*a*) the tangential force exerted by gear *D* on gear *E*, (*b*) the components of the reaction at shaft *O*.

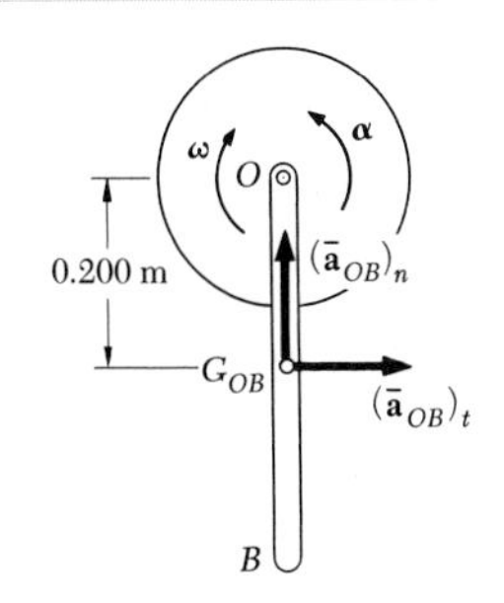

Solution. In determining the effective forces of the rigid body *AOB* we shall consider separately gear *E* and rod *OB*. Therefore, we shall first determine the components of the acceleration of the mass center G_{OB} of the rod:

$$(\bar{a}_{OB})_t = \bar{r}\alpha = (0.200\text{ m})(40\text{ rad/s}^2) = 8\text{ m/s}^2$$
$$(\bar{a}_{OB})_n = \bar{r}\omega^2 = (0.200\text{ m})(8\text{ rad/s})^2 = 12.8\text{ m/s}^2$$

Equations of Motion. Two sketches of the rigid body *AOB* have been drawn. The first shows the external forces consisting of the weight $\mathbf{W}_E$ of gear *E*, the weight $\mathbf{W}_{OB}$ of the rod *OB*, the force **F** exerted by gear *D*, and the components $\mathbf{R}_x$ and $\mathbf{R}_y$ of the reaction at *O*. The magnitudes of the weights are, respectively,

$$W_E = m_E g = (4\text{ kg})(9.81\text{ m/s}^2) = 39.2\text{ N}$$
$$W_{OB} = m_{OB} g = (3\text{ kg})(9.81\text{ m/s}^2) = 29.4\text{ N}$$

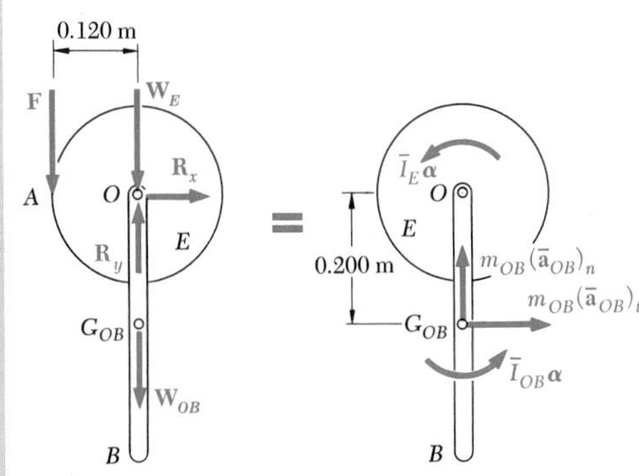

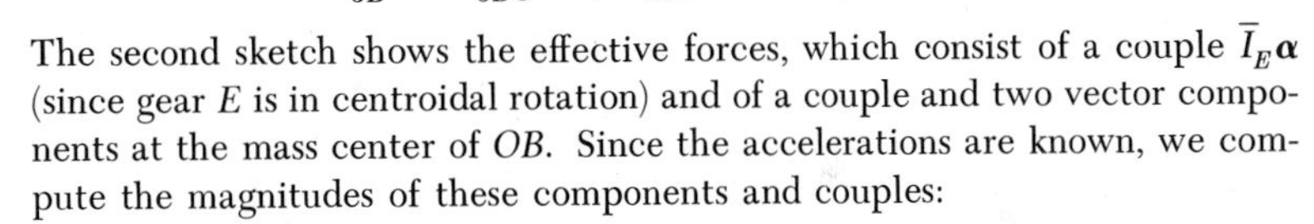

The second sketch shows the effective forces, which consist of a couple $\bar{I}_E\boldsymbol{\alpha}$ (since gear *E* is in centroidal rotation) and of a couple and two vector components at the mass center of *OB*. Since the accelerations are known, we compute the magnitudes of these components and couples:

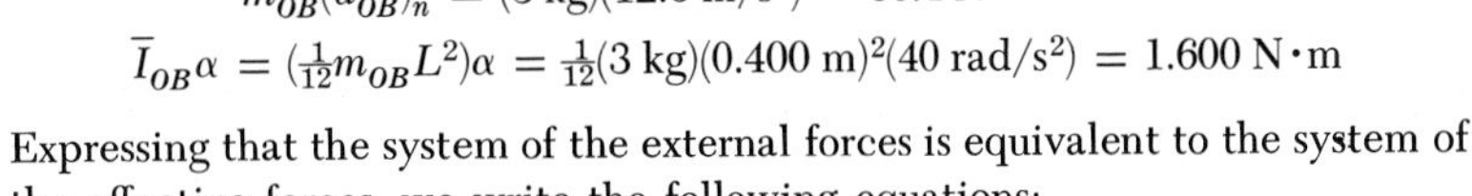

$$\bar{I}_E\alpha = m_E\bar{k}_E^2\alpha = (4\text{ kg})(0.085\text{ m})^2(40\text{ rad/s}^2) = 1.156\text{ N}\cdot\text{m}$$
$$m_{OB}(\bar{a}_{OB})_t = (3\text{ kg})(8\text{ m/s}^2) = 24.0\text{ N}$$
$$m_{OB}(\bar{a}_{OB})_n = (3\text{ kg})(12.8\text{ m/s}^2) = 38.4\text{ N}$$
$$\bar{I}_{OB}\alpha = (\tfrac{1}{12}m_{OB}L^2)\alpha = \tfrac{1}{12}(3\text{ kg})(0.400\text{ m})^2(40\text{ rad/s}^2) = 1.600\text{ N}\cdot\text{m}$$

Expressing that the system of the external forces is equivalent to the system of the effective forces, we write the following equations:

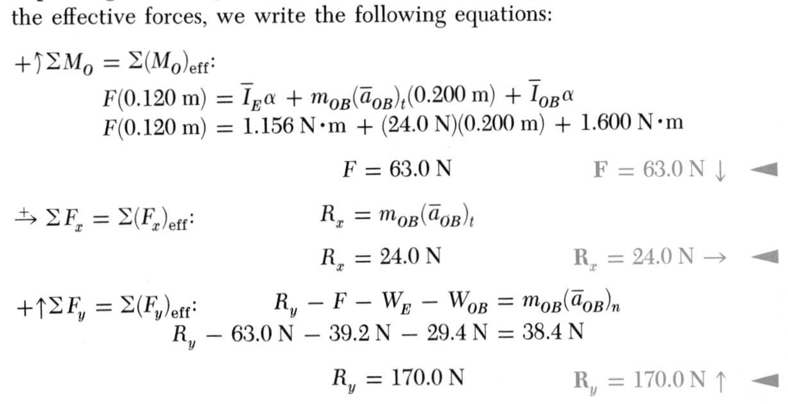

$+\circlearrowleft\Sigma M_O = \Sigma(M_O)_{\text{eff}}$:

$$F(0.120\text{ m}) = \bar{I}_E\alpha + m_{OB}(\bar{a}_{OB})_t(0.200\text{ m}) + \bar{I}_{OB}\alpha$$
$$F(0.120\text{ m}) = 1.156\text{ N}\cdot\text{m} + (24.0\text{ N})(0.200\text{ m}) + 1.600\text{ N}\cdot\text{m}$$

$$F = 63.0\text{ N} \qquad \mathbf{F} = 63.0\text{ N}\downarrow \blacktriangleleft$$

$\xrightarrow{+} \Sigma F_x = \Sigma(F_x)_{\text{eff}}$: $\quad R_x = m_{OB}(\bar{a}_{OB})_t$

$$R_x = 24.0\text{ N} \qquad \mathbf{R}_x = 24.0\text{ N}\rightarrow \blacktriangleleft$$

$+\uparrow\Sigma F_y = \Sigma(F_y)_{\text{eff}}$: $\quad R_y - F - W_E - W_{OB} = m_{OB}(\bar{a}_{OB})_n$

$$R_y - 63.0\text{ N} - 39.2\text{ N} - 29.4\text{ N} = 38.4\text{ N}$$

$$R_y = 170.0\text{ N} \qquad \mathbf{R}_y = 170.0\text{ N}\uparrow \blacktriangleleft$$

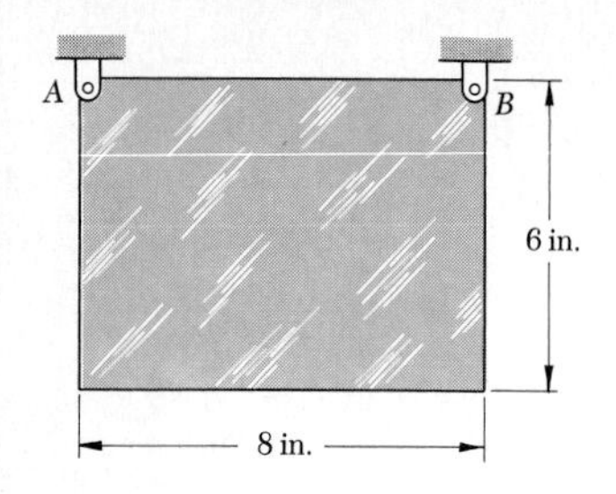

SAMPLE PROBLEM 16.7

A 6 × 8 in. rectangular plate weighing 60 lb is suspended from two pins A and B. If pin B is suddenly removed, determine (a) the angular acceleration of the plate, (b) the components of the reaction at pin A, immediately after pin B has been removed.

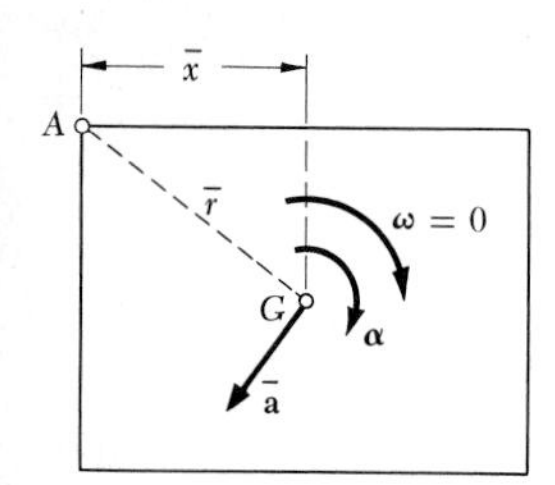

a. Angular Acceleration. We observe that as the plate rotates about point A, its mass center G describes a circle of radius $\bar{r}$ with center at A.

Since the plate is released from rest ($\omega = 0$), the normal component of the acceleration of G is zero. The magnitude of the acceleration $\bar{\mathbf{a}}$ of the mass center G is thus $\bar{a} = \bar{r}\alpha$. We draw the diagram shown to express that the external forces are equivalent to the effective forces:

$$+\circlearrowright \Sigma M_A = \Sigma(M_A)_{\text{eff}}: \qquad W\bar{x} = (m\bar{a})\bar{r} + \bar{I}\alpha$$

Since $\bar{a} = \bar{r}\alpha$, we have

$$W\bar{x} = m(\bar{r}\alpha)\bar{r} + \bar{I}\alpha \qquad \alpha = \frac{W\bar{x}}{\dfrac{W}{g}\bar{r}^2 + \bar{I}} \tag{1}$$

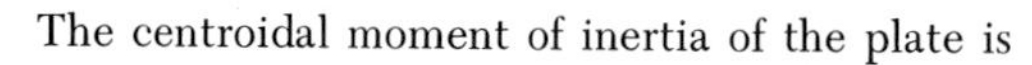

The centroidal moment of inertia of the plate is

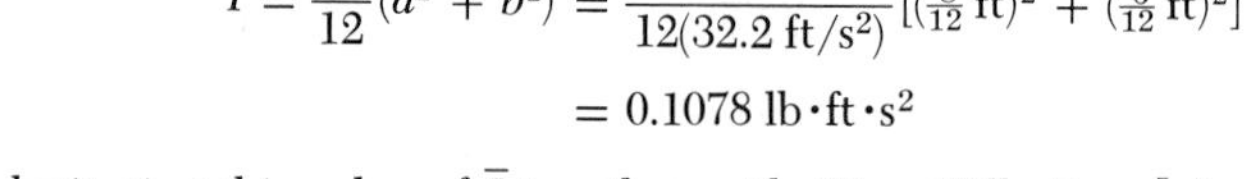

$$\bar{I} = \frac{m}{12}(a^2 + b^2) = \frac{60 \text{ lb}}{12(32.2 \text{ ft/s}^2)}[(\tfrac{8}{12} \text{ ft})^2 + (\tfrac{6}{12} \text{ ft})^2]$$
$$= 0.1078 \text{ lb}\cdot\text{ft}\cdot\text{s}^2$$

Substituting this value of $\bar{I}$ together with $W = 60$ lb, $\bar{r} = \frac{5}{12}$ ft, and $\bar{x} = \frac{4}{12}$ ft into Eq. (1), we obtain

$$\alpha = +46.4 \text{ rad/s}^2 \qquad \boldsymbol{\alpha} = 46.4 \text{ rad/s}^2 \circlearrowright$$

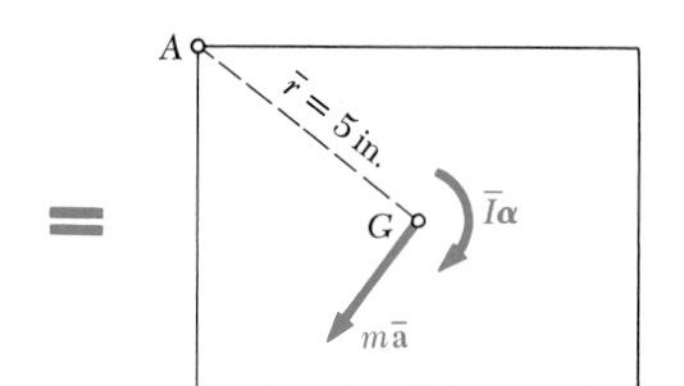

b. Reaction at A. Using the computed value of α, we determine the magnitude of the vector $m\bar{\mathbf{a}}$ attached at G,

$$m\bar{a} = m\bar{r}\alpha = \frac{60 \text{ lb}}{32.2 \text{ ft/s}^2}(\tfrac{5}{12} \text{ ft})(46.4 \text{ rad/s}^2) = 36.0 \text{ lb}$$

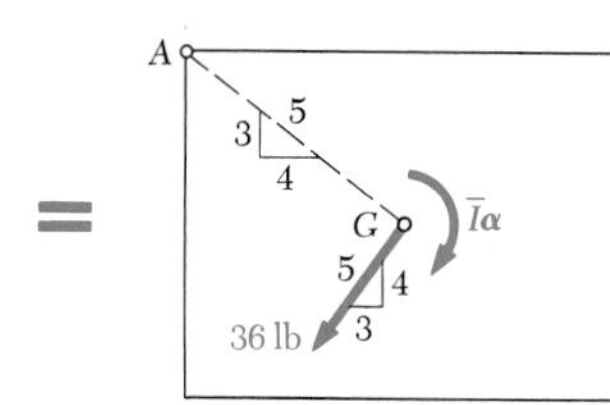

Showing this result on the diagram, we write the equations of motion

$$\xrightarrow{+} \Sigma F_x = \Sigma(F_x)_{\text{eff}}: \qquad A_x = -\tfrac{3}{5}(36 \text{ lb})$$
$$= -21.6 \text{ lb} \qquad \mathbf{A}_x = 21.6 \text{ lb} \leftarrow$$

$$+\uparrow \Sigma F_y = \Sigma(F_y)_{\text{eff}}: \qquad A_y - 60 \text{ lb} = -\tfrac{4}{5}(36 \text{ lb})$$
$$A_y = +31.2 \text{ lb} \qquad \mathbf{A}_y = 31.2 \text{ lb} \uparrow$$

The couple $\bar{I}\boldsymbol{\alpha}$ is not involved in the last two equations; nevertheless, it should be indicated on the diagram.

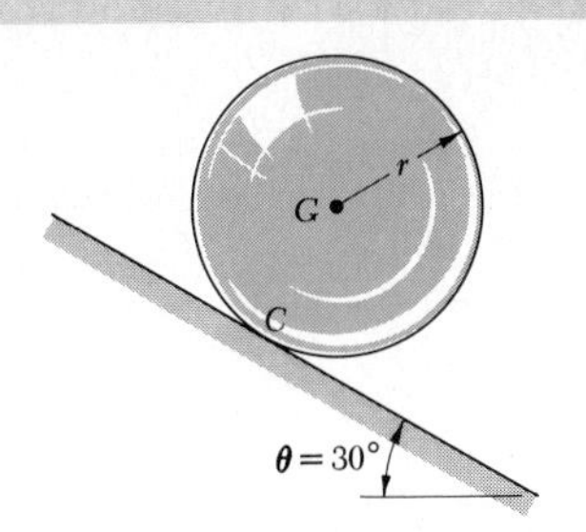

SAMPLE PROBLEM 16.8

A sphere of radius r and weight W is released with no initial velocity on the incline and rolls without slipping. Determine (*a*) the minimum value of the coefficient of static friction compatible with the rolling motion, (*b*) the velocity of the center G of the sphere after the sphere has rolled 10 ft, (*c*) the velocity of G if the sphere were to move 10 ft down a frictionless 30° incline.

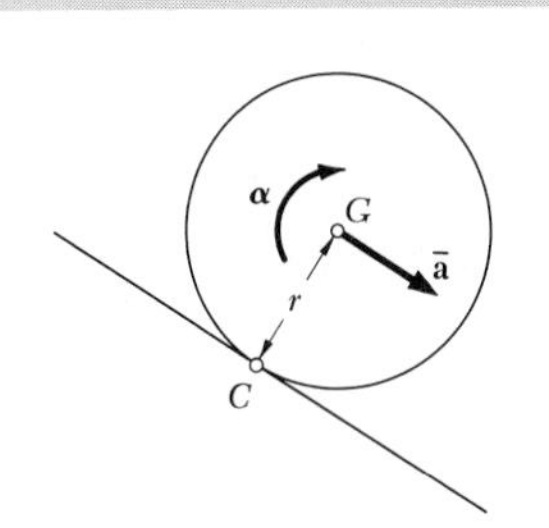

***a.* Minimum μ_s for Rolling Motion.** The external forces **W**, **N**, and **F** form a system equivalent to the system of effective forces represented by the vector $m\bar{\mathbf{a}}$ and the couple $\bar{I}\boldsymbol{\alpha}$. Since the sphere rolls without sliding, we have $\bar{a} = r\alpha$.

$$+\circlearrowright \Sigma M_C = \Sigma(M_C)_{\text{eff}}: \qquad (W \sin\theta)r = (m\bar{a})r + \bar{I}\alpha$$
$$(W \sin\theta)r = (mr\alpha)r + \bar{I}\alpha$$

Noting that $m = W/g$ and $\bar{I} = \frac{2}{5}mr^2$, we write

$$(W \sin\theta)r = \left(\frac{W}{g}r\alpha\right)r + \frac{2}{5}\frac{W}{g}r^2\alpha \qquad \alpha = +\frac{5g\sin\theta}{7r}$$

$$\bar{a} = r\alpha = \frac{5g\sin\theta}{7} = \frac{5(32.2\text{ ft/s}^2)\sin 30^\circ}{7} = 11.50\text{ ft/s}^2$$

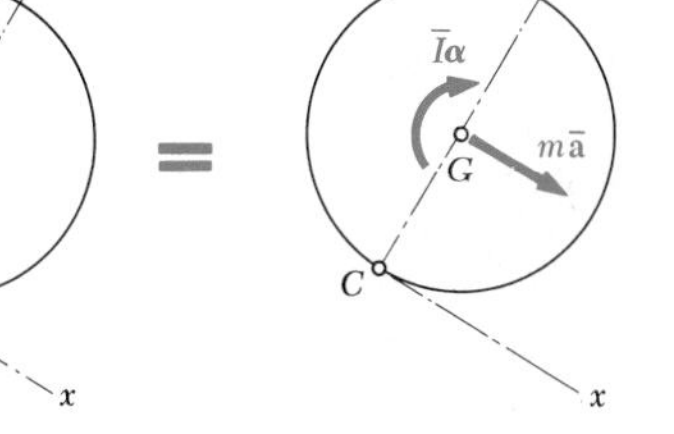

$$+\searrow \Sigma F_x = \Sigma(F_x)_{\text{eff}}: \qquad W\sin\theta - F = m\bar{a}$$
$$W\sin\theta - F = \frac{W}{g}\frac{5g\sin\theta}{7}$$

$$F = +\tfrac{2}{7}W\sin\theta = \tfrac{2}{7}W\sin 30^\circ \qquad \mathbf{F} = 0.143W \searrow 30^\circ$$

$$+\nearrow \Sigma F_y = \Sigma(F_y)_{\text{eff}}: \qquad N - W\cos\theta = 0$$
$$N = W\cos\theta = 0.866W \qquad \mathbf{N} = 0.866W \measuredangle 60^\circ$$

$$\mu_s = \frac{F}{N} = \frac{0.143W}{0.866W} \qquad \mu_s = 0.165 \blacktriangleleft$$

***b.* Velocity of Rolling Sphere.** We have uniformly accelerated motion:

$$\bar{v}_0 = 0 \qquad \bar{a} = 11.50\text{ ft/s}^2 \qquad \bar{x} = 10\text{ ft} \qquad \bar{x}_0 = 0$$
$$\bar{v}^2 = \bar{v}_0^2 + 2\bar{a}(\bar{x} - \bar{x}_0) \qquad \bar{v}^2 = 0 + 2(11.50\text{ ft/s}^2)(10\text{ ft})$$
$$\bar{v} = 15.17\text{ ft/s} \qquad \bar{\mathbf{v}} = 15.17\text{ ft/s} \searrow 30^\circ \blacktriangleleft$$

***c.* Velocity of Sliding Sphere.** Assuming now no friction, we have $F = 0$ and obtain

$$+\circlearrowright \Sigma M_G = \Sigma(M_G)_{\text{eff}}: \qquad 0 = \bar{I}\alpha \qquad \alpha = 0$$

$$+\searrow \Sigma F_x = \Sigma(F_x)_{\text{eff}}: \qquad W\sin 30^\circ = m\bar{a} \qquad 0.50W = \frac{W}{g}\bar{a}$$

$$\bar{a} = +16.1\text{ ft/s}^2 \qquad \bar{\mathbf{a}} = 16.1\text{ ft/s}^2 \searrow 30^\circ$$

Substituting $\bar{a} = 16.1\text{ ft/s}^2$ into the equations for uniformly accelerated motion, we obtain

$$\bar{v}^2 = \bar{v}_0^2 + 2\bar{a}(\bar{x} - \bar{x}_0) \qquad \bar{v}^2 = 0 + 2(16.1\text{ ft/s}^2)(10\text{ ft})$$
$$\bar{v} = 17.94\text{ ft/s} \qquad \bar{\mathbf{v}} = 17.94\text{ ft/s} \searrow 30^\circ \blacktriangleleft$$

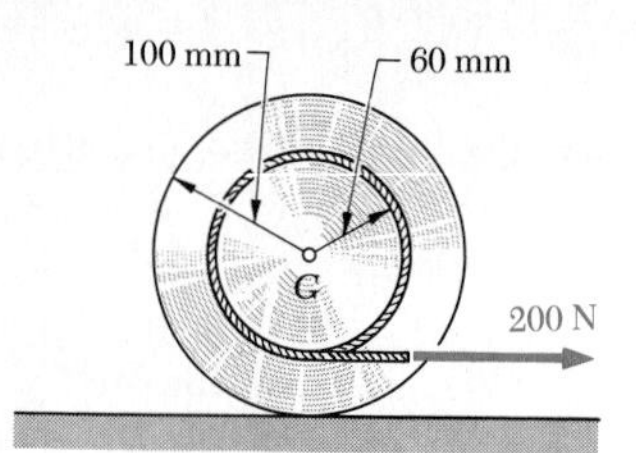

SAMPLE PROBLEM 16.9

A cord is wrapped around the inner drum of a wheel and pulled horizontally with a force of 200 N. The wheel has a mass of 50 kg and a radius of gyration of 70 mm. Knowing that $\mu_s = 0.20$ and $\mu_k = 0.15$, determine the acceleration of G and the angular acceleration of the wheel.

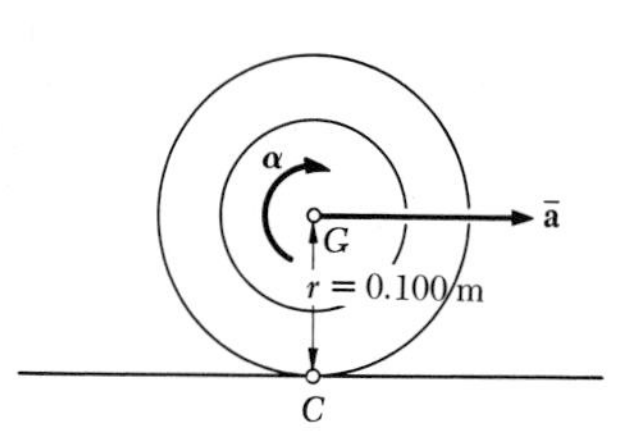

***a.* Assume Rolling without Sliding.** In this case, we have

$$\bar{a} = r\alpha = (0.100 \text{ m})\alpha$$

By comparing the friction force obtained with the maximum available friction force, we shall determine whether this assumption is justified. The moment of inertia of the wheel is

$$\bar{I} = m\bar{k}^2 = (50 \text{ kg})(0.070 \text{ m})^2 = 0.245 \text{ kg}\cdot\text{m}^2$$

Equations of Motion

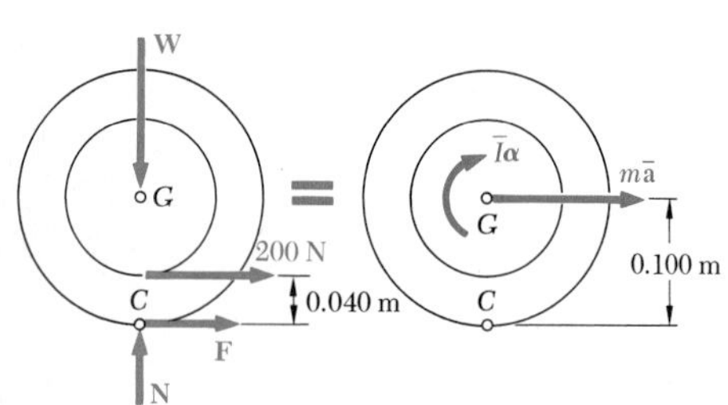

$$+\circlearrowright\Sigma M_C = \Sigma(M_C)_{\text{eff}}: \quad (200 \text{ N})(0.040 \text{ m}) = m\bar{a}(0.100 \text{ m}) + \bar{I}\alpha$$
$$8.00 \text{ N}\cdot\text{m} = (50 \text{ kg})(0.100 \text{ m})\alpha(0.100 \text{ m}) + (0.245 \text{ kg}\cdot\text{m}^2)\alpha$$
$$\alpha = +10.74 \text{ rad/s}^2$$
$$\bar{a} = r\alpha = (0.100 \text{ m})(10.74 \text{ rad/s}^2) = 1.074 \text{ m/s}^2$$

$$\xrightarrow{+} \Sigma F_x = \Sigma(F_x)_{\text{eff}}: \quad F + 200 \text{ N} = m\bar{a}$$
$$F + 200 \text{ N} = (50 \text{ kg})(1.074 \text{ m/s}^2)$$
$$F = -146.3 \text{ N} \qquad \mathbf{F} = 146.3 \text{ N} \leftarrow$$

$$+\uparrow\Sigma F_y = \Sigma(F_y)_{\text{eff}}:$$
$$N - W = 0 \qquad N - W = mg = (50 \text{ kg})(9.81 \text{ m/s}^2) = 490.5 \text{ N}$$
$$\mathbf{N} = 490.5 \text{ N} \uparrow$$

Maximum Available Friction Force

$$F_{\text{max}} = \mu_s N = 0.20(490.5 \text{ N}) = 98.1 \text{ N}$$

Since $F > F_{\text{max}}$, the assumed motion is impossible.

***b.* Rotating and Sliding.** Since the wheel must rotate and slide at the same time, we draw a new diagram, where $\bar{\mathbf{a}}$ and $\boldsymbol{\alpha}$ are independent and where

$$F = F_k = \mu_k N = 0.15(490.5 \text{ N}) = 73.6 \text{ N}$$

From the computation of part *a*, it appears that **F** should be directed to the left. We write the following equations of motion:

$$\xrightarrow{+} \Sigma F_x = \Sigma(F_x)_{\text{eff}}: \quad 200 \text{ N} - 73.6 \text{ N} = (50 \text{ kg})\bar{a}$$
$$\bar{a} = +2.53 \text{ m/s}^2 \qquad \bar{\mathbf{a}} = 2.53 \text{ m/s}^2 \rightarrow$$

$$+\circlearrowright\Sigma M_G = \Sigma(M_G)_{\text{eff}}:$$
$$(73.6 \text{ N})(0.100 \text{ m}) - (200 \text{ N})(0.060 \text{ m}) = (0.245 \text{ kg}\cdot\text{m}^2)\alpha$$
$$\alpha = -18.94 \text{ rad/s}^2 \qquad \boldsymbol{\alpha} = 18.94 \text{ rad/s}^2 \circlearrowleft$$

SAMPLE PROBLEM 16.10

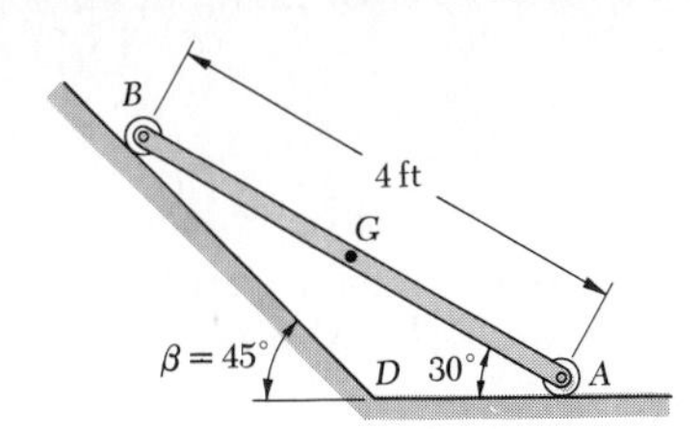

The extremities of a 4-ft rod, weighing 50 lb, may move freely and with no friction along two straight tracks as shown. If the rod is released with no velocity from the position shown, determine (*a*) the angular acceleration of the rod, (*b*) the reactions at *A* and *B*.

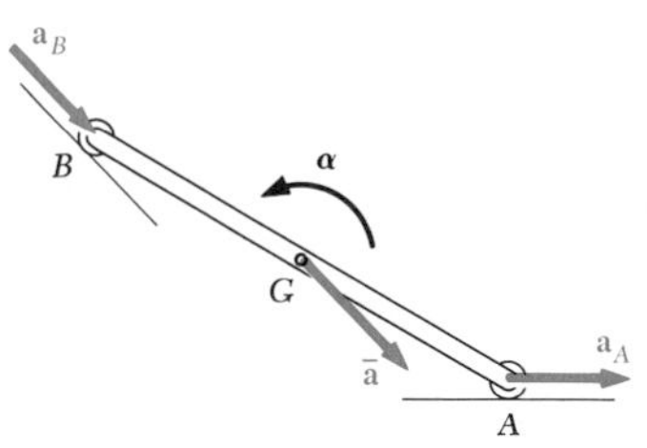

Kinematics of Motion. Since the motion is constrained, the acceleration of *G* must be related to the angular acceleration $\boldsymbol{\alpha}$. To obtain this relation, we shall first determine the magnitude of the acceleration $\mathbf{a}_A$ of point *A* in terms of α; assuming $\boldsymbol{\alpha}$ directed counterclockwise and noting that $a_{B/A} = 4\alpha$, we write

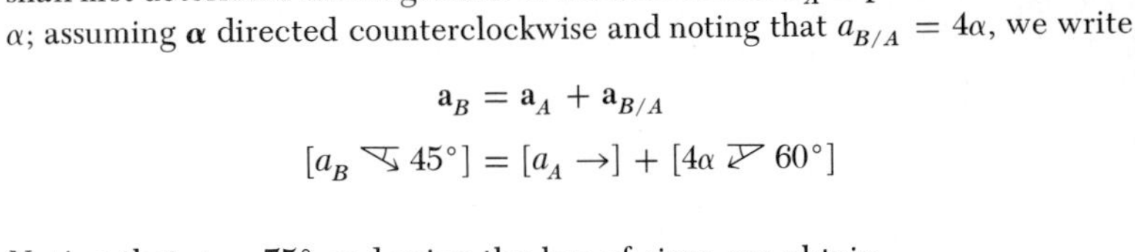

$$\mathbf{a}_B = \mathbf{a}_A + \mathbf{a}_{B/A}$$

$$[a_B \searrow 45°] = [a_A \rightarrow] + [4\alpha \nearrow 60°]$$

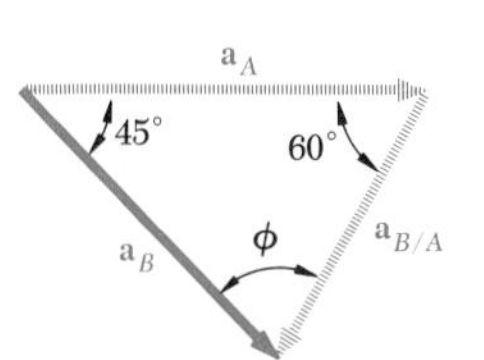

Noting that $\phi = 75°$ and using the law of sines, we obtain

$$a_A = 5.46\alpha \qquad a_B = 4.90\alpha$$

The acceleration of *G* is now obtained by writing

$$\bar{\mathbf{a}} = \mathbf{a}_G = \mathbf{a}_A + \mathbf{a}_{G/A}$$

$$\bar{\mathbf{a}} = [5.46\alpha \rightarrow] + [2\alpha \nearrow 60°]$$

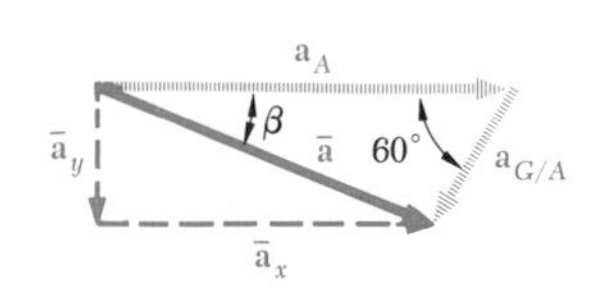

Resolving $\bar{\mathbf{a}}$ into *x* and *y* components, we obtain

$$\bar{a}_x = 5.46\alpha - 2\alpha \cos 60° = 4.46\alpha \qquad \bar{\mathbf{a}}_x = 4.46\alpha \rightarrow$$

$$\bar{a}_y = -2\alpha \sin 60° = -1.732\alpha \qquad \bar{\mathbf{a}}_y = 1.732\alpha \downarrow$$

Kinetics of Motion. We draw the two sketches shown to express that the system of external forces is equivalent to the system of effective forces represented by the vector of components $m\bar{\mathbf{a}}_x$ and $m\bar{\mathbf{a}}_y$ attached at *G* and the couple $\bar{I}\boldsymbol{\alpha}$. We compute the following magnitudes:

$$\bar{I} = \tfrac{1}{12}ml^2 = \frac{1}{12}\frac{50 \text{ lb}}{32.2 \text{ ft/s}^2}(4 \text{ ft})^2 = 2.07 \text{ lb}\cdot\text{ft}\cdot\text{s}^2 \qquad \bar{I}\alpha = 2.07\alpha$$

$$m\bar{a}_x = \frac{50}{32.2}(4.46\alpha) = 6.93\alpha \qquad m\bar{a}_y = -\frac{50}{32.2}(1.732\alpha) = -2.69\alpha$$

Equations of Motion

$+\circlearrowleft \Sigma M_E = \Sigma(M_E)_{\text{eff}}$:

$$(50)(1.732) = (6.93\alpha)(4.46) + (2.69\alpha)(1.732) + 2.07\alpha$$

$$\alpha = +2.30 \text{ rad/s}^2 \qquad \boldsymbol{\alpha} = 2.30 \text{ rad/s}^2 \circlearrowleft \blacktriangleleft$$

$\xrightarrow{+} \Sigma F_x = \Sigma(F_x)_{\text{eff}}$:

$$R_B \sin 45° = (6.93)(2.30) = 15.94$$

$$R_B = 22.5 \text{ lb} \qquad \mathbf{R}_B = 22.5 \text{ lb} \angle 45° \blacktriangleleft$$

$+\uparrow \Sigma F_y = \Sigma(F_y)_{\text{eff}}$:

$$R_A + R_B \cos 45° - 50 = -(2.69)(2.30)$$

$$R_A = -6.19 - 15.94 + 50 = 27.9 \text{ lb} \qquad \mathbf{R}_A = 27.9 \text{ lb} \uparrow \blacktriangleleft$$

Problems

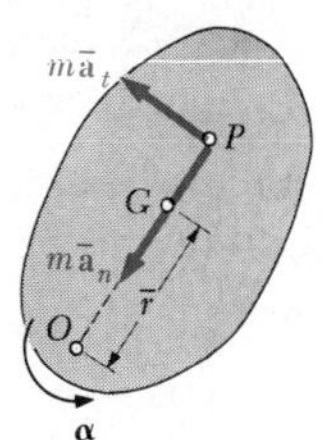

Fig. P16.78

16.78 Show that the couple $\bar{I}\boldsymbol{\alpha}$ of Fig. 16.14 may be eliminated by attaching the vectors $m\bar{\mathbf{a}}_t$ and $m\bar{\mathbf{a}}_n$ at a point P called the *center of percussion*, located on line OG at a distance $GP = \bar{k}^2/\bar{r}$ from the mass center of the body.

16.79 A uniform slender rod, of length $L = 900$ mm and mass $m = 2.5$ kg, hangs freely from a hinge at A. If a force $\mathbf{P}$ of magnitude 15 N is applied at B horizontally to the left ($h = L$), determine (a) the angular acceleration of the rod, (b) the components of the reaction at A.

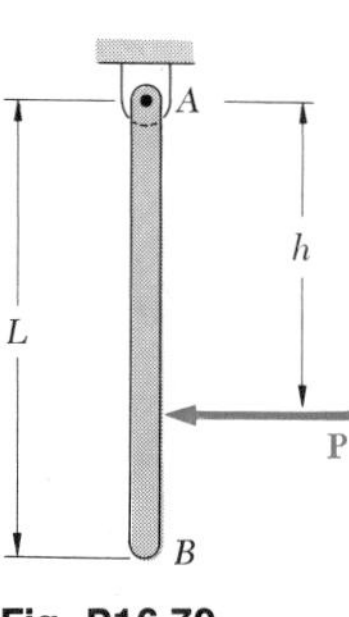

Fig. P16.79

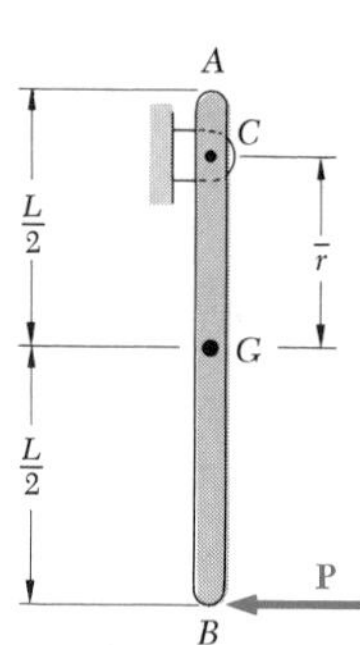

Fig. P16.80

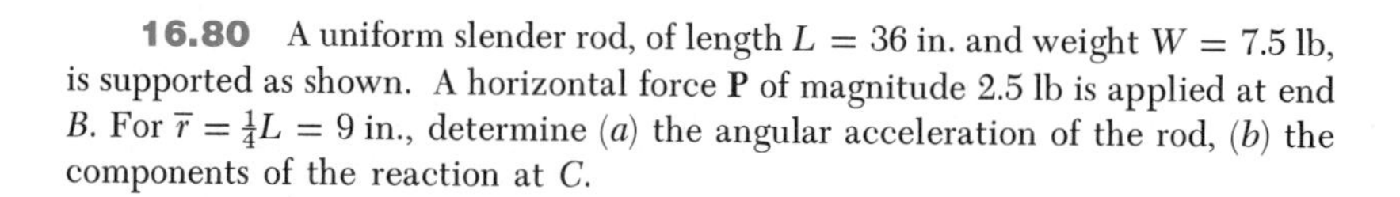

16.80 A uniform slender rod, of length $L = 36$ in. and weight $W = 7.5$ lb, is supported as shown. A horizontal force $\mathbf{P}$ of magnitude 2.5 lb is applied at end B. For $\bar{r} = \frac{1}{4}L = 9$ in., determine (a) the angular acceleration of the rod, (b) the components of the reaction at C.

16.81 In Prob. 16.80, determine (a) the distance $\bar{r}$ for which the horizontal component of the reaction at C is zero, (b) the corresponding angular acceleration of the rod.

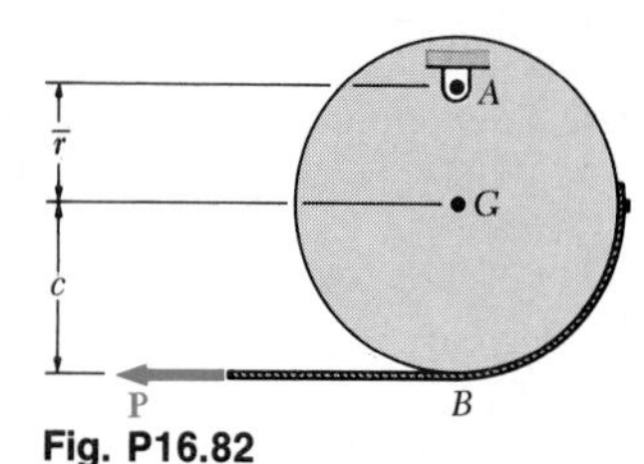

Fig. P16.82

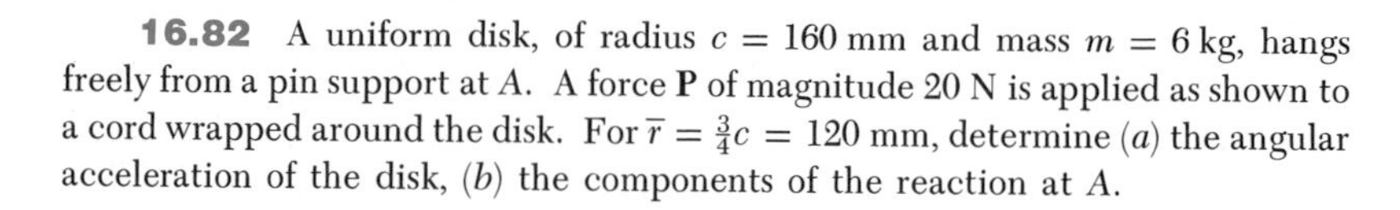

16.82 A uniform disk, of radius $c = 160$ mm and mass $m = 6$ kg, hangs freely from a pin support at A. A force $\mathbf{P}$ of magnitude 20 N is applied as shown to a cord wrapped around the disk. For $\bar{r} = \frac{3}{4}c = 120$ mm, determine (a) the angular acceleration of the disk, (b) the components of the reaction at A.

16.83 In Prob. 16.82, determine (a) the distance $\bar{r}$ for which the horizontal component of the reaction at A is zero, (b) the corresponding angular acceleration of the disk.

16.84 In Prob. 16.79, determine (a) the distance h for which the horizontal component of the reaction at A is zero, (b) the corresponding angular acceleration of the rod.

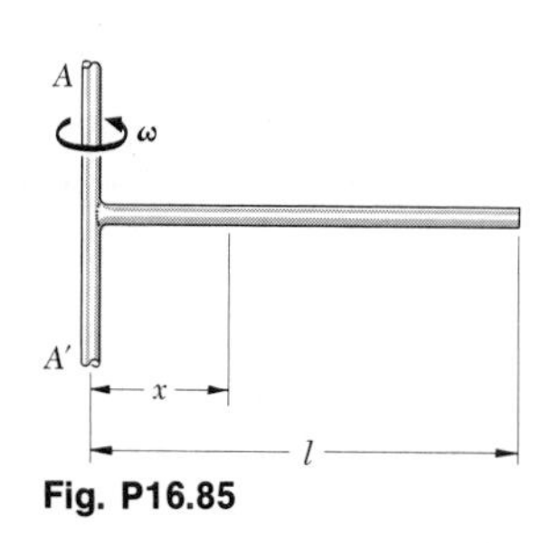

Fig. P16.85

16.85 A uniform slender rod of length l and mass m rotates about a vertical axis AA' at a constant angular velocity $\boldsymbol{\omega}$. Determine the tension in the rod at a distance x from the axis of rotation.

16.86 The rim of a flywheel has a mass of 1800 kg and a mean radius of 600 mm. As the flywheel rotates at a constant angular velocity of 360 rpm, radial forces are exerted on the rim by the spokes and internal forces are developed within the rim. Neglecting the weight of the spokes, determine (*a*) the internal forces in the rim, assuming the radial forces exerted by the spokes to be zero, (*b*) the radial force exerted by each spoke, assuming the tangential forces in the rim to be zero.

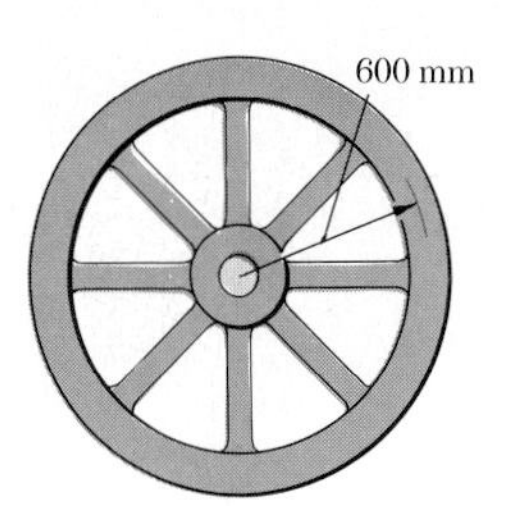

Fig. P16.86

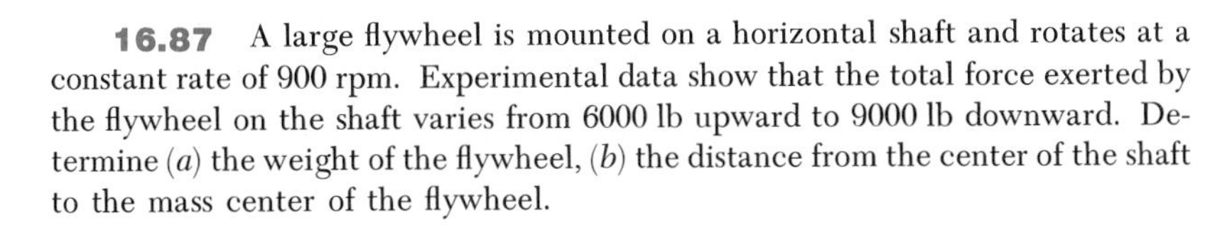

16.87 A large flywheel is mounted on a horizontal shaft and rotates at a constant rate of 900 rpm. Experimental data show that the total force exerted by the flywheel on the shaft varies from 6000 lb upward to 9000 lb downward. Determine (*a*) the weight of the flywheel, (*b*) the distance from the center of the shaft to the mass center of the flywheel.

16.88 A portion of a circular cylindrical shell forms a small vane which is welded to the vertical shaft *AB*. The vane and shaft rotate about the *y* axis with a constant angular velocity of 180 rpm counterclockwise. Knowing that the vane weighs 4 lb, determine the horizontal components of the reaction at *A*.

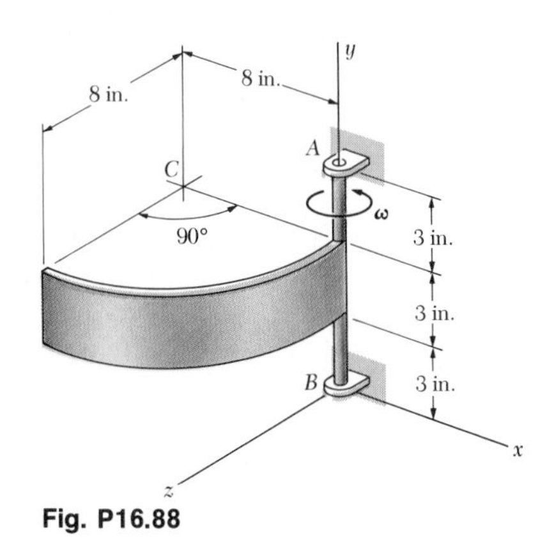

Fig. P16.88

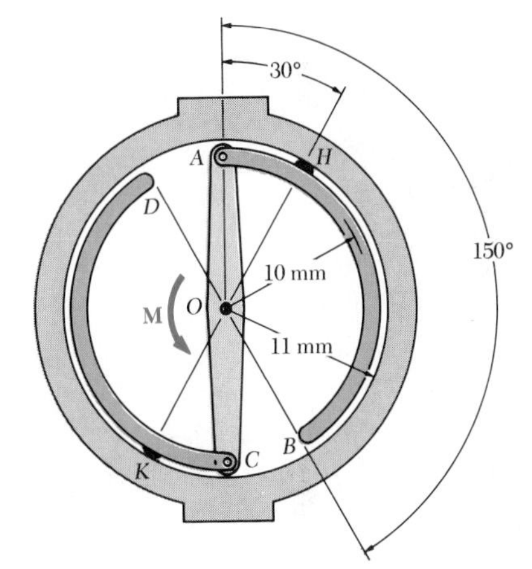

Fig. P16.89

16.89 Centrifugal clutches of the type shown are used to control the operating speed of equipment such as movie cameras and dial telephones. Thin curved members *AB* and *CD* are connected by pins at *A* and *C* to the arm *AC* which may rotate about a fixed point *O*. Each of the members *AB* and *CD* has a mass of 4.5 g and a radius of 10 mm. As the clutch rotates counterclockwise, knobs *H* and *K* slide on the inside of a fixed cylindrical surface of radius 11 mm. Knowing that the coefficient of kinetic friction at *H* and *K* is 0.35 and that the clutch is to have a constant angular velocity of 3000 rpm, determine the couple **M** which must be applied to arm *AC*.

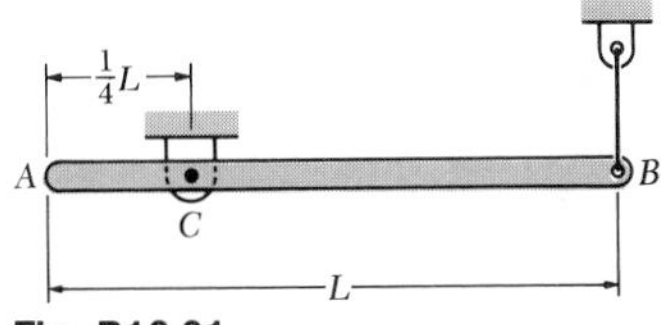

Fig. P16.91

16.90 For the centrifugal clutch of Prob. 16.89, determine the constant angular velocity which will result from the application of a couple **M** of constant moment 50 N·mm.

16.91 and 16.92 A uniform beam of length *L* and weight *W* is supported as shown. If the cable suddenly breaks, determine (*a*) the acceleration of end *B*, (*b*) the reaction at the pin support.

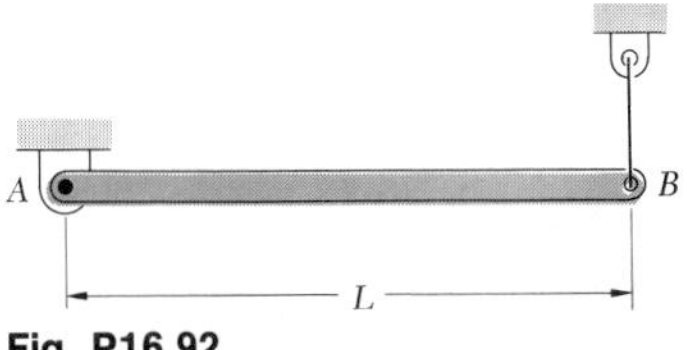

Fig. P16.92

16.93 A uniform square plate of weight W is supported as shown. If the cable suddenly breaks, determine (*a*) the angular acceleration of the plate, (*b*) the acceleration of corner C, (*c*) the reaction at A.

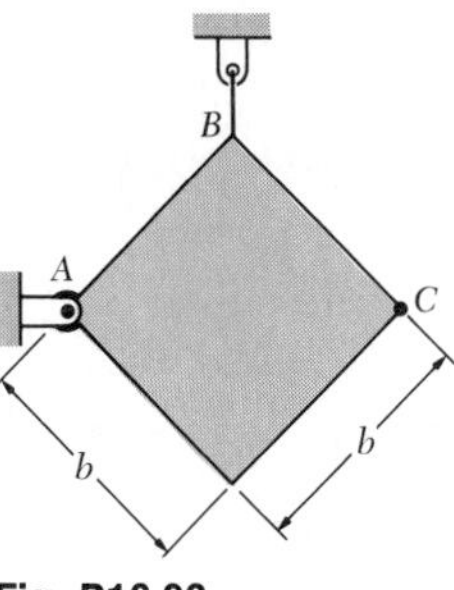

Fig. P16.93

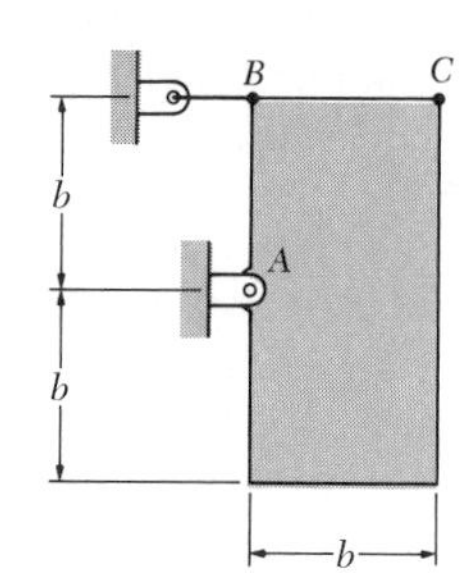

Fig. P16.94

16.94 A uniform rectangular plate of weight W is supported as shown. If the cable suddenly breaks, determine (*a*) the angular acceleration of the plate, (*b*) the acceleration of corner C, (*c*) the reaction at A.

16.95 An 8-lb uniform plate swings freely about A in a vertical plane. Knowing that $\mathbf{P} = 0$ and that in the position shown the plate has an angular velocity of 15 rad/s counterclockwise, determine (*a*) the angular acceleration of the plate, (*b*) the components of the reaction at A.

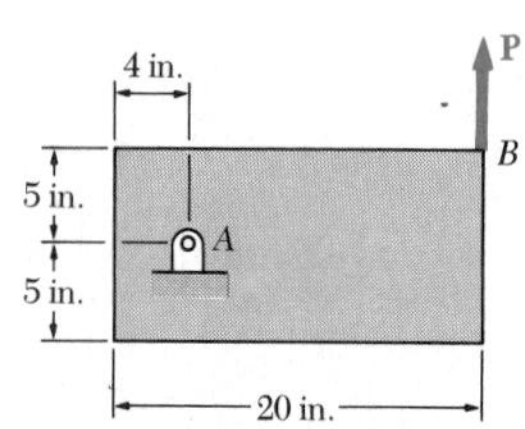

Fig. P16.95 and P16.96

16.96 An 8-lb uniform plate rotates about A in a vertical plane under the combined effect of gravity and of the vertical force $\mathbf{P}$. Knowing that at the instant shown the plate has an angular velocity of 25 rad/s and an angular acceleration of 20 rad/s^2 both counterclockwise, determine (*a*) the force $\mathbf{P}$, (*b*) the components of the reaction at A.

16.97 A 3-kg slender rod is welded to the edge of a 2-kg uniform disk as shown. The assembly rotates about A in a vertical plane under the combined effect of gravity and of the vertical force $\mathbf{P}$. Knowing that at the instant shown the assembly has an angular velocity of 12 rad/s and an angular acceleration of 24 rad/s^2 both counterclockwise, determine (*a*) the force $\mathbf{P}$, (*b*) the components of the reaction at A.

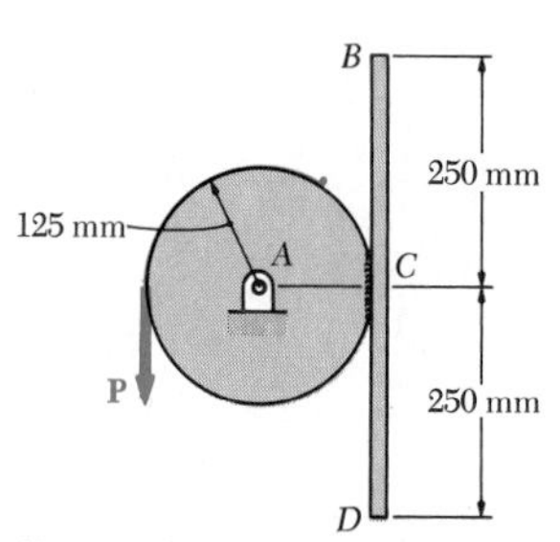

Fig. P16.97 and P16.98

16.98 A 3-kg slender rod is welded to the edge of a 2-kg uniform disk as shown. The assembly swings freely about A in a vertical plane. Knowing that $\mathbf{P} = 0$ and that in the position shown the assembly has an angular velocity of 16 rad/s counterclockwise, determine (*a*) the angular acceleration of the assembly, (*b*) the components of the reaction at A.

16.99 Two uniform rods, ABC of mass 3 kg and DCE of mass 4 kg, are connected by a pin at C and by two cords BD and BE. The T-shaped assembly rotates in a vertical plane under the combined effect of gravity and of a couple **M** which is applied to rod ABC. Knowing that at the instant shown the tension is 8 N in cord BD and zero in cord BE, determine (*a*) the angular acceleration of the assembly, (*b*) the couple **M**.

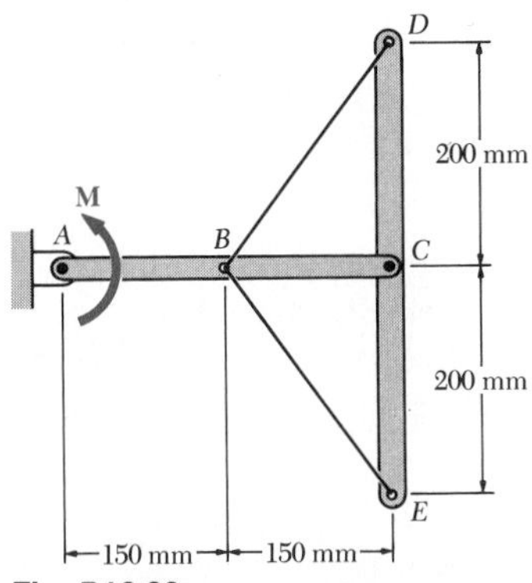

Fig. P16.99

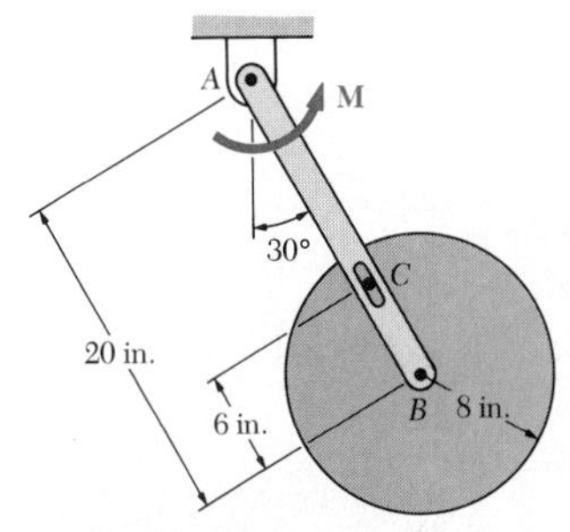

Fig. P16.100

16.100 A 16-lb uniform disk is attached to the 9-lb slender rod AB by means of frictionless pins at B and C. The assembly rotates in a vertical plane under the combined effect of gravity and of a couple **M** which is applied to rod AB. Knowing that at the instant shown the assembly has an angular velocity of 6 rad/s and an angular acceleration of 15 rad/s^2, both counterclockwise, determine (*a*) the couple **M**, (*b*) the force exerted by pin C on member AB.

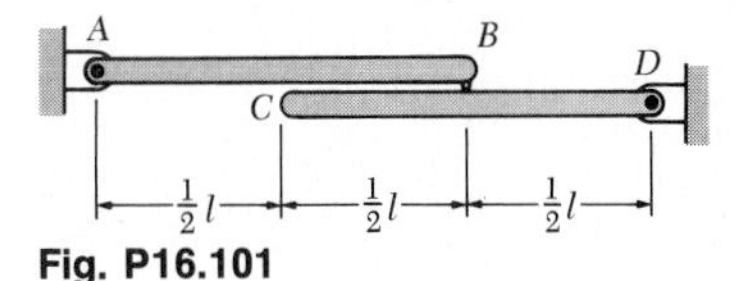

Fig. P16.101

16.101 Two slender rods, each of length l and mass m, are released from rest in the position shown. Knowing that a small frictionless knob at end B of rod AB bears on rod CD, determine immediately after release, (*a*) the acceleration of end C of rod CD, (*b*) the force exerted on knob B.

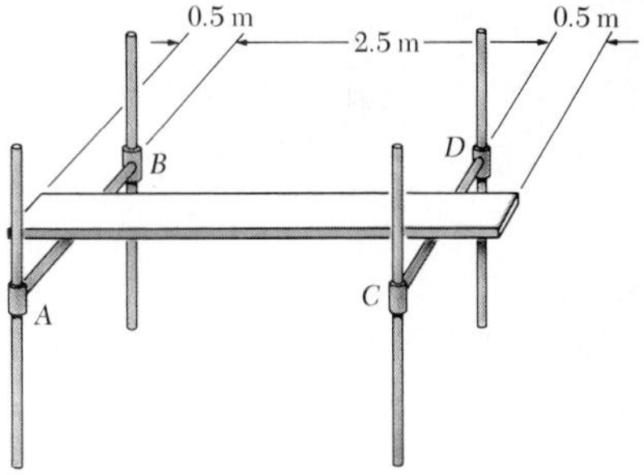

16.102 A 3.5-m plank of mass 30 kg rests on two horizontal pipes AB and CD of a scaffolding. The pipes are 2.5 m apart, and the plank overhangs 0.5 m at each end. A 70-kg worker is standing on the plank when pipe CD suddenly breaks. Determine the initial acceleration of the worker, knowing that he was standing (*a*) in the middle of the plank, (*b*) just above pipe CD.

Fig. P16.102

16.103 The uniform rod AB of mass m is released from rest when $\beta = 60°$. Assuming that the friction force between end A and the surface is large enough to prevent sliding, determine (*a*) the angular acceleration of the rod just after release, (*b*) the normal reaction and the friction force at A, (*c*) the minimum value of μ_s compatible with the described motion.

***16.104** Knowing that the coefficient of static friction between the rod and the floor is 0.30, determine the range of values of β for which the rod *will* slip immediately after being released from rest.

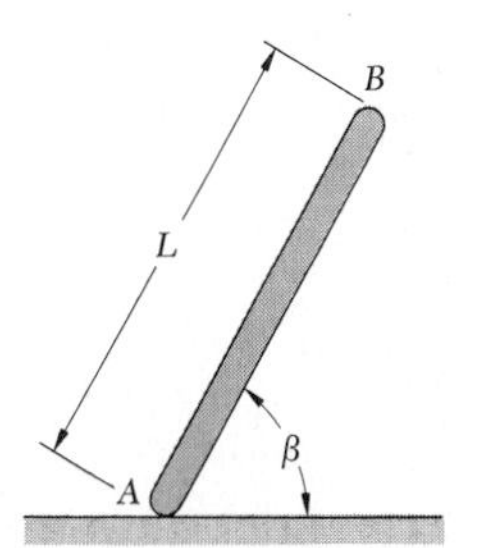

Fig. P16.103 and P16.104

16.105 Derive the equation $\Sigma M_C = I_C\alpha$ for the rolling disk of Fig. 16.16, where ΣM_C represents the sum of the moments of the external forces about the instantaneous center C and I_C the moment of inertia of the disk about C.

16.106 Show that in the case of an unbalanced disk, the equation derived in Prob. 16.105 is valid only when the mass center G, the geometric center O, and the instantaneous center C happen to lie in a straight line.

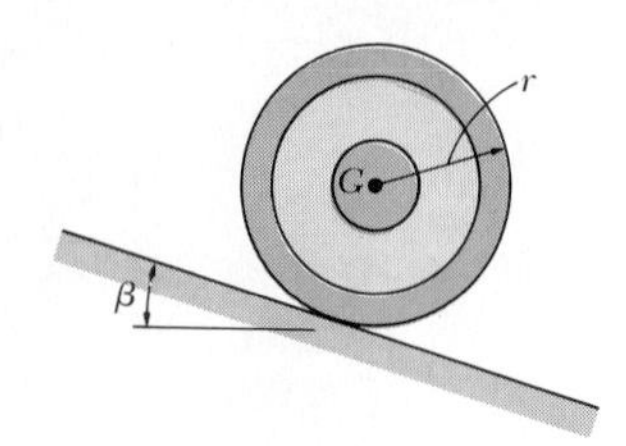

Fig. P16.107

16.107 A wheel of radius r and centroidal radius of gyration $\bar{k}$ is released from rest on the incline and rolls without slipping. Derive an expression for the acceleration of the center of the wheel in terms of r, $\bar{k}$, β, and g.

16.108 A flywheel is rigidly attached to a shaft of 50-mm radius which may roll along parallel rails as shown. When released from rest, the system rolls through a distance of 3.5 m in 20 s. Determine the centroidal radius of gyration of the system.

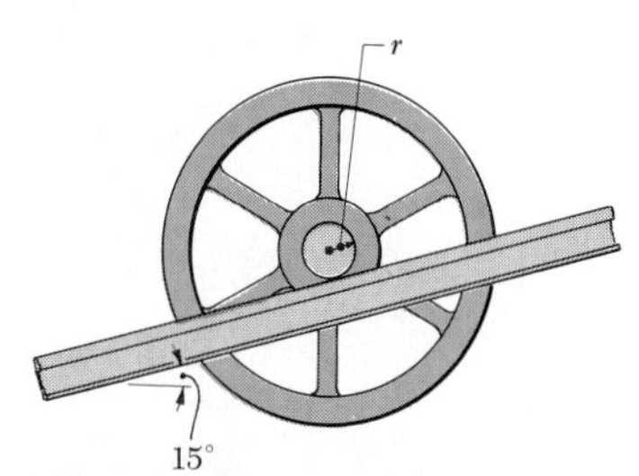

Fig. P16.108 and P16.109

16.109 A flywheel of centroidal radius of gyration $\bar{k} = 400$ mm is rigidly attached to a shaft of radius $r = 25$ mm which may roll along parallel rails. Knowing that the system is released from rest, determine the distance through which it will roll in 15 s.

16.110 A homogeneous cylinder C and a section of pipe P are in contact when they are released from rest. Knowing that both the cylinder and the pipe roll without slipping, determine the clear distance between them after 3 s.

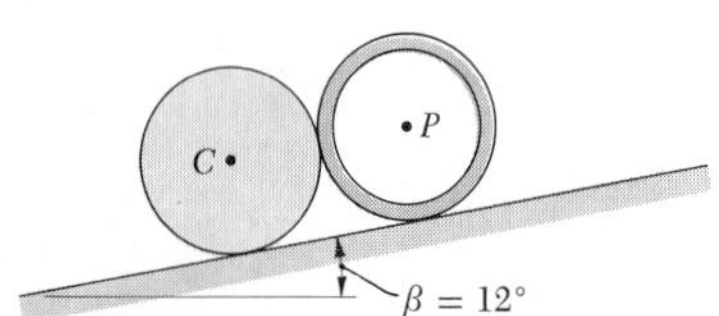

Fig. P16.110

16.111 through 16.114 A drum of 4-in. radius is attached to a disk of 8-in. radius. The disk and drum have a total weight of 10 lb and a radius of gyration of 6 in. A cord is attached as shown and pulled with a force **P** of magnitude 5 lb. Knowing that the disk rolls without sliding, determine (*a*) the angular acceleration of the disk and the acceleration of G, (*b*) the minimum value of the coefficient of static friction compatible with this motion.

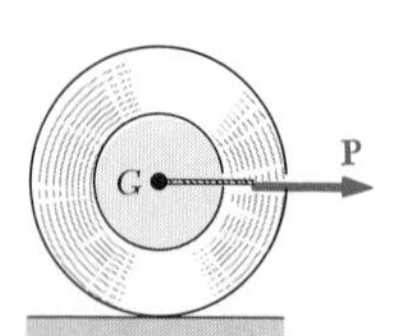

Fig. P16.111 and P16.115

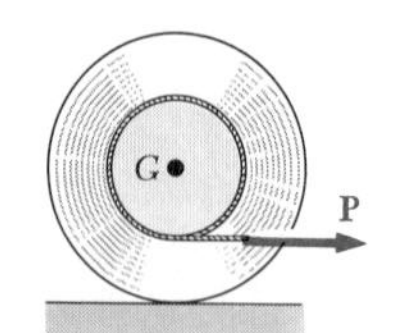

Fig. P16.112 and P16.116

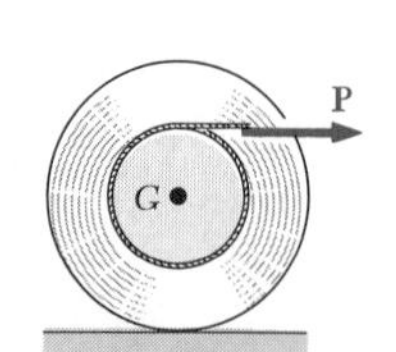

Fig. P16.113 and P16.117

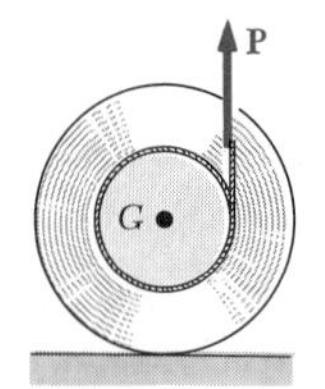

Fig. P16.114 and P16.118

16.115 through 16.118 A drum of 80-mm radius is attached to a disk of 160-mm radius. The disk and drum have a total mass of 5 kg and a radius of gyration of 120 mm. A cord is attached as shown and pulled with a force **P** of magnitude 18 N. Knowing that the coefficients of static and kinetic friction are $\mu_s = 0.20$ and $\mu_k = 0.15$, determine (*a*) whether or not the disk slides, (*b*) the angular acceleration of the disk and the acceleration of G.

16.119 and 16.120 The 9-kg carriage is supported as shown by two uniform disks each of mass 6 kg and radius 150 mm. Knowing that the disks roll without sliding, determine the acceleration of the carriage when a force of 30 N is applied to it.

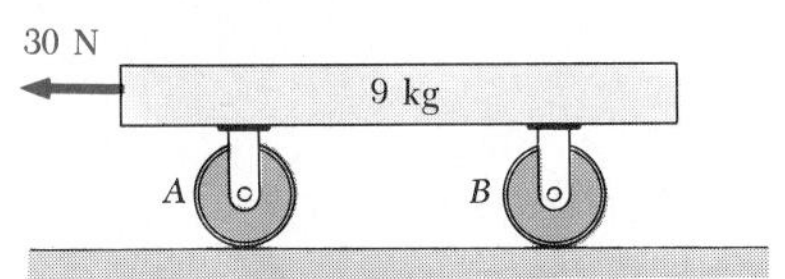

Fig. P16.119

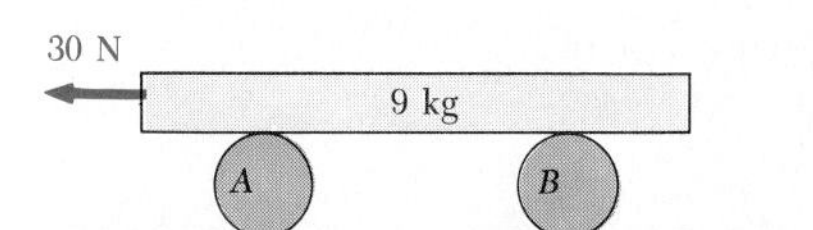

Fig. P16.120

16.121 and 16.122 Gear C has a mass of 3 kg and a centroidal radius of gyration of 75 mm. The uniform bar AB has a mass of 2.5 kg and gear D is stationary. If the system is released from rest in the position shown, determine (*a*) the angular acceleration of gear C, (*b*) the acceleration of point B.

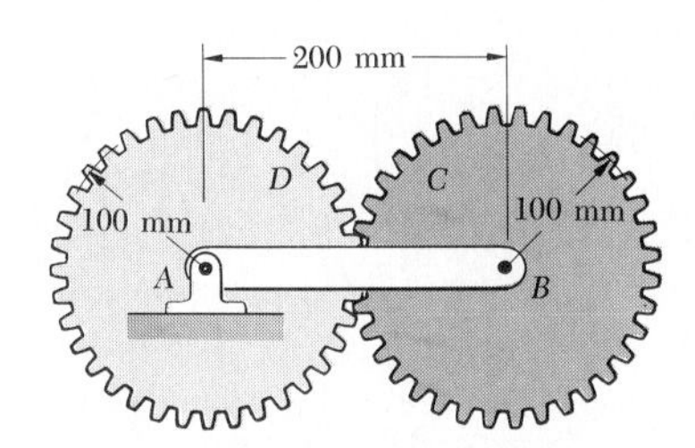

Fig. P16.121

16.123 A block B of mass m is attached to a cord wrapped around a cylinder of the same mass m and of radius r. The cylinder rolls without sliding on a horizontal surface. Determine the components of the accelerations of the center A of the cylinder and of the block B immediately after the system has been released from rest if (*a*) the block hangs freely, (*b*) the motion of the block is guided by a rigid member DAE, frictionless and of negligible mass, which is hinged to the cylinder at A.

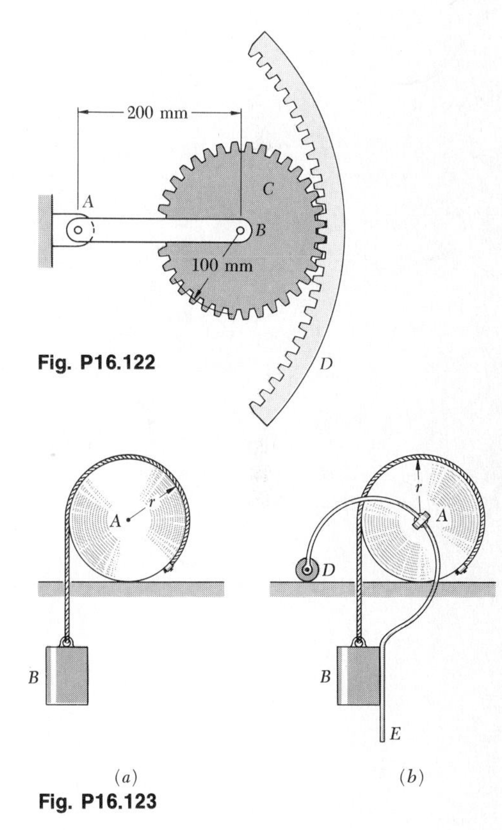

Fig. P16.122

Fig. P16.123

16.124 The disk-and-drum assembly A has a total weight of 15 lb and a centroidal radius of gyration of 6 in. A cord is attached as shown to the assembly A and to the 10-lb uniform disk B. Knowing that the disks roll without sliding, determine the acceleration of the center of each disk for $P = 4$ lb and $Q = 0$.

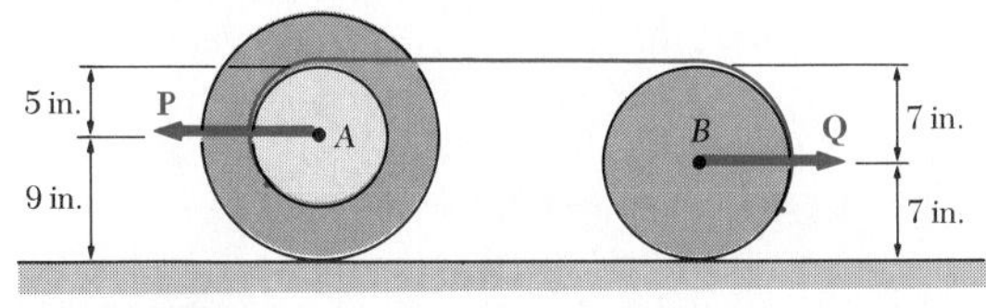

Fig. P16.124

16.125 Solve Prob. 16.124 for $P = 0$ and $Q = 4$ lb.

16.126 Two uniform disks A and B, each of weight 4 lb, are connected by a 3-lb rod CD as shown. A counterclockwise couple $\mathbf{M}$ of moment 1.5 lb·ft is applied to disk A. Knowing that the disks roll without sliding, determine (*a*) the acceleration of the center of each disk, (*b*) the horizontal component of the force exerted on disk B by pin D.

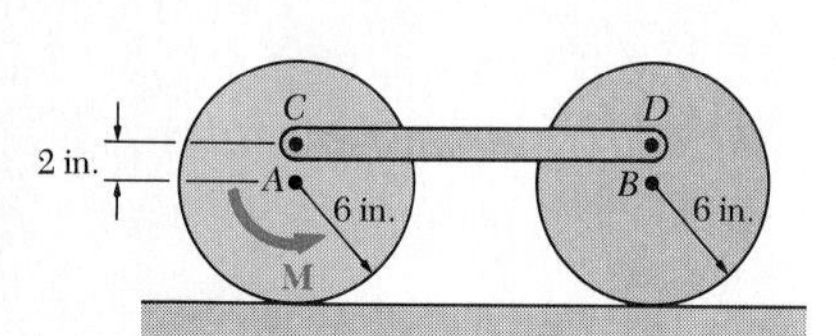

Fig. P16.126

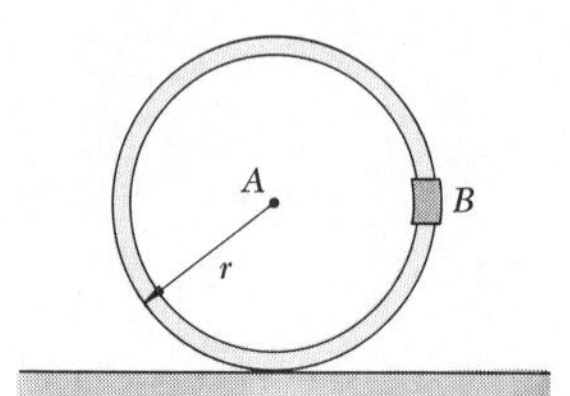

Fig. P16.127

16.127 A small block of mass m is attached at B to a hoop of mass m and radius r. The system is released from rest when B is directly above A and rolls without sliding. Knowing that at the instant shown the system has a clockwise angular velocity of magnitude $\omega = \sqrt{g/2r}$, determine (*a*) the angular acceleration of the hoop, (*b*) the acceleration of B.

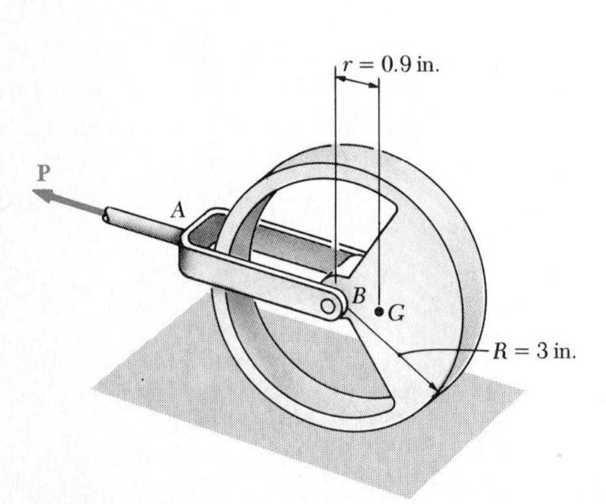

Fig. P16.128

16.128 The center of gravity G of a 2.5-lb unbalanced tracking wheel is located at a distance $r = 0.9$ in. from its geometric center B. The radius of the wheel is $R = 3$ in. and its centroidal radius of gyration is $\bar{k} = 2.2$ in. At the instant shown the center B of the wheel has a velocity of 1.5 ft/s and an acceleration of 5 ft/s^2, both directed to the left. Knowing that the wheel rolls without sliding and neglecting the weight of the driving yoke AB, determine the horizontal force $\mathbf{P}$ applied to the yoke.

16.129 For the wheel of Prob. 16.128, determine the horizontal force $\mathbf{P}$ applied to the yoke, knowing that at the instant shown the center B has a velocity of 1.5 ft/s and an acceleration of 5 ft/s^2, both directed to the right.

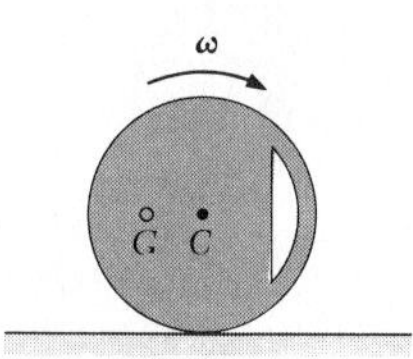

Fig. P16.130

16.130 The mass center G of a 5-kg wheel of radius $R = 300$ mm is located at a distance $r = 100$ mm from its geometric center C. The centroidal radius of gyration is $\bar{k} = 150$ mm. As the wheel rolls without sliding, its angular velocity varies and it is observed that $\omega = 8$ rad/s in the position shown. Determine the corresponding angular acceleration of the wheel.

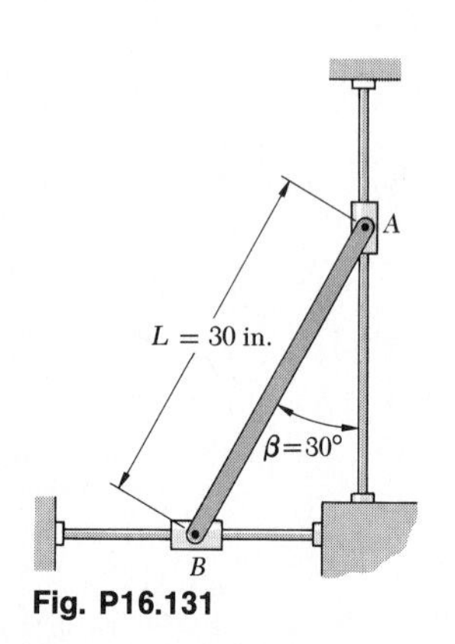

Fig. P16.131

16.131 The ends of the 6-lb slender rod AB are attached to collars of negligible weight which slide without friction along fixed rods. If rod AB is released from rest in the position shown, determine immediately after release (*a*) the angular acceleration of the rod, (*b*) the reaction at A, (*c*) the reaction at B.

16.132 The motion of the 3-kg uniform rod ACB is guided by two blocks of negligible mass which slide without friction in the slots shown. If the rod is released from rest in the position shown, determine immediately after release (*a*) the angular acceleration of the rod, (*b*) the reaction at A.

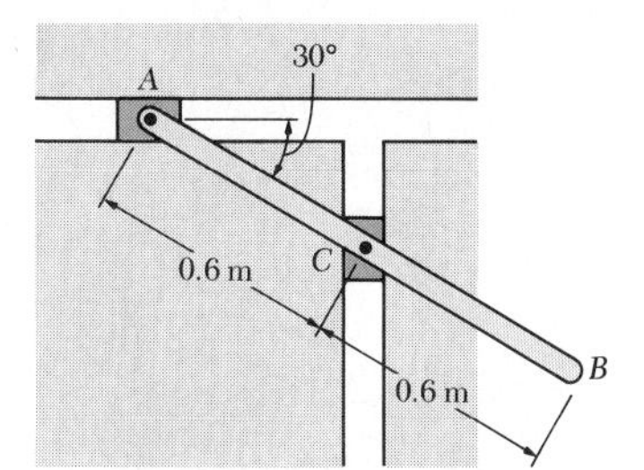

Fig. P16.132 and P16.133

16.133 The motion of the 3-kg uniform rod ACB is guided by two blocks of negligible mass which slide without friction in the slots shown. A horizontal force **P** is applied to block A, causing the rod to start from rest with a counterclockwise angular acceleration of 12 rad/s^2. Determine (*a*) the required force **P**, (*b*) the corresponding reaction at A.

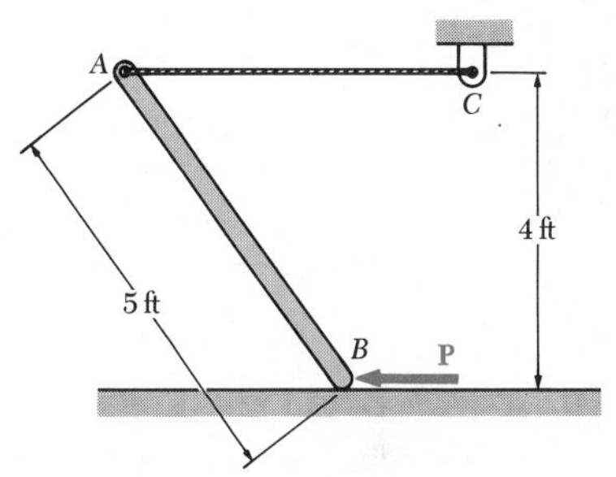

Fig. P16.134

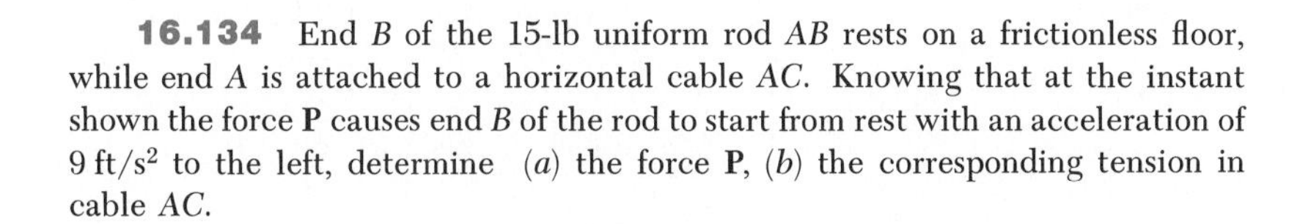

16.134 End B of the 15-lb uniform rod AB rests on a frictionless floor, while end A is attached to a horizontal cable AC. Knowing that at the instant shown the force **P** causes end B of the rod to start from rest with an acceleration of 9 ft/s^2 to the left, determine (*a*) the force **P**, (*b*) the corresponding tension in cable AC.

16.135 End A of the 5-kg uniform rod AB rests on the inclined surface, while end B is attached to a collar of negligible mass which may slide along the vertical rod shown. Knowing that the rod is released from rest when $\theta = 35°$ and neglecting the effect of friction, determine immediately after release (*a*) the angular acceleration of the rod, (*b*) the reaction at B.

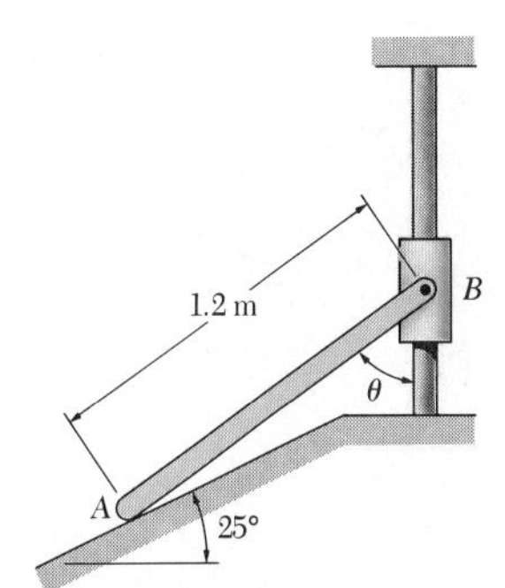

Fig. P16.135 and P16.138

16.136 Rod AB weighs 3 lb and is released from rest in the position shown. Assuming that the ends of the rod slide without friction, determine immediately after release (*a*) the angular acceleration of the rod, (*b*) the reaction at B.

16.137 Ends A and B of the 3-lb uniform rod slide without friction along the surfaces shown. When the rod is at rest a horizontal force **P** is applied at A, causing end A of the rod to start moving to the right with an acceleration of 8 ft/s^2. Determine (*a*) the force **P**, (*b*) the corresponding reaction at B.

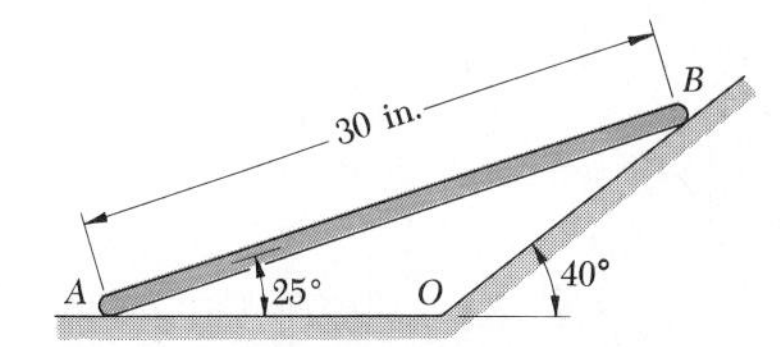

Fig. P16.136 and P16.137

16.138 End A of the 5-kg uniform rod AB rests on the inclined surface, while end B is attached to a collar of negligible mass which may slide along the vertical rod shown. When the rod is at rest a vertical force **P** is applied at B, causing end B of the rod to start moving upward with an acceleration of 4 m/s^2. Knowing that $\theta = 35°$, determine the force **P**.

16.139 The 5-kg uniform rod BD is connected as shown to crank AB and to a collar of negligible mass. Crank AB is rotated with a constant angular velocity of 20 rad/s counterclockwise. Neglecting the effect of friction, determine the reaction at D when $\theta = 0$.

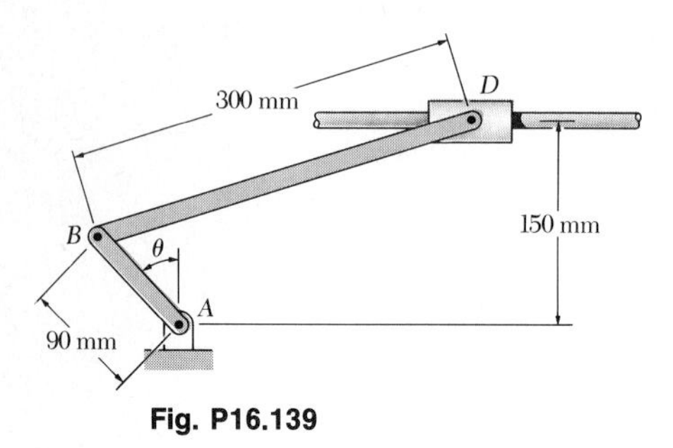

Fig. P16.139

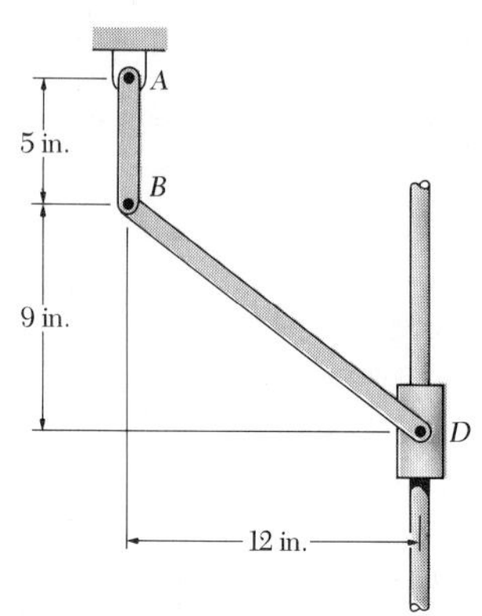

Fig. P16.140

16.140 The 1.5-lb uniform rod BD is connected to crank AB and to a collar of negligible weight. A couple (not shown) is applied to crank AB causing it to rotate with an angular velocity of 9 rad/s counterclockwise and an angular acceleration of 45 rad/s^2 clockwise at the instant shown. Neglecting the effect of friction, determine the reaction at D.

16.141 In Prob. 16.140, determine the reaction at D, knowing that in the position shown crank AB has an angular velocity of 9 rad/s and an angular acceleration of 45 rad/s^2, both counterclockwise.

16.142 Solve Prob. 16.139 when $\theta = 180°$.

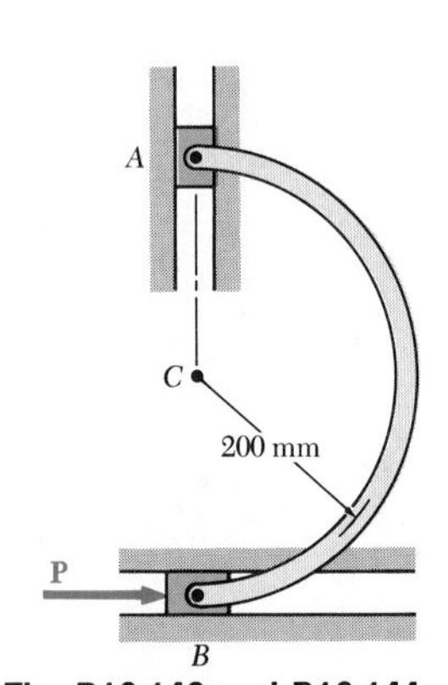

Fig. P16.143 and P16.144

16.143 The motion of a 1.5-kg semicircular rod is guided by two blocks of negligible mass which slide without friction in the slots shown. A horizontal and variable force **P** is applied at B, causing B to move to the right with a constant speed of 5 m/s. For the position shown, determine (*a*) the force **P**, (*b*) the reaction at B.

16.144 The motion of a 1.5-kg semicircular rod is guided by two blocks of negligible mass which slide without friction in the slots shown. Knowing that for the position shown $\mathbf{P} = 0$ and the speed of B is 5 m/s to the right, determine (*a*) the acceleration of B, (*b*) the reaction at B.

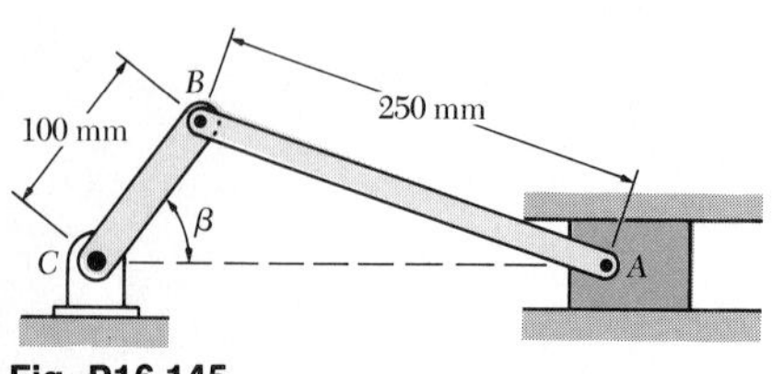

Fig. P16.145

16.145 The 1.8-kg sliding block is connected to crank BC by the uniform rod AB of mass 1.2 kg. Knowing that the crank is rotated with a constant angular velocity of 600 rpm, determine the forces exerted on the rod at A and B when $\beta = 0$. Assume that the motion takes place in a horizontal plane.

16.146 Solve Prob. 16.145 when $\beta = 180°$.

16.147 Two rods AB and BC, of mass m' per unit length, are connected as shown to a disk which is made to rotate in a vertical plane at a constant angular velocity $\boldsymbol{\omega}_0$. For the position shown, determine the components of the forces exerted at A and B on rod AB.

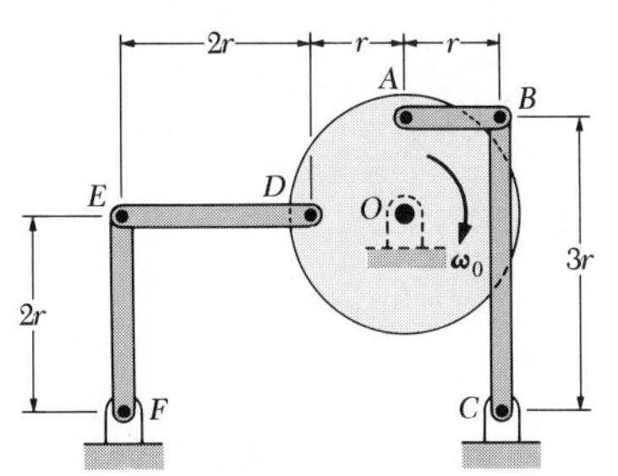

Fig. P16.147 and P16.148

16.148 Two rods DE and EF, of mass m' per unit length, are connected as shown to a disk which is made to rotate in a vertical plane at a constant angular velocity $\boldsymbol{\omega}_0$. For the position shown, determine the components of the forces exerted at D and E on rod DE.

***16.149 and *16.150** The 3-kg cylinder B and the 2-kg wedge A are held at rest in the position shown by the cord C. Assuming that the cylinder rolls without sliding on the wedge and neglecting friction between the wedge and the ground, determine (*a*) the acceleration of the wedge, (*b*) the angular acceleration of the cylinder, immediately after the cord C has been cut.

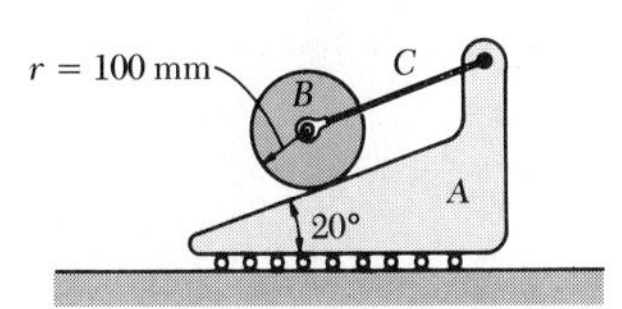

Fig. P16.149

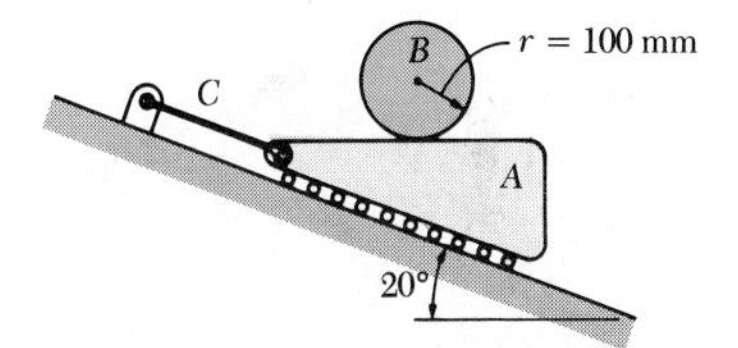

Fig. P16.150

***16.151 and *16.152** The uniform rod AB, of weight 20 lb and length 3 ft, is attached to the 30-lb cart C. Neglecting friction, determine immediately after the system has been released from rest, (*a*) the acceleration of the cart, (*b*) the angular acceleration of the rod.

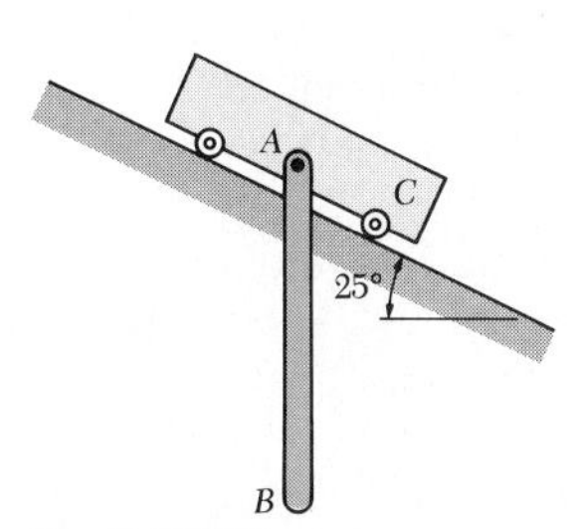

Fig. P16.151

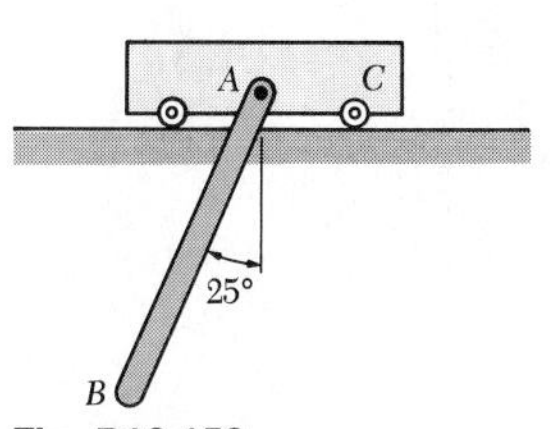

Fig. P16.152

*16.153 A uniform slender rod of length L and mass m is released from rest in the position shown. Derive an expression for (*a*) the angular acceleration of the rod, (*b*) the acceleration of end A, (*c*) the reaction at A, immediately after release. Neglect the mass and friction of the roller at A.

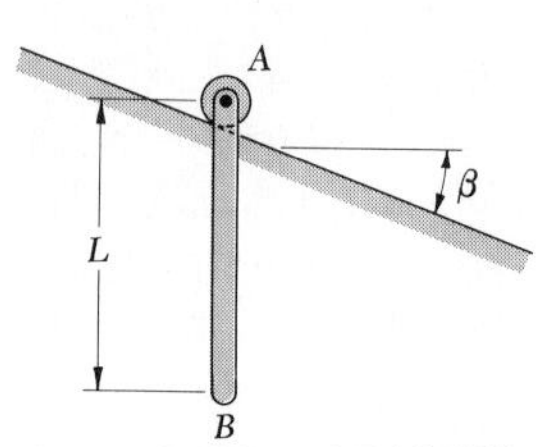

Fig. P16.153 and P16.154

*16.154 A uniform rod AB, of mass 3 kg and length $L = 1.2$ m, is released from rest in the position shown. Knowing that $\beta = 30°$, determine the values immediately after release of (*a*) the angular acceleration of the rod, (*b*) the acceleration of end A, (*c*) the reaction at A. Neglect the mass and friction of the roller at A.

*16.155 Each of the bars AB and BC is of length $L = 18$ in. and weight 3 lb. A horizontal force **P** of magnitude 4 lb is applied at C. Determine the angular acceleration of each bar.

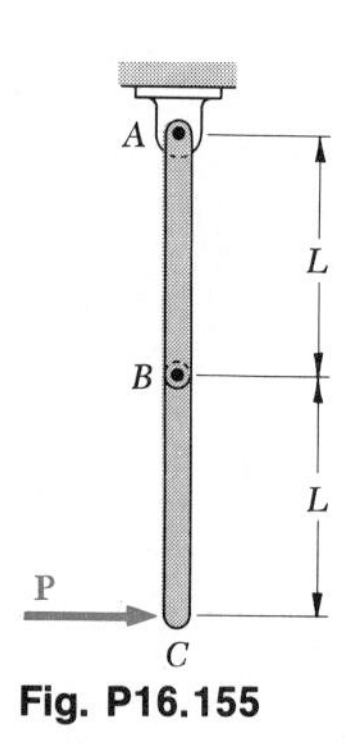

Fig. P16.155

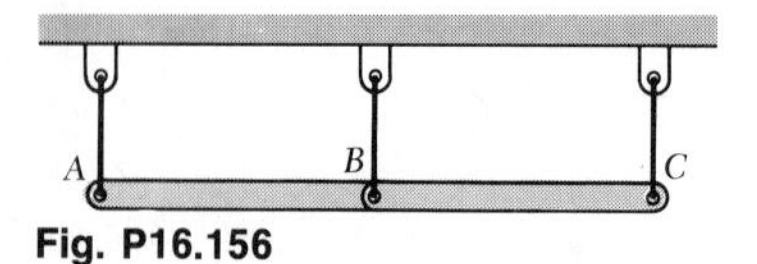

Fig. P16.156

*16.156 Two identical uniform rods are connected by a pin at B and are held in a horizontal position by three wires as shown. If the wires attached at A and B are cut simultaneously, determine at that instant the acceleration of (*a*) point A, (*b*) point B.

*16.157 (*a*) Determine the magnitude and the location of the maximum bending moment in the rod of Prob. 16.79. (*b*) Show that the answer to part *a* is independent of the mass of the rod.

*16.158 Draw the shear and bending-moment diagrams for the beam of Prob. 16.92 immediately after the cable at B breaks.

16.6. Rotation of a Three-Dimensional Body about a Fixed Axis. The rigid bodies considered in the preceding sections were assumed to be symmetrical with respect to the reference plane used to describe their motion. While such an assumption considerably simplifies the derivations involved, it also limits the range of application of the results obtained. One of the most important types of motion which *cannot* be analyzed directly by the methods considered so far is the rotation about a fixed axis of an unsymmetrical three-dimensional body.

We may, however, examine the particular case when the rotating body consists of several parallel slabs rigidly connected to a perpendicular shaft which is supported in bearings at A and B (Fig. 16.18). Since each slab

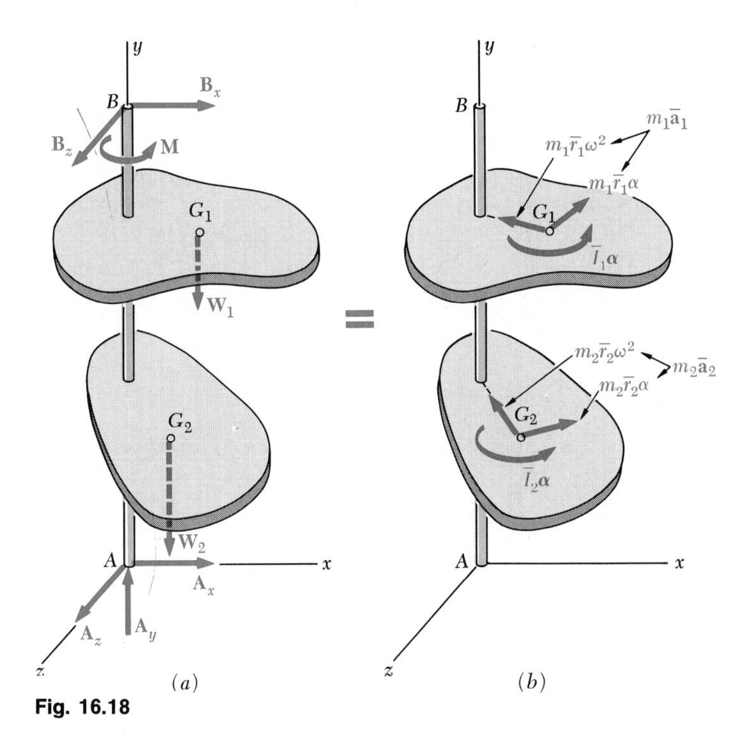

Fig. 16.18

rotates in its own plane, we may apply the method of Sec. 16.2 and reduce the effective forces of each slab to a force-couple system. By d'Alembert's principle the system of the external forces shown in part *a* of Fig. 16.18 must be equivalent to the system of the effective forces represented by the various force-couple systems shown in part *b* of the same figure. Equating components along, and moments about, each of the three coordinate axes, we write the following six equations:

$$\begin{aligned} \Sigma F_x &= \Sigma(F_x)_{\text{eff}} \qquad \Sigma F_y = \Sigma(F_y)_{\text{eff}} \qquad \Sigma F_z = \Sigma(F_z)_{\text{eff}} \\ \Sigma M_x &= \Sigma(M_x)_{\text{eff}} \qquad \Sigma M_y = \Sigma(M_y)_{\text{eff}} \qquad \Sigma M_z = \Sigma(M_z)_{\text{eff}} \end{aligned} \tag{16.17}$$

It should be noted that this method may also be used when the bodies attached to the shaft are three-dimensional, *as long as each body is symmetrical with respect to a plane perpendicular to the shaft.*

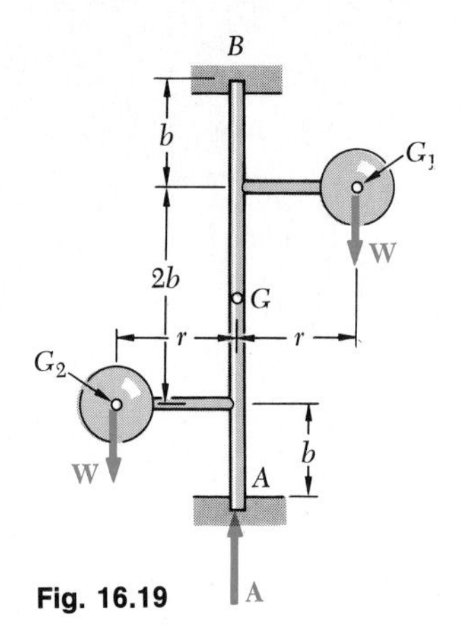

Fig. 16.19

As an example, we shall consider a shaft held in bearings at A and B and supporting two equal spheres of mass m and weight $W = mg$ (Fig. 16.19). We first observe that when the system is at rest, the shaft exerts no lateral thrust on its supports since the center of gravity G of the system is located directly above A. The system is said to be *statically balanced.* The reaction **A**, often referred to as a *static reaction,* is vertical, and its magnitude is $A = 2W = 2mg$. Let us now assume that the system rotates with a constant angular velocity $\boldsymbol{\omega}$. We shall apply d'Alembert's principle and express the fact that the system of the external forces, consisting of the weights and bearing reactions, is equivalent to the system of the effective forces. Since each sphere is symmetrical with respect to a plane perpendicular to the shaft and containing its own center, we may apply the method of Sec. 16.2 and replace the effective forces of each sphere by a force-couple system at its mass center. But in the present case the angular acceleration $\boldsymbol{\alpha}$ is zero and the force-couple systems reduce respectively to the vectors $m\bar{\mathbf{a}}_1$ and $m\bar{\mathbf{a}}_2$ directed toward the axis of rotation. Since these two vectors, shown in part b of Fig. 16.20, have the same magnitude $mr\omega^2$ and are located at a distance $2b$ from each other, we conclude that the system of the effective forces is equivalent to a counterclockwise couple of moment $2bmr\omega^2$. Additional bearing reactions $\mathbf{A}_x$ and $\mathbf{B}_x$, therefore, are

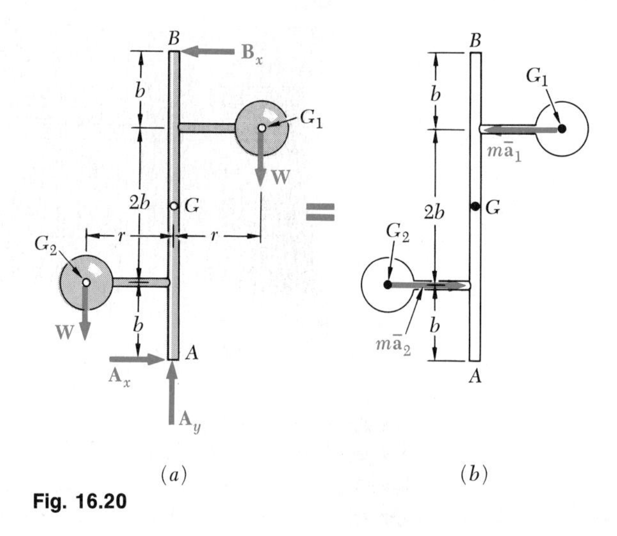

Fig. 16.20

required in part a of the figure. These reactions, called *dynamic reactions,* must also form a counterclockwise couple of moment $2bmr\omega^2$. They must be directed as shown and have a magnitude $A_x = B_x = \frac{1}{2}mr\omega^2$. Although the system is statically balanced, it is clearly *not dynamically balanced.* A system which is not dynamically balanced will have a tendency to tear away from its bearings when rotating at high speed, as indicated by the values obtained for the bearing reactions. Moreover, since the bearing reactions are contained in the plane defined by the shaft and the centers of the spheres, they rotate with the shaft and cause the structure supporting it to vibrate. *Rotating systems should therefore be dynamically as well as statically balanced.* This may be done by rearranging the distribution of mass around the shaft, or by adding corrective masses, until the system of the effective forces becomes equivalent to zero.

The analysis of the rotating shaft of Figs. 16.19 and 16.20 could also be carried out by the method of dynamic equilibrium. The effective forces shown on the right-hand side of the equals sign in Fig. 16.20 would then be replaced by equal and opposite inertia vectors on the left-hand side of the equals sign (Fig. 16.21). The problem is thus reduced to a statics problem, in which the inertia vectors represent the "centrifugal forces" which tend to "pull" the spheres away from the shaft.

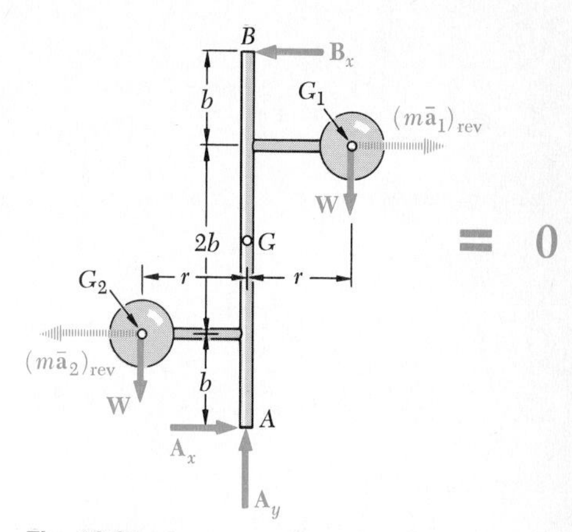

Fig. 16.21

In the general case of the rotation about a fixed axis of an unsymmetrical three-dimensional body, a typical element of mass should be isolated and its effective force should be determined. The sums of the components and moments of the effective forces may then be obtained by integration. While this approach is successfully used in Sample Prob. 16.12, it will lead, in general, to fairly involved computations. Other methods, beyond the scope of this chapter, may then be preferred. The approach indicated here shows clearly, however, that, even in as simple a motion as uniform rotation, the system of the effective forces *will not reduce to a single vector attached at the mass center of the rotating body* if the body does not possess a plane of symmetry perpendicular to the axis of rotation.

Consider, for example, an airplane making a turn of radius $\bar{r}$ in the banked position shown in Fig. 16.22. Since two symmetrical wing elements

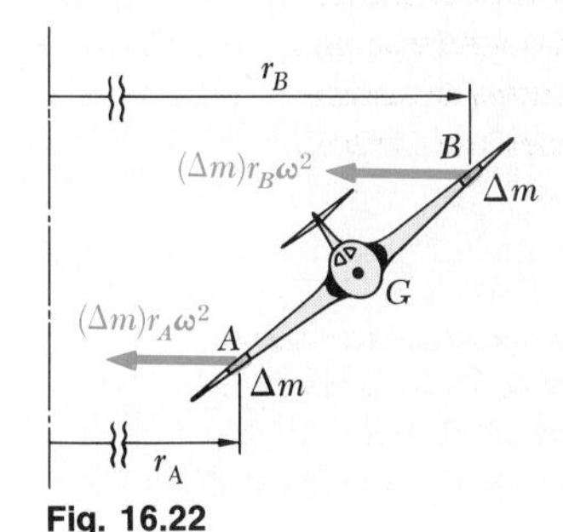

Fig. 16.22

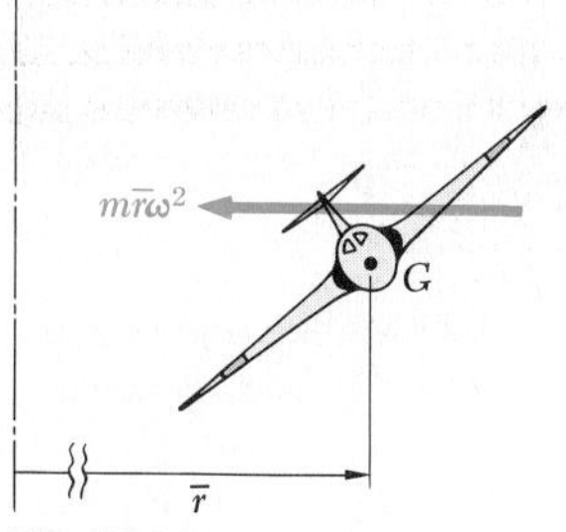

Fig. 16.23

A and B are not at the same distance from the axis of rotation, the effective forces associated with A and B are not equal. The system of the effective forces corresponding to the various elements of the airplane, therefore, will *not* reduce to a single vector at the mass center G of the airplane. It will reduce to a force-couple system at G which, in turn, may be replaced by a single vector located above G (Fig. 16.23). However, since the difference between the magnitudes of the effective forces corresponding to A and B is relatively small, the distance from G to the resultant of the effective forces of the airplane is also relatively small. It is therefore customary in such problems, and in problems dealing with the motion of cars along banked highway curves, to neglect the effect of the nonsymmetry of the three-dimensional body on the position of the resultant of its effective forces.

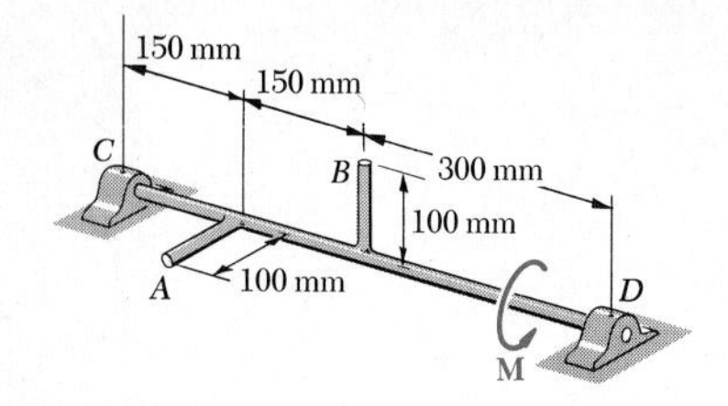

SAMPLE PROBLEM 16.11

Two 100-mm rods A and B, each of mass 300 g, are welded to the shaft CD which is supported by bearings at C and D. If a couple **M** of moment equal to $6 \text{ N} \cdot \text{m}$ is applied to the shaft, determine the components of the dynamic reactions at C and D at the instant when the shaft has reached an angular velocity of 1200 rpm. Neglect the moment of inertia of the shaft itself.

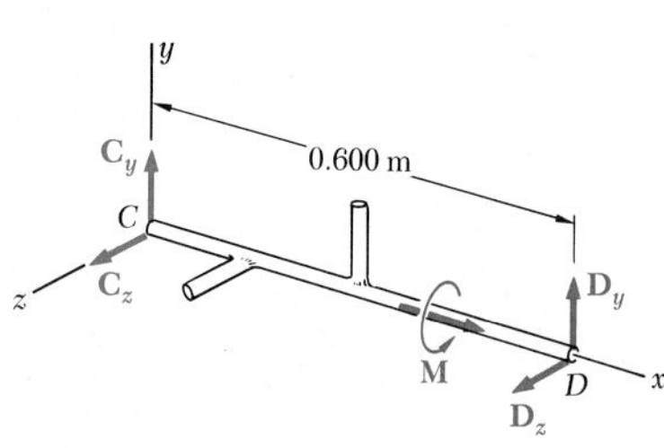

=

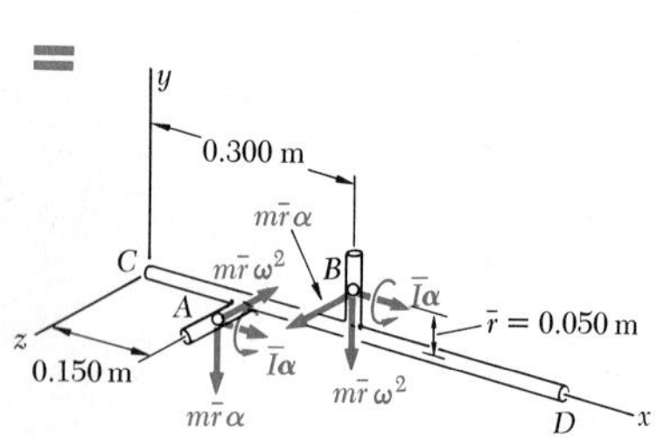

Solution. The external forces consist of the dynamic reactions at C and D. (The weights and the corresponding static reactions are ignored since they cancel each other.) The system of the effective forces of each rod reduces to a vector $m\bar{\mathbf{a}}$ at the mass center of the rod and a couple $\bar{I}\boldsymbol{\alpha}$ of axis parallel to the shaft. Each vector $m\bar{\mathbf{a}}$ is resolved into a tangential component of magnitude $m\bar{r}\alpha$ and a normal component of magnitude $m\bar{r}\omega^2$. The centroidal moment of inertia of each rod is

$$\bar{I} = \tfrac{1}{12}mL^2 = \tfrac{1}{12}(0.300 \text{ kg})(0.100 \text{ m})^2 = 0.25 \times 10^{-3} \text{ kg} \cdot \text{m}^2$$

Angular Acceleration of Shaft. Equating the moments about the x axis of the external and the effective forces, we first determine the angular acceleration of the shaft. The moments of the components of the vectors $m\bar{\mathbf{a}}$, as well as the moments of the couples $\bar{I}\boldsymbol{\alpha}$, must be taken into account.

$$+\circlearrowleft \Sigma M_x = \Sigma(M_x)_{\text{eff}}: \qquad M = 2(m\bar{r}\alpha)\bar{r} + 2\bar{I}\alpha$$

$$6 \text{ N} \cdot \text{m} = 2(0.300 \text{ kg})(0.050 \text{ m})^2\alpha + 2(0.25 \times 10^{-3} \text{ kg} \cdot \text{m}^2)\alpha$$

$$\alpha = 3000 \text{ rad/s}^2$$

Dynamic Reactions. Using the value obtained for α and noting that $\omega = 1200 \text{ rpm} = 125.7 \text{ rad/s}$, we compute the components of the vectors $m\bar{\mathbf{a}}$:

$$m\bar{r}\alpha = (0.300 \text{ kg})(0.050 \text{ m})(3000 \text{ rad/s}^2) = 45 \text{ N}$$

$$m\bar{r}\omega^2 = (0.300 \text{ kg})(0.050 \text{ m})(125.7 \text{ rad/s})^2 = 237 \text{ N}$$

Equating successively the moments about and the components along the y and z axes of the external and effective forces, we write

$$+\circlearrowleft \Sigma M_y = \Sigma(M_y)_{\text{eff}}:$$

$$-(0.600 \text{ m})D_z = (0.150 \text{ m})(m\bar{r}\omega^2) - (0.300 \text{ m})(m\bar{r}\alpha)$$

$$-(0.600 \text{ m})D_z = (0.150 \text{ m})(237 \text{ N}) - (0.300 \text{ m})(45 \text{ N})$$

$$D_z = -36.8 \text{ N} \quad ◄$$

$$+\circlearrowleft \Sigma M_z = \Sigma(M_z)_{\text{eff}}:$$

$$(0.600 \text{ m})D_y = -(0.150 \text{ m})(m\bar{r}\alpha) - (0.300 \text{ m})(m\bar{r}\omega^2)$$

$$(0.600 \text{ m})D_y = -(0.150 \text{ m})(45 \text{ N}) - (0.300 \text{ m})(237 \text{ N})$$

$$D_y = -129.8 \text{ N} \quad ◄$$

$$\Sigma F_y = \Sigma(F_y)_{\text{eff}}: \qquad C_y + D_y = -m\bar{r}\alpha - m\bar{r}\omega^2$$

$$C_y - 129.8 \text{ N} = -45 \text{ N} - 237 \text{ N} \qquad C_y = -152.2 \text{ N} \quad ◄$$

$$\Sigma F_z = \Sigma(F_z)_{\text{eff}}: \qquad C_z + D_z = -m\bar{r}\omega^2 + m\bar{r}\alpha$$

$$C_z - 36.8 \text{ N} = -237 \text{ N} + 45 \text{ N} \qquad C_z = -155.2 \text{ N} \quad ◄$$

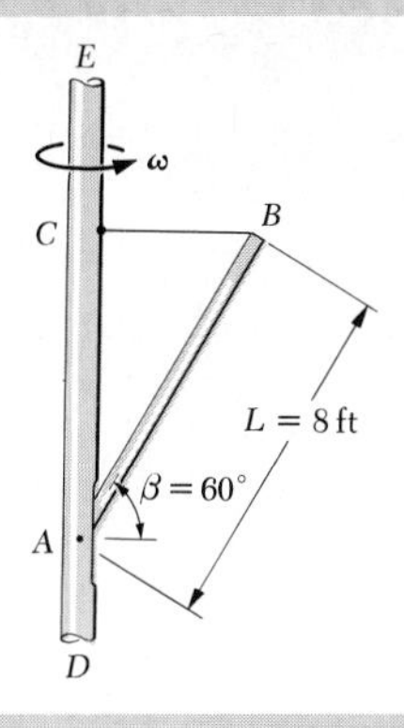

SAMPLE PROBLEM 16.12

A slender rod AB of length $L = 8$ ft and weight $W = 40$ lb is pinned at A to a vertical axle DE which rotates with a constant angular velocity $\omega = 15$ rad/s. The rod is maintained in position by means of a wire BC attached to the axle and to the end B of the rod. Determine the tension in the wire BC.

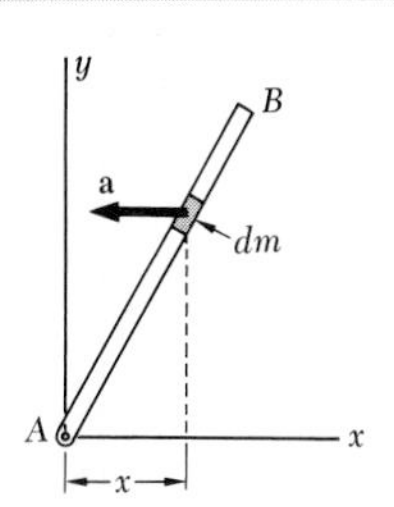

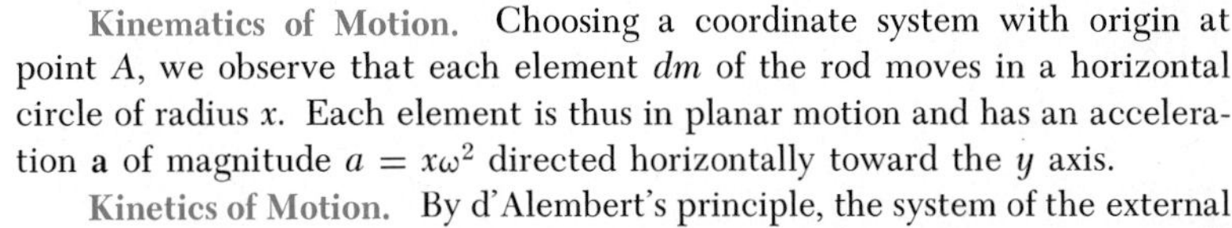

Kinematics of Motion. Choosing a coordinate system with origin at point A, we observe that each element dm of the rod moves in a horizontal circle of radius x. Each element is thus in planar motion and has an acceleration $\mathbf{a}$ of magnitude $a = x\omega^2$ directed horizontally toward the y axis.

Kinetics of Motion. By d'Alembert's principle, the system of the external forces, which consists of the weight $\mathbf{W}$, the force $\mathbf{T}$ exerted by the wire at B and the components $\mathbf{A}_x$ and $\mathbf{A}_y$ of the reaction at A, must be equivalent to the system of the effective forces $\mathbf{a}\,dm$ of the various elements. We write

$$+\circlearrowleft \Sigma M_A = \Sigma(M_A)_{\text{eff}}: \qquad T(L\sin\beta) - W(\tfrac{1}{2}L\cos\beta) = \int y(a\,dm) \tag{1}$$

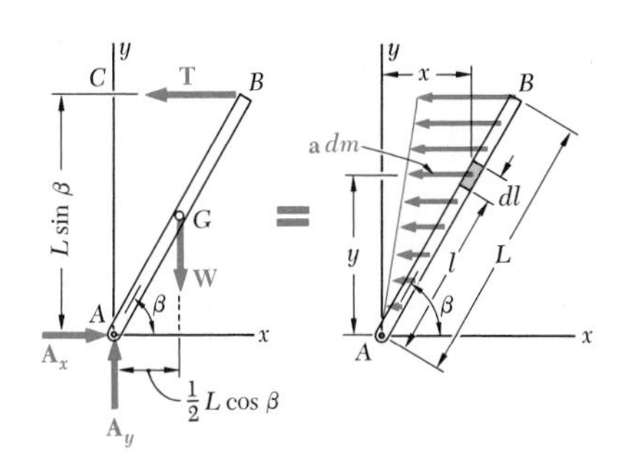

Denoting by l the distance from A to the element dm, we have

$$y = l\sin\beta \qquad a = x\omega^2 = (l\cos\beta)\omega^2 \qquad dm = \frac{m}{L}dl$$

Substituting these expressions into the integral and integrating from $l = 0$ to $l = L$, we obtain

$$\int y(a\,dm) = \int_0^L (l\sin\beta)(l\cos\beta)\omega^2\left(\frac{m}{L}dl\right) = \tfrac{1}{3}mL^2\omega^2\sin\beta\cos\beta \tag{2}$$

Substituting this expression into (1), we have

$$T(L\sin\beta) - W(\tfrac{1}{2}L\cos\beta) = \tfrac{1}{3}mL^2\omega^2\sin\beta\cos\beta$$
$$T = \tfrac{1}{2}W\cot\beta + \tfrac{1}{3}mL\omega^2\cos\beta \tag{3}$$

Substituting $W = 40$ lb, $m = 1.242$ lb · s²/ft, $L = 8$ ft, $\beta = 60°$, and $\omega = 15$ rad/s into (3), we obtain

$T = 384$ lb ◄

Note. The resultant of the effective forces $\mathbf{a}\,dm$ does not pass through the mass center of the rod. To show this, we first determine the magnitude of the resultant:

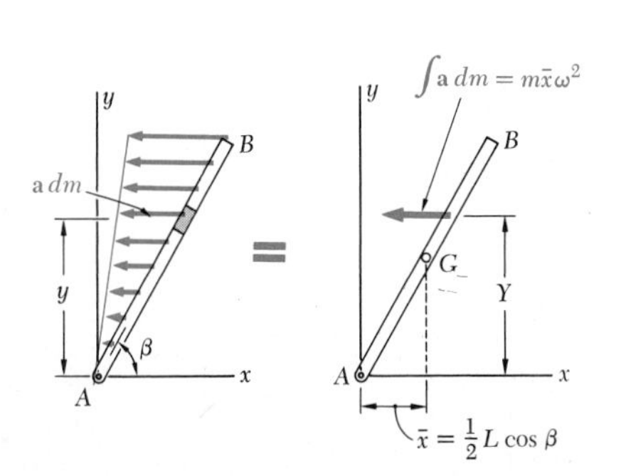

$$\int a\,dm = \int_0^L (l\cos\beta)\omega^2\left(\frac{m}{L}dl\right) = \tfrac{1}{2}mL\omega^2\cos\beta = m\bar{x}\omega^2 \tag{4}$$

Since the resultant must be equivalent to the effective forces,

$$+\circlearrowright \Sigma M_A = \Sigma M_A: \qquad \int y(a\,dm) = Y\int a\,dm \tag{5}$$

Substituting into (5) the expressions found in (2) and (4), we obtain

$$\tfrac{1}{3}mL^2\omega^2\sin\beta\cos\beta = Y(\tfrac{1}{2}mL\omega^2\cos\beta) \qquad Y = \tfrac{2}{3}L\sin\beta$$

Problems

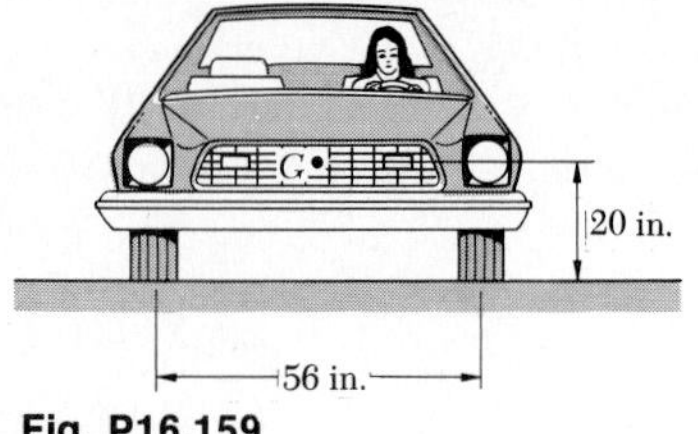

Fig. P16.159

16.159 The mass center of a 2400-lb automobile is 20 in. above the road, and the transverse distance between the wheels is 56 in. If the automobile takes a turn of 240-ft radius on an unbanked highway, determine (*a*) the maximum safe speed with regard to overturning, (*b*) the value of μ_s between the tires and pavement if the maximum safe speed with regard to skidding is equal to that found for overturning.

16.160 The mass center of a loaded freight car is located 2.4 m above the rails. Knowing that the gage (distance between the rails) is 1.435 m, determine the speed at which the car will overturn while rounding an unbanked curve of 600-m radius.

16.161 Solve Prob. 16.160, assuming that the rails are superelevated so that the outer rail is 6 mm higher than the inner rail.

16.162 The automobile described in Prob. 16.159 takes a turn of 240-ft radius on a highway banked at 18°. Determine the maximum safe speed with regard to overturning.

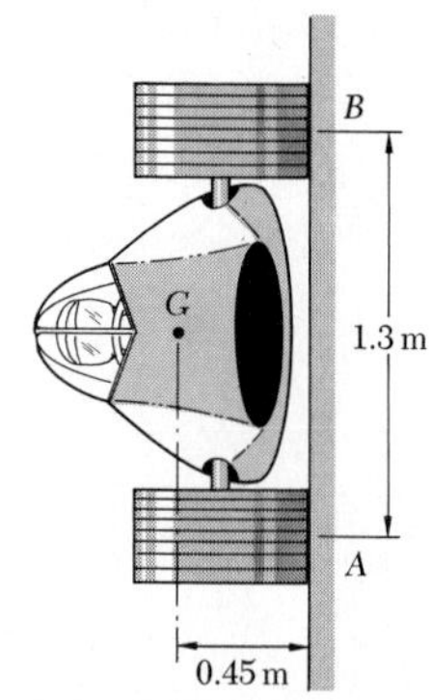

Fig. P16.163

16.163 A stunt driver drives a small automobile of mass 500 kg (driver included) on the vertical wall of a cylindrical pit of radius 12 m at a speed of 60 km/h. Knowing that this is the slowest speed at which he can perform this stunt, determine (*a*) the coefficient of friction between tires and wall, (*b*) the normal component of the reaction of the wall at each wheel.

16.164 If the coefficient of friction between the wall and the tires of the automobile of Prob. 16.163 is 0.60, determine (*a*) the lowest safe speed at which the stunt can be performed, (*b*) the normal component of the reaction of the wall at each wheel.

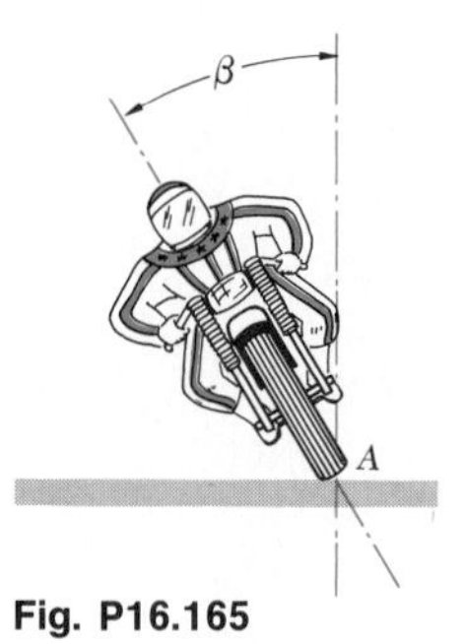

Fig. P16.165

16.165 A motorcyclist drives around an unbanked curve at a speed of 65 km/h. Assuming that the mass center of the driver and motorcycle travels in a circle of radius 50 m, determine (*a*) the angle between the plane of the motorcycle and the vertical, (*b*) the minimum safe value of μ_s.

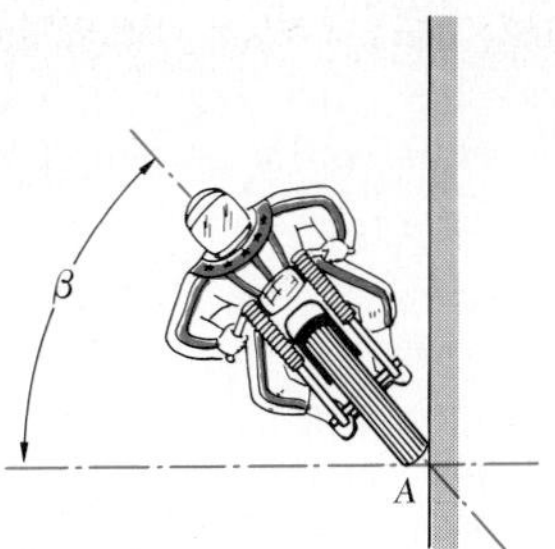

Fig. P16.166

16.166 A stunt driver rides a motorcycle at 75 km/h inside a vertical cylindrical wall. Assuming that the mass center of the driver and motorcycle travels in a circle of radius 12 m, determine (*a*) the angle β at which the motorcycle is inclined to the horizontal, (*b*) the minimum safe value of μ_s.

16.167 A mechanic has found that to statically balance the wheel shown (i.e., to insure that its mass center will be located on the axis of rotation XX'), he should use two 3-oz weights placed at A and B, respectively. If the wheel is attached to a machine which spins it at the rate of 900 rpm, determine (*a*) the force the wheel would have exerted on the machine before being statically balanced, (*b*) the moment of the additional couple exerted by the wheel on the machine after balancing if, to save time, the mechanic used a single 6-oz weight placed at A.

16.168 When the wheel shown is attached to a balancing machine and made to spin at the rate of 750 rpm, it is found that the wheel exerts on the machine a force-couple system located in the plane of the figure and consisting of a 36.2-lb force applied at C and directed upward, and of a 10.85-lb·ft clockwise couple. If only two weights are to be used to balance the wheel dynamically as well as statically, what should these weights be and at which of the points A, B, D, or E should they be placed?

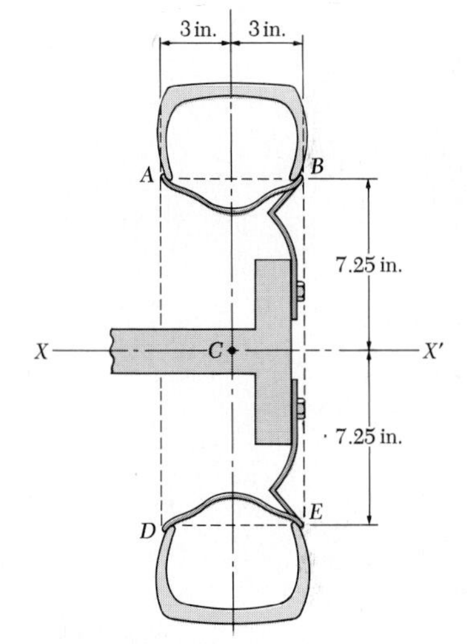

Fig. P16.167 and P16.168

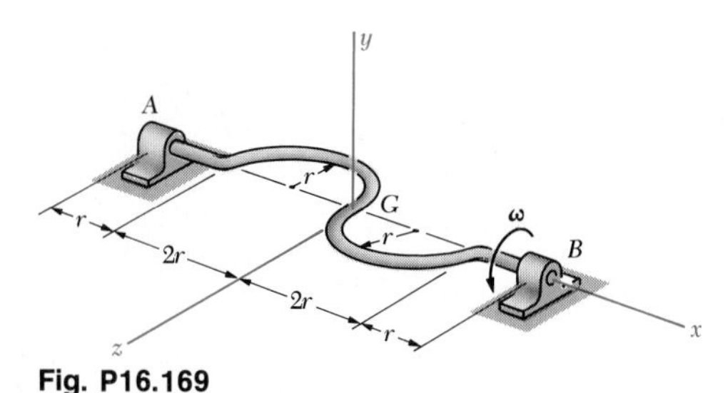

Fig. P16.169

16.169 A thin homogeneous rod of weight w per unit length is used to form the shaft shown. Knowing that the shaft rotates with a constant angular velocity ω, determine the dynamic reactions at A and B.

16.170 A thin homogeneous rod AB of mass m and length $2b$ is welded at its midpoint G to a vertical shaft GD. Knowing that the shaft rotates with a constant angular velocity $\boldsymbol{\omega}$, determine the couple exerted by the shaft on rod AB.

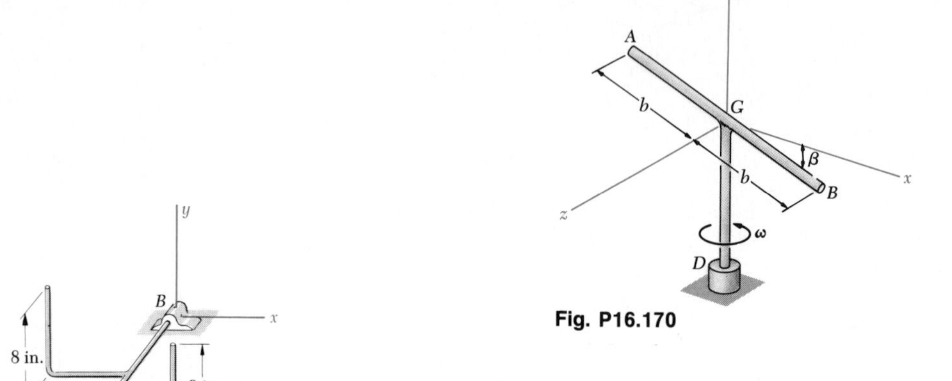

Fig. P16.170

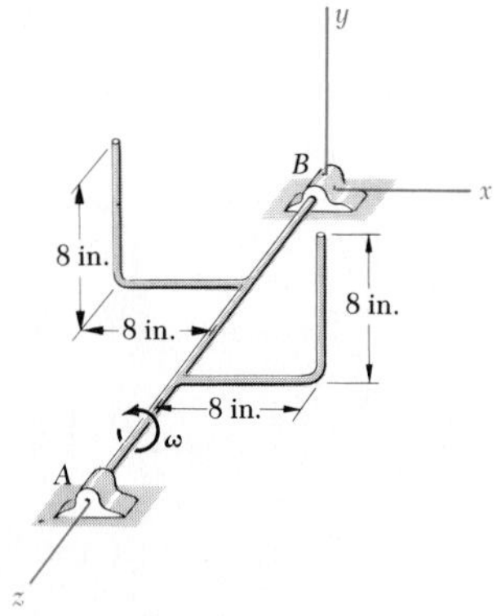

8 in.
8 in.
8 in.
8 in.

Fig. P16.171

16.171 Two L-shaped arms, each weighing 4 lb, are welded at the third points of a 2-ft horizontal shaft supported by bearings at A and B. Knowing that the shaft rotates at the constant rate $\omega = 240$ rpm, determine the dynamic reactions at A and B for the position shown.

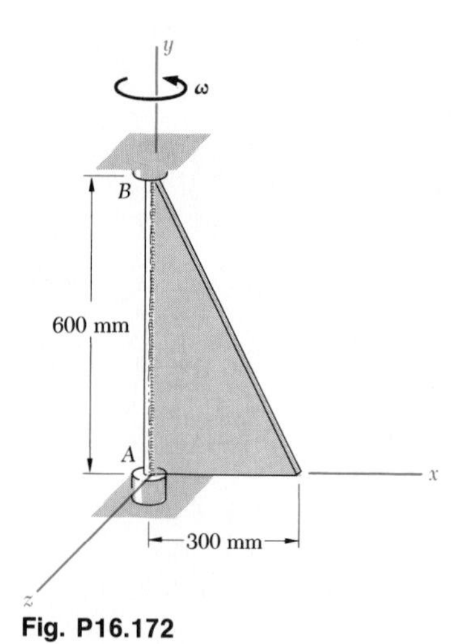

Fig. P16.172

16.172 A thin homogeneous triangular plate of mass 2.5 kg is welded to a light vertical axle supported by bearings at A and B. Knowing that the plate rotates at the constant rate $\omega = 8$ rad/s, determine the dynamic reactions at A and B. (*Hint.* Consider the effective forces of elements consisting of horizontal strips.)

16.173 Knowing that the shaft of Prob. 16.169 is originally at rest ($\omega = 0$), determine (*a*) the couple $\mathbf{M}_0$ required to impart to the shaft an angular acceleration $\boldsymbol{\alpha}$, (*b*) the dynamic reactions at A and B immediately after the couple $\mathbf{M}_0$ has been applied.

16.174 Knowing that the plate of Prob. 16.172 is initially at rest ($\omega = 0$) when a couple of moment $M_0 = 0.45$ N·m is applied to it, determine (*a*) the resulting angular acceleration of the plate, (*b*) the dynamic reactions at A and B immediately after the couple has been applied.

16.175 Knowing that the assembly of Prob. 16.171 is initially at rest ($\omega = 0$) when a couple of moment $M_0 = 20$ lb·in. is applied to the shaft, determine (*a*) the resulting angular acceleration of the shaft, (*b*) the dynamic reactions at A and B immediately after the couple has been applied.

16.176 Knowing that the shaft GD of Prob. 16.170 has an angular velocity $\boldsymbol{\omega}$ and an angular acceleration $\boldsymbol{\alpha}$, both counterclockwise as seen from above, determine for the position shown the components of the couple exerted by the shaft on rod AB.

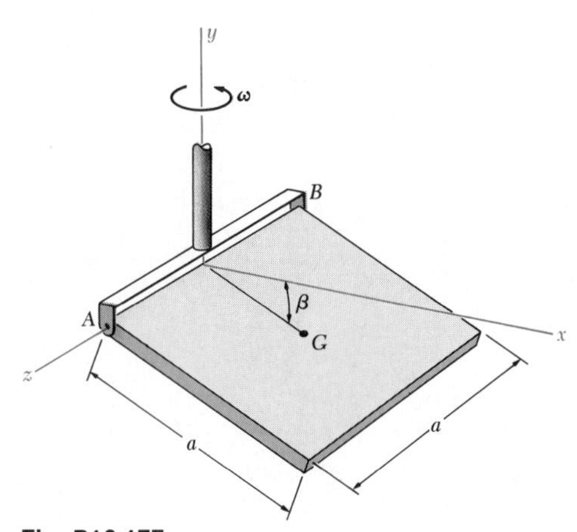

Fig. P16.177

16.177 A uniform square plate of side a and mass m is hinged at A and B to a clevis which rotates with a constant angular velocity $\boldsymbol{\omega}$. Determine (*a*) the value of ω for which the angle β maintains the constant value $\beta = 30°$, (*b*) the range of values of ω for which the plate remains vertical ($\beta = 90°$). (See hint of Prob. 16.172.)

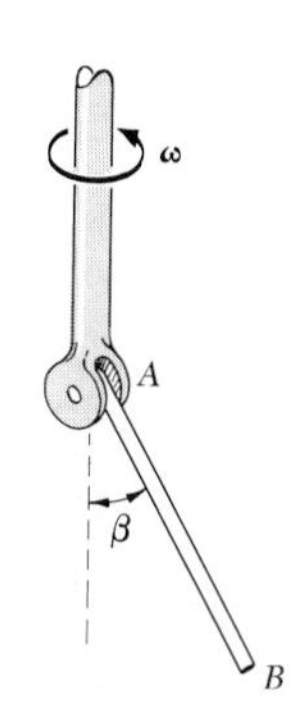

Fig. P16.178

16.178 A uniform rod AB of length l and mass m is attached to the pin of a clevis which rotates with a constant angular velocity $\boldsymbol{\omega}$. Determine (*a*) the constant angle β that the rod forms with the vertical, (*b*) the range of values of ω for which the rod remains vertical ($\beta = 0$).

16.7. Review and Summary. In this chapter, we considered the *kinetics of rigid bodies*. Our study was limited to the *plane motion* of rigid bodies and was based on the direct application of Newton's second law.

D'Alembert's principle

Considering first the case of a rigid slab—or, more generally, of a rigid body symmetrical with respect to the reference plane—we derived *d'Alembert's principle*, which states that *the external forces acting on a rigid body are equivalent to the effective forces of the various particles forming the rigid body* [Sec. 16.2].

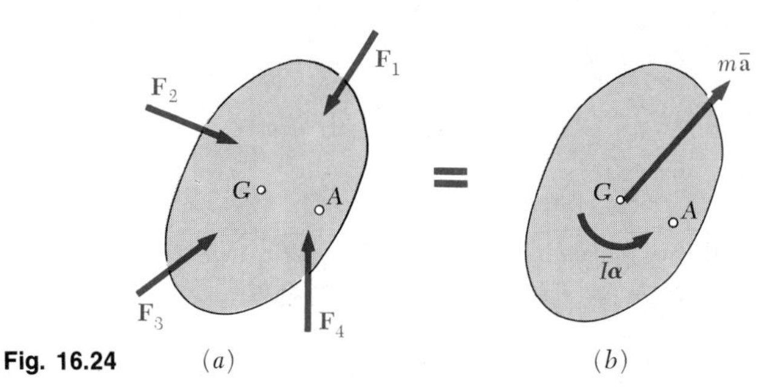

Fig. 16.24

Observing that in the case of a rigid slab, the effective forces reduce to a vector $m\bar{\mathbf{a}}$ attached at the mass center G of the slab and a couple $\bar{I}\boldsymbol{\alpha}$, we expressed d'Alembert's principle in the form of the vector diagram shown in Fig. 16.24. In the particular case of a slab in *translation*, the effective forces shown in part b of this figure reduce to the single vector $m\bar{\mathbf{a}}$, while in the particular case of a slab in *centroidal rotation*, they reduce to the single couple $\bar{I}\boldsymbol{\alpha}$; in any other case of plane motion, both the vector $m\bar{\mathbf{a}}$ and the couple $\bar{I}\boldsymbol{\alpha}$ should be included.

Free-body-diagram equation

Any problem involving the plane motion of a rigid slab may be solved by drawing a *free-body-diagram equation* similar to that of Fig. 16.24 [Sec. 16.3]. Three equations of motion may then be obtained by equating the x components, y components, or moments about an arbitrary point A, of the forces and vectors involved [Sample Probs. 16.1, 16.2, 16.4, and 16.5]. An alternative solution may also be obtained by adding to the external forces an inertia vector $(m\bar{\mathbf{a}})_{\text{rev}}$ of sense opposite to that of $\bar{\mathbf{a}}$, attached at G, and an *inertia couple* $(\bar{I}\boldsymbol{\alpha})_{\text{rev}}$ of sense opposite to that of $\boldsymbol{\alpha}$. The system obtained in this way is equivalent to zero, and the slab is said to be in *dynamic equilibrium*.

Connected rigid bodies

The method described above may also be used to solve problems involving the plane motion of several connected rigid bodies [Sec. 16.4]. A free-body-diagram equation is drawn for each part of the system and the equations of motion obtained are solved simultaneously. In some cases, however, a single diagram may be drawn for the entire system, including all the external forces as well as the vectors $m\bar{\mathbf{a}}$ and the couples $\bar{I}\boldsymbol{\alpha}$ associated with the various parts of the system [Sample Prob. 16.3].

Constrained plane motion

In the second part of the chapter, we were concerned with rigid bodies *moving under given constraints* [Sec. 16.5]. While the kinetic analysis of the constrained plane motion of a rigid slab is the same as above, it must be supplemented by a *kinematic analysis* which has for its object to express the components $\bar{a}_x$ and $\bar{a}_y$ of the acceleration of the mass center G of the slab in terms of its angular acceleration α. Problems solved in this way included the *noncentroidal rotation* of rods and plates [Sample Probs. 16.6 and 16.7], the *rolling motion* of spheres and wheels [Sample Probs. 16.8 and 16.9], and the plane motion of *various types of linkages* [Sample Prob. 16.10].

Rotation of a three-dimensional body

In the last part of the chapter, we considered the *rotation of a three-dimensional body about a fixed axis* [Sec. 16.6]. If the body consists of several rods or slabs contained in planes perpendicular to the axis of rotation—or, more generally, of bodies symmetrical with respect to such planes—the effective forces of each of these bodies may be reduced to a vector $m\bar{\mathbf{a}}$ attached at its mass center and a couple $\bar{I}\boldsymbol{\alpha}$. Equating, respectively, the x, y, and z components of the external and effective forces, as well as their moments about the x, y, and z axes, we obtained six equations which could be solved for the angular acceleration of the composite body and the reactions at its supports [Sample Prob. 16.11].

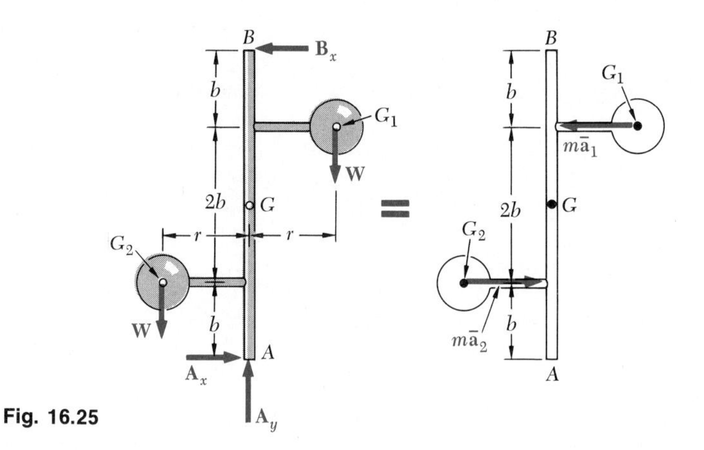

Fig. 16.25

Static and dynamic balancing

Using this approach, we considered a shaft supporting two equal spheres (Fig. 16.25) and observed that in addition to the *static reaction* $\mathbf{A}_y$, two equal and opposite *dynamic reactions* $\mathbf{A}_x$ and $\mathbf{B}_x$ develop at the bearings A and B when the shaft is made to rotate with a constant angular velocity $\boldsymbol{\omega}$. This led us to conclude that *rotating systems should be dynamically as well as statically balanced.*

Unsymmetrical rigid bodies

Finally, we noted that in the general case of the rotation about a fixed axis of an unsymmetrical three-dimensional body, a typical element of mass should be isolated and its effective force should be determined. The sums of the components and moments of the effective forces may then be obtained by integration [Sample Prob. 16.12].

Review Problems

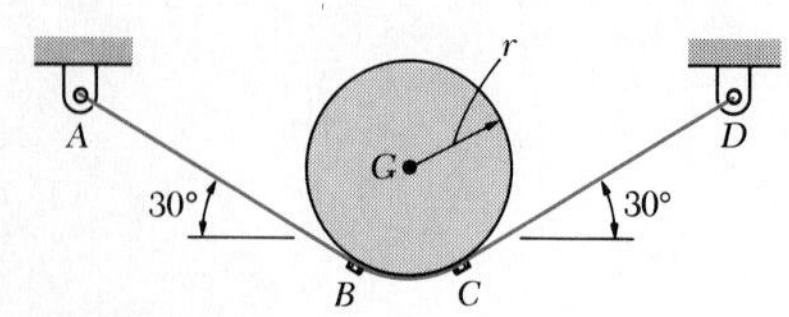

Fig. P16.179

16.179 A uniform disk of mass $m = 3$ kg and radius $r = 125$ mm is supported by the belt *ABCD* which is bolted to the disk at *B* and *C*. If the belt suddenly breaks at a point between *A* and *B*, determine at that instant (*a*) the acceleration of the center of the disk, (*b*) the tension in portion *CD* of the belt.

16.180 and 16.181 A uniform plate of mass *m* is suspended in each of the ways shown. For each case determine the acceleration of the center of the plate immediately after the connection at *B* has been released.

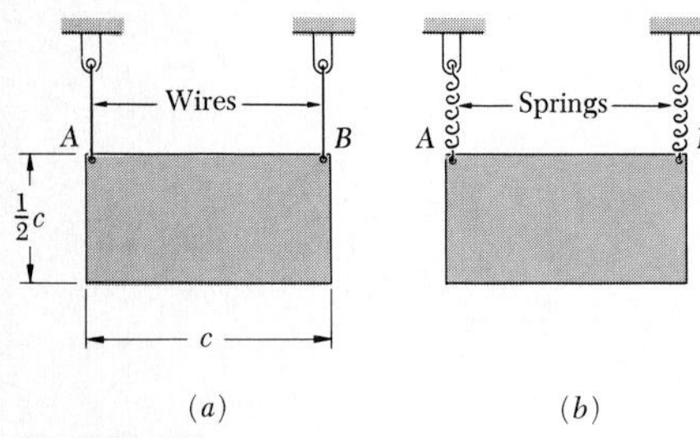

Fig. P16.180

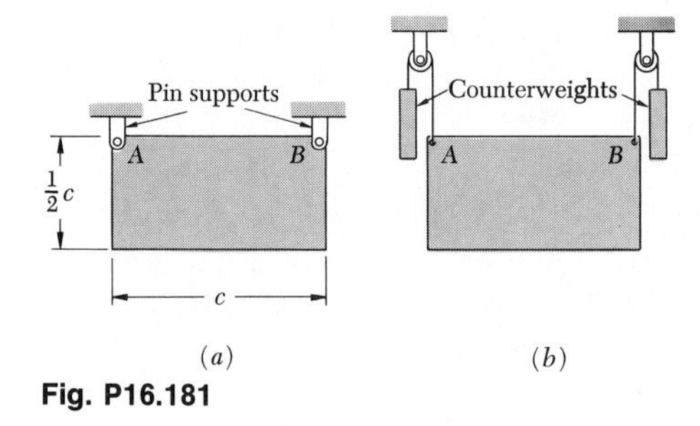

Fig. P16.181

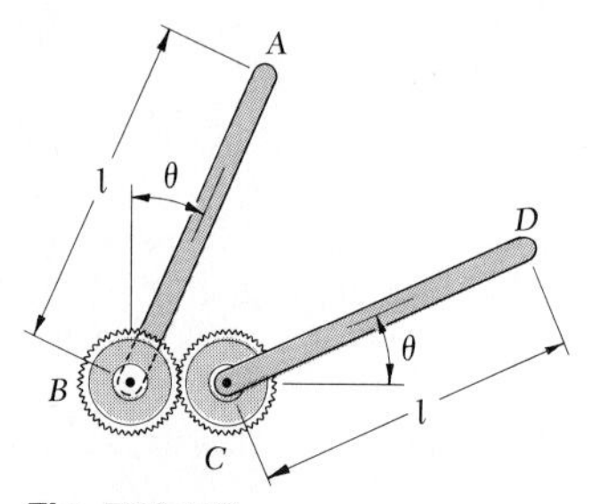

Fig. P16.183

16.182 A cyclist rides a bicycle at a speed of 20 mi/h. The distance between axles is 42 in., and the mass center of the cyclist and bicycle is located 26 in. behind the front axle and 40 in. above the ground. If the cyclist applies the brakes on the front wheel only, determine the shortest distance in which she can stop without being thrown over the front wheel.

16.183 Two uniform rods, each of mass *m*, are attached as shown to small gears of negligible mass. If the rods are released from rest in the position shown, determine the angular acceleration of rod *AB* immediately after release, assuming (*a*) $\theta = 0$, (*b*) $\theta = 30°$.

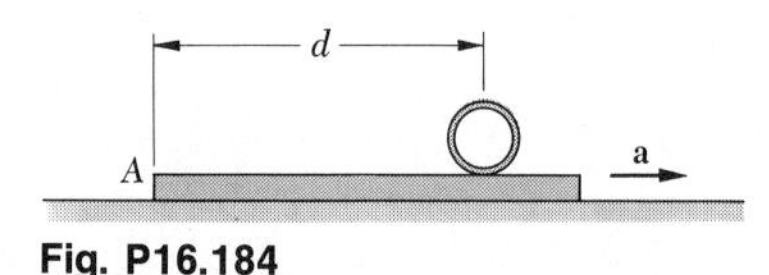

Fig. P16.184

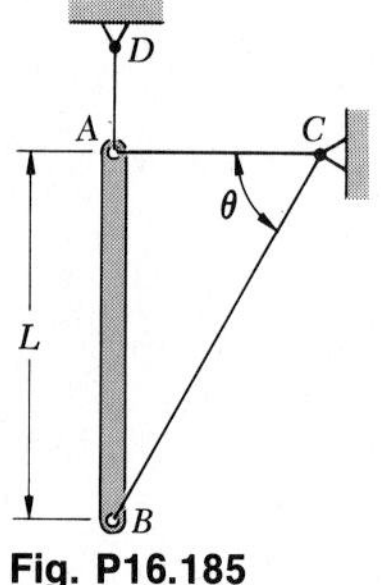

Fig. P16.185

16.184 A section of pipe rests on a plate. The plate is then given a constant acceleration **a** directed to the right. Assuming that the pipe rolls on the plate, determine (*a*) the acceleration of the pipe, (*b*) the distance through which the plate will move before the pipe reaches end *A*.

16.185 A uniform slender rod *AB*, of length $L = 1.2$ m and mass 3 kg, is held in the position shown by three wires. If $\theta = 60°$, determine the tension in wires *AC* and *BC* immediately after wire *AD* has been cut.

16.186 A 3-kg roll of paper of radius $r = 150$ mm is attached by a frictionless pin C to a 5-kg trolley which may roll freely along a horizontal pipe. Knowing that a horizontal force of magnitude $P = 12$ N is applied to the paper at D, determine (*a*) the acceleration of the trolley, (*b*) the acceleration of point D.

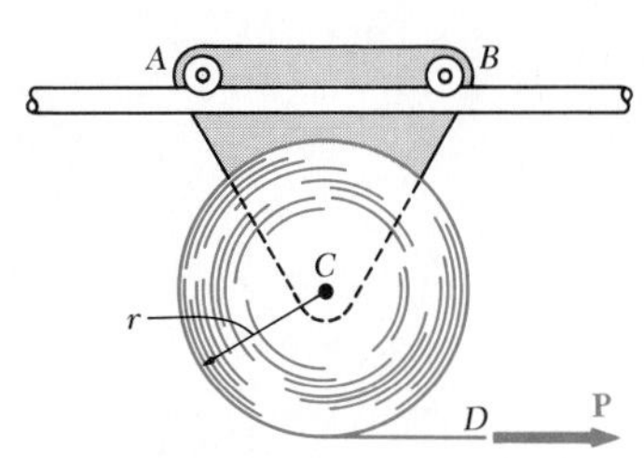

Fig. P16.186

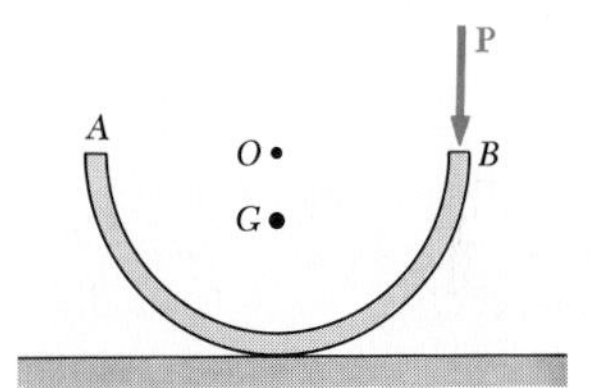

Fig. P16.187

16.187 A half section of pipe of mass m and radius r rests on a rough horizontal surface. A vertical force **P** is applied as shown. Assuming that the section rolls without sliding, derive an expression (*a*) for its angular acceleration, (*b*) for the minimum value of μ_s compatible with this motion. [*Hint.* Note that $OG = 2r/\pi$ and that, by the parallel-axis theorem, $\bar{I} = mr^2 - m(OG)^2$.]

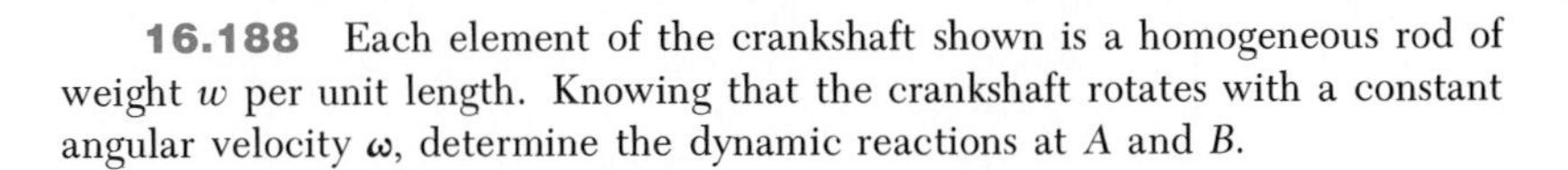

16.188 Each element of the crankshaft shown is a homogeneous rod of weight w per unit length. Knowing that the crankshaft rotates with a constant angular velocity $\boldsymbol{\omega}$, determine the dynamic reactions at A and B.

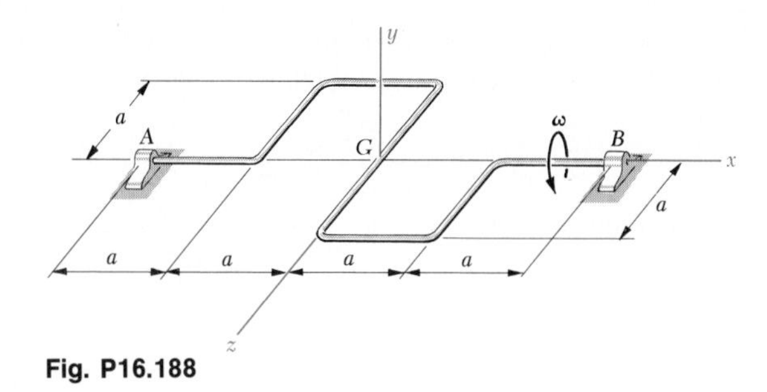

Fig. P16.188

16.189 The shaft of Prob. 16.188 is initially at rest ($\omega = 0$) and has an acceleration $\alpha = 80 \text{ rad/s}^2$. Knowing that $w = 5$ lb/ft and $a = 4$ in., determine (*a*) the couple **M** required to cause the acceleration, (*b*) the corresponding dynamic reactions at A and B.

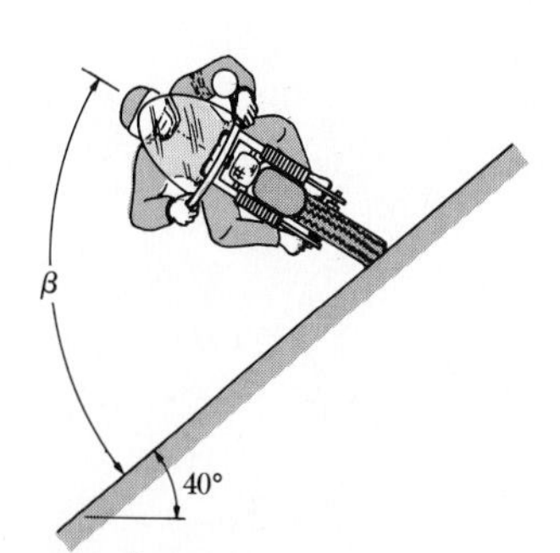

Fig. P16.190

16.190 A motorcyclist drives around a banked racing track at a constant speed v. The coefficient of static friction between the motorcycle tires and the track is $\mu_s = 0.50$. Assuming that the combined mass center of the motorcycle and driver travels in a horizontal circle of radius $r = 60$ ft, determine (*a*) the maximum and minimum safe values of the speed v with respect to skidding, (*b*) the corresponding values of the angle β.

The following problems are designed to be solved with a computer.

16.C1 A completely filled barrel and its contents have a combined mass of 125 kg. A cylinder is connected to the barrel as shown and it is known that the coefficients of friction between the barrel and the floor are $\mu_s = 0.35$ and $\mu_k = 0.30$. Write a computer program and use it to calculate the acceleration of the barrel and the range of values of h for which the barrel will not tip, for values of the mass of the cylinder from 25 to 75 kg at 5-kg intervals, and from 75 to 400 kg at 25-kg intervals.

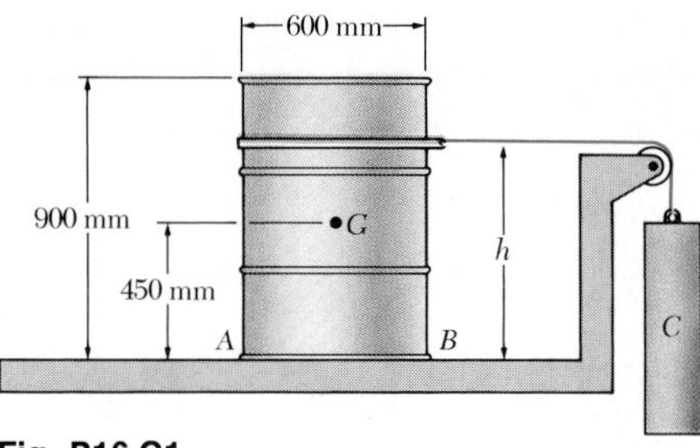

Fig. P16.C1

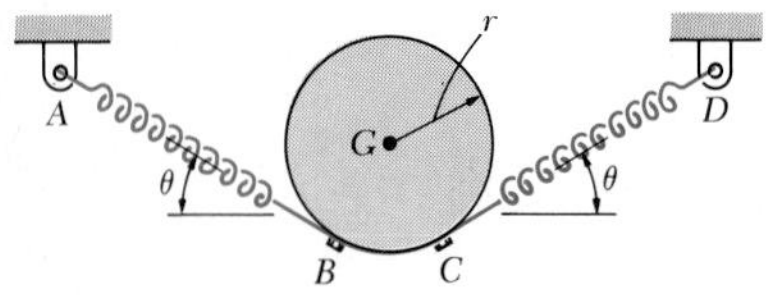

Fig. P16.C2

16.C2 A uniform disk is supported as shown by springs AB and CD. Write a computer program and use it to calculate, for values of θ from 30 to 90° at 5° intervals, the accelerations of the center G of the disk and of point C immediately after spring AB has broken.

16.C3 End A of the 6-kg uniform rod AB rests on the horizontal surface while end B is attached to a collar which is moved upward at a constant speed $v_B = 0.5$ m/s. Neglecting the effect of friction, write a computer program and use it to calculate the reaction at A for values of θ from 0 to 75° at 5° intervals.

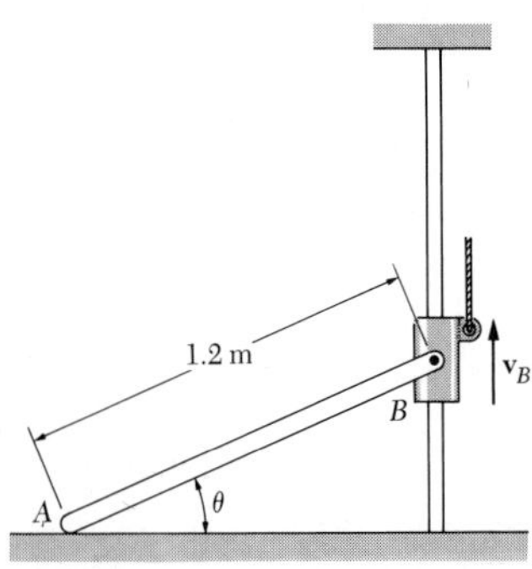

Fig. P16.C3

16.C4 In the engine system of Prob. 15.C4, the piston P weighs 3 lb and the connecting rod BD is assumed to be a 4-lb uniform slender rod. Knowing that during the test of the system no force is applied to the face of the piston, write a computer program and use it to calculate the horizontal and vertical components of the forces exerted on the connecting rod at B and D for values of θ from 0 to 180° at 10° intervals.

CHAPTER SEVENTEEN
Kinetics of Rigid Bodies: Work and Energy

17.1. Introduction. In this chapter, the method of work and energy will be used to analyze the motion of rigid bodies and of systems of rigid bodies. As was pointed out in Chap. 13, the method of work and energy is particularly well adapted to the solution of problems involving velocities and displacements. Its main advantage resides in the fact that the work of forces and the kinetic energy of particles are scalar quantities.

In Secs. 17.2 and 17.3, we shall derive the *principle of work and energy* for a rigid body and learn to compute the work of a force and the work of a couple acting on a rigid body. We shall then obtain an expression for the *kinetic energy of a rigid body in translation* (Sec. 17.4) and *in rotation about a fixed axis* (Sec. 17.5). The results obtained will be valid for any rigid body and any axis of rotation, regardless of the shape of the body or of the location of the axis of rotation. They will also be applicable to the solution of problems involving systems of rigid bodies, as long as the connection points move through equal distances (Sec. 17.6).

In the second part of the chapter, we shall determine the *kinetic energy of a rigid body in plane motion* (Sec. 17.7) and derive the *principle of conservation of energy* for a rigid body or a system of rigid bodies (Sec. 17.8). Finally, we shall extend the concept of power to rotating rigid bodies subjected to couples (Sec. 17.9).

17.2. Principle of Work and Energy for a Rigid Body. Consider a rigid body of mass m, and let P be a particle of the body, of mass Δm. The kinetic energy of the particle is $\Delta T = \frac{1}{2}(\Delta m)v^2$, where v is the speed of the particle. Consider now a displacement of the rigid body during which the particle P moves from a position P_1 to a position P_2, and denote by $\Delta U_{1\rightarrow 2}$ the work of all the forces acting on P during the displacement. We recall from Sec. 13.3 that the principle of work and energy for a particle states that

$$\Delta T_1 + \Delta U_{1\rightarrow 2} = \Delta T_2$$

where ΔT_1 = kinetic energy of the particle at P_1
ΔT_2 = kinetic energy of the particle at P_2

Similar relations may be written for all the particles forming the body. Adding the kinetic energy of all the particles, and the work of all the forces involved, we write

$$T_1 + U_{1\rightarrow 2} = T_2 \tag{17.1}$$

where T_1, T_2 = initial and final values of total kinetic energy of the particles forming the rigid body
$U_{1\rightarrow 2}$ = work of all forces acting on the various particles of the body

As indicated in Chap. 13, work and energy are expressed in joules (J) if SI units are used, and in ft · lb or in · lb if U.S. customary units are used.

The total kinetic energy

$$T = \Sigma\tfrac{1}{2}(\Delta m)v^2 \tag{17.2}$$

is obtained by adding positive scalar quantities and is itself a positive scalar quantity. We shall see later how T may be determined for various types of motion of a rigid body.

The expression $U_{1\rightarrow 2}$ in (17.1) represents the work of all the forces acting on the various particles of the body, whether these forces are internal or external. However, as we shall see presently, the total work of the internal forces holding together the particles of a rigid body is zero. Consider two particles A and B of a rigid body and the two equal and opposite forces **F** and **F**′ they exert on each other (Fig. 17.1). While, in general, small displacements $d\mathbf{r}$ and $d\mathbf{r}'$ of the two particles are different, the components of these displacements along AB must be equal; otherwise, the particles would not remain at the same distance from each other, and the body would not be rigid. Therefore, the work of **F** is equal in magnitude and opposite in sign to the work of **F**′, and their sum is zero. Thus, the total work of the internal forces acting on the particles of a rigid body is zero, and *the expression $U_{1\rightarrow 2}$ in Eq.* (17.1) *reduces to the work of the external forces* acting on the body during the displacement considered.

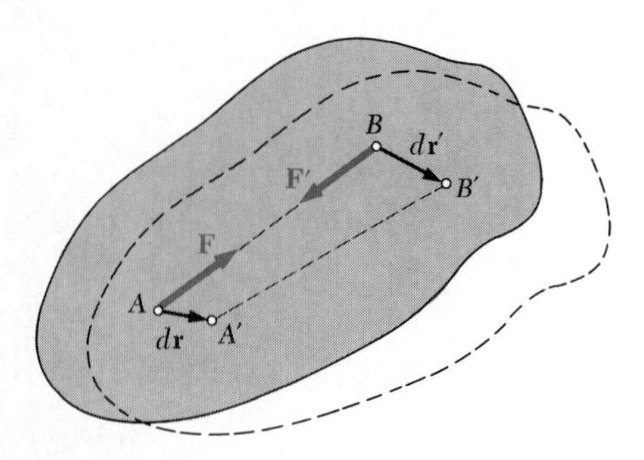

Fig. 17.1

17.3. Work of Forces Acting on a Rigid Body. We saw in Sec. 13.2 that the work of a force $\mathbf{F}$ during a displacement of its point of application from A_1 $(s = s_1)$ to A_2 $(s = s_2)$ is

$$U_{1\to2} = \int_{s_1}^{s_2} (F \cos \alpha)\, ds \tag{17.3}$$

where F is the magnitude of the force, α the angle it forms with the direction of motion of its point of application A, and s the variable of integration which measures the distance traveled by A along its path.

In computing the work of the external forces acting on a rigid body, it is often convenient to determine the work of a couple without considering separately the work of each of the two forces forming the couple. Consider the two forces $\mathbf{F}$ and $\mathbf{F}'$ forming a couple of moment $M = Fr$ and acting on a rigid body (Fig. 17.2). Any small displacement of the rigid body bringing

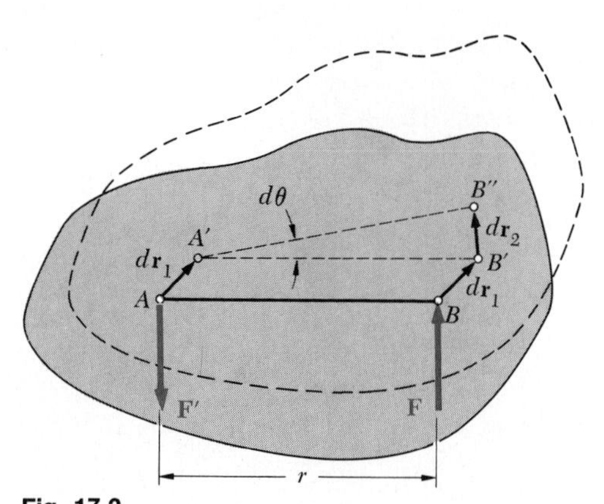

Fig. 17.2

A and B, respectively, into A' and B'' may be divided into two parts, one in which points A and B undergo equal displacements $d\mathbf{r}_1$, the other in which A' remains fixed while B' moves into B'' through a displacement $d\mathbf{r}_2$ of magnitude $ds_2 = r\, d\theta$. In the first part of the motion, the work of $\mathbf{F}$ is equal in magnitude and opposite in sign to the work of $\mathbf{F}'$ and their sum is zero. In the second part of the motion, only force $\mathbf{F}$ works, and its work is $dU = F\, ds_2 = Fr\, d\theta = M\, d\theta$. Thus, the work of a couple of moment M acting on a rigid body is

$$dU = M\, d\theta \tag{17.4}$$

where $d\theta$ is the small angle expressed in radians through which the body rotates. We again note that work should be expressed in units obtained by multiplying units of force by units of length. The work of the couple during a finite rotation of the rigid body is obtained by integrating both members of (17.4) from the initial value θ_1 of the angle θ to its final value θ_2. We write

$$U_{1\to2} = \int_{\theta_1}^{\theta_2} M\,d\theta \tag{17.5}$$

When the moment M of the couple is constant, formula (17.5) reduces to

$$U_{1\to2} = M(\theta_2 - \theta_1) \tag{17.6}$$

It was pointed out in Sec. 13.2 that a number of forces encountered in problems of kinetics *do no work.* They are forces applied to fixed points or acting in a direction perpendicular to the displacement of their point of application. Among the forces which do no work the following have been listed: the reaction at a frictionless pin when the body supported rotates about the pin, the reaction at a frictionless surface when the body in contact moves along the surface, the weight of a body when its center of gravity moves horizontally. We should also indicate now that *when a rigid body rolls without sliding on a fixed surface, the friction force* **F** *at the point of contact C does no work.* The velocity $\mathbf{v}_C$ of the point of contact C is zero, and the work of the friction force **F** during a small displacement of the rigid body is

$$dU = F\,ds_C = F(v_C\,dt) = 0$$

17.4. Kinetic Energy in Translation. Consider a rigid body in translation. Since, at a given instant, all the particles of the body have the same velocity as the mass center, we may substitute $v = \bar{v}$ in (17.2). We thus obtain the following expression for the kinetic energy of a rigid body in translation:

$$T = \Sigma\tfrac{1}{2}(\Delta m)\bar{v}^2 = \tfrac{1}{2}(\Sigma\,\Delta m)\bar{v}^2$$

$$T = \tfrac{1}{2}m\bar{v}^2 \tag{17.7}$$

This expression should be entered in formula (17.1) when the method of work and energy is used to analyze the motion of a rigid body in translation (either rectilinear or curvilinear).

We note that the expression obtained in (17.7) represents the kinetic energy of a particle of mass m moving with the velocity $\bar{\mathbf{v}}$ of the mass center. The approach used in Chap. 13, where bodies as large as automobiles or trains were considered as single particles, is therefore correct as long as the bodies considered are in translation.

17.5. Kinetic Energy in Rotation. Consider a rigid body rotating about a fixed axis intersecting the reference plane at O (Fig. 17.3). If the angular velocity of the body at a given instant is $\boldsymbol{\omega}$ (rad/s), the speed of a particle P located at a distance r from the axis of rotation is $v = r\omega$ at that instant. Substituting for v into (17.2), we obtain the following expression for the kinetic energy of the body:

$$T = \Sigma\tfrac{1}{2}(\Delta m)(r\omega)^2 = \tfrac{1}{2}\omega^2\Sigma r^2\,\Delta m$$

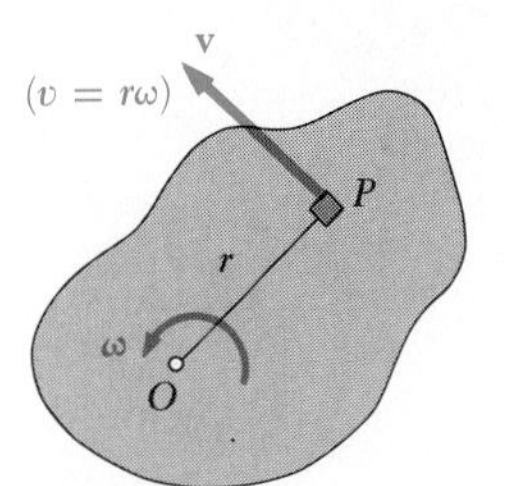

Fig. 17.3

Observing that the sum $\Sigma r^2\,\Delta m$ represents the moment of inertia I_O of the rigid body about the axis of rotation, we write

$$T = \tfrac{1}{2}I_O\omega^2 \tag{17.8}$$

This expression should be entered in formula (17.1) when the method of work and energy is used to analyze the motion of a rigid body rotating about a fixed axis.

We note that formula (17.8) is valid with respect to any fixed axis of rotation, whether or not this axis passes through the mass center of the body. Moreover, the result obtained is not limited to the rotation of plane slabs or to the rotation of bodies which are symmetrical with respect to the reference plane. Formula (17.8) may be applied to the rotation of any rigid body about a fixed axis, regardless of the shape of the body or of the location of the axis of rotation.

17.6. Systems of Rigid Bodies. When a problem involves several rigid bodies, each rigid body may be considered separately, and the principle of work and energy may be applied to each body. Adding the kinetic energies of all the particles and considering the work of all the forces involved, we may also write the equation of work and energy for the entire system. We have

$$T_1 + U_{1\to2} = T_2 \tag{17.9}$$

where T represents the arithmetic sum of the kinetic energies of the rigid bodies forming the system (all terms are positive) and $U_{1\to2}$ the work of all the forces acting on the various bodies, whether these forces are *internal* or *external* from the point of view of the system as a whole.

The method of work and energy is particularly useful in solving problems involving pin-connected members, or blocks and pulleys connected by inextensible cords, or meshed gears. In all these cases, the internal forces occur by pairs of equal and opposite forces, and the points of application of the forces in each pair *move through equal distances* during a small displacement of the system. As a result, the work of the internal forces is zero, and $U_{1\to2}$ reduces to the work of the *forces external to the system*.

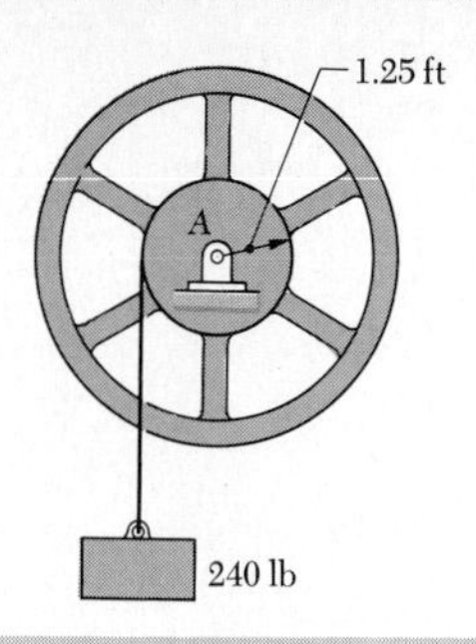

SAMPLE PROBLEM 17.1

A 240-lb block is suspended from an inextensible cable which is wrapped around a drum of 1.25-ft radius rigidly attached to a flywheel. The drum and flywheel have a combined centroidal moment of inertia $\bar{I} = 10.5 \text{ lb}\cdot\text{ft}\cdot\text{s}^2$. At the instant shown, the velocity of the block is 6 ft/s directed downward. Knowing that the bearing at A is poorly lubricated and that the bearing friction is equivalent to a couple **M** of moment 60 lb·ft, determine the velocity of the block after it has moved 4 ft downward.

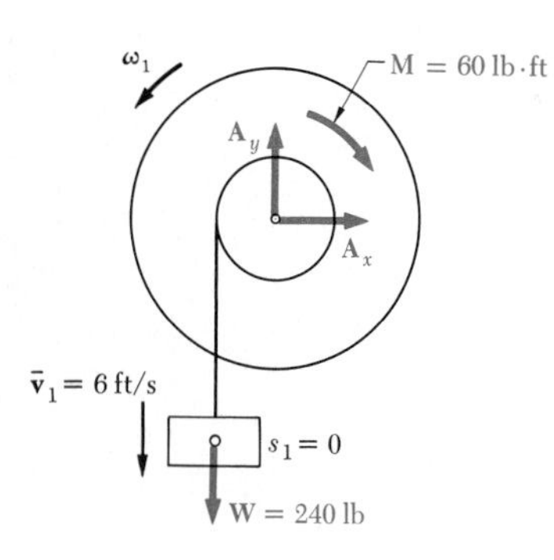

Solution. We consider the system formed by the flywheel and the block. Since the cable is inextensible, the work done by the internal forces exerted by the cable cancels. The initial and final positions of the system and the external forces acting on the system are as shown.

Kinetic Energy. *Position 1.*

Block: $\quad \bar{v}_1 = 6 \text{ ft/s}$

Flywheel: $\quad \omega_1 = \dfrac{\bar{v}_1}{r} = \dfrac{6 \text{ ft/s}}{1.25 \text{ ft}} = 4.80 \text{ rad/s}$

$$\begin{aligned} T_1 &= \tfrac{1}{2}m\bar{v}_1^2 + \tfrac{1}{2}\bar{I}\omega_1^2 \\ &= \frac{1}{2}\frac{240 \text{ lb}}{32.2 \text{ ft/s}^2}(6 \text{ ft/s})^2 + \tfrac{1}{2}(10.5 \text{ lb}\cdot\text{ft}\cdot\text{s}^2)(4.80 \text{ rad/s})^2 \\ &= 255 \text{ ft}\cdot\text{lb} \end{aligned}$$

Position 2. Noting that $\omega_2 = \bar{v}_2/1.25$, we write

$$\begin{aligned} T_2 &= \tfrac{1}{2}m\bar{v}_2^2 + \tfrac{1}{2}\bar{I}\omega_2^2 \\ &= \frac{1}{2}\frac{240}{32.2}(\bar{v}_2)^2 + (\tfrac{1}{2})(10.5)\left(\frac{\bar{v}_2}{1.25}\right)^2 = 7.09\bar{v}_2^2 \end{aligned}$$

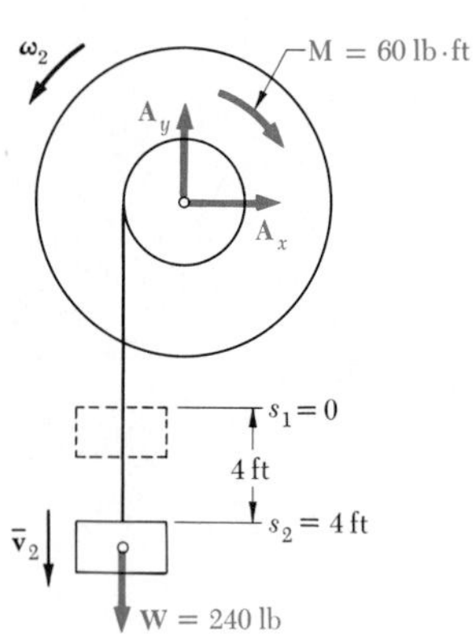

Work. During the motion, only the weight **W** of the block and the friction couple **M** do work. Noting that **W** does positive work and that the friction couple **M** does negative work, we write

$$s_1 = 0 \qquad s_2 = 4 \text{ ft}$$

$$\theta_1 = 0 \qquad \theta_2 = \frac{s_2}{r} = \frac{4 \text{ ft}}{1.25 \text{ ft}} = 3.20 \text{ rad}$$

$$\begin{aligned} U_{1\to2} &= W(s_2 - s_1) - M(\theta_2 - \theta_1) \\ &= (240 \text{ lb})(4 \text{ ft}) - (60 \text{ lb}\cdot\text{ft})(3.20 \text{ rad}) \\ &= 768 \text{ ft}\cdot\text{lb} \end{aligned}$$

Principle of Work and Energy

$$T_1 + U_{1\to2} = T_2$$
$$255 \text{ ft}\cdot\text{lb} + 768 \text{ ft}\cdot\text{lb} = 7.09\bar{v}_2^2$$
$$\bar{v}_2 = 12.01 \text{ ft/s} \qquad \bar{\mathbf{v}}_2 = 12.01 \text{ ft/s}\downarrow \blacktriangleleft$$

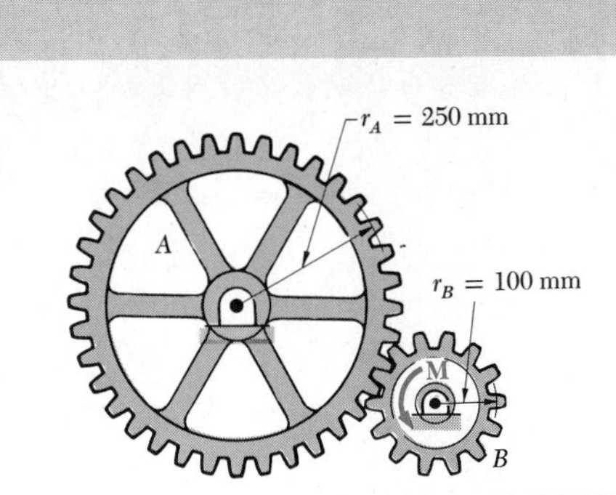

SAMPLE PROBLEM 17.2

Gear A has a mass of 10 kg and a radius of gyration of 200 mm, while gear B has a mass of 3 kg and a radius of gyration of 80 mm. The system is at rest when a couple **M** of moment 6 N·m is applied to gear B. Neglecting friction, determine (*a*) the number of revolutions executed by gear B before its angular velocity reaches 600 rpm, (*b*) the tangential force which gear B exerts on gear A.

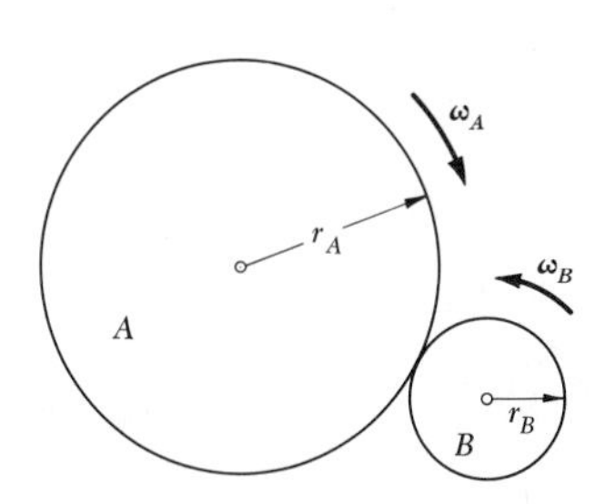

Motion of Entire System. Noting that the peripheral speeds of the gears are equal, we write

$$r_A\omega_A = r_B\omega_B \qquad \omega_A = \omega_B\frac{r_B}{r_A} = \omega_B\frac{100\text{ mm}}{250\text{ mm}} = 0.40\omega_B$$

For $\omega_B = 600$ rpm, we have

$$\omega_B = 62.8\text{ rad/s} \qquad \omega_A = 0.40\omega_B = 25.1\text{ rad/s}$$
$$\bar{I}_A = m_A\bar{k}_A^2 = (10\text{ kg})(0.200\text{ m})^2 = 0.400\text{ kg}\cdot\text{m}^2$$
$$\bar{I}_B = m_B\bar{k}_B^2 = (3\text{ kg})(0.080\text{ m})^2 = 0.0192\text{ kg}\cdot\text{m}^2$$

Kinetic Energy. Since the system is initially at rest, $T_1 = 0$. Adding the kinetic energies of the two gears when $\omega_B = 600$ rpm, we obtain

$$T_2 = \tfrac{1}{2}\bar{I}_A\omega_A^2 + \tfrac{1}{2}\bar{I}_B\omega_B^2$$
$$= \tfrac{1}{2}(0.400\text{ kg}\cdot\text{m}^2)(25.1\text{ rad/s})^2 + \tfrac{1}{2}(0.0192\text{ kg}\cdot\text{m}^2)(62.8\text{ rad/s})^2$$
$$= 163.9\text{ J}$$

Work. Denoting by θ_B the angular displacement of gear B, we have

$$U_{1\to2} = M\theta_B = (6\text{ N}\cdot\text{m})(\theta_B\text{ rad}) = (6\,\theta_B)\text{ J}$$

Principle of Work and Energy

$$T_1 + U_{1\to2} = T_2$$
$$0 + (6\,\theta_B)\text{ J} = 163.9\text{ J}$$
$$\theta_B = 27.32\text{ rad} \qquad \theta_B = 4.35\text{ rev} \blacktriangleleft$$

Motion of Gear A. ***Kinetic Energy.*** Initially, gear A is at rest, $T_1 = 0$. When $\omega_B = 600$ rpm, the kinetic energy of gear A is

$$T_2 = \tfrac{1}{2}\bar{I}_A\omega_A^2 = \tfrac{1}{2}(0.400\text{ kg}\cdot\text{m}^2)(25.1\text{ rad/s})^2 = 126.0\text{ J}$$

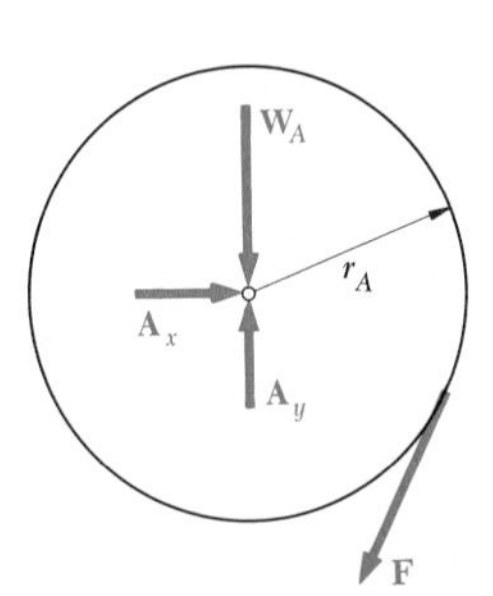

Work. The forces acting on gear A are as shown. The tangential force **F** does work equal to the product of its magnitude and of the length $\theta_A r_A$ of the arc described by the point of contact. Since $\theta_A r_A = \theta_B r_B$, we have

$$U_{1\to2} = F(\theta_B r_B) = F(27.3\text{ rad})(0.100\text{ m}) = F(2.73\text{ m})$$

Principle of Work and Energy

$$T_1 + U_{1\to2} = T_2$$
$$0 + F(2.73\text{ m}) = 126.0\text{ J}$$
$$F = +46.2\text{ N} \qquad \mathbf{F} = 46.2\text{ N} \swarrow \blacktriangleleft$$

Problems

17.1 A large flywheel weighs 6000 lb and has a radius of gyration of 36 in. It is observed that 1500 revolutions are required for the flywheel to coast from an angular velocity of 300 rpm to rest. Determine the average moment of the couple due to kinetic friction in the bearings.

17.2 The rotor of an electric motor has an angular velocity of 3600 rpm when the load and power are cut off. The 50-kg rotor, which has a centroidal radius of gyration of 180 mm, then coasts to rest. Knowing that the kinetic friction of the rotor produces a couple of moment 3.5 N·m, determine the number of revolutions that the rotor executes before coming to rest.

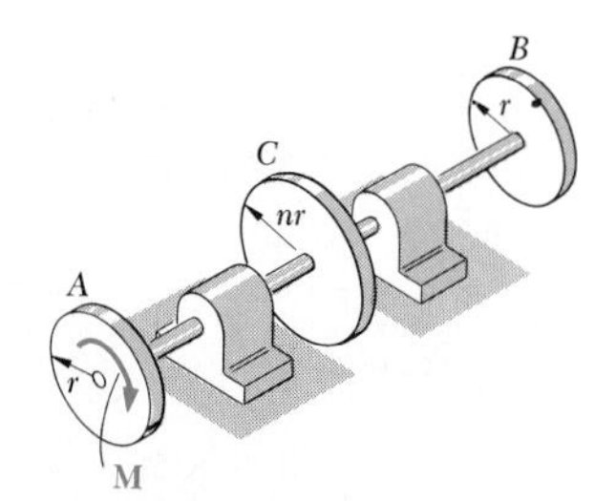

Fig. P17.3 and P17.4

17.3 Three disks of the same thickness and same material are attached to a shaft as shown. Disks A and B each have a radius r; disk C has a radius nr. A couple **M** of constant moment is applied when the system is at rest and is removed after the system has executed one revolution. Determine the radius of disk C which results in the largest final speed of a point on the rim of disk C.

17.4 Three disks of the same thickness and same material are attached to a shaft as shown. Disks A and B each have a mass of 8 kg and a radius $r = 240$ mm. A couple **M** of moment 40 N·m is applied to disk A when the system is at rest. Determine the radius nr of disk C if the angular velocity of the system is to be 900 rpm after 25 revolutions.

17.5 The flywheel of a punching machine has a mass of 300 kg and a radius of gyration of 600 mm. Each punching operation requires 2500 J of work. (*a*) Knowing that the speed of the flywheel is 300 rpm just before a punching, determine the speed immediately after the punching. (*b*) If a constant 25-N·m couple is applied to the shaft of the flywheel, determine the number of revolutions executed before the speed is again 300 rpm.

17.6 The flywheel of a small punch rotates at 300 rpm. It is known that 1800 ft·lb of work must be done each time a hole is punched. It is desired that the speed of the flywheel after one punching be not less than 90 percent of the original speed of 300 rpm. (*a*) Determine the required moment of inertia of the flywheel. (*b*) If a constant 25-lb·ft couple is applied to the shaft of the flywheel, determine the number of revolutions which must occur between each punching, knowing that the initial velocity is to be 300 rpm at the start of each punching.

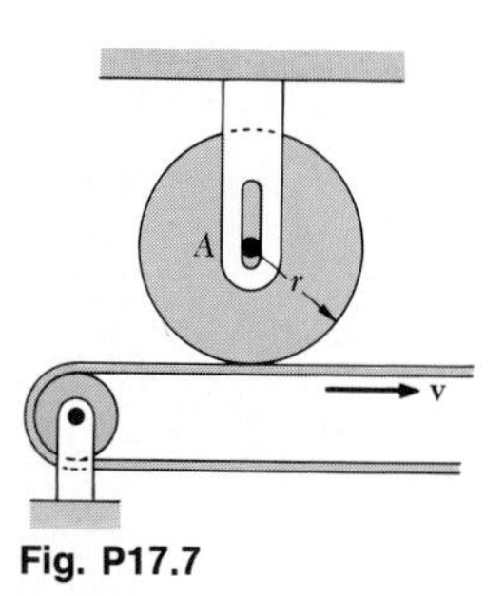

Fig. P17.7

17.7 A disk of uniform thickness and initially at rest is placed in contact with the belt, which moves with a constant velocity **v**. Denoting by μ_k the coefficient of kinetic friction between the disk and the belt, derive an expression for the number of revolutions executed by the disk before it reaches a constant angular velocity.

17.8 Disk A of Prob. 17.7 has a mass of 4 kg and a radius $r = 90$ mm; it is at rest when it is placed in contact with the belt, which moves with a constant speed $v = 15$ m/s. Knowing that $\mu_k = 0.25$ between the disk and the belt, determine the number of revolutions executed by the disk before it reaches a constant angular velocity.

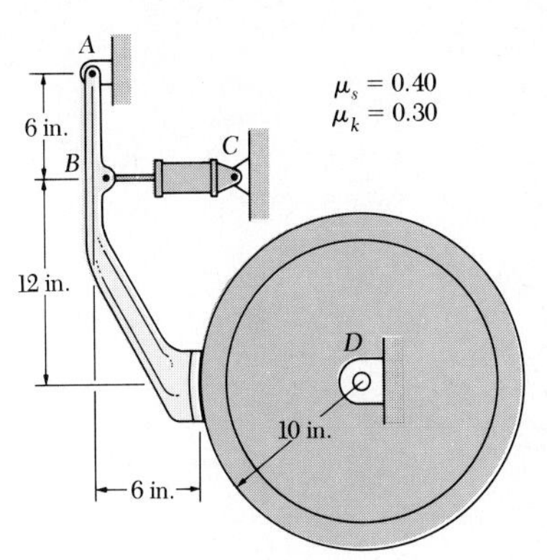

Fig. P17.9 and P17.10

17.9 The 10-in.-radius brake drum is attached to a larger flywheel which is not shown. The total mass moment of inertia of the flywheel and drum is 13.5 lb·ft·s^2. Knowing that the initial angular velocity is 180 rpm clockwise, determine the force which must be exerted by the hydraulic cylinder if the system is to stop in 50 revolutions.

17.10 Solve Prob. 17.9, assuming that the initial angular velocity of the flywheel is 180 rpm counterclockwise.

17.11 Using the principle of work and energy, solve Prob. 16.46*b*.

17.12 Gear G weighs 8 lb and has a radius of gyration of 2.4 in., while gear D weighs 30 lb and has a radius of gyration of 4.8 in. A 20-lb·in. couple **M** is applied to shaft CE as shown. Neglecting the mass of shafts CE and FH, determine (*a*) the number of revolutions of gear G required for its angular velocity to increase from 240 to 720 rpm, (*b*) the corresponding tangential force acting on gear G.

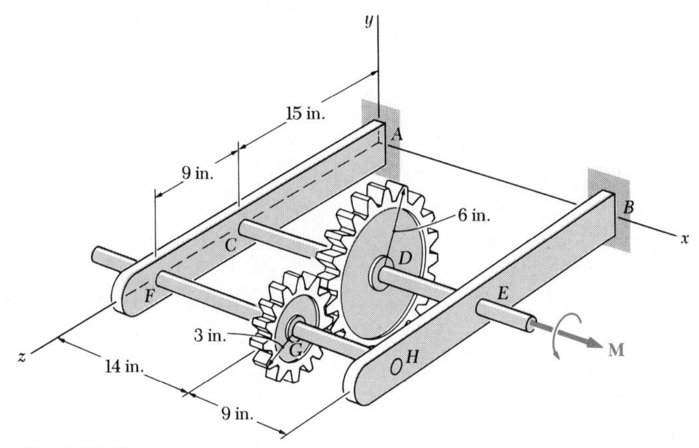

Fig. P17.12

17.13 Solve Prob. 17.12, assuming that the 20-lb·in. couple **M** is applied to shaft FH.

17.14 The double pulley shown has a total mass of 6 kg and a centroidal radius of gyration of 135 mm. Five collars, each of mass 1.2 kg, are attached to cords A and B as shown. When the system is at rest and in equilibrium, one collar is removed from cord A. Knowing that the bearing friction is equivalent to a couple **M** of moment 0.5 N·m, determine the velocity of cord A after it has moved 600 mm.

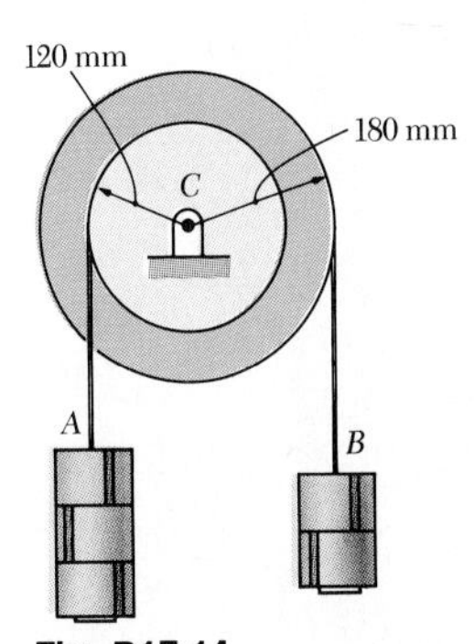

Fig. P17.14

17.15 Solve Prob. 17.14, assuming that one collar is removed from cord B.

17.16 The pulley shown has a mass of 5 kg and a radius of gyration of 150 mm. The 3-kg cylinder C is attached to a cord which is wrapped around the pulley as shown. A 1.8-kg collar B is then placed on the cylinder and the system is released from rest. After the cylinder has moved 300 mm, the collar is removed and the cylinder continues to move downward into a pit. Determine the velocity of cylinder C just before it strikes the bottom D of the pit.

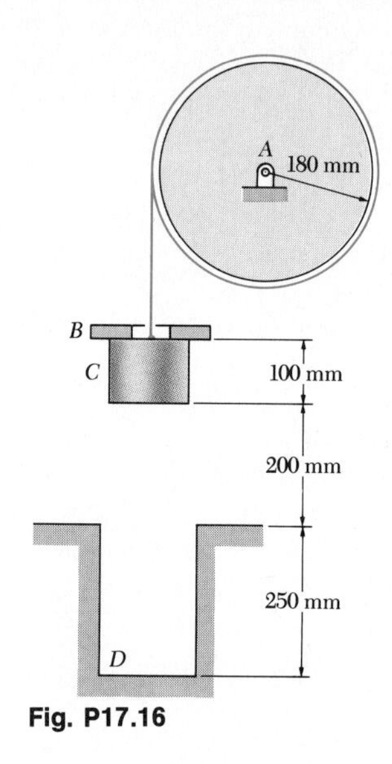

Fig. P17.16

17.17 and 17.18 A slender rod of length L and mass m is supported as shown. After the cable is cut the rod swings freely. (a) Determine the angular velocity of the rod as it first passes through a vertical position and the corresponding reaction at the pin support. (b) Solve part a for $m = 3$ kg and $L = 720$ mm.

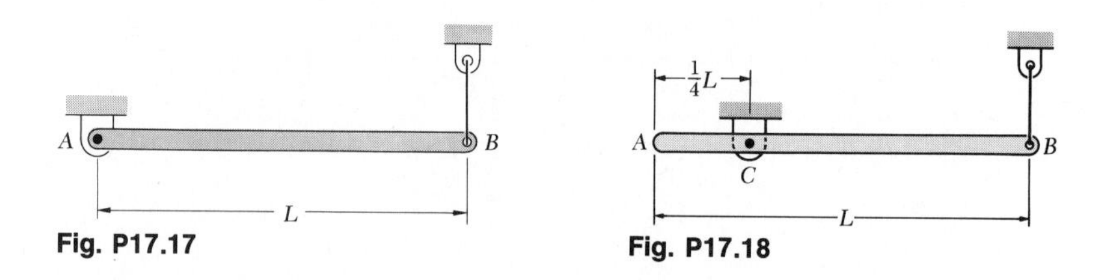

Fig. P17.17 Fig. P17.18

17.19 Two 6-lb slender rods are welded to the edge of an 8-lb uniform disk as shown. Knowing that in the position shown the assembly has an angular velocity of 9 rad/s clockwise, determine the largest and the smallest angular velocity of the assembly as it rotates about the pivot C.

17.20 Two 6-lb slender rods are welded to the edge of an 8-lb uniform disk as shown. The assembly is released from rest in the position shown and swings freely about the pivot C. Determine (a) the angular velocity of the assembly after it has rotated through 90°, (b) the maximum velocity attained by the assembly as it swings freely.

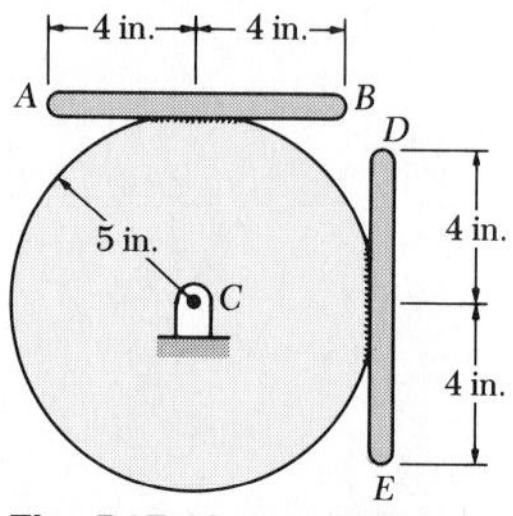

Fig. P17.19 and P17.20

17.7. Kinetic Energy in Plane Motion. Consider a rigid body in plane motion. We saw in Sec. 15.7 that at any given instant the velocities of the various particles of the body are the same as if the body were rotating about an axis perpendicular to the reference plane, called the *instantaneous axis of rotation*. We may therefore use the results obtained in Sec. 17.5 to express at any instant the kinetic energy of a rigid body in plane motion. Recalling that the instantaneous axis of rotation intersects the reference plane at a point C called the *instantaneous center of rotation* (Fig. 17.4), we denote by I_C the moment of inertia of the body about the instantaneous axis. If the body has an angular velocity $\boldsymbol{\omega}$ at the instant considered, its kinetic energy is therefore

$$T = \tfrac{1}{2} I_C \omega^2 \qquad (17.10)$$

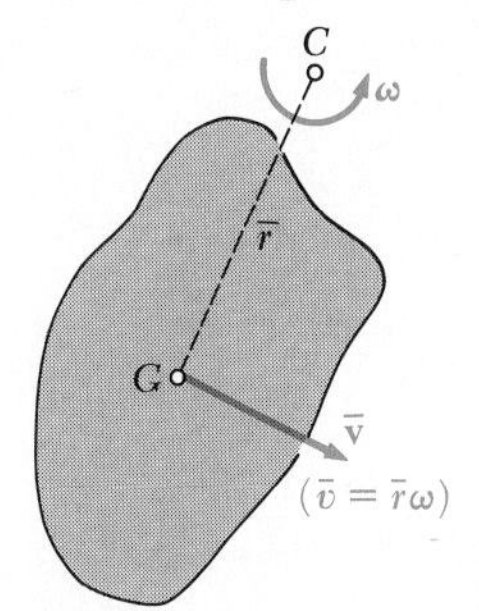

Fig. 17.4

But we know from the parallel-axis theorem (Sec. 9.12) that $I_C = m\bar{r}^2 + \bar{I}$, where $\bar{I}$ is the moment of inertia of the body about a centroidal axis perpendicular to the reference plane and $\bar{r}$ the distance from the instantaneous center C to the mass center G. Substituting for I_C into (17.10), we write

$$T = \tfrac{1}{2}(m\bar{r}^2 + \bar{I})\omega^2 = \tfrac{1}{2}m(\bar{r}\omega)^2 + \tfrac{1}{2}\bar{I}\omega^2$$

or, since $\bar{r}\omega$ represents the magnitude $\bar{v}$ of the velocity of the mass center,

$$T = \tfrac{1}{2}m\bar{v}^2 + \tfrac{1}{2}\bar{I}\omega^2 \qquad (17.11)$$

Either of the expressions (17.10) or (17.11) may be entered in formula (17.1) when the method of work and energy is used to analyze the plane motion of a rigid body.

We note that the expression given for the kinetic energy in (17.11) is separated into two distinct parts: (1) the kinetic energy $\tfrac{1}{2}m\bar{v}^2$ corresponding to the motion of the mass center G, and (2) the kinetic energy $\tfrac{1}{2}\bar{I}\omega^2$ corresponding to the rotation of the body about G (Fig. 17.5). This formula

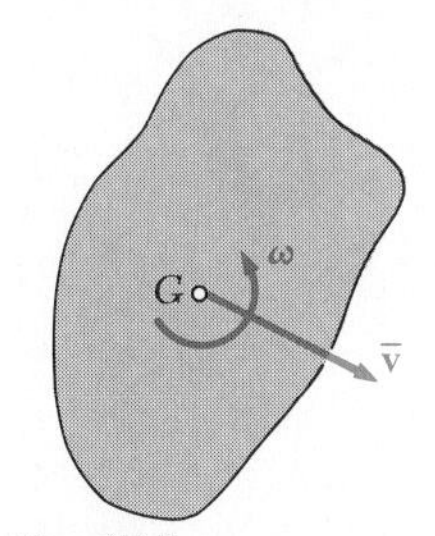

Fig. 17.5

thus conforms to the unified approach we have used so far, and we shall apply it in preference to (17.10) in the Sample Problems of this chapter.

We also note that the results obtained in this section are not limited to the motion of plane slabs or to the motion of bodies which are symmetrical with respect to the reference plane. They may be applied to the study of the plane motion of any rigid body, regardless of its shape.

17.8. Conservation of Energy. We saw in Sec. 13.6 that the work of conservative forces, such as the weight of a body or the force exerted by a spring, may be expressed as a change in potential energy. When a rigid body, or a system of rigid bodies, moves under the action of conservative forces, the principle of work and energy stated in Sec. 17.2 may be expressed in a modified form. Substituting for $U_{1\to2}$ from (13.15) into (17.1), we write

$$T_1 + V_1 = T_2 + V_2 \tag{17.12}$$

Formula (17.12) indicates that when a rigid body, or a system of rigid bodies, moves under the action of conservative forces, *the sum of the kinetic energy and of the potential energy of the system remains constant.* It should be noted that in the case of the plane motion of a rigid body, the kinetic energy of the body should include both the *translational* term $\frac{1}{2}m\bar{v}^2$ and the *rotational* term $\frac{1}{2}\bar{I}\omega^2$.

As an example of application of the principle of conservation of energy, we shall consider a slender rod AB, of length l and mass m, whose extremities are connected to blocks of negligible mass sliding along horizontal and vertical tracks. We assume that the rod is released with no initial velocity from a horizontal position (Fig. 17.6*a*), and we wish to determine its angular velocity after it has rotated through an angle θ (Fig. 17.6*b*).

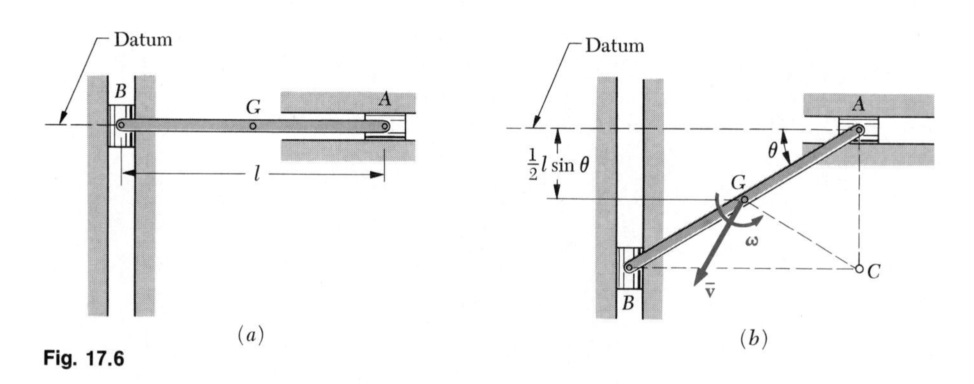

Fig. 17.6

Since the initial velocity is zero, we have $T_1 = 0$. Measuring the potential energy from the level of the horizontal track, we write $V_1 = 0$. After the rod has rotated through θ, the center of gravity G of the rod is at a distance $\frac{1}{2}l \sin\theta$ below the reference level and we have

$$V_2 = -\tfrac{1}{2}Wl\sin\theta = -\tfrac{1}{2}mgl\sin\theta$$

Observing that in this position, the instantaneous center of the rod is located at C, and that $CG = \frac{1}{2}l$, we write $\bar{v}_2 = \frac{1}{2}l\omega$ and obtain

$$T_2 = \tfrac{1}{2}m\bar{v}_2^2 + \tfrac{1}{2}\bar{I}\omega_2^2 = \tfrac{1}{2}m(\tfrac{1}{2}l\omega)^2 + \tfrac{1}{2}(\tfrac{1}{12}ml^2)\omega^2$$
$$= \frac{1}{2}\frac{ml^2}{3}\omega^2$$

Applying the principle of conservation of energy, we write

$$T_1 + V_1 = T_2 + V_2$$
$$0 = \frac{1}{2}\frac{ml^2}{3}\omega^2 - \tfrac{1}{2}mgl\sin\theta$$
$$\omega = \left(\frac{3g}{l}\sin\theta\right)^{1/2}$$

We recall that the advantages of the method of work and energy, as well as its shortcomings, were indicated in Sec. 13.4. In this connection, we wish to mention that the method of work and energy must be supplemented by the application of d'Alembert's principle when reactions at fixed axles, at rollers, or at sliding blocks are to be determined. For example, in order to compute the reactions at the extremities A and B of the rod of Fig. 17.6*b*, a diagram should be drawn to express that the system of the external forces applied to the rod is equivalent to the vector $m\bar{\mathbf{a}}$ and the couple $\bar{I}\boldsymbol{\alpha}$. The angular velocity $\boldsymbol{\omega}$ of the rod, however, is determined by the method of work and energy before the equations of motion are solved for the reactions. The complete analysis of the motion of the rod and of the forces exerted on the rod requires, therefore, the combined use of the method of work and energy and of the principle of equivalence of the external and effective forces.

17.9. Power. *Power* was defined in Sec. 13.8 as the time rate at which work is done. In the case of a body acted upon by a force $\mathbf{F}$, and moving with a velocity $\mathbf{v}$, the power was expressed as follows:

$$\text{Power} = \frac{dU}{dt} = \frac{(F\cos\alpha)\,ds}{dt} = (F\cos\alpha)v \qquad (13.18)$$

where α is the angle between $\mathbf{F}$ and $\mathbf{v}$. In the case of a rigid body acted upon by a couple $\mathbf{M}$ and rotating at an angular velocity $\boldsymbol{\omega}$, the power may be expressed in a similar way in terms of M and ω:

$$\text{Power} = \frac{dU}{dt} = \frac{M\,d\theta}{dt} = M\omega \qquad (17.13)$$

The various units used to measure power, such as the watt and the horsepower, were defined in Sec. 13.8.

SAMPLE PROBLEM 17.3

A sphere, a cylinder, and a hoop, each having the same mass and the same radius, are released from rest on an incline. Determine the velocity of each body after it has rolled through a distance corresponding to a change in elevation h.

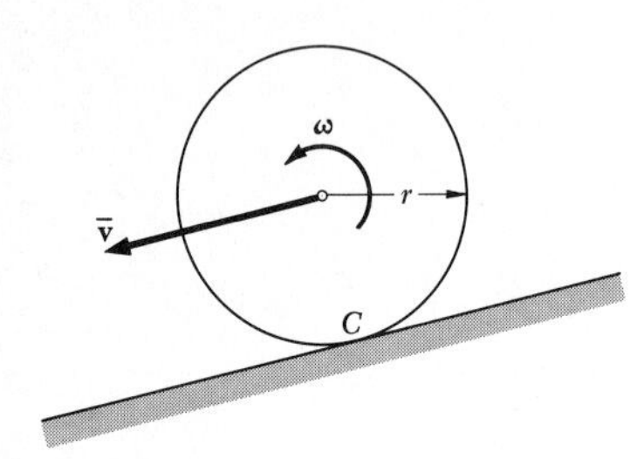

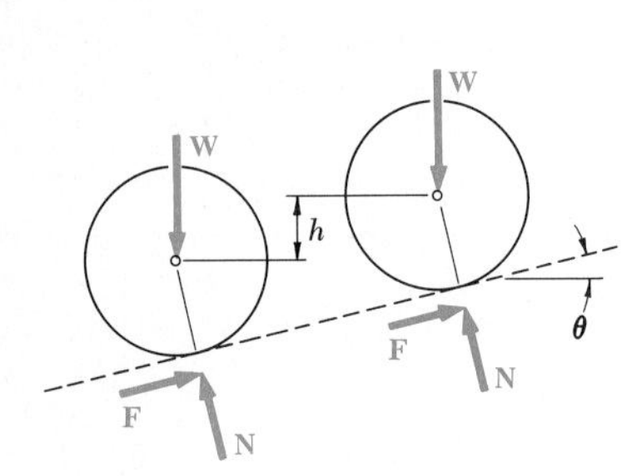

Solution. We shall first solve the problem in general terms and then find particular results for each body. We denote the mass by m, the centroidal moment of inertia by $\bar{I}$, the weight by W, and the radius by r.

Since each body rolls, the instantaneous center of rotation is located at C and we write

$$\omega = \frac{\bar{v}}{r}$$

Kinetic Energy

$$T_1 = 0$$

$$T_2 = \tfrac{1}{2}m\bar{v}^2 + \tfrac{1}{2}\bar{I}\omega^2$$

$$= \tfrac{1}{2}m\bar{v}^2 + \tfrac{1}{2}\bar{I}\left(\frac{\bar{v}}{r}\right)^2 = \tfrac{1}{2}\left(m + \frac{\bar{I}}{r^2}\right)\bar{v}^2$$

Work. Since the friction force $\mathbf{F}$ in rolling motion does no work,

$$U_{1\rightarrow 2} = Wh$$

Principle of Work and Energy

$$T_1 + U_{1\rightarrow 2} = T_2$$

$$0 + Wh = \tfrac{1}{2}\left(m + \frac{\bar{I}}{r^2}\right)\bar{v}^2 \qquad \bar{v}^2 = \frac{2Wh}{m + \bar{I}/r^2}$$

Noting that $W = mg$, we rearrange the result and obtain

$$\bar{v}^2 = \frac{2gh}{1 + \bar{I}/mr^2}$$

Velocities of Sphere, Cylinder, and Hoop. Introducing successively the particular expression for $\bar{I}$, we obtain

Sphere:	$\bar{I} = \frac{2}{5}mr^2$	$\bar{v} = 0.845\sqrt{2gh}$ ◄
Cylinder:	$\bar{I} = \frac{1}{2}mr^2$	$\bar{v} = 0.816\sqrt{2gh}$ ◄
Hoop:	$\bar{I} = mr^2$	$\bar{v} = 0.707\sqrt{2gh}$ ◄

Remark. We may compare the results with the velocity attained by a frictionless block sliding through the same distance. The solution is identical to the above solution except that $\omega = 0$; we find $\bar{v} = \sqrt{2gh}$.

Comparing the results, we note that the velocity of the body is independent of both its mass and radius. However, the velocity does depend upon the quotient $\bar{I}/mr^2 = \bar{k}^2/r^2$, which measures the ratio of the rotational kinetic energy to the translational kinetic energy. Thus the hoop, which has the largest $\bar{k}$ for a given radius r, attains the smallest velocity, while the sliding block, which does not rotate, attains the largest velocity.

SAMPLE PROBLEMS 17.4

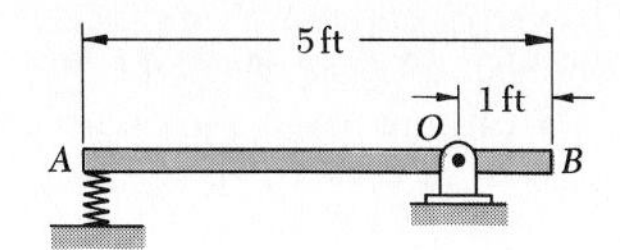

A 30-lb slender rod AB is 5 ft long and is pivoted about a point O which is 1 ft from end B. The other end is pressed against a spring of constant $k = 1800$ lb/in. until the spring is compressed 1 in. The rod is then in a horizontal position. If the rod is released from this position, determine its angular velocity and the reaction at the pivot O as the rod passes through a vertical position.

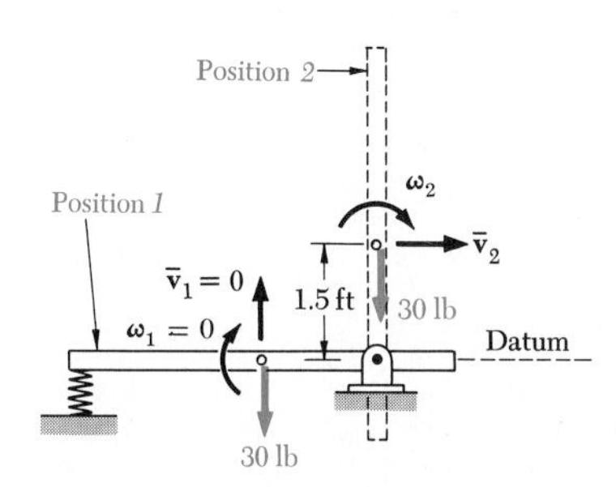

Position 1. ***Potential Energy.*** Since the spring is compressed 1 in., we have $x_1 = 1$ in.

$$V_e = \tfrac{1}{2}kx_1^2 = \tfrac{1}{2}(1800 \text{ lb/in.})(1 \text{ in.})^2 = 900 \text{ in}\cdot\text{lb}$$

Choosing the datum as shown, we have $V_g = 0$; therefore,

$$V_1 = V_e + V_g = 900 \text{ in}\cdot\text{lb} = 75 \text{ ft}\cdot\text{lb}$$

Kinetic Energy. Since the velocity in position 1 is zero, we have $T_1 = 0$.

Position 2. ***Potential Energy.*** The elongation of the spring is zero, and we have $V_e = 0$. Since the center of gravity of the rod is now 1.5 ft above the datum,

$$V_g = (30 \text{ lb})(+1.5 \text{ ft}) = 45 \text{ ft}\cdot\text{lb}$$
$$V_2 = V_e + V_g = 45 \text{ ft}\cdot\text{lb}$$

Kinetic Energy. Denoting by $\boldsymbol{\omega}_2$ the angular velocity of the rod in position 2, we note that the rod rotates about O and write $\bar{v}_2 = \bar{r}\omega_2 = 1.5\omega_2$.

$$\bar{I} = \tfrac{1}{12}ml^2 = \frac{1}{12}\frac{30 \text{ lb}}{32.2 \text{ ft/s}^2}(5 \text{ ft})^2 = 1.941 \text{ lb}\cdot\text{ft}\cdot\text{s}^2$$

$$T_2 = \tfrac{1}{2}m\bar{v}_2^2 + \tfrac{1}{2}\bar{I}\omega_2^2 = \frac{1}{2}\frac{30}{32.2}(1.5\omega_2)^2 + \tfrac{1}{2}(1.941)\omega_2^2 = 2.019\omega_2^2$$

Conservation of Energy

$$T_1 + V_1 = T_2 + V_2$$
$$0 + 75 \text{ ft}\cdot\text{lb} = 2.019\omega_2^2 + 45 \text{ ft}\cdot\text{lb}$$

$$\omega_2 = 3.86 \text{ rad/s} \curvearrowright \blacktriangleleft$$

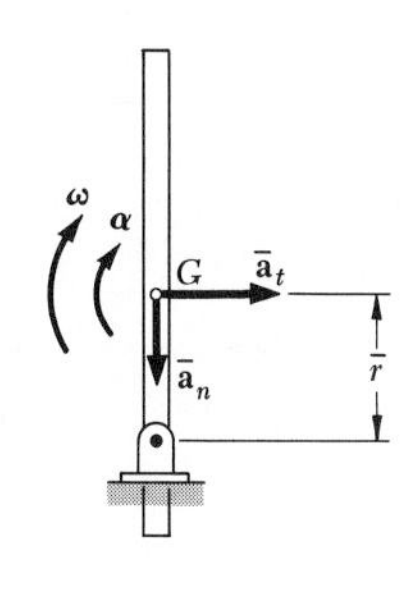

Reaction in Position 2. Since $\omega_2 = 3.86$ rad/s, the components of the acceleration of G as the rod passes through position 2 are

$$\bar{a}_n = \bar{r}\omega_2^2 = (1.5 \text{ ft})(3.86 \text{ rad/s})^2 = 22.3 \text{ ft/s}^2 \qquad \bar{\mathbf{a}}_n = 22.3 \text{ ft/s}^2 \downarrow$$
$$\bar{a}_t = \bar{r}\alpha \qquad \bar{\mathbf{a}}_t = \bar{r}\alpha \rightarrow$$

We express that the system of external forces is equivalent to the system of effective forces represented by the vector of components $m\bar{\mathbf{a}}_t$ and $m\bar{\mathbf{a}}_n$ attached at G and the couple $\bar{I}\boldsymbol{\alpha}$.

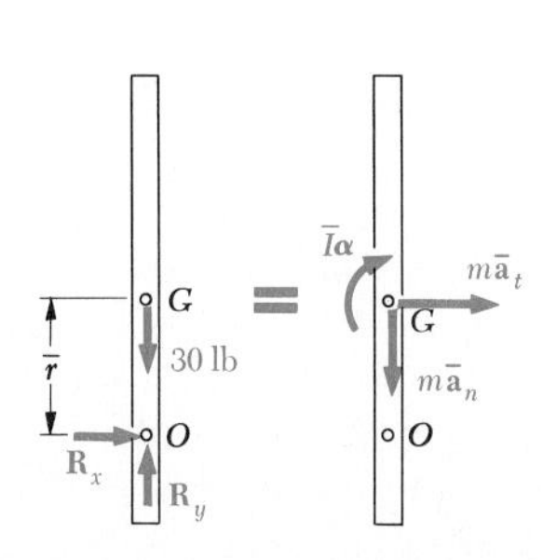

$$+\curvearrowright\Sigma M_O = \Sigma(M_O)_{\text{eff}}: \qquad 0 = \bar{I}\alpha + m(\bar{r}\alpha)\bar{r} \qquad \alpha = 0$$
$$\xrightarrow{+}\Sigma F_x = \Sigma(F_x)_{\text{eff}}: \qquad R_x = m(\bar{r}\alpha) \qquad R_x = 0$$
$$+\uparrow\Sigma F_y = \Sigma(F_y)_{\text{eff}}: \qquad R_y - 30 \text{ lb} = -m\bar{a}_n$$

$$R_y - 30 \text{ lb} = -\frac{30 \text{ lb}}{32.2 \text{ ft/s}^2}(22.3 \text{ ft/s}^2)$$

$$R_y = +9.22 \text{ lb} \qquad \mathbf{R} = 9.22 \text{ lb} \uparrow \blacktriangleleft$$

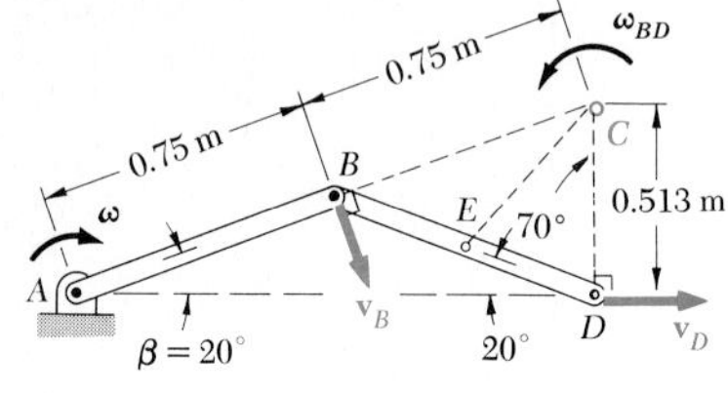

SAMPLE PROBLEM 17.5

Each of the two slender rods shown is 0.75 m long and has a mass of 6 kg. If the system is released from rest when $\beta = 60°$, determine (*a*) the angular velocity of rod *AB* when $\beta = 20°$, (*b*) the velocity of point *D* at the same instant.

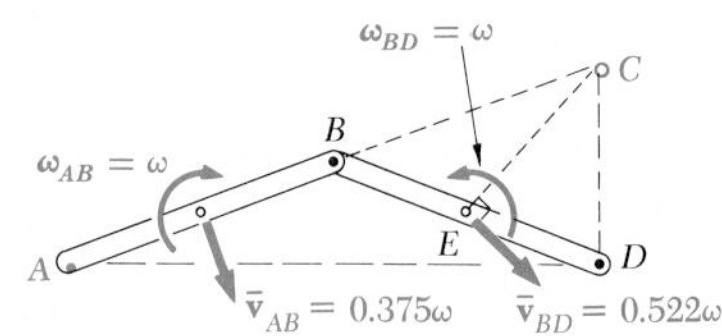

Kinematics of Motion When $\beta = 20°$. Since $\mathbf{v}_B$ is perpendicular to the rod *AB* and $\mathbf{v}_D$ is horizontal, the instantaneous center of rotation of rod *BD* is located at *C*. Considering the geometry of the figure, we obtain

$$BC = 0.75 \text{ m} \qquad CD = 2(0.75 \text{ m}) \sin 20° = 0.513 \text{ m}$$

Applying the law of cosines to triangle *CDE*, where *E* is located at the mass center of rod *BD*, we find $EC = 0.522$ m. Denoting by ω the angular velocity of rod *AB*, we have

$$\bar{v}_{AB} = (0.375 \text{ m})\omega \qquad \bar{\mathbf{v}}_{AB} = 0.375\omega \searrow$$
$$v_B = (0.75 \text{ m})\omega \qquad \mathbf{v}_B = 0.75\omega \searrow$$

Since rod *BD* seems to rotate about point *C*, we may write·

$$v_B = (BC)\omega_{BD} \qquad (0.75 \text{ m})\omega = (0.75 \text{ m})\omega_{BD} \qquad \omega_{BD} = \omega \circlearrowleft$$
$$\bar{v}_{BD} = (EC)\omega_{BD} = (0.522 \text{ m})\omega \qquad \bar{\mathbf{v}}_{BD} = 0.522\omega \searrow$$

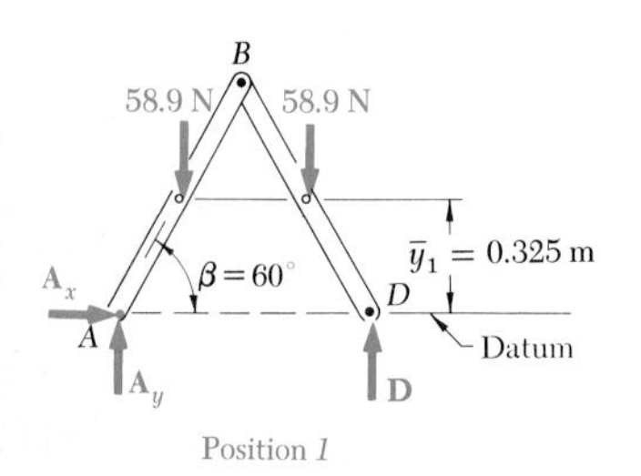

Position 1. *Potential Energy.* Choosing the datum as shown, and observing that $W = (6 \text{ kg})(9.81 \text{ m/s}^2) = 58.86$ N, we have

$$V_1 = 2W\bar{y}_1 = 2(58.86 \text{ N})(0.325 \text{ m}) = 38.26 \text{ J}$$

Kinetic Energy. Since the system is at rest, $T_1 = 0$.

Position 2. *Potential Energy*

$$V_2 = 2W\bar{y}_2 = 2(58.86 \text{ N})(0.1283 \text{ m}) = 15.10 \text{ J}$$

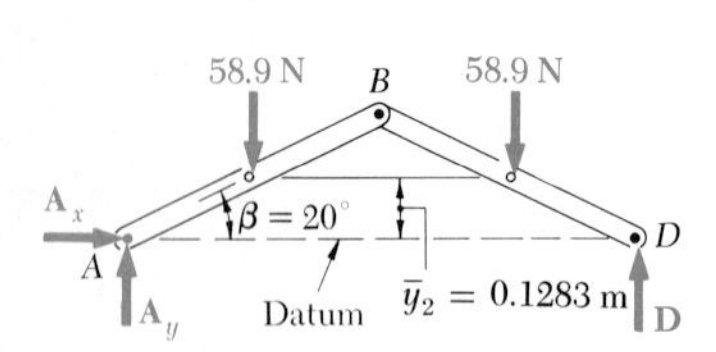

Kinetic Energy

$$\bar{I}_{AB} = \bar{I}_{BD} = \tfrac{1}{12}ml^2 = \tfrac{1}{12}(6 \text{ kg})(0.75 \text{ m})^2 = 0.281 \text{ kg}\cdot\text{m}^2$$
$$T_2 = \tfrac{1}{2}m\bar{v}_{AB}^2 + \tfrac{1}{2}\bar{I}_{AB}\omega_{AB}^2 + \tfrac{1}{2}m\bar{v}_{BD}^2 + \tfrac{1}{2}\bar{I}_{BD}\omega_{BD}^2$$
$$= \tfrac{1}{2}(6)(0.375\omega)^2 + \tfrac{1}{2}(0.281)\omega^2 + \tfrac{1}{2}(6)(0.522\omega)^2 + \tfrac{1}{2}(0.281)\omega^2$$
$$= 1.520\omega^2$$

Conservation of Energy

$$T_1 + V_1 = T_2 + V_2$$
$$0 + 38.26 \text{ J} = 1.520\omega^2 + 15.10 \text{ J}$$
$$\omega = 3.90 \text{ rad/s} \qquad \omega_{AB} = 3.90 \text{ rad/s} \circlearrowright \blacktriangleleft$$

Velocity of Point *D*

$$v_D = (CD)\omega = (0.513 \text{ m})(3.90 \text{ rad/s}) = 2.00 \text{ m/s}$$
$$\mathbf{v}_D = 2.00 \text{ m/s} \rightarrow \blacktriangleleft$$

Problems

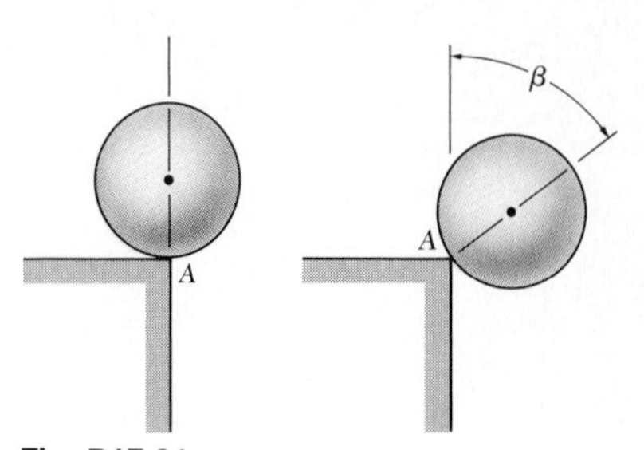

Fig. P17.21

17.21 A uniform sphere of radius r is placed at corner A and is given a slight clockwise motion. Assuming that the corner is sharp and becomes slightly embedded in the sphere, so that the coefficient of static friction at A is very large, determine (*a*) the angle β through which the sphere will have rotated when it loses contact with the corner, (*b*) the corresponding velocity of the center of the sphere.

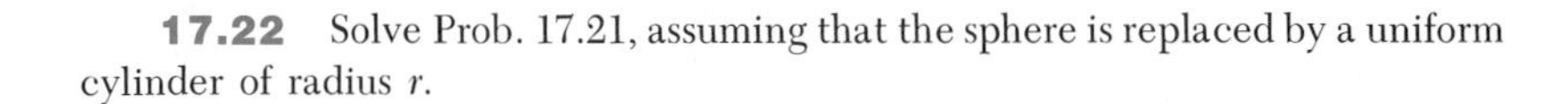

17.22 Solve Prob. 17.21, assuming that the sphere is replaced by a uniform cylinder of radius r.

17.23 The 10-lb slender rod AB is welded to the 6-lb uniform disk which rotates about a pivot at A. A spring of constant 0.625 lb/in. is attached to the disk and is unstretched when rod AB is horizontal. Knowing that the assembly is released from rest in the position shown, determine its angular velocity after it has rotated through 90°.

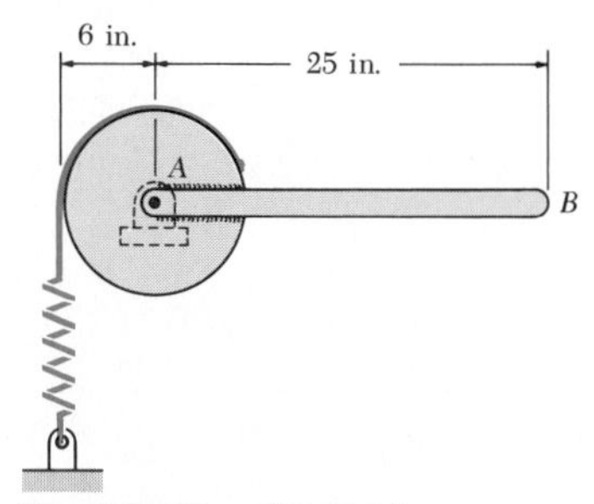

Fig. P17.23 and P17.24

17.24 The 10-lb slender rod AB is welded to the 6-lb uniform disk which rotates about a pivot at A. A spring of constant 0.625 lb/in. is attached to the disk and is unstretched when rod AB is horizontal. Determine the required angular velocity of the assembly when it is in the position shown, if its angular velocity is to be 8 rad/s after it has rotated through 90° clockwise.

17.25 A 4-kg slender rod rotates in a *vertical* plane about a pivot at B. A spring of constant $k = 200$ N/m and of unstretched length 150 mm is attached to the rod as shown. Knowing that in the position shown the rod has an angular velocity of 3 rad/s counterclockwise, determine the angular velocity of the rod after it has rotated through (*a*) 90°, (*b*) 180°.

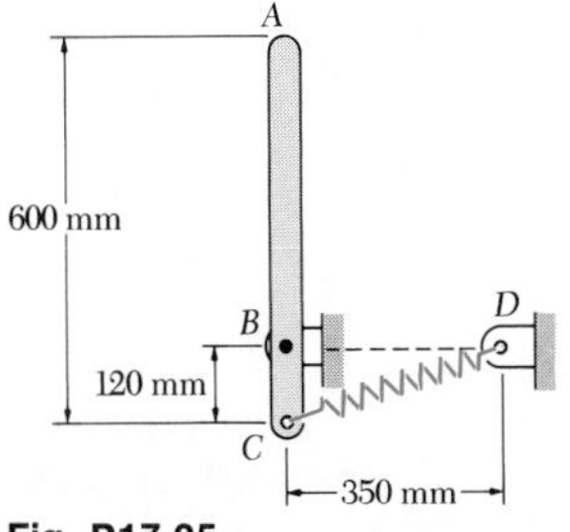

Fig. P17.25

17.26 Solve Prob. 17.25, assuming that the rod rotates in a *horizontal* plane.

17.27 A flywheel of centroidal radius of gyration $\bar{k} = 25$ in. is rigidly attached to a shaft of radius $r = 1.25$ in. which may roll along parallel rails. Knowing that the system is released from rest, determine the velocity of the center of the shaft after it has moved 10 ft.

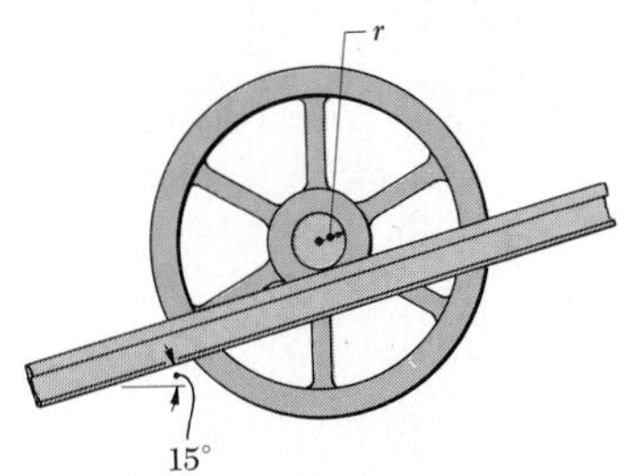

Fig. P17.27 and P17.28

17.28 A flywheel is rigidly attached to a 40-mm-radius shaft which rolls without sliding along parallel rails. The system is released from rest and attains a speed of 160 mm/s after moving 1.5 m along the rails. Determine the centroidal radius of gyration of the system.

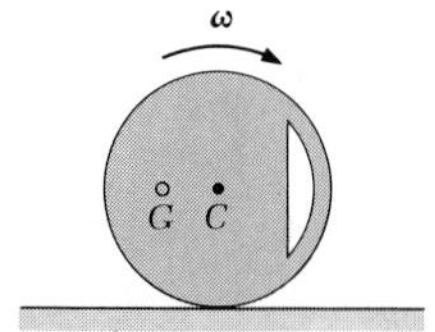

Fig. P17.29

17.29 The mass center G of a 5-kg wheel of radius $R = 300$ mm is located at a distance $r = 100$ mm from its geometric center C. The centroidal radius of gyration of the wheel is $\bar{k} = 150$ mm. As the wheel rolls without sliding, its angular velocity is observed to vary. Knowing that $\omega = 6$ rad/s in the position shown, determine (*a*) the angular velocity of the wheel when the mass center G is directly above the geometric center C, (*b*) the reaction at the horizontal surface at the same instant.

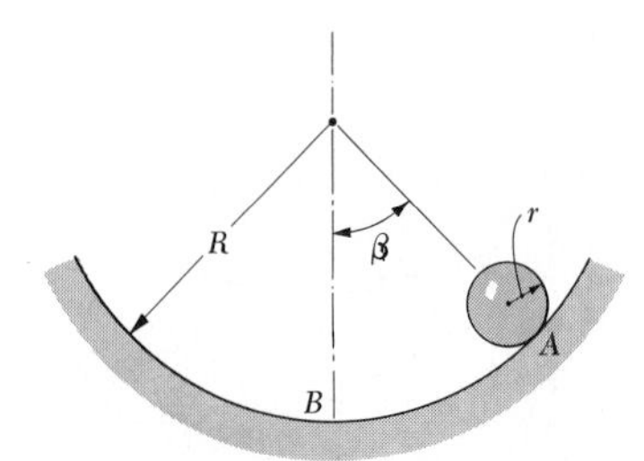

Fig. P17.30

17.30 A sphere of weight W and radius r rolls without slipping inside a curved surface of radius R. Knowing that the sphere is released from rest in the position shown, derive an expression (*a*) for the linear velocity of the sphere as it passes through B, (*b*) for the magnitude of the vertical reaction at that instant.

17.31 Solve Prob. 17.30, assuming that the sphere is replaced by a uniform cylinder of weight W and radius r.

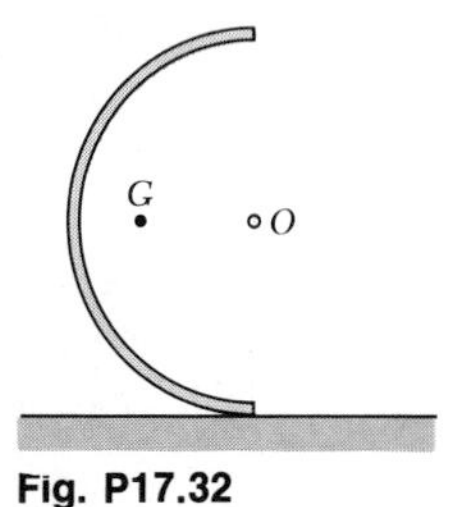

Fig. P17.32

17.32 A half section of pipe of mass m and radius r is released from rest in the position shown. Assuming that the pipe rolls without sliding, determine (*a*) its angular velocity after it has rolled through 90°, (*b*) the reaction at the horizontal surface at the same instant. [*Hint.* Note that $GO = 2r/\pi$ and that by the parallel-axis theorem, $\bar{I} = mr^2 - m(GO)^2$.]

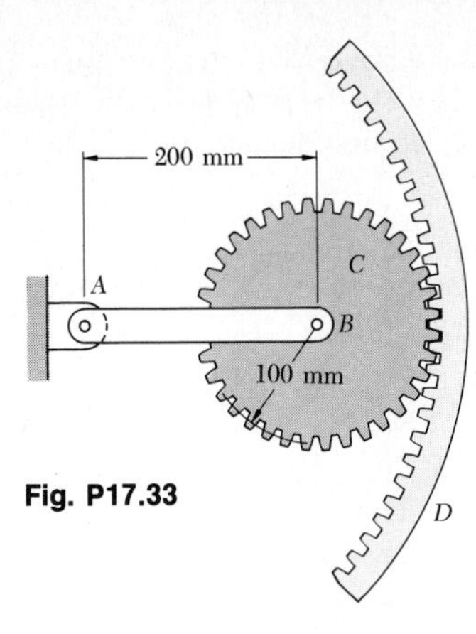

Fig. P17.33

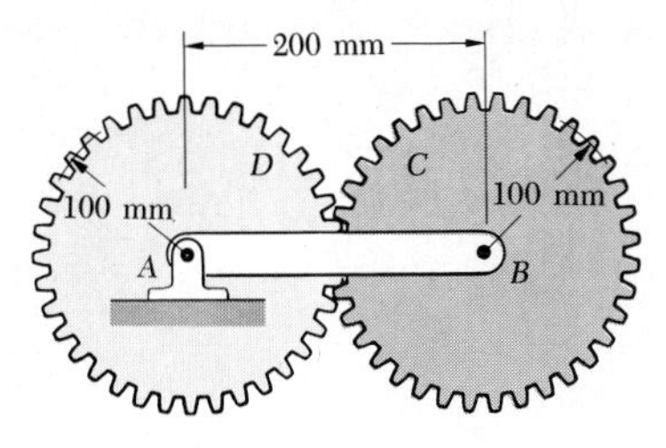

Fig. P17.34

17.33 and 17.34 Gear C has a mass of 3 kg and a centroidal radius of gyration of 75 mm. The uniform bar AB has a mass of 2.5 kg, and gear D is stationary. If the system is released from rest in the position shown, determine the velocity of point B after bar AB has rotated through 90°.

17.35 Two uniform cylinders, each of weight $W = 14$ lb and radius $r = 5$ in., are connected by a belt as shown. Knowing that at the instant shown the angular velocity of cylinder B is 30 rad/s clockwise, determine (*a*) the distance through which cylinder A will rise before the angular velocity of cylinder B is reduced to 5 rad/s, (*b*) the tension in the portion of belt connecting the two cylinders.

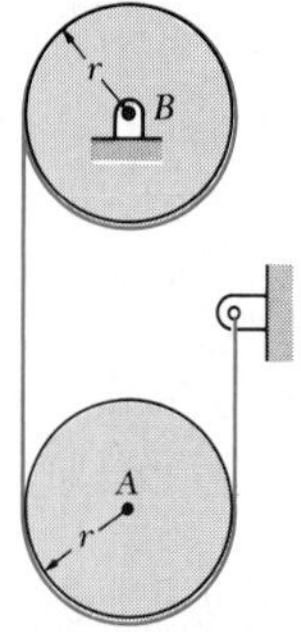

Fig. P17.35 and P17.36

17.36 Two uniform cylinders, each of weight $W = 14$ lb and radius $r = 5$ in., are connected by a belt as shown. If the system is released from rest, determine (*a*) the velocity of the center of cylinder A after it has moved through 3 ft, (*b*) the tension in the portion of belt connecting the two cylinders.

17.37 The 8-kg rod AB is attached by pins to two 5-kg uniform disks as shown. The assembly rolls without sliding on a horizontal surface. If the assembly is released from rest when $\theta = 60°$, determine (*a*) the angular velocity of the disks when $\theta = 180°$, (*b*) the force exerted by the surface on each disk at that instant.

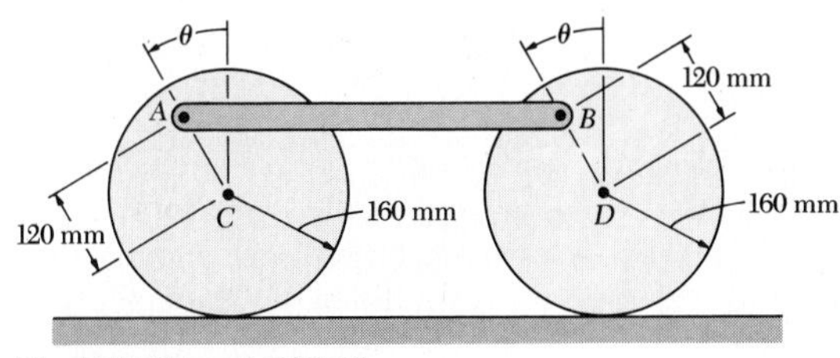

Fig. P17.37 and P17.38

17.38 The 8-kg rod AB is attached by pins to two 5-kg uniform disks as shown. The assembly rolls without sliding on a horizontal surface. Knowing that the velocity of rod AB is 700 mm/s to the left when $\theta = 0$, determine (*a*) the velocity of rod AB when $\theta = 180°$, (*b*) the force exerted by the surface on each disk at that instant.

17.39 The motion of the 5-lb slender bar AB is guided by collars of negligible weight which slide freely on the vertical and horizontal rods shown. Knowing that the bar is released from rest when $\theta = 30°$, determine the velocity of collars A and B when $\theta = 60°$.

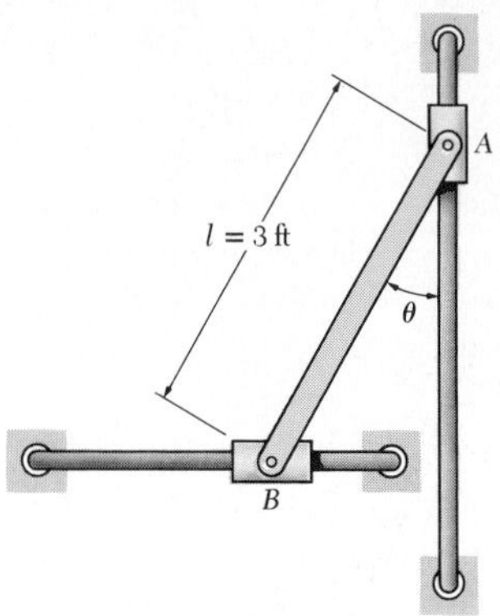

Fig. P17.39 and P17.40

17.40 The motion of the 5-lb slender bar AB is guided by collars of negligible weight which slide freely on the vertical and horizontal rods shown. Knowing that the bar is released from rest when $\theta = 45°$, determine the velocity of collar A when $\theta = 90°$.

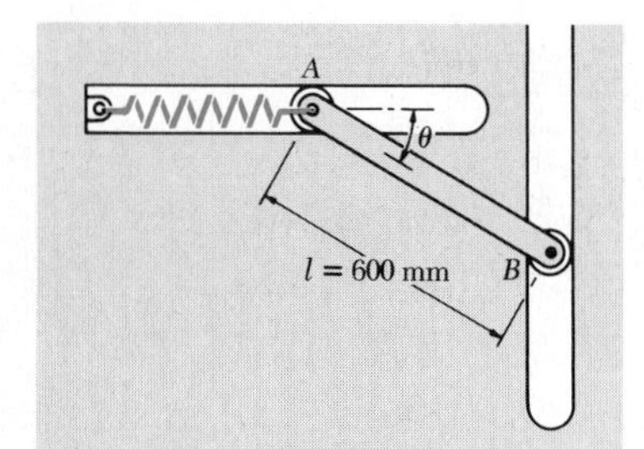

Fig. P17.41 and P17.42

17.41 The ends of a 4-kg rod AB are constrained to move along slots cut in a vertical plate as shown. A spring of constant 500 N/m is attached to end A in such a way that its tension is zero when $\theta = 0$. If the rod is released from rest when $\theta = 0$, determine the angular velocity of the rod and the velocity of end B when $\theta = 30°$.

17.42 The ends of a 4-kg rod AB are constrained to move along slots cut in a vertical plate as shown. A spring of constant 500 N/m is attached to end A in such a way that its tension is zero when $\theta = 0$. If the rod is released from rest when $\theta = 50°$, determine the angular velocity of the rod and the velocity of end B when $\theta = 0$.

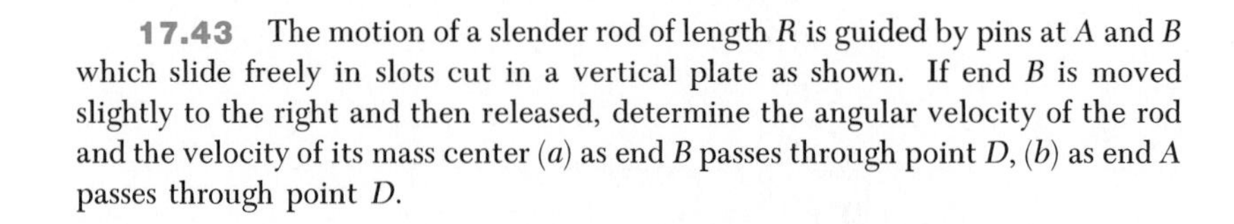

17.43 The motion of a slender rod of length R is guided by pins at A and B which slide freely in slots cut in a vertical plate as shown. If end B is moved slightly to the right and then released, determine the angular velocity of the rod and the velocity of its mass center (*a*) as end B passes through point D, (*b*) as end A passes through point D.

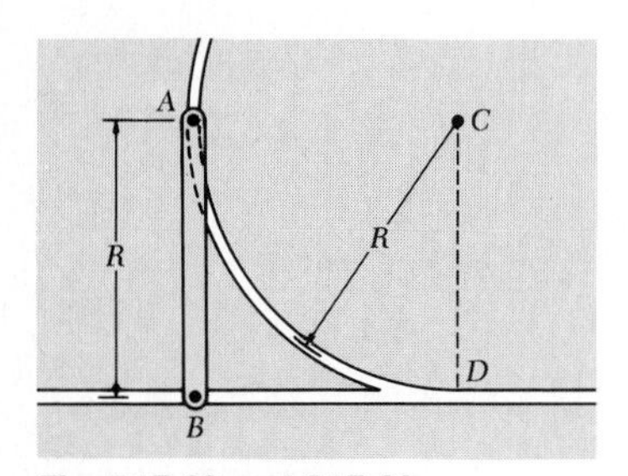

Fig. P17.43 and P17.44

17.44 The motion of a slender rod of length R is guided by pins at A and B which slide freely in slots cut in a vertical plate as shown. If end B is moved slightly to the left and then released, determine the angular velocity of the rod and the velocity of its mass center (*a*) at the instant when the velocity of end B is zero, (*b*) as end B passes through point D.

17.45 The uniform rods AB and BC are of mass 2 kg and 5 kg, respectively, and collar C has a mass of 3 kg. If the system is released from rest in the position shown, determine the velocity of point B after rod AB has rotated through 90°.

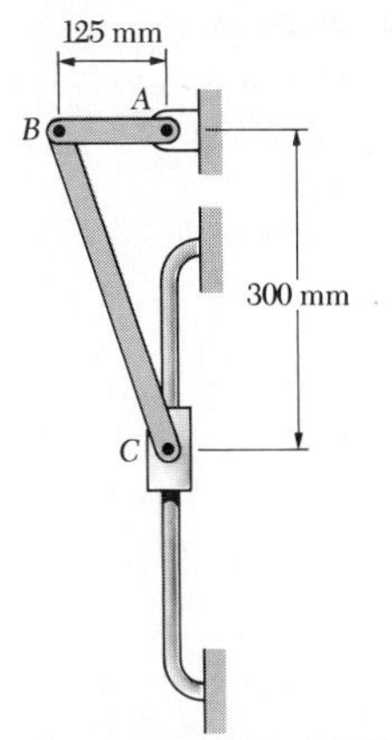

Fig. P17.45 and P17.46

17.46 The uniform rods AB and BC are of mass 2 kg and 5 kg, respectively, and collar C has a mass of 3 kg. Knowing that at the instant shown the velocity of collar C is 0.8 m/s downward, determine the velocity of point B after rod AB has rotated through 90°.

17.47 Two uniform rods, each of mass m and length l, are connected to form the linkage shown. End D of rod BD may slide freely in the vertical slot, while end A of rod AB is attached to a fixed pin support. If the system is released from rest in the position shown, determine the velocity of end D at the instant when (a) ends A and D are at the same elevation, (b) rod AB is horizontal.

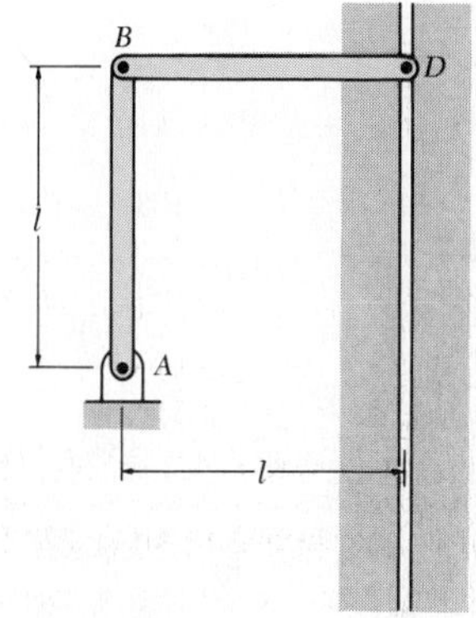

Fig. P17.47

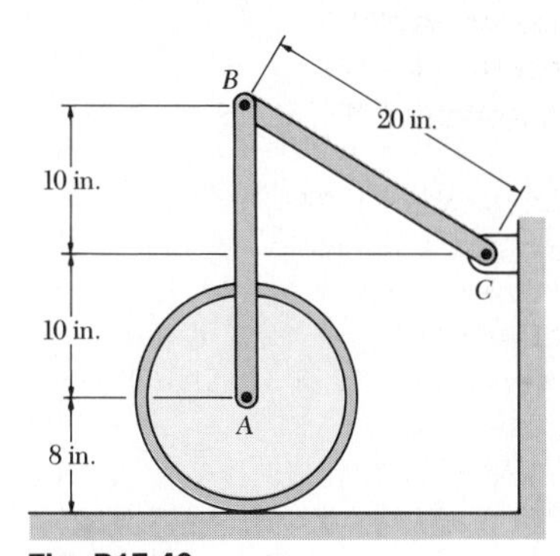

Fig. P17.48

17.48 Wheel A weighs 9 lb, has a centroidal radius of gyration of 6 in., and rolls without sliding on the horizontal surface. Each of the uniform rods AB and BC is 20 in. long and weighs 5 lb. If point A is moved slightly to the left and released, determine the velocity of point A as rod BC passes through a horizontal position.

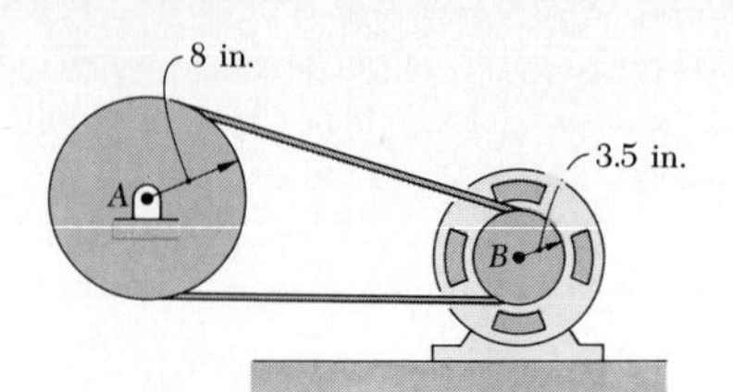

Fig. P17.49

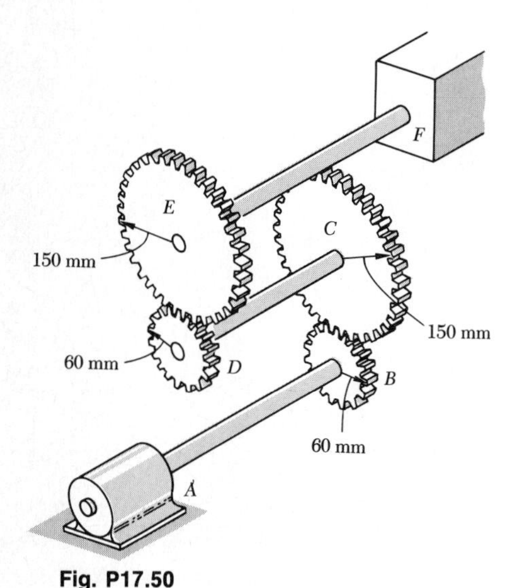

Fig. P17.50

17.49 The motor shown runs a machine attached to the shaft at A. The motor develops 5 hp and runs at a constant speed of 360 rpm. Determine the moment of the couple exerted (*a*) by the shaft on pulley A, (*b*) by the motor on pulley B.

17.50 Three shafts and four gears are used to form a gear train which will transmit 7.5 kW from the motor at A to a machine tool at F. (Bearings for the shafts are omitted in the sketch.) Knowing that the frequency of the motor is 30 Hz, determine the moment of the couple which is applied to shaft (*a*) AB, (*b*) CD, (*c*) EF.

17.51 Determine the moment of the couple which must be exerted by a motor to develop 250 W at a speed of (*a*) 3450 rpm, (*b*) 1150 rpm.

17.52 Knowing that the maximum allowable couple which can be applied to a shaft is 8 kip·in., determine the maximum horsepower which can be transmitted by the shaft (*a*) at 60 rpm, (*b*) at 300 rpm.

Principle of work and energy for a rigid body

17.10. Review and Summary. In this chapter, we used the method of work and energy to analyze the motion of rigid bodies and of systems of rigid bodies. In Sec. 17.2, we first expressed the principle of work and energy for a rigid body in the form

$$T_1 + U_{1\to2} = T_2 \qquad (17.1)$$

where T_1 and T_2 represent the initial and final values of the kinetic energy of the rigid body and $U_{1\to2}$ the work of the *external forces* acting on the rigid body.

Work of a force or a couple

In Sec. 17.3, we recalled the expression found in Chap. 13 for the work of a force $\mathbf{F}$ applied at a point A, namely

$$U_{1\to2} = \int_{s_1}^{s_2} (F\cos\alpha)\, ds \qquad (17.3)$$

where F was the magnitude of the force, α the angle it formed with the direction of motion of A, and s the variable of integration measuring the distance traveled by A along its path. We also derived the expression for the *work of a couple of moment* M applied to a rigid body during a rotation in θ of the rigid body:

$$U_{1\to2} = \int_{\theta_1}^{\theta_2} M\, d\theta \qquad (17.5)$$

We then derived an expression for the kinetic energy of a rigid body in translation [Sec. 17.4]. Observing that all the particles of the rigid body had the same velocity $\bar{v}$ as its mass center, we wrote

Kinetic energy in translation

$$T = \tfrac{1}{2}m\bar{v}^2 \tag{17.7}$$

For a rigid body rotating about a fixed axis through O with an angular velocity $\boldsymbol{\omega}$ [Sec. 17.5], we had

Kinetic energy in rotation

$$T = \tfrac{1}{2}I_O\omega^2 \tag{17.8}$$

where I_O was the moment of inertia of the body about the fixed axis. We noted that the result obtained was not limited to the rotation of plane slabs or of bodies symmetrical with respect to the reference plane. It is valid regardless of the shape of the body or of the location of the axis of rotation.

Systems of rigid bodies

Equation (17.1) may be applied to the motion of systems of rigid bodies [Sec. 17.6] as long as all the forces acting on the various bodies involved—internal as well as external to the system—are included in the computation of $U_{1\to2}$. However, in the case of systems consisting of pin-connected members, or blocks and pulleys connected by inextensible cords, or meshed gears, the points of application of the internal forces move through equal distances and the work of these forces cancels out [Sample Probs. 17.1 and 17.2].

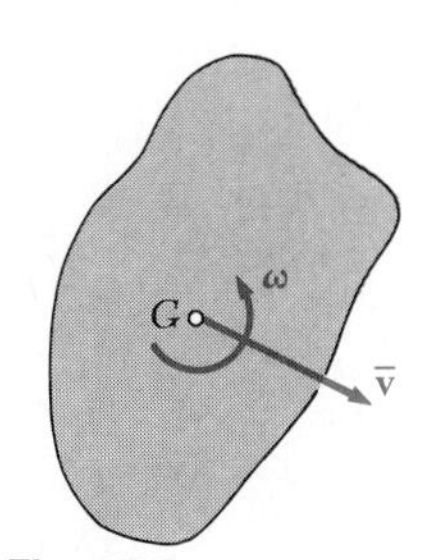

Fig. 17.7

Kinetic energy in plane motion

In the second part of the chapter, we determined the *kinetic energy of a rigid body in plane motion* [Sec. 17.7] and expressed it as

$$T = \tfrac{1}{2}m\bar{v}^2 + \tfrac{1}{2}\bar{I}\omega^2 \tag{17.11}$$

where $\bar{v}$ is the velocity of the mass center G of the body, ω the angular velocity of the body, and $\bar{I}$ its moment of inertia about an axis through G perpendicular to the plane of reference (Fig. 17.7) [Sample Prob. 17.3].

Conservation of energy

When a rigid body, or a system of rigid bodies, moves under the action of conservative forces, the principle of work and energy may be expressed in the form

$$T_1 + V_1 = T_2 + V_2 \tag{17.12}$$

which is referred to as the *principle of conservation of energy* [Sec. 17.8]. This principle may be used to solve problems involving conservative forces such as the force of gravity or the force exerted by a spring [Sample Probs.

17.4 and 17.5]. However, when a reaction is to be determined, the principle of conservation of energy must be supplemented by the application of d'Alembert's principle [Sample Prob. 17.4].

Power

Finally, in Sec. 17.9, we extended the concept of power to a rotating body subjected to a couple, writing

$$\text{Power} = \frac{dU}{dt} = \frac{M\,d\theta}{dt} = M\omega \tag{17.13}$$

where M is the moment of the couple and ω the angular velocity of the body.

Review Problems

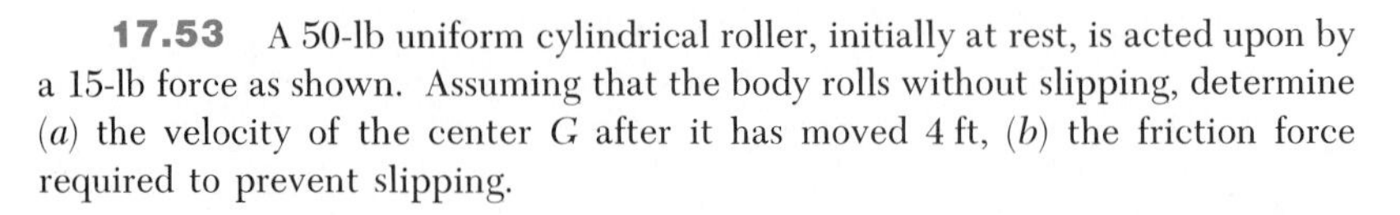

17.53 A 50-lb uniform cylindrical roller, initially at rest, is acted upon by a 15-lb force as shown. Assuming that the body rolls without slipping, determine (*a*) the velocity of the center G after it has moved 4 ft, (*b*) the friction force required to prevent slipping.

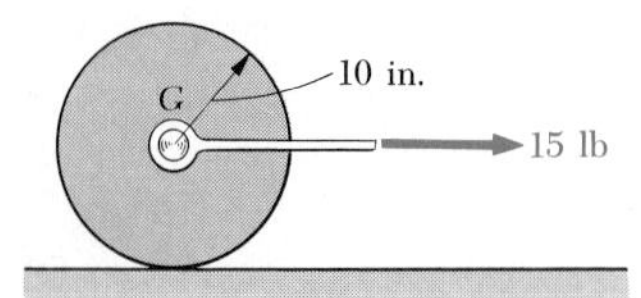

Fig. P17.53

17.54 A uniform rod of length L and weight W is attached to two wires, each of length b. The rod is released from rest when $\theta = 0$ and swings to the position $\theta = 90°$, at which time wire BD suddenly breaks. Determine the tension in wire AC (*a*) immediately before wire BD breaks, (*b*) immediately after wire BD breaks.

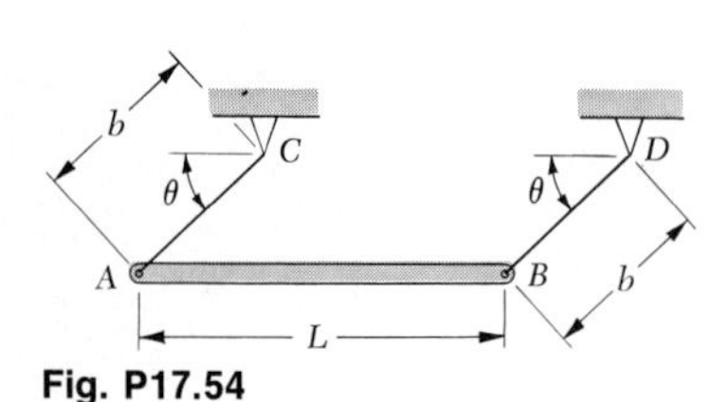

Fig. P17.54

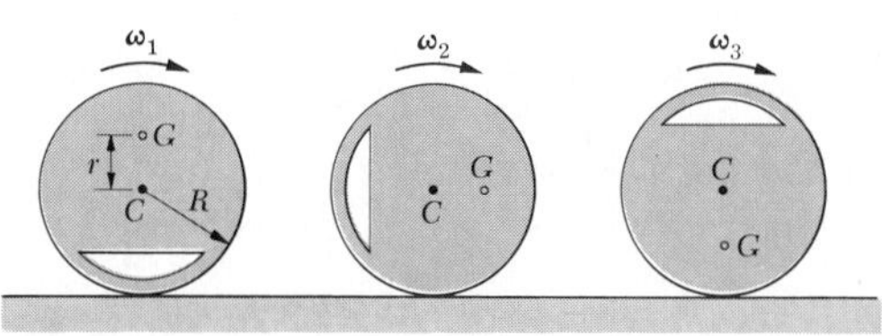

Fig. P17.55

17.55 The mass center G of a wheel of radius R is located at a distance r from its geometric center C. The centroidal radius of gyration of the wheel is denoted by $\bar{k}$. As the wheel rolls freely and without sliding on a horizontal plane, its angular velocity is observed to vary. Denoting by ω_1, ω_2, and ω_3, respectively, the angular velocity of the wheel when G is directly above C, level with C, and directly below C, show that ω_1, ω_2, and ω_3 satisfy the relation

$$\frac{\omega_2^2 - \omega_1^2}{\omega_3^2 - \omega_2^2} = \frac{g/R + \omega_1^2}{g/R + \omega_3^2}$$

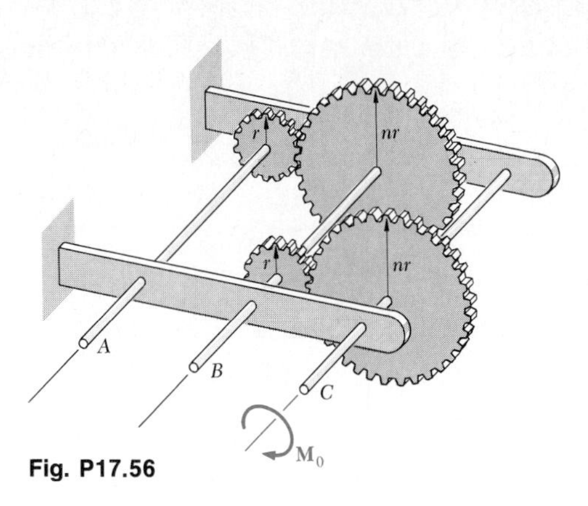

Fig. P17.56

17.56 The gear train shown consists of four gears of the same thickness and of the same material; two gears are of radius r, and the other two are of radius nr. The system is at rest when the couple $\mathbf{M}_0$ is applied to shaft C. Denoting by I_0 the moment of inertia of a gear of radius r, determine the angular velocity of shaft A if the couple $\mathbf{M}_0$ is applied for one revolution of shaft C.

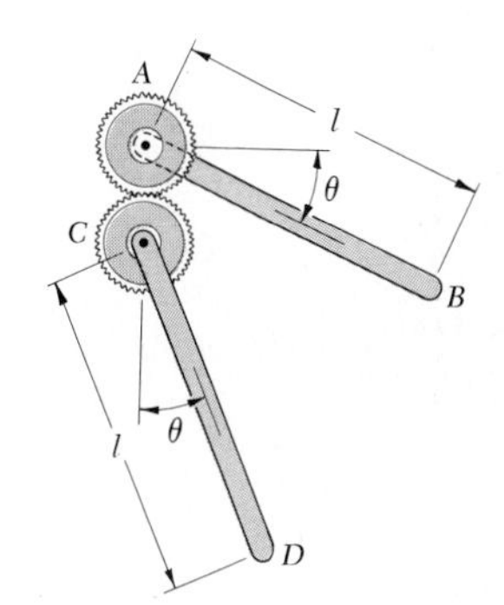

Fig. P17.57

17.57 Two uniform rods, each of mass m, are attached to two gears of the same radius as shown. The rods are released from rest in the position $\theta = 0$. Neglecting the mass of the gears, determine the angular velocity of the rods when (*a*) $\theta = 30°$, (*b*) $\theta = 45°$, (*c*) $\theta = 60°$.

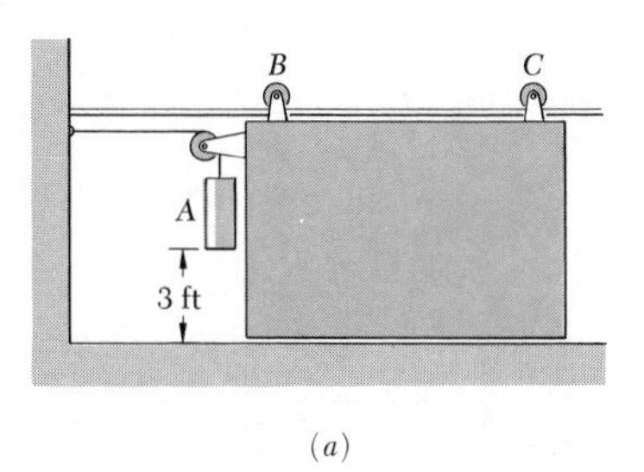

(*a*)

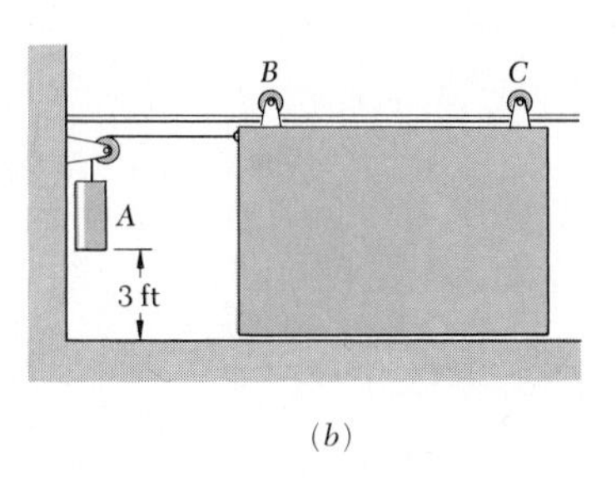

(*b*)

Fig. P17.58

17.58 The motion of a 20-lb sliding panel is guided by rollers at B and C. The counterweight A weighs 15 lb and is attached to a cable as shown. If the system is released from rest, determine for each case shown the velocity of the counterweight as it strikes the ground. Neglect the effect of friction.

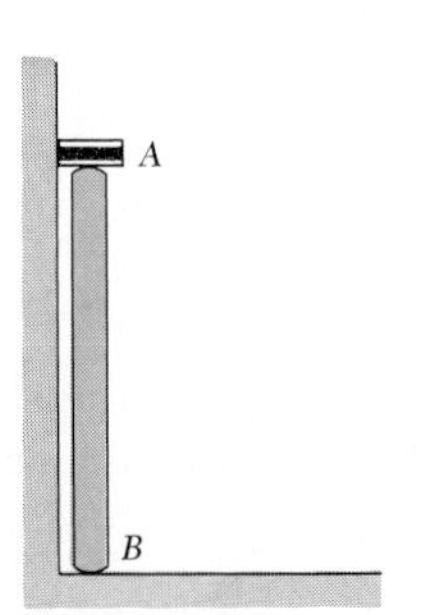

Fig. P17.59

17.59 A small matchbox is placed on top of the rod AB. End B of the rod is given a slight horizontal push, causing it to slide on the horizontal floor. Assuming no friction and neglecting the weight of the matchbox, determine the angle θ through which the rod will have rotated when the matchbox loses contact with the rod.

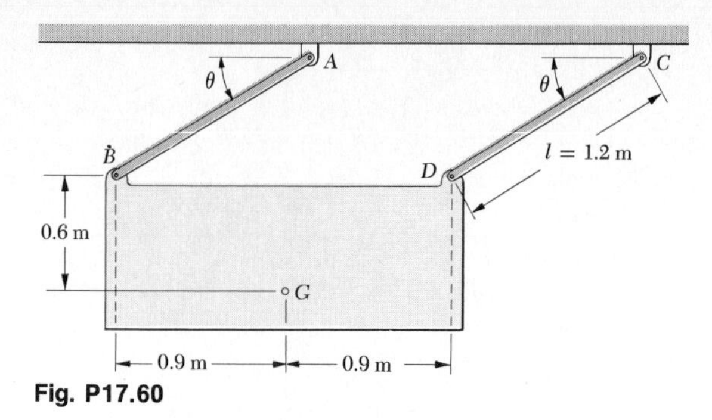

Fig. P17.60

17.60 A 100-kg bin is suspended from two rods AB and CD of negligible weight. The bin is released from rest in the position $\theta = 0$. As the bin passes through its lowest position, determine (*a*) the velocity of the bin, (*b*) the reactions at A and C. (*c*) Show that the answer to part *b* is independent of the length l of the rods.

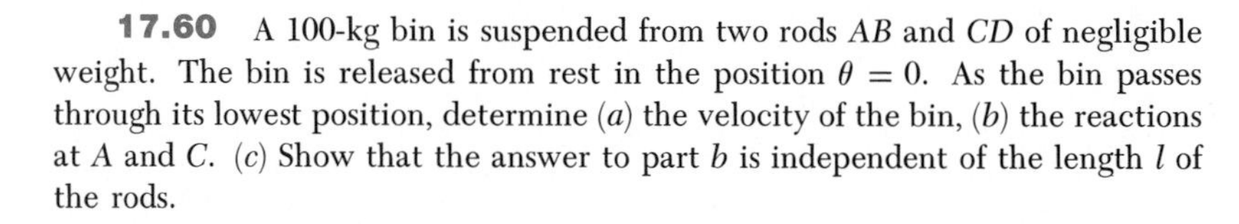

Fig. P17.61

17.61 A small collar of mass m is attached at B to the rim of a hoop of mass m and radius r. The hoop rolls without sliding on a horizontal plane. Find the angular velocity $\boldsymbol{\omega}_1$ of the hoop when B is directly above the center A in terms of g and r, knowing that the angular velocity of the loop is $3\boldsymbol{\omega}_1$ when B is directly below A.

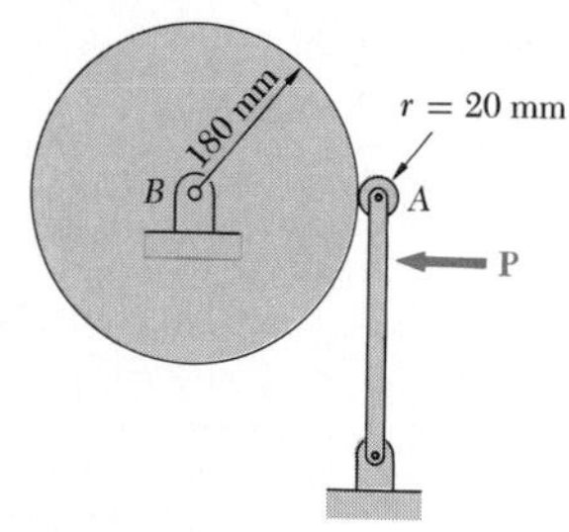

Fig. P17.62

17.62 A small disk A is driven at a constant angular velocity of 1200 rpm and is pressed against disk B, which is initially at rest. The normal force between disks is 50 N, and $\mu_k = 0.20$. Knowing that disk B has a mass of 20 kg, determine the number of revolutions executed by disk B before its speed reaches 120 rpm.

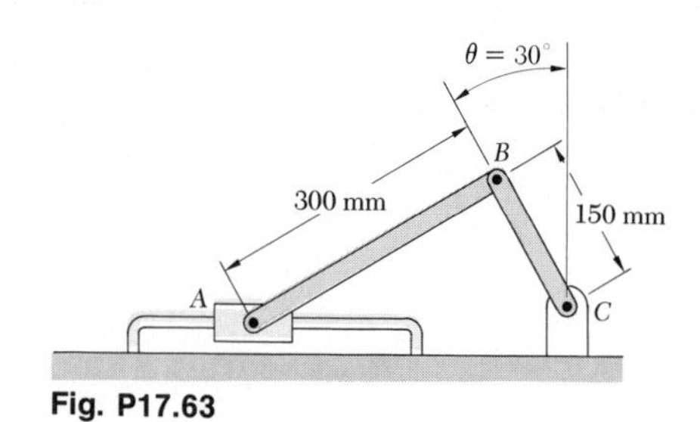

Fig. P17.63

17.63 Two uniform rods, AB of mass 2 kg and BC of mass 1 kg, are released from rest in the position shown. Neglecting both friction and the mass of the collar, determine the velocity of point B when $\theta = 90°$.

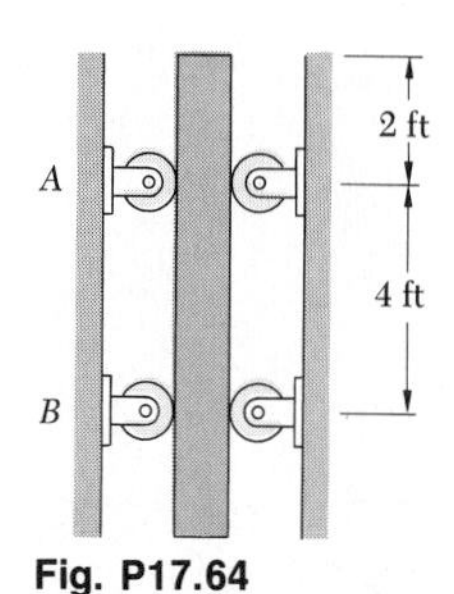

Fig. P17.64

17.64 A 10-lb bar is held between four disks as shown. Each disk weighs 5 lb and has a diameter of 8 in. The disks may rotate freely, and the normal reaction exerted by each disk on the bar is sufficient to prevent slipping. If the bar is released from rest, determine its velocity after it has dropped 2 ft.

The following problems are designed to be solved with a computer.

17.C1 The ends of a 4-kg rod AB are constrained to move along slots cut in a vertical plate as shown. A spring of constant 500 N/m is attached to end A in such a way that its tension is zero when $\theta = 0$. If the rod is released from rest when $\theta = 75°$, write a computer program and use it to calculate the velocity of end B for values of θ from 75° to 0 at 5° intervals.

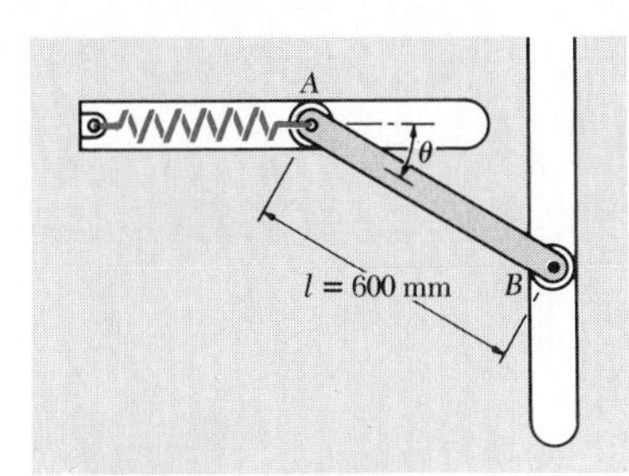

Fig. P17.C1

17.C2 Each of the two slender rods shown is 0.75 m long and has a mass of 6 kg. Knowing that the system is released from rest when $\beta = 60°$, write a computer program and use it to calculate the angular velocity of rod AB and the velocity of point D for values of β from 60° to 0 at 5° intervals.

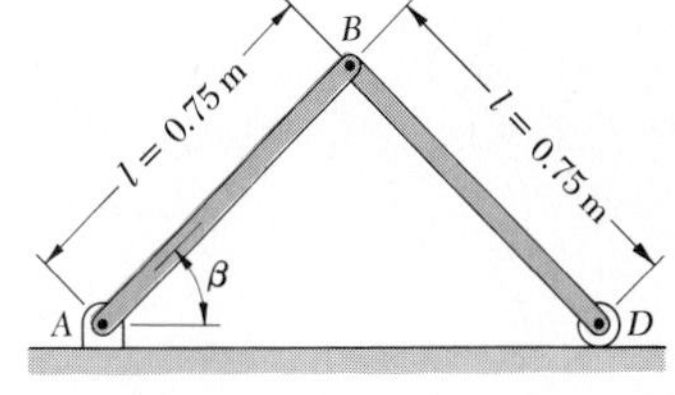

Fig. P17.C2

17.C3 A 5-lb uniform slender rod of length $L = 3$ ft is supported as shown. After the cable attached at B is cut the rod swings freely. (*a*) Write a computer program to determine the angular velocity $\boldsymbol{\omega}$ of the rod after it has rotated through an angle θ and use this program to calculate ω for values of θ from 0 to 90° at 5° intervals. (*b*) Expand the program written to calculate the approximate time required for the rod to swing from the horizontal position ($\theta = 0$) to the vertical position ($\theta = 90°$), using 1° intervals. [*Hint.* The time Δt_i required for a rigid body to rotate through an angle $\Delta\theta_i$ may be obtained by dividing $\Delta\theta_i$ expressed in radians by the average angular velocity $\frac{1}{2}(\omega_i + \omega_{i+1})$ of the body over Δt_i if the angular acceleration of the body is assumed to remain constant over Δt_i.]

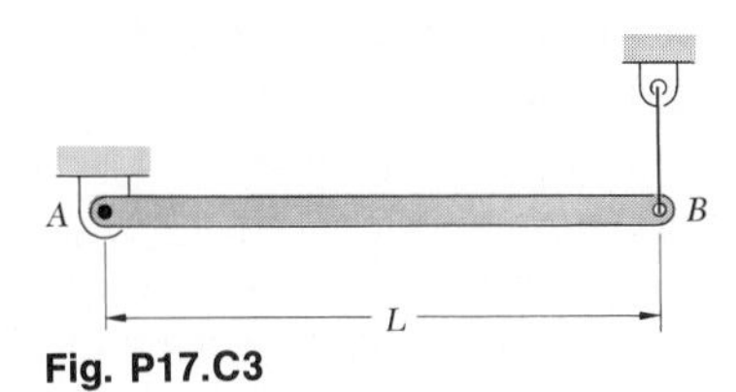

Fig. P17.C3

17.C4 A half section of pipe of mass m and radius r is released from rest in the position shown and rolls without sliding. (*a*) Write a computer program to determine the angular velocity of the pipe after it has rolled through an angle θ. (*b*) Use this program to calculate the angular velocity of a half section of pipe of radius $r = 8$ in. for values of θ from 0 to 90° at 5° intervals. [Note that $GO = 2r/\pi$ and that by the parallel-axis theorem, $\bar{I} = mr^2 - m(GO)^2$.] (*c*) Expand the program to determine the approximate time required for the half section to roll through 180°, using 1° intervals. (See hint of Prob. 17.C3.)

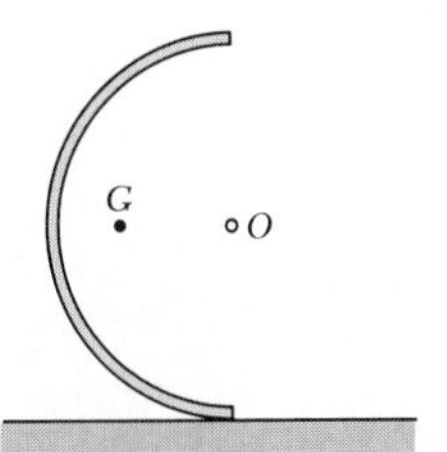

Fig. P17.C4

CHAPTER EIGHTEEN

Kinetics of Rigid Bodies: Impulse and Momentum

18.1. Introduction. In this chapter, the principle of impulse and momentum will be used to analyze the motion of rigid bodies and of systems of rigid bodies. As was pointed out in Chap. 14, the method of impulse and momentum is particularly well adapted to the solution of problems involving time and velocities. Moreover, the principle of impulse and momentum provides the only practicable method for the solution of problems of impact.

We shall first recall the principle of impulse and momentum as it was derived in Sec. 14.13 for a system of particles,

$$\textbf{Syst Momenta}_1 + \textbf{Syst Ext Imp}_{1\rightarrow 2} = \textbf{Syst Momenta}_2 \quad (14.45)$$

and apply it to the motion of a *rigid body* (Sec. 18.2). Next, we shall show that for a rigid slab in plane motion—or for a rigid body symmetrical with respect to the reference plane—the system of the momenta of the particles forming the body is equivalent to a *momentum vector* $m\bar{\mathbf{v}}$ attached at the mass center G of the body and a *momentum couple* $\bar{I}\omega$ (Sec. 18.3). The principle of impulse and momentum will then be applied to the solution of problems involving the plane motion of a rigid body (Sec. 18.4) and of a system of rigid bodies (Sec. 18.5). The same general approach will be used in problems involving conservation of angular momentum about a fixed point (Sec. 18.6).

In Secs. 18.7 and 18.8, we shall consider problems involving the *impulsive motion* and the *eccentric impact* of rigid bodies, i.e., an impact in which the mass centers of the two colliding bodies are *not* located on the line of impact (cf. Sec. 14.6). As was done in Chap. 14, where we considered the impact of particles, the coefficient of restitution between the colliding bodies will be used together with the principle of impulse and momentum in the solution of impact problems. It will also be shown that the method used is applicable not only when the colliding bodies move freely after the impact but also when the bodies are partially constrained in their motion.

In the last part of the chapter, we shall consider the *gyroscopic motion* of rigid bodies (Sec. 18.9). Our study will be limited to cases of *steady precession* where the axes of spin and precession are at a right angle to each other, or where the rate of spin is much larger than the rate of precession.

18.2. Principle of Impulse and Momentum for a Rigid Body. Since a rigid body of mass m may be considered as made of a large number of particles of mass Δm, Eq. (14.45), which was derived in Sec. 14.13 for a system of particles, may now be used to analyze the motion of a rigid body. We write

$$\text{Syst Momenta}_1 + \text{Syst Ext Imp}_{1\to2} = \text{Syst Momenta}_2 \qquad (18.1)$$

This equation relates the three systems of vectors shown in Fig. 18.1. It expresses that the system formed by the momenta of the particles of the rigid body at time t_1 and the system of the impulses of the external forces applied to the body from t_1 to t_2 are together equivalent to the system formed by the momenta of the particles at time t_2.

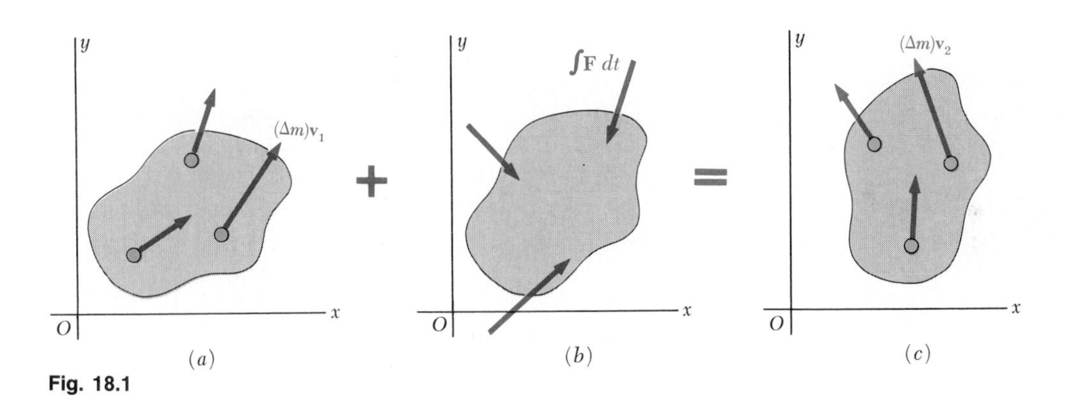

Fig. 18.1

We recall from Sec. 14.13 that, in the case of particles moving in the xy plane, three scalar equations are required to express the same relationship. The first two equations may be obtained by considering respectively the x and y components of the vectors shown in Fig. 18.1; they will relate the linear momenta of the particles and the linear impulses of the given forces in the x and y directions respectively. The third equation, which is obtained by computing the moments about O of the same vectors, will relate the angular momenta of the particles and the angular impulses of the forces.

Before we may conveniently apply the principle of impulse and momentum to the study of the plane motion of a rigid body, we must learn to express more simply the systems of vectors representing the momenta of the particles forming the rigid body. This is the object of the next section.

18.3. Momentum of a Rigid Body in Plane Motion. We shall see in this section that the system of the momenta of the particles forming a rigid body may be replaced at any given instant by an equivalent system consisting of a *momentum vector* attached at the mass center of the body and a *momentum couple*. This reduction of the system of momenta to a vector and a couple is similar to the reduction of the system of effective forces described in Chap. 16. While the method involved is quite general, our present analysis will be limited to the plane motion of rigid slabs and of rigid bodies which are symmetrical with respect to the reference plane.†

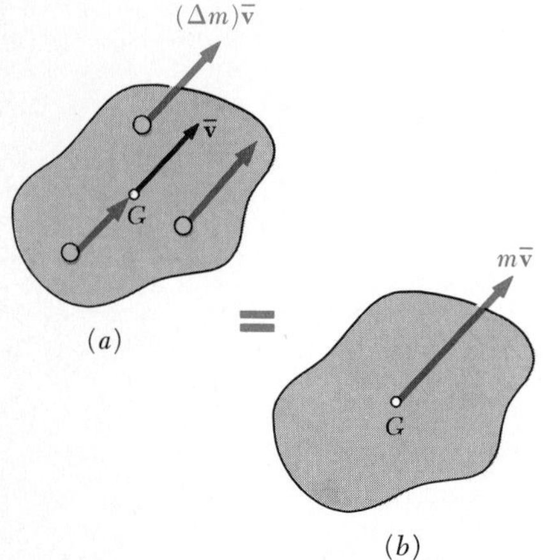

Fig. 18.2

Translation. Consider first the case of a rigid body in translation. Since each particle, at any given instant, has the same velocity $\overline{\mathbf{v}}$ as the mass center G of the body, the momentum of a particle of mass Δm is represented by the vector $(\Delta m)\overline{\mathbf{v}}$. The resultant of the momenta of the various particles of the body is thus

$$\Sigma(\Delta m)\overline{\mathbf{v}} = (\Sigma\,\Delta m)\overline{\mathbf{v}} = m\overline{\mathbf{v}}$$

where m is the mass of the body. On the other hand, a reasoning similar to that used in Sec. 16.2 for the effective forces of a rigid body in translation shows that the sum of the moments about G of the momenta $(\Delta m)\overline{\mathbf{v}}$ is zero. We thus conclude that, in the case of a rigid body in translation, *the system of the momenta* of the particles forming the body *is equivalent to a single momentum vector* $m\overline{\mathbf{v}}$ *attached at the mass center* G (Fig. 18.2). This vector is the *linear momentum* of the body.

Centroidal Rotation. When a rigid body rotates about a fixed centroidal axis perpendicular to the reference plane, the velocity $\overline{\mathbf{v}}$ of its mass center G is zero, since G is fixed, and it follows from Eqs. (14.10) of Sec. 14.4 that the sums of the x and y components of the momenta of the various particles of the body are zero:

$$\Sigma(\Delta m)v_x = (\Sigma\,\Delta m)\overline{v}_x = 0 \qquad \Sigma(\Delta m)v_y = (\Sigma\,\Delta m)\overline{v}_y = 0$$

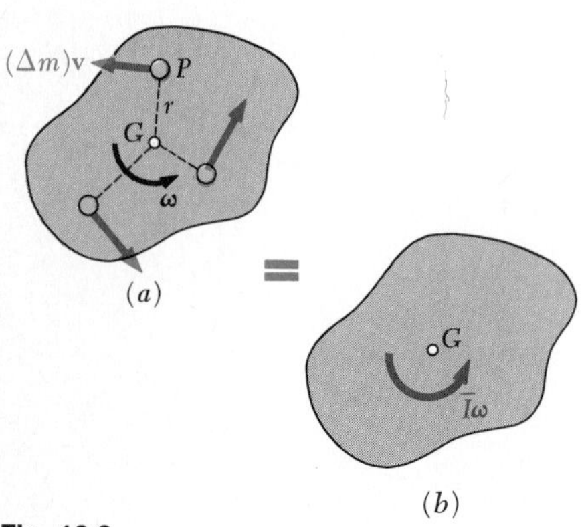

Fig. 18.3

Therefore, in the case of a rigid body in centroidal rotation, *the system of the momenta* of the particles forming the body *is equivalent to a single momentum couple*. If, as we have assumed in this section, the body is symmetrical with respect to the reference plane, the axis of the momentum couple coincides with the axis of rotation.

The moment of the momentum couple is obtained by summing the moments about G of the momenta $(\Delta m)\mathbf{v}$ of the various particles. This sum is denoted by H_G and referred to as the *angular momentum* of the body about the axis through G. Observing that the momentum $(\Delta m)\mathbf{v}$ of a particle P is perpendicular to the line GP (Fig. 18.3*a*) and that $v = r\omega$, where r is the distance from G to P and ω the angular velocity of the body at the instant considered, we write

$$+\circlearrowleft H_G = \Sigma r(\Delta m)v = \Sigma r(\Delta m)r\omega = \omega\Sigma r^2\,\Delta m$$

† Or, more generally, to the motion of rigid bodies which have a principal axis of inertia perpendicular to the reference plane.

Since $\Sigma r^2\,\Delta m$ represents the centroidal moment of inertia $\bar{I}$ of the body, we have

$$H_G = \bar{I}\omega \tag{18.2}$$

Recalling from Sec. 15.3 that the angular velocity of a rigid body rotating about a fixed axis may be represented by a vector $\boldsymbol{\omega}$ of magnitude ω directed along that axis, we note that the momentum couple may be represented by a couple vector $\bar{I}\boldsymbol{\omega}$ of magnitude $\bar{I}\omega$ directed along the axis of rotation. We therefore conclude that, in the case of the centroidal rotation of a rigid body symmetrical with respect to the reference plane, *the system of the momenta* of the particles of the body *is equivalent to a momentum couple* $\bar{I}\boldsymbol{\omega}$, where $\boldsymbol{\omega}$ denotes the angular velocity of the body (Fig. 18.3*b*).

General Plane Motion. In the general case of the plane motion of a rigid body, the velocity of any given particle P of the body may be resolved into the velocity $\bar{\mathbf{v}}$ of the mass center G and the relative velocity $\mathbf{v}'$ of P with respect to G. The momentum $(\Delta m)\mathbf{v}$ of each particle (Fig. 18.4*a*) may thus be resolved into the vectors $(\Delta m)\bar{\mathbf{v}}$ and $(\Delta m)\mathbf{v}'$ shown in Fig. 18.4*b*. Recalling the results obtained above in the particular cases of translation

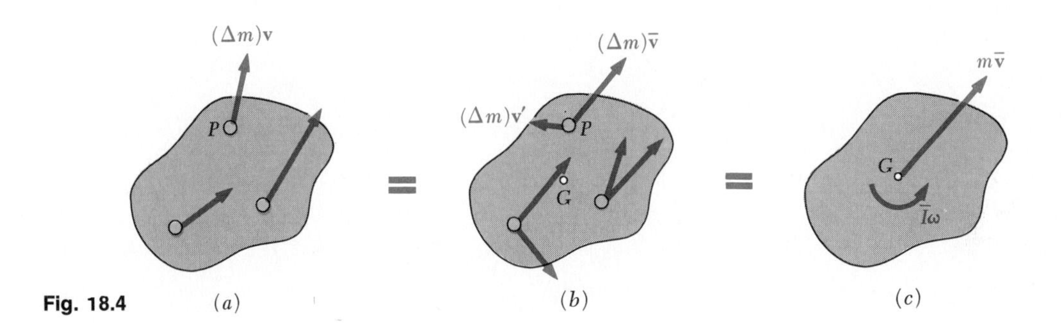

Fig. 18.4 (*a*) (*b*) (*c*)

and centroidal rotation, we find that the vectors $(\Delta m)\bar{\mathbf{v}}$ may be replaced by their resultant $m\bar{\mathbf{v}}$ attached at G and that the vectors $(\Delta m)\mathbf{v}'$ reduce to the couple $\bar{I}\boldsymbol{\omega}$. While this vector and this couple may in turn be replaced by a single vector,† it is generally found more convenient to represent the system of momenta by the momentum vector $m\bar{\mathbf{v}}$ attached at G and the momentum couple $\bar{I}\boldsymbol{\omega}$ (Fig. 18.4*c*). The momentum vector is associated with the translation of the body with G and represents the *linear momentum* of the body. The momentum couple corresponds to the rotation of the body about G and its moment $\bar{I}\omega$ represents the *angular momentum H_G of the body about an axis through G*. Recalling the definition given in Sec. 14.13 for the angular momentum of a system of particles, we note that the angular momentum H_A of the body about an axis through an arbitrary point A may be obtained by adding algebraically the moment about A of the momentum vector $m\bar{\mathbf{v}}$ and the moment $\bar{I}\omega$ of the momentum couple.‡

†See Probs. 18.13 and 18.14.

‡Note that, in general, H_A is *not* equal to $I_A\omega$ (see Prob. 18.15).

18.4. Application of the Principle of Impulse and Momentum to the Analysis of the Plane Motion of a Rigid Body.

We saw in the preceding section that, in the most general case of the plane motion of a rigid body, the system of the momenta of the particles of the body reduces to a vector $m\overline{\mathbf{v}}$ attached at G and a couple $\bar{I}\boldsymbol{\omega}$. Replacing the system of momenta in parts a and c of Fig. 18.1 by the equivalent momentum vector and momentum couple, we obtain the three diagrams shown in Fig. 18.5. This figure expresses graphically, in the case of the plane motion

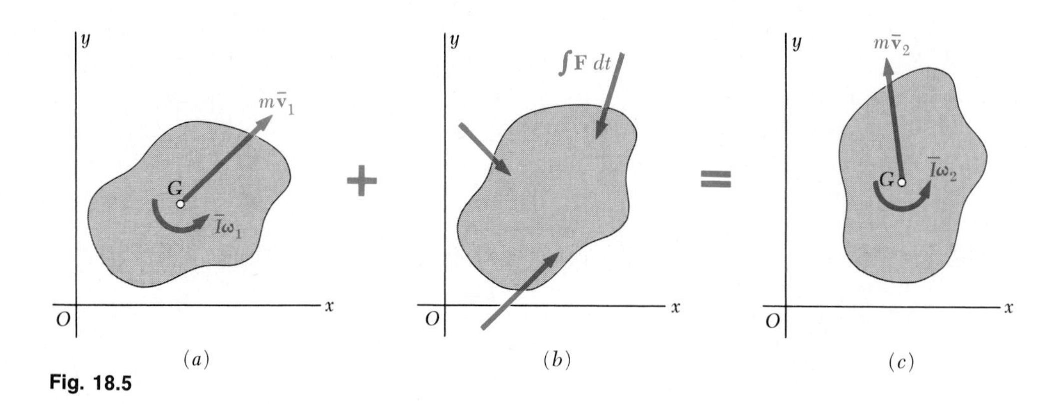

Fig. 18.5

of a rigid body symmetrical with respect to the reference plane, the fundamental relation.

$$\textbf{Syst Momenta}_1 + \textbf{Syst Ext Imp}_{1\to2} = \textbf{Syst Momenta}_2 \qquad (18.1)$$

Three equations of motion may be derived from Fig. 18.5. Two equations are obtained by summing and equating the x and y *components* of the momenta and impulses, and the third by summing and equating the *moments* of these vectors *about any given point.* The coordinate axes may be chosen fixed in space, or they may be allowed to move with the mass center of the body while maintaining a fixed direction. In either case, the point about which moments are taken should keep the same position relative to the coordinate axes during the interval of time considered.

In deriving the three equations of motion for a rigid body, care should be taken not to add indiscriminately linear and angular momenta. Confusion will be avoided if it is kept in mind that $m\bar{v}_x$ and $m\bar{v}_y$ represent the *components of a vector,* namely, the linear momentum vector $m\overline{\mathbf{v}}$, while $\bar{I}\omega$ represents the *moment of a couple,* namely, the momentum couple $\bar{I}\boldsymbol{\omega}$. Thus the angular momentum $\bar{I}\omega$ should be added only to the *moment* of the linear momentum $m\overline{\mathbf{v}}$, never to this vector itself nor to its components. All quantities involved will then be expressed in the same units, namely, $\mathrm{N \cdot m \cdot s}$ or $\mathrm{lb \cdot ft \cdot s}$.

As an example, we shall consider the motion of a wheel of radius r, mass m, and centroidal moment of inertia $\bar{I}$, which rolls down an incline without sliding. Assuming that the initial angular velocity ω_1 of the wheel is known, we propose to determine its angular velocity ω_2, after a time interval t, and the values of the components F and N of the reaction of the incline.

The system of momenta at times t_1 and t_2 are shown, respectively, in parts a and c of Fig. 18.6. They are represented in both cases by a vector

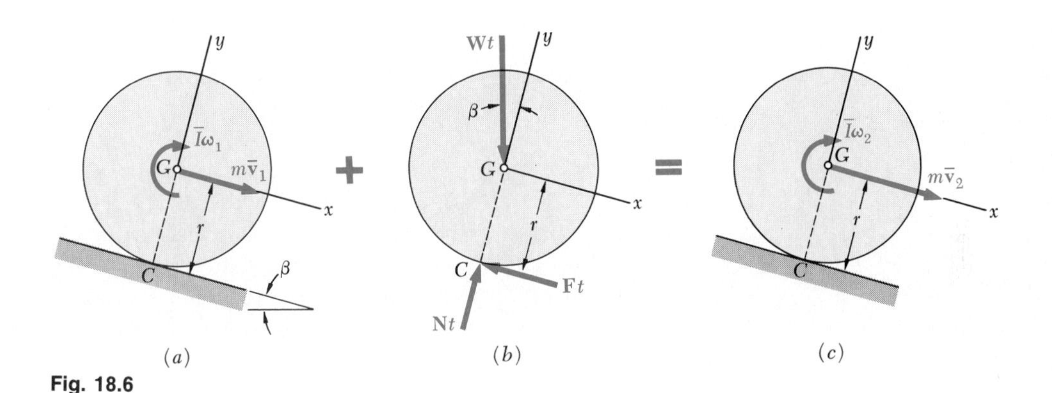

Fig. 18.6

$m\bar{\mathbf{v}}$ attached at the center G of the wheel and a couple $\bar{I}\boldsymbol{\omega}$. Choosing coordinate axes centered at G with the x and y axes parallel and perpendicular, respectively, to the incline, we note that the lines of action of the forces **W**, **F**, and **N** remain fixed with respect to the coordinate axes. Since these forces are also constant in magnitude, their impulses will be represented by the vectors $\mathbf{W}t$, $\mathbf{F}t$, and $\mathbf{N}t$ shown in part b of Fig. 18.6. Denoting by β the angle the incline forms with the horizontal, we add and equate successively the x components, y components, and moments about C (positive clockwise) of the momenta and impulses shown in Fig. 18.6. We write

$+\searrow x$ components:

$$m\bar{v}_1 - Ft + Wt \sin \beta = m\bar{v}_2 \tag{18.3}$$

$+\nearrow y$ components:

$$Nt - Wt \cos \beta = 0 \tag{18.4}$$

$+\downarrow$ moments about C:

$$\bar{I}\omega_1 + m\bar{v}_1 r + (Wt \sin \beta)r = \bar{I}\omega_2 + m\bar{v}_2 r \tag{18.5}$$

We note that all quantities involved in Eqs. (18.3) and (18.4) are expressed in the same units, namely, N · s or lb · s. Since, in Eq. (18.5), the magnitudes of the linear impulse $\mathbf{W}t \sin \beta$ and of the linear momenta $m\bar{\mathbf{v}}_1$ and $m\bar{\mathbf{v}}_2$ are multiplied by the radius r, all quantities involved in that equation are also expressed in the same units, namely, N · m · s or lb · ft · s.

Repeating the equations obtained on the preceding page,

$$m\bar{v}_1 - Ft + Wt \sin \beta = m\bar{v}_2 \tag{18.3}$$
$$Nt - Wt \cos \beta = 0 \tag{18.4}$$
$$\bar{I}\omega_1 + m\bar{v}_1 r + (Wt \sin \beta)r = \bar{I}\omega_2 + m\bar{v}_2 r \tag{18.5}$$

and observing that $\bar{v} = r\omega$, we write

$$mr\omega_1 - Ft + Wt \sin \beta = mr\omega_2 \tag{18.3'}$$
$$Nt - Wt \cos \beta = 0 \tag{18.4'}$$
$$(\bar{I} + mr^2)\omega_1 + (Wt \sin \beta)r = (\bar{I} + mr^2)\omega_2 \tag{18.5'}$$

Equation (18.5′) may be solved for ω_2 and Eq. (18.4′) for N; Eq. (18.3′) may be solved for F after substituting for ω_2.

Noncentroidal Rotation. In this particular case of plane motion, the magnitude of the velocity of the mass center of the body is $\bar{v} = \bar{r}\omega$, where $\bar{r}$ represents the distance from the mass center to the fixed axis of rotation and $\boldsymbol{\omega}$ the angular velocity of the body at the instant considered; the magnitude of the momentum vector attached at G is thus $m\bar{v} = m\bar{r}\omega$. Summing the moments about O of the momentum vector and momentum couple (Fig. 18.7) and using the parallel-axis theorem for moments of inertia, we find that the angular momentum of the body about O is

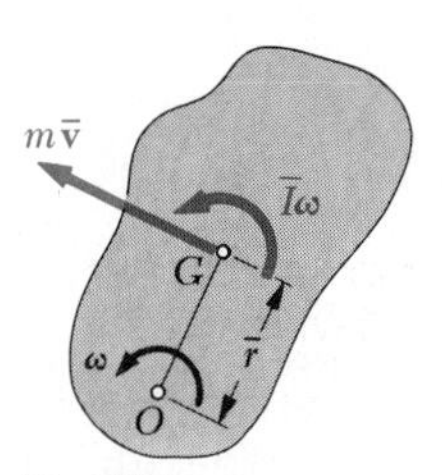

Fig. 18.7

$$\bar{I}\omega + (m\bar{r}\omega)\bar{r} = (\bar{I} + m\bar{r}^2)\omega = I_O\omega \tag{18.6}$$

Equating the moments about O of the momenta and impulses in (18.1), we write

$$I_O\omega_1 + \sum \int_{t_1}^{t_2} M_O\, dt = I_O\omega_2 \tag{18.7}$$

In the general case of plane motion of a rigid body symmetrical with respect to the reference plane, Eq. (18.7) may be used with respect to the instantaneous axis of rotation under certain conditions. It is recommended, however, that all problems of plane motion be solved by the general method described earlier in this section.

18.5. Systems of Rigid Bodies. The motion of several rigid bodies may be analyzed by applying the principle of impulse and momentum to each body separately (Sample Prob. 18.1).

However, in solving problems involving no more than three unknowns (including the impulses of unknown reactions), it is often found convenient to apply the principle of impulse and momentum to the system as a whole. The momentum and impulse diagrams are drawn for the entire system of bodies. The diagrams of momenta should include a momentum vector, a momentum couple, or both, for each moving part of the system. Impulses of forces internal to the system may be omitted from the impulse diagram since they occur in pairs of equal and opposite vectors. Summing and equating successively the x components, y components, and moments of all vectors involved, one obtains three relations which express that the momenta at time t_1 and the impulses of the external forces form a system equipollent to the system of the momenta at time t_2.† Again, care should be taken not to add indiscriminately linear and angular momenta; each equation should be checked to make sure that consistent units have been used. This approach has been used in Sample Prob. 18.3 and, further on, in Sample Probs. 18.4 and 18.5.

18.6. Conservation of Angular Momentum. When no external force acts on a rigid body or a system of rigid bodies, the impulses of the external forces are zero and the system of the momenta at time t_1 is equipollent to the system of the momenta at time t_2. Summing and equating successively the x components, y components, and moments of the momenta at times t_1 and t_2, we conclude that the total linear momentum of the system is conserved in any direction and that its total angular momentum is conserved about any point.

There are many engineering applications, however, in which *the linear momentum is not conserved* yet *the angular momentum* H_O of the system about a given point O is *conserved:*

$$(H_O)_1 = (H_O)_2 \tag{18.8}$$

Such cases occur when the lines of action of all external forces pass through O or, more generally, when the sum of the angular impulses of the external forces about O is zero.

Problems involving *conservation of angular momentum* about a point O may be solved by the general method of impulse and momentum, i.e., by drawing momentum and impulse diagrams as described in Secs. 18.4 and 18.5. Equation (18.8) is then obtained by summing and equating moments about O (Sample Prob. 18.3). As we shall see later in Sample Prob. 18.4, two additional equations may be written by summing and equating x and y components; these equations may be used to determine two unknown linear impulses, such as the impulses of the reaction components at a fixed point.

† Note that as in Sec. 16.4, we cannot speak of *equivalent* systems since we are not dealing with a single rigid body.

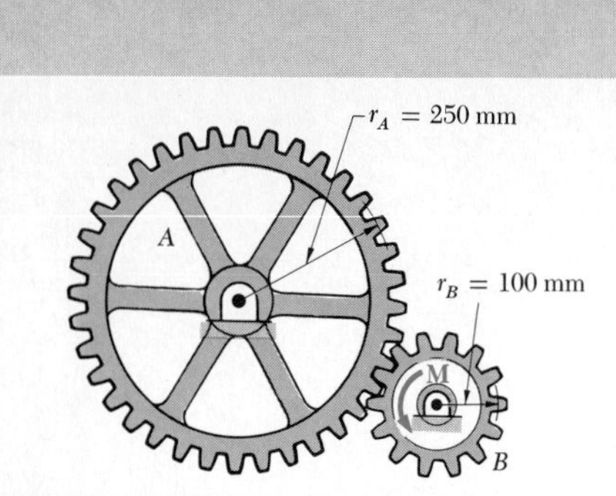

SAMPLE PROBLEM 18.1

Gear A has a mass of 10 kg and a radius of gyration of 200 mm, while gear B has a mass of 3 kg and a radius of gyration of 80 mm. The system is at rest when a couple **M** of moment 6 N·m is applied to gear B. Neglecting friction, determine (a) the time required for the angular velocity of gear B to reach 600 rpm, (b) the tangential force which gear B exerts on gear A. These gears have been previously considered in Sample Prob. 17.2.

Solution. We apply the principle of impulse and momentum to each gear separately. Since all forces and the couple are constant, their impulses are obtained by multiplying them by the unknown time t. We recall from Sample Prob. 17.2 that the centroidal moments of inertia and the final angular velocities are

$$\bar{I}_A = 0.400 \text{ kg·m}^2 \qquad \bar{I}_B = 0.0192 \text{ kg·m}^2$$
$$(\omega_A)_2 = 25.1 \text{ rad/s} \qquad (\omega_B)_2 = 62.8 \text{ rad/s}$$

Principle of Impulse and Momentum for Gear A. The systems of initial momenta, impulses, and final momenta are shown in three separate sketches.

Syst Momenta$_1$ + Syst Ext Imp$_{1\to2}$ = Syst Momenta$_2$

+↺moments about A: $\qquad 0 - Ftr_A = -\bar{I}_A(\omega_A)_2$

$$Ft(0.250 \text{ m}) = (0.400 \text{ kg·m}^2)(25.1 \text{ rad/s})$$
$$Ft = 40.2 \text{ N·s}$$

Principle of Impulse and Momentum for Gear B.

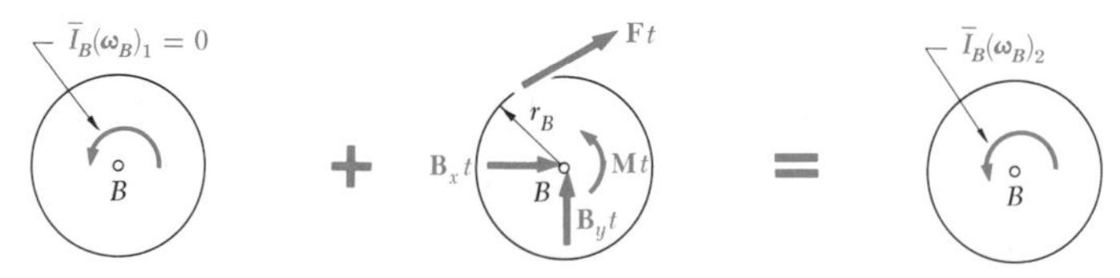

Syst Momenta$_1$ + Syst Ext Imp$_{1\to2}$ = Syst Momenta$_2$

+↺moments about B: $\qquad 0 + Mt - Ftr_B = \bar{I}_B(\omega_B)_2$

$$+(6 \text{ N·m})t - (40.2 \text{ N·s})(0.100 \text{ m}) = (0.0192 \text{ kg·m}^2)(62.8 \text{ rad/s})$$

$$t = 0.871 \text{ s} \quad ◀$$

Recalling that $Ft = 40.2$ N·s, we write

$$F(0.871 \text{ s}) = 40.2 \text{ N·s} \qquad F = +46.2 \text{ N}$$

Thus, the force exerted by gear B on gear A is $\qquad \mathbf{F} = 46.2 \text{ N} \swarrow \quad ◀$

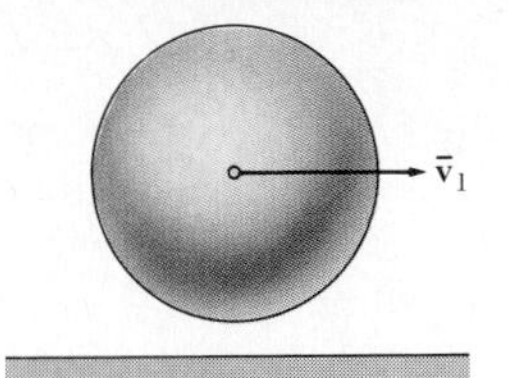

SAMPLE PROBLEM 18.2

A uniform sphere of mass m and radius r is projected along a rough horizontal surface with a linear velocity $\bar{\mathbf{v}}_1$ and no angular velocity. Denoting by μ_k the coefficient of kinetic friction between the sphere and the surface, determine (*a*) the time t_2 at which the sphere will start rolling without sliding, (*b*) the linear and angular velocities of the sphere at time t_2.

Solution. While the sphere is sliding relative to the surface, it is acted upon by the normal force **N**, the friction force **F**, and its weight **W** of magnitude $W = mg$.

Principle of Impulse and Momentum. We apply the principle of impulse and momentum to the sphere from the time $t_1 = 0$ when it is placed on the surface until the time $t_2 = t$ when it starts rolling without sliding.

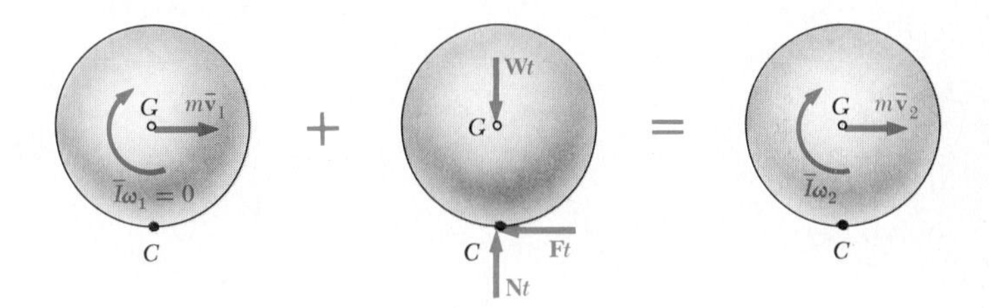

$$\textbf{Syst Momenta}_1 + \textbf{Syst Ext Imp}_{1\to 2} = \textbf{Syst Momenta}_2$$

$+\uparrow y$ components: $$Nt - Wt = 0 \tag{1}$$

$\xrightarrow{+} x$ components: $$m\bar{v}_1 - Ft = m\bar{v}_2 \tag{2}$$

$+\circlearrowright$ moments about G: $$Ftr = \bar{I}\omega_2 \tag{3}$$

From (1) we obtain $N = W = mg$. During the entire time interval considered, sliding occurs at point C and we have $F = \mu_k N = \mu_k mg$. Substituting for F into (2), we write

$$m\bar{v}_1 - \mu_k mgt = m\bar{v}_2 \qquad \bar{v}_2 = \bar{v}_1 - \mu_k gt \tag{4}$$

Substituting $F = \mu_k mg$ and $\bar{I} = \frac{2}{5}mr^2$ into (3),

$$\mu_k mgtr = \tfrac{2}{5}mr^2\omega_2 \qquad \omega_2 = \frac{5}{2}\frac{\mu_k g}{r}t \tag{5}$$

The sphere will start rolling without sliding when the velocity $\mathbf{v}_C$ of the point of contact is zero. At that time, point C becomes the instantaneous center of rotation, and we have $\bar{v}_2 = r\omega_2$. Substituting from (4) and (5), we write

$$\bar{v}_2 = r\omega_2 \qquad \bar{v}_1 - \mu_k gt = r\left(\frac{5}{2}\frac{\mu_k g}{r}t\right) \qquad t = \frac{2}{7}\frac{\bar{v}_1}{\mu_k g} \quad ◄$$

Substituting this expression for t into (5),

$$\omega_2 = \frac{5}{2}\frac{\mu_k g}{r}\left(\frac{2}{7}\frac{\bar{v}_1}{\mu_k g}\right) \qquad \omega_2 = \frac{5}{7}\frac{\bar{v}_1}{r} \qquad \omega_2 = \frac{5}{7}\frac{\bar{v}_1}{r} \circlearrowright \quad ◄$$

$$\bar{v}_2 = r\omega_2 \qquad \bar{v}_2 = r\left(\frac{5}{7}\frac{v_1}{r}\right) \qquad \bar{\mathbf{v}}_2 = \tfrac{5}{7}\bar{v}_1 \rightarrow \quad ◄$$

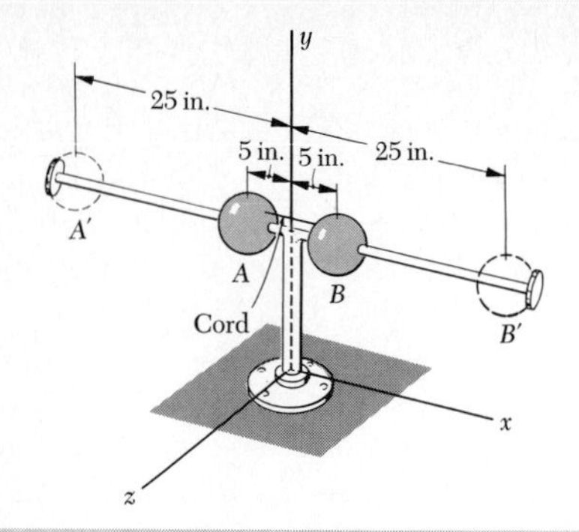

SAMPLE PROBLEM 18.3

Two solid spheres of radius 3 in., weighing 2 lb each, are mounted at A and B on the horizontal rod $A'B'$, which rotates freely about the vertical with a counterclockwise angular velocity of 6 rad/s. The spheres are held in position by a cord which is suddenly cut. Knowing that the centroidal moment of inertia of the rod and pivot is $\bar{I}_R = 0.25\ \text{lb}\cdot\text{ft}\cdot\text{s}^2$, determine (*a*) the angular velocity of the rod after the spheres have moved to positions A' and B', (*b*) the energy lost due to the plastic impact of the spheres and the stops at A' and B'.

***a.* Principle of Impulse and Momentum.** In order to determine the final angular velocity of the rod, we shall express that the initial momenta of the various parts of the system and the impulses of the external forces are together equipollent to the final momenta of the system.

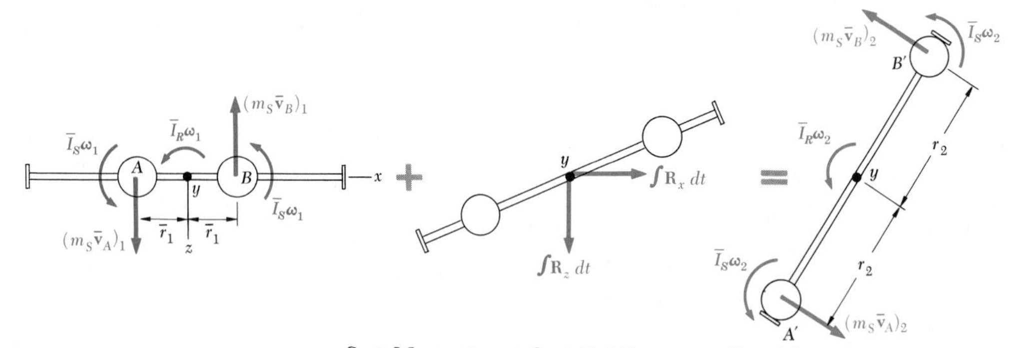

$$\textbf{Syst Momenta}_1 + \textbf{Syst Ext Imp}_{1\to2} = \textbf{Syst Momenta}_2$$

Observing that the external forces consist of the weights and the reaction at the pivot, which have no moment about the y axis, and noting that $\bar{v}_A = \bar{v}_B = \bar{r}\omega$, we write

$+\circlearrowleft$ moments about y axis:

$$2(m_S\bar{r}_1\omega_1)\bar{r}_1 + 2\bar{I}_S\omega_1 + \bar{I}_R\omega_1 = 2(m_S\bar{r}_2\omega_2)\bar{r}_2 + 2\bar{I}_S\omega_2 + \bar{I}_R\omega_2$$

$$(2m_S\bar{r}_1^2 + 2\bar{I}_S + \bar{I}_R)\omega_1 = (2m_S\bar{r}_2^2 + 2\bar{I}_S + \bar{I}_R)\omega_2 \qquad (1)$$

which expresses that *the angular momentum of the system about the y axis is conserved.* We now compute

$$\bar{I}_S = \tfrac{2}{5}m_Sa^2 = \frac{2}{5}\left(\frac{2\ \text{lb}}{32.2\ \text{ft/s}^2}\right)(\tfrac{3}{12}\ \text{ft})^2 = 0.00155\ \text{lb}\cdot\text{ft}\cdot\text{s}^2$$

$$m_S\bar{r}_1^2 = \frac{2}{32.2}\left(\frac{5}{12}\right)^2 = 0.0108 \qquad m_S\bar{r}_2^2 = \frac{2}{32.2}\left(\frac{25}{12}\right)^2 = 0.2696$$

Substituting these values and $\bar{I}_R = 0.25$, $\omega_1 = 6$ rad/s into (1):

$$0.275(6\ \text{rad/s}) = 0.792\omega_2 \qquad \omega_2 = 2.08\ \text{rad/s}\circlearrowleft \blacktriangleleft$$

***b.* Energy Lost.** The kinetic energy of the system at any instant is

$$T = 2(\tfrac{1}{2}m_S\bar{v}^2 + \tfrac{1}{2}\bar{I}_S\omega^2) + \tfrac{1}{2}\bar{I}_R\omega^2 = \tfrac{1}{2}(2m_S\bar{r}^2 + 2\bar{I}_S + \bar{I}_R)\omega^2$$

Recalling the numerical values found above, we have

$$T_1 = \tfrac{1}{2}(0.275)(6)^2 = 4.95\ \text{ft}\cdot\text{lb} \qquad T_2 = \tfrac{1}{2}(0.792)(2.08)^2 = 1.713\ \text{ft}\cdot\text{lb}$$

$$\Delta T = T_2 - T_1 = 1.71 - 4.95 \qquad \Delta T = -3.24\ \text{ft}\cdot\text{lb} \blacktriangleleft$$

Problems

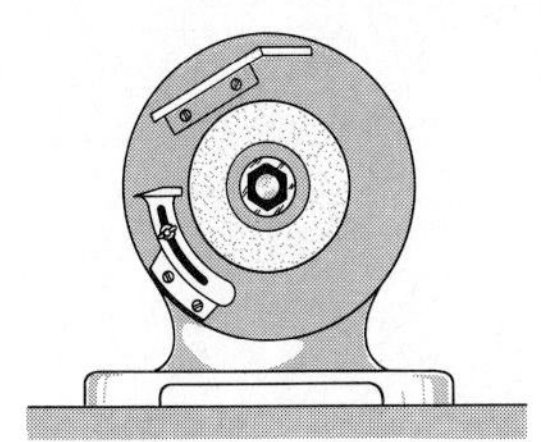
Fig. P18.1

18.1 A small grinding wheel is attached to the shaft of an electric motor which has a rated speed of 3600 rpm. When the power is turned off, the unit coasts to rest in 70 s. The grinding wheel and rotor have a combined weight of 6 lb and a combined radius of gyration of 2 in. Determine the average moment of the couple due to kinetic friction in the bearings of the motor.

18.2 A turbine-generator unit is shut off when its rotor is rotating at 3600 rpm; it is observed that the rotor coasts to rest in 7.10 min. Knowing that the 1850-kg rotor has a radius of gyration of 234 mm, determine the average moment of the couple due to bearing friction.

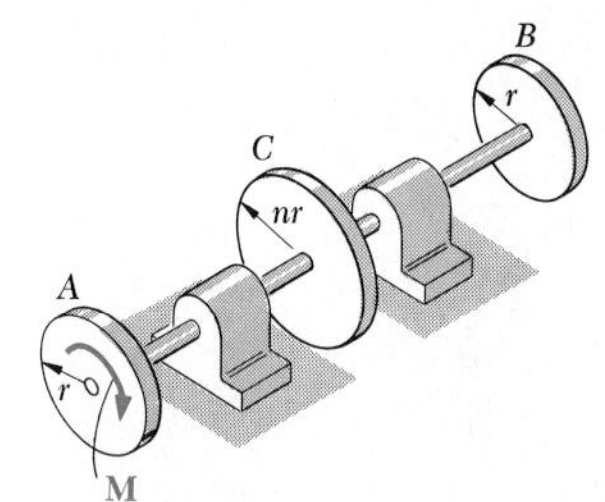

Fig. P18.3

18.3 Three disks of the same thickness and same material are attached to a shaft as shown. Disks A and B each have a mass of 8 kg and a radius $r = 240$ mm. A couple $\mathbf{M}$ of moment 40 N·m is applied to disk A when the system is at rest. Determine the radius nr of disk C if the angular velocity of the system is to be 900 rpm after 3 s.

18.4 A bolt located 2 in. from the center of an automobile wheel is tightened by applying the couple shown for 0.10 s. Assuming that the wheel is free to rotate and is initially at rest, determine the resulting angular velocity of the wheel. The wheel weighs 42 lb and has a radius of gyration of 11 in.

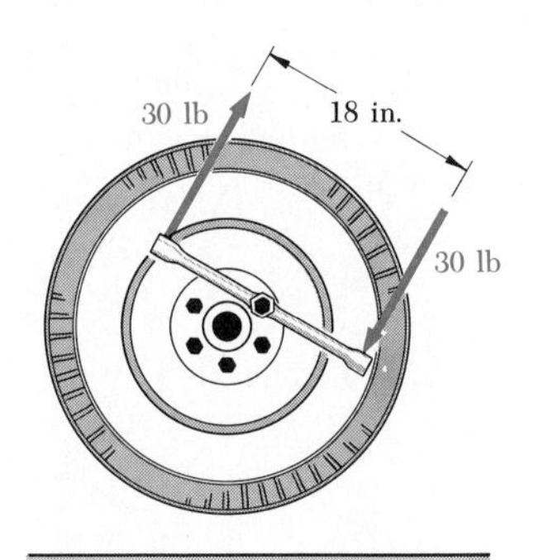

Fig. P18.4

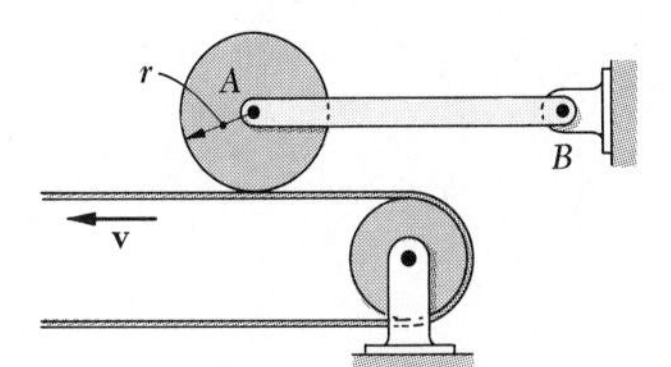

Fig. P18.5 and P18.6

18.5 A disk of constant thickness, initially at rest, is placed in contact with a belt which moves with a constant velocity $\mathbf{v}$. The link AB connecting the center of the disk to the support at B has a negligible weight. Denoting by μ_k the coefficient of kinetic friction between the disk and the belt, derive an expression for the time required for the disk to reach a constant angular velocity.

18.6 Disk A has a mass of 3 kg and a radius $r = 90$ mm; it is placed in contact with the belt, which moves with a constant speed $v = 18$ m/s. The link AB connecting the center of the disk to the support at B has a negligible weight. Knowing that $\mu_k = 0.20$ between the disk and the belt, determine the time required for the disk to reach a constant angular velocity.

18.7 Solve Prob. 18.6, assuming that the link AB has a mass of 750 g.

18.8 Each of the double pulleys shown has a centroidal mass moment of inertia of 0.10 kg·m², an inner radius of 80 mm, and an outer radius of 120 mm. Neglecting bearing friction, determine (*a*) the velocity of the cylinder 3 s after the system is released from rest, (*b*) the tension in the cord connecting the pulleys.

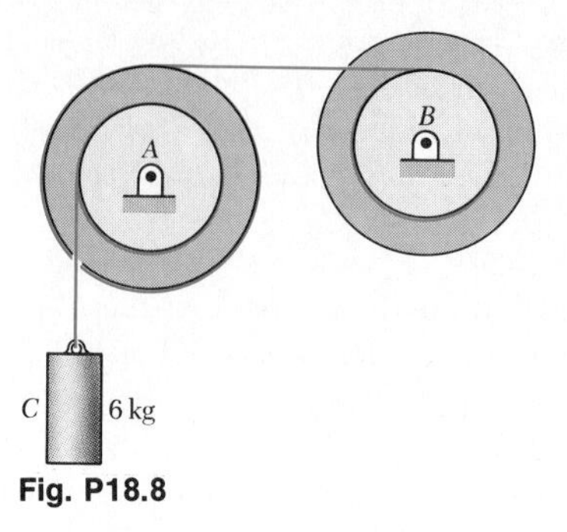

Fig. P18.8

18.9 Solve Prob. 18.8, assuming that the bearing friction at A and at B is equivalent to a couple of moment 0.25 N·m.

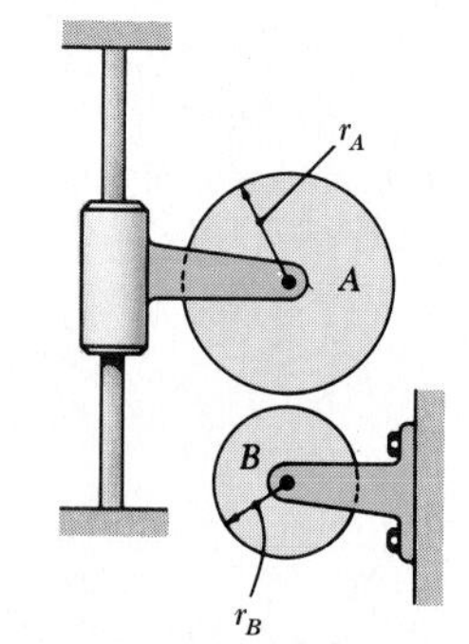

Fig. P18.10 and P18.11

18.10 Disk B has a mass $m_B = 3$ kg, a radius $r_B = 80$ mm, and an initial angular velocity $\omega_0 = 600$ rpm clockwise. Disk A has a mass $m_A = 5$ kg, a radius $r_A = 120$ mm, and is at rest when it is brought into contact with disk B. Neglecting friction in the bearings, determine (*a*) the final angular velocity of each disk, (*b*) the total impulse of the friction force which acted on disk A.

18.11 Disk A is at rest when it is brought into contact with disk B, which has an initial angular velocity $\boldsymbol{\omega}_0$. Show that the final angular velocity of disk B depends only on ω_0 and the ratio of the masses m_A and m_B of the two disks.

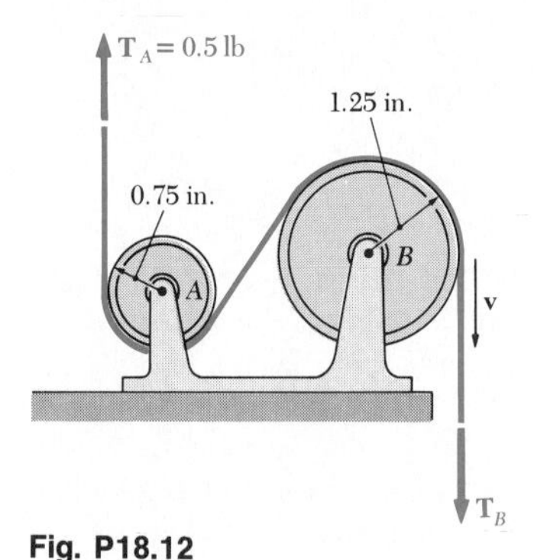

Fig. P18.12

18.12 A computer tape moves over the two drums shown. Drum A weighs 0.90 lb and has a radius of gyration of 0.70 in., while drum B weighs 1.8 lb and has a radius of gyration of 1.10 in. In the portion of tape above drum A the tension is constant and equal to $T_A = 0.5$ lb. Knowing that the tape is initially at rest, determine (*a*) the required constant tension T_B if the velocity of the tape is to be $v = 8$ ft/s after 0.2 s, (*b*) the corresponding tension in the portion of tape between the drums.

18.13 Show that the system of momenta for a rigid slab in plane motion reduces to a single vector, and express the distance from the mass center G to the line of action of this vector in terms of the centroidal radius of gyration $\bar{k}$ of the slab, the magnitude $\bar{v}$ of the velocity of G, and the angular velocity ω.

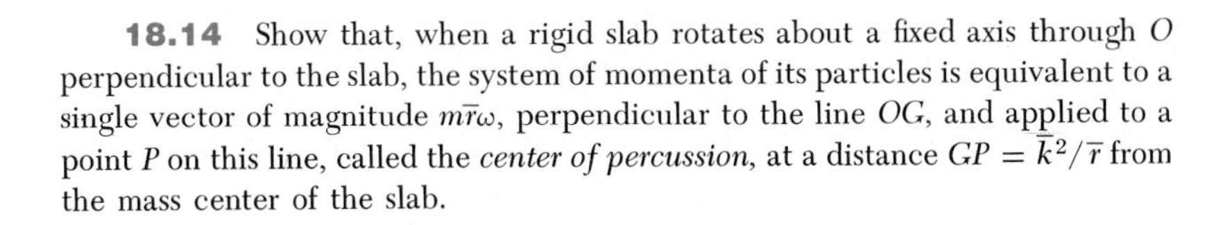

18.14 Show that, when a rigid slab rotates about a fixed axis through O perpendicular to the slab, the system of momenta of its particles is equivalent to a single vector of magnitude $m\bar{r}\omega$, perpendicular to the line OG, and applied to a point P on this line, called the *center of percussion*, at a distance $GP = \bar{k}^2/\bar{r}$ from the mass center of the slab.

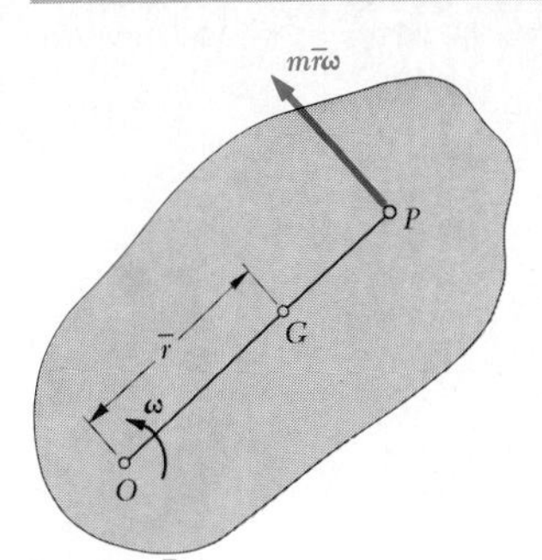

Fig. P18.14

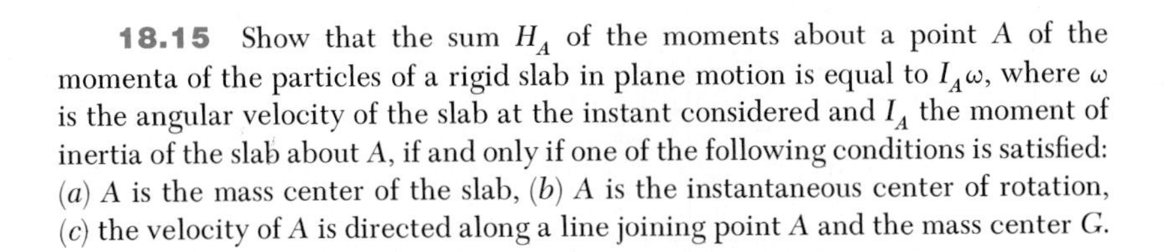

18.15 Show that the sum H_A of the moments about a point A of the momenta of the particles of a rigid slab in plane motion is equal to $I_A\omega$, where ω is the angular velocity of the slab at the instant considered and I_A the moment of inertia of the slab about A, if and only if one of the following conditions is satisfied: (*a*) A is the mass center of the slab, (*b*) A is the instantaneous center of rotation, (*c*) the velocity of A is directed along a line joining point A and the mass center G.

18.16 Consider a rigid slab initially at rest and subjected to an impulsive force **F** contained in the plane of the slab. We define the *center of percussion P* as the point of intersection of the line of action of **F** with the perpendicular drawn from G. (*a*) Show that the instantaneous center of rotation C of the slab is located on line GP at a distance $GC = \bar{k}^2/GP$ on the opposite side of G. (*b*) Show that if the center of percussion were located at C, the instantaneous center of rotation would be located at P.

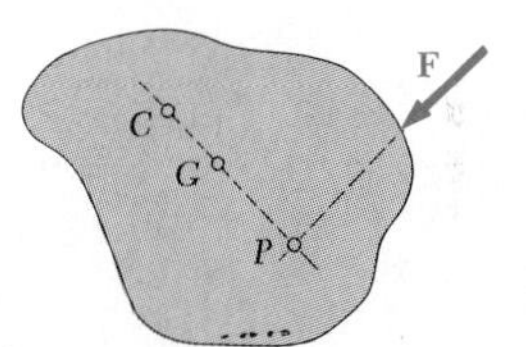

Fig. P18.16

18.17 A wheel of radius r and centroidal radius of gyration $\bar{k}$ is placed on an incline and released from rest at time $t = 0$. Assuming that the wheel rolls without slipping, determine (*a*) the velocity of the center at time t, (*b*) the coefficient of static friction required to prevent slipping.

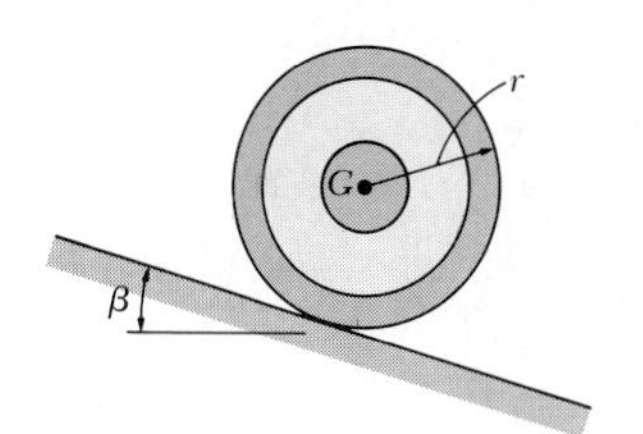

Fig. P18.17

18.18 A drum of 60-mm radius is attached to a disk of 120-mm radius. The disk and the drum have a total mass of 4 kg and a radius of gyration of 90 mm. A cord is wrapped around the drum and pulled with a force **P** of magnitude 15 N. Knowing that the disk is initially at rest, determine (*a*) the velocity of the center G after 2.5 s, (*b*) the friction force required to prevent slipping.

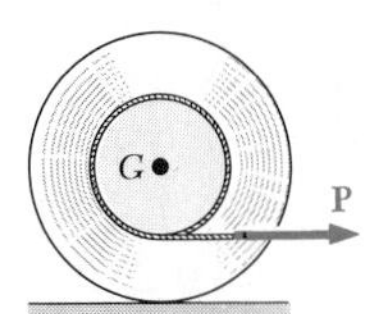

Fig. P18.18

18.19 Cords are wrapped around a thin-walled pipe and a solid cylinder as shown. Knowing that the pipe and the cylinder are each released from rest at time $t = 0$, determine at time t the velocity of the center of (*a*) the pipe, (*b*) the cylinder.

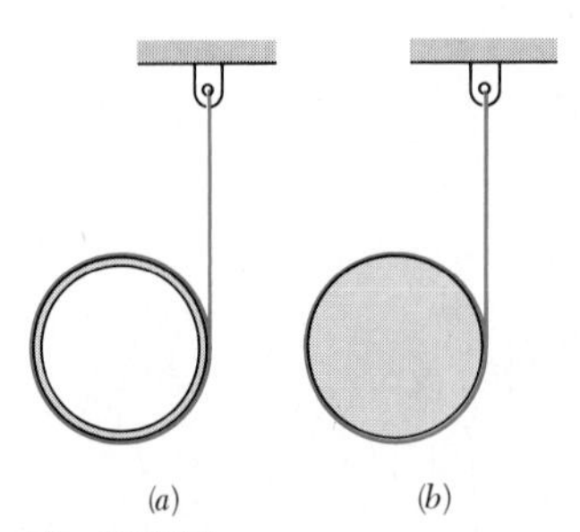

Fig. P18.19

18.20 The double pulley shown weighs 6 lb and has a centroidal radius of gyration of 3 in. The 2-in.-radius inner pulley is rigidly attached to the 4-in.-radius outer pulley, and the motion of the center A is guided by a smooth pin at A which slides in a vertical slot. When the pulley is at rest, a force **P** of magnitude 8 lb is applied to cord C. Determine the velocity of the center of the pulley after 5 s.

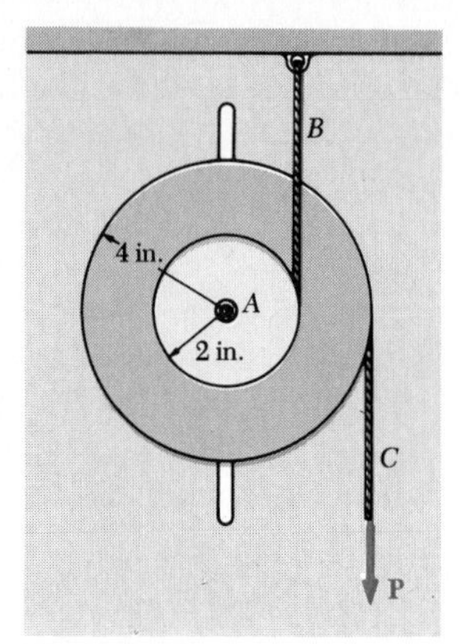

Fig. P18.20

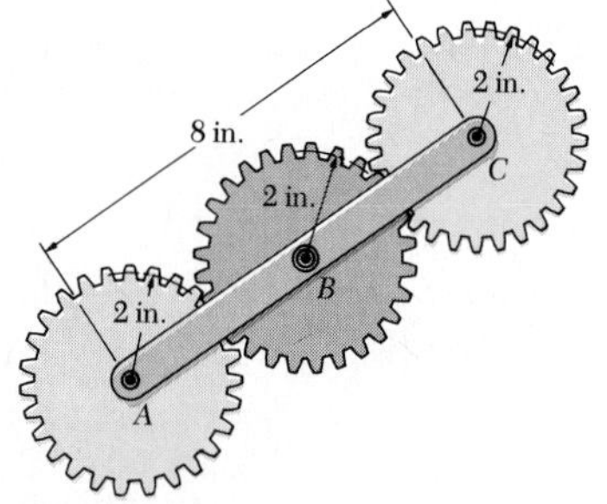

Fig. P18.22

18.21 Solve Prob. 18.20, assuming that the force **P** is replaced by an 8-lb weight attached to cord C.

18.22 In the gear arrangement shown, gears A and C are attached to the rod ABC, which is free to rotate about B, while the inner gear B is fixed. Knowing that the system is at rest, determine the moment of the couple **M** which must be applied to rod ABC, if 3 s later the angular velocity of the rod is to be 300 rpm clockwise. Gears A and C weigh 3 lb each and may be considered as disks of radius 2 in.; rod ABC weighs 5 lb.

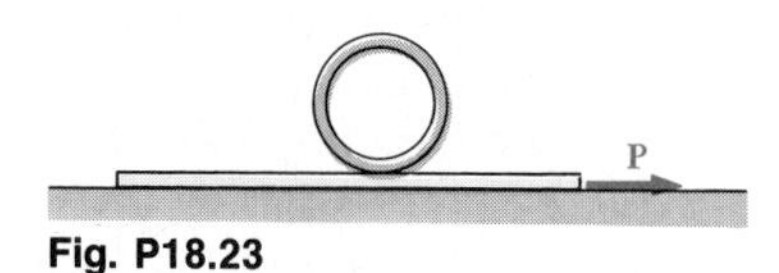

Fig. P18.23

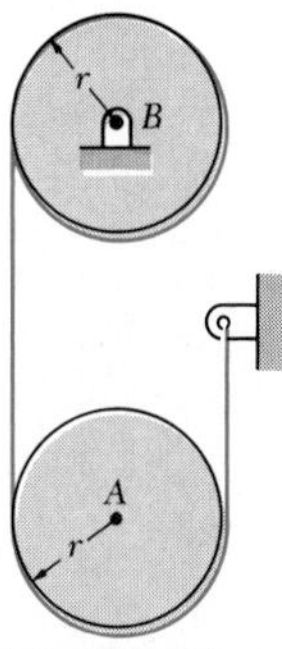

Fig. P18.24

18.23 A 240-mm-diameter pipe of mass 16 kg rests on a 4-kg plate. The pipe and plate are initially at rest when a force **P** of magnitude 80 N is applied for 0.50 s. Knowing that $\mu_s = 0.25$ and $\mu_k = 0.20$ between the plate and both the pipe and the floor, determine (*a*) whether the pipe slides with respect to the plate, (*b*) the resulting velocities of the pipe and of the plate.

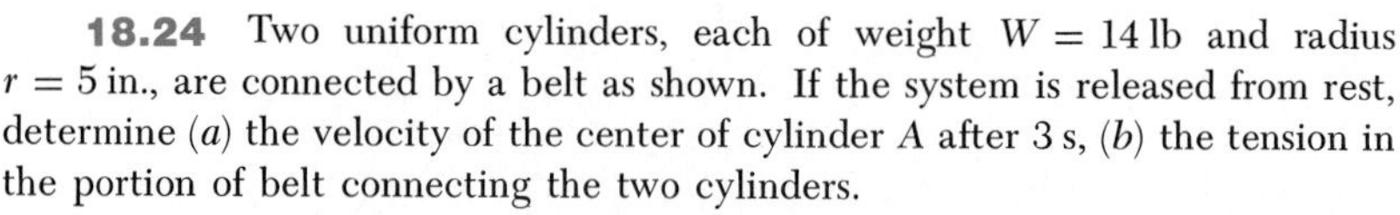

18.24 Two uniform cylinders, each of weight $W = 14$ lb and radius $r = 5$ in., are connected by a belt as shown. If the system is released from rest, determine (*a*) the velocity of the center of cylinder A after 3 s, (*b*) the tension in the portion of belt connecting the two cylinders.

18.25 and 18.26 The 6-kg carriage is supported as shown by two uniform disks, each having a mass of 4 kg and a radius of 75 mm. Knowing that the carriage is initially at rest, determine the velocity of the carriage 2.5 s after the 10-N force has been applied. Assume that the disks roll without sliding.

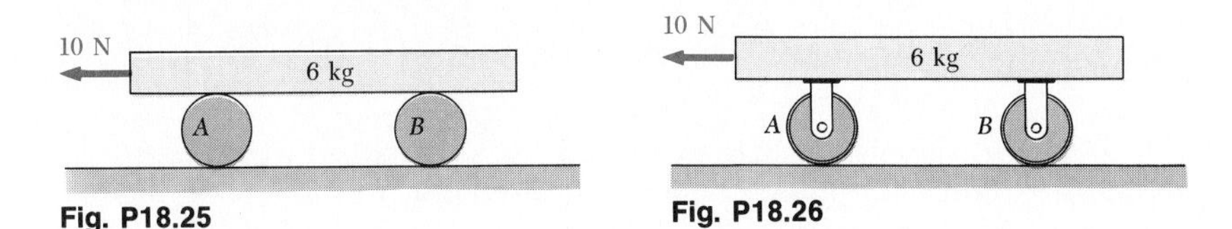

Fig. P18.25 **Fig. P18.26**

18.27 Solve Sample Prob. 18.2, assuming that the sphere is replaced by a uniform cylinder of mass m and radius r.

18.28 An 8-lb bar AB is attached by a pin at D to a 10-lb circular plate which may rotate freely about a vertical axis. Knowing that when the bar is vertical the angular velocity of the plate is 90 rpm, determine (a) the angular velocity of the plate after the bar has swung into a horizontal position and has come to rest against pin C, (b) the energy lost due to the plastic impact at C.

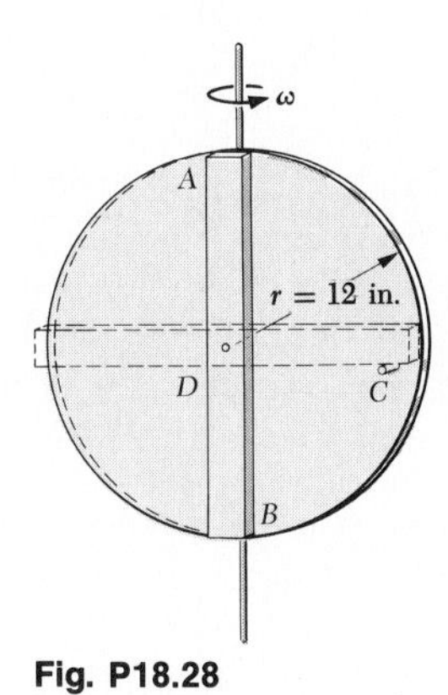

Fig. P18.28

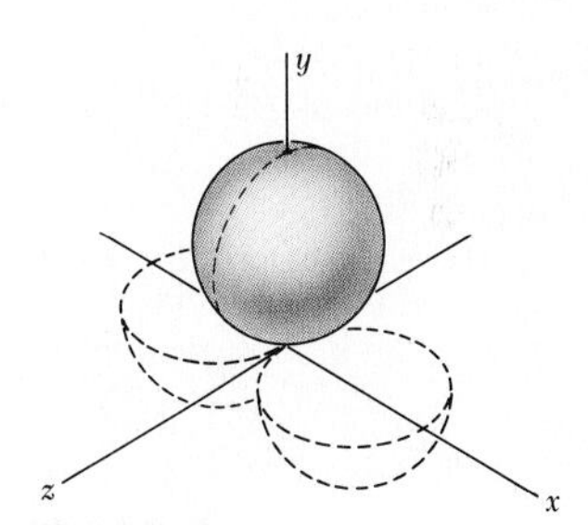

Fig. P18.29

18.29 A thin-walled spherical shell of mass m and radius r is used as a protective cover for a space shuttle experiment. In order to remove the cover, it is separated into two hemispherical shells which move to the final position shown. Knowing that during a test of the cover the initial angular velocity of the shell is ω_0 about the y axis, determine the angular velocity of the cover after it has reached its final position. (*Hint.* For a thin-walled shell, $I_{\text{diameter}} = \frac{2}{3}mr^2$.)

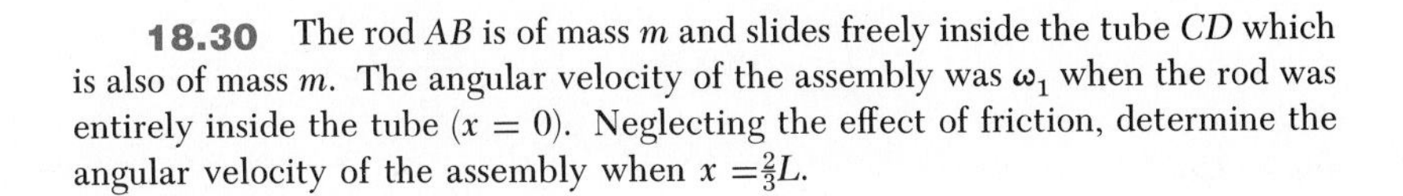

18.30 The rod AB is of mass m and slides freely inside the tube CD which is also of mass m. The angular velocity of the assembly was ω_1 when the rod was entirely inside the tube ($x = 0$). Neglecting the effect of friction, determine the angular velocity of the assembly when $x = \frac{2}{3}L$.

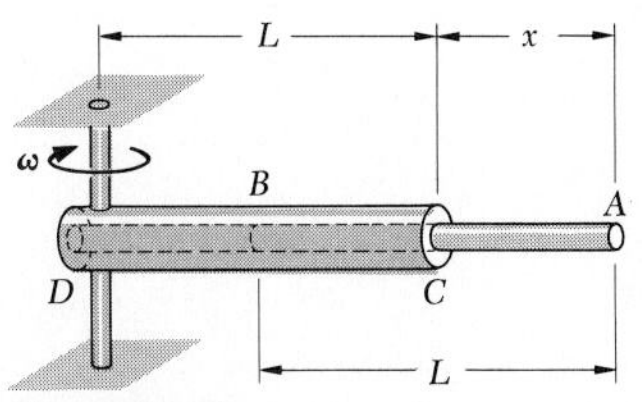

Fig. P18.30

18.31 A 1.5-kg tube AB may slide freely on rod DE which in turn may rotate freely in a horizontal plane. Initially the assembly is rotating with an angular velocity $\omega = 6$ rad/s and the tube is held in position by a cord. The moment of inertia of the rod and bracket about the vertical axis of rotation is $0.20 \text{ kg}\cdot\text{m}^2$ and the centroidal moment of inertia of the tube about a vertical axis is $0.0015 \text{ kg}\cdot\text{m}^2$. If the cord suddenly breaks, determine (*a*) the angular velocity of the assembly after the tube has moved to end E, (*b*) the energy lost during the plastic impact at E.

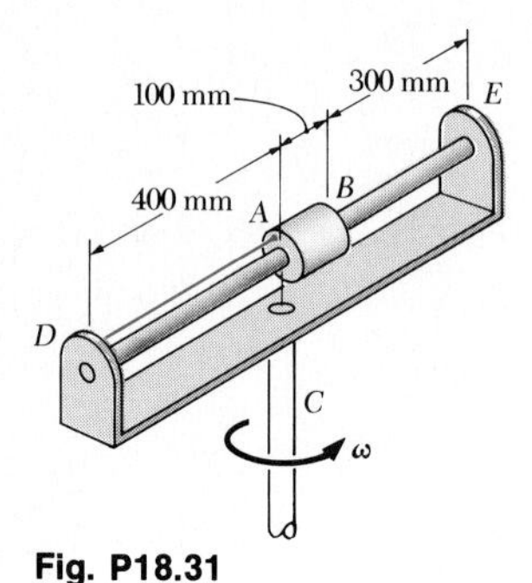

Fig. P18.31

18.32 Two square panels, each of side b, are attached with hinges to a rectangular plate and held by a wire in the position shown. The plate and the panels are made of the same material and have the same thickness. The entire assembly is rotating with an angular velocity ω_0 when the wire breaks. Determine the angular velocity of the assembly after the panels have come to rest in a horizontal position.

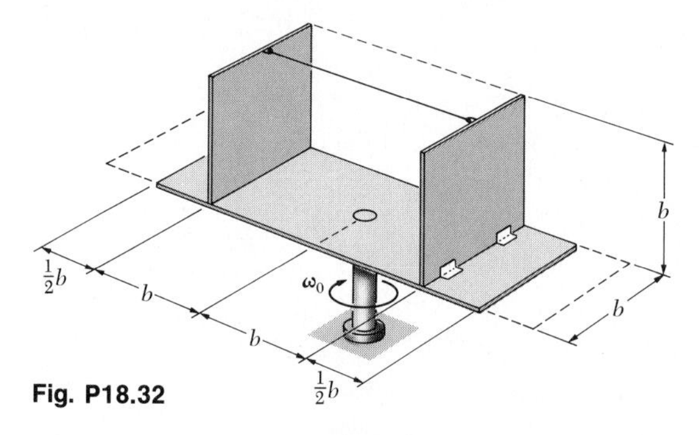

Fig. P18.32

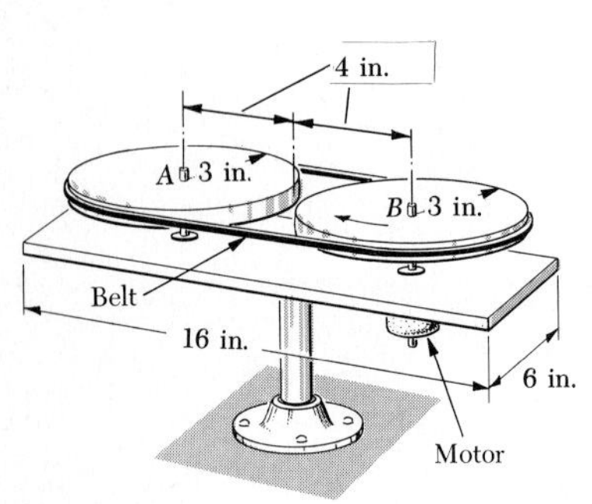

Fig. P18.33

18.33 Two 10-lb disks and a small motor are mounted on a 15-lb rectangular platform which is free to rotate about a central vertical spindle. The normal operating speed of the motor is 180 rpm. If the motor is started when the system is at rest, determine the angular velocity of all elements of the system after the motor has attained its normal operating speed. Neglect the mass of the motor and of the belt.

18.34 Solve Prob. 18.33, assuming that the belt is removed and that the bearing between disk A and the plate is (*a*) perfectly lubricated, (*b*) frozen.

18.35 In the helicopter shown, a vertical tail propeller is used to prevent rotation of the cab as the speed of the main blades is changed. Assuming that the tail propeller is not operating, determine the final angular velocity of the cab after the speed of the main blades has been changed from 200 to 300 rpm. The speed of the main blades is measured relative to the cab, which has a centroidal moment of inertia of $1200 \text{ kg}\cdot\text{m}^2$. Each of the four main blades is assumed to be a 5-m slender rod of mass 30 kg.

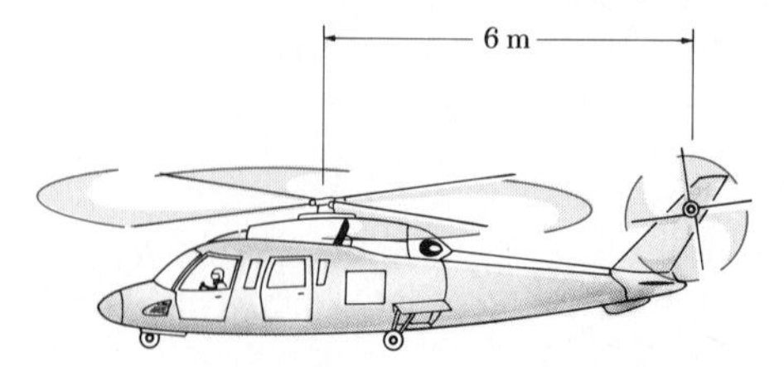

Fig. P18.35

18.36 Assuming that the tail propeller in Prob. 18.35 is operating and that the angular velocity of the cab remains zero, determine the final horizontal velocity of the cab when the speed of the main blades is changed from 200 to 300 rpm. The cab has a mass of 720 kg and is initially at rest. Also determine the force exerted by the tail propeller if this change in speed takes place uniformly in 12 s.

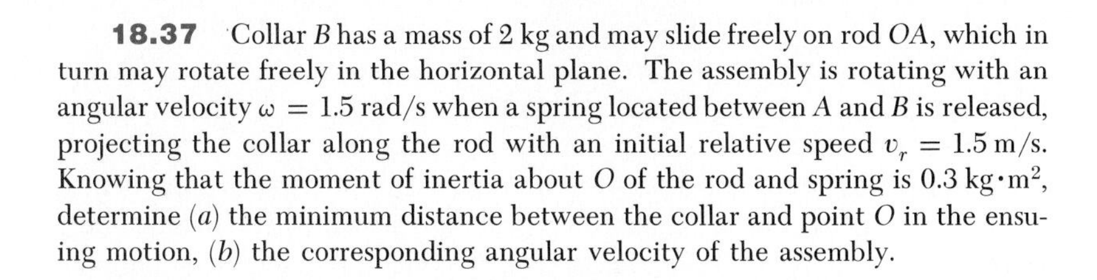

18.37 Collar B has a mass of 2 kg and may slide freely on rod OA, which in turn may rotate freely in the horizontal plane. The assembly is rotating with an angular velocity $\omega = 1.5$ rad/s when a spring located between A and B is released, projecting the collar along the rod with an initial relative speed $v_r = 1.5$ m/s. Knowing that the moment of inertia about O of the rod and spring is 0.3 kg·m^2, determine (a) the minimum distance between the collar and point O in the ensuing motion, (b) the corresponding angular velocity of the assembly.

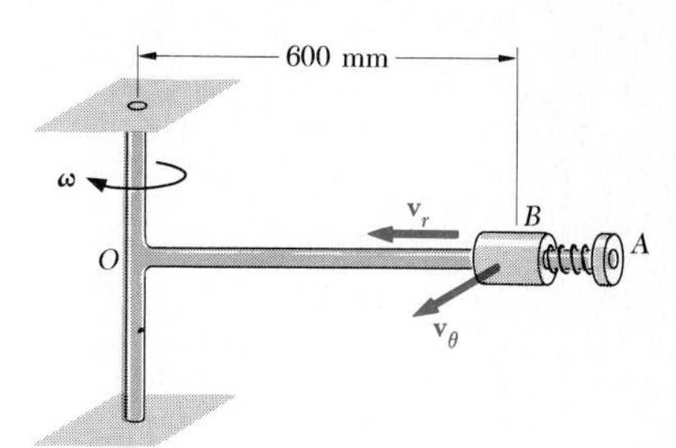

Fig. P18.37

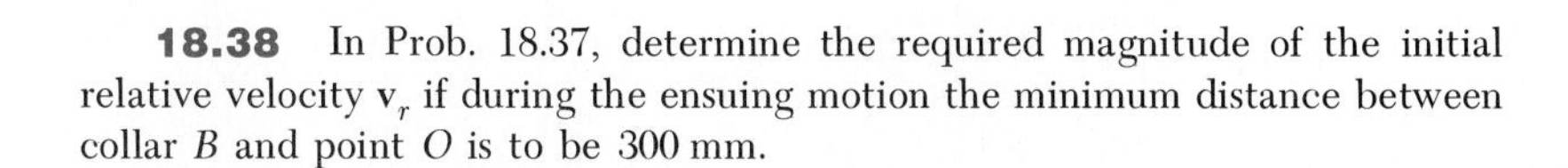

18.38 In Prob. 18.37, determine the required magnitude of the initial relative velocity $\mathbf{v}_r$ if during the ensuing motion the minimum distance between collar B and point O is to be 300 mm.

18.39 In Prob. 18.31, determine the velocity of the tube relative to the rod as the tube strikes end E of the assembly.

18.40 In Prob. 18.30, determine the velocity of the rod relative to the tube when $x = \frac{2}{3}L$.

18.41 In Prob. 18.30, determine the velocity of the rod relative to the tube when $x = \frac{1}{2}L$.

* **18.42** The uniform rod AB, of weight 20 lb and length 3 ft, is attached to the 30-lb cart C. Knowing that the system is released from rest in the position shown and neglecting friction, determine (a) the velocity of point B as the rod AB passes through a vertical position, (b) the corresponding velocity of the cart C.

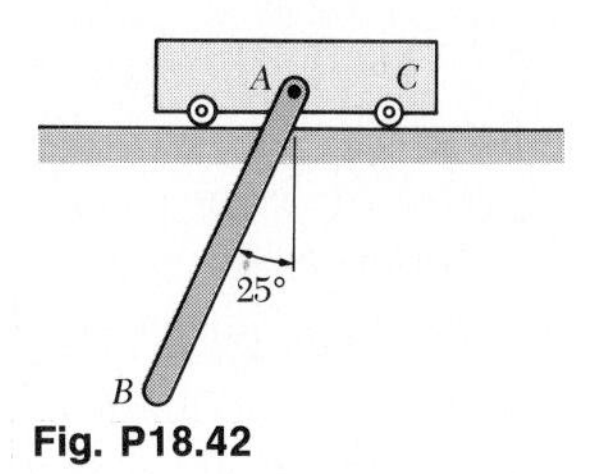

Fig. P18.42

* **18.43** The 3-kg cylinder B and the 2-kg wedge A are at rest in the position shown. The cord C connecting the cylinder and the wedge is then cut and the cylinder rolls without sliding on the wedge. Neglecting friction between the wedge and the ground, determine (a) the angular velocity of the cylinder after it has rolled 150 mm down the wedge, (b) the corresponding velocity of the wedge.

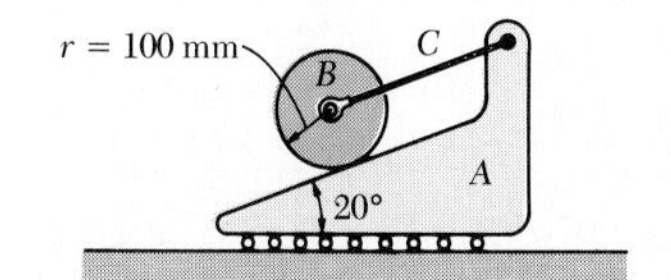

Fig. P18.43

18.7. Impulsive Motion. We saw in Chap. 14 that the method of impulse and momentum is the only practicable method for the solution of problems involving impulsive motion. Now we shall also find that compared with the various problems considered in the preceding sections, problems involving impulsive motion are particularly well adapted to a solution by the method of impulse and momentum. The computation of linear impulses and angular impulses is quite simple, since the time interval considered being very short, the bodies involved may be assumed to occupy the same position during that time interval.

18.8. Eccentric Impact. In Secs. 14.7 and 14.8, we learned to solve problems of *central impact,* i.e., problems in which the mass centers of the two colliding bodies are located on the line of impact. We shall now analyze the *eccentric impact* of two rigid bodies. Consider two bodies which collide, and denote by $\mathbf{v}_A$ and $\mathbf{v}_B$ the velocities before impact *of the two points of contact* A and B (Fig. 18.8a). Under the impact, the two

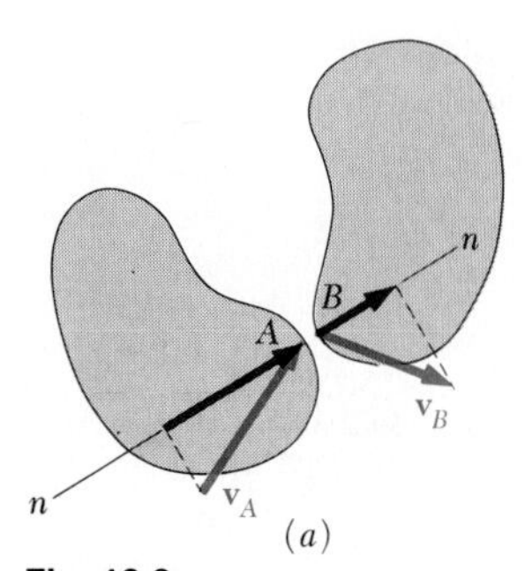

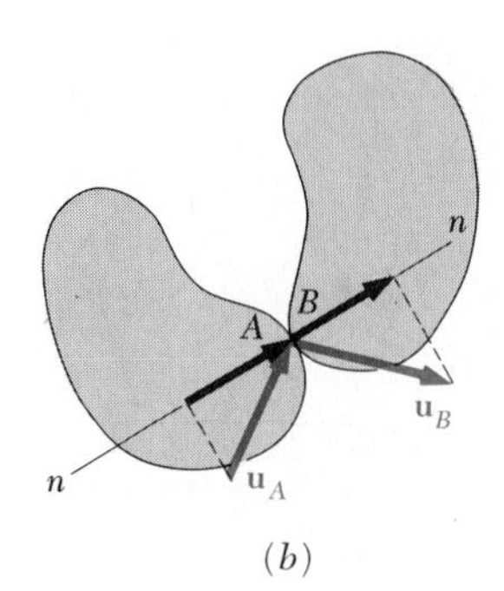

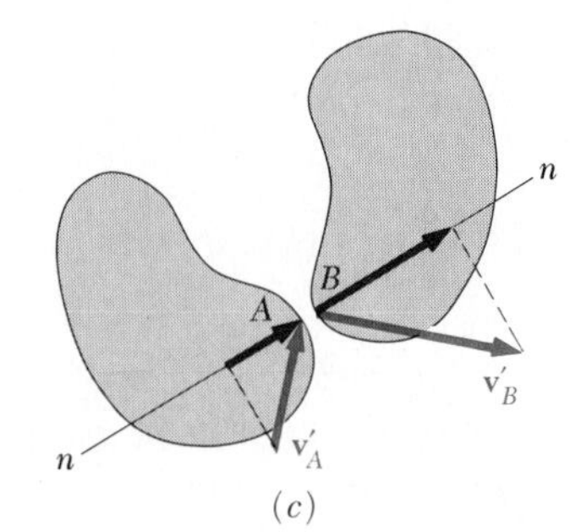

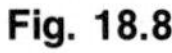
Fig. 18.8

bodies will *deform,* and at the end of the period of deformation, the velocities $\mathbf{u}_A$ and $\mathbf{u}_B$ of A and B will have equal components along the line of impact nn (Fig. 18.8b). A period of *restitution* will then take place, at the end of which A and B will have velocities $\mathbf{v}'_A$ and $\mathbf{v}'_B$ (Fig. 18.8c). Assuming that the bodies are frictionless, we find that the forces they exert on each other are directed along the line of impact. Denoting, respectively, by $\int P\,dt$ and $\int R\,dt$ the magnitude of the impulse of one of these forces during the period of deformation and during the period of restitution, we recall that the coefficient of restitution e is defined as the ratio

$$e = \frac{\int R\,dt}{\int P\,dt} \tag{18.9}$$

We propose to show that the relation established in Sec. 14.7 between the relative velocities of two particles before and after impact also holds between the components along the line of impact of the relative velocities of the two points of contact A and B. We propose to show, therefore, that

$$(v'_B)_n - (v'_A)_n = e[(v_A)_n - (v_B)_n] \tag{18.10}$$

We shall first assume that the motion of each of the two colliding bodies of Fig. 18.8 is unconstrained. Thus the only impulsive forces exerted on the bodies during the impact are applied at A and B, respectively. Consider the body to which point A belongs and draw the three momentum and impulse diagrams corresponding to the period of deformation (Fig. 18.9). We denote by $\bar{\mathbf{v}}$ and $\bar{\mathbf{u}}$, respectively, the velocity of the mass center

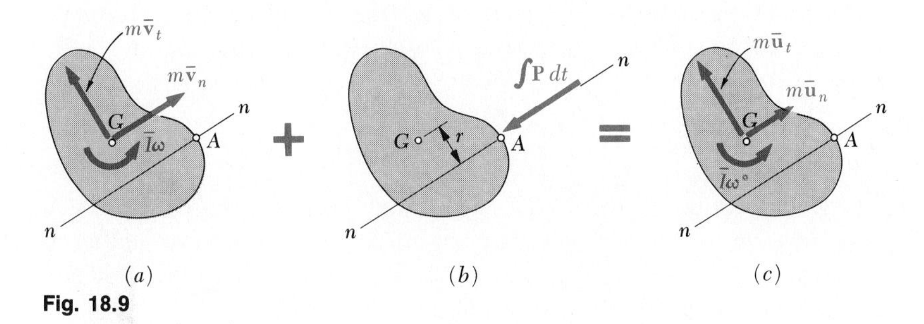

Fig. 18.9

at the beginning and at the end of the period of deformation, and by $\boldsymbol{\omega}$ and $\boldsymbol{\omega}^*$ the angular velocity of the body at the same instants. Summing and equating the components of the momenta and impulses along the line of impact nn, we write

$$m\bar{v}_n - \int P\,dt = m\bar{u}_n \tag{18.11}$$

Summing and equating the moments about G of the momenta and impulses, we also write

$$\bar{I}\omega - r\int P\,dt = \bar{I}\omega^* \tag{18.12}$$

where r represents the perpendicular distance from G to the line of impact. Considering now the period of restitution, we obtain in a similar way

$$m\bar{u}_n - \int R\,dt = m\bar{v}'_n \tag{18.13}$$

$$\bar{I}\omega^* - r\int R\,dt = \bar{I}\omega' \tag{18.14}$$

where $\bar{\mathbf{v}}'$ and $\boldsymbol{\omega}'$ represent, respectively, the velocity of the mass center and the angular velocity of the body after impact. Solving (18.11) and (18.13) for the two impulses and substituting into (18.9), and then solving (18.12) and (18.14) for the same two impulses and substituting again into (18.9), we obtain the following two alternative expressions for the coefficient of restitution:

$$e = \frac{\bar{u}_n - \bar{v}'_n}{\bar{v}_n - \bar{u}_n} \qquad e = \frac{\omega^* - \omega'}{\omega - \omega^*} \tag{18.15}$$

Multiplying by r the numerator and denominator of the second expression obtained for e, and adding respectively to the numerator and denominator of the first expression, we have

$$e = \frac{\bar{u}_n + r\omega^* - (\bar{v}'_n + r\omega')}{\bar{v}_n + r\omega - (\bar{u}_n + r\omega^*)} \tag{18.16}$$

Observing that $\bar{v}_n + r\omega$ represents the component $(v_A)_n$ along nn of the velocity of the point of contact A and that, similarly, $\bar{u}_n + r\omega^*$ and $\bar{v}'_n + r\omega'$ represent, respectively, the components $(u_A)_n$ and $(v'_A)_n$, we write

$$e = \frac{(u_A)_n - (v'_A)_n}{(v_A)_n - (u_A)_n} \tag{18.17}$$

The analysis of the motion of the second body leads to a similar expression for e in terms of the components along nn of the successive velocities of point B. Recalling that $(u_A)_n = (u_B)_n$, and eliminating these two velocity components by a manipulation similar to the one used in Sec. 14.7, we obtain relation (18.10).

If one or both of the colliding bodies is constrained to rotate about a fixed point O, as in the case of a compound pendulum (Fig. 18.10*a*), an impulsive reaction will be exerted at O (Fig. 18.10*b*). We shall verify that while their derivation must be modified, Eqs. (18.10) and (18.17) remain valid. Applying formula (18.7) to the period of deformation and to the period of restitution, we write

$$I_O\omega - r\int P\,dt = I_O\omega^* \tag{18.18}$$

$$I_O\omega^* - r\int R\,dt = I_O\omega' \tag{18.19}$$

where r represents the perpendicular distance from the fixed point O to the line of impact. Solving (18.18) and (18.19) for the two impulses and substituting into (18.9), and then observing that $r\omega$, $r\omega^*$, and $r\omega'$ represent the components along nn of the successive velocities of point A, we write

$$e = \frac{\omega^* - \omega'}{\omega - \omega^*} = \frac{r\omega^* - r\omega'}{r\omega - r\omega^*} = \frac{(u_A)_n - (v'_A)_n}{(v_A)_n - (u_A)_n}$$

and check that Eq. (18.17) still holds. Thus Eq. (18.10) remains valid when one or both of the colliding bodies is constrained to rotate about a fixed point O.

In order to determine the velocities of the two colliding bodies after impact, relation (18.10) should be used in conjunction with one or several other equations obtained by applying the principle of impulse and momentum (Sample Prob. 18.5).

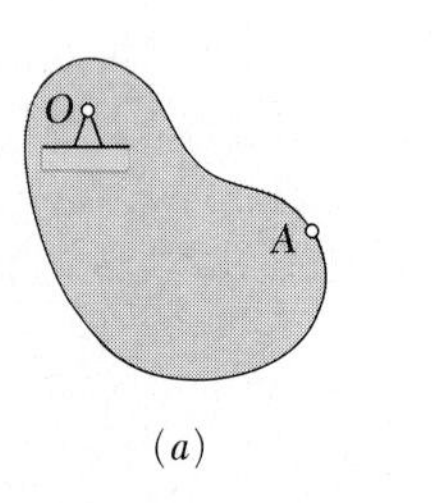

(*a*)

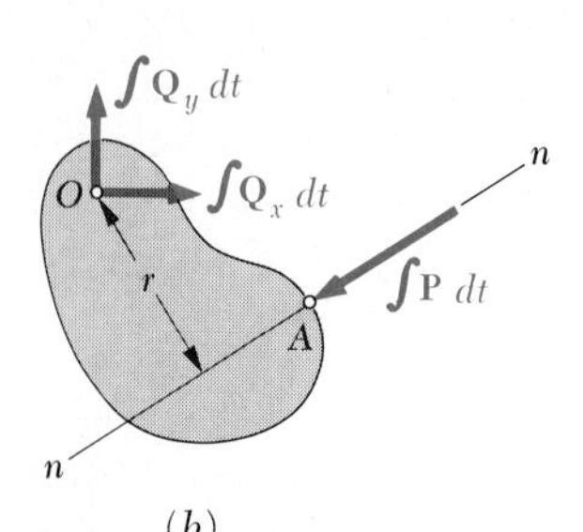

(*b*)

Fig. 18.10

SAMPLE PROBLEM 18.4

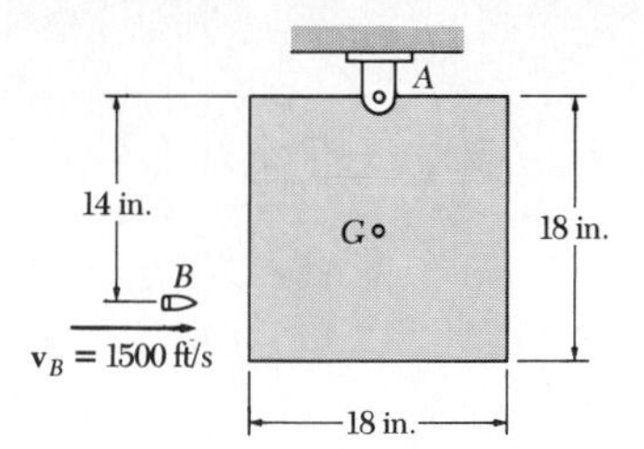

A 0.05-lb bullet B is fired with a horizontal velocity of 1500 ft/s into the side of a 20-lb square panel suspended from a hinge at A. Knowing that the panel is initially at rest, determine (*a*) the angular velocity of the panel immediately after the bullet becomes embedded, (*b*) the impulsive reaction at A, assuming that the bullet becomes embedded in 0.0006 s.

Solution. ***Principle of Impulse and Momentum.*** We consider the bullet and the panel as a single system and express that the initial momenta of the bullet and panel and the impulses of the external forces are together equipollent to the final momenta of the system. Since the time interval $\Delta t = 0.0006$ s is very short, we neglect all nonimpulsive forces and consider only the external impulses $\mathbf{A}_x \, \Delta t$ and $\mathbf{A}_y \, \Delta t$.

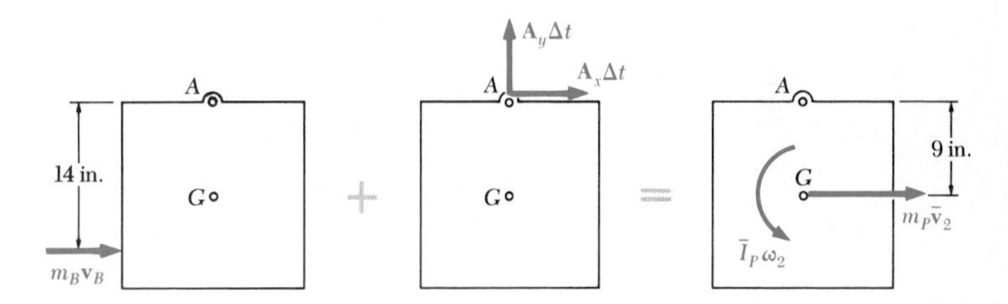

$$\textbf{Syst Momenta}_1 + \textbf{Syst Ext Imp}_{1\to2} = \textbf{Syst Momenta}_2$$

$+\circlearrowleft$ moments about A: $\quad m_B v_B(\frac{14}{12}\text{ ft}) + 0 = m_P \bar{v}_2(\frac{9}{12}\text{ ft}) + \bar{I}_P \omega_2 \quad (1)$

$\xrightarrow{+}$ x components: $\quad m_B v_B + A_x \, \Delta t = m_P \bar{v}_2 \quad (2)$

$+\uparrow y$ components: $\quad 0 + A_y \, \Delta t = 0 \quad (3)$

The centroidal mass moment of inertia of the square panel is

$$\bar{I}_P = \tfrac{1}{6} m_P b^2 = \frac{1}{6}\left(\frac{20\text{ lb}}{32.2}\right)(\tfrac{18}{12}\text{ ft})^2 = 0.2329\text{ lb}\cdot\text{ft}\cdot\text{s}^2$$

Substituting this value as well as the given data into (1) and noting that

$$\bar{v}_2 = (\tfrac{9}{12}\text{ ft})\omega_2$$

we write

$$\left(\frac{0.05}{32.2}\right)(1500)(\tfrac{14}{12}) = 0.2329\omega_2 + \left(\frac{20}{32.2}\right)(\tfrac{9}{12}\omega_2)(\tfrac{9}{12})$$

$$\omega_2 = 4.67\text{ rad/s} \qquad \omega_2 = 4.67\text{ rad/s}\circlearrowleft \blacktriangleleft$$

$$\bar{v}_2 = (\tfrac{9}{12}\text{ ft})\omega_2 = (\tfrac{9}{12}\text{ ft})(4.67\text{ rad/s}) = 3.50\text{ ft/s}$$

Substituting $\bar{v}_2 = 3.50$ ft/s and $\Delta t = 0.0006$ s into Eq. (2), we have

$$\left(\frac{0.05}{32.2}\right)(1500) + A_x(0.0006) = \left(\frac{20}{32.2}\right)(3.50)$$

$$A_x = -259\text{ lb} \qquad \mathbf{A}_x = 259\text{ lb} \leftarrow \blacktriangleleft$$

From Eq. (3), we find $\quad A_y = 0 \qquad \mathbf{A}_y = 0 \blacktriangleleft$

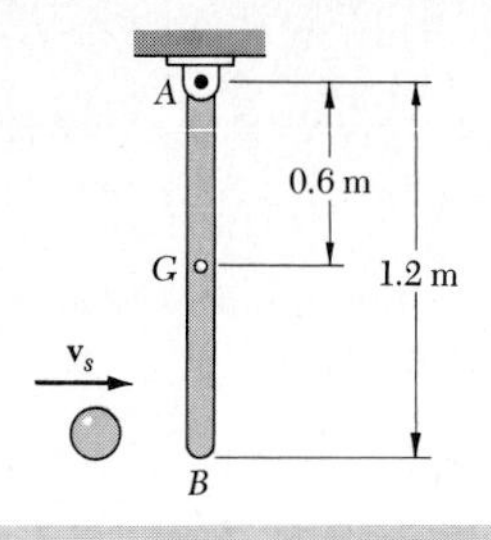

SAMPLE PROBLEM 18.5

A 2-kg sphere moving horizontally to the right with an initial velocity of 5 m/s strikes the lower end of an 8-kg rigid rod AB. The rod is suspended from a hinge at A and is initially at rest. Knowing that the coefficient of restitution between the rod and sphere is 0.80, determine the angular velocity of the rod and the velocity of the sphere immediately after the impact.

Principle of Impulse and Momentum. We consider the rod and sphere as a single system and express that the initial momenta of the rod and sphere and the impulses of the external forces are together equipollent to the final momenta of the system. We note that the only impulsive force external to the system is the impulsive reaction at A.

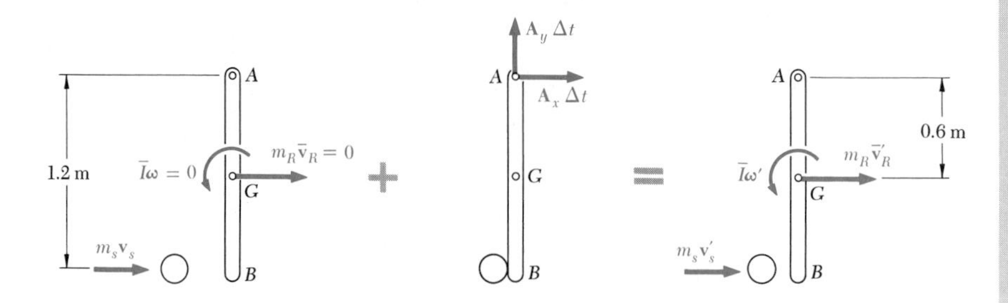

Syst Momenta$_1$ + Syst Ext Imp$_{1\to2}$ = Syst Momenta$_2$

$+\circlearrowleft$ moments about A:

$$m_s v_s(1.2 \text{ m}) = m_s v_s'(1.2 \text{ m}) + m_R \bar{v}_R'(0.6 \text{ m}) + \bar{I}\omega' \quad (1)$$

Since the rod rotates about A, we have $\bar{v}_R' = \bar{r}\omega' = (0.6 \text{ m})\omega'$. Also,

$$\bar{I} = \tfrac{1}{12}mL^2 = \tfrac{1}{12}(8 \text{ kg})(1.2 \text{ m})^2 = 0.96 \text{ kg}\cdot\text{m}^2$$

Substituting these values and the given data into Eq. (1), we have

$$(2 \text{ kg})(5 \text{ m/s})(1.2 \text{ m}) = (2 \text{ kg})v_s'(1.2 \text{ m}) + (8 \text{ kg})(0.6 \text{ m})\omega'(0.6 \text{ m}) + (0.96 \text{ kg}\cdot\text{m}^2)\omega'$$

$$12 = 2.4v_s' + 3.84\omega' \quad (2)$$

Relative Velocities. Choosing positive to the right, we write

$$v_B' - v_s' = e(v_s - v_B)$$

Substituting $v_s = 5$ m/s, $v_B = 0$, and $e = 0.80$, we obtain

$$v_B' - v_s' = 0.80(5 \text{ m/s}) \quad (3)$$

Again noting that the rod rotates about A, we write

$$v_B' = (1.2 \text{ m})\omega' \quad (4)$$

Solving Eqs. (2) to (4) simultaneously, we obtain

$$\omega' = 3.21 \text{ rad/s} \qquad \omega' = 3.21 \text{ rad/s}\circlearrowleft$$

$$v_s' = -0.143 \text{ m/s} \qquad \mathbf{v}_s' = 0.143 \text{ m/s} \leftarrow$$

SAMPLE PROBLEM 18.6

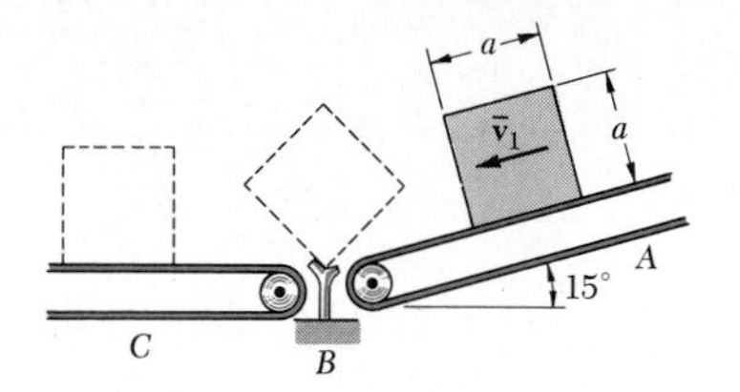

A square package of side a and mass m moves down a conveyor belt A with a constant velocity $\overline{\mathbf{v}}_1$. At the end of the conveyor belt, the corner of the package strikes a rigid support at B. Assuming that the impact at B is perfectly plastic, derive an expression for the smallest magnitude of the velocity $\overline{\mathbf{v}}_1$ for which the package will rotate about B and reach conveyor belt C.

Principle of Impulse and Momentum. Since the impact between the package and the support is perfectly plastic, the package rotates about B during the impact. We apply the principle of impulse and momentum to the package and note that the only impulsive force external to the package is the impulsive reaction at B.

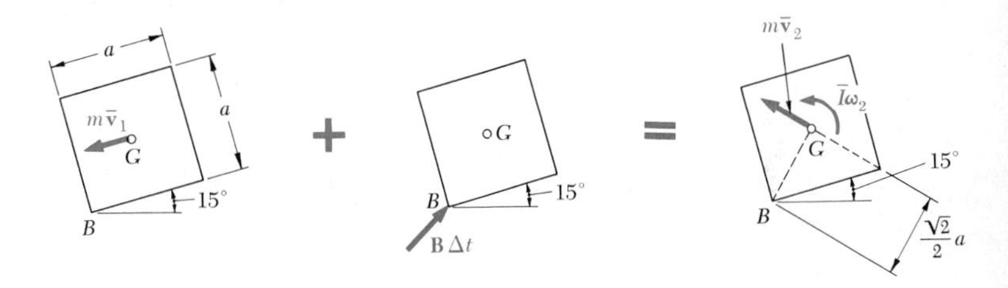

$$\textbf{Syst Momenta}_1 + \textbf{Syst Ext Imp}_{1\to 2} = \textbf{Syst Momenta}_2$$

$+\circlearrowleft$ moments about B: $\quad (m\bar{v}_1)(\tfrac{1}{2}a) + 0 = (m\bar{v}_2)(\tfrac{1}{2}\sqrt{2}a) + \bar{I}\omega_2 \qquad (1)$

Position 2

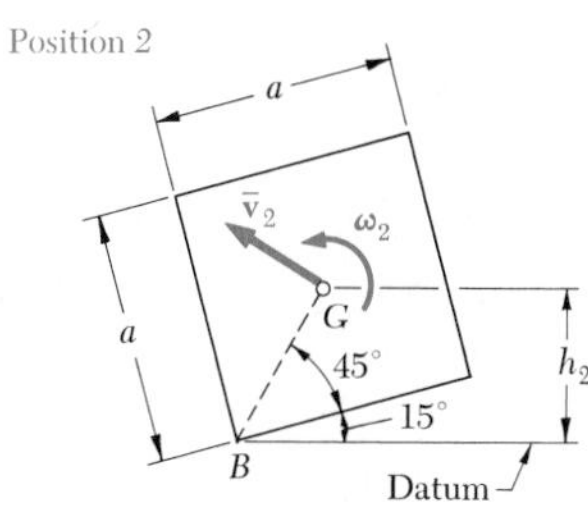

$GB = \frac{1}{2}\sqrt{2}a = 0.707a$
$h_2 = GB\sin(45° + 15°)$
$= 0.612a$

Since the package rotates about B, we have $\bar{v}_2 = (GB)\omega_2 = \frac{1}{2}\sqrt{2}a\omega_2$. We substitute this expression, together with $\bar{I} = \frac{1}{6}ma^2$, into Eq. (1):

$$(m\bar{v}_1)(\tfrac{1}{2}a) = m(\tfrac{1}{2}\sqrt{2}a\omega_2)(\tfrac{1}{2}\sqrt{2}a) + \tfrac{1}{6}ma^2\omega_2 \qquad \bar{v}_1 = \tfrac{4}{3}a\omega_2 \qquad (2)$$

Principle of Conservation of Energy. We apply the principle of conservation of energy between position 2 and position 3.

Position 2. $V_2 = Wh_2$. Recalling that $\bar{v}_2 = \frac{1}{2}\sqrt{2}a\omega_2$, we write

$$T_2 = \tfrac{1}{2}m\bar{v}_2^2 + \tfrac{1}{2}\bar{I}\omega_2^2 = \tfrac{1}{2}m(\tfrac{1}{2}\sqrt{2}a\omega_2)^2 + \tfrac{1}{2}(\tfrac{1}{6}ma^2)\omega_2^2 = \tfrac{1}{3}ma^2\omega_2^2$$

Position 3. Since the package must reach conveyor belt C, it must pass through position 3 where G is directly above B. Also, since we wish to determine the smallest velocity for which the package will reach this position, we choose $\bar{v}_3 = \omega_3 = 0$. Therefore $T_3 = 0$ and $V_3 = Wh_3$.

Conservation of Energy

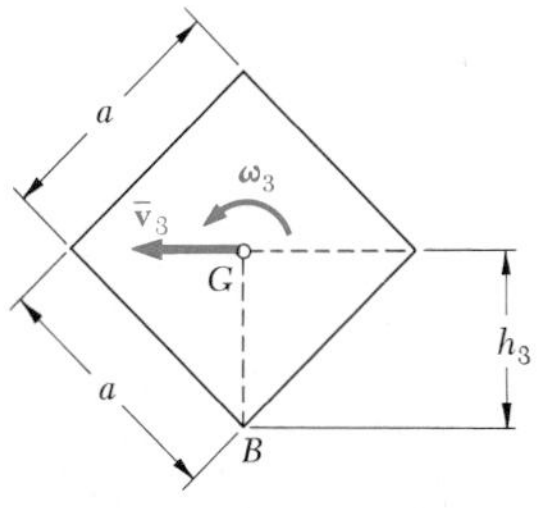

$h_3 = GB = 0.707a$

$$T_2 + V_2 = T_3 + V_3$$
$$\tfrac{1}{3}ma^2\omega_2^2 + Wh_2 = 0 + Wh_3$$
$$\omega_2^2 = \frac{3W}{ma^2}(h_3 - h_2) = \frac{3g}{a^2}(h_3 - h_2) \qquad (3)$$

Substituting the computed values of h_2 and h_3 into Eq. (3), we obtain

$$\omega_2^2 = \frac{3g}{a^2}(0.707a - 0.612a) = \frac{3g}{a^2}(0.095a) \qquad \omega_2 = \sqrt{0.285g/a}$$

$$\bar{v}_1 = \tfrac{4}{3}a\omega_2 = \tfrac{4}{3}a\sqrt{0.285g/a} \qquad \bar{v}_1 = 0.712\sqrt{ga} \blacktriangleleft$$

Problems

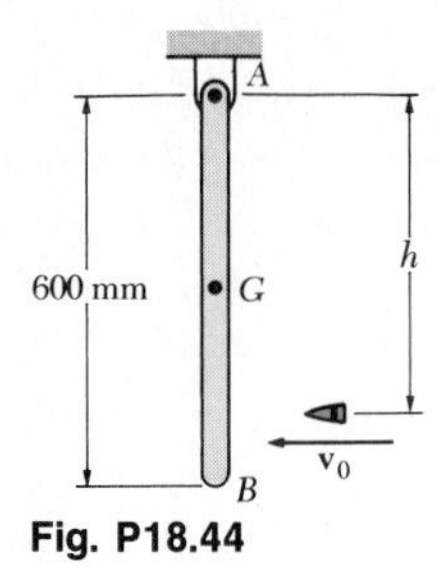

Fig. P18.44

18.44 A 32-g bullet is fired with a horizontal velocity of 450 m/s into a 5-kg wooden beam AB suspended from a pin support at A. Knowing that $h = 500$ mm and that the beam is initially at rest, determine (*a*) the velocity of the mass center G of the beam immediately after the bullet becomes embedded, (*b*) the impulsive reaction at A, assuming that the bullet becomes embedded in 1 ms.

18.45 In Prob. 18.44, determine (*a*) the required distance h if the impulsive reaction at A is to be zero, (*b*) the corresponding velocity of the mass center G of the beam after the bullet becomes embedded.

18.46 A 0.10-lb bullet is fired with a horizontal velocity of 1200 ft/s into a 20-lb wooden disk suspended from a pin support at A. Knowing that the disk is initially at rest, determine (*a*) the required distance h if the impulsive reaction at A is to be zero, (*b*) the corresponding velocity of the center G of the disk immediately after the bullet becomes embedded.

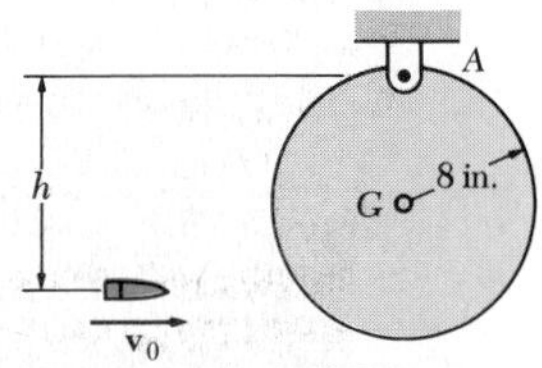

Fig. P18.46 and P18.47

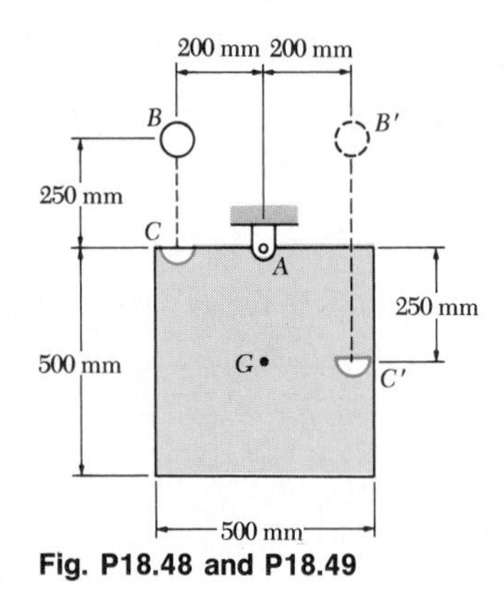

Fig. P18.48 and P18.49

18.47 A 0.10-lb bullet is fired with a horizontal velocity of 1200 ft/s into a 20-lb wooden disk suspended from a pin support at A. Knowing that $h = 10$ in. and that the disk is initially at rest, determine (*a*) the velocity of the center G of the disk immediately after the bullet becomes embedded, (*b*) the impulsive reaction at A, assuming that the bullet becomes embedded in 0.001 s.

18.48 An 8-kg wooden panel is suspended from a pin support at A and is initially at rest. A 2-kg metal sphere is released from rest at B and falls into a hemispherical cup C attached to the panel at a point located on its top edge. Assuming that the impact is perfectly plastic, determine the velocity of the mass center G of the panel immediately after the impact.

18.49 An 8-kg wooden panel is suspended from a pin support at A and is initially at rest. A 2-kg metal sphere is released from rest at B' and falls into a hemispherical cup C' attached to the panel at the same level as the mass center G. Assuming that the impact is perfectly plastic, determine the velocity of the mass center G of the panel immediately after the impact.

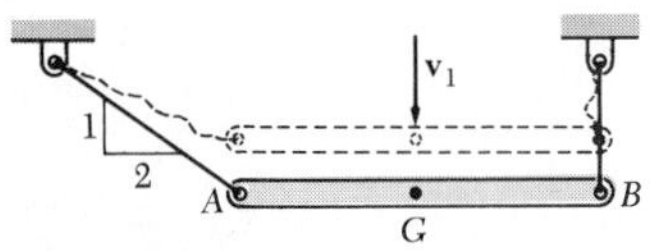

Fig. P18.50

18.50 A slender rod of length L is falling with a velocity $\mathbf{v}_1$ at the instant when the cords simultaneously become taut. Assuming that the impacts are perfectly plastic, determine the angular velocity of the rod and the velocity of its mass center immediately after the cords become taut.

18.51 A uniform square panel of side L and mass m is supported by a frictionless horizontal table. The panel is moving to the right with a velocity $\mathbf{v}_1$ when a hook located at the midpoint E of side AB engages the fixed pin F. Assuming that the impact is perfectly plastic, determine the angular velocity of the panel in its subsequent rotation about F.

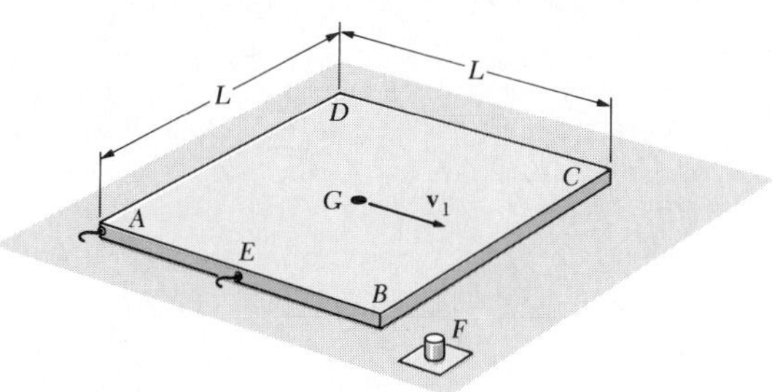

Fig. P18.51

18.52 Solve Prob. 18.51, assuming that the hook at E is removed and that a hook located at corner A engages the fixed pin F.

18.53 A uniform disk of radius r and mass m is supported by a frictionless horizontal table. Initially the disk is spinning freely about its mass center G with a constant angular velocity $\boldsymbol{\omega}_1$. Suddenly a latch B is moved to the right and is struck by a small stop A welded to the edge of the disk. Assuming that the impact of A and B is perfectly plastic, determine the angular velocity of the disk and the velocity of its mass center immediately after impact.

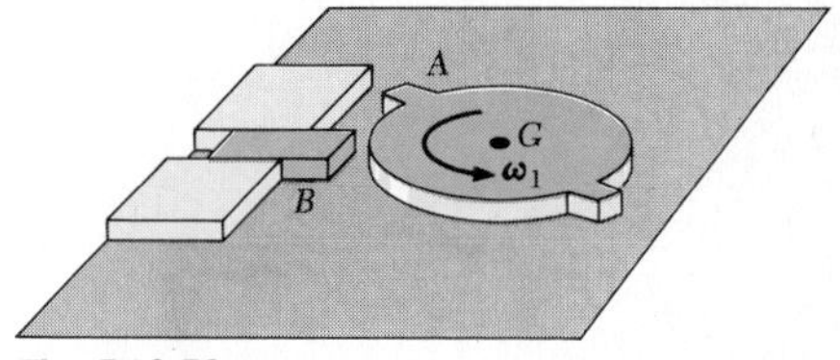

Fig. P18.53

18.54 A uniform slender rod AB is equipped at both ends with the hooks shown and is supported by a frictionless horizontal table. Initially the rod is hooked at A to a fixed pin C about which it rotates with the constant angular velocity $\boldsymbol{\omega}_1$. Suddenly end B of the rod hits and gets hooked to the pin D, causing end A to be released. Determine the magnitude of the angular velocity $\boldsymbol{\omega}_2$ of the rod in its subsequent rotation about D.

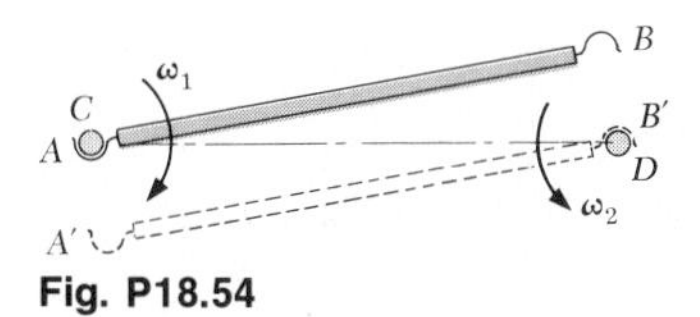

Fig. P18.54

18.55 A slender rod of length L and mass m is released from rest in the position shown and hits the edge D. Assuming perfectly plastic impact at D, determine for $b = 0.6L$, (*a*) the angular velocity of the rod immediately after the impact, (*b*) the maximum angle through which the rod will rotate after the impact.

18.56 A slender rod of mass m and length L is released from rest in the position shown and hits the edge D. Assuming perfectly elastic impact ($e = 1$) at D, determine the distance b for which the rod will rebound with no angular velocity.

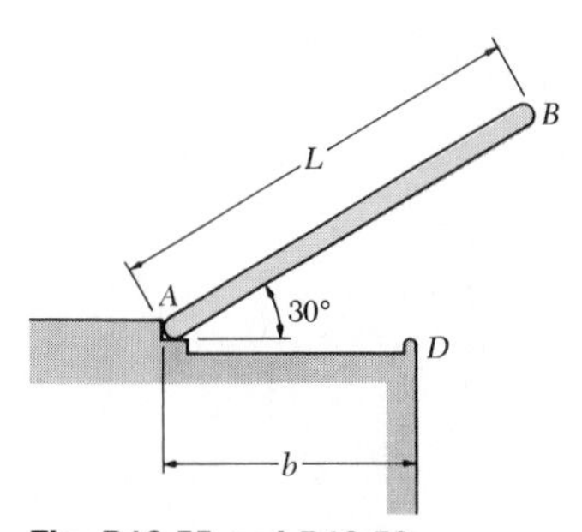

Fig. P18.55 and P18.56

18.57 Solve Prob. 18.56, assuming $e = 0.50$.

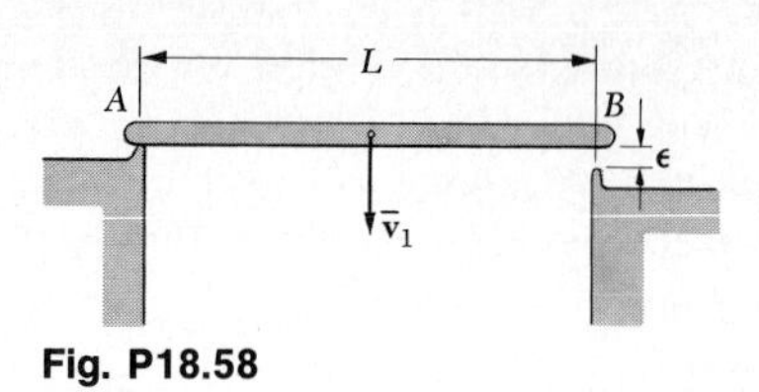

Fig. P18.58

18.58 A uniform slender rod of length L is dropped onto rigid supports at A and B. Immediately before striking A the velocity of the rod is $\overline{\mathbf{v}}_1$. Since support B is slightly lower than support A, the rod strikes A before it strikes B. Assuming perfectly elastic impact at both A and B, determine the angular velocity of the rod and the velocity of its mass center immediately after the rod (*a*) strikes support A, (*b*) strikes support B, (*c*) again strikes support A.

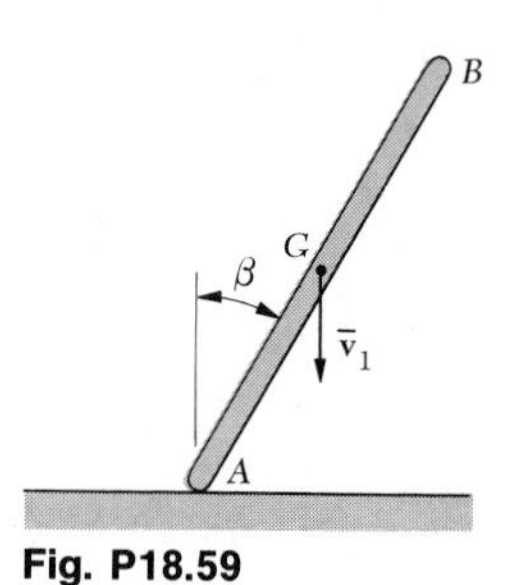

Fig. P18.59

18.59 A slender rod of length L forming an angle β with the vertical strikes a frictionless floor at A with a vertical velocity $\overline{\mathbf{v}}_1$ and no angular velocity. Assuming that the impact at A is perfectly elastic, derive an expression for the angular velocity of the rod immediately after impact.

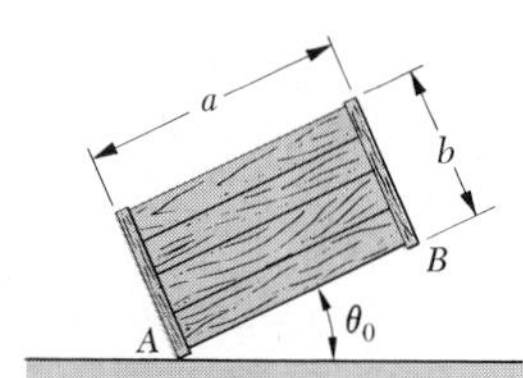

Fig. P18.60

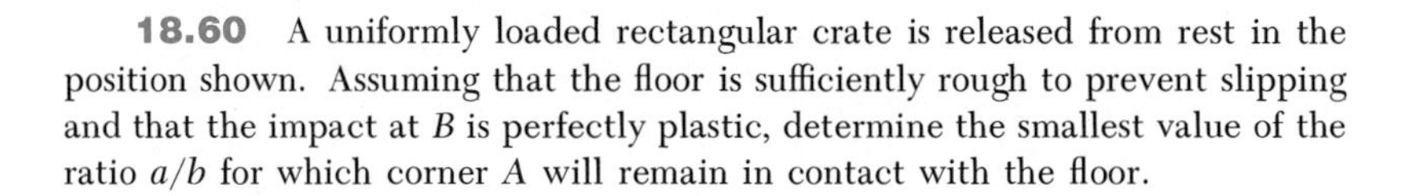

18.60 A uniformly loaded rectangular crate is released from rest in the position shown. Assuming that the floor is sufficiently rough to prevent slipping and that the impact at B is perfectly plastic, determine the smallest value of the ratio a/b for which corner A will remain in contact with the floor.

18.61 A uniformly loaded square crate is released from rest with its corner D directly above A; it rotates about A until its corner B strikes the floor, and then rotates about B. The floor is sufficiently rough to prevent slipping and the impact at B is perfectly plastic. Denoting by $\boldsymbol{\omega}_0$ the angular velocity of the crate immediately before B strikes the floor, determine (*a*) the angular velocity of the crate immediately after B strikes the floor, (*b*) the fraction of the kinetic energy of the crate lost during the impact, (*c*) the angle θ through which the crate will rotate after B strikes the floor.

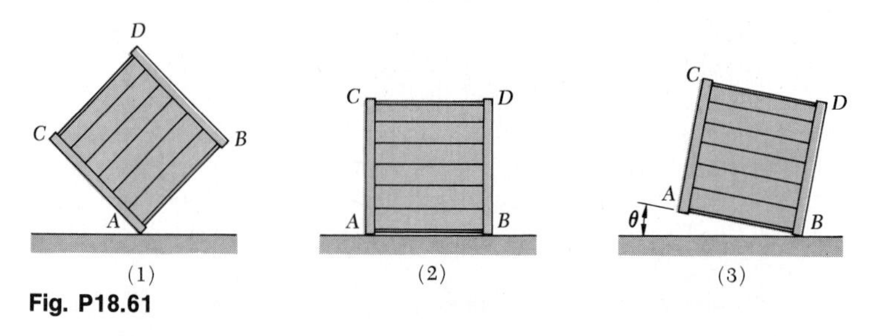

Fig. P18.61

18.62 Rod *AB*, of mass 1.5 kg and length $L = 0.5$ m, is suspended from a pin *A* which may move freely along a horizontal guide. If a 1.8-N·s horizontal impulse $\mathbf{Q}\,\Delta t$ is applied at *B*, determine the maximum angle θ_m through which the rod will rotate during its subsequent motion.

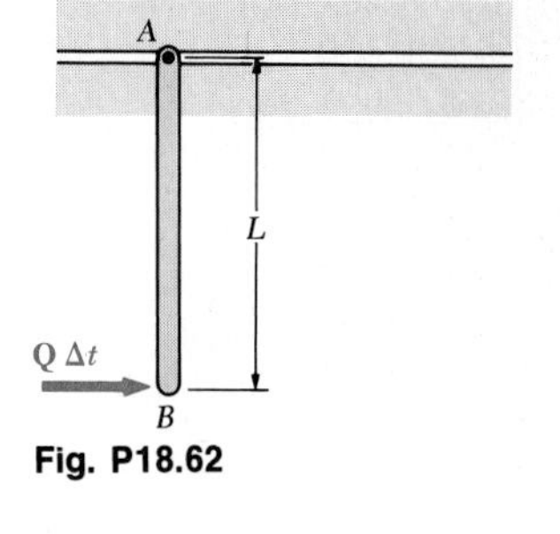

Fig. P18.62

18.63 For the rod of Prob. 18.62, determine the magnitude $Q\,\Delta t$ of the impulse for which the maximum angle through which the rod will rotate is (*a*) 90°, (*b*) 120°.

18.64 The plank *CDE* of mass m_P rests on top of a small pivot at *D*. A gymnast *A* of mass *m* stands on the plank at end *C*; a second gymnast *B* of the same mass *m* jumps from a height *h* and strikes the plank at *E*. Assuming perfectly plastic impact, determine the height to which gymnast *A* will rise. (Assume that gymnast *A* stands completely rigid.)

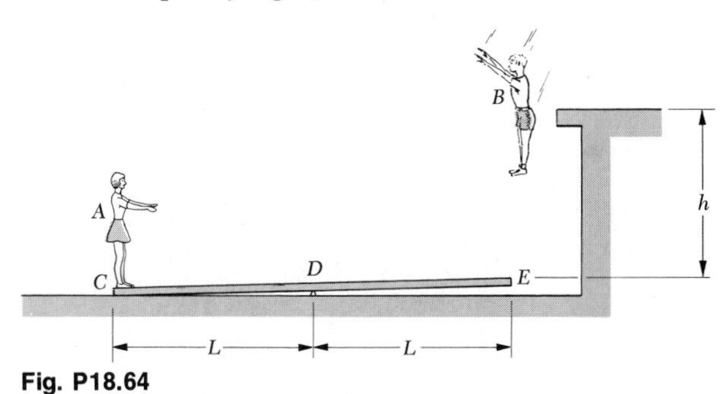

Fig. P18.64

18.65 A slender rod *CD*, of length *L* and mass *m*, is placed upright as shown on a frictionless horizontal surface. A second and identical rod *AB* is moving along the surface with a velocity $\bar{\mathbf{v}}_1$ when its end *B* strikes squarely end *C* of rod *CD*. Assuming that the impact is perfectly elastic, determine immediately after the impact, (*a*) the velocity of rod *AB*, (*b*) the angular velocity of rod *CD*, (*c*) the velocity of the mass center of rod *CD*.

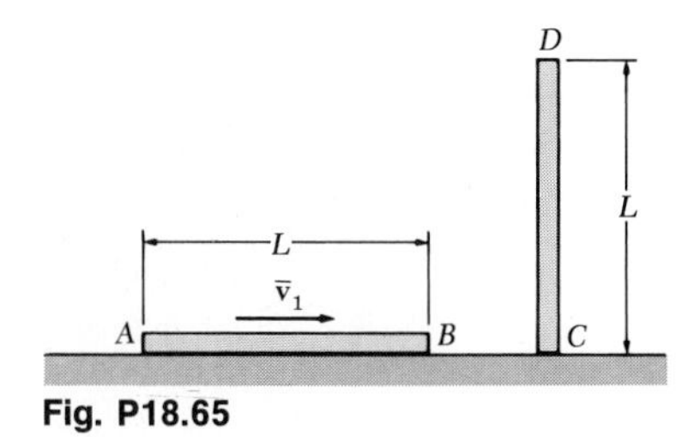

Fig. P18.65

18.66 Solve Prob. 18.65, assuming $e = 0.50$.

18.67 A sphere of mass *m* is dropped onto the end of a slender rod *BD* of length *L* and mass 2*m*. The rod is attached to a pin support at *C* which is located at a distance $c = \frac{1}{4}L$ from end *B*. Denoting by $\mathbf{v}_1$ the velocity of the sphere just before it strikes the rod and assuming perfectly elastic impact, determine immediately after the impact (*a*) the velocity of the sphere, (*b*) the angular velocity of the rod.

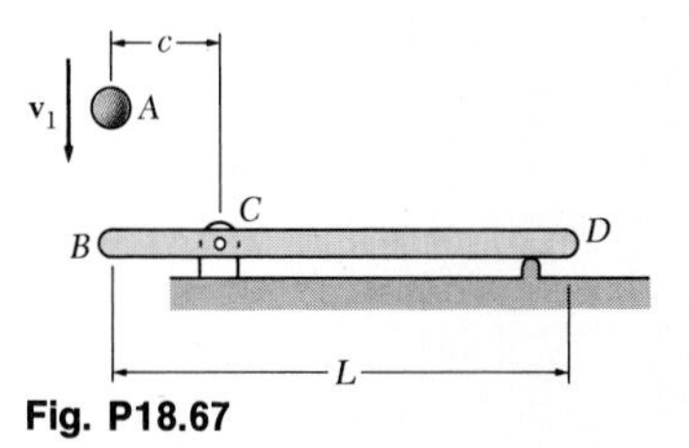

Fig. P18.67

18.68 Solve Prob. 18.67, assuming that $c = \frac{1}{2}L$.

18.69 A sphere of weight W_S is dropped from a height h and strikes at A the uniform slender plank AB of weight W which is held by two inextensible cords. Knowing that the impact is perfectly plastic and that the sphere remains attached to the plank, determine the velocity of the sphere immediately after impact.

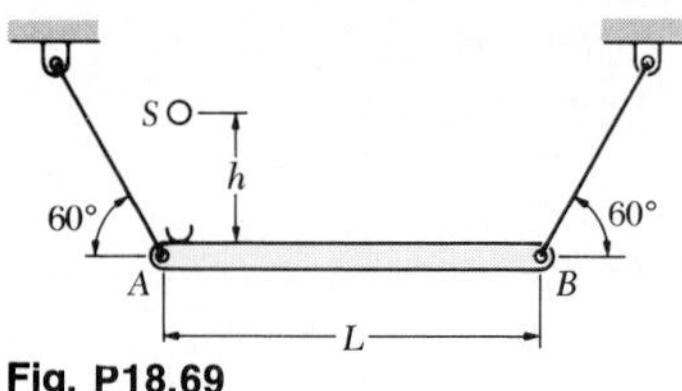

Fig. P18.69

18.70 Solve Prob. 18.69 when $W = 6$ lb, $W_S = 3$ lb, $L = 3$ ft, and $h = 4$ ft.

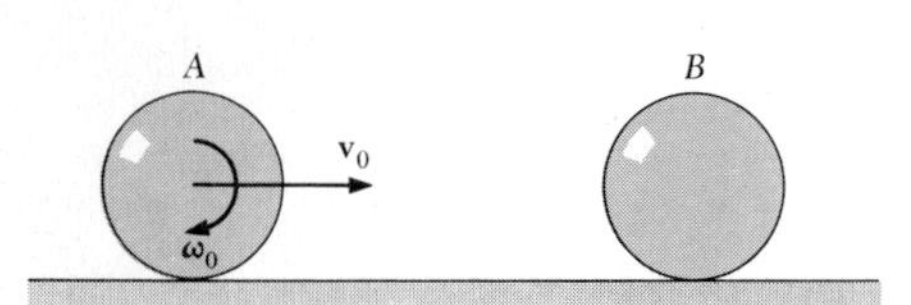

Fig. P18.71

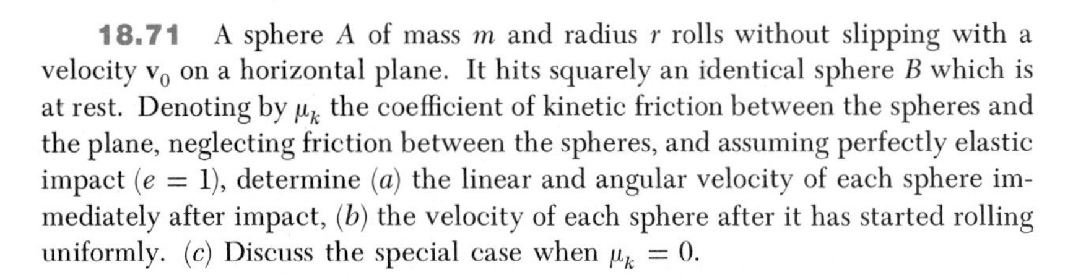

18.71 A sphere A of mass m and radius r rolls without slipping with a velocity $\mathbf{v}_0$ on a horizontal plane. It hits squarely an identical sphere B which is at rest. Denoting by μ_k the coefficient of kinetic friction between the spheres and the plane, neglecting friction between the spheres, and assuming perfectly elastic impact ($e = 1$), determine (a) the linear and angular velocity of each sphere immediately after impact, (b) the velocity of each sphere after it has started rolling uniformly. (c) Discuss the special case when $\mu_k = 0$.

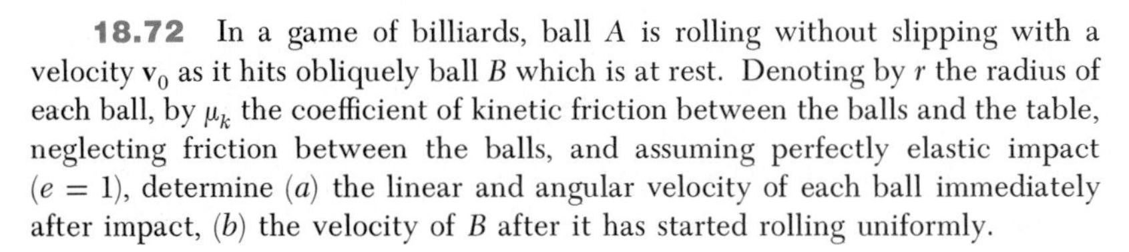

18.72 In a game of billiards, ball A is rolling without slipping with a velocity $\mathbf{v}_0$ as it hits obliquely ball B which is at rest. Denoting by r the radius of each ball, by μ_k the coefficient of kinetic friction between the balls and the table, neglecting friction between the balls, and assuming perfectly elastic impact ($e = 1$), determine (a) the linear and angular velocity of each ball immediately after impact, (b) the velocity of B after it has started rolling uniformly.

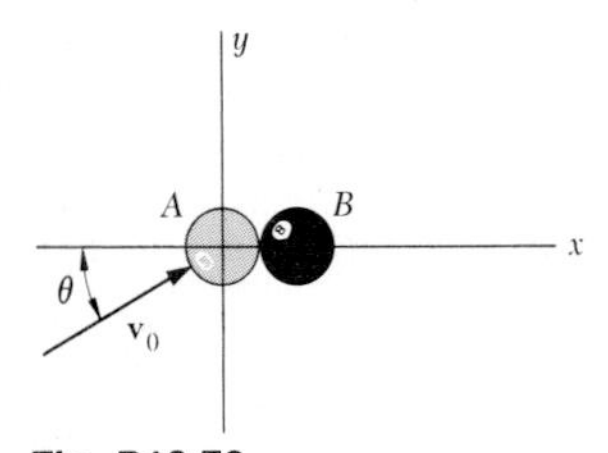

Fig. P18.72

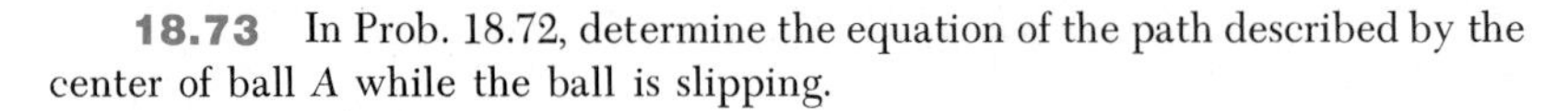

18.73 In Prob. 18.72, determine the equation of the path described by the center of ball A while the ball is slipping.

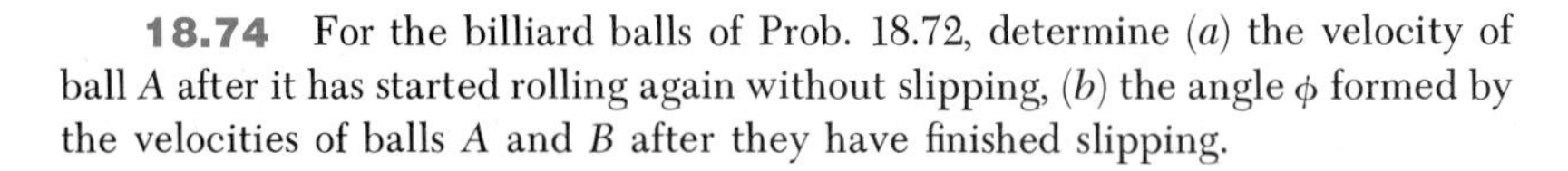

18.74 For the billiard balls of Prob. 18.72, determine (a) the velocity of ball A after it has started rolling again without slipping, (b) the angle ϕ formed by the velocities of balls A and B after they have finished slipping.

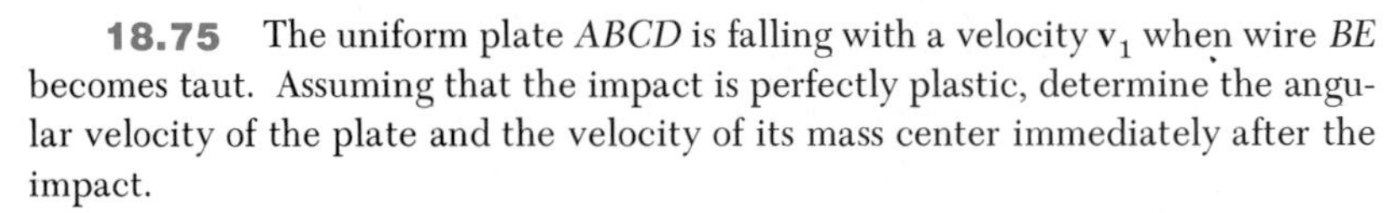

18.75 The uniform plate $ABCD$ is falling with a velocity $\mathbf{v}_1$ when wire BE becomes taut. Assuming that the impact is perfectly plastic, determine the angular velocity of the plate and the velocity of its mass center immediately after the impact.

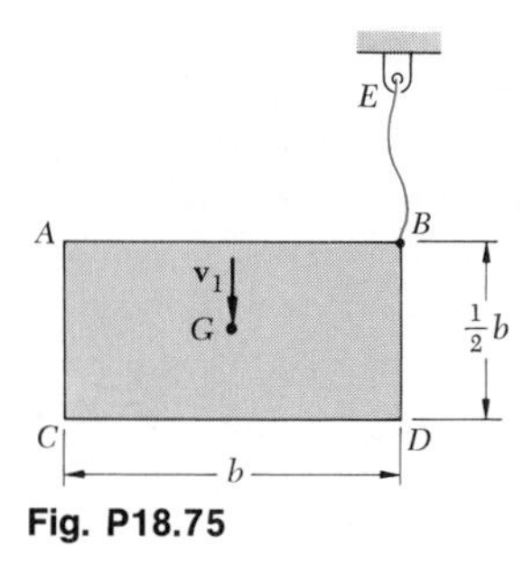

Fig. P18.75

18.76 A small rubber ball of radius r is thrown against a rough floor with a velocity $\mathbf{v}_A$ of magnitude v_0 and a "backspin" $\boldsymbol{\omega}_A$ of magnitude ω_0. It is observed that the ball bounces from A to B, then from B to A, then from A to B, etc. Assuming perfectly elastic impact, determine the required magnitude ω_0 of the backspin in terms of v_0 and r.

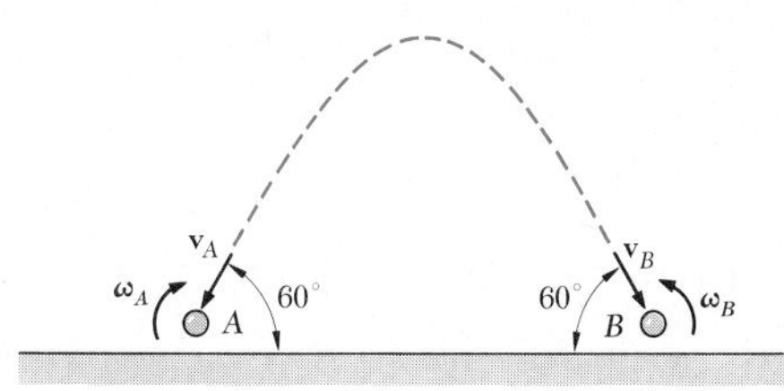

Fig. P18.76

18.77 A rectangular slab of mass m_S moves across a series of rollers, each of which is equivalent to a uniform disk of mass m_R and is initially at rest. Since the length of the slab is slightly less than three times the distance b between two adjacent rollers, the slab leaves a roller just before it reaches another one. Each time a new roller enters into contact with the slab, slipping occurs between the roller and the slab for a short period of time (less than the time needed for the slab to move through the distance b). Denoting by $\mathbf{v}_0$ the velocity of the slab in the position shown, determine the velocity of the slab after it has moved (a) a distance b, (b) a distance nb.

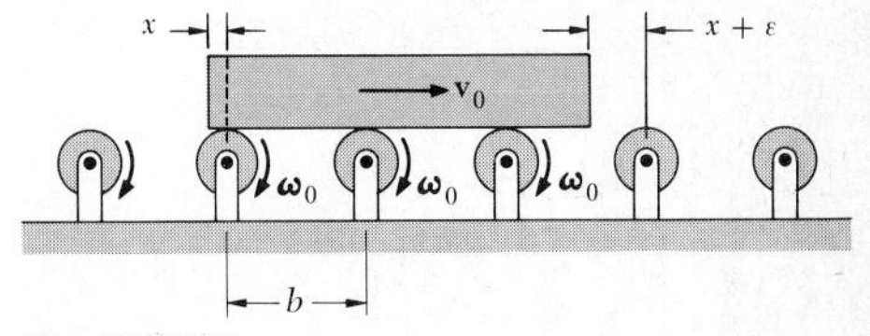

Fig. P18.77

18.78 Solve Prob. 18.77, assuming $v_0 = 6$ m/s, $m_S = 9$ kg, $m_R = 2$ kg, and $n = 8$.

* **18.79** Each of the bars AB and BC is of length $L = 15$ in. and weight $W = 2.5$ lb. Determine the angular velocity of each bar immediately after the 0.30-lb·s horizontal impulse $\mathbf{Q}\,\Delta t$ has been applied at C.

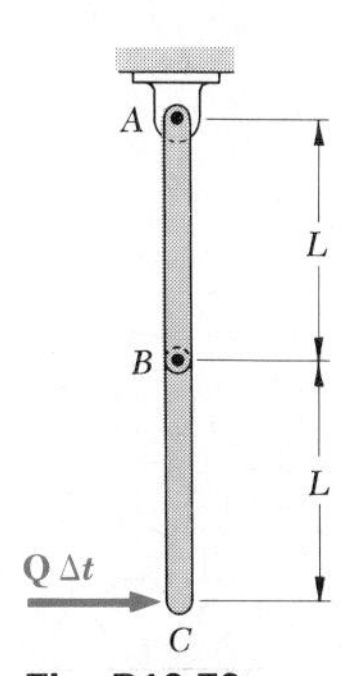

Fig. P18.79

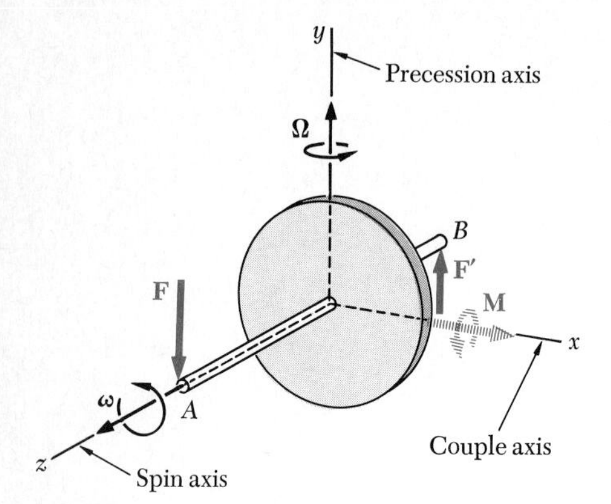

Fig. 18.11

***18.9. Gyroscopes.** Our study of kinetics is limited to the plane motion of rigid bodies. The analysis of the general motion of a rigid body in space is indeed beyond the scope of this text. Among the problems which we shall *not* attempt to solve with the means at our disposal are those involving the general motion of a top or of a gyroscope. However, we shall consider in this section a particular case of gyroscopic motion which requires only a very limited knowledge of the motion of a rigid body in space and yet has many useful engineering applications.

A *gyroscope* consists of a body of revolution mounted on an axle which coincides with its geometric axis. Consider a gyroscope which spins at a very large rate ω about an axle AB directed along the z axis (Fig. 18.11). This motion may be represented by a vector $\boldsymbol{\omega}$ of magnitude equal to the *rate of spin* ω of the gyroscope and directed along the *spin axis* AB in such a way that an observer located at the tip of the vector $\boldsymbol{\omega}$ will see the gyroscope rotate counterclockwise.

If we attempt to rotate the axle of the gyroscope about the x axis by applying two vertical forces $\mathbf{F}$ and $\mathbf{F}'$ at A and B, forming a couple of moment M, we find that the axle will resist the motion we try to impose and that *it will rotate about the y axis* with an angular velocity Ω. This motion of the gyroscope is called *precession;* it may be represented by a vector $\boldsymbol{\Omega}$ of magnitude equal to the *rate of precession* Ω of the gyroscope and directed along the y axis, or *precession axis,* in such a way that a person located at the tip of the vector will observe this motion as counterclockwise.

In order to explain this apparently odd behavior, we shall consider the gyroscope in two different positions and apply the principle of impulse and momentum. We first consider the gyroscope in its initial position at $t_1 = 0$. Denoting by $\bar{I}$ its centroidal moment of inertia, we find that the system of the momenta of its particles reduces to a couple of moment $\bar{I}\omega$ (Sec. 18.3). This momentum couple is contained in the xy plane and may therefore be represented by the couple vector $\bar{I}\boldsymbol{\omega}$ directed along the z axis (Fig. 18.12*a*). Next we consider the gyroscope at time $t_2 = \Delta t$, after its axle has rotated about the y axis through an angle $\Delta\phi$ as a result of our miscalculated effort. We may neglect the relatively small angular momentum of the gyroscope about the y axis and consider only the large momentum couple vector $\bar{I}\boldsymbol{\omega}'$ associated with the rotation of the gyroscope about its axle (Fig. 18.12*c*). The change in momentum during the time interval Δt is repre-

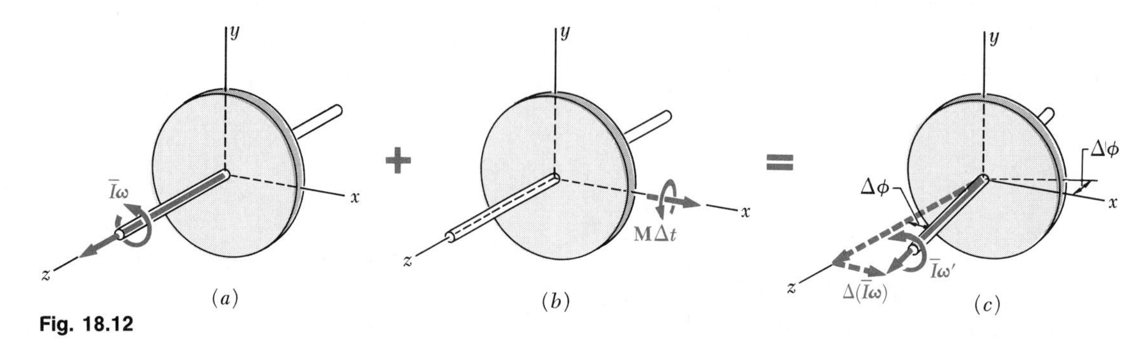

Fig. 18.12

sented by the couple vector $\Delta(\bar{I}\boldsymbol{\omega})$. This couple vector has the same direction as the couple vector $\mathbf{M}\,\Delta t$, which represents the angular impulse of the couple formed by the two forces applied to the axle (Fig. 18.12*b*).

Far from being strange, the behavior of the gyroscope is therefore entirely predictable. Noting that the couple vector $\Delta(\bar{I}\boldsymbol{\omega})$ has a magnitude equal to $\bar{I}\omega\,\Delta\phi$, we equate the angular impulse and the change in angular momentum and write

$$M\,\Delta t = \bar{I}\omega\,\Delta\phi$$

Dividing both members by Δt and letting Δt approach zero, we find that the moment M of the *gyroscopic couple* is equal to

$$M = \bar{I}\omega\frac{d\phi}{dt}$$

or

$$M = \bar{I}\omega\Omega \tag{18.20}$$

where ω = rate of spin
$\Omega = d\phi/dt$ = rate of precession of the axis of the gyroscope

We note that the spin axis, the couple axis (about which the force couple is applied), and the precession axis are at right angles to each other. The relative position of these axes may easily be found if we remember that the extremity of the vector couple $\bar{I}\boldsymbol{\omega}$ representing the momentum of the gyroscope tends to move in the direction defined by the vector $\mathbf{M}$ representing the gyroscopic couple.

If the motion just described is maintained for an appreciable length of time, the gyroscope is said to be in *steady precession*. Steady precession may be obtained only if the axis of the applied couple is maintained perpendicular to the spin axis and to the precession axis.

We shall now consider another case of steady precession, in which the axis of the gyroscope forms with the precession axis an angle β different from 90° (Fig. 18.13). Neglecting again the angular momentum associated with the motion of precession, we observe that the change in momentum during a time interval Δt is represented by the vector $\Delta(\bar{I}\boldsymbol{\omega})$ of magnitude $\bar{I}\omega \sin\beta\,\Delta\phi$. Equating angular impulse and change in angular momentum, we write

$$M\,\Delta t = \bar{I}\omega \sin\beta\,\Delta\phi$$

and obtain

$$M = \bar{I}\omega\Omega \sin\beta \tag{18.21}$$

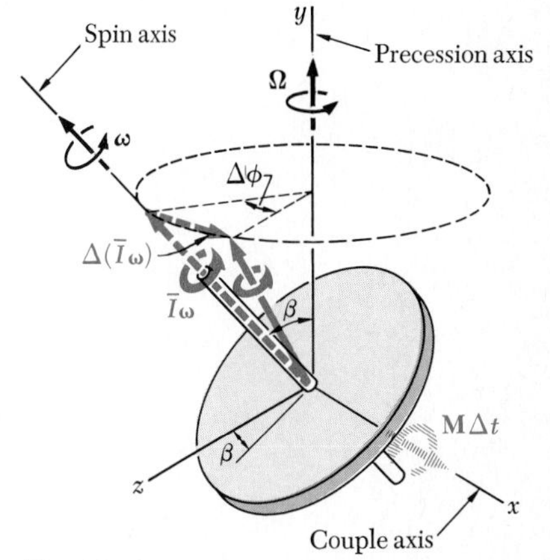

Fig. 18.13

where Ω represents the rate of precession $d\phi/dt$. Thus, the motion described will be obtained if a couple of moment M satisfying Eq. (18.21) is applied about an axis which is maintained perpendicular to the spin axis and to the precession axis. However, since the angular momentum associated with the motion of precession was neglected, Eq. (18.21) may be used only when the rate of spin is much larger than the rate of precession.†

An interesting example of gyroscopic motion is provided by a top supported at a fixed point O (Fig. 18.14). The weight $\mathbf{W}$ of the top and the normal component of the reaction at O form a couple of moment $M = Wc \sin \beta$. Assuming that the angle β the axis of the top forms with the vertical remains constant, we substitute for M in formula (18.21) and write

$$Wc = \bar{I}\omega\Omega \tag{18.22}$$

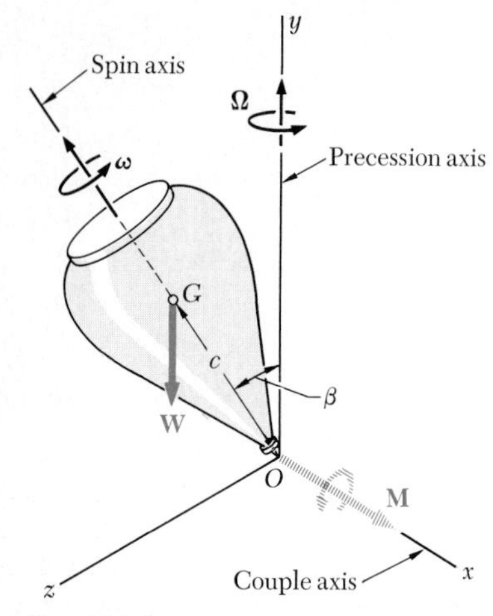

Fig. 18.14

We should note that the result obtained is valid only for a very large rate of spin and that the top will just fall if it does not spin. Besides, the top will move in *steady* precession, at a constant rate Ω and with its axis forming a constant angle β with the vertical, only if it is given initially a rate of spin ω and a rate of precession Ω which satisfy Eq. (18.22). Otherwise, *the precession will be unsteady* and will be accompanied by variations in β, referred to as *nutation*. The study of the general case of motion of the top is a difficult one which requires a full knowledge of the motion of rigid bodies in three-dimensional space.

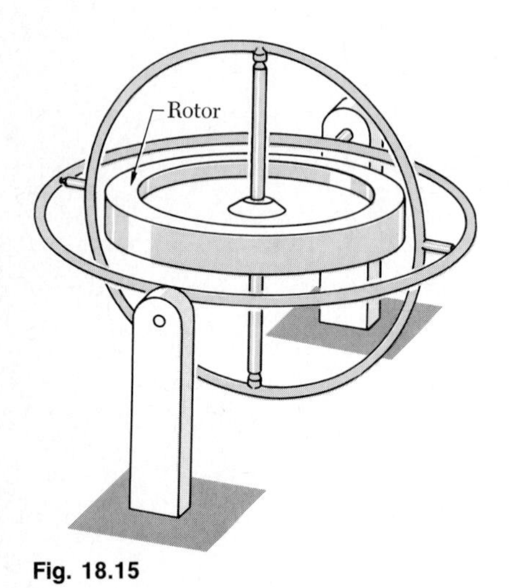

Fig. 18.15

A gyroscope mounted in a Cardan's suspension (Fig. 18.15) is free from any applied couple and thus will maintain a fixed direction with respect to a newtonian frame of reference. This was one of the facts used by the French physicist Jean Foucault (1819–1868) to prove that the earth rotates about its axis. The *gyrocompass* used to indicate the true north consists essentially of a gyroscope whose axis is maintained in a horizontal plane and oscillates symmetrically about the direction north-south. Other sensitive instruments based on gyroscopic motion include the inertial guidance systems of rockets, as well as the turn indicator, the artificial horizon, and the automatic pilot used on airplanes.

Because of the relatively large couples required to change the orientation of their axles, gyroscopes are used as stabilizers in torpedoes and ships. Spinning bullets and shells remain tangent to their trajectory because of gyroscopic action. And a bicycle is easier to keep balanced at high speeds because of the stabilizing effect of its spinning wheels. However, gyroscopic action is not always welcome and must be taken into account in the design of bearings supporting rotating shafts subjected to forced precession. The reactions exerted on an airplane by the rotating parts of its engines—and by its propellers, in the case of a propeller-driven airplane—when the airplane changes its direction of flight must also be taken into account and compensated for whenever possible.

† For a more comprehensive discussion of the steady precession of a gyroscope, see F. P. Beer and E. R. Johnston, Jr., *Vector Mechanics for Engineers: Dynamics*, 4th ed., McGraw-Hill, New York, 1984, sec. 18.10.

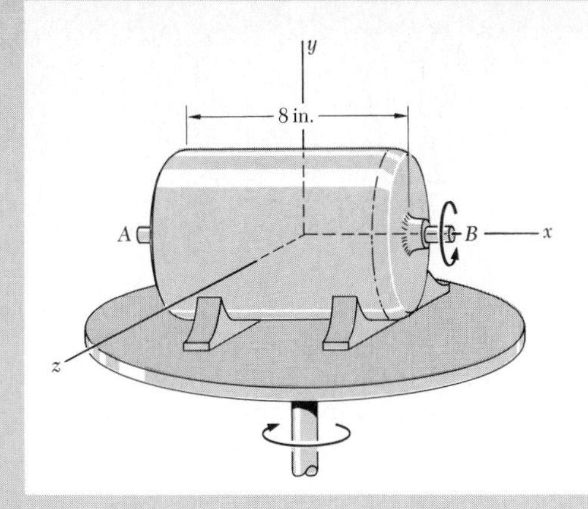

SAMPLE PROBLEM 18.7

The rotor of an electric motor weighs 6 lb and has a radius of gyration of 2 in. The angular velocity of the rotor is 3600 rpm counterclockwise as viewed from the positive x axis. Determine the reactions exerted by the bearings on the axle AB when the motor is rotated about the y axis clockwise as viewed from above and at a rate of 6 rpm.

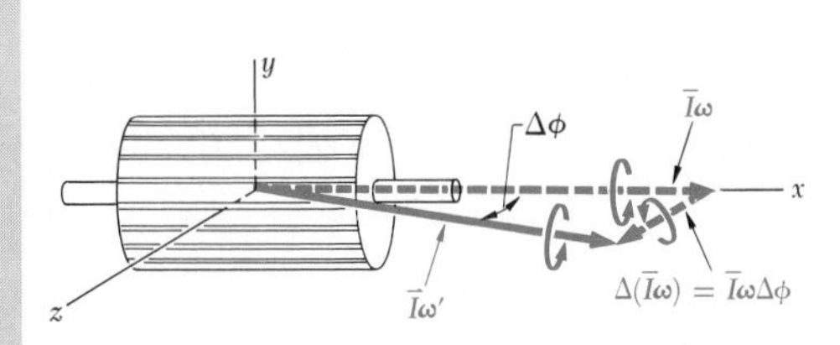

Solution. During the time interval Δt, the momentum couple vector $\bar{I}\boldsymbol{\omega}$ maintains a constant magnitude $\bar{I}\omega$ but is rotated in a horizontal plane through an angle $\Delta\phi$ into a new position $\bar{I}\boldsymbol{\omega}'$. Thus, the change in angular momentum is represented by the vector $\Delta(\bar{I}\boldsymbol{\omega})$ parallel to the z axis and of magnitude $\bar{I}\omega\,\Delta\phi$. Denoting by $\mathbf{M}_z$ the couple applied about the z axis, we write that the angular impulse $\mathbf{M}_z\,\Delta t$ must equal the change in angular momentum:

$$M_z\,\Delta t = \bar{I}\omega\,\Delta\phi \qquad M_z = \bar{I}\omega\frac{\Delta\phi}{\Delta t}$$

Denoting by $\boldsymbol{\Omega}$ the angular velocity about the vertical y axis, we write $\Omega = \Delta\phi/\Delta t$ and obtain

$$M_z = \bar{I}\omega\Omega \tag{1}$$

Using the given data, we have

$$\omega = 3600 \text{ rpm} = 377 \text{ rad/s}$$
$$\Omega = 6 \text{ rpm} = 0.628 \text{ rad/s}$$
$$\bar{I} = m\bar{k}^2 = \frac{6 \text{ lb}}{32.2 \text{ ft/s}^2}(\tfrac{2}{12}\text{ ft})^2 = 0.00518 \text{ lb}\cdot\text{ft}\cdot\text{s}^2$$

Substituting these values into (1), we obtain the moment of the gyroscopic couple:

$$M_z = (0.00518 \text{ lb}\cdot\text{ft}\cdot\text{s}^2)(377 \text{ rad/s})(0.628 \text{ rad/s})$$
$$= 1.226 \text{ lb}\cdot\text{ft}$$

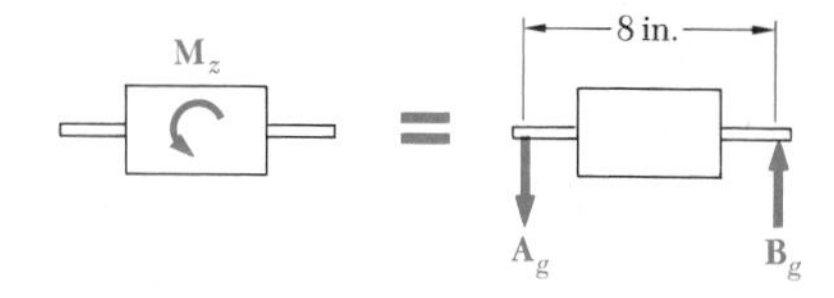

Reactions at A and B. The gyroscopic reactions $\mathbf{A}_g$ and $\mathbf{B}_g$ must be *equivalent* to the gyroscopic couple $\mathbf{M}_z$:

$$M_z = A_g(\tfrac{8}{12}\text{ ft})$$
$$1.226 \text{ lb}\cdot\text{ft} = A_g(\tfrac{8}{12}\text{ ft})$$
$$A_g = 1.84 \text{ lb} \qquad \mathbf{A}_g = 1.84 \text{ lb}\downarrow$$
$$B_g = 1.84 \text{ lb} \qquad \mathbf{B}_g = 1.84 \text{ lb}\uparrow$$

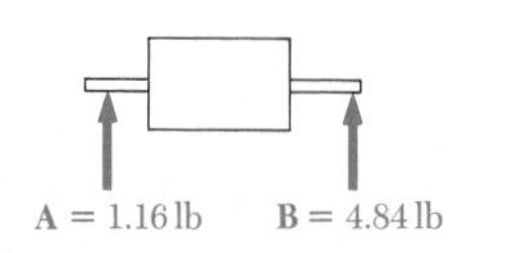

Since the static reactions are $\mathbf{A}_s = 3 \text{ lb}\uparrow$ and $\mathbf{B}_s = 3 \text{ lb}\uparrow$, the total reactions are

$$\mathbf{A} = 1.16 \text{ lb}\uparrow \qquad \mathbf{B} = 4.84 \text{ lb}\uparrow \blacktriangleleft$$

Problems

18.80 The major rotating elements of a given propeller-driven airplane rotate clockwise when viewed from behind. If the pilot turns the airplane to his right, determine in what direction the nose of the airplane will tend to move under the action of the gyroscopic couple exerted by the rotating elements.

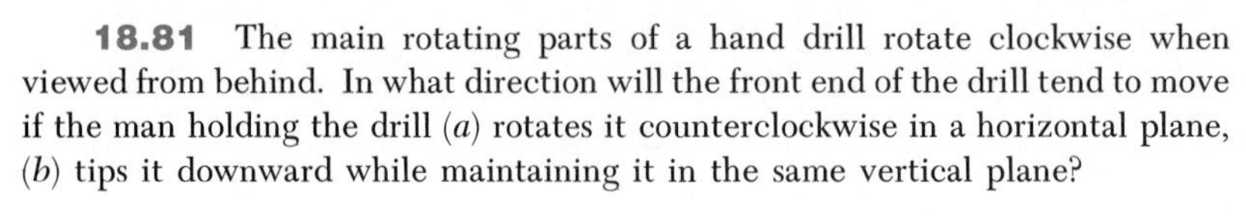

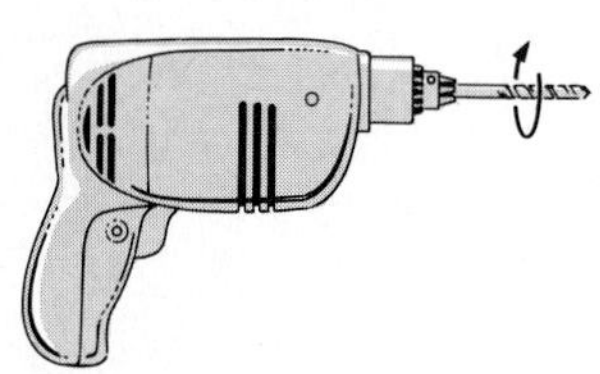

Fig. P18.81

18.81 The main rotating parts of a hand drill rotate clockwise when viewed from behind. In what direction will the front end of the drill tend to move if the man holding the drill (*a*) rotates it counterclockwise in a horizontal plane, (*b*) tips it downward while maintaining it in the same vertical plane?

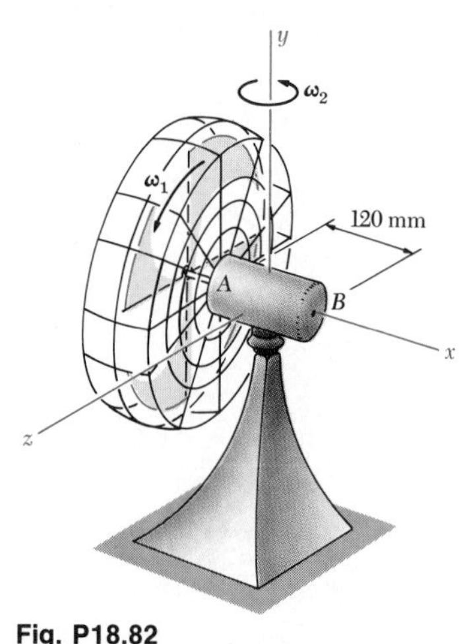

Fig. P18.82

18.82 The blade of an oscillating fan and the rotor of its motor have a total mass of 200 g and a combined radius of gyration of 75 mm. They are supported by bearings located at A and B, 120 mm apart, and have an angular velocity ω_1 of 2400 rpm at the high-speed setting. Determine the dynamic reactions at A and B when the motor casing has an angular velocity ω_2 of 0.5 rad/s as shown.

18.83 An airplane has a single four-bladed propeller which weighs 250 lb and has a radius of gyration of 32 in. Knowing that the propeller rotates at 1800 rpm clockwise as seen from the front, determine the magnitude of the couple exerted by the propeller on its shaft and the resulting effect on the airplane when the pilot executes a horizontal turn of 1500-ft radius to the left at a speed of 375 mi/h.

18.84 Each wheel of an automobile weighs 45 lb and has a radius of gyration of 10 in. The automobile travels around an unbanked curve of radius 600 ft at a speed of 60 mi/h. Knowing that each wheel is of 24-in. diameter and that the transverse distance between the wheels is 58 in., determine the additional normal reaction at each outside wheel due to the rotation of the car.

18.85 The rotor of a given turbine may be approximated by a 25-kg disk of 300-mm diameter. Knowing that the turbine rotates clockwise at 10 000 rpm as viewed from the positive x axis, determine the dynamic reactions at the bearings A and B if the instantaneous angular velocity of the turbine housing is 2 rad/s clockwise as viewed from (a) the positive x axis, (b) the positive y axis, (c) the positive z axis.

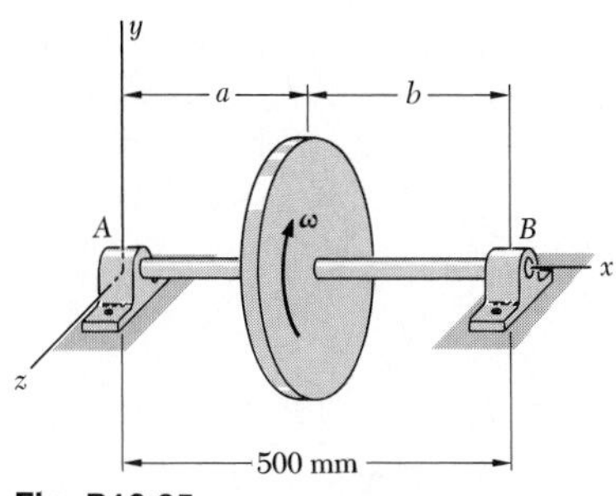

Fig. P18.85

18.86 The essential structure of a certain type of aircraft turn indicator is shown. Springs AC and BD are initially stretched and exert equal vertical forces at A and B when the airplane is traveling in a straight path. Each spring has a constant of 500 N/m and the uniform disk has a mass of 200 g and spins at the rate of 10 000 rpm. Determine the angle through which the yoke will rotate when the pilot executes a horizontal turn of 750-m radius to the right at a speed of 800 km/h. Indicate whether point A will move up or down.

Fig. P18.86

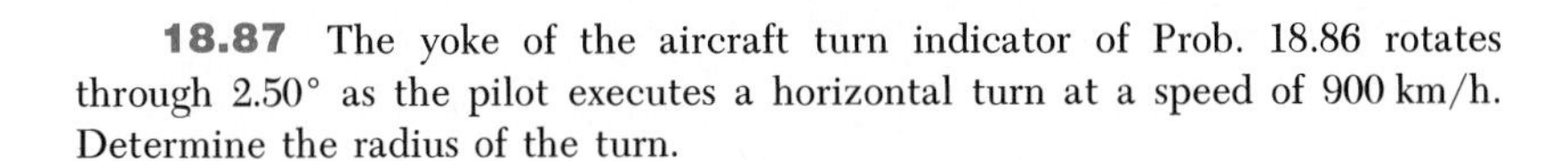

18.87 The yoke of the aircraft turn indicator of Prob. 18.86 rotates through 2.50° as the pilot executes a horizontal turn at a speed of 900 km/h. Determine the radius of the turn.

18.88 A thin homogeneous disk of mass m and radius a is held by a fork-ended horizontal rod ABC. The disk and the rod rotate with the angular velocities shown. Assuming that both ω_1 and ω_2 are constant, determine the dynamic reactions at A and B.

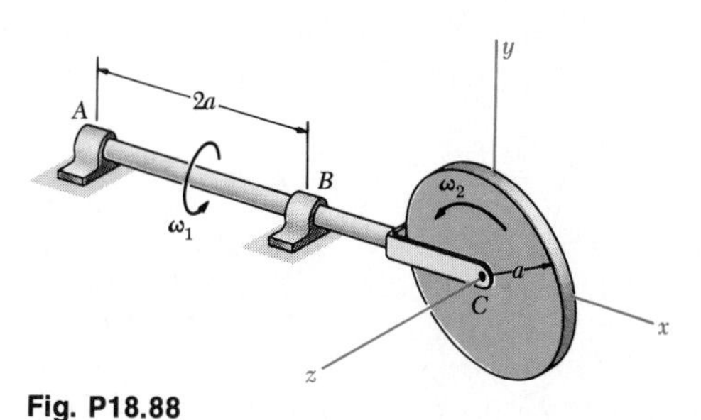

Fig. P18.88

18.89 Two disks, each of weight 10 lb, and radius 12 in., spin as shown at 1200 rpm about the rod AB, which is attached to shaft CD. The entire system is made to rotate about the z axis with an angular velocity $\mathbf{\Omega}$ of 60 rpm. (a) Determine the dynamic reactions at C and D due to gyroscopic action as the system passes through the position shown. (b) Solve part a assuming that the direction of spin of disk B is reversed.

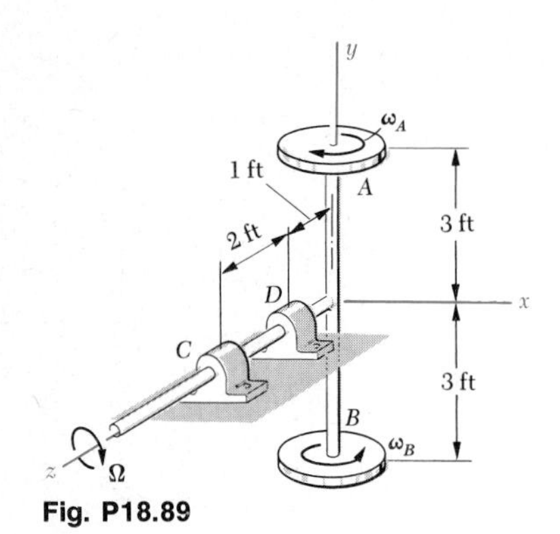

Fig. P18.89

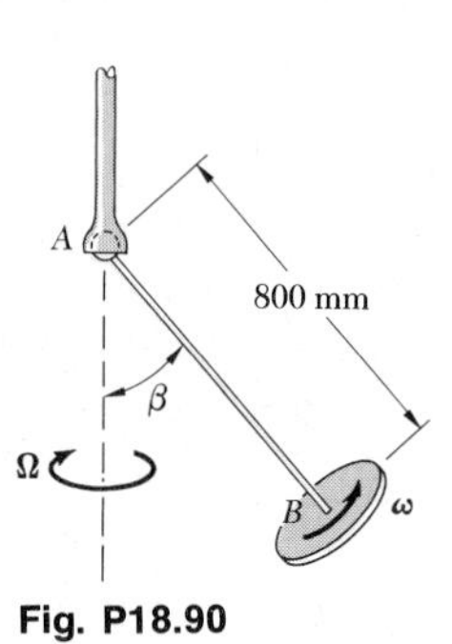

Fig. P18.90

18.90 A 2-kg disk of 200-mm diameter is attached to the end of a rod AB of negligible mass which is supported by a ball-and-socket joint at A. If the disk is observed to precess about the vertical in the sense indicated and at the constant rate of 6 rpm, determine the rate of spin ω of the disk about AB when $\beta = 60°$.

18.91 Determine the rate of precession of the disk of Prob. 18.90 when the disk spins at the rate of 18 000 rpm and $\beta = 50°$.

18.92 A toy top has a radius of gyration of 0.75 in. about its geometric axis and is set spinning at the rate of 1600 rpm. Knowing that $c = 1.50$ in. and assuming steady precession, determine the rate of precession (a) for $\beta = 10°$, (b) for $\beta = 20°$.

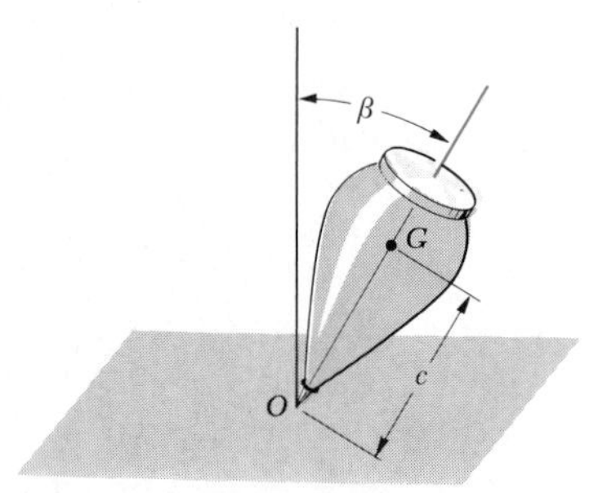

Fig. P18.92 and P18.93

18.93 A toy top having a radius of gyration of 0.75 in. about its geometric axis is set in motion as shown. It is observed that the top precesses through two complete turns in 1 s. Knowing that $c = 1.50$ in. and assuming steady precession, determine the rate at which the top spins about its geometric axis.

18.94 A thin disk of weight $W = 8$ lb rotates with an angular velocity $\boldsymbol{\omega}_2$ with respect to the arm OA, which itself rotates with an angular velocity $\boldsymbol{\omega}_1$ about the y axis. Knowing that $\omega_1 = 4$ rad/s and $\omega_2 = 12$ rad/s and that both are constant, determine the couple due to gyroscopic action exerted on the arm OA by the support at O.

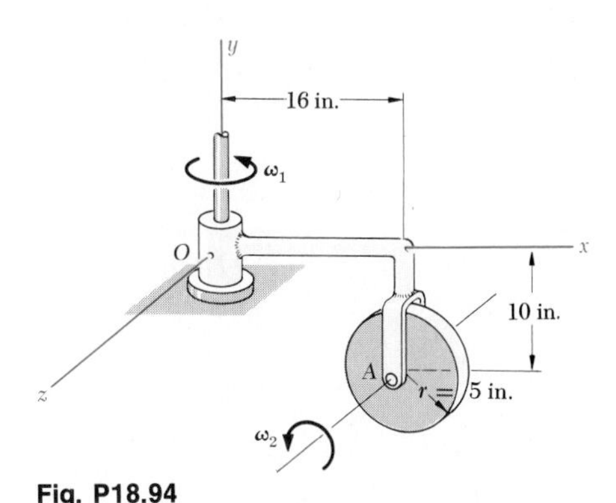

Fig. P18.94

18.95 A thin disk of mass $m = 5$ kg rotates with an angular velocity $\boldsymbol{\omega}_2$ with respect to the bent axle ABC, which itself rotates with an angular velocity $\boldsymbol{\omega}_1$ about the y axis. Knowing that $\omega_1 = 3$ rad/s and $\omega_2 = 8$ rad/s and that both are constant, determine the couple due to gyroscopic action exerted on the bent axle at the support at A.

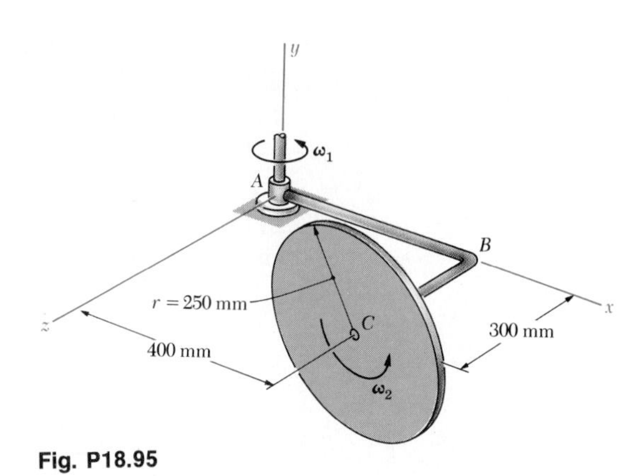

Fig. P18.95

* **18.96** For the assembly of Prob. 18.95, determine the force-couple system representing the total reaction at the support at A.

* **18.97** For the assembly of Prob. 18.94, determine the force-couple system representing the total reaction at the support at O.

18.10. Review and Summary. In this chapter, we used the method of impulse and momentum to analyze the motion of rigid bodies. Except for the last section, our study was limited to the plane motion of rigid slabs and rigid bodies symmetrical with respect to the reference plane.

Principle of impulse and momentum for a rigid body

We first recalled the *principle of impulse and momentum* as it was derived in Sec. 14.13 for a system of particles and applied it to the *motion of a rigid body* [Sec. 18.2]. We wrote

$$\textbf{Syst Momenta}_1 + \textbf{Syst Ext Imp}_{1\to 2} = \textbf{Syst Momenta}_2 \quad (18.1)$$

Next we showed that for a rigid slab or a rigid body symmetrical with respect to the reference plane, the system of the momenta of the particles forming the body is equivalent to a *momentum vector* $m\overline{\mathbf{v}}$ attached at the mass center G of the body and a *momentum couple* $\overline{I}\boldsymbol{\omega}$ (Fig. 18.16). The

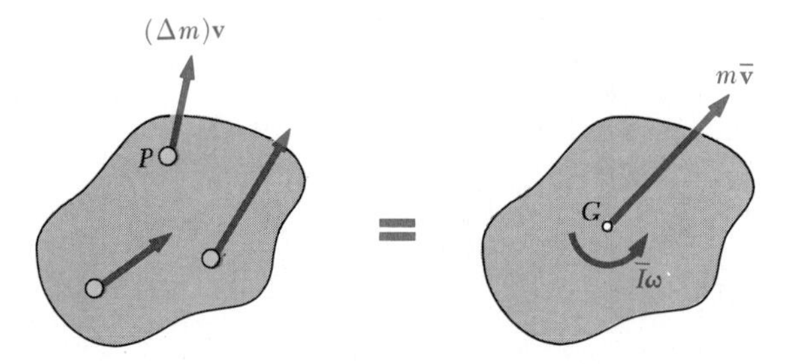

Fig. 18.16

momentum vector $m\overline{\mathbf{v}}$ is associated with the translation of the body with G and represents the *linear momentum* of the body, while the momentum couple $\overline{I}\boldsymbol{\omega}$ corresponds to the rotation of the body about G and its moment $\overline{I}\omega$ represents the *angular momentum* of the body about an axis through G [Sec. 18.3].

Equation (18.1) may be expressed graphically as shown in Fig. 18.17 by drawing three diagrams representing respectively the system of the initial momenta of the body, the impulses of the external forces acting on it, and the system of the final momenta of the body [Sec. 18.4]. Summing and

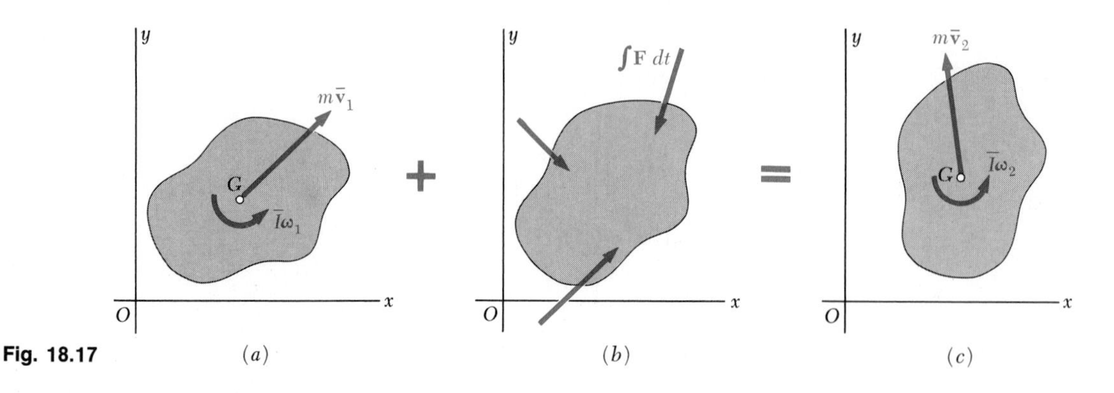

Fig. 18.17

equating respectively the x *components,* the y *components,* and the *moments about any given point* of the vectors shown in that figure, we obtain three equations of motion which may be solved for the desired unknowns [Sample Probs. 18.1 and 18.2].

Connected rigid bodies

In problems dealing with several connected rigid bodies [Sec. 18.5], each body may be considered separately [Sample Prob. 18.1] or, if no more than three unknowns are involved, the principle of impulse and momentum may be applied to the entire system, considering the impulses of the external forces only [Sample Prob. 18.3].

Conservation of angular momentum

When the lines of action of all the external forces acting on a system of rigid bodies pass through a given point O, the angular momentum of the system about O is conserved [Sec. 18.6]. It was suggested that problems involving conservation of angular momentum be solved by the general method described above [Sample Prob. 18.3].

Impulsive motion

The second part of the chapter was devoted to the *impulsive motion* and the *eccentric impact* of rigid bodies. In Sec. 18.7, we recalled that the method of impulse and momentum is the only practicable method for the solution of problems involving impulsive motion and that the computation of impulses in such problems is particularly simple [Sample Prob. 18.4].

Eccentric impact

In Sec. 18.8, we recalled that the eccentric impact of two rigid bodies is defined as an impact in which the mass centers of the colliding bodies are *not* located on the line of impact. It was shown that in such a situation a relation similar to that derived in Chap. 14 for the central impact of two particles and involving the coefficient of restitution e still holds, but that *the velocities of the points A and B where contact occurs during the impact should be used.* We have

$$(v'_B)_n - (v'_A)_n = e[(v_A)_n - (v_B)_n] \tag{18.10}$$

where $(v_A)_n$ and $(v_B)_n$ are the components along the line of impact of the velocities of A and B before the impact, and $(v'_A)_n$ and $(v'_B)_n$ their components after the impact (Fig. 18.18). Equation (18.10) is applicable not only

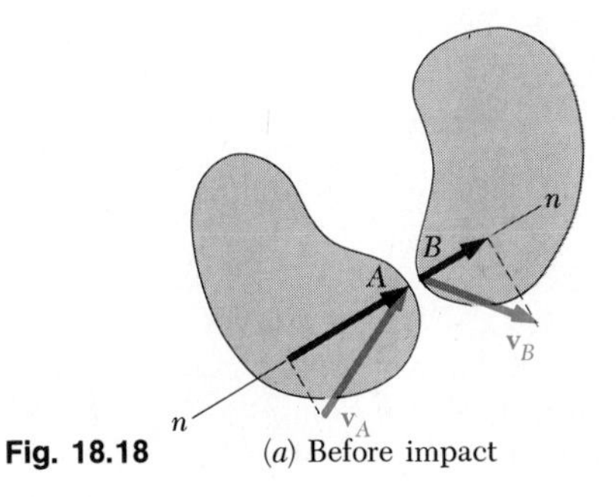

(a) Before impact

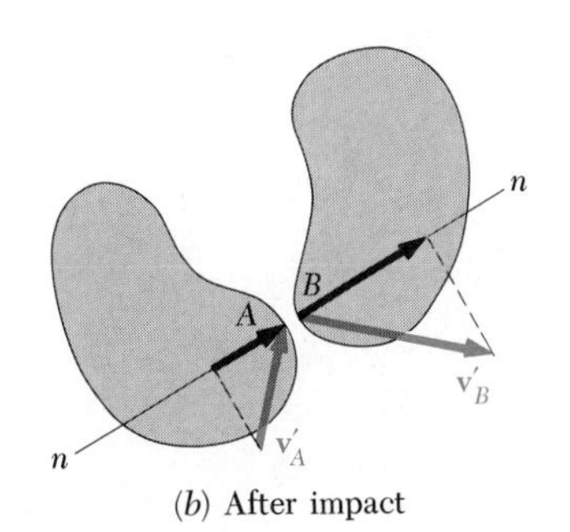

(b) After impact

Fig. 18.18

when the colliding bodies move freely after the impact but also when the bodies are partially constrained in their motion. It should be used in conjunction with one or several other equations obtained by applying the principle of impulse and momentum [Sample Prob. 18.5]. We also considered problems where the method of impulse and momentum and the method of work and energy may be combined [Sample Prob. 18.6].

Gyroscopic motion. Steady precession

In the last part of the chapter, we considered the *gyroscopic motion* of rigid bodies [Sec. 18.9]. We found that when a body of revolution which is spinning at the rate ω about its axis of symmetry (*spin axis*) is subjected to a couple of moment M about an axis (*couple axis*) perpendicular to the spin axis, the body—if free to do so—will *precess* at a rate Ω about an axis (*precession axis*) perpendicular to the other two. The sense of precession is as shown in Fig. 18.19, and the moment M of the couple, the rate of spin ω, and the rate of precession Ω must satisfy the relation

$$M = \bar{I}\omega\Omega \tag{18.20}$$

where $\bar{I}$ is the moment of inertia of the body about its axis of symmetry [Sample Prob. 18.7].

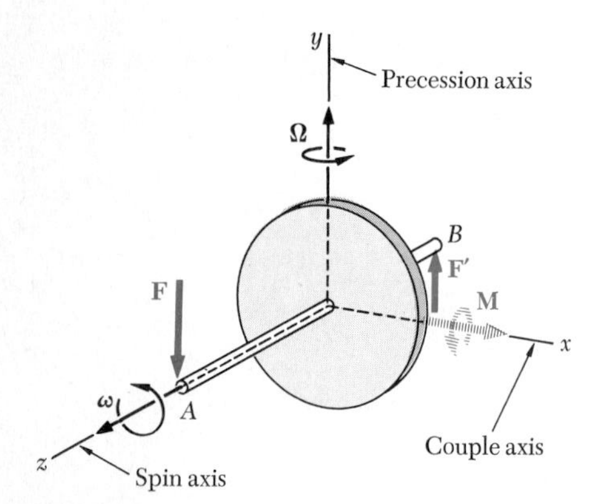

Fig. 18.19

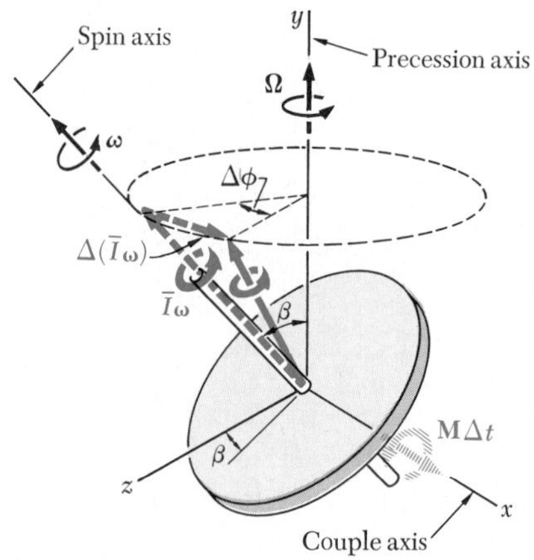

Fig. 18.20

General case of steady precession

In the more general case of steady precession where the spin axis and the precession axis form an angle β (Fig. 18.20), the moment M of the couple which should be applied about an axis perpendicular to the other two to maintain this motion was found to be

$$M = \bar{I}\omega\Omega \sin \beta \tag{18.21}$$

This formula, however, is approximate and should be used only when the rate of spin is much larger than the rate of precession.

Review Problems

18.98 A uniform sphere of radius R rolls to the right without slipping on a horizontal surface and strikes a step of height $\frac{1}{2}R$. Assuming that the impact is perfectly plastic and that no slipping occurs between the sphere and the corner of the step, determine the angular velocity of the sphere and the velocity of its center immediately after the impact.

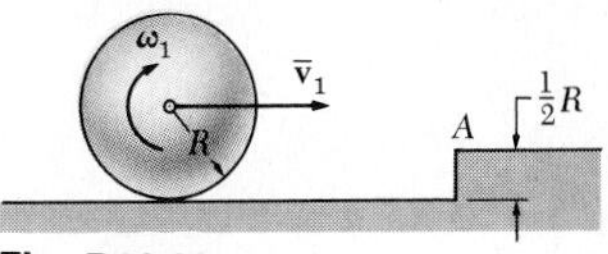

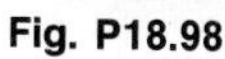

Fig. P18.98

18.99 The ends of a chain having a mass per unit length of 1.6 kg/m lie in piles at A and C. When released from rest at time $t = 0$, the chain moves over the pulley at B which has a mass of 2.5 kg, a radius of gyration of 160 mm, and an outside radius of 200 mm. It is known that $h = 300$ mm, and that the length of the chain connecting the two piles is $L = 4$ m. Neglecting friction, determine at $t = 2$ s (*a*) the speed v of the chain, (*b*) the length of chain which has been transferred from pile A to pile C. (Compare with Prob. 14.130 where the mass of the pulley was neglected.)

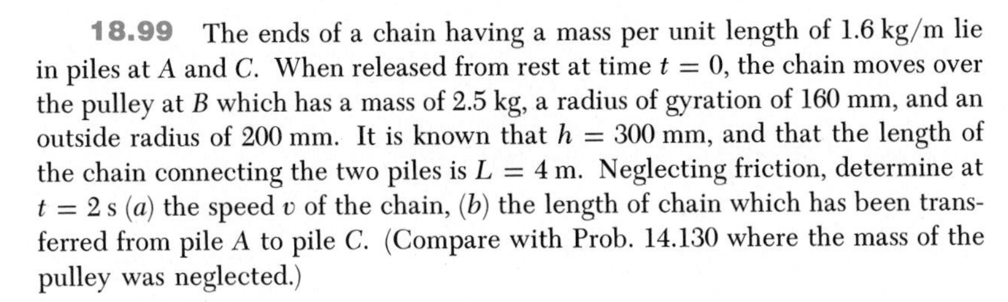

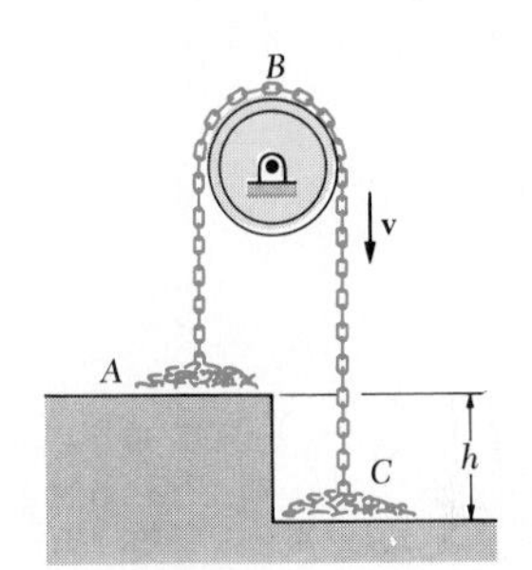

Fig. P18.99

18.100 The uniform rod AB of mass m is at rest on a frictionless horizontal surface when hook C of slider D engages a small pin at A. Knowing that the slider is pulled upward with a constant velocity $\mathbf{v}_0$, determine the impulse exerted on the rod (*a*) at A, (*b*) at B. Assume that the velocity of the slider is unchanged and that the impact is perfectly plastic.

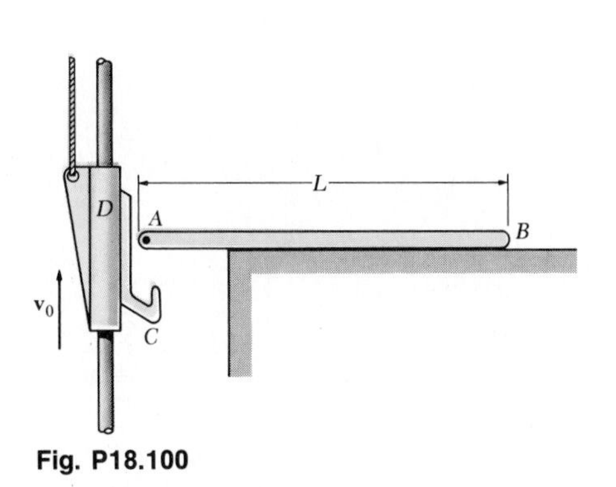

Fig. P18.100

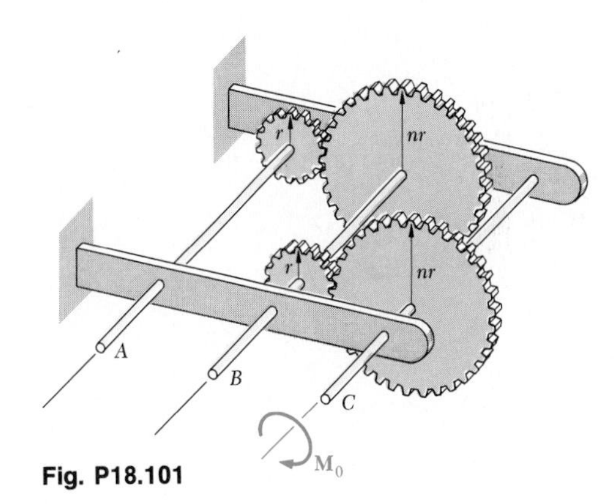

Fig. P18.101

18.101 The gear train shown consists of four gears of the same thickness and of the same material; two gears are of radius r, and the other two are of radius nr. The system is at rest when the couple $\mathbf{M}_0$ is applied to shaft C. Denoting by I_0 the moment of inertia of a gear of radius r, determine the angular velocity of shaft A if the couple $\mathbf{M}_0$ is applied for t seconds to shaft C.

18.102 A small 400-g ball may slide in a slender tube of length 1 m and of mass 1.2 kg which rotates freely about a vertical axis passing through its center C. If the angular velocity of the tube is 8 rad/s as the ball passes through C, determine the angular velocity of the tube (*a*) just before the ball leaves the tube, (*b*) just after the ball has left the tube.

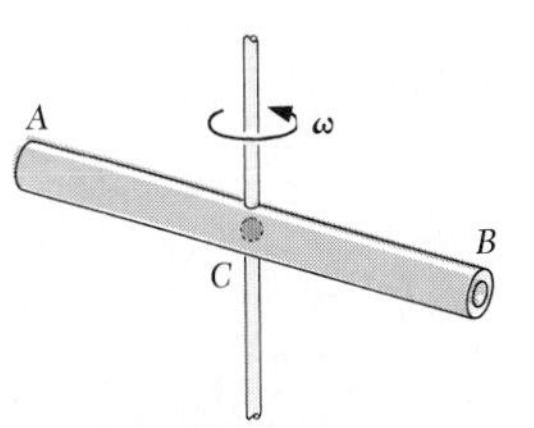

Fig. P18.102

18.103 A bullet of mass m is fired with a horizontal velocity $\mathbf{v}_0$ into the lower corner of a square panel of much larger mass M. The panel is held by two vertical wires as shown. Determine the velocity of the center G of the panel immediately after the bullet becomes embedded.

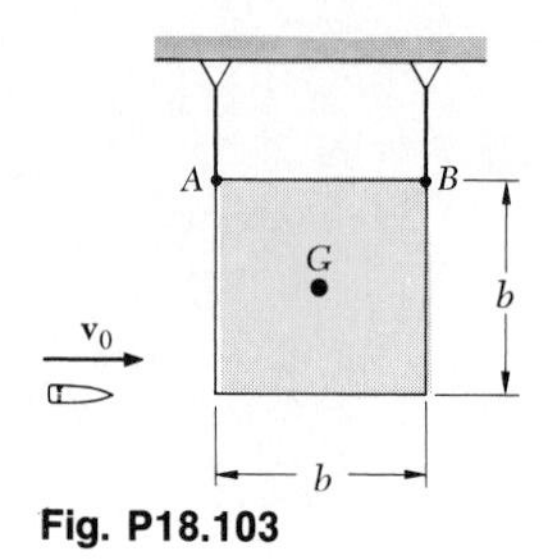

Fig. P18.103

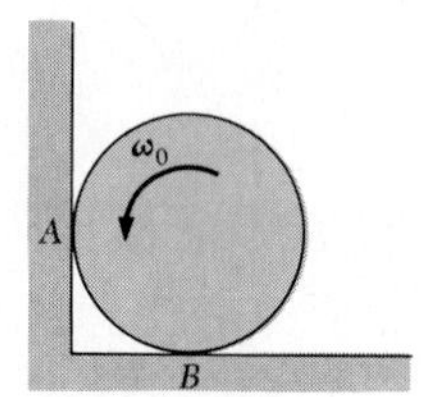

Fig. P18.104

18.104 A cylinder of radius r and mass m is placed in a corner with an initial counterclockwise angular velocity $\boldsymbol{\omega}_0$. Denoting by μ the coefficient of friction at A and B, derive an expression for the time required for the cylinder to come to rest.

18.105 Solve Prob. 18.104, assuming that the surface at A is frictionless.

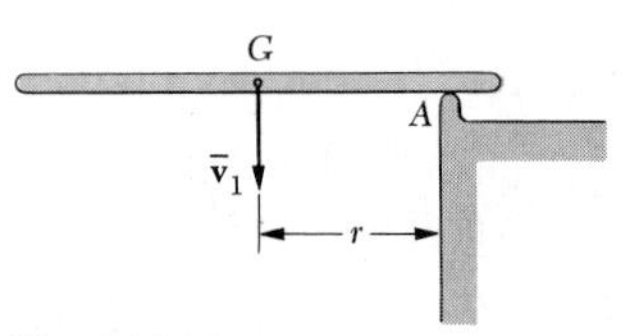

Fig. P18.106

18.106 A slender rigid rod of mass m and with a centroidal radius of gyration $\bar{k}$ strikes a rigid knob at A with a vertical velocity of magnitude $\bar{v}_1$ and no angular velocity. Assuming that the impact is perfectly elastic ($e = 1$), (*a*) show that the magnitude of the velocity of G after the impact is

$$\bar{v}_2 = \bar{v}_1 \frac{r^2 - \bar{k}^2}{r^2 + \bar{k}^2}$$

(*b*) derive an expression for the angular velocity of the rod after the impact.

18.107 The rotor shown spins at the rate of 12,000 rpm about the 6-in. shaft CD. At the same time the gimbal rotates about the vertical axis AB at the rate of 20 rpm. Knowing that the mass of the rotor is 1 oz and that it has a centroidal radius of gyration of 0.8 in., determine the magnitude of the reactions at C and D due to gyroscopic action.

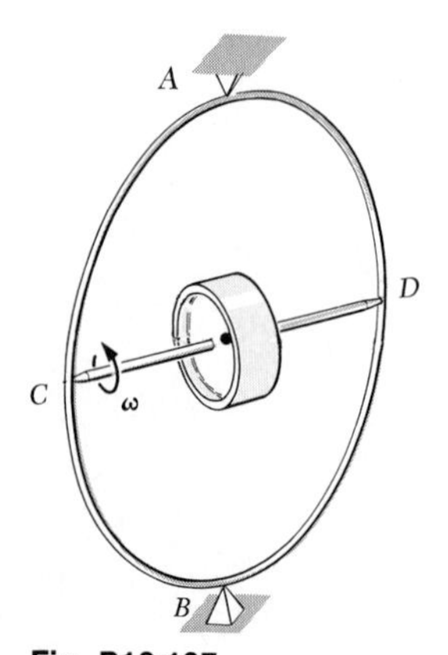

Fig. P18.107

18.108 A square block of mass m moves along a frictionless horizontal surface and strikes a small obstruction at B. Assuming that the impact between corner A and the obstruction B is perfectly plastic, determine the angular velocity of the block and the velocity of its mass center G immediately after the impact.

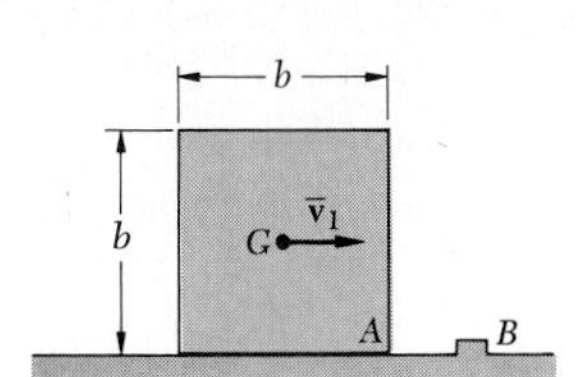

Fig. P18.108

18.109 Solve Prob. 18.108, assuming that the impact between corner A and the obstruction B is perfectly elastic ($e = 1$).

The following problems are designed to be solved with a computer.

18.C1 The 1.5-kg rod AB of length $L = 240$ mm slides freely inside the tube CD of mass m_t. Knowing that the angular velocity of the assembly was $\omega_1 = 8$ rad/s when the rod was entirely inside the tube ($x = 0$), write a computer program and use it to calculate the angular velocity of the assembly and the velocity of the rod relative to the tube for values of x from 40 to 200 mm at 40-mm intervals when (*a*) $m_t = 0.75$ kg, (*b*) $m_t = 1.5$ kg, (*c*) $m_t = 3$ kg.

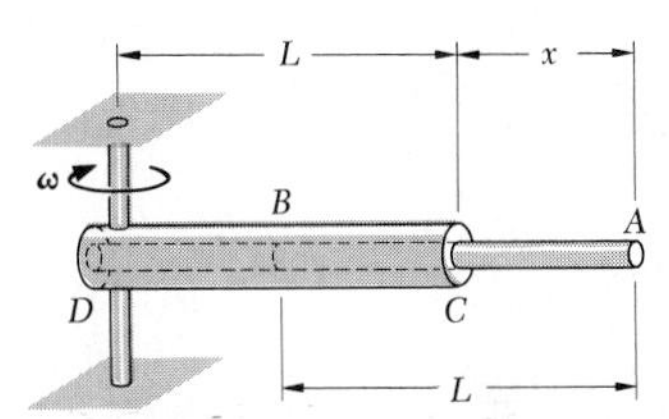

Fig. P18.C1

18.C2 A 2-kg sphere moving horizontally to the right with an initial velocity of 5 m/s strikes the lower end of an 8-kg rigid rod AB. The rod is suspended from a hinge at A and is initially at rest. Write a computer program and use it to calculate the angular velocity of the rod and the velocity of the sphere immediately after the impact, for values of the coefficient of restitution e from 0 to 1 at 0.1 intervals. Expand the program written to calculate the energy lost during the impact for the same values of e.

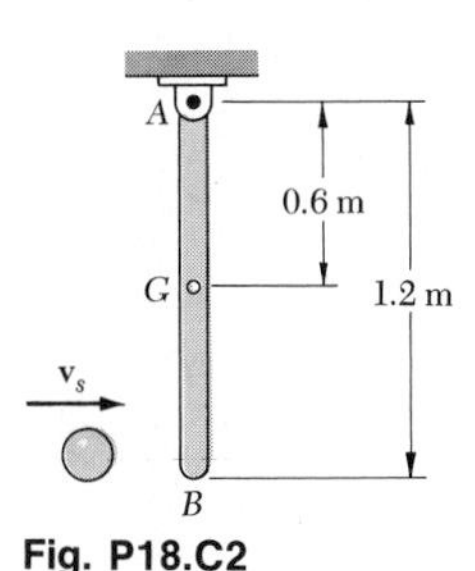

Fig. P18.C2

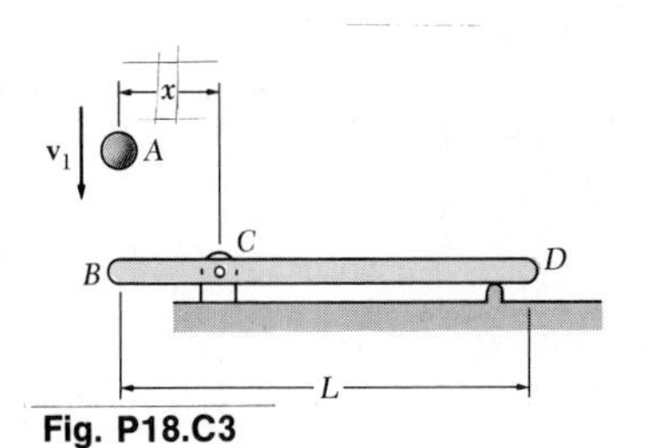

Fig. P18.C3

18.C3 A 0.8-kg sphere is dropped onto the end of a slender rod BD of length $L = 900$ mm and mass $m = 1.6$ kg. Knowing that the sphere strikes end B of the rod with a speed $v_1 = 10$ m/s and assuming perfectly elastic impact, write a computer program and use it to calculate the velocity of the sphere and the angular velocity of the rod immediately after impact, for values of x from 0 to 400 mm at 25-mm intervals.

18.C4 A 15-lb square package of side $a = 18$ in. moves down a conveyor belt A at a constant speed v_1. At the end of the conveyor belt, the corner of the package strikes a rigid support at B. Assuming perfectly plastic impact, write a computer program and use it to calculate for values of θ from 0 to 40° at 5° intervals (*a*) the smallest speed v_1 for which the package will reach conveyor belt C, (*b*) the energy lost in the impact.

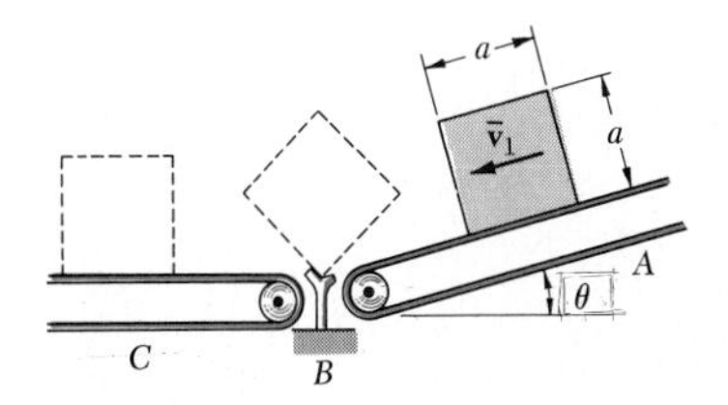

Fig. P18.C4

CHAPTER NINETEEN Mechanical Vibrations

19.1. Introduction. A *mechanical vibration* is the motion of a particle or a body which oscillates about a position of equilibrium. Most vibrations in machines and structures are undesirable because of the increased stresses and energy losses which accompany them. They should therefore be eliminated or reduced as much as possible by appropriate design. The analysis of vibrations has become increasingly important in recent years owing to the current trend toward higher-speed machines and lighter structures. There is every reason to expect that this trend will continue and that an even greater need for vibration analysis will develop in the future.

The analysis of vibrations is a very extensive subject to which entire texts have been devoted. We shall therefore limit our present study to the simpler types of vibrations, namely, the vibrations of a body or a system of bodies with one degree of freedom.

A mechanical vibration generally results when a system is displaced from a position of stable equilibrium. The system tends to return to this position under the action of restoring forces (either elastic forces, as in the case of a mass attached to a spring, or gravitational forces, as in the case of a pendulum). But the system generally reaches its original position with a certain acquired velocity which carries it beyond that position. Since the process can be repeated indefinitely, the system keeps moving back and forth across its position of equilibrium. The time interval required for the system to complete a full cycle of motion is called the *period* of the vibration. The number of cycles per unit time defines the *frequency*, and the maximum displacement of the system from its position of equilibrium is called the *amplitude* of the vibration.

When the motion is maintained by the restoring forces only, the vibration is said to be a *free vibration* (Secs. 19.2 to 19.6). When a periodic force is applied to the system, the resulting motion is described as a *forced vibration* (Sec. 19.7). When the effects of friction may be neglected, the vibrations are said to be *undamped*. However, all vibrations are actually *damped* to some degree. If a free vibration is only slightly damped, its amplitude slowly decreases until, after a certain time, the motion comes to a stop. But damping may be large enough to prevent any true vibration; the system then slowly regains its original position (Sec. 19.8). A damped forced vibration is maintained as long as the periodic force which produces the vibration is applied. The amplitude of the vibration, however, is affected by the magnitude of the damping forces (Sec. 19.9).

VIBRATIONS WITHOUT DAMPING

19.2. Free Vibrations of Particles. Simple Harmonic Motion. Consider a body of mass m attached to a spring of constant k (Fig. 19.1*a*). Since at the present time we are concerned only with the motion of its mass center, we shall refer to this body as a particle. When the particle is in static equilibrium, the forces acting on it are its weight $\mathbf{W}$ and the force $\mathbf{T}$ exerted by the spring, of magnitude $T = k\delta_{st}$, where δ_{st} denotes the elongation of the spring. We have, therefore,

$$W = k\delta_{st} \tag{19.1}$$

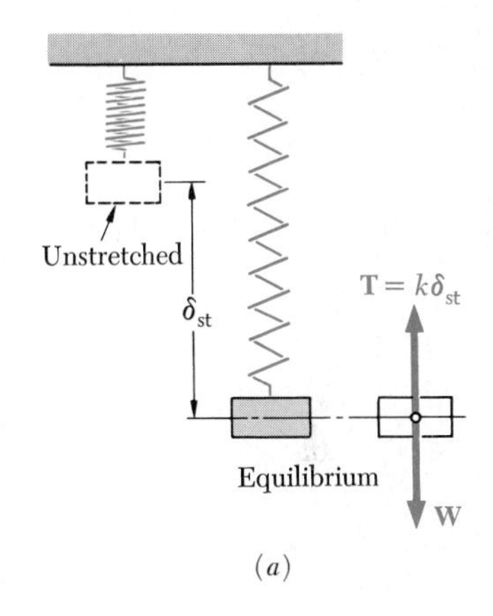

Suppose now that the particle is displaced through a distance x_m from its equilibrium position and released with no initial velocity. If x_m has been chosen smaller than δ_{st}, the particle will move back and forth through its equilibrium position; a vibration of amplitude x_m has been generated. Note that the vibration may also be produced by imparting a certain initial velocity to the particle when it is in its equilibrium position $x = 0$ or, more generally, by starting the particle from any given position $x = x_0$ with a given initial velocity $\mathbf{v}_0$.

To analyze the vibration, we shall consider the particle in a position P at some arbitrary time t (Fig. 19.1*b*). Denoting by x the displacement OP measured from the equilibrium position O (positive downward), we note that the forces acting on the particle are its weight $\mathbf{W}$ and the force $\mathbf{T}$ exerted by the spring which, in this position, has a magnitude $T = k(\delta_{st} + x)$. Recalling (19.1), we find that the magnitude of the resultant $\mathbf{F}$ of the two forces (positive downward) is

$$F = W - k(\delta_{st} + x) = -kx \tag{19.2}$$

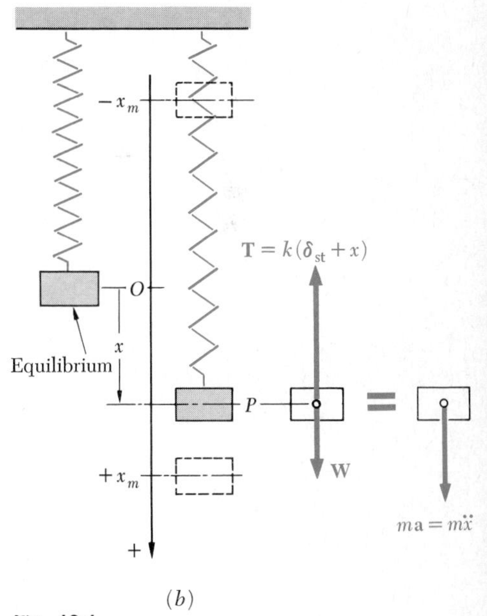

Fig. 19.1

Thus the *resultant* of the forces exerted on the particle is proportional to the displacement OP *measured from the equilibrium position.* Recalling the sign convention, we note that $\mathbf{F}$ is always directed *toward* the equilibrium position O. Substituting for F into the fundamental equation $F = ma$ and recalling that a is the second derivative $\ddot{x}$ of x with respect to t, we write

$$m\ddot{x} + kx = 0 \tag{19.3}$$

Note that the same sign convention should be used for the acceleration $\ddot{x}$ and for the displacement x, namely, positive downward.

Equation (19.3) is a linear differential equation of the second order. Setting

$$p^2 = \frac{k}{m} \tag{19.4}$$

we write (19.3) in the form

$$\ddot{x} + p^2 x = 0 \tag{19.5}$$

The motion defined by Eq. (19.5) is called *simple harmonic motion.* It is characterized by the fact that *the acceleration is proportional to the displacement and of opposite direction.* We note that each of the functions $x_1 = \sin pt$ and $x_2 = \cos pt$ satisfies (19.5). These functions, therefore, constitute two *particular solutions* of the differential equation (19.5). As we shall see presently, the *general solution* of (19.5) may be obtained by multiplying the two particular solutions by arbitrary constants A and B and adding. We write

$$x = Ax_1 + Bx_2 = A \sin pt + B \cos pt \tag{19.6}$$

Differentiating, we obtain successively the velocity and acceleration at time t:

$$v = \dot{x} = Ap \cos pt - Bp \sin pt \tag{19.7}$$

$$a = \ddot{x} = -Ap^2 \sin pt - Bp^2 \cos pt \tag{19.8}$$

Substituting from (19.6) and (19.8) into (19.5), we verify that the expression (19.6) provides a solution of the differential equation (19.5). Since this expression contains two arbitrary constants A and B, the solution obtained is the general solution of the differential equation. The values of the constants A and B depend upon the *initial conditions* of the motion. For example, we have $A = 0$ if the particle is displaced from its equilibrium position and released at $t = 0$ with no initial velocity, and we have $B = 0$ if P is started from O at $t = 0$ with a certain initial velocity. In general, substituting $t = 0$ and the initial values x_0 and v_0 of the displacement and velocity into (19.6) and (19.7), we find $A = v_0/p$ and $B = x_0$.

The expressions obtained for the displacement, velocity, and acceleration of a particle may be written in a more compact form if we observe that (19.6) expresses that the displacement $x = OP$ is the sum of the x components of two vectors **A** and **B**, respectively, of magnitude A and B, directed as shown in Fig. 19.2*a*. As t varies, both vectors rotate clockwise; we also note that the magnitude of their resultant $\overrightarrow{OQ}$ is equal to the maximum displacement x_m. The simple harmonic motion of P along the x axis may thus be obtained by projecting on this axis the motion of a point Q describing an *auxiliary circle* of radius x_m *with a constant angular velocity* p.

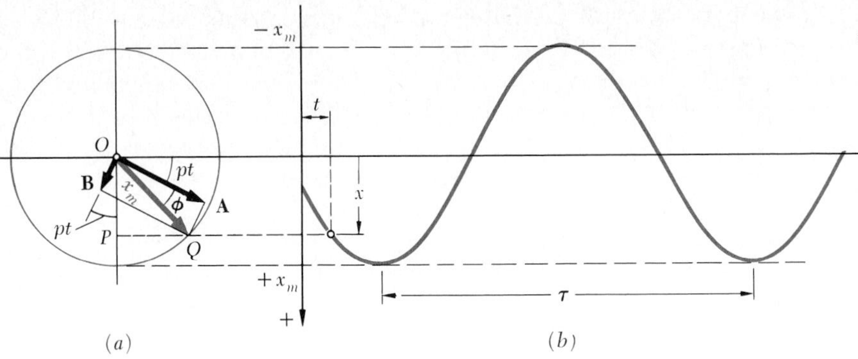

Fig. 19.2

Denoting by ϕ the angle formed by the vectors $\overrightarrow{OQ}$ and **A**, we write

$$OP = OQ \sin (pt + \phi) \tag{19.9}$$

which leads to new expressions for the displacement, velocity, and acceleration of P:

$$x = x_m \sin (pt + \phi) \tag{19.10}$$

$$v = \dot{x} = x_m p \cos (pt + \phi) \tag{19.11}$$

$$a = \ddot{x} = -x_m p^2 \sin (pt + \phi) \tag{19.12}$$

The displacement-time curve is represented by a sine curve (Fig. 19.2*b*), and the maximum value x_m of the displacement is called the *amplitude* of the vibration. The angular velocity p of the point Q which describes the auxiliary circle is known as the *circular frequency* of the vibration and is measured in rad/s, while the angle ϕ which defines the initial position of Q on the circle is called the *phase angle*. We note from Fig. 19.2 that a full *cycle* has been described after the angle pt has increased by 2π rad. The corresponding value of t, denoted by τ, is called the *period* of the vibration and is measured in seconds. We have

$$\text{Period} = \tau = \frac{2\pi}{p} \tag{19.13}$$

The number of cycles described per unit of time is denoted by f and is known as the *frequency* of the vibration. We write

$$\text{Frequency} = f = \frac{1}{\tau} = \frac{p}{2\pi} \tag{19.14}$$

The unit of frequency is a frequency of 1 cycle per second, corresponding to a period of 1 s. In terms of base units the unit of frequency is thus 1/s or s^{-1}. It is called a *hertz* (Hz) in the SI system of units. It also follows from Eq. (19.14) that a frequency of 1 s^{-1} or 1 Hz corresponds to a circular frequency of 2π rad/s. In problems involving angular velocities expressed in revolutions per minute (rpm), we have $1 \text{ rpm} = \frac{1}{60} \text{ s}^{-1} = \frac{1}{60} \text{ Hz}$, or $1 \text{ rpm} = (2\pi/60)$ rad/s.

Recalling that p was defined in (19.4) in terms of the constant k of the spring and the mass m of the particle, we observe that the period and the frequency are independent of the initial conditions and of the amplitude of the vibration. Note that τ and f depend on the *mass* rather than on the *weight* of the particle and thus are independent of the value of g.

The velocity-time and acceleration-time curves may be represented by sine curves of the same period as the displacement-time curve, but with different phase angles. From (19.11) and (19.12), we note that the maximum values of the magnitudes of the velocity and acceleration are

$$v_m = x_m p \qquad a_m = x_m p^2 \tag{19.15}$$

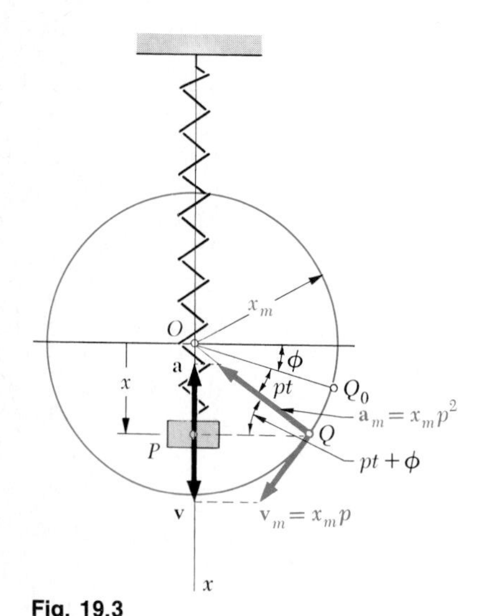

Fig. 19.3

Since the point Q describes the auxiliary circle, of radius x_m, at the constant angular velocity p, its velocity and acceleration are equal, respectively, to the expressions (19.15). Recalling Eqs. (19.11) and (19.12), we find, therefore, that the velocity and acceleration of P may be obtained at any instant by projecting on the x axis vectors of magnitudes $v_m = x_m p$ and $a_m = x_m p^2$ representing, respectively, the velocity and acceleration of Q at the same instant (Fig. 19.3).

The results obtained are not limited to the solution of the problem of a mass attached to a spring. They may be used to analyze the rectilinear motion of a particle *whenever the resultant* **F** *of the forces acting on the particle is proportional to the displacement x and directed toward O.* The fundamental equation of motion $F = ma$ may then be written in the form (19.5), which characterizes simple harmonic motion. Observing that the coefficient of x in (19.5) represents the square of the circular frequency p of the vibration, we easily obtain p and, after substitution into (19.13) and (19.14), the period τ and the frequency f of the vibration.

19.3. Simple Pendulum (Approximate Solution). Most of the vibrations encountered in engineering applications may be represented by a simple harmonic motion. Many others, although of a different type, may be *approximated* by a simple harmonic motion, provided that their amplitude remains small. Consider, for example, a *simple pendulum*, consisting of a bob of mass m attached to a cord of length l, which may oscillate in a vertical plane (Fig. 19.4*a*). At a given time t, the cord forms an angle θ with the vertical. The forces acting on the bob are its weight **W** and the force **T** exerted by the cord (Fig. 19.4*b*). Resolving the vector $m\mathbf{a}$ into tangential and normal components, with $m\mathbf{a}_t$ directed to the right, i.e., in

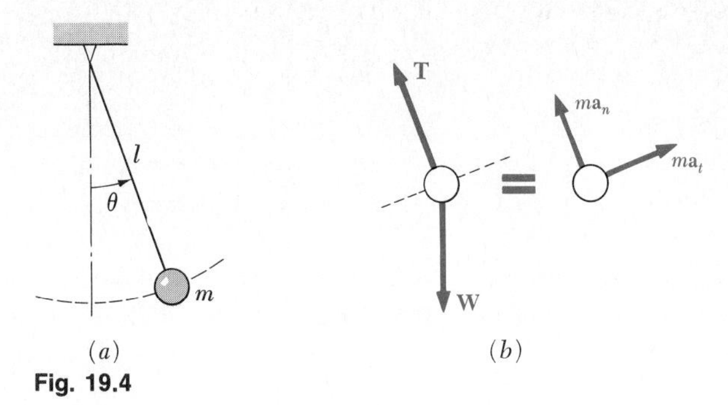

Fig. 19.4

the direction corresponding to increasing values of θ, and observing that $a_t = l\alpha = l\ddot{\theta}$, we write

$$\Sigma F_t = ma_t: \qquad -W \sin \theta = ml\ddot{\theta}$$

Noting that $W = mg$ and dividing through by ml, we obtain

$$\ddot{\theta} + \frac{g}{l} \sin \theta = 0 \tag{19.16}$$

For oscillations of small amplitude, we may replace $\sin \theta$ by θ, expressed in radians, and write

$$\ddot{\theta} + \frac{g}{l} \theta = 0 \tag{19.17}$$

Comparison with (19.5) shows that the equation obtained is that of a simple harmonic motion and that the circular frequency p of the oscillations is equal to $(g/l)^{1/2}$. Substitution into (19.13) yields the period of the small oscillations of a pendulum of length l:

$$\tau = \frac{2\pi}{p} = 2\pi \sqrt{\frac{l}{g}} \tag{19.18}$$

*** 19.4. Simple Pendulum (Exact Solution).** Formula (19.18) is only approximate. To obtain an exact expression for the period of the oscillations of a simple pendulum, we must return to (19.16). Multiplying both terms by $2\dot{\theta}$ and integrating from an initial position corresponding to the maximum deflection, that is, $\theta = \theta_m$ and $\dot{\theta} = 0$, we write

$$\dot{\theta}^2 = \frac{2g}{l} (\cos \theta - \cos \theta_m)$$

or

$$\left(\frac{d\theta}{dt}\right)^2 = \frac{2g}{l} (\cos \theta - \cos \theta_m)$$

Replacing $\cos\theta$ by $1 - 2\sin^2(\theta/2)$ and $\cos\theta_m$ by a similar expression, solving for dt, and integrating over a quarter period from $t = 0$, $\theta = 0$ to $t = \tau/4$, $\theta = \theta_m$, we have

$$\tau = 2\sqrt{\frac{l}{g}}\int_0^{\theta_m}\frac{d\theta}{\sqrt{\sin^2(\theta_m/2) - \sin^2(\theta/2)}}$$

The integral in the right-hand member is known as an *elliptic integral;* it cannot be expressed in terms of the usual algebraic or trigonometric functions. However, setting

$$\sin(\theta/2) = \sin(\theta_m/2)\sin\phi$$

we may write

$$\tau = 4\sqrt{\frac{l}{g}}\int_0^{\pi/2}\frac{d\phi}{\sqrt{1 - \sin^2(\theta_m/2)\sin^2\phi}} \qquad (19.19)$$

where the integral obtained, commonly denoted by K, may be found in *tables of elliptic integrals* for various values of $\theta_m/2$.† In order to compare the result just obtained with that of the preceding section, we write (19.19) in the form

$$\tau = \frac{2K}{\pi}\left(2\pi\sqrt{\frac{l}{g}}\right) \qquad (19.20)$$

Formula (19.20) shows that the actual value of the period of a simple pendulum may be obtained by multiplying the approximate value (19.18) by the correction factor $2K/\pi$. Values of the correction factor are given in Table 19.1 for various values of the amplitude θ_m. We note that for ordinary engineering computations the correction factor may be omitted as long as the amplitude does not exceed 10°.

Table 19.1 Correction Factor for the Period of a Simple Pendulum

θ_m	0°	10°	20°	30°	60°	90°	120°	150°	180°
K	1.571	1.574	1.583	1.598	1.686	1.854	2.157	2.768	∞
$2K/\pi$	1.000	1.002	1.008	1.017	1.073	1.180	1.373	1.762	∞

† See, for example, "Standard Mathematical Tables," Chemical Rubber Publishing Company, Cleveland, Ohio.

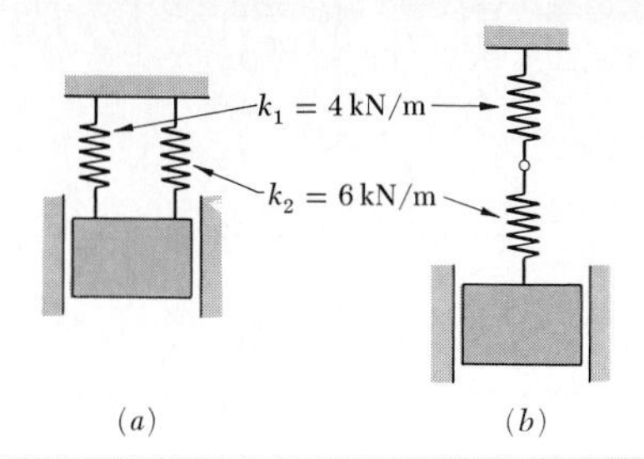

SAMPLE PROBLEM 19.1

A 50-kg block moves between vertical guides as shown. The block is pulled 40 mm down from its equilibrium position and released. For each spring arrangement, determine the period of the vibration, the maximum velocity of the block, and the maximum acceleration of the block.

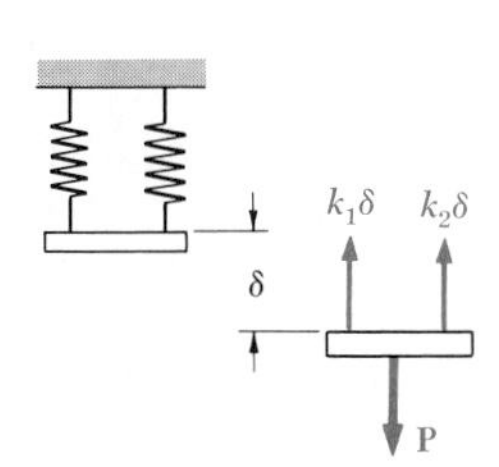

a. **Springs Attached in Parallel.** We first determine the constant k of a single spring equivalent to the two springs *by finding the magnitude of the force* **P** required to cause a given deflection δ. Since for a deflection δ the magnitudes of the forces exerted by the springs are, respectively, $k_1\delta$ and $k_2\delta$, we have

$$P = k_1\delta + k_2\delta = (k_1 + k_2)\delta$$

The constant k of the single equivalent spring is

$$k = \frac{P}{\delta} = k_1 + k_2 = 4\text{ kN/m} + 6\text{ kN/m} = 10\text{ kN/m} = 10^4\text{ N/m}$$

Period of Vibration: Since $m = 50$ kg, Eq. (19.4) yields

$$p^2 = \frac{k}{m} = \frac{10^4\text{ N/m}}{50\text{ kg}} \qquad p = 14.14\text{ rad/s}$$

$$\tau = 2\pi/p \qquad \tau = 0.444\text{ s} \blacktriangleleft$$

Maximum Velocity: $v_m = x_m p = (0.040\text{ m})(14.14\text{ rad/s})$

$$v_m = 0.566\text{ m/s} \qquad \mathbf{v}_m = 0.566\text{ m/s} \updownarrow \blacktriangleleft$$

Maximum Acceleration: $a_m = x_m p^2 = (0.040\text{ m})(14.14\text{ rad/s})^2$

$$a_m = 8.00\text{ m/s}^2 \qquad \mathbf{a}_m = 8.00\text{ m/s}^2 \updownarrow \blacktriangleleft$$

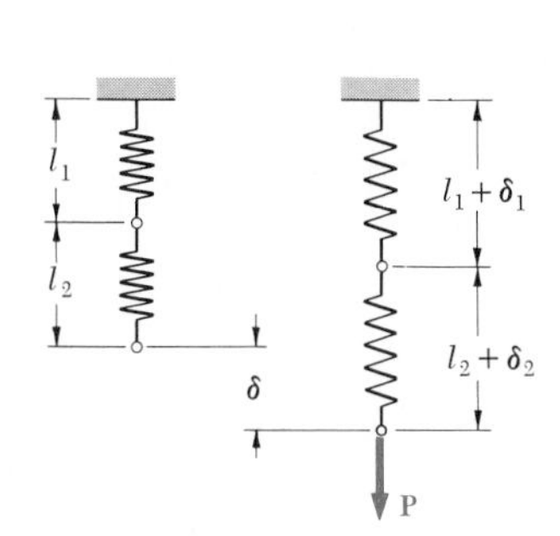

b. **Springs Attached in Series.** We first determine the constant k of a single spring equivalent to the two springs *by finding the total elongation* δ of the springs under a given static load **P**. To facilitate the computation, a static load of magnitude $P = 12$ kN is used.

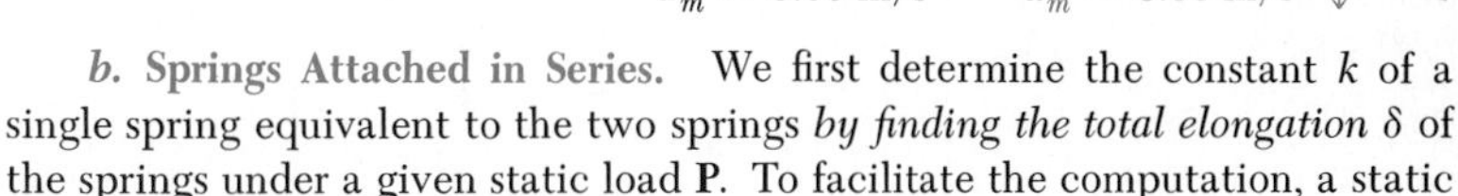

$$\delta = \delta_1 + \delta_2 = \frac{P}{k_1} + \frac{P}{k_2} = \frac{12\text{ kN}}{4\text{ kN/m}} + \frac{12\text{ kN}}{6\text{ kN/m}} = 5\text{ m}$$

$$k = \frac{P}{\delta} = \frac{12\text{ kN}}{5\text{ m}} = 2.4\text{ kN/m} = 2400\text{ N/m}$$

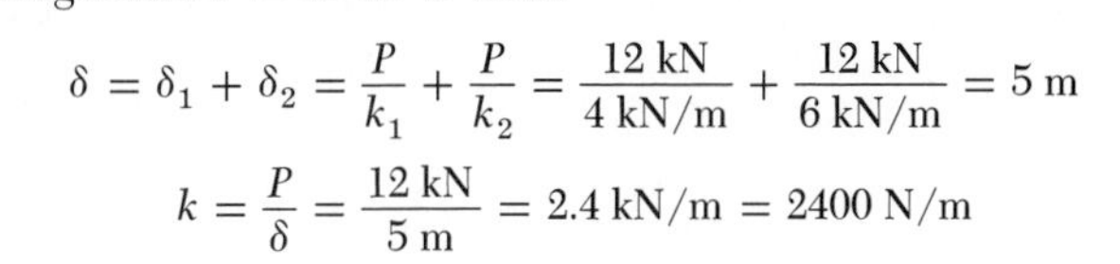

Period of Vibration: $p^2 = \dfrac{k}{m} = \dfrac{2400\text{ N/m}}{50\text{ kg}} \qquad p = 6.93\text{ rad/s}$

$$\tau = \frac{2\pi}{p} \qquad \tau = 0.907\text{ s} \blacktriangleleft$$

Maximum Velocity: $v_m = x_m p = (0.040\text{ m})(6.93\text{ rad/s})$

$$v_m = 0.277\text{ m/s} \qquad \mathbf{v}_m = 0.277\text{ m/s} \updownarrow \blacktriangleleft$$

Maximum Acceleration: $a_m = x_m p^2 = (0.040\text{ m})(6.93\text{ rad/s})^2$

$$a_m = 1.920\text{ m/s}^2 \qquad \mathbf{a}_m = 1.920\text{ m/s}^2 \updownarrow \blacktriangleleft$$

Problems

19.1 A particle is known to move with a simple harmonic motion. The maximum acceleration is 3 m/s^2, and the maximum velocity is 150 mm/s. Determine the amplitude and the frequency of the motion.

19.2 Determine the maximum velocity and maximum acceleration of a particle which moves in simple harmonic motion with an amplitude of 150 mm and a period of 0.9 s.

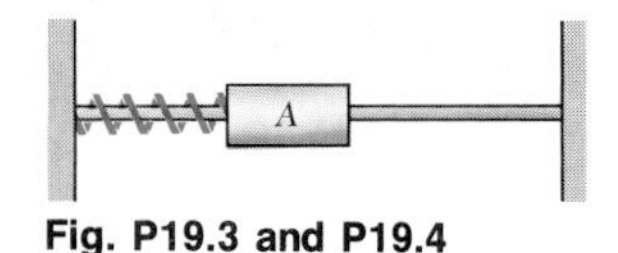

Fig. P19.3 and P19.4

19.3 A 5-lb collar is attached to a spring of constant 3 lb/in. and may slide without friction on a horizontal rod. If the collar is moved 4 in. from its equilibrium position and released, determine the maximum velocity and the maximum acceleration of the collar during the resulting motion.

19.4 A 3-lb collar is attached to a spring of constant 4 lb/in. and may slide without friction on a horizontal rod. The collar is at rest when it is struck with a mallet and given an initial velocity of 50 in./s. Determine the amplitude and the maximum acceleration of the collar during the resulting motion.

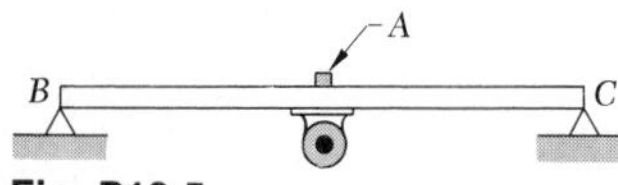

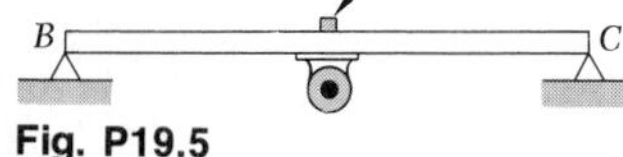

Fig. P19.5

19.5 A variable-speed motor is rigidly attached to the beam BC. The rotor is slightly unbalanced and causes the beam to vibrate with a frequency equal to the motor speed. When the speed of the motor is less than 450 rpm or more than 900 rpm, a small object placed at A is observed to remain in contact with the beam. For speeds between 450 and 900 rpm the object is observed to "dance" and actually to lose contact with the beam. Determine the amplitude of the motion of A when the speed of the motor is (a) 450 rpm, (b) 900 rpm. Give answers in both SI and U.S. customary units.

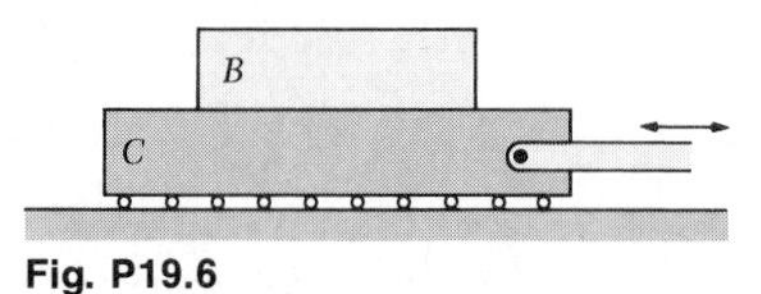

Fig. P19.6

19.6 An instrument package B is placed on the shaking table C as shown. The table is made to move horizontally in simple harmonic motion with a frequency of 3 Hz. Knowing that the coefficient of static friction is $\mu_s = 0.40$ between the package and the table, determine the largest allowable amplitude of the motion if the package is not to slip on the table. Give answers in both SI and U.S. customary units.

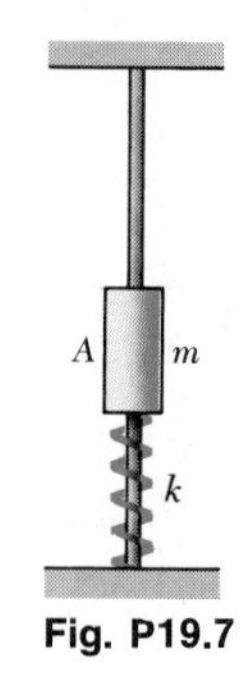

Fig. P19.7

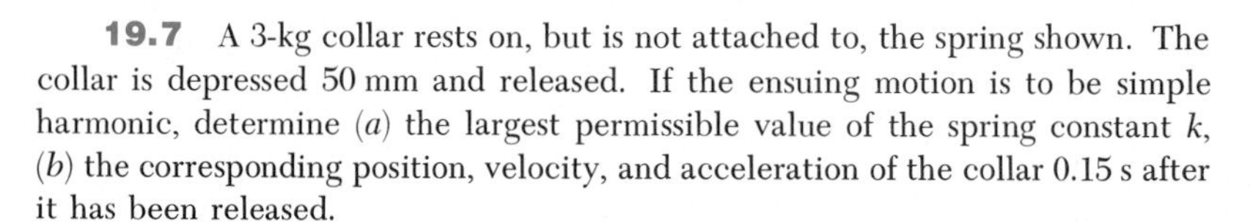

19.7 A 3-kg collar rests on, but is not attached to, the spring shown. The collar is depressed 50 mm and released. If the ensuing motion is to be simple harmonic, determine (a) the largest permissible value of the spring constant k, (b) the corresponding position, velocity, and acceleration of the collar 0.15 s after it has been released.

19.8 A 4-kg collar is attached to a spring of constant $k = 800$ N/m as shown. If the collar is given a displacement of 40 mm downward from its equilibrium position and released, determine (*a*) the time required for the collar to move 60 mm upward, (*b*) the corresponding velocity and acceleration of the collar.

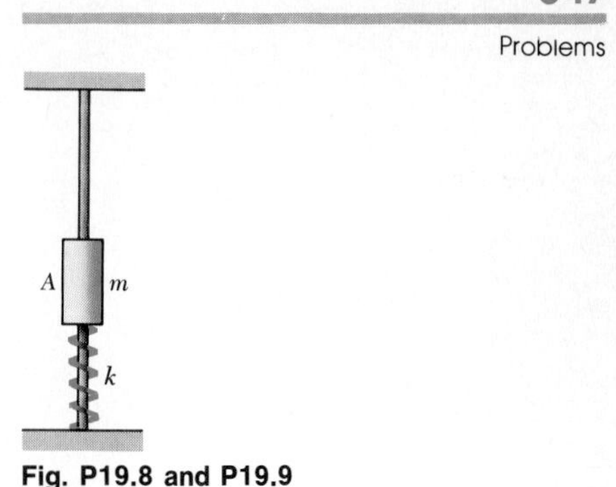

Fig. P19.8 and P19.9

19.9 A 3-lb collar is attached to a spring of constant $k = 5$ lb/in. as shown. If the collar is given a displacement of 2.5 in. downward from its equilibrium position and released, determine (*a*) the time required for the collar to move 2 in. upward, (*b*) the corresponding velocity and acceleration of the collar.

19.10 In Prob. 19.9, determine the position, velocity, and acceleration of the collar 0.20 s after it has been released.

19.11 and 19.12 A 40-lb block is supported by the spring arrangement shown. If the block is moved vertically downward from its equilibrium position and released, determine (*a*) the period and frequency of the resulting motion, (*b*) the maximum velocity and acceleration of the block if the amplitude of the motion is 1.25 in.

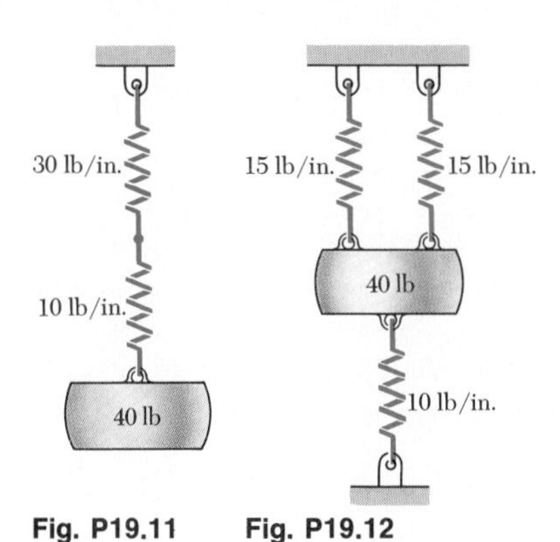

Fig. P19.11 Fig. P19.12

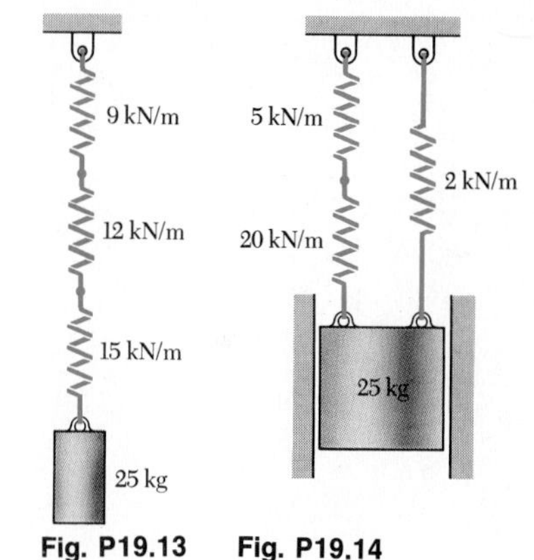

Fig. P19.13 Fig. P19.14

19.13 and 19.14 A 25-kg block is supported by the spring arrangement shown. If the block is moved vertically downward from its equilibrium position and released, determine (*a*) the period and frequency of the resulting motion, (*b*) the maximum velocity and acceleration of the block if the amplitude of the motion is 25 mm.

19.15 The period of vibration of the system shown is observed to be 0.4 s. After cylinder *B* has been removed, the period is observed to be 0.3 s. Determine (*a*) the weight of cylinder *A*, (*b*) the constant of the spring.

19.16 The period of vibration of the system shown is observed to be 1.5 s. After cylinder *B* has been removed and replaced with a 4-lb cylinder, the period is observed to be 1.6 s. Determine (*a*) the weight of cylinder *A*, (*b*) the constant of the spring.

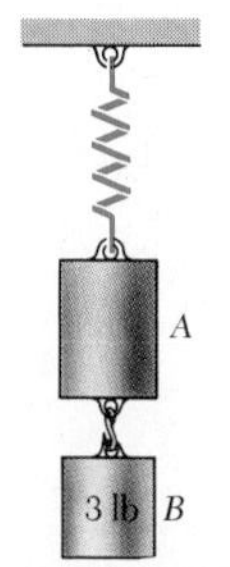

Fig. P19.15 and P19.16

19.17 A tray of mass m is attached to three springs as shown. The period of vibration of the empty tray is 0.5 s. After a 1.5-kg block has been placed in the center of the tray, the period is observed to be 0.6 s. Knowing that the amplitude of the vibration is small, determine the mass m of the tray.

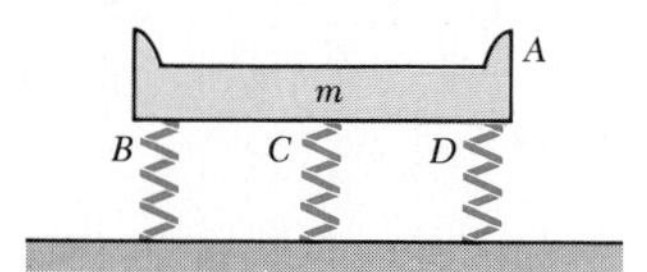

Fig. P19.17 and P19.18

19.18 A tray of mass m is attached to three springs as shown. The period of vibration of the empty tray is 0.75 s. After the center spring C has been removed, the period is observed to be 0.90 s. Knowing that the constant of spring C is 100 N/m, determine the mass m of the tray.

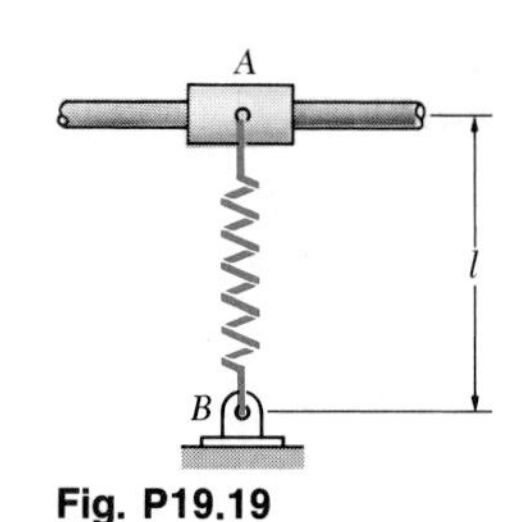

Fig. P19.19

19.19 A collar of mass m slides without friction on a horizontal rod and is attached to a spring AB of constant k. (*a*) If the unstretched length of the spring is just equal to l, show that the collar does not execute simple harmonic motion even when the amplitude of the oscillations is small. (*b*) If the unstretched length of the spring is less than l, show that the motion may be approximated by a simple harmonic motion for small oscillations.

19.20 The rod AB is attached to a hinge at A and to two springs each of constant k. If $h = 600$ mm, $d = 250$ mm, and $m = 25$ kg, determine the value of k for which the period of small oscillations is (*a*) 1 s, (*b*) infinite. Neglect the mass of the rod and assume that each spring can act in either tension or compression.

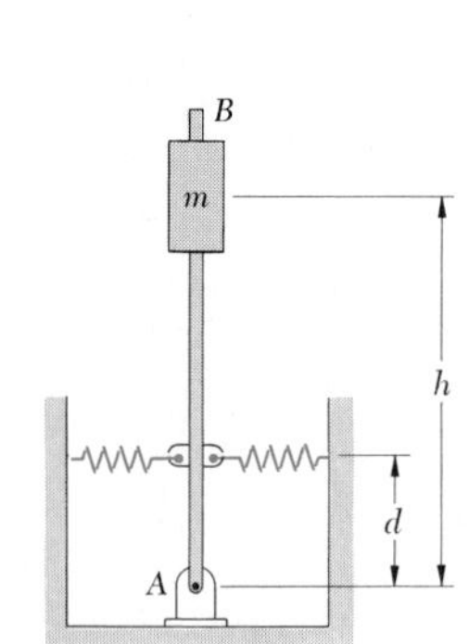

Fig. P19.20 and P19.21

19.21 If $h = 600$ mm and $d = 400$ mm and each spring has a constant $k = 700$ N/m, determine the mass m for which the period of small oscillations is (*a*) 0.50 s, (*b*) infinite. Neglect the mass of the rod and assume that each spring can act in either tension or compression.

19.22 Denoting by δ_{st} the static deflection of a beam under a given load, show that the frequency of vibration of the load is

$$f = \frac{1}{2\pi}\sqrt{\frac{g}{\delta_{st}}}$$

Neglect the mass of the beam, and assume that the load remains in contact with the beam.

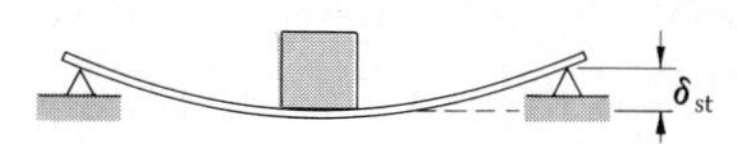

Fig. P19.22

* **19.23** Expanding the integrand in (19.19) into a series of even powers of $\sin\phi$ and integrating, show that the period of a simple pendulum of length l may be approximated by the formula

$$\tau = 2\pi\sqrt{\frac{l}{g}}\left(1 + \tfrac{1}{4}\sin^2\frac{\theta_m}{2}\right)$$

where θ_m is the amplitude of the oscillations.

* **19.24** Using the formula given in Prob. 19.23, determine the amplitude θ_m for which the period of a simple pendulum is 1 percent longer than the period of the same pendulum for small oscillations.

* **19.25** Using the data of Table 19.1, determine the period of a simple pendulum of length $l = 750$ mm (*a*) for small oscillations, (*b*) for oscillations of amplitude $\theta_m = 60°$, (*c*) for oscillations of amplitude $\theta_m = 90°$.

* **19.26** Using a table of elliptic integrals, determine the period of a simple pendulum of length $l = 750$ mm if the amplitude of the oscillations is $\theta_m = 50°$.

19.5. Free Vibrations of Rigid Bodies. The analysis of the vibrations of a rigid body or of a system of rigid bodies possessing a single degree of freedom is similar to the analysis of the vibrations of a particle. An appropriate variable, such as a distance x or an angle θ, is chosen to define the position of the body or system of bodies, and an equation relating this variable and its second derivative with respect to t is written. If the equation obtained is of the same form as (19.5), i.e., if we have

$$\ddot{x} + p^2x = 0 \qquad \text{or} \qquad \ddot{\theta} + p^2\theta = 0 \tag{19.21}$$

the vibration considered is a simple harmonic motion. The period and frequency of the vibration may then be obtained by identifying p and substituting into (19.13) and (19.14).

In general, a simple way to obtain one of Eqs. (19.21) is to express that the system of the external forces is equivalent to the system of the effective forces by drawing a diagram of the body for an arbitrary value of the variable and writing the appropriate equation of motion. We recall that our goal should be *the determination of the coefficient* of the variable x or

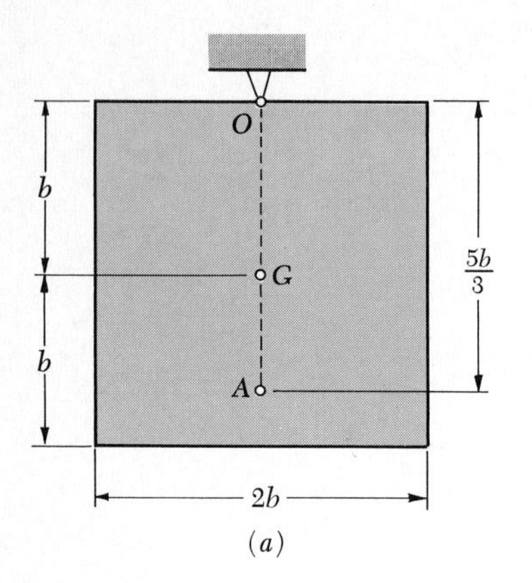

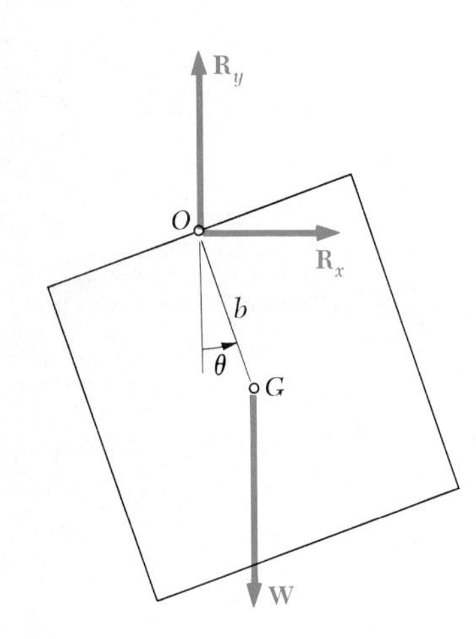

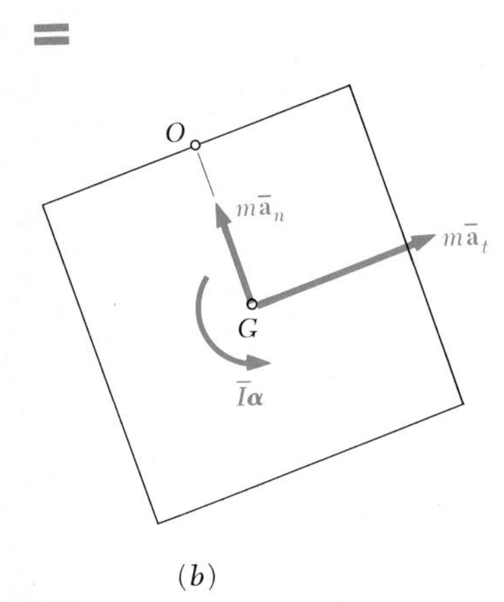

Fig. 19.5

θ, *not* the determination of the variable itself or of the derivative $\ddot{x}$ or $\ddot{\theta}$. Setting this coefficient equal to p^2, we obtain the circular frequency p, from which τ and f may be determined.

The method we have outlined may be used to analyze vibrations which are truly represented by a simple harmonic motion, or vibrations of small amplitude which can be *approximated* by a simple harmonic motion. As an example, we shall determine the period of the small oscillations of a square plate of side $2b$ which is suspended from the midpoint O of one side (Fig. 19.5*a*). We consider the plate in an arbitrary position defined by the angle θ that the line OG forms with the vertical and draw a diagram to express that the weight $\mathbf{W}$ of the plate and the components $\mathbf{R}_x$ and $\mathbf{R}_y$ of the reaction at O are equivalent to the vectors $m\mathbf{a}_t$ and $m\mathbf{a}_n$ and to the couple $\bar{I}\boldsymbol{\alpha}$ (Fig. 19.5*b*). Since the angular velocity and angular acceleration of the plate are equal, respectively, to $\dot{\theta}$ and $\ddot{\theta}$, the magnitudes of the two vectors are, respectively, $mb\ddot{\theta}$ and $mb\dot{\theta}^2$, while the moment of the couple is $\bar{I}\ddot{\theta}$. In previous applications of this method (Chap. 16), we tried whenever possible to assume the correct sense for the acceleration. Here, however, we must assume the same positive sense for θ and $\ddot{\theta}$ in order to obtain an equation of the form (19.21). Consequently, the angular acceleration $\ddot{\theta}$ will be assumed positive counterclockwise, even though this assumption is obviously unrealistic. Equating moments about O, we write

$$+\circlearrowleft \qquad -W(b \sin\theta) = (mb\ddot{\theta})b + \bar{I}\ddot{\theta}$$

Noting that $\bar{I} = \frac{1}{12}m[(2b)^2 + (2b)^2] = \frac{2}{3}mb^2$ and $W = mg$, we obtain

$$\ddot{\theta} + \frac{3}{5}\frac{g}{b}\sin\theta = 0 \tag{19.22}$$

For oscillations of small amplitude, we may replace $\sin\theta$ by θ, expressed in radians, and write

$$\ddot{\theta} + \frac{3}{5}\frac{g}{b}\theta = 0 \tag{19.23}$$

Comparison with (19.21) shows that the equation obtained is that of a simple harmonic motion and that the circular frequency p of the oscillations is equal to $(3g/5b)^{1/2}$. Substituting into (19.13), we find that the period of the oscillations is

$$\tau = \frac{2\pi}{p} = 2\pi\sqrt{\frac{5b}{3g}} \tag{19.24}$$

The result obtained is valid only for oscillations of small amplitude. A more accurate description of the motion of the plate is obtained by comparing Eqs. (19.16) and (19.22). We note that the two equations are identical if we choose l equal to $5b/3$. This means that the plate will oscillate as a simple pendulum of length $l = 5b/3$, and the results of Sec. 19.4 may be used to correct the value of the period given in (19.24). The point A of the plate located on line OG at a distance $l = 5b/3$ from O is defined as the *center of oscillation* corresponding to O (Fig. 19.5*a*).

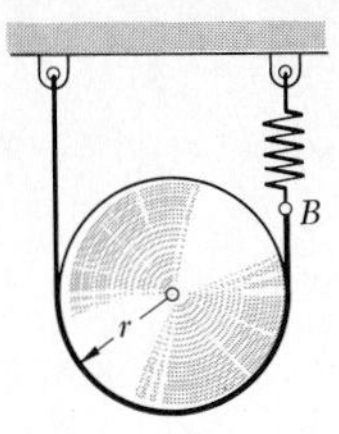

SAMPLE PROBLEM 19.2

A cylinder of weight W and radius r is suspended from a looped cord as shown. One end of the cord is attached directly to a rigid support, while the other end is attached to a spring of constant k. Determine the period and frequency of vibration of the cylinder.

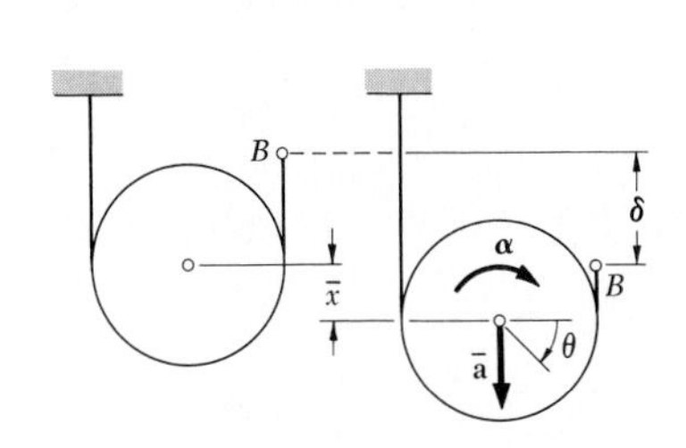

Kinematics of Motion. We express the linear displacement and the acceleration of the cylinder in terms of the angular displacement θ. Choosing the positive sense clockwise and measuring the displacements from the equilibrium position, we write

$$\bar{x} = r\theta \qquad \delta = 2\bar{x} = 2r\theta$$

$$\boldsymbol{\alpha} = \ddot{\theta} \downarrow \qquad \bar{a} = r\alpha = r\ddot{\theta} \qquad \bar{\mathbf{a}} = r\ddot{\theta} \downarrow \tag{1}$$

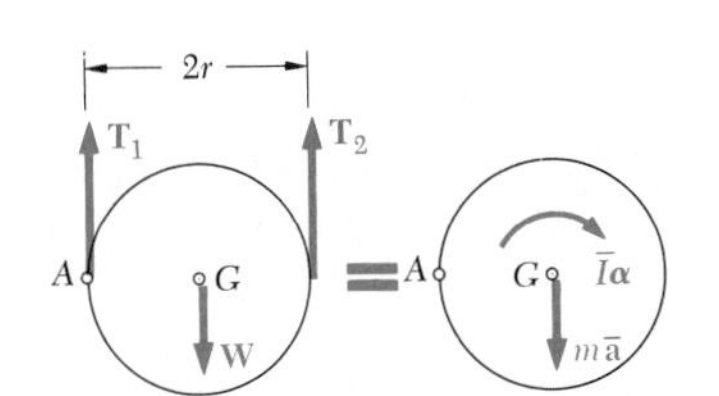

Equations of Motion. The system of external forces acting on the cylinder consists of the weight $\mathbf{W}$ and of the forces $\mathbf{T}_1$ and $\mathbf{T}_2$ exerted by the cord. We express that this system is equivalent to the system of effective forces represented by the vector $m\bar{\mathbf{a}}$ attached at G and the couple $\bar{I}\boldsymbol{\alpha}$.

$$+\downarrow\Sigma M_A = \Sigma(M_A)_{\text{eff}}: \qquad Wr - T_2(2r) = m\bar{a}r + \bar{I}\alpha \tag{2}$$

When the cylinder is in its position of equilibrium, the tension in the cord is $T_0 = \frac{1}{2}W$. We note that for an angular displacement θ, the magnitude of $\mathbf{T}_2$ is

$$T_2 = T_0 + k\delta = \tfrac{1}{2}W + k\delta = \tfrac{1}{2}W + k(2r\theta) \tag{3}$$

Substituting from (1) and (3) into (2), and recalling that $\bar{I} = \frac{1}{2}mr^2$, we write

$$Wr - (\tfrac{1}{2}W + 2kr\theta)(2r) = m(r\ddot{\theta})r + \tfrac{1}{2}mr^2\ddot{\theta}$$

$$\ddot{\theta} + \frac{8}{3}\frac{k}{m}\theta = 0$$

The motion is seen to be simple harmonic, and we have

$$p^2 = \frac{8}{3}\frac{k}{m} \qquad p = \sqrt{\frac{8}{3}\frac{k}{m}}$$

$$\tau = \frac{2\pi}{p} \qquad \tau = 2\pi\sqrt{\frac{3}{8}\frac{m}{k}} \quad ◀$$

$$f = \frac{p}{2\pi} \qquad f = \frac{1}{2\pi}\sqrt{\frac{8}{3}\frac{k}{m}} \quad ◀$$

SAMPLE PROBLEM 19.3

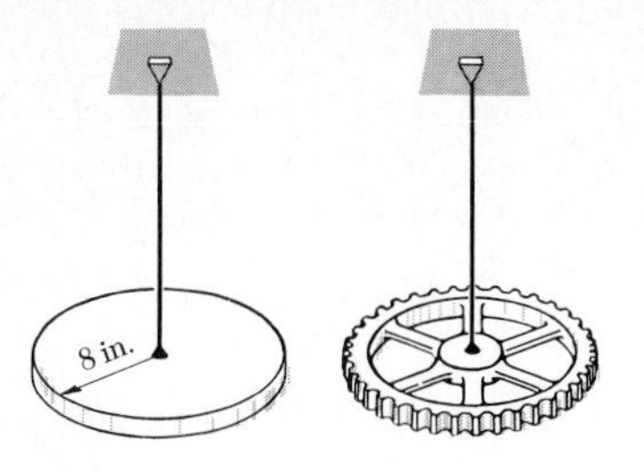

A circular disk, weighing 20 lb and of radius 8 in., is suspended from a wire as shown. The disk is rotated (thus twisting the wire) and then released; the period of the torsional vibration is observed to be 1.13 s. A gear is then suspended from the same wire, and the period of torsional vibration for the gear is observed to be 1.93 s. Assuming that the moment of the couple exerted by the wire is proportional to the angle of twist, determine (*a*) the torsional spring constant of the wire, (*b*) the centroidal moment of inertia of the gear, (*c*) the maximum angular velocity reached by the gear if it is rotated through 90° and released.

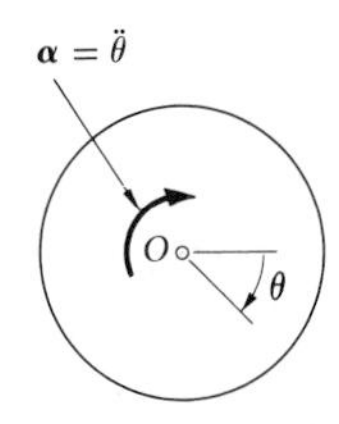

***a.* Vibration of Disk.** Denoting by θ the angular displacement of the disk, we express that the moment of the couple exerted by the wire is $M = K\theta$, where K is the torsional spring constant of the wire. Since this couple must be equivalent to the couple $\bar{I}\boldsymbol{\alpha}$ representing the effective forces of the disk, we write

$$+\circlearrowleft \Sigma M_O = \Sigma(M_O)_{\text{eff}}: \qquad +K\theta = -\bar{I}\ddot{\theta}$$

$$\ddot{\theta} + \frac{K}{\bar{I}}\theta = 0$$

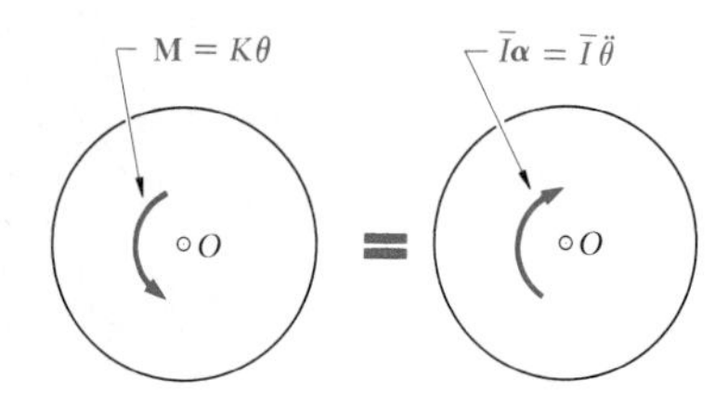

The motion is seen to be simple harmonic, and we have

$$p^2 = \frac{K}{\bar{I}} \qquad \tau = \frac{2\pi}{p} \qquad \tau = 2\pi\sqrt{\frac{\bar{I}}{K}} \tag{1}$$

For the disk, we have

$$\tau = 1.13 \text{ s} \qquad \bar{I} = \tfrac{1}{2}mr^2 = \frac{1}{2}\left(\frac{20 \text{ lb}}{32.2 \text{ ft/s}^2}\right)\left(\frac{8}{12}\text{ ft}\right)^2 = 0.138 \text{ lb}\cdot\text{ft}\cdot\text{s}^2$$

Substituting into (1), we obtain

$$1.13 = 2\pi\sqrt{\frac{0.138}{K}} \qquad K = 4.27 \text{ lb}\cdot\text{ft/rad} \blacktriangleleft$$

***b.* Vibration of Gear.** Since the period of vibration of the gear is 1.93 s and $K = 4.27$ lb·ft/rad, Eq. (1) yields

$$1.93 = 2\pi\sqrt{\frac{\bar{I}}{4.27}} \qquad \bar{I}_{\text{gear}} = 0.403 \text{ lb}\cdot\text{ft}\cdot\text{s}^2 \blacktriangleleft$$

***c.* Maximum Angular Velocity of the Gear.** Since the motion is simple harmonic, we have

$$\theta = \theta_m \sin pt \qquad \omega = \theta_m p \cos pt \qquad \omega_m = \theta_m p$$

Recalling that $\theta_m = 90° = 1.571$ rad and $\tau = 1.93$ s, we write

$$\omega_m = \theta_m p = \theta_m\left(\frac{2\pi}{\tau}\right) = (1.571 \text{ rad})\left(\frac{2\pi}{1.93 \text{ s}}\right)$$

$$\omega_m = 5.11 \text{ rad/s} \blacktriangleleft$$

Problems

19.27 The 3-kg uniform rod shown is attached to a spring of constant $k = 900$ N/m. If end A of the rod is depressed 25 mm and released, determine (a) the period of vibration, (b) the maximum velocity of end A.

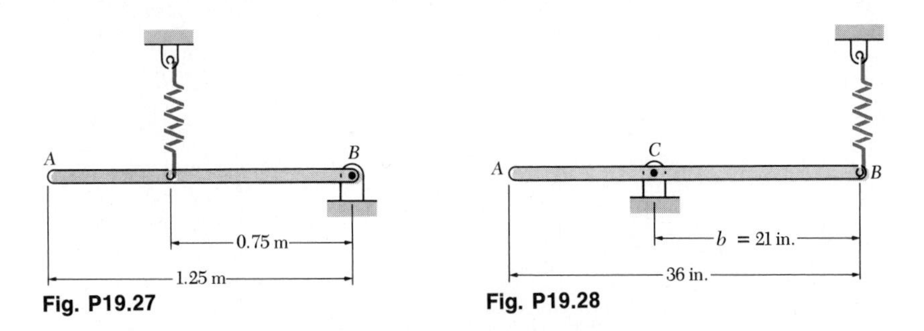

Fig. P19.27 **Fig. P19.28**

19.28 The uniform rod shown weighs 12 lb and is attached to a spring of constant $k = 3$ lb/in. If end B of the rod is depressed 0.5 in. and released, determine (a) the period of vibration, (b) the maximum velocity of end B.

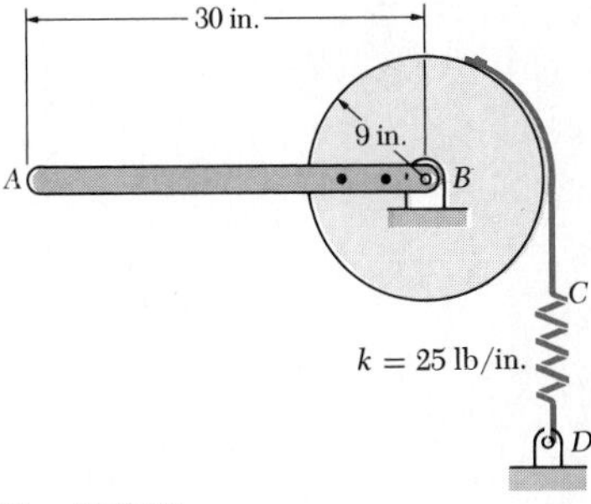

Fig. P19.29

19.29 A 12-lb slender rod AB is riveted to a 10-lb uniform disk as shown. A belt is attached to the rim of the disk and to a spring which holds the rod at rest in the position shown. If end A of the rod is moved 1.5 in. down and released, determine (a) the period of vibration, (b) the maximum velocity of end A.

19.30 A 5-kg uniform cylinder rolls without sliding on an incline and is attached to a spring AB as shown. If the center of the cylinder is moved 10 mm down the incline from its equilibrium position and released, determine (a) the period of vibration, (b) the maximum velocity of the center of the cylinder.

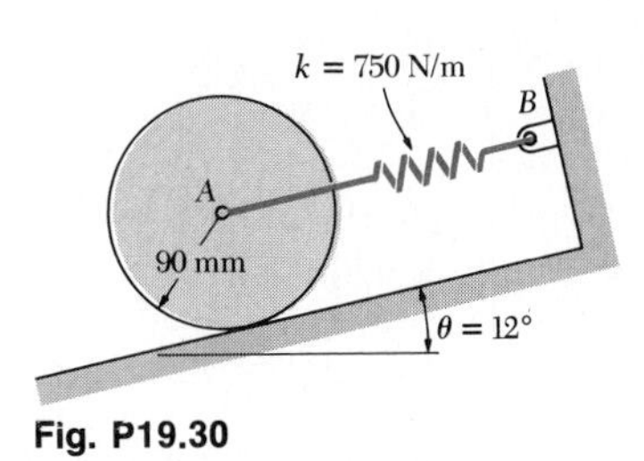

Fig. P19.30

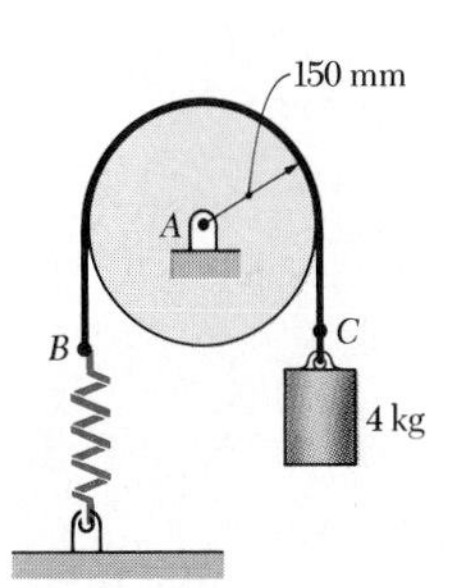

Fig. P19.31

19.31 A belt is placed over the rim of a 12-kg disk as shown and then attached to a 4-kg cylinder and to a spring of constant $k = 500$ N/m. If the cylinder is moved 75 mm down from its equilibrium position and released, determine (a) the period of vibration, (b) the maximum velocity of the cylinder. Assume friction is sufficient to prevent the belt from slipping on the rim.

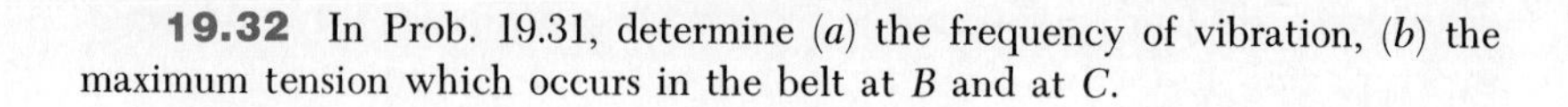

19.32 In Prob. 19.31, determine (*a*) the frequency of vibration, (*b*) the maximum tension which occurs in the belt at B and at C.

19.33 In Prob. 19.28, determine (*a*) the value of b for which the smallest period of vibration occurs, (*b*) the corresponding period of vibration.

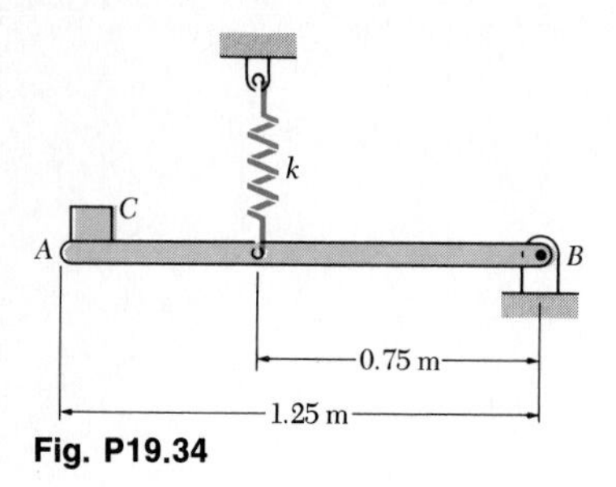

Fig. P19.34

19.34 The 3-kg uniform rod AB is attached as shown to a spring of constant $k = 900$ N/m. A small 0.5-kg block C is placed on the rod at A. (*a*) If end A of the rod is then moved down through a small distance δ_0 and released, determine the period of the vibration. (*b*) Determine the largest allowable value of δ_0 if block C is to remain at all times in contact with the rod.

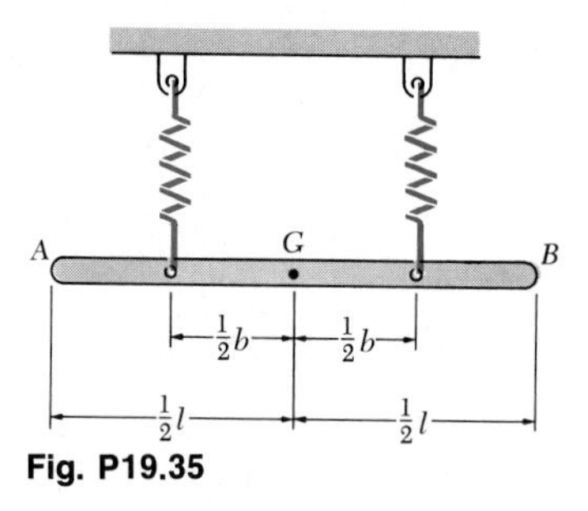

Fig. P19.35

19.35 A slender rod of mass m and length l is held by two springs, each of constant k. Determine the frequency of the resulting vibration if the rod is (*a*) given a small vertical displacement and released, (*b*) rotated through a small angle about a horizontal axis through G and released. (*c*) Determine the ratio b/l for which the frequencies found in parts *a* and *b* are equal.

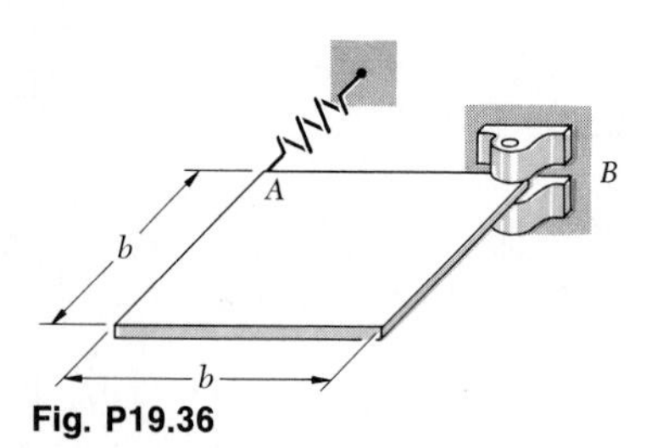

Fig. P19.36

19.36 A uniform square plate of mass m is supported in a horizontal plane by a vertical pin at B and is attached at A to a spring of constant k. If corner A is given a small displacement and released, determine the period of the resulting motion.

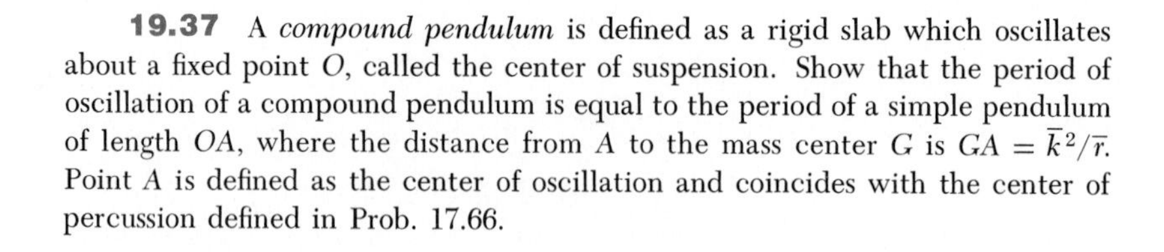

19.37 A *compound pendulum* is defined as a rigid slab which oscillates about a fixed point O, called the center of suspension. Show that the period of oscillation of a compound pendulum is equal to the period of a simple pendulum of length OA, where the distance from A to the mass center G is $GA = \bar{k}^2/\bar{r}$. Point A is defined as the center of oscillation and coincides with the center of percussion defined in Prob. 17.66.

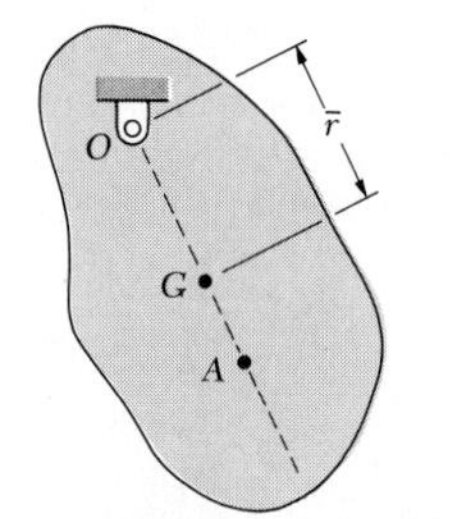

Fig. P19.37 and P19.39

19.38 Show that if the compound pendulum of Prob. 19.37 is suspended from A instead of O, the period of oscillation is the same as before and the new center of oscillation is located at O.

19.39 A rigid slab oscillates about a fixed point O. Show that the smallest period of oscillation occurs when the distance $\bar{r}$ from point O to the mass center G is equal to $\bar{k}$.

19.40 If either a simple or a compound pendulum is used to determine experimentally the acceleration of gravity g, difficulties are encountered. In the simple pendulum, the string is not truly weightless, while in the compound pendulum, the exact location of the mass center is difficult to establish. In the case of a compound pendulum, the difficulty may be eliminated by using a reversible, or Kater, pendulum. Two knife-edges A and B are placed so that they are obviously not at the same distance from the mass center G, and the distance l is measured with great precision. The position of a counterweight D is then adjusted so that the period of oscillation τ is the same when either knife-edge is used. Show that the period τ obtained is equal to that of a true simple pendulum of length l and that $g = 4\pi^2 l/\tau^2$.

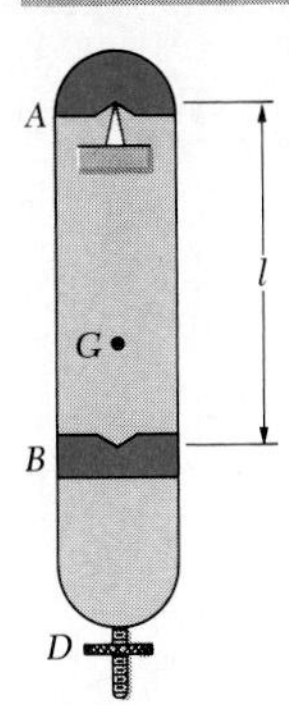

Fig. P19.40

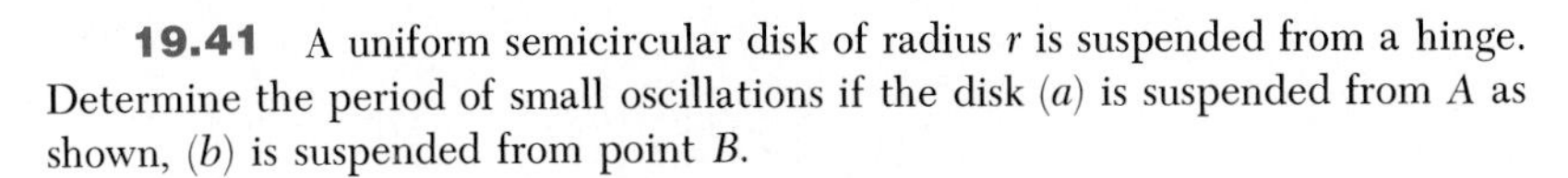

19.41 A uniform semicircular disk of radius r is suspended from a hinge. Determine the period of small oscillations if the disk (a) is suspended from A as shown, (b) is suspended from point B.

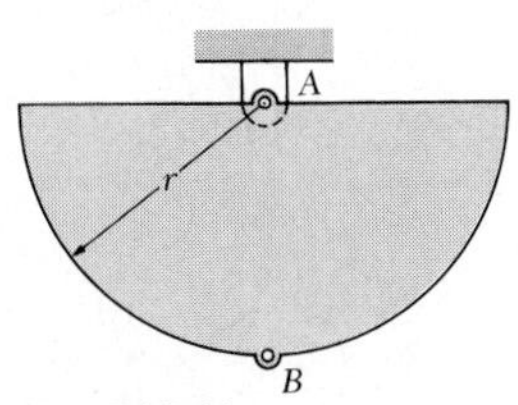

Fig. P19.41

19.42 A uniform bar of length l may oscillate about a hinge A located at a distance c from the mass center G of the bar. (a) Determine the frequency of small oscillations if $c = \frac{1}{2}l$. (b) Determine a second value of c for which the frequency of small oscillations is the same as that found in part a.

19.43 A uniform bar of length l may oscillate about a hinge A located at a distance c from the mass center G of the bar. Determine (a) the distance c for which the period of oscillation is minimum, (b) the corresponding value of the period.

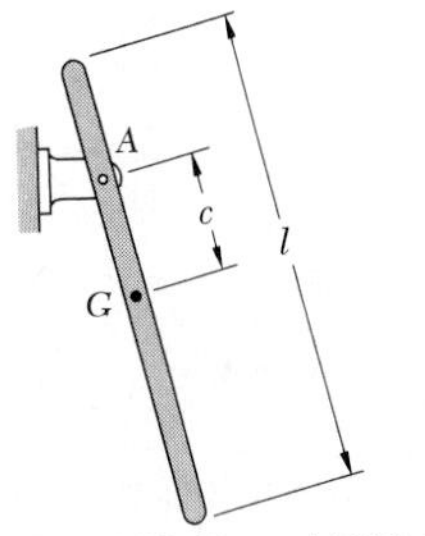

Fig. P19.42 and P19.43

19.44 A thin homogeneous wire is bent into the shape of an equilateral triangle of side $l = 250$ mm. Determine the period of small oscillations if the wire (a) is suspended as shown, (b) is suspended from a pin located at the midpoint of one side.

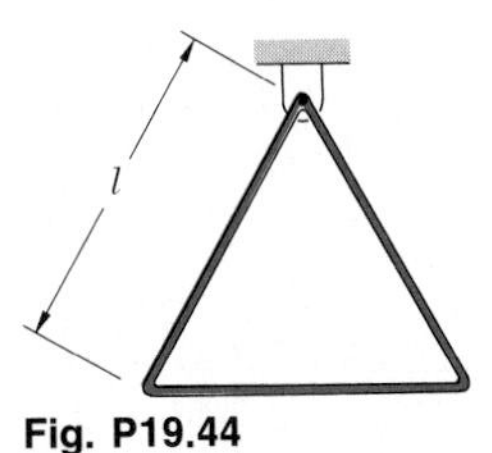

Fig. P19.44

19.45 Two uniform rods, each of mass m and length l, are welded together to form the T-shaped assembly shown. Determine the frequency of small oscillations of the assembly.

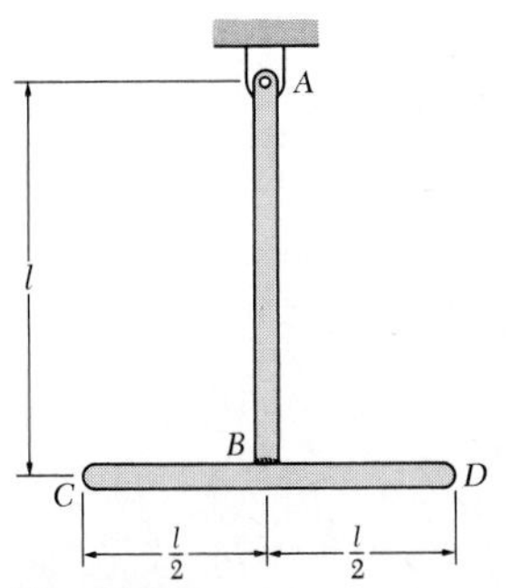

Fig. P19.45

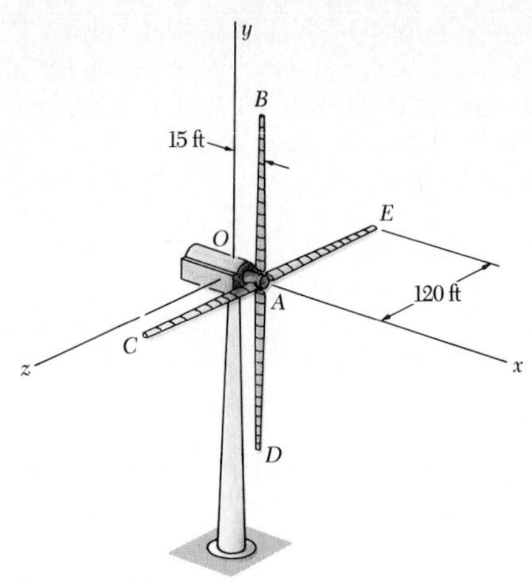

Fig. P19.46

19.46 Blade AB of the wind-turbine generator shown is to be temporarily removed. Motion of the turbine generator about the y axis is prevented, but the remaining three blades may oscillate as a unit about the x axis. Assuming that each blade is equivalent to a 120-ft slender rod, determine the period of small oscillations of the blades in the absence of wind.

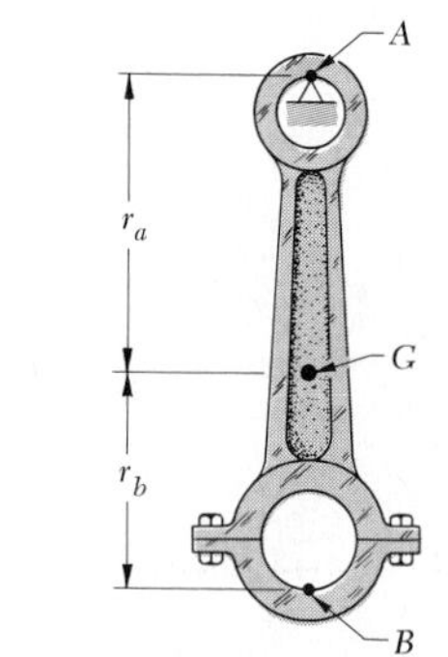

Fig. P19.47 and P19.48

19.47 The period of small oscillations about A of a connecting rod is observed to be 1.06 s. Knowing that the distance r_a is 170 mm, determine the centroidal radius of gyration of the connecting rod.

19.48 A connecting rod is supported by a knife-edge at point A; the period of small oscillations is observed to be 0.895 s. The rod is then inverted and supported by a knife-edge at point B and the period of small oscillations is observed to be 0.805 s. Knowing that $r_a + r_b = 270$ mm, determine (*a*) the location of the mass center G, (*b*) the centroidal radius of gyration $\bar{k}$.

19.49 and 19.50 A thin disk of radius r may oscillate about the axis AB located as shown at a distance b from the mass center G. (*a*) Determine the period of small oscillations if $b = r$. (*b*) Determine a second value of b for which the period of oscillation is the same as that found in part *a*.

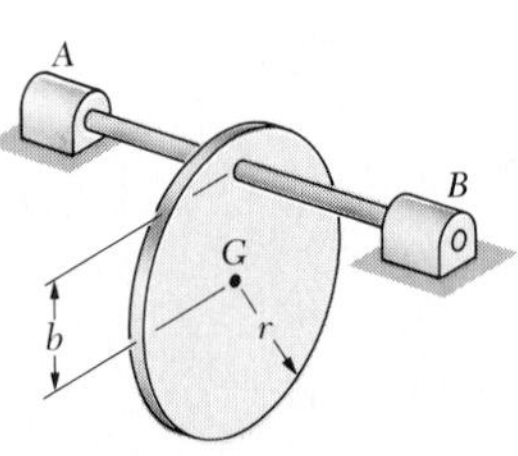

Fig. P19.49

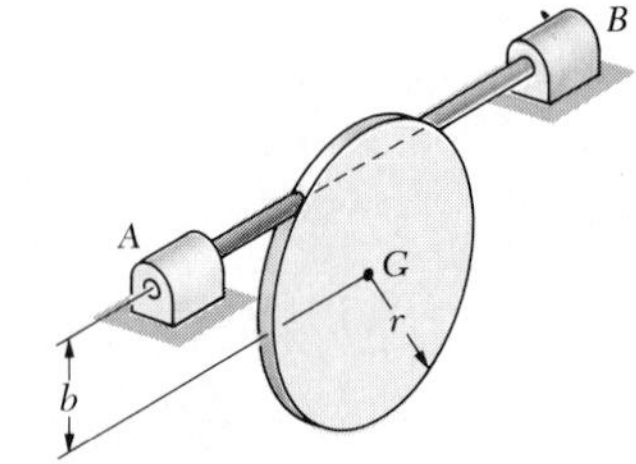

Fig. P19.50

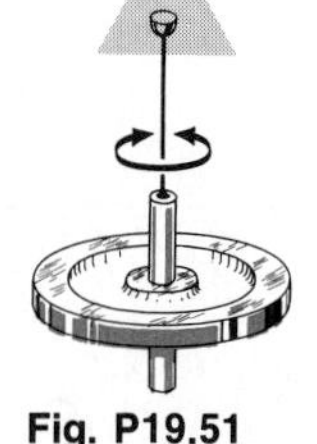

Fig. P19.51

19.51 A period of 3.60 s is observed for the angular oscillations of a 750-g gyroscope rotor suspended from a wire as shown. Knowing that a period of 5.10 s is obtained when a 60-mm-diameter steel sphere is suspended in the same fashion, determine the centroidal radius of gyration of the rotor. (Density of steel = 7850 kg/m^3.)

19.52 A 6-kg slender rod is suspended from a steel wire which is known to have a torsional spring constant $K = 1.75$ N·m/rad. If the rod is rotated through 180° about the vertical and then released, determine (*a*) the period of oscillation, (*b*) the maximum velocity of end *A* of the rod.

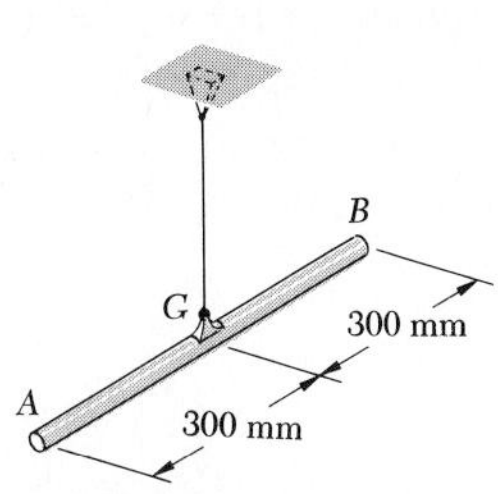

Fig. P19.52

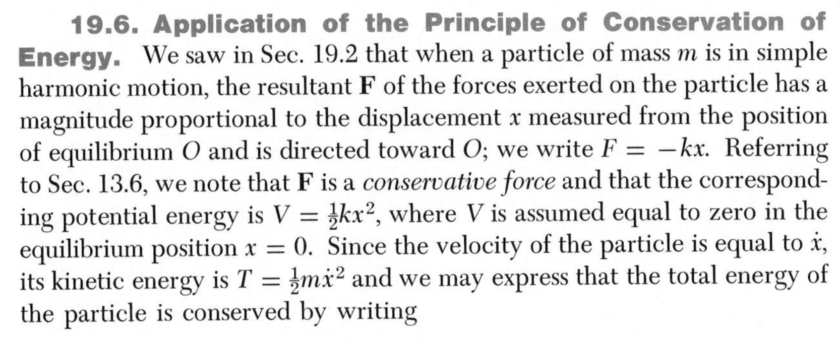

Fig. P19.53

19.53 A thin circular plate of radius *r* is suspended from three vertical wires of length *h* equally spaced around the perimeter of the plate. Determine the period of oscillation when (*a*) the plate is rotated through a small angle about a vertical axis passing through its mass center and released, (*b*) the plate is given a small horizontal translation and released.

19.54 Solve Prob. 19.53, assuming that $r = 750$ mm and $h = 600$ mm.

19.55 A uniform disk of radius 10 in. and weighing 18 lb is attached to a vertical shaft which is rigidly held at *B*. It is known that the disk rotates through 3° when a 40-lb·in. static couple is applied to the disk. If the disk is rotated through 8° and then released, determine (*a*) the period of the resulting vibration, (*b*) the maximum velocity of a point on the rim of the disk.

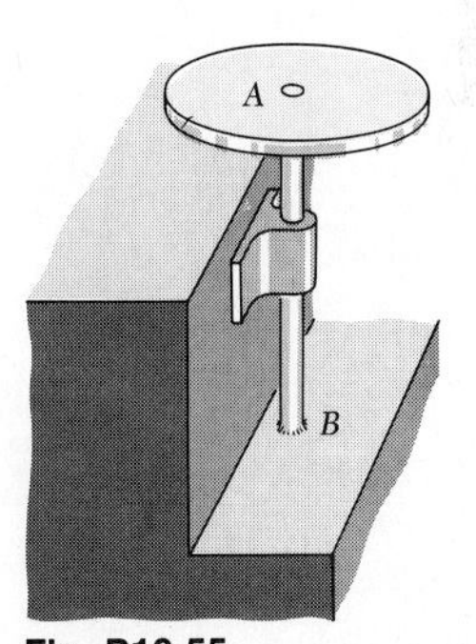

Fig. P19.55

19.56 A steel casting is rigidly bolted to the disk of Prob. 19.55. Knowing that the period of torsional vibration of the disk and casting is 0.90 s, determine the moment of inertia of the casting with respect to the shaft *AB*.

19.6. Application of the Principle of Conservation of Energy. We saw in Sec. 19.2 that when a particle of mass m is in simple harmonic motion, the resultant **F** of the forces exerted on the particle has a magnitude proportional to the displacement x measured from the position of equilibrium O and is directed toward O; we write $F = -kx$. Referring to Sec. 13.6, we note that **F** is a *conservative force* and that the corresponding potential energy is $V = \frac{1}{2}kx^2$, where V is assumed equal to zero in the equilibrium position $x = 0$. Since the velocity of the particle is equal to $\dot{x}$, its kinetic energy is $T = \frac{1}{2}m\dot{x}^2$ and we may express that the total energy of the particle is conserved by writing

$$T + V = \text{constant} \qquad \tfrac{1}{2}m\dot{x}^2 + \tfrac{1}{2}kx^2 = \text{constant}$$

Setting $p^2 = k/m$ as in (19.4), where p is the circular frequency of the vibration, we have

$$\dot{x}^2 + p^2x^2 = \text{constant} \tag{19.25}$$

Equation (19.25) is characteristic of simple harmonic motion; it may be obtained directly from (19.5) by multiplying both terms by $2\dot{x}$ and integrating.

The principle of conservation of energy provides a convenient way for determining the period of vibration of a rigid body or of a system of rigid bodies possessing a single degree of freedom, once it has been established that the motion of the system is a simple harmonic motion, or that it may be approximated by a simple harmonic motion. Choosing an appropriate variable, such as a distance x or an angle θ, we consider two particular positions of the system:

1. *The displacement of the system is maximum;* we have $T_1 = 0$, and V_1 may be expressed in terms of the amplitude x_m or θ_m (choosing $V = 0$ in the equilibrium position).
2. *The system passes through its equilibrium position;* we have $V_2 = 0$, and T_2 may be expressed in terms of the maximum velocity $\dot{x}_m$ or $\dot{\theta}_m$.

We then express that the total energy of the system is conserved and write $T_1 + V_1 = T_2 + V_2$. Recalling from (19.15) that for simple harmonic motion the maximum velocity is equal to the product of the amplitude and of the circular frequency p, we find that the equation obtained may be solved for p.

As an example, we shall consider again the square plate of Sec. 19.5. In the position of maximum displacement (Fig. 19.6*a*), we have

$$T_1 = 0 \qquad V_1 = W(b - b\cos\theta_m) = Wb(1 - \cos\theta_m)$$

or, since $1 - \cos\theta_m = 2\sin^2(\theta_m/2) \approx 2(\theta_m/2)^2 = \theta_m^2/2$ for oscillations of small amplitude,

$$T_1 = 0 \qquad V_1 = \tfrac{1}{2}Wb\theta_m^2 \tag{19.26}$$

As the plate passes through its position of equilibrium (Fig. 19.6*b*), its velocity is maximum and we have

$$T_2 = \tfrac{1}{2}m\bar{v}_m^2 + \tfrac{1}{2}\bar{I}\omega_m^2 = \tfrac{1}{2}mb^2\dot{\theta}_m^2 + \tfrac{1}{2}\bar{I}\dot{\theta}_m^2 \qquad V_2 = 0$$

or, recalling from Sec. 19.5 that $\bar{I} = \frac{2}{3}mb^2$,

$$T_2 = \tfrac{1}{2}(\tfrac{5}{3}mb^2)\dot{\theta}_m^2 \qquad V_2 = 0 \tag{19.27}$$

Substituting from (19.26) and (19.27) into $T_1 + V_1 = T_2 + V_2$, and noting that the maximum velocity $\dot{\theta}_m$ is equal to the product $\theta_m p$, we write

$$\tfrac{1}{2}Wb\theta_m^2 = \tfrac{1}{2}(\tfrac{5}{3}mb^2)\theta_m^2 p^2 \tag{19.28}$$

which yields $p^2 = 3g/5b$ and

$$\tau = \frac{2\pi}{p} = 2\pi\sqrt{\frac{5b}{3g}} \tag{19.29}$$

as previously obtained.

(*a*)

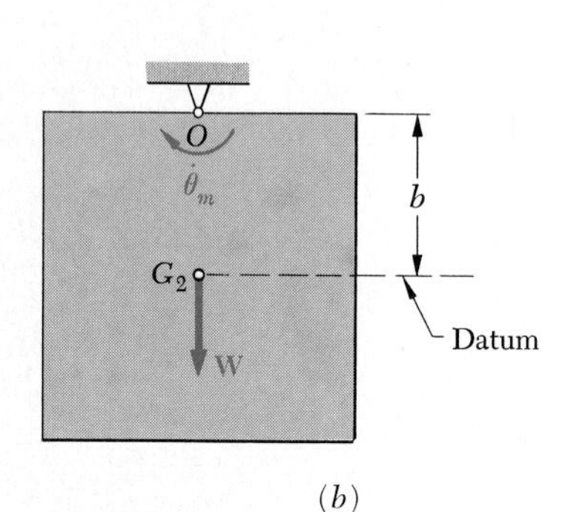

(*b*)

Fig. 19.6

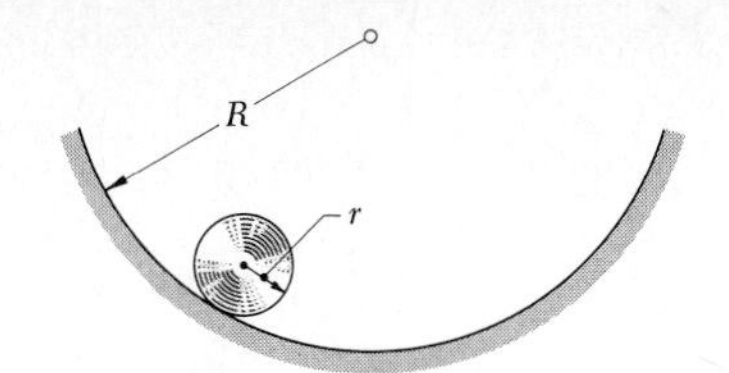

SAMPLE PROBLEM 19.4

Determine the period of small oscillations of a cylinder of radius r which rolls without slipping inside a curved surface of radius R.

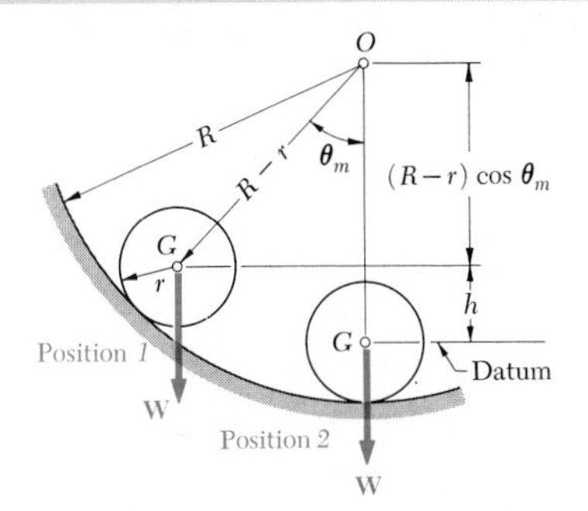

Solution. We denote by θ the angle which line OG forms with the vertical. Since the cylinder rolls without slipping, we may apply the principle of conservation of energy between position *1*, where $\theta = \theta_m$, and position 2, where $\theta = 0$.

Position *1.* ***Kinetic Energy.*** Since the velocity of the cylinder is zero, we have $T_1 = 0$.

Potential Energy. Choosing a datum as shown and denoting by W the weight of the cylinder, we have

$$V_1 = Wh = W(R - r)(1 - \cos\theta)$$

Noting that for small oscillations $(1 - \cos\theta) = 2\sin^2(\theta/2) \approx \theta^2/2$, we have

$$V_1 = W(R - r)\frac{\theta_m^2}{2}$$

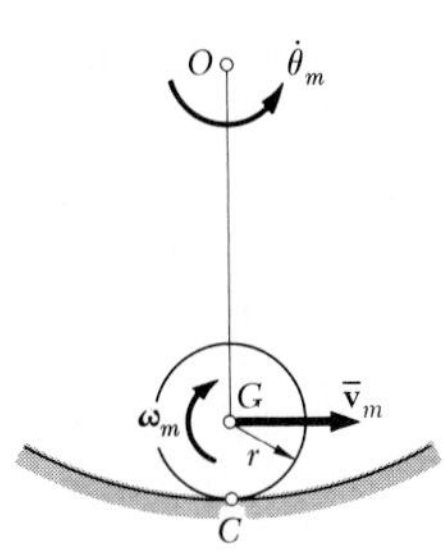

Position 2. Denoting by θ_m the angular velocity of line OG as the cylinder passes through position 2, and observing that point C is the instantaneous center of rotation of the cylinder, we write

$$\bar{v}_m = (R - r)\dot{\theta}_m \qquad \omega_m = \frac{\bar{v}_m}{r} = \frac{R - r}{r}\dot{\theta}_m$$

Kinetic Energy

$$\begin{aligned} T_2 &= \tfrac{1}{2}m\bar{v}_m^2 + \tfrac{1}{2}\bar{I}\omega_m^2 \\ &= \tfrac{1}{2}m(R - r)^2\dot{\theta}_m^2 + \tfrac{1}{2}(\tfrac{1}{2}mr^2)\left(\frac{R - r}{r}\right)^2\dot{\theta}_m^2 \\ &= \tfrac{3}{4}m(R - r)^2\dot{\theta}_m^2 \end{aligned}$$

Potential Energy $\qquad V_2 = 0$

Conservation of Energy

$$T_1 + V_1 = T_2 + V_2$$

$$0 + W(R - r)\frac{\theta_m^2}{2} = \tfrac{3}{4}m(R - r)^2\dot{\theta}_m^2 + 0$$

Since $\dot{\theta}_m = p\theta_m$ and $W = mg$, we write

$$mg(R - r)\frac{\theta_m^2}{2} = \tfrac{3}{4}m(R - r)^2(p\theta_m)^2 \qquad p^2 = \frac{2}{3}\frac{g}{R - r}$$

$$\tau = \frac{2\pi}{p} \qquad \tau = 2\pi\sqrt{\frac{3}{2}\frac{R - r}{g}} \quad \blacktriangleleft$$

Problems

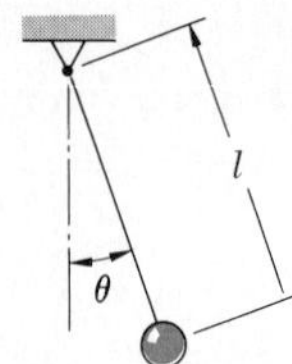

Fig. P19.57

19.57 Using the method of Sec. 19.6, determine the period of a simple pendulum of length l.

19.58 The springs of an automobile are observed to expand 200 mm to an undeformed position as the body is lifted by several jacks. Assuming that each spring carries an equal portion of the weight of the automobile, determine the frequency of the free vertical vibrations of the body.

19.59 Using the method of Sec. 19.6, solve Prob. 19.3.

19.60 Using the method of Sec. 19.6, solve Prob. 19.4.

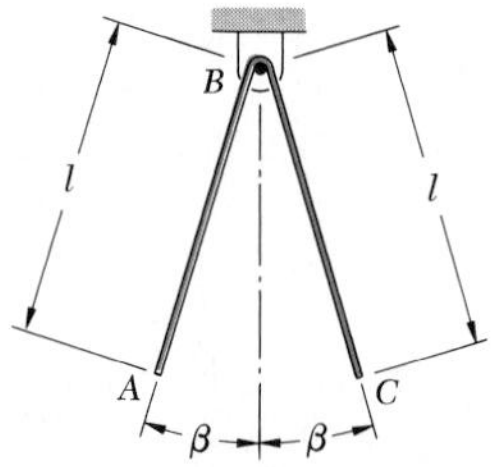

Fig. P19.61 and P19.62

19.61 A homogeneous wire of length $2l$ is bent as shown and allowed to oscillate about a frictionless pin at B. Denoting by τ_0 the period of small oscillations when $\beta = 0$, determine the angle β for which the period of small oscillations is $2\tau_0$.

19.62 Knowing that $l = 750$ mm and $\beta = 40°$, determine the period of small oscillations of the bent homogeneous wire shown.

19.63 A homogeneous wire is bent to form a square of side l which is supported by a ball-and-socket joint at A. Determine the period of small oscillations of the square (*a*) in the plane of the square, (*b*) in a direction perpendicular to the square.

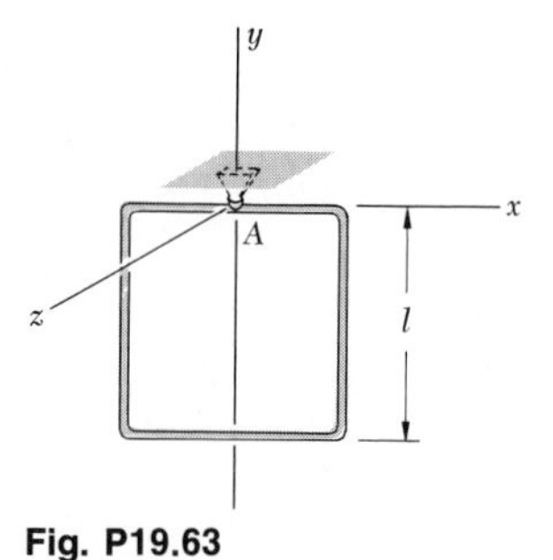

Fig. P19.63

19.64 Solve Prob. 19.63, assuming that the square is supported by a ball-and-socket joint located at one corner of the square.

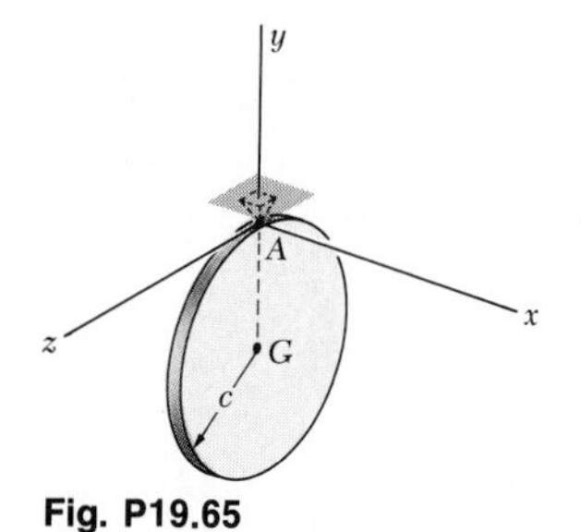

Fig. P19.65

19.65 A uniform disk of radius c is supported by a ball-and-socket joint at A. Determine the frequency of small oscillations of the disk (*a*) in the plane of the disk, (*b*) in a direction perpendicular to the disk.

19.66 It is observed that when an 8-lb weight is attached to the rim of a 6-ft-diameter flywheel, the period of small oscillations of the flywheel is 22 s. Neglecting axle friction, determine the centroidal moment of inertia of the flywheel.

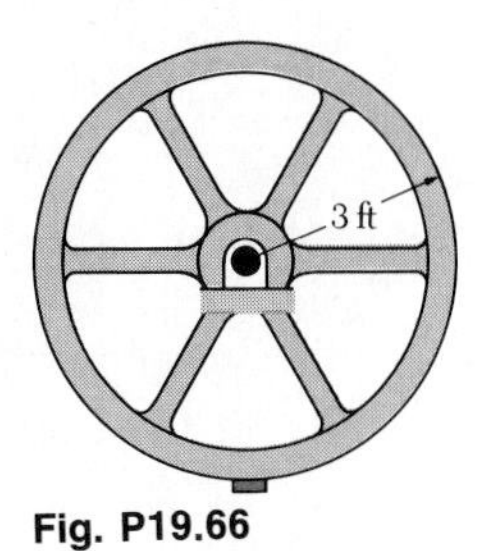

Fig. P19.66

19.67 Using the method of Sec. 19.6, solve Prob. 19.45.

19.68 The uniform rod *ABC* weighs 5 lb and is attached to two springs as shown. If end *C* is given a small displacement and released, determine the frequency of vibration of the rod.

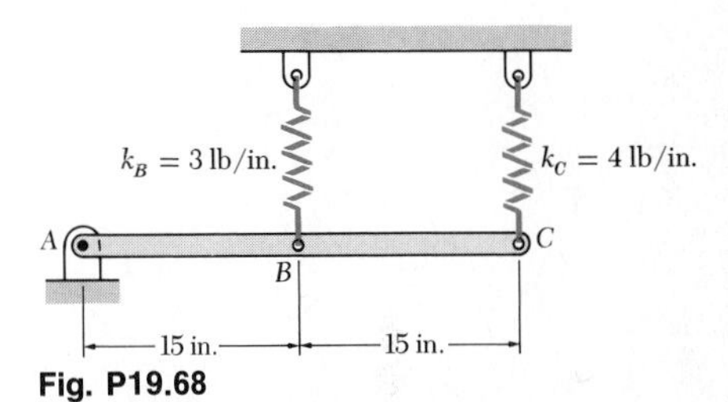

Fig. P19.68

19.69 For the rod of Prob. 19.68, determine the frequency of vibration of the rod if the springs are interchanged so that $k_B = 4$ lb/in. and $k_C = 3$ lb/in.

19.70 The 5-kg slender rod *AB* is welded to the 8-kg uniform disk. A spring of constant 450 N/m is attached to the disk and holds the rod at rest in the position shown. If end *B* of the rod is given a small displacement and released, determine the period of vibration of the rod.

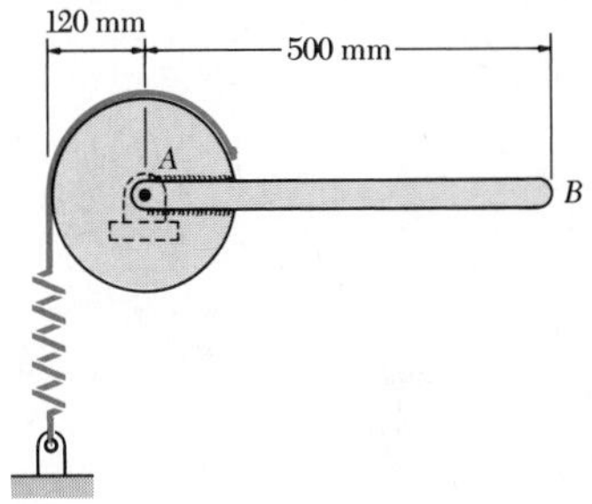

Fig. P19.70

19.71 For the rod and disk of Prob. 19.70, determine the constant of the spring for which the period of vibration of the rod is 1.5 s.

19.72 The slender rod *AB* of mass *m* is attached to two collars of negligible mass. Knowing that the system lies in a horizontal plane and is in equilibrium in the position shown, determine the period of vibration if collar *A* is given a small displacement and released.

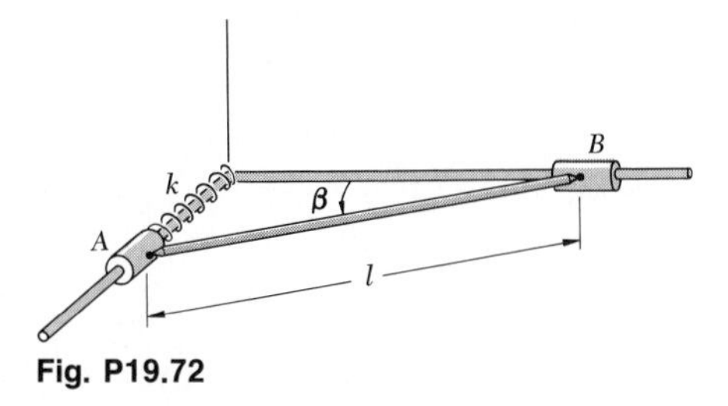

Fig. P19.72

19.73 Disks *A* and *B* are of mass 3 kg and 8 kg, respectively, and a small block *C* of mass 750 g is attached to the rim of disk *B*. Assuming that no slipping occurs between the disks, determine the period of small oscillations of the system.

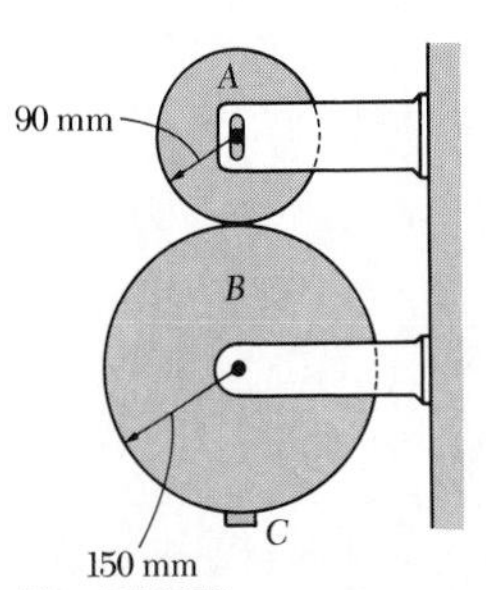

Fig. P19.73

19.74 Two 5-kg uniform disks are attached to the 8-kg rod AB as shown. Knowing that the constant of the spring is 4 kN/m and that the disks roll without sliding, determine the frequency of vibration of the system.

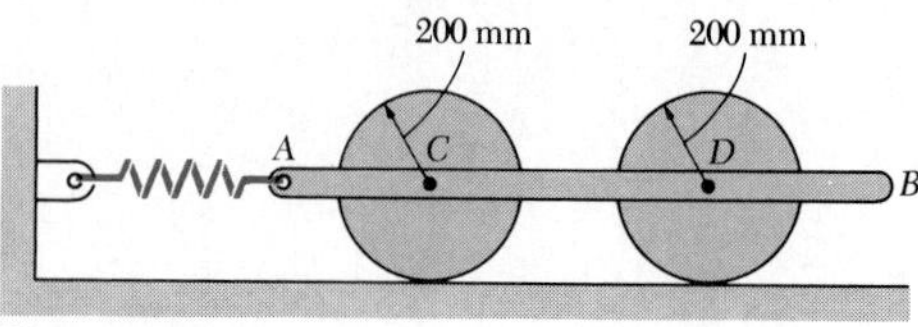

Fig. P19.74

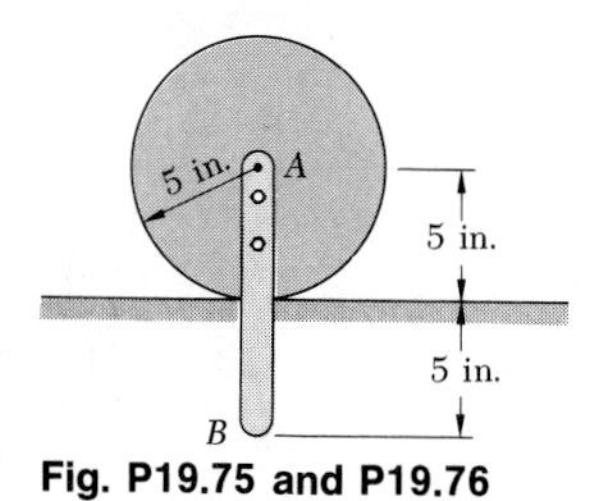

Fig. P19.75 and P19.76

19.75 The 4-lb rod AB is bolted to the 6-lb disk. Knowing that the disk rolls without sliding, determine the period of small oscillations of the system.

19.76 The 4-lb rod AB is bolted to the 5-in.-radius disk as shown. Knowing that the disk rolls without sliding, determine the weight of the disk for which the period of small oscillations of the system is 1.5 s.

19.77 Three identical rods are connected as shown. If $b = \frac{3}{4}l$, determine the frequency of small oscillations of the system.

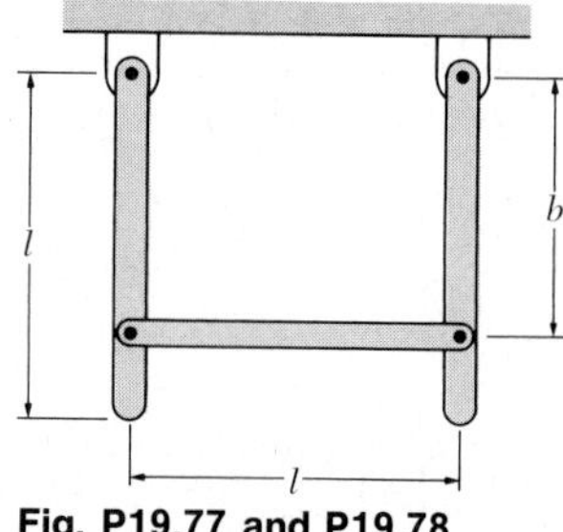

Fig. P19.77 and P19.78

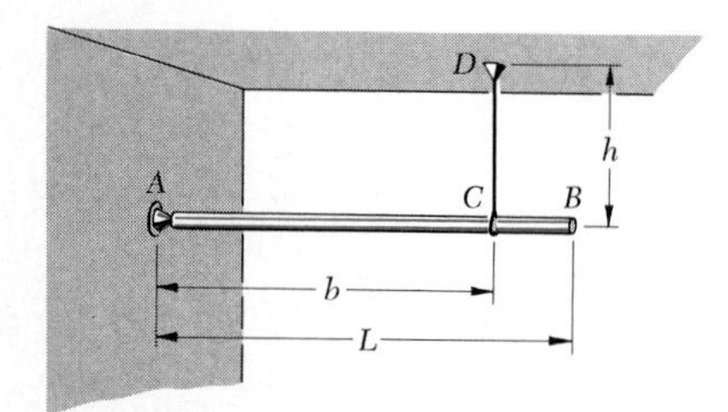

Fig. P19.79

19.78 Three identical rods are connected as shown. Determine (*a*) the distance b for which the frequency of oscillation is maximum, (*b*) the corresponding maximum frequency.

19.79 A uniform rod of length L is supported by a ball-and-socket joint at A and by the vertical wire CD. Derive an expression for the period of oscillation of the rod if end B is given a small horizontal displacement and then released.

19.80 Solve Prob. 19.79, assuming that $L = 3$ m, $b = 2.5$ m, and $h = 2$ m.

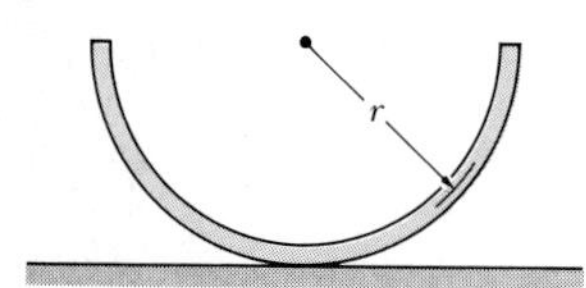

Fig. P19.81

19.81 A half section of pipe is placed on a horizontal surface, rotated through a small angle, and then released. Assuming that the pipe section rolls without sliding, determine the period of oscillation.

19.82 A slender rod of length l is suspended from two vertical wires of length h, each located a distance $\frac{1}{2}b$ from the mass center G. Determine the period of oscillation when (*a*) the rod is rotated through a small angle about a vertical axis passing through G and released, (*b*) the rod is given a small horizontal translation along AB and released.

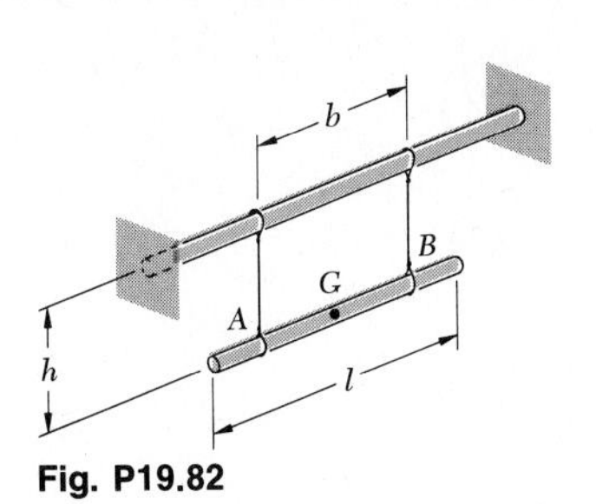

Fig. P19.82

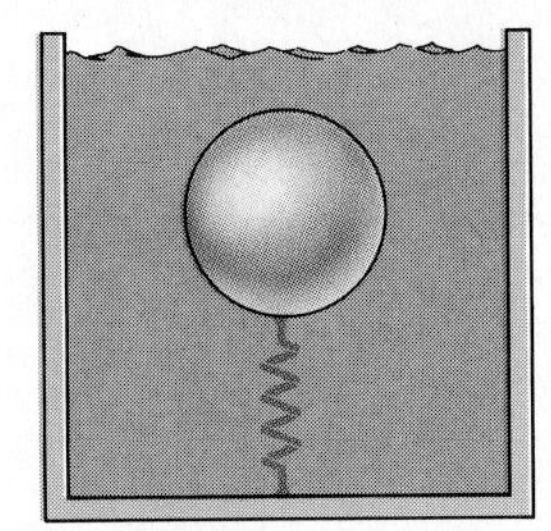

Fig. P19.83

* **19.83** As a submerged body moves through a fluid, the particles of the fluid flow around the body and thus acquire kinetic energy. In the case of a sphere moving in an ideal fluid, the total kinetic energy acquired by the fluid is $\frac{1}{4}\rho V v^2$, where ρ is the density of the fluid, V the volume of the sphere, and v the velocity of the sphere. Consider a 500-g hollow spherical shell of radius 75 mm which is held submerged in a tank of water by a spring of constant 600 N/m. (*a*) Neglecting fluid friction, determine the period of vibration of the shell when it is displaced vertically and then released. (*b*) Solve part *a*, assuming that the tank is accelerated upward at the constant rate of 3 m/s^2.

* **19.84** A thin plate of length l rests on a half cylinder of radius r. Derive an expression for the period of small oscillations of the plate.

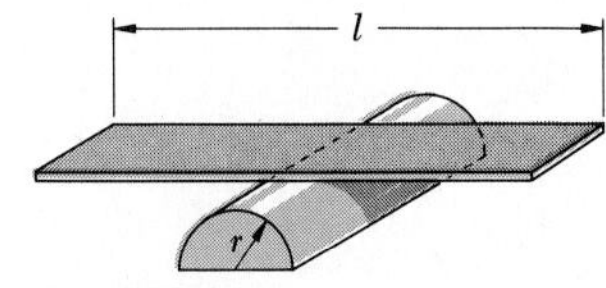

Fig. P19.84

19.7. Forced Vibrations.

The most important vibrations from the point of view of engineering applications are the *forced vibrations* of a system. These vibrations occur when a system is subjected to a periodic force or when it is elastically connected to a support which has an alternating motion.

Consider first the case of a body of mass m suspended from a spring and subjected to a periodic force $\mathbf{P}$ of magnitude $P = P_m \sin \omega t$ (Fig. 19.7). This force may be an actual external force applied to the body, or it may be a centrifugal force produced by the rotation of some unbalanced part of the body (see Sample Prob. 19.5). Denoting by x the displacement of the body measured from its equilibrium position, we write the equation of motion,

$$+\downarrow \Sigma F = ma: \qquad P_m \sin \omega t + W - k(\delta_{st} + x) = m\ddot{x}$$

Recalling that $W = k\delta_{st}$, we have

$$m\ddot{x} + kx = P_m \sin \omega t \qquad (19.30)$$

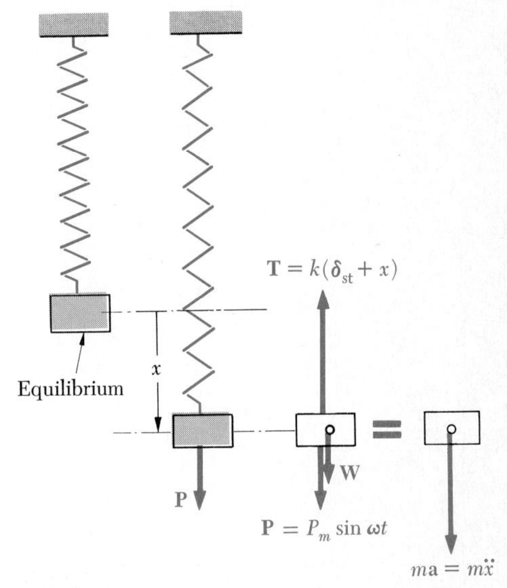

Fig. 19.7

Next we consider the case of a body of mass m suspended from a spring attached to a moving support whose displacement δ is equal to $\delta_m \sin \omega t$ (Fig. 19.8). Measuring the displacement x of the body from the position of static equilibrium corresponding to $\omega t = 0$, we find that the total elongation of the spring at time t is $\delta_{st} + x - \delta_m \sin \omega t$. The equation of motion is thus

$$+\downarrow \Sigma F = ma: \qquad W - k(\delta_{st} + x - \delta_m \sin \omega t) = m\ddot{x}$$

Recalling that $W = k\delta_{st}$, we have

$$m\ddot{x} + kx = k\delta_m \sin \omega t \tag{19.31}$$

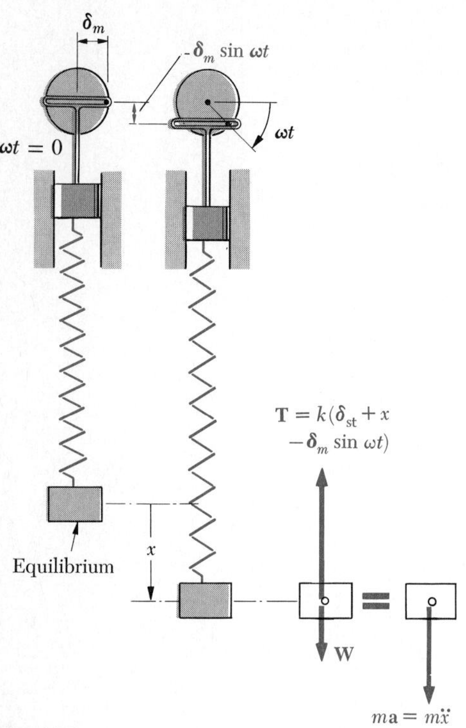

Fig. 19.8

We note that Eqs. (19.30) and (19.31) are of the same form and that a solution of the first equation will satisfy the second if we set $P_m = k\delta_m$.

A differential equation such as (19.30) or (19.31), possessing a right-hand member different from zero, is said to be *nonhomogeneous*. Its general solution is obtained by adding a particular solution of the given equation to the general solution of the corresponding *homogeneous* equation (with right-hand member equal to zero). A *particular solution* of (19.30) or (19.31) may be obtained by trying a solution of the form

$$x_{part} = x_m \sin \omega t \tag{19.32}$$

Substituting x_{part} for x into (19.30), we find

$$-m\omega^2 x_m \sin \omega t + kx_m \sin \omega t = P_m \sin \omega t$$

which may be solved for the amplitude,

$$x_m = \frac{P_m}{k - m\omega^2}$$

Recalling from (19.4) that $k/m = p^2$, where p is the circular frequency of the free vibration of the body, we write

$$x_m = \frac{P_m/k}{1 - (\omega/p)^2} \tag{19.33}$$

Substituting from (19.32) into (19.31), we obtain in a similar way

$$x_m = \frac{\delta_m}{1 - (\omega/p)^2} \tag{19.33'}$$

The homogeneous equation corresponding to (19.30) or (19.31) is Eq. (19.3), defining the free vibration of the body. Its general solution, called the *complementary function*, was found in Sec. 19.2,

$$x_{\text{comp}} = A \sin pt + B \cos pt \tag{19.34}$$

Adding the particular solution (19.32) and the complementary function (19.34), we obtain the *general solution* of Eqs. (19.30) and (19.31):

$$x = A \sin pt + B \cos pt + x_m \sin \omega t \tag{19.35}$$

We note that the vibration obtained consists of two superposed vibrations. The first two terms in (19.35) represent a free vibration of the system. The frequency of this vibration, called the *natural frequency* of the system, depends only upon the constant k of the spring and the mass m of the body, and the constants A and B may be determined from the initial conditions. This free vibration is also called a *transient* vibration, since in actual practice, it will soon be damped out by friction forces (Sec. 19.9).

The last term in (19.35) represents the *steady-state* vibration produced and maintained by the impressed force or impressed support movement. Its frequency is the *forced frequency* imposed by this force or movement, and its amplitude x_m, defined by (19.33) or (19.33′), depends upon the *frequency ratio* ω/p. The ratio of the amplitude x_m of the steady-state vibration to the static deflection P_m/k caused by a force P_m, or to the amplitude δ_m of the support movement, is called the *magnification factor*. From (19.33) and (19.33′), we obtain

$$\text{Magnification factor} = \frac{x_m}{P_m/k} = \frac{x_m}{\delta_m} = \frac{1}{1 - (\omega/p)^2} \tag{19.36}$$

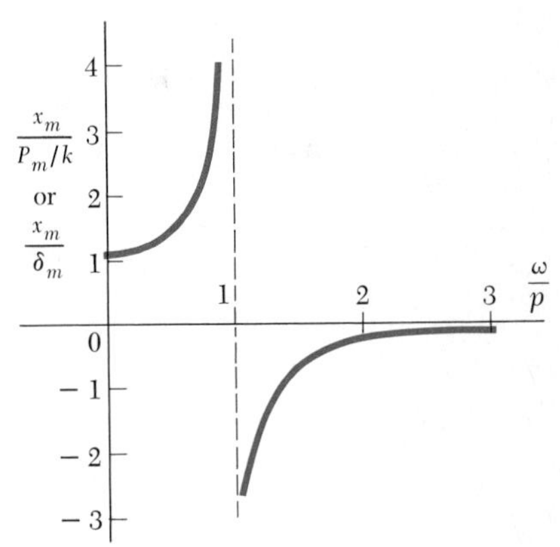

Fig. 19.9

The magnification factor has been plotted in Fig. 19.9 against the frequency ratio ω/p. We note that when $\omega = p$, the amplitude of the forced vibration becomes infinite. The impressed force or impressed support movement is said to be in *resonance* with the given system. Actually, the amplitude of the vibration remains finite because of damping forces (Sec. 19.9); nevertheless, such a situation should be avoided, and the forced frequency should not be chosen too close to the natural frequency of the system. We also note that for $\omega < p$ the coefficient of $\sin \omega t$ in (19.35) is positive, while for $\omega > p$ this coefficient is negative. In the first case the forced vibration is *in phase* with the impressed force or impressed support movement, while in the second case it is 180° *out of phase*.

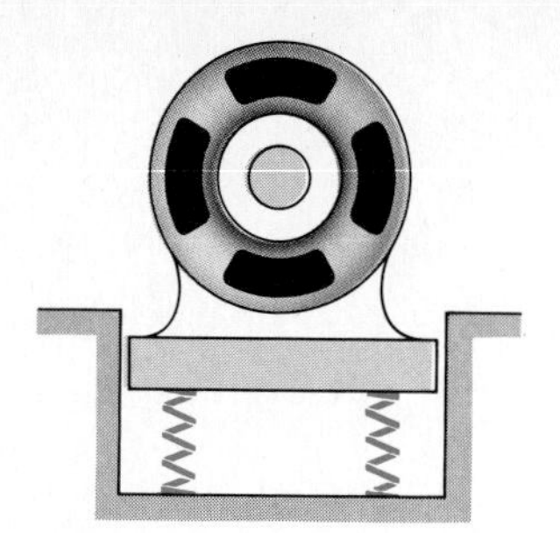

SAMPLE PROBLEM 19.5

A motor weighing 350 lb is supported by four springs, each having a constant of 750 lb/in. The unbalance of the rotor is equivalent to a weight of 1 oz located 6 in. from the axis of rotation. Knowing that the motor is constrained to move vertically, determine (*a*) the speed in rpm at which resonance will occur, (*b*) the amplitude of the vibration of the motor at a speed of 1200 rpm.

***a*. Resonance Speed.** The resonance speed is equal to the circular frequency (in rpm) of the free vibration of the motor. The mass of the motor and the equivalent constant of the supporting springs are

$$m = \frac{350 \text{ lb}}{32.2 \text{ ft/s}^2} = 10.87 \text{ lb}\cdot\text{s}^2/\text{ft}$$

$$k = 4(750 \text{ lb/in.}) = 3000 \text{ lb/in.} = 36{,}000 \text{ lb/ft}$$

$$p = \sqrt{\frac{k}{m}} = \sqrt{\frac{36{,}000}{10.87}} = 57.5 \text{ rad/s} = 549 \text{ rpm}$$

Resonance speed = 549 rpm ◄

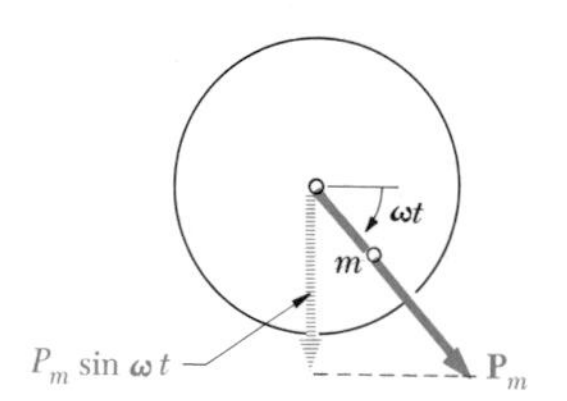

***b*. Amplitude of Vibration at 1200 rpm.** The angular velocity of the motor and the mass of the equivalent 1-oz weight are

$$\omega = 1200 \text{ rpm} = 125.7 \text{ rad/s}$$

$$m = (1 \text{ oz})\frac{1 \text{ lb}}{16 \text{ oz}}\frac{1}{32.2 \text{ ft/s}^2} = 0.00194 \text{ lb}\cdot\text{s}^2/\text{ft}$$

The magnitude of the centrifugal force due to the unbalance of the rotor is

$$P_m = ma_n = mr\omega^2 = (0.00194 \text{ lb}\cdot\text{s}^2/\text{ft})(\tfrac{6}{12} \text{ ft})(125.7 \text{ rad/s})^2 = 15.3 \text{ lb}$$

The static deflection that would be caused by a constant load P_m is

$$\frac{P_m}{k} = \frac{15.3 \text{ lb}}{3000 \text{ lb/in.}} = 0.00510 \text{ in.}$$

Substituting the value of P_m/k together with the known values of ω and p into Eq. (19.33), we obtain

$$x_m = \frac{P_m/k}{1 - (\omega/p)^2} = \frac{0.00510 \text{ in.}}{1 - (125.7/57.5)^2} = -0.00135 \text{ in.}$$

$x_m = 0.00135$ in. (out of phase) ◄

Note. Since $\omega > p$, the vibration is 180° out of phase with the centrifugal force due to the unbalance of the rotor. For example, when the unbalanced mass is directly below the axis of rotation, the position of the motor is $x_m =$ 0.00135 in. above the position of equilibrium.

Problems

19.85 A 5-kg cylinder is suspended from a spring of constant 320 N/m and is acted upon by a vertical periodic force of magnitude $P = P_m \sin \omega t$, where $P_m = 14$ N. Determine the amplitude of the motion of the cylinder if (*a*) $\omega = 6$ rad/s, (*b*) $\omega = 12$ rad/s.

Fig. P19.85 and P19.86

19.86 A cylinder of mass m is suspended from a spring of constant k and is acted upon by a vertical periodic force of magnitude $P = P_m \sin \omega t$. Determine the range of values of ω for which the amplitude of the vibration exceeds twice the static deflection caused by a constant force of magnitude P_m.

19.87 In Prob. 19.86, determine the range of values of ω for which the amplitude of the vibration is less than the static deflection caused by a constant force of magnitude P_m.

19.88 A simple pendulum of length l is suspended from a collar C which is forced to move horizontally according to the relation $x_C = \delta_m \sin \omega t$. Determine the range of values of ω for which the amplitude of the motion of the bob is less than δ_m. (Assume that δ_m is small compared with the length l of the pendulum.)

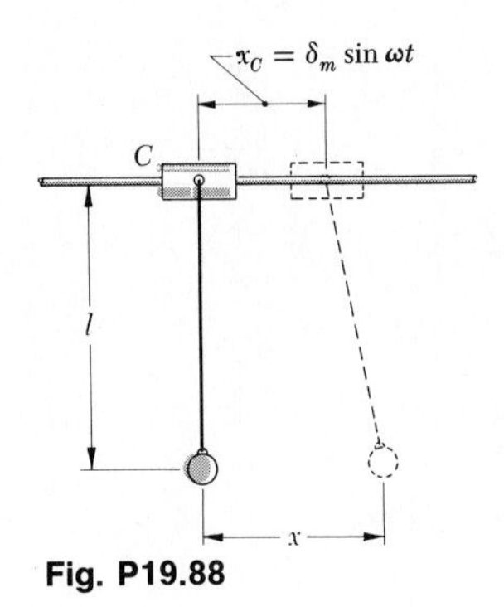

Fig. P19.88

19.89 In Prob. 19.88, determine the range of values of ω for which the amplitude of the motion of the bob exceeds $3\delta_m$.

19.90 A 125-kg motor is supported by a light horizontal beam. The unbalance of the rotor is equivalent to a mass of 25 g located 200 mm from the axis of rotation. Knowing that the static deflection of the beam due to the mass of the motor is 6.9 mm, determine (*a*) the speed (in rpm) at which resonance will occur, (*b*) the amplitude of the steady-state vibration of the motor at a speed of 720 rpm.

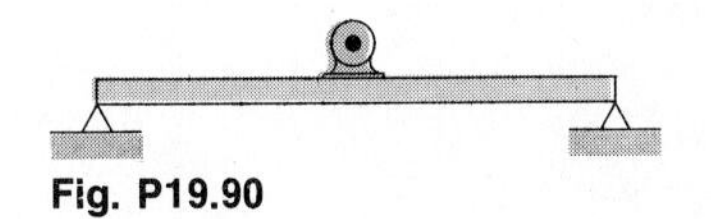
Fig. P19.90

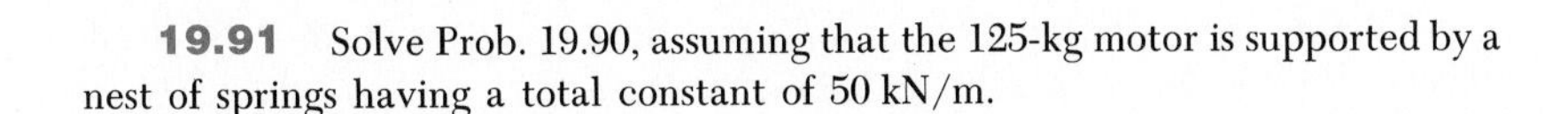
19.91 Solve Prob. 19.90, assuming that the 125-kg motor is supported by a nest of springs having a total constant of 50 kN/m.

19.92 As the speed of a spring-supported motor is slowly increased from 300 to 400 rpm, the amplitude of the vibration due to the unbalance of the rotor is observed to decrease continuously from 0.075 to 0.040 in. Determine the speed at which resonance will occur.

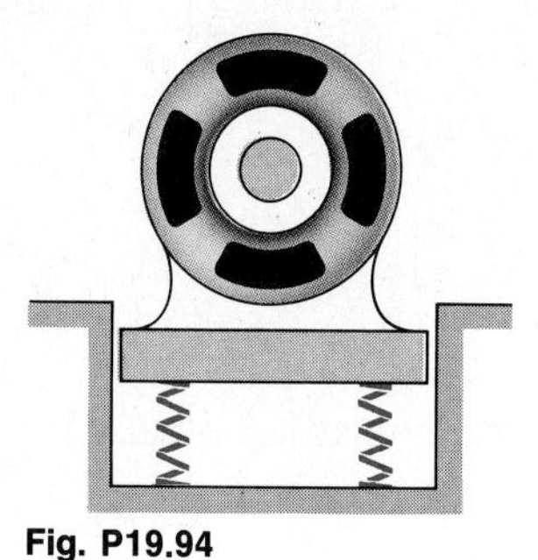

Fig. P19.94

19.93 For the spring-supported motor of Prob. 19.92, determine the speed of the motor for which the amplitude of the vibration is 0.100 in.

19.94 A motor of mass 9 kg is supported by four springs, each of constant 20 kN/m. The motor is constrained to move vertically, and the amplitude of its motion is observed to be 1.2 mm at a speed of 1200 rpm. Knowing that the mass of the rotor is 2.5 kg, determine the distance between the mass center of the rotor and the axis of the shaft.

19.95 In Prob. 19.94, determine the amplitude of the vertical motion of the motor at a speed of (*a*) 450 rpm, (*b*) 1600 rpm, (*c*) 900 rpm.

19.96 Rod AB is rigidly attached to the frame of a motor running at a constant speed. When a collar of mass m is placed on the spring, it is observed to vibrate with an amplitude of 10 mm. When two collars, each of mass m, are placed on the spring, the amplitude is observed to be 12 mm. What amplitude of vibration should be expected when three collars, each of mass m, are placed on the spring? (Obtain two answers.)

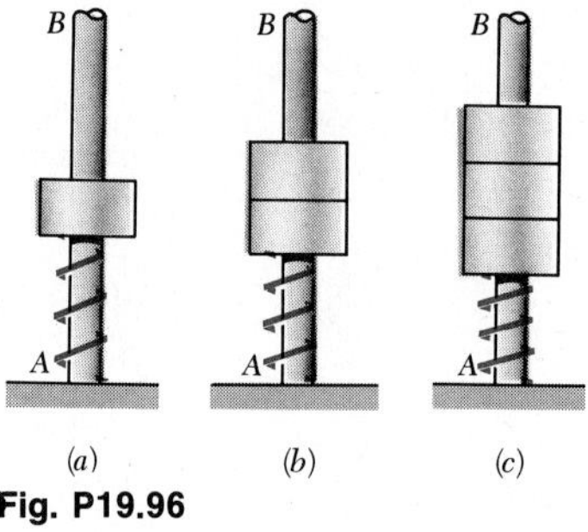

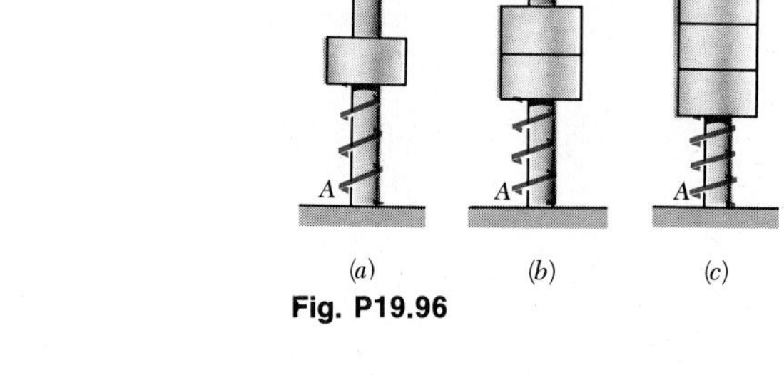

Fig. P19.96

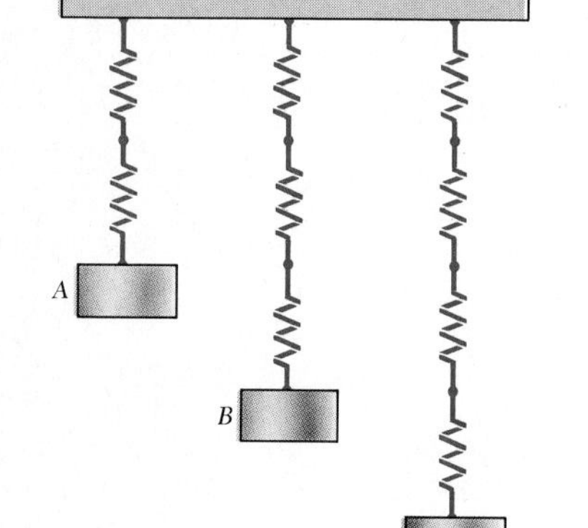

Fig. P19.98

19.97 Solve Prob. 19.96, assuming that the speed of the motor is changed and that one collar has an amplitude of 12 mm and two collars have an amplitude of 4 mm.

19.98 Three identical cylinders A, B, and C are suspended from a bar DE by two or more identical springs as shown. Bar DE is known to move vertically according to the relation $y = \delta_m \sin \omega t$. Knowing that the amplitudes of vibration of cylinders A and B are, respectively, 1.5 in. and 0.75 in., determine the expected amplitude of vibration of cylinder C. (Obtain two answers.)

19.99 Solve Prob. 19.98, assuming that the amplitudes of vibration of cylinders A and B are, respectively, 0.8 in. and 1.2 in.

19.100 A variable-speed motor is rigidly attached to the beam BC. When the speed of the motor is less than 750 rpm or more than 1500 rpm, a small object placed at A is observed to remain in contact with the beam. For speeds between 750 and 1500 rpm the object is observed to "dance" and actually to lose contact with the beam. Determine the speed at which resonance will occur.

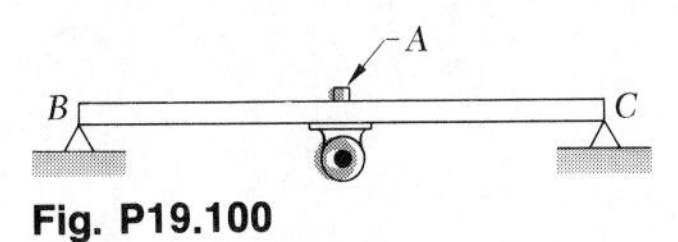

Fig. P19.100

19.101 A disk of mass m is attached to the midpoint of the vertical shaft AB, which revolves at a constant angular velocity ω. Denoting by k the spring constant of the system for a horizontal movement of the disk and by e the eccentricity of the disk with respect to the shaft, show that the deflection of the center of the shaft may be written in the form

$$r = \frac{e(\omega/p)^2}{1 - (\omega/p)^2}$$

where $p = \sqrt{k/m}$.

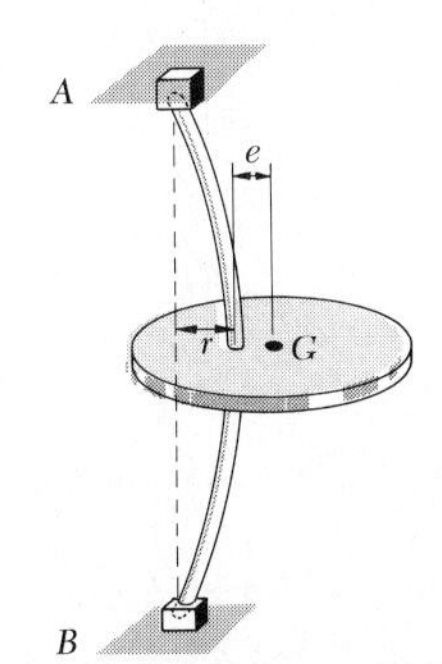

Fig. P19.101 and P19.102

19.102 A disk of mass 20 kg is attached with an eccentricity $e = 0.25$ mm to the midpoint of the vertical shaft AB, which revolves at a constant angular velocity ω. Knowing that a static force of 150 N will deflect the shaft 0.4 mm, determine (a) the angular velocity ω at which resonance occurs, (b) the deflection r of the shaft when $\omega = 1200$ rpm.

19.103 The amplitude of the motion of the pendulum bob shown is observed to be 60 mm, when the amplitude of the motion of collar C is 15 mm. Knowing that the length of the pendulum is $l = 0.90$ m, determine the two possible values of the frequency of the horizontal motion of collar C.

19.104 A pendulum is suspended from collar C as shown. Knowing that the amplitude of the motion of the bob is 12 mm for $l = 750$ mm and 17 mm for $l = 500$ mm, determine the frequency and amplitude of the horizontal motion of collar C.

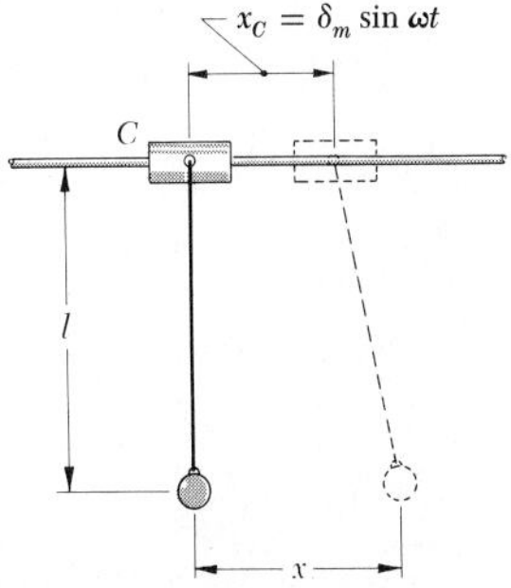

Fig. P19.103 and P19.104

19.105 A small trailer weighing 600 lb with its load is supported by two springs, each of constant 100 lb/in. The trailer is pulled over a road, the surface of which may be approximated by a sine curve of amplitude $1\frac{1}{2}$ in. and of period 16 ft (i.e., the distance between two successive crests is 16 ft, and the vertical distance from a crest to a trough is 3 in.). Determine (*a*) the speed at which resonance will occur, (*b*) the amplitude of the vibration of the trailer at a speed of 40 mi/h.

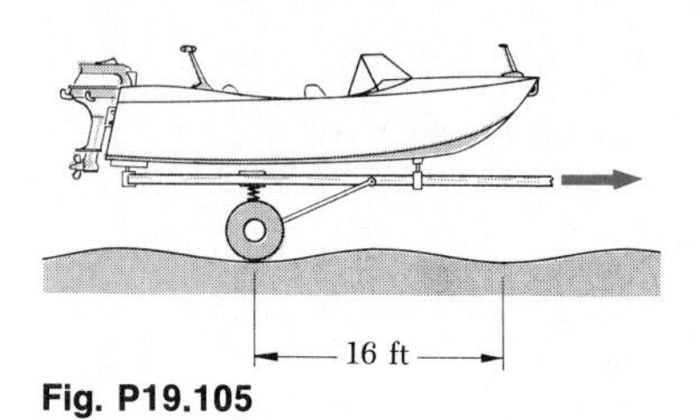

Fig. P19.105

19.106 Knowing that the amplitude of the vibration of the trailer of Prob. 19.105 is not to exceed $\frac{3}{4}$ in., determine the smallest speed at which the trailer can be pulled over the road.

DAMPED VIBRATIONS

*** 19.8. Damped Free Vibrations.** The vibrating systems considered in the first part of this chapter were assumed free of damping. Actually all vibrations are damped to some degree by friction forces. These forces may be caused by *dry friction*, or *Coulomb friction*, between rigid bodies, by *fluid friction* when a rigid body moves in a fluid, or by *internal friction* between the molecules of a seemingly elastic body.

A type of damping of special interest is the *viscous damping* caused by fluid friction at low and moderate speeds. Viscous damping is characterized by the fact that the friction force is *directly proportional to the speed* of the moving body. As an example, we shall consider again a body of mass m suspended from a spring of constant k, and we shall assume that the body is attached to the plunger of a dashpot (Fig. 19.10). The magnitude of the friction force exerted on the plunger by the surrounding fluid is equal to $c\dot{x}$, where the constant c, expressed in N·s/m or lb·s/ft and known as the *coefficient of viscous damping*, depends upon the physical properties of the fluid and the construction of the dashpot. The equation of motion is

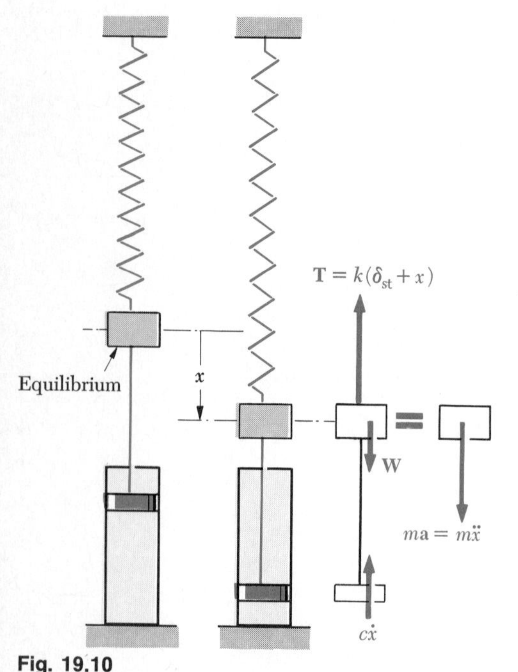

Fig. 19.10

$$+\downarrow\Sigma F = ma: \qquad W - k(\delta_{st} + x) - c\dot{x} = m\ddot{x}$$

Recalling that $W = k\delta_{st}$, we write

$$m\ddot{x} + c\dot{x} + kx = 0 \tag{19.37}$$

Substituting $x = e^{\lambda t}$ into (19.37) and dividing through by $e^{\lambda t}$, we write the *characteristic equation*

$$m\lambda^2 + c\lambda + k = 0 \tag{19.38}$$

and obtain the roots

$$\lambda = -\frac{c}{2m} \pm \sqrt{\left(\frac{c}{2m}\right)^2 - \frac{k}{m}} \tag{19.39}$$

Defining the *critical damping coefficient* c_c as the value of c which makes the radical in (19.39) equal to zero, we write

$$\left(\frac{c_c}{2m}\right)^2 - \frac{k}{m} = 0 \qquad c_c = 2m\sqrt{\frac{k}{m}} = 2mp \tag{19.40}$$

where p is the circular frequency of the system in the absence of damping. We may distinguish three different cases of damping, depending upon the value of the coefficient c.

1. *Heavy damping:* $c > c_c$. The roots λ_1 and λ_2 of the characteristic equation (19.38) are real and distinct, and the general solution of the differential equation (19.37) is

$$x = Ae^{\lambda_1 t} + Be^{\lambda_2 t} \tag{19.41}$$

This solution corresponds to a nonvibratory motion. Since λ_1 and λ_2 are both negative, x approaches zero as t increases indefinitely. However, the system actually regains its equilibrium position after a finite time.

2. *Critical damping:* $c = c_c$. The characteristic equation has a double root $\lambda = -c_c/2m = -p$, and the general solution of (19.37) is

$$x = (A + Bt)e^{-pt} \tag{19.42}$$

The motion obtained is again nonvibratory. Critically damped systems are of special interest in engineering applications since they regain their equilibrium position in the shortest possible time without oscillation.

3. *Light damping:* $c < c_c$. The roots of (19.38) are complex and conjugate, and the general solution of (19.37) is of the form

$$x = e^{-(c/2m)t}(A \sin qt + B \cos qt) \tag{19.43}$$

where q is defined by the relation

$$q^2 = \frac{k}{m} - \left(\frac{c}{2m}\right)^2$$

Substituting $k/m = p^2$ and recalling (19.40), we write

$$q = p\sqrt{1 - \left(\frac{c}{c_c}\right)^2} \tag{19.44}$$

where the constant c/c_c is known as the *damping factor*. A substitution similar to the one used in Sec. 19.2 enables us to write the general solution of (19.37) in the form

$$x = x_m e^{-(c/2m)t} \sin (qt + \phi) \tag{19.45}$$

The motion defined by (19.45) is vibratory with diminishing amplitude (Fig. 19.11). Although this motion does not actually repeat itself, the time interval $\tau = 2\pi/q$, corresponding to two successive points where the curve (19.45) touches one of the limiting curves shown in Fig. 19.11, is commonly referred to as the period of the damped vibration. Recalling (19.44), we observe that τ is larger than the period of vibration of the corresponding undamped system.

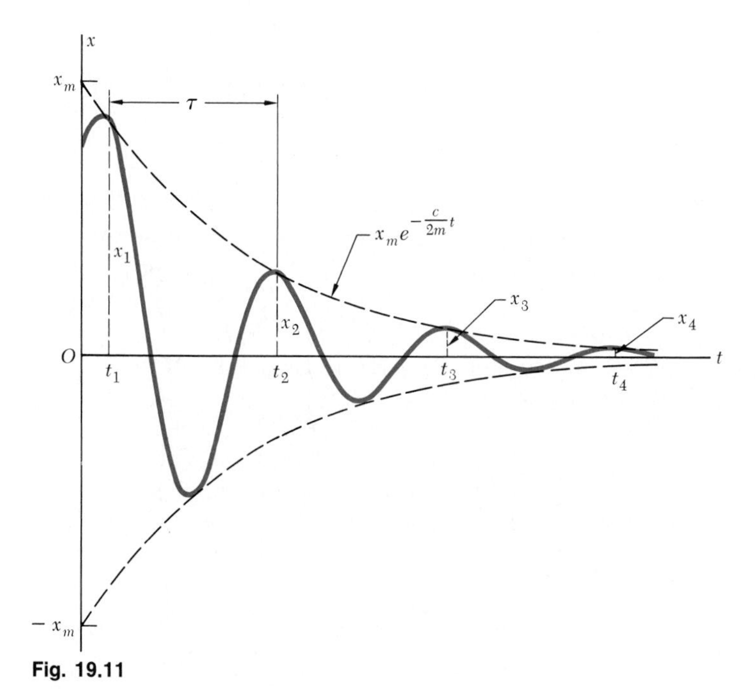

Fig. 19.11

***19.9. Damped Forced Vibrations.** If the system considered in the preceding section is subjected to a periodic force **P** of magnitude $P = P_m \sin \omega t$, the equation of motion becomes

$$m\ddot{x} + c\dot{x} + kx = P_m \sin \omega t \qquad (19.46)$$

The general solution of (19.46) is obtained by adding a particular solution of (19.46) to the complementary function or general solution of the homogeneous equation (19.37). The complementary function is given by (19.41), (19.42), or (19.43), depending upon the type of damping considered. It represents a *transient* motion which is eventually damped out.

Our interest in this section is centered on the steady-state vibration represented by a particular solution of (19.46) of the form

$$x_{\text{part}} = x_m \sin (\omega t - \varphi) \qquad (19.47)$$

Substituting x_{part} for x into (19.46), we obtain

$$-m\omega^2 x_m \sin (\omega t - \varphi) + c\omega x_m \cos (\omega t - \varphi) + kx_m \sin (\omega t - \varphi) = P_m \sin \omega t$$

Making $\omega t - \varphi$ successively equal to 0 and to $\pi/2$, we write

$$c\omega x_m = P_m \sin \varphi \qquad (19.48)$$

$$(k - m\omega^2)x_m = P_m \cos \varphi \qquad (19.49)$$

Squaring both members of (19.48) and (19.49) and adding, we have

$$[(k - m\omega^2)^2 + (c\omega)^2]x_m^2 = P_m^2 \qquad (19.50)$$

Solving (19.50) for x_m and dividing (19.48) and (19.49) member by member, we obtain, respectively,

$$x_m = \frac{P_m}{\sqrt{(k - m\omega^2)^2 + (c\omega)^2}} \qquad \tan \varphi = \frac{c\omega}{k - m\omega^2} \qquad (19.51)$$

Recalling from (19.4) that $k/m = p^2$, where p is the circular frequency of the undamped free vibration, and from (19.40) that $2mp = c_c$, where c_c is the critical damping coefficient of the system, we write

$$\frac{x_m}{P_m/k} = \frac{x_m}{\delta_m} = \frac{1}{\sqrt{[1 - (\omega/p)^2]^2 + [2(c/c_c)(\omega/p)]^2}} \qquad (19.52)$$

$$\tan \varphi = \frac{2(c/c_c)(\omega/p)}{1 - (\omega/p)^2} \qquad (19.53)$$

Formula (19.52) expresses the magnification factor in terms of the frequency ratio ω/p and damping factor c/c_c. It may be used to determine the amplitude of the steady-state vibration produced by an impressed force of magnitude $P = P_m \sin \omega t$ or by an impressed support movement $\delta = \delta_m \sin \omega t$. Formula (19.53) defines in terms of the same parameters the *phase difference* φ between the impressed force or impressed support movement and the resulting steady-state vibration of the damped system. The magnification factor has been plotted against the frequency ratio in Fig. 19.12 for various values of the damping factor. We observe that the amplitude of a forced vibration may be kept small by choosing a large coefficient of viscous damping c or by keeping the natural and forced frequencies far apart.

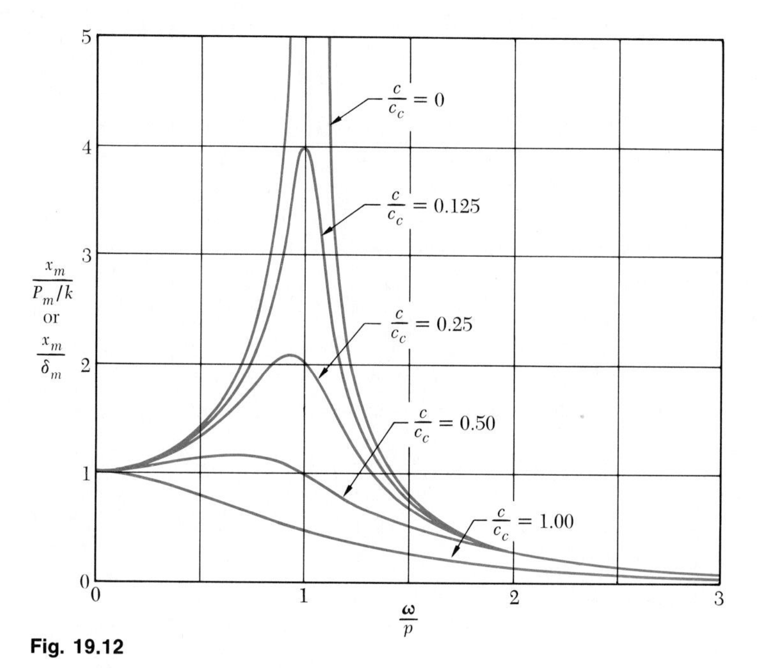

Fig. 19.12

*** 19.10. Electrical Analogues.** Oscillating electrical circuits are characterized by differential equations of the same type as those obtained in the preceding sections. Their analysis is therefore similar to that of a mechanical system, and the results obtained for a given vibrating system may be readily extended to the equivalent circuit. Conversely, any result obtained for an electrical circuit will also apply to the corresponding mechanical system.

Consider an electrical circuit consisting of an inductor of inductance L, a resistor of resistance R, and a capacitor of capacitance C, connected in series with a source of alternating voltage $E = E_m \sin \omega t$ (Fig. 19.13). It is recalled from elementary circuit theory† that if i denotes the current in the circuit and q the electric charge on the capacitor, the drop in potential is $L(di/dt)$ across the inductor, Ri across the resistor, and q/C across the capacitor. Expressing that the algebraic sum of the applied voltage and of the drops in potential around the circuit loop is zero, we write

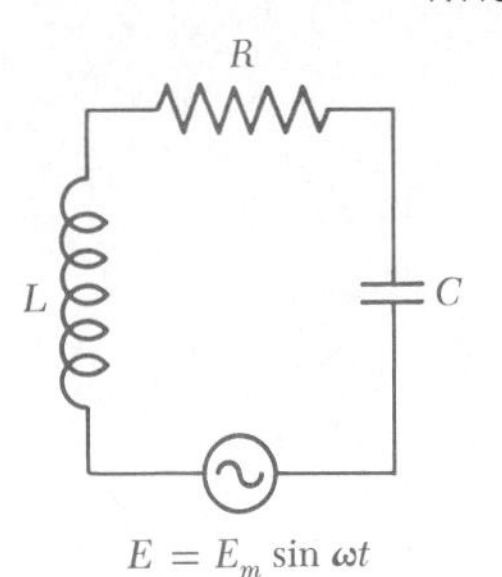

Fig. 19.13

$$E_m \sin \omega t - L\frac{di}{dt} - Ri - \frac{q}{C} = 0 \qquad (19.54)$$

Rearranging the terms and recalling that at any instant the current i is equal to the rate of change $\dot{q}$ of the charge q, we have

$$L\ddot{q} + R\dot{q} + \frac{1}{C}q = E_m \sin \omega t \qquad (19.55)$$

We verify that Eq. (19.55), which defines the oscillations of the electrical circuit of Fig. 19.13, is of the same type as Eq. (19.46), which characterizes the damped forced vibrations of the mechanical system of Fig. 19.10. By comparing the two equations, we may construct a table of the analogous mechanical and electrical expressions.

Table 19.2 may be used to extend to their electrical analogues the results obtained in the preceding sections for various mechanical systems. For instance, the amplitude i_m of the current in the circuit of Fig. 19.13 may be obtained by noting that it corresponds to the maximum value v_m of the velocity in the analogous mechanical system. Recalling that $v_m = \omega x_m$,

Table 19.2 **Characteristics of a Mechanical System and of Its Electrical Analogue**

Mechanical System		Electrical Circuit	
m	Mass	L	Inductance
c	Coefficient of viscous damping	R	Resistance
k	Spring constant	$1/C$	Reciprocal of capacitance
x	Displacement	q	Charge
v	Velocity	i	Current
P	Applied force	E	Applied voltage

†See A. E. Fitzgerald, D. E. Higginbotham, and A. Grabel, *Basic Electrical Engineering,* 5th ed., McGraw-Hill, New York, 1981; or J. W. Nilsson, *Electric Circuits,* Addison-Wesley, Reading, Mass., 1983.

substituting for x_m from Eq. (19.51), and replacing the constants of the mechanical system by the corresponding electrical expressions, we have

$$i_m = \frac{\omega E_m}{\sqrt{\left(\frac{1}{C} - L\omega^2\right)^2 + (R\omega)^2}}$$

$$= \frac{E_m}{\sqrt{R^2 + \left(L\omega - \frac{1}{C\omega}\right)^2}} \quad (19.56)$$

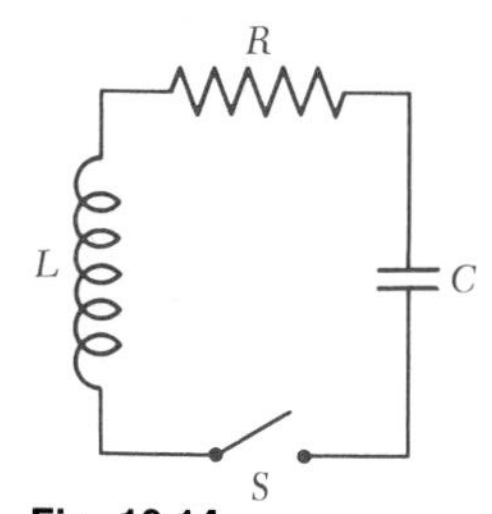

Fig. 19.14

The radical in the expression obtained is known as the *impedance* of the electrical circuit.

The analogy between mechanical systems and electrical circuits holds for transient as well as steady-state oscillations. The oscillations of the circuit shown in Fig. 19.14, for instance, are analogous to the damped free vibrations of the system of Fig. 19.10. As far as the initial conditions are concerned, we may note that closing the switch S when the charge on the capacitor is $q = q_0$ is equivalent to releasing the mass of the mechanical system with no initial velocity from the position $x = x_0$. We should also observe that if a battery of constant voltage E is introduced in the electrical circuit of Fig. 19.14, closing the switch S will be equivalent to suddenly applying a force of constant magnitude P to the mass of the mechanical system of Fig. 19.10.

The discussion above would be of questionable value if its only result were to make it possible for mechanics students to analyze electrical circuits without learning the elements of circuit theory. It is hoped that this discussion will instead encourage students to apply to the solution of problems in mechanical vibrations the mathematical techniques they may learn in later courses in circuit theory. The chief value of the concept of electrical analogue, however, resides in its application to *experimental methods* for the determination of the characteristics of a given mechanical system. Indeed, an electrical circuit is much more easily constructed than is a mechanical model, and the fact that its characteristics may be modified by varying the inductance, resistance, or capacitance of its various components makes the use of the electrical analogue particularly convenient.

To determine the electrical analogue of a given mechanical system, we shall focus our attention on each moving mass in the system and observe which springs, dashpots, or external forces are applied directly to it. An equivalent electrical loop may then be constructed to match each of the mechanical units thus defined; the various loops obtained in that way will form together the desired circuit. Consider, for instance, the mechanical system of Fig. 19.15. We observe that the mass m_1 is acted upon by two springs of constants k_1 and k_2 and by two dashpots characterized by the coefficients of viscous damping c_1 and c_2. The electrical circuit should therefore include a loop consisting of an inductor of inductance L_1 proportional to m_1, of two capacitors of capacitance C_1 and C_2 inversely proportional to k_1 and k_2, respectively, and of two resistors of resistance R_1 and R_2, proportional to c_1 and c_2, respectively. Since the mass m_2 is acted upon by the spring k_2 and the dashpot c_2, as well as by the force $P = P_m \sin \omega t$, the circuit should also include a loop containing the capacitor C_2, the resistor R_2, the new inductor L_2, and the voltage source $E = E_m \sin \omega t$ (Fig. 19.16).

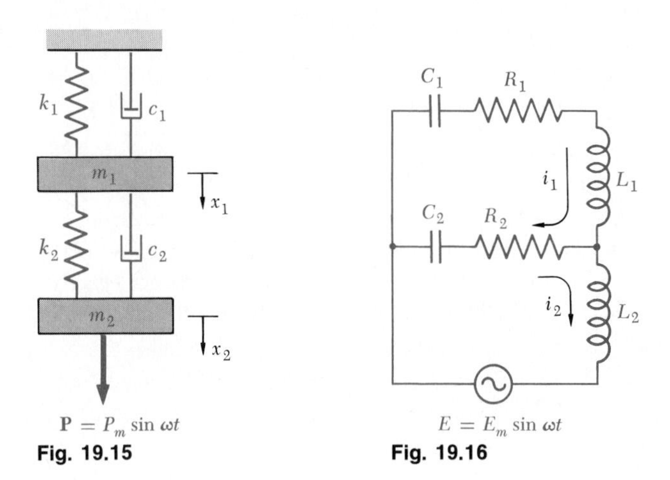

$\mathbf{P} = P_m \sin \omega t$
Fig. 19.15

$E = E_m \sin \omega t$
Fig. 19.16

To check that the mechanical system of Fig. 19.15 and the electrical circuit of Fig. 19.16 actually satisfy the same differential equations, we shall first derive the equations of motion for m_1 and m_2. Denoting, respectively, by x_1 and x_2 the displacements of m_1 and m_2 from their equilibrium positions, we observe that the elongation of the spring k_1 (measured from the equilibrium position) is equal to x_1, while the elongation of the spring k_2 is equal to the relative displacement $x_2 - x_1$ of m_2 with respect to m_1. The equations of motion for m_1 and m_2 are therefore

$$m_1\ddot{x}_1 + c_1\dot{x}_1 + c_2(\dot{x}_1 - \dot{x}_2) + k_1x_1 + k_2(x_1 - x_2) = 0 \qquad (19.57)$$

$$m_2\ddot{x}_2 + c_2(\dot{x}_2 - \dot{x}_1) + k_2(x_2 - x_1) = P_m \sin \omega t \qquad (19.58)$$

Consider now the electrical circuit of Fig. 19.16; we denote, respectively, by i_1 and i_2 the current in the first and second loops, and by q_1 and q_2 the integrals $\int i_1\,dt$ and $\int i_2\,dt$. Noting that the charge on the capacitor C_1 is q_1, while the charge on C_2 is $q_1 - q_2$, we express that the sum of the potential differences in each loop is zero:

$$L_1\ddot{q}_1 + R_1\dot{q}_1 + R_2(\dot{q}_1 - \dot{q}_2) + \frac{q_1}{C_1} + \frac{q_1 - q_2}{C_2} = 0 \qquad (19.59)$$

$$L_2\ddot{q}_2 + R_2(\dot{q}_2 - \dot{q}_1) + \frac{q_2 - q_1}{C_2} = E_m \sin \omega t \qquad (19.60)$$

We easily check that Eqs. (19.59) and (19.60) reduce to (19.57) and (19.58), respectively, when the substitutions indicated in Table 19.2 are performed.

Problems

19.107 Show that in the case of heavy damping $(c > c_c)$, a body never passes through its position of equilibrium O (*a*) if it is released with no initial velocity from an arbitrary position or (*b*) if it is started from O with an arbitrary initial velocity.

19.108 Show that in the case of heavy damping $(c > c_c)$, a body released from an arbitrary position with an arbitrary initial velocity cannot pass more than once through its equilibrium position.

19.109 In the case of light damping, the displacements x_1, x_2, x_3, and so forth, shown in Fig. 19.11 may be assumed equal to the maximum displacements. Show that the ratio of any two successive maximum displacements x_n and x_{n+1} is a constant and that the natural logarithm of this ratio, called the *logarithmic decrement,* is

$$\ln \frac{x_n}{x_{n+1}} = \frac{2\pi(c/c_c)}{\sqrt{1 - (c/c_c)^2}}$$

19.110 In practice, it is often difficult to determine the logarithmic decrement of a system with light damping defined in Prob. 19.109 by measuring two successive maximum displacements. Show that the logarithmic decrement may also be expressed as $(1/k) \ln (x_n/x_{n+k})$, where k is the number of cycles between readings of the maximum displacement.

19.111 In a system with light damping $(c < c_c)$, the period of vibration is commonly defined as the time interval $\tau = 2\pi/q$ corresponding to two successive points where the displacement-time curve touches one of the limiting curves shown in Fig. 19.11. Show that the interval of time (*a*) between a maximum positive displacement and the following maximum negative displacement is $\frac{1}{2}\tau$, (*b*) between two successive zero displacements is $\frac{1}{2}\tau$, (*c*) between a maximum positive displacement and the following zero displacement is greater than $\frac{1}{4}\tau$.

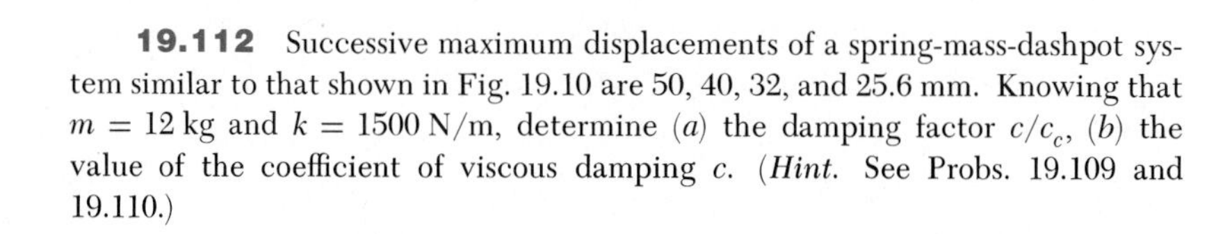

19.112 Successive maximum displacements of a spring-mass-dashpot system similar to that shown in Fig. 19.10 are 50, 40, 32, and 25.6 mm. Knowing that $m = 12$ kg and $k = 1500$ N/m, determine (*a*) the damping factor c/c_c, (*b*) the value of the coefficient of viscous damping c. (*Hint.* See Probs. 19.109 and 19.110.)

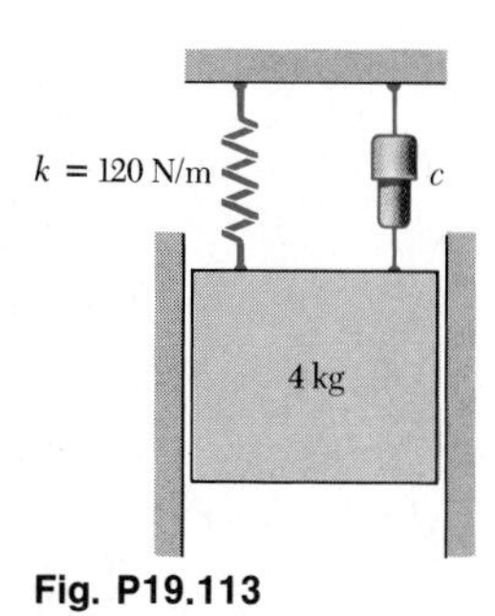

Fig. P19.113

19.113 The block shown is depressed 20 mm from its equilibrium position and released. Knowing that after eight cycles the maximum displacement of the block is 12 mm, determine (*a*) the damping factor c/c_c, (*b*) the value of the coefficient of viscous damping. (*Hint.* See Probs. 19.109 and 19.110.)

19.114 The barrel of a field gun weighs 1400 lb and is returned into firing position after recoil by a recuperator of constant $k = 10{,}000$ lb/ft. Determine (*a*) the value of the coefficient of damping of the recuperator which causes the barrel to return into firing position in the shortest possible time without oscillation, (*b*) the time needed for the barrel to move from its maximum-recoil position halfway back to its firing position.

19.115 Assuming that the barrel of the gun of Prob. 19.114 is modified, with a resulting increase in weight of 400 lb, determine (*a*) the constant k which should be used for the recuperator if the barrel is to remain critically damped, (*b*) the time needed for the modified barrel to move from its maximum-recoil position halfway back to its firing position.

19.116 In the case of the forced vibration of a system with a given damping factor c/c_c, determine the frequency ratio ω/p for which the amplitude of the vibration is maximum.

19.117 Show that for a small value of the damping factor c/c_c, (*a*) the maximum amplitude of a forced vibration occurs when $\omega \approx p$, (*b*) the corresponding value of the magnification factor is approximately $\frac{1}{2}(c_c/c)$.

19.118 A 30-lb motor is directly supported by a light horizontal beam which has a static deflection of 0.05 in. due to the weight of the motor. Knowing that the unbalance of the rotor is equivalent to a weight of 1 oz located 7.5 in. from the axis of rotation, determine the amplitude of vibration of the motor at a speed of 900 rpm, assuming (*a*) that no damping is present, (*b*) that the damping factor c/c_c is equal to 0.075.

19.119 A motor weighing 50 lb is supported by four springs, each having a constant of 1000 lb/in. The unbalance of the rotor is equivalent to a weight of 1 oz located 5 in. from the axis of rotation. Knowing that the motor is constrained to move vertically, determine the amplitude of the steady-state vibration of the motor at a speed of 1800 rpm, assuming (*a*) that no damping is present, (*b*) that the damping factor c/c_c is equal to 0.125.

19.120 Solve Prob. 19.94, assuming that a dashpot having a coefficient of damping $c = 200$ N·s/m has been connected to the motor and to the ground.

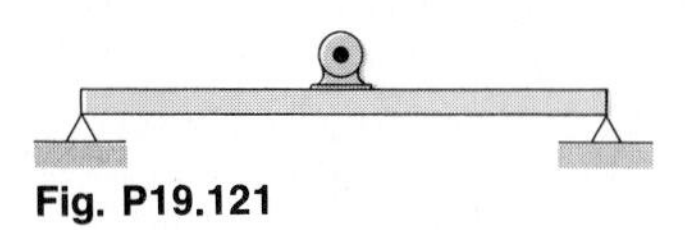

Fig. P19.121

19.121 A 50-kg motor is directly supported by a light horizontal beam which has a static deflection of 6 mm due to the weight of the motor. The unbalance of the rotor is equivalent to a mass of 100 g located 75 mm from the axis of rotation. Knowing that the amplitude of the vibration of the motor is 0.9 mm at a speed of 400 rpm, determine (*a*) the damping factor c/c_c, (*b*) the coefficient of damping c.

19.122 A machine element having a mass of 400 kg is supported by two springs, each having a constant of 38 kN/m. A periodic force of maximum magnitude equal to 135 N is applied to the element with a frequency of 2.5 Hz. Knowing that the coefficient of damping is 1400 N·s/m, determine the amplitude of the steady-state vibration of the element.

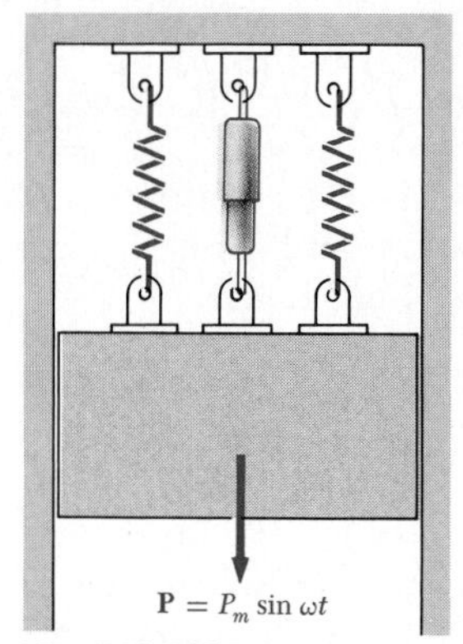

Fig. P19.122

19.123 In Prob. 19.122, determine the required value of the coefficient of damping if the amplitude of the steady-state vibration of the element is to be 3.5 mm.

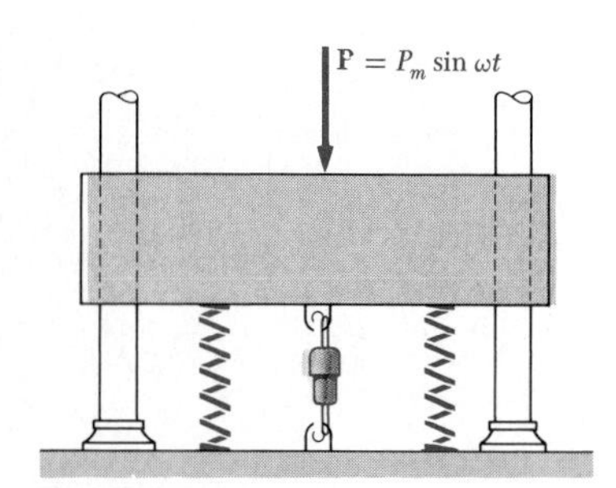

Fig. P19.124

19.124 A platform of weight 200 lb, supported by two springs each of constant $k = 250$ lb/in., is subjected to a periodic force of maximum magnitude equal to 125 lb. Knowing that the coefficient of damping is 10 lb·s/in., determine (*a*) the natural frequency in rpm of the platform if there were no damping, (*b*) the frequency in rpm of the periodic force corresponding to the maximum value of the magnification factor, assuming damping, (*c*) the amplitude of the actual motion of the platform for each of the frequencies found in parts *a* and *b*.

19.125 Solve Prob. 19.124, assuming that the coefficient of damping is increased to 15 lb·s/in.

* **19.126** The suspension of an automobile may be approximated by the simplified spring-and-dashpot system shown. (*a*) Write the differential equation defining the absolute motion of the mass m when the system moves at a speed v over a road of sinusoidal cross section as shown. (*b*) Derive an expression for the amplitude of the absolute motion of m.

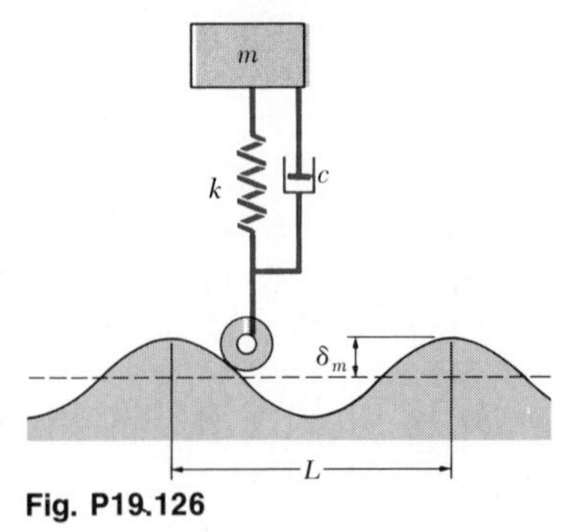

Fig. P19.126

* **19.127** Two blocks A and B, each of mass m, are suspended as shown by means of five springs of the same constant k and are connected by a dashpot of coefficient of damping c. Block B is subjected to a force of magnitude $P = P_m \sin \omega t$. Write the differential equations defining the displacements x_A and x_B of the two blocks from their equilibrium positions.

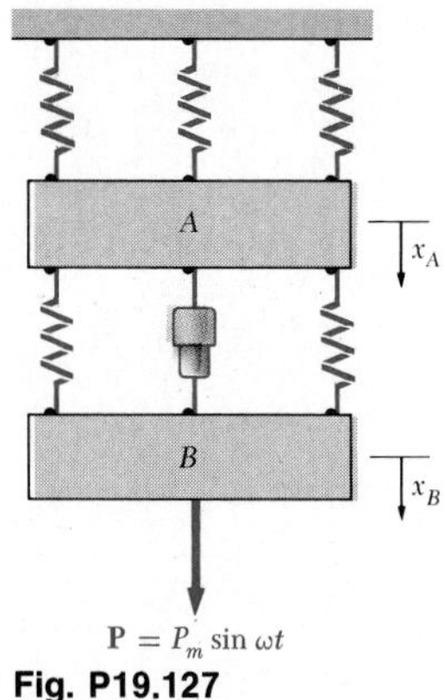

Fig. P19.127

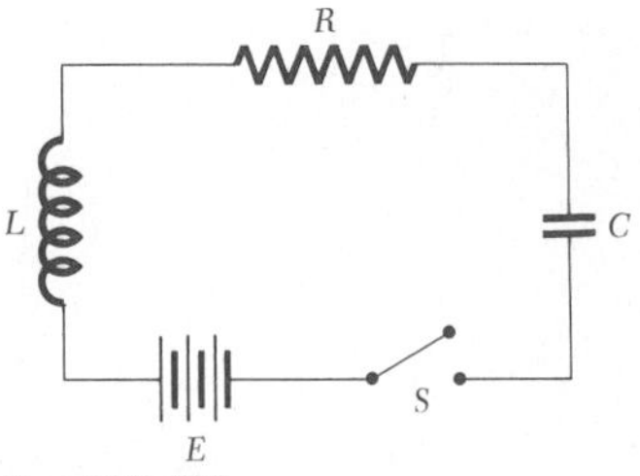

Fig. P19.128

19.128 Determine the range of values of the resistance R for which oscillations will take place in the circuit shown when the switch S is closed.

19.129 Consider the circuit of Prob. 19.128 when the capacitor C is removed. If the switch S is closed at time $t = 0$, determine (*a*) the final value of the current in the circuit, (*b*) the time t at which the current will have reached $(1 - 1/e)$ times its final value. (The desired value of t is known as the *time constant* of the circuit.)

19.130 and 19.131 Draw the electrical analogue of the mechanical system shown. (*Hint.* Draw the loops corresponding to the free bodies m and A.)

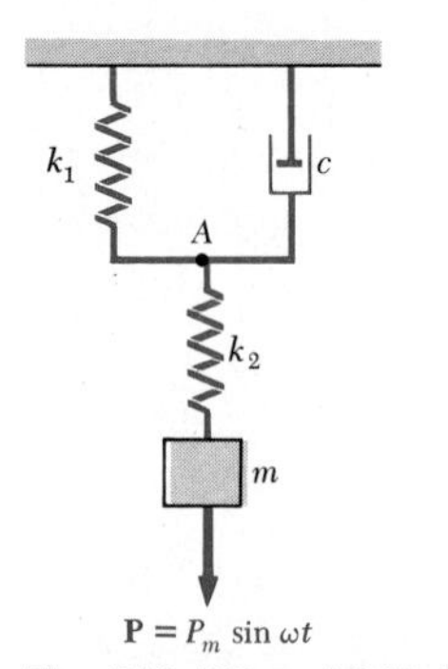

Fig. P19.130 and P19.132

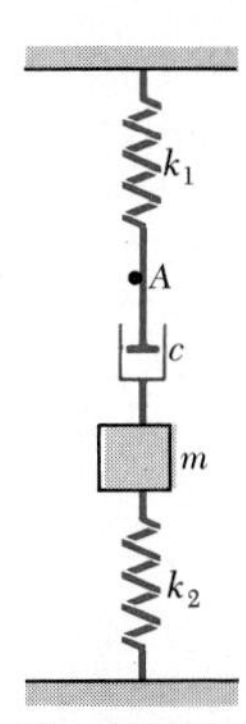

Fig. P19.131 and P19.133

19.132 and 19.133 Write the differential equations defining (*a*) the displacements of mass m and point A, (*b*) the currents in the corresponding loops of the electrical analogue.

19.134 and 19.135 Draw the electrical analogue of the mechanical system shown.

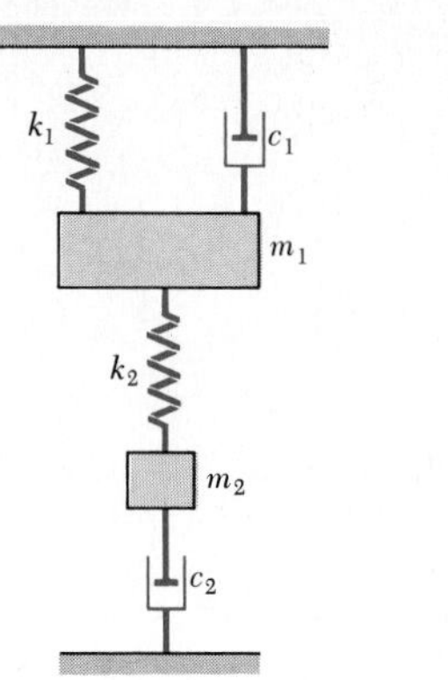

Fig. P19.134 and P19.136

$\mathbf{P} = P_m \sin \omega t$

Fig. P19.135 and P19.137

19.136 and 19.137 Write the differential equations defining (*a*) the displacements of the masses m_1 and m_2, (*b*) the currents in the corresponding loops of the electrical analogue.

19.11. Review and Summary. This chapter was devoted to the study of *mechanical vibrations*, i.e., to the analysis of the motion of particles and rigid bodies oscillating about a position of equilibrium. In the first part of the chapter [Secs. 19.2 through 19.7], we considered *vibrations without damping*, while the second part was devoted to *damped vibrations* [Secs. 19.8 through 19.10].

Free vibrations of a particle

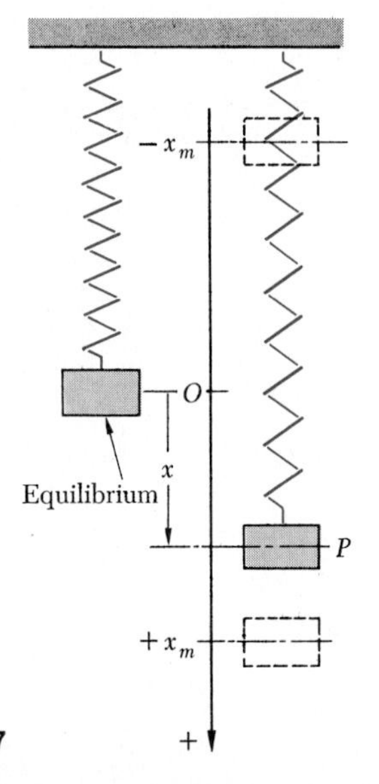

Fig. 19.17

In Sec. 19.2, we considered the *free vibrations of a particle*, i.e., the motion of a particle P subjected to a restoring force proportional to the displacement of the particle—such as the force exerted by a spring. If the displacement x of the particle P is measured from its equilibrium position O (Fig. 19.17), the resultant $\mathbf{F}$ of the forces acting on P (including its weight) has a magnitude kx and is directed toward O. Applying Newton's second law $F = ma$ and recalling that $a = \ddot{x}$, we wrote the differential equation

$$m\ddot{x} + kx = 0 \qquad (19.3)$$

or, setting $p^2 = k/m$,

$$\ddot{x} + p^2 x = 0 \qquad (19.5)$$

The motion defined by this equation is called a *simple harmonic motion*.

The solution of Eq. (19.5), which represents the displacement of the particle P, was expressed as

$$x = x_m \sin (pt + \phi) \qquad (19.10)$$

where x_m = amplitude of the vibration
$p = \sqrt{k/m}$ = circular frequency
ϕ = phase angle

The *period of the vibration* (i.e., the time required for a full cycle) and its *frequency* (i.e., the number of cycles per second) were expressed as

$$\text{Period} = \tau = \frac{2\pi}{p} \qquad (19.13)$$

$$\text{Frequency} = f = \frac{1}{\tau} = \frac{p}{2\pi} \qquad (19.14)$$

The velocity and acceleration of the particle were obtained by differentiating Eq. (19.10), and their maximum values were found to be

$$v_m = x_m p \qquad a_m = x_m p^2 \qquad (19.15)$$

Since all the above parameters depend directly upon the circular frequency p and thus upon the ratio k/m, it is essential in any given problem to calculate the value of the constant k; this may be done by determining the relation between the restoring force and the corresponding displacement of the particle [Sample Prob. 19.1].

It was also shown that the oscillatory motion of the particle P may be represented by the projection on the x axis of the motion of a point Q describing an auxiliary circle of radius x_m with the constant angular velocity p (Fig. 19.18). The instantaneous values of the velocity and acceleration of P may then be obtained by projecting on the x axis the vectors $\mathbf{v}_m$ and $\mathbf{a}_m$ representing, respectively, the velocity and acceleration of Q.

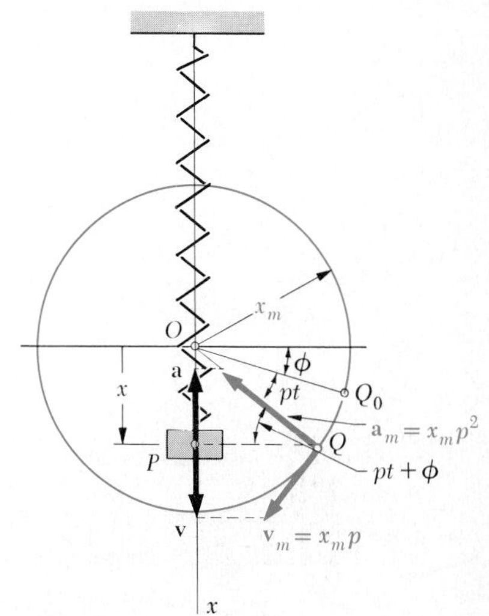

Fig. 19.18

Simple pendulum

While the motion of a *simple pendulum* is not truly a simple harmonic motion, the formulas derived above may be used to calculate the period and frequency of the *small oscillations* of a simple pendulum [Sec. 19.3]. Large-amplitude oscillations of a simple pendulum were discussed in Sec. 19.4.

Free vibrations of a rigid body

The *free vibrations of a rigid body* may be analyzed by choosing an appropriate variable, such as a distance x or an angle θ, to define the position of the body, drawing a diagram expressing the equivalence of the external and effective forces, and writing an equation relating the selected variable and its second derivative [Sec. 19.5]. If the equation obtained is of the form

$$\ddot{x} + p^2 x = 0 \qquad \text{or} \qquad \ddot{\theta} + p^2\theta = 0 \qquad (19.21)$$

the vibration considered is a simple harmonic motion and its period and frequency may be obtained *by identifying* p and substituting in (19.13) and (19.14) [Sample Probs. 19.2 and 19.3].

Using the principle of conservation of energy

The *principle of conservation of energy* may be used as an alternative method for the determination of the period and frequency of the simple harmonic motion of a particle or rigid body [Sec. 19.6]. Choosing again an appropriate variable, such as θ, to define the position of the system, we express that the total energy of the system is conserved, $T_1 + V_1 = T_2 + V_2$, between the position of maximum displacement ($\theta_1 = \theta_m$) and the position of maximum velocity ($\dot{\theta}_2 = \dot{\theta}_m$). If the motion considered is simple harmonic, the two members of the equation obtained consist of homogeneous quadratic expressions in θ_m and $\dot{\theta}_m$, respectively.† Substituting $\dot{\theta}_m = \theta_m p$ in this equation, we may factor out θ_m^2 and solve for the circular frequency p [Sample Prob. 19.4].

Forced vibrations

In Sec. 19.7, we considered the *forced vibrations* of a mechanical system. These vibrations occur when the system is subjected to a periodic force (Fig. 19.19) or when it is elastically connected to a support which has an alternating motion (Fig. 19.20). In the first case, the motion of the system is defined by the differential equation

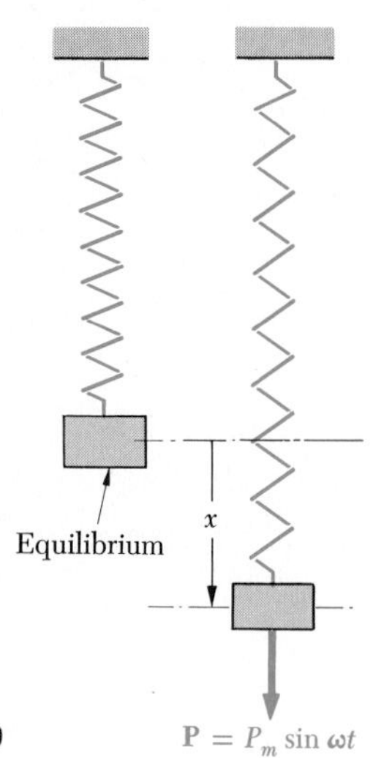

Fig. 19.19

$$m\ddot{x} + kx = P_m \sin \omega t \tag{19.30}$$

and in the second case by the differential equation

$$m\ddot{x} + kx = k\delta_m \sin \omega t \tag{19.31}$$

The general solution of these equations is obtained by adding a particular solution of the form

$$x_{\text{part}} = x_m \sin \omega t \tag{19.32}$$

to the general solution of the corresponding homogeneous equation. The particular solution (19.32) represents a *steady-state vibration* of the system, while the solution of the homogeneous equation represents a *transient free vibration* which may generally be neglected.

Dividing the amplitude x_m of the steady-state vibration by P_m/k in the case of a periodic force, or by δ_m in the case of an oscillating support, we defined the *magnification factor* of the vibration and found that

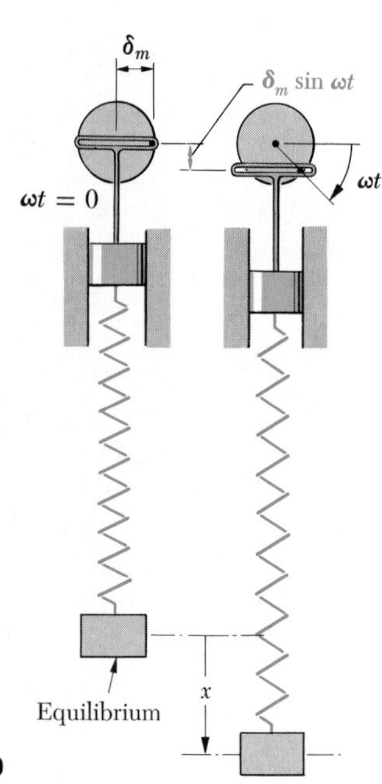

Fig. 19.20

$$\text{Magnification factor} = \frac{x_m}{P_m/k} = \frac{x_m}{\delta_m} = \frac{1}{1-(\omega/p)^2} \tag{19.36}$$

According to Eq. (19.36), the amplitude x_m of the forced vibration *becomes infinite when* $\omega = p$, i.e., *when the forced frequency is equal to the natural frequency of the system*. The impressed force or impressed support movement is then said to be in *resonance* with the system [Sample Prob. 19.5]. Actually the amplitude of the vibration remains finite, due to damping forces.

† If the motion considered may only be *approximated* by a simple harmonic motion, such as for the small oscillations of a body under gravity, the potential energy must be approximated by a quadratic expression in θ_m.

Damped free vibrations

In the last part of the chapter, we considered the *damped vibrations* of a mechanical system. First, we analyzed the *damped free vibrations* of a system with *viscous damping* [Sec. 19.8]. We found that the motion of such a system was defined by the differential equation

$$m\ddot{x} + c\dot{x} + kx = 0 \qquad (19.37)$$

where c is a constant called the *coefficient of viscous damping*. Defining the *critical damping coefficient* c_c as

$$c_c = 2m\sqrt{\frac{k}{m}} = 2mp \qquad (19.40)$$

where p is the circular frequency of the system in the absence of damping, we distinguished three different cases of damping, namely, (1) *heavy damping* when $c > c_c$, (2) *critical damping* when $c = c_c$, and (3) *light damping* when $c < c_c$. In the first two cases, the system when disturbed tends to regain its equilibrium position without any oscillation. In the third case, the motion is vibratory with diminishing amplitude.

Damped forced vibrations

In Sec. 19.9, we considered the *damped forced vibrations* of a mechanical system. These vibrations occur when a system with viscous damping is subjected to a periodic force **P** of magnitude $P = P_m \sin \omega t$ or when it is elastically connected to a support with an alternating motion $\delta = \delta_m \sin \omega t$. In the first case, the motion of the system was defined by the differential equation

$$m\ddot{x} + c\dot{x} + kx = P_m \sin \omega t \qquad (19.46)$$

and in the second case by a similar equation obtained by replacing P_m by $k\,\delta_m$ in (19.46).

The *steady-state vibration* of the system is represented by a particular solution of Eq. (19.46) of the form

$$x_{\text{part}} = x_m \sin(\omega t - \varphi) \qquad (19.47)$$

Dividing the amplitude x_m of the steady-state vibration by P_m/k in the case of a periodic force, or by δ_m in the case of an oscillating support, we obtained the following expression for the magnification factor:

$$\frac{x_m}{P_m/k} = \frac{x_m}{\delta_m} = \frac{1}{\sqrt{[1 - (\omega/p)^2]^2 + [2(c/c_c)(\omega/p)]^2}} \qquad (19.52)$$

where $p = \sqrt{k/m}$ = natural circular frequency of undamped system
$c_c = 2mp$ = critical damping coefficient
c/c_c = damping factor

We also found that the *phase difference* φ between the impressed force or support movement and the resulting steady-state vibration of the damped system was defined by the relation

$$\tan \varphi = \frac{2(c/c_c)(\omega/p)}{1 - (\omega/p)^2} \qquad (19.53)$$

Electrical analogues

The chapter ended with a discussion of *electrical analogues* [Sec. 19.10], in which it was shown that the vibrations of mechanical systems and the oscillations of electrical circuits are defined by the same differential equations. Electrical analogues of mechanical systems may therefore be used to study or predict the behavior of these systems.

Review Problems

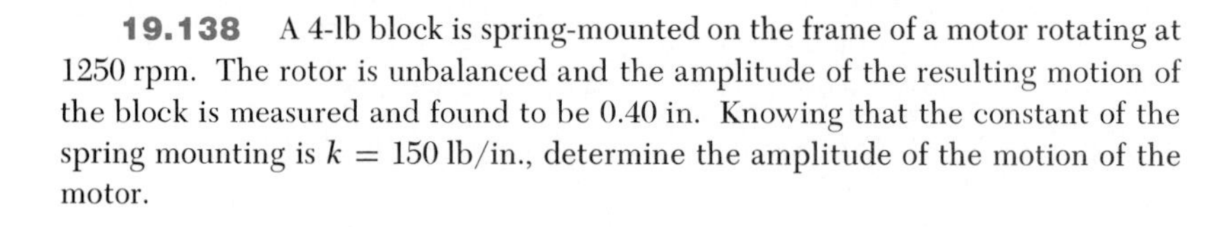

19.138 A 4-lb block is spring-mounted on the frame of a motor rotating at 1250 rpm. The rotor is unbalanced and the amplitude of the resulting motion of the block is measured and found to be 0.40 in. Knowing that the constant of the spring mounting is $k = 150$ lb/in., determine the amplitude of the motion of the motor.

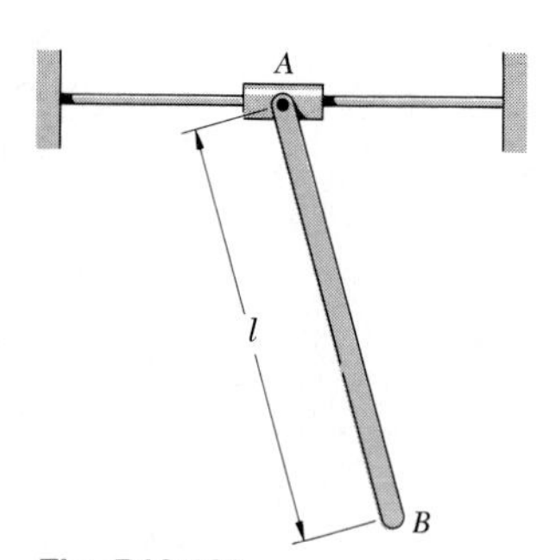

Fig. P19.139

19.139 A slender bar of length l is attached by a smooth pin A to a collar of negligible mass. Determine the period of small oscillations of the bar, assuming that the coefficient of friction between the collar and the horizontal rod (*a*) is sufficient to prevent any movement of the collar, (*b*) is zero.

19.140 The 10-kg rod AB is attached to 4-kg disks as shown. Knowing that the disks roll without sliding, determine the frequency of small oscillations of the system.

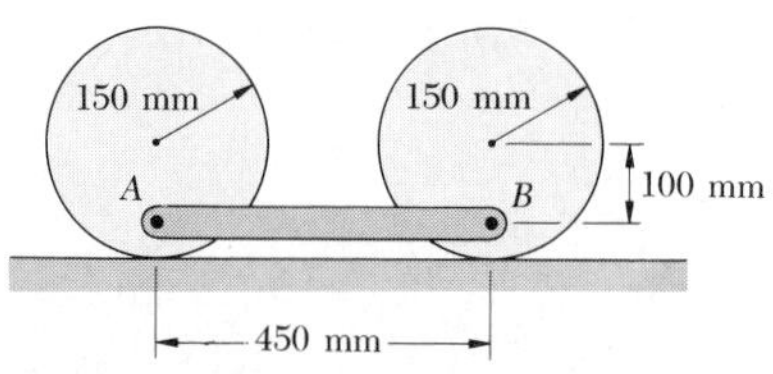

Fig. P19.140

19.141 A particle is placed with no initial velocity on a frictionless *plane* tangent to the surface of the earth. (*a*) Show that the particle will theoretically execute simple harmonic motion with a period of oscillation equal to that of a simple pendulum of length equal to the radius of the earth. (*b*) Compute the theoretical period of oscillation and show that it is equal to the periodic time of an earth satellite describing a low-altitude circular orbit. [*Hint.* See Eq. (12.35).]

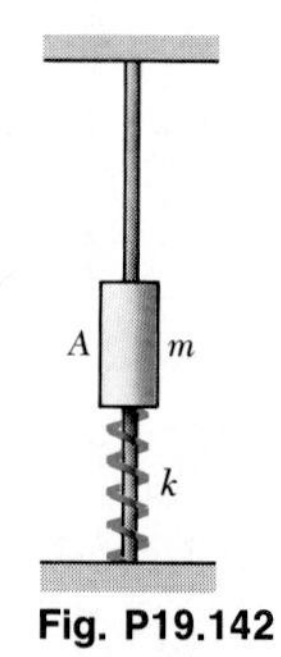

Fig. P19.142

19.142 A 1.5-kg collar is attached to a spring of constant $k = 750$ N/m as shown. If the collar is depressed 20 mm and released, determine (*a*) the maximum velocity of the resulting motion, (*b*) the position and velocity of the collar 0.08 s after release.

19.143 A rod of mass m and length L rests on the two pulleys A and B which rotate in the directions shown. Denoting by μ_k the coefficient of kinetic friction between the rod and the pulleys, determine the frequency of vibration if the rod is given a small displacement to the right and released.

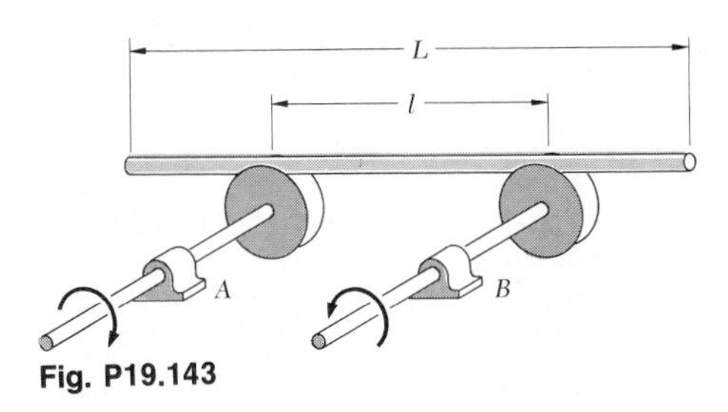

Fig. P19.143

19.144 A torsion pendulum may be used to determine experimentally the moment of inertia of a given object. The horizontal platform P is held by several rigid bars connected to a vertical wire. The period of oscillation of the platform is found equal to τ_0 when the platform is empty and to τ_A when an object of known moment of inertia I_A is placed on the platform so that its mass center is directly above the center of the plate. (*a*) Show that the moment of inertia I_0 of the platform and its supports may be expressed as $I_0 = I_A\tau_0^2/(\tau_A^2 - \tau_0^2)$. (*b*) If a period of oscillation τ_B is observed when an object B of unknown moment of inertia I_B is placed on the platform, show that $I_B = I_A(\tau_B^2 - \tau_0^2)/(\tau_A^2 - \tau_0^2)$.

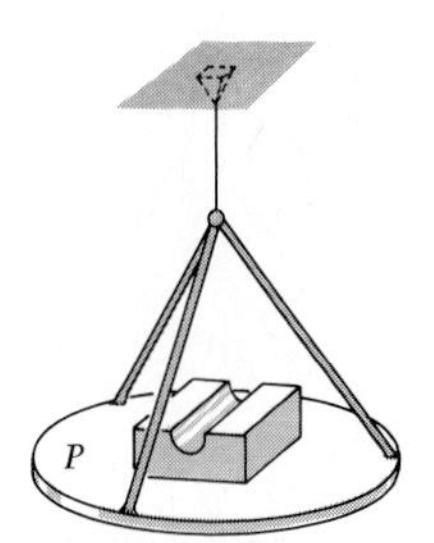

Fig. P19.144

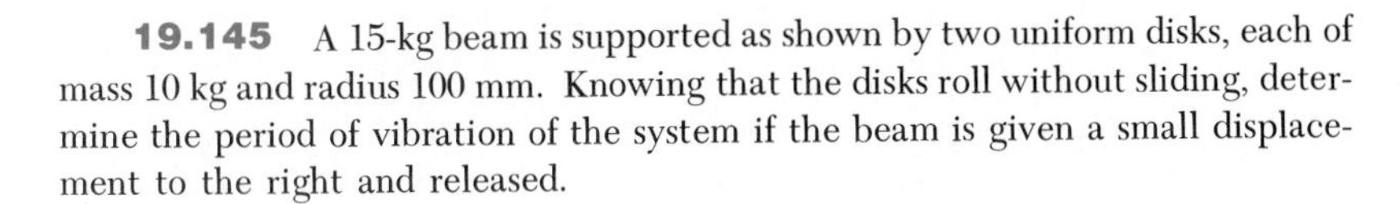

19.145 A 15-kg beam is supported as shown by two uniform disks, each of mass 10 kg and radius 100 mm. Knowing that the disks roll without sliding, determine the period of vibration of the system if the beam is given a small displacement to the right and released.

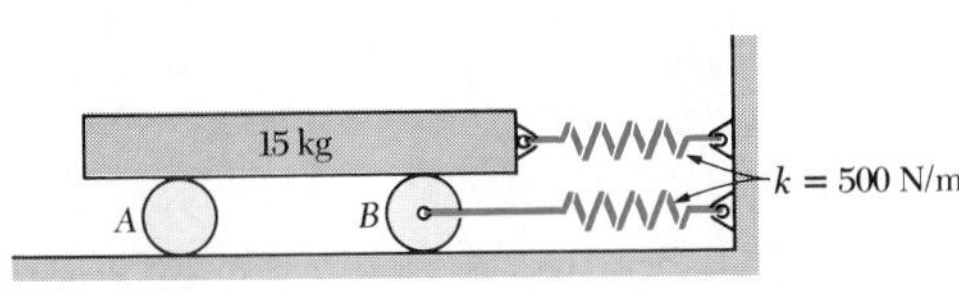

Fig. P19.145

19.146 Solve Prob. 19.145, assuming that the spring attached to the beam is removed.

19.147 A certain vibrometer used to measure vibration amplitudes consists essentially of a box containing a slender rod to which a mass m is attached; the natural frequency of the mass-rod system is known to be 8 Hz. When the box is rigidly attached to the casing of a motor rotating at 960 rpm, the mass is observed to vibrate with an amplitude of 0.08 in. relative to the box. Determine the amplitude of the vertical motion of the motor.

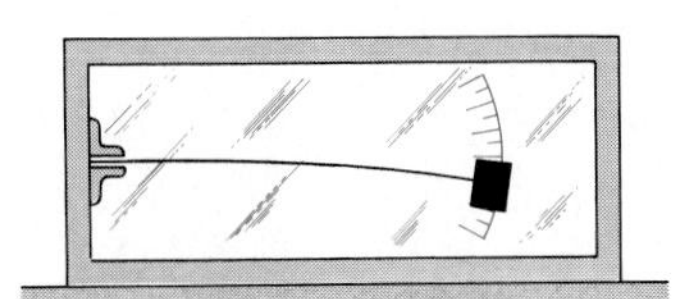

Fig. P19.147

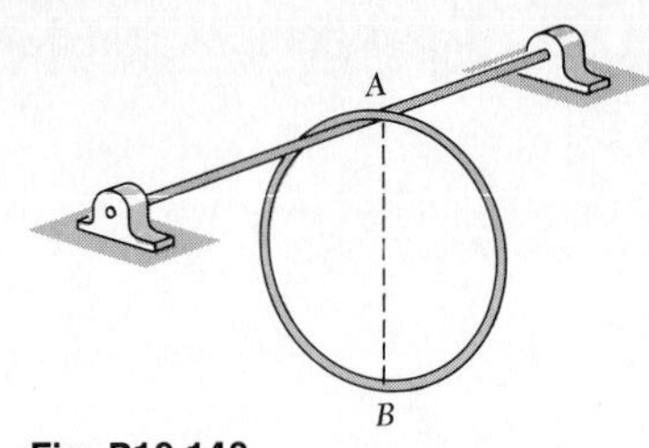

Fig. P19.148

19.148 A thin hoop of radius r and mass m is suspended from a rough rod as shown. Determine the frequency of small oscillations of the hoop (*a*) in the plane of the hoop, (*b*) in a direction perpendicular to the plane of the hoop. Assume that friction is sufficiently large to prevent slipping at A.

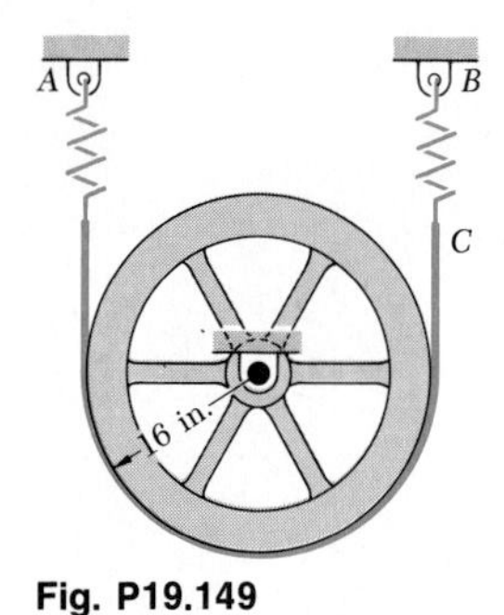

Fig. P19.149

19.149 A 400-lb flywheel has a diameter of 32 in. and a radius of gyration of 14 in. A belt is placed around the rim and attached to two springs, each of constant $k = 60$ lb/in. The initial tension in the belt is sufficient to prevent slipping. If the end C of the belt is pulled 1.25 in. down and released, determine (*a*) the period of vibration, (*b*) the maximum angular velocity of the flywheel.

The following problems are designed to be solved with a computer.

19.C1 By expanding the integrand in Eq. (19.19) into a series of even powers of $\sin\phi$ and integrating, it may be shown that the period of a simple pendulum of length l may be approximated by the expression

$$\tau = 2\pi\sqrt{\frac{l}{g}}\left[1 + \left(\frac{1}{2}\right)^2 c^2 + \left(\frac{1 \times 3}{2 \times 4}\right)^2 c^4 + \left(\frac{1 \times 3 \times 5}{2 \times 4 \times 6}\right)^2 c^6 + \cdots\right]$$

where $c = \sin\frac{1}{2}\theta_m$ and θ_m is the amplitude of the oscillations. Write a computer program and use it to calculate the sum of the series in brackets using successively 1, 2, 4, 8, and 16 terms for values of the amplitude θ_m from 30 to 120° at 30° intervals. Express the results with five significant figures.

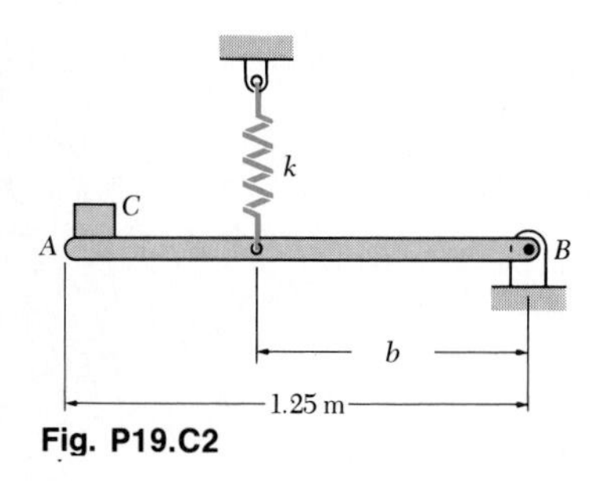

Fig. P19.C2

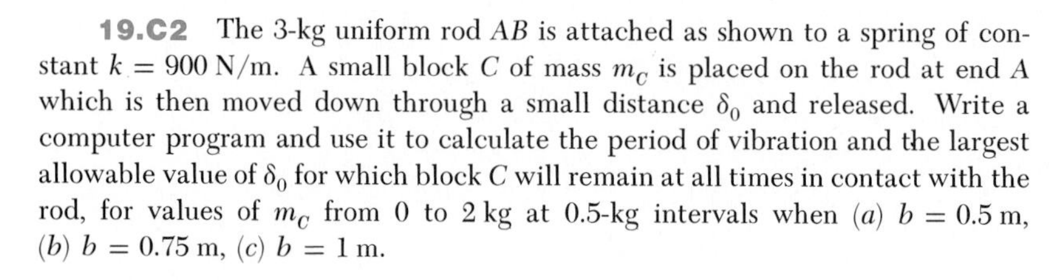

19.C2 The 3-kg uniform rod AB is attached as shown to a spring of constant $k = 900$ N/m. A small block C of mass m_C is placed on the rod at end A which is then moved down through a small distance δ_0 and released. Write a computer program and use it to calculate the period of vibration and the largest allowable value of δ_0 for which block C will remain at all times in contact with the rod, for values of m_C from 0 to 2 kg at 0.5-kg intervals when (*a*) $b = 0.5$ m, (*b*) $b = 0.75$ m, (*c*) $b = 1$ m.

19.C3 For the motor of Sample Prob. 19.5, write a computer program and use it to determine the amplitude of the vibration and the maximum acceleration of the motor base for values of the motor speed from 300 to 1200 rpm at 50-rpm intervals.

19.C4 Solve Prob. 19.C3, assuming that a dashpot having a coefficient of damping $c = 75$ lb·s/ft has been connected to the motor base and to the ground. Also determine the phase difference between the motion of the motor base and that of the rotor for the various values of the motor speed considered.

APPENDIX

Moments of Inertia of Masses

MOMENTS OF INERTIA OF MASSES*

9.11. Moment of Inertia of a Mass. Consider a small mass Δm mounted on a rod of negligible mass which may rotate freely about an axis AA' (Fig. 9.20*a*). If a couple is applied to the system, the rod and mass, assumed initially at rest, will start rotating about AA'. The details of this motion will be studied later in dynamics. At present, we wish only to indicate that the time required for the system to reach a given speed of rotation is proportional to the mass Δm and to the square of the distance r. The product $r^2\,\Delta m$ provides, therefore, a measure of the *inertia* of the system, i.e., of the resistance the system offers when we try to set it in motion. For this reason, the product $r^2\,\Delta m$ is called the *moment of inertia* of the mass Δm with respect to the axis AA'.

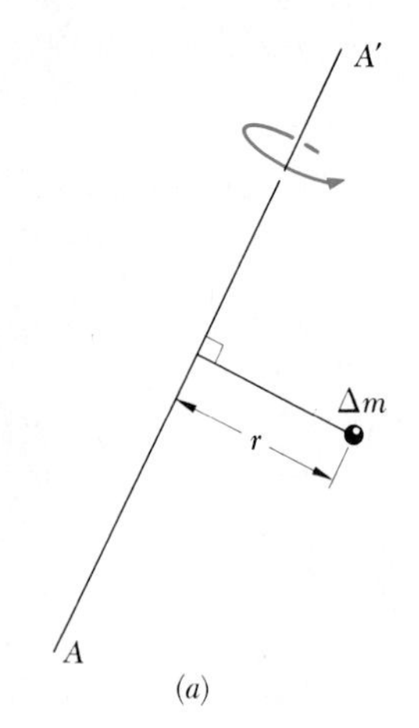

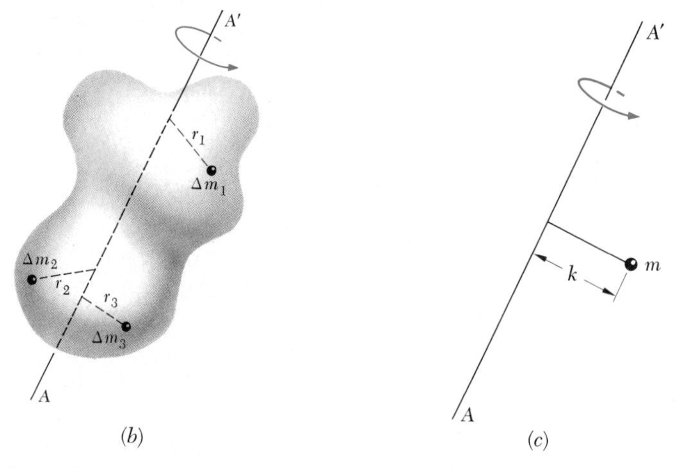

Fig. 9.20 (*a*) (*b*) (*c*)

Consider now a body of mass m which is to be rotated about an axis AA' (Fig. 9.20*b*). Dividing the body into elements of mass Δm_1, Δm_2, etc., we find that the resistance offered by the body is measured by the sum $r_1^2\,\Delta m_1 + r_2^2\,\Delta m_2 + \cdots$. This sum defines, therefore, the moment of inertia of the body with respect to the axis AA'. Increasing the number of elements, we find that the moment of inertia is equal, at the limit, to the integral

$$I = \int r^2\,dm \tag{9.28}$$

The *radius of gyration* k of the body with respect to the axis AA' is defined by the relation

$$I = k^2 m \qquad \text{or} \qquad k = \sqrt{\frac{I}{m}} \tag{9.29}$$

The radius of gyration k represents, therefore, the distance at which the entire mass of the body should be concentrated if its moment of inertia with respect to AA' is to remain unchanged (Fig. 9.20*c*). Whether it is kept in its

*This repeats Secs. 9.11 through 9.17 of the volume on statics.

original shape (Fig. 9.20*b*) or whether it is concentrated as shown in Fig. 9.20*c*, the mass m will react in the same way to a rotation, or *gyration*, about AA'.

If SI units are used, the radius of gyration k is expressed in meters and the mass m in kilograms. The moment of inertia of a mass, therefore, will be expressed in $\text{kg}\cdot\text{m}^2$. If U.S. customary units are used, the radius of gyration is expressed in feet and the mass in slugs, i.e., in $\text{lb}\cdot\text{s}^2/\text{ft}$. The moment of inertia of a mass, then, will be expressed in $\text{lb}\cdot\text{ft}\cdot\text{s}^2$.†

The moment of inertia of a body with respect to a coordinate axis may easily be expressed in terms of the coordinates x, y, z of the element of mass dm (Fig. 9.21). Noting, for example, that the square of the distance r from

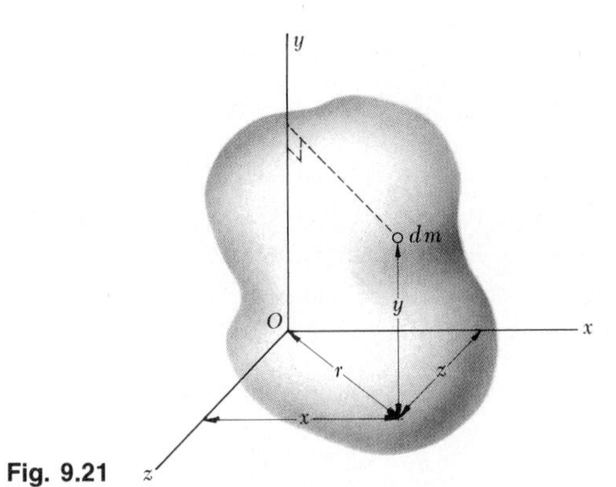

Fig. 9.21

the element dm to the y axis is $z^2 + x^2$, we express the moment of inertia of the body with respect to the y axis as

$$I_y = \int r^2\,dm = \int (z^2 + x^2)\,dm$$

Similar expressions may be obtained for the moments of inertia with respect to the x and z axes. We write

$$\begin{aligned} I_x &= \int (y^2 + z^2)\,dm \\ I_y &= \int (z^2 + x^2)\,dm \\ I_z &= \int (x^2 + y^2)\,dm \end{aligned} \qquad (9.30)$$

†It should be kept in mind when converting the moment of inertia of a mass from U.S. customary units to SI units that the base unit pound used in the derived unit $\text{lb}\cdot\text{ft}\cdot\text{s}^2$ is a unit of force (*not* of mass) and should therefore be converted into newtons. We have

$$1\ \text{lb}\cdot\text{ft}\cdot\text{s}^2 = (4.45\ \text{N})(0.3048\ \text{m})(1\ \text{s})^2 = 1.356\ \text{N}\cdot\text{m}\cdot\text{s}^2$$

or, since $\text{N} = \text{kg}\cdot\text{m/s}^2$,

$$1\ \text{lb}\cdot\text{ft}\cdot\text{s}^2 = 1.356\ \text{kg}\cdot\text{m}^2$$

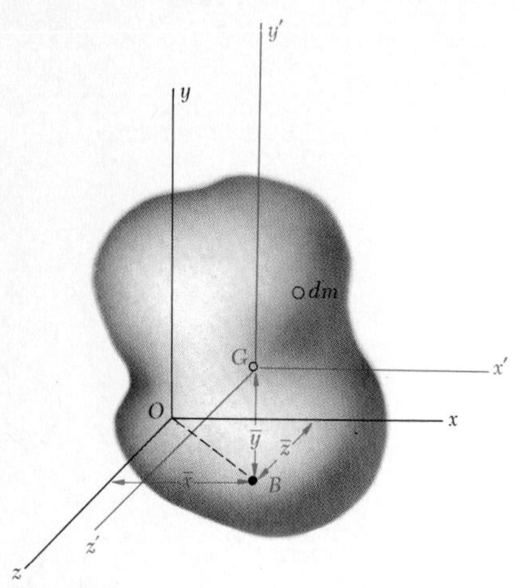

Fig. 9.22

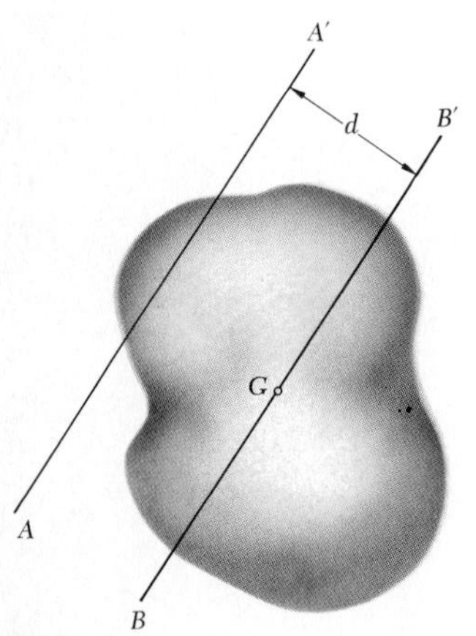

Fig. 9.23

9.12. Parallel-Axis Theorem. Consider a body of mass m. Let $Oxyz$ be a system of rectangular coordinates with origin at an arbitrary point O, and $Gx'y'z'$ a system of parallel *centroidal axes,* i.e., a system with origin at the center of gravity G of the body† and with axes x', y', z', respectively, parallel to x, y, z (Fig. 9.22). Denoting by $\bar{x}$, $\bar{y}$, $\bar{z}$ the coordinates of G with respect to $Oxyz$, we write the following relations between the coordinates x, y, z of the element dm with respect to $Oxyz$ and its coordinates x', y', z' with respect to the centroidal axes $Gx'y'z'$:

$$x = x' + \bar{x} \qquad y = y' + \bar{y} \qquad z = z' + \bar{z} \tag{9.31}$$

Referring to Eqs. (9.30), we may express the moment of inertia of the body with respect to the x axis as follows:

$$\begin{aligned} I_x &= \int (y^2 + z^2)\, dm = \int [(y' + \bar{y})^2 + (z' + \bar{z})^2]\, dm \\ &= \int (y'^2 + z'^2)\, dm + 2\bar{y} \int y'\, dm + 2\bar{z} \int z'\, dm + (\bar{y}^2 + \bar{z}^2) \int dm \end{aligned}$$

The first integral in the expression obtained represents the moment of inertia $\bar{I}_{x'}$ of the body with respect to the centroidal axis x'; the second and third integrals represent the first moment of the body with respect to the $z'x'$ and $x'y'$ planes, respectively, and, since both planes contain G, the two integrals are zero; the last integral is equal to the total mass m of the body. We write, therefore,

$$I_x = \bar{I}_{x'} + m(\bar{y}^2 + \bar{z}^2) \tag{9.32}$$

and, similarly,

$$I_y = \bar{I}_{y'} + m(\bar{z}^2 + \bar{x}^2) \qquad I_z = \bar{I}_{z'} + m(\bar{x}^2 + \bar{y}^2) \tag{9.32'}$$

We easily verify from Fig. 9.22 that the sum $\bar{z}^2 + \bar{x}^2$ represents the square of the distance OB between the y and y' axis. Similarly, $\bar{y}^2 + \bar{z}^2$ and $\bar{x}^2 + \bar{y}^2$ represent the squares of the distances between the x and x' axes, and the z and z' axes, respectively. Denoting by d the distance between an arbitrary axis AA' and a parallel centroidal axis BB' (Fig. 9.23), we may, therefore, write the following general relation between the moment of inertia I of the body with respect to AA' and its moment of inertia $\bar{I}$ with respect to BB':

$$I = \bar{I} + md^2 \tag{9.33}$$

Expressing the moments of inertia in terms of the corresponding radii of gyration, we may also write

$$k^2 = \bar{k}^2 + d^2 \tag{9.34}$$

where k and $\bar{k}$ represent the radii of gyration about AA' and BB', respectively.

†Note that the term centroidal is used to define an axis passing through the center of gravity G of the body, whether or not G coincides with the centroid of the volume of the body.

9.13. Moments of Inertia of Thin Plates. Consider a thin plate of uniform thickness t, made of a homogeneous material of density ρ (density = mass per unit volume). The mass moment of inertia of the plate with respect to an axis AA' *contained in the plane* of the plate (Fig. 9.24a) is

$$I_{AA',\text{mass}} = \int r^2\,dm$$

Since $dm = \rho t\,dA$, we write

$$I_{AA',\text{mass}} = \rho t \int r^2\,dA$$

But r represents the distance of the element of area dA to the axis AA'; the

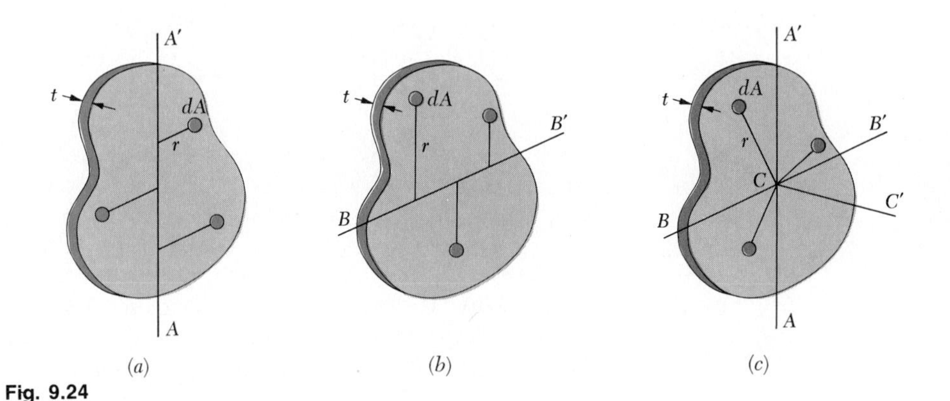

Fig. 9.24

integral is therefore equal to the moment of inertia of the area of the plate with respect to AA'. We have

$$I_{AA',\text{mass}} = \rho t I_{AA',\text{area}} \tag{9.35}$$

Similarly, we have with respect to an axis BB' perpendicular to AA' (Fig. 9.24b)

$$I_{BB',\text{mass}} = \rho t I_{BB',\text{area}} \tag{9.36}$$

Considering now the axis CC' *perpendicular* to the plate through the point of intersection C of AA' and BB' (Fig. 9.24c), we write

$$I_{CC',\text{mass}} = \rho t J_{C,\text{area}} \tag{9.37}$$

where J_C is the *polar* moment of inertia of the area of the plate with respect to point C.

Recalling the relation $J_C = I_{AA'} + I_{BB'}$ existing between polar and rectangular moments of inertia of an area, we write the following relation between the mass moments of inertia of a thin plate:

$$I_{CC'} = I_{AA'} + I_{BB'} \tag{9.38}$$

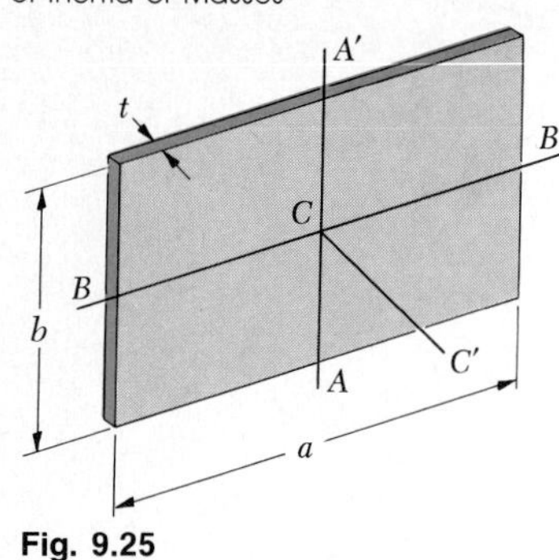

Fig. 9.25

Rectangular Plate. In the case of a rectangular plate of sides a and b (Fig. 9.25), we obtain the following mass moments of inertia with respect to axes through the center of gravity of the plate:

$$I_{AA',\text{mass}} = \rho t I_{AA',\text{area}} = \rho t(\tfrac{1}{12}a^3b)$$
$$I_{BB',\text{mass}} = \rho t I_{BB',\text{area}} = \rho t(\tfrac{1}{12}ab^3)$$

Observing that the product ρabt is equal to the mass m of the plate, we write the mass moments of inertia of a thin rectangular plate as follows:

$$I_{AA'} = \tfrac{1}{12}ma^2 \qquad I_{BB'} = \tfrac{1}{12}mb^2 \tag{9.39}$$
$$I_{CC'} = I_{AA'} + I_{BB'} = \tfrac{1}{12}m(a^2 + b^2) \tag{9.40}$$

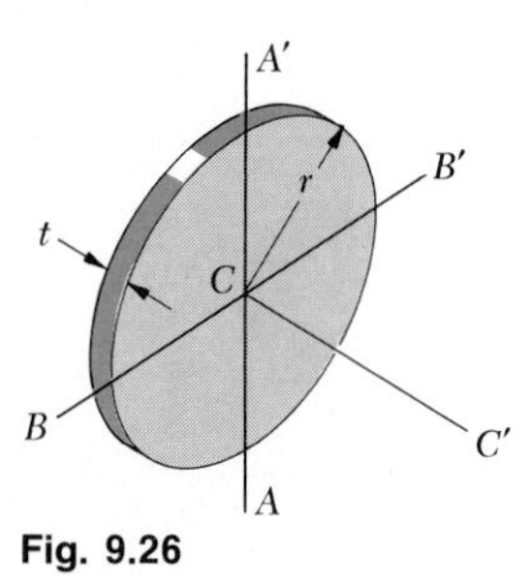

Fig. 9.26

Circular Plate. In the case of a circular plate, or disk, of radius r (Fig. 9.26), we write

$$I_{AA',\text{mass}} = \rho t I_{AA',\text{area}} = \rho t(\tfrac{1}{4}\pi r^4)$$

Observing that the product $\rho\pi r^2 t$ is equal to the mass m of the plate and that $I_{AA'} = I_{BB'}$, we write the mass moments of inertia of a circular plate as follows:

$$I_{AA'} = I_{BB'} = \tfrac{1}{4}mr^2 \tag{9.41}$$
$$I_{CC'} = I_{AA'} + I_{BB'} = \tfrac{1}{2}mr^2 \tag{9.42}$$

9.14. Determination of the Moment of Inertia of a Three-Dimensional Body by Integration. The moment of inertia of a three-dimensional body is obtained by computing the integral $I = \int r^2\,dm$. If the body is made of a homogeneous material of density ρ, we have $dm = \rho\,dV$ and write $I = \rho\int r^2\,dV$. This integral depends only upon the shape of the body. Thus, in order to compute the moment of inertia of a three-dimensional body, it will generally be necessary to perform a triple, or at least a double, integration.

However, if the body possesses two planes of symmetry, it is usually possible to determine the body's moment of inertia through a single integration by choosing as an element of mass dm the mass of a thin slab perpendicular to the planes of symmetry. In the case of bodies of revolution, for example, the element of mass should be a thin disk (Fig. 9.27). Using formula (9.42), the moment of inertia of the disk with respect to the axis of revolution may be readily expressed as indicated in Fig. 9.27. Its moment of inertia with respect to each of the other two axes of coordinates will be obtained by using formula (9.41) and the parallel-axis theorem. Integration of the expressions obtained will yield the desired moment of inertia of the body of revolution.

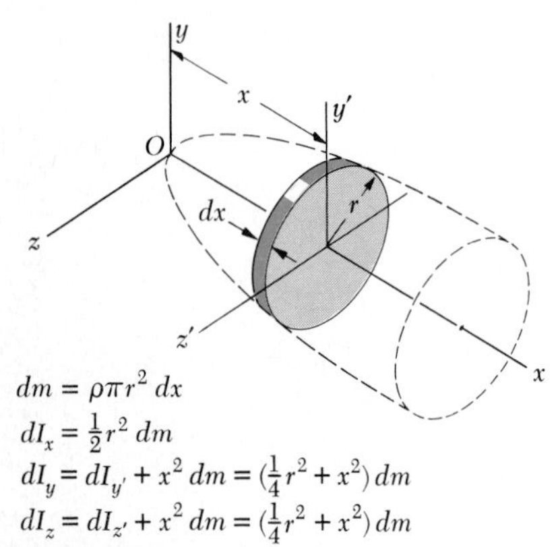

Fig. 9.27 Determination of the moment of inertia of a body of revolution.

9.15. Moments of Inertia of Composite Bodies. The moments of inertia of a few common shapes are shown in Fig. 9.28. The moment of inertia with respect to a given axis of a body made of several of these simple shapes may be obtained by computing the moments of inertia of its component parts about the desired axis and adding them together. We should note, as we have already noted in the case of areas, that the radius of gyration of a composite body *cannot* be obtained by adding the radii of gyration of its component parts.

Slender rod	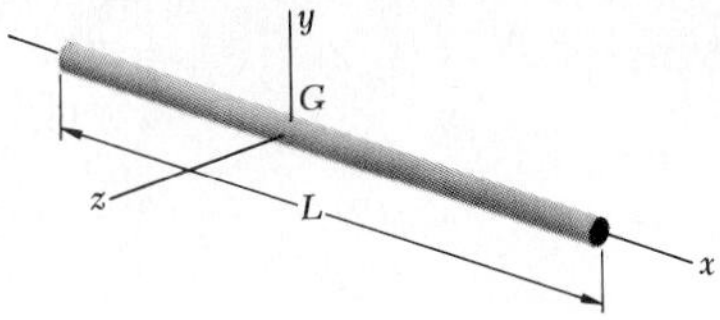	$I_y = I_z = \frac{1}{12}mL^2$
Thin rectangular plate	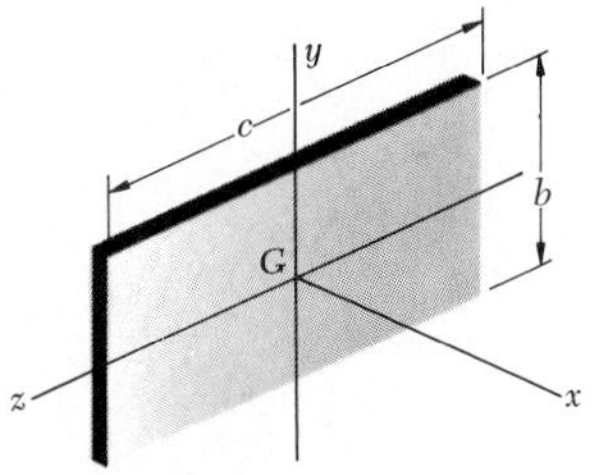	$I_x = \frac{1}{12}m(b^2 + c^2)$ $I_y = \frac{1}{12}mc^2$ $I_z = \frac{1}{12}mb^2$
Rectangular prism	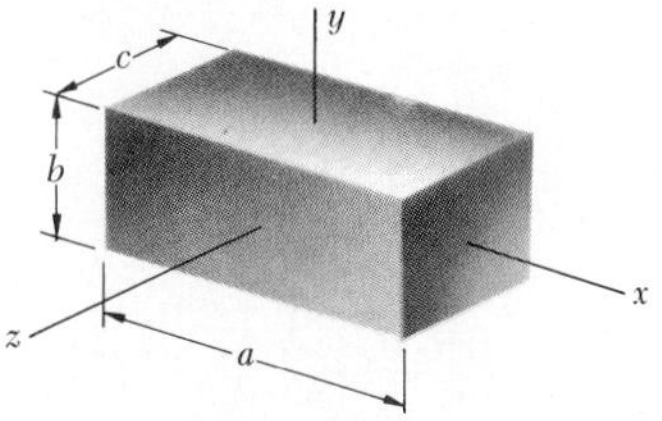	$I_x = \frac{1}{12}m(b^2 + c^2)$ $I_y = \frac{1}{12}m(c^2 + a^2)$ $I_z = \frac{1}{12}m(a^2 + b^2)$
Thin disk	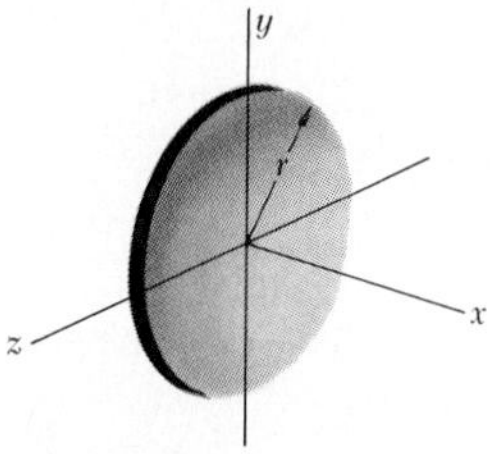	$I_x = \frac{1}{2}mr^2$ $I_y = I_z = \frac{1}{4}mr^2$
Circular cylinder	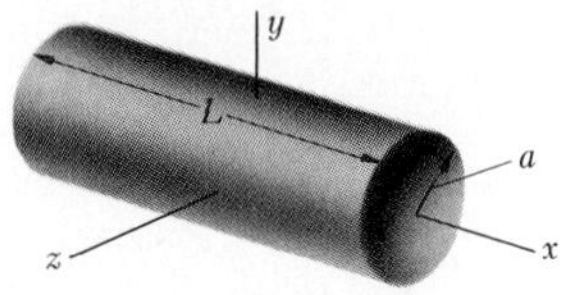	$I_x = \frac{1}{2}ma^2$ $I_y = I_z = \frac{1}{12}m(3a^2 + L^2)$
Circular cone	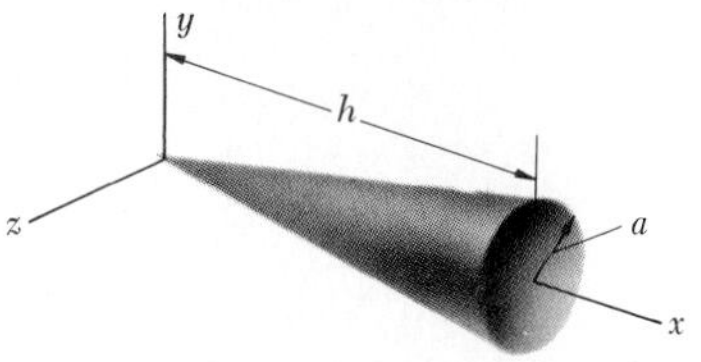	$I_x = \frac{3}{10}ma^2$ $I_y = I_z = \frac{3}{5}m(\frac{1}{4}a^2 + h^2)$
Sphere	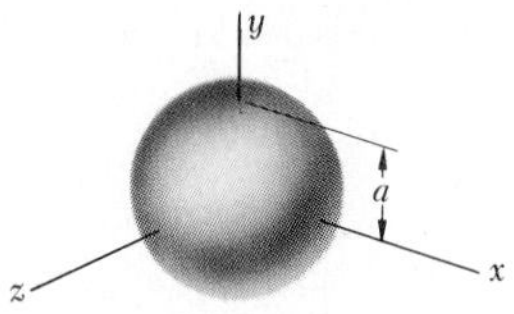	$I_x = I_y = I_z = \frac{2}{5}ma^2$

Fig. 9.28 Mass moments of inertia of common geometric shapes.

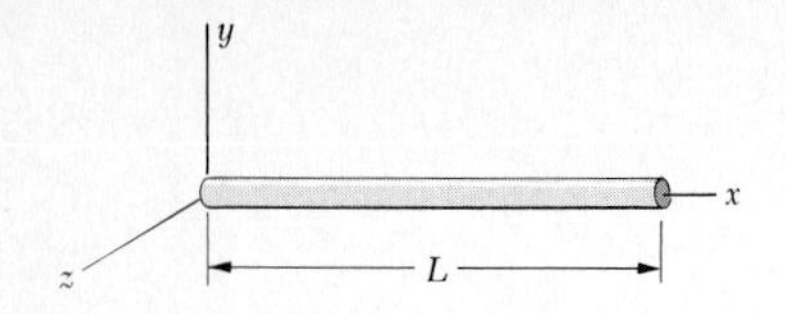

SAMPLE PROBLEM 9.9

Determine the mass moment of inertia of a slender rod of length L and mass m with respect to an axis perpendicular to the rod and passing through one end of the rod.

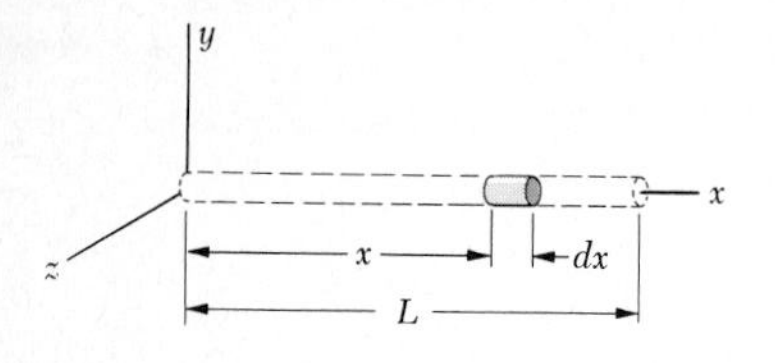

Solution. Choosing the differential element of mass shown, we write

$$dm = \frac{m}{L}\,dx$$

$$I_y = \int x^2\,dm = \int_0^L x^2 \frac{m}{L}\,dx = \left[\frac{m}{L}\frac{x^3}{3}\right]_0^L \qquad I_y = \tfrac{1}{3}mL^2 \blacktriangleleft$$

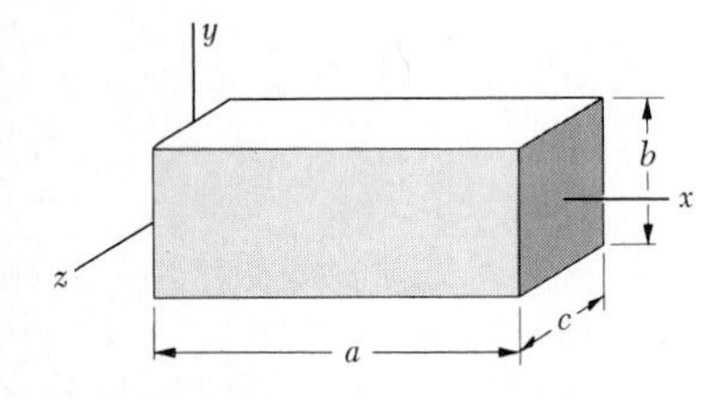

SAMPLE PROBLEM 9.10

Determine the mass moment of inertia of the homogeneous rectangular prism shown with respect to the z axis.

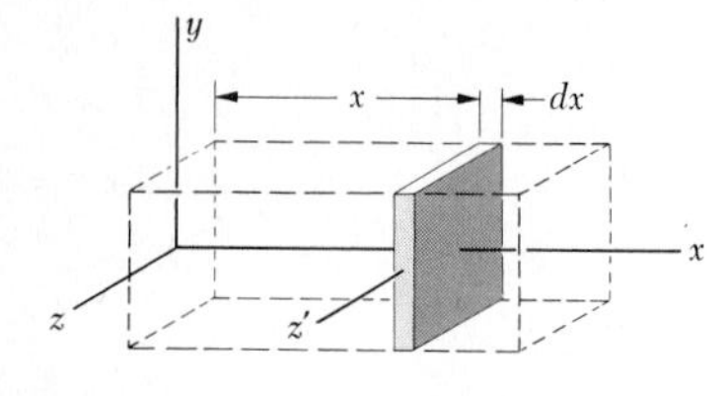

Solution. We choose as a differential element of mass the thin slab shown for which

$$dm = \rho bc\,dx$$

Referring to Sec. 9.13, we find that the moment of inertia of the element with respect to the z' axis is

$$dI_{z'} = \tfrac{1}{12}b^2\,dm$$

Applying the parallel-axis theorem, we obtain the mass moment of inertia of the slab with respect to the z axis.

$$dI_z = dI_{z'} + x^2\,dm = \tfrac{1}{12}b^2\,dm + x^2\,dm = (\tfrac{1}{12}b^2 + x^2)\,\rho bc\,dx$$

Integrating from $x = 0$ to $x = a$, we obtain

$$I_z = \int dI_z = \int_0^a (\tfrac{1}{12}b^2 + x^2)\,\rho bc\,dx = \rho abc\,(\tfrac{1}{12}b^2 + \tfrac{1}{3}a^2)$$

Since the total mass of the prism is $m = \rho abc$, we may write

$$I_z = m(\tfrac{1}{12}b^2 + \tfrac{1}{3}a^2) \qquad I_z = \tfrac{1}{12}m(4a^2 + b^2) \blacktriangleleft$$

We note that if the prism is slender, b is small compared to a and the expression for I_z reduces to $\frac{1}{3}ma^2$, which is the result obtained in Sample Prob. 9.9 when $L = a$.

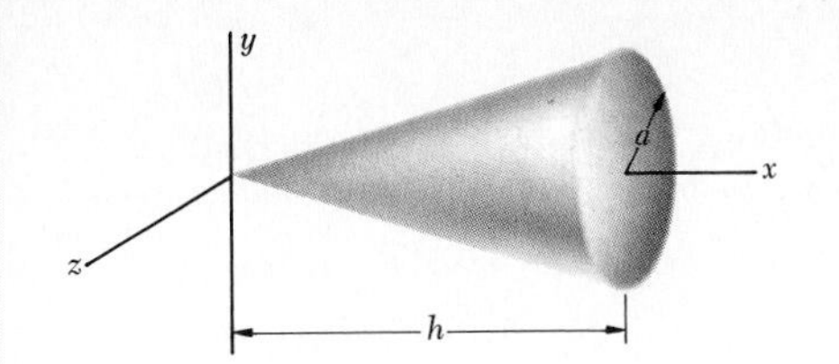

SAMPLE PROBLEM 9.11

Determine the mass moment of inertia of a right circular cone with respect to (*a*) its longitudinal axis, (*b*) an axis through the apex of the cone and perpendicular to its longitudinal axis, (*c*) an axis through the centroid of the cone and perpendicular to its longitudinal axis.

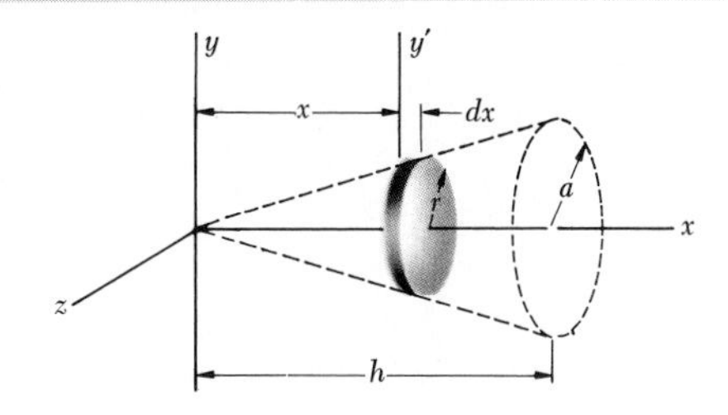

Solution. We choose the differential element of mass shown.

$$r = a\frac{x}{h} \qquad dm = \rho\pi r^2\,dx = \rho\pi\frac{a^2}{h^2}x^2\,dx$$

a. Moment of Inertia I_x. Using the expression derived in Sec. 9.13 for a thin disk, we compute the mass moment of inertia of the differential element with respect to the x axis.

$$dI_x = \tfrac{1}{2}r^2\,dm = \tfrac{1}{2}\left(a\frac{x}{h}\right)^2\left(\rho\pi\frac{a^2}{h^2}x^2\,dx\right) = \tfrac{1}{2}\rho\pi\frac{a^4}{h^4}x^4\,dx$$

Integrating from $x = 0$ to $x = h$, we obtain

$$I_x = \int dI_x = \int_0^h \tfrac{1}{2}\rho\pi\frac{a^4}{h^4}x^4\,dx = \tfrac{1}{2}\rho\pi\frac{a^4}{h^4}\frac{h^5}{5} = \tfrac{1}{10}\rho\pi a^4 h$$

Since the total mass of the cone is $m = \tfrac{1}{3}\rho\pi a^2 h$, we may write

$$I_x = \tfrac{1}{10}\rho\pi a^4 h = \tfrac{3}{10}a^2(\tfrac{1}{3}\rho\pi a^2 h) = \tfrac{3}{10}ma^2 \qquad I_x = \tfrac{3}{10}ma^2 \blacktriangleleft$$

b. Moment of Inertia I_y. The same differential element will be used. Applying the parallel-axis theorem and using the expression derived in Sec. 9.13 for a thin disk, we write

$$dI_y = dI_{y'} + x^2\,dm = \tfrac{1}{4}r^2\,dm + x^2\,dm = (\tfrac{1}{4}r^2 + x^2)\,dm$$

Substituting the expressions for r and dm, we obtain

$$dI_y = \left(\frac{1}{4}\frac{a^2}{h^2}x^2 + x^2\right)\left(\rho\pi\frac{a^2}{h^2}x^2\,dx\right) = \rho\pi\frac{a^2}{h^2}\left(\frac{a^2}{4h^2} + 1\right)x^4\,dx$$

$$I_y = \int dI_y = \int_0^h \rho\pi\frac{a^2}{h^2}\left(\frac{a^2}{4h^2} + 1\right)x^4\,dx = \rho\pi\frac{a^2}{h^2}\left(\frac{a^2}{4h^2} + 1\right)\frac{h^5}{5}$$

Introducing the total mass of the cone m, we rewrite I_y as follows:

$$I_y = \tfrac{3}{5}(\tfrac{1}{4}a^2 + h^2)\tfrac{1}{3}\rho\pi a^2 h \qquad I_y = \tfrac{3}{5}m(\tfrac{1}{4}a^2 + h^2) \blacktriangleleft$$

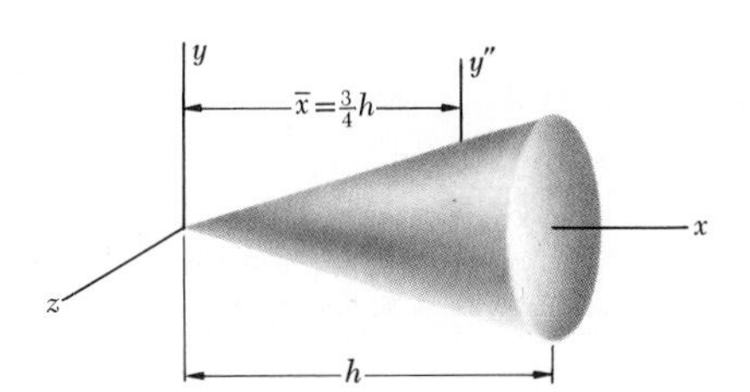

c. Moment of Inertia $\bar{I}_{y''}$. We apply the parallel-axis theorem and write

$$I_y = \bar{I}_{y''} + m\bar{x}^2$$

Solving for $\bar{I}_{y''}$ and recalling that $\bar{x} = \tfrac{3}{4}h$, we have

$$\bar{I}_{y''} = I_y - m\bar{x}^2 = \tfrac{3}{5}m(\tfrac{1}{4}a^2 + h^2) - m(\tfrac{3}{4}h)^2$$

$$\bar{I}_{y''} = \tfrac{3}{20}m(a^2 + \tfrac{1}{4}h^2) \blacktriangleleft$$

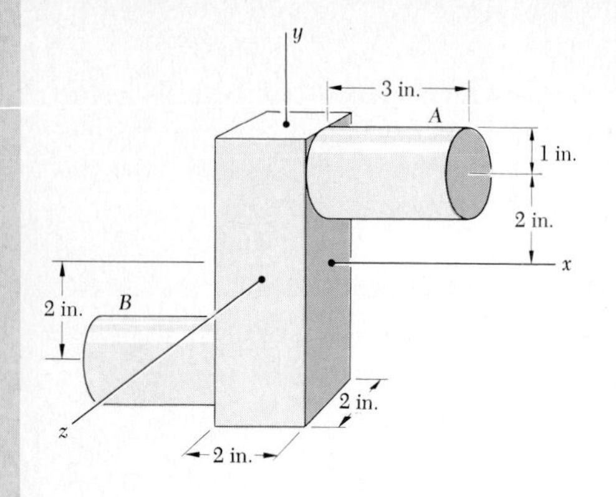

SAMPLE PROBLEM 9.12

A steel forging consists of a rectangular prism $6 \times 2 \times 2$ in. and of two cylinders of diameter 2 in. and length 3 in., as shown. Determine the mass moments of inertia with respect to the coordinate axes. (Specific weight of steel = 490 lb/ft³.)

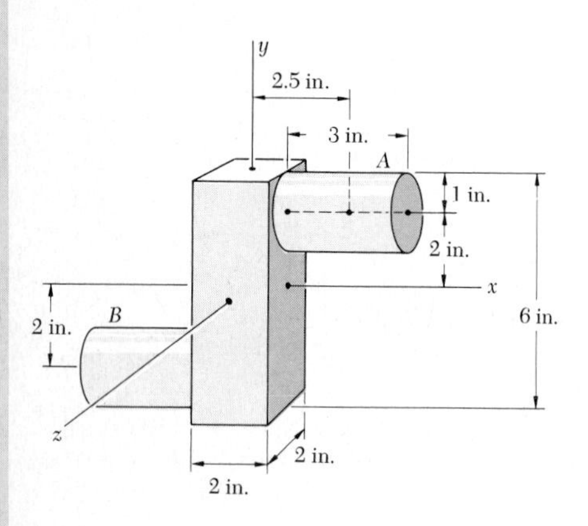

Computation of Masses

Prism

$$V = (2 \text{ in.})(2 \text{ in.})(6 \text{ in.}) = 24 \text{ in}^3$$

$$W = \frac{(24 \text{ in}^3)(490 \text{ lb/ft}^3)}{1728 \text{ in}^3/\text{ft}^3} = 6.81 \text{ lb}$$

$$m = \frac{6.81 \text{ lb}}{32.2 \text{ ft/s}^2} = 0.211 \text{ lb}\cdot\text{s}^2/\text{ft}$$

Each Cylinder

$$V = \pi(1 \text{ in.})^2(3 \text{ in.}) = 9.42 \text{ in}^3$$

$$W = \frac{(9.42 \text{ in}^3)(490 \text{ lb/ft}^3)}{1728 \text{ in}^3/\text{ft}^3} = 2.67 \text{ lb}$$

$$m = \frac{2.67 \text{ lb}}{32.2 \text{ ft/s}^2} = 0.0829 \text{ lb}\cdot\text{s}^2/\text{ft}$$

Mass Moments of Inertia. The mass moments of inertia of each component are computed from Fig. 9.28, using the parallel-axis theorem when necessary. Note that all lengths should be expressed in feet.

Prism

$$I_x = I_z = \tfrac{1}{12}(0.211 \text{ lb}\cdot\text{s}^2/\text{ft})[(\tfrac{6}{12} \text{ ft})^2 + (\tfrac{2}{12} \text{ ft})^2] = 4.88 \times 10^{-3} \text{ lb}\cdot\text{ft}\cdot\text{s}^2$$
$$I_y = \tfrac{1}{12}(0.211 \text{ lb}\cdot\text{s}^2/\text{ft})[(\tfrac{2}{12} \text{ ft})^2 + (\tfrac{2}{12} \text{ ft})^2] = 0.977 \times 10^{-3} \text{ lb}\cdot\text{ft}\cdot\text{s}^2$$

Each Cylinder

$$I_x = \tfrac{1}{2}ma^2 + m\bar{y}^2 = \tfrac{1}{2}(0.0829 \text{ lb}\cdot\text{s}^2/\text{ft})(\tfrac{1}{12} \text{ ft})^2 + (0.0829 \text{ lb}\cdot\text{s}^2/\text{ft})(\tfrac{2}{12} \text{ ft})^2 = 2.59 \times 10^{-3} \text{ lb}\cdot\text{ft}\cdot\text{s}^2$$
$$I_y = \tfrac{1}{12}m(3a^2 + L^2) + m\bar{x}^2 = \tfrac{1}{12}(0.0829 \text{ lb}\cdot\text{s}^2/\text{ft})[3(\tfrac{1}{12} \text{ ft})^2 + (\tfrac{3}{12} \text{ ft})^2] + (0.0829 \text{ lb}\cdot\text{s}^2/\text{ft})(\tfrac{2.5}{12} \text{ ft})^2 = 4.17 \times 10^{-3} \text{ lb}\cdot\text{ft}\cdot\text{s}^2$$
$$I_z = \tfrac{1}{12}m(3a^2 + L^2) + m(\bar{x}^2 + \bar{y}^2) = \tfrac{1}{12}(0.0829)[3(\tfrac{1}{12})^2 + (\tfrac{3}{12})^2] + (0.0829)[(\tfrac{2.5}{12})^2 + (\tfrac{2}{12})^2] = 6.48 \times 10^{-3} \text{ lb}\cdot\text{ft}\cdot\text{s}^2$$

Entire Body. Adding the values obtained,

$I_x = 4.88 \times 10^{-3} + 2(2.59 \times 10^{-3})$ $\qquad I_x = 10.06 \times 10^{-3} \text{ lb}\cdot\text{ft}\cdot\text{s}^2$ ◀

$I_y = 0.977 \times 10^{-3} + 2(4.17 \times 10^{-3})$ $\qquad I_y = 9.32 \times 10^{-3} \text{ lb}\cdot\text{ft}\cdot\text{s}^2$ ◀

$I_z = 4.88 \times 10^{-3} + 2(6.48 \times 10^{-3})$ $\qquad I_z = 17.84 \times 10^{-3} \text{ lb}\cdot\text{ft}\cdot\text{s}^2$ ◀

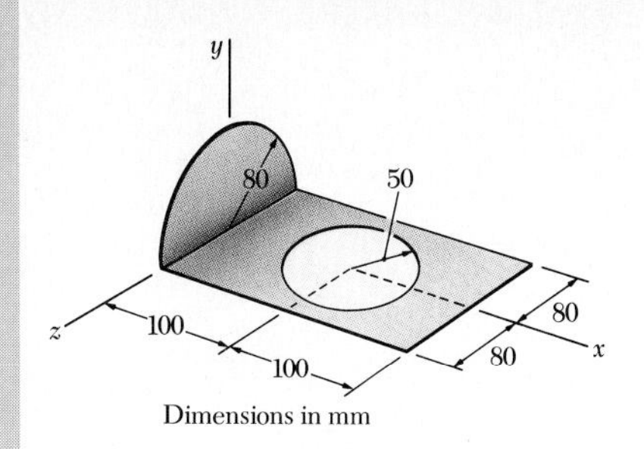

Dimensions in mm

SAMPLE PROBLEM 9.13

A thin steel plate 4 mm thick is cut and bent to form the machine part shown. Knowing that the density of steel is 7850 kg/m^3, determine the mass moment of inertia of the machine part with respect to the coordinate axes.

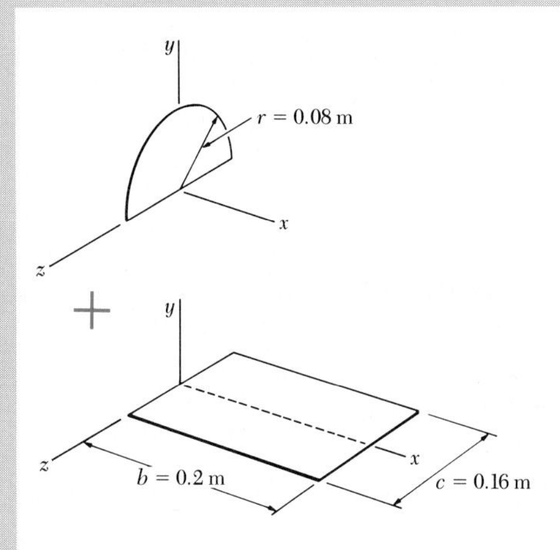

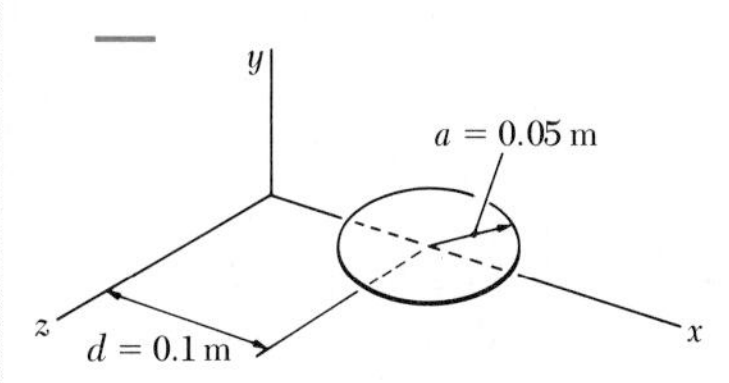

Solution. We observe that the machine part consists of a semicircular plate, plus a rectangular plate, minus a circular plate.

Computation of Masses. *Semicircular Plate*

$$V_1 = \tfrac{1}{2}\pi r^2 t = \tfrac{1}{2}\pi(0.08 \text{ m})^2(0.004 \text{ m}) = 40.21 \times 10^{-6} \text{ m}^3$$
$$m_1 = \rho V_1 = (7.85 \times 10^3 \text{ kg/m}^3)(40.21 \times 10^{-6} \text{ m}^3) = 0.3156 \text{ kg}$$

Rectangular Plate

$$V_2 = (0.200 \text{ m})(0.160 \text{ m})(0.004 \text{ m}) = 128 \times 10^{-6} \text{ m}^3$$
$$m_2 = \rho V_2 = (7.85 \times 10^3 \text{ kg/m}^3)(128 \times 10^{-6} \text{ m}^3) = 1.005 \text{ kg}$$

Circular Plate

$$V_3 = \pi a^2 t = \pi(0.050 \text{ m})^2(0.004 \text{ m}) = 31.42 \times 10^{-6} \text{ m}^3$$
$$m_3 = \rho V_3 = (7.85 \times 10^3 \text{ kg/m}^3)(31.42 \times 10^{-6} \text{ m}^3) = 0.2466 \text{ kg}$$

Mass Moments of Inertia. Using the method presented in Sec. 9.13, we compute the mass moment of inertia of each component.

Semicircular Plate. From Fig. 9.28, we observe that for a circular plate of mass m and radius r

$$I_x = \tfrac{1}{2}mr^2 \qquad I_y = I_z = \tfrac{1}{4}mr^2$$

Because of symmetry, we note that for a semicircular plate

$$I_x = \tfrac{1}{2}(\tfrac{1}{2}mr^2) \qquad I_y = I_z = \tfrac{1}{2}(\tfrac{1}{4}mr^2)$$

Since the mass of the semicircular plate is $m_1 = \tfrac{1}{2}m$, we have

$$I_x = \tfrac{1}{2}m_1 r^2 = \tfrac{1}{2}(0.3156 \text{ kg})(0.08 \text{ m})^2 = 1.010 \times 10^{-3} \text{ kg}\cdot\text{m}^2$$
$$I_y = I_z = \tfrac{1}{4}(\tfrac{1}{2}mr^2) = \tfrac{1}{4}m_1 r^2 = \tfrac{1}{4}(0.3156 \text{ kg})(0.08 \text{ m})^2 = 0.505 \times 10^{-3} \text{ kg}\cdot\text{m}^2$$

Rectangular Plate

$$I_x = \tfrac{1}{12}m_2 c^2 = \tfrac{1}{12}(1.005 \text{ kg})(0.16 \text{ m})^2 = 2.144 \times 10^{-3} \text{ kg}\cdot\text{m}^2$$
$$I_z = \tfrac{1}{3}m_2 b^2 = \tfrac{1}{3}(1.005 \text{ kg})(0.2 \text{ m})^2 = 13.400 \times 10^{-3} \text{ kg}\cdot\text{m}^2$$
$$I_y = I_x + I_z = (2.144 + 13.400)(10^{-3}) = 15.544 \times 10^{-3} \text{ kg}\cdot\text{m}^2$$

Circular Plate

$$I_x = \tfrac{1}{4}m_3 a^2 = \tfrac{1}{4}(0.2466 \text{ kg})(0.05 \text{ m})^2 = 0.154 \times 10^{-3} \text{ kg}\cdot\text{m}^2$$
$$I_y = \tfrac{1}{2}m_3 a^2 + m_3 d^2$$
$$= \tfrac{1}{2}(0.2466 \text{ kg})(0.05 \text{ m})^2 + (0.2466 \text{ kg})(0.1 \text{ m})^2 = 2.774 \times 10^{-3} \text{ kg}\cdot\text{m}^2$$
$$I_z = \tfrac{1}{4}m_3 a^2 + m_3 d^2 = \tfrac{1}{4}(0.2466 \text{ kg})(0.05 \text{ m})^2 + (0.2466 \text{ kg})(0.1 \text{ m})^2$$
$$= 2.620 \times 10^{-3} \text{ kg}\cdot\text{m}^2$$

Entire Machine Part

$$I_x = (1.010 + 2.144 - 0.154)(10^{-3}) \text{ kg}\cdot\text{m}^2 \qquad I_x = 3.00 \times 10^{-3} \text{ kg}\cdot\text{m}^2 \blacktriangleleft$$
$$I_y = (0.505 + 15.544 - 2.774)(10^{-3}) \text{ kg}\cdot\text{m}^2 \qquad I_y = 13.28 \times 10^{-3} \text{ kg}\cdot\text{m}^2 \blacktriangleleft$$
$$I_z = (0.505 + 13.400 - 2.620)(10^{-3}) \text{ kg}\cdot\text{m}^2 \qquad I_z = 11.29 \times 10^{-3} \text{ kg}\cdot\text{m}^2 \blacktriangleleft$$

Problems

9.87 A thin semicircular plate has a radius a and a mass m. Determine the mass moment of inertia of the plate with respect to (a) the centroidal axis BB', (b) the centroidal axis CC'.

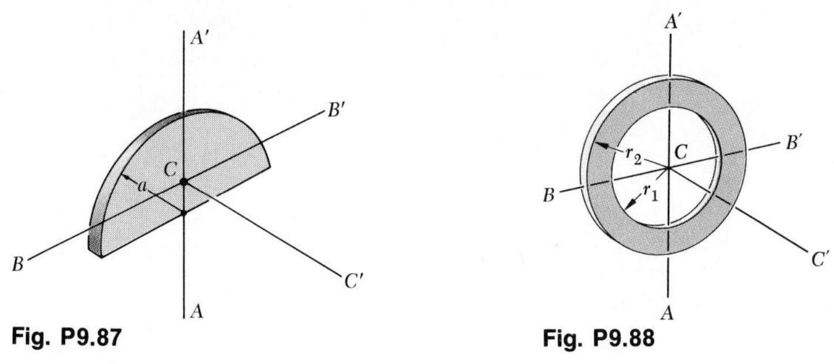

Fig. P9.87 **Fig. P9.88**

9.88 Determine the mass moment of inertia of a ring of mass m, cut from a thin uniform plate, with respect to (a) the diameter AA' of the ring, (b) the axis CC' perpendicular to the plane of the ring.

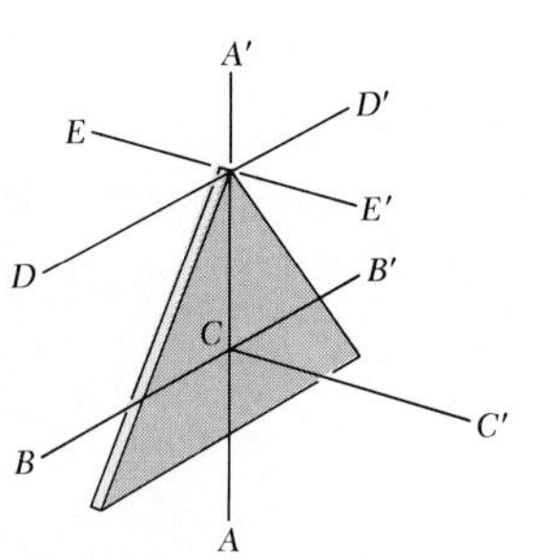

Fig. P9.89

9.89 A thin plate of mass m is cut in the shape of an equilateral triangle of side a. Determine the mass moment of inertia of the plate with respect to (a) the centroidal axes AA' and BB' in the plane of the plate, (b) the centroidal axis CC' perpendicular to the plate.

9.90 Determine the mass moments of inertia of the plate of Prob. 9.89 with respect to the axes DD' and EE' parallel to the centroidal axes BB' and CC', respectively.

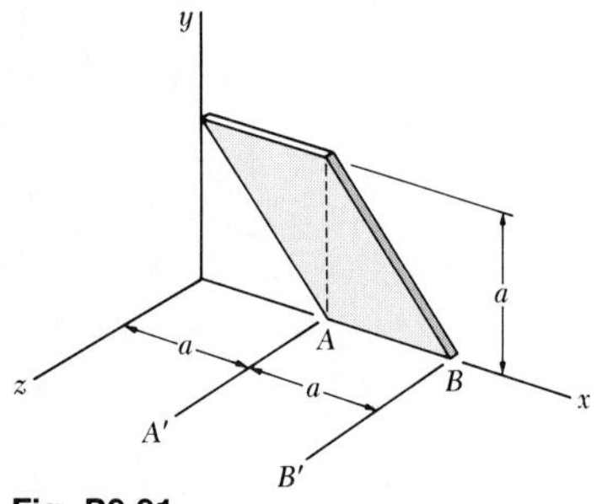

Fig. P9.91

9.91 A thin plate of mass m is cut in the shape of a parallelogram as shown. Determine the mass moment of inertia of the plate with respect to (a) the x axis, (b) the axis BB' perpendicular to the plate.

9.92 Determine the mass moment of inertia of the plate of Prob. 9.91 with respect to (a) the y axis, (b) the axis AA' perpendicular to the plate.

9.93 The area shown is revolved about the x axis to form a homogeneous solid of revolution of mass m. Express the mass moment of inertia of the solid with respect to the x axis in terms of m, a, and n. The expression obtained may be used to verify (a) the value given in Fig. 9.28 for a cone (with $n = 1$), (b) the answer to Prob. 9.95 (with $n = \frac{1}{2}$).

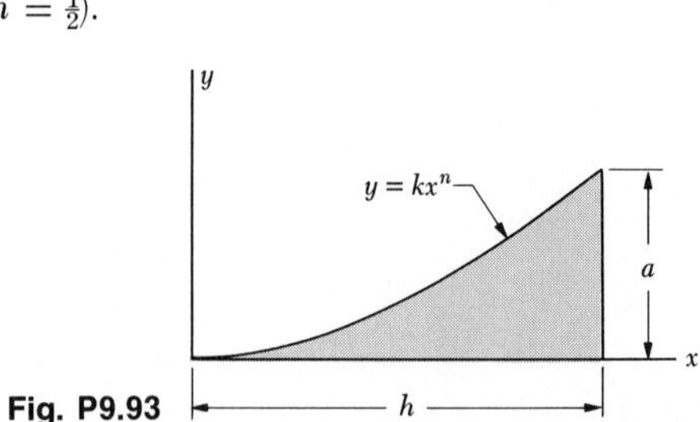

Fig. P9.93

9.94 Determine by direct integration the mass moment of inertia with respect to the y axis of the right circular cylinder shown, assuming a uniform density and a mass m.

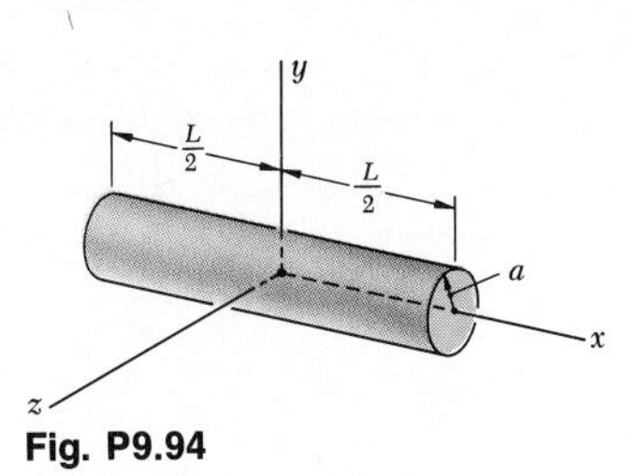

Fig. P9.94

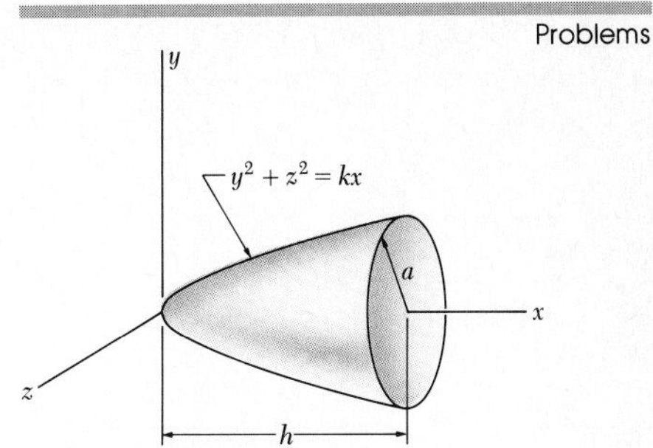

Fig. P9.95 and P9.96

9.95 Determine by direct integration the mass moment of inertia and the radius of gyration with respect to the x axis of the paraboloid shown, assuming a uniform density and a mass m.

9.96 Determine by direct integration the mass moment of inertia and the radius of gyration with respect to the y axis of the paraboloid shown, assuming a uniform density and a mass m.

9.97 Determine by direct integration the mass moment of inertia with respect to the x axis of the pyramid shown, assuming a uniform density and a mass m.

9.98 Determine by direct integration the mass moment of inertia with respect to the y axis of the pyramid shown, assuming a uniform density and a mass m.

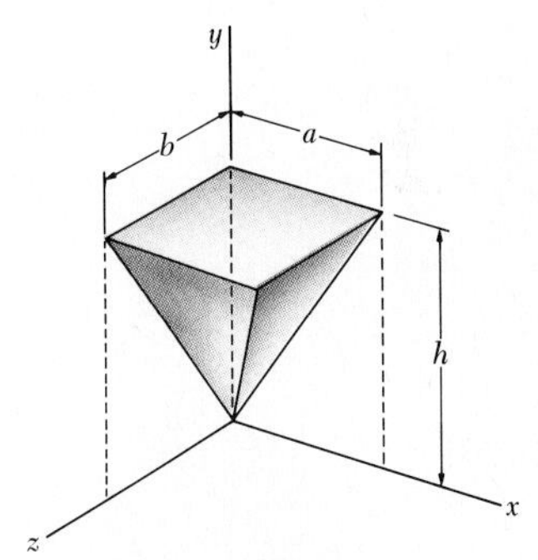

Fig. P9.97 and P9.98

9.99 A thin steel wire is bent into the shape of a circular arc as shown. Denoting by m' the mass per unit length of the wire, determine the mass moments of inertia of the wire with respect to the coordinate axes.

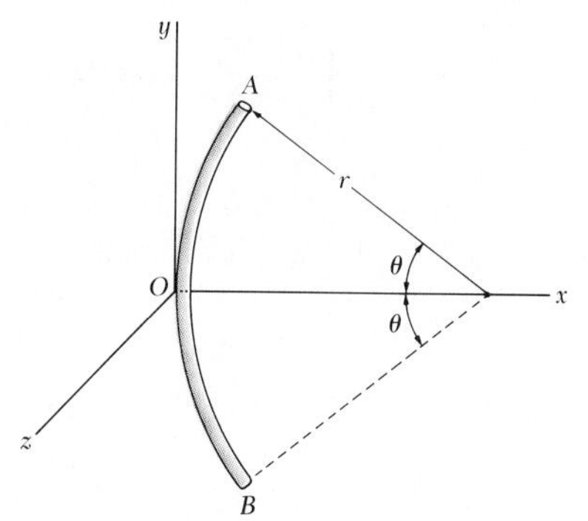

Fig. P9.99

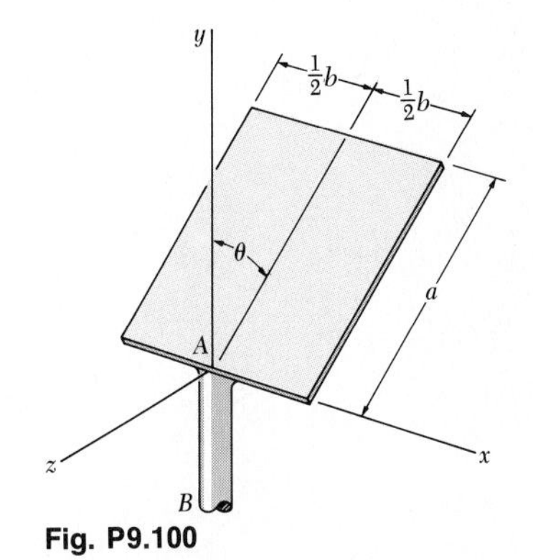

Fig. P9.100

9.100 A thin rectangular plate of mass m is welded to a vertical shaft AB with which it forms an angle θ. Determine by direct integration the mass moment of inertia with respect to (a) the y axis, (b) the z axis.

9.101 The cross section of a small flywheel is shown. The rim and hub are connected by eight spokes (two of which are shown in the cross section). Each spoke has a cross-sectional area of 160 mm^2. Determine the mass moment of inertia and radius of gyration of the flywheel with respect to the axis of rotation. (Density of steel $= 7850$ kg/m^3.)

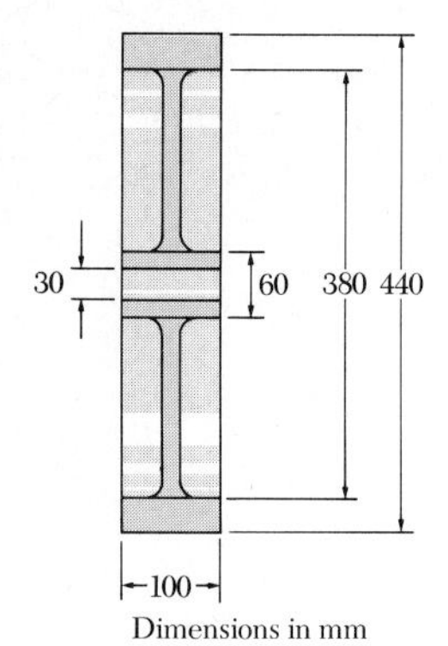

Fig. P9.101

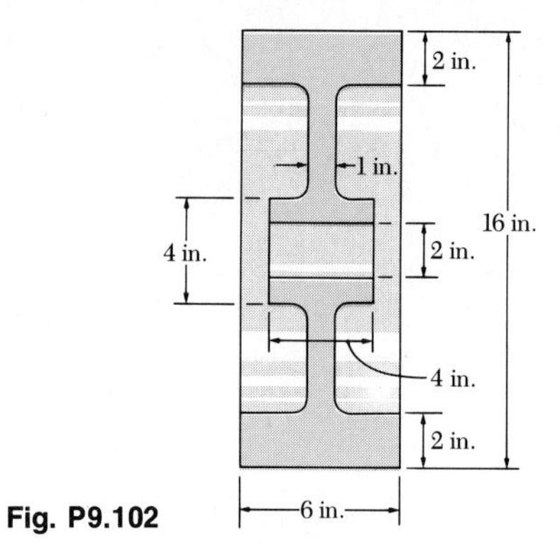

Fig. P9.102

9.102 Determine the mass moment of inertia and the radius of gyration of the steel flywheel shown with respect to the axis of rotation. The web of the flywheel consists of a solid plate 1 in. thick. (Specific weight of steel $=$ 490 lb/ft^3.)

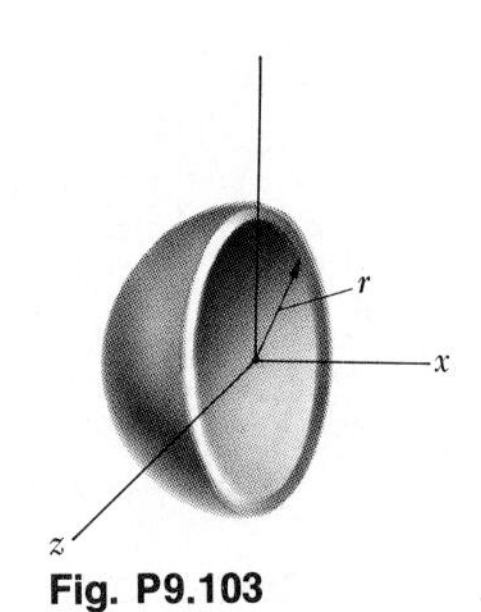

Fig. P9.103

9.103 Knowing that the thin hemispherical shell shown is of mass m and thickness t, determine the mass moment of inertia and the radius of gyration of the shell with respect to the x axis. (*Hint.* Consider the shell as formed by removing a hemisphere of radius r from a hemisphere of radius $r + t$; then neglect the terms containing t^2 and t^3 and keep those terms containing t.)

9.104 The machine part shown is formed by machining a conical surface into a circular cylinder. For $b = \frac{1}{2}h$, determine the mass moment of inertia and the radius of gyration of the machine part with respect to the y axis.

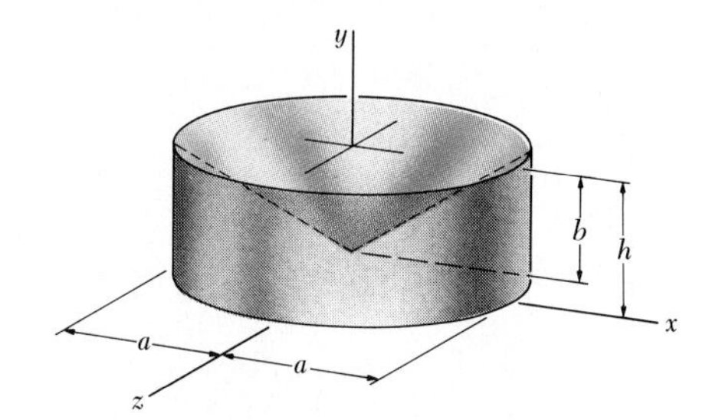

Fig. P9.104

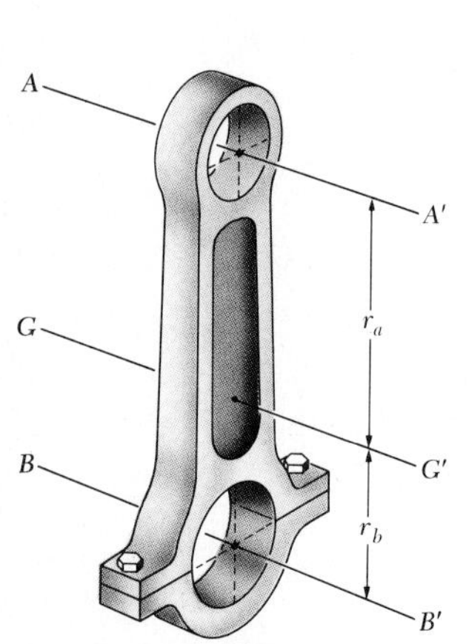

Fig. P9.105 and P9.106

9.105 For the 1.8-kg connecting rod shown, it has been experimentally determined that the mass moments of inertia of the rod with respect to the centerline axes of the bearings AA' and BB' are, respectively, $I_{AA'} = 65$ g$\cdot$m^2 and $I_{BB'} = 33$ g$\cdot$m^2. Knowing that $r_a + r_b = 275$ mm, determine (*a*) the location of the centroidal axis GG', (*b*) the radius of gyration with respect to axis GG'.

9.106 Knowing that for the 3.1-kg connecting rod shown the mass moment of inertia with respect to axis AA' is 160 g$\cdot$m^2, determine the mass moment of inertia with respect to axis BB', for $r_a = 204$ mm and $r_b = 126$ mm.

9.107 A circular hole of radius r is to be drilled through the center of a rectangular steel plate to form the machine component shown. Denoting the density of the steel by ρ, determine (*a*) the mass moment of inertia of the component with respect to the axis BB', (*b*) the value of r for which, given a and h, $I_{BB'}$ is maximum, (*c*) the corresponding value of $I_{BB'}$ and of the radius of gyration $k_{BB'}$.

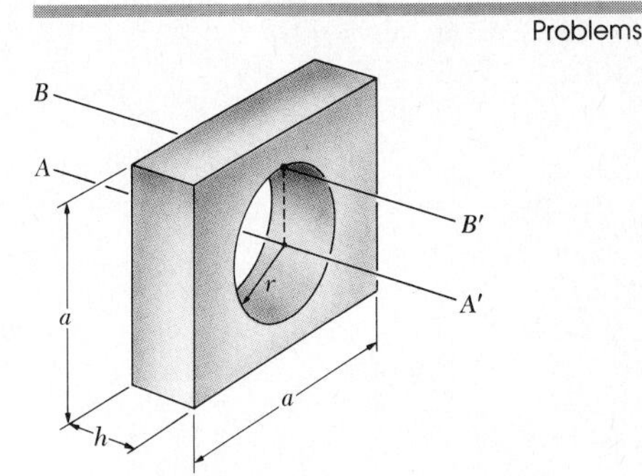

Fig. P9.107 and P9.108

9.108 Determine the mass moment of inertia and the radius of gyration of the steel machine component shown with respect to the axis BB' when $a = 8$ in., $h = 4$ in., and $r = 3$ in. (Specific weight of steel = 490 lb/ft^3.)

9.109 and 9.110 A section of sheet steel 2 mm thick is cut and bent into the machine component shown. Knowing that the density of steel is 7850 kg/m^3, determine the mass moment of inertia of the component with respect to (*a*) the x axis, (*b*) the y axis, (*c*) the z axis.

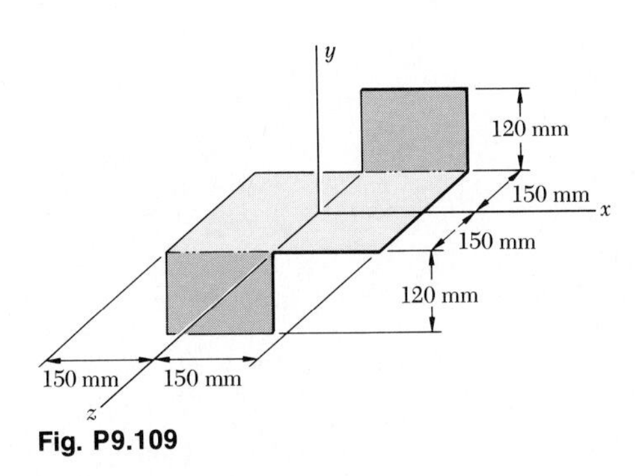

Fig. P9.109

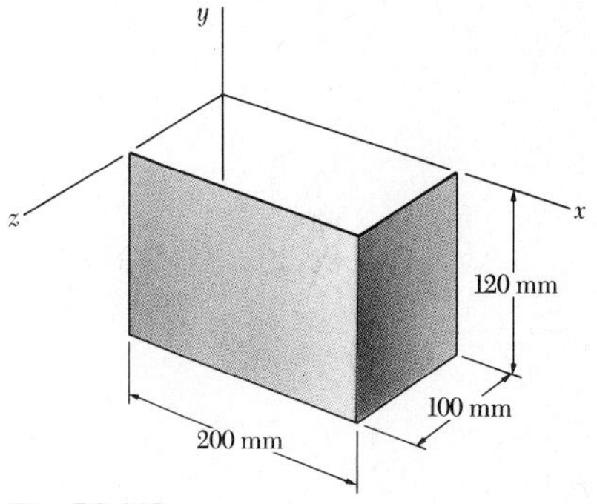

Fig. P9.110

9.111 A corner reflector for tracking by radar has two sides in the shape of a quarter circle of radius 15 in. and one side in the shape of a triangle. Each part of the reflector is formed from aluminum plate of uniform 0.05-in. thickness. Knowing that the specific weight of the aluminum used is 170 lb/ft^3, determine the mass moment of inertia of the reflector with respect to each of the coordinate axes.

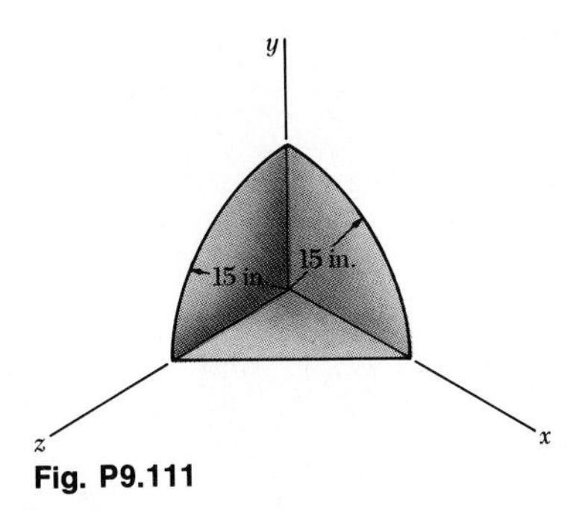

Fig. P9.111

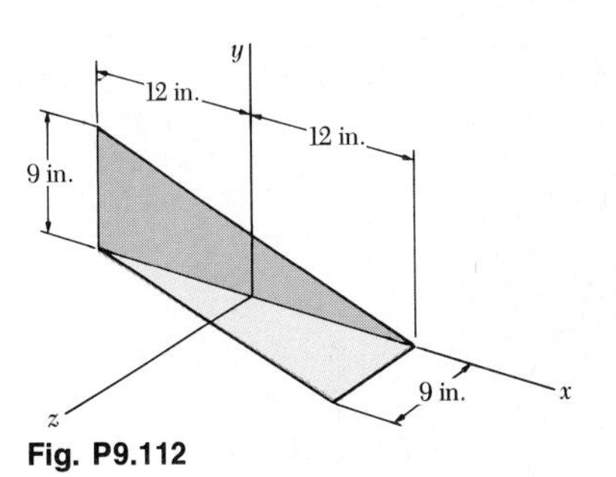

Fig. P9.112

9.112 A section of sheet steel 0.03 in. thick is cut and bent into the sheet-metal machine component shown. Knowing that the specific weight of steel is 490 lb/ft^3, determine the mass moment of inertia of the component with respect to each of the coordinate axes.

9.113 and 9.114 Determine the mass moment of inertia and the radius of gyration of the steel machine element shown with respect to the x axis. (Specific weight of steel = 490 lb/ft^3; density of steel = 7850 kg/m^3.)

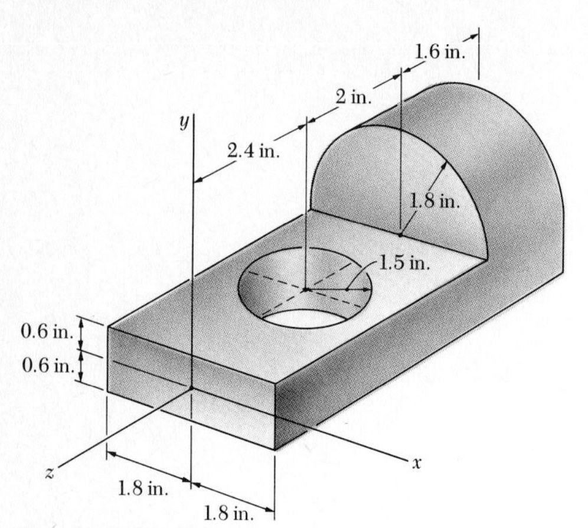

Fig. P9.114 and P9.115

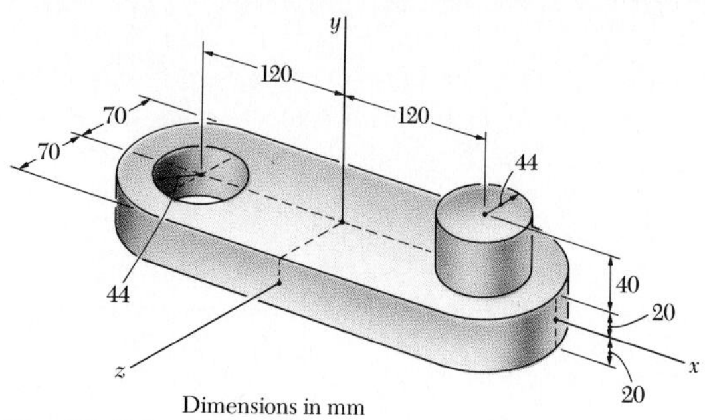

Fig. P9.113 and P9.116

9.115 and 9.116 Determine the mass moment of inertia and the radius of gyration of the steel machine element shown with respect to the y axis. (Specific weight of steel = 490 lb/ft^3; density of steel = 7850 kg/m^3.)

9.117 A homogenous wire with a mass per unit length of 1.5 kg/m is used to form the figure shown. Determine the mass moment of inertia of the wire figure with respect to (a) the x axis, (b) the y axis.

9.118 In Prob. 9.117, determine the mass moment of inertia of the wire figure with respect to the z axis.

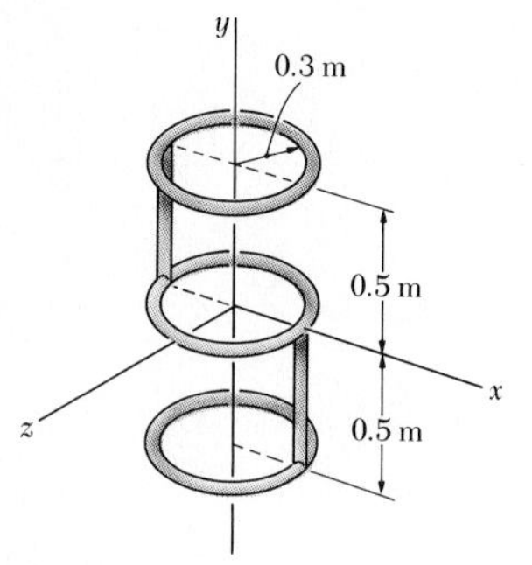

Fig. P9.117

Index

Index

Absolute acceleration, 682–684
Absolute motion of a particle, 469
Absolute system of units, 6, 501
Absolute velocity, 666–668, 676
Acceleration, 437, 466
  absolute, 682–684
  angular, 656, 658
  components of: normal, 479–481
    radial, 481–483
    rectangular, 466–467
    tangential, 479–481
    transverse, 481–483
  Coriolis, 693–695
  of gravity, 3, 447, 518
  in plane motion, 682–684
  relative: of a particle: 664–666
      in plane motion, 682–684
      with respect to axes in translation, 468
      with respect to rotating axes, 694
    of two particles, 449, 468–469
  in rotation, 656
Acceleration-time curve, 439, 456–457
Accuracy, numerical, 13
Action and reaction, 3, 207
Addition:
  of couples: in a plane, 73
    in space, 120–121
  of forces: concurrent: in a plane, 18, 26
      in space, 43
    nonconcurrent: in a plane, 80–81
      in space, 128–129
  of vectors, 16
Amplitude, 838, 841
Analogue, electrical, 874–877
Angle:
  of kinetic friction, 304
  lead, 319
  phase, 841
  of repose, 304
  of static friction, 304
Angular acceleration, 656, 658
Angular coordinate, 656
Angular impulse, 622
Angular momentum:
  conservation of, 620, 801
  of a particle, 620
  of a rigid body, 796
  of a system of particles, 662
Angular velocity, 656, 657, 668
Apogee, 532
Archimedes, 2
Areal velocity, 517
Aristotle, 2
Associative property for sums of vectors, 18
Auxiliary circle, 840
Axioms of mechanics, 2–5
Axis:
  of precession, 824
  of rotation, instantaneous, 654, 674
  of spin, 824
  of symmetry, 158
  of a wrench, 129
Axle friction, 325–327

Balancing of rotating shafts, 754–755
Ball-and-socket supports, 138
Ball supports, 138
Ballistic missiles, 532
Band brakes, 335
Banking of curves, 519
Beams, 265–280
  combined, 265
  loading of, 265
  span of, 265
  supports of, 265–266
  types of, 265
Bearings, 138, 325–328
  collar, 327
  end, 327
  journal, 325–327
  thrust, 327–328
Belt drive, 335
Belt friction, 334–335
Bending, 260
Bending moment, 261, 266–280
Bending-moment diagram, 268
Bernoulli, Jean, 402
Body centrode, 676
Borda's mouthpiece, 642

Cables:
  with concentrated loads, 281–282
  with distributed loads, 282–283
  parabolic, 283–284
  reactions at, 137
  span of, 284, 292
Calculators:
  accuracy, 13
  use of, 20, 21, 26*n*, 28, 34, 41*n*, 44
Cardan's suspension, 826
Catenary, 290–292
Cathode-ray tube, 524
Center:
  of gravity, 155, 188–189, 506
  of oscillation, 850

Center (*Cont.*):
- of percussion, 742, 807
- of pressure, 182, 367, 368
- of rotation, instantaneous, 674–676
- of symmetry, 158
- (*See also* Mass center)

Centimeter, 7
Central force, 517
Central impact, 598–604
Centrifugal force, 517, 735
Centrode, 676
Centroidal axes, principal, 372
Centroidal rotation:
- effective forces in, 713
- kinetic energy in, 771
- momentum in, 796

Centroids, 154–161, 170–173, 181–182, 188–191
- of areas and lines, 156–161
- of common shapes: of areas, 159
  - of lines, 160
  - of volumes, 190
- of composite areas and lines, 161
- of composite volumes, 191
- determination of, by integration, 170–171, 191
- of volumes, 188–191

Circular frequency, 841
Circular orbit, 532
Coefficient:
- of critical damping, 871
- of damping, 870
- of kinetic friction, 302–303
- of restitution, 599, 812
- of rolling resistance, 329
- of static friction, 302
- of viscous damping, 870

Collar bearings, 327
Complementary acceleration, 694
Complementary function, 865
Complete constraints, 92
Components:
- of acceleration (*see* Acceleration, components of)
- of force, 19, 24–25, 39–42

Components (*Cont.*):
- of velocity (*see* Velocity, components of)

Composite areas:
- centroids of, 161
- moments of inertia of, 358–359

Composite bodies:
- centroids of, 191
- moments of inertia of, 384

Composition of forces (*see* Addition, of forces)
Compound pendulum, 854
Compound trusses, 221
Compression, 63, 209, 260
Computers, xii
Concurrent forces, 18
Conic section, 530
Conservation:
- of angular momentum, 620, 801
- of energy: for particles, 566–567
  - for rigid bodies, 778–779
  - in vibrations, 857–858
- of momentum, 590–591, 620, 801

Conservative force, 421, 566, 857
Constrained plane motion, 733–736
Constraining forces, 88
Constraints, 407
- complete, 92
- improper, 92–93
- partial, 92–93, 138

Coordinate:
- angular, 656
- position, 436

Coplanar forces, 18
Coplanar vectors, 18
Coriolis acceleration, 693–695
Coulomb friction, 301, 870
Counters, 227
Couple vector, 120
Couples, 70–74, 120–121
- gyroscopic, 824
- inertia, 715
- momentum, 796
- in a plane, 70–74
  - addition of, 73
  - equivalent, 72–73

Couples (*Cont.*):
- in space, 120–121
  - addition of, 121
  - equivalent, 120

Critical damping, coefficient of, 871
Curvature, radius of, 480
Curvilinear motion of a particle:
- kinematics of, 464–483
- kinetics of, 503–533

Curvilinear translation, 653
Customary units, U.S., 8–13

D'Alembert, Jean, 2
D'Alembert's principle, 710
Damped vibrations (*see* Vibrations, damped)
Damping:
- coefficient of, 870
- critical, 871
- heavy, 871
- light, 871–872
- viscous, 870

Damping coefficient, 870
Damping factor, 872
Deceleration, 438
Decimeter, 7
Decrement, logarithmic, 878
Deformation, period of, 598–599, 812
Degrees of freedom, 421, 423
Density, 156*n*, 383
Dependent motions, 450
Determinate reactions, 92
Determinate structures, 231
Determinate trusses, 222
Diagram:
- acceleration-time, 439, 456
- bending-moment, 268
- displacement-time, 439, 456
- free-body, 32–33, 137
- shear, 266, 268
- velocity-displacement, 457
- velocity-time, 439, 456

Differential elements:
- for centroids: of areas, 171
  - of volumes, 191

Differential elements (*Cont.*):
for moments of inertia: of areas, 348–351
of masses, 384
Direct central impact, 598–601
Direction cosines, 40
Direction of a force, 14
Disk clutches, 327
Disk friction, 327–328
Displacement, 403
virtual, 405
Distance, 439
Distributed forces, 154, 347
Distributed loads, 181, 265
Dry friction, 301, 870
Dynamic balancing, 754
Dynamic equilibrium:
of a particle, 503
of a rigid body: in noncentroidal rotation, 734
in plane motion, 714
Dynamics, definition of, 1, 435

Earth satellites, 529
Eccentric impact, 812–814
Eccentricity, 530, 869
Effective forces, 504
for a rigid body: in plane motion, 710–714
in three dimensions, 753
Efficiency, 408–409, 574–575
Elastic impact, 600–601
Electrical analogue, 874–877
Elliptic orbit, 530
End bearings, 327
Energy:
conservation of (*see* Conservation of energy)
kinetic: of a particle, 550–552
of a rigid body: in plane motion, 777
in rotation, 771
in translation, 770
potential, 420–422, 564–566, 778
total mechanical, 566
Equations:
of equilibrium: for a particle, 32, 48
for a rigid body: in a plane, 87
in space, 137
of motion: for a particle, 503
for a system of particles, 504
for a three-dimensional body, 753
for a two-dimensional body:
in noncentroidal rotation, 735
in plane motion, 713, 715
Equilibrium:
dynamic (*see* Dynamic equilibrium)
equations (*see* Equations of equilibrium)
neutral, 422–423
of a particle: in a plane, 31–32
in space, 48
of a rigid body: in a plane, 87–93, 104–105
in space, 137–139
stability of, 422–423
Equipollence of external forces and effective forces, 505
Equipollent systems of vectors, 505–506
Equivalence of external forces and effective forces for a rigid body, 710, 753
Equivalent forces, 61–63
Equivalent systems of forces:
in a plane, 59–86
in space, 117–136
Escape velocity, 447, 531
External forces, 59–61

Fan, 632–633
First moments:
of areas and lines, 157–158
of volumes, 189
Fixed supports, 88, 138
Flexible cords (*see* Cables)
Fluid flow, 632
Fluid friction, 870
Focus of conic section, 530
Foot, 8
Force, 2
central, 517
centrifugal, 517, 735
conservative, 421, 566, 857
effective (*see* Effective forces)
external, 60–61, 504
gravitational, 517–518, 548, 549, 564, 620
impulsive, 588
inertia, 503
internal, 60–61, 504
in a member, 211
nonconservative, 567
nonimpulsive, 588
on a particle: in a plane, 14–27
in space, 39–48
reversed effective (*see* Inertia vector)
on a rigid body: in a plane, 59–109
in space, 117–148
Force-couple system, 74
Force systems, 59–86, 117–136
Forced frequency, 865
Forced vibrations:
damped, 873–874
undamped, 863–865
Frame of reference, 500
Frames, 228–231
Free-body diagram:
of a particle, 32–33
of a rigid body, 137
Free vibrations:
damped, 870–872
undamped, 839–870
Freedom, degrees of, 421, 423, 450
Freely falling body, 449
Frequency, 838, 841
circular, 841
forced, 865
natural, 865
Frequency ratio, 865
Friction, 301–340
angles of, 304

Friction (*Cont.*):
belt, 334–335
circle of, 327
coefficient of, 302–303
Coulomb, 301, 870
dry, 301–306, 870
fluid, 870
kinetic, 302–303
laws of, 302–303
static, 302–304
wheel, 328–329
Frictionless surface, 88, 138

Gears:
analysis of, 668, 684
planetary, 672
Geneva mechanism, 696–697
Geometric instability, 93
Gram, 6
Graphical methods for solution of rectilinear-motion problems, 456–457
Gravitation:
constant of, 3, 518
Newton's law of, 3–4, 517–518
Gravitational forces, 517–518, 548, 549, 564, 620
Gravitational potential energy, 564
Gravitational system of units, 8
Gravity:
acceleration of, 3, 447, 518
center of, 155, 188–189
Guldinus, theorems of, 172–173
Gun, recoil, 597
Gyration, radius of, 352, 380
Gyrocompass, 826
Gyroscope, 824–826
Gyroscopic couple, 824

Hamilton, Sir William R., 2
Harmonic motion, simple, 839–842
Helicopter, 641, 810
Hertz (unit), 842
Hinges, 138
Horsepower, 575
Hydraulic jump, 642
Hydrostatic forces, 182
Hyperbolic trajectory, 530

Ideal machines, 408–409
Impact, 598
central: direct, 598–601
oblique, 601–604
eccentric, 812–814
elastic, 600–601
line of, 598
plastic, 600
Improper constraints, 92–93
Impulse:
angular, 622
linear, 587
and momentum, principle (*see* Principle of impulse and momentum)
Impulsive force, 588
Impulsive motion, 588, 812
Inch, 9
Inclined axes, moments of inertia, 369
Indeterminate reactions, 92–93
Indeterminate structures, 231
Indeterminate trusses, 222
Inertia:
moments of (*see* Moments of inertia)
principal axes of, 369–372
product of, 368–369
parallel-axis theorem for, 369
Inertia couple, 715
Inertia force, 503
Inertia vector:
for a particle, 503
for a rigid body in plane motion, 715
Inertial system, 500
Initial conditions, 440
Input forces, 243
Input power, 575
Input work, 408–409
Instantaneous axis of rotation, 654, 674
Instantaneous center of rotation, 674–676
Internal forces, 60–61, 504
in members, 211, 260–261
in structures, 206
International system of units, 5–8

Jacks, 318–319
Jerk, 462
Jet engine, 632
Joints, method of, 210–215
Joule (unit), 403
Journal bearings, 325–327

Kater pendulum, 855
Kepler, Johann, 533
Kepler's laws, 533
Kilogram, 5
Kilometer, 6
Kilonewton, 6
Kilopound, 9
Kilowatt, 575
Kinematics, 435
of particles: in curvilinear motion, 464–483
in rectilinear motion, 436–457
in relative motion, 449–450, 693–695
of rigid bodies: in plane motion, 653–655
in rotation, 656–658
in translation, 655
Kinetic energy (*see* Energy, kinetic)
Kinetic friction, 302–303
Kinetics, 435
of particles, 498–652
of rigid bodies, 709–837
Kip, 9

Lagrange, J. L., 2
Laws:
of friction, 301–303
Kepler's, 533
Newton's (*see* Newton's law)

Lead angle, 319
Lead of a screw, 319
Line of action, 15, 61
Linear impulse, 587
Linear momentum:
  of a particle, 586
  of a rigid body, 796
Links, 88
Liter, 7
Loading of beams, 265
Logarithmic decrement, 878

Machines, 228, 243
  ideal, 408–409
  real, 408–409
Magnification factor, 865, 874
Magnitude of a force, 14
Mass, 2, 499
Mass center:
  of a rigid body, 714
  of a system of particles, 506
Mass moments of inertia, 380–385
Maxwell's diagram, 213
Mechanical efficiency, 408–409, 575
Mechanical energy, 566–567
Mechanics:
  definition of, 1
  Newtonian, 2
  principles of, 2–5
Megagram, 6
Meter, 5
Metric ton, 6*n*
Metric units, 5–7
Mile, 9
Millimeter, 6
Mohr's circle, 372–373
Moment:
  bending, 261, 266–280
  of a couple, 70–71
  first (*see* First moments)
  of a force, 64–65
  of momentum (*see* Angular momentum)
  second, 348–351
Moment-area method, 457
Moments of inertia, 347–385, 890–895
Moments of inertia (*Cont.*):
  of areas, 347–372
    parallel-axis theorem for, 357–358
  of common geometric shapes, 359, 895
  of composite areas, 358–359
  of composite bodies, 894
  determination of, by integration, 360–361, 894
  inclined axes, 369
  of masses, 890–895
    parallel-axis theorem for, 892
  polar, 351
  principal, 371
  rectangular, 350–351, 892
  of thin plates, 893–894
Momentum:
  angular (*see* Angular momentum)
  conservation of, 590–591, 620, 801
  linear (*see* Linear momentum)
  of a particle, 586–588
  of a rigid body, 795–800
Momentum couple, 796
Momentum vector, 796
Motion:
  absolute, 469
  under a central force, 517
  curvilinear (*see* Curvilinear motion of a particle)
  equations of (*see* Equations of motion)
  harmonic, simple, 839–842
  impulsive, 588, 812
  about mass center, 714
  of mass center, 506–507, 714
  Newton's laws of (*see* Newton's law, of motion)
  of a particle, 436–652
  plane (*see* Plane motion)
  rectilinear (*see* Rectilinear motion of a particle)
  relative (*see* Relative motion)
  of a rigid body, 653–837
  rolling, 735–736
  of a system of particles, 504–506
Motion curves, 439, 456–457
Multiforce members, 228, 261

Natural frequency, 865
Neutral equilibrium, 422–423
Newton, Sir Isaac, 2, 3
Newton (unit), 5
Newtonian frame of reference, 500
Newtonian mechanics, 2
Newton's law:
  of gravitation, 3–4, 517–518
  of motion: first, 3, 32
    second, 3, 499–500
    third, 3–4, 207
Noncentroidal rotation:
  dynamic equilibrium in, 734
  effective forces in, 734
  equations of motion in, 734–735
  kinetic energy in, 771
  momentum in, 800
Nonconservative force, 567
Nonimpulsive force, 588
Nonrigid truss, 222
Normal component of acceleration, 479–481
Numerical accuracy, 13
Nutation, 826

Oblique central impact, 601–604
Orbit, 532
Oscillation, center of, 850
Oscillations:
  of a rigid body, 850, 858
  of a simple pendulum, 842–844
Output forces, 243
Output power, 575
Output work, 408
Overrigid trusses, 222

Pappus, theorems of, 172–173
Parabolic cable, 283–284
Parabolic trajectory, 467, 530

Parallel-axis theorem:
for moments of inertia: of areas, 357–358
of masses, 892
for products of inertia, 369
Parallelogram law, 3, 15, 16–17
Partial constraints, 92–93, 138
Particles, 3, 14
equilibrium of: in a plane, 31–32
in space, 48
free-body diagram of, 32–33
kinematics of, 436–497
kinetics of, 498–652
moving on a slab, 693–695
relative motion of, 449, 468, 693–695
systems of (*see* Systems of particles)
vibrations of (*see* Vibrations)
Pascal (unit), 182*n*
Pendulum:
compound, 854
Kater, or reversible, 855
simple, 842–844
Percussion, center of, 742, 807
Perigee, 532
Period:
of deformation, 598–599, 812
of restitution, 598–599, 812
of vibration, 838
damped, 872
undamped, 841
Periodic time, 532–533
Phase angle, 841
Phase difference, 874
Pile driver, 616
Pin-and-bracket supports, 138
Pins, 88, 210
Pitch:
of a thread, 319
of a wrench, 129
Plane of symmetry, 191
Plane motion, 653–655
constrained, 733–736
dynamic equilibrium in, 714
effective forces in, 710–714
equations of motion in, 713, 715
kinematics of, 653–655
Plane motion (*Cont.*):
kinetic energy in, 777
momentum in, 795
Planetary gears, 672
Planetary motion, 533
Plastic impact, 600
Point of application of a force, 14, 61
Polar coordinates, 481
Polar moment of inertia, 351
Pole, 351
Polygon rule, 18
Position coordinate, 436
relative, 449
Position vector, 73, 464
relative, 468
Potential energy, 420–422, 564–566, 778
Pound, 8
Pound force, 8
Pound mass, 11
Power, 574–575, 779
Precession, 824–826
steady, 825
unsteady, 826
Pressure, center of, 155, 188–189
Principal axes of inertia, 369–372
Principal moments of inertia, 371
Principle:
of impulse and momentum: for a particle, 586–588
for a rigid body, 795
for a system of particles, 589
of transmissibility, 3, 59, 61–63
of virtual work, 402, 405–406
of work and energy: for a particle, 550–553
for a rigid body, 768
for a system of particles, 553
Principles of mechanics, 2–5
Problem solution, method of, 11–13
Product of inertia, 368–369
Projectile, 467, 516
Propeller, 633

Radial component:
of acceleration, 481–483
Radial component (*Cont.*):
of velocity, 481–483
Radius:
of curvature, 480
of gyration, 352, 890
Rated speed, 519
Reactions at supports and connections, 88–90, 137–139
Real machines, 408–409
Rectangular components:
of acceleration, 466–467
of force, 24, 39–41
of velocity, 464
Rectilinear motion of a particle, 435–457
uniform, 448
uniformly accelerated, 448–449
Rectilinear-motion problems, solution of: analytical, 440–441
graphical, 456–457
Rectilinear translation, 653
Reduction of a system of forces:
in a plane, 80–81
in space, 128–129
Redundant members, 222
Reference frame, 500
Relative acceleration (*see* Acceleration, relative)
Relative motion:
of a particle: with respect to axes in translation, 468
with respect to rotating axes, 693–695
of two particles, 449, 468–469
Relative position, 449, 468
Relative velocity:
of a particle: in plane motion, 666–668
with respect to axes in translation, 468
with respect to rotating axes, 693
of two particles, 449, 468
Relativity, theory of, 2
Repose, angle of, 304
Resolution of a force:
into components: in a plane, 19, 24–25
in space, 39–41

Resolution of a force (*Cont.*):
into a force and a couple: in a plane, 73–74
in space, 127
Resonance, 865
Restitution:
coefficient of, 599, 812
period of, 598–599, 812
Resultant of forces, 3, 14–15, 43, 128
(*See also* Addition, of forces; Addition, of vectors)
Reversed effective force (*see* Inertia vector)
Reversible pendulum, 855
Revolution:
body of, 172, 384
surface of, 172
Right-hand rule, 120
Rigid body, 2, 59
equilibrium of: in a plane, 87–93, 104–105
in space, 137
free-body diagram of, 137
kinematics of, 653–708
kinetics of, 709–837
vibrations of, 849–850
Rigid truss, 210
Rocket, 634
Rollers, 88, 138
Rolling motion, 735–736
Rolling resistance, 328–329
coefficient of, 329
Rotating shafts, 754
Rotation, 654, 656–658, 753–755
centroidal (*see* Centroidal rotation)
dynamic equilibrium in, 734
effective forces in, 713, 734
equations of motion in, 734, 753
instantaneous axis of, 654, 674
instantaneous center of, 674–676
kinematics of, 656–658
kinetic energy in, 771
momentum in, 796–797
noncentroidal (*see* Noncentroidal rotation)
Rotation (*Cont.*):
of a three-dimensional body, 753–755
uniform, 657
uniformly accelerated, 657–658
Rough surfaces, 88, 138

Sag, 284, 292
Satellites, 529, 534, 625
Scalar components, 24
Scalars, 15
Screws, 318–319
Second, 5, 8
Second moment, 348–351
Sections, method of, 220–221
Self-locking screws, 319
Semimajor axis, 532–533
Semiminor axis, 532–533
Sense of a force, 15
Shafts, rotating, 754
Shear, 261, 266–268
Shear diagram, 266, 268
SI units, 5–8
Significant figures, 13
Simple harmonic motion, 839–842
Simple pendulum, 842–844
Simple trusses, 210, 215
Slipstream, 632
Slug, 9
Space, 2
Space centrode, 676
Space mechanics, 529–533
Space shuttle, 536, 537, 627, 628, 644
Space truss, 215
Space vehicles, 529, 620
Specific weight, 156*n*, 188
Speed, 437, 465
rated, 519
Spin, 824
Spring:
force exerted by, 419, 549
potential energy, 420, 565
Spring constant, 419, 549
Square-threaded screws, 318–319
Stable equilibrium, 422–423
Static friction, 302–304
angle of, 304
coefficient of, 302
Statically determinate reactions, 92
Statically determinate structures, 231
Statically determinate trusses, 222
Statically indeterminate reactions, 92–93
Statically indeterminate structures, 231
Statically indeterminate trusses, 222
Statics, definition of, 1, 435
Steady precession, 825
Steady-state vibrations, 865, 873
Stream of particles, 630–633
Structural shapes, properties of, 360–361
Structures:
analysis of, 206–243
determinate, 231
indeterminate, 231
internal forces in, 206
two-dimensional, 63
Submerged surfaces, forces on, 182, 349
Supports:
ball, 138
ball-and-socket, 138
of beams, 265–266
reactions at, 88–90, 137–139
Surface:
frictionless, 88, 138
of revolution, 172
rough, 88, 138
submerged, forces on, 182, 349
Suspension bridges, 283
Symmetry:
axis of, 158
center of, 158
plane of, 191
Systems:
of forces, 59–86, 117–136
of particles: angular momentum of, 622
equations of motion for, 504
impulse-momentum principle for, 589

Systems, of particles (*Cont.*):
kinetic energy of, 553
mass center of, 506
variable, 630
work-energy principle for, 553
of units, 5–13

Tangential component of acceleration, 479–481
Tension, 63, 209, 260
Three-force body, 105
Thrust, 632
Thrust bearings, 327–328
Time, 2
Toggle vise, analysis of, 406–408
Ton:
metric, 6*n*
U.S., 9
Top, 826
Torsional vibrations, 852
Trajectory:
of projectile, 467, 532
of space vehicle, 530–533
Transfer formula (*see* Parallel-axis theorem)
Transient vibrations, 865, 873
Translation, 653, 655
curvilinear, 653
effective forces in, 713
kinematics of, 655
kinetic energy in, 770
momentum in, 796
rectilinear, 653
Transmissibility, principle of, 3, 59, 61–63
Transverse component:
of acceleration, 481–483
of velocity, 481–483
Triangle rule, 17
Trusses, 207–215, 220–222
compound, 221
determinate, 222
indeterminate, 222
overrigid, 222
rigid, 210
simple, 210, 215

Trusses (*Cont.*):
space, 215
typical, 209
Two-dimensional structures, 63
Two-force body, 104–105

Unbalanced disk, 736
Uniform rectilinear motion, 448
Uniform rotation, 657
Uniformly accelerated rectilinear motion, 448–449
Uniformly accelerated rotation, 657–658
U.S. customary units, 8–13
Units, 5–13
(*See also specific systems of units*)
Universal joints, 138
Unstable equilibrium, 422
Unstable rigid bodies, 92*n*
Unsteady precession, 826

V belts, 335
Variable systems of particles, 630
Varignon's theorem, 65
Vector addition, 16
Vector components, 24
Vectors, 15–16
bound, fixed, 15
coplanar, 18
couple, 120
free, 15
inertia, 503, 715
momentum, 796
sliding, 16, 62
Velocity, 437, 464
absolute, 666–668, 676
angular, 656, 657, 668
areal, 517
components of: radial, 481–483
rectangular, 464
transverse, 481–483
escape, 447, 531
in plane motion, 666–668
relative (*see* Relative velocity)

Velocity (*Cont.*):
in rotation, 656
Velocity-displacement curve, 457
Velocity-time curve, 439, 456–457
Vibrations, 838–888
damped: forced, 873–874
free, 870–872
forced, 863–865, 873–874
free, 839–872
frequency of, 838, 841
period of, 838, 841, 872
of rigid bodies, 849–850
steady-state, 865, 873
torsional, 852
transient, 865, 873
undamped: forced, 863–865
free, 839–870
Vibrometer, 887
Virtual displacement, 405
Virtual work, 405–408
principle of, 402, 405–406
Viscous damping, 870

Watt (unit), 575
Wedges, 318
Weight, 4, 518, 548
Wheel friction, 328–329
Wheels, 137, 328, 735
Work:
of a couple, 404–405, 418, 770
and energy, principle of (*see* Principle, of work and energy)
of a force, 403, 418, 547–550
of force exerted by spring, 419
of forces on a rigid body, 404–405, 769–770
of gravitational force, 549–550
input and output, 408
virtual, 405–408
of a weight, 418–419, 548
Wrench, 129

Zero-force member, 214

Answers to Even-Numbered Problems

Answers to Even-Numbered Problems†

CHAPTER 11

11.2 $x = 3$ in., $v = 28$ in./s, $a = 54$ in./s^2.
11.4 $t = 1$ s, $x = 25$ m, $a = -12$ m/s^2; $t = 5$ s, $x = -7$ m, $a = +12$ m/s^2.
11.6 (*a*) 1 s, 4 s. (*b*) +1.5 m, 24.5 m.
11.8 $a = 2t$; $v = t^2 - 9$; $x = \frac{1}{3}t^3 - 9t + 30$.
11.10 (*a*) 3 ft/s^4. (*b*) $a = 3t^2$; $v = t^3 - 24$; $x = \frac{1}{4}t^4 - 24t + 50$.
11.12 (*a*) ± 2.21 m/s. (*b*) 1.219 m.
11.14 (*a*) ± 20 in./s. (*b*) +6.71 in. (*c*) 3.87 in.
11.16 (*a*) 7.05 m/s. (*b*) 10.95 m/s. (*c*) 25.3 m/s.
11.18 (*a*) 241 ft. (*b*) Infinite.
11.20 412 m.
11.22 (*b*) 4 m/s; $\frac{1}{6}$ s.
11.24 (*a*) -2.43×10^6 ft/s^2. (*b*) 1.366×10^{-3} s.
11.26 $\sqrt{2gR}$.
11.30 (*a*) 0.496 m/s^2. (*b*) 32.7 km/h.
11.32 (*a*) 70.6 ft. (*b*) 68.5 ft/s.
11.34 (*a*) 5.61 s. (*b*) 60.2 km/h.
11.36 (*b*) 1.529 s; 11.47 m.
11.38 (*a*) $t = 16$ s; $x = 960$ ft (*b*) $v_A = 51.8$ mi/h; $v_B = 28.6$ mi/h
11.40 (*a*) 1.25 m/s^2 $\uparrow$. (*b*) 2.5 m/s^2 $\downarrow$. (*c*) 7.5 m/s $\uparrow$.
11.42 (*a*) $\mathbf{a}_A = 4.5$ in./s^2 $\leftarrow$; $\mathbf{a}_B = 3$ in./s^2 $\leftarrow$. (*b*) 22.5 in./s $\leftarrow$; 56.3 in. $\leftarrow$.
11.44 (*a*) 300 mm/s $\rightarrow$. (*b*) 600 mm/s $\leftarrow$. (*c*) 450 mm/s $\leftarrow$.
11.46 (*a*) 2.5 mm/s^2 $\uparrow$; 20 mm/s $\uparrow$. (*b*) 225 mm $\uparrow$.
11.48 (*a*) 2 s. (*b*) 1.500 in. $\downarrow$.
11.50 $\mathbf{v}_A = 125$ mm/s $\uparrow$; $\mathbf{v}_B = 75$ mm/s $\downarrow$; $\mathbf{v}_C = 175$ mm/s $\downarrow$.
11.52 +15 ft/s; +91.5 ft; 143.5 ft.
11.54 (*a*) 44 m. (*b*) 4 s, 18 s.
11.56 Accelerate for 11.84 s; 76.7 km/h.
11.58 15 s.
11.60 (*a*) -1.464 m/s^2; 0.75 m/s^2. (*b*) 32.7 km/h.
11.62 8.04 s.
11.64 (*a*) and (*b*) 23.3 ft/s^2.
11.66 (*a*) 3 s. (*b*) 1000 m.
11.68 (*a*) 25 s; 300 ft. (*b*) 100 s; 1800 ft.
11.70 (*a*) 8 s. (*b*) 24 m. (*c*) 3 m/s.
11.72 (*a*) 272 m/s. (*b*) 679 m.
11.74 (*a*) -7.56 in./s^2. (*b*) -8.80 in./s^2.
11.76 (*a*) 2.7 s. (*b*) 12.15 m.
11.78 51.5 ft.
11.80 (*a*) 5.66 ft/s $\angle$45°; 12.65 ft/s^2 $\angle$71.6°. (*b*) 6.12 ft/s $\angle$11.2°; 4.65 ft/s^2 $\angle$30.7°.
11.82 (*a*) 80 mm/s $\uparrow$; 307 mm/s^2 $\angle$15.1°. (*b*) 247 mm/s $\angle$40.3°; 80 mm/s^2 $\uparrow$.
11.84 (*a*) $\mathbf{v} = 4\pi$ in./s $\leftarrow$; $\mathbf{a} = 4\pi^2$ in./s^2 $\uparrow$.
11.86 (*a*) $\pm 90°$. (*b*) Ap. (*c*) Circle of radius A with center at the origin.
11.88 (*a*) 46.3 ft/s. (*b*) 44.0 ft/s $\leq v_0 \leq$ 56.7 ft/s.
11.90 4.70 m/s $\leq v_0 \leq$ 7.23 m/s.
11.92 5.15 ft $\leq h \leq$ 10.92 ft.
11.94 4.12 m from B.
11.96 (*a*) 214 m/s. (*b*) 231 m/s.
11.98 14.91 ft/s $\leq v_0 \leq$ 21.5 ft/s.
11.100 (*a*) 213 ft/s. (*b*) 29.5°.
11.102 51.6° or 69.3°.
11.104 4.34 m.
11.106 54.0 km/h $\angle$31.2°.
11.108 (*a*) 91.0 ft/s $\angle$63.0°. (*b*) 273 ft $\angle$63.0°. (*c*) 434 ft.
11.110 (*a*) 80.7 mm/s^2 $\angle$68.3°. (*b*) 323 mm $\angle$68.3°.
11.112 (*a*) 9.91 in./s $\angle$66.2°. (*b*) 14.87 in./s^2 $\angle$66.2°.

† Partial answers are provided or referenced for *all* problems designed to be solved with a computer.

11.114 (*a*) 63.9°. (*b*) 1195 ft/s ↑; 32.2 ft/s^2 ↓.
11.116 2.45 m/s ↗ 35.3°.
11.118 23.7 mi/h.
11.120 44.1 km/h.
11.122 (*a*) 5 ft/s. 240 ft/s^2.
11.124 93.8 m; 17.68 s.
11.126 43.8 s.
11.128 (*a*) 60.1 km/h ∠8.6°.
(*b*) 3.21 m/s^2 ↘5.9°.
11.130 22,800 ft; 58,100 ft.
11.132 (*a*) 14.01 m/s. (*b*) 12.80 m.
11.134 17,390 mi/h.
11.136 11 060 km/h.
11.138 3670 km/h.
11.140 (*a*) $v = 1.118$ m/s, $\beta = 259.7°$.
(*b*) $a = 6.67$ m/s^2, $\gamma = 197.8°$.
(*c*) $a_{B/OA} = 2.90$ m/s^2, $\gamma = 180°$. Where β and γ are measured counterclockwise from the rod OA.
11.142 (*a*) $v = 10$ in./s, $\beta = 0$; $a = 125.7$ in./s^2, $\gamma = 90°$. (*b*) $v = 18.62$ in./s, $\beta = 147.5°$; $a = 159.8$ in./s^2, $\gamma = 218.1°$. Where β and γ are measured counterclockwise from the axis $\theta = 0$.
11.144 (*a*) $v = 62.8$ mm/s, $\beta = 270°$;
$a = 112.8$ mm/s^2, $\gamma = 0$.
(*b*) $v = 137.5$ mm/s, $\beta = 323.1°$;
$a = 520$ mm/s^2, $\gamma = 355.2°$. Where β and γ are measured counterclockwise from the axis $\theta = 0$.
11.146 $(v_0/h)\sin^2\theta$.
11.148 $(2v_0^2/h^2)\sin^3\theta\cos\theta$.
11.150 437 mi/h ↘5.9°.
11.152 $v = e^{b\theta}\sqrt{1+b^2}\,\dot{\theta}$.
11.154 $a = (1+b^2)\omega^2 e^{b\theta}$.
11.156 (*b*) Toward A.
11.158 (*a*) 63.7 mm/s. (*b*) 127.3 mm.
(*c*) 31.8 mm/s.
11.160 88.4 ft.
11.162 13.59 m.
11.164 (*a*) 86.4 m; 4.20 s. (*b*) 166.6 m; 5.83 s.
11.166 (*a*) 11.09 s. (*b*) 688 ft.
11.168 80 mm/s ↓.
11.C1 $v_0 = 250$ *m/s:* $h = 1432$ m, $v_f = 122.2$ m/s.
11.C2 $v_0 = 6$ *m/s:* Strikes ground 3.76 m from firefighter; $v_0 = 10$ *m/s:* Strikes building 5.02 m from ground; $v_0 = 13$ *m/s:* Strikes roof 6.74 m from B; $v_0 = 15$ *m/s:* Strikes ground 20.5 m from firefighter.
11.C3 (*a*) Range = 44,000 ft; $t = 30$ *s:*
$x = 27{,}600$ ft, $y = 8650$ ft.
(*b*) Range = 41,300 ft; $t = 30$ *s:*
$x = 26{,}700$ ft, $y = 8220$ ft.
(*c*) Range = 33,500 ft; $t = 30$ *s:*
$x = 24{,}000$ ft, $y = 6750$ ft.
(*d*) Range = 27,400 ft; $t = 30$ *s:*
$x = 21{,}300$ ft, $y = 5340$ ft.
11.C4 38.4 s.

CHAPTER 12

12.2 (*a*) 4.987 lb at 0°; 5.000 lb at 45°; 5.013 lb at 90°. (*b*) 5.000 lb at all latitudes.
(*c*) 0.1554 lb·s^2/ft at all latitudes.
12.4 2.84×10^6 kg·m/s.
12.6 (*a*) 78.1 km/h. (*b*) 63.8 km/h.
12.8 1.610 ft/s^2.
12.10 (*a*) $\mathbf{a}_A = 8.40$ ft/s^2 ↗; $\mathbf{a}_B = 4.20$ ft/s^2 ↓.
(*b*) 152.2 lb.
12.12 (*a*) 3.60 m/s^2 ←. (*b*) 180 N →.
12.14 (*a*) 5.21 m/s^2 ↙. (*b*) 215 N.
12.16 (*a*) 1.2 m/s →. (*b*) 0.6 m/s ←.
12.18 (*a*) 8.05 ft/s^2 ↓; 12.88 ft/s^2 ↓; 1.342 ft/s^2 ↓.
(*b*) 16.10 ft/s ↓; 25.8 ft/s ↓; 2.68 ft/s ↓.
(*c*) 11.35 ft/s ↓; 14.36 ft/s ↓; 4.63 ft/s ↓.
12.20 (*a*) Load will shift. (*b*) 10.64 ft/s ←.
12.22 (*a*) 3.37 m/s. (*b*) 1.372 s.
12.24 1.981 km.
12.26 $x = Pt/k - (Pm/k^2)(1 - e^{-kt/m})$.
12.28 (*a*) 1.656 lb. (*b*) 20.83 lb.
12.30 (*a*) 22.4 N. (*b*) 1.121 m/s^2 →.
12.32 (*a*) 6.17 m/s^2 ↘25°. (*b*) 144.0 N.
12.34 (*a*) 8.94 ft/s^2 ←; 18.06 lb.
(*b*) and (*c*) 12.38 ft/s^2 ←; 15.38 lb.
12.36 (*a*) 24.2°. (*b*) 32.3 N.
12.38 27.1 ft/s.
12.40 2.82 m/s $\leq v \leq$ 3.15 m/s.
12.42 2.61 m/s $\leq v \leq$ 3.36 m/s.
12.44 (*a*) 0.395W. (*b*) 0.766W.
12.46 (*a*) $a_t = g\cos\theta$; $v = \sqrt{2gl\sin\theta}$. (*b*) 41.8°.
12.48 (*a*) 0.303 lb ↑. (*b*) 9.01°.
12.50 $-58.2° \leq \theta \leq 238.2°$.
12.52 0.45.
12.54 (*a*) 43.7°. (*b*) 0.68. (*c*) 56.1 km/h.
12.56 (*a*) 0.455 N ∠70°; does not slide.
(*b*) 0.556 N ↗30°; slides.
12.58 (*a*) 1.010 s. (*b*) 6.06 ft/s.

12.60 (a) 1.207 m/s. (b) 21.8° and 158.2°.
12.62 $\delta = eVlL/mv_0^2 d$.
12.64 Length l should be doubled.
12.66 $F_r = +0.400$ N, $F_\theta = +3.20$ N.
12.68 (a) $F_r = -(11.65 \text{ lb}) \cos\theta$, $F_\theta = -(11.65 \text{ lb}) \sin\theta$. (b) $\mathbf{P} = 0$; $\mathbf{Q} = 11.65$ lb ⦪ 2θ.
12.70 (a) $a_r = -72$ m/s², $a_\theta = +43.2$ m/s². (b) $\ddot{r} = -14.40$ m/s². (c) $F_\theta = +8.64$ N.
12.72 (a) $v_r = -3.00$ m/s, $v_\theta = +3.60$ m/s. (b) $a_r = -62.1$ m/s², $a_\theta = -72.0$ m/s². (c) $F_\theta = -14.40$ N.
12.74 $v_r = 2v_0 \sin 2\theta$, $v_\theta = v_0 \cos 2\theta$.
12.76 (b) $3mv_0^2/r_0$.
12.78 383×10^3 km, 238,000 mi.
12.80 (a) 1.90×10^{27} kg. (b) 422×10^3 km.
12.82 (a) 448 m/s. (b) 2290 m/s.
12.84 4800 ft/s.
12.86 (a) 5.74 in. (b) 15.69 ft/s.
12.88 (a) 2.45 m/s. (b) -12.25 rad/s².
12.94 18.82 km/s.
12.96 1.150.
12.98 (a) 26.0×10^3 ft/s. (b) 173.3 ft/s.
12.100 $\beta = \sqrt{2/(2+\alpha)}$.
12.102 (a) 125×10^3 km. (b) 128.5 m/s; 2370 m/s.
12.104 19 h 10 min 7 s.
12.106 (a) 1 h 29 min 14 s. (b) 43 min 43 s.
12.108 3.29×10^9 mi.
12.110 123.0 m/s.
12.112 28.8 mi.
12.114 (a) $\varepsilon = (r_1 - r_0)/(r_1 + r_0)$. (b) $\varepsilon_1 = 0.778$; $\varepsilon_2 = 0.852$.
12.118 3.50 m/s.
12.120 For $P < 18.75$ lb: $a = 0.429P$; For $P \geq 18.75$ lb: $a = 1.073(P - 9 \text{ lb})$.
12.122 $m_B = 2.59m_A$; $m_C = 1.337m_A$.
12.124 (a) 1.804 μN. (b) 190.9 s.
12.126 (a) 18,180 mi/h. (b) 16,900 mi.
12.128 (a) $\mathbf{a}_A = 3.46$ m/s² ←; $\mathbf{a}_B = 2.31$ m/s² ←. (b) 19.23 N.
12.C1 (a) $\theta = 30°$: 2.66 m/s, 2.50 m. (b) $\theta = 30°$: 1.657 m/s, 0.970 m. (c) $\theta = 30°$: 1.521 m/s, 0.817 m.
12.C2 $\mu = 0.05$: $\mathbf{a}_A = 2.76$ ft/s² →, $\mathbf{a}_{B/A} = 17.16$ ft/s² ⦪ 30°; $\mu = 0.20$: $\mathbf{a}_A = 0$, $\mathbf{a}_{B/A} = 10.52$ ft/s² ⦪ 30°; $\mu = 0.60$: $\mathbf{a}_A = \mathbf{a}_{B/A} = 0$.
12.C3 33.8°.
12.C4 $\beta = 0.8$: 18.0°; $\beta = 0.98$: 70.5°.

CHAPTER 13

13.2 (a) 224 ft·lb; 55.9 ft. (b) 224 ft·lb; 339 ft.
13.4 100.9 km/h.
13.6 3.39 m/s.
13.8 (a) 14.81 ft. (b) 15.17 ft/s. (c) 107.4 ft·lb.
13.10 (a) 322 ft. (b) $F_{AB} = 5400$ lb C; $F_{BC} = 12{,}600$ lb C.
13.12 (a) 156.25 m. (b) 2100 N →.
13.16 (a) 5.26 m/s. (b) 27.2 m.
13.18 (a) 2.21 m/s ↓. (b) $\mathbf{F}_A = 91.5$ N ↑; $\mathbf{F}_B = 98.2$ N ↑.
13.20 (a) $\mathbf{F}_A = 149.8$ lb ↑; $\mathbf{F}_D = 240.4$ lb ↑. (b) 0.15. (c) 453 ft·lb.
13.22 (a) 6.95 ft/s ↓. (b) 5.54 ft/s ↓.
13.24 $v_1 = 3.43$ m/s; $v_2 = 4.72$ m/s.
13.26 1.180 m/s ↓.
13.28 (a) 3.2 in. (b) 28.7 in./s.
13.30 134.7 mm.
13.32 A: 5.66 in.; B: 7.5 in.
13.34 (a) 32.7 mm; 98.1 N. (b) 30.4 mm; 104.9 N.
13.36 5.38 m/s.
13.38 (a) 77.2 km. (b) 80.7 km.
13.40 867 mi.
13.42 (a) 3.13 m/s. (b) 3.50 m/s.
13.44 (a) 1120 lb ↑; 160 lb ↓. (b) 80 ft.
13.46 (a) 0.75r. (b) 0.1676r.
13.48 (a) 440 mm; 4.80 m/s. (b) 440 mm; 2.40 m/s.
13.50 12.04 ft/s.
13.52 44.0 ft/s.
13.54 (a) 4.58 m/s. (b) $x = 160$ mm, $y = 300$ mm, $z = 120$ mm.
13.56 (a) 37.1 ft/s. (b) 21.3 ft.
13.58 1.546 m.
13.60 (a) 2.48 m/s. (b) 1.732 m/s.
13.62 51.4 N ↑.
13.64 2.33 lb/in.
13.66 $3l/5$.
13.68 (a) $mgR[1 - (R/r)]$. (b) $\frac{1}{2}mgR^2/r$. (c) $mgR[1 - (R/2r)]$.
13.70 (a) 2.81 MJ/kg. (b) 1.330 MJ/kg.
13.74 2.40 MW.
13.76 (a) and (b) 63.3 kW.
13.78 (a) 55.2 kW. (b) 241 kW.
13.80 For $t \leq 50$ s: $P = (132.5 \text{ kW/s})t$. At $t = 50$ s: $P = 6.63$ MW. For $t > 50$ s: $P = 375$ kW.

13.82 (*a*) 36.4 hp. (*b*) 60.1 hp.

13.84 (*a*) 20 s; 207 m. (*b*) 33.3 s; 563 m.

13.86 4.20 km/h.

13.88 5.08 m/s.

13.90 3.43 m/s $\uparrow$.

13.92 103.1 mi/h.

13.94 4.95 m/s $\rightarrow$.

13.96 (*a*) 8.79 ft/s. (*b*) 0.

13.98 (*a*) 5.18 ft/s. (*b*) 0.375 lb.

13.C1 $F_m = 16$ *lb:* (*a*) 4.27 ft, (*b*) 3.91 ft;
$F_m = 24$ *lb:* (*a*) 6.40 ft, (*b*) 6.89 ft.

13.C2 $k = 0.5$ *lb/in.:* 8.31 ft/s;
$k = 1.5$ *lb/in.:* 4.59 ft/s.

13.C3 (*a*) *200 mm from B:* 1.530 m/s;
300 mm from B: 1.655 m/s.
(*b*) 0.335 s.

13.C4 $v_0^2 = 3gr$*:* $x = -0.649r$;
$v_0^2 = 4gr$*:* $x = 0.296r$.

CHAPTER 14

14.2 (*a*) 3.40 s. (*b*) 25.5 s.

14.4 (*a*) and (*b*) 30.5 kips.

14.8 (*a*) 22.8 s. (*b*) $F_{AB} = 12{,}600$ lb *C*;
$F_{BC} = 5400$ lb *C*.

14.12 (*a*) 15.19 m/s. (*b*) 16.80 m/s.

14.14 (*a*) 38.6 ft/s $\rightarrow$. (*b*) 12.08 ft/s $\rightarrow$.

14.16 24 kN.

14.18 33.0 kips $\measuredangle$ 79.6°.

14.20 (*a*) 6.17 m/s. (*b*) 0.01235.

14.22 (*a*) 1.333 m/s $\leftarrow$. (*b*) 1.467 m/s $\leftarrow$.
(*c*) 1.456 m/s $\leftarrow$.

14.24 (*a*) Car *B*. (*b*) 52.9 mi/h.

14.26 (*a*) 4.80 N·s; 14.40 J. (*b*) 4.00 N·s; 12.00 J.

14.28 (*a*) 0.6 m/s $\rightarrow$. (*b*) 0.75.

14.30 (*a*) $\mathbf{v}'_A = 1$ ft/s $\leftarrow$; $\mathbf{v}'_B = 15$ ft/s $\rightarrow$.
(*b*) 3.35 ft·lb.

14.32 $\mathbf{v}'_A = 0.209$ m/s $\rightarrow$; $\mathbf{v}'_B = 0.235$ m/s $\rightarrow$;
$\mathbf{v}'_C = 1.156$ m/s $\rightarrow$.

14.34 $\mathbf{v}'_A = 5.27$ ft/s $\measuredangle$ 27.1°;
$\mathbf{v}'_B = 7.17$ ft/s $\measuredangle$ 50°.

14.36 (*a*) 45°. (*b*) $(1 + e)v_0/4$.

14.40 $0.425h$.

14.42 (*a*) $h_1 = d_1 = 6.40$ in.
(*b*) $h_2 = 4.10$ in; $d_2 = 5.12$ in.

14.44 (*a*) 0.80. (*b*) 283 mm.

14.46 $\mathbf{v}'_A = 0.4v_0 \measuredangle 60°$; $\mathbf{v}'_B = 0.7v_0 \leftarrow$.

14.48 $\mathbf{v}'_A = 2.28$ m/s $\measuredangle$ 18.9°; $\mathbf{v}'_B = 0.719$ m/s $\leftarrow$.

14.50 (*a*) 37.8°. (*b*) 40.9°.

14.52 (*a*) 0.50. (*b*) 29.0°.

14.54 (*a*) 2.90 m. (*b*) 75.6%.

14.56 439 m/s.

14.58 79.0 kN $\uparrow$.

14.60 (*a*) 0.80. (*b*) 12 oz.

14.62 $\mathbf{v}_A = 0.990$ m/s $\leftarrow$; $\mathbf{v}_B = 1.485$ m/s $\rightarrow$.

14.64 $v_A = 1.690$ m/s; $v_B = 1.532$ m/s;
$v_C = 3.29$ m/s.

14.66 (*a*) 11.09 ft/s $\measuredangle$ 30°. (*b*) 2.74 ft/s $\rightarrow$.

14.68 6.37 in.

14.72 (*a*) 0.919 m/s. (*b*) 8.27 m/s.

14.74 7.35 m/s.

14.76 15.61 in.

14.82 (*a*) 7960 ft/s. (*b*) 4820 ft/s.

14.84 3190 m/s.

14.86 17,980 ft/s.

14.88 67.2°.

14.92 $(1 \pm \sin\alpha)r_0$.

14.96 4.50 kN $\measuredangle$ 30°.

14.98 (*a*) 4100 lb. (*b*) 6450 lb.

14.100 540 N.

14.102 (*a*) 36.9°. (*b*) 82.9 lb $\downarrow$.

14.104 $\mathbf{C} = 31.0$ lb $\leftarrow$; $\mathbf{D}_x = 58.7$ lb $\leftarrow$,
$\mathbf{D}_y = 51.8$ lb $\downarrow$.

14.108 $\mathbf{C}_x = 90$ N $\rightarrow$, $\mathbf{C}_y = 2360$ N $\uparrow$;
$\mathbf{D} = 2900$ N $\uparrow$.

14.110 75 kg/s.

14.112 37.2 ft/s.

14.114 5500 kg.

14.116 (*a*) 847 km/h. (*b*) 638 km/h.

14.118 (*a*) 47.4 kJ/s. (*b*) 0.316.

14.120 (*a*) $\frac{1}{2}v_A \rightarrow$. (*b*) $\frac{1}{4}A\rho(1 - \cos\theta)v_A^3$.
(*c*) $2(V/v_A)[1 - (V/v_A)](1 - \cos\theta)$.

14.124 9.21 m^3/s.

14.126 (*a*) $\frac{1}{2}mg + mv^2/l$. (*b*) $\frac{1}{2}mg$.

14.128 $h = v^2/g$.

14.130 (*a*) 1.192 m/s. (*b*) 1.320 m.

14.132 (*a*) m_0e^{qL/m_0v_0}. (*b*) v_0e^{-qL/m_0v_0}.

14.134 51.6 ft/s^2.

14.136 894 kips.

14.138 (*a*) 31.9 m/s^2 $\uparrow$. (*b*) 240 m/s^2 $\uparrow$.

14.140 1890 kg.

14.142 27,200 ft/s.

14.144 709,000 ft.

14.146 136.9 km.

14.150 (*a*) 0.4 m/s $\rightarrow$. (*b*) 0.8 m/s $\rightarrow$.

14.152 $h = d/(1 - e^2)$.

14.154 (*a*) 23.1 s. (*b*) 444 lb *C*.

14.156 (*a*) 1.048 kN·s; 2.75 kJ.
(*b*) 0.749 kN·s; 1.962 kJ.

14.158 (*a*) $\mathbf{v}_A = \mathbf{v}_B = \mathbf{v}_C = 0.50$ km/h →.
(*b*) $\mathbf{v}_A = \mathbf{v}_B = 2.75$ km/h ←;
$\mathbf{v}_C = 7.00$ km/h →.

14.160 6.31 lb.

14.C1 $W =$ *120 lb:* (*a*) 4.80 s, (*b*) 56.5 ft/s,
(*c*) 18.47 s; $W =$ *180 lb:* (*a*) 7.20 s,
(*b*) 8.90 ft/s, (*c*) 13.06 s.

14.C2 $\phi_0 =$ *75°* and *105°*: $h_{min} = 1504$ km,
$h_{max} = 17\,390$ km;
$\phi_0 =$ *90°*: $h_{min} = 2400$ km,
$h_{max} = 16\,490$ km.

14.C3 $\theta_0 =$ *0.15°*: (*a*) $v'_A = 0.1448$ m/s,
$\theta'_A = 45.0°$ ∠;
(*b*) $v'_A = 0.226$ m/s, $\theta'_A = 26.3°$ ∠;
(*c*) $v'_A = 0.609$ m/s, $\theta'_A = 8.6°$ ∠.
$\theta_0 =$ *4°*: (*a*) $v'_A = 2.791$ m/s,
$\theta'_A = 48.3°$ ∠;
(*b*) $v'_A = 2.794$ m/s, $\theta'_A = 46.8°$ ∠;
(*c*) $v'_A = 2.82$ m/s, $\theta'_A = 41.0°$ ∠.

14.C4 $\theta =$ *30°*: (*a*) $\mathbf{v}'_A = 2.66$ m/s ∠ 25.7°,
$\mathbf{v}'_B = 0.799$ m/s ←;
(*b*) $\mathbf{v}'_A = 2.28$ m/s ∠ 18.9°,
$\mathbf{v}'_B = 0.719$ m/s ←.
$\theta =$ *60°*: (*a*) $\mathbf{v}'_A = 2.75$ m/s ∠ 40.9°,
$\mathbf{v}'_B = 0.693$ m/s ←;
(*b*) $\mathbf{v}'_A = 2.68$ m/s ∠ 45.7°,
$\mathbf{v}'_B = 0.624$ m/s ←.

CHAPTER 15

15.2 (*a*) 2 s. (*b*) 4.8 rad; -3.6 rad/s^2.

15.4 (*a*) 0; 0.2 rad/s; -0.05 rad/s^2.
(*b*) 0.506 rad; 0.0736 rad/s;
-0.01839 rad/s^2. (*c*) 0.8 rad; 0; 0.

15.6 (*a*) 75 rev. (*b*) 1350 rev.

15.8 (*a*) -2.71 rad/s^2. (*b*) 27,900 rev.

15.10 (*a*) 2.40 rev. (*b*) Infinite. (*c*) 1.842 s.

15.12 (*a*) 1525 ft/s; 0.1112 ft/s^2.
(*b*) 1168 ft/s; 0.0852 ft/s^2. (*c*) 0; 0.

15.14 ± 6.52 rad/s^2.

15.16 (*a*) 100 mm/s^2. (*b*) 111.8 mm/s^2.
(*c*) 224 mm/s^2.

15.18 (*a*) 67.7 ft/s ←; 24,500 ft/s^2 ↓.
(*b*) 6.77 ft/s ←; 245 ft/s^2 ∠ 86.8°.

15.20 (*a*) 9.6 rad/s^2 ↻. (*b*) 16.04 rev ↻.

15.22 (*a*) 0.509 rad/s^2. (*b*) 20.6 s.

15.24 (*a*) 15.28 rev. (*b*) 10.14 s.

15.26 (*a*) 9.75 s.
(*b*) $\boldsymbol{\omega}_A = 279$ rpm ↺; $\boldsymbol{\omega}_B = 419$ rpm ↻.

15.28 (*a*) $\boldsymbol{\alpha}_A = 4.19$ rad/s^2 ↻; $\boldsymbol{\alpha}_B = 2.36$ rad/s^2 ↻.
(*b*) 6.00 s

15.30 $\mathbf{v}_B = 132$ ft/s →, $\mathbf{v}_C = 0$,
$\mathbf{v}_D = 127.5$ ft/s ∠ 15°,
$\mathbf{v}_E = 93.3$ ft/s ∠ 45°.

15.32 (*a*) 1.570 rad/s ↺. (*b*) 1.703 m/s ∠ 25°.

15.34 (*a*) 6.03 rad/s ↺. (*b*) 2.94 m/s ∠ 60°.

15.36 (*a*) 1.677 rad/s ↻. (*b*) 21.7 in./s →.

15.38 (*a*) $\frac{1}{2}\omega_A$ ↺. (*b*) $\frac{1}{4}\omega_A$ ↻.

15.40 (*a*) 240 rpm ↺. (*b*) Zero.

15.42 $\boldsymbol{\omega}_B = 144$ rpm ↻; $\boldsymbol{\omega}_C = 80$ rpm ↺.

15.44 (*a*) $\boldsymbol{\omega}_{BD} = 0$; $\mathbf{v}_D = 1.885$ m/s ←.
(*b*) $\boldsymbol{\omega}_{BD} = 7.26$ rad/s ↺; $\mathbf{v}_D = 1.088$ m/s ←.
(*c*) $\boldsymbol{\omega}_{BD} = 0$; $\mathbf{v}_D = 1.885$ m/s →.

15.46 $\mathbf{v}_P = 12.40$ ft/s →; $\boldsymbol{\omega}_{BD} = 20.6$ rad/s ↺.

15.48 $\boldsymbol{\omega}_{BD} = 0$; $\boldsymbol{\omega}_{DE} = 1.6$ rad/s ↻.

15.50 $\boldsymbol{\omega}_{BD} = 1.5$ rad/s ↻; $\boldsymbol{\omega}_{DE} = 3.75$ rad/s ↻.

15.52 (*a*) $\mathbf{v}_A = 3$ m/s →; $\boldsymbol{\omega} = 0$.
(*b*) $\mathbf{v}_A = 1.113$ m/s →; $\boldsymbol{\omega} = 1.291$ rad/s ↻.

15.54 $\omega_C = (r_A/r_C)\omega_A + [1 - (r_A/r_C)]\omega_{ABC}$.

15.56 (*a*) 3 in. below *D*. (*b*) 16 in./s →.

15.58 (*a*) 10 mm to the right of *A*. (*b*) 40 mm/s ↓.
(*c*) Outer pulley: unwrapped, 240 mm/s;
inner pulley: unwrapped, 120 mm/s.

15.60 (*a*) 6.93 rad/s ↺. (*b*) 1.039 m/s ←.
(*c*) 2.11 m/s ∠ 34.7°.

15.62 (*a*) 1.5 rad/s ↺. (*b*) 18 in./s ↑.
(*c*) 11.25 in./s ∠ 53.1°.

15.64 (*a*) 1.867 rad/s ↺. (*b*) 1.029 m/s ∠ 76.5°.

15.66 (*a*) $\boldsymbol{\omega}_{AB} = 1.4$ rad/s ↺; $\boldsymbol{\omega}_{BD} = 4$ rad/s ↺.
(*b*) 0.7 m/s ∠ 36.9°.

15.68 (*a*) 24.9 in./s ∠ 60°. (*b*) 20.9 in./s →.

15.70 (*a*) 64.8°. (*b*) 60.6 in./s ←.

15.72 (*a*) $\boldsymbol{\omega}_{AB} = 1.2$ rad/s ↺; $\boldsymbol{\omega}_{DE} = 0.80$ rad/s ↺.
(*b*) 140 mm/s ↑.

15.74 Space centrode: horizontal surface. Body centrode: circumference of cylinder.

15.80 $\mathbf{a}_A = 2.6$ m/s^2 ↑; $\mathbf{a}_B = 0.2$ m/s^2 ↑.

15.82 (*a*) 2.5 in. from *B* on *AB*. (*b*) 2 in. from *A* on *AB*.

15.84 14.04 m/s^2 ∠ 36.7°.

15.86 6.63 ft/s.

15.88 2.30 m/s^2 ∠ 43.2°.

15.90 $\mathbf{a}_D = 7400$ ft/s^2 ∠ 45°;
$\mathbf{a}_E = 1529$ ft/s^2 ∠ 45°.

15.92 $32.0 \text{ m/s}^2 \searrow$.
15.94 $\boldsymbol{\omega}_{DE} = \frac{1}{2}\omega_0 \circlearrowleft$; $\boldsymbol{\alpha}_{DE} = 0$;
$\boldsymbol{\omega}_{EF} = 0$; $\boldsymbol{\alpha}_{EF} = \frac{3}{4}\omega_0^2 \circlearrowright$.
15.96 (*a*) $6.90 \text{ rad/s}^2 \circlearrowleft$. (*b*) $5.52 \text{ rad/s}^2 \circlearrowleft$.
15.98 (*a*) $1.925 \text{ rad/s}^2 \circlearrowright$. (*b*) $3.86 \text{ m/s}^2 \measuredangle 40.3°$.
15.100 (*a*) $1.974 \text{ rad/s}^2 \circlearrowleft$. (*b*) $10.03 \text{ ft/s}^2 \measuredangle 70.3°$.
15.102 $\omega = (v_B \sin \beta)/(l \cos \theta)$.
15.104 (*a*) $\mathbf{v}_B = v_0 \cos \frac{1}{2}\theta \nwarrow$.
(*b*) $\mathbf{a}_B = \frac{1}{2}(v_0^2/b) \sin \frac{1}{2}\theta \searrow$.
15.106 (*a*) $v_A = b\omega \cos \theta$.
(*b*) $a_A = b\alpha \cos \theta - b\omega^2 \sin \theta$.
15.108 $\boldsymbol{\omega} = v_0/(2l \sin \theta) \circlearrowright$;
$\boldsymbol{\alpha} = (v_0/2l)^2 \cos \theta / \sin^3 \theta \circlearrowright$.
15.110 (*a*) $\boldsymbol{\omega} = (v_0/b) \sin^2 \theta \circlearrowright$.
(*b*) $(\mathbf{v}_A)_x = (v_0 l/b) \sin^2 \theta \cos \theta \rightarrow$.
$(\mathbf{v}_A)_y = (v_0 l/b) \sin^3 \theta \uparrow$.
15.112 $v_x = v[1 - \cos (vt/r)]$, $v_y = v \sin (vt/r)$.
15.114 (*a*) $1.815 \text{ rad/s} \circlearrowright$. (*b*) $16.42 \text{ in./s} \measuredangle 20°$.
15.116 $\boldsymbol{\omega}_A = 4.21 \text{ rad/s} \circlearrowleft$; $\boldsymbol{\omega}_B = 1.274 \text{ rad/s} \circlearrowleft$.
15.118 $706 \text{ mm/s} \measuredangle 40.9°$.
15.120 $\mathbf{a}_1 = [r\omega^2 \rightarrow] + [2u\omega \downarrow]$,
$\mathbf{a}_2 = [r\omega^2 \downarrow] + [2u\omega \rightarrow]$,
$\mathbf{a}_3 = [r\omega^2 \leftarrow] + [u^2/r \leftarrow] + [2u\omega \leftarrow]$,
$\mathbf{a}_4 = [r\omega^2 \uparrow] + [2u\omega \uparrow]$.
15.122 (*a*) $25.3 \text{ in./s} \measuredangle 78.4°$.
(*b*) $2.31 \text{ in./s}^2 \measuredangle 3.7°$.
15.124 (*a*) $240 \text{ m/s}^2 \measuredangle 51.9°$.
(*b*) $351 \text{ m/s}^2 \measuredangle 32.5°$.
15.126 (*a*) $742 \text{ mm/s} \measuredangle 76.0°$.
(*b*) $876 \text{ mm/s}^2 \measuredangle 9.5°$.
15.128 (*a*) $24.7 \text{ in./s} \measuredangle 76.0°$.
(*b*) $74.2 \text{ in./s}^2 \measuredangle 14.0°$.
15.130 $\boldsymbol{\omega}_S = 4.77 \text{ rad/s} \circlearrowright$; $\boldsymbol{\alpha}_S = 127.2 \text{ rad/s}^2 \circlearrowright$.
15.132 $3.61 \text{ rad/s}^2 \circlearrowleft$.
15.134 $317 \text{ ft/s}^2 \measuredangle 41.8°$.
15.136 (*a*) $0.291 \text{ rad/s} \circlearrowleft$. (*b*) $0.1206 \text{ rad/s}^2 \circlearrowleft$.
15.138 $5.45 \text{ m/s}^2 \measuredangle 51.6°$.
15.140 (*a*) $3.61 \text{ rad/s} \circlearrowleft$. (*b*) $86.6 \text{ in./s} \measuredangle 30°$.
(*c*) $563 \text{ in./s}^2 \measuredangle 46.1°$.
15.142 (*a*) $450 \text{ mm/s}^2 \uparrow$. (*b*) $759 \text{ mm/s}^2 \measuredangle 71.6°$.
(*c*) $1265 \text{ mm/s}^2 \measuredangle 34.7°$.
15.144 (*a*) $2 \text{ rad/s} \circlearrowleft$. (*b*) $6 \text{ rad/s} \circlearrowleft$.
15.146 (*a*) 2.88 rad/s^2. (*b*) $a_n = 1.791 \text{ m/s}^2$;
$a_t = 0.432 \text{ m/s}^2$. (*c*) 1.843 m/s^2.
15.148 (*a*) $1.540 \text{ m/s} \measuredangle 83.1°$.
(*b*) $0.414 \text{ m/s}^2 \measuredangle 5.02°$.
15.150 (*a*) $4 \text{ rad/s} \circlearrowright$. (*b*) $108.2 \text{ in./s} \measuredangle 73.9°$.
15.152 (*a*) $1.260 \text{ m/s} \uparrow$. (*b*) $1.250 \text{ rad/s} \circlearrowleft$.

15.C1 $\theta = 50°$: $\boldsymbol{\omega} = 2.34 \text{ rad/s} \circlearrowright$,
$\mathbf{v}_A = 0.938 \text{ m/s} \rightarrow$; $\theta = 130°$:
$\boldsymbol{\omega} = 4.03 \text{ rad/s} \circlearrowright$, $\mathbf{v}_A = 0.552 \text{ m/s} \rightarrow$.
15.C2 $\theta = 30°$: (*a*) $11.00 \text{ in./s} \measuredangle 52.2°$;
(*b*) $6.58 \text{ in./s} \measuredangle 46.8°$;
(*c*) $4.41 \text{ in./s} \measuredangle 40.1°$.
$\theta = 60°$: (*a*) $3.97 \text{ in./s} \measuredangle 10.9°$;
(*b*) $3.00 \text{ in./s} \measuredangle 30.0°$;
(*c*) $2.70 \text{ in./s} \measuredangle 43.9°$.
15.C3 $\theta = 30°$: $\boldsymbol{\omega}_{BE} = 3.23 \text{ rad/s} \circlearrowright$,
$\boldsymbol{\alpha}_{BE} = 4.19 \text{ rad/s}^2 \circlearrowleft$.
$\theta = 60°$: $\boldsymbol{\omega}_{BE} = 2.86 \text{ rad/s} \circlearrowright$,
$\boldsymbol{\alpha}_{BE} = 10.60 \text{ rad/s}^2 \circlearrowleft$.
$\theta = 150°$: $\boldsymbol{\omega}_{BE} = 4.77 \text{ rad/s} \circlearrowleft$,
$\boldsymbol{\alpha}_{BE} = 127.2 \text{ rad/s}^2 \circlearrowleft$.
15.C4 $\theta = 60°$: (*a*) $41.5 \text{ rad/s} \circlearrowleft$, $14,470 \text{ rad/s}^2 \circlearrowright$;
(*b*) $54.3 \text{ ft/s} \rightarrow$, $3440 \text{ ft/s}^2 \rightarrow$. $\theta = 120°$:
(*a*) $41.5 \text{ rad/s} \circlearrowright$, $14,470 \text{ rad/s}^2 \circlearrowright$;
(*b*) $36.4 \text{ ft/s} \rightarrow$, $7530 \text{ ft/s}^2 \leftarrow$.

CHAPTER 16

16.2 (*a*) 9.24 lb. (*b*) $18.59 \text{ ft/s}^2 \leftarrow$.
16.4 (*a*) $3.33 \text{ m/s}^2 \rightarrow$.
(*b*) $\mathbf{A} = 7.55 \text{ N} \uparrow$; $\mathbf{B} = 7.16 \text{ N} \uparrow$.
16.6 (*a*) $7.85 \text{ m/s}^2 \leftarrow$. (*b*) $3.89 \text{ m/s}^2 \leftarrow$.
(*c*) $3.95 \text{ m/s}^2 \leftarrow$.
16.8 35.6 N.
16.10 (*a*) $0.882g$. (*b*) 4.
16.12 At each wheel:
$\mathbf{A} = 15.63 \text{ kN} \uparrow$; $\mathbf{B} = 2.52 \text{ kN} \uparrow$.
16.14 (*a*) $16.1 \text{ ft/s}^2 \rightarrow$. (*b*) $10 \text{ in.} \leq h \leq 46 \text{ in.}$
16.16 (*a*) $1.193 \text{ m/s}^2 \rightarrow$. (*b*) $h \leq 841 \text{ mm}$.
16.18 $F_{AD} = 41.5 \text{ N } T$; $F_{BE} = 4.41 \text{ N } T$.
16.20 (*a*) $27.9 \text{ ft/s}^2 \measuredangle 60°$. (*b*) $1.708 \text{ lb} \uparrow$.
16.22 $F_{AB} = 36.1 \text{ N } T$; $F_{DE} = 71.2 \text{ N } T$.
16.24 54.0 lb.
16.26 (*a*) $16.1 \text{ ft/s}^2 \measuredangle 30°$. (*b*) $\mathbf{a}_B = 20.1 \text{ ft/s}^2 \downarrow$;
$\mathbf{a}_P = 40.2 \text{ ft/s}^2 \measuredangle 30°$.
16.28 $V_D = -13.24 \text{ N}$; $M_D = -2.65 \text{ N} \cdot \text{m}$.
16.30 (*a*) $8.50 \text{ m/s}^2 \measuredangle 60°$. (*b*) $V_B = -6.13 \text{ N}$;
$M_B = -3.07 \text{ N} \cdot \text{m}$.
16.32 3660 rev.
16.34 (*a*) $62.8 \text{ rad/s}^2 \circlearrowright$. (*b*) $69.8 \text{ rad/s}^2 \circlearrowright$.
16.36 382 mm.
16.38 82.5 lb.
16.40 26.6 s.
16.42 (*a*) $6.37 \text{ rad/s}^2 \circlearrowright$. (*b*) $1.912 \text{ m/s} \uparrow$.

16.44 63.4 lb·ft·s^2.

16.46 (*a*) 2.07 m/s^2 $\downarrow$. (*b*) 3.52 m/s $\downarrow$.

16.48 $I_R = \left(n + \frac{1}{n}\right)^2 I_0 + n^4 I_C$.

16.50 (*a*) $\boldsymbol{\alpha}_A = 19.32$ rad/s^2 $\circlearrowleft$; $\boldsymbol{\alpha}_B = 7.25$ rad/s^2 $\circlearrowleft$.
(*b*) $\boldsymbol{\omega}_A = 300$ rpm $\circlearrowright$; $\boldsymbol{\omega}_B = 225$ rpm $\circlearrowleft$.

16.52 (*a*) $\boldsymbol{\alpha}_A = 11.77$ rad/s^2 $\circlearrowleft$; $\boldsymbol{\alpha}_B = 49.1$ rad/s^2 $\circlearrowleft$.
(*b*) $\mathbf{C} = 41.2$ N $\uparrow$; $\mathbf{M}_C = 1.130$ N·m $\circlearrowleft$.

16.54 $\boldsymbol{\alpha} = \dfrac{2\mu_k g \sin\phi}{r(\sin 2\phi - \mu_k \cos 2\phi)} \circlearrowleft$.

16.56 (*a*) 4.91 m/s^2 $\downarrow$. (*b*) 2.71 m/s $\downarrow$.

16.58 (*a*) 19.32 ft/s^2 $\rightarrow$. (*b*) 6.44 ft/s^2 $\leftarrow$.

16.60 (*a*) 300 mm from *A*.
(*b*) $\boldsymbol{\alpha} = 5.33$ rad/s^2 $\circlearrowleft$; $\bar{\mathbf{a}} = 2.4$ m/s^2 $\rightarrow$.

16.62 $\mathbf{a}_A = 0.430$ m/s^2 $\uparrow$; $\mathbf{a}_B = 1.966$ m/s^2 $\uparrow$.

16.64 $T_A = 593$ lb; $T_B = 505$ lb.

16.66 (*a*) *W*. (*b*) $rg/\bar{k}^2$ $\circlearrowleft$.

16.68 (*a*) 0.431 rad/s^2 $\circlearrowleft$. (*b*) 0.575 ft/s^2 $\uparrow$.

16.70 (*a*) $3g/L$ $\circlearrowright$. (*b*) g $\uparrow$. (*c*) $2g$ $\downarrow$.

16.72 (*a*) g/L $\circlearrowright$. (*b*) 0. (*c*) g $\downarrow$.

16.74 (*a*) $t_1 = 2\omega_0 r/7\,\mu_k g$.
(*b*) $\bar{\mathbf{v}}_1 = 2r\omega_0/7 \rightarrow$; $\boldsymbol{\omega}_1 = 2\omega_0/7$ $\circlearrowright$.

16.76 (*a*) $t_1 = 2v_1/7\,\mu_k g$.
(*b*) $\bar{\mathbf{v}} = 2v_1/7 \rightarrow$; $\boldsymbol{\omega} = 5v_1/7r$ $\circlearrowleft$.

16.80 (*a*) 18.40 rad/s^2 $\circlearrowright$.
(*b*) $\mathbf{C}_x = 0.714$ lb $\leftarrow$; $\mathbf{C}_y = 7.5$ lb $\uparrow$.

16.82 (*a*) 34.3 rad/s^2 $\circlearrowright$.
(*b*) $\mathbf{A}_x = 4.71$ N $\leftarrow$; $\mathbf{A}_y = 58.9$ N $\uparrow$.

16.84 (*a*) 600 mm. (*b*) 13.33 rad/s^2 $\circlearrowright$.

16.86 (*a*) 244 kN. (*b*) 187.0 kN.

16.88 $A_x = 6.64$ lb; $A_z = -11.63$ lb.

16.90 3280 rpm.

16.92 (*a*) $\frac{3}{2}g$ $\downarrow$. (*b*) $\frac{1}{4}W$ $\uparrow$.

16.94 (*a*) $\frac{3}{4}g/b$ $\circlearrowright$. (*b*) $1.061g$ $\measuredangle 45°$. (*c*) $\frac{5}{8}W$ $\uparrow$.

16.96 (*a*) 5.01 lb $\uparrow$. (*b*) $\mathbf{A}_x = 77.6$ lb $\leftarrow$; $\mathbf{A}_y =$ 5.47 lb $\uparrow$.

16.98 (*a*) 29.4 rad/s^2 $\circlearrowright$.
(*b*) $\mathbf{A}_x = 96.0$ N $\leftarrow$; $\mathbf{A}_y = 38.0$ N $\uparrow$.

16.100 (*a*) 43.3 lb·ft $\circlearrowleft$. (*b*) 3.31 lb $\measuredangle 30°$.

16.102 (*a*) 8.20 m/s^2 $\downarrow$. (*b*) 9.81 m/s^2 $\downarrow$.

16.104 $6.0° < \beta < 67.3°$.

16.108 600 mm.

16.110 5.02 ft or 1.530 m.

16.112 (*a*) 7.73 rad/s^2 $\circlearrowright$; 5.15 ft/s^2 $\rightarrow$. (*b*) 0.340.

16.114 (*a*) 7.73 rad/s^2 $\circlearrowleft$; 5.15 ft/s^2 $\leftarrow$. (*b*) 0.320.

16.116 (*a*) Slides. (*b*) 3.65 rad/s^2 $\circlearrowleft$; 2.13 m/s^2 $\rightarrow$.

16.118 (*a*) Does not slide.
(*b*) 7.2 rad/s^2 $\circlearrowleft$; 1.152 m/s^2 $\leftarrow$.

16.120 2.22 m/s^2 $\leftarrow$.

16.122 (*a*) 75.5 rad/s^2 $\circlearrowleft$. (*b*) 7.55 m/s^2 $\downarrow$.

16.124 $\mathbf{a}_A = 4.19$ ft/s^2 $\leftarrow$; $\mathbf{a}_B = 3.26$ ft/s^2 $\leftarrow$.

16.126 (*a*) $\bar{\mathbf{a}}_A = \bar{\mathbf{a}}_B = 5.57$ ft/s^2 $\leftarrow$. (*b*) 0.779 lb $\leftarrow$.

16.128 1.592 lb $\leftarrow$.

16.130 23.7 rad/s^2 $\circlearrowleft$.

16.132 (*a*) 13.07 rad/s^2 $\circlearrowright$. (*b*) 9.06 N $\uparrow$.

16.134 (*a*) 7.81 lb $\leftarrow$. (*b*) 5.71 lb.

16.136 (*a*) 2.98 rad/s^2 $\circlearrowright$. (*b*) 1.396 lb $\measuredangle 50°$.

16.138 49.6 N $\uparrow$.

16.140 3.50 lb $\rightarrow$.

16.142 52.1 N $\downarrow$.

16.144 (*a*) 33.5 m/s^2 $\rightarrow$. (*b*) 16.14 N $\downarrow$.

16.146 $\mathbf{A} = 426$ N $\leftarrow$; $\mathbf{B} = 805$ N $\rightarrow$.

16.148 $\mathbf{D}_x = \frac{7}{2}m'r^2\omega_0^2 \rightarrow$, $\mathbf{D}_y = \mathbf{E}_y = m'rg \uparrow$,
$\mathbf{E}_x = m'r^2\omega_0^2 \leftarrow$.

16.150 (*a*) 5.19 m/s^2 $\measuredangle 20°$. (*b*) 32.5 rad/s^2 $\circlearrowleft$.

16.152 (*a*) 4.91 ft/s^2 $\leftarrow$. (*b*) 9.03 rad/s^2 $\circlearrowleft$.

16.154 (*a*) 12.14 rad/s^2 $\circlearrowright$. (*b*) 11.21 m/s^2 $\measuredangle 30°$.
(*c*) 14.56 N $\measuredangle 60°$.

16.156 (*a*) $\frac{6}{7}g$ $\downarrow$. (*b*) $\frac{9}{7}g$ $\downarrow$.

16.158 $M_{\max} = WL/27$ at $\frac{1}{3}L$ from *A*.

16.160 151.0 km/h.

16.162 106.6 mi/h.

16.164 (*a*) 49.5 km/h. (*b*) $\mathbf{A}_n = 2.89$ kN $\leftarrow$,
$\mathbf{B}_n = 1.195$ kN $\leftarrow$.

16.166 (*a*) 15.18°. (*b*) 0.271.

16.168 0.5 oz at *B* and 5.5 oz at *D*.

16.170 $M_x = M_y = 0$, $M_z = -\frac{1}{6}(mb^2\omega^2 \sin 2\beta)$.

16.172 $A_x = -12.00$ N, $A_y = A_z = 0$;
$B_x = -4.00$ N, $B_y = B_z = 0$.

16.174 (*a*) 12.00 rad/s^2. (*b*) $A_x = A_y = 0$,
$A_z = -2.25$ N; $B_x = B_y = 0$, $B_z = -0.75$ N.

16.176 $M_x = \frac{1}{6}(mb^2\alpha \sin 2\beta)$,
$M_y = \frac{1}{6}mb^2\alpha(1 + \cos 2\beta)$,
$M_z = -\frac{1}{6}(mb^2\omega^2 \sin 2\beta)$.

16.178 (*a*) $\cos^{-1}(3g/2l\omega^2)$. (*b*) $0 \le \omega \le \sqrt{3g/2l}$

16.180 (*a*) $12g/17$ $\downarrow$. (*b*) $\frac{1}{2}g$ $\downarrow$.

16.182 20.6 ft.

16.184 (*a*) $\frac{1}{2}a$ $\rightarrow$. (*b*) $2d$ $\rightarrow$.

16.186 (*a*) 1.5 m/s^2 $\rightarrow$. (*b*) 9.5 m/s^2 $\rightarrow$.

16.188 $A_x = A_y = 0$, $A_z = \frac{1}{2}(w/g)a^2\omega^2$;
$B_x = B_y = 0$, $B_z = -\frac{1}{2}(w/g)a^2\omega^2$.

16.190 (*a*) 66.8 ft/s; 21.5 ft/s. (*b*) 63.4°; 116.6°.

16.C1 *m = 40 kg:* $\mathbf{a} = 0$, $h \le 937$ mm;
m = 200 kg: $\mathbf{a} = 4.91$ m/s^2 $\rightarrow$, $h \le 656$ mm;
m = 400 kg: $\mathbf{a} = 6.77$ m/s^2 $\rightarrow$,
10.8 mm $\le h \le$ 617 mm.

16.C2 $\theta = 40°$: $\bar{\mathbf{a}} = 0.778g$ ⦪ 40°,
$\mathbf{a}_C = 1.856g$ ⦩ 15.6°;
$\theta = 70°$: $\bar{\mathbf{a}} = 0.532g$ ⦪ 70°,
$\mathbf{a}_C = 0.740g$ ⦩ 42.5°.

16.C3 $\theta = 30°$: $\mathbf{A} = 29.1$ N ↑;
$\theta = 60°$: $\mathbf{A} = 23.7$ N ↑.

16.C4 $\theta = 60°$: $\mathbf{B}_x = 874$ lb →, $\mathbf{B}_y = 195.4$ lb ↓;
$\mathbf{D}_x = 320$ lb ←, $\mathbf{D}_y = 394$ lb ↓.
$\theta = 120°$: $\mathbf{B}_x = 1510$ lb ←, $\mathbf{B}_y = 780$ lb ↓;
$\mathbf{D}_x = 701$ lb →, $\mathbf{D}_y = 190.3$ lb ↑.

CHAPTER 17

17.2 5230 rev.

17.4 342 mm.

17.6 (*a*) 19.20 lb·ft·s². (*b*) 11.46 rev.

17.8 40.6 rev.

17.10 100.8 lb.

17.12 (*a*) 22.8 rev. (*b*) 0.702 lb ↓.

17.14 770 mm/s ↑.

17.16 2.19 m/s ↓.

17.18 (*a*) $\sqrt{24g/7L}$ ↻; $13mg/7$ ↑.
(*b*) 6.83 rad/s ↻; 54.7 N ↑.

17.20 (*a*) 10.00 rad/s ↻. (*b*) 10.98 rad/s.

17.22 (*a*) 55.2°. (*b*) $0.756\sqrt{gr}$ ⦪ 55.2°.

17.24 5.45 rad/s ↻.

17.26 (*a*) 6.53 rad/s ↺. (*b*) 3.00 rad/s ↺.

17.28 689 mm.

17.30 (*a*) $\sqrt{\frac{10}{7}g(R-r)(1-\cos\beta)}$.
(*b*) $\frac{1}{7}W(17 - 10\cos\beta)$.

17.32 (*a*) $1.324\sqrt{g/r}$ ↺. (*b*) $2.12mg$ ↑.

17.34 1.738 m/s ←.

17.36 (*a*) 7.43 ft/s ↓. (*b*) 4.00 lb.

17.38 (*a*) 421 mm/s ←. (*b*) 141.5 N ↑.

17.40 14.32 ft/s ↓.

17.42 $\boldsymbol{\omega} = 3.21$ rad/s ↺: $\mathbf{v}_B = 1.923$ m/s ↑.

17.44 (*a*) $\boldsymbol{\omega} = 1.225\sqrt{g/R}$ ↻;
$\bar{\mathbf{v}} = 0.612\sqrt{gR}$ ⦪ 60°.
(*b*) $\boldsymbol{\omega} = 0$; $\bar{\mathbf{v}} = \sqrt{gR}$ →.

17.46 3.60 m/s →.

17.48 3.23 ft/s ←.

17.50 (*a*) 39.8 N·m. (*b*) 99.5 N·m.
(*c*) 249 N·m.

17.52 (*a*) 7.62 hp. (*b*) 38.1 hp.

17.54 (*a*) $\frac{3}{2}W$. (*b*) $\frac{3}{4}W$.

17.56 $\dfrac{2n}{n^2+1}\sqrt{\dfrac{\pi M_0}{I_0}}$.

17.58 (*a*) 10.77 ft/s ⦨ 45°. (*b*) 9.10 ft/s ↓.

17.60 (*a*) 4.85 m/s →. (*b*) $\mathbf{A} = \mathbf{C} = 1472$ N ↑.

17.62 2.26 rev.

17.64 8.02 ft/s ↓.

17.C1 $\theta = 50°$: 4.70 m/s ↑; $\theta = 25°$: 7.22 m/s ↑.

17.C2 $\theta = 30°$: $\boldsymbol{\omega}_{AB} = 2.86$ rad/s ↻,
$\mathbf{v}_D = 2.15$ m/s →;
$\theta = 10°$: $\boldsymbol{\omega}_{AB} = 4.99$ rad/s ↻,
$\mathbf{v}_D = 1.300$ m/s →.

17.C3 (*a*) $\theta = 30°$: $\boldsymbol{\omega} = 4.01$ rad/s ↻;
$\theta = 60°$: $\boldsymbol{\omega} = 5.28$ rad/s ↻.
(*b*) $t = 0.462$ s.

17.C4 (*b*) $\theta = 30°$: $\boldsymbol{\omega} = 4.75$ rad/s ↺;
$\theta = 60°$: $\boldsymbol{\omega} = 7.70$ rad/s ↺.
(*c*) $t = 0.785$ s.

CHAPTER 18

18.2 89.7 N·m.

18.4 4.11 rad/s ↻.

18.6 4.59 s.

18.8 (*a*) 3.11 m/s ↓. (*b*) 24.3 N.

18.10 (*a*) $\boldsymbol{\omega}_A = 150$ rpm ↺; $\boldsymbol{\omega}_B = 225$ rpm ↻.
(*b*) 4.71 N·s →.

18.12 (*a*) 3.21 lb. (*b*) 1.474 lb.

18.18 (*a*) 3 m/s →. (*b*) 10.20 N ←.

18.20 16.51 ft/s ↑.

18.22 0.385 lb·ft ↻.

18.24 (*a*) 27.6 ft/s ↓. (*b*) 4.00 lb.

18.26 1.389 m/s ←.

18.28 (*a*) 43.5 rpm. (*b*) 1.780 ft·lb.

18.30 $3\omega_1/8$.

18.32 $7\omega_0/11$.

18.34 (*a*) $\boldsymbol{\omega}_A = 0$; $\boldsymbol{\omega}_B = 168.9$ rpm ↻;
$\boldsymbol{\omega}_P = 11.10$ rpm ↺.
(*b*) $\boldsymbol{\omega}_A = \boldsymbol{\omega}_P = 10.45$ rpm ↺;
$\boldsymbol{\omega}_B = 169.5$ rpm ↻.

18.36 2.08 m/s; 145.4 N

18.38 1.136 m/s.

18.40 $0.645L\omega_1$.

18.42 (*a*) 4.98 ft/s →. (*b*) 1.246 ft/s ←.

18.44 (*a*) 3.6 m/s ←. (*b*) 3600 N ←.

18.46 (*a*) 12 in. (*b*) 6 ft/s →.

18.48 242 mm/s →.

18.50 $6v_1/7L$ ↺; $3\sqrt{2}v_1/7$ ⦨ 45°.

18.52 $3v_1/4L$ ↻.

18.54 $\omega_2 = \frac{1}{2}\omega_1$.

18.56 $L/\sqrt{3}$.

18.58 (*a*) $3\bar{v}_1/L$ ↻; $\frac{1}{2}\bar{v}_1$ ↓. (*b*) $3\bar{v}_1/L$ ↺; $\frac{1}{2}\bar{v}_1$ ↑.
(*c*) 0; $\bar{v}_1$ ↑.

18.60 $\sqrt{2}$.
18.62 83.2°.
18.64 $h\left(\frac{3m}{6m + m_P}\right)^2$
18.66 (*a*) $0.7\bar{v}_1 \rightarrow$. (*b*) $1.8\bar{v}_1/L$ ↺. (*c*) $0.3\bar{v}_1 \rightarrow$.
18.68 (*a*) $v_1/5 \downarrow$. (*b*) $12v_1/5L$ ↺.
18.70 3.01 ft/s ⦩ 30°.
18.72 (*a*) $\bar{\mathbf{v}}_A = v_0 \sin\theta \uparrow$, $\boldsymbol{\omega}_A = v_0/r$ ↻ θ;
$\bar{\mathbf{v}}_B = v_0 \cos\theta \rightarrow$, $\boldsymbol{\omega}_B = 0$.
(*b*) $\bar{\mathbf{v}}_B = \frac{5}{7}v_0 \cos\theta \rightarrow$.
18.74 (*a*) $(\bar{\mathbf{v}}_A)_x = \frac{2}{7}v_0 \cos\theta \rightarrow$, $(\bar{\mathbf{v}}_A)_y = v_0 \sin\theta \uparrow$.
(*b*) $\tan\phi = \frac{7}{2}\tan\theta$.
18.76 $5v_0/4r$.
18.78 (*a*) 5.5 m/s →. (*b*) 2.99 m/s →.
18.80 Nose of plane tends to move down.
18.82 $A_y = 1.178$ N, $A_z = 0$;
$B_y = -1.178$ N, $B_z = 0$.
18.84 5.18 lb ↑.
18.86 1.14°; Point *A* will move up.
18.88 $A_y = 0$, $A_z = -\frac{1}{4}ma\omega_1\omega_2$;
$B_y = 0$, $B_z = \frac{1}{4}ma\omega_1\omega_2$.
18.90 23 900 rpm.
18.92 (*a*) and (*b*) 58.7 rpm.
18.94 $M_x = 1.035$ lb·ft, $M_y = M_z = 0$.
18.96 $A_x = -18.00$ N, $A_y = 49.05$ N,
$A_z = -13.50$ N; $M_x = -10.96$ N·m,
$M_y = 0$, $M_z = 19.62$ N·m.
18.98 $9\omega_1/14$ ↻; $9\bar{v}_1/14$ ⦟ 60°.
18.100 (*a*) $\frac{1}{3}mv_0 \uparrow$. (*b*) $\frac{1}{6}mv_0 \uparrow$.
18.102 (*a*) and (*b*) 4.00 rad/s.
18.104 $t = \frac{1 + \mu^2}{\mu + \mu^2}\frac{r\omega_0}{2g}$.
18.106 (*b*) $\omega_2 = 2\bar{v}_1 r/(r^2 + \bar{k}^2)$.
18.108 $\omega = \frac{3}{4}\bar{v}_1/b$ ↻; $\bar{\mathbf{v}} = \frac{3}{8}\sqrt{2}\bar{v}_1$ ⦟ 45°.
18.C1 *x = 120 mm:* (*a*) 3.20 rad/s, 1.052 m/s;
(*b*) 3.76 rad/s, 1.141 m/s; (*c*) 4.57 rad/s,
1.257 m/s.
18.C2 *e = 0.5:* 2.68 rad/s ↺, 0.714 m/s →, 10.71 J.
e = 0.9: 3.39 rad/s ↺, 0.429 m/s ←, 2.71 J.
18.C3 *x = 150 mm:* 8.67 m/s ↑, 8.89 rad/s ↺;
x = 300 mm: 3.33 m/s ↑, 22.2 rad/s ↺.
18.C4 *θ = 10°:* (*a*) 5.74 ft/s; (*b*) 4.80 ft·lb;
θ = 30°: (*a*) 2.49 ft/s; (*b*) 0.904 ft·lb.

CHAPTER 19

19.2 1.047 m/s; 7.31 m/s^2.
19.4 2.20 in.; 94.6 ft/s^2.
19.6 11.04 mm or 0.435 in.
19.8 (*a*) 0.1481 s. (*b*) 0.490 m/s ↑; 4 m/s^2 ↓.
19.10 0.888 in. below equilibrium position;
4.94 ft/s ↓; 47.6 ft/s^2 ↑.
19.12 (*a*) 0.320 s; 3.13 Hz. (*b*) 2.05 ft/s; 40.3 ft/s^2.
19.14 (*a*) 0.406 s; 2.47 Hz. (*b*) 0.387 m/s; 6 m/s^2.
19.16 (*a*) 4.26 lb. (*b*) 3.96 lb/ft.
19.18 4.66 kg.
19.20 (*a*) 4.02 kN/m. (*b*) 1.777 kN/m.
19.24 23.1°.
19.26 1.824 s.
19.28 (*a*) 0.329 s. (*b*) 0.795 ft/s.
19.30 (*a*) 0.628 s. (*b*) 100.0 mm/s.
19.32 (*a*) 1.125 Hz. (*b*) $T_B = 76.7$ N; $T_C = 54.2$ N.
19.34 (*a*) 0.428 s. (*b*) 45.4 mm.
19.36 (*a*) $\tau = 2\pi\sqrt{2m/3k}$.
19.42 (*a*) $f = (1/2\pi)\sqrt{3g/2l}$. (*b*) $\frac{1}{6}l$.
19.44 (*a*) and (*b*) 0.933 s.
19.46 17.15 s.
19.48 (*a*) $r_a = 163.5$ mm. (*b*) 76.2 mm.
19.50 (*a*) $\tau = 2\pi\sqrt{5r/4g}$. (*b*) $\frac{1}{4}r$.
19.52 (*a*) 2.015 s. (*b*) 2.94 m/s.
19.54 (*a*) 1.099 s. (*b*) 1.554 s.
19.56 1.112 lb·ft·s^2.
19.58 1.115 Hz.
19.62 1.621 s.
19.64 (*a*) $6.82\sqrt{l/g}$. (*b*) $6.10\sqrt{l/g}$.
19.66 292 lb·ft·s^2.
19.68 5.28 Hz.
19.70 1.700 s.
19.72 $\tau = 2\pi\sqrt{m/3k\cos^2\beta}$.
19.74 2.10 Hz.
19.76 10.86 lb.
19.78 (*a*) 0.291 *l*. (*b*) $0.209\sqrt{g/l}$.
19.80 2.54 s.
19.82 (*a*) $\tau = (2\pi l/b)\sqrt{h/3g}$.
(*b*) $\tau = 2\pi\sqrt{h/g}$.
19.84 $\tau = \pi l/\sqrt{3gr}$.
19.86 $\sqrt{k/2m} < \omega < \sqrt{3k/2m}$.
19.88 $\omega > \sqrt{2g/l}$.
19.90 (*a*) 360 rpm. (*b*) 53.3 μm.
19.92 245 rpm.
19.94 1.888 mm.
19.96 15 mm or 3.75 mm.
19.98 0.5 in. or 0.3 in.
19.100 813 rpm.
19.102 (*a*) 1308 rpm. (*b*) 1.334 mm.
19.104 0.642 Hz; 2.91 mm.
19.106 34.1 mi/h.
19.112 (*a*) 0.0355. (*b*) 9.53 N·s/m.

19.114 (*a*) 1319 lb·s/ft. (*b*) 0.1107 s.
19.116 $\sqrt{1 - 2(c/c_c)^2}$.
19.118 (*a*) 0.1202 in. (*b*) 0.0818 in.
19.120 2.04 mm.
19.122 4.28 mm.
19.124 (*a*) 297 rpm. (*b*) 267 rpm. (*c*) 0.402 in.; 0.423 in.
19.126 (*a*) $m\dfrac{d^2x}{dt^2} + c\dfrac{dx}{dt} + kx = (k \sin \omega t + c\omega \cos \omega t)\delta_m$.
(*b*) $x_m = \dfrac{\sqrt{k^2 + c^2\omega^2}\,\delta_m}{\sqrt{(k - m\omega^2)^2 + (c\omega)^2}}$, with $\omega = 2\pi v/L$.
19.128 $R < 2\sqrt{L/C}$.
19.132 (*a*) $m\dfrac{d^2x_m}{dt^2} + k_2(x_m - x_A) = P_m \sin \omega t$,
$c\dfrac{dx_A}{dt} + k_1x_A + k_2(x_A - x_m) = 0$.
(*b*) $L\dfrac{d^2q_m}{dt^2} + \dfrac{(q_m - q_A)}{C_2} = E_m \sin \omega t$,
$R\dfrac{dq_A}{dt} + \dfrac{q_A}{C_1} + \dfrac{(q_A - q_m)}{C_2} = 0$.
19.136 (*a*) $m_1\dfrac{d^2x_1}{dt^2} + c_1\dfrac{dx_1}{dt} + k_1x_1 + k_2(x_1 - x_2) = 0$,
$m_2\dfrac{d^2x_2}{dt^2} + c_2\dfrac{dx_2}{dt} + k_2(x_2 - x_1) = 0$.
(*b*) $L_1\dfrac{d^2q_1}{dt^2} + R_1\dfrac{dq_1}{dt} + \dfrac{q_1}{C_1} + \dfrac{(q_1 - q_2)}{C_2} = 0$,
$L_2\dfrac{d^2q_2}{dt^2} + R_2\dfrac{dq_2}{dt} + \dfrac{(q_2 - q_1)}{C_2} = 0$.
19.138 0.0730 in.
19.140 0.918 Hz.
19.142 (*a*) 447 mm/s. (*b*) 4.33 mm above the equilibrium position; 437 mm/s ↑.
19.146 2.67 s.
19.148 (*a*) $(1/2\pi)\sqrt{g/2r}$. (*b*) $(1/2\pi)\sqrt{2g/3r}$.
19.C1 $\theta = 60°$: 1 term, 1.0625; 2 terms, 1.0713; 4 terms, 1.0731; 8 terms, 1.0732; 16 terms, 1.0732.
19.C2 $m_C = 0.5$ *kg:* (*a*) 0.641 s, 102.2 mm; (*b*) 0.428 s, 45.4 mm; (*c*) 0.321 s, 25.6 mm. $m_C = 1.5$ *kg:* (*a*) 0.828 s, 170.3 mm; (*b*) 0.552 s, 75.7 mm; (*c*) 0.414 s, 42.6 mm.
19.C3 $\omega = 500$ *rpm:* $x_m = 0.00515$ in. (in phase); $a_m = 1.176$ ft/s². $\omega = 1000$ *rpm:* $x_m = 0.001535$ in. (out of phase); $a_m = 1.403$ ft/s².
19.C4 $\omega = 500$ *rpm:* $x_m = 0.00435$ in.; $a_m = 0.994$ ft/s²; $\varphi = 32.3°$. $\omega = 1000$ *rpm:* $x_m = 0.001528$ in.; $a_m = 1.397$ ft/s²; $\varphi = -5.39°$.

APPENDIX

9.88 (*a*) $\frac{1}{4}m(r_1^2 + r_2^2)$. (*b*) $\frac{1}{2}m(r_1^2 + r_2^2)$.
9.90 $I_{DD'} = 3ma^2/8$; $I_{EE'} = 5ma^2/12$.
9.92 (*a*) $7ma^2/6$. (*b*) $\frac{1}{2}ma^2$.
9.94 $m(3a^2 + L^2)/12$.
9.96 $m(a^2 + 3h^2)/6$; $\sqrt{(a^2 + 3h^2)/6}$.
9.98 $m(a^2 + b^2)/5$.
9.100 (*a*) $m(b^2 + 4a^2 \sin^2 \theta)/12$. (*b*) $m(b^2 + 4a^2 \cos^2 \theta)/12$.
9.102 1.743 lb·ft·s², 6.54 in.
9.104 $27ma^2/50$; $0.735a$.
9.106 80.2 g·m².
9.108 0.215 lb · ft · s²; 4.95 in.
9.110 (*a*) 7.11 g·m². (*b*) 16.96 g·m². (*c*) 15.27 g·m².
9.112 $I_x = 5.35 \times 10^{-3}$ lb·ft·s²; $I_y = I_z = 21.7 \times 10^{-3}$ lb·ft·s².
9.114 30.5×10^{-3} lb·ft·s².
9.116 183.8 g · m².
9.118 2.06 kg·m².

SI Prefixes

Multiplication Factor	Prefix†	Symbol
1 000 000 000 000 = 10^{12}	tera	T
1 000 000 000 = 10^{9}	giga	G
1 000 000 = 10^{6}	mega	M
1 000 = 10^{3}	kilo	k
100 = 10^{2}	hecto‡	h
10 = 10^{1}	deka‡	da
0.1 = 10^{-1}	deci‡	d
0.01 = 10^{-2}	centi‡	c
0.001 = 10^{-3}	milli	m
0.000 001 = 10^{-6}	micro	μ
0.000 000 001 = 10^{-9}	nano	n
0.000 000 000 001 = 10^{-12}	pico	p
0.000 000 000 000 001 = 10^{-15}	femto	f
0.000 000 000 000 000 001 = 10^{-18}	atto	a

† The first syllable of every prefix is accented so that the prefix will retain its identity. Thus, the preferred pronunciation of kilometer places the accent on the first syllable, not the second.

‡ The use of these prefixes should be avoided, except for the measurement of areas and volumes and for the nontechnical use of centimeter, as for body and clothing measurements.

Principal SI Units Used in Mechanics

Quantity	Unit	Symbol	Formula
Acceleration	Meter per second squared	. . .	m/s^2
Angle	Radian	rad	†
Angular acceleration	Radian per second squared	. . .	rad/s^2
Angular velocity	Radian per second	. . .	rad/s
Area	Square meter	. . .	m^2
Density	Kilogram per cubic meter	. . .	kg/m^3
Energy	Joule	J	N·m
Force	Newton	N	$kg \cdot m/s^2$
Frequency	Hertz	Hz	s^{-1}
Impulse	Newton-second	. . .	kg·m/s
Length	Meter	m	‡
Mass	Kilogram	kg	‡
Moment of a force	Newton-meter	. . .	N·m
Power	Watt	W	J/s
Pressure	Pascal	Pa	N/m^2
Stress	Pascal	Pa	N/m^2
Time	Second	s	‡
Velocity	Meter per second	. . .	m/s
Volume, solids	Cubic meter	. . .	m^3
Liquids	Liter	L	10^{-3} m^3
Work	Joule	J	N·m

† Supplementary unit (1 revolution = 2π rad = 360°).

‡ Base unit.